LINEAR CONTROL SYSTEM ANALYSIS AND DESIGN

Conventional and Modern

McGraw-Hill Electrical Engineering Series

Consulting Editor

Stephen W. Director, *Carnegie-Mellon University*

CIRCUITS AND SYSTEMS
COMMUNICATIONS AND SIGNAL PROCESSING
CONTROL THEORY
ELECTRONICS AND ELECTRONIC CIRCUITS
POWER AND ENERGY
ELECTROMAGNETICS
COMPUTER ENGINEERING
INTRODUCTORY
RADAR AND ANTENNAS
VLSI

Previous Consulting Editors

Ronald N. Bracewell, Colin Cherry, James F. Gibbons, Willis W. Harman, Hubert Heffner, Edward W. Herold, John G. Linvill, Simon Ramo, Ronald A. Rohrer, Anthony E. Siegman, Charles Susskind, Frederick E. Terman, John G. Truxal, Ernest Weber, and John R. Whinnery

CONTROL THEORY

Consulting Editor

Stephen W. Director, *Carnegie-Mellon University*

Athans and Falb: *Optimal Controls*
Auslander, Rabins and Takahashi: *Introducing Systems and Control*
D'Azzo and Houpis: *Feedback Control System Analysis and Synthesis*
D'Azzo and Houpis: *Linear Control System Analysis and Design: Conventional and Modern*
Emanuel and Leff: *Introduction to Feedback Control Systems*
Eveleigh: *Introduction to Control Systems Design*
Friedland: *Control System Design: An Introduction to State Space Methods*
Houpis and Lamont: *Digital Control Systems: Theory, Hardware, Software*
Raven: *Automatic Control Engineering*
Schultz and Melsa: *State Functions and Linear Control Systems*

LINEAR CONTROL SYSTEM ANALYSIS AND DESIGN:
Conventional and Modern

Third Edition

John J. D'Azzo
Constantine H. Houpis

Air Force Institute of Technology
Wright-Patterson Air Force Base

McGraw-Hill Book Company

New York St. Louis San Francisco Auckland Bogotá Caracas
Colorado Springs Hamburg Lisbon London Madrid Mexico
Milan Montreal New Delhi Oklahoma City Panama Paris
San Juan São Paulo Singapore Sydney Tokyo Toronto

This book was set in Times Roman by Science Typographers, Inc.
The editor was Alar E. Elken;
the production supervisors were Diane Renda and Louise Karam.
The cover was designed by Kao & Kao Associates.
Project supervision was done by Science Typographers, Inc.
Arcata Graphics/Halliday was printer and binder.

LINEAR CONTROL SYSTEM ANALYSIS AND DESIGN
Conventional and Modern

1 2 3 4 5 6 7 8 9 0 H A L H A L 8 9 3 2 1 0 9 8

ISBN 0-07-016186-0

Library of Congress Cataloging-in-Publication Data

D'Azzo, John Joachim.
Linear control system analysis and design.

(McGraw-Hill series in electrical engineering. Control theory)
Includes index.
1. Automatic control. 2. Control theory.
I. Houpis, Constantine H. II. Title. III. Series.
TJ213.D33 1988 629.8'32 87-15909
ISBN 0-07-016186-0

CONTENTS

PREFACE

This textbook is intended to provide a clear, understandable, and motivated account of the subject which spans both conventional and modern control theory. The authors have tried to exert meticulous care with explanations, diagrams, calculations, tables, and symbols. They have tried to ensure that the student is made aware that rigor is necessary for advanced control work. Also stressed is the importance of clearly understanding the concepts which provide the rigorous foundations of modern control theory. The text provides a strong, comprehensive, and illuminating account of those elements of conventional control theory which have relevance in the design and analysis of control systems. The presentation of a variety of different techniques contributes to the development of the student's working understanding of what A. T. Fuller has called "the enigmatic control system." To provide a coherent development of the subject, an attempt is made to eschew formal proofs and lemmas with an organization that draws the perceptive student steadily and surely onto the demanding theory of multivariable control systems. It is the opinion of the authors that a student who has reached this point is fully equipped to undertake with confidence the challenges presented by more advanced control theories as typified by Chapters 18 through 22. The importance and necessity of making extensive use of computers is emphasized by references to comprehensive computer-aided-design (CAD) programs.

The establishment of appropriate differential equations to describe the performance of physical systems, networks, and devices is set forth in Chapter 2, which also introduces some elementary matrix algebra, the block diagram, and the transfer function. The essential concept of modern control theory, the state space, is dealt with also. The approach used is the simultaneous derivation of the state-vector differential equation with the single-input single-output differential equation for a chosen physical system. The relationship of the transfer function to the state equation of the system is deferred until Chapter 4. The derivation of a mathematical description of a physical system by using Lagrange equations is also given.

The first half of Chapter 3 deals with the classical method of solving differential equations and with the nature of the resulting response. This serves the individual who needs to learn this material or needs a reference. Once the state-variable equation has been introduced, careful account is given of its solution. The central importance of the state transition matrix is brought out, and the state transition equation is derived. The idea of an eigenvalue is next explained and this theory is used with the Cayley-Hamilton and Sylvester theorems to evaluate the state transition matrix.

The early part of Chapter 4 presents a comprehensive account of Laplace transform methods and pole-zero maps. Some further aspects of matrix algebra are dealt with before dealing with the solution of the state equation by the use of Laplace transforms. Finally the evaluation of transfer matrices is clearly explained.

Chapter 5 begins with system representation by the conventional block-diagram approach. It is followed by a straightforward account of simulation diagrams and the determination of the state transition equation by the use of signal flow graphs. By deriving parallel state diagrams from system transfer functions, the advantages of having the state equation in uncoupled form are established. This is followed by the methods of diagonalizing the system matrix. A feature of this chapter is the clear treatment of how to transform an **A** matrix which has complex eigenvalues into a suitable alternative block diagonal form and the transformation to companion form.

In Chapter 6 the system characteristics are introduced. This includes the relationship between system type and the ability of the system to follow or track polynomial inputs.

Chapters 7 to 11 are updated versions of the same chapters in the previous edition and present substantially the same material, with a greater emphasis on CAD packages. In Chapter 7 the details of the root-locus method of analysis are presented. Then the frequency-response method of analysis is given in Chapters 8 and 9, using both the log and the polar plots. These chapters include the following topics: Nyquist stability criterion; correlation between the s plane, frequency domain, and time domain; and gain setting to achieve a desired output response peak value. Chapters 10 and 11 describe the possible improvements in system performance, along with examples of the technique for applying cascade and feedback compensators. Both the root-locus and frequency-response methods of designing compensators are covered.

The concept of modeling a desired control ratio which has figures of merit that satisfy the system performance specifications is developed in Chapter 12. The system inputs generally fall into two categories: (1) a desired input which the system output is to track (a tracking system) and (2) a disturbance input for which the system output is to be minimal (a disturbance-rejection system). Desired control ratios for both types of systems are synthesized by the proper placement of its poles and inclusion of zeros if required. Chapter 12 also includes the Guillemin-Truxal design procedure for designing a tracking control system and a design procedure for a disturbance-rejection control system.

The technique of achieving desired system characteristics by using complete state-variable feedback are developed thoroughly and carefully in Chapter 13. The very important concepts of modern control theory—controllability and observability—are treated in a simple, straightforward, and correct manner. Although the treatment is brief, it provides sufficient coverage of these topics for the requirements of the remainder of the book. This provides a useful foundation for the work of Chapters 16 to 22.

Chapter 14 includes a presentation of the sensitivity concepts of Bode for the variation of system parameters. Also included is the method of using feedback transfer functions to form estimates of inaccessible states for use in state feedback.

Additional matrix algebra is presented in Chapter 15 with particular emphasis on quadratic forms. This material is used in the presentation of a short account of some of the important aspects of stability considered from the Liapunov point of view. An account of trajectories in the state space and some associated phase-plane techniques is also given. A feature of this approach is that the arguments are extended to nonlinear systems. The treatment of such systems by linearization is presented. The use of a Liapunov function is presented for the determination of system stability and instability. It serves as an introduction to its use in establishing a performance index, as shown in Chapter 16.

Chapter 16 starts with a careful and comprehensive treatment of the use of performance indexes for the parameter-optimization methods for single-input single-output systems. The chapter goes on to deal with the nature of the problem of optimal control. The solution to the infinite-time linear quadratic problem is deal with *in extenso* in terms of the algebraic Riccati equation, but the results are derived using the Liapunov function approach rather than the calculus of variations of the method proposed by Pontryagin et al. The heuristic advantages of this approach for beginning students are very great. A treatment of the same problem from the standpoints of the Bode diagram and root-square locus underscores the essential unity of the subject and the mutuality of the modern and conventional control theories.

Chapter 17 presents some methods of optimal linear system design. The relationship of these methods to the conventional methods is stressed and evaluated. Although the account is limited, for heuristic convenience, to single-input systems, the subsequent correlation provides the student with the opportunity to develop an invaluable insight into the nature of the linear quadratic optimal control problem.

Chapters 18 and 19 provide a thorough presentation of the principles and techniques of entire eigenstructure assignment for multiple-input multiple-output systems by means of state feedback. This extends the use of eigenvalue assignment to include the simultaneous assignment of the associated eigenvectors and provides the means for shaping the output response to meet design specifications.

There are many worthwhile control-system design techniques available in the technical literature which are based on both modern and conventional control theory. Each technique may have limited applicability to certain classes of design problems. The control engineer must have a sufficiently broad perspective to be able to apply the right technique to the right design problem. For some techniques the designer is assisted by available computer-aided-design (CAD) packages. Design techniques like LQR, LQG, LTR, etc. are thoroughly covered in the literature.

Two additional design techniques have definite design characteristics that warrant their inclusion in a textbook for both the beginning and the practicing engineer. Thus, this third edition has been expanded to include, in Chapter 20, an output feedback state-space technique based on singular perturbation theory and, in Chapter 21, a conventional control technique based on quantitative feedback theory (QFT). The authors feel that these methods have proven their applicability to the design of practical multiple-input multiple-output control systems. These chapters are intended to further strengthen the fundamentals presented earlier in the text and to "whet the appetite" of the budding control engineer. Application of the singular perturbation theory is applied to achieve robust output feedback high-gain systems in Chapter 20. This leads to the design principles developed by Professor Brian Porter which incorporate proportional plus integral controllers. Chapter 21 presents an introduction to and lays the foundation for the quantitative feedback theory (QFT) developed by Professor Isaac M. Horowitz. This technique incorporates the concept of designing a robust control system that maintains the desired system performance over a prescribed region of plant parameter uncertainty.

Chapter 22 presents an introduction to digital control systems. The advances in digital computers and microprocessors have made their use very attractive as components in control systems. The effectiveness of digital compensation is clearly demonstrated. The concept of a pseudo-continuous-time (PCT) model of a digital system permits the use of continuous-time methods for the design of digital control systems.

The authors have tried to provide students of control engineering with a clear, unambiguous, and relevant account of appropriate, contemporary, and state-of-the-art control theory. It is suitable as an introductory and bridging text for undergraduate and graduate students.

The text is arranged so that it can be used for self-study by the engineer in practice. Included are as many examples of feedback control systems in various areas of practice (electrical, aeronautical, mechanical, etc.) as space permits while maintaining a strong basic feedback control text that can be used for study in any of the various branches of engineering. To make the text meaningful and valuable to all engineers, the authors have attempted to unify the treatment of physical control systems through use of mathematical and block-diagram models common to all. The text has been class-tested, thus enhancing its value for classroom and self-study use. There are many computer-aided-design (CAD) packages available to assist a control engineer in the analysis, design, and simulation of control systems. Some of these are listed in Appendix B.

The authors express their thanks to the students who have used this book and to the faculty who have reviewed it for their helpful comments and corrections, especially, Theodore Bernstein, University of Wisconsin; Robert Fenton, Ohio State University; Frank Kern, University of Missouri-Rolla; A. A. Moezzi, Florida Institute of Technology; E. Noges, University of Washington; Floyd Patterson, North Dakota State University; David Tsui, Northeastern University; and Carlin Weimer, Ohio State University. Appreciation is expressed to Dr. R. E. Fontana, Professor Emeritus of Electrical Engineering, Air Force Institute of Technology, for the encouragement he has given, and Dr. T. J. Higgins, Professor Emeritus of Electrical Engineering, University of Wisconsin, for his thorough review of the earlier manuscript.

Especial appreciation is expressed to Dr. Donald McLean, Professor at the University of Southampton, England, formerly a visiting Professor at the Air Force Institute of Technology. His perception and insight have contributed extensively to the clarity and rigor of the presentation. Our association with him has been an enlightening and refreshing experience. Important advanced concepts are based on collaborative work with Professors Brian Porter, University of Salford, England and Isaac M. Horowitz, Weizmann Institute of Science, Rehovat, Israel. Extensive reference to their work is given in appropriate chapters. The personal relationship with them has been a source of inspiration and deep respect.

John J. D'Azzo
Constantine H. Houpis

LINEAR CONTROL SYSTEM ANALYSIS AND DESIGN

Conventional and Modern

CHAPTER 1

INTRODUCTION

1.1 INTRODUCTION

Automatic control systems permeate life in all advanced societies today. Such systems act as a catalyst in promoting progress and development. The automatic toaster, the thermostat, the washer and dryer, the computer, the microprocessor, the space vehicles, the robots, and the control systems that have speeded up the production and quality of manufactured goods—all influence our way of life. Control systems are an integral component of any industrial society and are necessary for the production of goods required by an increasing world population. Technological developments have made it possible to travel to the moon and to explore outer space. The successful operation of space vehicles, the space shuttle, space stations, and reconfigurable flight control systems depend on the proper functioning of the large number of control systems used in such ventures.

1.2 INTRODUCTION TO CONTROL SYSTEMS

Assume that the toaster in Fig. 1.1*a* is set for the desired darkness of the toasted bread. The setting of the "darkness," or timer, knob represents the input quantity, and the degree of darkness of the toast produced is the output quantity. If the degree of darkness is not satisfactory, because of the condition of the bread or some similar reason, this condition can in no way automatically alter the length of time that heat is applied. Thus, the output quantity has no influence on the input quantity. The heater portion of the toaster, excluding the timer unit, represents the dynamic part of the overall system.

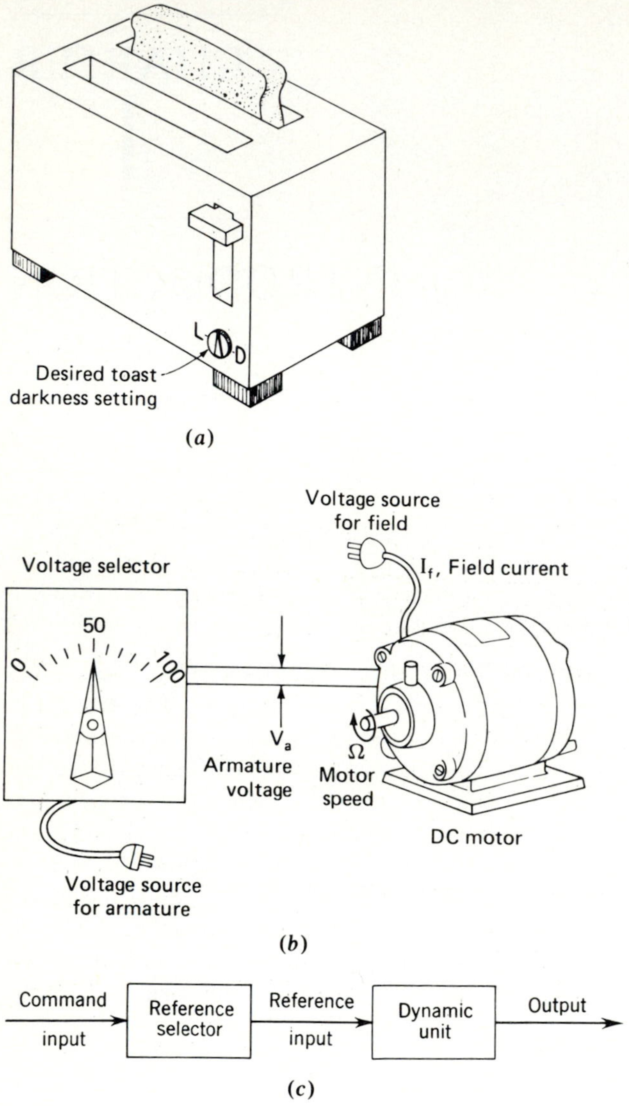

FIGURE 1.1
Open-loop control systems: (*a*) automatic toaster; (*b*) electric motor; (*c*) functional block diagram.

Another example is the dc shunt motor of Fig. 1.1*b*. For a given value of field current, a certain value of voltage is applied to the armature to produce the desired value of motor speed. In this case the motor is the dynamic part of the system, the applied armature voltage is the input quantity, and the speed of the shaft is the output quantity. A variation of the speed from the desired value, due to a change of mechanical load on the shaft, can in no way cause a change in the value of the applied armature voltage to maintain the desired speed. In this example it can also be said that the output quantity has no influence on the input quantity.

Systems in which the output quantity has no effect upon the input quantity are called *open-loop control systems*. The two examples just cited can be represented symbolically by a functional block diagram, as shown in Fig. 1.1*c*. The desired darkness of the toast or the desired speed of the motor is the command input; the selection of the value of time on the toaster timer or the value of voltage applied to the motor armature is represented by the reference-selector block; and the output of this block is identified as the reference input. The reference input is applied to the dynamic unit that performs the desired control function, and the output of this block is the desired output.

A person can be involved in the controlling actions of the systems above to sense the actual value of the output and compare it with the command input. If the output does not have the desired value, a person can alter the reference-selector position to achieve this value. Introducing the person provides a means through which the output is fed back and is compared with the input. Any necessary change is then made in order to cause the output to equal the desired value. The *feedback* action therefore controls the input to the dynamic unit. Systems in which the output has an effect upon the input quantity are called *closed-loop control systems*.

To improve the performance of the closed-loop system so that the output quantity is maintained as close to the desired quantity as possible, the person can be replaced by a mechanical, electrical, or other form of comparison unit. The functional block diagram of a *single-input single-output* (SISO) closed-loop control system is illustrated in Fig. 1.2. Comparison between the reference input and the feedback signals results in an actuating signal that is the difference between these two quantities. The actuating signal acts to maintain the output at the desired value.

This system may now be properly called a *closed-loop control system*. The designation *closed-loop* implies the action resulting from the comparison between the output and input quantities in order to maintain the output at the desired value. Thus, the output is controlled in order to maintain the desired value.

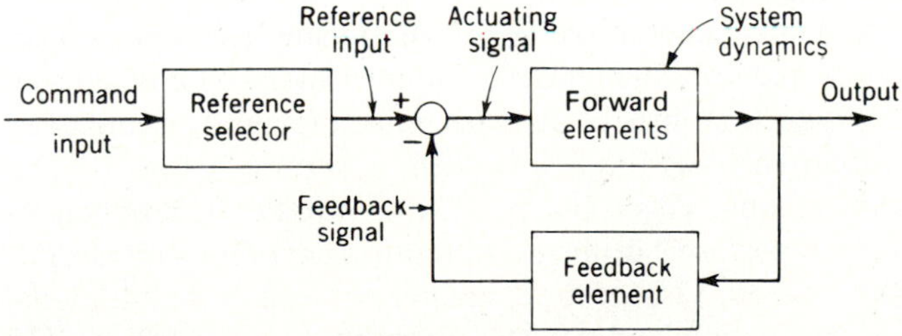

FIGURE 1.2
Functional block diagram of a closed-loop control system.

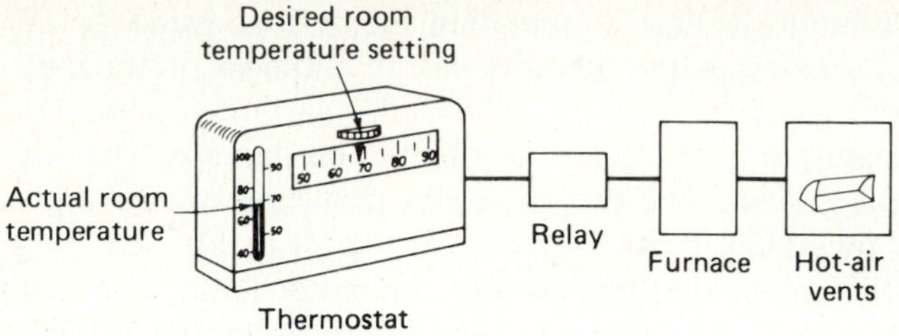

FIGURE 1.3
Home heating control system.

Examples of closed-loop control systems are illustrated in Figs. 1.3 to 1.6. A person selects the desired room temperature (command input) and adjusts the temperature setting on the thermostat in Fig. 1.3 (reference selector). A bimetallic coil in the thermostat is affected by the actual room temperature (output) and the reference-selector setting. If the room temperature is lower than the desired temperature, the coil strip alters its shape and causes a mercury switch to operate a relay, which in turn activates the furnace to produce heat in the room. When the room temperature reaches the desired temperature, the shape of the coil strip is again altered so that the mercury switch opens. This deactivates the relay and in turn shuts off the furnace. In this example, the bimetallic coil performs the function of a comparator since the output (room temperature) is fed back directly to the comparator. The switch, relay, and furnace are the dynamic elements of this closed-loop control system.

A closed-loop control system of great importance to all multistory buildings is the automatic elevator of Fig. 1.4. A person in the elevator presses the button corresponding to the desired floor. This produces an actuating signal which indicates the desired floor and activates the forward elements (control devices and elevator). As the elevator approaches the desired floor, the actuating signal decreases in value and, with the proper switching sequences, the elevator stops at the desired floor and the actuating signal is reset to zero. The closed-loop control system for the express elevator in the Sears Tower building in Chicago is designed so that it ascends or descends the 103 floors in just under 1 min with maximum passenger comfort.

The Air Force Materials Laboratory, Wright-Patterson Air Force Base, Ohio, has developed robots under its Integrated Computer-Aided Manufacturing program for use on the production lines of aerospace manufacturing companies. The first "true" robot, referred to as the 6 CM Arm, is shown in Fig. 1.5 and looks like a gigantic artificial limb. This robot performs a number of production-line functions. Specifically, it performs drilling and routing operations on aircraft access panels ranging in size from 12 in to 1 yd square. It can also do additional jobs like painting, welding, manipulating parts, assembly, and handling. The development of small reliable, versatile minicomputers has permitted the practi-

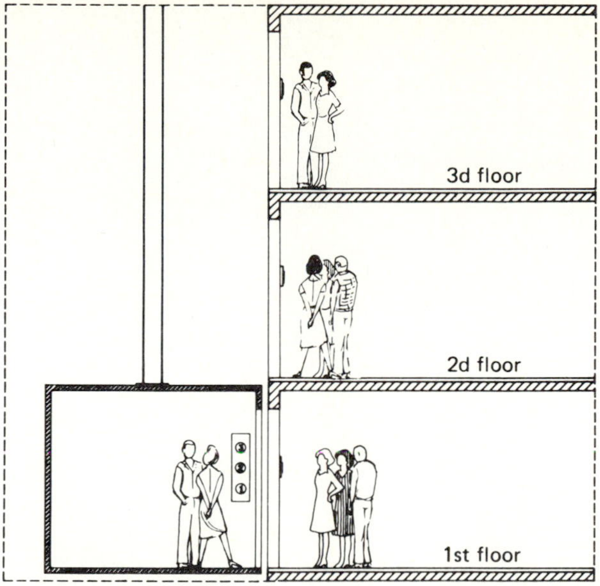

FIGURE 1.4
Automatic elevator.

FIGURE 1.5
Production-line robot.

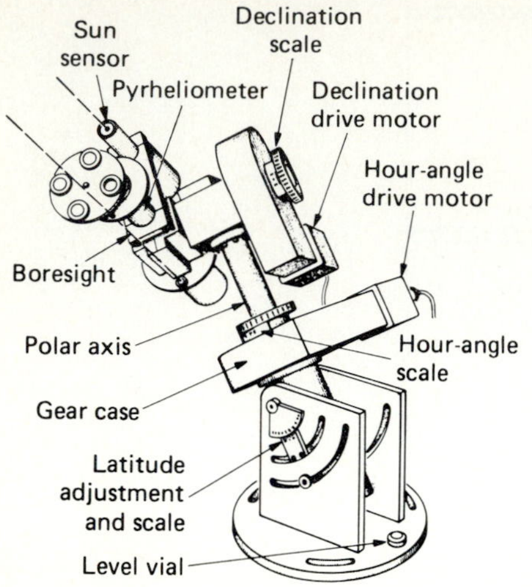

FIGURE 1.6
Sun-following pyrheliometer.

cal development of robots which have an increasingly important role in manufacturing.[12]†

The emphasis and urgency for practical large-scale solar-energy systems has spurred the development of sun-following control systems. Figure 1.6 shows a polar sun-following pyrheliometer.[11] The movement of the sun in relation to the earth is nonuniform and often obscured by clouds. This sun follower has been designed to point at the sun within 1 degree accuracy and to compensate for day-to-day changes in the sun's motion. This sun follower has essentially an open-loop drive actuated by a clock. There is a periodic sensing of the correct sun position, and corrections are made to reduce the sun-sensor error. The use of a microprocessor permits accurate sun following with trouble-free operation. This system can be used to control sun pointing of photovoltaic arrays, solar concentrators, or other devices that need to follow the sun in order to receive the maximum energy.

Desired performance can often be achieved for many systems that are structured as indicated in Fig. 1.2. The design methods for SISO control systems are covered in Chaps. 6 to 17. Some systems require a precision in their performance that cannot be achieved by the structure of Fig. 1.2. Also, systems exist for which there are multiple inputs and/or multiple outputs as discussed in Chaps. 18 to 21. The design methods for such systems are often based on a

† References are indicated by superscript numbers and will be found at the end of the chapter.

representation of the system in terms of *state variables*. For example, position, velocity, and acceleration may represent the state variables of a position control system. The definition of state variables and their use in representing systems are contained in Chaps. 2, 3, and 5. The use of state-variable methods to achieve improved or sometimes optimal performance is presented in Chaps. 13 to 20. The use of a digital computer to assist the engineer in the design process is emphasized throughout this book and a sampling of available computer-aided-design (CAD) programs is given in App. B.

1.3 DEFINITIONS

From the preceding discussion the following definitions are evolved, based in part on the standards of the IEEE.[1]

System. A combination of components that act together to perform a function not possible with any of the individual parts. The word *system* as used herein is interpreted to include physical, biological, organizational, and other entities, and combinations thereof, which can be represented through a common mathematical symbolism. The formal name *systems engineering* can also be assigned to this definition of the word *system*. Thus, the study of feedback control systems is essentially a study of an important aspect of systems engineering and its application.

Command input. The motivating input signal to the system, which is independent of the output of the system and exercises complete control over it (if the system is completely controllable).

Reference selector (*reference input element*). The unit that establishes the value of the reference input. The reference selector is calibrated in terms of the desired value of the system output.

Reference input. The reference signal produced by the reference selector, i.e., the command expressed in a form directly usable by the system. It is the actual signal input to the control system.

Disturbance input. A disturbance input signal to the system that has an unwanted effect on the system output.

Forward element (*system dynamics*). The unit that reacts to an actuating signal to produce a desired output. This unit does the work of controlling the output and thus may be a power amplifier.

Output (*controlled variable*). The quantity that must be maintained at a prescribed value, i.e., following the command input without responding to disturbance inputs.

Open-loop control system. A system in which the output has no effect upon the input signal.

Feedback element. The unit that provides the means for feeding back the output quantity, or a function of the output, in order to compare it with the reference input.

Actuating signal. The signal that is the difference between the reference input and the feedback signal. It actuates the control unit in order to cause the output to have the desired value.

Closed-loop control system. A system in which the output has an effect upon the input quantity in such a manner as to maintain the desired output value.

Note that the fundamental difference between the open- and closed-loop systems is the *feedback action*, which may be continuous or discontinuous. Continuous control implies that the output is continuously fed back and compared with the reference input. In one form of discontinuous control the input and output quantities are periodically sampled and compared, i.e., the control action is discontinuous in time. This type is commonly called a *discrete-data* or *sampled-data* feedback control system. A discrete-data control system may incorporate a digital computer which improves the performance achievable by the system. In another form of discontinuous control system the actuating signal must reach a prescribed value before the system dynamics reacts to it; i.e., the control action is discontinuous in amplitude rather than in time. This type of discontinuous control system is commonly called an *on-off* or *relay* feedback control system. Both forms may be present in a system. In this text continuous control systems are considered in detail since they lend themselves readily to a basic understanding of feedback control systems. Digital control systems are introduced in Chap. 22.

With the above introductory material, it is proper to state a definition[1] of a feedback control system: "A control system that operates to achieve prescribed relationships between selected system variables by comparing functions of these variables and using the comparison to effect control." In other books and papers on this subject the following terms may also be used.

Servomechanism (often abbreviated as servo). The term is often used to refer to a mechanical system in which the steady-state error is zero for a constant input signal. Sometimes, by generalization, it is used to refer to any feedback control system.

Regulator. This term is used to refer to systems in which there is a constant steady-state output for a constant signal. The name is derived from the early speed and voltage controls, called speed and voltage regulators.

The reader is cautioned to ascertain the meaning of a particular author. Throughout this text an attempt is made to conform to the IEEE definitions.[1]

1.4 HISTORICAL BACKGROUND[2]

The action of steering an automobile to maintain a prescribed direction of movement satisfies the definition of a feedback control system. In Fig. 1.7, the prescribed direction is the reference input. The eyes perform the function of comparing the actual direction of movement with the prescribed direction, the desired output. The eyes transmit a signal to the brain, which interprets this signal and transmits a signal to the arms to turn the steering wheel, adjusting the actual direction of movement to bring it in line with the desired direction. Thus, steering an automobile constitutes a feedback control system.

One of the earliest open-loop control systems was Hero's device for opening the doors of a temple. The command input to the system (see Fig. 1.8) was lighting a fire upon the altar. The expanding hot air under the fire drove the water from the container into the bucket. As the bucket became heavier, it descended and turned the door spindles by means of ropes, causing the counterweight to rise. The door could be closed by dousing the fire. As the air in the container

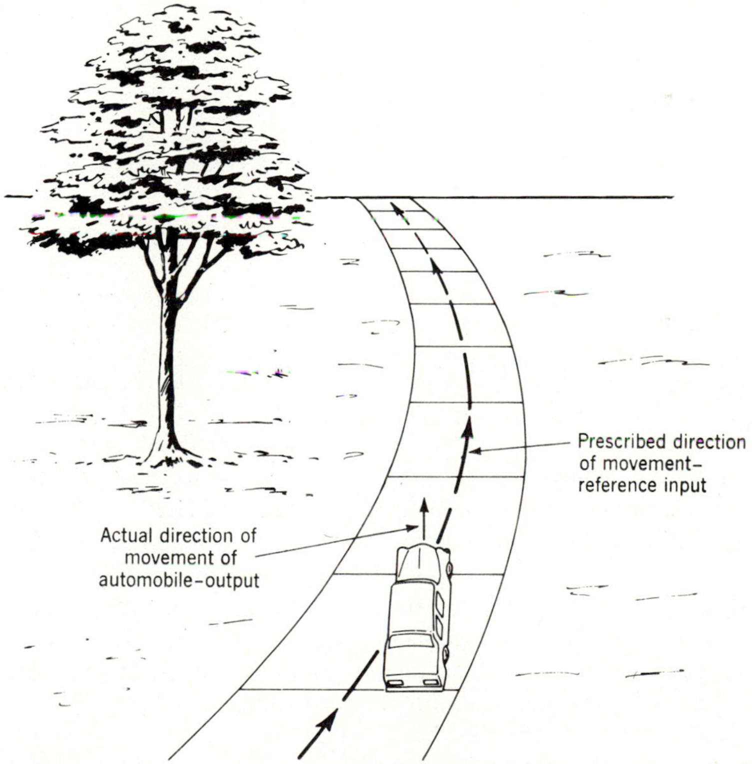

FIGURE 1.7
A pictorial demonstration of an automobile as a feedback control system.

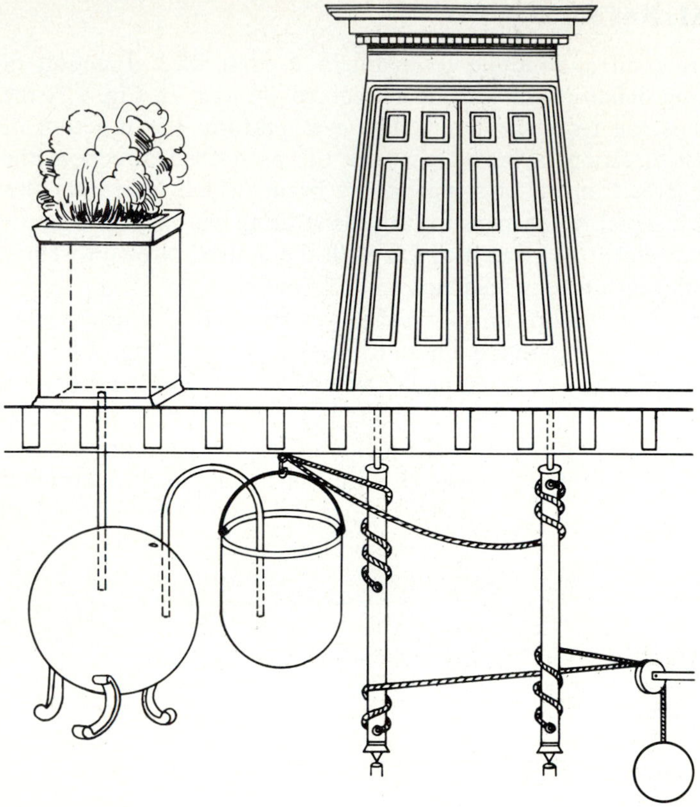

FIGURE 1.8
Hero's device for opening temple doors.

cooled and the pressure was thereby reduced, the water from the bucket siphoned back into the storage container. Thus, the bucket became lighter and the counterweight, being heavier, moved down, thereby closing the door. This occurs as long as the bucket is higher than the container. The device was probably actuated when the ruler and his entourage started to ascend the temple steps. The system for opening the door was not visible or known to the masses. Thus, it created an air of mystery and demonstrated the power of the Olympian gods.

James Watt's flyball governor for controlling speed, developed in 1788, can be considered the first widely used feedback control system not involving a human being. Maxwell, in 1868, made an analytic study of the stability of the flyball governor. This was followed by a more detailed solution of the stability of a third-order flyball governor in 1876 by the Russian engineer Wischnegradsky.[3] Minorsky made one of the earlier deliberate applications of nonlinear elements in closed-loop systems in his study of automatic ship steering about 1922.[4]

A significant data in the history of automatic feedback control systems is 1934, when Hazen's paper Theory of Servomechanisms was published in the

Journal of the Franklin Institute, marking the beginning of the very intense modern interest in this new field. It was in this paper that the word *servomechanism* originated, from the words *servant* (or slave) and *mechanism*. The word *servomechanism* thus implies a slave mechanism. It is interesting to note that in the same year Black's important paper on feedback amplifiers appeared.[5] During the following 6 years, further basic work was accomplished. Owing to World War II security restrictions, the developments in the period 1940 to 1945 were obscured, delaying rapid progress in this field. During this time three important laboratories organized at the Massachusetts Institute of Technology—the Servomechanisms, Radiation, and Instrumentation Laboratories—contributed much to the advancement of the control field. The research done by many companies during this period also helped to strengthen the foundation of this new science. After the lifting of wartime security restrictions in 1945, rapid progress was made in the control field. Since then many books and thousands of articles and technical papers have been written, and the application of control systems in the industrial and military fields has been extensive. This rapid growth of feedback control systems was accelerated by the equally rapid development and widespread use of computers.

An early example of a military application of a feedback control system is the antiaircraft radar tracking control system shown in Fig. 1.9. The radar antenna detects the position and velocity of the target airplane, and the computer

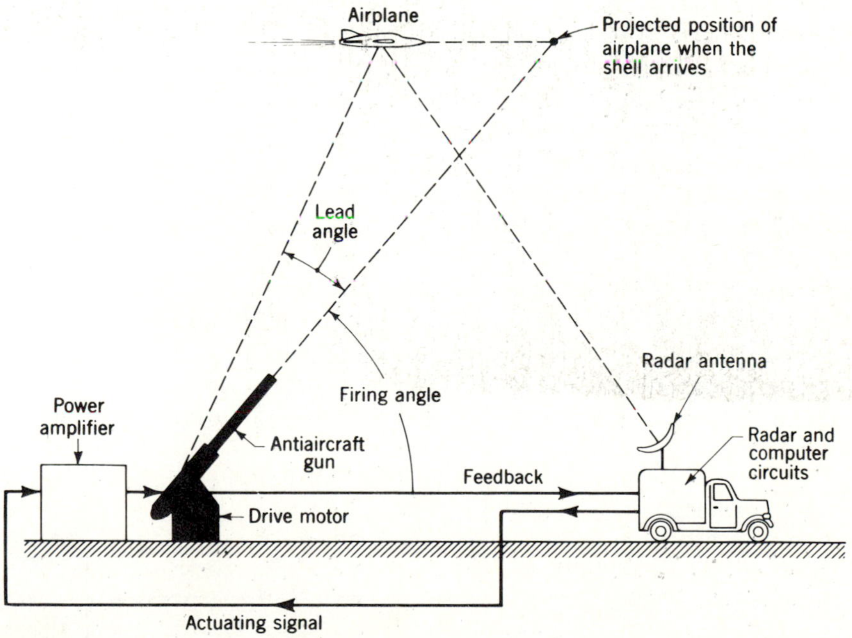

FIGURE 1.9
Antiaircraft radar tracking control system.

takes this information and determines the correct firing angle for the gun. This angle includes the necessary lead angle so that the shell reaches the projected position at the same time as the airplane. The output signal of the computer, a function of the firing angle, is fed into an amplifier which provides power for the drive motor. The motor then aims the gun at the necessary firing angle. A feedback signal proportional to the gun position ensures correct alignment with the position determined by the computer. Since the gun must be positioned both horizontally and vertically, this system has two drive motors, which are parts of two coordinated feedback loops.

The advent of the nuclear reactor was a milestone in the advancement of science and technology. For proper operation the power level of the reactor must be maintained at a desired value or must vary in a prescribed manner. This must be accomplished automatically with minimum human supervision. Figure 1.10 is a simplified block diagram of a feedback control system for controlling the power output level of a reactor. If the power output level differs from the reference input value, the actuating signal produces a signal at the output of the control elements. This, in turn, moves the regulating rod in the proper direction to achieve the desired power level of the nuclear reactor. The position of the regulating rod determines the rate of nuclear fission and therefore the total power generated. This output nuclear power can be converted into steam power, for example, which is then used for generating electric energy.

The control theory developed through the late 1950s may be categorized as *conventional* control theory. Conventional control theory is effectively applied to many control-design problems, especially to systems with a single input signal and single output signal. The control theory that has been developed since the late 1950s for the design of more complicated systems and for multiple-input multiple-output systems is called *modern* control theory. The advances that have been made in space travel were possible only because of the advent of modern control theory. Areas such as trajectory optimization and minimum-time and/or minimum-fuel problems, which are very important in space travel, can be readily handled only by modern control theory.

The advent of the steam engine and the subsequent industrial revolution provided human beings with larger quantities of controlled power than they had had at their command before. This created the need for controlling large amounts of power by means of low-power input signals and paved the way for the development of feedback control system analysis. The desire to span the universe

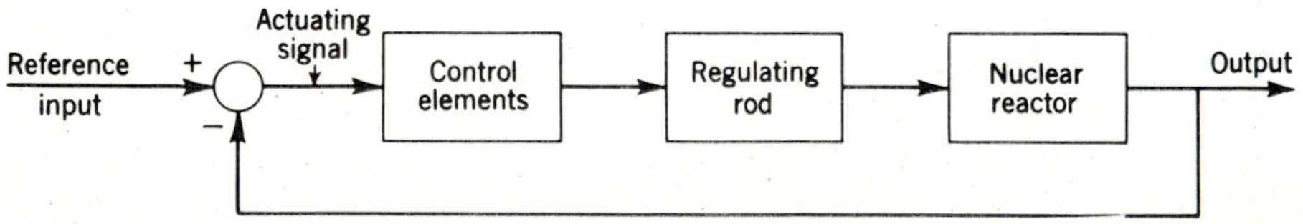

FIGURE 1.10
A nuclear-reactor power-level control system.

has given an impetus to the development of modern control theory. These same principles applied to many industrial processes have revolutionized manufacturing methods, in terms of quality, quantity, and technical sophistication of the goods produced. The result has been to change the way of life completely in the industrialized nations.

The introduction of microprocessors as control elements, i.e., performing control functions in contrast to being used solely as computational tools, has had an enormous impact on the design of feedback control systems and achieving desired control-system specifications.

As is frequently true of a science, the large number of people with many different backgrounds who have worked in this field has resulted in a large number of terms and definitions. An effort has been made by several major engineering societies to eliminate this confusion and to establish a set of definitions that can serve as a standard.[1]

The development of the control concept in the engineering field has been extended to the realm of human engineering. The basic concept of feedback control has been used extensively in the field of business management. The field of medicine is also one to which the principles of control systems and systems engineering are being applied extensively. Thus, standards of optimum performance are established in all areas of endeavor: the actual performance is compared with the standard, and any difference between the two is used to bring them into closer agreement.

1.5 DIGITAL CONTROL DEVELOPMENT[13]

With the decreasing cost of digital hardware, economical digital control implementation is now used extensively. Applications include process control, automatic aircraft stabilization and control, guidance and control of aerospace vehicles, and numeric control of manufacturing machines. The development of digital control systems is illustrated by the following example of a digital flight system.

Aircraft Digital Control Development

Modern technology has brought about numerous changes in aircraft flight control systems. Initially, flight control systems were purely mechanical, which was ideal for smaller, slow-speed, low-performance aircraft because they were easy to maintain. However, more control-surface force is required in modern high-performance airplanes. Thus, a hydraulic boost system was added to the mechanical control. This modification still maintains the direct mechanical linkage between the pilot and the control surface. As aircraft became larger, faster, and heavier, and had increased performance, they became harder to control because the pilot could not provide the necessary power to directly operate the control surface. Thus, the entire effort of moving the control surface must be provided by the actuator. Although planes were originally designed to be statically stable, under certain flight conditions large changes in the longitudinal stability began to occur. At this point, the basic airframe could no longer provide all the required flight stability and the flight control system was called upon to aid in performing this function. A stability augmentation system (SAS) was added to the hydraulically boosted mechanical regulator system to make the aircraft flyable under all flight configurations. Motion sensors were used to detect aircraft perturbations and to provide electric signals to a SAS computer, which, in turn, calculated the proper amount of servo actuator force required.

When a higher-authority SAS was required, as with advanced aircraft, both series- and parallel-pitch axis dampers were installed. This so-called command augmentation system (CAS) allowed greater flexibility in control because the parallel damper could provide full-authority travel without restricting the pilot's stick movements. The next step in the evolution of flight control systems was the use of a fly-by-wire (FBW) control system shown in Fig. 1.11. In this design, all pilot commands were transmitted to the control-surface actuators through electric wires. Thus, all mechanical linkages forward of the servo actuators were removed from the aircraft. The FBW system offered the advantages of reduced weight, improved survivability, and decreased maintenance. Its major disadvantage was the pilot's sense of insecurity with the realization that there was no mechanical backup to a totally electric system. However, the increased survivability was provided by using redundancy throughout the entire flight control system. Individual component reliability was also increased by replacing the older analog circuitry with newer digital hardware. These updated systems were referred to as digital flight control systems (DFCS).

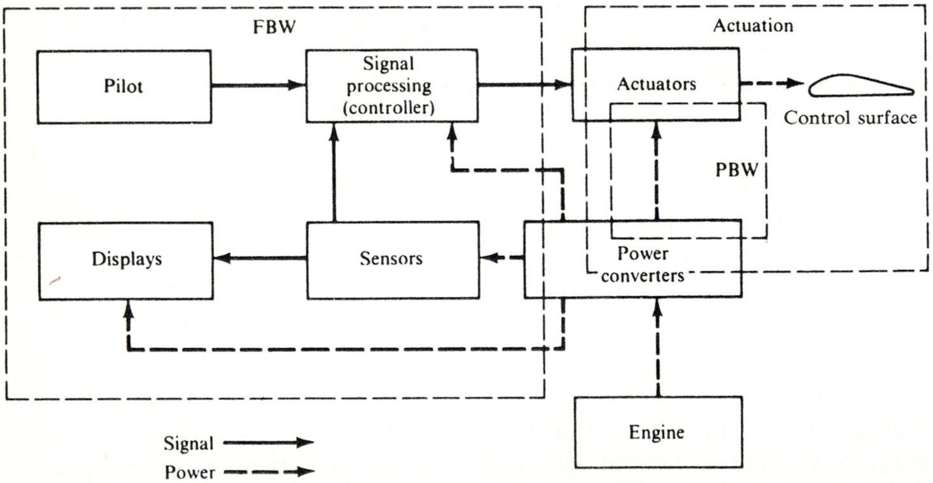

FIGURE 1.11
Fly-by-wire (FBW) and power-by-wire (PBW) control systems. (*Control Systems Development Branch, Flight Dynamics Laboratory, Wright-Patterson Air Force Base, Ohio.*)

FIGURE 1.12
An aircraft with quad-redundant three-axis fly-by-wire flight control system.

The use of airborne digital processors further reduced the cost, weight, and maintenance of modern aircraft. Other advantages associated with newer digital equipment included greater accuracy, increased modification flexibility through the use of software changes, improved in-flight reconfiguration techniques, and more reliable preflight and postflight maintenance testing. An example of a modern high-performance aircraft is shown in Fig. 1.12. This aircraft has a quad-redundant three-axis fly-by-wire flight control system. It also includes a digital built-in task computer which runs through all the preflight tests to make sure that all equipment is functioning properly.

1.6 MATHEMATICAL BACKGROUND

The early studies of control systems were based upon the solution of differential equations by classical means. Other than for simple systems, the analysis in this approach is tedious and does not readily indicate what changes should be made to improve system performance. Use of the Laplace transform simplifies this analysis somewhat. Nyquist's paper[6] published in 1932 dealt with the application of steady-state frequency-response techniques to feedback amplifier design. This work was extended by Black[5] and Bode.[7] Hall[8] and Harris[9] applied frequency-response analysis in the study of feedback control systems, which furthered the development of control theory as a whole.

Another advance occurred in 1948, when Evans[10] presented his root-locus theory. This theory affords a graphical display of the stability properties of a system and permits the graphical evaluation of the frequency response. Laplace transform theory and network theory are joined in the root-locus calculation. In the conventional control-theory portion of this text the reader learns to appreciate the simplicity and value of this phase of analysis.

Laplace transform theory and linear-algebra theory are utilized in the application of modern control theory to system analysis and design. The nth-order differential equation describing the system can be converted into a set of n first-order differential equations expressed in terms of the state variables. These equations can be written in matrix notation for simpler mathematical manipulation. The matrix equations lend themselves very well to computer computation. It is this characteristic that has enabled modern control theory to solve many problems, such as nonlinear and optimization problems, that could not be solved by conventional control theory.

Throughout the various phases of linear analysis presented in this text, mathematical models are used. Once a physical system has been described by a set of mathematical equations, they are manipulated to achieve an appropriate mathematical format. When this has been done, the subsequent method of analysis is independent of the nature of the physical system; i.e., it does not matter whether the system is electrical, mechanical, etc. This technique helps the designer to spot similarities from previous experience.

As readers cover the various phases of conventional control-theory analysis presented here, they should keep in mind that no single method is intended to be used to the exclusion of the others. Depending upon the known factors and the simplicity or complexity of a control-system problem, a designer may use one method exclusively or a combination. With experience in the design of feedback control systems comes the ability to use the advantages of each method to a greater extent.

The modern control theory presented in this text is intended as an introduction to the area of system performance optimization. In addition, one widely used method of system optimization, by solution of the algebraic Riccati equation, is presented in detail. This is extended to incorporate the manner of achieving some specified values of the conventional control-theory figures of merit along with an optimal performance. Synthesis methods are also presented for the assignment of the complete eigenstructure. This consists of both the closed-loop eigenvalue spectrum and the associated set of eigenvectors. It is an extension of previous methods which assigned only the closed-loop eigenvalues. The eigenvectors can be selected from identifiable subspaces which are associated with the system plant and control matrices. This method provides great potential for shaping the system output response to meet desired performance standards.

Two state-of-the-art design techniques for *multiple-input multiple-output* (MIMO) controls systems are presented that bring together many of the fundamentals presented earlier in the text. These chapters present the concepts of designing a robust control system in which the plant parameters may vary over specified ranges during the entire operating regime.

There is a considerable range of capabilities of digital computers, extending from the hand-operated programmable calculators, to microcomputers (PCs), to large-scale computer systems. A control engineer must be proficient in the use of available digital-computer programs provided for the programmable calculators[11] and those comprehensive programs similar to TOTAL (App. B),[12] which is a control-system computer-aided-design program. There are several CAD packages for personal computers (PCs) and mainframe computers (see App. B) available commercially. The appropriate CAD package secured depends on the computer at the designer's disposal. The use of a CAD package enhances the designer's control-system design proficiency. These programs minimize and expedite the tedious and repetitive calculations involved in the synthesis of a satisfactory control-system design. To understand and use a computer-aided analysis and design package, one must first achieve a conceptual understanding of the theory and processes involved in the analysis and synthesis of control systems. Once the conceptual understanding is achieved by direct calculations, the reader is urged to use all available computer aids. For complicated design problems engineers must write their own digital-computer program especially geared to help achieve a satisfactory system performance.

1.7 GENERAL NATURE OF THE ENGINEERING CONTROL PROBLEM

In general, a control problem can be divided into the following steps:

1. A set of performance specifications is established.
2. As a result of the performance specifications a control problem exists.
3. A set of differential equations that describe the physical system is formulated.
4. Using the conventional control-theory approach aided by available[12] or specially written computer programs:
   a. The performance of the basic (original or uncompensated) system is determined by application of one of the available methods of analysis (or a combination of them).
   b. If the performance of the original system does not meet the required specifications, cascade or feedback compensation must be added to improve the response.
5. Using the modern control-theory approach, the designer specifies an optimal performance index for the system. With the help of computer programs, the design yields the necessary structure to minimize the specified performance index, thus producing an optimal system.

6. An alternate modern control-theory approach is the method of entire eigenstructure assignment. First the desired closed-loop eigenvalue spectrum is selected. Then the desired contribution of each mode to each state and output response is selected. The eigenvector spaces are identified, and the eigenvectors are assigned which best meet the selected modal composition of the states and outputs.

Design of the system to obtain the desired performance is the control problem. The necessary basic equipment is then assembled into a system to perform the desired control function. To a varying extent, most systems are nonlinear. In many cases the nonlinearity is small enough to be neglected, or the limits of operation are small enough to allow a linear analysis to be made. In this textbook linear systems or those which can be approximated as linear systems are considered. Because of the relative simplicity and straightforwardness of this approach, the reader can obtain a thorough understanding of linear systems. After mastering the terminology, definitions, and methods of analysis for linear control systems, the engineer will find it easier to undertake a study of nonlinear systems. A method of linearizing a nonlinear system is included in Chap. 15.

A basic system has the minimum amount of equipment necessary to accomplish the control function. The differential equations that describe the physical system are derived, and an analysis of the basic system is made. If the analysis indicates that the desired performance has not been achieved with this basic system, additional equipment must be inserted into the system. Generally this analysis also indicates the characteristics for the additional equipment that are necessary to achieve the desired performance. After the system is synthesized to achieve the desired performance, based upon a linear analysis, final adjustments can be made on the actual system to take into account the nonlinearities that were neglected. It should be noted that a computer is generally used in the design depending upon the complexity of the system.

1.8 OUTLINE OF TEXT

The first few chapters deal with the mathematics underlying the analysis of control systems. In conjunction with this presentation, various basic physical units are discussed. Once the technique of writing the system equations (and, in turn, the Laplace transforms) that describe the performance of a dynamic system has been mastered, the ideas of block and simulation diagrams and transfer functions are developed. When physical systems are described in terms of block diagrams and transfer functions, they exhibit basic servo characteristics. These characteristics are described and discussed. The concept of state is introduced, and the system equations are developed in the standard matrix format. The necessary linear algebra required to manipulate the matrix equations is included. Then follows a presentation of the various methods of analysis that can be used in the study of feedback control systems. Single-input single-output (SISO) systems are used initially to facilitate an understanding of the synthesis methods.

These methods rely on root-locus and steady-state frequency-response analysis. How a basic system can be improved by using compensators if it does not meet the desired specifications is then presented. These compensators can be designed by analyzing and synthesizing a desired open-loop transfer function or by synthesizing a desired closed-loop transfer function that produces the desired overall system performance. Chapter 12 is devoted to modeling a desired closed-loop transfer function for tracking a desired input or for rejecting (not responding to) disturbance inputs.

The next portion of the text deals with single-input single-output system design using modern control theory. Topics such as controllability and observability, pole placement via state-variable feedback, parameter sensitivity, phase-plane analysis, linearization of nonlinear systems, Liapunov stability concepts, the use of performance indexes for parameter optimization, the solution of the infinite-time linear quadratic problem (LQP), and the method of entire eigenstructure assignment, etc., are presented. Chapters 16 and 17 present some methods of optimal linear system design. These methods indicate how conventional control figures of merit can be achieved while satisfying the LQP. Chapters 18 and 19 present the importance of the eigenvector on system response for multi-input multi-output systems. Then the subspaces are identified in which the eigenvectors must be located. The selected eigenvectors are used to compute the required feedback matrix. There are many worthwhile control-system design techniques available in the technical literature involving both modern and conventional control theory. Each technique has its advantages and disadvantages and may be applicable to only certain classes of design problems. Ease of design is directly dependent upon the available control-system computer-aided-design (CAD) packages. Chapters 20 and 21 present an introduction to two multivariable control-system design techniques that are applicable to the design of practical control systems. These two design methods yield robust control systems that maintain the desired system performance over a prescribed region of plant parameter uncertainty. The fundamentals of sampled-data control-system analysis and design are presented in Chap. 22. The use of a digital computer as a compensator is stressed. Appendix A gives a table of Laplace transform pairs. A description of computer-aided-design packages that are useful to a feedback control system engineer are presented in App. B.

It seems appropriate to point out that feedback control engineers are essentially "system engineers," i.e., people whose primary concern is with the design and synthesis of an overall system. To an extent depending on their own background and experience, they rely on, and work closely with, engineers in the various recognized branches of engineering (electrical, mechanical, chemical, aeronautical, etc.) to furnish them with the transfer functions and/or system equations of various portions of a control system. In the biomedical area a control engineer works closely with the medical profession in modeling biological functions, man-machine interface problems, etc.

In closing this introductory chapter it is important to stress that both conventional and optimal theory may be used for designing practical control systems. The two theories produce system designs that have different performance and require different implementation in terms of cost and equipment. The following design policy includes factors that are worthy of consideration:

1. Use proven design methods.
2. Select the system design which has the minimum complexity.
3. Use minimum specifications or requirements that yield a satisfactory system response. Compare the cost with the performance and select the fully justified system implementation.
4. Perform a complete and adequate simulation and testing of the system.

REFERENCES

1. *IEEE Standard Dictionary of Electrical and Electronics Terms*, Wiley-Interscience, New York, 1972.
2. Mayr, O.: *Origins of Feedback Control*, M.I.T. Press, Cambridge, Mass., 1971.
3. Trinks, W.: *Governors and the Governing of Prime Movers*, Van Nostrand, Princeton, N.J., 1919.
4. Minorsky, N.: "Directional Stability and Automatically Steered Bodies," *J. Am. Soc. Nav. Eng.*, vol. 34, p. 280, 1922.
5. Black, H. S.: "Stabilized Feedback Amplifiers," *Bell Syst. Tech. J.*, 1934.
6. Nyquist, H.: "Regeneration Theory," *Bell Syst. Tech. J.*, 1932.
7. Bode, H. W.: *Network Analysis and Feedback Amplifier Design*, Van Nostrand, Princeton, N.J., 1945.
8. Hall, A. C.: "Application of Circuit Theory to the Design of Servomechanisms," *J. Franklin Inst.*, 1946.
9. Harris, H.: "The Frequency Response of Automatic Control System," *Trans. AIEE*, vol. 65, pp. 539–546, 1946.
10. Evans, W. R.: "Graphical Analysis of Control Systems," *Trans. AIEE*, vol. 67, pt. II, pp. 547–551, 1948.
11. Much, C. H., P. L. Rossini, and D. G. Stuart: "A Microprocessor-Controlled Autonomous Sun-Following Mount for Solar Photovoltaic Energy Planning," *5th Annu. Conf. Ind. Control Appl. Microprocessors*, Philadelphia, March 1979.
12. Allan, R.: "Busy Robots Spur Productivity," *IEEE Spectrum*, vol. 16, pp. 31–36, 1979.
13. Houpis, C. H., and G. B. Lamont: *Digital Control Systems Theory, Hardward, Software*, McGraw-Hill, New York, 1985.

CHAPTER 2

WRITING SYSTEM EQUATIONS

2.1 INTRODUCTION

In order to analyze a dynamic system, an accurate mathematical model that describes the system completely must be determined. The derivation of this model can be based upon the fact that the dynamic system can be completely described by known differential equations (Chap. 2) or by experimental test data (Sec. 8.9). Thus the ability to analyze the system and determine its performance depends on how well the characteristics can be expressed mathematically. Numerous exact techniques are available for solving linear differential equations with constant coefficients. However, when the equations are time-varying or nonlinear, it is more difficult and often impossible to solve them analytically. In general, except for some low-order equations, the solution of a time-varying or nonlinear equation often requires a numerical, graphic, or computer procedure. Sections 15.2 and 15.3 present graphical and linearization techniques, respectively, for analyzing nonlinear systems. The systems considered in this chapter are those which are described completely by a set of linear *constant-coefficient* differential equations. Such systems are said to be *linear time-invariant* (LTI),[1] i.e., the relationship between the system input and system output is independent of time. Since the system does not change with time, the output is independent of the time at which the input is applied. There are regions of operation for which the linear representation works very well. Linear methods are used because there is extensive and elegant mathematics for solving linear equations.

This chapter presents methods for writing the differential and state equations for a variety of electrical, mechanical, thermal, and hydraulic systems.[2] This is the first step that must be mastered by the control-systems engineer. The basic physical laws are given for each system, and the associated parameters are defined. Examples are included to show the application of the basic laws to physical equipment. The result is a differential equation or a set of differential equations that describes the system. The equations derived are limited to linear systems or to those systems which can be represented by linear equations over their useful operating range. The important concepts of system state and of state variables are also introduced. The system equations, expressed in terms of state variables, are called *state equations*. The analytical tools from linear algebra are introduced as they are needed in the development of the state equations and their solutions. The solutions of both the differential and the state equations are covered in Chap. 3.

The analysis of system behavior and the development of the system equations are enhanced by using the block-diagram representation of the system. Control systems generally have many components. Complete drawings of all the detailed parts are frequently too congested to show the specific functions that are performed. It is common to use a block diagram in which each function in the system is represented by a block to simplify the picture of the complete system. Each block is labeled with the name of the component, and the blocks are appropriately interconnected by line segments. This type of diagram removes excess detail from the picture and shows the functional operation of the system. A block diagram[3] represents the flow of information and the functions performed by each component in the system.

A primary concern is the dynamic behavior of a complete control system. The use of a block diagram provides a simple means by which the functional relationship of the various components can be shown and reveals the operation of the system more readily than observation of the physical system itself. The simple functional block diagram shows clearly that apparently different physical systems can be analyzed by the same techniques. Since a block diagram is involved not with the physical characteristics of the system but only with the functional relationship between various points in the system, it can reveal the similarity between apparently unrelated physical systems.

A further step taken to increase the information supplied by the block diagram is to label the input quantity into each block and the output quantity from each block. Arrows are used to show the direction of the flow of information. The block represents the function or dynamic characteristics of the component and is represented by a transfer function (Sec. 2.5). The complete block diagram shows how the functional components are connected and the mathematical equations that determine the response of each component. Examples of block diagrams are shown throughout this chapter.

In general, variables which are functions of time are represented by lower-case letters. These are sometimes indicated by the form $x(t)$, but more often this is written just as x. There are some exceptions, because of established convention in the use of certain symbols.

To simplify the writing of differential equations, the D operator notation is used.[4] The symbols D and $1/D$ are defined by

$$Dy \equiv \frac{dy(t)}{dt} \qquad D^2y \equiv \frac{d^2y(t)}{dt^2} \tag{2.1}$$

$$D^{-1}y \equiv \frac{1}{D}y \equiv \int_0^t y(\tau)\,d\tau + \int_{-\infty}^0 y(\tau)\,d\tau = \int_0^t y(\tau)\,d\tau + Y_0 \tag{2.2}$$

where Y_0 represents the value of the integral at time $t = 0$, that is, the *initial value* of the integral.

2.2 ELECTRIC CIRCUITS AND COMPONENTS[5]

The equations for an electric circuit obey Kirchhoff's laws, which state:

1. The algebraic sum of the potential differences around a closed path equals zero. This can be restated as follows: in traversing any closed loop the sum of the voltage rises equals the sum of the voltage drops.
2. The algebraic sum of the currents entering (or leaving) a node is equal to zero. In other words, the sum of the currents entering the junction equals the sum of the currents leaving the junction.

Both these laws are used in examples in this chapter.

The voltage sources are generators. The usual direct-current (dc) voltage source is a battery or dc generator. The voltage drops appear across the three basic electrical elements: resistors, inductors, and capacitors. These elements have constant component values.

The voltage drop across a resistor is given by Ohm's law, which states that the voltage drop across a resistor is equal to the product of the current through the resistor and its resistance. Resistors absorb energy from the system. Symbolically, this voltage is written as

$$v_R = Ri \tag{2.3}$$

The voltage drop across an inductor is given by Faraday's law, which is written

$$v_L = L\frac{di}{dt} \equiv L\,Di \tag{2.4}$$

This equation states that the voltage drop across an inductor is equal to the product of the inductance and the time rate of increase of current. A positive-valued derivative Di implies an increasing current and thus a positive voltage drop; and a negative-valued derivative implies a decreasing current and thus a negative voltage drop.

TABLE 2.1
Electrical symbols and units

Symbol	Quantity	Units
e or v	Voltage	Volts
i	Current	Amperes
L	Inductance	Henrys
C	Capacitance	Farads
R	Resistance	Ohms

The positively directed voltage drop across a capacitor is defined as the ratio of the magnitude of the positive electric charge on its positive plate to the value of its capacitance. Its direction is from the positive plate to the negative plate. The charge on a capacitor plate is equal to the time integral of the current entering the plate from the initial instant to the arbitrary time t, plus the initial value of the charge. The capacitor voltage is written in the form

$$v_C = \frac{q}{C} = \frac{1}{C}\int_0^t i\,d\tau + \frac{Q_0}{C} = \frac{i}{CD} \tag{2.5}$$

The mks units for these electrical quantities in the practical system are given in Table 2.1.

Series Resistor-Inductor Circuit

The voltage source e in Fig. 2.1 is a function of time. When the switch is closed, setting the voltage rise equal to the sum of the voltage drops produces

$$v_R + v_L = e$$

$$Ri + L\frac{di}{dt} = Ri + L\,Di = e \tag{2.6}$$

The equation can be solved for the current i as shown in Chap. 3 when the applied voltage e is known. This equation can also be written in terms of the voltage v_L across the inductor in the following manner. The voltage across the inductor is

$$v_L = L\,Di$$

The current through the inductor is therefore

$$i = \frac{1}{LD}v_L$$

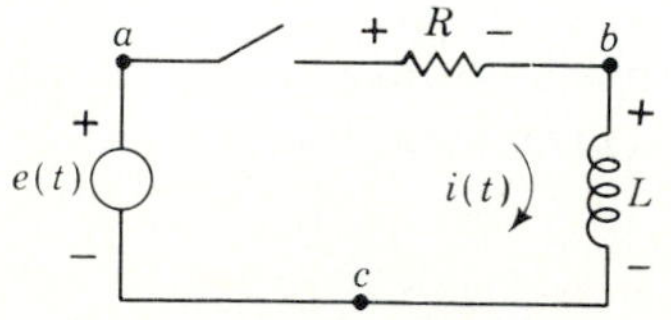

FIGURE 2.1
Series resistor-inductor circuit.

Substituting this value into the original equation gives

$$\frac{R}{LD}v_L + v_L = e \tag{2.7}$$

The node method is also convenient for writing the system equations directly in terms of the voltages. The junctions of any two elements are called *nodes*. This circuit has three nodes, labeled a, b, and c (see Fig. 2.1). One node is used as a reference point; in this circuit node c is used as the reference. The voltages at the other node are all considered with respect to the reference node. Thus v_{ac} is the voltage drop from node a to node c, and v_{bc} is the voltage drop from node b to the reference node c. For simplicity, these voltages are written just as v_a and v_b.

The source voltage $v_a = e$ is known; therefore there is only one unknown voltage, v_b, and only one node equation is necessary. Kirchhoff's second law, that the algebraic sum of the currents entering (or leaving) a node must equal zero, is applied to node b.

The current from node b to node a through the resistor R is $(v_b - v_a)/R$. The current from node b to node c through the inductor L is $(1/LD)v_b$. The sum of these currents must equal zero:

$$\frac{v_b - v_a}{R} + \frac{1}{LD}v_b = 0 \tag{2.8}$$

Rearranging terms gives

$$\left(\frac{1}{R} + \frac{1}{LD}\right)v_b - \frac{1}{R}v_a = 0 \tag{2.9}$$

Except for the use of different symbols for the voltages, this is the same as Eq. (2.7). Note that the node method requires writing only one equation.

Series Resistor-Inductor-Capacitor Circuit

For the series RLC circuit shown in Fig. 2.2, the applied voltage is equal to the sum of the voltage drops when the switch is closed:

$$v_L + v_R + v_C = e \qquad LDi + Ri + \frac{1}{CD}i = e \tag{2.10}$$

The circuit equation can be written in terms of the voltage drop across any circuit element. For example, in terms of the voltage across the resistor, $v_R = Ri$, Eqs. (2.10) become

$$v_R + \frac{L}{R}Dv_R + \frac{1}{RCD}v_R = e \tag{2.11}$$

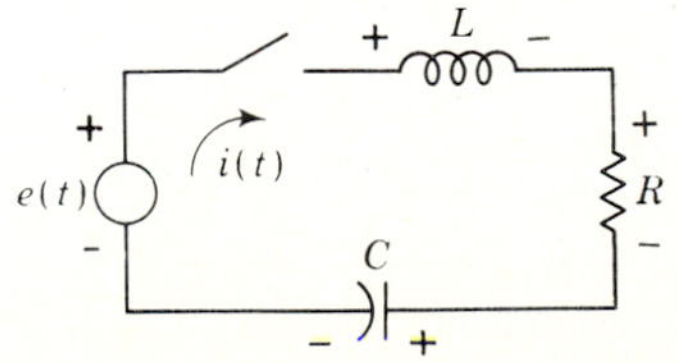

FIGURE 2.2
Series RLC circuit.

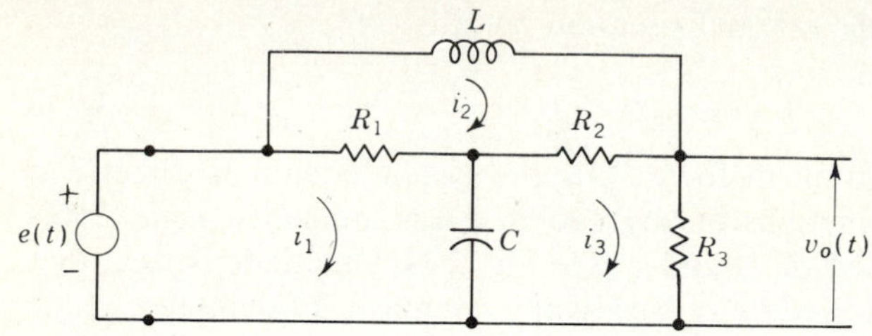

FIGURE 2.3
Multiloop network.

Multiloop Electric Circuits

Multiloop electric circuits (see Fig. 2.3) can be solved by either loop or nodal equations. The following example illustrates both methods. The problem is to solve for the output voltage v_o.

LOOP METHOD. A loop current is drawn in each closed loop; then Kirchhoff's voltage equation is written for each loop:

$$\left(R_1 + \frac{1}{CD}\right)i_1 - R_1 i_2 - \frac{1}{CD}i_3 = e \tag{2.12}$$

$$-R_1 i_1 + (R_1 + R_2 + LD)i_2 - R_2 i_3 = 0 \tag{2.13}$$

$$-\frac{1}{CD}i_1 - R_2 i_2 + \left(R_2 + R_3 + \frac{1}{CD}\right)i_3 = 0 \tag{2.14}$$

The output voltage is

$$v_o = R_3 i_3 \tag{2.15}$$

These four equations must be solved simultaneously to obtain $v_o(t)$ in terms of the input voltage $e(t)$ and the circuit parameters.

NODE METHOD. The junctions, or nodes, are labeled by letters in Fig. 2.4. Kirchhoff's current equations are written for each node in terms of the node voltages, where node d is taken as reference. The voltage v_{bd} is the voltage of node b with reference to node d. For simplicity, the voltage v_{bd} is written just as v_b. Since there is one known node voltage $v_a = e$ and two unknown voltages v_b

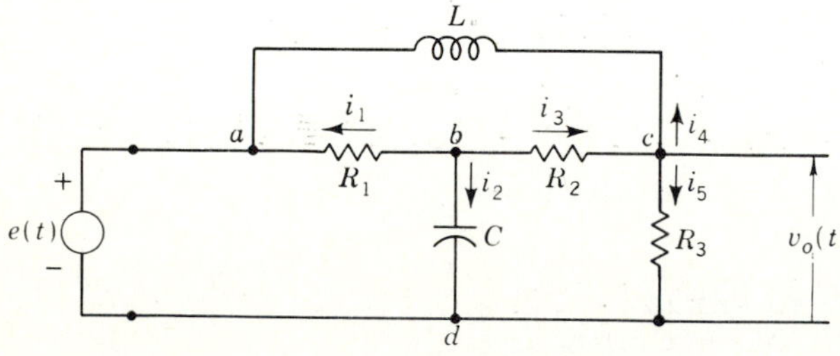

FIGURE 2.4
Multinode network.

and v_o, only two equations are required:

For node b: $$i_1 + i_2 + i_3 = 0 \tag{2.16}$$

For node c: $$-i_3 + i_4 + i_5 = 0 \tag{2.17}$$

In terms of the node voltages, these equations are

$$\frac{v_b - v_a}{R_1} + C D v_b + \frac{v_b - v_o}{R_2} = 0 \tag{2.18}$$

$$\frac{v_o - v_b}{R_2} + \frac{v_o}{R_3} + \frac{1}{LD}(v_o - e) = 0 \tag{2.19}$$

Rearranging the terms to systematize the form of the equations gives

$$\left(\frac{1}{R_1} + CD + \frac{1}{R_2}\right) v_b - \frac{1}{R_2} v_o = \frac{1}{R_1} e \tag{2.20}$$

$$-\frac{1}{R_2} v_b + \left(\frac{1}{R_2} + \frac{1}{R_3} + \frac{1}{LD}\right) v_o = \frac{1}{LD} e \tag{2.21}$$

For this example, only two nodal equations are needed to solve for the potential at node c. An additional equation must be used if the current in R_3 is required. With the loop method, three equations must be solved simultaneously to obtain the current in any branch; an additional equation must be used if the voltage across R_3 is required. The method that requires the solution of the fewest equations should be used. This varies with the circuit.

The rules for writing the node equations in a straight forward systematic manner can be summarized as follows:

1. The number of equations required is equal to the number of unknown node voltages.
2. An equation is written for each node.
3. Each equation includes:
   a. The node voltage multiplied by the sum of all the admittances that are connected to this node. This term is positive.
   b. The node voltage at the other end of each branch multiplied by the admittance connected between the two nodes. This term is negative.

The reader should learn to apply these rules so that Eqs. (2.20) and (2.21) can be written directly from the circuit of Fig. 2.4.

2.3 BASIC MATRIX LINEAR ALGEBRA[6]

Before proceeding with a discussion of other physical systems it is appropriate to introduce some basic matrix concepts needed in the development of the state method of describing and analyzing physical systems.

MATRIX. A *matrix* is a rectangular arrav of elements. The elements may be real or complex numbers or variables of time or frequency. A matrix with α rows and

β columns is called an $\alpha \times \beta$ matrix or is said to be of *order* $\alpha \times \beta$. The matrix is also said to have size or dimension $\alpha \times \beta$. If $\alpha = \beta$, the matrix is called a *square matrix*. Boldface capital letters are used to denote rectangular matrices, and boldface small letters are used to denote column matrices. A general expression for an $\alpha \times \beta$ matrix is

$$\mathbf{M} = \begin{bmatrix} m_{11} & m_{12} & \cdots & m_{1\beta} \\ m_{21} & m_{22} & \cdots & m_{2\beta} \\ \cdots & \cdots & \cdots & \cdots \\ m_{\alpha 1} & m_{\alpha 2} & \cdots & m_{\alpha\beta} \end{bmatrix} = \left[m_{ij} \right] \tag{2.22}$$

The elements are denoted by lowercase letters. A double-subscript notation is used to denote the location of the element in the matrix; thus, the element m_{ij} is located in the ith row and the jth column.

TRANSPOSE. The transpose of a matrix $\mathbf{M}$ is denoted by $\mathbf{M}^T$. The matrix $\mathbf{M}^T$ is obtained by interchanging the rows and columns of the matrix $\mathbf{M}$. In general, with $\mathbf{M} = [m_{ij}]$, the transpose matrix is $\mathbf{M}^T = [m_{ji}]$.

VECTOR. A vector is a matrix which has either one row or one column. An $\alpha \times 1$ matrix is called a *column vector*, written

$$\mathbf{x} = \begin{bmatrix} x_1 \\ x_2 \\ \vdots \\ x_\alpha \end{bmatrix} \tag{2.23}$$

and a $1 \times \beta$ matrix is called a *row vector*, written

$$\mathbf{x}^T = \begin{bmatrix} x_1 & x_2 & \cdots & x_\beta \end{bmatrix} \tag{2.24}$$

Thus, the transpose of a column vector $\mathbf{x}$ yields the row vector $\mathbf{x}^T$.

ADDITION AND SUBTRACTION OF MATRICES. The sum or difference of two matrices $\mathbf{M}$ and $\mathbf{N}$, both of order $\alpha \times \beta$, is a matrix $\mathbf{W}$ of order $\alpha \times \beta$. The element w_{ij} of $\mathbf{W} = \mathbf{M} \pm \mathbf{N}$ is

$$w_{ij} = m_{ij} \pm n_{ij} \tag{2.25}$$

These operations are commutative and associative; i.e.,

Commutative: $\mathbf{M} \pm \mathbf{N} = \pm\mathbf{N} + \mathbf{M}$

Associative: $(\mathbf{M} + \mathbf{N}) + \mathbf{Q} = \mathbf{M} + (\mathbf{N} + \mathbf{Q})$

Example 1 Addition of matrices

$$\begin{bmatrix} 3 & 1 & 2 \\ 1 & 0 & -4 \\ 0 & 5 & 7 \end{bmatrix} + \begin{bmatrix} 1 & -3 & 1 \\ 2 & 1 & 5 \\ -4 & -1 & 0 \end{bmatrix} = \begin{bmatrix} 4 & -2 & 3 \\ 3 & 1 & 1 \\ -4 & 4 & 7 \end{bmatrix}$$

MULTIPLICATION OF MATRICES. The multiplication of two matrices $\mathbf{MN}$ can be performed *if and only if* (iff) they *conform*. If the orders of $\mathbf{M}$ and $\mathbf{N}$ are $\alpha \times \beta$ and $\gamma \times \delta$, respectively, they are conformable iff $\beta = \gamma$; that is, the number of columns of $\mathbf{M}$ must be equal to the number of rows of $\mathbf{N}$. Under this condition

$\mathbf{MN} = \mathbf{W}$, where the elements of $\mathbf{W}$ are defined by

$$w_{ij} = \sum_{k=1}^{\beta} m_{ik} n_{kj} \tag{2.26}$$

That is, each element of the ith row of $\mathbf{M}$ is multiplied by the corresponding element of the jth column of $\mathbf{N}$, and these products are summed to yield the ijth element of $\mathbf{W}$. The dimension or order of the resulting matrix $\mathbf{W}$ is $\alpha \times \delta$.

Matrix multiplication operations are summarized as follows:

1. An $\alpha \times \beta$ $\mathbf{M}$ matrix times a $\beta \times \gamma$ $\mathbf{N}$ matrix yields an $\alpha \times \gamma$ $\mathbf{W}$ matrix; that is, $\mathbf{MN} = \mathbf{W}$.
2. An $\alpha \times \beta$ $\mathbf{M}$ matrix times a $\beta \times \alpha$ $\mathbf{N}$ matrix yields an $\alpha \times \alpha$ $\mathbf{W}$ square matrix; that is, $\mathbf{MN} = \mathbf{W}$. Note that although $\mathbf{NM} = \mathbf{Y}$ is also a square matrix, it is of order $\beta \times \beta$.
3. When $\mathbf{M}$ is of order $\alpha \times \beta$ and $\mathbf{N}$ is of order $\beta \times \alpha$, each of the products $\mathbf{MN}$ and $\mathbf{NM}$ is conformable. However, in general,

$$\mathbf{MN} \neq \mathbf{NM}$$

   Thus the product of $\mathbf{M}$ and $\mathbf{N}$ is said to be *noncommutable*, that is, the order of multiplication cannot be interchanged.
4. The product $\mathbf{MN}$ can be referred to as $\mathbf{N}$ *premultiplied* by $\mathbf{M}$ or as $\mathbf{M}$ *postmultiplied* by $\mathbf{N}$. In other words, premultiplication or postmultiplication is used to indicate whether one matrix is multiplied by another from the left or from the right.
5. A $1 \times \alpha$ row vector times an $\alpha \times \beta$ matrix yields a $1 \times \beta$ row vector; that is, $\mathbf{x}^T\mathbf{M} = \mathbf{y}^T$.
6. A $\beta \times \alpha$ matrix times an $\alpha \times 1$ column vector yields a $\beta \times 1$ column vector; that is, $\mathbf{Mx} = \mathbf{y}$.
7. A $1 \times \alpha$ row vector times an $\alpha \times 1$ column vector yields a 1×1 matrix, that is, $\mathbf{x}^T\mathbf{y} = \mathbf{w}$, which contains a single scalar element.
8. The k-fold multiplication of a square matrix $\mathbf{M}$ by itself is indicated by $\mathbf{M}^k$.

These operations are illustrated by the following examples.

Example 2 Matrix multiplication

$$\begin{bmatrix} 2 & 1 & 3 \\ 1 & 4 & 1 \end{bmatrix} \begin{bmatrix} 2 & 1 & 0 \\ 0 & 3 & 1 \\ 1 & 2 & 1 \end{bmatrix} = \begin{bmatrix} 7 & 11 & 4 \\ 3 & 15 & 5 \end{bmatrix} \qquad \begin{bmatrix} 2 & 1 & 3 \\ 1 & 4 & 1 \end{bmatrix} \begin{bmatrix} 2 & 1 \\ 0 & 3 \\ 1 & 2 \end{bmatrix} = \begin{bmatrix} 7 & 11 \\ 3 & 15 \end{bmatrix}$$

$$\begin{bmatrix} 2 & 1 \\ 0 & 3 \\ 1 & 2 \end{bmatrix} \begin{bmatrix} 2 & 1 & 3 \\ 1 & 4 & 1 \end{bmatrix} = \begin{bmatrix} 5 & 6 & 7 \\ 3 & 12 & 3 \\ 4 & 9 & 5 \end{bmatrix} \qquad \begin{bmatrix} 2 & 1 \\ 1 & 1 \end{bmatrix} \begin{bmatrix} 1 & 0 \\ 2 & 3 \end{bmatrix} = \begin{bmatrix} 4 & 3 \\ 3 & 3 \end{bmatrix}$$

$$\begin{bmatrix} 1 & 0 \\ 2 & 3 \end{bmatrix} \begin{bmatrix} 2 & 1 \\ 1 & 1 \end{bmatrix} = \begin{bmatrix} 2 & 1 \\ 7 & 5 \end{bmatrix} \qquad \begin{bmatrix} 1 & 3 \end{bmatrix} \begin{bmatrix} 2 & 1 \\ 0 & 3 \end{bmatrix} = \begin{bmatrix} 2 & 10 \end{bmatrix}$$

$$\begin{bmatrix} 2 & 1 \\ 0 & 3 \end{bmatrix} \begin{bmatrix} 1 \\ 3 \end{bmatrix} = \begin{bmatrix} 5 \\ 9 \end{bmatrix} \qquad \begin{bmatrix} 1 & 3 \end{bmatrix} \begin{bmatrix} 2 \\ 1 \end{bmatrix} = \begin{bmatrix} 5 \end{bmatrix}$$

MULTIPLICATION OF A MATRIX BY A SCALAR. *The multiplication of matrix* **M** *by a scalar k* is effected by multiplying each element m_{ij} by k, that is,

$$k\mathbf{M} = \mathbf{M}k = \left[km_{ij}\right] \tag{2.27}$$

Example 3

$$2\begin{bmatrix} 1 & 0 \\ 5 & -7 \end{bmatrix} = \begin{bmatrix} 1 & 0 \\ 5 & -7 \end{bmatrix} 2 = \begin{bmatrix} 2 & 0 \\ 10 & -14 \end{bmatrix}$$

UNIT OR IDENTITY MATRIX. A unit matrix, denoted **I**, is a diagonal matrix in which each element on the principal diagonal is unity. Sometimes the notation $\mathbf{I}_n$ is used to indicate an identity matrix of order n. An example of a unit matrix is

$$\mathbf{I} = \begin{bmatrix} 1 & 0 & 0 & 0 \\ 0 & 1 & 0 & 0 \\ 0 & 0 & 1 & 0 \\ 0 & 0 & 0 & 1 \end{bmatrix}$$

The premultiplication or postmultiplication of a matrix **M** by the unit matrix **I** leaves the matrix unchanged.

$$\mathbf{MI} = \mathbf{IM} = \mathbf{M}$$

DIFFERENTIATION OF A MATRIX. The differentiation of a matrix with respect to a scalar is effected by differentiating each element of the matrix with respect to the indicated variable:

$$\frac{d}{dt}[\mathbf{M}(t)] = \dot{\mathbf{M}}(t) = \left[\dot{m}_{ij}\right] = \begin{bmatrix} \dot{m}_{11} & \dot{m}_{12} & \cdots & \dot{m}_{1\beta} \\ \dot{m}_{21} & \dot{m}_{22} & \cdots & \dot{m}_{2\beta} \\ \cdots & \cdots & \cdots & \cdots \\ \dot{m}_{\alpha 1} & \dot{m}_{\alpha 2} & \cdots & \dot{m}_{\alpha\beta} \end{bmatrix}$$

The derivative of a product of matrices follows rules similar to those for the derivative of a scalar product, with preservation of order. Thus, typically,

$$\frac{d}{dt}[\mathbf{M}(t)\mathbf{N}(t)] = \mathbf{M}(t)\dot{\mathbf{N}}(t) + \dot{\mathbf{M}}(t)\mathbf{N}(t)$$

INTEGRATION OF A MATRIX. The integration of a matrix is effected by integrating each element of the matrix with respect to the indicated variable:

$$\int \mathbf{M}(t)\,dt = \left[\int m_{ij}\,dt\right]$$

2.4 STATE CONCEPTS

With basic matrix concepts, the system state concept and the method of writing and solving the state equations can be presented. The state of a system (henceforth referred to only as state) is defined by Kalman[7] as follows:

STATE. The *state* of a system is a mathematical structure containing a set of n variables $x_1(t), x_2(t), \ldots, x_i(t), \ldots, x_n(t)$, called the *state variables*, such that *the initial values* $x_i(t_0)$ of this set and the system inputs $u_j(t)$ are sufficient to uniquely describe the system's future response of $t \geq t_0$. There is a minimum set of state variables which is required to represent the system accurately. The m inputs, $u_1(t), u_2(t), \ldots, u_j(t), \ldots, u_m(t)$, are deterministic; i.e., they have specific values for all values of time $t \geq t_0$.

Generally the initial starting time t_0 is taken to be zero. The state variables need not be physically observable and measurable quantities; they may be purely mathematical quantities. As a consequence of the definition of state, the following additional definitions are generated.

STATE VECTOR. The set of state variables $x_i(t)$ represents the elements or components of the n-dimensional state vector $\mathbf{x}(t)$; that is,

$$\mathbf{x}(t) \equiv \begin{bmatrix} x_1(t) \\ x_2(t) \\ \vdots \\ x_n(t) \end{bmatrix} = \begin{bmatrix} x_1 \\ x_2 \\ \vdots \\ x_n \end{bmatrix} \equiv \mathbf{x} \tag{2.28}$$

The order of the system characteristic equation is n, and the state equation representation of the system consists of n first-order differential equations. When all the inputs $u_j(t)$ to a given system are specified for $t > t_0$, the resulting state vector uniquely determines the system behavior for any $t > t_0$.

STATE SPACE. State space is defined as the n-dimensional space in which the components of the state vector represent its coordinate axes.

STATE TRAJECTORY. State trajectory is defined as the path produced in the state space by the state vector $\mathbf{x}(t)$ as it changes with the passage of time. State space and state trajectory in the two-dimensional case are referred to as the *phase plane* and *phase trajectory*, respectively. Examples of state trajectories are presented in Chap. 15.

The first step in applying these definitions to a physical system is the selection of the system variables that are to represent the state of the system. It should be clearly understood that there is no unique way of making this selection. The three common representations for expressing the system state are the *physical*, *phase*, and *canonical state variables*. The physical state-variable representation is introduced in this chapter. The other two representations are introduced in later chapters.

The selection of the state variables for the physical-variable method is based upon the energy-storage elements of the system. Table 2.2 lists some common energy-storage elements that exist in physical systems and the corresponding energy equations. The physical variable in the energy equation for each energy-storage element *can* be selected as a state variable of the system. Only independent physical variables are chosen to be state variables. *Independent state variables* are those state variables which cannot be expressed in terms of the remaining assigned state variables. In some systems it may be necessary to identify more state variables than just the energy-storage variables. This is

TABLE 2.2
Energy-storage elements

Element	Energy	Physical variable
Capacitor C	$\frac{Cv^2}{2}$	Voltage v
Inductor L	$\frac{Li^2}{2}$	Current i
Mass M	$\frac{Mv^2}{2}$	Translational velocity v
Moment of inertia J	$\frac{J\omega^2}{2}$	Rotational velocity ω
Spring K	$\frac{Kx^2}{2}$	Displacement x
Fluid compressibility $\frac{V}{K_B}$	$\frac{VP_L^2}{2K_B}$	Pressure P_L
Fluid capacitor $C = \rho A$	$\frac{\rho A h^2}{2}$	Height h
Thermal capacitor C	$\frac{C\theta^2}{2}$	Temperature θ

illustrated in some of the following examples where velocity is a state variable. When position, the integral of this state variable, is of interest, it must also be assigned as a state variable.

Example 1 Series resistor-inductor circuit (Fig. 2.1). This circuit contains only one energy-storage element, the inductor; thus only one state variable is required. From Table 2.2, the state variable is $x_1 = i$. The equation desired is one which contains the first derivative of the state variable. The loop equation, Eq. (2.6), can be rewritten, letting $u = e$, as

$$Rx_1 + L\dot{x}_1 = u$$

$$\dot{x}_1 = -\frac{R}{L}x_1 + \frac{1}{L}u \tag{2.29}$$

The letter u is the standard notation for the input forcing function and is called the *control variable*. Equation (2.29) is called the *state equation* of the system. There is only one state equation because this is a first-order system, $n = 1$.

Example 2 Series resistor-inductor-capacitor circuit (Fig. 2.2). This circuit contains two energy-storage elements, the inductor and capacitor. From Table 2.2, the two assigned state variables are immediately identified as $x_1 = v_c$ (the voltage across the capacitor) and $x_2 = i$ (the current in the inductor). Thus two state equations are required.

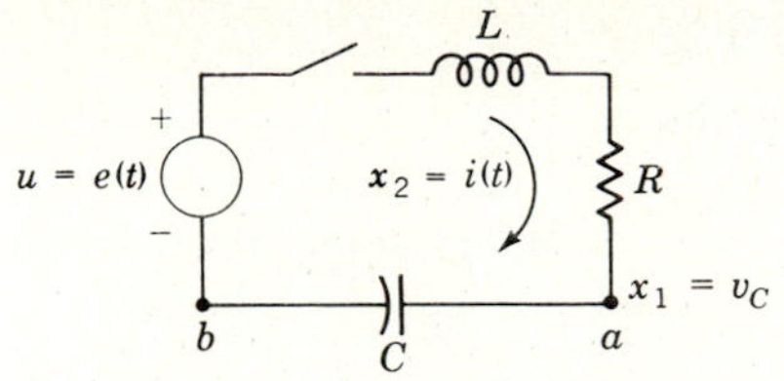

FIGURE 2.5
Series *RLC* circuit.

A loop or branch equation is written to obtain an equation containing the derivative of the current in the inductor. A node equation is written to obtain an equation containing the derivative of the capacitor voltage. The number of loop equations that must be written is equal to the number of state variables representing currents in inductors. The number of equations involving node voltages that must be written is equal to the number of state variables representing voltages across capacitors. These are usually, but not always, node equations. These loop and node equations are written in terms of the inductor *branch* currents and capacitor branch voltages. From these equations, it is necessary to determine which of the assigned physical variables are independent. When a variable of interest is not an energy physical variable, then this variable is identified as an *augmented state variable.*

Figure 2.2 is redrawn in Fig. 2.5 with node b as the reference node. The node equation for node a and the loop equation are, respectively,

$$C\dot{x}_1 = x_2 \tag{2.30}$$

and

$$L\dot{x}_2 + Rx_2 + x_1 = u \tag{2.31}$$

Rearranging terms yields

$$\dot{x}_1 = \frac{1}{C}x_2 \tag{2.32}$$

$$\dot{x}_2 = -\frac{1}{L}x_1 - \frac{R}{L}x_2 + \frac{1}{L}u \tag{2.33}$$

Equations (2.32) and (2.33) represent the state equations of the system containing two independent state variables. Note that they are *first-order* linear differential equations and are $n = 2$ in number. They are the minimum number of state equations required to represent the system's future performance.

The following definition is based upon these two examples.

STATE EQUATION. The state equations of a system are a set of n first-order differential equations, where n is the number of independent states.

The state equations represented by Eqs. (2.32) and (2.33) are expressed in matrix notation as

$$\begin{bmatrix} \dot{x}_1 \\ \dot{x}_2 \end{bmatrix} = \begin{bmatrix} 0 & \frac{1}{C} \\ -\frac{1}{L} & -\frac{R}{L} \end{bmatrix} \begin{bmatrix} x_1 \\ x_2 \end{bmatrix} + \begin{bmatrix} 0 \\ \frac{1}{L} \end{bmatrix} [u] \tag{2.34}$$

It can be expressed in a more compact form as

$$\dot{\mathbf{x}} = \mathbf{A}\mathbf{x} + \mathbf{b}\mathbf{u} \tag{2.35}$$

where $\dot{\mathbf{x}} = \begin{bmatrix} \dot{x}_1 \\ \dot{x}_2 \end{bmatrix}$ an $n \times 1$ column vector

$\mathbf{A} = \begin{bmatrix} a_{11} & a_{12} \\ a_{21} & a_{22} \end{bmatrix} = \begin{bmatrix} 0 & \frac{1}{C} \\ -\frac{1}{L} & -\frac{R}{L} \end{bmatrix}$ an $n \times n$ plant coefficient matrix

$\mathbf{x} = \begin{bmatrix} x_1 \\ x_2 \end{bmatrix}$ an $n \times 1$ state vector

$\mathbf{b} = \begin{bmatrix} b_1 \\ b_2 \end{bmatrix} = \begin{bmatrix} 0 \\ \frac{1}{L} \end{bmatrix}$ an $n \times 1$ control matrix

and, in this case, $\mathbf{u} = [u]$ is a one-dimensional control vector. In Eq. (2.34) matrices **A** and **x** are conformable.

If the output quantity $y(t)$ for the circuit of Fig. 2.2 is the voltage across the capacitor v_C, then

$$y(t) = v_C = x_1$$

Thus the matrix *system output* equation for this example is

$$\mathbf{y}(t) = \mathbf{c}^T\mathbf{x} + \mathbf{d}\mathbf{u} = [1 \quad 0]\begin{bmatrix} x_1 \\ x_2 \end{bmatrix} \tag{2.36}$$

where

$$\mathbf{c}^T = [1 \quad 0]$$

is a 1×2 row vector, $\mathbf{y} = [y]$ is a one-dimensional output vector, and $\mathbf{d} = \mathbf{0}$.

Equations (2.35) and (2.36) are for a single-input single-output (SISO) system. For a multiple-input multiple-output (MIMO) system, with m inputs and l outputs, these equations become

$$\dot{\mathbf{x}} = \mathbf{A}\mathbf{x} + \mathbf{B}\mathbf{u} \tag{2.37}$$

$$\mathbf{y} = \mathbf{C}\mathbf{x} + \mathbf{D}\mathbf{u} \tag{2.38}$$

where $\mathbf{B} = n \times m$ control matrix
$\mathbf{C} = l \times n$ output matrix
$\mathbf{D} = l \times m$ feedforward matrix
$\mathbf{u} = m$-dimensional control vector
$\mathbf{y} = l$-dimensional output vector

Example 3. Obtain the state equation for the circuit of Fig. 2.6, where i_2 is considered to be the output of this system. The assigned state variables are $x_1 = i_1$, $x_2 = i_2$, and $x_3 = v_C$. Thus two loop and one node equations are written

$$R_1x_1 + L_1\dot{x}_1 + x_3 = u \tag{2.39}$$

$$-x_3 + L_2\dot{x}_2 + R_2x_2 = 0 \tag{2.40}$$

$$-x_1 + x_2 + C\dot{x}_3 = 0 \tag{2.41}$$

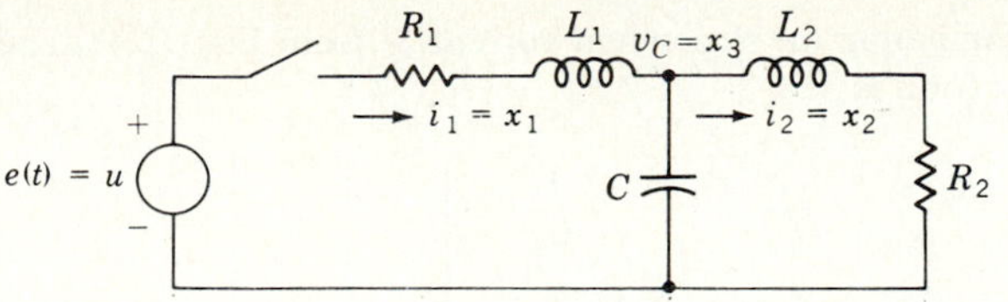

FIGURE 2.6
An electric circuit.

The three state variables are independent, and the system state and output equations are

$$\dot{\mathbf{x}} = \begin{bmatrix} -\dfrac{R_1}{L_1} & 0 & -\dfrac{1}{L_1} \\ 0 & -\dfrac{R_2}{L_2} & \dfrac{1}{L_2} \\ \dfrac{1}{C} & -\dfrac{1}{C} & 0 \end{bmatrix} \mathbf{x} + \begin{bmatrix} \dfrac{1}{L_1} \\ 0 \\ 0 \end{bmatrix} \mathbf{u} \tag{2.42}$$

$$\mathbf{y} = [0 \quad 1 \quad 0]\mathbf{x} \tag{2.43}$$

Example 4. Obtain the state equations for the circuit of Fig. 2.7. The output is the voltage v_1. The input or control variable is a current source $i(t)$. The assigned state variables are i_1, i_2, i_3, v_1, and v_2. Three loop equations and two node equations are written:

$$v_1 = L_1\, Di_1 \tag{2.44}$$

$$v_2 = L_2\, Di_2 + v_1 \tag{2.45}$$

$$v_2 = L_3\, Di_3 \tag{2.46}$$

$$i_2 = C_1\, Dv_1 + i_1 \tag{2.47}$$

$$i = i_3 + C_2\, Dv_2 + i_2 \tag{2.48}$$

Substituting from Eqs. (2.44) and (2.46) into Eq. (2.45), or writing the loop equation through L_1, L_2, and L_3 and then integrating (multiplying by $1/D$), gives

$$L_3 i_3 = L_2 i_2 + L_1 i_1 + K \tag{2.49}$$

where K is a function of the initial conditions. This equation reveals that one inductor current is dependent upon the other two inductor currents. Thus, this circuit has only *four* independent physical state variables, two inductor currents and two capacitor voltages. The four independent state variables are designated as $x_1 = v_1$, $x_2 = v_2$, $x_3 = i_1$, and $x_4 = i_2$, and the control variable is $u = i$. Three state equations are obtainable from Eqs. (2.44), (2.45), and (2.47). The fourth

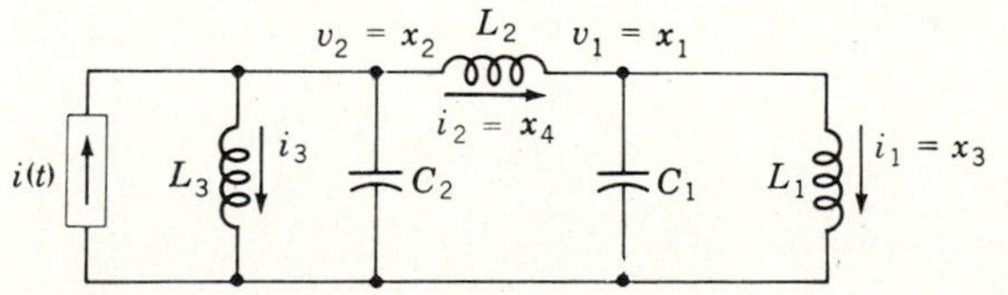

FIGURE 2.7
Electric circuit.

equation is obtained by eliminating the dependent current i_3 from Eqs. (2.48) and (2.49). The result in matrix form is

$$\dot{\mathbf{x}} = \begin{bmatrix} 0 & 0 & -\dfrac{1}{C_1} & \dfrac{1}{C_1} \\ 0 & 0 & -\dfrac{L_1}{L_3C_2} & -\dfrac{L_2 + L_3}{L_3C_2} \\ \dfrac{1}{L_1} & 0 & 0 & 0 \\ -\dfrac{1}{L_2} & \dfrac{1}{L_2} & 0 & 0 \end{bmatrix} \mathbf{x} + \begin{bmatrix} 0 \\ \dfrac{1}{C_2} \\ 0 \\ 0 \end{bmatrix} \mathbf{u} \tag{2.50}$$

$$\mathbf{y} = [1 \quad 0 \quad 0 \quad 0]\mathbf{x} \tag{2.51}$$

The dependence of i_3 on i_1 and i_2 as shown by Eq. (2.49) may not be readily observed. In that case the matrix state equation for this example would be written with five state variables.

The examples considered in this section are fairly straightforward. In general it is necessary to write more than just the number of state equations because other system variables appear in them. These equations are solved simultaneously to eliminate all internal variables in the circuit except for the state variables. This is necessary in some of the problems for this chapter. For more complex circuits it is possible to introduce more generalized and systematized methods using linear graphs for obtaining the state equations. They are not included in this book.

2.5 TRANSFER FUNCTION AND BLOCK DIAGRAM

A quantity that plays a very important role in control theory is the system *transfer function*, defined as follows:

TRANSFER FUNCTION. If the differential equation which describes the system is known and is linear, the ratio of the output variable to the input variable, where the variables are expressed as functions of the D operator, is called the transfer function.

Consider the system output $v_C = y$ in the RLC circuit of Fig. 2.2. Substituting $i = C\,Dv_C$ into Eq. (2.10) yields

$$(LCD^2 + RCD + 1)v_C(t) = e(t) \tag{2.52}$$

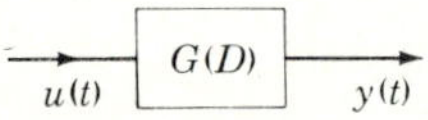

FIGURE 2.8
Block-diagram representation of Fig. 2.2.

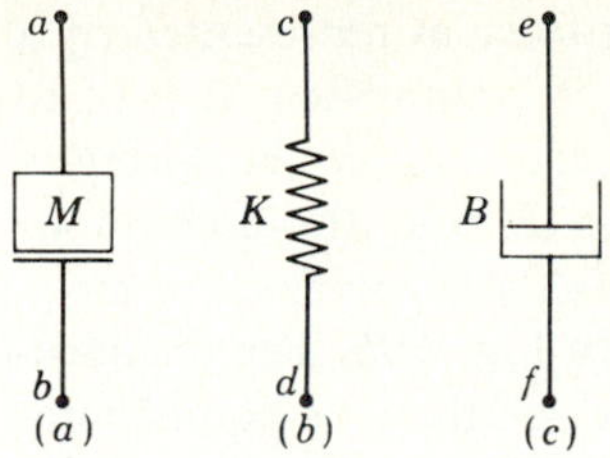

FIGURE 2.9
Network elements of mechanical translation systems.

The system transfer function is

$$G(D) = \frac{y(t)}{u(t)} = \frac{v_C(t)}{e(t)} = \frac{1}{LCD^2 + RCD + 1} \tag{2.53}$$

The notation $G(D)$ is used to denote a transfer function when it is expressed in terms of the D operator. It may also be written simply as G.

The *block-diagram* representation of this system (Fig. 2.8) represents the mathematical operation $G(D)u(t) = y(t)$; that is, the transfer function times the input is equal to the output of the block. The resulting equation is the differential equation of the system.

2.6 MECHANICAL TRANSLATION SYSTEMS

Mechanical systems obey the basic law that the sum of the forces equals zero. This is known as Newton's law and may be restated as follows: the sum of the applied forces must be equal to the sum of the reactive forces. The following analysis includes only linear functions. The three qualities characterizing elements in a mechanical translation† system are mass, elastance, and damping. Static friction, coulomb friction, and other nonlinear friction terms are not included. Basic elements entailing these qualities are represented as network elements,[8] and a mechanical network is drawn for each mechanical system to facilitate writing the differential equations.

The mass M is the inertial element. A force applied to a mass produces an acceleration of the mass. The reaction force f_M is equal to the product of mass and acceleration and is opposite in direction to the applied force. In terms of displacement x, velocity v, and acceleration a, the force equation is

$$f_M = Ma = MDv = MD^2x \tag{2.54}$$

The network representation of mass is shown in Fig. 2.9*a*. One terminal, *a*, has the motion of the mass; and the other terminal, *b*, is considered to have the motion of the reference. The reaction force f_M is a function of time and acts "through" M.

† Translation means motion in a straight line.

The elastance, or stiffness, K provides a restoring force as represented by a spring. Thus, if stretched, the string tries to contract; if compressed, it tries to expand to its normal length. The reaction force f_K on each end of the spring is the same and is equal to the product of the stiffness K and the amount of deformation of the spring.

The network representation of a spring is shown in Fig. 2.9*b*. The displacement of each end of the spring is measured from the original or equilibrium position. End c has a position x_c, and end d has a position x_d, measured from the respective equilibrium positions. The force equation, in accordance with Hooke's law, is

$$f_K = K(x_c - x_d) \tag{2.55}$$

If the end d is stationary, then $x_d = 0$ and the above equation reduces to

$$f_K = Kx_c \tag{2.56}$$

The plot f_K vs. x_c for a spring is not usually a straight line because the spring characteristic is nonlinear. However, over a limited region of operation, the linear approximation, i.e., a constant value for K, gives satisfactory results.

The damping, or viscous friction, B characterizes the element that absorbs energy. The damping force is proportional to the difference in velocity of two bodies, and the assumption is made that the viscous friction is linear. This assumption simplifies the solution of the dynamic equation. The network representation of damping action is a dashpot, as shown in Fig. 2.9*c*. It should be realized that damping may either be intentional or occur unintentionally and is present because of physical construction.

The reaction damping force f_B is approximated by the product of damping B and the relative velocity of the two ends of the dashpot. The direction of this force, given by Eq. (2.57), depends on the relative magnitudes and directions of the velocities Dx_e and Dx_f:

$$f_B = B(v_e - v_f) = B(Dx_e - Dx_j) \tag{2.57}$$

Damping may be added to a system by use of a dashpot. The basic operation of a dashpot, in which the housing is filled with an incompressible fluid, is shown in Fig. 2.10. If a force f is applied to the shaft, the piston presses against the fluid, increasing the pressure of side b and decreasing the pressure on side a. As a result, the fluid flows around the piston from side b to side a. If necessary, a small hold can be drilled through the piston to provide a positive

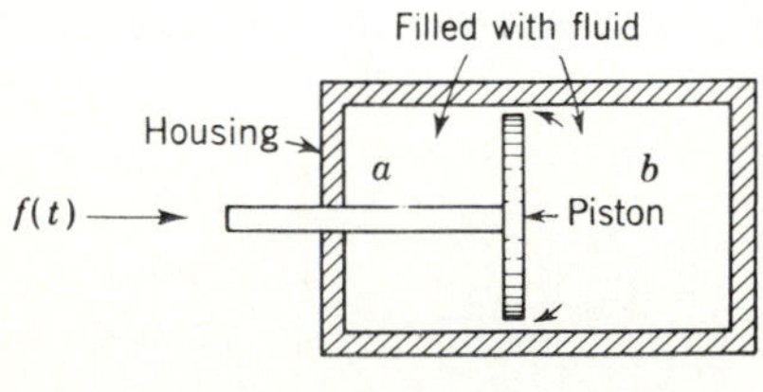

FIGURE 2.10
Dashpot construction.

TABLE 2.3
Mechanical translation symbols and units

Symbol	Quantity	U.S. customary units	Metric units
f	Force	Pounds	Newtons
x	Distance	Feet	Meters
v	Velocity	Feet/second	Meters/second
a	Acceleration	Feet/second^2	Meters/second^2
M†	Mass	Slugs $= \dfrac{\text{pound-seconds}^2}{\text{foot}}$	Kilograms
K	Stiffness coefficient	Pounds/foot	Newtons/meter
B	Damping coefficient	Pounds/(foot/second)	Newtons/(meter/second)

† Mass M in the U.S. customary system above has the dimensions of slugs. Sometimes it is given in units of pounds. If so, then in order to use the consistent set of units above, the mass must be expressed in slugs by using the conversion factor 1 slug = 32.2 lb.

path for the flow of fluid. The force required to move the piston inside the housing is given by Eq. (2.57), where the damping B depends on the dimensions and the fluid used.

Before writing the differential equations of a complete system, the first step is to draw the mechanical network. This is done by connecting the terminals of elements that have the same displacement. Then the force equation is written for each node or position by equating the sum of the forces at each position to zero. The equations are similar to the node equations in an electric circuit, with force analogous to current, velocity analogous to voltage, and the mechanical elements with their appropriate operators analogous to admittance. Several examples are shown in the following pages. The reference position in all cases should be taken from the static equilibrium positions. The force of gravity therefore does not appear in the system equations. The U.S. customary and metric systems of units are shown in Table 2.3.

Simple Mechanical Translation System

The system shown in Fig. 2.11 is initially at rest. The end of the spring and the mass have positions denoted as the reference positions, and any displacements from these reference positions are labeled x_a and x_b, respectively. A force f applied at the end of the spring must be balanced by a compression of the spring. The same force is also transmitted through the spring and acts at point x_b.

To draw the mechanical network, the points x_a and x_b and the reference are located. The network elements are then connected between these points. For example, one end of the spring has the position x_a, and the other end has the position x_b. Therefore the spring is connected between these points. The complete mechanical network is drawn in Fig. 2.11*b*.

The displacements x_a and x_b are nodes of the circuit. At each node the sum of the forces must add to zero. Accordingly, the equations may be written

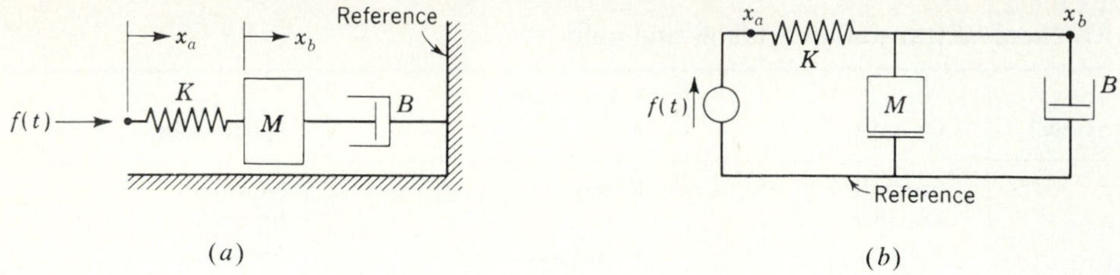

FIGURE 2.11
(*a*) Simple mass-spring-damper mechanical system; (*b*) corresponding mechanical network.

for nodes *a* and *b* as

$$f = f_K = K(x_a - x_b) \tag{2.58}$$

$$f_K = f_M + f_B = MD^2x_b + BDx_b \tag{2.59}$$

These two equations can be solved for the two displacements x_a and x_b and their respective velocities $v_a = Dx_a$ and $v_b = Dx_b$.

It is possible to obtain one equation relating x_a to f, x_b to x_a, or x_b to f by combining Eqs. (2.58) and (2.59):

$$K(MD^2 + BD)x_a = (MD^2 + BD + K)f \tag{2.60}$$

$$(MD^2 + BD + K)x_b = Kx_a \tag{2.61}$$

$$(MD^2 + BD)x_b = f \tag{2.62}$$

The solution of Eq. (2.61) shows the motion x_b resulting from a given motion x_a. Also, the solutions of Eqs. (2.60) and (2.62) show the motions x_a and x_b, respectively, resulting from a given force f. From each of these three equations the following transfer functions are obtained:

$$G_1 = \frac{x_a}{f} = \frac{MD^2 + BD + K}{K(MD^2 + BD)} \tag{2.63}$$

$$G_2 = \frac{x_b}{x_a} = \frac{K}{MD^2 + BD + K} \tag{2.64}$$

$$G = \frac{x_b}{f} = \frac{1}{MD^2 + BD} \tag{2.65}$$

Note that the last equation is equal to the product of the first two, i.e.,

$$G = G_1G_2 = \frac{x_a}{f}\frac{x_b}{x_a} = \frac{x_b}{f} \tag{2.66}$$

The block diagram representing the mathematical operation of Eq. (2.66) is shown in Fig. 2.12. Figure 2.12*a* is a detailed representation which indicates all variables in the system. The two blocks G_1 and G_2 are said to be in *cascade*. Figure 2.12*b*, called the *overall block-diagram* representation, shows only the

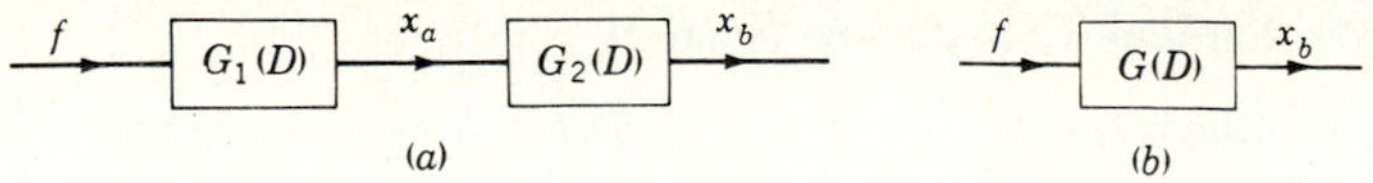

FIGURE 2.12
(*a*) Detailed and (*b*) overall block-diagram representation of Fig. 2.11.

input f and the output x_b, where x_b is considered the output variable of the system of Fig. 2.11.

The multiplication of transfer functions, as in Eq. (2.66), is valid as long as there is no coupling or loading between the two blocks in Fig. 2.12*a*. The signal x_a is unaffected by the presence of the block having the transfer function G_2; thus the multiplication is valid. When electric circuits are coupled, the transfer functions may not be independent unless they are isolated by an electronic amplifier with a very high input impedance.

Example 1. Determine the state equations for Eq. (2.59).

Equation (2.59) involves only one energy-storage element, the mass M, whose energy variable is v_b. The output quantity in this system is the position $y = x_b$. Since this quantity is not one of the physical or energy-related state variables, it is necessary to increase the number of state variables to 2, i.e., $x_b = x_1$ is an augmented state variable. Equation (2.59) is of second order, which confirms that two state variables are required. Note that the spring constant does not appear in this equation since the force f is transmitted through the spring to the mass. With $x_1 = x_b$, $x_2 = v_b = \dot{x}_1$, $u = f$, and $y = x_1$, the state and output equations are

$$\dot{\mathbf{x}} = \begin{bmatrix} 0 & 1 \\ 0 & -\dfrac{B}{M} \end{bmatrix}\mathbf{x} + \begin{bmatrix} 0 \\ \dfrac{1}{M} \end{bmatrix}\mathbf{u} \qquad \mathbf{y} = [1 \quad 0]\mathbf{x} \tag{2.67}$$

Example 2. Determine the state equations for Eq. (2.61).

Equation (2.61) involves two energy-storage elements, K and M, whose energy variables are x_b and v_b, respectively. Note that the spring constant K does appear in this equation since x_a is the input u. Therefore a state variable must be associated with this energy element. Thus the assigned state variables are x_b and $v_b = Dx_b$, which are independent state variables of this system. Let $x_1 = x_b = y$ and $x_2 = v_b = \dot{x}_1$. Equation (2.61) converted into state-variable form yields the two equations

$$\dot{x}_1 = x_2 \qquad \dot{x}_2 = -\frac{K}{M}x_1 - \frac{B}{M}x_2 + \frac{K}{M}u \tag{2.68}$$

or, in matrix form,

$$\dot{\mathbf{x}} = \mathbf{Ax} + \mathbf{bu}$$

where

$$\mathbf{A} = \begin{bmatrix} 0 & 1 \\ -\dfrac{K}{M} & -\dfrac{B}{M} \end{bmatrix} \qquad \mathbf{b} = \begin{bmatrix} 0 \\ \dfrac{K}{M} \end{bmatrix}$$

The output equation expressed in terms of the state vector is

$$\mathbf{y} = [y] = [x_1] = [1 \quad 0]\begin{bmatrix} x_1 \\ x_2 \end{bmatrix} = \mathbf{c}^T\mathbf{x} \tag{2.69}$$

Multiple-Element Mechanical Translation System

A force $f(t)$ is applied to the mass M_1 of Fig. 2.13*a*. The sliding friction between the masses M_1 and M_2 and the surface is indicated by the viscous friction coefficients B_1 and B_2. The system equations can be written in terms of the two displacements x_a and x_b. The mechanical network is drawn by connecting the terminals of the elements that have the same displacement (see Fig. 2.13*b*). Since the forces at each node must add to zero, the equations are written according to the rules for nodes equations:

For node a: $$\left(M_1D^2 + B_1D + B_3D + K_1\right)x_a - \left(B_3D\right)x_b = f. \qquad (2.70)$$

For node b: $$-\left(B_3D\right)x_a + \left(M_2D^2 + B_2D + B_3D + K_2\right)x_b = 0 \qquad (2.71)$$

A definite pattern to these equations can be detected. Observe that K_1, M_1, B_1, and B_3 are connected to node a and that Eq. (2.70), for node a, contains all four of these terms as coefficients of x_a. Notice that element B_3 is also connected to node b and that the term $-B_3$ appears as a coefficient of x_b. When this pattern is used, Eq. (2.71) can be written directly. Thus, since K_2, M_2, B_2, and B_3 are connected to node b, they appear as coefficients of x_b. B_3 is also connected to node a, and $-B_3$ appears as the coefficient of x_a. Each term in the equation must be a force.

The node equations for a mechanical system follow directly from the mechanical network of Fig. 2.13*b*. They are similar in form to the node equations for an electric circuit and follow the same rules.

The block diagram representing the system in Fig. 2.13 is also given by Fig. 2.12, where again x_b is considered to be the system output. The transfer functions G_1 and G_2 are obtained by solving the equations for this system.

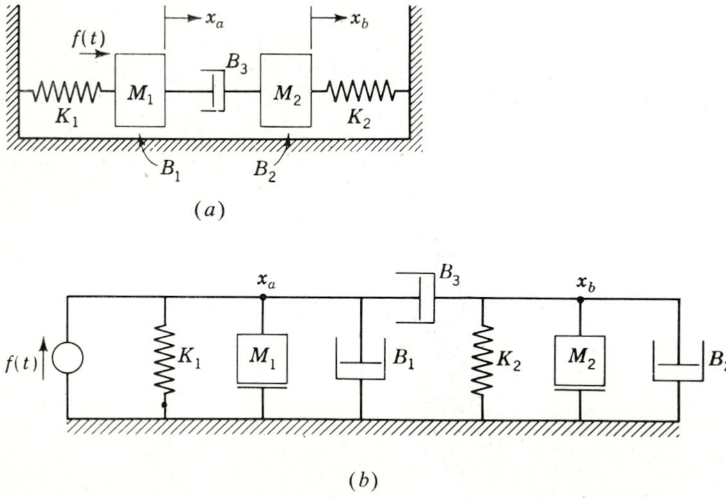

FIGURE 2.13
(*a*) Multiple-element mechanical system; (*b*) corresponding mechanical network.

Example 3. Obtain the state equations for the system of Fig. 2.13, where x_b is the system output. There are four energy-storage elements, thus the four assigned state variables are x_a, x_b, Dx_a, and Dx_b. Analyzing Eqs. (2.70), and (2.71) shows that this is a fourth-order system. Let

$$x_1 = x_b \quad \text{for spring } K_2 \qquad x_2 = \dot{x}_1 = v_b \quad \text{for mass } M_2$$

$$x_3 = x_a \quad \text{for spring } K_1 \qquad x_4 = \dot{x}_3 = v_a \quad \text{for mass } M_1$$

$$u = f \qquad y = x_b = x_1$$

The four state variables are independent, i.e., no one state variable can be expressed in terms of the remaining state variables. The system equations are

$$\dot{\mathbf{x}} = \mathbf{A}\mathbf{x} + \mathbf{b}u \qquad y = \mathbf{c}^T\mathbf{x} \tag{2.72}$$

where
$$\mathbf{A} = \begin{bmatrix} 0 & 1 & 0 & 0 \\ -\dfrac{K_2}{M_2} & -\dfrac{B_2 + B_3}{M_2} & 0 & \dfrac{B_3}{M_2} \\ 0 & 0 & 0 & 1 \\ 0 & \dfrac{B_3}{M_1} & -\dfrac{K_1}{M_1} & -\dfrac{B_1 + B_3}{M_1} \end{bmatrix}, \qquad \mathbf{b} = \begin{bmatrix} 0 \\ 0 \\ 0 \\ \dfrac{1}{M_1} \end{bmatrix}$$

$$\mathbf{c}^T = [1 \quad 0 \quad 0 \quad 0]$$

2.7 ANALOGOUS CIRCUITS

Analogous circuits represent systems for which the differential equations have the same form. The corresponding variables and parameters in two circuits represented by equations of the same form are called *analogs*. An electric circuit can be drawn that looks like the mechanical circuit and is represented by node equations that have the same mathematical form as the mechanical equations. The analogs are listed in Table 2.4.

In this table the force f and the current i are analogs and are classified as "through" variables. There is a physical similarity since a measuring instrument must be placed in series in both cases; i.e., an ammeter and a force indicator must be placed in series with the system. Also, the velocity "across" a mechanical element is analogous to voltage across an electrical element. Again, the physical

TABLE 2.4
Electrical and mechanical analogs

Mechanical translation element		Electrical element	
Symbol	**Quantity**	**Symbol**	**Quantity**
f	Force	i	Current
$v = Dx$	Velocity	e or v	Voltage
M	Mass	C	Capacitance
K	Stiffness coefficient	$1/L$	Reciprocal inductance
B	Damping coefficient	$G = 1/R$	Conductance

similarity is present since a measuring instrument must be placed across the system in both cases. A voltmeter must be placed across a circuit to measure voltage; it must have a point of reference. A velocity indicator must also have a point of reference. Nodes in the mechanical network are analogous to nodes in the electric network.

2.8 MECHANICAL ROTATIONAL SYSTEMS

The equations characterizing rotational systems are similar to those for translation systems. Writing torque equations parallels the writing of force equations, with the displacement, velocity, and acceleration terms now angular quantities. The applied torque is equal to the sum of the reaction torques. The three elements in a rotational system are inertia, the spring, and the dashpot. The mechanical-network representation of these elements is shown in Fig. 2.14.

The torque applied to a body having a moment of inertia J produces an angular acceleration. The reaction torque T_J is opposite to the direction of the applied torque and is equal to the product of moment of inertia and acceleration. In terms of angular displacement θ, angular velocity ω, or angular acceleration α, the torque equation is

$$T_J = J\alpha = J\,D\omega = J\,D^2\theta \tag{2.73}$$

When a torque is applied to a spring, the spring is twisted by an angle θ. The applied torque is transmitted through the spring and appears at the other end. The reaction spring torque T_K that is produced is equal to the product of the *stiffness*, or *elastance*, K of the spring and the angle of twist. By denoting the positions of the two ends of the spring, measured from the neutral position, as θ_c and θ_d, the reaction torque is

$$T_K = K(\theta_c - \theta_d) \tag{2.74}$$

Damping occurs whenever a body moves through a fluid, which may be a liquid or a gas such as air. To produce motion of the body a torque must be applied to overcome the reaction damping torque. The damping is represented as a dashpot with a *viscous friction coefficient* B. The damping torque T_B is equal to the product of damping B and the relative angular velocity of the ends of the

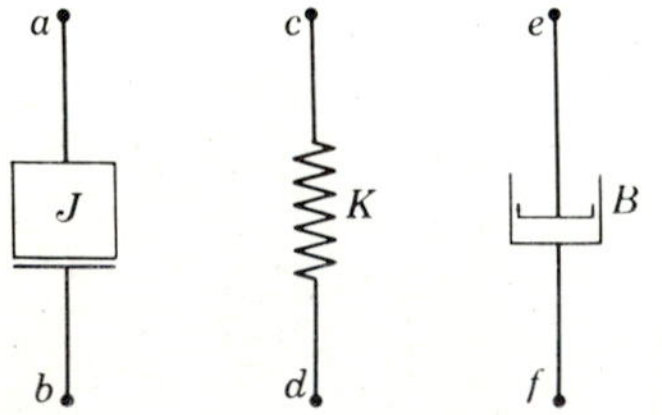

FIGURE 2.14
Network elements of mechanical rotational systems.

TABLE 2.5
Mechanical rotational symbols and units

Symbol	Quantity	U.S. customary units	Metric units
T	Torque	Pound-feet	Newton-meters
θ	Angle	Radians	Radians
ω	Angular velocity	Radians/second	Radians/second
α	Angular acceleration	Radians/second^2	Radians/second^2
J†	Moment of inertia	Slug-feet^2 (or pound-feet-seconds^2)	Kilogram-meters^2
K	Stiffness coefficient	Pound-feet/radian	Newton-meters/radian
B	Damping coefficient	Pound-feet/(radian/second)	Newton-meters/(radians/second)

† The moment of inertia J has the dimensions of mass-distance^2, which have the units slug-feet^2 in the U.S. customary system. Sometimes the units are given as pound-feet^2. To use the consistent set of units above, the moment of inertia must be expressed in slug-feet^2 by using the conversion factor 1 slug = 32.2 lb.

dashpot. The reaction torque of a damper is

$$T_B = B(\omega_e - \omega_f) = B(D\theta_e - D\theta_f) \tag{2.75}$$

The work of writing the differential equations can be simplified by first drawing the mechanical network for the system. This is done by first designating nodes that correspond to each angular displacement. Then each element is connected between the nodes that correspond to the two motions at each end of that element. The inertia elements are connected from the reference node to the node representing its position. The spring and dashpot elements are connected to the two nodes that represent the position of each end of the element. Then the torque equation is written for each node by equating the sum of the torques at each node to zero. These equations are similar to those for mechanical translation and are analogous to those for electric circuits. Several examples are shown in the following pages. The units for these elements are given in Table 2.5.

An electrical analog can be obtained for a mechanical rotational system in the manner described in Sec. 2.7. The changes to Table 2.4 are due only to the fact that rotational quantities are involved. Thus torque is the analog of current, and the moment of inertia is the analog of capacitance.

Simple Mechanical Rotational System

The system shown in Fig. 2.15 has a mass with a moment of inertia J immersed in a fluid. A torque T is applied to the mass. The wire produces a reactive torque proportional to its stiffness K and to the angle of twist. The fins moving through the fluid have a damping B, which requires a torque proportional to the rate at which they are moving. The mechanical network is drawn in Fig. 2.15b. There is

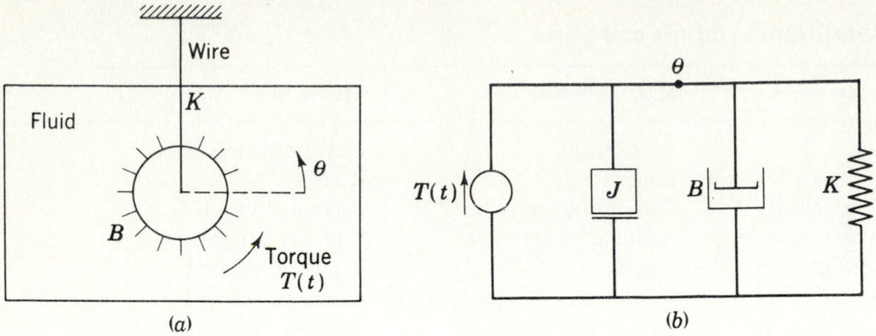

FIGURE 2.15
(*a*) Simple rotational system; (*b*) mechanical network.

one node having a displacement θ; therefore, only one equation is necessary:

$$J D^2\theta + B D\theta + K\theta = T(t) \tag{2.76}$$

Multiple-Element Mechanical Rotational System

The system represented by Fig. 2.16*a* has two disks which have damping between them and also between each of them and the frame. The mechanical network is drawn in Fig. 2.16*b* by connecting the terminals of those elements which have the same angular displacement. The torques at each node must add to zero.

The equations are written directly in systematized form:

Node 1: $$K_1\theta_1 - K_1\theta_2 = T(t) \tag{2.77}$$

Node 2: $$-K_1\theta_1 + [J_1D^2 + (B_1 + B_3)D + K_1]\theta_2 - (B_3D)\theta_3 = 0 \tag{2.78}$$

Node 3: $$-(B_3D)\theta_2 + [J_2D^2 + (B_2 + B_3)D + K_2]\theta_3 = 0 \tag{2.79}$$

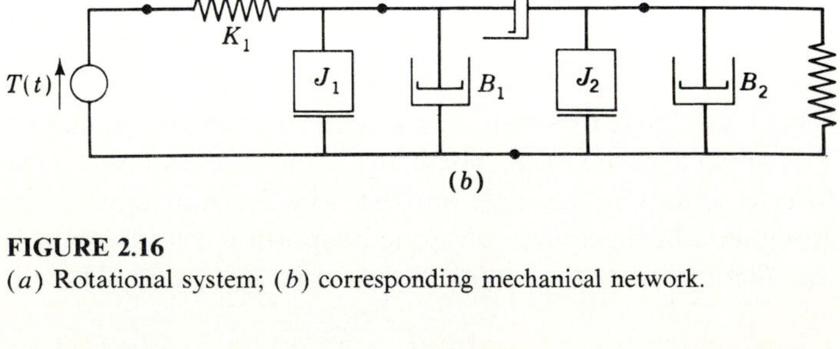

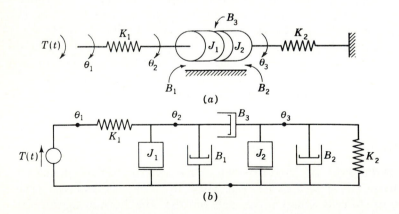

FIGURE 2.16
(*a*) Rotational system; (*b*) corresponding mechanical network.

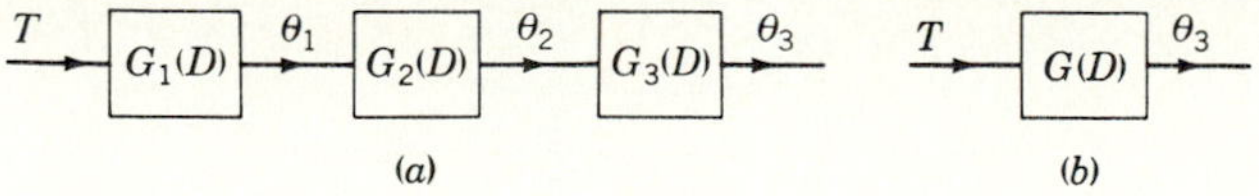

FIGURE 2.17
Detailed and overall block-diagram representations of Fig. 2.16.

These three equations can be solved simultaneously for θ_1, θ_2, and θ_3 as a function of the applied torque. If θ_3 is the system output, these equations can be solved for the following four transfer functions:

$$G_1 = \frac{\theta_1}{T} \qquad G_2 = \frac{\theta_2}{\theta_1} \qquad G_3 = \frac{\theta_3}{\theta_2} \qquad G = G_1G_2G_3 = \frac{\theta_1}{T}\frac{\theta_2}{\theta_1}\frac{\theta_3}{\theta_2} = \frac{\theta_3}{T} \tag{2.80}$$

The detailed and overall block-diagram representations of Fig. 2.16 are shown in Fig. 2.17. The overall transfer function G given by Eqs. (2.80) is the product of the transfer functions, which are said to be *in cascade*. This is general and applies to any number of elements in cascade when there is no loading between the blocks.

Example. Determine the state variables for the system of Fig. 2.16, where θ_3 is the system output y and torque T is the system input u. The spring K_1 is effectively not in the system because the torque is transmitted through the spring. Since there are three energy-storage elements J_1, J_2 and K_2, the assigned state variables are $D\theta_2$, $D\theta_3$, and θ_3. Analyzing Eqs. (2.77) to (2.79) reveals that these variables are independent state variables. Therefore, let $x_1 = \theta_3$, $x_2 = D\theta_3$, and $x_3 = D\theta_2$. If the system input is $u = \theta_1$, then the spring K_1 is included, resulting in four state variables (see Prob. 2.19).

Effective Moment of Inertia and Damping of a Gear Train

When the load is coupled to a drive motor through a gear train, the moment of inertia and damping relative to the motor are important. Since the shaft length between gears is very short, the stiffness may be considered infinite. A simple representation of a gear train is shown in Fig. 2.18a. The following definitions are used:

N = number of teeth on each gear
$\omega = D\theta$ = velocity of each gear
n_a = ratio of (speed of driving shaft)/(speed of driven shaft)
θ = angular position

Typically, gearing may not mesh perfectly, and the teeth of one gear may not fill the full space between the gear teeth of the matching gear. As a consequence, there can be times when the drive gear is not in contact with the outer gear. This nonlinear characteristic is called *dead zone*. When the dead zone is very small, it is neglected for the linearized gearing model.

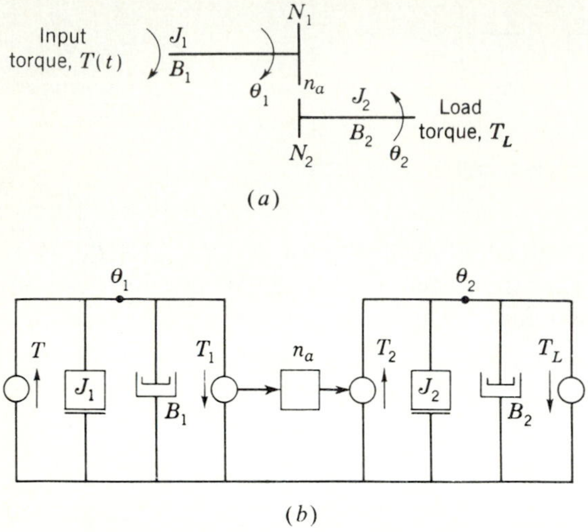

FIGURE 2.18
(a) Representation of a gear train; (b) corresponding mechanical network.

The mechanical network for the gear train is shown in Fig. 2.18b. At each gear pair two torques are produced. For example, a restraining torque T_1 is produced on gear 1 by the rest of the gear train. There is also produced a driving torque T_2 on gear 2. T_1 is the load on gear 1 produced by the rest of the gear train. T_2 is the torque transmitted to gear 2 to drive the rest of the gear train. These torques are inversely proportional to the speeds of the respective gears. The block labeled n_a between T_1 and T_2 is used to show the relationship between them; that is, $T_2 = n_a T_1$. A transformer may be considered as the electrical analog of the gear train, with angular velocity analogous to voltage and torque analogous to current. The equations describing the system are

$$J_1 D^2\theta_1 + B_1 D\theta_1 + T_1 = T \qquad J_2 D^2\theta_2 + B_2 D\theta_2 + T_L = T_2$$
$$n_a = \frac{\omega_1}{\omega_2} = \frac{\theta_1}{\theta_2} = \frac{N_2}{N_1} \qquad \theta_2 = \frac{\theta_1}{n_a} \qquad T_2 = n_a T_1 \tag{2.81}$$

The equations can be combined to produce

$$J_1 D^2\theta_1 + B_1 D\theta_1 + \frac{1}{n_a}\left(J_2 D^2\theta_2 + B_2 D\theta_2 + T_L\right) = T \tag{2.82}$$

This equation can be expressed in terms of the torque and the input position θ_1 only:

$$\left(J_1 + \frac{J_2}{n_a^2}\right) D^2\theta_1 + \left(B_1 + \frac{B_2}{n_a^2}\right) D\theta_1 + \frac{T_L}{n_a} = T \tag{2.83}$$

Equation (2.83) represents the system performance as a function of a single dependent variable. An equivalent system is one having an equivalent moment of inertia and damping equal to

$$J_{eq} = J_1 + \frac{J_2}{n_a^2} \tag{2.84}$$

$$B_{eq} = B_1 + \frac{B_2}{n_a^2} \tag{2.85}$$

If the solution for θ_2 is wanted, the equation can be altered by the substitution of $\theta_1 = n_a\theta_2$. This system can be generalized for any number of gear stages. When the gear-reduction ratio is large, the load moment of inertia may contribute a negligible value to the equivalent moment of inertia. When a motor rotor is geared to a long shaft that provides translational motion, then the elastance or stiffness K of the shaft must be taken into account in deriving the system's differential equations (see Prob. 2.27).

2.9 THERMAL SYSTEMS[9]

A limited number of thermal systems can be represented by linear differential equations. The basic requirement is that the temperature of a body be considered uniform. This approximation is valid when the body is small. Also, when the region consists of a body of air or liquid, the temperature can be considered uniform if there is perfect mixing of the fluid. The necessary condition of equilibrium requires that the heat added to the system equal the heat stored plus the heat carried away. This can also be expressed in terms of rate of heat flow.

The symbols shown in Table 2.6 are used for thermal systems. A thermal-system network is drawn for each system in which the thermal capacitance and thermal resistance are represented by network elements. From the thermal network the differential equations can be written.

The additional heat stored in a body whose temperature is raised from θ_1 to θ_2 is given by

$$h = \frac{q}{D} = C(\theta_2 - \theta_1) \tag{2.86}$$

TABLE 2.6
Thermal symbols and units

Symbol	Quantity	U.S. customary units	Metric units
q	Rate of heat flow	Btu/minute	Joules/second
M	Mass	Pounds	Kilograms
S	Specific heat	Btu/(pounds)(°F)	Joules/(kilogram)(°C)
C	Thermal capacitance $C = MS$	Btu/°F	Joules/°C
K	Thermal conductance	Btu/(minute)(°F)	Joules/(second)(°C)
R	Thermal resistance	Degrees/(Btu/minute)	Degrees/(joule/second)
θ	Temperature	°F	°C
h	Heat energy	Btu	Joules

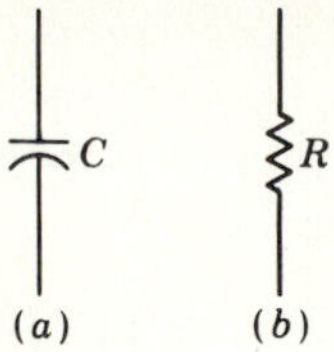

FIGURE 2.19
Network elements of thermal systems.

In terms of rate of heat flow, this equation can be written as

$$q = CD(\theta_2 - \theta_1)$$

The thermal capacitance determines the amount of heat stored in a body. It is analogous to the electric capacitance of a capacitor in an electric circuit, which determines the amount of charge stored. The network representation of thermal capacitance is shown in Fig. 2.19*a*.

Rate of heat flow through a body in terms of the two boundary temperatures θ_3 and θ_4 is

$$q = \frac{\theta_3 - \theta_4}{R} \tag{2.87}$$

The thermal resistance determines the rate of heat flow through the body. This is analogous to the resistance of a resistor in an electric circuit, which determines the current flow. The network representation of thermal resistance is shown in Fig. 2.19*b*. In the thermal network the temperature is analogous to potential.

Simple Mercury Thermometer

Consider a thin glass-walled thermometer filled with mercury which has stabilized at a temperature θ_1. It is plunged into a bath of temperature θ_0 at $t = 0$. In its simplest form, the thermometer can be considered to have a capacitance C which stores heat and a resistance R which limits the heat flow. The temperature of the mercury is θ_m. The flow of heat into the thermometer is

$$q = \frac{\theta_0 - \theta_m}{R}$$

The heat entering the thermometer is stored in the thermal capacitance and is given by

$$h = \frac{q}{D} = C(\theta_m - \theta_1)$$

These equations can be combined to form

$$\frac{\theta_0 - \theta_m}{RD} = C(\theta_m - \theta_1) \tag{2.88}$$

Differentiating Eq. (2.88) and rearranging terms gives

$$RC\,D\theta_m + \theta_m = \theta_0 \tag{2.89}$$

The thermal network is drawn in Fig. 2.20. The node equation for this circuit, with the temperature considered as a voltage, gives Eq. (2.89) directly. From this equation the transfer function $G = \theta_m/\theta_0$ may be obtained. Since there is one energy-storage element, the independent state variable is $x_1 = \theta_m$ and the input is

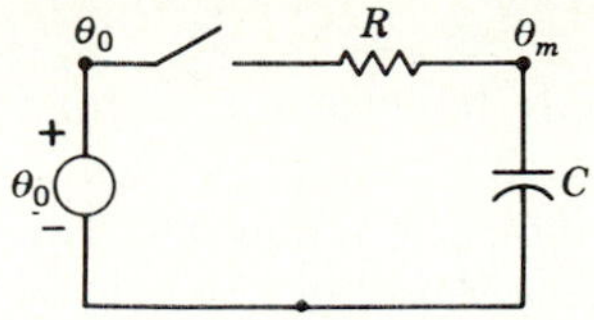

FIGURE 2.20
Simple network representation of a thermometer.

$u = \theta_0$. Thus the state equation is

$$\dot{x}_1 = -\frac{1}{RC}x_1 + \frac{1}{RC}u$$

More Exact Analysis of a Mercury Thermometer

A more exact analysis of the thermometer takes into account both the resistance R_g of the glass and the resistance R_m of the mercury. Also, the glass has a capacitance C_g, and the mercury has a capacitance C_m. The thermal network can be represented by Fig. 2.21, where θ_s represents the temperature at the inner surface between the glass case and the mercury. The temperature of the mercury is θ_m. The node equations can be written from the thermal network:

For node s:
$$\left(\frac{1}{R_g} + \frac{1}{R_m} + C_g D\right)\theta_s - \frac{1}{R_m}\theta_m = \frac{\theta_0}{R_g} \tag{2.90}$$

For node m:
$$-\frac{1}{R_m}\theta_s + \left(\frac{1}{R_m} + C_m D\right)\theta_m = 0 \tag{2.91}$$

The two equations can be combined to eliminate θ_s:

$$\left[C_g C_m D^2 + \left(\frac{C_g}{R_m} + \frac{C_m}{R_g} + \frac{C_m}{R_m}\right)D + \frac{1}{R_g R_m}\right]\theta_m = \frac{1}{R_m R_g}\theta_0 \tag{2.92}$$

Equations (2.90) to (2.92) yield the transfer functions $G_1 = \theta_s/\theta_0$, $G_2 = \theta_m/\theta_s$, and $G = \theta_m/\theta_0$, which can be represented, with the appropriate change in variable notation, by the block diagrams of Fig. 2.12.

The two energy-storage elements in Fig. 2.21 yield the assigned state variables θ_s and θ_m, which are independent. With $y = x_1 = \theta_m$, $x_2 = \theta_s$, and $u = \theta_0$, the state equation is

$$\dot{\mathbf{x}} = \begin{bmatrix} -\dfrac{1}{R_m C_m} & \dfrac{1}{R_m C_m} \\ \dfrac{1}{R_m C_g} & -\dfrac{1}{R_g C_g} - \dfrac{1}{R_m C_g} \end{bmatrix}\mathbf{x} + \begin{bmatrix} 0 \\ \dfrac{1}{R_g C_g} \end{bmatrix}\mathbf{u}$$

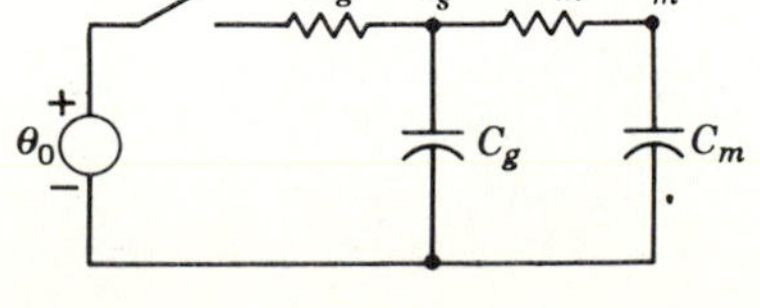

FIGURE 2.21
More exact network representation of a thermometer.

Simple Heat-Transfer System

The electric water heater used in many homes to supply hot water is a good example of a simple heat-transfer problem. A sketch of such a heater is shown in Fig. 2.22. The tank is insulated to reduce heat loss to the surrounding air. The electrical heating element is turned on and off by a thermostatic switch to maintain a reference temperature. A demand from any faucet in the house causes hot water to leave and cold water to enter the tank. The necessary simplifying assumptions are as follows:

1. There is no heat storage in the insulation. This is valid since the specific heat of the insulation is small and the water-temperature variation is small.
2. All the water in the tank is at a uniform temperature. This requires perfect mixing of the water.

Definitions of the system parameters and variations are as follows:

q = rate of heat flow of heating element
q_t = rate of heat flow into water in tank
q_0 = rate of heat flow carried out by hot water leaving tank
q_i = rate of heat flow carried in by cold water entering tank
q_e = rate of heat flow through tank insulation
θ = temperature of water in tank
θ_i = temperature of water entering tank
θ_a = temperature of air surrounding tank
C = thermal capacitance of water in tank
R = thermal resistance of insulation
n = water flow from tank
S = specific heat of water

There are three temperatures, θ, θ_i, and θ_a; therefore the thermal network must have three nodes. The network can be drawn first or after the equations are

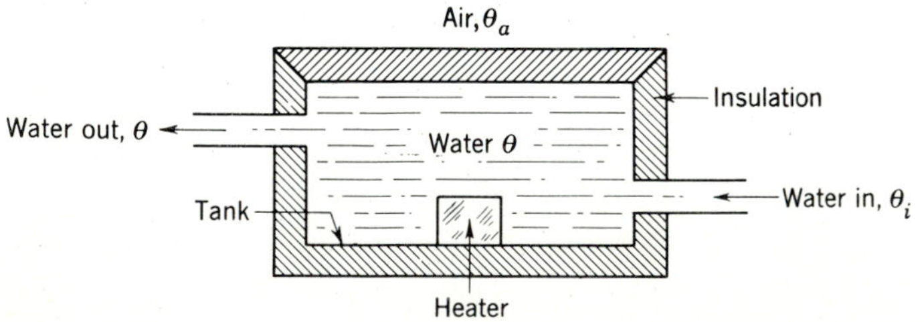

FIGURE 2.22
Electric water heater.

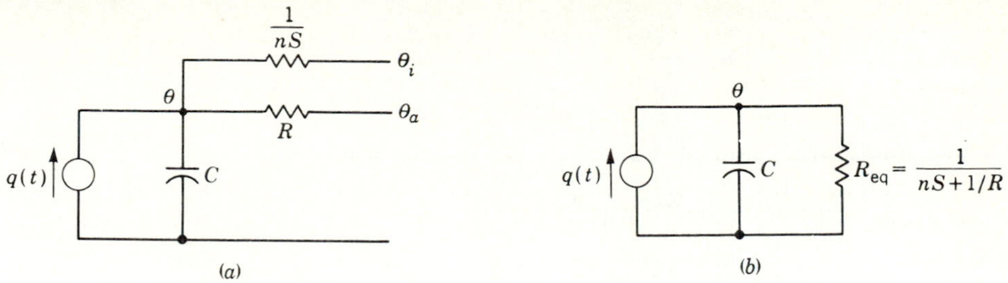

FIGURE 2.23
(a) Thermal network of an electric water heater; (b) simplified network.

written. The equilibrium equation for rate of heat flow is

$$q_t + q_0 - q_i + q_e = q \tag{2.93}$$

where $\quad q_t = C\,D\theta \qquad q_0 = nS\theta \qquad q_i = nS\theta_i \qquad q_e = \dfrac{\theta - \theta_a}{R}$

Combining these equations gives

$$C\,D\theta + nS(\theta - \theta_i) + \frac{\theta - \theta_a}{R} = q \tag{2.94}$$

The thermal network is shown in Fig. 2.23a. The heat loss, due to the difference $\theta - \theta_i$, can be represented as occurring through the resistor of value $1/nS$. In this thermal network the rate of heat flow from the heater is analogous to the current from a constant-current source in an electric circuit.

There are four variables in this system, θ, θ_i, θ_a, and n. Three of them must be specified in order to solve the problem. For the special case in which n is a constant and $\theta_a = \theta_i$, Eq. (2.94) can be simplified. In terms of θ, which now is the temperature above the reference θ_a, the equation is

$$C\,D\theta + \left(nS + \frac{1}{R}\right)\theta = q \tag{2.95}$$

The thermal network is drawn in Fig. 2.23b.

2.10 HYDRAULIC LINEAR ACTUATOR

The valve-controlled hydraulic actuator is used in many applications as a force amplifier. Very little force is required to position the valve, but a large output force is controlled. The hydraulic unit is relatively small, which makes its use very attractive. Figure 2.24 shows a simple hydraulic actuator in which motion of the valve regulates the flow of oil to either side of the main cylinder. An input motion x of a few thousandths of an inch results in a large change of oil flow. The resulting difference in pressure on the main piston causes motion of the output shaft. The oil flowing in is supplied by a source which maintains a constant high

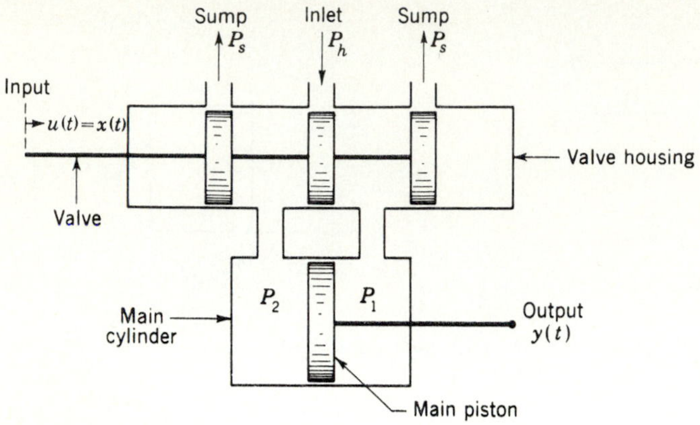

FIGURE 2.24
Hydraulic actuator.

pressure P_h, and the oil on the opposite side of the piston flows into the drain (sump) at low pressure P_s. The load-induced pressure P_L is the difference between the pressures on each side of the main piston:

$$P_L = P_1 - P_2 \tag{2.96}$$

The flow of fluid through an inlet orifice is given by[10]

$$q = ca\sqrt{2g\frac{\Delta p}{w}} \tag{2.97}$$

where c = orifice coefficient
a = orifice area
w = specific weight of fluid
Δp = pressure drop across orifice
g = gravitational acceleration constant
q = rate of flow of fluid

Simplified Analysis

As a first-order approximation, it can be assumed that the orifice coefficient and the pressure drop across the orifice are constant and independent of valve position. Also, the orifice area is proportional to the valve displacement x. Equation (2.97), which gives the rate of flow of hydraulic fluid through the valve, can be rewritten as

$$q = C_x x \tag{2.98}$$

where x is the displacement of the valve. The displacement of the main piston is directly proportional to the volume of fluid that enters the main cylinder, or, equivalently, the rate of flow of fluid is proportional to the rate of displacement of the output. When the compressibility of the fluid and the leakage around the valve and main piston are neglected, the equation of motion of the main piston is

$$q = C_b Dy \tag{2.99}$$

Combining the two equations gives

$$Dy = \frac{C_x}{C_b} x = C_1 x \tag{2.100}$$

This analysis is essentially correct when the load reaction is small.

More Complete Analysis

When the load reaction is not negligible, a more complete analysis should take into account the pressure drop across the orifice, the leakage of oil around the piston, and the compressibility of the oil.

The pressure drop Δp across the orifice is a function of the source pressure P_h and the load pressure P_L. Since P_h is assumed constant, the flow equation for q is a function of valve displacement x and load pressure P_L:

$$q = f(x, P_L) \tag{2.101}$$

The differential rate of flow dq, expressed in terms of partial derivatives, is

$$dq = \frac{\partial q}{\partial x} dx + \frac{\partial q}{\partial P_L} dP_L \tag{2.102}$$

If q, x, and P_L are measured from zero values as reference points, and if the partial derivatives are constant at the values they have at zero, the integration of Eq. (2.102) gives

$$q = \left(\frac{\partial q}{\partial x}\right)_0 x + \left(\frac{\partial q}{\partial P_L}\right)_0 P_L \tag{2.103}$$

By defining

$$C_x \equiv \left(\frac{\partial q}{\partial x}\right)_0 \qquad \text{and} \qquad C_p \equiv \left(\frac{-\partial q}{\partial P_L}\right)_0$$

the flow equation for fluid entering the main cylinder can be written as

$$q = C_x x - C_p P_L \tag{2.104}$$

Both C_x and C_p have positive values. A comparison with Eq. (2.98) shows that the load pressure reduces the flow into the main cylinder. The flow of fluid into the cylinder must satisfy the continuity conditions of equilibrium. This flow is equal to the sum of the components:

$$q = q_0 + q_l + q_c \tag{2.105}$$

where q_0 = incompressible component (causes motion of piston)
q_l = leakage component
q_c = compressible component

The component q_0, which produces a motion y of the main piston, is

$$q_0 = C_b\, Dy \tag{2.106}$$

The compressible component is derived in terms of the bulk modulus of elasticity of the fluid, which is defined as the ratio of incremental stress to incremental strain. Thus

$$K_B = \frac{\Delta P_L}{\Delta V/V}$$

Solving for ΔV and dividing both sides of the equation by Δt gives

$$\frac{\Delta V}{\Delta t} = \frac{V}{K_B}\frac{\Delta P_L}{\Delta t}$$

Taking the limit as Δt approaches zero and letting $q_c = dV/dt$ gives

$$q_c = \frac{V}{K_B}\, DP_L \tag{2.107}$$

where V is the effective volume of fluid under compression and K_B is the bulk modulus of the hydraulic oil. The volume V at the middle position of the piston stroke is often used in order to linearize the differential equation.

The leakage component is

$$q_l = LP_L \tag{2.108}$$

where L is the leakage coefficient of the whole system.

Combining these equations gives

$$q = C_x x - C_p P_L = C_b\, Dy + \frac{V}{K_B}\, DP_L + LP_L \tag{2.109}$$

and rearranging terms gives

$$C_b\, Dy + \frac{V}{K_B}\, DP_L + (L + C_p)P_L = C_x x \tag{2.110}$$

The force developed by the main piston is

$$F = n_F A P_L = C P_L \tag{2.111}$$

where n_F is the force conversion efficiency of the unit and A is the area of the main actuator piston.

An example of a specific type of load consisting of a mass and a dashpot is shown in Fig. 2.25. The equation for this system is obtained by equating the force produced by the piston, which is given by Eq. (2.111) to the reactive load forces:

$$F = M D^2 y + B\, Dy = C P_L \tag{2.112}$$

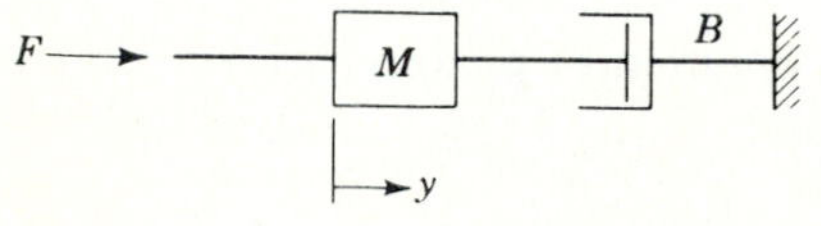

FIGURE 2.25
Load on a hydraulic piston.

Substituting the value of P_L from Eq. (2.112) into Eq. (2.110) gives the equation relating the input motion x to the response y:

$$\frac{MV}{CK_B} D^3y + \left[\frac{BV}{CK_B} + \frac{M}{C}(L + C_p)\right] D^2y + \left[C_b + \frac{B}{C}(L + C_p)\right] Dy = C_x x \tag{2.113}$$

The analysis above is based on perturbations about the reference set of values $x = 0$, $q = 0$, $P_L = 0$. For the entire range of motion x of the valve, the quantities $\partial q/\partial x$ and $-\partial q/\partial P_L$ can be determined experimentally. Although they are not constant at values equal to the values C_x and C_p at the zero reference point, average values can be assumed in order to simulate the system by linear equations. For conservative design the volume V is determined for the main piston at the midpoint.

To write the state equation for the hydraulic actuator and load of Figs. 2.24 and 2.25 the energy-related variables must be determined. The mass M yields one energy-storage variable which is the output velocity Dy. The compressible component q_c also represents an energy-storage element in a hydraulic system. The compression of a fluid produces stored energy, just as in the compression of a spring. The equation for hydraulic energy is

$$E(t) = \int_0^t P(\tau)q(\tau)\, d\tau \tag{2.114}$$

where $P(t)$ is the pressure and $q(t)$ is the rate of flow of fluid. The energy storage in a compressed fluid is obtained in terms of the bulk modulus of elasticity K_B. Combining Eq. (2.107) with Eq. (2.114) for a constant volume yields

$$E_c(P_L) = \int_0^{P_L} \frac{V}{K_B} P_L\, dP_L = \frac{V}{2K_B} P_L^2 \tag{2.115}$$

The stored energy in a compressed fluid is proportional to the pressure P_L squared; thus P_L may be used as a physical state variable.

Since the output quantity in this system is the position y, it is necessary to increase the state variables to three. Further evidence of the need for three state variables is the fact that Eq. (2.113) is a third-order equation. Therefore, in this example, let $x_1 = y$, $x_2 = Dy = \dot{x}_1$, $x_3 = P_L$, and $u = x$. Then, from Eqs. (2.110) and (2.112), the state and output equations are

$$\dot{\mathbf{x}} = \begin{bmatrix} 0 & 1 & 0 \\ 0 & -\dfrac{B}{M} & \dfrac{C}{M} \\ 0 & -\dfrac{C_b K_B}{V} & -\dfrac{K_B(L + C_p)}{V} \end{bmatrix} \mathbf{x} + \begin{bmatrix} 0 \\ 0 \\ \dfrac{K_B C_X}{V} \end{bmatrix} \mathbf{u} \tag{2.116}$$

$$\mathbf{y} = [y] = [1 \quad 0 \quad 0]\mathbf{x} = [x_1] \tag{2.117}$$

The effect of augmenting the state variables by adding the piston displacement $x_1 = y$ is to produce a singular system; that is, $|\mathbf{A}| = 0$. This property does not appear if a spring is added to the load, as shown in Prob. 2.21. In that case $x_1 = y$ is an independent state variable.

2.11 POSITIVE-DISPLACEMENT ROTATIONAL HYDRAULIC TRANSMISSION[11]

When a large torque is required in a control device, it is possible to use a hydraulic transmission. The transmission contains a variable-displacement pump driven at constant speed. It pumps a quantity of oil that is proportional to a control stroke and independent of back pressure. The direction of fluid flow is determined by the direction of displacement of the control stroke. The hydraulic motor has an angular velocity proportional to the volumetric flow rate and in the direction of the oil flow from the pump.

The assumption is made that over a limited range of operation the hydraulic transmission is linear. A schematic picture of the system is shown in Fig. 2.26.

The following symbols are used:

q_p = total volumetric flow rate from pump
q_m = volumetric flow rate through motor
q_l = volumetric leakage flow rate of both pump and motor
q_c = compressibility flow rate
x = control stroke (x varies from 0 to ± 1)
ω_p = angular velocity of pump shaft (constant)
ω_m = angular velocity of motor shaft (variable)
θ_m = angular position of motor shaft
d_p = volumetric displacement (at $x = 1$) per unit angular displacement
d_m = volumetric motor displacement per unit angular displacement
L = leakage coefficient of complete system, $(\text{ft}^3/\text{s})/(\text{lb}/\text{ft}^2)$
V = total volume of liquid under compression, ft^3
K_B = bulk modulus of oil, lb/ft^2
P_L = load-induced pressure drop across motor, lb/ft^2
C = motor torque constant, ft^3
n_T = torque conversion efficiency of motor

The basic equation is based on the fact that the fluid flow rate from the pump equals the sum of the flow rates in the system. This is given by

$$q_p = q_m + q_l + q_c \tag{2.118}$$

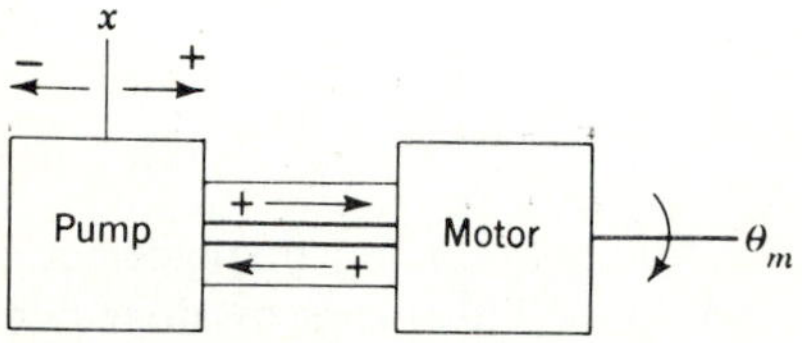

FIGURE 2.26
Hydraulic transmission.

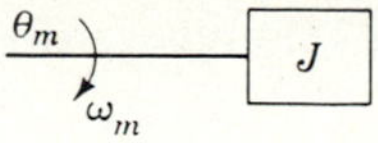

FIGURE 2.27
Inertia load on a hydraulic transmission.

where $q_p = x\,d_p\omega_p \qquad q_m = d_m\omega_m \qquad q_l = LP_L \qquad q_c = DV = \dfrac{V}{K_B}\,DP_L$

Combining these flow rates into the original equation produces

$$d_m\omega_m + LP_L + \frac{V}{K_B}\,DP_L = d_p\omega_p x \tag{2.119}$$

The torque produced at the motor shaft is

$$T = n_T d_m P_L = CP_L \tag{2.120}$$

Since the torque required depends on the load, two cases are considered.

Case 1: Inertia Load

Figure 2.27 shows an inertia load on a hydraulic transmission. Equating the generated torque to the load reaction torque gives

$$T = J\,D^2\theta_m = CP_L \tag{2.121}$$

When P_L from Eq. (2.121) is inserted in Eq. (2.119), the result is

$$\frac{VJ}{K_B C}\,D^3\theta_m + \frac{LJ}{C}\,D^2\theta_m + d_m\,D\theta_m = d_p\omega_p x \tag{2.122}$$

This equation can be solved for the motor position θ_m in terms of the stroke position x.

Based upon the analogs for the linear hydraulic actuator, the three state variables for the rotational hydraulic system are $x_1 = \theta_m$, $x_2 = \omega_m = D\theta_m = \dot{x}_1$, and $x_3 = P_L$. The state equation can be obtained in the usual manner.

Case 2: Inertia Load Coupled Through a Spring

Figure 2.28 shows a spring and inertia load on a hydraulic transmission. Equating the torque generated by the hydraulic motor to the load torque and then solving for P_L in terms of θ_m gives

$$T = K(\theta_m - \theta_L) = J\,D^2\theta_L = CP_L \tag{2.123}$$

$$P_L = \frac{KJ\,D^2\theta_m}{(J\,D^2 + K)C} \tag{2.124}$$

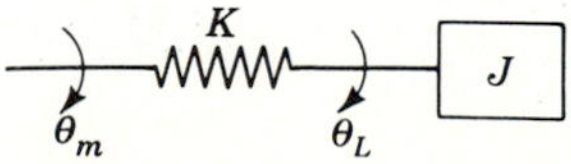

FIGURE 2.28
Spring and inertia load on a hydraulic transmission.

Using the value of P_L from Eq. (2.124) in the system equation (2.119) relates the motor output position θ_m to the stroke input x by

$$\left[\left(\frac{J}{K} + \frac{VJ}{K_B C d_m}\right) D^3 + \frac{LJ}{Cd_m} D^2 + D\right]\theta_m = \frac{d_p \omega_p}{d_m}\left(\frac{J}{K} D^2 + 1\right)x \quad (2.125)$$

The solution of these differential equations is considered in the next chapter. It will be found that the leakage L is essential for stable response. Without leakage the system would have a sustained steady-state oscillation. Therefore, rather than rely on accidental leakage, the designer provides for a positive and finite leakage of hydraulic fluid around the pistons of the motor. This leakage can also serve to lubricate the piston joints.

The transfer function for this system, obtained from Eq. (2.125), is

$$G = \frac{(d_p \omega_p / d_m)(J D^2/K + 1)}{(J/K + VJ/K_B C d_m) D^3 + (LJ/Cd_m) D^2 + D} \quad (2.126)$$

Note that this transfer function contains the operator D in the numerator. In such systems the physical variables may not always be used as the state variables. This problem is discussed in more detail in Chap. 5 in which the phase and canonical state variables are introduced.

2.12 LIQUID-LEVEL SYSTEM[12]

Figure 2.29*a* represents a two-tank liquid-level control system. Definitions of the system parameters are

q_i, q_1, q_2 = rates of flow of fluid
R_1, R_2 = flow resistance
h_1, h_2 = heights of fluid level
A_1, A_2 = cross-sectional tank areas

The following basic linear relationships hold for this system:

$$q = \frac{h}{R} = \text{rate of flow through orifice} \quad (2.127)$$

$$q_n = (\text{tank input rate of flow}) - (\text{tank output rate of flow})$$

$$= \text{net tank rate of flow} = A\,Dh \quad (2.128)$$

Applying Eq. (2.128) to tanks 1 and 2 yields, respectively,

$$q_{n_1} = A_1\,Dh_1 = q_i - q_1 = q_i - \frac{h_1 - h_2}{R_1} \quad (2.129)$$

$$q_{n_2} = A_2\,Dh_2 = q_1 - q_2 = \frac{h_1 - h_2}{R_1} - \frac{h_2}{R_2} \quad (2.130)$$

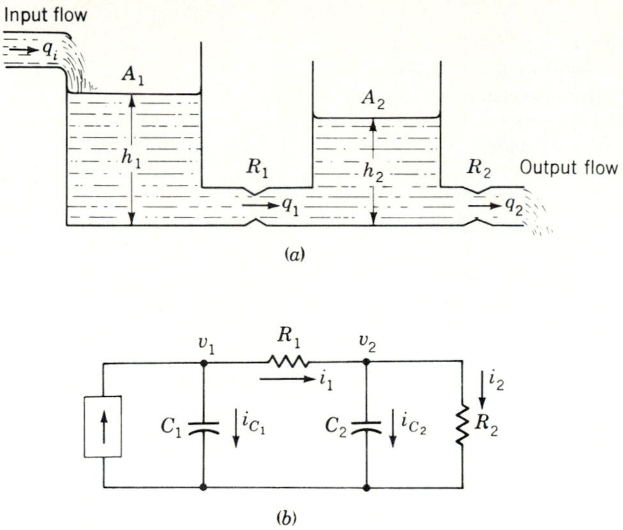

FIGURE 2.29
Liquid-level system and its equivalent electrical analog.

These equations can be solved simultaneously to obtain the transfer functions h_1/q_i and h_2/q_i.

The energy stored in each tank represents potential energy, which is equal to $\rho A h^2/2$, where ρ is the fluid density coefficient. Since there are two tanks, the system has two energy-storage elements, whose energy-storage variables are h_1 and h_2. Letting $x_1 = h_1$, $x_2 = h_2$, and $u = q_i$ in Eqs. (2.129) and (2.130) reveals that x_1 and x_2 are independent state variables. Thus the state equation is

$$\dot{\mathbf{x}} = \begin{bmatrix} -\dfrac{1}{R_1A_1} & \dfrac{1}{R_1A_1} \\ \dfrac{1}{R_1A_2} & -\dfrac{1}{R_1A_2} - \dfrac{1}{R_2A_2} \end{bmatrix} \mathbf{x} + \begin{bmatrix} \dfrac{1}{A_1} \\ 0 \end{bmatrix} \mathbf{u} \tag{2.131}$$

The levels of the two tanks are the outputs of the system. Letting $y_1 = x_1 = h_1$ and $y_2 = x_2 = h_2$ yields

$$\mathbf{y} = \begin{bmatrix} 1 & 0 \\ 0 & 1 \end{bmatrix} \mathbf{x} \tag{2.132}$$

The potential energy of a tank can be represented as a capacitor whose stored potential energy is $Cv_c^2/2$; thus the electrical analog of h is v_C. As a consequence, an analysis of Eqs. (2.129) and (2.130) yields the analogs between the hydraulic and electrical quantities listed in Table 2.7. The analogous electrical

TABLE 2.7
Hydraulic and electrical analogs

Hydraulic element		Electrical element	
Symbol	**Quantity**	**Symbol**	**Quantity**
q_i	Input flow rate	i_i	Current source
h	Height	v_C	Capacitor voltage
A	Tank area	C	Capacitance
R	Flow resistance	R	Resistance

equations are

$$C_1 Dv_1 = i_i - \frac{v_1 - v_2}{R_1} \tag{2.133}$$

$$C_2 Dv_2 = \frac{v_1 - v_2}{R_1} - \frac{v_2}{R_2} \tag{2.134}$$

These two equations yield the analogous electric circuit of Fig. 2.29*b*.

2.13 ROTATING POWER AMPLIFIERS[13,14]

A dc generator can be used as a power amplifier in which the power required to excite the field circuit is lower than the power output rating of the armature circuit. The voltage e_g induced in the armature circuit is directly proportional to the product of the magnetic flux ϕ set up by the field and the speed of rotation ω of the armature. This is expressed by

$$e_g = K_1 \phi \omega \tag{2.135}$$

The flux is a function of field current and the type of iron used in the field. A typical magnetization curve showing flux as a function of field current is given in Fig. 2.30. Up to saturation the relation is approximately linear, and the flux is directly proportional to field current:

$$\phi = K_2 i_f \tag{2.136}$$

Combining these equations gives

$$e_g = K_1 K_2 \omega i_f \tag{2.137}$$

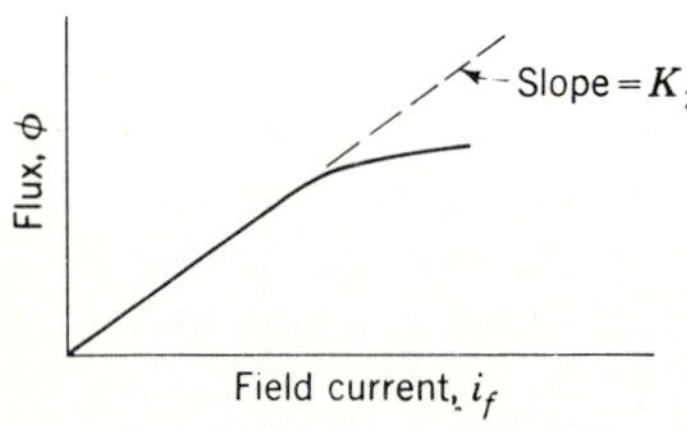

FIGURE 2.30
Magnetization curve.

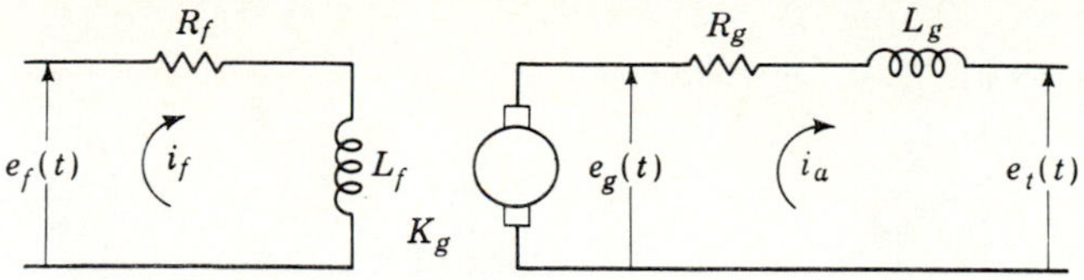

FIGURE 2.31
Schematic diagram of a generator.

When used as a power amplifier the armature is driven at constant speed, and this equation becomes

$$e_g = K_g i_f \tag{2.138}$$

A generator is represented schematically in Fig. 2.31, in which L_f and R_f and L_g and R_g are the inductance and resistance of the field and armature circuits, respectively. The equations for the generator are

$$e_f = (L_f D + R_f) i_f \qquad e_g = K_g i_f \qquad e_t = e_g - (L_g D + R_g) i_a \tag{2.139}$$

The armature current depends on the load connected to the generator terminals. Combining the first two equations gives

$$(L_f D + R_f) e_g = K_g e_f \tag{2.140}$$

Equation (2.140) relates the generated voltage e_g to the input field voltage e_f. The power output $p_o = e_t i_a$ is much larger than the power input $p_i = e_f i_f$. Thus, the ratio p_o/p_i represents the power gain.

2.14 DC SERVOMOTOR

A current-carrying conductor located in a magnetic field experiences a force proportional to the magnitude of the flux, the current, the length of the conductor, and the sine of the angle between the conductor and the direction of the flux. When the conductor is a fixed distance from an axis about which it can rotate, a torque is produced that is proportional to the product of the force and the radius. In a motor the resultant torque is the sum of the torques produced by each conductor. For any given motor the only two adjustable quantities are the flux and armature current. The developed torque can be expressed as

$$T(t) = K_3 \phi i_m \tag{2.141}$$

In this case, to avoid confusion with t for time, the capital letter T is used to indicate torque. It may denote either a constant or a function that varies with time. There are two modes of operation of a servomotor. In one mode the field current is held constant, and an adjustable voltage is applied to the armature. In the second mode the armature current is held constant, and an adjustable voltage is applied to the field. These methods of operation are considered separately.

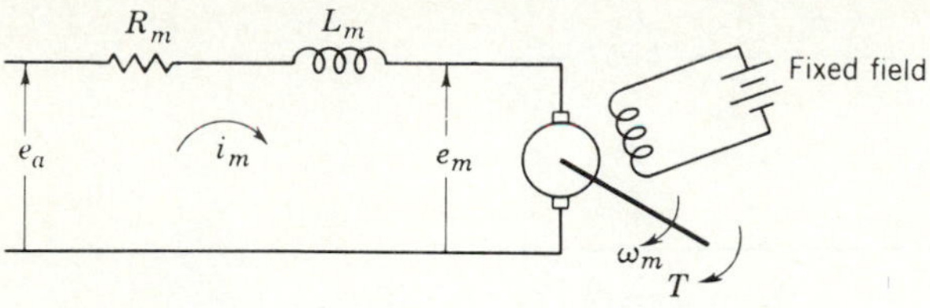

FIGURE 2.32
Circuit diagram of a dc motor.

Armature Control

A constant field current is obtained by separately exciting the field from a fixed dc source. The flux is produced by the field current and is therefore constant. Thus the torque is proportional only to the armature current and Eq. (2.141) becomes

$$T(t) = K_T i_m \tag{2.142}$$

When the motor armature is rotating, there is induced a voltage e_m which is proportional to the product of flux and speed. Because the polarity of this voltage opposes the applied voltage e_a, it is commonly called the *back emf*. Since the flux is held constant, the induced voltage e_m is directly proportional to the speed ω_m:

$$e_m = K_1 \phi \omega_m = K_b \omega_m = K_b \, D\theta_m \tag{2.143}$$

The torque constant K_T and the generator constant K_b are the same for mks units. Control of the motor speed is obtained by adjusting the voltage applied to the armature. Its polarity determines the direction of the armature current and therefore the direction of the torque generated. This, in turn, determines the direction of rotation of the motor. A circuit diagram of the armature-controlled dc motor is shown in Fig. 2.32. The armature inductance and resistance are labeled L_m and R_m. The voltage equation of the armature circuit is

$$L_m \, Di_m + R_m i_m + e_m = e_a \tag{2.144}$$

The current in the armature produces the required torque according to Eq. (2.142). The required torque depends on the load connected to the motor shaft. If the load consists only of a moment of inertia and damper (friction), as shown in Fig. 2.33, the torque equation can be written:

$$J \, D\omega_m + B\omega_m = T(t) \tag{2.145}$$

The required armature current i_m can be obtained by equating the generated torque of Eq. (2.142) to the required load torque of Eq. (2.145). Inserting this current and the back emf from Eq. (2.143) into Eq. (2.144) produces the system

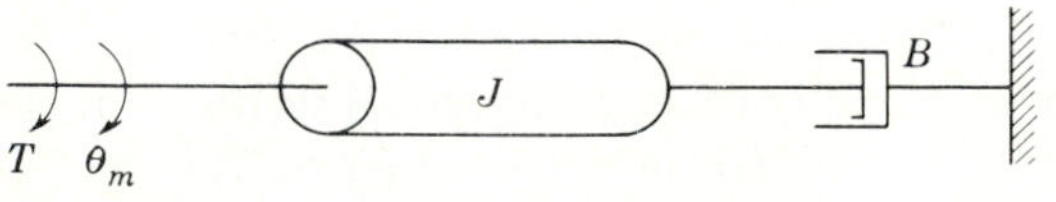

FIGURE 2.33
Inertia and friction as a motor load.

equation in terms of the velocity ω_m:

$$\frac{L_m J}{K_T} D^2\omega_m + \frac{L_m B + R_m J}{K_T} D\omega_m + \left(\frac{R_m B}{K_T} + K_b\right)\omega_m = e_a \qquad (2.146)$$

This equation can also be written in terms of motor position θ_m:

$$\frac{L_m J}{K_T} D^3\theta_m + \frac{L_m B + R_m J}{K_T} D^2\theta_m + \frac{R_m B + K_b K_T}{K_T} D\theta_m = e_a \qquad (2.147)$$

There are two energy-storage elements, J and L_m, for the system represented by Figs. 2.32 and 2.33. Designating the independent state variables as $x_1 = \omega_m$ and $x_2 = i_m$ and the input as $u = e_a$ yields the state equation

$$\dot{\mathbf{x}} = \begin{bmatrix} -\dfrac{B}{J} & \dfrac{K_T}{J} \\ -\dfrac{K_b}{L_m} & -\dfrac{R_m}{L_m} \end{bmatrix}\mathbf{x} + \begin{bmatrix} 0 \\ \dfrac{1}{L_m} \end{bmatrix}\mathbf{u} \qquad (2.148)$$

If motor position θ_m is the output, another differential equation is required; that is, $\dot{\theta}_m = \omega$. If the solution of θ_m is required, then the system is of third order. A third state variable, $x_3 = \theta_m$, must therefore be added.

The transfer functions ω_m/e_a and θ_m/e_a can be obtained from Eqs. (2.146) and (2.147). The armature inductance is small and can usually be neglected, $L_m \approx 0$. Equation (2.147) is thus reduced to a second-order equation. The corresponding transfer function has the form

$$G = \frac{\theta_m}{e_a} = \frac{K_M}{D(T_m D + 1)} \qquad (2.149)$$

Field Control

If the armature current i_m is constant, the torque $T(t)$ is proportional only to the flux ϕ. In the unsaturated region (see Fig. 2.30) the flux is directly proportional to the field current [see Eq. (2.136)]. Therefore the torque equation is written as

$$T(t) = K_3\phi i_m = K_3 K_2 i_m i_f = K_f i_f \qquad (2.150)$$

Control of the motor speed is obtained by adjusting the voltage applied to the field. Its magnitude and polarity determine the magnitude of the torque and the direction of rotation. The circuit diagram of the field-controlled dc motor is shown in Fig. 2.34. The field winding inductance and resistance are labeled L_f and R_f. The voltage equation of the field circuit is

$$L_f\, Di_f + R_f i_f = e_f \qquad (2.151)$$

If the load consists of moment of inertia and viscous friction, as shown in Fig. 2.33, the torque required to drive the load is given by Eq. (2.145). By combining Eqs. (2.145), (2.150), and (2.151), the system equation in terms of the motor

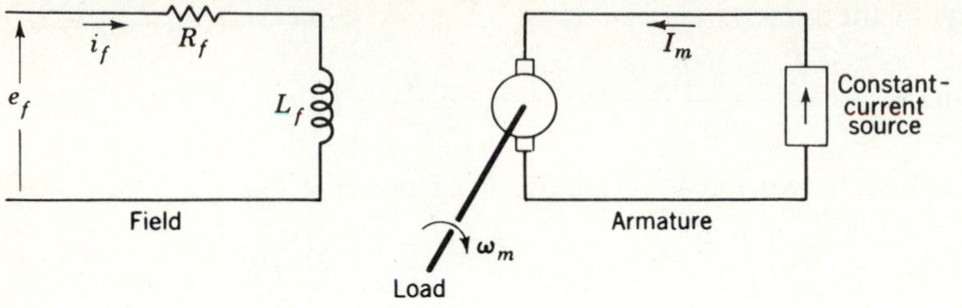

FIGURE 2.34
Circuit diagram of a field-controlled dc motor.

velocity is

$$(L_f D + R_f)(JD + B)\omega_m = K_f e_f \tag{2.152}$$

An advantage of field control over armature control is that the power required by the field is much smaller than that required by the armature. Since this power is usually supplied by an amplifier, smaller power capacity is required in the amplifier.

Although a constant armature current i_m is specified in the derivation above, this is not easily achieved. If the armature is connected to a fixed dc voltage source E_a, the armature current depends on the back emf, as shown by

$$i_m = \frac{E_a - e_m}{R_m} = \frac{E_a - K_2\phi\omega_m}{R_m}$$

For the range of speed that the motor experiences, the armature circuit may be designed so that $E_a \gg e_m$. Thus

$$i_m \approx \frac{E_a}{R_m} \tag{2.153}$$

A practical circuit for field control of a dc motor with a split field is shown in Fig. 2.35. The field is energized by a balanced amplifier which has the input e. When this input is zero, the current i_1 and i_2 are equal. Since they flow in opposite directions, the net flux is zero, thus the torque is zero and the motor is stationary. If e is not zero, one of the currents increases and the other decreases in proportion to the magnitude of e. The resulting flux is proportional to the magnitude of e, and its direction depends on the polarity of e. The size and

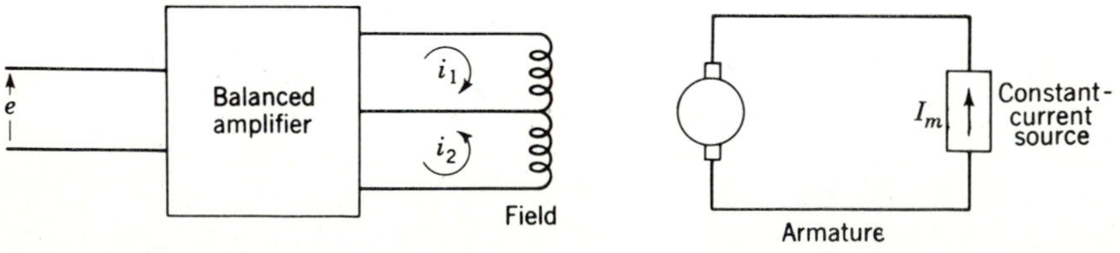

FIGURE 2.35
Circuit diagram of a split-field-controlled dc motor.

direction of the generated torque and the resulting speed therefore respond to the magnitude and polarity of the input e.

In this example, using the independent state variables $x_1 = \omega_m$ and $x_2 = i_f$, the state equation can be readily obtained.

2.15 AC SERVOMOTOR[15]

An ac servomotor is basically a two-phase induction motor that has two stator field coils placed 90 electrical degrees apart, as shown in Fig. 2.36a. In a two-phase motor the ac voltages e and e_c are equal in magnitude and separated by a phase angle of 90°. A two-phase induction motor runs at a speed slightly below the synchronous speed and is essentially a constant-speed motor. The synchronous speed n_s is determined by the number of poles P produced by the stator windings and the frequency f of the voltage applied to the stator windings where $n_s = 120f/P$ revolutions per minute.

When the unit is used as a servomotor, the speed must be proportional to an input voltage. The two-phase motor can be used as a servomotor by applying an ac voltage e of fixed amplitude to one of the motor windings. When the other voltage e_c is varied, the torque and speed are a function of this voltage. Figure 2.36c shows a set of torque-speed curves for various control voltages.

It is important to note that the curve for zero control-field voltage goes through the origin and that the slope is negative. This means that when the control-field voltage becomes zero, the motor develops a decelerating torque, causing it to stop. The curves show a large torque at zero speed. This is a requirement for a servomotor in order to provide rapid acceleration. It is accomplished in an induction motor by building the rotor with a high resistance.

The torque-speed curves are not straight lines. Therefore, a linear differential equation cannot be used to represent the exact motor characteristics. Sufficient accuracy may be obtained by approximating the characteristics by straight lines. The following analysis is based on this approximation.

The torque generated is a function of both the speed ω and the control-field voltage E_c. In terms of partial derivatives, the torque equation is approximated

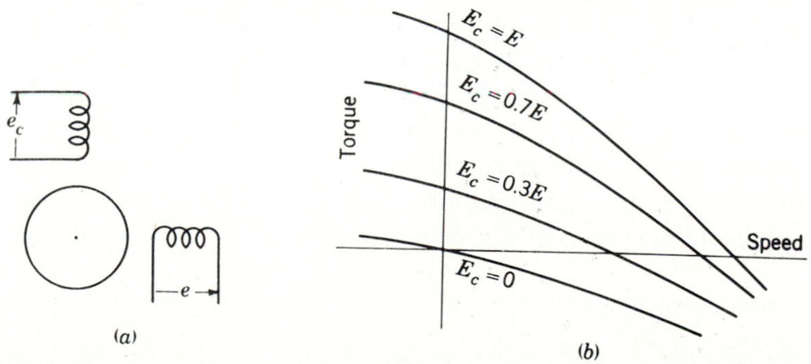

FIGURE 2.36
(a) Schematic diagram of a two-phase induction motor; (b) servomotor characteristics.

by effecting a double Taylor series expansion of $T(E_c, \omega)$ about the origin and keeping only the linear terms:

$$\left.\frac{\partial T}{\partial E_c}\right|_{\text{origin}} E_c + \left.\frac{\partial T}{\partial \omega}\right|_{\text{origin}} \omega = T(E_c, \omega) \tag{2.154}$$

If the torque-speed motor curves are approximated by parallel straight lines, the partial-derivative coefficients of Eq. (2.154) are constants which can be evaluated from the graph. Let

$$\frac{\partial T}{\partial E_c} = K_c \qquad \text{and} \qquad \frac{\partial T}{\partial \omega} = K_\omega \tag{2.155}$$

For a load consisting of a moment of inertia and damping, the load torque required is

$$T_L = J D\omega + B\omega \tag{2.156}$$

Since the generated and load torques must be equal, Eqs. (2.154) and (2.156) are equated:

$$K_c E_c + K_\omega \omega = J D\omega + B\omega$$

Rearranging terms gives

$$J D\omega + (B - K_\omega)\omega = K_c E_c \tag{2.157}$$

In terms of position θ, this equation can be written as

$$J D^2\theta + (B - K_\omega)\, D\theta = K_c E_c. \tag{2.158}$$

In order for the system to be stable (see Chap. 3) the coefficient $B - K_\omega$ must be positive. Observation of the motor characteristics shows that $K_\omega = \partial T/\partial \omega$ is negative; therefore the stability requirement is satisfied.

Analyzing Eq. (2.157) reveals that this system has only one energy-storage element J. Thus the state equation, where $x_1 = \omega$ and $u = E_c$, is

$$\dot{x}_1 = -\frac{B - K_\omega}{J} x_1 + \frac{K_c}{J} u \tag{2.159}$$

2.16 LAGRANGE'S EQUATION

Previous sections show the application of Kirchhoff's laws for writing the differential equations of electric networks and the application of Newton's laws for writing the equations of motion of mechanical systems. These laws can be applied, depending on the complexity of the system, with relative ease. In many instances there are systems that contain combinations of electric and mechanical components. The use of Lagrange's equation provides a systematic unified approach for handling a broad class of physical systems, no matter how complex their structure.[16]

Lagrange's equation is given by

$$\frac{d}{dt}\left(\frac{\partial T}{\partial \dot{q}_n}\right) - \frac{\partial T}{\partial q_n} + \frac{\partial D}{\partial \dot{q}_n} + \frac{\partial V}{\partial q_n} = Q_n \qquad n = 1, 2, 3, \ldots \tag{2.160}$$

where T = total kinetic energy of system
D = dissipation function of system
V = total potential energy of system
Q_n = generalized applied force at the coordinate n
q_n = generalized coordinate
$\dot{q}_n = dq_n/dt$ (generalized velocity)

and $n = 1, 2, 3, \ldots$ denote the number of independent coordinates or degrees of freedom which exist in the system. The total kinetic energy T includes all energy terms, regardless of whether they are electrical or mechanical. The dissipation function D represents one-half the rate at which energy is dissipated as heat; dissipation is produced by friction in mechanical systems and by resistance in electric circuits. The total potential energy stored in the system is designated by V. The forcing functions applied to a system are designated by Q_n; they take the form of externally applied forces or torques in mechanical systems and appear as voltage or current sources in electric circuits. Examples of these quantities are illustrated for electrical and mechanical systems. Kinetic energy is associated with the generalized velocity and, for an inductor and mass, is given by $T_L = L\dot{q}^2/2 = Li^2/2$ and $T_M = M\dot{x}^2/2 = Mv^2/2$, respectively. Potential energy is associated with the generalized position and, for a capacitor and spring, is given by $V_C = q^2/2C$ and $V_K = Kx^2/2$, respectively. The dissipation function is always a function of the generalized velocity and, for an electrical resistor and mechanical friction, is given by $D_R = R\dot{q}^2/2 = Ri^2/2$ and $D_B = B\dot{x}^2/2 = Bv^2/2$, respectively. The generalized applied force for an electric circuit is the electromotive force of a generator, thus $Q_q = E$. For the earth's gravitational force applied to a mass, the generalized applied force is $Q_f = Mg$. Similar relationships can be developed for other physical systems. The application of Lagrange's equation is illustrated by the following example.

Example Electromechanical system with capacitive coupling. Figure 2.37 shows a system in which mechanical motion is converted into electric energy. This represents the action which takes place in a capacitor microphone. Plate a of the capacitor is fastened rigidly to the frame. Sound waves impinge upon and exert a force on plate b of mass M, which is suspended from the frame by a spring K and which has damping B. The output voltage which appears across the resistor R is intended to reproduce electrically the sound-wave patterns which strike the plate b.

At equilibrium, with no external force extended on plate b, there is a charge q_0 on the capacitor. This produces a force of attraction between the plates so that the spring is stretched by an amount x_1 and the space between the plates is x_0. When sound waves exert a force on plate b there will be a resulting motion x which is measured from the equilibrium position. The distance between the plates will then be $x_0 - x$, and the charge on the plates will be $q_0 + q$.

The capacitance is approximated by

$$C = \frac{\varepsilon A}{x_0 - x} \qquad \text{and} \qquad C_0 = \frac{\varepsilon A}{x_0}$$

where ε is the dielectric constant for air and A is the area of the plate.

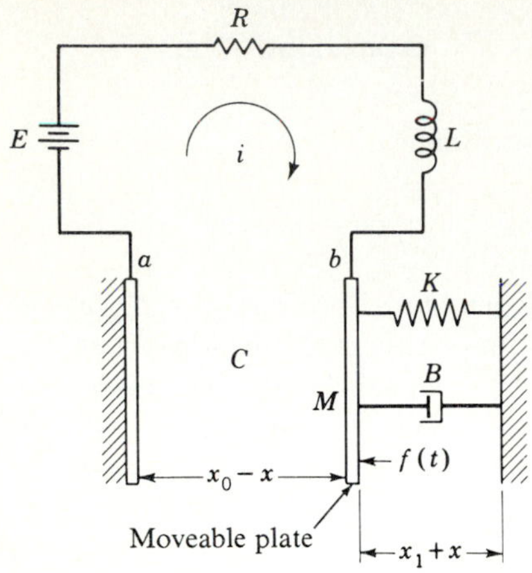

FIGURE 2.37
Electromechanical system with capacitive coupling.

The energy expressions for this system are

$$T = \tfrac{1}{2}L\dot{q}^2 + \tfrac{1}{2}M\dot{x}^2$$

$$D = \tfrac{1}{2}R\dot{q}^2 + \tfrac{1}{2}B\dot{x}^2$$

$$V = \frac{1}{2C}(q_0 + q)^2 + \tfrac{1}{2}K(x_1 + x)^2$$

$$= \frac{1}{2\varepsilon A}(x_0 - x)(q_0 + q)^2 + \tfrac{1}{2}K(x_1 + x)^2$$

The method is simple and direct. It is merely necessary to include all the energy terms, whether electrical or mechanical. The electromechanical coupling in this example appears in the potential energy. Here, the presence of charge on the plates of the capacitor exerts a force on the mechanical system. Also, motion of the mechanical system produces an equivalent emf in the electric circuit.

The two degrees of freedom are the displacement x of plate b and the charge q on the capacitor. Applying Lagrange's equation twice gives

$$M\ddot{x} + B\dot{x} - \frac{1}{2\varepsilon A}(q_0 + q)^2 + K(x_1 + x) = f(t) \tag{2.161}$$

$$L\ddot{q} + R\dot{q} + \frac{1}{\varepsilon A}(x_0 - x)(q_0 - q) = E \tag{2.162}$$

These equations are nonlinear. However, a good linear approximation can be obtained when it is realized that x and q are very small quantities and therefore the x^2, q^2, and xq terms can be neglected. This gives

$$(q_0 + q)^2 \approx q_0^2 + 2q_0 q$$

$$(x_0 - x)(q_0 - q) \approx x_0 q_0 - q_0 x + x_0 q$$

With these approximations the system equations become

$$M\ddot{x} + Kx_1 + Kx - \frac{q_0^2}{2\varepsilon A} - \frac{2q_0 q}{2\varepsilon A} + B\dot{x} = f(t) \tag{2.163}$$

$$L\ddot{q} + \frac{x_0 q_0}{\varepsilon A} - \frac{q_0 x}{\varepsilon A} + \frac{x_0 q}{\varepsilon A} + R\dot{q} = E \tag{2.164}$$

TABLE 2.8
Identification of energy functions for electromechanical elements

Definition	Element and symbol	Kinetic energy T
x = relative motion i = current in coil $U = l\beta N_c$ (electromechanical coupling constant) β = flux density produced by permanent magnet l = length of coil N_c = no. of turns in coil	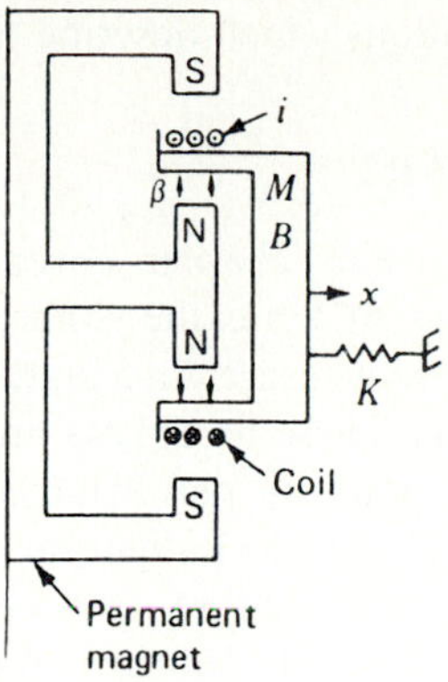	Uxi

From the first equation, by setting $f(t) = 0$ and taking steady-state conditions, the result is

$$Kx_1 - \frac{q_0^2}{2\varepsilon A} = 0$$

This simply equates the force on the spring and the force due to the charges at the equilibrium condition. Similarly, in the second equation at equilibrium

$$\frac{x_0 q_0}{\varepsilon A} = \frac{q_0}{C_0} = E$$

Therefore the two system equations can be written in linearized form as

$$M\ddot{x} + B\dot{x} + Kx - \frac{q_0}{\varepsilon A} q = f(t) \tag{2.165}$$

$$L\ddot{q} + R\dot{q} + \frac{q}{C_0} - \frac{q_0}{\varepsilon A} x = 0 \tag{2.166}$$

These equations show that $q_0/\varepsilon A$ is the coupling factor between the electrical and mechanical portions of the system.

Another form of electromechanical coupling exists when current in a coil produces a force which is exerted on a mechanical system and, simultaneously, motion of a mass induces an emf in an electric circuit. The electromagnetic coupling is represented in Table 2.8. In that case the kinetic energy includes a term

$$T = l\beta N_c xi = Uxi \tag{2.167}$$

The energy for this system, which is shown in Table 2.8, may also be considered potential energy. This influences the sign on the corresponding term in the differential equation.

The main advantage of Lagrange's equation is the use of a single systematic procedure, eliminating the need to consider separately Kirchhoff's laws for the electrical aspects and Newton's law for the mechanical aspects of the system in formulating the statements of equilibrium. Once this procedure is mastered, the differential equations which describe the system are readily obtained.

2.17 SUMMARY

The examples in this chapter cover many of the basic elements of control systems. In order to write the differential and state equations, the basic laws governing performance are first stated for electrical, mechanical, thermic, and hydraulic systems. These basic laws are then applied to specific devices, and their differential equations of performance are obtained. The basic matrix, state, transfer-function, and block-diagram concepts are introduced. Lagrange's equation has been introduced to provide a systematized method for writing the differential equations of electrical, mechanical, and electromechanical systems. This chapter constitutes a reference for the reader who wants to review the fundamental concepts involved in writing differential equations of performance.

REFERENCES

1. DeRusso, P. M., et al.: *State Variables for Engineers*, Wiley, New York, 1965.
2. Blackburn, J. F. (ed.): *Components Handbook*, McGraw-Hill, New York, 1948.
3. Stout, T. M.: "A Block Diagram Approach to Network Analysis," *Trans. AIEE*, vol. 71, pp. 255–260, 1952.
4. Wylie, C. R., Jr.: *Advanced Engineering Mathematics, 4th ed., McGraw-Hill, New York*, 1975.
5. Kinariwala, B., et al.: *Linear Circuits and Computation*, Wiley, New York, 1973.
6. Gantmacher, F. R.: *Applications of the Theory of Matrices*, Wiley-Interscience, New York, 1959.
7. Kalman, R. E.: "Mathematical Description of Linear Dynamical Systems," *J. Soc. Ind. Appl. Math., ser. A, Control*, vol. 1, no. 2, 1963.
8. Gardner, M. F., and J. L. Barnes: *Transients in Linear Systems*, Wiley, New York, 1942, chap. 2.
9. Hornfeck, A. J.: "Response Characteristics of Thermometer Elements," *Trans. ASME*, vol. 71, pp. 121–132, 1949.
10. *Flow Meters: Their Theory and Application*, American Society of Mechanical Engineers, New York, 1937.
11. Newton, C. G., Jr.: "Hydraulic Variable Speed Transmissions as Servomotors," *J. Franklin Inst.*, vol. 243, no. 6, pp. 439–469, June 1947.
12. Takahashi, Y., et al.: *Control and Dynamic Systems*, Addison-Wesley, Reading, Mass., 1970.
13. Saunders, R. M.: "The Dynamo Electric Amplifier: Class A Operation," *Trans. AIEE*, vol. 68, pp. 1368–1373, 1949.
14. Litman, B.: "An Analysis of Rotating Amplifiers," *Trans. AIEE*, vol. 68, pt. II, pp. 1111–1117, 1949.
15. Hopkin, A. M.: "Transient Response of Small Two-Phase Servomotors," *Trans. AIEE*, vol. 70, pp. 881–886, 1951.
16. Ogar, G. W., and J. J. D'Azzo: "A Unified Procedure for Deriving the Differential Equations of Electrical and Mechanical Systems," *IRE Trans. Educ.*, vol. E-5, no. 1, pp. 18–26, March 1962.

CHAPTER 3

SOLUTION OF DIFFERENTIAL EQUATIONS

3.1 INTRODUCTION

The general solution of a linear differential equation[1,2] is the sum of two components, the particular integral and the complementary function. Often the particular integral is the steady-state component of the solution of the differential equation; and the complementary function, which is the solution of the corresponding homogeneous equation, is the transient component of the solution. Often the steady-state component of the response has the same form as the driving function. In this book the particular integral is called the steady-state solution even when it is not periodic. The form of the transient component of the response depends only on the roots of the characteristic equation and sometimes on the initial or boundary conditions. The instantaneous value of the transient component depends on the boundary conditions, the roots of the characteristic equation, and the instantaneous value of the steady-state component.

This chapter covers methods of determining the steady-state and the transient components of the solution. These components are first determined separately and then added to form the complete solution. Analysis of the transient solution should develop in the student a feel for the solution to be expected.

The method of solution is next applied to the matrix state equation. A general format is obtained for the complementary solution in terms of the state transition matrix (STM). Then the complete solution is obtained as a function of the STM and the input forcing function. The discrete representation of the state equation is also developed. This form is useful for systems which use digital controllers as described in Chap. 22. It is also useful for solution of the state equation using a digital computer.

3.2 INPUTS TO CONTROL SYSTEMS

For some control systems the input has a specific form which may be represented either by an analytical expression or as a specific curve. An example of the latter is the pattern used in a machining operation where the cutting tool is required to follow the path indicated by the pattern outline.

For other control systems the input may be random in shape. In this case it cannot be expressed analytically and is not repetitive. An example is the camera platform used in a photographic airplane. The airplane flies at a fixed altitude and speed, and the camera takes a series of pictures of the terrain below it, which are then fitted together to form one large picture of the area. This requires that the camera platform remain level regardless of the motion of the airplane. Since the attitude of the airplane varies with wind gusts and depends on the stability of the airplane itself, the input to the camera platform is obviously a random function.

It is important to have a basis of comparison for various systems. One way of doing this is by comparing the response with a standardized input. The input or inputs used as a basis of comparison must be determined from the required response of the system and the actual form of its input. The following standard inputs, with unit amplitude, are often used in checking the response of a system:

1. Sinusoidal function $r = \cos \omega t$
2. Power-series function $r = a_0 + a_1 t + a_2 \frac{t^2}{2} + \cdots$
3. Step function $r = u_{-1}(t)$
4. Ramp (step velocity) function $r = u_{-2}(t) = t u_{-1}(t)$
5. Parabolic (step acceleration) function $r = u_{-3}(t) = \frac{t^2}{2} u_{-1}(t)$
6. Impulse function $r = u_0(t)$

Functions 3 to 6 are called *singularity functions* (Fig. 3.1). The singularity functions can be obtained from one another by successive differentiation or integration. For example, the derivative of the parabolic function is the ramp function, the derivative of the ramp function is the step function, and the derivative of the step function is the impulse function.

For each of these inputs a complete solution of the differential equation is determined in this chapter. First, generalized methods are developed to determine the steady-state output $c(t)_{ss}$ for each type of input. These methods are applicable to linear differential equations of any order. Next, the method of evaluating the transient component of response $c(t)_t$ is determined, and it is shown that the

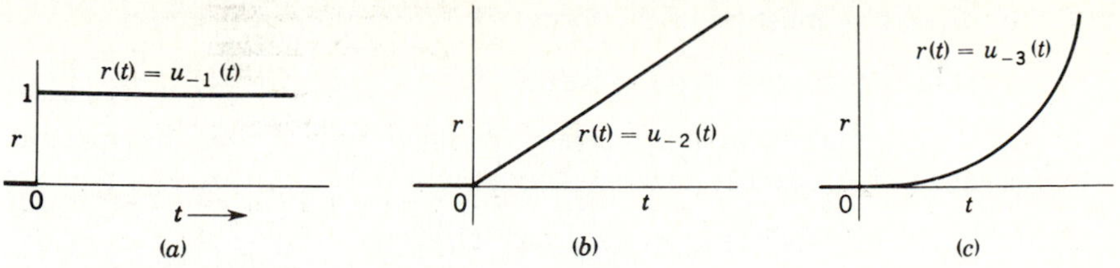

FIGURE 3.1
Singularity functions: (*a*) step function, $u_{-1}(t)$; (*b*) ramp function, $u_{-2}(t)$; (*c*) parabolic function, $u_{-3}(t)$.

form of the transient component of response depends on the characteristic equation. Addition of the steady-state component and the transient component gives the complete solution, that is $c(t) = c(t)_{ss} + c(t)_t$. The coefficients of the transient terms are determined by the instantaneous value of the steady-state component, the roots of the characteristic equation, and the initial conditions. Several examples are used to illustrate these principles.

The solution of the differential equation with a pulse input is postponed until the next chapter, where the Laplace transform is used.

3.3 STEADY-STATE RESPONSE: SINUSOIDAL INPUT

The input quantity r is assumed to be a sinusoidal function of the form

$$r(t) = R\cos(\omega t + \alpha) \tag{3.1}$$

The general integrodifferential equation to be solved is of the form

$$A_v D^v c + A_{v-1} D^{v-1} c + \cdots + A_0 D^0 c + A_{-1} D^{-1} c + \cdots + A_{-w} D^{-w} c = r \tag{3.2}$$

The steady-state solution can be obtained directly by use of Euler's identity,

$$e^{j\omega t} = \cos\omega t + j\sin\omega t$$

The input can then be written

$$r = R\cos(\omega t + \alpha) = \text{real part of } (Re^{j(\omega t+\alpha)})$$
$$= \text{Re}(Re^{j(\omega t+\alpha)}) = \text{Re}(Re^{j\alpha}e^{j\omega t}) = \text{Re}(\mathbf{R}e^{j\omega t}) \tag{3.3}$$

For simplicity, the phrase *real part of* or its symbolic equivalent *Re* is often omitted, but it must be remembered that the real part is intended. The quantity $\mathbf{R} = Re^{j\alpha}$ is the *phasor* representation of the input; i.e., it has both a magnitude R and an angle α. The magnitude R represents the maximum value of the input quantity $r(t)$. For simplicity the angle $\alpha = 0°$ usually is chosen for $\mathbf{R}$. The input r from Eq. (3.3) is inserted in Eq. (3.2). Then, in order for the expression to be an equality, the response c must be of the form

$$c(t)_{ss} = C\cos(\omega t + \phi) = \mathrm{Re}(Ce^{j\phi}e^{j\omega t}) = \mathrm{Re}(\mathbf{C}e^{j\omega t}) \tag{3.4}$$

where $\mathbf{C} = Ce^{j\phi}$ is a phasor quantity having the magnitude C and the angle ϕ. The nth derivative of c_{ss} with respect to time is

$$D^n c(t)_{ss} = \mathrm{Re}\left[(j\omega)^n \mathbf{C}e^{j\omega t}\right] \tag{3.5}$$

Inserting c_{ss} and its derivatives from Eqs. (3.4) and (3.5) into Eq. (3.2) gives

$$\mathrm{Re}\left[A_v(j\omega)^v \mathbf{C}e^{j\omega t} + A_{v-1}(j\omega)^{v-1}\mathbf{C}e^{j\omega t} + \cdots + A_{-w}(j\omega)^{-w}\mathbf{C}e^{j\omega t}\right] = \mathrm{Re}(\mathbf{R}e^{j\omega t}) \tag{3.6}$$

Canceling $e^{j\omega t}$ from both sides of the equation and solving for $\mathbf{C}$ gives

$$\mathbf{C} = \frac{\mathbf{R}}{A_v(j\omega)^v + A_{v-1}(j\omega)^{v-1} + \cdots + A_0 + A_{-1}(j\omega)^{-1} + \cdots + A_{-w}(j\omega)^{-w}} \tag{3.7}$$

where $\mathbf{C}$ is the phasor representation of the output; i.e., it has a magnitude C and an angle ϕ. Since the values of C and ϕ are functions of the frequency ω, it may be written as $\mathbf{C}(j\omega)$ to show this relationship. Similarly, $\mathbf{R}(j\omega)$ denotes the fact that the input is sinusoidal and may be a function of frequency.

When Eqs. (3.2) and (3.6) are compared, it can be seen that one equation can be determined easily from the other. Substituting $j\omega$ for D, $\mathbf{C}(j\omega)$ for c, and $\mathbf{R}(j\omega)$ for r in Eq. (3.2) results in Eq. (3.6). The reverse is also true and is independent of the order of the equation. It should be realized that this is simply a *rule of thumb* which yields the desired expression.

The time response can be obtained directly from the phasor response. The output is

$$c(t)_{ss} = \mathrm{Re}(\mathbf{C}e^{j\omega t}) = |\mathbf{C}|\cos(\omega t + \phi) \tag{3.8}$$

As an example consider the rotational hydraulic transmission described in Sec. 2.11, which has an input x and an output angular position θ_m. Equation (2.122), which relates input to output in terms of system parameters, is repeated below:

$$\frac{VJ}{K_B C} D^3\theta_m + \frac{LJ}{C} D^2\theta_m + d_m D\theta_m = d_p\omega_p x \tag{3.9}$$

When the input is the sinusoid, $x = X\sin\omega t$, the corresponding phasor equation can be obtained from Eq. (3.9) by replacing $x(t)$ by $\mathbf{X}(j\omega)$, θ_m by $\mathbf{\Theta}_m(j\omega)$, and D by $j\omega$. The ratio of phasor output to phasor input is termed the *frequency transfer function*, often designated by $\mathbf{G}(j\omega)$. This ratio, in terms of the steady-state sinusoidal phasors, is

$$\mathbf{G}(j\omega) = \frac{\mathbf{\Theta}_m(j\omega)}{\mathbf{X}(j\omega)} = \frac{d_p\omega_p/d_m}{j\omega\left[(VJ/K_B C d_m)(j\omega)^2 + (LJ/Cd_m)(j\omega) + 1\right]} \tag{3.10}$$

3.4 STEADY-STATE RESPONSE: POLYNOMIAL INPUT

A general development of the steady-state solution of a differential equation with a general polynomial input is first given. Particular cases of the power series are then covered in detail.

The general differential equation is repeated here:

$$A_v D^v c + A_{v-1} D^{v-1} c + \cdots + A_0 c + A_{-1} D^{-1} c + \cdots + A_{-w} D^{-w} c = r \tag{3.11}$$

The polynomial input is of the form

$$r(t) = R_0 + R_1 t + \frac{R_2 t^2}{2!} + \cdots + \frac{R_k t^k}{k!} \tag{3.12}$$

where the highest-order term in the input is $R_k t^k / k!$. For $t < 0$ the value of $r(t)$ is zero. The problem is to find the steady-state or particular solution of the dependent variable c. The method used is to assume a polynomial solution of the form

$$c(t)_{ss} = b_0 + b_1 t + \frac{b_2 t^2}{2!} + \cdots + \frac{b_q t^q}{q!} \tag{3.13}$$

where determination of the value of q is given below.

The assumed solution is then substituted into the differential equation. The coefficients $b_0, b_1, b_2, \ldots$ of the polynomial solution are evaluated by equating the coefficients of like powers of t on both sides of the equation. The highest power of t on the right side of Eq. (3.11) is k; therefore t^k must also appear on the left side of this equation. The highest power of t on the left side of the equation is produced by the lowest-order derivative term $D^{-w}c$ and is equal to q minus the order of the lowest derivative. With this information, the value of the highest-order exponent of t to use in the assumed solution of Eq. (3.11) is

$$q = k + X \qquad q \geq 0 \tag{3.14}$$

where k is the highest exponent appearing in the input and X is the order of the lowest derivative appearing in the differential equation. When integral terms are present, X is a negative number. X is the lowest-order exponent of the differential operator D appearing in the differential equation. For the general differential equation (3.11), the value of X is equal to $-w$. Equation (3.14) is valid only for positive values of q. For each of the examples the response is a polynomial since the input is of that form. However, the highest power in the response may not be the same as that of the input. When the lowest-order derivative is zero, that is, $w = 0$ in Eq. (3.11), then $q = k$.

Step-Function Input

A convenient input $r(t)$ to a system is an abrupt change represented by a unit step function, as shown in Fig. 3.1a. This type of input cannot always be put into a system since it may take a definite length of time to make the change in input, but it represents a good mathematical input for checking system response.

The servomotor described in Sec. 2.14 is used as an example. The response of motor velocity ω_m in terms of the voltage e_a applied to the armature, as given by Eq. (2.146), is of the form

$$A_2 D^2x + A_1 Dx + A_0x = r \tag{3.15}$$

The unit step function $r = u_{-1}(t)$ is a polynomial in which the highest exponent of t is $k = 0$. When the input is a unit step function, the method of solution for a polynomial can therefore be used, that is, the steady-state response is also a polynomial. The lowest-order derivative in Eq. (3.15) is $X = 0$; therefore $q = 0$ and the steady-state response has only one term of the form

$$x_{ss} = b_0 \tag{3.16}$$

The derivatives are $Dx_{ss} = 0$ and $D^2x_{ss} = 0$. Inserting these values into Eq. (3.15) yields

$$x(t)_{ss} = b_0 = \frac{1}{A_0} \tag{3.17}$$

Ramp-Function Input (Step Function of Velocity)

The ramp function is a fixed rate of change of a variable as a function of time. This input and its rate of change are shown in Fig. 3.1*b*. The input is expressed mathematically as

$$r(t) = u_{-2}(t) = tu_{-1}(t) \qquad Dr = u_{-1}(t)$$

A ramp input is a polynomial input where the highest power of t is $k = 1$. When the input $r(t)$ in Eq. (3.15) is the ramp function $u_{-2}(t)$, the highest power of t in the polynomial output is $q = k + X = 1$. The output is therefore

$$x(t)_{ss} = b_0 + b_1t \tag{3.18}$$

The derivatives are $Dx_{ss} = b_1$ and $D^2x_{ss} = 0$. Inserting these values into Eq. (3.15) and equating coefficients of t raised to the same power yields

$$t^0: \qquad A_1b_1 + A_0b_0 = 0 \qquad b_0 = -\frac{A_1b_1}{A_0} = -\frac{A_1}{A_0^2}$$

$$t^1: \qquad A_0b_1 = 1 \qquad b_1 = \frac{1}{A_0}$$

Thus, the steady-state solution is

$$x(t)_{ss} = -\frac{A_1}{A_0^2} + \frac{1}{A_0}t \tag{3.19}$$

Parabolic-Function Input (Step Function of Acceleration)

A parabolic function has a constant second derivative, which means that the first derivative is a ramp. The input $r(t)$ is expressed mathematically as

$$r = u_{-3}(t) = \frac{t^2}{2}u_{-1}(t) \qquad Dr = u_{-2}(t) = tu_{-1}(t) \qquad D^2r = u_{-1}(t)$$

A parabolic input is a polynomial input where the highest power of t is $k = 2$. A steady-state solution is found in the conventional manner for a polynomial input. Consider Eq. (3.15) with a parabolic input. The order of the lowest derivative in the system equation is $X = 0$. The value of q is therefore equal to 2, and the steady-state response is of the form

$$x(t)_{ss} = b_0 + b_1 t + \frac{b_2 t^2}{2} \tag{3.20}$$

The derivatives of x are $Dx_{ss} = b_1 + b_2 t$ and $D^2 x_{ss} = b_2$. Inserting these values into Eq. (3.15) and equating the coefficients of t raised to the same power yields $b_2 = 1/A_0$, $b_1 = -A_1/A_0^2$, and $b_0 = A_1^2/A_0^3 - A_2/A_0^2$.

3.5 TRANSIENT RESPONSE: CLASSICAL METHOD

The classical method of solving for the complementary function or transient response of a differential equation requires, first, the writing of the homogeneous equation. The general differential equation has the form

$$b_v D^v c + b_{v-1} D^{v-1} c + \cdots + b_0 D^0 c + b_{-1} D^{-1} c + \cdots + b_{-w} D^{-w} c = r \tag{3.21}$$

where r is the forcing function and c is the response.

The homogeneous equation is formed by letting the right-hand side of the differential equation equal zero:

$$b_v D^v c_t + b_{v-1} D^{v-1} c_t + \cdots + b_0 c_t + b_{-1} D^{-1} c_t + \cdots + b_{-w} D^{-w} c_t = 0 \tag{3.22}$$

where c_t is the transient component of the general solution.

The general expression for the transient response, which is the solution of the homogeneous equation, is obtained by assuming a solution of the form

$$c_t = A_m e^{mt} \tag{3.23}$$

where m is a constant yet to be determined. Substituting this value of c_t into Eq. (3.22) and factoring $A_m e^{mt}$ from all terms gives

$$A_m e^{mt}\left(b_v m^v + b_{v-1} m^{v-1} + \cdots + b_0 + \cdots + b_{-w} m^{-w}\right) = 0 \tag{3.24}$$

Equation (3.24) must be satisfied for $A_m e^{mt}$ to be a solution. Since e^{mt} cannot be zero for all values of time t, it is necessary that

$$Q(m) = b_v m^v + b_{v-1} m^{v-1} + \cdots + b_0 + \cdots + b_{-w} m^{-w} = 0 \tag{3.25}$$

This is purely an algebraic equation and is termed the *characteristic equation*. There are $v + w$ roots, or *eigenvalues*, of the characteristic equation; therefore the complete transient solution contains the same number of terms of the form $A_m e^{mt}$ if all the roots are simple. Thus the transient component, where there are no multiple roots, is

$$c_t = A_1 e^{m_1 t} + A_2 e^{m_2 t} + \cdots + A_k e^{m_k t} + \cdots + A_{v+w} e^{m_{v+w} t} \tag{3.26}$$

where each $e^{m_k t}$ is described as a *mode* of the system. If there is a root m_q of multiplicity p, the transient includes corresponding terms of the form

$$A_{q1} e^{m_q t} + A_{q2} t e^{m_q t} + \cdots + A_{qp} t^{p-1} e^{m_q t} \tag{3.27}$$

Instead of using the detailed procedure outlined above, the characteristic equation is usually obtained directly from the homogeneous equation by substituting m for Dc_t, m^2 for $D^2 c_t$, etc.

Since the coefficients of the transient solution must be determined from the initial conditions, there must be $v + w$ known initial conditions. These conditions are values of the variable c and of its derivatives which are known at specific times. The $v + w$ initial conditions are used to set up $v + w$ simultaneous equations of c and its derivatives. The value of c includes both the steady-state and transient components. Since determination of the coefficients includes consideration of the steady-state component, the input affects the value of the coefficient of each exponential term.

Complex Roots

If all values of m_k are real, the transient terms can be evaluated as indicated above. Frequently, some values of m_k are complex. When this happens, they always occur in pairs that are complex conjugates and are of the form

$$m_k = \sigma + j\omega_d \qquad m_{k+1} = \sigma - j\omega_d \tag{3.28}$$

where σ is called the *damping coefficient* and ω_d is called the *damped natural frequency*. The transient terms corresponding to these values of m are

$$A_k e^{(\sigma + j\omega_d)t} + A_{k+1} e^{(\sigma - j\omega_d)t} \tag{3.29}$$

These terms are combined to a more useful form by factoring the term $e^{\sigma t}$:

$$e^{\sigma t}\left(A_k e^{j\omega_d t} + A_{k+1} e^{-j\omega_d t}\right) \tag{3.30}$$

By using the Euler identity $e^{\pm j\omega_d t} = \cos \omega_d t \pm j \sin \omega_d t$ and then combining terms, expression (3.30) can be put in the form

$$e^{\sigma t}(B_1 \cos \omega_d t + B_2 \sin \omega_d t) \tag{3.31}$$

This can be converted into the form

$$A e^{\sigma t} \sin(\omega_d t + \phi) \tag{3.32}$$

where $A = \sqrt{B_1^2 + B_2^2}$ and $\phi = \tan^{-1}(B_1/B_2)$. This is a very convenient form

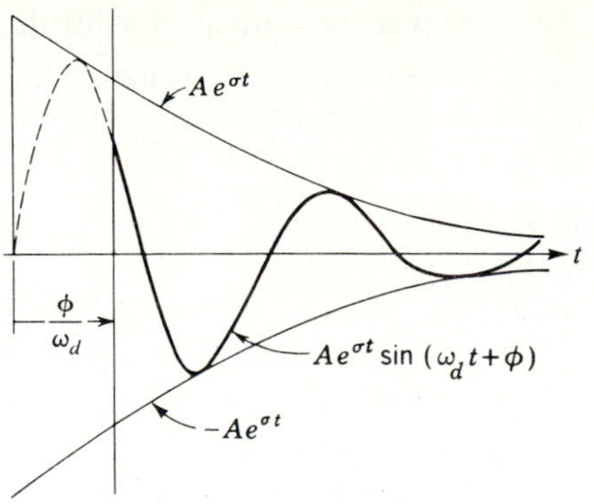

FIGURE 3.2
Sketch of an exponentially damped sinusoid.

for plotting the transient response. The student must learn to use this equation directly without deriving it each time. Often the constants in the transient term can be evaluated more readily from the initial conditions by using the form of Eq. (3.31).

This transient term is called an *exponentially damped sinusoid*; it consists of a sine wave of frequency ω_d whose magnitude is $Ae^{\sigma t}$; that is, it is decreasing exponentially with time if σ is negative. It has the form shown in Fig. 3.2, in which the curves $\pm Ae^{\sigma t}$ constitute the *envelope*. The plot of the time solution always remains betwen the two branches of the envelope.

For the complex roots given in Eq. (3.28), where σ is negative, the transient decays with time and eventually dies out. This represents a stable system. When σ is positive, the transient increases with time and will destroy the equipment unless otherwise restrained. This represents the undesirable case of an unstable system. Control systems must be designed so that they are always stable.

Damping Ratio ζ and Undamped Natural Frequency ω_n

When the characteristic equation has a pair of complex-conjugate roots, it has a quadratic factor of the form $b_2m^2 + b_1m + b_0$. The roots of this factor are

$$m_{1,2} = -\frac{b_1}{2b_2} \pm j\sqrt{\frac{4b_2b_0 - b_1^2}{4b_2^2}} = \sigma \pm j\omega_d \qquad (3.33)$$

The real part σ is recognized as the exponent of e, and ω_d is the frequency of the oscillatory portion of the component stemming from this pair of roots, as given by expression (3.32).

The quantity b_1 represents the effective damping constant of the system. If the numerator under the square root in Eq. (3.33) is zero, then b_1 has the value

$2\sqrt{b_2 b_0}$ and the two roots $m_{1,2}$ are equal. This represents the critical value of the damping constant and is written $b_1' = 2\sqrt{b_2 b_0}$. The *damping ratio* ζ is defined as the ratio of the actual damping constant to the critical value of the damping constant:

$$\zeta = \frac{\text{actual damping constant}}{\text{critical damping constant}} = \frac{b_1}{b_1'} = \frac{b_1}{2\sqrt{b_2 b_0}} \tag{3.34}$$

When ζ is positive and less than unity, the roots are complex and the transient is a damped sinusoid of the form of expression (3.32). When ζ is less than unity, the response is said to be *underdamped*. When ζ is greater than unity, the roots are real and the response is *overdamped*; i.e., the transient solution consists of two exponential terms with real exponents.

The undamped natural frequency ω_n is defined as the frequency of the sustained oscillation of the transient if the damping is zero:

$$\omega_n = \sqrt{\frac{b_0}{b_2}} \tag{3.35}$$

The case of zero damping constant, $b_1 = 0$, means that the transient response does not die out; it is a sine wave of constant amplitude.

The quadratic factors are frequently written in terms of the damping ratio and the undamped natural frequency. Underdamped systems are generally analyzed in terms of these two parameters. After factoring b_0, the quadratic factor of the characteristic equation is

$$\frac{b_2}{b_0}m^2 + \frac{b_1}{b_0}m + 1 = \frac{1}{\omega_n^2}m^2 + \frac{2\zeta}{\omega_n}m + 1 \tag{3.36}$$

When it is multiplied through by ω_n^2, the quadratic appears in the form

$$m^2 + 2\zeta\omega_n m + \omega_n^2 \tag{3.37}$$

The two forms given by Eqs. (3.36) and (3.37) are called the *standard forms* of the quadratic factor, and the corresponding roots are

$$m_{1,2} = \sigma \pm j\omega_d = -\zeta\omega_n \pm j\omega_n\sqrt{1 - \zeta^2} \tag{3.38}$$

The transient response of Eq. (3.32) for the underdamped case, written in terms of ζ and ω_n, is

$$Ae^{\sigma t}\sin(\omega_d t + \phi) = Ae^{-\zeta\omega_n t}\sin\left(\omega_n\sqrt{1 - \zeta^2}\,t + \phi\right) \tag{3.39}$$

From this expression the effect on the transient of the terms ζ and ω_n can readily be seen. The larger the product $\zeta\omega_n$, the faster the transient will decay. These terms also affect the damped natural frequency of oscillation of the transient, $\omega_d = \omega_n\sqrt{1 - \zeta^2}$, which varies directly as the undamped natural frequency and decreases with an increase in the damping ratio.

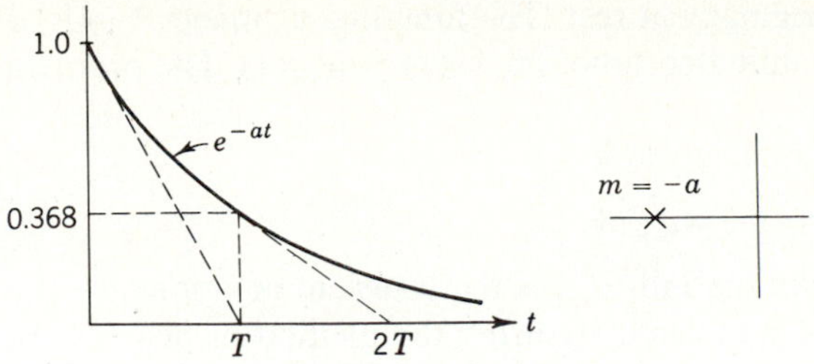

FIGURE 3.3
Plot of the exponential e^{-at} and the root location.

3.6 DEFINITION OF TIME CONSTANT

The transient terms have the exponential form Ae^{mt}. When $m = -a$ is real and negative, the plot of Ae^{-at} has the form shown in Fig. 3.3. The value of time that makes the exponent of e equal to -1 is called the *time constant* T. Thus

$$-aT = -1 \qquad \text{and} \qquad T = \frac{1}{a} \tag{3.40}$$

In a duration of time equal to one time constant the exponential e^{-at} decreases from the value 1 to the value 0.368. Geometrically the tangent drawn to the curve Ae^{-at} at $t = 0$ intersects the time axis at the value of time equal to the time constant T.

When $m = \sigma \pm j\omega_d$ is a complex quantity, the transient has the form $Ae^{\sigma t}\sin(\omega_d t + \phi)$. A plot of this function is shown in Fig. 3.2. In the case of the damped sinusoid the time constant is defined in terms of the parameter σ that characterizes the envelope $Ae^{\sigma t}$. Thus the time constant T is equal to

$$T = \frac{1}{|\sigma|} \tag{3.41}$$

In terms of the damping ratio and the undamped natural frequency, the time constant is $T = 1/\zeta\omega_n$. Therefore, the larger the product $\zeta\omega_n$, the greater the instantaneous rate of decay of the transient.

3.7 EXAMPLE: SECOND-ORDER SYSTEM —MECHANICAL

The simple mechanical system of Sec. 2.6 is used as an example and is shown in Fig. 3.4. Equation (2.61) relates the displacement x_b to x_a:

$$MD^2x_b + BDx_b + Kx_b = Kx_a \tag{3.42}$$

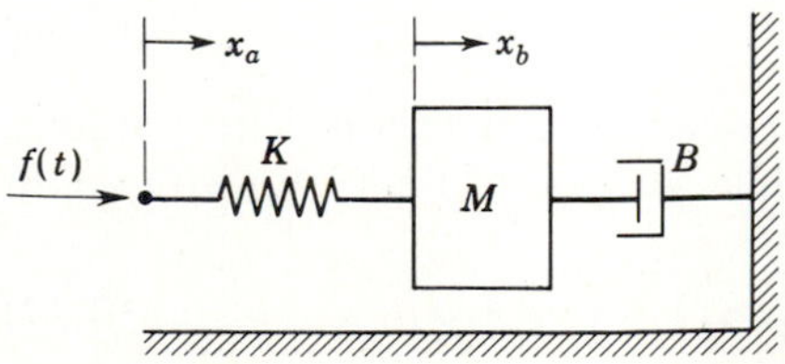

FIGURE 3.4
Simple mechanical system.

The system is considered to be originally at rest. The function x_a moves 1 unit at time $t = 0$; that is, the input is a unit step function $x_a(t) = u_{-1}(t)$. The problem is to find the motion $x_b(t)$.

The displacement of the mass is given by

$$x_b(t) = x_{b,\mathrm{ss}} + x_{b,t}$$

where $x_{b,\mathrm{ss}}$ is the steady-state solution and $x_{b,t}$ is the transient solution.

The steady-state solution is found first by using the method of Sec. 3.4. In this example it may be easier to consider the following:

1. Since the input x_a is a constant, the response x_b must reach a fixed steady-state position.
2. When x_b reaches a constant value, the velocity and acceleration become zero.

By putting $D^2x_b = Dx_b = 0$ into Eq. (3.42), the final or steady-state value of x_b is $x_{b,\mathrm{ss}} = x_a$. When the steady-state solution has been found, the transient solution is determined. The characteristic equation is

$$Mm^2 + Bm + K = M\left(m^2 + \frac{B}{M}m + \frac{K}{M}\right) = 0$$

Putting this in terms of ζ and ω_n gives

$$m^2 + 2\zeta\omega_n m + \omega_n^2 = 0 \tag{3.43}$$

for which the roots are $m_{1,2} = -\zeta\omega_n \pm \omega_n\sqrt{\zeta^2 - 1}$. The transient solution depends on whether the damping ratio ζ is (1) greater than unity, (2) equal to unity, or (3) smaller than unity. For ζ greater than unity the roots are real and have the values $m_1 = -a$ and $m_2 = -b$, and the transient response is

$$\begin{aligned} x_{b,t} &= A_1 \exp\left[\left(-\zeta + \sqrt{\zeta^2 - 1}\right)\omega_n t\right] + A_2 \exp\left[\left(-\zeta - \sqrt{\zeta^2 - 1}\right)\omega_n t\right] \\ &= A_1 \exp(-at) + A_2 \exp(-bt) \end{aligned} \tag{3.44}$$

For ζ equal to unity, the roots are real and equal; that is, $m_1 = m_2 = -\zeta\omega_n$. Since there are multiple roots, the transient response is

$$x_{b,t} = A_1 e^{-\zeta\omega_n t} + A_2 t e^{-\zeta\omega_n t} \tag{3.45}$$

For ζ less than unity, the roots are complex,

$$m_{1,2} = -\zeta\omega_n \pm j\omega_n\sqrt{1 - \zeta^2}$$

and the transient solution, as outlined in Sec. 3.5, is

$$x_{b,t} = Ae^{-\zeta\omega_n t}\sin\left(\omega_n\sqrt{1 - \zeta^2}\,t + \phi\right) \tag{3.46}$$

The complete solution is the sum of the steady-state and transient solutions. For the underdamped case, $\zeta < 1$, the complete solution of Eq. (3.42) with a unit step input is

$$x_b(t) = 1 + Ae^{-\zeta\omega_n t}\sin\left(\omega_n\sqrt{1 - \zeta^2}\,t + \phi\right) \tag{3.47}$$

The two constants A and ϕ must next be determined from the initial conditions. In this example the system was initially at rest; therefore $x_b(0) = 0$. The energy stored in a mass is $W = \frac{1}{2}Mv^2$. From the principle of conservation of energy, the *velocity of a system with mass cannot change instantaneously*; thus $Dx_b(0) = 0$. Two equations are necessary, one for $x_b(t)$ and one for $Dx_b(t)$. Differentiating Eq. (3.47) yields

$$\begin{aligned} Dx_b(t) = & -\zeta\omega_n A e^{-\zeta\omega_n t} \sin\left(\omega_n\sqrt{1-\zeta^2}\,t + \phi\right) \\ & + \omega_n\sqrt{1-\zeta^2}\,A e^{-\zeta\omega_n t} \cos\left(\omega_n\sqrt{1-\zeta^2}\,t + \phi\right) \end{aligned} \tag{3.48}$$

Inserting in Eqs. (3.47) and (3.48) the initial conditions

$$x_b(0) = 0, \qquad Dx_b(0) = 0 \qquad t = 0$$

yields

$$0 = 1 + A\sin\phi \qquad 0 = -\zeta\omega_n A\sin\phi + \omega_n\sqrt{1-\zeta^2}\,A\cos\phi$$

These equations are then solved for A and ϕ:

$$A = \frac{-1}{\sqrt{1-\zeta^2}} \qquad \phi = \tan^{-1}\frac{\sqrt{1-\zeta^2}}{\zeta} = \cos^{-1}\zeta$$

Thus the complete solution is

$$x_b(t) = 1 - \frac{e^{-\zeta\omega_n t}}{\sqrt{1-\zeta^2}} \sin\left(\omega_n\sqrt{1-\zeta^2}\,t + \cos^{-1}\zeta\right) \tag{3.49}$$

When the complete solution has been obtained, it should be checked to see that it satisfies the known conditions. For example, putting $t = 0$ into Eq. (3.49) gives $x_b(0) = 0$; therefore the solution checks. In a like manner, the constants can be evaluated for the other two cases of damping. Equation (3.49) shows that the steady-state value $x_{b_{ss}}$ is equal to x_a. Thus, the output *tracks* the input.

3.8 EXAMPLE: SECOND-ORDER SYSTEM —ELECTRICAL

The electric circuit of Fig. 3.5 is used to further illustrate the determination of initial conditions. The circuit, as shown, is in the steady state. At time $t = 0$ the switch is closed. The problem is to solve for the current $i_2(t)$ through the inductor for $t > 0$.

Two loop equations are written for this circuit:

$$10 = 20i_1 - 10i_2 \tag{3.50}$$

$$0 = -10i_1 + \left(D + 25 + \frac{100}{D}\right)i_2 \tag{3.51}$$

Eliminating i_1 from these equations yields

$$10 = \left(2D + 40 + \frac{200}{D}\right)i_2 \tag{3.52}$$

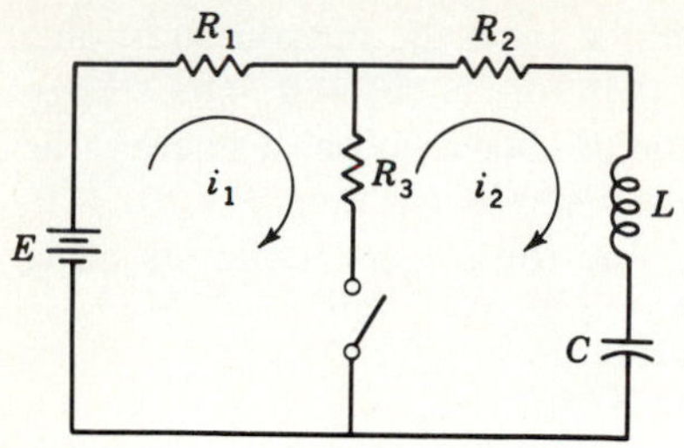

FIGURE 3.5
Electric circuit: $E = 10$ V, $R_1 = 10\ \Omega$, $R_2 = 15\ \Omega$, $R_3 = 10\ \Omega$, $L = 1$ H, $C = 0.01$ F.

Differentiating Eq. (3.52) yields

$$(2D^2 + 40D + 200)i_2 = 0$$

The steady-state solution is found by using the method of Sec. 3.4. Since the input is a step function of value zero, the steady-state output is

$$i_{2,\,ss} = 0 \tag{3.53}$$

This can also be deduced from an inspection of the circuit. When a branch contains a capacitor, the steady-state current is always zero for a dc source.

Next the transient solution is determined. The characteristic equation is $m^2 + 20m + 100 = 0$, for which the roots are $m_{1,2} = -10$. Thus the circuit is critically damped, and the current through the inductor can be expressed by

$$i_2(t) = i_2(t)_t = A_1 e^{-10t} + A_2 t e^{-10t} \tag{3.54}$$

Two equations and two initial conditions are necessary to evaluate the two constants A_1 and A_2. Equation (3.54) and its derivative,

$$Di_2(t) = -10A_1 e^{-10t} + A_2(1 - 10t)e^{-10t} \tag{3.55}$$

are utilized in conjunction with the initial conditions $i_2(0^+)$ and $Di_2(0^+)$. In this example the currents just before the switch is closed are $i_1(0^-) = i_2(0^-) = 0$. The energy stored in the magnetic field of a single inductor is $W = \frac{1}{2}Li^2$. Since this energy cannot change instantly, the current through an inductor cannot change instantly either. Therefore $i_2(0^+) = i_2(0^-) = 0$. $Di_2(t)$ is found from the original circuit equation, Eq. (3.51):

$$Di_2(t) = 10i_i(t) - 25i_2(t) - v_c(t) \tag{3.56}$$

To determine $Di_2(0^+)$, it is necessary first to determine $i_1(0^+)$, $i_2(0^+)$, and $v_c(0^+)$. Since $i_2(0^+) = 0$, Eq. (3.50) yields $i_1(0^+) = 0.5$ A. Since the energy $W = \frac{1}{2}Cv^2$ stored in a capacitor cannot change instantly, the voltage across the capacitor cannot change instantly either. The steady-state value of capacitor voltage for $t < 0$ is 10 V; thus

$$v_c(0^-) = v_c(0^+) = 10 \text{ V} \tag{3.57}$$

Inserting these values into Eq. (3.56) yields $Di_2(0^+) = -5$. Substituting the initial conditions into Eqs. (3.54) and (3.55) results in $A_1 = 0$ and $A_2 = -5$.

Therefore the current through the inductor for $t \geq 0$ is

$$i_2(t) = -5te^{-10t} \tag{3.58}$$

3.9 SECOND-ORDER TRANSIENTS[2]

The response to a unit step-function input is usually used as a means of evaluating the response of a system. The example of Sec. 3.7 is used as an illustrative second-order system. The differential equation given by Eq. (3.42) can be expressed in the form

$$\frac{D^2c}{\omega_n^2} + \frac{2\zeta}{\omega_n} Dc + c = r$$

This is defined as a *simple* second-order equation because there are no derivatives of r on the right side of the equation. The underdamped response ($\zeta < 1$) to a unit step input, subject to zero initial conditions, is derived in Sec. 3.7 and is given by

$$c(t) = 1 - \frac{e^{-\zeta\omega_n t}}{\sqrt{1-\zeta^2}} \sin\left(\omega_n\sqrt{1-\zeta^2}\,t + \cos^{-1}\zeta\right) \tag{3.59}$$

A family of curves representing this equation is shown in Fig. 3.6, where the abscissa is the dimensionless variable $\omega_n t$. The curves are thus a function only of the damping ratio ζ.

These curves show that the amount of overshoot depends on the damping ratio ζ. For the overdamped and critically damped case, $\zeta \geq 1$, there is no

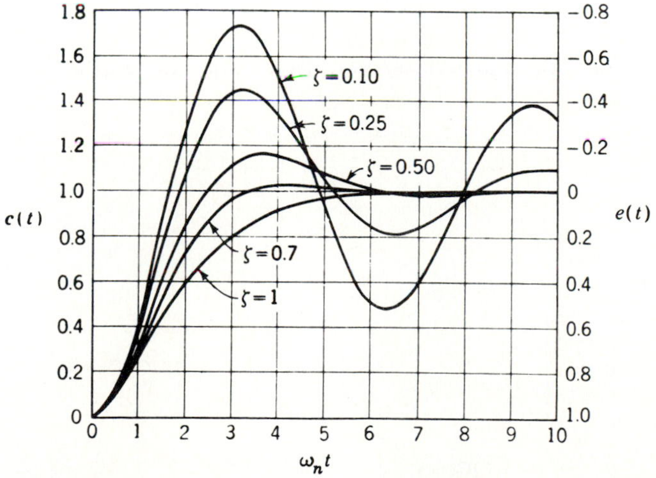

FIGURE 3.6
Simple second-order transients.

overshoot. For the underdamped case, $\zeta < 1$, the system oscillates around the final value. The oscillations decrease with time, and the system response approaches the final value. The peak overshoot for the underdamped system is the first overshoot. The time at which the peak overshoot occurs, t_p, can be found by differentiating $c(t)$ from Eq. (3.59) with respect to time and setting this derivative equal to zero:

$$\frac{dc}{dt} = \frac{\zeta\omega_n e^{-\zeta\omega_n t}}{\sqrt{1-\zeta^2}} \sin\left(\omega_n\sqrt{1-\zeta^2}\,t + \cos^{-1}\zeta\right) - \omega_n e^{-\zeta\omega_n t} \cos\left(\omega_n\sqrt{1-\zeta^2}\,t + \cos^{-1}\zeta\right) = 0$$

This derivative is zero at $\omega_n\sqrt{1-\zeta^2}\,t = 0, \pi, 2\pi, \ldots$. The peak overshoot occurs at the first value after zero, provided there are zero initial conditions; therefore

$$t_p = \frac{\pi}{\omega_n\sqrt{1-\zeta^2}} \tag{3.60}$$

Inserting this value of time in Eq. (3.59) gives the peak overshoot as

$$M_p = c_p = 1 + \exp\left(-\frac{\zeta\pi}{\sqrt{1-\zeta^2}}\right) \tag{3.61}$$

The per unit overshoot M_o as a function of damping ratio is shown in Fig. 3.7, where

$$M_o = \frac{c_p - c_{ss}}{c_{ss}} \tag{3.62}$$

The variation of the frequency of oscillation of the transient with variation of damping ratio is also of interest. In order to represent this variation by one curve, the quantity ω_d/ω_n is plotted against ζ in Fig. 3.8. If the scales of ordinate and abscissa are equal, the curve is an arc of a circle. Note that this curve has been plotted for $\zeta \le 1$. Values of damped natural frequency for $\zeta > 1$ are mathematical only, not physical.

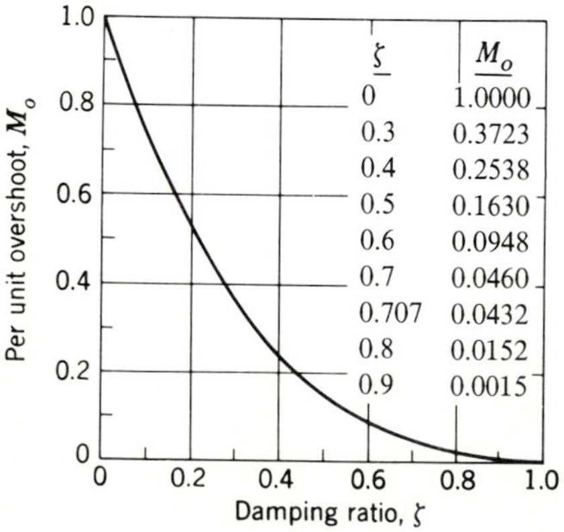

FIGURE 3.7
Peak overshoot vs. damping ratio for a simple second-order equation
$D^2 c/\omega_n^2 + (2\zeta/\omega_n) Dc + c = r(t)$.

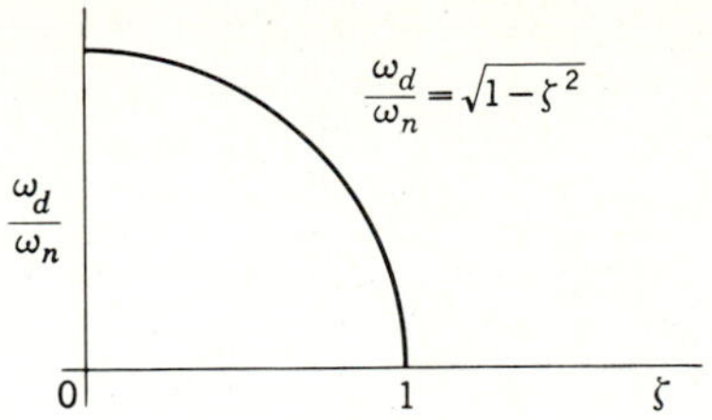

FIGURE 3.8
Frequency of oscillation vs. damping ratio.

The error in the system is the input minus the output; thus the error equation is

$$e = r - c = \frac{e^{-\zeta\omega_n t}}{\sqrt{1-\zeta^2}} \sin\left(\omega_n\sqrt{1-\zeta^2}\,t + \cos^{-1}\zeta\right) \tag{3.63}$$

The variation of error with time is sometimes plotted. These curves can be obtained from Fig. 3.6 by realizing that the curves start at $e(0) = +1$ and have the final value $e(\infty) = 0$.

Response Characteristics

The transient-response curves for the second-order system show a number of significant characteristics.

The overdamped system is slow-acting and does not oscillate about the final position. For some applications the absence of oscillations may be necessary. For example, an elevator cannot be allowed to oscillate at each stop. But for systems where a fast response is necessary, the slow response of an overdamped system cannot be tolerated.

The underdamped system reaches the final value faster than the overdamped system, but the response oscillates about this final value. If this oscillation can be tolerated, the underdamped system is faster-acting. The amount of permissible overshoot determines the desirable value of the damping ratio. A damping ratio $\zeta = 0.4$ has an overshoot of 25.4 percent, and a damping ratio $\zeta = 0.8$ has an overshoot of 1.52 percent.

The settling time is the time required for the oscillations to decrease to a specified absolute percentage of the final value and thereafter remains less than this value. Errors of 2 or 5 percent are common values used to determine settling time. For second-order systems the value of the transient component at any time is equal to or less than the exponential $e^{-\zeta\omega_n t}$. The value of this term is given in Table 3.1 for several values of t expressed in a number of time constants T.

The settling time for a 2 percent error criterion is approximately 4 time constants; for a 5 percent error criterion, it is 3 time constants. The percent error criterion used must be determined from the response desired for the system. The

TABLE 3.1
Exponential values

t	$e^{-\zeta\omega_n t}$	Error, %
$1T$	0.368	36.8
$2T$	0.135	13.5
$3T$	0.050	5.0
$4T$	0.018	1.8
$5T$	0.007	0.7

time for the envelope of the transient to die out is

$$T_s = \frac{\text{number of time constants}}{\zeta\omega_n} \tag{3.64}$$

Since ζ must be determined and adjusted for the permissible overshoot, the undamped natural frequency determines the settling time. When $c(t)_{ss} = 0$, then t_s may be based upon ± 2 percent of $c(t_p)$.

3.10 TIME-RESPONSE SPECIFICATIONS[3]

The desired performance characteristics of a tracking system of any order may be specified in terms of the transient response to a unit step-function input. The performance of a system may be evaluated in terms of the following quantities, as shown in Fig. 3.9.

1. Peak overshoot c_p is the magnitude of the largest overshoot and often occurs at the first overshoot. This may also be expressed in percent of the final value.
2. Time to maximum overshoot t_p is the time required to reach the maximum overshoot.
3. Time to first zero error t_0 is the time required to reach the final value the first time. It is often referred to as *duplicating time*.
4. Settling time t_s is the time required for the output response first to reach and thereafter remain within a prescribed percentage of the final value. This percentage must be specified in the individual case. Common values used for settling time are 2 and 5 percent. As commonly used, the 2 or 5 percent is applied to the envelope which yields T_s. The actual t_s may be smaller than T_s.
5. Rise time t_r is defined as the time for the response, on its initial rise, to go from 0.1 to 0.9 times the steady-state value.
6. Frequency of oscillation of the transient ω_d.

If the system has an initial constant output $c_1(t)_{ss}$ and then the input is changed to a new constant value $c_2(t)_{ss}$, the transient response may have the under-

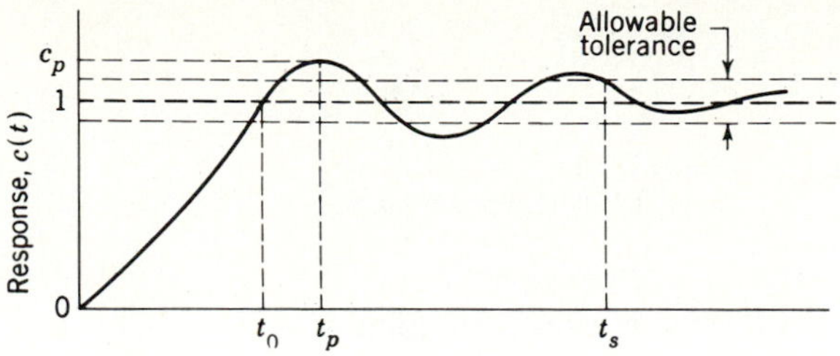

FIGURE 3.9
Typical underdamped response to a step input.

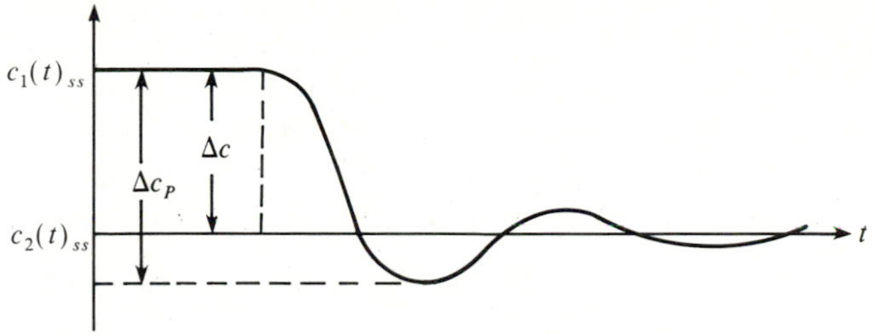

FIGURE 3.10
Underdamped transient.

damped characteristic shown in Fig. 3.10. The per unit overshoot may then be redefined from the expression in Eq. (3.62) to the new value

$$M_o = \frac{\text{maximum overshoot}}{\text{signal transition}} = \frac{|\Delta c_p| - |\Delta c|}{|\Delta c|} \tag{3.65}$$

The time response differs for each set of initial conditions. Therefore, to compare the time response of various systems it is necessary to start with standard initial conditions. The most practical standard is to start with the system at rest. Then the response characteristics, such as maximum overshoot and settling time, can be compared significantly.

For some systems these specifications are also applied for a ramp input. In such cases the plot of error with time is used with the definitions. For systems subject to shock inputs the response due to an impulse is used as a criterion of performance.

3.11 STATE-VARIABLE EQUATIONS[4-7]

The differential equations and the corresponding state equations of various physical systems are derived in Chap. 2. The state variables selected in Chap. 2 are restricted to the energy-storage variables. In Chap. 5 different formulations of the state variables are presented. In large-scale systems, with many inputs and outputs, the state-variable approach can have distinct advantages over conventional methods, especially when digital computers are used to obtain the solutions. While this text is restricted to linear time-invariant systems, the state-variable approach is applicable to nonlinear and to time-varying systems. In these cases a computer is a practical method for obtaining the solution. A feature

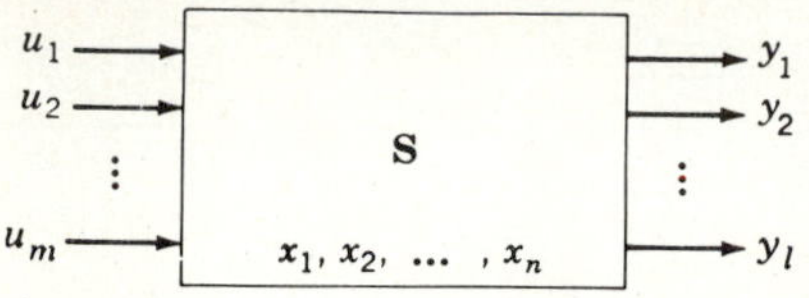

FIGURE 3.11
General system representation.

of the state-variable method is that it decomposes a complex system into a set of smaller systems which can be normalized to have a minimum interaction and which can be solved individually. Also, it provides a unified approach that is used extensively in modern control theory.

The block diagram of Fig. 3.11 represents a system **S** which has m inputs, l outputs, and n state variables. The coefficients in the equations representing a linear time-invariant system are constants. The matrix state and output equations are then

$$\dot{\mathbf{x}}(t) = \mathbf{A}\mathbf{x}(t) + \mathbf{B}\mathbf{u}(t) \tag{3.66}$$

$$\mathbf{y}(t) = \mathbf{C}\mathbf{x}(t) + \mathbf{D}\mathbf{u}(t) \tag{3.67}$$

The variables $\mathbf{x}(t)$, $\mathbf{u}(t)$, and $\mathbf{y}(t)$ are column vectors, and **A**, **B**, **C**, and **D** are matrices having constant elements. Equation (3.66) is solved for the state vector $\mathbf{x}(t)$. This result is then used in Eq. (3.67) to determine the output $\mathbf{y}(t)$.

Homogeneous Solution (State Transition Matrix)

The homogeneous state equation, with the input $\mathbf{u}(t) = \mathbf{0}$, is

$$\dot{\mathbf{x}} = \mathbf{A}\mathbf{x} \tag{3.68}$$

where **A** is a constant $n \times n$ matrix and **x** is an $n \times 1$ column vector.

For the scalar first-order equation $\dot{x} = ax$, the solution, in terms of the initial conditions at time $t = 0$, is

$$x(t) = e^{at}x(0) \tag{3.69}$$

For any other initial condition, at time $t = t_0$, the solution is

$$x(t) = e^{a(t-t_0)}x(t_0) \tag{3.70}$$

Comparing the scalar and the state equations shows the solution of Eq. (3.68) to be analogous to the solution given by Eq. (3.70); it is

$$\mathbf{x}(t) = \exp\left[\mathbf{A}(t - t_0)\right]\mathbf{x}(t_0) \tag{3.71}$$

The exponential function of a scalar which appears in Eq. (3.69) can be expressed as the infinite series

$$e^{at} = \exp[at] = 1 + \frac{at}{1!} + \frac{(at)^2}{2!} + \frac{(at)^3}{3!} + \cdots + \frac{(at)^k}{k!} + \cdots \tag{3.72}$$

The analogous exponential function of a square matrix **A** which appears in

Eq. (3.71), with $t_0 = 0$, is

$$e^{\mathbf{A}t} = \exp[\mathbf{A}t] = \mathbf{I} + \frac{\mathbf{A}t}{1!} + \frac{(\mathbf{A}t)^2}{2!} + \frac{(\mathbf{A}t)^3}{3!} + \cdots + \frac{(\mathbf{A}t)^k}{k!} + \cdots \tag{3.73}$$

Thus $\exp[\mathbf{A}t]$ is a square matrix of the same order as **A**. It is more useful when the infinite series of Eq. (3.73) is put in closed form. This is done in Sec. 3.13. It is common to call this the *state transition matrix* or the *fundamental matrix* of the system and to denote it by

$$\mathbf{\Phi}(t) = e^{\mathbf{A}t} = \exp[\mathbf{A}t] \tag{3.74}$$

The term *state transition matrix* (STM) is descriptive of the unforced or natural response and is the expression preferred by engineers. The STM has the following properties:[10]

1. $\mathbf{\Phi}(t_2 - t_1)\mathbf{\Phi}(t_1 - t_0) = \mathbf{\Phi}(t_2 - t_0)$ for any t_0, t_1, t_2 (3.75)
2. $\mathbf{\Phi}(t)\mathbf{\Phi}(t)\cdots\mathbf{\Phi}(t) = \mathbf{\Phi}^q(t) = \mathbf{\Phi}(qt)$ q = positive integer (3.76)
3. $\mathbf{\Phi}^{-1}(t) = \mathbf{\Phi}(-t)$ (3.77)
4. $\mathbf{\Phi}(0) = \mathbf{I}$ unity matrix (3.78)
5. $\mathbf{\Phi}(t)$ is nonsingular for all finite values of t (3.79)

3.12 CHARACTERISTIC VALUES

Consider a system of equations represented by

$$\dot{\mathbf{x}} = \mathbf{A}\mathbf{x} \tag{3.80}$$

The signals $\dot{\mathbf{x}}$ and $\mathbf{x}$ are column vectors, and **A** is a square matrix of order n. One case for which a solution of this equation exists is if $\mathbf{x}$ and $\dot{\mathbf{x}}$ have the same direction in the state space but differ only in magnitude by a scalar proportionality factor λ. The solution must therefore have the form $\dot{\mathbf{x}} = \lambda\mathbf{x}$. Inserting this into Eq. (3.80) and rearranging terms yields

$$[\lambda\mathbf{I} - \mathbf{A}]\mathbf{x} = \mathbf{0}$$

This equation has a nontrivial solution only if $\mathbf{x}$ is not zero. It is therefore required that the matrix $[\lambda\mathbf{I} - \mathbf{A}]$ must not have full rank; therefore the determinant of the coefficients of $\mathbf{x}$ must be zero:

$$Q(\lambda) \equiv |\lambda\mathbf{I} - \mathbf{A}| = 0 \tag{3.81}$$

When **A** is of order n, the resulting polynomial equation is the *characteristic equation*

$$Q(\lambda) = \lambda^n + a_{n-1}\lambda^{n-1} + \cdots + a_1\lambda + a_0 = 0 \tag{3.82}$$

The roots λ_i of the characteristic equation are called the *characteristic values* or *eigenvalues* of **A**. The roots may be distinct (simple) or repeated with a multiplicity p. Also, a root may be real or complex. Complex roots must appear in conjugate pairs and the set of roots containing the complex-conjugate roots is said to be *self-conjugate*. The polynomial $Q(\lambda)$ may be written in factored form as

$$Q(\lambda) = (\lambda - \lambda_1)(\lambda - \lambda_2)\cdots(\lambda - \lambda_n) \tag{3.83}$$

The product of *eigenvalues* of a matrix **A** is equal to its determinant, that is, $\lambda_1\lambda_2\cdots\lambda_n = |\mathbf{A}|$. Also, the sum of the *eigenvalues* is equal to the sum of the elements on the main diagonal (the trace), of **A**, that is,

$$\sum_{l}^{n}\lambda_i = \sum_{i=1}^{n} a_{ii} \equiv \text{trace}\,\mathbf{A}$$

3.13 EVALUATING THE STATE TRANSITION MATRIX

There are several methods for evaluating the STM $\boldsymbol{\Phi}(t) = \exp[\mathbf{A}t]$ in closed form for a given matrix **A**. The method illustrated below is based on the Cayley-Hamilton theorem. The Laplace transform method is covered in the next chapter, and a state transformation method is covered in Chap. 5. Consider a general polynomial of the form

$$N(\lambda) = \lambda^m + C_{m-1}\lambda^{m-1} + \cdots + C_1\lambda + C_0 \qquad (3.84a)$$

When the polynomial $N(\lambda)$ is divided by the characteristic polynomial $Q(\lambda)$, the result is

$$\frac{N(\lambda)}{Q(\lambda)} = F(\lambda) + \frac{R(\lambda)}{Q(\lambda)}$$

or

$$N(\lambda) = F(\lambda)Q(\lambda) + R(\lambda) \qquad (3.84b)$$

The function $R(\lambda)$ is the remainder and it is a polynomial whose maximum order is $n - 1$, or 1 less than the order of $Q(\lambda)$. For $\lambda = \lambda_i$ the value $Q(\lambda_i) = 0$; thus

$$N(\lambda_i) = R(\lambda_i) \qquad (3.85)$$

The matrix polynomial corresponding to Eq. (3.84a), using **A** as the variable, is

$$\mathbf{N}(\mathbf{A}) = \mathbf{A}^m + C_{m-1}\mathbf{A}^{m-1} + \cdots + C_1\mathbf{A} + C_0\mathbf{I} \qquad (3.86)$$

Since the characteristic equation $Q(\lambda) = 0$ has n roots, there are n equations $Q(\lambda_1) = 0$, $Q(\lambda_2) = 0, \ldots, Q(\lambda_n) = 0$. The analogous matrix equation is

$$Q(\mathbf{A}) = \mathbf{A}^n + a_{n-1}\mathbf{A}^{n-1} + \cdots + a_1\mathbf{A} + a_0\mathbf{I} = \mathbf{0}$$

where **0** indicates a null matrix of the same order as $Q(\mathbf{A})$. This equation implies the Cayley-Hamilton theorem, which is sometimes expressed as "every square matrix **A** satisfies its own characteristic equation." The matrix polynomial corresponding to Eqs. (3.84b) and (3.85) is therefore

$$\mathbf{N}(\mathbf{A}) = \mathbf{F}(\mathbf{A})\mathbf{Q}(\mathbf{A}) + \mathbf{R}(\mathbf{A}) = \mathbf{R}(\mathbf{A}) \qquad (3.87)$$

Equations (3.85) and (3.87) are valid when $N(\lambda)$ is a polynomial of any order (or even an infinite series) as long as it is analytic. The exponential function $N(\lambda) = e^{\lambda t}$ is an analytic function which can be represented by an infinite series as shown in Eq. (3.72). Since this function converges in the region of analyticity, it can be expressed in closed form by a polynomial in λ of degree $n - 1$. Thus, for each eigenvalue, from Eq. (3.85)

$$e^{\lambda_i t} = R(\lambda_i) = \alpha_0(t) + \alpha_1(t)\lambda_i + \cdots + \alpha_k(t)\lambda_i^k + \cdots + \alpha_{n-1}(t)\lambda_i^{n-1} \quad (3.88)$$

Inserting the n distinct roots λ_i into Eq. (3.88) yields n equations which can be solved simultaneously for the coefficients α_k. These coefficients may be inserted in the corresponding matrix equation

$$e^{\mathbf{A}t} = N(\mathbf{A}) = R(\mathbf{A}) = \alpha_0(t)\mathbf{I} + \alpha_1(t)\mathbf{A} + \cdots + \alpha_k(t)\mathbf{A}^k + \cdots + \alpha_{n-1}(t)\mathbf{A}^{n-1} \quad (3.89)$$

For the STM this yields

$$\mathbf{\Phi}(t) = \exp[\mathbf{A}t] = \sum_{k=0}^{n-1} \alpha_k(t)\mathbf{A}^k \quad (3.90)$$

For the case of multiple roots, refer to a text on linear algebra[4] for the evaluation of $\mathbf{\Phi}(t)$.

Example. Find $\exp[\mathbf{A}t]$ when

$$\mathbf{A} = \begin{bmatrix} 0 & 6 \\ -1 & -5 \end{bmatrix}$$

The characteristic equation is

$$Q(\lambda) = |\lambda\mathbf{I} - \mathbf{A}| = \begin{vmatrix} \lambda & -6 \\ 1 & \lambda + 5 \end{vmatrix} = \lambda^2 + 5\lambda + 6 = 0$$

The roots of this equation are $\lambda_1 = -2$ and $\lambda_2 = -3$. Since $\mathbf{A}$ is a second-order matrix, the remainder polynomial given by Eq. (3.88) is of first order:

$$N(\lambda_i) = R(\lambda_i) = \alpha_0 + \alpha_1\lambda_i \quad (3.91)$$

Substituting $N(\lambda_i) = e^{\lambda_i t}$ and the two roots λ_1 and λ_2 into Eq. (3.91) yields

$$e^{-2t} = \alpha_0 - 2\alpha_1 \qquad \text{and} \qquad e^{-3t} = \alpha_0 - 3\alpha_1$$

Solving these equations for the coefficients α_0 and α_1 yields

$$\alpha_0 = 3e^{-2t} - 2e^{-3t} \qquad \text{and} \qquad \alpha_1 = e^{-2t} - e^{-3t}$$

The STM obtained by using Eq. (3.90) is

$$\mathbf{\Phi}(t) = \exp[\mathbf{A}t] = \alpha_0\mathbf{I} + \alpha_1\mathbf{A} = \begin{bmatrix} \alpha_0 & 0 \\ 0 & \alpha_0 \end{bmatrix} + \begin{bmatrix} 0 & 6\alpha_1 \\ -\alpha_1 & -5\alpha_1 \end{bmatrix}$$

$$= \begin{bmatrix} 3e^{-2t} - 2e^{-3t} & 6e^{-2t} - 6e^{-3t} \\ -e^{-2t} + e^{-3t} & -2e^{-2t} + 3e^{-3t} \end{bmatrix} \quad (3.92)$$

Another procedure for evaluating the STM, $\mathbf{\Phi}(t)$, is by use of the Sylvester expansion. The format in this method makes it especially useful for use with a digital computer. The method is presented here without proof for the case when $\mathbf{A}$ is a square matrix with n distinct eigenvalues. The polynomial $\mathbf{N(A)}$ can be written

$$e^{\mathbf{A}t} = \mathbf{N}(\mathbf{A}) = \sum_{i=1}^{n} N(\lambda_i)\mathbf{Z}_i(\lambda) \tag{3.93}$$

where

$$\mathbf{Z}_i(\lambda) = \frac{\sum\limits_{\substack{j=1\\ j\neq i}}^{n} (\mathbf{A} - \lambda_j \mathbf{I})}{\sum\limits_{\substack{j=1\\ j\neq i}}^{n} (\lambda_i - \lambda_j)} \tag{3.94}$$

For the previous example there are two eigenvalues $\lambda_1 = -2$ and $\lambda_2 = -3$. The two matrices $\mathbf{Z}_1$ and $\mathbf{Z}_2$ are evaluated from Eq. (3.94):

$$\mathbf{Z}_1 = \frac{\mathbf{A} - \lambda_2 \mathbf{I}}{\lambda_1 - \lambda_2} = \frac{\begin{bmatrix} 0+3 & 6 \\ -1 & -5+3 \end{bmatrix}}{-2-(-3)} = \begin{bmatrix} 3 & 6 \\ -1 & -2 \end{bmatrix}$$

$$\mathbf{Z}_2 = \frac{\mathbf{A} - \lambda_1 \mathbf{I}}{\lambda_2 - \lambda_1} = \frac{\begin{bmatrix} 0+2 & 6 \\ -1 & -5+2 \end{bmatrix}}{-3-(-2)} = \begin{bmatrix} -2 & -6 \\ 1 & 3 \end{bmatrix}$$

Using these values in Eq. (3.93), where $N(\lambda_i) = e^{\lambda_i t}$, yields

$$\mathbf{\Phi}(t) = \exp[\mathbf{A}t] = e^{\lambda_1 t}\mathbf{Z}_1 + e^{\lambda_2 t}\mathbf{Z}_2$$

Equation (3.92) contains the value of $\mathbf{\Phi}(t)$.

The matrices $\mathbf{Z}_i(\lambda)$ are called the *constituent matrices* of **A** and have the following properties:

1. Orthogonality:

$$\mathbf{Z}_i\mathbf{Z}_j = \mathbf{0} \qquad i \neq j \tag{3.95}$$

2. The sum of the complete set is equal to the unit matrix:

$$\sum_{i=1}^{n} \mathbf{Z}_i = \mathbf{I} \tag{3.96}$$

3. Idempotency:

$$\mathbf{Z}_i^r = \mathbf{Z}_i \tag{3.97}$$

where $r > 0$ is any positive integer

Equation (3.93) can be conveniently applied to evaluate other functions of **A**. For example, to obtain $\mathbf{A}^r$ use the form

$$\mathbf{A}^r = \lambda_1^r\mathbf{Z}_1 + \lambda_2^r\mathbf{Z}_2 + \cdots + \lambda_n^r\mathbf{Z}_n = \sum_{i=1}^{n} \lambda_i^r \mathbf{Z}_i \tag{3.98}$$

The operations involved in computing Eq. (3.98) consist primarily of addition, which is simpler than multiplying **A** by itself r times.

3.14 COMPLETE SOLUTION OF THE STATE EQUATION

When an input $\mathbf{u}(t)$ is present, the complete solution for $\mathbf{x}(t)$ is obtained from Eq. (3.66). The starting point for obtaining the solution is the following equality, obtained by applying the rule for the derivative of the product of two matrices (see Sec. 2.3):

$$\frac{d}{dt}\left[e^{-\mathbf{A}t}\mathbf{x}(t)\right] = e^{-\mathbf{A}t}\left[\dot{\mathbf{x}}(t) - \mathbf{A}\mathbf{x}(t)\right] \tag{3.99}$$

Utilizing Eq. (3.66), $\dot{\mathbf{x}} = \mathbf{A}\mathbf{x} + \mathbf{B}\mathbf{u}$, in the right side of Eq. (3.99) yields

$$\frac{d}{dt}\left[e^{-\mathbf{A}t}\mathbf{x}(t)\right] = e^{-\mathbf{A}t}\mathbf{B}\mathbf{u}(t)$$

Integrating this equation between 0 and t gives

$$e^{-\mathbf{A}t}\mathbf{x}(t) - \mathbf{x}(0) = \int_0^t e^{-\mathbf{A}\tau}\mathbf{B}\mathbf{u}(\tau)\,d\tau \tag{3.100}$$

Multiplying by $e^{\mathbf{A}t}$ and rearranging terms produces

$$\mathbf{x}(t) = e^{\mathbf{A}t}\mathbf{x}(0) + \int_0^t e^{\mathbf{A}(t-\tau)}\mathbf{B}\mathbf{u}(\tau)\,d\tau = \boldsymbol{\Phi}(t)\mathbf{x}(0) + \int_0^t \boldsymbol{\Phi}(t-\tau)\mathbf{B}\mathbf{u}(\tau)\,d\tau \tag{3.101}$$

Using the STM and generalizing for initial conditions at time $t = t_0$ gives the solution to the state-variable equation with an input $\mathbf{u}(t)$ as

$$\mathbf{x}(t) = \boldsymbol{\Phi}(t-t_0)\mathbf{x}(t_0) + \int_{t_0}^t \boldsymbol{\Phi}(t-\tau)\mathbf{B}\mathbf{u}(\tau)\,d\tau \qquad t > t_0 \tag{3.102}$$

This equation is called the *state transition equation*; i.e., it describes the change of state relative to the initial conditions $\mathbf{x}(t_0)$ and the input $\mathbf{u}(t)$. It may be more convenient to change the variables in the integral of Eq. (3.101) by letting $\beta = t - \tau$, which gives

$$\mathbf{x}(t) = \boldsymbol{\Phi}(t)\mathbf{x}(0) + \int_0^t \boldsymbol{\Phi}(\beta)\mathbf{B}\mathbf{u}(t-\beta)\,d\beta \tag{3.103}$$

In modern terminology the state solution $\mathbf{x}(t)$ as given by Eqs. (3.101) to (3.103) is described by

$$\mathbf{x}(t) = \mathbf{x}_{zi}(t) + \mathbf{x}_{zs}(t) \tag{3.104}$$

where $\mathbf{x}_{zi}(t)$ is called the *zero-input* response, that is, $\mathbf{u}(t) = \mathbf{0}$, and $\mathbf{x}_{zs}(t)$ is called the *zero-state* response, that is, $\mathbf{x}(t_0) = \mathbf{0}$.

Example. Solve the following state equation for a unit step-function scalar input, $u(t) = u_{-1}(t)$:

$$\dot{\mathbf{x}} = \begin{bmatrix} 0 & 6 \\ -1 & -5 \end{bmatrix}\mathbf{x} + \begin{bmatrix} 0 \\ 1 \end{bmatrix}u(t)$$

The STM for this equation is given by Eq. (3.92). Thus, the total time solution is obtained by substituting this value of $\boldsymbol{\Phi}(t)$ into Eq. (3.103). Since $u(t-\beta)$ has the value of unity for $0 < \beta < t$, Eq. (3.103) yields

$$\mathbf{x}(t) = \mathbf{\Phi}(t)\mathbf{x}(0) + \int_0^t \begin{bmatrix} 6e^{-2\beta} - 6e^{-3\beta} \\ -2e^{-2\beta} + 3e^{-3\beta} \end{bmatrix} d\beta$$

$$= \begin{bmatrix} 3e^{-2t} - 2e^{-3t} & 6e^{-2t} - 6e^{-3t} \\ -e^{-2t} + e^{-3t} & -2e^{-2t} + 3e^{-3t} \end{bmatrix} \begin{bmatrix} x_1(0) \\ x_2(0) \end{bmatrix} + \begin{bmatrix} 1 - 3e^{-2t} + 2e^{-3t} \\ e^{-2t} - e^{-3t} \end{bmatrix} \tag{3.105}$$

The integral of a matrix is the integral of each element in the matrix with respect to the indicated scalar variable. This property is used in evaluating the particular integral in Eq. (3.105). Inserting the initial conditions produces the final solution for the state transition equation. The Laplace transform method presented in Chap. 4 is a more direct method for finding this solution. An additional method is shown in Chap. 5.

3.15 DISCRETE STATE EQUATION

The solution of the state equation in the continuous-time domain is given by Eq. (3.103). The evaluation of $\mathbf{x}(t)$ at equally spaced intervals of time t_k can be obtained from a discrete-time representation of this equation. The times at which the solution is to be obtained are $t = T, 2T, 3T, 4T, \ldots,$ that is,

$$t_k = kT \qquad k = 1, 2, \ldots \tag{3.106}$$

where $t_k - t_{k-1} = T =$ constant. With $t = kT$ and $t_0 = (k-1)T$, Eq. (3.103) becomes

$$\mathbf{x}(kT) = e^{\mathbf{A}T}\mathbf{x}[(k-1)T] + \int_{(k-1)T}^{kT} e^{\mathbf{A}\tau}\mathbf{B}\mathbf{u}(kT - \tau)\, d\tau \tag{3.107}$$

If $u(kT - \tau)$ is assumed to be constant over the interval $0 < \tau \le T$ and has the value $\mathbf{u}[(k-1)T]$, then, since $\mathbf{B}$ is a constant matrix, Eq. (3.107) becomes

$$\mathbf{x}(kT) = e^{\mathbf{A}T}\mathbf{x}[(k-1)T] + \left(\int_0^T e^{\mathbf{A}\tau}\, d\tau\, \mathbf{B}\right)\mathbf{u}[(k-1)T]$$

$$= \mathbf{A}(T)\mathbf{x}[(k-1)T] + \mathbf{B}(T)\mathbf{u}[(k-1)T] \qquad k = 1, 2, \ldots \tag{3.108}$$

where the matrices

$$\mathbf{A}(T) = e^{\mathbf{A}T} \tag{3.109}$$

$$\mathbf{B}(T) = \int_0^T e^{\mathbf{A}\tau}\, d\tau\, \mathbf{B} \tag{3.110}$$

must be evaluated accurately. Then Eq. (3.108) is a simple recursive equation for obtaining accurate values of $\mathbf{x}(kt)$. The matrices $\mathbf{A}(T)$ and $\mathbf{B}(T)$ can be evaluated by first forming the partitioned matrix[9]

$$\overline{\mathbf{A}} = \begin{bmatrix} \dfrac{\mathbf{A}}{2} & \mathbf{0} \\ \mathbf{I} & -\dfrac{\mathbf{A}}{2} \end{bmatrix} \tag{3.111}$$

The matrix $e^{\overline{\mathbf{A}}t}$ is partitioned into subblocks of corresponding sizes in the form

$$e^{\overline{\mathbf{A}}t} = \begin{bmatrix} \mathbf{\Phi}_{11}(t) & \mathbf{\Phi}_{12}(t) \\ \mathbf{\Phi}_{21}(t) & \mathbf{\Phi}_{22}(t) \end{bmatrix} \tag{3.112}$$

where, for $t = 0$

$$e^{\overline{\mathbf{A}}t}\Big|_{t=0} = \begin{bmatrix} \mathbf{I} & \mathbf{0} \\ \mathbf{0} & \mathbf{I} \end{bmatrix} \tag{3.113}$$

The derivative of the right side of Eq. (3.112) is

$$\frac{d}{dt} e^{\overline{\mathbf{A}}t} = \begin{bmatrix} \dot{\mathbf{\Phi}}_{11}(t) & \dot{\mathbf{\Phi}}_{12}(t) \\ \dot{\mathbf{\Phi}}_{21}(t) & \dot{\mathbf{\Phi}}_{22}(t) \end{bmatrix} \tag{3.114}$$

The derivative obtained from the left side of Eq. (3.112) is

$$\frac{d}{dt} e^{\overline{\mathbf{A}}t} = \overline{\mathbf{A}} e^{\overline{\mathbf{A}}t} = \begin{bmatrix} \dfrac{\mathbf{A}}{2} & \mathbf{0} \\ \mathbf{I} & -\dfrac{\mathbf{A}}{2} \end{bmatrix} \begin{bmatrix} \mathbf{\Phi}_{11}(t) & \mathbf{\Phi}_{12}(t) \\ \mathbf{\Phi}_{21}(t) & \mathbf{\Phi}_{22}(t) \end{bmatrix}$$

$$= \begin{bmatrix} \dfrac{\mathbf{A}\mathbf{\Phi}_{11}(t)}{2} & \dfrac{\mathbf{A}\mathbf{\Phi}_{12}(t)}{2} \\ -\dfrac{\mathbf{A}\mathbf{\Phi}_{21}(t)}{2} + \mathbf{\Phi}_{11}(t) & \mathbf{\Phi}_{12} - \dfrac{\mathbf{A}\mathbf{\Phi}_{22}(t)}{2} \end{bmatrix} \tag{3.115}$$

Equating the corresponding submatrices of Eqs. (3.114) and (3.115) and solving for $\mathbf{\Phi}_{11}(t)$ and $\mathbf{\Phi}_{21}(t)$, with the boundary conditions of Eqs. (3.113), yields

$$\mathbf{\Phi}_{11}(t) = e^{\mathbf{A}t/2} \tag{3.116}$$

$$\mathbf{\Phi}_{21}(t) = e^{-\mathbf{A}t/2} \int_0^t e^{\mathbf{A}\tau}\, d\tau \tag{3.117}$$

Therefore, the matrices $\mathbf{A}(T)$ and $\mathbf{B}(T)$ given by Eqs. (3.109) and (3.110), respectively, are evaluated from Eqs. (3.116) and (3.117), respectively, by letting $t = T$, that is,

$$\mathbf{A}(T) = e^{\mathbf{A}T} = \mathbf{\Phi}_{11}^2(T) \tag{3.118}$$

$$\mathbf{B}(T) = \int_0^T e^{\mathbf{A}\tau}\, d\tau\, \mathbf{B} = \mathbf{\Phi}_{11}(T)\mathbf{\Phi}_{21}(T)\mathbf{B} \tag{3.119}$$

The matrix $e^{\overline{\mathbf{A}}T}$ can be evaluated by truncating the infinite series of Eq. (3.72) to include the terms

$$e^{\overline{\mathbf{A}}T} = \mathbf{I} + \frac{\overline{\mathbf{A}}T}{1!} + \frac{\overline{\mathbf{A}}^2T^2}{2!} + \cdots + \frac{\overline{\mathbf{A}}^LT^L}{L!} \tag{3.120}$$

This summation converges rapidly for small values of T. (T is selected to be much smaller than the reciprocal of the magnitude of the real part of the eigenvalue closest to the imaginary axis for stable systems.) A simple test is available[10] for determining L in Eq. (3.120) which must be used to achieve a desired accuracy. The resulting evaluation of $e^{\bar{\mathbf{A}}T}$ yields the submatrices $\mathbf{\Phi}_{11}(T)$ and $\mathbf{\Phi}_{21}(T)$, which are used to compute $\mathbf{A}(T)$ and $\mathbf{B}(T)$ in accordance with Eqs. (3.118) and (3.119). The approximate evaluation of $e^{\bar{\mathbf{A}}T}$ in accordance with Eq. (3.120) is easily accomplished on a digital computer since it involves only the operations of matrix multiplication and addition.

In Eq. (3.108) the index k can be replaced by $k+1$ and yields the *difference equation*

$$\mathbf{x}[(k+1)T] = \mathbf{A}(T)\mathbf{x}(kT) + \mathbf{B}(T)\mathbf{u}(kT) \qquad k = 0, 1, 2, \ldots \tag{3.121}$$

This is the discrete-state-equation representation of the system in the discrete-time domain. It is useful in the synthesis of control systems which incorporate digital controllers. This is introduced in Chap. 22 and is an area for further study.[8,10]

3.16 SUMMARY

This chapter has established the manner of solving differential equations. The steady-state and transient solutions are determined separately, and then the constants of the transient component of the solution are evaluated to satisfy the initial conditions. The ability to anticipate the form of the response is very important. The solution of differential equations has been extended to include solution of the matrix state and output equations. This method is also applicable to multiple-input multiple-output, time-varying, and nonlinear systems. The matrix formulation of these systems lends itself to digital-computer solutions. Obtaining the complete solution by means of the Laplace transform is covered in the next chapter.

REFERENCES

1. Wylie, C. R., Jr.: *Advanced Engineering Mathematics*, 4th ed., McGraw-Hill, New York, 1975.
2. Kinariwala, B., et al.: *Linear Circuits and Computation*, Wiley, New York, 1973.
3. D'Azzo, J. J., and C. H. Houpis: *Feedback Control System Analysis and Synthesis*, 2nd ed., McGraw-Hill, New York, 1966.
4. DeRusso, P. M., et al.: *State Variables for Engineers*, Wiley, New York, 1965.
5. Schwartz, R. J., and B. Friedland: *Linear Systems*, McGraw-Hill, New York, 1965.
6. Ward, J. R., and R. D. Strum: *State Variable Analysis*, Prentice-Hall, Englewood Cliffs, N.J., 1970.
7. Wiberg, D. M.: *State Space and Linear Systems*, Schaum's Outline Series, McGraw-Hill, New York, 1971.
8. Kuo, B. C.: *Discrete-Data Control Systems*, Prentice-Hall, Englewood Cliffs, N.J., 1970.
9. Reid, J. G.: "Linear Systems Analysis and Digital Computation," unpublished notes, Air Force Institute of Technology, Wright-Patterson Air Force Base, Ohio, 1979.
10. Cadzow, J. A., and H. R. Martens: *Discrete-Time and Computer Control Systems*, Prentice-Hall, Englewood Cliffs, N.J., 1970.

CHAPTER 4

LAPLACE TRANSFORM

4.1 INTRODUCTION

Solution of differential equations with discontinuous inputs or of higher than second order is laborious by the classical method. Also, inserting the initial conditions to evaluate the constants of integration requires the simultaneous solution of a number of algebraic equations equal to the order of the differential equation. To facilitate and systematize the solution of ordinary constant-coefficient differential equations, the Laplace transform method is used extensively.[1-3] The advantages of this modern transform method for the analysis of feedback systems are:

1. It includes the boundary or initial conditions.
2. The work involved in the solution is simple algebra.
3. The work is systematized.
4. The use of a table of transforms reduces the labor required.
5. Discontinuous inputs can be treated.
6. The transient and steady-state components of the solution are obtained simultaneously.

The disadvantage of transform methods is that if they are used mechanically, without knowledge of the actual theory involved, they sometimes yield erroneous results. Also, a particular equation can sometimes be solved more simply and with less work by the classical method. Although an understanding of the Laplace transform method is essential, it should be emphasized that the solutions of differential equations are readily obtained by using digital-computer programs (see App. B)[8,11].

Laplace transforms are also applied to the solution of system equations which are in matrix state-variable format. The method for using the state and output equations to obtain the system transfer function is presented.

4.2 DEFINITION OF THE LAPLACE TRANSFORM

The direct Laplace transformation of a function of time $f(t)$ is given by

$$\mathscr{L}[f(t)] = \int_0^\infty f(t)e^{-st}\,dt = F(s) \tag{4.1}$$

where $\mathscr{L}[f(t)]$ is a shorthand notation for the Laplace integral. Evaluation of the integral results in a function $F(s)$ which has s as the parameter. This parameter s is a complex quantity of the form $\sigma + j\omega$. It is to be noted that as the limits of integration are zero and infinity, it is immaterial what value $f(t)$ has for negative or zero time.

There are limitations on the functions $f(t)$ that are Laplace-transformable. Basically, the requirement is that the Laplace integral converge, which means that this integral has a definite functional value. To meet this requirement[3] the function $f(t)$ must be (1) piecewise continuous over every finite interval $0 \le t_1 \le t \le t_2$ and (2) of exponential order. A function is piecewise continuous in a finite interval if that interval can be divided into a finite number of subintervals over each of which the function is continuous and at the ends of each of which $f(t)$ possesses finite right- and left-hand limits. A function $f(t)$ is of exponential order if there exists a constant a such that the product $e^{-at}|f(t)|$ is bounded for all values of t greater than some finite value T. This imposes the restriction that σ, the real part of s, must be greater than a lower bound σ_a for which the product $e^{-\sigma_a t}|f(t)|$ is of exponential order. A linear differential equation with constant coefficients and with a finite number of terms is Laplace-transformable if the driving function is Laplace-transformable.

All cases covered in this book are Laplace-transformable. The basic purpose in using the Laplace transform is to obtain a method of solving differential equations which involves only simple algebraic operations in conjunction with a table of transforms.

4.3 DERIVATION OF LAPLACE TRANSFORMS OF SIMPLE FUNCTIONS

A number of examples are presented to show the derivation of the Laplace transform of several time functions. A list of common transform pairs is given in App. A.

Step Function $u_{-1}(t)$

The Laplace transform of the unit step function $u_{-1}(t)$ is

$$\mathscr{L}[u_{-1}(t)] = \int_0^\infty u_{-1}(t)e^{-st}\,dt = U_{-1}(s) \tag{4.2}$$

Since $u_{-1}(t)$ has the value 1 over the limits of integration,

$$U_{-1}(s) = \int_0^\infty e^{-st}\,dt = -\left.\frac{e^{-st}}{s}\right|_0^\infty = \frac{1}{s} \qquad \text{if } \sigma > 0 \tag{4.3}$$

The step function is undefined at $t = 0$; but this is immaterial, for it is noted that the integral is defined by a limit process,

$$\int_0^\infty f(t)e^{-st}\,dt = \lim_{\substack{T\to\infty \\ \varepsilon\to 0}} \int_\varepsilon^T f(t)e^{-st}\,dt \tag{4.4}$$

and the explicit value at $t = 0$ does not affect the value of the integral. The value of the integral obtained by taking limits is implied in each case but is not written out explicitly.

Decaying Exponential $e^{-\alpha t}$

The exponent α is a positive real number.

$$\mathscr{L}[e^{-\alpha t}] = \int_0^\infty e^{-\alpha t}e^{-st}\,dt = \int_0^\infty e^{-(s+\alpha)t}\,dt$$

$$= -\left.\frac{e^{-(s+\alpha)t}}{s+\alpha}\right|_0^\infty = \frac{1}{s+\alpha} \qquad \sigma > -\alpha \tag{4.5}$$

Sinusoid $\cos \omega t$

Here ω is a positive real number.

$$\mathscr{L}[\cos \omega t] = \int_0^\infty \cos \omega t\, e^{-st}\,dt \tag{4.6}$$

Expressing $\cos \omega t$ in exponential form gives

$$\cos \omega t = \frac{e^{j\omega t} + e^{-j\omega t}}{2}$$

Then $$\mathscr{L}[\cos \omega t] = \frac{1}{2}\left(\int_0^\infty e^{(j\omega - s)t}\,dt + \int_0^\infty e^{(-j\omega - s)t}\,dt\right)$$

$$= \frac{1}{2}\left[\frac{e^{(j\omega - s)t}}{j\omega - s} + \frac{e^{(-j\omega - s)t}}{-j\omega - s}\right]_0^\infty$$

$$= \frac{1}{2}\left(-\frac{1}{j\omega - s} - \frac{1}{-j\omega - s}\right) = \frac{s}{s^2 + \omega^2} \qquad \sigma > 0 \tag{4.7}$$

Ramp Function $u_{-2}(t) = tu_{-1}(t)$

$$\mathscr{L}[t] = \int_0^\infty te^{-st}\,dt \qquad \sigma > 0 \tag{4.8}$$

This is integrated by parts by using

$$\int_a^b u\,dv = uv\Big|_a^b - \int_a^b v\,du$$

Let $u = t$ and $dv = e^{-st}\,dt$. Then $du = dt$ and $v = -e^{-st}/s$. Thus

$$\int_0^\infty te^{-st}\,dt = -\frac{te^{-st}}{s}\Big|_0^\infty - \int_0^\infty \left(-\frac{e^{-st}}{s}\right) dt$$

$$= 0 - \frac{e^{-st}}{s^2}\Big|_0^\infty = \frac{1}{s^2} \qquad \sigma > 0 \tag{4.9}$$

4.4 LAPLACE TRANSFORM THEOREMS

Several theorems that are useful in applying the Laplace transform are presented in this section. In general, they are helpful in evaluating transforms.

Theorem 1 Linearity. If a is a constant or is independent of s and t, and if $f(t)$ is transformable, then

$$\mathscr{L}[af(t)] = a\mathscr{L}[f(t)] = aF(s) \tag{4.10}$$

Theorem 2 Superposition. If $f_1(t)$ and $f_2(t)$ are both Laplace-transformable, the principle of superposition applies:

$$\mathscr{L}[f_1(t) \pm f_2(t)] = \mathscr{L}[f_1(t)] \pm \mathscr{L}[f_2(t)] = F_1(s) \pm F_2(s) \tag{4.11}$$

Theorem 3 Translation in time. If the Laplace transform of $f(t)$ is $F(s)$ and a is a positive real number, the Laplace transform of the translated function $f(t-a)u_{-1}(t-a)$ is

$$\mathscr{L}[f(t-a)u_{-1}(t-a)] = e^{-as}F(s) \tag{4.12}$$

Translation in the positive t direction in the real domain becomes multiplication by the exponential e^{-as} in the s domain.

Theorem 4 Complex differentiation. If the Laplace transform of $f(t)$ is $F(s)$, then

$$\mathscr{L}[tf(t)] = -\frac{d}{ds}F(s) \tag{4.13}$$

Multiplication by time in the real domain entails differentiation with respect to s in the s domain.

Example 1.

$$\mathscr{L}[t\cos\omega t] = -\frac{d}{ds}\left[\frac{s}{s^2+\omega^2}\right] = \frac{s^2-\omega^2}{(s^2+\omega^2)^2}$$

Example 2.

$$\mathscr{L}[te^{-\alpha t}] = -\frac{d}{ds}\mathscr{L}[e^{-\alpha t}] = -\frac{d}{ds}\left(\frac{1}{s+\alpha}\right) = \frac{1}{(s+\alpha)^2}$$

Theorem 5 Translation in the s **domain.** If the Laplace transform of $f(t)$ is $F(s)$ and a is either real or complex, then

$$\mathscr{L}[e^{at}f(t)] = F(s-a) \tag{4.14}$$

Multiplication of e^{at} in the real domain becomes translation in the s domain.

Example 3. Starting with $\mathscr{L}[\sin \omega t] = \omega/(s^2 + \omega^2)$ and applying Theorem 5 gives

$$\mathscr{L}[e^{-\alpha t}\sin \omega t] = \frac{\omega}{(s+\alpha)^2 + \omega^2}$$

Theorem 6 Real differentiation. If the Laplace transform of $f(t)$ is $F(s)$, and if the first derivative of $f(t)$ with respect to time $Df(t)$ is transformable, then

$$\mathscr{L}[Df(t)] = sF(s) - f(0^+) \tag{4.15}$$

The term $f(0^+)$ is the value of the right-hand limit of the function $f(t)$ as the origin, $t = 0$, is approached from the right side (thus through positive values of time). This includes functions, such as the step function, that may be undefined at $t = 0$. For simplicity, the plus sign following the zero is usually omitted although its presence is implied.

The transform of the second derivative $D^2f(t)$ is

$$\mathscr{L}[D^2f(t)] = s^2F(s) - sf(0) - Df(0) \tag{4.16}$$

where $Df(0)$ is the value of the limit of the derivative of $f(t)$ as the origin, $t = 0$, is approached from the right side.

The transform of the nth derivative $D^nf(t)$ is

$$\mathscr{L}[D^nf(t)] = s^nF(s) - s^{n-1}f(0) - s^{n-2}Df(0) - \cdots - sD^{n-2}f(0) - D^{n-1}f(0) \tag{4.17}$$

Note that the transform includes the initial conditions, whereas in the classical method of solution the initial conditions are introduced separately to evaluate the coefficients of the solution of the differential equation. When all initial conditions are zero, the Laplace transform of the nth derivative of $f(t)$ is simply $s^nF(s)$.

Theorem 7 Real integration. If the Laplace transform of $f(t)$ is $F(s)$, its integral

$$D^{-1}f(t) = \int_0^t f(t)\,dt + D^{-1}f(0^+)$$

is transformable and the value of its transform is

$$\mathscr{L}[D^{-1}f(t)] = \frac{F(s)}{s} + \frac{D^{-1}f(0^+)}{s} \tag{4.18}$$

The term $D^{-1}f(0^+)$ is the constant of integration and is equal to the value of the integral as the origin is approached from the positive, or right, side. The plus sign is omitted in the remainder of this text.

The transform of the double integral $D^{-2}f(t)$ is

$$\mathscr{L}[D^{-2}f(t)] = \frac{F(s)}{s^2} + \frac{D^{-1}f(0)}{s^2} + \frac{D^{-2}f(0)}{s} \tag{4.19}$$

The transform of the nth-order integral $D^{-n}f(t)$ is

$$\mathscr{L}[D^{-n}f(t)] = \frac{F(s)}{s^n} + \frac{D^{-1}f(0)}{s^n} + \cdots + \frac{D^{-n}f(0)}{s} \tag{4.20}$$

Theorem 8 Final value. If $f(t)$ and $Df(t)$ are Laplace-transformable, if the Laplace transform of $f(t)$ is $F(s)$, and if the limit $f(t)$ as $t \to \infty$ exists, then

$$\lim_{s \to 0} sF(s) = \lim_{t \to \infty} f(t) \tag{4.21}$$

This theorem states that the behavior of $f(t)$ in the neighborhood of $t = \infty$ is related to the behavior of $sF(s)$ in the neighborhood of $s = 0$. If $sF(s)$ has poles [values of s for which $|sF(s)|$ becomes infinite] on the imaginary axis (excluding the origin) or in the right-half s plane, there is no finite final value of $f(t)$ and the theorem cannot be used. If $f(t)$ is sinusoidal, the theorem is invalid, since $\mathscr{L}[\sin \omega t]$ has poles at $s = \pm j\omega$ and $\lim_{t \to \infty} \sin \omega t$ does not exist. However, for poles of $sF(s)$ at the origin, $s = 0$, this theorem gives the final value of $f(\infty) = \infty$. This correctly describes the behavior of $f(t)$ as $t \to \infty$.

Theorem 9 Initial value. If the function $f(t)$ and its first derivative are Laplace-transformable, if the Laplace transform of $f(t)$ is $F(s)$, and if $\lim_{s \to \infty} sF(s)$ exists, then

$$\lim_{s \to \infty} sF(s) = \lim_{t \to 0} f(t) \tag{4.22}$$

This theorem states that the behavior of $f(t)$ in the neighborhood of $t = 0$ is related to the behavior of $sF(s)$ in the neighborhood of $|s| = \infty$. There are no limitations on the locations of the poles of $sF(s)$.

Theorem 10 Complex integration. If the Laplace transform of $f(t)$ is $F(s)$ and if $f(t)/t$ has a limit as $t \to 0^+$, then

$$\mathscr{L}\left[\frac{f(t)}{t}\right] = \int_s^{\infty} F(s)\, ds \tag{4.23}$$

This theorem states that division by the variable in the real domain entails integration with respect to s in the s domain.

4.5 APPLICATION OF THE LAPLACE TRANSFORM TO DIFFERENTIAL EQUATIONS

The Laplace transform is now applied to the solution of the differential equation for the simple mechanical system which is solved by the classical method in

Sec. 3.7. The differential equation of the system is repeated below:

$$M D^2 x_2 + B D x_2 + K x_2 = K x_1 \tag{4.24}$$

The position $x_1(t)$ undergoes a unit step displacement. This is the input and is called the *driving* function. The unknown quantity for which the equation is to be solved is the output displacement $x_2(t)$, called the *response function*. The Laplace transform of Eq. (4.24) is

$$\mathscr{L}[K x_1] = \mathscr{L}\left[M D^2 x_2 + B D x_2 + K x_2\right] \tag{4.25}$$

The transform of each term is

$$\begin{aligned}
\mathscr{L}[K x_1] &= K X_1(s) \\
\mathscr{L}[K x_2] &= K X_2(s) \\
\mathscr{L}[B D x_2] &= B\left[s X_2(s) - x_2(0)\right] \\
\mathscr{L}\left[M D^2 x_2\right] &= M\left[s^2 X_2(s) - s x_2(0) - D x_2(0)\right]
\end{aligned}$$

Substituting these terms into Eq. (4.25) and collecting terms gives

$$K X_1(s) = (M s^2 + B s + K) X_2(s) - \left[M s x_2(0) + M D x_2(0) + B x_2(0)\right] \tag{4.26}$$

Equation (4.26) is the transform equation and shows how the initial conditions—the initial position $x_2(0)$ and the initial velocity $Dx_2(0)$—are incorporated into the equation. The function $X_1(s)$ is called the *driving transform*; the function $X_2(s)$ is called the *response transform*. The coefficient of $X_2(s)$, which is $Ms^2 + Bs + K$, is called the *characteristic function*. The equation formed by setting the characteristic function equal to zero is called the *characteristic equation* of the system. Solving for $X_2(s)$ gives

$$X_2(s) = \frac{K}{M s^2 + B s + K} X_1(s) + \frac{M s x_2(0) + B x_2(0) + M D x_2(0)}{M s^2 + B s + K} \tag{4.27}$$

The coefficient of $X_1(s)$ is defined as the *system transfer function*. The second terms on the right side of the equation is called the *initial condition component*. Combining the terms of Eq. (4.27) yields

$$X_2(s) = \frac{K X_1(s) + M s x_2(0) + B x_2(0) + M D x_2(0)}{M s^2 + B s + K} \tag{4.28}$$

Finding the function $x_2(t)$ whose transform is given by Eq. (4.28) is symbolized by the inverse transform operator $\mathscr{L}^{-1}$; thus

$$\begin{aligned}
x_2(t) &= \mathscr{L}^{-1}[X_2(s)] \\
&= \mathscr{L}^{-1}\left[\frac{K X_1(s) + M s x_2(0) + B x_2(0) + M D x_2(0)}{M s^2 + B s + K}\right]
\end{aligned} \tag{4.29}$$

The function $x_2(t)$ can be found in the table of App. A after inserting numerical values into Eq. (4.29). For this example the system is initially at rest. From the principle of conservation of energy the initial conditions are $x_2(0) = 0$ and $Dx_2(0) = 0$. Assume, as in Sec. 3.7, that the damping ratio ζ is less than unity.

Since $x_1(t)$ is a step function, $X_1(s) = 1/s$ and

$$x_2(t) = \mathscr{L}^{-1}\left[\frac{K/M}{s(s^2 + Bs/M + K/M)}\right] = \mathscr{L}^{-1}\left[\frac{\omega_n^2}{s(s^2 + 2\zeta\omega_n s + \omega_n^2)}\right] \tag{4.30}$$

where $\omega_n = \sqrt{K/M}$ and $\zeta = B/2\sqrt{KM}$. Reference to transform pair 27*a* in App. A provides the solution directly as

$$x_2(t) = 1 - \frac{e^{-\zeta\omega_n t}}{\sqrt{1 - \zeta^2}} \sin\left(\omega_n\sqrt{1 - \zeta^2}\,t + \cos^{-1}\zeta\right)$$

4.6 INVERSE TRANSFORMATION

The application of the Laplace transforms to a differential equation yields an algebraic equation. From the algebraic equation the transform of the response function is readily found. To complete the solution the inverse transform must be found. In some cases the inverse-transform operation

$$f(t) = \mathscr{L}^{-1}[F(s)] \tag{4.31}$$

can be performed by direct reference to transform tables or by use of a digital-computer program (see App. B). The linearity and translation theorems are useful in extending the tables. When the response transform cannot be found in the tables, the general procedure is to express $F(s)$ as the sum of partial fractions with constant coefficients. The partial fractions have a first-order or quadratic factor in the denominator and are readily found in the table of transforms. The complete inverse transform is the sum of the inverse transforms of each fraction.

The response transform $F(s)$ can be expressed, in general, as the ratio of two polynomials $P(s)$ and $Q(s)$. Consider that these polynomials are of degree w and n, respectively, and are arranged in descending order of the powers of the variable s; thus,

$$F(s) = \frac{P(s)}{Q(s)} = \frac{a_w s^w + a_{w-1}s^{w-1} + \cdots + a_1 s + a_0}{s^n + b_{n-1}s^{n-1} + \cdots + b_1 s + b_0} \tag{4.32}$$

The a's and b's are real constants, and the coefficient of the highest power of s in the denominator has been made equal to unity. Only those $F(s)$ which are proper fractions are considered, i.e., those in which n is greater than w.† The first step is to factor $Q(s)$ into first-order and quadratic factors with real coefficients:

$$F(s) = \frac{P(s)}{Q(s)} = \frac{P(s)}{(s - s_1)(s - s_2)\cdots(s - s_k)\cdots(s - s_n)} \tag{4.33}$$

† If $n = w$, first divide $P(s)$ by $Q(s)$ to obtain $F(s) = a_w + P_1(s)/Q(s) = a_w + F_1(s)$. Then express $F_1(s)$ as the sum of partial fractions with constant coefficients.

The values $s_1, s_2, \ldots, s_n$ in the finite plane that make the denominator equal to zero are called the *zeros* of the denominator. These values of s, which may be either real or complex, also make $|F(s)|$ infinite, and so they are called *poles* of $F(s)$. Therefore, the values $s_1, s_2, \ldots, s_n$ are referred to as zeros of the denominator or poles of the complete function in the finite plane, i.e., there are n poles of $F(s)$. Methods of factoring polynomials exist in the literature. Digital-computer programs are available to perform this operation (see App. B).[8,11]

The transform $F(s)$ can be expressed as a series of fractions. If the poles are simple (nonrepeated), the number of fractions is equal to n, the number of poles of $F(s)$. In such case the function $F(s)$ can be expressed as

$$F(s) = \frac{P(s)}{Q(s)} = \frac{A_1}{s - s_1} + \frac{A_2}{s - s_2} + \cdots + \frac{A_k}{s - s_k} + \cdots + \frac{A_n}{s - s_n} \tag{4.34}$$

The problem is to evaluate the constants $A_1, A_2, \ldots, A_n$ corresponding to the poles $s_1, s_2, \ldots, s_n$. The coefficients $A_1, A_2, \ldots$ are termed the *residues*† of $F(s)$ at the corresponding poles. Cases of repeated factors and complex factors are treated separately. Several ways of evaluating the constants are taken up in the following section.

4.7 HEAVISIDE PARTIAL-FRACTION EXPANSION THEOREMS

The technique of partial-fraction expansion is set up to take care of all cases systematically. There are four classes of problems, depending on the denominator $Q(s)$. Each of these cases is illustrated separately.

Case 1: $F(s)$ has first-order real poles.

Case 2: $F(s)$ has repeated first-order real poles.

Case 3: $F(s)$ has a pair of complex-conjugate poles (a quadratic factor in the denominator).

Case 4: $F(s)$ has repeated pairs of complex-conjugate poles (a repeated quadratic factor in the denominator).

Case 1: First-Order Real Poles

The position of three real poles of $F(s)$ in the s plane is shown in Fig. 4.1. The poles may be positive, zero, or negative, and they lie on the real axis in the s plane. In this example, s_1 is positive, s_0 is zero, and s_2 is negative. For the poles shown in Fig. 4.1 the transform $F(s)$ and its partial fractions are

$$F(s) = \frac{P(s)}{Q(s)} = \frac{P(s)}{s(s - s_1)(s - s_2)} = \frac{A_0}{s} + \frac{A_1}{s - s_1} + \frac{A_2}{s - s_2} \tag{4.35}$$

† More generally the residue is the coefficient of the $(s - s_i)^{-1}$ term in the Laurent expansion of $F(s)$ about $s = s_i$.

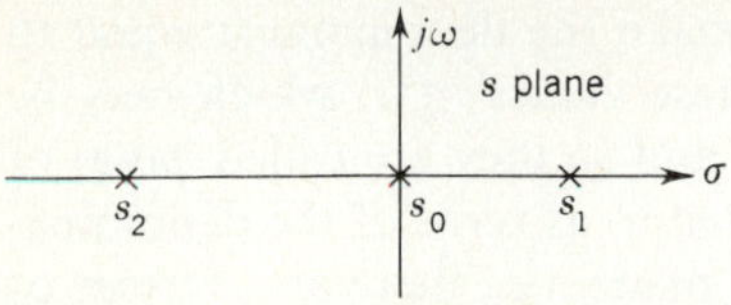

FIGURE 4.1
Location of real poles in the s plane.

There are as many fractions as there are factors in the denominator of $F(s)$. Since $s_0 = 0$, the factor $s - s_0$ is written simply as s. The inverse transform of $F(s)$ is

$$f(t) = A_0 + A_1 e^{s_1 t} + A_2 e^{s_2 t} \tag{4.36}$$

The pole s_1 is positive; therefore the term $A_1 e^{s_1 t}$ is an increasing exponential and the system is unstable. The pole s_2 is negative, and the term $A_2 e^{s_2 t}$ is a decaying exponential with a final value of zero. Therefore, for a system to be stable, all real poles that contribute to the complementary solution must be in the left half of the s plane.

To evaluate a typical coefficient A_k, multiply both sides of Eq. (4.34) by the factor $s - s_k$. The result is

$$\begin{aligned}(s - s_k)F(s) &= (s - s_k)\frac{P(s)}{Q(s)} \\ &= A_1 \frac{s - s_k}{s - s_1} + A_2 \frac{s - s_k}{s - s_2} + \cdots + A_k + \cdots + A_n \frac{s - s_k}{s - s_n}\end{aligned} \tag{4.37}$$

The multiplying factor $s - s_k$ on the left side of the equation and the same factor of $Q(s)$ should be divided out. By letting $s = s_k$, each term on the right side of the equation is zero except A_k. Thus, a general rule for evaluating the constants for single-order real poles is

$$A_k = \left[(s - s_k)\frac{P(s)}{Q(s)}\right]_{s=s_k} = \left[\frac{P(s)}{Q'(s)}\right]_{s=s_k} \tag{4.38}$$

where $Q'(s_k) = [dQ(s)/ds]_{s=s_k} = [Q(s)/(s - s_k)]_{s=s_k}$. The coefficients A_k are the *residues* of $F(s)$ at the corresponding poles. For the case of

$$F(s) = \frac{s + 2}{s(s + 1)(s + 3)} = \frac{A_0}{s} + \frac{A_1}{s + 1} + \frac{A_2}{s + 3}$$

the constants are

$$A_0 = [sF(s)]_{s=0} = \left[\frac{s + 2}{(s + 1)(s + 3)}\right]_{s=0} = \frac{2}{3}$$

$$A_1 = [(s + 1)F(s)]_{s=-1} = \left[\frac{s + 2}{s(s + 3)}\right]_{s=-1} = -\frac{1}{2}$$

$$A_2 = [(s + 3)F(s)]_{s=-3} = \left[\frac{s + 2}{s(s + 1)}\right]_{s=-3} = -\frac{1}{6}$$

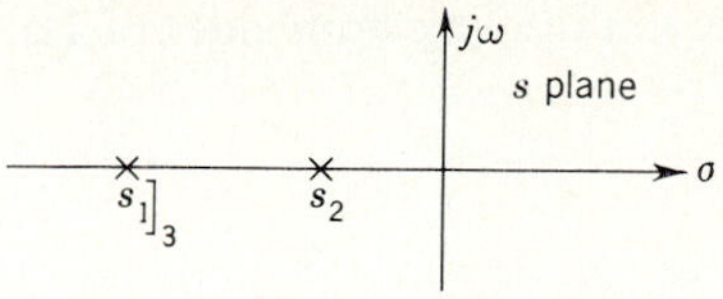

FIGURE 4.2
Location of real poles in the s plane.

The solution as a function of time is

$$f(t) = \frac{2}{3} - \frac{e^{-t}}{2} - \frac{1}{6}e^{-3t} \tag{4.39}$$

Case 2: Multiple-Order Real Poles

The position of real poles of $F(s)$, some of which are repeated, is shown in Fig. 4.2. The notation $]_r$ indicates a pole of order r. All real poles lie on the real axis of the s plane. For the poles shown in Fig. 4.2 the transform $F(s)$ and its partial fractions are

$$F(s) = \frac{P(s)}{Q(s)} = \frac{P(s)}{(s - s_1)^3(s - s_2)}$$

$$= \frac{A_{13}}{(s - s_1)^3} + \frac{A_{12}}{(s - s_1)^2} + \frac{A_{11}}{s - s_1} + \frac{A_2}{s - s_2} \tag{4.40}$$

The order of $Q(s)$ in this case is 4, and there are four fractions. Note that the multiple pole s_1, which is of order 3, has resulted in three fractions on the right side of Eq. (4.40). To designate the constants in the partial fractions, a single subscript is used for a first-order pole. For multiple-order poles a double-subscript notation is used. The first subscript designates the pole, and the second subscript designates the order of the pole in the partial fraction. The constants associated with first-order denominators in the partial-fraction expansion are called residues: therefore only the constants A_{11} and A_2 are residues of Eq. (4.40).

The inverse transform of $F(s)$ is

$$f(t) = A_{13}\frac{t^2}{2}e^{s_1 t} + A_{12}te^{s_1 t} + A_{11}e^{s_1 t} + A_2 e^{s_2 t} \tag{4.41}$$

For the general transform with repeated real roots,

$$F(s) = \frac{P(s)}{Q(s)} = \frac{P(s)}{(s - s_q)^r(s - s_1)\cdots}$$

$$= \frac{A_{qr}}{(s - s_q)^r} + \frac{A_{q(r-1)}}{(s - s_q)^{r-1}} + \cdots + \frac{A_{q(r-k)}}{(s - s_q)^{r-k}} + \cdots$$

$$+ \frac{A_{q1}}{s - s_q} + \frac{A_1}{s - s_1} + \cdots \tag{4.42}$$

The constant A_{qr} can be evaluated by simply multiplying both sides of Eq. (4.42) by $(s - s_q)^r$, giving

$$\begin{aligned}(s - s_q)^r F(s) &= \frac{(s - s_q)^r P(s)}{Q(s)} \\ &= \frac{P(s)}{(s - s_1)\cdots} \\ &= A_{qr} + A_{q(r-1)}(s - s_q) + \cdots \\ &\quad + A_{q1}(s - s_q)^{r-1} + A_1 \frac{(s - s_q)^r}{s - s_1} + \cdots\end{aligned} \tag{4.43}$$

Note that the factor $(s - s_q)^r$ is divided out of the left side of the equation.

For $s = s_q$, all terms on the right side of the equation are zero except A_{qr}:

$$A_{qr} = \left[(s - s_q)^r \frac{P(s)}{Q(s)}\right]_{s=s_q} \tag{4.44}$$

Evaluation of $A_{q(r-1)}$ cannot be performed in a similar manner. Multiplying both sides of Eq. (4.42) by $(s - s_q)^{r-1}$ and letting $s = s_q$ would result in both sides being infinite, which leaves $A_{q(r-1)}$ indeterminate. If the term A_{qr} were eliminated from Eq. (4.43), $A_{q(r-1)}$ could be evaluated. This can be done by differentiating Eq. (4.43) with respect to s:

$$\frac{d}{ds}\left[(s - s_q)^r \frac{P(s)}{Q(s)}\right] = A_{q(r-1)} + 2A_{q(r-2)}(s - s_q) + \cdots \tag{4.45}$$

Letting $s = s_q$ gives

$$A_{q(r-1)} = \left\{\frac{d}{ds}\left[(s - s_q)^r \frac{P(s)}{Q(s)}\right]\right\}_{s=s_q} \tag{4.46}$$

Repeating the differentiation gives the coefficient $A_{q(r-2)}$ as

$$A_{q(r-2)} = \left\{\frac{1}{2}\frac{d^2}{ds^2}\left[(s - s_q)^r \frac{P(s)}{Q(s)}\right]\right\}_{s=s_q} \tag{4.47}$$

This process can be repeated until each constant is determined. A general formula for finding these coefficients associated with the repeated real pole of order r is

$$A_{q(r-k)} = \left\{\frac{1}{k!}\frac{d^k}{ds^k}\left[(s - s_q)^r \frac{P(s)}{Q(s)}\right]\right\}_{s=s_q} \tag{4.48}$$

For the case of

$$F(s) = \frac{1}{(s+2)^3(s+3)} = \frac{A_{13}}{(s+2)^3} + \frac{A_{12}}{(s+2)^2} + \frac{A_{11}}{s+2} + \frac{A_2}{s+3} \tag{4.49}$$

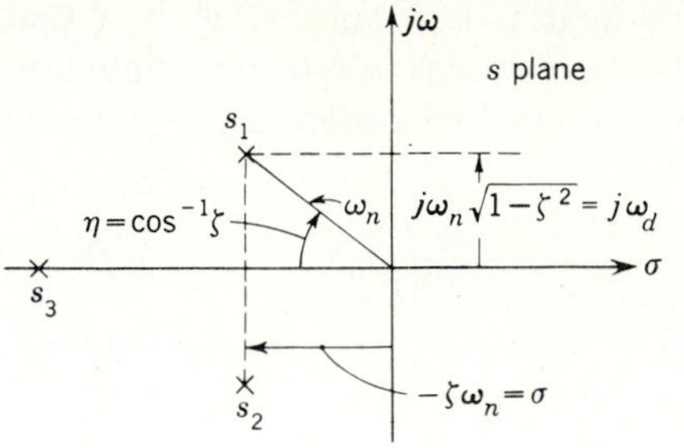

FIGURE 4.3
Location of complex-conjugate poles in the s plane.

the constants are

$$A_{13} = \left[(s+2)^3 F(s)\right]_{s=-2} = 1 \qquad A_{12} = \left\{\frac{d}{ds}\left[(s+2)^3 F(s)\right]\right\}_{s=-2} = -1$$

$$A_{11} = \left\{\frac{d^2}{2ds^2}\left[(s+2)^3 F(s)\right]\right\}_{s=-2} = 1 \qquad A_2 = \left[(s+3)F(s)\right]_{s=-3} = -1$$

and the solution as a function of time is

$$f(t) = \frac{t^2}{2}e^{-2t} - te^{-2t} + e^{-2t} - e^{-3t} \tag{4.50}$$

A recursive algorithm for computing the partial fraction expansion of rational functions having any number of multiple poles is readily programmed on a digital computer and is given in Ref. 12.

Case 3: Complex-Conjugate Poles

The position of complex poles of $F(s)$ in the s plane is shown in Fig. 4.3. Complex poles always are present in complex-conjugate pairs; their real part may be either positive or negative. For the poles shown in Fig. 4.3 the transform $F(s)$ and its partial fractions are

$$\begin{aligned} F(s) = \frac{P(s)}{Q(s)} &= \frac{P(s)}{(s^2 + 2\zeta\omega_n s + \omega_n^2)(s - s_3)} \\ &= \frac{A_1}{s - s_1} + \frac{A_2}{s - s_2} + \frac{A_3}{s - s_3} \\ &= \frac{A_1}{s + \zeta\omega_n - j\omega_n\sqrt{1-\zeta^2}} + \frac{A_2}{s + \zeta\omega_n + j\omega_n\sqrt{1-\zeta^2}} + \frac{A_3}{s - s_3} \end{aligned} \tag{4.51}$$

The inverse transform of $F(s)$ is

$$\begin{aligned} f(t) = A_1 \exp\left[\left(-\zeta\omega_n + j\omega_n\sqrt{1-\zeta^2}\right)t\right] \\ + A_2 \exp\left[\left(-\zeta\omega_n - j\omega_n\sqrt{1-\zeta^2}\right)t\right] + A_3 e^{s_3 t} \end{aligned} \tag{4.52}$$

Since the poles s_1 and s_2 are complex conjugates, and since $f(t)$ is a real quantity, the coefficients A_1 and A_2 must also be complex conjugates. Equation (4.52) can be written with the first two terms combined to a more useful damped sinusoidal form:

$$
\begin{aligned}
f(t) &= 2|A_1|e^{-\zeta\omega_n t}\sin\left(\omega_n\sqrt{1-\zeta^2}\,t+\phi\right)+A_3e^{s_3t} \\
&= 2|A_1|e^{\sigma t}\sin\left(\omega_d t+\phi\right)+A_3e^{s_3t}
\end{aligned}
\tag{4.53}
$$

where $\phi = \text{angle of } A_1 + 90°$

The values of A_1 and A_3, as found in the manner shown previously, are

$$A_1 = [(s-s_1)F(s)]_{s=s_1} \qquad \text{and} \qquad A_3 = [(s-s_3)F(s)]_{s=s_3}$$

Since s_1 is complex, the constant A_1 is also complex. *Remember that A_1 is associated with the complex pole with the positive imaginary part.*

In Fig. 4.3 the complex poles have a negative real part, $\sigma = -\zeta\omega_n$, where the damping ratio ζ is positive. For this case the corresponding transient response is known as a damped sinusoid and is shown in Fig. 3.2. Its final value is zero. The angle η shown in Fig. 4.3 is measured from the negative real axis and is related to the damping ratio by

$$\cos\eta = \zeta$$

If the complex pole has a positive real part, the time response increases exponentially with time and the system is unstable. If the complex roots are in the right half of the s plane, the damping ratio ζ is negative. The angle η for this case is measured from the positive real axis and is given by

$$\cos\eta = |\zeta|$$

For the case of

$$F(s) = \frac{10}{(s^2+6s+25)(s+2)} = \frac{A_1}{s+3-j4} + \frac{A_2}{s+3+j4} + \frac{A_3}{s+2} \tag{4.54}$$

the constants are

$$
\begin{aligned}
A_1 &= \left[(s+3-j4)\frac{10}{(s^2+6s+25)(s+2)}\right]_{s=-3+j4} \\
&= \left[\frac{10}{(s+3+j4)(s+2)}\right]_{s=-3+j4} = 0.303\underline{/-194°} \\
A_3 &= \left[(s+2)\frac{10}{(s^2+6s+25)(s+2)}\right]_{s=-2} \\
&= \left(\frac{10}{s^2+6s+25}\right)_{s=-2} = 0.59
\end{aligned}
$$

The solution is

$$f(t) = 0.6e^{-3t}\sin(4t-104°) + 0.59e^{-2t} \tag{4.55}$$

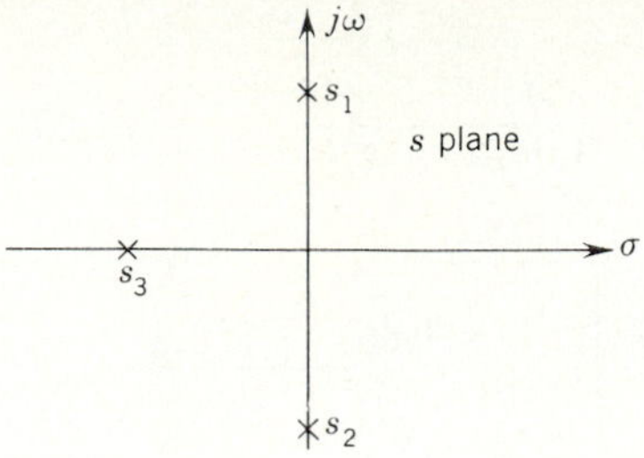

FIGURE 4.4
Poles of $F(s)$ containing imaginary conjugate poles in the s plane.

The particular function $F(s)$ of Eq. (4.54) appears in App. A as transform pair 29. With the notation of App. A, the phase angle in the damped sinusoidal term is

$$\phi = \tan^{-1}\frac{b}{c-a} = \tan^{-1}\frac{4}{2-3} = \tan^{-1}\frac{4}{-1} = 104° \qquad (4.56)$$

It is important to note that $\tan^{-1}[4/(-1)] \neq \tan^{-1}(-4/1)$. To get the correct value for the angle ϕ, it is useful to draw a sketch in order to avoid ambiguity and ensure that ϕ is evaluated correctly.

IMAGINARY POLES. The position of imaginary poles of $F(s)$ in the s plane is shown in Fig. 4.4 As the real part of the poles is zero, the poles lie on the imaginary axis. This situation is a special case of complex poles, i.e., the damping ratio $\zeta = 0$. For the poles shown in Fig. 4.4 the transform $F(s)$ and its partial fractions are

$$F(s) = \frac{P(s)}{Q(s)} = \frac{P(s)}{(s^2+\omega_n^2)(s-s_3)} = \frac{A_1}{s-s_1} + \frac{A_2}{s-s_2} + \frac{A_3}{s-s_3} \qquad (4.57)$$

The quadratic can be factored in terms of the poles s_1 and s_2; thus,

$$s^2 + \omega_n^2 = (s - j\omega_n)(s + j\omega_n) = (s - s_1)(s - s_2)$$

The inverse transform of Eq. (4.57) is

$$f(t) = A_1 e^{j\omega_n t} + A_2 e^{-j\omega_n t} + A_3 e^{s_3 t} \qquad (4.58)$$

As $f(t)$ is a real quantity, the coefficients A_1 and A_2 are complex conjugates. The terms in Eq. (4.58) can be combined into the more useful form

$$f(t) = 2|A_1|\sin(\omega_n t + \phi) + A_3 e^{s_3 t} \qquad (4.59)$$

Note that since there is no damping term multiplying the sinusoid, this term represents a steady-state value. The angle

$$\phi = \text{angle of } A_1 + 90°$$

The values of A_1 and A_3, found in the conventional manner, are

$$A_1 = [(s-s_1)F(s)]_{s=s_1} \qquad A_3 = [(s-s_3)F(s)]_{s=s_3}$$

For the case where

$$F(s) = \frac{100}{(s^2 + 25)(s + 2)} = \frac{A_1}{s - j5} + \frac{A_2}{s + j5} + \frac{A_3}{s + 2} \tag{4.60}$$

the values of the coefficients are

$$A_1 = [(s - j5)F(s)]_{s=j5} = \left[\frac{100}{(s + j5)(s + 2)}\right]_{s=j5} = 1.86\underline{/-158.2^\circ}$$

$$A_3 = [(s + 2)F(s)]_{s=-2} = \left(\frac{100}{s^2 + 25}\right)_{s=-2} = 3.45$$

The solution is

$$f(t) = 3.72 \sin(5t - 68.2^\circ) + 3.45e^{-2t} \tag{4.61}$$

Case 4: Multiple-Order Complex Poles

Multiple-order complex-conjugate poles are rare. They can be treated in much the same fashion as repeated real poles.

4.8 PARTIAL-FRACTION SHORTCUTS

Some shortcuts can be used to simplify the partial-fraction expansion procedures previously described. They are most useful for transform functions that have multiple poles or complex poles. With multiple poles, the evaluation of the constants by the process of repeated differentiation given by Eq. (4.48) can be very tedious, particularly when a factor containing s appears in the numerator of $F(s)$. The modified procedure is to evaluate only those coefficients for real poles which can be obtained readily without differentiation. The corresponding partial fractions are then subtracted from the original function. The resultant function is easier to work with to get the additional partial fractions than the original function $F(s)$.

Example 1. For the function $F(s)$ of Eq. (4.49), which contains a pole $s_1 = -2$ of multiplicity 3, the coefficient $A_{13} = 1$ is readily found. Subtracting the associated fraction from $F(s)$ and putting the result over a common denominator yields

$$F_x(s) = \frac{1}{(s + 2)^3(s + 3)} - \frac{1}{(s + 2)^3} = \frac{-1}{(s + 2)^2(s + 3)}$$

Note that the remainder $F_x(s)$ contains the pole $s_1 = -2$ with a multiplicity of only 2. Thus the coefficient A_{12} is now easily obtained from $F_x(s)$ without differentiation. The procedure is repeated to obtain all the coefficients associated with the multiple pole.

Example 2. Consider the function

$$F(s) = \frac{20}{(s^2 + 6s + 25)(s + 1)} = \frac{As + B}{s^2 + 6s + 25} + \frac{C}{s + 1} \tag{4.62}$$

which has a pair of complex poles. Instead of partial fractions for each pole, leave the quadratic factor in one fraction. Note that the numerator of the fraction containing the quadratic is a polynomial in s and is one degree lower than the denominator. Residue C is $C = [(s+1)F(s)]_{s=-1} = 1$. Subtracting this fraction from $F(s)$ gives

$$\frac{20}{(s^2+6s+25)(s+1)} - \frac{1}{s+1} = -\frac{-s^2-6s-5}{(s^2+6s+25)(s+1)}$$

Since the pole at $s = -1$ has been removed, the numerator of the remainder must be divisible by $s + 1$. The simplified function must be equal to the remaining partial fraction; i.e.,

$$-\frac{s+5}{s^2+6s+25} = \frac{As+B}{s^2+6s+25} = \frac{-(s+5)}{(s+3)^2+4^2}$$

The inverse transform for this fraction can be obtained directly from App. A, pair 26. The complete inverse transform of $F(s)$ is

$$f(t) = e^{-t} - 1.12e^{-3t}\sin(4t + 63.4^\circ) \tag{4.63}$$

Leaving the complete quadratic in one fraction has resulted in real constants for the numerator. The chances for error are probably less than evaluating the residues, which are complex numbers, at the complex poles.

A rule presented by Hazony and Riley[4] is very useful for evaluating the coefficients of partial-fraction expansions. For a normalized ratio of polynomials:

1. If the denominator is one degree higher than the numerator, the sum of the residues is 1.
2. If the denominator is two or more degrees higher than the numerator, the sum of the residues is 0.

Equation (4.32) with a_w factored from the numerator is a normalized ratio of polynomials. These rules are applied to the ratio $F(s)/a_w$. It should be noted that *residue* refers only to the coefficients of terms in a partial-fraction expansion with first-degree denominators. Coefficients of terms with higher-degree denominators are referred to only as coefficients.

These rules can be used to simplify the work involved in evaluating the coefficients of partial-fraction expansions, particularly when the original function has a multiple-order pole. For example, in Eq. (4.49) it can immediately be written that $A_{11} + A_2 = 0$. Since $A_2 = -1$, the value of $A_{11} = 1$ is obtained directly.

Digital-computer programs (see App. B) are readily available for evaluating the partial-fraction coefficients and obtaining a tabulation and plot of $f(t)$.[8,11]

4.9 GRAPHICAL INTERPRETATION OF PARTIAL-FRACTION COEFFICIENTS[5]

The preceding sections describe the analytical evaluation of the partial-fraction coefficients. These constants are directly related to the pole-zero pattern of the function $F(s)$ and can be determined graphically, whether the poles and zeros

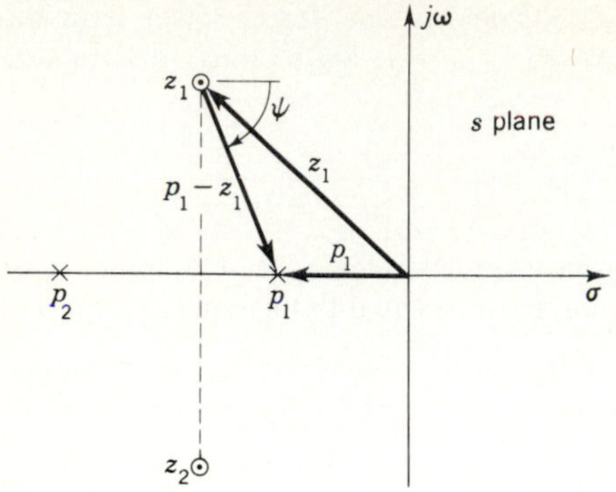

FIGURE 4.5
Directed line segments drawn on a pole-zero diagram of a function $F(s)$.

are real or in complex-conjugate pairs. As long as $P(s)$ and $Q(s)$ are in factored form, the coefficients can be determined graphically by inspection. Rewriting Eq. (4.32) with the numerator and denominator factored and $a_w = K$ gives

$$F(s) = \frac{P(s)}{Q(s)} = \frac{K(s - z_1)(s - z_2)\cdots(s - z_m)\cdots(s - z_w)}{(s - p_1)(s - p_2)\cdots(s - p_k)\cdots(s - p_n)}$$

$$= K\frac{\prod_{m=1}^{w}(s - z_m)}{\prod_{k=1}^{n}(s - p_k)} \tag{4.64}$$

The zeros of this function, $s = z_m$, are those values of s for which the function is zero; that is, $F(z_m) = 0$. Zeros are indicated by a small circle on the s plane. The poles of this function, $s = p_k$, are those values of s for which the function is infinite; that is, $|F(p_k)| = \infty$. Poles are indicated by a small x on the s plane. Poles are also known as singularities† of the function. When $F(s)$ has only simple poles (first-order poles), it can be expanded into partial fractions of the form

$$F(s) = \frac{A_1}{s - p_1} + \frac{A_2}{s - p_2} + \cdots + \frac{A_k}{s - p_k} + \cdots + \frac{A_n}{s - p_n} \tag{4.65}$$

The coefficients are obtained from Eq. (4.38) and are given by

$$A_k = [(s - p_k)F(s)]_{s=p_k} \tag{4.66}$$

The first coefficient is

$$A_1 = \frac{K(p_1 - z_1)(p_1 - z_2)\cdots(p_1 - z_w)}{(p_1 - p_2)(p_1 - p_3)\cdots(p_1 - p_n)} \tag{4.67}$$

Figure 4.5 shows the poles and zeros of a function $F(s)$. The quantity $s = p_1$ is drawn as an arrow from the origin of the s plane to the pole p_1. This

† A singularity of a function $F(s)$ is a point where $F(s)$ does not have a derivative.

arrow is called a *directed line segment*. It has a magnitude equal to $|p_1|$ and an angle of 180°. Similarly, the directed line segment $s = z_1$ is drawn as an arrow from the origin to the zero z_1 and has a corresponding magnitude and angle. A directed line segment is also drawn from the zero z_1 to the pole p_1. By the rules of vector addition, this directed line segment is equal to $p_1 - z_1$; in polar form it has a magnitude equal to its length and an angle ψ, as shown. By referring to Eq. (4.67), it is seen that $p_1 - z_1$ appears in the numerator of A_1. While this quantity can be evaluated analytically, it can also be measured from the pole-zero diagram. In a similar fashion, each factor in Eq. (4.67) can be obtained graphically; the directed line segments are drawn *from* the zeros and each of the other poles *to* the pole p_1. The angle from a zero is indicated by the symbol ψ, and the angle from a pole is indicated by the symbol θ.

The general rule for evaluating the coefficients in the partial-fraction expansion is quite simple. The value of A_k is the product of K and the directed distances from each zero to the pole p_k divided by the product of the directed distances from each of the other poles to the pole p_k. Each of these directed distances is characterized by a magnitude and an angle. This statement can be written in equation form as

$$A_k = K\frac{\text{product of directed distances from each zero to pole } p_k}{\text{product of directed distances from all other poles to pole } p_k}$$

$$= K\frac{\sum_{m=1}^{w} |p_k - z_m| \;\angle \sum_{m=1}^{w} \psi(p_k - z_m)}{\sum_{\substack{c=1\\c\neq k}}^{n} |p_k - p_c| \;\angle \sum_{c=1}^{n} \theta(p_k - p_c)} \tag{4.68}$$

Equations (4.67) and (4.68) readily reveal that as the pole p_k comes closer to the zero z_m, the corresponding coefficient A_k becomes smaller (approaches a zero value). This characteristic is very important in analyzing system response characteristics. For real poles the values of A_k must be real but can be either positive or negative. A_k is positive if the total number of real poles and real zeros to the right of p_k is even, and it is negative if the total number of real poles and real zeros to the right if p_k is odd. For complex poles the values of A_k are complex. If $F(s)$ has no finite zeros, the numerator of Eq. (4.68) is equal to K. A repeated factor of the form $(s + a)^r$ in either the numerator or the denominator is included r times in Eq. (4.68), where the values of $|s + a|$ and $\angle s + a$ are obtained graphically. The application of the graphical technique is illustrated by the following examples:

Example. *A function $F(s)$ with complex poles* (step function response)

$$F(s) = \frac{K(s+\alpha)}{s\left[(s+a)^2 + b^2\right]} = \frac{K(s+\alpha)}{s(s+a-jb)(s+a+jb)}$$

$$= \frac{K(s-z_1)}{s(s-p_1)(s-p_2)} = \frac{A_0}{s} + \frac{A_1}{s+a-jb} + \frac{A_2}{s+a+jb} \tag{4.69}$$

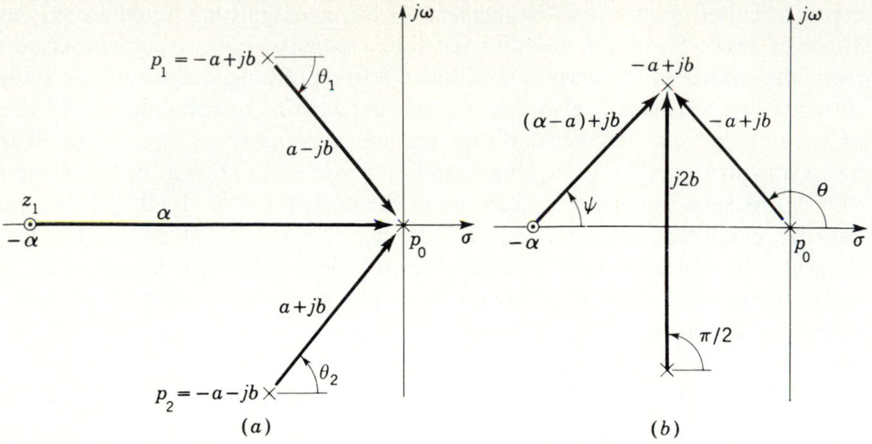

FIGURE 4.6
Directed line segments for evaluating (*a*) A_0 of Eq. (4.70) and (*b*) A_1 of Eq. (4.71).

The coefficient A_0 is obtained by use of the directed line segments shown in Fig. 4.6*a*:

$$A_0 = \frac{K(\alpha)}{(a-jb)(a+jb)} = \frac{K\alpha}{\sqrt{a^2+b^2}\,e^{j\theta_1}\sqrt{a^2+b^2}\,e^{j\theta_2}} = \frac{K\alpha}{a^2+b^2} \tag{4.70}$$

Note that the angles θ_1 and θ_2 are equal in magnitude but opposite in sign. The coefficient for a real pole is therefore always a real number.

The coefficient A_1 is obtained by use of the directed line segments shown in Fig. 4.6*b*:

$$A_1 = \frac{K[(\alpha-a)+jb]}{(-a+jb)(j2b)} = \frac{K\sqrt{(\alpha-a)^2+b^2}\,e^{j\psi}}{\sqrt{a^2+b^2}\,e^{j\theta}2be^{j\pi/2}}$$

$$= \frac{K}{2b}\sqrt{\frac{(\alpha-a)^2+b^2}{a^2+b^2}}\,e^{j(\psi-\theta-\pi/2)} \tag{4.71}$$

where $\qquad \psi = \tan^{-1}\frac{b}{\alpha-a} \qquad$ and $\qquad \theta = \tan^{-1}\frac{b}{-a}$

Since the constants A_1 and A_2 are complex conjugates, it is not necessary to evaluate A_2. It should be noted that zeros in $F(s)$ affect both the magnitude and the angle of the coefficients in the time function.

The response as a function of time is therefore

$$f(t) = A_0 + 2|A_1|e^{-at}\sin\left(bt + \angle A_1 + 90°\right)$$

$$= \frac{K\alpha}{a^2+b^2} + \frac{K}{b}\sqrt{\frac{(\alpha-a)^2+b^2}{a^2+b^2}}\,e^{-at}\sin(bt+\phi) \tag{4.72}$$

where $\qquad \phi = \psi - \theta = \tan^{-1}\frac{b}{\alpha-a} - \tan^{-1}\frac{b}{-a} = \text{angle of } A_1 + 90° \tag{4.73}$

This agrees with transform pair 28 in App. A. A_0 and A_1 can be evaluated by measuring the directed line segments directly on the pole-zero diagram. It is interesting to note that, for a given ζ, as the magnitude $\omega_n = \sqrt{a^2 + b^2}$ becomes larger, the coefficients A_0 and A_1 approach zero as revealed by an analysis of Eqs. (4.70) and (4.71), respectively. (The limiting value is $\lim_{\omega_n \to \infty}[A_1] = K/2b$.) This is an important characteristic when it is desired that $f(t)$ be very small. The application of this characteristic is used in disturbance rejection (Chap. 12). Also see Prob. 4.17 for further illustration.

4.10 FREQUENCY RESPONSE FROM THE POLE-ZERO DIAGRAM

The frequency response of a system is described as the steady-state response with a sine-wave forcing function for all values of frequency. This information is often presented graphically, using two curves. One curve shows the ratio of output amplitude to input amplitude M and the other curve the phase angle of the output α, where both are plotted as a function of frequency, often on a logarithmic scale.

Consider the input to a system as sinusoidal and given by

$$x_1(t) = X_1 \sin \omega t \tag{4.74}$$

and the output is

$$x_2(t) = X_2 \sin(\omega t + \alpha) \tag{4.75}$$

The magnitude of the frequency response for the function given in Eq. (4.64) is

$$M = \left|\frac{\mathbf{X}_2(j\omega)}{\mathbf{X}_1(j\omega)}\right| = \left|\frac{\mathbf{P}(j\omega)}{\mathbf{Q}(j\omega)}\right| = \left|\frac{K(j\omega - z_1)(j\omega - z_2)\cdots}{(j\omega - p_1)(j\omega - p_2)\cdots}\right| \tag{4.76}$$

and the angle is

$$\begin{aligned}\alpha &= \angle \mathbf{P}(j\omega) - \angle \mathbf{Q}(j\omega) \\ &= \angle \mathbf{K} + \angle j\omega - z_1 + \angle j\omega - z_2 + \cdots - \angle j\omega - p_1 - \angle j\omega - p_2 - \cdots\end{aligned} \tag{4.77}$$

Figure 4.7 shows a possible pole-zero diagram and the directed line segments for evaluating M and α corresponding to a frequency ω_1. As ω increases, each of the directed line segments changes in both magnitude and angle. Several characteristics can be noted by observing the change in the magnitude and angle of each directed line segment as the frequency ω increases from 0 to ∞. The magnitude and angle of the frequency response can be obtained graphically from the pole-zero diagram.

The magnitude M and the angle α are a composite of all the effects of all the poles and zeros. In particular, for a system function which has all poles and zeros in the left half of the s plane, the following characteristics of the frequency

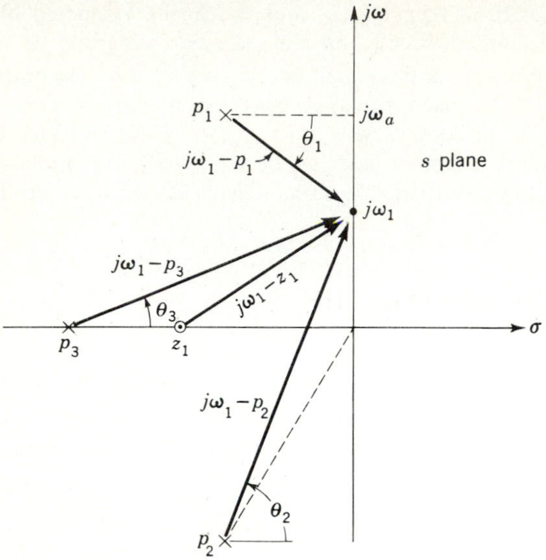

FIGURE 4.7
Pole-zero diagram of Eq. (4.76) showing the directed line segments for evaluating the frequency response $M\angle\alpha$.

response are noted:

1. At $\omega = 0$ the magnitude is a finite value, and the angle is 0°.
2. If there are more poles than zeros, as $\omega \to \infty$, the magnitude of M approaches zero and the angle is $-90°$ times the difference between the number of poles and zeros.
3. The magnitude can have a peak value M_m only if there are complex poles fairly close to the imaginary axis. It is shown later that the damping ratio ζ of the complex poles must be less than 0.707 to obtain a peak value for M. Of course, the presence of zeros could counteract the effect of the complex poles, and so it is possible that no peak would be present even if ζ of the poles were less than 0.707.

Typical frequency-response characteristics are illustrated for two common functions. As a first example,

$$\frac{X_2(s)}{X_1(s)} = \frac{P(s)}{Q(s)} = \frac{a}{s+a} \tag{4.78}$$

The frequency response starts with a magnitude of unity and an angle of 0°. As ω increases, the magnitude decreases and approaches zero while the angle approaches $-90°$. At $\omega = a$ the magnitude is 0.707 and the angle is $-45°$. The frequency $\omega = a$ is called the *corner frequency* ω_{cf} or the break frequency. Figure 4.8 shows the pole location, the frequency-response magnitude M, and the frequency-response angle α.

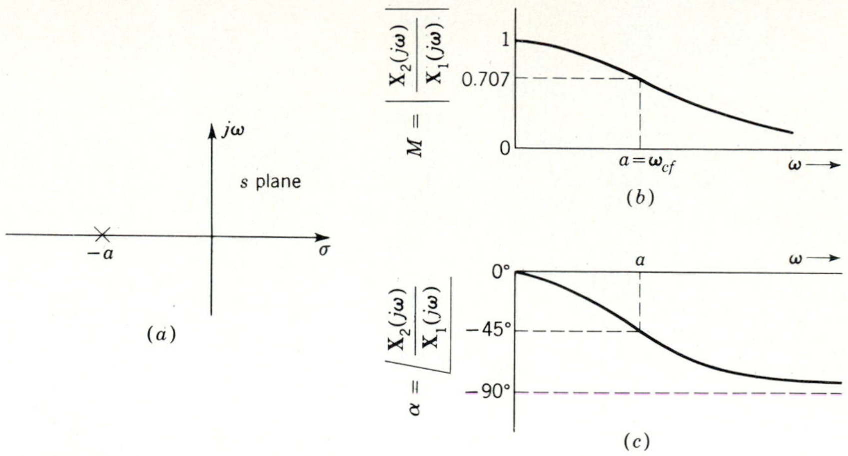

FIGURE 4.8
(*a*) Pole location of Eq. (4.78); (*b*) M vs. ω; (*c*) α vs. ω.

The second example is a system transform with a pair of conjugate poles:

$$\frac{X_2(s)}{X_1(s)} = \frac{\omega_n^2}{s^2 + 2\zeta\omega_n s + \omega_n^2} = \frac{\omega_n^2}{(s - p_1)(s - p_2)}$$

$$= \frac{\omega_n^2}{\left(s + \zeta\omega_n - j\omega_n\sqrt{1 - \zeta^2}\right)\left(s + \zeta\omega_n + j\omega_n\sqrt{1 - \zeta^2}\right)} \tag{4.79}$$

The location of the poles of Eq. (4.79) and the directed line segments for $s = j\omega$ are shown in Fig. 4.9*a*. At $\omega = 0$ the values are $M = 1$ and $\alpha = 0°$. As ω increases, the magnitude M first increases because $j\omega - p_1$ is decreasing faster than $j\omega - p_2$ is increasing. By differentiating the magnitude of $|\mathbf{M}(j\omega)|$, obtained from Eq. (4.79), with respect to ω, it can be shown (see Sec. 9.3) that the maximum value M_m and the frequency ω_m at which it occurs are given by

$$M_m = \frac{1}{2\zeta\sqrt{1 - \zeta^2}} \tag{4.80}$$

$$\omega_m = \omega_n\sqrt{1 - 2\zeta^2} \tag{4.81}$$

A circle drawn on the s plane in Fig. 4.9*a* with the poles p_1 and p_2 as the diameter intersects the imaginary axis at the value ω_m given by Eq. (4.81). Both Eq. (4.81) and the geometrical construction of the circle show that the M curve has a maximum value, other than at $\omega = 0$, only if $\zeta < 0.707$. The angle α becomes more negative as ω increases. At $\omega = \omega_n$ the angle is equal to $-90°$. This is the corner frequency for a quadratic factor with complex roots. As $\omega \to \infty$, the angle α approaches $-180°$. The magnitude and angle of the frequency response for $\zeta < 0.707$ are shown in Fig. 4.9. The M and α curves are continuous for all linear systems. The M vs. ω plot of Fig. 4.9*b* describes the frequency response characteristics for the undamped simple second-order system

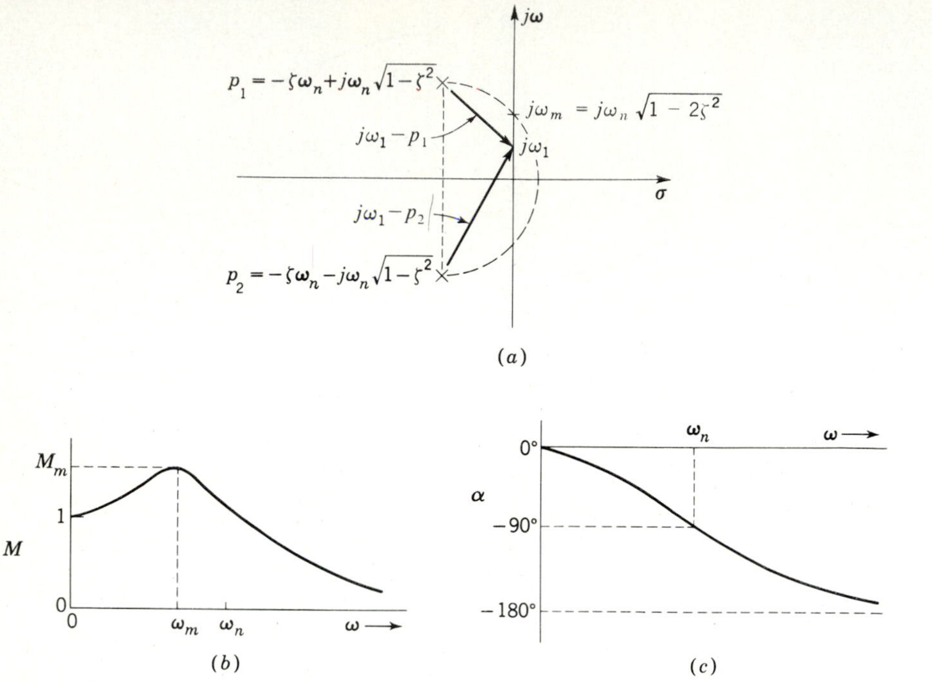

FIGURE 4.9
(*a*) Pole-zero diagram, $\zeta < 0.707$; (*b*) magnitude of frequency response; (*c*) angle of frequency response.

of Eq. (4.79). In later chapters this characteristic is used as the standard reference for analyzing and synthesizing higher-order systems.

4.11 LOCATION OF POLES AND STABILITY

The stability and the corresponding response of a system can be determined from the location of the poles of the response transform $F(s)$ in the s plane. The possible positions of the poles are shown in Fig. 4.10, and the responses are given in Table 4.1. These poles are the roots of the characteristic equation.

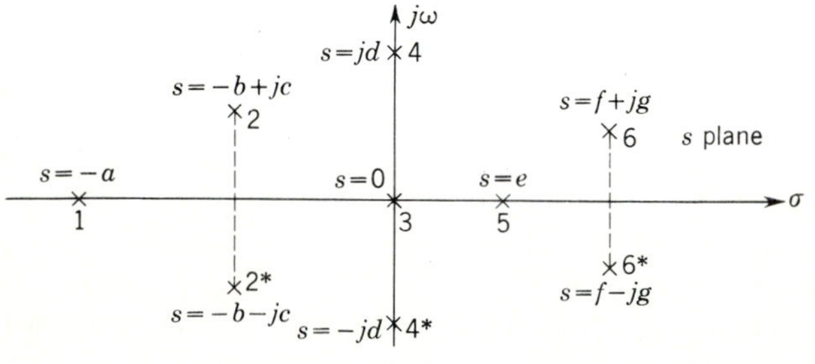

FIGURE 4.10
Location of poles in the s plane. (Numbers are used to identify the poles.)

TABLE 4.1
Relation of response to location of poles

Position of pole	Form of response	Characteristics
1	Ae^{-at}	Damped exponential
2–2*	$Ae^{-bt}\sin(ct+\phi)$	Exponentially damped sinusoid
3	A	Constant
4–4*	$A\sin(dt+\phi)$	Constant-sinusoid
5	Ae^{et}	Increasing exponential (unstable)
6–6*	$Ae^{ft}\sin(gt+\phi)$	Exponentially increasing sinusoid (unstable)

Poles of the response transform at the origin or on the imaginary axis that are not contributed by the forcing function result in a continuous output. These outputs are undesirable in a control system. Poles in the right-half s plane result in transient terms that increase with time. Such performance characterizes an unstable system; therefore poles in the right-half s plane are not desirable, in general.

4.12 LAPLACE TRANSFORM OF THE IMPULSE FUNCTION

Figure 4.11 shows a rectangular pulse of amplitude $1/a$ and of duration a. The analytical expression for this pulse is

$$f(t) = \frac{u_{-1}(t) - u_{-1}(t-a)}{a} \tag{4.82}$$

and its Laplace transform is

$$F(s) = \frac{1 - e^{-as}}{as} \tag{4.83}$$

If a is decreased, the amplitude increases and the duration of the pulse decreases but the area under the pulse, and thus its strength, remains unity. The limit of $f(t)$ as $a \to 0$ is termed a unit impulse and is designated by $\delta(t) = u_0(t)$. The Laplace transform of the unit impulse, as evaluated by use of L'Hospital's theorem, is $\Delta(s) = 1$.

When any function has a jump discontinuity, the derivative of the function has an impulse at the discontinuity. Some systems are subjected to shock inputs.

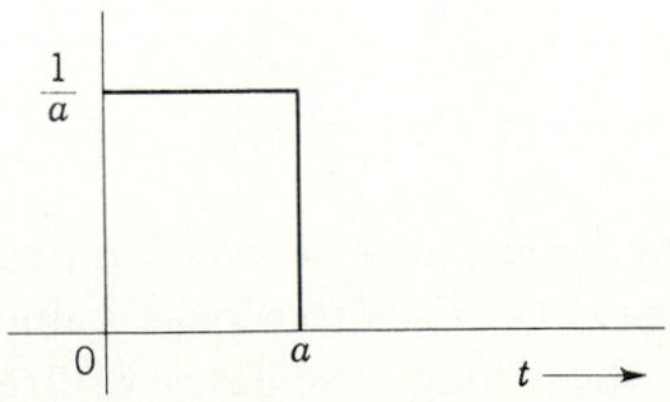

FIGURE 4.11
Rectangular pulse of unit area.

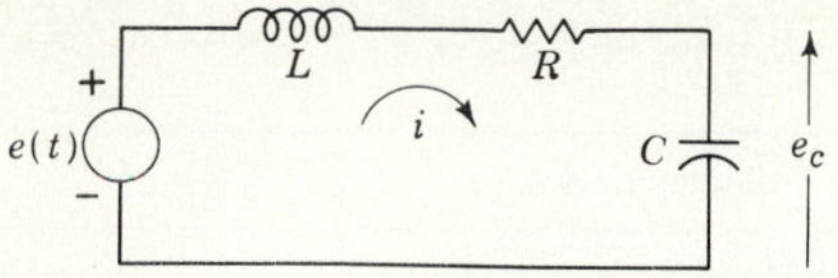

FIGURE 4.12
RLC series circuit.

For example, an airplane in flight may be jolted by a gust of wind of short duration. A gun that is fired has a reaction force of large magnitude and very short duration, and so does a steel ball bouncing off a steel plate. When the duration of a disturbance or input to a system is very short compared with the natural periods of the system, the response can often be well approximated by considering the input to be an impulse of proper strength. An impulse of infinite magnitude and zero duration does not occur in nature; but if the pulse duration is much smaller than the time constants of the system, a representation of the input by an impulse is a good approximation. The shape of the impulse is unimportant as long as the strength of the equivalent impulse is equal to the strength of the actual pulse.

Since the Laplace transform of an impulse is defined, the approximate response of a system to a pulse is often found more easily by this method than by the classical method. Consider the *RLC* circuit shown in Fig. 4.12 when the input voltage is an impulse. The problem is to find the voltage e_c across the capacitor as a function of time. The differential equation relating e_c to the input voltage e and the impedances is written by the node method:

$$\left(CD + \frac{1}{R + LD}\right)e_c - \frac{1}{R + LD}e = 0 \tag{4.84}$$

Rationalizing the equation yields

$$(LCD^2 + RCD + 1)e_c = e \tag{4.85}$$

Taking the Laplace transform of this equation gives

$$(s^2LC + sRC + 1)E_c(s) - sLCe_c(0) - LCDe_c(0) - RCe_c(0) = E(s) \tag{4.86}$$

Consider an initial condition $e_c(0^-) = 0$. The voltage across the capacitor is related to the current in the series circuit by the expression $De_c = i/C$. Since the current cannot change instantaneously through an inductor, $De_c(0^-) = 0$. These initial conditions which exist at $t = 0^-$ are used in the Laplace transformed equation when the forcing function is an impulse. The solution for the current shows a discontinuous step change at $t = 0$.

The input voltage e is an impulse; therefore $E(s) = 1$. Solving for $E_c(s)$ gives

$$E_c(s) = \frac{1/LC}{s^2 + (R/L)s + 1/LC} = \frac{\omega_n^2}{s^2 + 2\zeta\omega_n s + \omega_n^2} \tag{4.87}$$

Depending on the relative sizes of the parameters, the circuit may be overdamped or underdamped. In the overdamped case the voltage $e_c(t)$ rises to a peak value and then decays to zero. In the underdamped case the voltage oscillates with a frequency $\omega_d = \omega_n\sqrt{1 - \zeta^2}$ and eventually decays to zero. The form of the voltage $e_c(t)$ for several values of damping is shown in the next section.

The stability of a system is revealed by the impulse response. With an impulse input the response transform contains only poles and zeros contributed by the system parameters. The impulse response is therefore the inverse transform of the system transfer function and is called the *weighting function*:

$$g(t) = \mathscr{L}^{-1}[G(s)] \tag{4.88}$$

The impulse response provides one method of evaluating the system function; although the method is not simple, it is possible to take the impulse response which is a function of time and convert it into the transfer function $G(s)$ which is a function of s.

The system response $g(t)$ to an impulse input $\delta(t)$ can be used to determine the response $c(t)$ to any input $r(t)$.[6] The response $c(t)$ is determined by the use of the convolution or superposition integral

$$c(t) = \int_0^t r(\tau)g(t - \tau)\,d\tau \tag{4.89}$$

This method is normally used for inputs which are not sinusoidal or a power series.

4.13 SECOND-ORDER SYSTEM WITH IMPULSE EXCITATION

The response of a second-order system to a step-function input is shown in Fig. 3.6. The impulse function is the derivative of the step function; therefore the response to an impulse is the derivative of the response to a step function. Figure 4.13 shows the response for several values of damping ratio ζ for the second-order function of Eq. (4.87).

The first zero of the impulse response occurs at t_p, which is the time at which the maximum value of the step-function response occurs.

For the underdamped case the impulse response is

$$f(t) = \frac{\omega_n}{\sqrt{1 - \zeta^2}} e^{-\zeta\omega_n t} \sin \omega_n\sqrt{1 - \zeta^2}\,t \tag{4.90}$$

By setting the derivative of $f(t)$ with respect to time equal to zero, it can be shown that the maximum overshoot occurs at

$$t_m = \frac{\cos^{-1}\zeta}{\omega_n\sqrt{1 - \zeta^2}} \tag{4.91}$$

The maximum value of $f(t)$ is

$$f(t_m) = \omega_n \exp\left(-\frac{\zeta\cos^{-1}\zeta}{\sqrt{1 - \zeta^2}}\right) \tag{4.92}$$

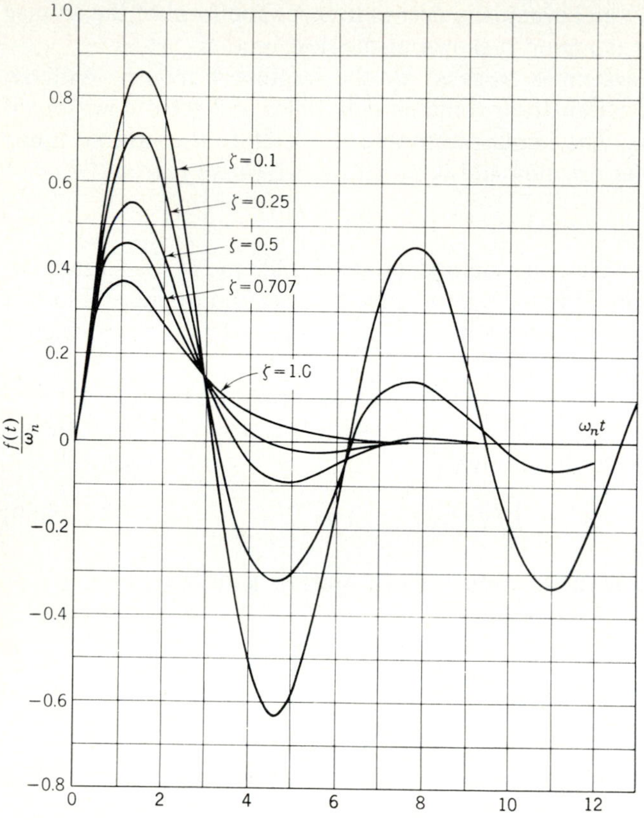

FIGURE 4.13
Response to an impulse for $F(s) = \omega_n^2/(s^2 + 2\zeta\omega_n s + \omega_n^2)$.

The impulse response $g(t)$ can be measured after applying an input pulse which has a short duration compared with the largest system time constant. Then it can be used in a deconvolution technique to determine the transfer function $G(s)$. It can also be used to determine the disturbance-rejection characteristics of control systems.

4.14 ADDITIONAL MATRIX OPERATIONS AND PROPERTIES

Additional characteristics of matrices required to solve matrix state equations by using the Laplace transform are presented in this section.

PRINCIPAL DIAGONAL. The principal diagonal of a square matrix $\mathbf{M} = [m_{ij}]$ consists of the elements m_{ii}.

DIAGONAL MATRIX. A diagonal matrix is a square matrix in which all the elements off the principal diagonal are zero. This can be expressed as $m_{ij} = 0$ for $i \neq j$ and the principal diagonal elements $m_{ii} = m_i$ are not all zero. When the diagonal elements are all equal, the matrix is called a *scalar* matrix.

TRACE. The *trace* of a square matrix $\mathbf{M}$ of order n is the sum of all the elements along the principal diagonal; that is,

$$\text{trace}\,\mathbf{M} = \sum_{i=1}^{n} m_{ii}$$

DETERMINANT. The determinant of a square matrix $\mathbf{M}$ of order n is the sum of all possible signed products of n elements containing one and only one element from every row and column in the matrix. The determinant of $\mathbf{M}$ may be denoted by $|\mathbf{M}|$, or $\det \mathbf{M}$, or Δ_M. It is assumed that the reader knows how to evaluate determinants

Example 1.

$$\mathbf{M} = \begin{bmatrix} 3 & 1 & 2 \\ 1 & 0 & -4 \\ 0 & 5 & 7 \end{bmatrix} \qquad |\mathbf{M}| = \begin{vmatrix} 3 & 1 & 2 \\ 1 & 0 & -4 \\ 0 & 5 & 7 \end{vmatrix} = 63$$

Characteristics of determinants are as follows:

1. The determinant of a unit matrix is $|\mathbf{I}| = 1$.
2. The determinant of a matrix is zero if (a) any row or column contains all zeros or (b) the elements of any two rows (or columns) have a common ratio.

SINGULAR MATRIX. A square matrix is said to be *singular* if the value of its determinant is zero. If the value is not zero, the matrix is *nonsingular*.

MINOR. The minor M_{ij} of a square matrix $\mathbf{M}$ of order n is the determinant formed after the ith row and jth column are deleted from $\mathbf{M}$.

PRINCIPAL MINOR. A principal minor is a minor M_{ii} whose diagonal elements are also the diagonal elements of the square matrix $\mathbf{M}$.

COFACTOR. A cofactor is a *signed* minor and is given by

$$C_{ij} = \Delta_{ij} = (-1)^{i+j}\mathbf{M}_{ij} \tag{4.93}$$

Example 2. For the matrix $\mathbf{M}$ of Example 1, the cofactor C_{21} is obtained by deleting the second row and first column, giving

$$C_{21} = (-1)^{2+1}\begin{vmatrix} 1 & 2 \\ 5 & 7 \end{vmatrix} = +3$$

ADJOINT MATRIX. The adjoint of square matrix **M**, denoted as adj **M**, is the transpose of the cofactor matrix. The cofactor matrix is formed by replacing each element of **M** by its cofactor.

$$\operatorname{adj}\mathbf{M} = [C_{ji}] = [C_{ij}]^T = \begin{bmatrix}\text{array of}\\ \text{cofactors}\end{bmatrix}^T \tag{4.94}$$

Example 3. The adj **M** for the matrix **M** of Example 1 is

$$\operatorname{adj}\mathbf{M} = \begin{bmatrix} 20 & -7 & 5 \\ 3 & 21 & -15 \\ -4 & 14 & -1 \end{bmatrix}^T = \begin{bmatrix} 20 & 3 & -4 \\ -7 & 21 & 14 \\ 5 & -15 & -1 \end{bmatrix}$$

INVERSE MATRIX. The product of a matrix and its adjoint is a scalar matrix, as illustrated by the following example.

Example 4. The product **M** adj **M**, using values in Examples 1 and 3, is

$$\mathbf{M}\operatorname{adj}\mathbf{M} = \begin{bmatrix} 3 & 1 & 2 \\ 1 & 0 & -4 \\ 0 & 5 & 7 \end{bmatrix}\begin{bmatrix} 20 & 3 & -4 \\ -7 & 21 & 14 \\ 5 & -15 & -1 \end{bmatrix}$$

$$= \begin{bmatrix} 63 & 0 & 0 \\ 0 & 63 & 0 \\ 0 & 0 & 63 \end{bmatrix} = 63\begin{bmatrix} 1 & 0 & 0 \\ 0 & 1 & 0 \\ 0 & 0 & 1 \end{bmatrix} = |\mathbf{M}|\mathbf{I} \tag{4.95}$$

The inverse of a square matrix **M** is denoted by $\mathbf{M}^{-1}$ and has the property

$$\mathbf{M}\mathbf{M}^{-1} = \mathbf{M}^{-1}\mathbf{M} = \mathbf{I} \tag{4.96}$$

The inverse matrix $\mathbf{M}^{-1}$ is defined from the result of Example 4 as

$$\mathbf{M}^{-1} = \frac{\operatorname{adj}\mathbf{M}}{|\mathbf{M}|} \tag{4.97}$$

The inverse exists only if $|\mathbf{M}| \neq 0$, that is, the matrix **M** is nonsingular.

INVERSE OF A PRODUCT OF MATRICES. The inverse of a product of matrices is equal to the product of the inverses of the individual matrices in reverse order:

$$[\mathbf{AB}]^{-1} = \mathbf{B}^{-1}\mathbf{A}^{-1}$$

PRODUCT OF DETERMINANTS. The determinant of the product of square matrices is equal to the product of the individual determinants:

$$|ABC| = |A| \cdot |B| \cdot |C| \tag{4.98}$$

RANK OF A MATRIX. The rank r of a matrix **M**, not necessarily square, is the order of the largest square array contained in **M** which has a nonzero determinant.

Example 5.

$$\mathbf{M} = \begin{bmatrix} 1 & 2 & 3 \\ 2 & 3 & 4 \\ 3 & 5 & 7 \end{bmatrix}$$

The determinant $|\mathbf{M}| = 0$; that is, $\mathbf{M}$ is a singular matrix. The square array obtained by deleting the first row and second column has a nonzero determinant:

$$\begin{vmatrix} 2 & 4 \\ 3 & 7 \end{vmatrix} = 2$$

Therefore, $\mathbf{M}$ has a rank of 2.

DEGENERACY (OR NULLITY) OF A MATRIX. When a matrix $\mathbf{M}$ of order n has rank r, there are $q = n - r$ rows or columns which are linear combinations of the r rows or columns. The matrix $\mathbf{M}$ is then said to have *degeneracy* q. In Example 5 the matrix $\mathbf{M}$ is of order $n = 3$, the rank is $r = 2$, and the degeneracy is $q = 1$ (simple degeneracy). Note that the third row of $\mathbf{M}$ is the sum of the first two rows. Also, the third column is two times the second column minus the first column.

SYMMETRIC MATRIX. A square matrix containing only real elements is symmetric if it is equal to its transpose, that is, $\mathbf{A} = \mathbf{A}^T$. The relationship between the elements is

$$a_{ij} = a_{ji}$$

A symmetric matrix is symmetrical about its principal diagonal.

TRANSPOSE OF A PRODUCT OF MATRICES. The transpose of a product of matrices is the product of the transposed matrices in reverse order:

$$(\mathbf{AB})^T = \mathbf{B}^T\mathbf{A}^T \tag{4.99}$$

This relationship can be described by the statement that the product of transposed matrices is equal to the transpose of the product in reverse order of the original matrices.

HERMITE NORMAL FORM. A number of properties of a matrix can be determined by transforming the matrix to *Hermite normal form* (HNF). These properties include the determination of the rank of a matrix, the inverse of a matrix of full rank, the characteristic polynomial of a matrix, the eigenvectors associated with the eigenvalue, the controllability of a system, and the observability of a system. The transformation is accomplished on a matrix $\mathbf{M}$ by performing the following basic or *elementary* operations:

1. Interchanging the ith and jth rows
2. Multiplying the ith row by a nonzero scalar k
3. Adding to the elements of the ith row the corresponding elements of the jth row multiplied by a scalar k

The purpose of these operations is to produce a unit element as the first nonzero element in each row (from the left), with the remaining elements of the column all zero. These operations can be accomplished by premultiplying by *elementary matrices*, where the elementary matrices are formed by performing the desired basic operations on a unit matrix. For example, for a third-order matrix **M**, premultiplying by the following elementary matrices $\mathbf{E}_i$ produces the results indicated:

$$\underset{\substack{\text{Interchanging}\\ \text{rows 1 and 2}}}{\begin{bmatrix} 0 & 1 & 0 \\ 1 & 0 & 0 \\ 0 & 0 & 1 \end{bmatrix}} \quad \underset{\substack{\text{Multiplying}\\ \text{row 3 by } k}}{\begin{bmatrix} 1 & 0 & 0 \\ 0 & 1 & 0 \\ 0 & 0 & k \end{bmatrix}} \quad \underset{\substack{\text{Adding } k \text{ times}\\ \text{row 3 to row 2}}}{\begin{bmatrix} 1 & 0 & 0 \\ 0 & 1 & k \\ 0 & 0 & 1 \end{bmatrix}} \tag{4.100}$$

Example 6. The matrix of Example 5 is reduced to HNF by the following operations:

1. Subtract 2 times row 1 from row 2 and 3 times row 1 from row 3. Premultiplying by the corresponding elementary matrix $\mathbf{E}_1$ produces

$$\mathbf{E}_1\mathbf{M} = \begin{bmatrix} 1 & 0 & 0 \\ -2 & 1 & 0 \\ -3 & 0 & 1 \end{bmatrix}\begin{bmatrix} 1 & 2 & 3 \\ 2 & 3 & 4 \\ 3 & 5 & 7 \end{bmatrix} = \begin{bmatrix} 1 & 2 & 3 \\ 0 & -1 & -2 \\ 0 & -1 & -2 \end{bmatrix} = \mathbf{M}_1$$

2. Add 2 times row 2 to row 1, subtract row 2 from row 3, and multiply row 2 by -1. Premultiplying by the corresponding elementary matrix $\mathbf{E}_2$ produces

$$\mathbf{E}_2\mathbf{M}_1 = \begin{bmatrix} 1 & 2 & 0 \\ 0 & -1 & 0 \\ 0 & -1 & 1 \end{bmatrix}\begin{bmatrix} 1 & 2 & 3 \\ 0 & -1 & -2 \\ 0 & -1 & -2 \end{bmatrix}\begin{bmatrix} 1 & 0 & -1 \\ 0 & 1 & 2 \\ 0 & 0 & 0 \end{bmatrix} = \mathbf{M}_2$$

The matrix $\mathbf{M}_2$ is in Hermite normal form and has the same rank as the matrix **M**. Since $\mathbf{M}_2$ has one zero row, it has a rank of 2, which agrees with the result of Example 5.

MATRIX INVERSION BY ROW OPERATIONS. The inverse $\mathbf{M}^{-1}$ of a nonsingular matrix **M** can be obtained as follows:

1. Form the augmented matrix $[\mathbf{M} \,\vdots\, \mathbf{I}]$.
2. Perform row operations on the augmented matrix so that the left portion **M** becomes the identity matrix **I**. When the left portion is so transformed, the right portion **I** is transformed into the matrix inverse $\mathbf{M}^{-1}$. Thus, the transformed augmented matrix becomes $\{\mathbf{I} \,\vdots\, \mathbf{M}^{-1}]$.

Example 7. For the matrix of Example 1, the augmented matrix is

$$[\mathbf{M} \,\vdots\, \mathbf{I}] = \left[\begin{array}{ccc|ccc} 3 & 1 & 2 & 1 & 0 & 0 \\ 1 & 0 & -4 & 0 & 1 & 0 \\ 0 & 5 & 7 & 0 & 0 & 1 \end{array}\right]$$

A suitable set of basic row operations includes subtracting 3 times row 2 from row 1 and then interchanging rows 1 and 2, yielding

$$\left[\begin{array}{ccc|ccc} 1 & 0 & -4 & 0 & 1 & 0 \\ 0 & 1 & 14 & 1 & -3 & 0 \\ 0 & 5 & 7 & 0 & 0 & 1 \end{array}\right]$$

Subtracting 5 times row 2 from row 3 yields

$$\left[\begin{array}{ccc|ccc} 1 & 0 & -4 & 0 & 1 & 0 \\ 0 & 1 & 14 & 1 & -3 & 0 \\ 0 & 0 & -63 & -5 & 15 & 1 \end{array}\right]$$

Dividing row 3 by -63, adding 4 times the new row 3 to row 1, and subtracting 14 times the new row 3 from row 2 yields

$$\left[\begin{array}{ccc|ccc} 1 & 0 & 0 & 20/63 & 3/63 & -4/63 \\ 0 & 1 & 0 & -7/63 & 21/63 & 14/63 \\ 0 & 0 & 1 & 5/63 & -15/63 & -1/63 \end{array}\right] = [\mathbf{I} \vdots \mathbf{M}^{-1}]$$

The inverse matrix $\mathbf{M}^{-1}$ agrees with the value obtained from Eq. (4.97) and the results of Examples 1 and 3.

EVALUATION OF THE CHARACTERISTIC POLYNOMIAL. The characteristic polynomial for the matrix **A** appearing in the state equation

$$\dot{\mathbf{x}} = \mathbf{A}\mathbf{x} + \mathbf{b}u \tag{4.101}$$

is

$$Q(\lambda) = |\lambda \mathbf{I} - \mathbf{A}| = \lambda^n + a_{n-1}\lambda^{n-1} + \cdots + a_1\lambda + a_0 \tag{4.102}$$

When the $n \times (n+1)$ matrix

$$\mathbf{M}_c = [\mathbf{b} \quad \mathbf{Ab} \quad \mathbf{A}^2\mathbf{b} \quad \cdots \quad \mathbf{A}^n\mathbf{b}] \tag{4.103}$$

is reduced to HNF, which must be full rank, the coefficients of the characteristic polynomial appear in the last column,[10] i.e.,

$$\begin{bmatrix} 1 & 0 & 0 & \cdots & 0 & 0 & -a_0 \\ 0 & 1 & 0 & \cdots & 0 & 0 & -a_1 \\ \cdots & \cdots & \cdots & \cdots & \cdots & \cdots & \cdots \\ 0 & 0 & 0 & \cdots & 0 & 1 & -a_{n-1} \end{bmatrix} \tag{4.104}$$

This procedure may be simpler than forming the determinant $Q(\lambda) = |\lambda \mathbf{I} - \mathbf{A}|$.

Example 8.

$$\mathbf{A} = \begin{bmatrix} 1 & 6 & -3 \\ -1 & -1 & 1 \\ -2 & 2 & 0 \end{bmatrix} \qquad \mathbf{b} = \begin{bmatrix} 1 \\ 1 \\ 1 \end{bmatrix}$$

$$\mathbf{M}_c = \begin{bmatrix} 1 & 4 & -2 & 10 \\ 1 & -1 & -3 & -5 \\ 1 & 0 & -10 & -2 \end{bmatrix} \tag{4.105}$$

Reducing $\mathbf{M}_c$ to HNF yields

$$\begin{bmatrix} 1 & 0 & 0 & -2 \\ 0 & 1 & 0 & 3 \\ 0 & 0 & 1 & 0 \end{bmatrix} \begin{matrix} \leftarrow -a_0 \\ \leftarrow -a_1 \\ \leftarrow -a_2 \end{matrix}$$

The characteristic polynomial is therefore

$$Q(\lambda) = \lambda^3 - 3\lambda + 2$$

LINEAR INDEPENDENCE. A set of vectors $\mathbf{v}_1, \mathbf{v}_2, \ldots, \mathbf{v}_k, \ldots, \mathbf{v}_n$ are linearly independent provided there is no set of scalars α_i ($i = 1, 2, \ldots, n$), not all zero, which satisfies

$$\alpha_1\mathbf{v}_1 + \cdots + \alpha_k\mathbf{v}_k + \cdots + \alpha_n\mathbf{v}_n = \mathbf{0} \tag{4.106}$$

If this equation is satisfied with the α_i not all zero, then the vectors are *linearly dependent*. Linear independence can be checked by forming the matrix

$$\begin{bmatrix} \mathbf{v}_1^T \\ \mathbf{v}_2^T \\ \vdots \\ \mathbf{v}_n^T \end{bmatrix}$$

If this matrix has rank n, the vectors are linearly independent; otherwise they are linearly dependent. The rank of this matrix can be determined by reducing it to Hermite normal form.

4.15 SOLUTION OF STATE EQUATION[7]

The homogeneous state equation is

$$\dot{\mathbf{x}} = \mathbf{A}\mathbf{x} \tag{4.107}$$

The solution of this equation can be obtained by taking the Laplace transform

$$s\mathbf{X}(s) - \mathbf{x}(0) = \mathbf{A}\mathbf{X}(s)$$

Grouping of the terms containing $\mathbf{X}(s)$ yields

$$[s\mathbf{I} - \mathbf{A}]\mathbf{X}(s) = \mathbf{x}(0)$$

The unit matrix $\mathbf{I}$ is introduced so that all terms in the equations are proper matrices. Premultplying both sides of the equation by $[s\mathbf{I} - \mathbf{A}]^{-1}$ yields

$$\mathbf{X}(s) = [s\mathbf{I} - \mathbf{A}]^{-1}\mathbf{x}(0) \tag{4.108}$$

The inverse Laplace transform of Eq. (4.108) gives

$$\mathbf{x}(t) = \mathscr{L}^{-1}\{[s\mathbf{I} - \mathbf{A}]^{-1}\}\mathbf{x}(0) \tag{4.109}$$

Comparing this solution with Eq. (3.70) yields the following expression for the state transition matrix:

$$\mathbf{\Phi}(t) = \mathscr{L}^{-1}[s\mathbf{I} - \mathbf{A}]^{-1} \tag{4.110}$$

The *resolvent matrix* is designated by $\mathbf{\Phi}(s)$ and is defined by†

$$\mathbf{\Phi}(s) = [s\mathbf{I} - \mathbf{A}]^{-1} \tag{4.111}$$

† When $\mathbf{A}$ is a companion matrix, a direct algorithm for evaluating $[s\mathbf{I} - \mathbf{A}]^{-1}$ and $\mathbf{\Phi}(t)$ is presented in Ref. 9.

Example 1. Solve for $\mathbf{\Phi}(t)$ (see the example of Sec. 3.13) for

$$\mathbf{A} = \begin{bmatrix} 0 & 6 \\ -1 & -5 \end{bmatrix}$$

Using Eq. (4.111) gives

$$\mathbf{\Phi}(s) = \left[\begin{bmatrix} s & 0 \\ 0 & s \end{bmatrix} - \begin{bmatrix} 0 & 6 \\ -1 & -5 \end{bmatrix}\right]^{-1} = \begin{bmatrix} s & -6 \\ 1 & s+5 \end{bmatrix}^{-1}$$

Using Eq. (4.97) gives

$$\mathbf{\Phi}(s) = \frac{1}{s^2 + 5s + 6}\begin{bmatrix} s+5 & 6 \\ -1 & s \end{bmatrix} = \begin{bmatrix} \dfrac{s+5}{(s+2)(s+3)} & \dfrac{6}{(s+2)(s+3)} \\ \dfrac{-1}{(s+2)(s+3)} & \dfrac{s}{(s+2)(s+3)} \end{bmatrix} \tag{4.112}$$

Using transform pairs 12 to 14 in App. A yields the inverse transform

$$\mathbf{\Phi}(t) = \begin{bmatrix} 3e^{-2t} - 2e^{-3t} & 6e^{-2t} - 6e^{-3t} \\ -e^{-2t} + e^{-3t} & -2e^{-2t} + 3e^{-3t} \end{bmatrix}$$

This is the same as the value obtained in Eq. (3.92) and is an alternate method of obtaining the same result.

The state equation with an input present is

$$\dot{\mathbf{x}} = \mathbf{A}\mathbf{x} + \mathbf{B}\mathbf{u}$$

Solving by means of the Laplace transform yields

$$\mathbf{X}(s) = [s\mathbf{I} - \mathbf{A}]^{-1}\mathbf{x}(0) + [s\mathbf{I} - \mathbf{A}]^{-1}\mathbf{B}\mathbf{U}(s) = \mathbf{\Phi}(s)\mathbf{x}(0) + \mathbf{\Phi}(s)\mathbf{B}\mathbf{U}(s) \tag{4.113}$$

Example 2. Solve for the state equation in the example of Sec. 3.14 for a step-function input. The function $\mathbf{\Phi}(s)$ is given by Eq. (4.112) in Example 1 and $\mathbf{U}(s) = 1/s$. Inserting these values into Eq. (4.113) gives

$$\mathbf{X}(s) = \mathbf{\Phi}(s)\mathbf{x}(0) + \begin{bmatrix} \dfrac{6}{s(s+2)(s+3)} \\ \dfrac{1}{(s+2)(s+3)} \end{bmatrix}$$

Applying the transform pairs 10 and 11 of App. A yields the same result for $\mathbf{x}(t)$ as in Eq. (3.105). The Laplace transform method is a direct method of solving the state equation.

The possibility exists in the elements of $\mathbf{\Phi}(s)$ that a numerator factor of the form $s + a$ will cancel a similar term in the denominator. In that case the transient mode e^{-at} does not appear in the corresponding element of $\mathbf{\Phi}(t)$. This has a very important theoretic significance since it affects the *controllability and observability* of a control system. These properties are described in Sec. 13.7. Also, as described in Chap. 15, the ability to achieve an optimal control system requires that the system be observable and controllable.

4.16 EVALUATION OF THE TRANSFER-FUNCTION MATRIX

The transfer-function matrix of a system with multiple inputs and outputs is evaluated from the state and output equations

$$\dot{\mathbf{x}} = \mathbf{A}\mathbf{x} + \mathbf{B}\mathbf{u} \qquad \mathbf{y} = \mathbf{C}\mathbf{x} + \mathbf{D}\mathbf{u}$$

Taking the Laplace transform of these equations and solving for the output $\mathbf{Y}(s)$ in terms of the input $\mathbf{U}(s)$ yields

$$\mathbf{Y}(s) = \mathbf{C}\boldsymbol{\Phi}(s)\mathbf{x}(0) + \mathbf{C}\boldsymbol{\Phi}(s)\mathbf{B}\mathbf{U}(s) + \mathbf{D}\mathbf{U}(s)$$

Since the transfer-function relationship is $\mathbf{Y}(s) = \mathbf{G}(s)\mathbf{U}(s)$ and is defined for zero initial conditions, the system transfer-function matrix is given by analogy as

$$\mathbf{G}(s) = \mathbf{C}\boldsymbol{\Phi}(s)\mathbf{B} + \mathbf{D} \tag{4.114}$$

For multiple-input multiple-output systems $\mathbf{G}(s)$ is a matrix in which the elements are the transfer functions between each output and each input of the system.

Example. Determine the transfer functions and draw a block diagram for the two-input two-output system represented by

$$\dot{\mathbf{x}} = \begin{bmatrix} 0 & 1 \\ -2 & -3 \end{bmatrix}\mathbf{x} + \begin{bmatrix} 1 & 1 \\ 0 & -2 \end{bmatrix}\mathbf{u} \qquad \mathbf{y} = \begin{bmatrix} 0 & -2 \\ 1 & 0 \end{bmatrix}\mathbf{x}$$

The characteristic polynomial is

$$|s\mathbf{I} - \mathbf{A}| = (s+1)(s+2)$$

From Eq. (4.111), the resolvent matrix is

$$\boldsymbol{\Phi}(s) = \frac{1}{(s+1)(s+2)}\begin{bmatrix} s+3 & 1 \\ -2 & s \end{bmatrix}$$

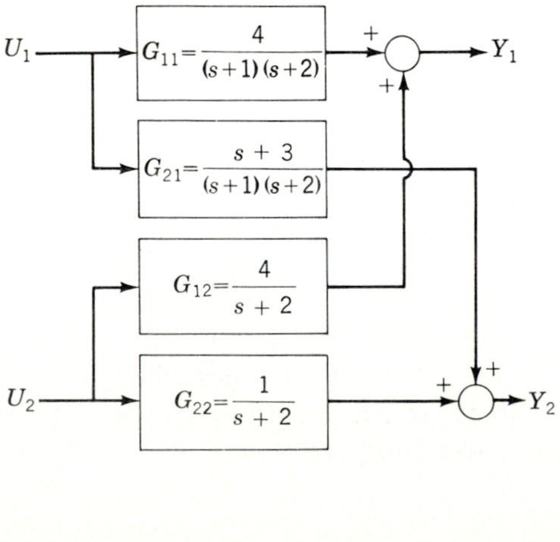

FIGURE 4.14
Block diagram for Eq. (4.115).

Then, from Eq. (4.114)

$$\mathbf{G}(s) = \mathbf{C}\boldsymbol{\Phi}(s)\mathbf{B} = \begin{bmatrix} \dfrac{4}{(s+1)(s+2)} & \dfrac{4}{s+2} \\ \dfrac{s+3}{(s+1)(s+2)} & \dfrac{1}{s+2} \end{bmatrix} = \begin{bmatrix} G_{11}(s) & G_{12}(s) \\ G_{21}(s) & G_{22}(s) \end{bmatrix} \tag{4.115}$$

The block diagram for the system is shown in Fig. 4.14.

4.17 SUMMARY

This chapter discusses the important characteristics and the use of the Laplace transform, which is employed extensively with differential equations because it systematizes their solution. Also it is used extensively in feedback-system synthesis. The pole-zero pattern has been introduced to represent a system function. The pole-zero pattern is significant because it determines the amplitudes of all the time-response terms. The frequency response has also been shown to be a function of the pole-zero pattern. The solution of linear time-invariant state equations can be obtained by means of the Laplace transform. The procedure is readily adapted for the use of a digital computer.[8,11] This is advantageous for multiple-input multiple-output systems.

Later chapters cover feedback-system analysis and synthesis by three methods. The first of these is the root-locus method, which locates the poles and zeros of the system in the s plane. Knowing the poles and zeros permits an exact determination of the time response. The second method is based on the frequency response. Since the frequency response is a function of the pole-zero pattern, the two methods are complementary and give equivalent information in different forms. The root-locus and frequency-response methods are primarily applicable to single-input single-output (SISO) systems and rely on use of transfer functions to represent the system. They can be adapted for the synthesis of multiple-input multiple-output (MIMO) systems. The state-feedback method is readily used for both SISO and MIMO systems. The necessary linear-algebra operations for state-feedback synthesis are presented in this chapter in preparation for their use in later chapters.

System stability requires that all roots of the characteristic equation be located in the left half of the s plane. They can be identified on the pole-zero diagram and are the poles of the overall transfer function. The transfer function can be obtained from the overall differential equation relating an input to an output. The transfer function can also be obtained from the state-equation formulation, as shown in this chapter.

REFERENCES

1. Churchill, R. V.: *Operational Mathematics*, 3d ed., McGraw-Hill, New York, 1972.
2. Thomson, W. T.: *Laplace Transformation*, 2d ed., Prentice-Hall, Englewood Cliffs, N.J., 1966.

3. Wylie, C. R., Jr.: *Advanced Engineering Mathematics*, 4th ed., McGraw-Hill, New York, 1975.
4. Hazony, D., and I. Riley: Simplified Technique for Evaluating Residues in the Presence of High Order Poles, paper presented at the Western Electric Show and Convention, August 1959.
5. Aseltine, J. A.: *Transform Method in Linear System Analysis*, McGraw-Hill, New York, 1958.
6. Kinariwala, B., et al.: *Linear Circuits and Computation*, Wiley, New York, 1973.
7. Ward, J. R., and R. D. Strum: *State Variable Analysis*, Prentice-Hall, Englewood Cliffs, N.J., 1970.
8. Larimer, S. J.: "An Interactive Computer-Aided Design Program for Digital and Continuous System Analysis and Synthesis (TOTAL)," M.S. thesis, GE/GGC/EE/78-2, School of Engineering. Air Force Institute of Technology, Wright-Patterson Air Force Base, Dayton, Ohio, 1978. [available from Defense Documentation Center (DDC), Cameron Station, Alexandria, VA 22314.]
9. Taylor, F. J.: "A Novel Inversion of ($s\mathbf{I} - \mathbf{A}$)," *Int. J. Systems Science*, vol. 5, no. 2, pp. 153–160, Feb. 1974.
10. Reid, J. G.: "Linear Systems Analysis and Digital Computation," unpublished notes, Air Force Institute of Technology, Wright-Patterson Air Force Base, Dayton, Ohio, 1979.
11. Thompson, Peter M.: *USER's Guide to Program CC, Version 3*, Systems Technology, Inc., Hawthorne, CA, March, 1985.
12. Bongiorno, J. J., Jr.: "A Recursive Algorithm for Computing the Partial Fraction Expansion of Rational Functions Having Multiple Poles," *IEEE Trans. Autom. Control*, vol. AC-29, pp. 650–652, 1984.

CHAPTER 5

SYSTEM REPRESENTATION

5.1 INTRODUCTION

This chapter introduces the basic principles of system representation. From the concepts introduced in earlier chapters, a number of systems are represented in block-diagram form. The individual functions are represented by transfer function which are used to describe the blocks of the system. Feedback is included in these systems in order to achieve the desired performance. Also, the standard symbols and definitions are presented. These are extensively used in the technical literature on control systems and form a common basis for understanding and clarity. While block diagrams simplify the representation of functional relationships within a system, the use of signal flow graphs (SFG) provides further simplification for larger systems which contain intercoupled relationships. Simulation diagrams are presented as a means of representing a system for which the overall differential equation is known. Then the inclusion of initial conditions in the SFG leads to the state-diagram representation.

To provide flexibility in the method used by the designer for system analysis and design, the system representation may be in either the transfer-function or the state-equation form. The procedures for converting from one representation to the other are presented in Chap. 4 and in this chapter. When the state-variable format is used, it is often desirable to transform the mathematical equations so that the states are uncoupled. This simplifies the equations, leading to an **A** matrix which is diagonal. This is called the *normal* or *canonical form* and is useful for evaluating observability and controllability (see Chap. 13). It also considerably simplifies the computer programs required to solve these equations. The techniques for transforming the state vector are presented.

5.2 BLOCK DIAGRAMS

The representation of physical components by blocks is shown in Chap. 2. For each block the transfer function provides the dynamic mathematical relationship between the input and output quantities. Also, Chap. 1 and Fig. 1.2 describe the concept of feedback, which is used to achieve a better response of a control system to a command input. This section presents several examples of control systems and their representation by block diagrams. The blocks represent the functions performed rather than the components of the system.

Example 1. A temperature-control system. The first example is an industrial-process temperature-control system, shown in Fig. 5.1*a*. The voltage r, obtained from a potentiometer, is calibrated in terms of the desired temperature θ_{comm} to be produced within the tank. This voltage represents the input quantity to the control system. The actual temperature θ, the output quantity, is measured by means of a thermocouple immersed in the tank. The voltage e_{th} produced in the thermocouple is proportional to θ. The voltage e_{th} is amplified to produce a voltage b, which is the feedback quantity. The voltage $e = r - b$ is the actuating signal and is amplified to produce the solenoid voltage e_1. The current i_s in the solenoid, which results from applying e_1, produces a proportional force f_s which acts on the solenoid armature and valve to control the valve position x. The valve position in turn controls the flow q of hot steam into the heating coil in the tank. The resulting temperature θ of the tank is directly proportional to the steam flow with a time delay which depends on the specific heat of the fluid and the mixing rate. The

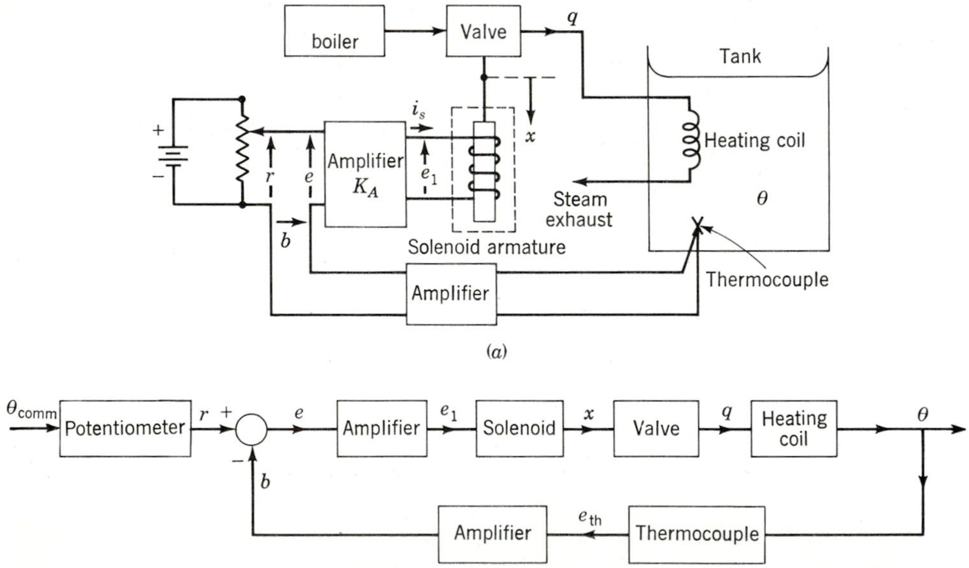

FIGURE 5.1
An industrial-process temperature-control system.

block-diagram representation for this system (Fig. 5.1b) shows the functions of each unit and the signal flow through the system.

To show the operation of the system consider that an increase in the tank temperature is required. The voltage r is increased to a value that represents the value of the desired temperature. This change in r causes an increase in the actuating signal e. This increase in e, through its effect on the solenoid and valve, causes an increase in the amount of hot steam flowing through the heating coil in the tank. The temperature θ therefore increases in proportion to the steam flow. When the output temperature rises to a value essentially equal to the desired temperature, the feedback voltage b is equal to the reference input r, $b \approx r$. The flow of steam through the heating coil is stabilized at a steady-state value that maintains θ at the desired value.

Example 2. Command guidance interceptor system. A more complex system is a command guidance system which directs the flight of a missile in space in order to intercept a moving target. The target may be an enemy bomber whose aim is to drop bombs at some position. The defense uses the missile with the objective of intercepting and destroying the bomber before it launches its bombs. A sketch of a generalized command guidance interceptor system is shown in Fig. 5.2. The target-tracking radar is used first for detection and then for tracking the target. It supplies information on target range and angle and their rates of change (time derivatives). This information is continuously fed into the computer, which calculates a predicted course for the target. The missile-tracking radar supplies similar information which is used by the computer to determine its flight path. The computer compares the two flight paths and determines the necessary change in missile flight path to produce a collision course. The necessary flight-path changes are supplied to the radio command link, which transmits this information to the missile. This electrical information containing corrections in flight path is used by a control system in the missile. The missile control system converts the error signals into mechanical displacements of the missile airframe control surfaces by means of actuators. The missile responds to the positions of the aerodynamic control surfaces to follow the prescribed flight path, which is intended to produce a collision with the target. Monitoring of the

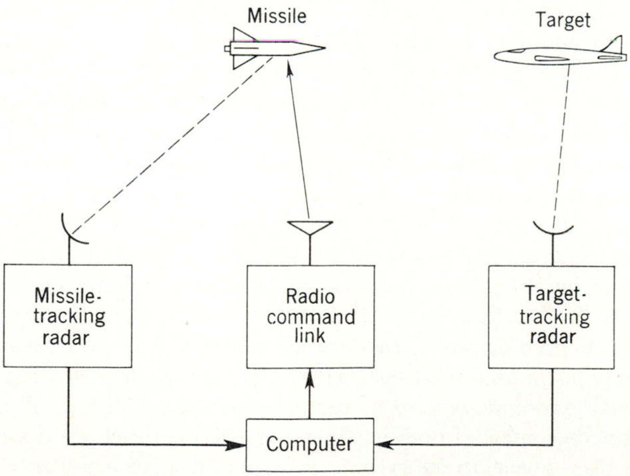

FIGURE 5.2
Command guidance interceptor system.

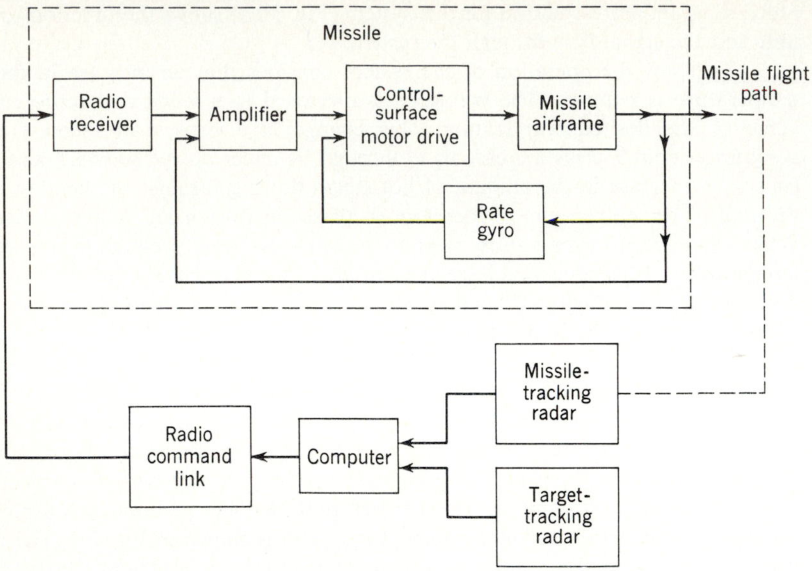

FIGURE 5.3
Block diagram of a generalized command guidance interceptor system.

target is continuous so that changes in the missile course can be corrected up to the point of impact. A block diagram depicting the functions of this command guidance system is shown in Fig. 5.3. Many individual functions are performed within each block. Some of the components of the missile control system are shown within the block representing the missile.

Example 3. Aircraft control system.[1] The feedback control system used to keep an airplane on a predetermined course or heading is necessary for the navigation of commercial airliners. Despite poor weather conditions and lack of visibility the airplane must maintain a specified heading and altitude in order to reach its destination safely. In addition, in spite of rough air, the trip must be made as smooth and comfortable as possible for the passengers and crew. The problem is considerably complicated by the fact that the airplane has six degrees of freedom. This makes control more difficult than the control of a ship, whose motion is limited to the surface of the water. A *flight controller* is used to control the aircraft motion.

Two typical signals to the system are the correct flight path, which is set by the pilot, and the level position (attitude) of the airplane. The ultimately controlled variable is the actual course and position of the airplane. The output of the control system, the controlled variable, is the aircraft heading. In conventional aircraft there are three primary control surfaces used to control the physical three-dimensional attitude of the airplane, the elevators, rudder, and ailerons. The axes used for an airplane and the motions produced by the control surfaces are shown in Fig. 5.4.

The directional gyroscope is used as the error-measuring device. Two gyros must be used to provide control of both heading and attitude (level position) of the airplane. The error that appears in the gyro as an angular displacement between the

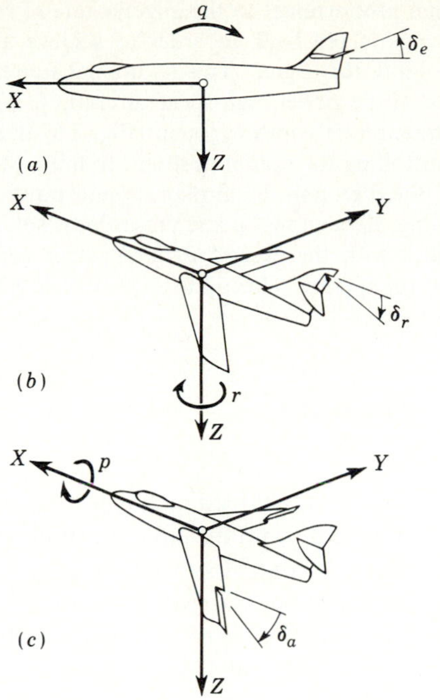

FIGURE 5.4
Airplane control surfaces: (*a*) elevator deflection produces pitching velocity q; (*b*) rudder deflection produces yawing velocity r; (*c*) aileron deflection produces rolling velocity p.

rotor and case is translated into a voltage by various methods, including the use of transducers such as potentiometers, synchros, transformers, or microsyns. Selection of the method used depends on the preference of the gyro manufacturer and the sensitivity required.

Additional stabilization for the aircraft can be provided in the control system by rate feedback. In other words, in addition to the primary feedback, which is the

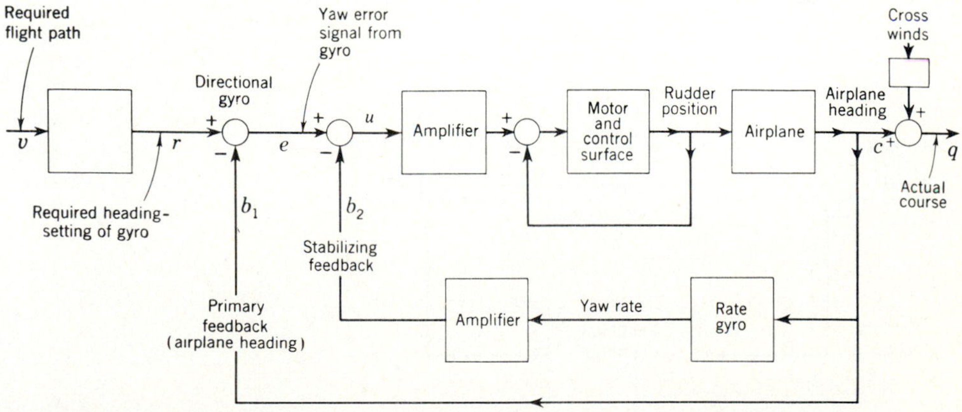

FIGURE 5.5
Airplane directional-control system.

position of the airplane, another signal proportional to the angular rate of rotation of the airplane around the vertical axis is fed back in order to achieve a stable response. A "rate" gyro is used to supply this signal. This additional stabilization may be absolutely necessary for some of the newer high-speed aircraft.

A typical block diagram of the aircraft control system (Fig. 5.5) illustrates control of the airplane heading by controlling the rudder position. In this system the airplane heading is controlled and is the direction the airplane would travel in still air. The pilot must correct this heading, depending on the crosswinds, so that the actual course of the airplane coincides with the desired path. Another control is included in the complete airplane control system to keep the airplane in level flight. It controls the ailerons and elevators.

5.3 DETERMINATION OF THE OVERALL TRANSFER FUNCTION

The block diagram of a control system with negative feedback can be simplified to the form shown in Fig. 5.6, where the standard symbols and definitions used in feedback systems are indicated. In this feedback control system the output is the controlled variable C. This output is measured by a feedback element H to produce the primary feedback signal B which is then compared with the reference input R. The difference E, between the reference input R and the feedback signal B, is the input to the controlled system G and is referred to as the *actuating signal*. The actuating signal is sometimes referred to as the error signal, but it is actually the difference $R - C$ only for unity-feedback systems where $H = 1$. The transfer functions of the forward and feedback components of the system are G and H, respectively.

In using the block diagram to represent a linear feedback control system where the transfer functions of the components are known, the letter symbol is capitalized, indicating that it is a transformed quantity; i.e., it is a function of the operator D, the complex parameter s, or the frequency parameter $j\omega$. This applies for the transfer function, where G is used to represent $G(D)$, $G(s)$, or $\mathbf{G}(j\omega)$. It also applies to all variable quantities, such as C, which represents

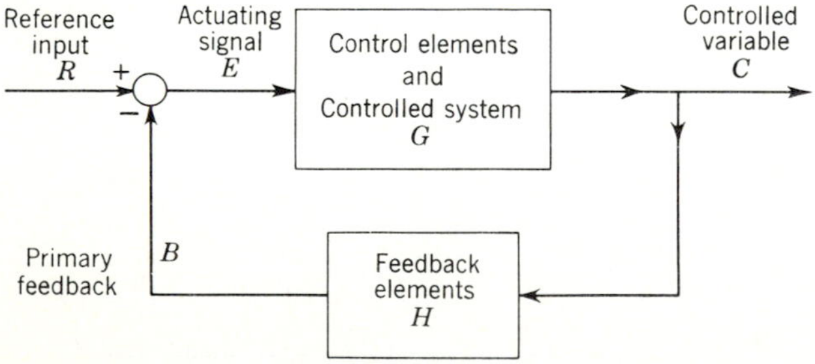

FIGURE 5.6
Block diagram of a feedback system.

$C(D)$, $C(s)$, or $\mathbf{C}(j\omega)$. Lowercase symbols are used, as in Fig. 5.5, to represent any function in the time domain. For example, the symbol c represents $c(t)$.

The important characteristic of such a system is the *overall transfer function*, which is the ratio of the transform of the controlled variable C to the transform of the reference input R. This ratio may be expressed in operational, Laplace transform, or frequency (phasor) form. The overall transfer function is also referred to as the *control ratio*. The terms are used interchangeably in this text.

The equations describing this system in terms of the transform variable are

$$C(s) = G(s)E(s) \tag{5.1}$$

$$B(s) = H(s)C(s) \tag{5.2}$$

$$E(s) = R(s) - B(s) \tag{5.3}$$

Combining these equations produces the *control ratio*, or *overall transfer function*,

$$\frac{C(s)}{R(s)} = \frac{G(s)}{1 + G(s)H(s)} \tag{5.4}$$

The characteristic equation of the closed-loop system is

$$1 + G(s)H(s) = 0 \tag{5.5}$$

This is obtained from the denominator of the control ratio. The stability and response of the closed-loop system, as determined by analysis of the characteristic equation, are discussed more fully in later chapters.

For simplified systems where the feedback is unity, that is, $H(s) = 1$, the actuating signal, given by Eq. (5.3), is now the error present in the system, i.e., *reference input minus the controlled variable*, expressed by

$$E(s) = R(s) - C(s) \tag{5.6}$$

The control ratio with unity feedback is

$$\frac{C(s)}{R(s)} = \frac{G(s)}{1 + G(s)} \tag{5.7}$$

The *open-loop transfer function* is defined as the ratio of the output of the feedback path $B(s)$ to the actuating signal $E(s)$ for any given feedback loop. In terms of Fig. 5.6, the open-loop transfer function is

$$\frac{B(s)}{E(s)} = G(s)H(s) \tag{5.8}$$

The *forward transfer function* is defined as the ratio of the controlled variable $C(s)$ to the actuating signal $E(s)$. For the system shown in Fig. 5.5 the forward transfer function is

$$\frac{C(s)}{E(s)} = G(s) \tag{5.9}$$

In the case of unity feedback, where $H(s) = 1$, the open-loop and the forward transfer functions are the same.

The forward transfer function $G(s)$ may be made up not only of elements in cascade but also may contain internal, or *minor*, feedback loops. The algebra of combining these internal feedback loops is similar to that used above. An example of a controlled system with an internal feedback loop is shown in Fig. 5.5.

It is often useful to express the actuating signal E in terms of the input R. Solving from Eqs. (5.1) to (5.3) gives

$$\frac{E(s)}{R(s)} = \frac{1}{1 + G(s)H(s)} \tag{5.10}$$

The concept of system error y_e is important. It is defined as the ideal or desired system value minus the actual system output. The ideal value establishes the desired performance of the system. For unity-feedback systems the actuating signal is an actual measure of the error and is directly proportional to the system error (see Fig. 5.7).

5.4 STANDARD BLOCK-DIAGRAM TERMINOLOGY[2]

Figure 5.7 shows a block-diagram representation of a feedback control system containing the basic elements. Figure 5.8 shows the block diagram and symbols of a more complicated system with multiple paths. Numerical subscripts are used to distinguish between blocks of similar functions in the circuit. For example, in Fig. 5.7 the control elements are designated by G_1 and the controlled system by G_2. In Fig. 5.8 the control elements are divided into two blocks, G_1 and G_2, to aid in representing relations between parts of the system. Also, in this figure the primary feedback is represented by H_1. Additional minor feedback loops might be designated by H_2, H_3, and so forth.

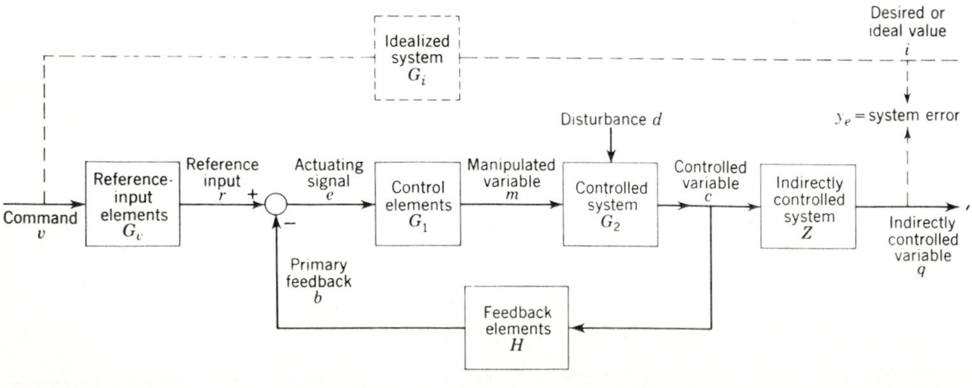

FIGURE 5.7
Block diagram of feedback control system containing all basic elements.

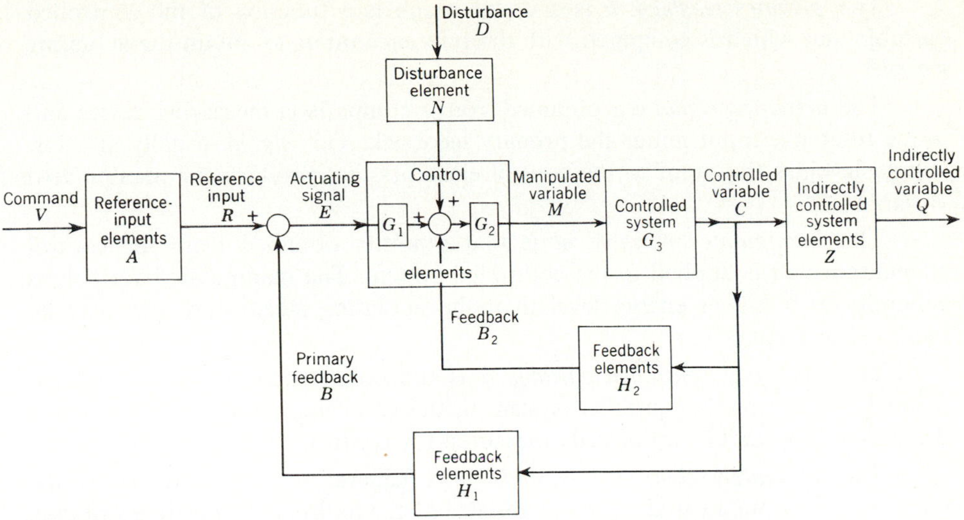

FIGURE 5.8
Block diagram of representative feedback control system showing multiple feedback loops (all uppercase letters denote transformation).

The idealized system represented by the block enclosed with dashed lines in Fig. 5.7 can be understood to show the relation between the basic input to the system and the performance of the system in terms of the desired output. This would be the system agreed upon to establish the ideal value for the output of the system. In systems where the command is actually the desired value or ideal value, the idealized system would be represented by unity. The arrows and block associated with the idealized system, the ideal value, and the system error are shown in dashed lines on the block diagram because they do not exist physically in any feedback control system. For any specific problem, it represents the system (conceived in the mind of the designer) that will give the nearest approach, when considered as a perfect system, to the desired output or ideal value.

Definitions: Variables in the System

The *command* v is the input which is established by some means external to, and independent of, the feedback control system.

The *reference input* r is derived from the command and is the actual signal input to the system.

The *controlled variable* c is the quantity that is directly measured and controlled. It is the output of the controlled system.

The *primary feedback* b is a signal which is a function of the controlled variable and which is compared with the reference input to obtain the actuating signal.

The *actuating signal* e is obtained from a comparison measuring device and is the reference input minus the primary feedback. This signal, usually at a low energy level, is the input to the control elements that produce the manipulated variable.

The *manipulated variable* m is that quantity obtained from the control elements which is applied to the controlled system. The manipulated variable is generally at a higher energy level than the actuating signal and may also be modified in form.

The *indirectly controlled variable* q is the output quantity and is related through the indirectly controlled system to the controlled variable. It is outside the closed loop and is not directly measured for control.

The *ultimately controlled variable* is a general term that refers to the indirectly controlled variable. In the absence of the indirectly controlled variable, it refers to the controlled variable.

The *ideal value* i is the value of the ultimately controlled variable that would result from an idealized system operating from the command as the actual system.

The *system error* y_e is the ideal value minus the value of the ultimately controlled variable.

The *disturbance* d is the unwanted signal that tends to affect the controlled variable. The disturbance may be introduced into the system at many places.

Definitions: System Components

The *reference input elements* G_v produce a signal r proportional to the command.

The *control elements* G produce the manipulated variable m from the actuating signal.

The *controlled system* G is the device that is to be controlled. This is frequently a high-power element.

The *feedback elements* H produce the primary feedback b from the controlled variable. This is generally a proportionality device but may also modify the characteristics of the controlled variable.

The *indirectly controlled system* Z relates the indirectly controlled variable q to the controlled quantity c. This component is outside the feedback loop.

The *idealized system* G_i is one whose performance is agreed upon to define the relationship between the ideal value and the command. This is often called the *model* or *desired* system.

The *disturbance element* N denotes the functional relationship between the variable representing the disturbance and its effect on the control system.

5.5 POSITION-CONTROL SYSTEM

Figure 5.9 shows a simplified block diagram of an angular-position-control system. The reference selector and the sensor, which produce the reference input $R = \theta_R$ and the controlled output position $C = \theta_o$, respectively, consist of rotational potentiometers. The combination of these units to produce a rotational comparison unit to generate the actuating signal E for the position-control system is shown in Fig. 5.10*a*, where K_θ, in volts per radian, is the potentiometer-sensitivity constant. The symbolic comparator for this system is shown in Fig. 5.10*b*.

The transfer function of the motor-generator control is obtained by writing the equations for the schematic diagram shown in Fig. 5.11*a*. This figure shows a dc motor which has a constant field excitation and drives an inertia and friction load. The armature voltage for the motor is furnished by the generator which is driven at constant speed by a prime mover. The generator voltage e_g is determined by the voltage e_f applied to the generator field. The generator is acting as a power amplifier for the signal voltage e_f. The equations for this system are

$$e_f = L_f Di_f + R_f i_f \tag{5.11}$$

$$e_g = K_g i_f \tag{5.12}$$

$$e_g - e_m = (L_g + L_m)\, Di_m + (R_g + R_m) i_m \tag{5.13}$$

$$e_m = K_b D\theta_o \tag{5.14}$$

$$T = K_T i_m = J D^2\theta_o + B D\theta_o \tag{5.15}$$

The complete block diagram drawn in Fig. 5.11*b* is based on the performance represented by these equations. Starting with the input quantity e_f, Eq. (5.11) shows that a current i_f is produced. Therefore, a block is drawn with e_f as the input and i_f as the output. Equation (5.12) shows that a voltage e_g is generated as a function of the current i_f. Therefore, a second block is drawn with the current i_f as the input and e_g as the output. Equation (5.13) relates the current i_m in the motor to the difference of two voltages, $e_g - e_m$. To obtain this difference a summation point is introduced. The quantity e_g from the previous block enters this summation point. To obtain the quantity $e_g - e_m$ there must be

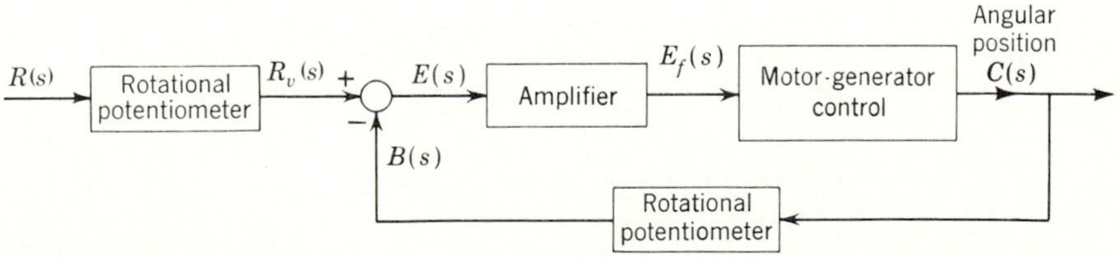

FIGURE 5.9
Position-control system.

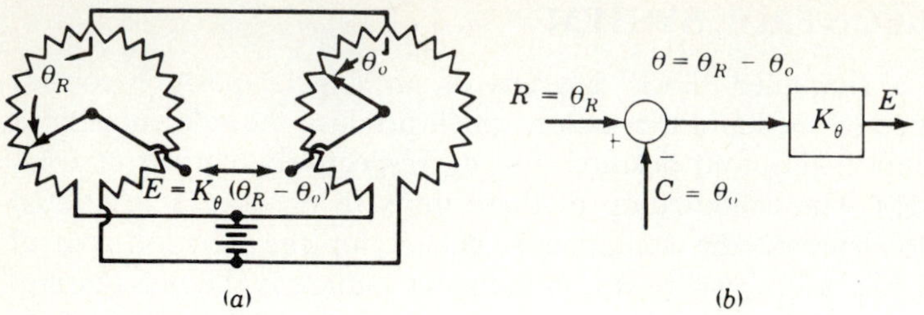

FIGURE 5.10
Rotational position comparison and its block-diagram representation.

added the quantity e_m entering the summation point with a minus sign. Up to this point the manner in which e_m is obtained has not yet been determined. The output of this summation point is used as the input to the next block from which the current i_m is the output. In the same manner, the block with current as the input and the generated torque as the output and the block with the torque input and the resultant motor position as the output are drawn. There must be no loose ends in the complete diagram; i.e., every dependent variable must be connected through a block or blocks into the system. Therefore e_m is obtained from Eq. (5.14), and a block representing this relationship is drawn with θ_o as the input and e_m as the output. Using this procedure, the block diagram is completed and the functional relationships in the system are described. Note that the generator and the motor are no longer separately distinguishable. The input and the output of each block are not in the same units, as both electrical and mechanical quantities are included.

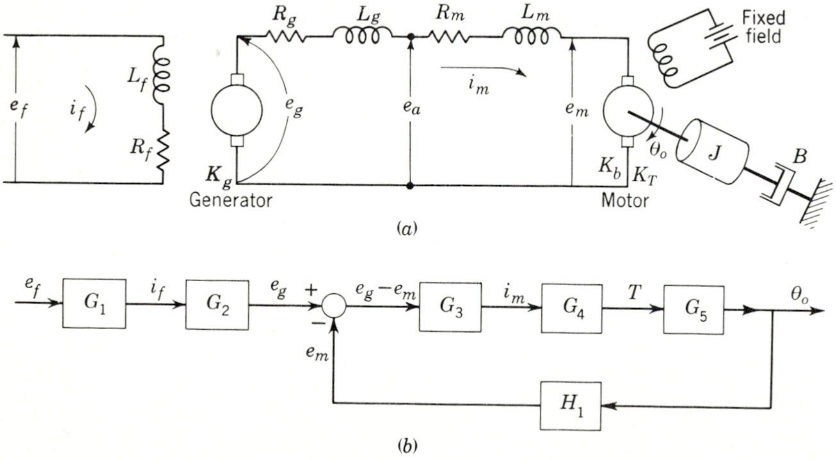

FIGURE 5.11
(a) Motor-generator control: (b) block diagram.

The transfer functions of each block, as determined in terms of the pertinent Laplace transforms, are

$$G_1(s) = \frac{I_f(s)}{E_f(s)} = \frac{1/R_f}{1 + (L_f/R_f)s} = \frac{1/R_f}{1 + T_f s} \tag{5.16}$$

$$G_2(s) = \frac{E_g(s)}{I_f(s)} = K_g \tag{5.17}$$

$$G_3(s) = \frac{I_m(s)}{E_g(s) - E_m(s)} = \frac{1/(R_g + R_m)}{1 + [(L_g + L_m)/(R_g + R_m)]s}$$

$$= \frac{1/R_{gm}}{1 + (L_{gm}/R_{gm})s} = \frac{1/R_{gm}}{1 + T_{gm}s} \tag{5.18}$$

$$G_4(s) = \frac{T(s)}{I_m(s)} = K_T \tag{5.19}$$

$$G_5(s) = \frac{\Theta_o(s)}{T(s)} = \frac{1/B}{s[1 + (J/B)s]} = \frac{1/B}{s(1 + T_n s)} \tag{5.20}$$

$$H_1(s) = \frac{E_m(s)}{\Theta_o(s)} = K_b s \tag{5.21}$$

The block diagram can be simplified, as shown in Fig. 5.12, by combining the blocks in cascade. The block diagram is further simplified, as shown in Fig. 5.13, by evaluating an equivalent block from $E_g(s)$ to $\Theta_o(s)$, using the principle of Eq. (5.4). The final simplification results in Fig. 5.14. The overall transfer

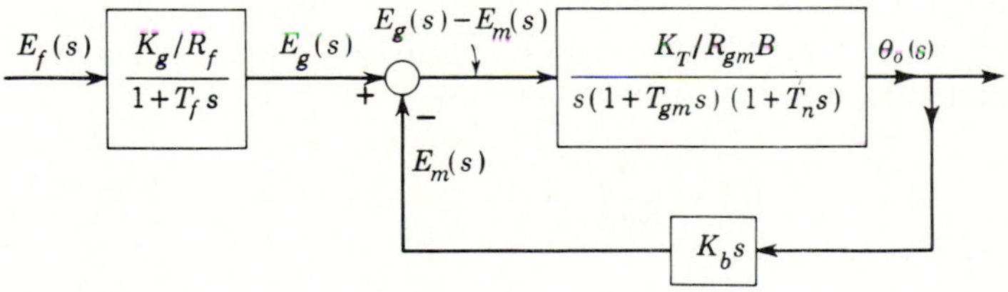

FIGURE 5.12
Simplified block diagram.

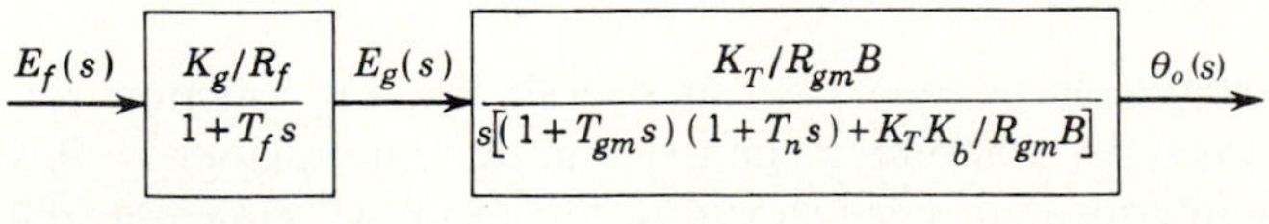

FIGURE 5.13
Reduced block diagram.

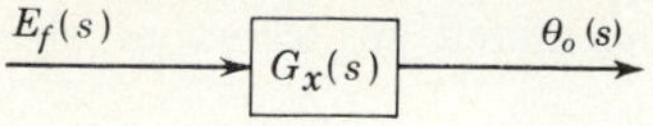

FIGURE 5.14
Simplified block diagram for Fig. 5.11.

function $G_x(s)$ is

$$G_x(s) = \frac{\Theta_o(s)}{E_f(s)} = \frac{K_g K_T / R_f B R_{gm}}{s(1 + T_f s)\left[(1 + K_T K_b / B R_{gm}) + (T_{gm} + T_n)s + T_{gm} T_n s^2\right]} \tag{5.22}$$

This expression is the exact transfer function for the entire motor and generator combination. Certain approximations, if valid, may be made to simplify this expression. The first approximation is that the inductance of the generator and motor armatures is very small ($T_{gm} \approx 0$). With this approximation, the transfer function reduces to

$$\frac{\Theta_o(s)}{E_f(s)} = G_x(s) \approx \frac{K_x}{s(1 + T_f s)(1 + T_m s)} \tag{5.23}$$

where

$$K_x = \frac{K_g K_T}{R_f (B R_{gm} + K_T K_b)} \tag{5.24}$$

$$T_f = \frac{L_f}{R_f} \tag{5.25}$$

$$T_m = \frac{J R_{gm}}{B R_{gm} + K_T K_b} \tag{5.26}$$

If the frictional effect of the load is very small, the approximation can be made that $B \approx 0$. With this additional approximation, the transfer function remains of the same form as Eq. (5.23), but the constants are now

$$K_x = \frac{K_g}{R_f K_b} \tag{5.27}$$

$$T_f = \frac{L_f}{R_f} \tag{5.28}$$

$$T_m = \frac{J R_{gm}}{K_T K_b} \tag{5.29}$$

For simple components, as in this case, an overall transfer function can often be derived more easily by combining the original system equations. This can be done without the intermediate steps shown in this example. However, the purpose of this example is to show the representation of dynamic components by individual blocks and the combination of blocks.

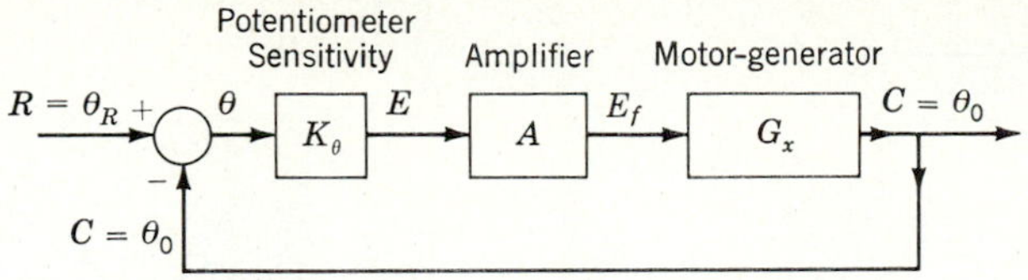

FIGURE 5.15
Equivalent representation of Fig. 5.9.

A new block diagram representing the position-control system of Fig. 5.9 is redrawn in Fig. 5.15. Since the transfer function of the amplifier of Fig. 5.9 is $E_f(s)/E(s) = A$, the forward transfer function of the position-control system of Fig. 5.15 is

$$G(s) = \frac{\Theta_o(s)}{\Theta(s)} = \frac{AK_xK_\theta}{s(1 + T_fs)(1 + T_ms)} \tag{5.30}$$

The overall transfer function (control ratio) for this system is

$$\frac{C(s)}{R(s)} = \frac{\Theta_o(s)}{\Theta_R(s)} = \frac{G(s)}{1 + G(s)H(s)} = \frac{AK_xK_\theta}{s(1 + T_fs)(1 + T_ms) + AK_xK_\theta} \tag{5.31}$$

where $H(s) = 1$.

5.6 SIMULATION DIAGRAMS[11,12]

A block diagram is often used to represent the dynamic equations of a system. The simulation may show physical variables which appear in the system, or it may show variables that are used purely for mathematical convenience. In either case the overall response of the system is the same. The simulation diagram is similar to the diagram used to represent the system on an analog computer. The basic elements used are ideal integrators, ideal amplifiers, and ideal summers, shown in Fig. 5.16. Additional elements such as multipliers and dividers may be used for nonlinear systems.

One of the methods used to obtain a simulation diagram includes the following steps:

1. Start with the differential equation.
2. On the left side of the equation put the highest-order derivative of the *dependent* variable. A first-order or higher-order derivative of the input may appear in the equation. In this case the highest-order derivative of the input is also placed on the left side of the equation. All other terms are put on the right side.
3. Start the diagram by assuming that the signal, represented by the terms on the left side of the equation, is available. Then integrate it as many times as needed to obtain all the lower-order derivatives. It may be necessary to add a summer in the simulation diagram to obtain the dependent variable explicitly.

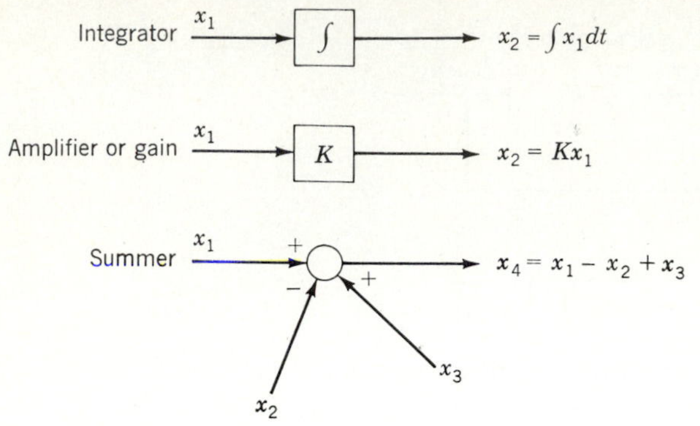

FIGURE 5.16
Elements used in a simulation diagram.

4. Complete the diagram by feeding back the appropriate outputs of the integrators to a summer to generate the original signal of step 2. Include the input function if it is required.

Example. Draw the simulation diagram for the series *RLC* circuit of Fig. 2.2 in which the output is the voltage across the capacitor.

Step 1. When $y = v_c$ and $u = e$ are used, Eq. (2.10) becomes

$$LC\ddot{y} + RC\dot{y} + y = u \tag{5.32}$$

Step 2. Rearrange terms to the form

$$\ddot{y} = bu - a\dot{y} - by \qquad \begin{aligned} a &= R/L \\ b &= 1/LC \end{aligned} \tag{5.33}$$

Step 3. The signal $\ddot{y}$ is integrated twice, as shown in Fig. 5.17*a*.

Step 4. The block or simulation diagram is completed as shown in Fig. 5.17*b* in order to satisfy Eq. (5.33).

The state variables are often selected as the outputs of the integrators in the simulation diagram. In this case they are $x_1 = y$ and $x_2 = \dot{x}_1 = \dot{y}$. The state and output equations are therefore

$$\begin{bmatrix} \dot{x}_1 \\ \dot{x}_2 \end{bmatrix} = \begin{bmatrix} 0 & 1 \\ -\dfrac{1}{LC} & -\dfrac{R}{L} \end{bmatrix} \begin{bmatrix} x_1 \\ x_2 \end{bmatrix} + \begin{bmatrix} 0 \\ \dfrac{1}{LC} \end{bmatrix} u \tag{5.34}$$

$$\mathbf{y} = [1 \quad 0] \begin{bmatrix} x_1 \\ x_2 \end{bmatrix} + \mathbf{0}u \tag{5.35}$$

It is common in a physical system for the **D** matrix to be zero, as in this example. These equations are different from Eqs. (2.30) and (2.32), yet they represent the same system. This illustrates that state variables are not unique. When the state variables are the dependent variable and the derivatives of the

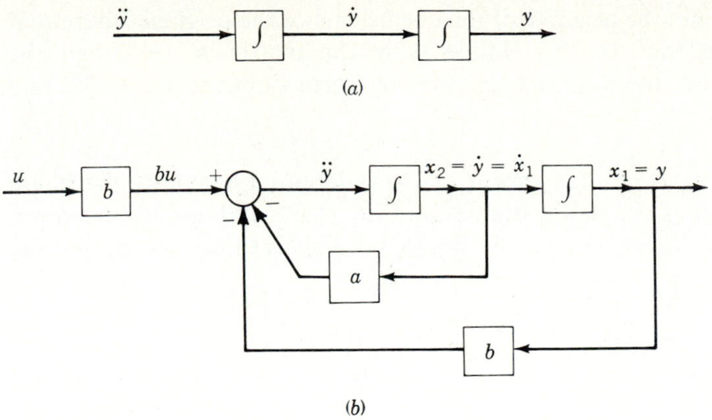

FIGURE 5.17
Simulation diagram for Eq. (5.32).

dependent variable, as in this example, they are called *phase variables*. The phase variables are applicable to differential equations of any order and can be used without drawing the simulation diagram.

Case 1. The general differential equation which contains no derivatives of the input is

$$D^n y + a_{n-1} D^{n-1} y + \cdots + a_1 Dy + a_o y = u \tag{5.36}$$

The state variables are selected as the *phase variables* defined by $x_1 = y$, $x_2 = Dx_1 = \dot{x}_1 = Dy$, $x_3 = \dot{x}_2 = D^2 y, \ldots, x_n = \dot{x}_{n-1} = D^{n-1} y$. Thus, $D^n y = \dot{x}_n$. When these state variables are used, Eq. (5.36) becomes

$$\dot{x}_n + a_{n-1} x_n + a_{n-2} x_{n-1} + \cdots + a_1 x_2 + a_0 x_1 = u \tag{5.37}$$

The resulting state and output equations using phase variables are

$$\dot{\mathbf{x}} = \begin{bmatrix} 0 & 1 & & & & & \\ & 0 & 1 & & & \mathbf{0} & \\ & & 0 & 1 & & & \\ & \mathbf{0} & & 0 & \ddots & & \\ & & & & \ddots & & \\ & & & & & 0 & 1 \\ -a_0 & -a_1 & -a_2 & -a_3 & \cdots & -a_{n-2} & -a_{n-1} \end{bmatrix} \mathbf{x} + \begin{bmatrix} 0 \\ 0 \\ \vdots \\ \\ 0 \\ 1 \end{bmatrix} \mathbf{u}$$

$$= \mathbf{A}_c \mathbf{x} + \mathbf{b}_c \mathbf{u} \tag{5.38}$$

$$\mathbf{y} = [x_1] = [1 \quad 0 \quad 0 \quad \cdots]\mathbf{x} = \mathbf{c}_c^T \mathbf{x} \tag{5.39}$$

This form of the state and output equations can be written directly from the original differential equation (5.36). The plant matrix $\mathbf{A}_c$ contains the number 1 in the *superdiagonal* and the negative of the coefficients of the original differential equation in the nth row. In this simple form the matrix $\mathbf{A}_c$ is called the *companion matrix*. Also, the **B** matrix takes on the form shown in Eq. (5.38) and is indicated by $\mathbf{b}_c$.

Case 2. When derivatives of $u(t)$ appear in the differential equation, the phase variables may be used as the state variables and Eq. (5.38) still applies; however, the output equation is no longer given by Eq. (5.39). This is shown by considering the differential equation

$$\left(D^n + a_{n-1}D^{n-1} + \cdots + a_1 D + a_0\right)y = \left(c_w D^w + c_{w-1}D^{w-1} + \cdots + c_1 D + c_0\right)u \qquad w \le n \tag{5.40}$$

The output y is specified as

$$y = \left(c_w D^w + c_{w-1}D^{w-1} + \cdots + c_1 D + c_0\right)x_1 \tag{5.41}$$

In terms of the state variables this output equation becomes

$$y = c_w \dot{x}_w + c_{w-1}x_w + \cdots + c_1 x_2 + c_0 x_1 \tag{5.42}$$

Substituting Eq. (5.41) into Eq. (5.40) and cancelling common terms on both sides of the equation yields

$$\left(D^n + a_{n-1}D^{n-1} + \cdots + a_1 D + a_0\right)x_1 = u \tag{5.43}$$

This equation has the same form as Eq. (5.36). Thus, in terms of phase variables, Eq. (5.43) can be represented in the form $\dot{\mathbf{x}} = \mathbf{A}_c\mathbf{x} + \mathbf{b}_c\mathbf{u}$ where the matrices $\mathbf{A}_c$ and $\mathbf{b}_c$ are the same as in Eq. (5.38).†

(1) For $w = n$, the output equation (5.42) becomes

$$y = c_n \dot{x}_n + c_{n-1}x_n + \cdots + c_1 x_2 + c_0 x_1$$

and Eq. (5.37) is used to eliminate $\dot{x}_n$. Thus

$$\begin{aligned}\mathbf{y} &= \left[(c_0 - a_0 c_n) \quad (c_1 - a_1 c_n) \quad \cdots \quad (c_{n-1} - a_{n-1}c_n)\right]\mathbf{x} + [c_n]\mathbf{u} \\ &= \mathbf{c}_c^T\mathbf{x} + \mathbf{d}_c\mathbf{u}\end{aligned} \tag{5.44}$$

(2) When $w < n$, $\dot{x}_w = x_{w+1}$, and $c_{w+i} = 0$ for $i = 1, 2, \ldots, n - w$. For this case Eq. (5.44) reduces to

$$\mathbf{y} = \begin{bmatrix} c_0 & c_1 & \cdots & c_w & 0 & \cdots & 0 \end{bmatrix}\mathbf{x} \tag{5.45}$$

Figure 5.18 shows the simulation diagram which represents the system of Eq. (5.40) in terms of Eqs. (5.38) and (5.45). It has the desirable feature that differentiation of the input signal $u(t)$ is not required to satisfy the differential equation. The simulation diagram has the advantage of requiring only two summers, regardless of the order of the system, one at the input and one at the output. This representation of a differential equation is considered again in

† When u is replaced by $K_G u$ in Eqs. (5.38) and (5.40), the state equations can be expressed as in Eq. (13.17). Then in Fig. 5.18 the signal u is replaced by $K_G u$.

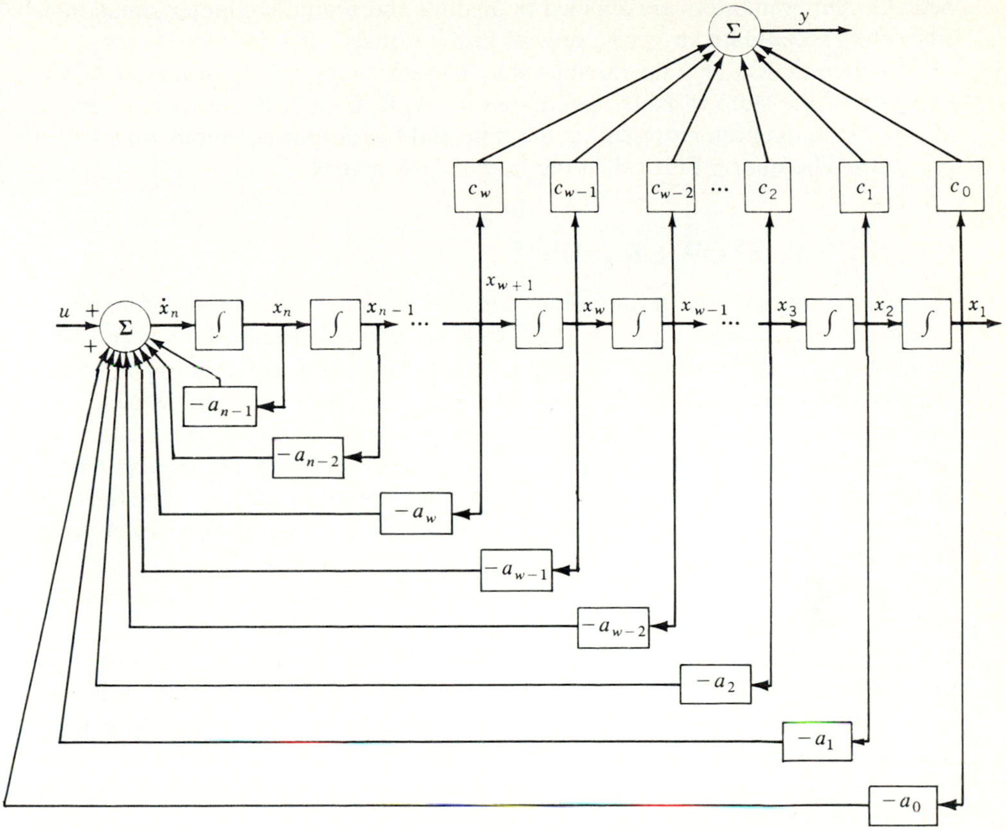

FIGURE 5.18
Simulation diagram representing the system of Eq. (5.40) in terms of Eqs. (5.38) and (5.45).

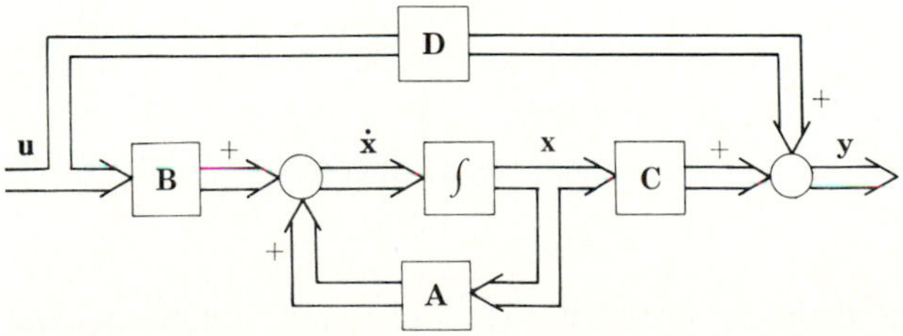

FIGURE 5.19
General matrix block diagram representing the state and output equations.

Sec. 5.8. Differentiators are avoided in analog- and digital-computer simulations since they accentuate any noise present in the signals.

Different sets of state variables may be selected to represent a system. The selection of the state variables determines the **A**, **B**, **C**, and **D** matrices. A general matrix block diagram representing the state and the output equations is shown in Fig. 5.19. The matrix **D** is called the *feedforward* matrix.

5.7 SIGNAL FLOW GRAPHS[3,4]

The block diagram is a useful tool for simplifying the representation of a system. The block diagrams of Figs. 5.9 and 5.15 have only one feedback loop and may be categorized as simple block diagrams. The system represented in Fig. 5.5 has a total of three feedback loops and is no longer a simple system. When intercoupling exists between feedback loops, and when a system has more than one input and one output, the control system and block diagram are more complex. Having the block diagram simplifies the analysis of a complex system. Such an analysis can be further simplified by using a *signal flow graph* (SFG), which looks like a simplified block diagram.

An SFG is a diagram which represents a set of simultaneous equations. It consists of a *graph* in which *nodes* are connected by directed *branches*. The nodes represent each of the system variables. A branch connected between two nodes acts as a one-way signal multiplier: the direction of signal flow is indicated by an arrow placed on the branch, and the multiplication factor (transmittance or transfer function) is indicated by a letter placed near the arrow. Thus, in Fig. 5.20, the branch transmits the signal x_1 from left to right and multiplies it by the quantity a in the process. The quantity a is the transmittance, or transfer function. It may also be indicated by $a = t_{12}$, where the subscripts show that the signal flow is from node 1 to node 2.

Flow-Graph Definitions

A *node* performs two functions:

1. *Addition* of the signals on all incoming branches
2. *Transmission* of the total node signal (the sum of all incoming signals) to all outgoing branches

These functions are illustrated in the graph of Fig. 5.21, which represents the equations

$$w = au + bv \qquad x = cw \qquad y = dw \tag{5.46}$$

FIGURE 5.20
Signal flow graph for $x_2 = ax_1$.

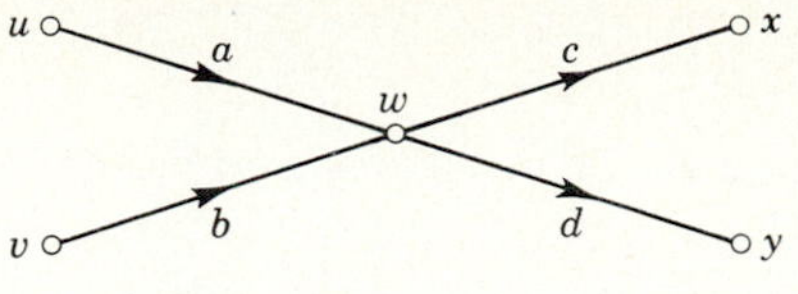

FIGURE 5.21
Signal flow graph for Eqs. (5.46).

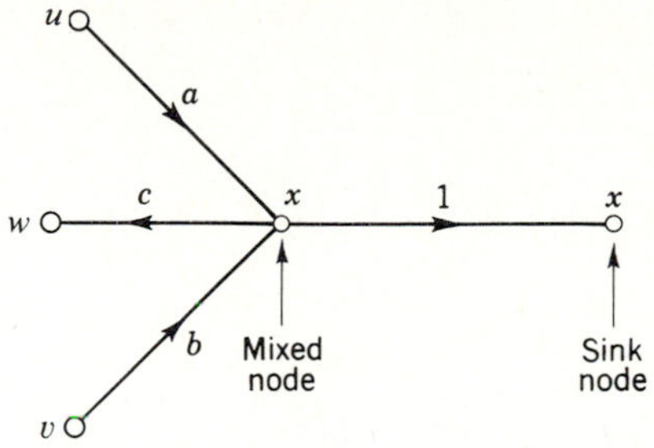

FIGURE 5.22
Mixed and sink nodes for a variable.

There are three types of nodes that are of particular interest:

Source nodes (*independent nodes*). These represent independent variables and have only outgoing branches. In Fig. 5.21, nodes u and v are source nodes.

Sink nodes (*dependent nodes*). These represent dependent variables and have only incoming branches. In Fig. 5.21, nodes x and y are sink nodes.

Mixed nodes (*general nodes*). These have both incoming and outgoing branches. In Fig. 5.21, node w is a mixed node. A mixed node may be treated as a sink node by adding an outgoing branch of unity transmittance, as shown in Fig. 5.22, for the equation $x = au + bv + cw$.

A *path* is any connected sequence of branches whose arrows are in the same direction.

A *forward path* between two nodes is one which follows the arrows of successive branches and in which a node appears only once. In Fig. 5.21 the path uwx is a forward path between the nodes u and x.

Flow-Graph Algebra

The following rules are useful for simplifying a signal flow graph.

Series paths (*cascade nodes*). Series paths can be combined into a single path by multiplying the transmittances as shown in Fig. 5.23*a*.

Path gain. The product of the transmittances in a series path.

Parallel paths. Parallel paths can be combined by adding the transmittances as shown in Fig. 5.23*b*.

Node absorption. A node representing a variable other than a source or sink can be eliminated as shown in Fig. 5.23*c*.

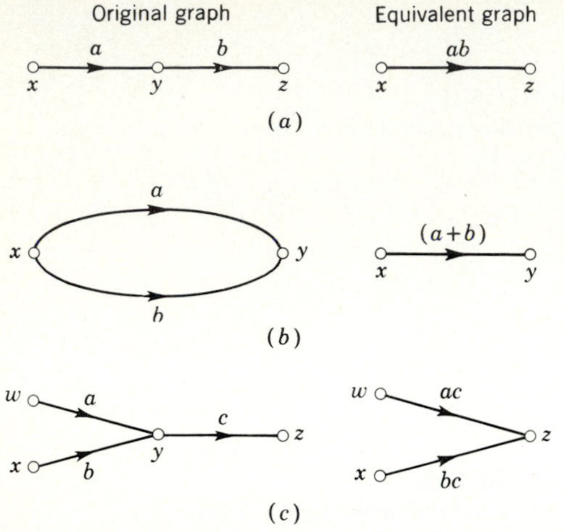

FIGURE 5.23
Flow-graph simplifications.

Feedback loop. A closed path which starts at a node and ends at the same node.

Loop gain. The product of the transmittances of a feedback loop.

The equations for the feedback system of Fig. 5.6 are:

$$C = GE \tag{5.47}$$

$$B = HC \tag{5.48}$$

$$E = R - B \tag{5.49}$$

Note that an equation is written for each dependent variable. The corresponding signal flow graph is shown in Fig. 5.24a. The node B can be eliminated to produce Fig. 5.24b. The node E can be eliminated to produce Fig. 5.24c, which

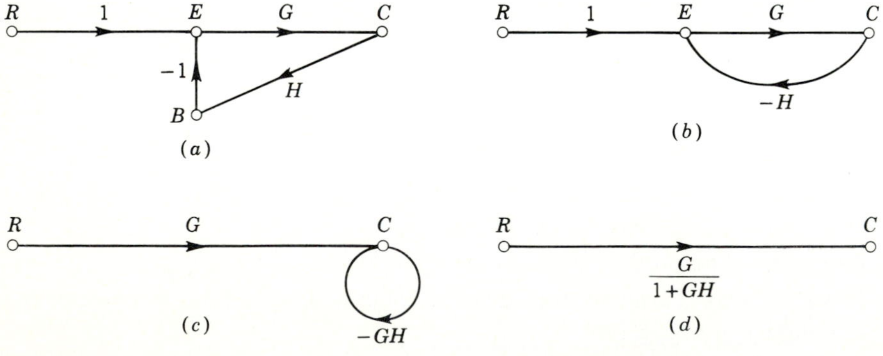

FIGURE 5.24
Successive reduction of the flow graph for the feedback system of Fig. 5.6.

has a *self-loop* of value $-GH$. The final simplification is to eliminate the self-loop to produce the overall transmittance from the input R to the output C. This is obtained by summing signals at node C in Fig. 5.24c, yielding $C = GR - GHC$. Solving for C produces Fig. 5.24d.

General Flow-Graph Analysis

If all the source nodes are brought to the left and all the sink nodes are brought to the right, the SFG for an arbitrarily complex system can be represented by Fig. 5.25a. The effect of the *internal* nodes can be factored out by ordinary algebraic processes to yield the equivalent graph represented by Fig. 5.25b. This simplified graph is represented by

$$y_1 = T_a x_1 + T_d x_2 \tag{5.50}$$

$$y_2 = T_b x_1 + T_e x_2 \tag{5.51}$$

$$y_3 = T_c x_1 + T_f x_2 \tag{5.52}$$

The T's, called overall graph transmittances, are the overall transmittances from a specified source node to a specified dependent node. For linear systems the principle of superposition can be used to "solve" the graph. This means that the sources can be considered one at a time. Then the output signal is equal to the sum of the contributions produced by each input.

The overall transmittances can be found by the ordinary processes of linear algebra, i.e., by the solution of the set of simultaneous equations representing the

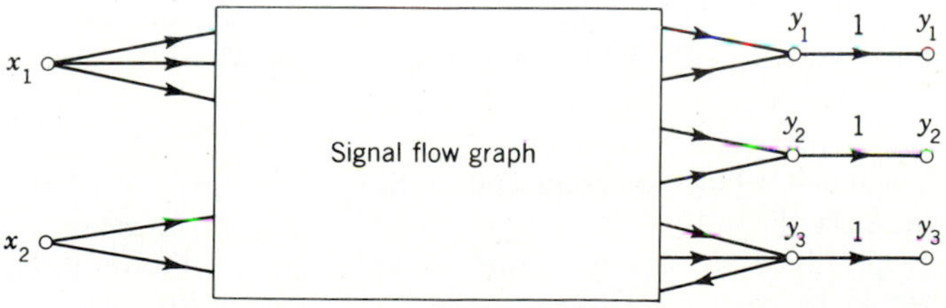

(a)

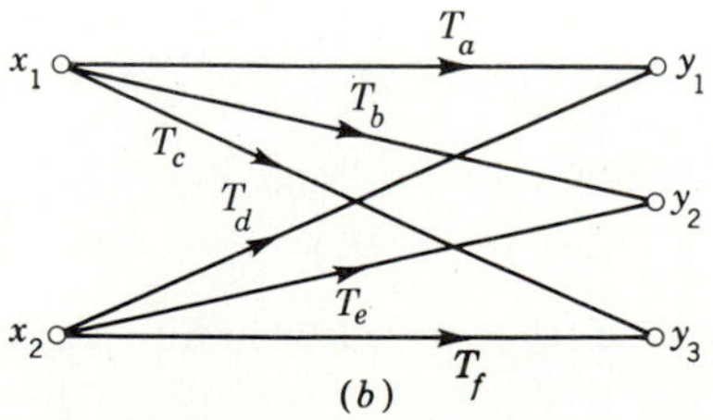

(b)

FIGURE 5.25
Equivalent signal flow graphs.

system. However, the same results can be obtained directly from the signal flow graph. The fact that they can produce answers to large sets of linear equations *by inspection* gives the signal flow graphs their power and usefulness.

The Mason Gain Rule

The overall transmittance can be obtained from the formula developed by S. J. Mason.[3] The formula and definitions are followed by an example to show its application. The overall transmittance is given by

$$T = \frac{\sum T_n \Delta_n}{\Delta} \tag{5.53}$$

where T_n is the transmittance of each forward path between a source and a sink node and Δ is the graph determinant found from

$$\Delta = 1 - \sum L_1 + \sum L_2 - \sum L_3 + \cdots \tag{5.54}$$

In this equation L_1 is the transmittance of each closed path, and ΣL_1 is the sum of the transmittances of all closed paths in the graph. L_2 is the product of the transmittances of two nontouching loops. Loops are nontouching if they do not have any common nodes. ΣL_2 is the sum of the product of transmittances of all possible combinations of nontouching loops taken two at a time. L_3 is the product of the transmittances of three nontouching loops. ΣL_3 is the sum of the product of transmittances of all possible combinations of nontouching loops taken three at a time.

In Eq. (5.53) Δ_n is the cofactor of T_n. It is the determinant of the remaining subgraph when the forward path which produces T_n is removed. Thus, Δ_n does not include any loops which touch the forward path in question. Δ_n is equal to unity when the forward path touches all the loops in the graph or when the graph contains no loops. Δ_n has the same form as Eq. (5.54).

Example. Figure 5.26 shows a block diagram and its SFG. Note that not all the variables are shown. Since $E_1 = M_1 - B_1 = G_1 E - H_1 M_2$, it is not necessary to show M_1 and B_1 explicitly. Since this is a fairly complex system, the resulting equation is expected to be complex. However, the application of Mason's rule produces the resulting overall transmittance in a systematic manner. This system has four loops whose transmittances are $-G_2 H_1$, $-G_5 H_2$, $-G_1 G_2 G_3 G_5$, and $-G_2 G_3 G_5 H_3$. Therefore

$$\sum L_1 = -G_2 H_1 - G_5 H_2 - G_1 G_2 G_3 G_5 - G_1 G_2 G_5 H_3 \tag{5.55}$$

Only two loops are nontouching; therefore

$$\sum L_2 = (-G_2 H_1)(-G_5 H_2) \tag{5.56}$$

Although there are four loops, there is no set of three loops which are nontouching; therefore

$$\sum L_3 = 0 \tag{5.57}$$

The system determinant can therefore be obtained from Eq. (5.54).

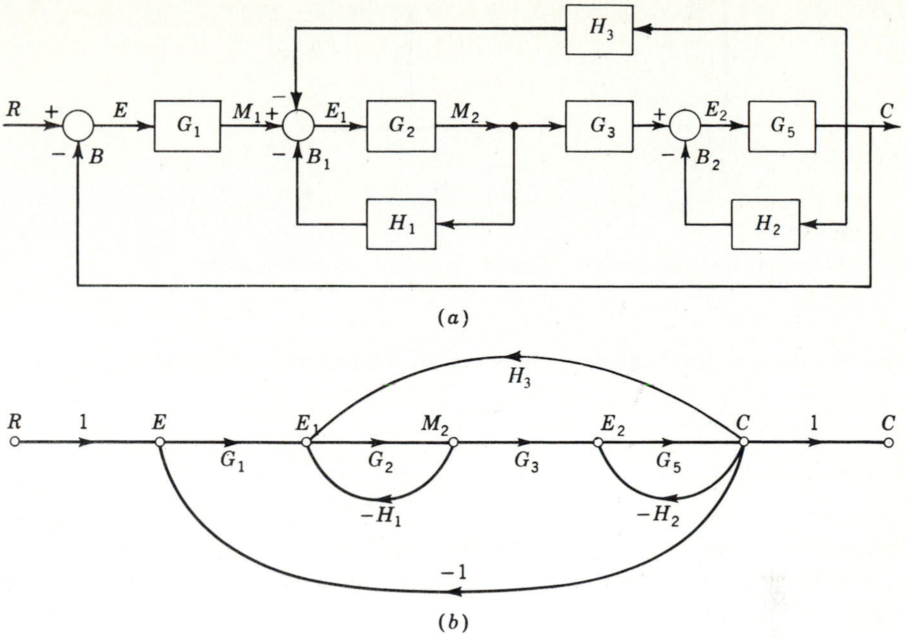

FIGURE 5.26
Block diagram and its signal flow graph.

There is one forward path between R and C. The corresponding forward transmittance is $G_1G_2G_3G_5$. If this path, with its corresponding nodes, is removed from the graph, the remaining subgraph has no loops. The cofactor Δ_n is therefore equal to unity. The complete overall transmittance from R to C, obtained from Eq. (5.53), is

$$T = \frac{G_1G_2G_3G_5}{1 + G_2H_1 + G_5H_2 + G_1G_2G_3G_5 + G_2G_3G_5H_3 + G_2G_5H_1H_2} \tag{5.58}$$

The overall transmittance has been obtained by inspection from the signal flow graph. This is much simpler than solving the five simultaneous equations which represent this system.

5.8 STATE TRANSITION SIGNAL FLOW GRAPH[5]

The state transition SFG or, more simply, the *state diagram*, is a simulation diagram for a system of equations and includes the initial conditions of the states. Since the state diagram in the Laplace domain satisfies the rules of Mason's SFG, it can be used to obtain the transfer function of the system and the state transition equation. As described in Sec. 5.6, the basic elements used in a simulation diagram are a gain, a summer, and an integrator. The signal-flow

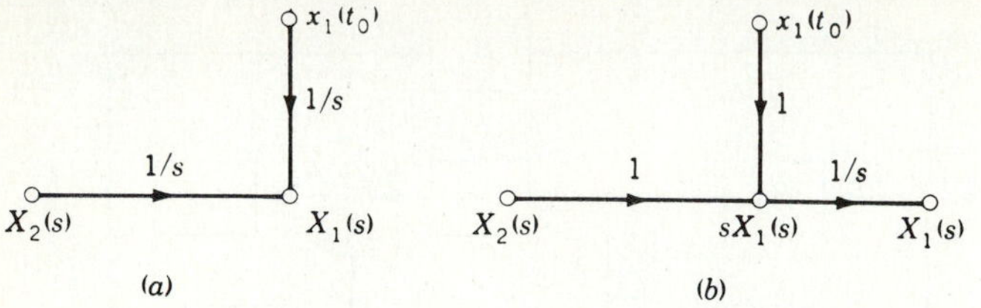

FIGURE 5.27
Representations of an integrator in the Laplace domain in a signal flow graph.

representation in the Laplace domain for an integrator is obtained as follows:

$$\dot{x}_1(t) = x_2(t)$$

$$X_1(s) = \frac{X_2(s)}{s} + \frac{x_1(t_0)}{s} \tag{5.59}$$

Equation (5.59) may be represented either by Fig. 5.27*a* or Fig. 5.27*b*.

A differential equation which contains no derivatives of the input, as given by Eq. (5.36), is repeated here:

$$D^n y + a_{n-1}D^{n-1}y + \cdots + a_1 Dy + a_0 y = u \tag{5.60}$$

In terms of the phase variables, with $x_1 = y$ and $\dot{x}_i = x_{i+1}$, this equation becomes

$$D\dot{x}_n + a_{n-1}x_n + \cdots + a_1 x_2 + a_0 x_1 = u \tag{5.60a}$$

The state diagram is drawn in Fig. 5.28 with phase variables. It is obtained by first drawing the number of integrator branches $1/s$ equal to the order of the differential equation. The outputs of the integrators are designated as the state variables, and the initial conditions of the n states are included as inputs in accordance with Fig. 5.27*a*. Taking the Laplace transform of Eq. (5.60*a*) yields,

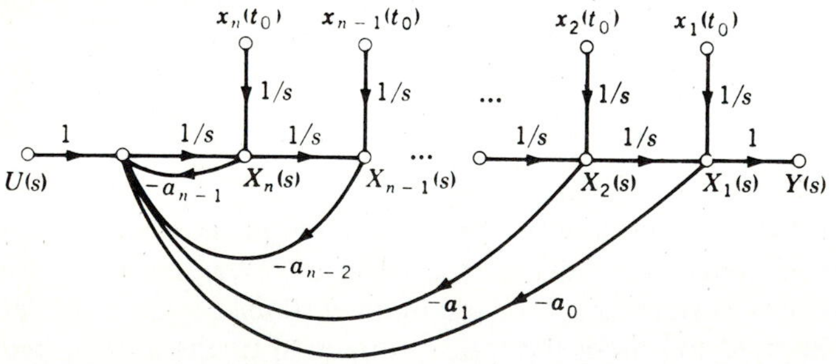

FIGURE 5.28
State diagram for Eq. (5.60).

after dividing by s,

$$X_n(s) = \frac{1}{s}[-a_0 X_1(s) - a_1 X_2(s) - \cdots - a_{n-1} X_n(s) + U(s)] + \frac{1}{s} x_n(t_0)$$

The node $X_n(s)$ in Fig. 5.28 satisfies this equation. The signal at the unlabeled node between $U(s)$ and $X_n(s)$ is equal to $sX_n(s) - x_n(t_0)$.

The overall transfer function $Y(s)/U(s)$ is defined with all initial values of the states equal to zero. Using this condition and applying the Mason gain formula given by Eq. (5.53) yields

$$G(s) = \frac{Y(s)}{U(s)} = \frac{s^{-n}}{\Delta(s)} = \frac{s^{-n}}{1 + a_{n-1}s^{-1} + \cdots + a_1 s^{-(n-1)} + a_0 s^{-n}}$$

$$= \frac{1}{s^n + a_{n-1}s^{n-1} + \cdots + a_1 s + a_0} \tag{5.61}$$

The state transition equation of the system can be obtained from the state diagram by applying the Mason gain formula and considering each initial condition as a source. This is illustrated by the following example.

Example 1. Equation (5.32) can be expressed as

$$\ddot{y} + \frac{R}{L}\dot{y} + \frac{1}{LC}y = \frac{1}{LC}u$$

(a) Draw the state diagram. (b) Determine the state transition equation.

(a) The state diagram, Fig. 5.29, includes two integrators since this is a second-order equation. The state variables are selected as the phase variables which are the outputs of the integrators, that is, $x_1 = y$ and $x_2 = \dot{x}_1$.

(b) The state transition equations are obtained by applying the Mason gain formula with the three inputs u, $x_1(t_0)$, and $x_2(t_0)$:

$$X_1(s) = \frac{s^{-1}(1 + s^{-1}R/L)}{\Delta(s)} x_1(t_0) + \frac{s^{-2}}{\Delta(s)} x_2(t_0) + \frac{s^{-2}/LC}{\Delta(s)} U(s)$$

$$X_2(s) = \frac{-s^{-2}/LC}{\Delta(s)} x_1(t_0) + \frac{s^{-1}}{\Delta(s)} x_2(t_0) + \frac{s^{-1}/LC}{\Delta(s)} U(s)$$

$$\Delta(s) = 1 + \frac{s^{-1}R}{L} + \frac{s^{-2}}{LC}$$

After simplification these equations become

$$\mathbf{X}(s) = \begin{bmatrix} X_1(s) \\ X_2(s) \end{bmatrix}$$

$$= \frac{1}{s^2 + (R/L)s + 1/LC} \left\{ \begin{bmatrix} s + \frac{R}{L} & 1 \\ \frac{-1}{LC} & s \end{bmatrix} \begin{bmatrix} x_1(t_0) \\ x_2(t_0) \end{bmatrix} + \begin{bmatrix} \frac{1}{LC} \\ \frac{s}{LC} \end{bmatrix} U(s) \right\}$$

This equation can be recognized as being of the form given by Eq. (4.113), i.e.,

$$\mathbf{X}(s) = \mathbf{\Phi}(s)\mathbf{x}(t_0) + \mathbf{\Phi}(s)\mathbf{B}U(s)$$

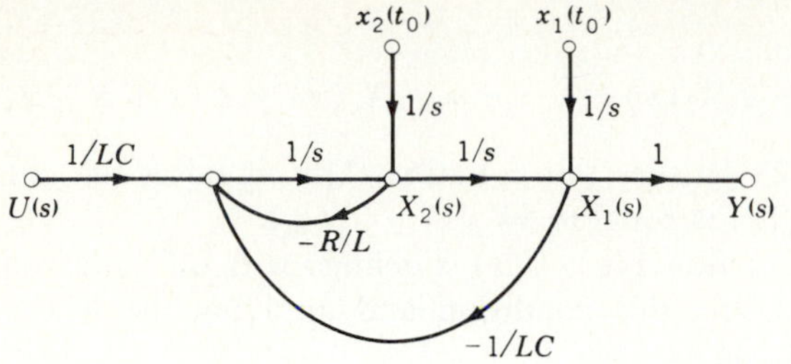

FIGURE 5.29
State diagram.

Thus the resolvant matrix $\mathbf{\Phi}(s)$ is readily identified and by use of the inverse Laplace transform yields the state transition matrix $\mathbf{\Phi}(t)$. The elements of the resolvent matrix can be obtained directly from the SFG with $\mathbf{U}(s) = 0$ and only one $x_j(t_0)$ considered at a time, i.e, all other initial conditions set to zero. Each elements of $\mathbf{\Phi}(s)$ is

$$\phi_{ij}(s) = \frac{X_i(s)}{x_j(t_0)}$$

For example,

$$\phi_{11}(s) = \frac{X_1(s)}{x_1(t_0)} = \frac{s^{-1}(1 + s^{-1}R/L)}{\Delta(s)} = \frac{s + R/L}{s^2\,\Delta(s)}$$

The complete state transition equation $\mathbf{x}(t)$ is obtainable through use of the state diagram. Therefore, $\mathbf{\Phi}(s)$ is obtained without performing the inverse operation $[s\mathbf{I} - \mathbf{A}]^{-1}$. Since $\mathbf{x}(t)$ represents phase variables, the system output is $y(t) = x_1(t)$.

Example 2. A differential equation containing derivatives of the input, given by Eq. (5.40), is repeated below.

$$\left(D^n + a_{n-1}D^{n-1} + \cdots + a_1 D + a_0\right)y$$
$$= \left(c_w D^w + c_{w-1}D^{w-1} + \cdots + c_1 D + c_0\right)u \qquad w \le n \tag{5.62}$$

Phase variables can be specified as the state variables, provided the output is identified by Eq. (5.41) as

$$y = \left(c_w D^w + c_{w-1}D^{w-1} + \cdots + c_1 D + c_0\right)x_1 \tag{5.63}$$

Equation (5.62) then reduces to

$$\left(D^n + a_{n-1}D^{n-1} + \cdots + a_1 D + a_0\right)x_1 = u \tag{5.64}$$

The resulting state diagram for Eqs. (5.63) and (5.64) is shown in Fig. 5.30 for $w = n$. The state equations obtained from the state diagram are given by Eq. (5.38). The output equation in matrix form is readily obtained from Fig. 5.30 in terms of the transformed variables $X(s)$ and the input $U(s)$ as

$$\begin{aligned}\mathbf{Y}(s) &= [(c_0 - a_0 c_n) \quad (c_1 - a_1 c_n) \cdots \\ &\qquad (c_{n-2} - a_{n-2}c_n) \quad (c_{n-1} - a_{n-1}c_n)]\mathbf{X}(s) + c_n\mathbf{U}(s) \\ &= \mathbf{c}_c^T\mathbf{X}(s) + \mathbf{d}_c\mathbf{U}(s)\end{aligned} \tag{5.65}$$

Equation (5.65) is expressed in terms of the state $\mathbf{X}(s)$ and the input $\mathbf{U}(s)$. In order to obtain Eq. (5.65) from Fig. 5.30, delete all the branches having a transmittance $1/s$ and all the initial conditions in Fig. 5.30. The Mason gain formula is then used

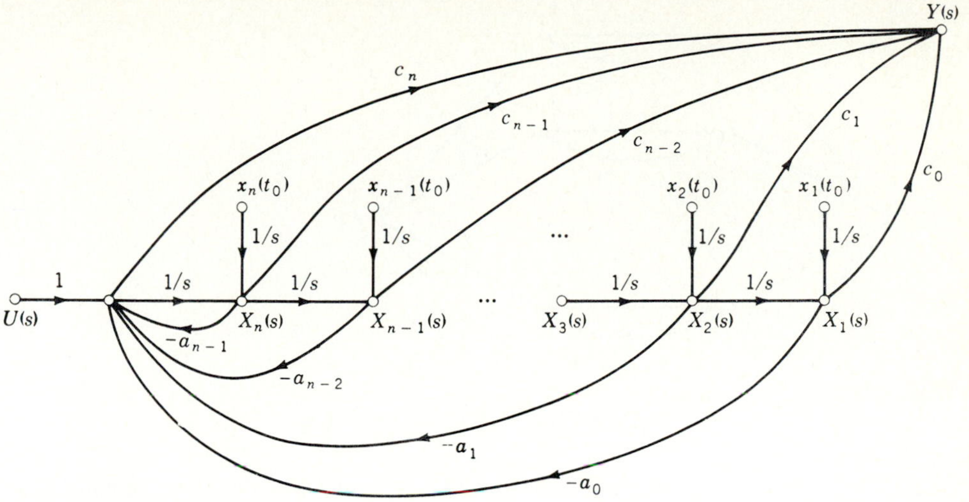

FIGURE 5.30
State diagram for Eq. (5.62) using phase variables, for $w = n$.

to obtain the output in terms of the state variables and the input. Note that there are two forward paths from each state variable to the output when $w = n$. If $w < n$, Eq. (5.65) becomes

$$\mathbf{Y}(s) = [c_0 \quad c_1 \quad \cdots \quad c_w \quad 0 \quad \cdots \quad 0]\mathbf{X}(s) \tag{5.66}$$

5.9 PARALLEL STATE DIAGRAMS FROM TRANSFER FUNCTIONS

A single-input single-output system represented by the differential equation (5.62) may also be represented by an overall transfer function of the form

$$\frac{Y(s)}{U(s)} = G(s) = \frac{c_w s^w + c_{w-1}s^{w-1} + \cdots + c_1 s + c_0}{s^n + a_{n-1}s^{n-1} + \cdots + a_1 s + a_0} \qquad w \le n \tag{5.67}$$

An alternate method of determining a simulation diagram and a set of state variables is to factor the denominator and to express $G(s)$ in partial fractions. When there are no repeated roots and $w = n$, the form is

$$G(s) = \frac{c_w s^w + c_{w-1}s^{w-1} + \cdots + c_1 s + c_0}{(s-\lambda_1)(s-\lambda_2)\cdots(s-\lambda_n)}$$

$$= c_n + \frac{f_1}{s-\lambda_1} + \frac{f_2}{s-\lambda_2} + \cdots + \frac{f_n}{s-\lambda_n} = c_n + \sum_{i=1}^{n} G_i(s)$$

where $$G_i(s) = \frac{f_i Z_i(s)}{U(s)} = \frac{f_i}{s-\lambda_i}$$

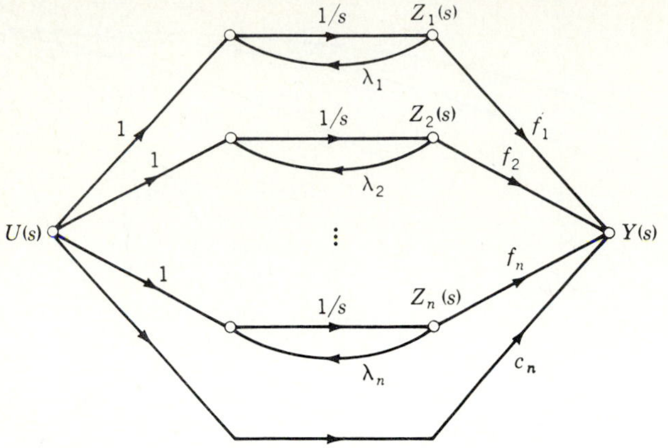

FIGURE 5.31
Simulation of Eq. (5.67) by parallel decomposition for $w = n$ (Jordan diagram for distinct roots).

The symbol z_i and its Laplace transform $Z_i(s)$ are used for the state variables in order to distinguish the form of the associated diagonal matrix. The output $Y(s)$ produced by the input $U(s)$ is therefore

$$\begin{aligned} Y(s) &= c_n U(s) + \frac{f_1 U(s)}{s - \lambda_1} + \frac{f_2 U(s)}{s - \lambda_2} + \cdots \\ &= c_n U(s) + f_1 Z_1(s) + f_2 Z_2(s) + \cdots \end{aligned} \tag{5.68}$$

The state variables $Z_i(s)$ are selected to satisfy this equation. Each fraction represents a first-order differential equation of the form

$$\dot{z}_i - \lambda_i z_i = u \tag{5.69}$$

This can be simulated by an integrator with a feedback path of gain equal to λ_i, followed by a gain f_i. Therefore the complete simulation diagram is drawn in Fig. 5.31. The reader may add the initial conditions of the states, $z_i(t_0)$, to Fig. 5.31. The $z_i(t_0)$ are the inputs to additional branches having transmittances of value $1/s$ and terminating at the nodes $Z_i(s)$. The term $c_n U(s)$ in Eq. (5.68) is satisfied in Fig. 5.31 by the feedforward path of gain c_n. This term appears only when the numerator and denominator of $G(s)$ have the same degree. The output of each integrator is defined as a state variable. The state equations are therefore of the form

$$\dot{\mathbf{z}} = \begin{bmatrix} \lambda_1 & & & \mathbf{0} \\ & \lambda_2 & & \\ & & \ddots & \\ \mathbf{0} & & & \lambda_n \end{bmatrix} \mathbf{z} + \begin{bmatrix} 1 \\ 1 \\ \vdots \\ 1 \end{bmatrix} \mathbf{u} = \Lambda \mathbf{z} + \mathbf{b}_n \mathbf{u} \tag{5.70}$$

$$\mathbf{y} = [f_1 \quad f_2 \quad \cdots \quad f_n] \mathbf{z} + c_n \mathbf{u} = \mathbf{c}_n^T \mathbf{z} + \mathbf{d}_n \mathbf{u} \tag{5.71}$$

An important feature of these equations is that the **A** matrix appears in diagonal or *normal form*. This diagonal form is indicated by Λ (or $\mathbf{A}^*$), and the corresponding state variables are often called *canonical variables*. The elements of the **B** vector (indicated by $\mathbf{b}_n$) are all unity, and the elements of the **C** row matrix (indicated by $\mathbf{c}_n^T$) are the coefficients of the partial fractions. The state diagram of Fig. 5.31 and the state equation of Eq. (5.70) represent the Jordan form for the system with distinct eigenvalues. The state and output equations for a multiple-input multiple-output system can be expressed as

$$\dot{\mathbf{z}} = \Lambda\mathbf{z} + \mathbf{B}_n\mathbf{u} \tag{5.72}$$

$$\mathbf{y} = \mathbf{C}_n\mathbf{z} + \mathbf{D}_n\mathbf{u} \tag{5.73}$$

The diagonal matrix $\mathbf{A} = \Lambda$ means that each state equation is *uncoupled*; i.e., each state z_i can be solved independently of the other states. This simplifies the procedure for finding the state transition matrix $\mathbf{\Phi}(t)$. This form is also useful for studying the observability and controllability of a system, as discussed in Chap. 13. When the state equations are expressed in normal form, the state variables are often denoted by z_i. Transfer functions with repeated roots are not encountered very frequently and are not considered in this text. If they should occur, the reader is referred to more extensive books on linear algebra.

Example. Draw the parallel state diagram and determine the state equation. A partial-fraction expansion is performed on $G(s)$, and the state diagram is drawn in Fig. 5.32.

$$G(s) = \frac{4s^2 + 15s + 13}{s^2 + 3s + 2} = 4 + \frac{1}{s+2} + \frac{2}{s+1}$$

$$\dot{\mathbf{z}}(t) = \begin{bmatrix} -2 & 0 \\ 0 & -1 \end{bmatrix}\mathbf{z}(t) + \begin{bmatrix} 1 \\ 1 \end{bmatrix}\mathbf{u}(t)$$

$$\mathbf{y}(t) = [1 \quad 2]\mathbf{z}(t) + 4\mathbf{u}(t)$$

5.10 DIAGONALIZING THE A MATRIX[6,7]

In Sec. 5.9 the method of partial-fraction expansion is shown to lead to the desirable normal form of the state equation in which the **A** matrix is diagonal. The partial-fraction method is not convenient for multiple-input multiple-output systems or when the system equations are already given in state form. Therefore, this section presents a more general method for converting the state equation by

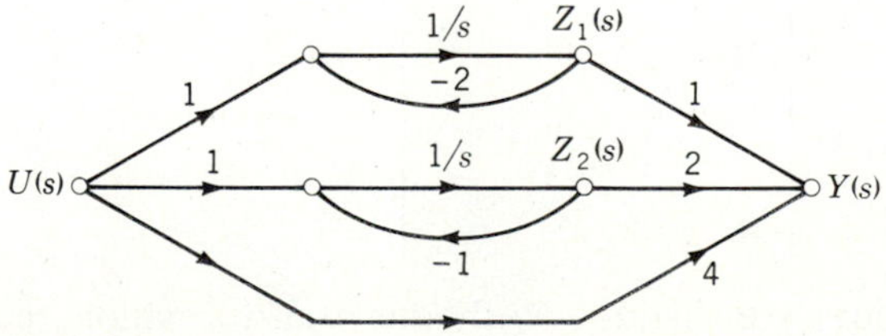

FIGURE 5.32
Simulation diagram for example of Sec. 5.9.

means of a linear *similarity transformation*. Since the state variables are not unique, the intention is to transform the state vector $\mathbf{x}$ to a new state vector $\mathbf{z}$ by means of a constant, square, nonsingular transformation matrix $\mathbf{T}$ so that

$$\mathbf{x} = \mathbf{Tz} \tag{5.74}$$

Since $\mathbf{T}$ is a constant matrix, the differentiation of this equation yields

$$\dot{\mathbf{x}} = \mathbf{T}\dot{\mathbf{z}} \tag{5.75}$$

Substituting these values into the state equation $\dot{\mathbf{x}} = \mathbf{Ax} + \mathbf{Bu}$ produces

$$\mathbf{T}\dot{\mathbf{z}} = \mathbf{ATz} + \mathbf{Bu} \tag{5.76}$$

Premultiplying by $\mathbf{T}^{-1}$ gives

$$\dot{\mathbf{z}} = \mathbf{T}^{-1}\mathbf{ATz} + \mathbf{T}^{-1}\mathbf{Bu} \tag{5.77}$$

The corresponding output equation is

$$\mathbf{y} = \mathbf{CTz} + \mathbf{Du} \tag{5.78}$$

It is easily shown below that the eigenvalues are the same for the original and for the transformed equations. From Eq. (5.77) the new characteristic equation is

$$|\lambda\mathbf{I} - \mathbf{T}^{-1}\mathbf{AT}| = 0 \tag{5.79}$$

In this equation the unit matrix can be replaced by $\mathbf{I} = \mathbf{T}^{-1}\mathbf{T}$. Then, after prefactoring $\mathbf{T}^{-1}$ and postfactoring $\mathbf{T}$, the characteristic equation becomes

$$|\mathbf{T}^{-1}(\lambda\mathbf{I} - \mathbf{A})\mathbf{T}| = 0 \tag{5.80}$$

From the property that the determinant of the product of matrices is equal to the product of the determinants of the individual matrices, Eq. (5.80) is equal to

$$|\mathbf{T}^{-1}| \cdot |\lambda\mathbf{I} - \mathbf{A}| \cdot |\mathbf{T}| = |\mathbf{T}^{-1}| \cdot |\mathbf{T}| \cdot |\lambda\mathbf{I} - \mathbf{A}| = 0$$

Since $|\mathbf{T}^{-1}| \cdot |\mathbf{T}| = |\mathbf{I}| = 1$, the characteristic equation of the transformed equation as given by Eq. (5.79) is equal to the characteristic equation of the original system, i.e.,

$$|\lambda\mathbf{I} - \mathbf{A}| = 0 \tag{5.81}$$

Therefore, it is concluded that the eigenvalues are invariant in a linear transformation given by Eq. (5.74).

The matrix $\mathbf{T}$ is called the *modal* matrix when it is selected so that $\mathbf{T}^{-1}\mathbf{AT}$ is diagonal, i.e.,

$$\mathbf{T}^{-1}\mathbf{AT} = \mathbf{\Lambda} = \begin{bmatrix} \lambda_1 & & & \mathbf{0} \\ & \lambda_2 & & \\ & & \ddots & \\ \mathbf{0} & & & \lambda_n \end{bmatrix} \tag{5.82}$$

This supposes that the eigenvalues are distinct. With this transformation the system equations are

$$\dot{\mathbf{z}} = \mathbf{\Lambda z} + \mathbf{B'u} \tag{5.83}$$

$$\mathbf{y} = \mathbf{C'z} + \mathbf{D'u} \tag{5.84}$$

where $\quad \mathbf{\Lambda} = \mathbf{T}^{-1}\mathbf{AT} \quad \mathbf{B'} = \mathbf{T}^{-1}\mathbf{B} \quad \mathbf{C'} = \mathbf{CT} \quad \mathbf{D'} = \mathbf{D}$

When $\mathbf{T}$ is selected so that $\mathbf{B'} = \mathbf{T}^{-1}\mathbf{B}$ contains only unit elements, then it is denoted as $\mathbf{B}_n$ as in Eq. (5.70) and in method 3. When $\mathbf{x} = \mathbf{Tz}$ is used to transform the equations into any other state-variable representation $\mathbf{T}^{-1}\mathbf{AT} = \mathbf{A'}$, then $\mathbf{T}$ is just called a *transformation matrix*. The case where $\mathbf{A'}$ is the companion matrix $\mathbf{A}_c$ is covered in Sec. 5.13.

Four methods are presented for obtaining the modal matrix $\mathbf{T}$ for the case where the matrix $\mathbf{A}$ has distinct eigenvalues. When $\mathbf{A}$ has multiple eigenvalues, this matrix can be transformed into diagonal form only if the number of independent eigenvectors is equal to the multiplicity of the eigenvalues. In such cases, the columns of $\mathbf{T}$ may be determined by methods of 2, 3, and 4.

Method 1. When there are distinct eigenvalues $\lambda_1, \lambda_2, \ldots, \lambda_n$ for the matrix $\mathbf{A}$ and it is in companion form ($\mathbf{A} = \mathbf{A}_c$), the Vandermonde matrix, which is easily obtained, is the modal matrix. The Vandermonde matrix is defined by

$$\mathbf{T} = \begin{bmatrix} 1 & 1 & \cdots & 1 \\ \lambda_1 & \lambda_2 & \cdots & \lambda_n \\ \lambda_1^2 & & \cdots & \lambda_n^2 \\ \cdots & \cdots & \cdots & \cdots \\ \lambda_1^{n-1} & \lambda_2^{n-1} & \cdots & \lambda_n^{n-1} \end{bmatrix} \tag{5.85}$$

Example 1. Transform the state variables in the following equation in order to uncouple the states:

$$\dot{\mathbf{x}} = \begin{bmatrix} 0 & 1 & 0 \\ 0 & 0 & 1 \\ -24 & -26 & -9 \end{bmatrix} \mathbf{x} + \begin{bmatrix} 1 \\ 0 \\ 2 \end{bmatrix} \mathbf{u} \tag{5.86}$$

$$\mathbf{y} = [3 \quad 3 \quad 1]\mathbf{x} \tag{5.87}$$

The characteristic equation $|\lambda\mathbf{I} - \mathbf{A}_c| = 0$ yields the roots $\lambda_1 = -2$, $\lambda_2 = -3$, and $\lambda_3 = -4$. Since these eigenvalues are distinct and the $\mathbf{A}$ matrix is in companion form, the Vandermonde matrix can be used as the modal matrix. Using Eq. (5.85) yields

$$\mathbf{T} = \begin{bmatrix} 1 & 1 & 1 \\ -2 & -3 & -4 \\ 4 & 9 & 16 \end{bmatrix}$$

A matrix inversion routine (using a programmable calculator, computer, or hand procedure) yields

$$\mathbf{T}^{-1} = \frac{1}{2}\begin{bmatrix} 12 & 7 & 1 \\ -16 & -12 & -2 \\ 6 & 5 & 1 \end{bmatrix}$$

In order to show that this does uncouple the states, the matrix $\mathbf{T}^{-1}\mathbf{AT}$ is evaluated and found to be

$$\mathbf{T}^{-1}\mathbf{A}\mathbf{T} = \begin{bmatrix} -2 & 0 & 0 \\ 0 & -3 & 0 \\ 0 & 0 & -4 \end{bmatrix} = \mathbf{\Lambda} \tag{5.88}$$

The matrix system equations, in terms of the new state vector **z**, are obtained from Eqs. (5.77) and (5.78) as

$$\dot{\mathbf{z}} = \begin{bmatrix} -2 & 0 & 0 \\ 0 & -3 & 0 \\ 0 & 0 & -4 \end{bmatrix}\mathbf{z} + \begin{bmatrix} 7 \\ -10 \\ 4 \end{bmatrix}\mathbf{u} \tag{5.89}$$

$$\mathbf{y} = [1 \quad 3 \quad 7]\mathbf{z} \tag{5.90}$$

Method 2. The second method does not require the **A** matrix to be in companion form. Premultiplying Eq. (5.82) by **T**, where the elements of the modal matrix are defined by $\mathbf{T} = [v_{ij}]$, yields

$$\mathbf{AT} = \mathbf{T\Lambda} = \begin{bmatrix} v_{11} & v_{12} & \cdots & v_{1n} \\ v_{21} & v_{22} & \cdots & v_{2n} \\ \cdots & \cdots & \cdots & \cdots \\ v_{n1} & v_{n2} & \cdots & v_{nn} \end{bmatrix} \begin{bmatrix} \lambda_1 & & & 0 \\ & \lambda_2 & & \\ & & \ddots & \\ 0 & & & \lambda_n \end{bmatrix}$$

$$= \begin{bmatrix} \lambda_1 v_{11} & \lambda_2 v_{12} & \cdots & \lambda_n v_{1n} \\ \lambda_1 v_{21} & \lambda_2 v_{22} & \cdots & \lambda_n v_{2n} \\ \cdots & \cdots & \cdots & \cdots \\ \lambda_1 v_{n1} & \lambda_2 v_{n2} & \cdots & \lambda_n v_{nn} \end{bmatrix} \tag{5.91}$$

The column vectors $\mathbf{v}_i$ of the modal matrix

$$\mathbf{T} = [\mathbf{v}_1 \quad \mathbf{v}_2 \quad \cdots \quad \mathbf{v}_n] \tag{5.92}$$

are called the *eigenvectors*. In terms of the $\mathbf{v}_i$, Eq. (5.91) is written as

$$[\mathbf{A}\mathbf{v}_1 \quad \mathbf{A}\mathbf{v}_2 \quad \cdots \quad \mathbf{A}\mathbf{v}_n] = [\lambda_1\mathbf{v}_1 \quad \lambda_2\mathbf{v}_2 \quad \cdots \quad \lambda_n\mathbf{v}_n] \tag{5.93}$$

Equating columns on the left and right sides of Eq. (5.93) yields

$$\mathbf{A}\mathbf{v}_i = \lambda_i\mathbf{v}_i \tag{5.94}$$

This equation can be put in the form

$$[\lambda_i\mathbf{I} - \mathbf{A}]\mathbf{v}_i = \mathbf{0} \tag{5.95}$$

Equation (5.95) identifies that the eigenvectors† $\mathbf{v}_i$ are in the *null space* of the matrix $[\lambda_i\mathbf{I} - \mathbf{A}]$. Using the property $\mathbf{M}\,\text{adj}\,\mathbf{M} = |\mathbf{M}|\mathbf{I}$ (see Sec. 4.14) and letting $\mathbf{M} = \lambda_i\mathbf{I} - \mathbf{A}$ yields $[\lambda_i\mathbf{I} - \mathbf{A}]\,\text{adj}\,[\lambda_i\mathbf{I} - \mathbf{A}] = |\lambda_i\mathbf{I} - \mathbf{A}|\mathbf{I}$. Since $|\lambda_i\mathbf{I} - \mathbf{A}|$ is the

† These are sometimes called the right eigenvector because $[\lambda_i\mathbf{I} - \mathbf{A}]$ is multiplied on the right by $\mathbf{v}_i$ in Eq. (5.95).

characteristic polynomial and λ_i is an eigenvalue, this equation becomes

$$[\lambda_i \mathbf{I} - \mathbf{A}] \operatorname{adj} [\lambda_i \mathbf{I} - \mathbf{A}] = \mathbf{0} \tag{5.96}$$

A comparison of Eqs. (5.95) and (5.96) shows that the eigenvector

$$\mathbf{v}_i \text{ is proportional to any nonzero column of adj } [\lambda_i \mathbf{I} - \mathbf{A}] \tag{5.97}$$

Since Eq. (5.77) contains $\mathbf{T}^{-1}$, a necessary condition is that $\mathbf{T}$ must be nonsingular. This is satisfied if the $\mathbf{v}_i$ are linearly independent. Since Eq. (5.95) is a homogeneous equation whose rank is $n - 1$ when the eigenvalues are distinct, each value of λ_i yields only one eigenvector $\mathbf{v}_i$. When the eigenvalue λ_i has multiplicity r the *rank deficiency* α of Eq. (5.95) is $1 \le \alpha \le r$. Rank deficiency denotes that the rank of Eq. (5.95) is $n - \alpha$. Then the linearly independent eigenvectors $\mathbf{v}_i$ which satisfy Eq. (5.95) are α in number. If $\alpha < r$, then the matrix $\mathbf{A}$ cannot be diagonalized.

Example 2. Use adj $[\lambda_i \mathbf{I} - \mathbf{A}]$ to obtain the modal matrix for the equations

$$\dot{\mathbf{x}} = \begin{bmatrix} -9 & 1 & 0 \\ -26 & 0 & 1 \\ -24 & 0 & 0 \end{bmatrix} \mathbf{x} + \begin{bmatrix} 2 \\ 5 \\ 0 \end{bmatrix} u \tag{5.98}$$

$$\mathbf{y} = [1 \quad 2 \quad -1]\mathbf{x} \tag{5.99}$$

The characteristic equation $|\lambda \mathbf{I} - \mathbf{A}| = 0$ yields[9] the roots $\lambda_1 = -2$, $\lambda_2 = -3$, and $\lambda_3 = -4$. These eigenvalues are distinct, but $\mathbf{A}$ is not in companion form. Therefore, the Vandermonde matrix is not the modal matrix. Using the adjoint method gives

$$\operatorname{adj} [\lambda \mathbf{I} - \mathbf{A}] = \operatorname{adj} \begin{bmatrix} \lambda + 9 & -1 & 0 \\ 26 & \lambda & -1 \\ 24 & 0 & \lambda \end{bmatrix}$$

$$= \begin{bmatrix} \lambda^2 & \lambda & 1 \\ -26\lambda - 24 & \lambda^2 + 9\lambda & \lambda + 9 \\ -24\lambda & -24 & \lambda^2 + 9\lambda + 26 \end{bmatrix}$$

$$\text{For } \lambda_1 = -2: \quad \operatorname{adj} [-2\mathbf{I} - \mathbf{A}] = \begin{bmatrix} 4 & -2 & 1 \\ 28 & -14 & 7 \\ 48 & -24 & 12 \end{bmatrix} \qquad \mathbf{v}_1 = \begin{bmatrix} 1 \\ 7 \\ 12 \end{bmatrix}$$

$$\text{For } \lambda_2 = -3: \quad \operatorname{adj} [-3\mathbf{I} - \mathbf{A}] = \begin{bmatrix} 9 & -3 & 1 \\ 54 & -18 & 6 \\ 72 & -24 & 8 \end{bmatrix} \qquad \mathbf{v}_2 = \begin{bmatrix} 1 \\ 6 \\ 8 \end{bmatrix}$$

$$\text{For } \lambda_3 = -4: \quad \operatorname{adj} [-4\mathbf{I} - \mathbf{A}] = \begin{bmatrix} 16 & -4 & 1 \\ 80 & -20 & 5 \\ 96 & -24 & 6 \end{bmatrix} \qquad \mathbf{v}_3 = \begin{bmatrix} 1 \\ 5 \\ 6 \end{bmatrix}$$

In each case the columns of adj $[\lambda_i \mathbf{I} - \mathbf{A}]$ are linearly related; i.e., they are proportional. The $\mathbf{v}_i$ may be multiplied by a constant and are selected to contain the smallest integers; often the leading term is reduced to 1. In practice, it is necessary to calculate only one column of the adjoint matrix. The modal matrix is

$$\mathbf{T} = [\mathbf{v}_1 \quad \mathbf{v}_2 \quad \mathbf{v}_3] = \begin{bmatrix} 1 & 1 & 1 \\ 7 & 6 & 5 \\ 12 & 8 & 6 \end{bmatrix}$$

The inverse is

$$\mathbf{T}^{-1} = -\frac{1}{2}\begin{bmatrix} -4 & 2 & -1 \\ 18 & -6 & 2 \\ -16 & 4 & -1 \end{bmatrix}$$

The matrix equations, in terms of the new state vector **z**, are

$$\dot{\mathbf{z}} = \begin{bmatrix} -2 & 0 & 0 \\ 0 & -3 & 0 \\ 0 & 0 & -4 \end{bmatrix}\mathbf{z} + \begin{bmatrix} -1 \\ -3 \\ 6 \end{bmatrix}\mathbf{u} = \mathbf{\Lambda z} + \mathbf{B'u} \tag{5.100}$$

$$\mathbf{y} = [3 \quad 5 \quad 5]\mathbf{z} = \mathbf{C'z} \tag{5.101}$$

Once each eigenvalue is assigned the designation λ_1, λ_2, and λ_3, they appear in this order along the diagonal of $\mathbf{\Lambda}$. Note that the systems of Examples 1 and 2 have the same eigenvalues, but different modal matrices are required to uncouple the states.

Method 3. An alternate method of evaluating the n elements of each $\mathbf{v}_i$ is to form a set of n equations from the matrix equation, Eq. (5.94). This is illustrated by the following example.

Example 3. Rework Example 2 using Eq. (5.94). The matrices formed by using the **A** matrix of Eq. (5.98) are

$$\begin{bmatrix} -9 & 1 & 0 \\ -26 & 0 & 1 \\ -24 & 0 & 0 \end{bmatrix}\begin{bmatrix} v_{1i} \\ v_{2i} \\ v_{3i} \end{bmatrix} = \lambda_i \begin{bmatrix} v_{1i} \\ v_{2i} \\ v_{3i} \end{bmatrix} \tag{5.102}$$

Performing the multiplication yields

$$\begin{bmatrix} -9v_{1i} + v_{2i} \\ -26v_{1i} + v_{3i} \\ -24v_{1i} \end{bmatrix} = \begin{bmatrix} \lambda_i v_{1i} \\ \lambda_i v_{2i} \\ \lambda_i v_{3i} \end{bmatrix} \tag{5.103}$$

Each value of λ_i is inserted in this matrix, and the corresponding elements are equated to form three equations. For $\lambda_1 = -2$, the equations are

$$\begin{aligned} -9v_{11} + v_{21} &= -2v_{11} \\ -26v_{11} + v_{31} &= -2v_{21} \\ -24v_{11} &= -2v_{31} \end{aligned} \tag{5.104}$$

Only two of these equations are independent. This can be demonstrated by inserting v_{21} from the first equation and v_{31} from the third equation into the second equation. The result is the identity $v_{11} = v_{11}$. This merely confirms the fact that $|\lambda_i \mathbf{I} - \mathbf{A}|$ is of rank $n - 1 = 2$. Since there are three elements (v_{1i}, v_{2i}, and v_{3i}) in the eigenvector $\mathbf{v}_i$ and only two independent equations are obtained from Eq. (5.102), the procedure is to arbitrarily set one of the elements equal to unity.

Thus, after letting $v_{11} = 1$, these equations yield

$$\mathbf{v}_1 = \begin{bmatrix} v_{11} \\ v_{21} \\ v_{31} \end{bmatrix} = \begin{bmatrix} v_{11} \\ 7v_{11} \\ 12v_{11} \end{bmatrix} = \begin{bmatrix} 1 \\ 7 \\ 12 \end{bmatrix}$$

This is the same result as obtained in Example 2. The procedure is repeated to evaluate $\mathbf{v}_2$ and $\mathbf{v}_3$.

The need to select one of the elements in each column vector v_i arbitrarily can be eliminated by specifying that the desired matrix $\mathbf{B}' = \mathbf{B}_n$. In Sec. 5.9 the partial-fraction expansion of the transfer function leads to a $\mathbf{b}_n$ matrix containing all 1's, that is,

$$\mathbf{b}_n = [1 \quad 1 \quad \cdots \quad 1]^T \tag{5.105}$$

In order to achieve this form for $\mathbf{b}_n$ there is imposed the restriction that $\mathbf{b} = \mathbf{T}\mathbf{b}_n$. This produces n equations which are added to the n^2 equations formed by equating the elements of $\mathbf{AT} = \mathbf{T\Lambda}$. Only n^2 of these equations are independent, and they can be solved simultaneously for the n^2 elements of $\mathbf{T}$. The restriction that $\mathbf{b} = \mathbf{T}\mathbf{b}_n$ can be satisfied only if the system is controllable. The determination and definition of controllability are presented in Sec. 13.7.

Method 4. The eigenvectors† $\mathbf{v}_i$ which satisfy the equation

$$[\lambda_i \mathbf{I} - \mathbf{A}]\mathbf{v}_i = \mathbf{0} \tag{5.106}$$

are said to lie in the *null* space of the matrix $[\lambda_i \mathbf{I} - \mathbf{A}]$. These eigenvectors can be computed by the following procedure:[10]

1. Use elementary row operations (Sec. 4.14) to transform $[\lambda_i \mathbf{I} - \mathbf{A}]$ into HNF.
2. Rearrange the rows of this matrix if necessary so that the leading unit elements appear on the principal diagonal.
3. Identify the column of this matrix which contains a zero on the principal diagonal. After replacing the zero on the principal diagonal by -1, this column vector is identified as the basis vector for the eigenvector space.

Example 4. Rework Example 2 using method 4.

$$[\lambda_i \mathbf{I} - \mathbf{A}] = \begin{bmatrix} \lambda_i + 9 & -1 & 0 \\ 26 & \lambda_i & -1 \\ 24 & 0 & \lambda_i \end{bmatrix} \tag{5.107}$$

For $\lambda_i = -2$ this matrix and its HNF normal form are

$$\begin{bmatrix} 7 & -1 & 0 \\ 26 & -2 & -1 \\ 24 & 0 & -2 \end{bmatrix} \sim \begin{bmatrix} 1 & 0 & -\frac{1}{12} \\ 0 & 1 & -\frac{7}{12} \\ 0 & 0 & 0 \end{bmatrix}$$

† The eigenvectors $\mathbf{v}_i$ are not unique since they can be formed by any combination of vectors in the null space and can be multiplied by any nonzero constant.

where ~ indicates the result of elementary operations. The eigenvector $\mathbf{v}_1$ lies in the space defined by the third column after inserting -1 on the principal diagonal. This is written as

$$\mathbf{v}_1 \in \text{span}\left\{\begin{bmatrix} -\frac{1}{12} \\ -\frac{7}{12} \\ -1 \end{bmatrix}\right\} \tag{5.108}$$

Similarly, for $\lambda_2 = -3$

$$\begin{bmatrix} 6 & -1 & 0 \\ 26 & -3 & -1 \\ 24 & 0 & -3 \end{bmatrix} \sim \begin{bmatrix} 1 & 0 & -\frac{1}{8} \\ 0 & 1 & -\frac{6}{8} \\ 0 & 0 & 0 \end{bmatrix}$$

The eigenvector $\mathbf{v}_2$ lies in the space defined by the third column after inserting -1 on the principal diagonal, i.e.,

$$\mathbf{v}_2 \in \text{span}\left\{\begin{bmatrix} -\frac{1}{8} \\ -\frac{6}{8} \\ -1 \end{bmatrix}\right\} \tag{5.109}$$

Also, for $\lambda_3 = -4$

$$\begin{bmatrix} 5 & -1 & 0 \\ 26 & -4 & -1 \\ 24 & 0 & -4 \end{bmatrix} \sim \begin{bmatrix} 1 & 0 & -\frac{1}{6} \\ 0 & 1 & -\frac{5}{6} \\ 0 & 0 & 0 \end{bmatrix}$$

The eigenvector $\mathbf{v}_3$ lies in the space defined by the third column after inserting -1 on the principal diagonal, i.e.,

$$\mathbf{v}_3 \in \text{span}\left\{\begin{bmatrix} -\frac{1}{6} \\ -\frac{5}{6} \\ -1 \end{bmatrix}\right\} \tag{5.110}$$

The eigenvectors are selected from the one-dimensional spaces indicated in Eqs. (5.108) to (5.110). These vectors are linearly independent. For convenience, the basic vectors can be multiplied by a negative constant such that the elements are positive integers. Therefore, one selection for the modal matrix is

$$\mathbf{T} = [\mathbf{v}_1 \quad \mathbf{v}_2 \quad \mathbf{v}_3] = \begin{bmatrix} 1 & 1 & 1 \\ 7 & 6 & 5 \\ 12 & 8 & 6 \end{bmatrix} \tag{5.111}$$

which coincides with the results of method 2.

When the matrix **A** has *multiple eigenvalues*, the number of independent eigenvectors associated with the eigenvalue may be equal to the multiplicity of the eigenvalue. This property can be determined by using method 4. The number of columns of the modified HNF of $[\lambda_i \mathbf{I} - \mathbf{A}]$ containing zeros on the principal diagonal may be equal to r, the multiplicity of λ_i. When this occurs, the independent eigenvectors are used to form the modal matrix **T** which results in $\mathbf{T}^{-1}\mathbf{A}\mathbf{T} = \mathbf{\Lambda}$ being a diagonal matrix (see Prob. 5.22).

5.11 USE OF STATE TRANSFORMATION FOR THE STATE EQUATION SOLUTION

The usefulness of the transformation of a matrix to diagonal form is now illustrated in solving the state equation. In Sec. 3.14 it is shown that the state equation and its solution are given, respectively, by

$$\dot{\mathbf{x}}(t) = \mathbf{A}\mathbf{x}(t) + \mathbf{B}\mathbf{u}(t) \tag{5.112}$$

and

$$\dot{x}(t) = e^{\mathbf{A}t}\mathbf{x}(0) + \int_0^t e^{\mathbf{A}(t-\tau)}\mathbf{B}\mathbf{u}(\tau)\, d\tau \tag{5.113}$$

The evaluation of the state transition matrix $e^{\mathbf{A}t}$ is presented in Sec. 3.13 using the Cayley-Hamilton theorem. This procedure can be simplified by expressing the matrix $\mathbf{A}$ in terms of the diagonalized matrix Λ and the modal matrix $\mathbf{T}$. Thus, from the development in Sec. 5.10

$$\mathbf{A} = \mathbf{T}\Lambda\mathbf{T}^{-1} \tag{5.114}$$

where Λ is the diagonal matrix given in Eq. (5.82) and $\mathbf{T}$ is the modal matrix given in Eq. (5.92). Using the series representation of $e^{\mathbf{A}t}$ given in Eq. (3.72) and then using Eq. (5.114) yields

$$\begin{aligned} e^{\mathbf{A}t} &= \mathbf{I} + \mathbf{A}t + \frac{\mathbf{A}^2t^2}{2!} + \frac{\mathbf{A}^3t^3}{3!} + \cdots \\ &= \mathbf{I} + (\mathbf{T}\Lambda\mathbf{T}^{-1})t + \frac{(\mathbf{T}\Lambda\mathbf{T}^{-1})^2t^2}{2!} + \frac{(\mathbf{T}\Lambda\mathbf{T}^{-1})^3t^3}{3!} + \cdots \\ &= \mathbf{T}\left(\mathbf{I} + \Lambda t + \frac{\Lambda^2t^2}{2!} + \frac{\Lambda^3t^3}{3!} + \cdots\right)\mathbf{T}^{-1} = \mathbf{T}e^{\Lambda t}\mathbf{T}^{-1} \end{aligned} \tag{5.115}$$

Since Λ is a diagonal matrix, the matrix $e^{\Lambda t}$ is also diagonal and has the form

$$e^{\Lambda t} = \begin{bmatrix} e^{\lambda_1 t} & & & \mathbf{0} \\ & e^{\lambda_2 t} & & \\ & & \ddots & \\ \mathbf{0} & & & e^{\lambda_n t} \end{bmatrix} \tag{5.116}$$

The work of finding the modal matrix $\mathbf{T}$ is therefore balanced by the ease of obtaining $e^{\Lambda t}$. This is illustrated in the following example.

Example. Evaluate $e^{\mathbf{A}t}$ for the example of Sec. 3.13. The plant matrix $\mathbf{A}$ is

$$\mathbf{A} = \begin{bmatrix} 0 & 6 \\ -1 & -5 \end{bmatrix}$$

and the eigenvalues are $\lambda_1 = -2$ and $\lambda_2 = -3$. Since $\mathbf{A}$ is not in companion form, the modal matrix is evaluated using method 2, 3, or 4 of Sec. 5.10:

$$\mathbf{T} = [\mathbf{v}_1 \quad \mathbf{v}_2] = \begin{bmatrix} 3 & 2 \\ -1 & -1 \end{bmatrix}$$

Now, using Eq. (5.115) yields

$$e^{\mathbf{A}T} = \mathbf{T}e^{\Lambda}\mathbf{T}^{-1} = \mathbf{T}\begin{bmatrix} e^{\lambda_1 t} & 0 \\ 0 & e^{\lambda_2 t} \end{bmatrix}\mathbf{T}^{-1}$$

$$= \begin{bmatrix} 3 & 2 \\ -1 & -1 \end{bmatrix}\begin{bmatrix} e^{-2t} & 0 \\ 0 & e^{-3t} \end{bmatrix}\begin{bmatrix} 1 & 2 \\ -1 & -3 \end{bmatrix}$$

$$= \begin{bmatrix} 3e^{-2t} - 2e^{-3t} & 6e^{-2t} - 6e^{-3t} \\ -e^{-2t} + e^{-3t} & -2e^{-2t} + 3e^{-3t} \end{bmatrix}$$

This is the same result given in Eq. (3.92).

5.12 TRANSFORMING A MATRIX WITH COMPLEX EIGENVALUES

The transformation of variables described in Sec. 5.10 simplifies the system representation mathematically by removing any coupling between the system modes. This diagonalizing of the **A** matrix produces the type of system representation shown in Fig. 5.31. Each state equation corresponding to the normal form of the state equation, Eq. (5.83), has the form

$$\dot{z}_i(t) = \lambda_i z_i(t) + b_i u \tag{5.117}$$

Since only one state variable z_i appears in this equation, it can be solved independently of all the other states.

When the eigenvalues are complex, the matrix **T** which produces the diagonal matrix Λ contains complex numbers. This is illustrated for the differential equation

$$\ddot{x} - 2\sigma\dot{x} + (\sigma^2 + \omega_d^2)x = u(t) \tag{5.118}$$

where $\omega_n^2 = \sigma^2 + \omega_d^2$ and the eigenvalues are $\lambda_{1,2} = \sigma \pm j\omega_d$. Using the phase variables $x_1 = x$ and $x_2 = \dot{x}_1$, the state and output equations are

$$\dot{\mathbf{x}} = \begin{bmatrix} 0 & 1 \\ -\omega_n^2 & 2\sigma \end{bmatrix}\mathbf{x} + \begin{bmatrix} 0 \\ 1 \end{bmatrix}\mathbf{u} \tag{5.119}$$

$$\mathbf{y} = [1 \quad 0]\mathbf{x} \tag{5.120}$$

The modes can be uncoupled, but this is undesirable, as shown below. Since the eigenvalues are distinct and **A** is a companion matrix, the modal matrix **T** is the Vandermonde matrix

$$\mathbf{T} = \begin{bmatrix} 1 & 1 \\ \sigma + j\omega_d & \sigma - j\omega_d \end{bmatrix} = [\mathbf{v}_1 \quad \mathbf{v}_2] \tag{5.121}$$

When the transformation $\mathbf{x} = \mathbf{Tz}$ is used,[8] the normal form of the state and output equations obtained from Eqs. (5.83) and (5.84) are

$$\dot{\mathbf{z}} = \begin{bmatrix} \sigma + j\omega_d & 0 \\ 0 & \sigma - j\omega_d \end{bmatrix}\mathbf{z} + \begin{bmatrix} \dfrac{1}{j2\omega_d} \\ -\dfrac{1}{j2\omega_d} \end{bmatrix}\mathbf{u} \tag{5.122}$$

$$\mathbf{y} = [1 \quad 1]\mathbf{z} \tag{5.123}$$

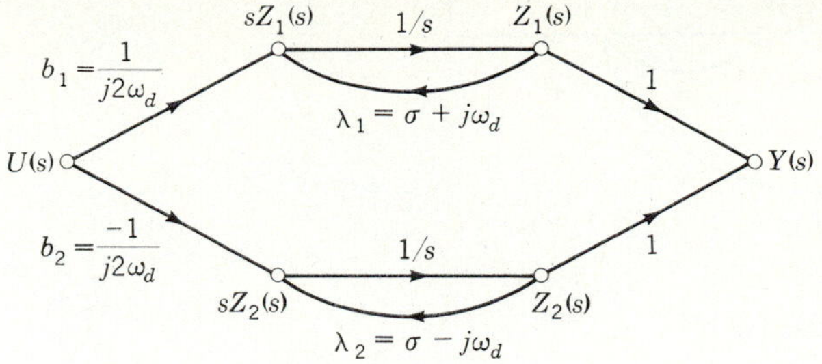

FIGURE 5.33
Simulation diagram for Eqs. (5.122) and (5.123).

The simulation diagram based on these equations is shown in Fig. 5.33. The parameters in this diagram are complex quantities, which increases the difficulty when obtaining the mathematical solution. Thus, it may be desirable to perform another transformation in order to obtain a simulation diagram which contains only real quantities. The pair of complex-conjugate eigenvalues λ_1 and λ_2 jointly contribute to one transient mode which has the form of a damped sinusoid. The additional transformation is $\mathbf{z} = \mathbf{Qw}$, where

$$\mathbf{Q} = \begin{bmatrix} \frac{1}{2} & -\frac{j}{2} \\ \frac{1}{2} & \frac{j}{2} \end{bmatrix} \tag{5.124}$$

The new state and output equations are

$$\dot{\mathbf{w}} = \mathbf{Q}^{-1}\mathbf{\Lambda}\mathbf{Q}\mathbf{w} + \mathbf{Q}^{-1}\mathbf{T}^{-1}\mathbf{B}\mathbf{u} = \mathbf{\Lambda}_m\mathbf{w} + \mathbf{B}_m\mathbf{u} \tag{5.125}$$

$$\mathbf{y} = \mathbf{CTQw} + \mathbf{Du} = \mathbf{C}_m\mathbf{w} + \mathbf{D}_m\mathbf{u} \tag{5.126}$$

For this example these equations† are in the modified canonical form

$$\dot{\mathbf{w}} = \begin{bmatrix} \sigma & \omega_d \\ -\omega_d & \sigma \end{bmatrix}\mathbf{w} + \begin{bmatrix} 0 \\ \frac{1}{\omega_d} \end{bmatrix}\mathbf{u} \tag{5.127}$$

$$\mathbf{y} = [1 \quad 0]\mathbf{w} \tag{5.128}$$

The simulation diagram for these equations is shown in Fig. 5.34. Note that only real quantities appear. Also, the two modes have not been isolated, but this is an advantage for complex-conjugate roots.

The effect of the two transformations $\mathbf{x} = \mathbf{Tz}$ and $\mathbf{z} = \mathbf{Qw}$ can be considered as one transformation given by

$$\mathbf{x} = \mathbf{TQw} = \begin{bmatrix} 1 & 0 \\ \sigma & \omega_d \end{bmatrix}\mathbf{w} = \mathbf{T}_m\mathbf{w} \tag{5.129}$$

Comparison of the modified modal matrix $\mathbf{T}_m$ in Eq. (5.129) with the original

† Although the trajectories in the state space are not drawn here, the result of this transformation produces trajectories which are symmetrical with respect to the **w** axes; see Sec. 15.2.

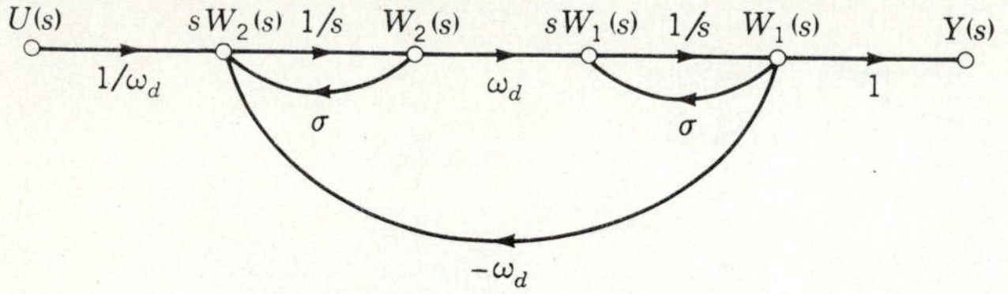

FIGURE 5.34
Simulation diagram for Eqs. (5.127) and (5.128).

matrix in Eq. (5.121) shows that

$$\mathbf{T}_m = [\mathrm{Re}(\mathbf{v}_1) \quad \mathrm{Im}(\mathbf{v}_1)] \tag{5.130}$$

Therefore, $\mathbf{T}_m$ can be obtained directly from $\mathbf{T}$ without using the $\mathbf{Q}$ transformation.

In the general case containing both complex and real eigenvalues the modified block diagonal matrix Λ_m and the resulting matrices $\mathbf{B}_m$ and $\mathbf{C}_m$ have the form

$$\Lambda_m = \begin{bmatrix} \sigma_1 & \omega_1 & & & & & \\ -\omega_1 & \sigma_1 & & & & \mathbf{0} & \\ & & \sigma_2 & \omega_2 & & & \\ & & -\omega_2 & \sigma_2 & & & \\ & & & & \lambda_5 & & \\ & \mathbf{0} & & & & \ddots & \\ & & & & & & \lambda_n \end{bmatrix} \qquad \mathbf{B}_m = \mathbf{T}_m^{-1}\mathbf{B} \tag{5.131}$$

$$\mathbf{C}_m = \mathbf{C}\mathbf{T}_m \tag{5.132}$$

With this modified matrix Λ_m the two oscillating modes produced by the two sets of complex eigenvalues $\sigma_1 \pm j\omega_1$ and $\sigma_2 \pm j\omega_2$ are uncoupled. The column vectors contained in $\mathbf{T}$ for the conjugate eigenvalues are also conjugates. Thus

$$\mathbf{T} = [\mathbf{v}_1 \quad \mathbf{v}_1^* \quad \mathbf{v}_3 \quad \mathbf{v}_3^* \quad \mathbf{v}_5 \quad \cdots \quad \mathbf{v}_n] \tag{5.133}$$

The modified modal matrix $\mathbf{T}_m$ can then be obtained directly by using

$$\mathbf{T}_m = [\mathrm{Re}(\mathbf{v}_1) \quad \mathrm{Im}(\mathbf{v}_1) \quad \mathrm{Re}(\mathbf{v}_3) \quad \mathrm{Im}(\mathbf{v}_3) \quad \mathbf{v}_5 \quad \cdots \quad \mathbf{v}_n] \tag{5.134}$$

5.13 TRANSFORMING AN A MATRIX INTO COMPANION FORM

The synthesis of feedback control systems and the analysis of their response characteristics is often considerably facilitated when the plant and control matrices in the matrix state equation have particular forms. The transformation from a physical or a general state variable $\mathbf{x}_p$ to the phase variable $\mathbf{x}$ is presented in this section.[10,13] In this case the state equation in terms of the general state variable is

$$\dot{\mathbf{x}}_p = \mathbf{A}_p\mathbf{x}_p + \mathbf{b}_p\mathbf{u} \tag{5.135}$$

and the transformed state equation in terms of the phase variables is

$$\dot{\mathbf{x}} = \mathbf{A}_c\mathbf{x} + \mathbf{b}_c\mathbf{u} \tag{5.136}$$

where the companion matrix $\mathbf{A}_c$ and the vector $\mathbf{b}_c$ have the respective forms

$$\mathbf{A}_c = \begin{bmatrix} 0 & 1 & & & & \\ & 0 & 1 & \ddots & & \mathbf{0} \\ & & 0 & & & \\ & & & \ddots & & \\ \mathbf{0} & & & \cdots & 0 & 1 \\ -a_0 & -a_1 & -a_2 & \cdots & -a_{n-2} & -a_{n-1} \end{bmatrix} \qquad \mathbf{b}_c = \begin{bmatrix} 0 \\ 0 \\ \vdots \\ 0 \\ 1 \end{bmatrix} \tag{5.137}$$

The state equation given by Eq. (5.136) is often described in the literature as being in the *control canonical form*. The general state and phase variables are related by

$$\mathbf{x}_p = \mathbf{T}\mathbf{x} \tag{5.138}$$

so that

$$\mathbf{A}_c = \mathbf{T}^{-1}\mathbf{A}_p\mathbf{T} \qquad \mathbf{b}_c = \mathbf{T}^{-1}\mathbf{b}_p \tag{5.139}$$

Since the characteristic polynomial is invariant under a similarity transformation, it can be obtained from $\mathbf{Q}(\lambda) = |\lambda\mathbf{I} - \mathbf{A}_p|$ or by the method of Sec. 4.14 and has the form

$$\mathbf{Q}(\lambda) = \lambda^n + a_{n-1}\lambda^{n-1} + \cdots + a_1\lambda + a_0 \tag{5.140}$$

Thus, by using the coefficients of Eq. (5.140), the companion matrix can readily be formed.

The controllability matrix [see Eq. (13.68)] for the phase-variable state equation is

$$\mathbf{M}_{cc} = \begin{bmatrix}\mathbf{b}_c & \mathbf{A}_c\mathbf{B}_c & \cdots & \mathbf{A}_c^{n-1}\mathbf{b}_c\end{bmatrix}$$

$$= \begin{bmatrix} & \mathbf{0} & & & & 1 \\ & & & 0 & \cdots & \\ & & 0 & 1 & \cdots & \\ & 0 & 1 & -a_{n-1} & \cdots & \\ 0 & 1 & -a_{n-1} & a_{n-1}^2 - a_{n-2} & \cdots & \\ 1 & -a_{n-1} & a_{n-1}^2 - a_{n-2} & -a_{n-1}^3 + 2a_{n-1}a_{n-2} - a_{n-3} & \cdots & \end{bmatrix} \tag{5.141}$$

The matrix $\mathbf{M}_{cc}$† in Eq. (5.141) can be formed since the coefficients of the characteristic polynomial $\mathbf{Q}(\lambda)$ are known. The matrix $\mathbf{M}_{cc}$ expressed in terms of the transformed matrices of Eq. (5.139) is

$$\begin{aligned}\mathbf{M}_{cc} &= \left[\mathbf{T}^{-1}\mathbf{b}_p \quad \mathbf{T}^{-1}\mathbf{A}_p\mathbf{b}_p \quad \mathbf{T}^{-1}\mathbf{A}_p^2\mathbf{b}_p \quad \cdots \quad \mathbf{T}^{-1}\mathbf{A}_p^{n-1}\mathbf{b}_p\right] \\ &= \mathbf{T}^{-1}\left[\mathbf{b}_p \quad \mathbf{A}_p\mathbf{b}_p \quad \mathbf{A}_p^2\mathbf{b}_p \quad \cdots \quad \mathbf{A}_p^{n-1}\mathbf{b}_p\right] = \mathbf{T}^{-1}\mathbf{M}_{cp} \qquad (5.142)\end{aligned}$$

The matrix $\mathbf{M}_{cp}$ is obtained by using $\mathbf{b}_p$ and $\mathbf{A}_p$ of the general state-variable equation (5.135). Equation (5.142) yields

$$\mathbf{T} = \mathbf{M}_{cp}\mathbf{M}_{cc}^{-1} \qquad (5.143)$$

An alternate procedure is to determine $\mathbf{T}^{-1}$ as follows:

1. Form the augmented matrix $[\mathbf{M}_{cp} \quad \mathbf{I}]$.
2. Perform elementary row operations (see Sec. 4.14) until this matrix has the form $[\mathbf{M}_{cc} \quad \mathbf{T}^{-1}]$.
3. The right half of this matrix identifies the required transformation matrix.

Example. A general state-variable equation contains the matrices

$$\mathbf{A}_p = \begin{bmatrix} 1 & 6 & -3 \\ -1 & -1 & 1 \\ -2 & 2 & 0 \end{bmatrix} \qquad \mathbf{b}_p = \begin{bmatrix} 1 \\ 1 \\ 1 \end{bmatrix} \qquad (5.144)$$

for which the characteristic polynomial is evaluated in Sec. 4.14 as

$$\mathbf{Q}(\lambda) = \lambda^3 - 3\lambda + 2 = (\lambda - 1)^2(\lambda + 2) \qquad (5.145)$$

The matrix $\mathbf{M}_{cc}$ obtained in accordance with Eq. (5.141) is

$$\mathbf{M}_{cc} = \begin{bmatrix} 0 & 0 & 1 \\ 0 & 1 & 0 \\ 1 & 0 & 3 \end{bmatrix} \qquad (5.146)$$

The augmented matrix obtained in accordance with step 1 of the procedure is

$$\left[\mathbf{M}_{cp} \quad \mathbf{I}\right] = \left[\begin{array}{ccc|ccc} 1 & 4 & -2 & 1 & 0 & 0 \\ 1 & -1 & -3 & 0 & 1 & 0 \\ 1 & 0 & -10 & 0 & 0 & 1 \end{array}\right] \qquad (5.147)$$

Performing row operations until this matrix has the form $[\mathbf{M}_{cc} \quad \mathbf{T}^{-1}]$ yields

$$\left[\begin{array}{ccc|ccc} 0 & 0 & 1 & 1/36 & 4/36 & -5/36 \\ 0 & 1 & 0 & 7/36 & -8/36 & 1/36 \\ 1 & 0 & 3 & 13/36 & 52/36 & -29/36 \end{array}\right] = \left[\mathbf{M}_{cc} \quad \mathbf{T}^{-1}\right] \qquad (5.148)$$

The matrix $\mathbf{T}^{-1}$ from Eq. (5.148) yields the transformation matrix

$$\mathbf{T} = \begin{bmatrix} -5 & 4 & 1 \\ -6 & -1 & 1 \\ -13 & 0 & 1 \end{bmatrix} \qquad (5.149)$$

---

† The matrix $\mathbf{M}_{cc}$ has *striped* form, that is, the elements comprising the secondary diagonal (from the lower left to the upper right) are all equal. The same is true of all elements in the lines parallel to it.

which produces the desired forms of the matrices

$$\mathbf{A}_c = \mathbf{T}^{-1}\mathbf{A}_p\mathbf{T} = \begin{bmatrix} 0 & 1 & 0 \\ 0 & 0 & 1 \\ -2 & 3 & 0 \end{bmatrix} \tag{5.150}$$

and

$$\mathbf{b}_c = \mathbf{T}^{-1}\mathbf{b}_p = \begin{bmatrix} 0 \\ 0 \\ 1 \end{bmatrix} \tag{5.151}$$

The transformation from the general state variable $\mathbf{x}_p$ to the phase variable $\mathbf{x}$ can also be accomplished by first obtaining the transfer function $G(s) = \mathbf{c}_p^T[s\mathbf{I} - \mathbf{A}_p]^{-1}\mathbf{b}_p$. Then the state and output equations can be obtained directly by using Eqs. (5.38) and (5.44), respectively. The advantage of the method presented in this section is that it avoids having to obtain the inverse of the *polynomial matrix* $[s\mathbf{I} - \mathbf{A}_p]^{-1}$. Also, the method is generalized in Chap. 18 for multivariable systems.

5.14 SUMMARY

The foundation has been completed in Chaps. 2 to 5 for analyzing the performance of feedback control systems. The analysis of such systems and the adjustments of parameters to achieve the desired performance is covered in the following chapters. In order to provide clarity in understanding the principles and to avoid ambiguity, the systems considered in Chaps. 6 to 17 are restricted primarily to those which have a single input and a single output (SISO). The procedures can be extended to multiple-input multiple-output (MIMO) systems, and some synthesis techniques for multivariable systems are presented in Chaps. 18 through 21. The block-diagram representation containing a single feedback loop is the basic unit used to develop the various techniques of analysis.

Since state variables are not unique, there are several ways of selecting them. One method is the use of physical variables, as in Chap. 2. The selection of the state variables as phase or canonical variables is demonstrated in this chapter through the medium of the simulation diagram. This diagram is similar to an analog-computer diagram and uses integrators, amplifiers, and summers. The outputs of the integrators in the simulation diagram are defined as the state variables.

Preparation for overall system analysis has been accomplished by writing the conventional differential and state equations in Chap. 2, obtaining the solution of these equations in Chap. 3, introducing the Laplace transform in Chap. 4, and incorporating this into the representation of a complete system in Chap. 5. Matrix methods are also presented in these chapters in preparation for system synthesis via the use of modern control theory.

REFERENCES

1. Etkin, B.: *Dynamics of Atmospheric Flight*, Wiley, New York, 1972.
2. *IEEE Standard Dictionary of Electrical and Electronics Terms*, Wiley-Interscience, New York, 1972.

3. Mason, S. J.: "Feedback Theory: Further Properties of Signal Flow Graphs," *Proc. IRE*, vol. 44, no. 7, pp. 920–926, July 1956.
4. Mason, S. J., and H. J. Zimmerman: *Electronic Circuits, Signals, and Systems*, M.I.T. Press, Cambridge, Mass., 1960.
5. Kuo, B. C.: *Linear Networks and Systems*, McGraw-Hill, New York, 1967.
6. Trimmer, J. D.: *Response of Physical Systems*, Wiley, New York, 1950, p. 17.
7. Langholz, G., and S. Frankenthal: "Reduction to Normal Form of a State Equation in the Presence of Input Derivatives," *Intern. J. Systems Science*, vol. 5, no. 7, July 1974.
8. Crossley, T. R., and B. Porter: "Inversion of Complex Matrices," *Electronics Letters*, vol. 6, no. 4, Feb. 1970.
9. *Master Library, TI Programmable 58/59*, Texas Instruments, Inc., Dallas, Texas, 1977.
10. Reid, J. G.: *Linear System Fundamentals: Continuous and Discrete, Classic and Modern*, McGraw-Hill, New York, 1983.
11. DeRusso, P. M., et al.: *State Variables for Engineers*, Wiley, New York, 1965.
12. Wiberg, D. M.: *State Space and Linear Systems*, Schaum's Outline Series, McGraw-Hill, New York, 1971.
13. Franklin, G. F., and J. D. Powell: *Digital Control of Dynamic Systems*, Addison-Wesley, Reading, Mass., 1980, chap. 6.

CHAPTER 6

CONTROL-SYSTEM CHARACTERISTICS

6.1 INTRODUCTION

The transfer functions of both open- and closed-loop systems are developed for various types of controlled quantities in Chaps. 2 and 4. These transfer functions have certain basic characteristics that permit transient and steady-state analyses of the feedback-controlled system. Five factors of prime importance in feedback-control systems are *stability, the existence and magnitude of the steady-state error, controllability, observability, and parameter sensitivity*. The stability characteristic of a linear time-invariant system is determined from the system's characteristic equation. Routh's stability criterion provides a means for determining stability without evaluating the roots of this equation. The steady-state characteristics are obtainable from the forward transfer function for unity-feedback systems (or equivalent unity-feedback systems), yielding figures of merit and a ready means for classifying systems. The first two properties are developed in this chapter. The remaining properties are presented in Chaps. 12 through 14.

6.2 ROUTH'S STABILITY CRITERION[1,3,8,9]

The *response transform* $X_2(s)$ has the general form given by Eq. (4.32), which is repeated here in slightly modified form. $X_1(s)$ is the driving transform.

$$X_2(s) = \frac{P(s)}{Q(s)}X_1(s) = \frac{P(s)X_1(s)}{b_n s^n + b_{n-1}s^{n-1} + b_{n-2}s^{n-2} + \cdots + b_1 s + b_0} \tag{6.1}$$

Sections 4.6 and 4.9 describe the methods used to evaluate the inverse transform $\mathscr{L}^{-1}[F(s)] = f(t)$. However, before the inverse transformation can be performed, the polynomial $Q(s)$ must be factored. Computer and calculator programs are readily available for obtaining the roots of a polynomial.[5,6] Section 4.11 shows that stability of the response $x_2(t)$ requires that all zeros of $Q(s)$ have negative real parts. Since it is usually not necessary to find the exact solution when the response is unstable, a simple procedure to determine the existence of zeros with positive real parts is needed. If such zeros of $Q(s)$ with positive real parts are found, the system is unstable and must be modified. Routh's criterion is a simple method of determining the number of zeros with positive real parts without actually solving for the zeros of $Q(s)$. It should be noted that zeros of $Q(s)$ are poles of $X_2(s)$.

The characteristic equation is

$$Q(s) = b_n s^n + b_{n-1}s^{n-1} + b_{n-2}s^{n-2} + \cdots + b_1 s + b_0 = 0 \tag{6.2}$$

If the b_0 term is zero, divide by s to obtain the equation in the form of Eq. (6.2). The b's are real coefficients, and all powers of s from s^n to s^0 must be present in the characteristic equation. A necessary but not sufficient condition for stable roots is that all the coefficients in Eq. (6.2) be positive. If any coefficients other than b_0 are zero, or if the coefficients do not all have the same sign, then there are pure imaginary roots or roots with positive real parts and the system is unstable. It is therefore unnecessary to continue if only stability or instability is to be determined. When all the coefficients are present and positive, the system may or may not be stable because there still may be roots on the imaginary axis or in the right-half s plane. Routh's criterion is mainly used to determine stability. In special situations it may be necessary to determine the actual number of roots in the right-half s plane. For these situations the procedure described in this section can be used.

The coefficients of the characteristic equation are arranged in the pattern shown in the first two rows of the following Routhian array. These coefficients are then used to evaluate the rest of the constants to complete the array.

$$\begin{array}{c|ccccc}
s^n & b_n & b_{n-2} & b_{n-4} & b_{n-6} & \cdots \\
s^{n-1} & b_{n-1} & b_{n-3} & b_{n-5} & b_{n-7} & \cdots \\
s^{n-2} & c_1 & c_2 & c_3 & \cdots & \\
s^{n-3} & d_1 & d_2 & \cdots & & \\
\cdots & \cdots & \cdots & \cdots & \cdots & \cdots \\
s^1 & j_1 & & & & \\
s^0 & k_1 & & & &
\end{array}$$

The constants c_1, c_2, c_3, etc., in the third row are evaluated as follows:

$$c_1 = \frac{b_{n-1}b_{n-2} - b_n b_{n-3}}{b_{n-1}} \tag{6.3}$$

$$c_2 = \frac{b_{n-1}b_{n-4} - b_n b_{n-5}}{b_{n-1}} \tag{6.4}$$

$$c_3 = \frac{b_{n-1}b_{n-6} - b_n b_{n-7}}{b_{n-1}} \tag{6.5}$$

This pattern is continued until the rest of the c's are all equal to zero. Then the d row is formed by using the s^{n-1} and s^{n-2} row. The constants are

$$d_1 = \frac{c_1 b_{n-3} - b_{n-1} c_2}{c_1} \tag{6.6}$$

$$d_2 = \frac{c_1 b_{n-5} - b_{n-1} c_3}{c_1} \tag{6.7}$$

$$d_3 = \frac{c_1 b_{n-7} - b_{n-1} c_4}{c_1} \tag{6.8}$$

This is continued until no more d terms are present. The rest of the rows are formed in this way down to the s^0 row. The complete array is triangular, ending with the s^0 row. Notice that the s^1 and s^0 rows contain only one term each. Once the array has been found, *Routh's criterion states that the number of roots of the characteristic equation with position real parts is equal to the number of changes of sign of the coefficients in the first column*. Therefore the system is stable if all terms in the first column have the same sign.†

The following example illustrates this criterion:

$$Q(s) = s^5 + s^4 + 10s^3 + 72s^2 + 152s + 240 \tag{6.9}$$

The Routhian array is formed by using the procedure described above:

s^5	1	10	152
s^4	1	72	240
s^3	−62	−88	
s^2	70.6	240	
s^1	122.6		
s^0	240		

In the first column there are two changes of sign, from 1 to −62 and from −62 to 70.6; therefore $Q(s)$ has two roots in the right-half s plane. Note that this criterion gives the number of roots with positive real parts but does not tell the values of the roots. If Eq. (6.9) is factored, the roots are $s_1 = -3$, $s_{2,3} = -1 \pm j\sqrt{3}$, and $s_{4,5} = +2 \pm j4$. This confirms that there are two roots with positive real parts. The Routh criterion does not distinguish between real and complex roots.

Theorem 1 Division of a row. The coefficients of any row may be multiplied or divided by a positive number without changing the signs of the first column. The labor of evaluating the coefficients in Routh's array can be reduced by multiplying or dividing any row by a constant. This may result, for example, in reducing the size of the coefficients and therefore simplifying the evaluation of the remaining coefficients.

† Reference 7 shows that a system is unstable if there is a negative element in any position in any row.

The following example illustrates this theorem:

$$Q(s) = s^6 + 3s^5 + 2s^4 + 9s^3 + 5s^2 + 12s + 20 \tag{6.10}$$

The Routhian array is

s^6	1	2	5	20	
s^5	~~3~~	~~9~~	~~12~~		
	1	3	4		(after dividing by 3)
s^4	−1	1	20		
s^3	~~4~~	~~24~~			
	1	6			(after dividing by 4)
s^2	7	20			
s^1	22				(after multiplying by 7)
s^0	20				

Notice that the size of the numbers has been reduced by dividing the s^5 row by 3 and the s^3 row by 4. The result is unchanged; i.e., there are two changes of signs in the first column and therefore there are two roots with positive real parts.

Theorem 2 A zero coefficient in the first column. When the first term in a row is zero but not all the other terms are zero, the following methods can be used:

1. Substitute in the original equation $s = 1/x$; then solve for the roots of x with positive real parts. The number of roots x with positive real parts will be the same as the number of s roots with positive real parts.
2. Multiply the original polynomial by the factor $(s + 1)$ which introduces an additional negative root. Then form the Routhian array for the new polynomial.

Both of these methods are illustrated in the following example:

$$Q(s) = s^4 + s^3 + 2s^2 + 2s + 5 \tag{6.11}$$

The Routhian array is

s^4	1	2	5
s^3	1	2	
s^2	0	5	

The zero in the first column prevents completion of the array. The following methods overcome this problem.

Method 1. Letting $s = 1/x$ and rearranging the polynomial gives

$$Q(x) = 5x^4 + 2x^3 + 2x^2 + x + 1 \tag{6.12}$$

The new Routhian array is

x^4	5	2	1
x^3	2	1	
x^2	−1	2	
x^1	5		
x^0	2		

There are two changes of sign; therefore there are two roots of x in the right-half s plane. The number of roots s with positive real parts is also two. This method does not work when the coefficients of $Q(s)$ and of $Q(x)$ are identical.

Method 2

$$Q_1(s) = Q(s)(s + 1) = s^5 + 2s^4 + 3s^3 + 4s^2 + 7s + 5$$

s^5	1	3	7
s^4	2	4	5
s^3	2	9	
s^2	−10	10	
s^1	11		
s^0	10		

The same result is obtained by both methods. There are two changes of sign in the first column, so there are two zeros of $Q(s)$ with positive real parts. An additional method is described in Ref. 5.

Theorem 3 A zero row. When all the coefficients of one row are zero, the procedure is as follows:

1. The auxiliary equation can be formed from the preceding row, as shown below.
2. The Routhian array can be completed by replacing the all-zero row with the coefficients obtained by differentiating the auxiliary equation.
3. The roots of the auxiliary equation are also roots of the original equation. These roots occur in pairs and are the negative of each other. Therefore these roots may be imaginary (complex conjugates) or real (one positive and one negative), may lie in quadruplets (two pairs of complex-conjugate roots), etc.

Consider the system which has the characteristic equation

$$Q(s) = s^4 + 2s^3 + 11s^2 + 18s + 18 = 0 \tag{6.13}$$

The Routhian array is

s^4	1	11	18
s^3	~~2~~	~~18~~	
	1	9	(after dividing by 2)
s^2	~~2~~	~~18~~	
	1	9	(after dividing by 2)
s^1	0		

The presence of a zero row indicates that there are roots that are the negatives of each other. The next step is to form the auxiliary equation from the preceding row, which is the s^2 row. The highest power of s is s^2, and only even powers of s appear. Therefore the auxiliary equation is

$$s^2 + 9 = 0 \tag{6.14}$$

The roots of this equation are

$$s = \pm j3$$

These are also roots of the original equation. The presence of imaginary roots indicates that the output includes a sinusoidally oscillating component.

To complete the Routhian array, the auxiliary equation is differentiated and is

$$2s + 0 = 0 \tag{6.15}$$

The coefficients of this equation are inserted in the s^1 row, and the array is then completed:

$$\begin{array}{c|c} s^1 & 2 \\ s^0 & 9 \end{array}$$

Since there are no changes of sign in the first column, there are no roots with positive real parts.

In feedback systems, covered in detail in the following chapters, the ratio of the output to the input does not have an explicitly factored denominator. An example of such a function is

$$\frac{X_2(s)}{X_1(s)} = \frac{P(s)}{Q(s)} = \frac{K(s+2)}{s(s+5)(s^2+2s+5) + K(s+2)} \tag{6.16}$$

The value of K is an adjustable parameter in the system and may be positive or negative. The value of K determines the location of the poles and therefore the stability of the system. It is important to know the range of values of K for which the system is stable. This information must be obtained from the characteristic equation, which is

$$Q(s) = s^4 + 7s^3 + 15s^2 + (25 + K)s + 2K = 0 \tag{6.17}$$

The coefficients must all be positive in order for the zeros of $Q(s)$ to lie in the left half of the s plane, but this it not a sufficient condition for stability. The Routhian array permits evaluation of precise boundaries for K:

$$\begin{array}{c|ccc} s^4 & 1 & 15 & 2K \\ s^3 & 7 & 25 + K & \\ s^2 & 80 - K & 14K & \\ s^1 & \dfrac{(80 - K)(25 + K) - 98K}{80 - K} & & \\ s^0 & 14K & & \end{array}$$

The term $80 - K$ for the s^2 row imposes the restriction $K < 80$, and the s^0 row requires $K > 0$. The numerator of the first term in the s^1 row is equal to $-K^2 - 43K + 2000$ and this function must be positive for a stable system. By use of the quadratic formula the zeros of this function are $K = 28.1$ and $K = -71.1$. The combined restrictions on K for stability of the system are therefore $0 < K < 28.1$. For the value $K = 28.1$ the characteristic equation has imaginary roots which can be evaluated by applying Theorem 3 to form the auxiliary equation. Also, for $K = 0$, it can be seen from Eq. (6.16) that there is no output. The methods for selecting the "best" value of K in the range between 0 and 28.1 are contained in later chapters. It is important to note that the Routh criterion has provided useful but restricted information.

Another method of determining stability is Hurwitz's criterion,[1,4] which establishes the necessary conditions in terms of the system determinants. This method is more amenable to digital-computer system-stability evaluation.

6.3 MATHEMATICAL AND PHYSICAL FORMS

In various systems the controlled variable C shown in Fig. 6.1 may have the physical form of position, speed, temperature, rate of change of temperature, voltage, rate of flow, pressure, etc. To generalize the study of feedback systems the controlled variable has been labeled C. Once the blocks in the diagram are related to transfer functions, it is immaterial to the analysis of the system what the physical form of the controlled variable may be.

Generally, the important quantities are the controlled quantity c, its rate of change Dc, and its second derivative D^2c, that is, the first several derivatives of c, including the zeroth derivative. For any specific control system each of these "mathematical" functions has a definite "physical" meaning. For example, if the controlled variable c is position, then Dc is velocity and D^2c is acceleration. As a second example, if the controlled variable c is velocity, then Dc is acceleration and D^2c is the rate of change of acceleration.

Often the input signal to a system has an irregular form, such as that shown in Fig. 6.2, which cannot be expressed by any simple equation. This prevents a straightforward analysis of system response. One notes, though, that the signal form shown in Fig. 6.2 may be considered to be composed of three basic forms of known types of input signals, i.e., a step in the region cde, a ramp in the region $0b$, and a parabola in the region ef. Thus, if the given linear system is analyzed separately for each of these types of input signals, there is then established a fair measure of performance with the irregular input. For example, consider the

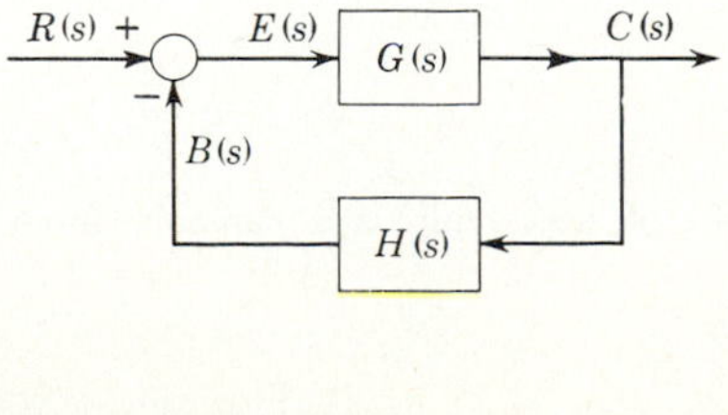

FIGURE 6.1
Simple block diagram.

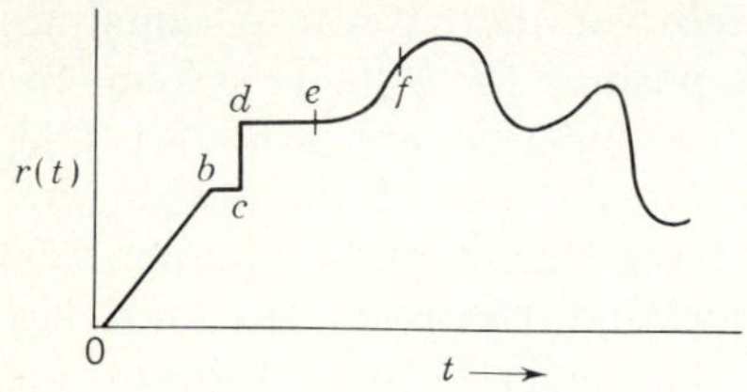

FIGURE 6.2
Input signal to a system.

control system for a missile which receives its command input from an interceptor guidance system (see Fig. 5.3). This input changes slowly compared with the control-system dynamics. Thus, step and ramp inputs adequately approximate most command inputs. Therefore, the three standard inputs not only approximate most command inputs but provide a means for comparing the performance of different systems.

Consider that the system shown in Fig. 6.1 is a position-control system. Feedback control systems are often analyzed on the basis of a unit step input signal (see Fig. 6.3). This system can first be analyzed on the basis that the unit step input signal $r(t)$ represents position. Since this gives only a limited idea of how the system responds to the actual input signal, the system can then be analyzed on the basis that the unit step signal represents a constant velocity $Dr(t) = u_{-1}(t)$. This in reality gives an input position signal of the form of a ramp (Fig. 6.3) and thus a closer idea of how the system responds to the actual input signal. In the same manner the unit step input signal can represent a constant acceleration, $D^2r(t) = u_{-1}(t)$, to obtain the system's performance to a parabolic position input signal. The curves shown in Fig. 6.3 then represent acceleration, velocity, and position.

6.4 TYPES OF FEEDBACK SYSTEMS

A simple closed-loop feedback system with unity feedback is shown in Fig. 6.4. This system may be called a *tracker* since the output $c(t)$ is expected to track or follow the input $r(t)$. The open-loop transfer function for this system is $G(s) = C(s)/E(s)$, which is determined by the components of the actual control system. Several examples are derived in the preceding chapter. Generally the

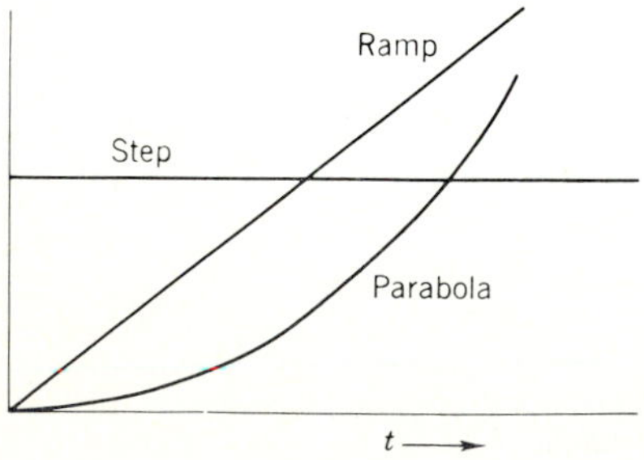

FIGURE 6.3
Graphical forms of step, ramp, and parabolic input functions.

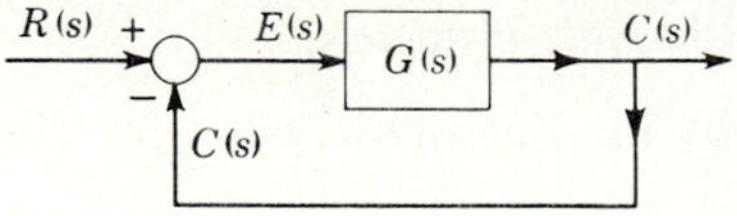

FIGURE 6.4
Unity-feedback control system.

transfer function has one of the following mathematical forms:

$$G(s) = \frac{K_0(1 + T_1 s)(1 + T_2 s)\cdots}{(1 + T_a s)(1 + T_b s)\cdots} \tag{6.18}$$

$$G(s) = \frac{K_1(1 + T_1 s)(1 + T_2 s)\cdots}{s(1 + T_a s)(1 + T_b s)\cdots} \tag{6.19}$$

$$G(s) = \frac{K_2(1 + T_1 s)(1 + T_2 s)\cdots}{s^2(1 + T_a s)(1 + T_b s)\cdots} \tag{6.20}$$

Note that the constant term in each factor is equal to unity. The above equations are expressed in a more generalized manner by defining the standard form of the transfer function as

$$G(s) = \frac{K_m\left(1 + b_1 s + b_2 s^2 + \cdots + b_w s^w\right)}{s^m\left(1 + a_1 s + a_2 s^2 + \cdots + a_u s^u\right)} = K_m G'(s) \tag{6.21}$$

where $a_1, a_2, \ldots$ = constant coefficients
$b_1, b_2, \ldots$ = constant coefficients
K_m = gain constant of the transfer function $G(s)$
$m = 0, 1, 2, \ldots$ denotes the transfer function type
$G'(s)$ = forward transfer function with unity gain

The degree of the denominator is $n = m + u$. For a unity-feedback system, E and C have the same units. Therefore, K_0 is nondimensional, K_1 has the units of seconds^{-1}, and K_2 has the units of seconds^{-2}.

In order to analyze each control system, a "type" designation is introduced. The designation is based upon the value of the exponent m of s in Eq. (6.21). Thus, when $m = 0$, the system represented by this equation is called a Type 0 system; when $m = 1$ it is called a Type 1 system; when $m = 2$ it is called a Type 2 system; etc. See Prob. 6.21 for a Type -1 ($m = -1$) system. Once a physical system has been expressed mathematically, the analysis is independent of the nature of the physical system. It is immaterial whether the system is electrical, mechanical, hydraulic, thermal, or a combination of these. The most common feedback control systems have Type 0, 1, or 2 open-loop transfer functions, as shown in Eqs. (6.18) to (6.20). It is important to analyze each type thoroughly and to relate it as closely as possible to its transient and steady-state solution.

The various types exhibit the following steady-state properties:

Type 0: A constant actuating signal results in a constant value for the controlled variable.

Type 1: A constant actuating signal results in a constant rate of change (constant velocity) of the controlled variable.

Type 2: A constant actuating signal results in a constant second derivative (constant acceleration) of the controlled variable.

Type 3: A constant actuating signal results in a constant rate of change of acceleration of the controlled variable.

These classifications lend themselves to definition in terms of the differential equations of the system and to identification in terms of the forward transfer function. For all classifications the degree of the denominator of the $G(s)H(s)$ function usually is equal to or greater than the degree of the numerator because of the physical nature of feedback control systems. That is, in every physical system there are energy-storage and dissipative elements such that there can be no instantaneous transfer of energy from the input to the output. However, exceptions do occur.

6.5 ANALYSIS OF SYSTEM TYPES

The properties presented in the preceding section are now examined in detail for each type of $G(s)$ appearing in stable unity-feedback tracking systems. First, remember the following theorems:

Final-value theorem:
$$\lim_{t\to\infty} f(t) = \lim_{s\to 0} sF(s) \tag{6.22}$$

Differentiation theorem:
$$\mathscr{L}[D^m c(t)] = s^m C(s) \qquad \text{when all initial conditions are zero} \tag{6.23}$$

It should also be remembered that the steady-state output of a stable closed-loop system that has *unity feedback* has the same form as the input when the input is a polynomial. Therefore, if the input is a ramp function, the steady-state output must also include a ramp function plus a constant, and so forth.

From the preceding section it is seen that the forwarded transfer function defines the system type. In deriving the transfer function it is generally in the factored form

$$G(s) = \frac{C(s)}{E(s)} = \frac{K_m(1 + T_1 s)(1 + T_2 s)\cdots}{s^m(1 + T_a s)(1 + T_b s)(1 + T_c s)\cdots} \tag{6.24}$$

Solving this equation for $E(s)$ yields

$$E(s) = \frac{(1 + T_a s)(1 + T_b s)(1 + T_c s)\cdots}{K_m(1 + T_1 s)(1 + T_2 s)\cdots} s^m C(s) \tag{6.25}$$

Thus, applying the final-value theorem gives

$$e(t)_{ss} = \lim_{s\to 0}[sE(s)] = \lim_{s\to 0}\left[\frac{s(1 + T_a s)(1 + T_b s)(1 + T_c s)\cdots}{K_m(1 + T_1 s)(1 + T_s 2)\cdots} s^m C(s)\right]$$
$$= \lim_{s\to 0} \frac{s[s^m C(s)]}{K_m} \tag{6.26}$$

But applying the final-value theorem to Eq. (6.23) gives

$$\lim_{s \to 0} s[s^m C(s)] = D^m c(t)_{ss} \tag{6.27}$$

Therefore, Eq. (6.26) can be written as

$$e(t)_{ss} = \frac{D^m c(t)_{ss}}{K_m} \tag{6.28}$$

or

$$K_m e(t)_{ss} = D^m c(t)_{ss} \tag{6.29}$$

This equation relates a derivative of the output to the error; it is most useful for the case where $D^m c(t)_{ss}$ = constant. Then $e(t)_{ss}$ must also equal a constant, that is, $e(t)_{ss} = E_0$, and Eq. (6.29) may be expressed as

$$K_m E_0 = D^m c(t)_{ss} = \text{constant} = C_m \tag{6.30}$$

Note that $C(s)$ has the form

$$C(s) = \frac{G(s)}{1 + G(s)} R(s)$$

$$= \frac{K_m[(1 + T_1 s)(1 + T_2 s) \cdots]}{s^m(1 + T_a s)(1 + T_b s) \cdots + K_m(1 + T_1 s)(1 + T_2 s) \cdots} R(s) \tag{6.31}$$

The expression for $E(s)$ in terms of the input $R(s)$ is obtained as follows:

$$E(s) = \frac{C(s)}{G(s)} = \frac{1}{G(s)} \frac{G(s)R(s)}{1 + G(s)} = \frac{R(s)}{1 + G(s)} \tag{6.32}$$

With $G(s)$ given by Eq. (6.24), the expression for $E(s)$ is

$$E(s) = \frac{s^m(1 + T_a s)(1 + T_b s) \cdots R(s)}{s^m(1 + T_a s)(1 + T_b s) \cdots + K_m(1 + T_1 s)(1 + T_2 s) \cdots} \tag{6.33}$$

Applying the final-value theorem to Eq. (6.33) yields

$$e(t)_{ss} = \lim_{s \to 0} s \left[\frac{s^m(1 + T_a s)(1 + T_b s) \cdots R(s)}{s^m(1 + T_a s)(1 + T_b s) \cdots + K_m(1 + T_1 s)(1 + T_2 s) \cdots} \right] \tag{6.34}$$

Equation (6.34) relates the steady-state error to the input; it is now analyzed for various system types and for step, ramp, and parabolic inputs.

Case 1: $m = 0$ (Type 0 System)

STEP INPUT $r(t) = R_0 u_{-1}(t)$, $R(s) = R_0/s$. From Eqs. (6.34)

$$e(t)_{ss} = \frac{R_0}{1 + K_0} = \text{constant} = E_0 \neq 0 \tag{6.35}$$

Applying the final-value theorem to Eq. (6.31) yields

$$c(t)_{ss} = \frac{K_0}{1 + K_0} R_0 \tag{6.36}$$

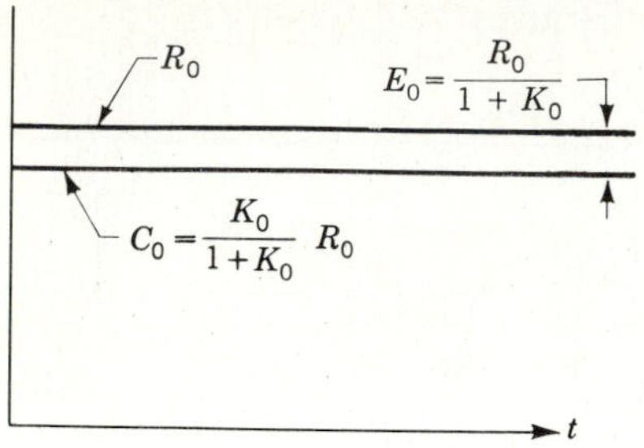

FIGURE 6.5
Steady-state response of a Type 0 system with a step input.

From Eq. (6.35) it is seen that a Type 0 system with a constant input produces a constant value of the output and a constant actuating signal. The same results can be obtained by applying Eq. (6.30). This means that in a Type 0 system a fixed error E_0 is required to produce a desired constant output C_0; that is, $K_0 E_0 = c(t)_{ss} = \text{constant} = C_0$. For steady-state conditions

$$e(t)_{ss} = r(t)_{ss} - c(t)_{ss} = R_0 - C_0 = E_0 \tag{6.37}$$

Differentiating the above equation yields $Dr(t)_{ss} = Dc(t)_{ss} = 0$. Figure 6.5 illustrates the results obtained above.

RAMP INPUT $r(t) = R_1 t u_{-1}(t)$, $R(s) = R_1/s^2$. From Eq. (6.34)

$$e(t)_{ss} = \infty$$

Also, by the use of the Heaviside partial-fraction expansion, the particular solution of $e(t)$ obtained from Eq. (6.33) contains the term $[R_1/(1 + K_0)]t$. Therefore the conclusion is that a Type 0 system with a ramp-function input produces a ramp output with a smaller slope; thus there is an error which increases with time and approaches a value of infinity. This means that a Type 0 system cannot follow a ramp input.

Similarly, it can be shown that a Type 0 system cannot follow a parabolic input

$$r(t) = \frac{R_2 t^2}{2} u_{-1}(t) \tag{6.38}$$

that is, $e(t)_{ss} = r(t)_{ss} - c(t)_{ss}$ approaches a value of infinity.

Case 2: $m = 1$ (Type 1 System)

STEP INPUT $R(s) = R_0/s$. From Eq. (6.34)

$$e(t)_{ss} = 0 \tag{6.39}$$

Therefore it can be stated that a Type 1 system with a constant input produces a steady-state constant output of value identical with the input. This means that for a Type 1 system there is zero steady-state error between the output and input for a step input, i.e.,

$$e(t)_{ss} = r(t)_{ss} - c(t)_{ss} = 0 \tag{6.40}$$

The above analysis is in agreement with Eq. (6.30). That is, for a constant input

$r(t) = R_0u_{-1}(t)$, the steady-state output must also be a constant $c(t) = C_0u_{-1}(t)$ so that

$$Dc(t)_{ss} = 0 = K_1E_0 \tag{6.41}$$

or

$$E_0 = 0 \tag{6.42}$$

RAMP INPUT $R(s) = R_1/s^2$. From Eq. (6.34)

$$e(t)_{ss} = \frac{R_1}{K_1} = \text{constant} = E_0 \neq 0 \tag{6.43}$$

From Eq. (6.43) it is seen that a Type 1 system with a ramp input produces a constant actuating signal. That is, in a Type 1 system a fixed error E_0 is required to produce a ramp output. This result can also be obtained from Eq. (6.30), that is, $K_1E_0 = Dc(t)_{ss} = \text{constant} = C_1$. For steady-state conditions

$$e(t)_{ss} = r(t)_{ss} - c(t)_{ss} = E_0 \tag{6.44}$$

For the ramp input

$$r(t) = R_1tu_{-1}(t) \tag{6.45}$$

the output has the form of a power series

$$c(t)_{ss} = C_0 + C_1t \tag{6.46}$$

Substituting Eqs. (6 .45) and (6.46) into Eq. (6.44) yields

$$E_0 = R_1t - C_0 - C_1t \tag{6.47}$$

This can occur only with

$$R_1 = C_1 \tag{6.48}$$

This result signifies that the slope of the ramp input and the ramp output are equal. This, of course, is a necessary condition if the difference between input and output is a constant. Figure 6.6 illustrates the results obtained above. The delay of the steady-state output ramp shown in Fig. 6.6 is equal to

$$\text{delay} = \frac{e(t)_{ss}}{Dr(t)} = \frac{1}{K_1} \tag{6.49}$$

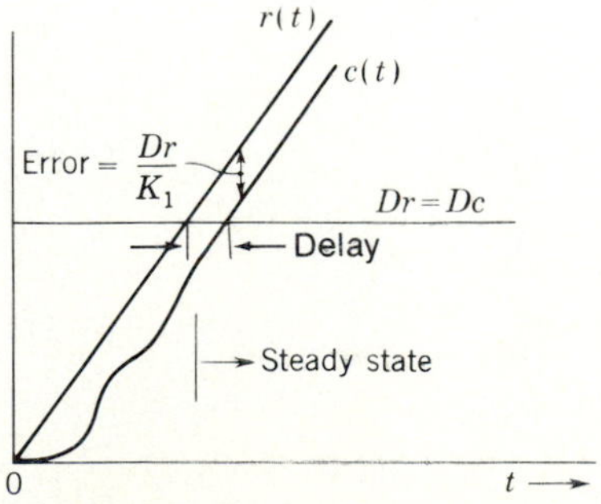

FIGURE 6.6
Steady-state response of a Type 1 system with a ramp input.

PARABOLIC INPUT $r(t) = (R_2 t^2/2)u_{-1}(t)$, $R(s) = R_2/s^3$. From Eq. (6.34), $e(t)_{ss} = \infty$. Also, by use of the partial-fraction expansion, the particular solution of $e(t)$ obtained from Eq. (6.33) contains the term

$$e(t) = \frac{R_2 t}{K_1} \tag{6.50}$$

Therefore, it can be stated that a Type 1 system with a parabolic input produces a parabolic output but with an error which increases with time and approaches a value of infinity. This means that a Type 1 system cannot follow a parabolic input.

Case 3: $m = 2$ (Type 2 System)

STEP INPUT $R(s) = R_0/s$. Performing a corresponding analysis in the same manner as for Cases 1 and 2 reveals that a Type 2 system can follow a step input with zero steady-state error, i.e., $c(t)_{ss} = r(t) = R_0 u_{-1}(t)$ and

$$E_0 = 0 \tag{6.51}$$

RAMP INPUT $R(s) = R_1/s^2$. Performing a corresponding analysis in the same manner as for Cases 1 and 2 reveals that a Type 2 system can follow a ramp input with zero steady-state error, i.e., $c(t)_{ss} = r(t) = R_1 t$ and

$$E_0 = 0 \tag{6.52}$$

PARABOLIC INPUT $R(s) = R_2/s^3$. From Eq. (6.34)

$$e(t)_{ss} = \frac{R_2}{K_2} = \text{constant} = E_0 \neq 0 \tag{6.53}$$

From Eq. (6.53) it is seen that a Type 2 system with a parabolic input produces a parabolic output with a constant actuating signal. That is, in a Type 2 system a fixed error E_0 is required to produce a parabolic output. This result is further confirmed by applying Eq. (6.30), i.e.,

$$K_2 E_0 = D^2 c(t)_{ss} = \text{constant} = C_2 \tag{6.54}$$

Thus, for steady-state conditions

$$e(t)_{ss} = r(t)_{ss} - c(t)_{ss} = E_0 \tag{6.55}$$

For a parabolic input given by

$$r(t) = \frac{R_2 t^2}{2} u_{-1}(t) \tag{6.56}$$

the particular solution of the output must be given by

$$c(t)_{ss} = \frac{C_2}{2} t^2 + C_1 t + C_0 \tag{6.57}$$

Substituting Eqs. (6.56) and (6.57) into Eq. (6.55) yields

$$E_0 = \frac{R_2 t^2}{2} - \frac{C_2 t^2}{2} - C_1 t - C_0 \tag{6.58}$$

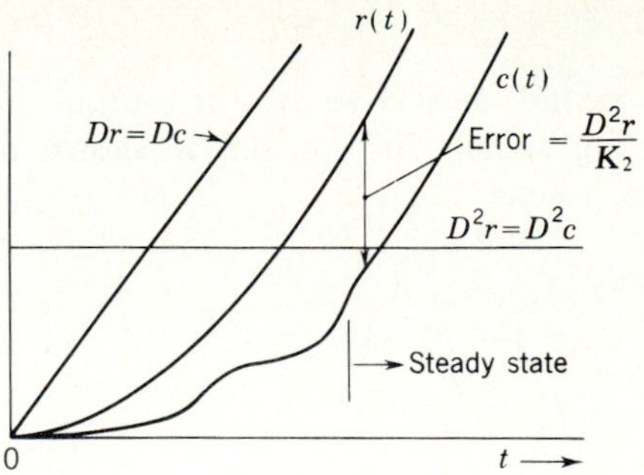

FIGURE 6.7
Steady-state response of a Type 2 system with a parabolic input.

Differentiating the above equation twice results in

$$C_1 = 0 \tag{6.59}$$

and

$$R_2 = C_2 \tag{6.60}$$

Therefore, $E_0 = -C_0$ and the input and output curves have the same shape but are displaced by a constant error. Figure 6.7 illustrates the results obtained above.

The results determined in this section verify the properties stated in the previous section for the system types. The steady-state response characteristics for Type 0, 1, and 2 unity-feedback systems are given in Table 6.1 and apply only to stable systems.

TABLE 6.1
Steady-state response characteristics for stable unity-feedback systems

System type m	$r(t)_{ss}$	$c(t)_{ss}$	$e(t)_{ss}$	$e(\infty)$	Derivatives
0	$R_0 u_{-1}(t)$	$\frac{K_0}{1+K_0} R_0$	$\frac{R_0}{1+K_0}$	$\frac{R_0}{1+K_0}$	$Dr = Dc = 0$
	$R_1 t u_{-1}(t)$	$\frac{K_0 R_1}{1+K_0} t + C_0$	$\frac{R_1}{1+K_0} t - C_0$	∞	$Dr \neq Dc$
	$\frac{R_2 t^2}{2} u_{-1}(t)$	$\frac{K_0 R_2}{2(1+K_0)} t^2 + C_1 t + C_0$	$\frac{R_2}{2(1+K_0)} t^2 - C_1 t - C_0$	∞	$Dr \neq Dc$
1	$R_0 u_{-1}(t)$	R_0	0	0	$Dr = Dc = 0$
	$R_1 t u_{-1}(t)$	$R_1 t - \frac{R_1}{K_1}$	$\frac{R_1}{K_1}$	$\frac{R_1}{K_1}$	$Dr = Dc = R_1$
	$\frac{R_2 t^2}{2} u_{-1}(t)$	$\frac{R_2 t^2}{2} - \frac{R_2}{K_1} t + C_0$	$\frac{R_2}{K_1} t - C_0$	∞	$Dr \neq Dc$
2	$R_0 u_{-1}(t)$	R_0	0	0	$Dr = Dc = 0$
	$R_1 t u_{-1}(t)$	$R_1 t$	0	0	$Dr = Dc = R_1$
	$\frac{R_2 t^2}{2} u_{-1}(t)$	$\frac{R_2 t^2}{2} - \frac{R_2}{K_2}$	$\frac{R_2}{K_2}$	$\frac{R_2}{K_2}$	$D^2 r = D^2 c = R_2$ $Dr = Dc = R_2 t$

6.6 EXAMPLE: TYPE 2 SYSTEM

A Type 2 system is one in which the second derivative of the output is maintained constant by a constant actuating signal. In Fig. 6.8 is shown a positioning system which is a Type 2 system, where

K_b = motor back emf constant, V/(rad/s)
K_x = potentiometer constant, V/rad
K_T = motor torque constant, lb · ft/A
A = integrator amplifier gain, s^{-1}
K_g = generator constant, V/A
$T_m = JR_{gm}/(K_bK_T + R_{gm}B)$ = motor mechanical constant, s
T_f = generator field constant, s
$K_M = K_T/(BR_{gm} + K_TK_b)$ = overall motor constant, rad/V · s

The motor-generator control unit is described in Secs. 2.13 and 2.14. It is assumed that the inductance L_{gm} of the generator and motor armatures is negligible. The transfer function of this unit, with the inductance neglected, is given by Eq. (2.148). Figure 6.9 is the block-diagram representation of the Type 2 position-control system shown in Fig. 6.8. Thus the open-loop transfer function is

$$G(s) = \frac{\theta_o(s)}{E(s)} = \frac{K_xAK_gK_M/R_f}{s^2(1 + T_fs)(1 + T_ms)} \tag{6.61}$$

or

$$G(s) = \frac{K_2}{s^2(1 + T_fs)(1 + T_ms)} \tag{6.62}$$

where K_2 has the units of seconds^{-2}.

This Type 2 control system is unstable. As shown in Chap. 10, an appropriate cascade compensator can be added to the forward path to produce a

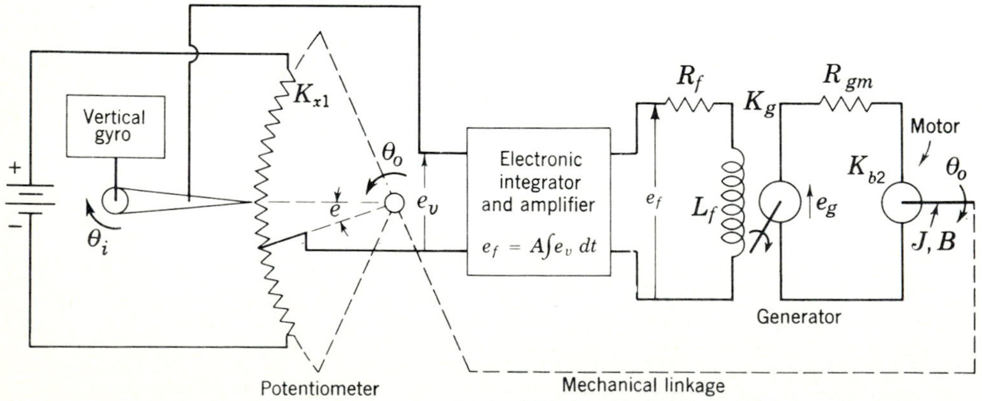

FIGURE 6.8
Position control of a space-vehicle camera (Type 2 system).

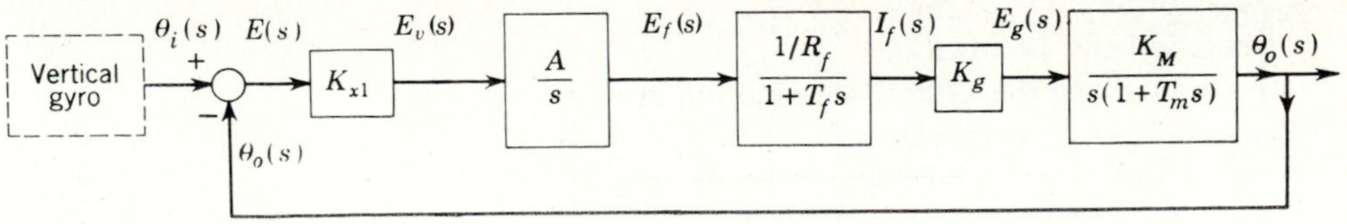

FIGURE 6.9
Block-diagram representation of the system of Fig. 6.8.

stable system. The new transfer function with $T_1 > T_2$, has the form

$$G_0(s) = G_c(s)G(s) = \frac{K_2'(1 + T_1 s)}{s^2(1 + T_f s)(1 + T_m s)(1 + T_2 s)} \tag{6.63}$$

For a constant input $r(t) = R_0 u_{-1}(t)$ the steady-state value of $c(t)_{ss}$ is

$$c(t)_{ss} = \lim_{s\to 0}\left[s\frac{C(s)}{R(s)}\frac{R_0}{s}\right] = \lim_{s\to 0}\left[\frac{G_0(s)}{1 + G_0(s)}R_0\right] = R_0 \tag{6.64}$$

Thus
$$E_0 = r(t)_{ss} - c(t)_{ss} = R_0 - R_0 = 0 \tag{6.65}$$

Therefore, as expected, a Type 2 system follows a step-function input with no error.

6.7 STEADY-STATE ERROR COEFFICIENTS[2]

System types are defined in the preceding sections of this chapter. This is the first step toward establishing a set of standard characteristics that permit the engineer to obtain as much information as possible about a given system with a minimum amount of calculation. Also, these standard characteristics must point the direction in which a given system must be modified to meet a given set of performance specifications.

An item of importance is the ability of a system to achieve the desired steady-state output with a minimum error. Thus, in this section are defined system error coefficients that are a measure of a unity-feedback control system's steady-state accuracy for a given desired output that is relatively constant or slowly varying.

In Eq. (6.30) it is shown that when the derivative of the output is constant, a constant actuating signal exists. This derivative is proportional to the actuating signal E_0 and to a constant K_m, which is the gain of the forward transfer function. The conventional names of these constants for the Type 0, 1, and 2 systems are *position*, *velocity*, and *acceleration error coefficients*, respectively. Since the conventional names were originally selected for application to position-control systems (servomechanisms), these names referred to the actual physical form of $c(t)$ or $r(t)$, which is position, as well as to the mathematical form of $c(t)$, that is, c Dc, and D^2c. These names are ambiguous when the

TABLE 6.2
Correspondence between the conventional and the author's designation of steady-state error coefficients

Symbol	Conventional designation of error coefficients	Authors' designation of error coefficients
K_p	Position	Step
K_v	Velocity	Ramp
K_a	Acceleration	Parabolic

analysis is extended to cover control of temperature, velocity, etc. To avoid this ambiguity of terminology and to define general terms that are universally applicable, the authors have selected the terminology *step*, *ramp*, and *parabolic steady-state error coefficients*. Table 6.2 lists the conventional and the author's designation of the error coefficients.

The *following derivatives of the error coefficients are independent of the system type. They apply to any system type and are defined for specific forms of the input, i.e., for a step, ramp, or parabolic input. These error coefficients are also defined and are applicable only for stable unity-feedback systems.* The results are summarized in Table 6.3.

Steady-State Step Error Coefficient

The step error coefficient is defined as

$$\text{Step error coefficient} = \frac{\text{steady-state value of output } c(t)_{ss}}{\text{steady-state actuating signal } e(t)_{ss}} = K_p \tag{6.66}$$

and applies only for a step input $r(t) = R_0 u_{-1}(t)$. The steady-state value of the

TABLE 6.3
Definitions of steady-state error coefficients for stable unity-feedback systems

Error coefficient	Definition of error coefficient	Value of error coefficient	Form of input signal $r(t)$
Step (K_p)	$\frac{c(t)_{ss}}{e(t)_{ss}}$	$\lim_{s \to 0} G(s)$	$R_0 u_{-1}(t)$
Ramp (K_v)	$\frac{(Dc)_{ss}}{e(t)_{ss}}$	$\lim_{s \to 0} sG(s)$	$R_1 t u_{-1}(t)$
Parabolic (K_a)	$\frac{(D^2 c)_{ss}}{e(t)_{ss}}$	$\lim_{s \to 0} s^2 G(s)$	$\frac{R_2 t^2}{2} u_{-1}(t)$

output is obtained by applying the final-value theorem to Eq. (6.31):

$$c(t)_{ss} = \lim_{s\to 0} sC(s) = \lim_{s\to 0}\left[\frac{sG(s)}{1+G(s)}\frac{R_0}{s}\right] = \lim_{s\to 0}\left[\frac{G(s)}{1+G(s)}R_0\right] \tag{6.67}$$

Similarly, from Eq. (6.32), for a unity-feedback system

$$e(t)_{ss} = \lim_{s\to 0}\left[s\frac{1}{1+G(s)}\frac{R_0}{s}\right] = \lim_{s\to 0}\left[\frac{1}{1+G(s)}R_0\right] \tag{6.68}$$

Substituting Eqs. (6.67) and (6.68) into Eq. (6.66) yields

$$\text{Step error coefficient} = \frac{\lim_{s\to 0}\left[\dfrac{G(s)}{1+G(s)}R_0\right]}{\lim_{s\to 0}\left[\dfrac{1}{1+G(s)}R_0\right]} \tag{6.69}$$

Since both the numerator and the denominator of Eq. (6.69) in the limit can never be zero or infinity simultaneously, where $K_m \neq 0$, the indeterminate forms $0/0$ and ∞/∞ never occur. Thus, this equation reduces to

$$\text{Step error coefficient} = \lim_{s\to 0} G(s) = K_p \tag{6.70}$$

Therefore, applying Eq. (6.70) to each type system yields

$$K_p = \begin{cases} \lim_{s\to 0} \dfrac{K_0(1+T_1 s)(1+T_2 s)\cdots}{(1+T_a s)(1+T_b s)(1+T_c s)\cdots} = K_0 & \text{Type 0 system} \quad (6.71) \\ \infty & \text{Type 1 system} \quad (6.72) \\ \infty & \text{Type 2 system} \quad (6.73) \end{cases}$$

Steady-State Ramp Error Coefficient

The ramp error coefficient is defined as

$$\text{Ramp error coefficient} = \frac{\text{steady-state value of derivative of output } Dc(t)_{ss}}{\text{steady-state actuating signal } e(t)_{ss}} = K_v \tag{6.74}$$

and applies only for a ramp input $r(t) = R_1 t u_{-1}(t)$. The first derivative of the output is given by

$$\mathcal{L}[Dc] = sC(s) = \frac{sG(s)}{1+G(s)}R(s) \tag{6.75}$$

The steady-state value of the derivative of the output is obtained by using the final-value theorem:

$$Dc(t)_{ss} = \lim_{s\to 0} s[sC(s)] = \lim_{s\to 0}\left[\frac{s^2G(s)}{1+G(s)}\frac{R_1}{s^2}\right] = \lim_{s\to 0}\left[\frac{G(s)}{1+G(s)}R_1\right] \tag{6.76}$$

Similarly, from Eq. (6.32), for a unity-feedback system

$$e(t)_{ss} = \lim_{s\to 0}\left[s\frac{1}{1+G(s)}\frac{R_1}{s^2}\right] = \lim_{s\to 0}\left[\frac{1}{1+G(s)}\frac{R_1}{s}\right] \tag{6.77}$$

Substituting Eqs. (6.76) and (6.77) into Eq. (6.74) yields

$$\text{Ramp error coefficient} = \frac{\lim_{s\to 0}\left[\dfrac{G(s)}{1+G(s)}R_1\right]}{\lim_{s\to 0}\left[\dfrac{1}{1+G(s)}\dfrac{R_1}{s}\right]} \tag{6.78}$$

Since the above equation never has the indeterminate form 0/0 or ∞/∞, it can be simplified to

$$\text{Ramp error coefficient} = \lim_{s\to 0} sG(s) = K_v \tag{6.79}$$

Therefore, applying Eq. (6.79) to each type system yields

$$K_v = \begin{cases} \lim_{s\to 0} \dfrac{sK_0(1+T_1 s)(1+T_2 s)\cdots}{(1+T_a s)(1+T_b s)(1+T_c s)\cdots} = 0 & \text{Type 0 system} \quad (6.80) \\ K_1 & \text{Type 1 system} \quad (6.81) \\ \infty & \text{Type 2 system} \quad (6.82) \end{cases}$$

Steady-State Parabolic Error coefficient

The parabolic error coefficient is defined as

Parabolic error coefficient

$$= \frac{\text{steady-state value of second derivative of output } D^2c(t)_{ss}}{\text{steady-state actuating signal } e(t)_{ss}} = K_a \tag{6.83}$$

and applies only for a parabolic input $r(t) = (R_2 t^2/2)u_{-1}(t)$. The second derivative of the output is given by

$$\mathscr{L}[D^2c] = s^2C(s) = \frac{s^2G(s)}{1+G(s)}R(s) \tag{6.84}$$

The steady-state value of the second derivative of the output is obtained by using the final-value theorem:

$$D^2c(t)_{ss} = \lim_{s\to 0} s[s^2C(s)] = \lim_{s\to 0}\left[\frac{s^3G(s)}{1+G(s)}\frac{R_2}{s^3}\right] = \lim_{s\to 0}\left[\frac{G(s)}{1+G(s)}R_2\right] \tag{6.85}$$

Similarly, from Eq. (6.32), for a unity-feedback system

$$e(t)_{ss} = \lim_{s\to 0}\left[s\frac{1}{1+G(s)}\frac{R_2}{s^3}\right] = \lim_{s\to 0}\left[\frac{1}{1+G(s)}\frac{R_2}{s^2}\right] \tag{6.86}$$

Substituting Eqs. (6.85) and (6.86) into Eq. (6.83) yields

$$\text{Parabolic error coefficient} = \frac{\lim_{s\to 0}\left[\frac{G(s)}{1+G(s)}R_2\right]}{\lim_{s\to 0}\left[\frac{1}{1+G(s)}\frac{R_2}{s^2}\right]} \tag{6.87}$$

Since the above equation never has the indeterminate form 0/0 or ∞/∞, it can be simplified to

$$\text{Parabolic error coefficient} = \lim_{s\to 0} s^2 G(s) = K_a \tag{6.88}$$

$$K_a = \begin{cases} \lim_{s\to 0}\dfrac{s^2K_0(1+T_1s)(1+T_2s)\cdots}{(1+T_as)(1+T_bs)(1+T_cs)\cdots} = 0 & \text{Type 0 system} \quad (6.89)\\ 0 & \text{Type 1 system} \quad (6.90)\\ K_2 & \text{Type 2 system} \quad (6.91)\end{cases}$$

Based upon the static error coefficient definitions and using the corresponding Figs. 6.5 through 6.7, the values of K_0, K_1, and K_2 can also be determined from the computer simulation of the control system. For example, from Fig. 6.6 and using Eq. (6.49), the gain is $K_1 = 1/\text{delay}$.

6.8 USE OF STEADY-STATE ERROR COEFFICIENTS

The use of steady-state error coefficients is discussed for stable unity-feedback systems, as illustrated in Fig. 6.10. It is important to stress that the steady-state error coefficients *can only be used* to find errors for unity-feedback systems which are *also* stable.

Type 1 System

For a Type 1 system considered at steady state, the value of $(Dc)_{ss}$ is

$$Dc(t)_{ss} = K_1E_0 \tag{6.92}$$

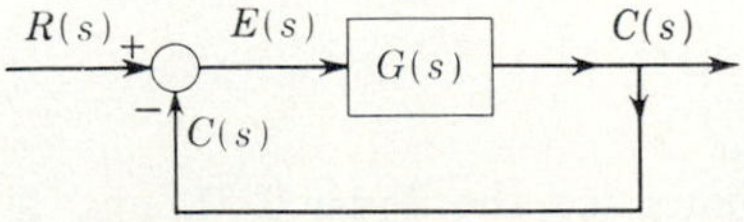

FIGURE 6.10
Unity-feedback system.

Thus, the larger K_1, the smaller the size of the actual signal necessary to maintain a constant rate of change of the output. From the standpoint of trying to maintain $c(t) = r(t)$ at all times, a larger K_1 results in a more sensitive system. In other words, a larger K_1 results in a greater speed of response of the system to a given actuating signal $e(t)$. Therefore, K_1 is another standard characteristic of a system's performance. The maximum value of K_1 is limited by stability considerations and is discussed in later chapters.

For the Type 1 system the step error coefficient is equal to infinity, and the steady-state error is zero for a step input. Therefore, the steady-state output $c(t)_{ss}$ for a Type 1 system is equal to the input when $r(t) =$ constant.

Consider now a ramp input $r(t) = R_1 t u_{-1}(t)$. The steady-state value of $Dc(t)_{ss}$ is found by using the final-value theorem:

$$Dc(t)_{ss} = \lim_{s \to 0} s[sC(s)] = \lim_{s \to 0} \left[s \frac{sG(s)}{1 + G(s)} R(s) \right] \tag{6.93}$$

where $R(s) = R_1/s^2$ and

$$G(s) = \frac{K_1(1 + T_1 s)(1 + T_2 s) \cdots (1 + T_w s)}{s(1 + T_a s)(1 + T_b s) \cdots (1 + T_u s)}$$

Inserting these values in Eq. (6.93) gives

$$Dc(t)_{ss} = \lim_{s \to 0} \left[s \frac{sK_1(1 + T_1 s)(1 + T_2 s) \cdots (1 + T_w s)}{s(1 + T_a s)(1 + T_b s) \cdots (1 + T_u s) + K_1(1 + T_1 s)(1 + T_2 s) \cdots (1 + T_w s)} \frac{R_1}{s^2} \right] = R_1$$

Therefore,

$$(Dc)_{ss} = (Dr)_{ss} \tag{6.94}$$

The magnitude of the steady-state error is found by using the ramp error coefficient. From Eq. (6.79),

$$K_v = \lim_{s \to 0} sG(s) = K_1$$

From the definition of ramp error coefficient, the steady-state error is

$$e(t)_{ss} = \frac{Dc(t)_{ss}}{K_1} \tag{6.95}$$

Since $Dc(t)_{ss} = Dr(t)$,

$$e(t)_{ss} = \frac{Dr}{K_1} = \frac{R_1}{K_1} = E_0 \tag{6.96}$$

Therefore a Type 1 system follows a ramp input with a constant error E_0. Figure 6.11 illustrates these conditions graphically.

Table of Steady-State Error Coefficients

Table 6.4 gives the values of the error coefficients for the Type 0, 1, and 2 systems. These values are determined from Table 6.3. The reader should be able to make ready use of Table 6.4 for evaluating the appropriate error coefficient.

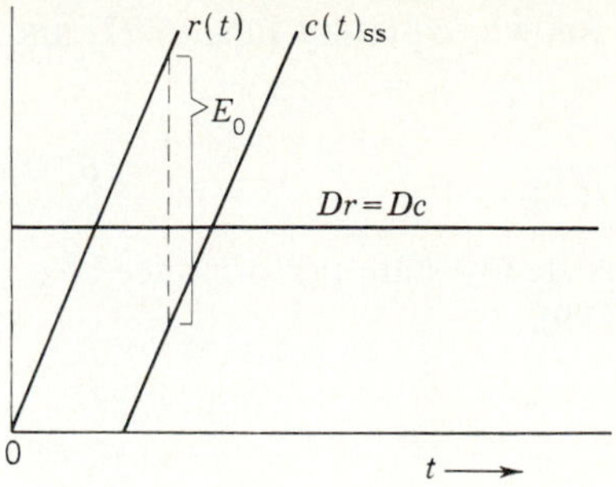

FIGURE 6.11
Steady-state response of a Type 1 system for Dr = constant.

TABLE 6.4
Steady-state error coefficients for stable systems

System type	Step error coefficient K_p	Ramp error coefficient K_v	Parabolic error coefficient K_a
0	K_0	0	0
1	∞	K_1	0
2	∞	∞	K_2

The error coefficient is then used with the definitions given in Table 6.3 to evaluate the magnitude of the steady-state error.

Polynomial Input: t^{m+1}

Note that a Type m system can follow an input of the form t^{m-1} with zero steady-state error. It can follow an input t^m, but there is a constant steady-state error. It cannot follow an input t^{m+1} because the steady-state error approaches infinity. However, for this case the input may be present only for a finite length of time. Thus the error is also finite. Then the error can be evaluated by taking the inverse Laplace transform of Eq. (6.33) and inserting the value of time. The maximum permissible error limits the time ($0 < t < t_1$) that an input t^{m+1} can be applied to a control system.

6.9 NONUNITY-FEEDBACK SYSTEM

The nonunity-feedback system of Fig. 6.1 may be mathematically converted to an equivalent unity-feedback system from which its "effective" system type and static error coefficient can be determined. The control ratio for Fig. 6.1 is

$$\frac{C(s)}{R(s)} = \frac{G(s)}{1 + G(s)H(s)} = \frac{N(s)}{D(s)} \tag{6.97}$$

and for the equivalent unity-feedback control system of the form of Fig. 6.4, the control ratio is

$$\frac{C(s)}{R(s)} = \frac{G_{eq}(s)}{1 + G_{eq}(s)} \tag{6.98}$$

Since the transfer functions $G(s)$ and $H(s)$ are known, equating Eqs. (6.97) and (6.98) yields

$$G_{eq}(s) = \frac{N(s)}{D(s) - N(s)} \tag{6.99}$$

When the nonunity-feedback system is stable, its steady-state performance characteristics can be determined based upon Eq. (6.99).

6.10 SUMMARY

In this chapter an attempt is made to clarify the distinction between the physical forms that the reference input and the controlled variable may have and the mathematical formulations of these quantities. In a position-control system the error always represents position, regardless of whether the input is $r(t) = u_{-1}(t)$, $u_{-2}(t)$, or $u_{-3}(t)$. In a system that controls another physical quantity, such as temperature, the error represents the variable which is being controlled. Since, in general, the forward transfer functions of most unity feedback control systems fall into three categories, they can be identified as Type 0, 1, and 2 systems, with the corresponding definitions of the steady-state error coefficients. These error coefficients are indicative of a stable system's steady-state performance. Thus, a start has been made in developing a set of standard characteristics. A nonunity-feedback system can be represented by an equivalent unity-feedback system having the forward transfer function $G_{eq}(s)$ (see Sec. 6.9). Then the steady-state performance can be evaluated on the basis of the system type and the error coefficients determined from $G_{eq}(s)$.

REFERENCES

1. Guillemin, E. A.: *The Mathematics of Circuit Analysis*, Wiley, New York, 1949.
2. Chestnut, H., and R. W. Mayer: Servomechanisms and Regulating System Design, 2d ed., vol. 1, Wiley, New York, 1959.
3. Singh, V.: "Comments on the Routh-Hurwitz Criterion," *IEEE Trans. Autom. Control*, vol. AC-26, p. 612, 1981.
4. Porter, Brian: *Stability Criteria for Linear Dynamical Systems*, Academic, New York, 1968.
5. Gantmacher, F. R.: *Applications of the Theory of Matrices*, Wiley Interscience, New York, 1959.
6. *Master Library*, *TI Programmable 58/59*, Texas Instruments, Inc., Dallas, Texas, 1977.
7. Krishnamurthi, V.: "Implications of Routh Stability Criteria," *IEEE Trans. Autom. Control*, vol. AC-25, pp. 554–555, 1980.
8. Khatwani, K. J.: "On Routh-Hurwitz Criterion," *IEEE Trans. Autom. Control*, vol. AC-26, pp. 583–584, 1981.
9. Pillai, S. K.: "On the ε-Method of the Routh-Hurwitz Criterion," *IEEE Trans. Autom. Control*, vol. AC-26, p. 584, 1981.

CHAPTER 7

ROOT LOCUS

7.1 INTRODUCTION

A designer can determine whether his design of a control system meets the specifications if he knows the desired time response of the controlled variable. An accurate solution of the system's performance can be obtained by deriving the differential equations for the control system and solving them. However, if the response does not meet the specifications, it is not easy to determine from this solution just what physical parameters in the system should be changed to improve the response.

A designer wishes to be able to predict a system's performance by an analysis that does not require the actual solution of the differential equations. Also, he would like this analysis to indicate readily the manner or method by which this system must be adjusted or compensated to produce the desired performance characteristics.

The first thing that a designer wants to know about a given system is whether or not it is stable. This can be determined by examining the roots obtained from the characteristic equation $1 + G(s)H(s) = 0$. By applying Routh's criterion to the characteristic equation it is possible in short order to determine whether the system is stable or unstable. Yet this does not satisfy the designer because it does not indicate the degree of stability of the system, i.e., the amount of overshoot and the settling time of the controlled variable. Not only must the system be stable, but the overshoot must be maintained within prescribed limits and transients must die out in a sufficiently short time. The graphical methods described in this text not only indicate whether a system is stable or unstable but, for a stable system, also show the degree of stability.

There are two basic methods available to a designer. He can choose to analyze and interpret the steady-state sinusoidal response of the transfer function of the system to obtain an idea of the system's response. This method is based upon the interpretation of a Nyquist plot, discussed in more detail in Chaps. 8 and 9. Although this frequency-response approach does not yield an exact quantitative prediction of the system's performance, i.e., the poles of the control ratio $C(s)/R(s)$ cannot be determined, enough information can be obtained to indicate whether the system needs to be adjusted or compensated. Also, the analysis indicates how the system should be compensated.

This chapter deals with the second method, *the root-locus method*,[1,2] devised by W. R. Evans, which incorporates the more desirable features of both the classical method and the frequency-response method. *The root locus is a plot of the roots of the characteristic equation of the closed-loop system as a function of the gain*. This graphical approach yields a clear indication of the effect of gain adjustment with relatively small effort compared with other methods. The underlying principle is that the poles of $C(s)/R(s)$ (transient-response modes) are related to the zeros and poles of the open-loop transfer function $G(s)H(s)$ and also to the gain. An important advantage of the root-locus method is that the roots of the characteristic equation of the system can be obtained directly; this results in a complete and accurate solution of the transient and steady-state response of the controlled variable. Another important feature is that an approximate solution can be obtained with a reduction of the work required. As with any other design technique, a person can readily obtain proficiency with the root-locus method. With the help of a digital-computer program, it is possible to synthesize a compensator, if one is required, with relative ease.

7.2 PLOTTING ROOTS OF A CHARACTERISTIC EQUATION

To give a better insight into the root-locus plots, consider the position-control system shown in Fig. 7.1. The motor produces a torque $T(s)$ which is proportional to the actuating signal $E(s)$. The load consists of the combined motor and

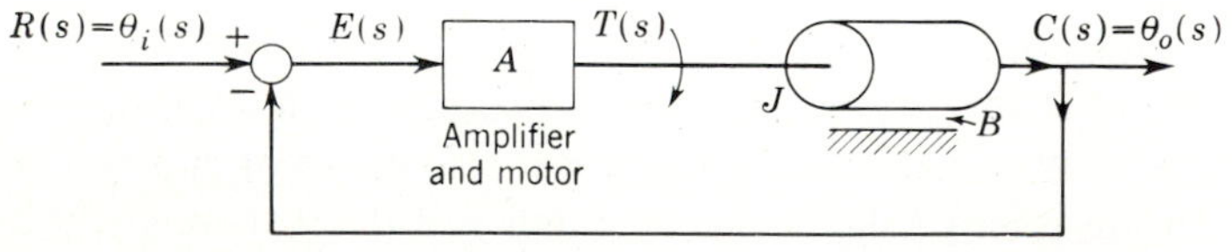

FIGURE 7.1
A position-control system.

load inertia J and viscous friction B. The forward transfer function is

$$G(s) = \frac{\theta_o(s)}{E(s)} = \frac{A/J}{s(s + B/J)} = \frac{K}{s(s + a)} \tag{7.1}$$

where $K = A/J$ and $a = B/J$. Assume that $a = 2$. Thus

$$G(s) = \frac{C(s)}{E(s)} = \frac{K}{s(s + 2)} \tag{7.2}$$

When the transfer function is expressed with the coefficients of the highest powers of s in both the numerator and the denominator equal to unity, the value of K is defined as the *static loop sensitivity*. The control ratio (closed-loop transfer function) is

$$\frac{C(s)}{R(s)} = \frac{K}{s(s + 2) + K} = \frac{K}{s^2 + 2s + K} = \frac{\omega_n^2}{s^2 + 2\zeta\omega_n s + \omega_n^2} \tag{7.3}$$

where $\omega_n = \sqrt{K}$, $\zeta = 1/\sqrt{K}$, and K is considered to be adjustable from zero to an infinite value.

The problem is to determine the roots of the characteristic equation for all values of K and to plot these roots in the s plane. The roots of the characteristic equation are given by

$$s_{1,2} = -1 \pm \sqrt{1 - K} \tag{7.4}$$

For $K = 0$, the roots are $s_1 = 0$ and $s_2 = -2$, which also are the poles of the open-loop transfer function given by Eq. (7.2). When $K = 1$, then $s_{1,2} = -1$. Thus when $0 < K < 1$, the roots $s_{1,2}$ are real and lie on the negative real axis of the s plane between -2 and -1 and 0 to -1, respectively. For the case where $K > 1$, the roots are complex and are given by

$$s_{1,2} = \sigma \pm j\omega_d = -\zeta\omega_n \pm j\omega_n\sqrt{1 - \zeta^2} = -1 \pm j\sqrt{K - 1} \tag{7.5}$$

Note that the real part of all the roots is constant for all values of $K > 1$.

The roots of the characteristic equation $s^2 + 2s + K = 0$ are determined for a number of values of K (see Table 7.1) and are plotted in Fig. 7.2. Curves are drawn through these plotted points. On these curves, containing two branches, lie all possible roots of the characteristic equation for all values of K from zero to infinity. Note that each branch is calibrated with K as a parameter and the values of K at points on the locus are underlined; the arrows show the direction of increasing values of K. *These curves are defined as the root-locus plot of Eq.* (7.3). Once this plot is obtained, the roots that best fit the system performance specifications can be selected. Corresponding to the selected roots there is a required value of K which can be determined from the plot. When the roots have been selected, the time response can be obtained. Since this process of finding the root locus by calculating the roots for various values of K becomes tedious for characteristic equations of order higher than second, a simpler method of obtaining the root locus is desired. The graphical methods for

TABLE 7.1
Location of roots for the characteristic equation
$s^2 + 2s + K = 0$

K	s_1	s_2
0	$-0 + j0$	$-2.0 - j0$
0.5	$-0.293 + j0$	$-1.707 - j0$
0.75	$-0.5 + j0$	$-1.5 - j0$
1.0	$-1.0 + j0$	$-1.0 - j0$
2.0	$-1.0 + j1.0$	$-1.0 - j1.0$
3.0	$-1.0 + j1.414$	$-1.0 - j1.414$
50.0	$-1.0 + j7.0$	$-1.0 - j7.0$

determining the root-locus plot are the subject of this chapter. Digital-computer methods[3] are used to facilitate obtaining the root locus.

The value of K is normally considered to be positive. However, it is possible for K to be negative. For the example in this section, if the value of K is negative, Eq. (7.4) gives only real roots. Thus the entire locus lies on the real axis, that is, $0 \le s_1 < +\infty$ and $-2 \ge s_2 > -\infty$ for $0 \ge K > -\infty$. For any negative value of K there is always a root in the right half of the s plane, and the system is unstable.

Once the root locus has been obtained for a control system, it is possible to determine the variation in system performance with respect to a variation in

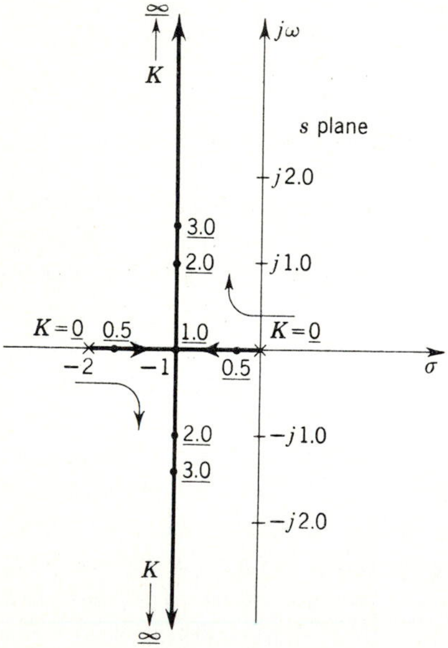

FIGURE 7.2
Plot of all roots of the characteristic equation $s^2 + 2s + K = 0$ for $0 \le K < \infty$. Values of K are underlined.

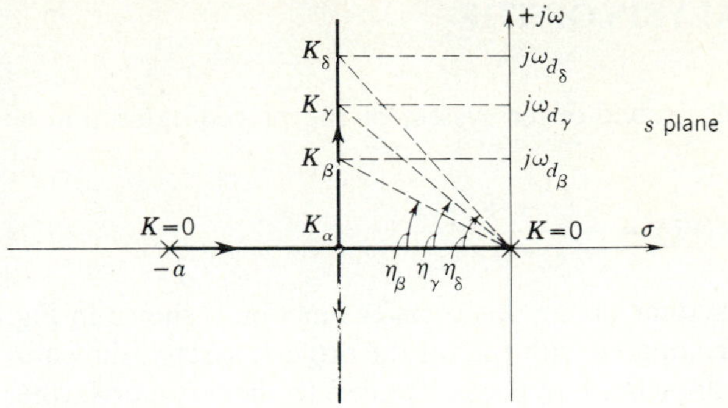

FIGURE 7.3
Root-locus plot of the position-control system of Fig. 7.1. $\eta = \cos^{-1}\zeta$.

sensitivity K. For the example of Fig. 7.1, the control ratio is written in terms of its roots, for $K > 1$, as

$$\frac{C(s)}{R(s)} = \frac{K}{(s - \sigma - j\omega_d)(s - \sigma + j\omega_d)} \tag{7.6}$$

Note, as defined in Fig. 4.3, that a root with a damping ratio ζ lies on a line making the angle $\eta = \cos^{-1}\zeta$ with the negative real axis. The damping ratio of several roots is indicated in Fig. 7.3. Analysis of the root locus reveals the following characteristics for an increase in the gain of the system:

1. A decrease in the damping ratio ζ. This increases the overshoot of the time response.
2. An increase in the undamped natural frequency ω_n. The value of ω_n is the distance from the origin to the complex root.
3. An increase in the damped natural frequency ω_d. The value of ω_d is the imaginary component of the complex root and is equal to the frequency of the transient response.
4. No effect on the rate of decay σ; that is, it remains constant for all values of gain equal to or greater than K_a. For higher-order systems this is not the case.
5. The root locus is a vertical line for $K \geq K_a$, and $\sigma = -\zeta\omega_n$ is constant. This means that no matter how much the gain is increased in a linear *simple* second-order system, the system can never become unstable. The time response of this system with a step-function input, for $\zeta < 1$, is of the form

$$c(t) = A_0 + A_1 e^{-\zeta\omega_n t} \sin(\omega_d t + \phi)$$

The root locus of each control system can be analyzed in a similar manner to obtain an idea of the variation in its time response which results from a variation in its loop sensitivity K.

7.3 QUALITATIVE ANALYSIS OF THE ROOT LOCUS

A zero is added to the simple second-order system of the preceding section so that the transfer function is

$$G(s) = \frac{K(s + 1/T_2)}{s(s + 1/T_1)} \tag{7.7}$$

The root locus of the control system having this transfer function is shown in Fig. 7.4*b*. When this root locus is compared with that of the original system, shown in Fig. 7.4*a*, it is seen that the branches have been "pulled to the left," or farther from the imaginary axis. For values of static loop sensitivity greater than K_α, the roots are farther to the left than for the original system. Therefore the transients will decay faster, yielding a more stable system.

If a pole, instead of a zero, is added to the original system the resulting transfer function is

$$G(s) = \frac{K}{s(s + 1/T_1)(s + 1/T_3)} \tag{7.8}$$

Figure 7.4*c* shows the root locus of the control system having this transfer function. Note that the addition of a pole has "pulled the locus to the right" so that two branches cross the imaginary axis. For values of static loop sensitivity greater than K_β, the roots are closer to the imaginary axis than for the original system. Therefore the transients will decay more slowly, yielding a less stable system. Also, for values of $K > K_\gamma$, two of the three roots lie in the right half of the s plane, resulting in an unstable system. The addition of the pole has resulted in a less stable system, compared with the original second-order system. Thus the following general conclusions can be drawn:

1. The addition of a zero to a system has the general effect of pulling the root locus to the left, tending to make it a more stable and faster-responding system (shorter settling time T_s).
2. The addition of a pole to a system has the effect of pulling the root locus to the right, tending to make it a less stable and slower-responding system.

Figure 7.5 illustrates the root-locus configurations for negative-feedback control systems having the following transfer functions:

$$G(s)H(s) = \frac{K(s + 1/T_2)(s + 1/T_4)}{s(s + 1/T_1)(s + 1/T_3)} \tag{7.9}$$

$$G(s)H(s) = \frac{K(s + 1/T_2)(s + 1/T_4)}{(s + 1/T_1)(s + 1/T_3)(s + 1/T_5)} \tag{7.10}$$

$$G(s)H(s) = \frac{K(s + 1/T_2)}{s(s - 1/T)(s^2 + 2\zeta\omega_n s + \omega_n^2)} \tag{7.11}$$

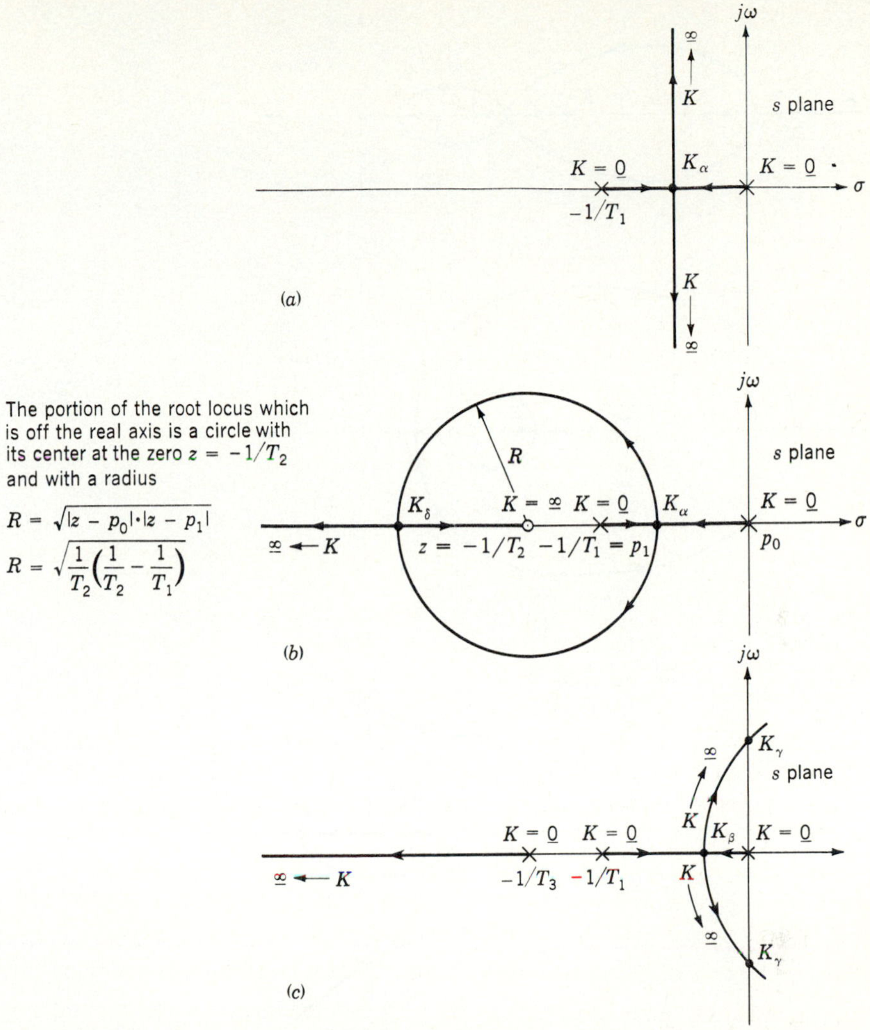
jω
∞
K
s plane
K = 0
Kα
K = 0
σ
−1/T₁
K
∞
(a)
The portion of the root locus which is off the real axis is a circle with its center at the zero $z = -1/T_2$ and with a radius
$R = \sqrt{|z - p_0| \cdot |z - p_1|}$
$R = \sqrt{\frac{1}{T_2}\left(\frac{1}{T_2} - \frac{1}{T_1}\right)}$
jω
R
s plane
Kδ
K = ∞
K = 0
Kα
K = 0
σ
∞ ← K
z = −1/T₂
−1/T₁ = p₁
p₀
(b)
jω
Kγ
∞
s plane
K
Kβ
K = 0
K = 0
K = 0
∞ ← K
−1/T₃
−1/T₁
K
∞
Kγ
(c)

FIGURE 7.4
Various root-locus configurations. (a) Root locus of basic transfer function:

$$G(s) = \frac{K}{s(s + 1/T_1)} \qquad H(s) = 1$$

(b) Root locus with additional zero:

$$G(s) = \frac{K(s + 1/T_2)}{s(s + 1/T_1)} \qquad H(s) = 1$$

(c) Root locus with additional pole:

$$G(s) = \frac{K}{s(s + 1/T_1)(s + 1/T_3)} \qquad H(s) = 1$$

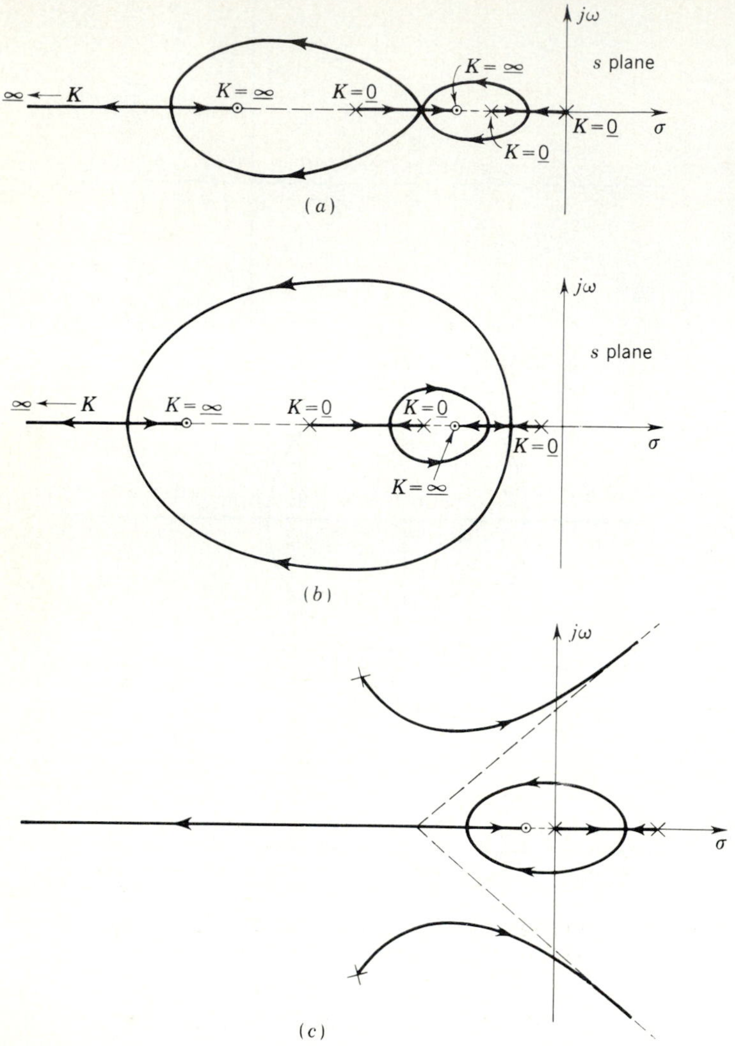

FIGURE 7.5
Various root-locus configurations:

(a) $$G(s)H(s) = \frac{K(s + 1/T_2)(s + 1/T_4)}{s(s + 1/T_1)(s + 1/T_3)}$$

(b) $$G(s)H(s) = \frac{K(s + 1/T_2)(s + 1/T_4)}{(s + 1/T_1)(s + 1/T_3)(s + 1/T_5)}$$

(c) $$G(s)H(s) = \frac{K(s + 1/T_2)}{s(s - 1/T)(s^2 + 2\zeta\omega_n s + \omega_n^2)}$$

These figures are not drawn to scale. Several other root-locus shapes are possible for a given pole-zero arrangement, depending on the specific values of the poles and zeros. Variations of the possible root-locus plots for a given pole-zero arrangement are shown in Ref. 9.

Note that the third system contains a pole in the right half of the s plane. It represents the performance of an airplane with an autopilot in the longitudinal mode.

The root-locus method is a graphical technique for readily determining the location of all possible roots of a characteristic equation as the gain is varied from zero to infinity. Also, how the locus should be altered in order to improve the system's performance can be readily determined, based upon the knowledge of the effect of the addition of poles or zeros.

7.4 PROCEDURE OUTLINE

To help the reader visualize the order of the root-locus approach, the procedure to be followed in applying this method is first outlined. This procedure is modified when a computer program[3,10] is used to obtain the root locus. Such a program can provide the desired data for the root locus in plotted or tabular form. The procedure outlined below and discussed in the following sections is intended to establish firmly for the reader the fundamentals of the root-locus method and to enhance the interpretation of the data obtained from the computer program.

Step 1. Derive the open-loop transfer function $G(s)H(s)$ of the system.

Step 2. Factor the numerator and denominator of the transfer function into linear factors of the form $s + a$, where a may be real or complex.

Step 3. Plot the zeros and poles of the open-loop transfer function in the $s = \sigma + j\omega$ plane.

Step 4. The plotted zeros and poles of the open-loop function determine the roots of the characteristic equation of the closed-loop system $[1 + G(s)H(s) = 0]$. By use of the geometrical shortcuts or a digital-computer program,[3] determine the locus that describes the roots of the closed-loop characteristic equation.

Step 5. Calibrate the locus in terms of the loop sensitivity K. If the gain of the open-loop system is predetermined, the locations of the exact roots of $1 + G(s)H(s)$ are immediately known. If the location of the roots is specified, the required value of K can be determined.

Step 6. Once the roots have been found, in step 5, the system's time response can be calculated by taking the inverse Laplace transform of $C(s)$, either manually or by use of a computer program.

Step 7. If the response does not meet the desired specifications, determine the shape that the root locus must have to meet these specifications.

Step 8. Synthesize the network that must be inserted into the system, if other than gain adjustment is required, to make the required modification on the original locus. This process, called *compensation*, is described in later chapters.

7.5 OPEN-LOOP TRANSFER FUNCTION

In securing the open-loop transfer function, keep the terms in the factored form of $s + a$ or $s^2 + 2\zeta\omega_n s + \omega_n^2$. For unity feedback the open-loop function is equal to the forward transfer function $G(s)$. For nonunity feedback it also includes the transfer function of the feedback path. This open-loop transfer function is of the form

$$G(s)H(s) = \frac{K(s+a_1)\cdots(s+a_h)\cdots(s+a_w)}{s^m(s+b_1)(s+b_2)\cdots(s+b_c)\cdots(s+b_u)} \tag{7.12}$$

where a_h and b_c may be real or complex numbers and may lie in either the left-half or right-half s plane. The value of K may be either positive or negative. For example, consider

$$G(s)H(s) = \frac{K(s+a_1)}{s(s+b_1)(s+b_2)} \tag{7.13}$$

When the transfer function is in this form (with the coefficients of s all equal to unity), the K is defined as the *loop sensitivity*. By inspection it can be seen that for this example a zero of the open-loop transfer function exists at $s = -a_1$ and the poles are at $s = 0$, $s = -b_1$, and $s = -b_2$. Now let the zeros and poles of $G(s)H(s)$ in Eq. (7.12) be denoted by the letters z and p, respectively, i.e.,

$$z_1 = -a_1 \qquad z_2 = -a_2 \qquad \cdots \qquad z_w = -a_w$$

$$p_1 = -b_1 \qquad p_2 = -b_2 \qquad \cdots \qquad p_u = -b_u$$

Then Eq. (7.12) can be rewritten as

$$G(s)H(s) = \frac{K(s-z_1)\cdots(s-z_w)}{s^m(s-p_1)\cdots(s-p_u)} = \frac{K\prod_{h=1}^{w}(s-z_h)}{s^m\prod_{c=1}^{u}(s-p_c)} \tag{7.14}$$

where $\prod$ indicates a product of terms. The degree of the numerator is w, and that of the denominator is $m + u = n$.

7.6 POLES OF THE CONTROL RATIO $C(s)/R(s)$

The underlying principle of the root-locus method is that the poles of the control ratio $C(s)/R(s)$ are related to the zeros and poles of the $G(s)H(s)$ function and to the loop sensitivity K. This can be shown as follows. Let

$$G(s) = \frac{N_1(s)}{D_1(s)} \tag{7.15}$$

and

$$H(s) = \frac{N_2(s)}{D_2(s)} \tag{7.16}$$

Then

$$G(s)H(s) = \frac{N_1 N_2}{D_1 D_2} \tag{7.17}$$

Thus

$$\frac{C(s)}{R(s)} = M(s) = \frac{A(s)}{B(s)} = \frac{G(s)}{1 + G(s)H(s)} = \frac{N_1/D_1}{1 + N_1N_2/D_1D_2} \tag{7.18}$$

where

$$B(s) \equiv 1 + G(s)H(s) = 1 + \frac{N_1N_2}{D_1D_2} = \frac{D_1D_2 + N_1N_2}{D_1D_2} \tag{7.19}$$

Rationalizing Eq. (7.18) gives

$$\frac{C(s)}{R(s)} = M(s) = \frac{N_1D_2}{D_1D_2 + N_1N_2} = \frac{P(s)}{Q(s)} \tag{7.20}$$

From Eqs. (7.19) and (7.20) it is seen that the zeros of $B(s)$ are the poles of $M(s)$ and determine the form of the system's transient response. In terms of $G(s)H(s)$, given by Eq. (7.14), the degree of $B(s)$ is equal to $m + u$; therefore $B(s)$ has $n = m + u$ finite zeros. As shown in Chap. 4, Laplace Transforms, the factors of $Q(s)$ produce transient components of $c(t)$ which fall into the categories shown in Table 7.2. The numerator $P(s)$ of Eq. (7.20) merely modifies the constant multipliers of these transient components. The roots of $B(s) = 0$, which is the characteristic equation of the system, must satisfy the equation:

$$B(s) \equiv 1 + G(s)H(s) = 0 \tag{7.21}$$

These roots must also satisfy the equation:

$$G(s)H(s) = \frac{K(s - z_1)\cdots(s - z_w)}{s^m(s - p_1)\cdots(s - p_u)} = -1 \tag{7.22}$$

Thus, as the loop sensitivity K assumes values from zero to infinity, the transfer

TABLE 7.2
Time response terms in $c(t)$

Denominator factor of $C(s)$	Corresponding inverse	Form
s	$u_{-1}(t)$	Step function
$s + \frac{1}{T}$	$e^{-t/T}$	Decaying exponential
$s^2 + 2\zeta\omega_n s + \omega_n^2$	$e^{-\zeta\omega_n t}\sin(\omega_n\sqrt{1-\zeta^2}\,t + \phi)$ where $\zeta < 1$	Damped sinusoid

function $G(s)H(s)$ must always be equal to -1. The corresponding values of s which satisfy Eq. (7.22), for any value of K, are the poles of $M(s)$. The plots of these values of s are defined as the root locus of $M(s)$.

Conditions that determine the root locus for *positive values* of loop sensitivity are now determined. The general form of $G(s)H(s)$ for *any* value of s is

$$G(s)H(s) = Fe^{-j\beta}$$

The right side of Eq. (7.22), -1, can be written as

$$-1 = e^{j(1+2h)\pi} \qquad h = 0, \pm 1, \pm 2, \ldots$$

Equation (7.22) is satisfied *only* for those values of s for which

$$Fe^{-j\beta} = e^{j(1+2h)\pi}$$

where $$F = |G(s)H(s)| = 1$$

and $$-\beta = (1 + 2h)\pi \tag{7.23}$$

From the above it can be concluded that the magnitude of $G(s)H(s)$, a function of the complex variable s, must always be unity and its phase angle must be an odd multiple of π if the particular value of s is to be a zero of $B(s) = 1 + G(s)H(s)$. Consequently, the following two conditions are formalized for the root locus for all positive values of K from zero to infinity:

For $K > 0$:

Magnitude condition: $$|G(s)H(s)| = 1 \tag{7.24}$$

Angle condition:

$$\angle G(s)H(s) = (1 + 2h)180^\circ \qquad \text{for } h = 0, \pm 1, \pm 2, \ldots \tag{7.25}$$

In a similar manner, the conditions for *negative* values of loop sensitivity ($-\infty < K < 0$) can be determined. [This corresponds to positive feedback, $e(t) = r(t) + b(t)$, and positive values of K.] The root locus must satisfy the conditions:

For $K < 0$:

Magnitude condition: $$|G(s)H(s)| = 1 \tag{7.26}$$

Angle condition: $$\angle G(s)H(s) = h360^\circ \qquad \text{for } h = 0, \pm 1, \pm 2, \ldots \tag{7.27}$$

Thus the root-locus method provides a plot of the variation of each of the poles of $C(s)/R(s)$ in the complex s plane as the loop sensitivity is varied from $K = 0$ to $K = \pm\infty$.

7.7 APPLICATION OF THE MAGNITUDE AND ANGLE CONDITIONS

Once the open-loop transfer function $G(s)H(s)$ has been determined and put into the proper form, the poles and zeros of this function are plotted in the

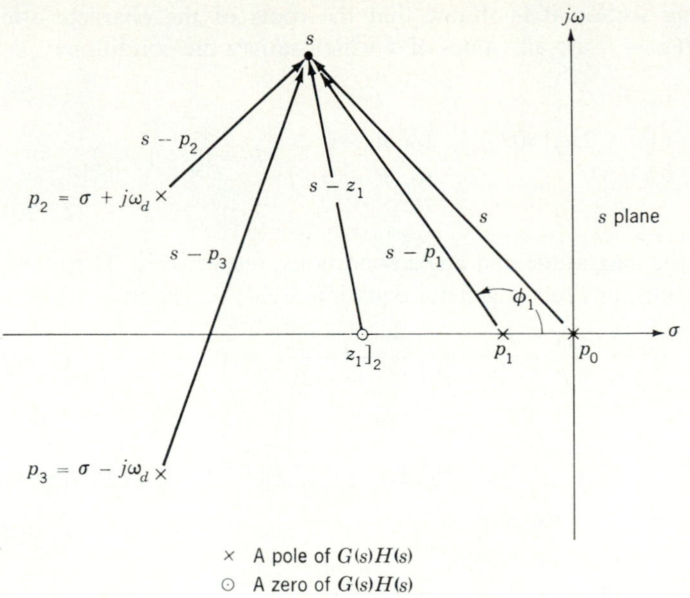

FIGURE 7.6
Pole-zero diagram for Eq. (7.28).

$s = \sigma + j\omega$ plane. As an example, consider

$$G(s)H(s) = \frac{K(s + 1/T_1)^2}{s(s + 1/T_2)(s^2 + 2\zeta\omega_n s + \omega_n^2)} = \frac{K(s - z_1)^2}{s(s - p_1)(s - p_2)(s - p_3)} \tag{7.28}$$

For the quadratic factor $s^2 + 2\zeta\omega_n s + \omega_n^2$ with the damping ratio $\zeta < 1$, the complex-conjugate poles of Eq. (7.28) are

$$p_{2,3} = -\zeta\omega_n \pm j\omega_n\sqrt{1 - \zeta^2} = \sigma \pm j\omega_d$$

The plot of the poles and zeros of Eq. (7.28) is shown in Fig. 7.6. Remember that complex poles or zeros always occur in conjugate pairs, that σ is the damping constant, and ω_d is the damped natural frequency of oscillation. A multiple pole or zero is indicated on the pole-zero diagram by $x]_q$ or $\odot]_q$, where $q = 1, 2, 3, \ldots$ is the order of the pole or zero. With the open-loop poles and zeros plotted, they are now used in the graphical construction of the locus of the poles (the roots of the characteristic equation) of the closed-loop control ratio.

For any particular value (real or complex) of s the terms $s, s - p_1, s - p_2, s - z_1, \ldots$ are complex numbers designating *directed line segments*. For example, if $s = -4 + j4$ and $p_1 = -1$, then $s - p_1 = -3 + j4$ or

$$|s - p_1| = 5$$

and $$\phi_1 = \angle s - p_1 = 126.8° \qquad \text{(see Fig. 7.6)}$$

In the preceding section it is shown that the roots of the characteristic equation $1 + G(s)H(s) = 0$ are all values of s which satisfy the conditions

$$|G(s)H(s)| = 1 \tag{7.29}$$

$$\underline{/G(s)H(s)} = \begin{cases} (1 + 2h)180° & \text{for } K > 0 \\ h360° & \text{for } K < 0 \end{cases} \qquad h = 0, \pm 1, \pm 2, \ldots \tag{7.30}$$

These are labeled as the magnitude and angle conditions, respectively. Therefore, applying these two conditions to the general equation (7.14) results in

$$\frac{|K| \cdot |s - z_1| \cdots |s - z_w|}{|s^m| \cdot |s - p_1| \cdot |s - p_2| \cdots |s - p_u|} = 1 \tag{7.31}$$

and

$$-\beta = \underline{/s - z_1} + \cdots + \underline{/s - z_w} - m\underline{/s} - \underline{/s - p_1} - \cdots - \underline{/s - p_u}$$
$$= \begin{cases} (1 + 2h)180° & \text{for } K > 0 \\ h360° & \text{for } K < 0 \end{cases} \tag{7.32}$$

Rewriting these equations gives

$$|K| = \frac{|s^m| \cdot |s - p_1| \cdot |s - p_2| \cdots |s - p_u|}{|s - z_1| \cdots |s - z_w|} = \text{loop sensitivity} \tag{7.33}$$

and

$$-\beta = \sum(\text{angles of numerator terms}) - \sum(\text{angles of denominator terms})$$
$$= \begin{cases} 1 + 2h)180° & \text{for } K > 0 \\ h360° & \text{for } K < 0 \end{cases} \tag{7.34}$$

All angles are considered positive, measured in the counterclockwise sense. Since $G(s)H(s)$ usually has more poles than zeros, it is convenient to multiply Eq. (7.34) by -1. Since h may be a positive or negative integer, rearranging terms yields

$$\beta = \underline{/\text{denominator}} - \underline{/\text{numerator}}$$
$$= \begin{cases} (1 + 2h)180° & \text{for } K > 0 \\ h360° & \text{for } K < 0 \end{cases} \tag{7.35}$$

Equations (7.33) and (7.35) are in the form generally used in the graphical construction of the roots locus. In other words, there are particular values of s for which $G(s)H(s)$ satisfies the angle condition. For a given loop sensitivity only certain of these values of s simultaneously satisfy the magnitude condition. Those values of s which satisfy both the angle and the magnitude conditions are the roots of the characteristic equation and are $n = m + u$ in number. Thus, corresponding to step 4 in Sec. 7.4, the locus of all possible roots is obtained by applying the angle condition. This root locus can be calibrated in terms of the loop sensitivity K by using the magnitude condition.

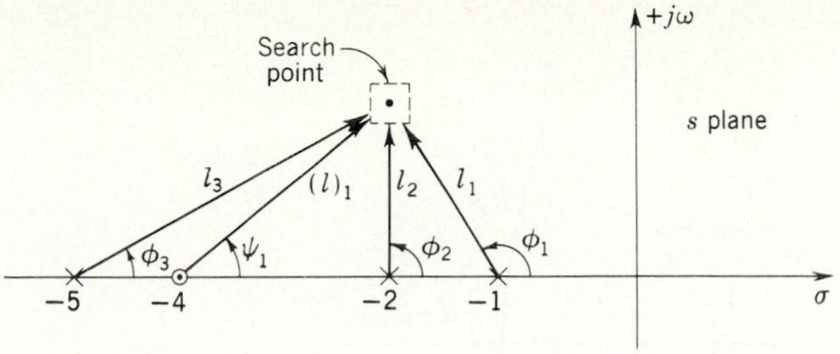

FIGURE 7.7
Construction of the root locus.

Example

$$G(s)H(s) = \frac{K_0(1 + 0.25s)}{(1 + s)(1 + 0.5s)(1 + 0.2s)} \tag{7.36}$$

Determine the locus of all possible closed-loop poles of $C(s)/R(s)$.

$$G(s)H(s) = \frac{K_0(0.25)}{(0.5)(0.2)} \frac{s + 4}{(s + 1)(s + 2)(s + 5)}$$

$$G(s)H(s) = \frac{K(s + 4)}{(s + 1)(s + 2)(s + 5)} \qquad K = 2.5K_0 \tag{7.37}$$

Step 1. The poles and zeros are plotted in Fig. 7.7.

Step 2. In Fig. 7.7 the ϕ's are denominator angles and ψ's are numerator angles. Also, the l's are the lengths of the directed segments stemming from the denominator factors, and (l)'s are the lengths of the directed segments stemming from the numerator factors. After plotting the poles and zeros of the open-loop transfer functions, arbitrarily choose a search point. To this point, draw directed line segments from all the open-loop poles and zeros and label as indicated. For this search point to be a point on the locus, the following angle condition must be true:

$$\beta = \phi_1 + \phi_2 + \phi_3 - \psi_1 = \begin{cases} (1 + 2h)180° & \text{for } K > 0 \\ h360° & \text{for } K < 0 \end{cases} \tag{7.38}$$

If this equation is not satisfied, select another search point until it is satisfied. Locate a sufficient number of points in the s plane that satisfy the angle condition. In the next section additional information is given that lessens the work involved in this trial-and-error approach.

Step 3. Once the complete locus has been determined, the locus can be calibrated in terms of the loop sensitivity for any root s_1 as follows:

$$|K| = \frac{l_1 l_2 l_3}{(l)_1} \tag{7.39}$$

where $l_1 = |s_1 + 1| \qquad l_2 = |s_1 + 2| \qquad l_3 = |s_1 + 5| \qquad (l)_1 = |s_1 + 4|$

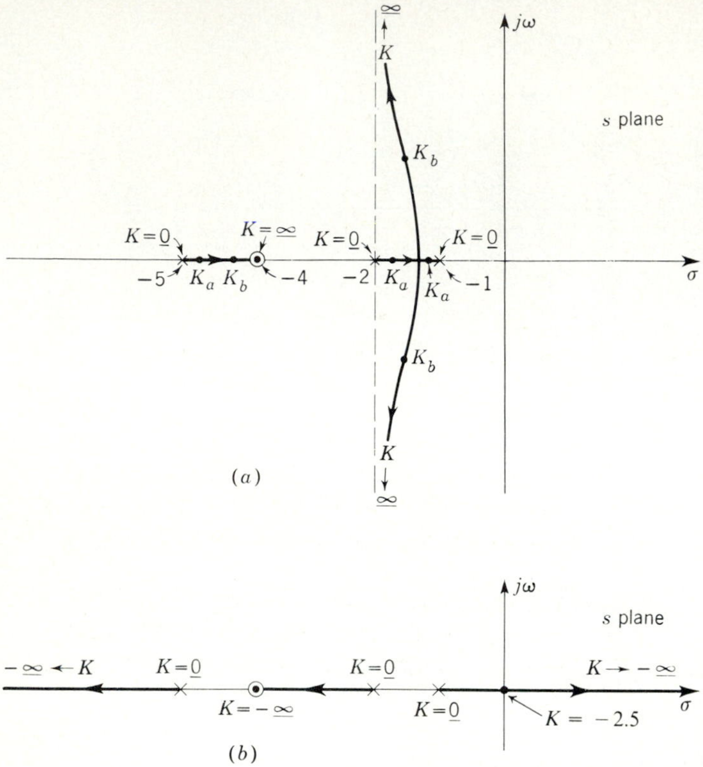

FIGURE 7.8
The complete root locus of Eq. (7.36); (a) for $K > 0$; (b) for $K < 0$.

In other words, the values of l_1, l_2, l_3, and $(l)_1$ are known for a given point s_1 that satisfies the angle condition; thus the value of $|K|$ for this point can be calculated. The appropriate sign must be given to the magnitude of K compatible with the particular angle condition which is utilized to obtain the root locus. Note that since complex roots must occur in conjugate pairs, the locus is symmetrical about the real axis. Thus the bottom half of the locus can be drawn once the locus above the real axis has been determined. The root locus for this system is shown in Fig. 7.8. Note that the three branches of the root locus, for negative values of K, lie in the left half plane for $K > -2.5$.

$W(s)$ Plane

From Eqs. (7.37) and (7.22)

$$W(s) = u_x + jv_y = \frac{(s+1)(s+2)(s+5)}{s+4} = -K \tag{7.40}$$

The line $u_x = -K$ in the $W(s)$ plane maps into the curves indicated in Fig. 7.8. That is, for each value of u_x in the $W(s)$ plane there is a particular value or a set of values of s in the s plane.

7.8 GEOMETRICAL PROPERTIES (CONSTRUCTION RULES)

To facilitate the application of the root-locus method, the following rules are established for $K > 0$. These rules are based upon the interpretation of the angle condition and an analysis of the characteristic equation. The reader should be able to extend these rules for the case where $K < 0$. The rules for both $K > 0$ and $K < 0$ are listed in Sec. 7.14 for easy reference.

The rules presented aid in obtaining the root locus by expediting the manual plotting of the locus. The root locus can also be obtained by using the digital computer.[3] These rules provide *checkpoints* to ensure that the computer solution is correct. They also permit rapid sketching of the root locus, which provides a qualitative idea of achievable system performance.

RULE 1: NUMBER OF BRANCHES OF THE LOCUS. The characteristic equation $B(s) = 0$ is of degree $n = m + u$; therefore there are n roots, which are continuous functions of the open-loop sensitivity K. As K is varied from zero to infinity, each root traces a continuous curve. Since there are n roots, there are the same number of curves or branches in the complete root locus. Since the degree of the polynomial $B(s)$ is determined by the poles of the open-loop transfer function, *the number of branches of the root locus is equal to the number of poles of the open-loop transfer function.*

RULE 2: REAL-AXIS LOCUS. In Fig. 7.9 are shown a number of open-loop poles and zeros. If the angle condition is applied to any search point such as s_1 on the real axis, the angular contribution of all the poles and zeros on the real axis to the left of this point is zero. The angular contribution of the complex-conjugate poles to this point is 360°. (This is also true for complex-conjugate zeros.) Finally, the poles and zeros on the real axis to the right of this point each contribute 180° (with the appropriate sign included). From Eq. (7.35) the angle of $G(s)H(s)$ to

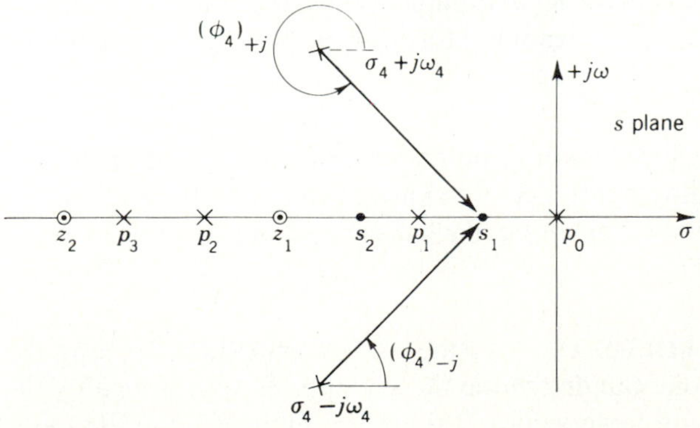

FIGURE 7.9
Determination of the real-axis locus.

the point s_1 is given by

$$\phi_0 + \phi_1 + \phi_2 + \phi_3 + (\phi_4)_{+j} + (\phi_4)_{-j} - (\psi_1 + \psi_2) = (1 + 2h)180° \quad (7.41)$$

or

$$180° + 360° = (1 + 2h)180° \quad (7.42)$$

Therefore s_1 is a point on a locus. Similarly, it can be shown that the point s_2 is not a point on the locus. The poles and zeros to the left of a point s on the real axis and the 360° contributed by the complex-conjugate poles or zeros do not affect the odd-multiple-of-180° requirement. Thus, *if the total number of real poles and zeros to the right of a search point s on the real axis is odd, this point lies on the locus*. In Fig. 7.9 the root locus exists on the real axis from p_0 to p_1, z_1 to p_2, and p_3 to z_2.

All points on the real axis between z_1 and p_2 in Fig. 7.9 satisfy the angle condition and are therefore points on the root locus. However, there is no guarantee that this section of the real axis is part of just one branch. Figure 7.5*b* and Prob. 7.4 illustrate the situation where the part of the real axis between a pole and a zero is divided into three sections which are parts of three different branches.

RULE 3: LOCUS END POINTS. The magnitude of the loop sensitivity which satisfies the magnitude condition is given by Eq. (7.33) and has the general form

$$|W(s)| = K = \frac{\prod_{c=1}^{n} |s - p_c|}{\prod_{h=1}^{w} |s - z_h|} \quad (7.43)$$

Since the numerator and denominator factors of Eq. (7.43) locate the poles and zeros, respectively, of the open-loop transfer function, the following conclusions can be drawn:

1. When $s = p_c$, the loop sensitivity K is zero.
2. When $s = z_h$, the loop sensitivity K is infinite. When the numerator of Eq. (7.43) is of higher order than the denominator, $s = \infty$ also makes K infinite, thus being equivalent in effect to a zero.

Thus it can be said that the locus starting points ($K = 0$) are at the open-loop poles and that the locus ending points ($K = \infty$) are at the open-loop zeros (the point at infinity being considered as an equivalent zero of multiplicity equal to $n - w$).

RULE 4: ASYMPTOTES OF LOCUS AS s APPROACHES INFINITY. Plotting the locus is greatly facilitated if one can determine the asymptotes approached by the various branches as s takes on large values. Taking the limit of $G(s)H(s)$ as s

approaches infinity, based on Eqs. (7.14) and (7.22), yields

$$\lim_{s\to\infty} G(s)H(s) = \lim_{s\to\infty}\left[K\frac{\prod_{h=1}^{w}(s-z_h)}{\prod_{c=1}^{n}(s-p_c)}\right] = \lim_{s\to\infty}\frac{K}{s^{n-w}} = -1 \tag{7.44}$$

It must be remembered that K in Eq. (7.44) is still a variable in the manner prescribed previously, thus allowing the magnitude condition to be met. Therefore, as $s \to \infty$,

$$-K = s^{n-w} \tag{7.45}$$

$$|-K| = |s^{n-w}| \qquad \text{magnitude condition} \tag{7.46}$$

$$\angle -K = \angle s^{n-w} = (1+2h)180° \qquad \text{angle condition} \tag{7.47}$$

Rewriting Eq. (7.47) gives

$$(n-w)\angle s = (1+2h)180°$$

or

$$\gamma = \frac{(1+2h)180°}{n-w} \qquad \text{as } s \to \infty \tag{7.48}$$

There are $n - w$ asymptotes of the root locus and their angles are given by

$$\gamma = \frac{(1+2h)180°}{[\text{number of poles of } G(s)H(s)] - [\text{number of zeros of } G(s)H(s)]} \tag{7.49}$$

Equation (7.49) reveals that, no matter what magnitude s may have, after a sufficiently large value has been reached, the argument (angle) of s on the root locus remains constant. For a search point that has a sufficiently large magnitude the open-loop poles and zeros appear to it as if they had collapsed into a single point. Therefore the branches are asymptotic to straight lines whose slopes and directions are given by Eq. (7.49) (see Fig. 7.10) These asymptotes usually do not go through the origin. The correct real-axis intercept of the asymptotes is obtained from Rule 5.

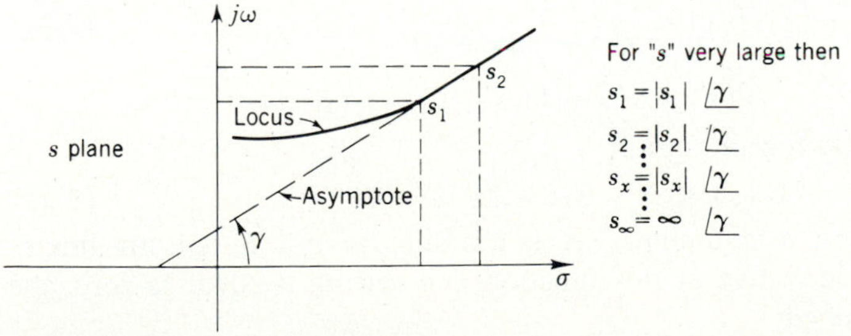

FIGURE 7.10
Asymptotic condition for large values of s.

RULE 5: REAL-AXIS INTERCEPT OF THE ASYMPTOTES. The real-axis crossing σ_o of the asymptotes can be obtained by applying the theory of equations. The result is

$$\sigma_o = \frac{\sum_{c=1}^{n} \mathrm{Re}(p_c) - \sum_{h=1}^{w} \mathrm{Re}(z_h)}{n - w} \tag{7.50}$$

The asymptotes are not dividing lines, and a locus may cross its asymptote. It may be valuable to know from which side the root locus approaches its asymptote. Lorens and Titsworth[4] present a method for obtaining this information. The locus lies exactly along the asymptote if the pole-zero pattern is symmetric about the asymptote line extended through the point σ_o.

RULE 6: BREAKAWAY POINT ON THE REAL AXIS.[11] The branches of the root locus start at the open-loop poles where $K = 0$ and end at the finite open-loop zeros or at $s = \infty$. Consider now the case where the root locus has branches on the real axis between two poles (between p_0 and p_1 in Fig. 7.11(a-3) and between p_2 and p_3 in Fig. 7.11(b-2)). There must be a point at which the two branches break away from the real axis and enter the complex region of the s plane in order to approach zeros or the point at infinity. For two finite zeros [see Fig. 7.11(b-1)] or one finite zero and one at infinity [see Fig. 7.11(a-1)] the branches are coming from the complex region and enter the real axis. In Fig. 7.11(a-3) between two poles there is a point s_a for which the loop sensitivity K_z is greater than for points on either side of s_a on the real axis. In other words, since K starts with a value of zero at the poles and increases in value as the locus moves away from the poles, there is a point somewhere in between where the K's for the two branches simultaneously reach a maximum value. This point is called the *breakaway point*. Plots of K vs. σ utilizing Eq. (7.33) are shown in Fig. 7.11 for the portions of the root locus which exist on the real axis for $K > 0$. The point s_b for which the value of K is a minimum between two zeros is called the *break-in point*. The breakaway and break-in points can easily be calculated for an open-loop pole-zero combination for which the derivative of $W(s) = -K$ is of the second order.

As an example, if

$$G(s)H(s) = \frac{K}{s(s+1)(s+2)} \tag{7.51}$$

then

$$W(s) = s(s+1)(s+2) = -K$$

Multiplying the factors together gives

$$W(s) = s^3 + 3s^2 + 2s = -K \tag{7.52}$$

When $s^3 + 3s^2 + 2s$ is a minimum, $-K$ is a minimum and K is a maximum. Thus by taking the derivative of this function and setting it equal to zero, the points can be determined:

$$\frac{dW(s)}{ds} = 3s^2 + 6s + 2 = 0 \tag{7.53}$$

or

$$s_{a,b} = -1 \pm 0.5743 = -0.4257, -1.5743$$

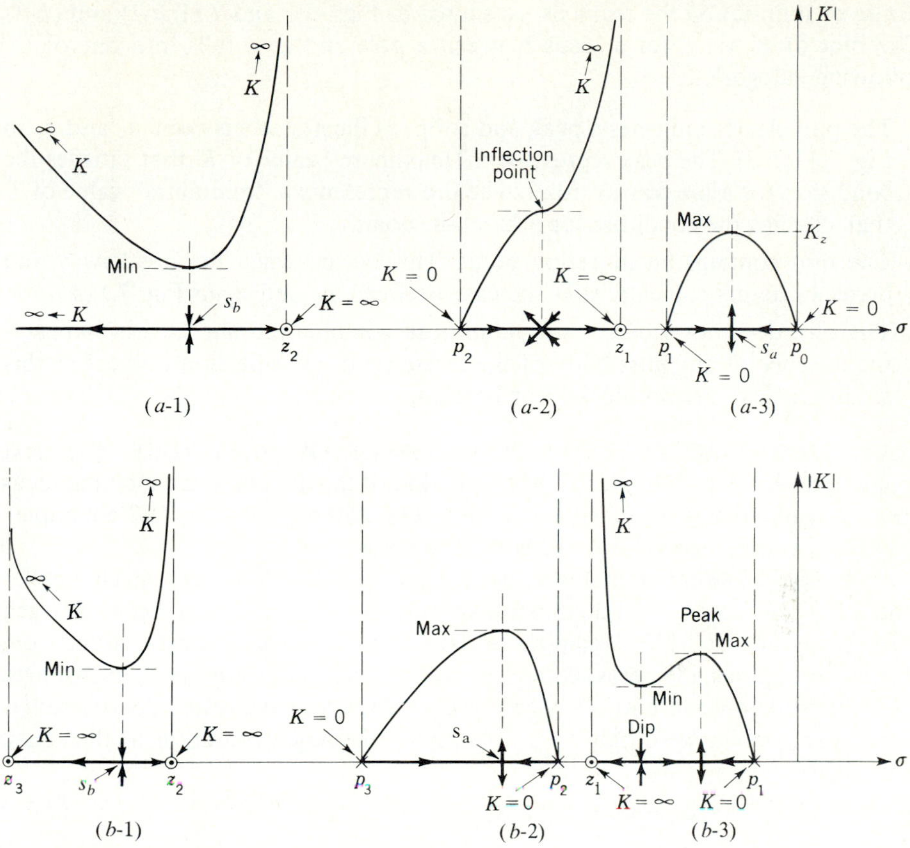

FIGURE 7.11
Plots of K and the corresponding real-axis locus for (a) Fig. 7.5a and (b) Fig. 7.5b.

Since the breakaway points s_a for $K > 0$ must lie between $s = 0$ and $s = -1$ in order to satisfy the angle condition, the value is

$$s_a = -0.4257 \tag{7.54}$$

The other point, $s_b = -1.5743$, is the break-in point on the root locus for $K < 0$. Substituting $s_a = -0.4257$ into Eq. (7.52) gives the value of K at the breakaway point for $K > 0$ as

$$\begin{aligned} -K &= (-0.426)^3 + (3)(-0.426)^2 + (2)(-0.426) \\ K &= 0.385 \end{aligned} \tag{7.55}$$

When the derivative of $W(s)$ is of higher order than 2, a digital-computer program can be used to calculate the roots of the numerator polynomial of $dW(s)/ds$; these roots locate the breakaway and break-in points. Note that it is possible to have both a breakaway and a break-in point between a pole and zero (finite or infinite) on the real axis, as shown in Figs. 7.5 and 7.11(a-2) and (b-3). The plot of K vs. σ for a locus between a pole and zero falls into one of the following categories:

1. The plot clearly indicates a peak and a dip, as illustrated between p_1 and z_1 in Fig. 7.11(b-3). The peak represents a "maximum" value of K that satisfies the condition for a breakaway point. The dip represents a "minimum" value of K that satisfies the condition for a break-in point.
2. The plot contains an inflection point. This occurs when the breakaway and break-in points coincide, as is the case between p_2 and z_1 in Fig. 7.11a.
3. The plot does not indicate a dip-and-peak combination and clearly indicates the absence of any possibility of the existence of an inflection point. For this situation there are no break-in or breakaway points.

RULE 7: COMPLEX POLE (OR ZERO): ANGLE OF DEPARTURE. The next geometrical shortcut is the rapid determination of the direction in which the locus leaves a complex pole or enters a complex zero. Although in Fig. 7.12 a complex pole is considered, the results also hold for a complex zero.

In Fig. 7.12a, an area about p_2 is chosen so that l_2 is very much smaller than l_0, l_1, l_3, and $(l)_1$. For illustrative purposes, this area has been enlarged many times in Fig. 7.12b. Under these conditions the angular contributions from all the other poles and zeros, except p_2, to a search point anywhere in this area are approximately constant. They can be considered to have values determined as if the search point were right at p_2. Applying the angle condition to this small area yields

$$\phi_0 + \phi_1 + \phi_2 + \phi_3 - \psi_1 = (1 + 2h)180° \tag{7.56}$$

or the departure angle is

$$\phi_{2D} = (1 + 2h)180° - (\phi_0 + \phi_1 + 90° - \psi_1)$$

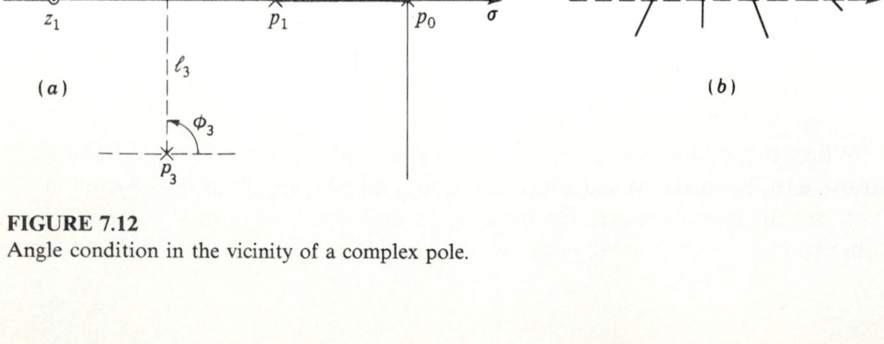

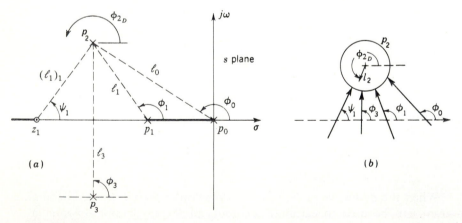

FIGURE 7.12
Angle condition in the vicinity of a complex pole.

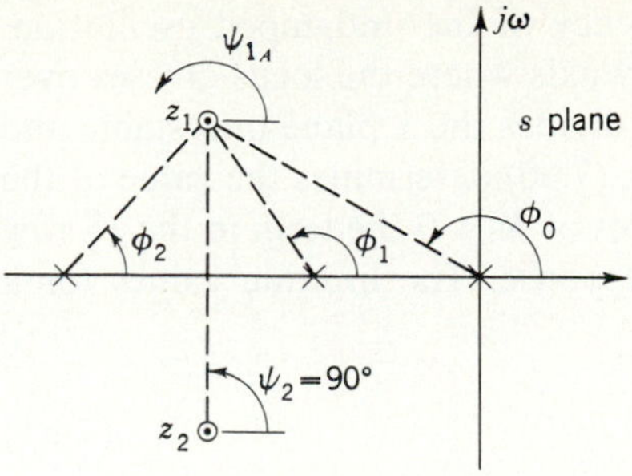

FIGURE 7.13
Angle condition in the vicinity of a complex zero.

In a similar manner the angle of approach to a complex zero can be determined. For an open-loop transfer function having the pole-zero arrangement shown in Fig. 7.13, the angle of approach ψ_1 to z_1 is given by

$$\psi_{1A} = (\phi_0 + \phi_1 + \phi_2 - 90°) - (1 + 2h)180° \tag{7.57}$$

In other words, the direction of the locus as it leaves a pole or approaches a zero can be determined by adding up, according to the angle condition, all the angles of all vectors from all the other poles and zeros to the pole or zero in question. Subtracting this sum from $(1 + 2h)180°$ gives the required direction.

RULE 8: IMAGINARY-AXIS CROSSING POINT. In cases where the locus crosses the imaginary axis into the right-half s plane, the crossover point can usually be determined by Routh's method or by similar means. For example, if the closed-loop characteristic equation $D_1D_2 + N_1N_2 = 0$ is of the form

$$s^3 + bs^2 + cs + Kd = 0$$

the Routhian array is

$$\begin{array}{c|ll} s^3 & 1 & c \\ s^2 & b & Kd \\ s^1 & (bc - Kd)/b & \\ s^0 & Kd & \end{array}$$

An undamped oscillation may exist if the s^{-1} row in the array equals zero. For this condition the auxiliary equation obtained from the s^2 row is

$$bs^2 + Kd = 0 \tag{7.58}$$

and its roots are

$$s_{1,2} = \pm j\sqrt{\frac{Kd}{b}} = \pm j\omega_n \tag{7.59}$$

The loop sensitivity term K is determined by setting the s^1 row to zero:

$$K = \frac{bc}{d} \tag{7.60}$$

For $K > 0$, Eq. (7.59) gives the natural frequency of the undamped oscillation. This corresponds to the point on the imaginary axis where the locus crosses over into the right-half s plane. The imaginary axis divides the s plane into stable and unstable regions. Also, the value of K from Eq. (7.60) determines the value of the loop sensitivity at the crossover point. For values of $K < 0$ the term in the s^0 row is negative, thus characterizing an unstable response. The limiting values for a stable response are therefore

$$0 < K < \frac{bc}{d} \tag{7.61}$$

In like manner, the crossover point can be determined for higher-order characteristic equations. For these higher-order systems care must be exercised in *analyzing all terms in the first column that contain the term* K in order to obtain the correct range of values of gain for stability.

RULE 9: NONINTERSECTION OR INTERSECTION OF ROOT-LOCUS BRANCHES.[5] By utilizing the theory of complex variables, the following properties are evolved:

1. A value of s which satisfies the angle condition of Eq. (7.35) is a point on the root locus. If $dW(s)/ds \neq 0$ at this point, there is one and only one branch of the root locus through the point.
2. If the first $y - 1$ derivatives of $W(s)$ vanish at a given point on the root locus, there are y branches approaching and y branches leaving this point. Thus, it can be said that there are root-locus intersections at this point. The angle between two adjacent *approaching* branches is given by

$$\lambda_y = \pm \frac{360°}{y} \tag{7.62}$$

Also, the angle between a branch *leaving* and an adjacent branch that is *approaching* the same point is given by

$$\theta_y = \pm \frac{180°}{y} \tag{7.63}$$

Figure 7.14 illustrates these angles.

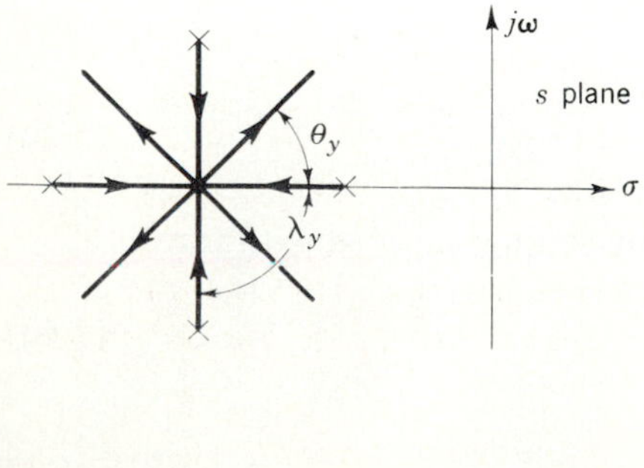

FIGURE 7.14
Root locus for

$$G(s)H(s) = \frac{K}{(s+2)(s+4)(s^2+6s+10)}$$

RULE 10: CONSERVATION OF THE SUM OF THE SYSTEM ROOTS.[6] The technique described by this rule aids in the determination of the general shape of the root locus. Consider the general open-loop transfer function in the form

$$G(s)H(s) = \frac{K\prod_{h=1}^{w}(s - z_h)}{s^m \prod_{c=1}^{u}(s - p_c)} \tag{7.64}$$

By recalling that for physical systems $w \le n = u + m$, the denominator of $C(s)/R(s)$ can be written

$$B(s) = 1 + G(s)H(s) = \frac{\prod_{j=1}^{n}(s - r_j)}{s^m \prod_{c=1}^{u}(s - p_c)} \tag{7.65}$$

where r_j are the roots described by the root locus.

Substituting from Eq. (7.64) into Eq. (7.65) and equating numerators on each side of the resulting equation yields

$$s^m \prod_{c=1}^{u}(s - p_c) + K\prod_{h=1}^{w}(s - z_h) = \prod_{j=1}^{n}(s - r_j) \tag{7.66}$$

Expanding both sides of this equation gives

$$\left(s^n - \sum_{c=1}^{u} p_c s^{n-1} + \cdots\right) + K\left(s^w - \sum_{h=1}^{w} z_h s^{w-1} + \cdots\right) = s^n - \sum_{j=1}^{n} r_j s^{n-1} + \cdots \tag{7.67}$$

For those open-loop transfer functions which satisfy the condition $w \le n - 2$, the following is obtained by equating the coefficients of s^{n-1} of Eq. (7.67):

$$\sum_{c=1}^{u} p_c = \sum_{j=1}^{n} r_j$$

Since m open-loop poles have values of zero, this equation can also be written as

$$\sum_{j=1}^{n} p_j = \sum_{j=1}^{n} r_j \tag{7.68}$$

where p_j now represents all the open-loop poles, including those at the origin, and r_j are the roots of the characteristic equation. This equation reveals that as the system gain is varied from zero to infinity, the sum of the system roots is constant. In other words, the sum of the system roots is conserved and is independent of K. When a system has several root-locus branches which go to infinity (as $K \to \infty$), the directions of the branches are such that the sum of the roots is constant. A branch going to the right therefore requires that there will be a branch going to the left. The root locus of Fig. 7.2 satisfies the conservancy law for the root locus. The sum of the roots is a constant for all values of K.

For a unity-feedback system, Rao[7] has shown that the closed-loop pole and zero locations of

$$\frac{C(s)}{R(s)} = \frac{K(s-z_1)\cdots(s-z_w)}{(s-p_1)\cdots(s-p_n)} \tag{7.69}$$

satisfy the relation

$$\sum_{i=1}^{w}(-z_i)^{-q} = \sum_{j=1}^{n}(-p_j)^{-q} \qquad q = 1, 2, \ldots, m-1 \tag{7.70}$$

for $m > 1$, that is, for a Type 2 (or higher) system. This serves as a check on the accuracy of the root determination.

RULE 11: DETERMINATION OF ROOTS ON THE ROOT LOCUS. After the root locus has been plotted, the specifications for system performance are used to determine the dominant roots. The root locus must be analyzed to determine the *dominant branch*. This is the branch that yields the root which has the largest influence on the time response. When this branch yields complex dominant roots, the time response is oscillatory and the figures of merit (see Sec. 3.10) are the peak overshoot M_p, the time t_p at which the peak overshoot occurs, and the settling time t_s. An additional figure of merit, the gain K_m, significantly affects the steady-state error (see Chap. 6). All of these quantities can be used to select the *dominant roots*. Thus, the designer may use the damping ratio ζ, the undamped natural frequency ω_n, the damped natural frequency ω_d, the damping coefficient σ, or the gain K_m to select the dominant roots. When the dominant roots are selected, the required loop sensitivity can be determined by applying the magnitude condition, as shown in Eq. (7.33). The remaining roots on each of the other branches can be determined by any of the following methods:

Method 1. Determine the point on each branch of the locus that satisfies the same value of loop sensitivity as for the dominant roots.

Method 2. If all except one real or a complex pair of roots are known, either of the following procedures can be used.

PROCEDURE 1. Divide the characteristic equation by the factors representing the known roots. The remainder gives the remaining roots.

PROCEDURE 2. Equation (7.68), known as *Grant's rule*, can be used to find some of the roots. A necessary condition is that the denominator of $G(s)H(s)$ be at least of degree 2 higher than the numerator. If all the roots except one real root are known, application of Eq. (7.68) yields directly the value of the real root. However, for complex roots of the form $r = \sigma \pm j\omega_d$ it yields only the value of the real component σ.

Digital-computer root-locus programs[1] are available that yield the loop sensitivity and the roots of the characteristic equation of the system. When the damping ratio is specified for the dominant roots, these roots determine the value of K and all the remaining roots.

7.9 EXAMPLES

Example 1. Given:

$$G(s) = \frac{K_1}{s(s^2/2600 + s/26 + 1)} \qquad \text{and} \qquad H(s) = \frac{1}{0.04s + 1}$$

Rearranging gives

$$G(s) = \frac{2600K_1}{s(s^2 + 100s + 2600)} = \frac{N_1}{D_1} \qquad \text{and} \qquad H(s) = \frac{25}{s + 25} = \frac{N_2}{D_2}$$

Thus

$$G(s)H(s) = \frac{65{,}000K_1}{s(s + 25)(s^2 + 100s + 2600)} = \frac{K}{s^4 + 125s^3 + 5100s^2 + 6500s}$$

where $K = 65{,}000K_1$.

Find $C(s)/R(s)$ with $\zeta = 0.5$ for the dominant roots (roots closest to the imaginary axis).

1. The poles of $G(s)H(s)$ are plotted on the s plane in Fig. 7.15; the values of these poles are $s = 0, -25, -50 + j10, -50 - j10$. The system is completely unstable for $K < 0$. Therefore this example is solved only for the condition $K > 0$.
2. There are four branches of the root locus.
3. The locus exists on the real axis between 0 and -25.
4. The angles of the asymptotes are

$$\gamma = \frac{(1 + 2h)180°}{4} = \pm 45°, \pm 135°$$

5. The real-axis intercept of the asymptotes is

$$\sigma_o = \frac{0 - 25 - 50 - 50}{4} = -31.25$$

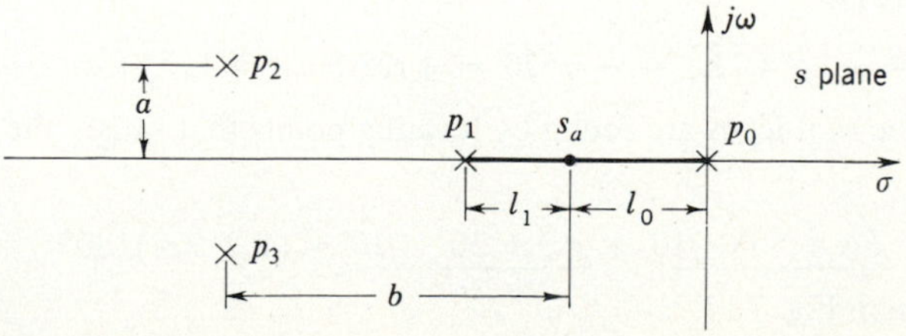

FIGURE 7.15
Location of the breakaway point.

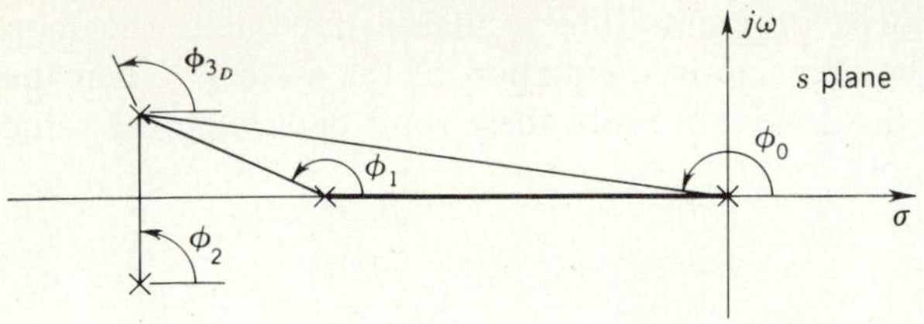

FIGURE 7.16
Determination of the departure angle.

6. The breakaway point s_a (see Fig. 7.15) on the real axis between 0 and -25 is found by solving $dW(s)ds = 0$ [see Eq. (7.40)]:

$$-K = s^4 + 125s^3 + 5100s^2 + 6500s$$
$$\frac{d(-K)}{ds} = 4s^3 + 375s^2 + 10200s + 65000 = 0$$
$$s_a = 9.15$$

7. The angle of departure ϕ_{3_D} (see Fig. 7.16) from the pole $-50 + j10$ is obtained from

$$\phi_0 + \phi_1 + \phi_2 + \phi_{3_D} = (1 + 2h)180°$$
$$168.7° + 158.2° + 90° + \phi_{3_D} = (1 + 2h)180°$$
$$\phi_{3_D} = 123.1°$$

Similarly, the angle of departure from the pole $-50 - j10$ is $-123.1°$.

8. The imaginary-axis intercepts are obtained from

$$\frac{C(s)}{R(s)} = \frac{2600K_1(s + 25)}{s^4 + 125s^3 + 5100s^2 + 65{,}000s + 65{,}000K_1} \tag{7.71}$$

The Routhian array for the denominator of $C(s)/R(s)$, which is the characteristic polynomial, is

s^4	1	5100	$65{,}000K_1$
s^3	1	520 (after division by 125)	
s^2	1	$14.2K_1$ (after division by 4580)	
s^1	$520 - 14.2K_1$		
s^0	$14.2K_1$		

Pure imaginary roots exist when the s^1 row is zero. This occurs when $K_1 = 520/14.2 = 36.6$. The auxiliary equation is

$$s^2 + 14.2K_1 = 0$$

and the imaginary roots are

$$s = \pm j\sqrt{14.2K_1} = \pm j\sqrt{520} = \pm j22.8$$

9. Additional points on the root locus are found by locating points that satisfy the angle condition

$$\angle s + \angle s + 25 + \angle s + 50 - j10 + \angle s + 50 + j10 = (1 + 2m)180°$$

The root locus is shown in Fig. 7.17.

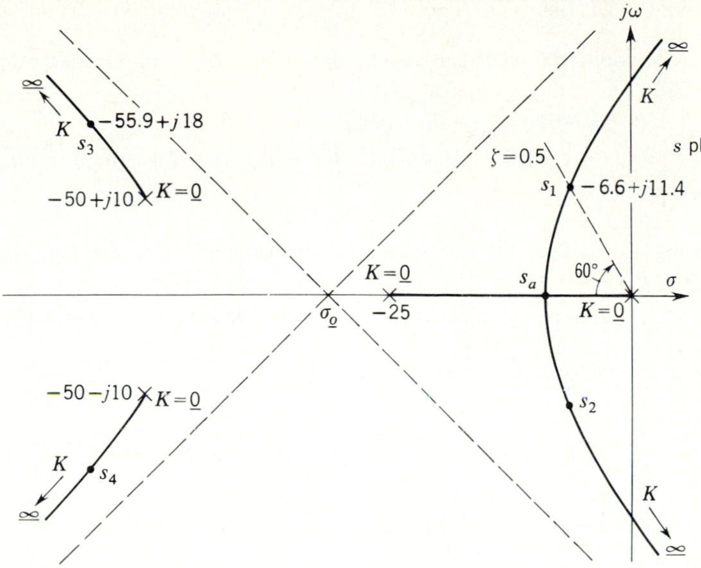

FIGURE 7.17
Root locus for

$$G(s)H(s) = \frac{65{,}000K_1}{s(s+25)(s^2+100s+2600)}$$

10. The radial line for $\zeta = 0.5$ is drawn on the graph of Fig. 7.17 at the angle (see Fig. 4.3 for definition of η)

$$\eta = \cos^{-1} 0.5 = 60°$$

The dominant roots obtained from the graph are

$$s_{1,2} = -6.6 \pm j11.4$$

11. The gain is obtained from the expression

$$K = 65{,}000K_1 = |s| \cdot |s+25| \cdot |s+50-j10| \cdot |s+50+j10|$$

by using a calculator. For $s = -6.6 + j11.4$,

$$K = 65{,}000K_1 = 598{,}800$$

$$K_1 = 9.25$$

12. The other roots are evaluated to satisfy the magnitude condition $K = 598{,}800$. The other roots of the characteristic equation are

$$s_{3,4} = -55.9 \pm j18.0$$

13. *Alternative method*. With $K_1 = 9.25$ the characteristic equation is

$$s^4 + 125s^3 + 5100s^2 + 65{,}000s + 598{,}800 = 0$$

The quadratic factor representing the dominant roots is

$$(s + 6.6 - j11.4)(s + 6.6 + j11.4) = s^2 + 13.2s + 173.5$$

Dividing the characteristic equation by this quadratic factor and ignoring the remainder leaves the quadratic representing the other roots as

$$s^2 + 112s + 3450$$

The other roots obtained from this factor are

$$s_{3,4} = -56 \pm j18$$

The additional roots can also be determined by using Grant's rule. From Eq. (7.68):

$$0 - 25 + (-50 + j10) + (-50 - j10)$$
$$= (-6.6 + j11.4) + (-6.6 - j11.4) + (\sigma + j\omega_d) + (\sigma - j\omega_d)$$

This gives

$$\sigma = -55.9$$

By using this value, the roots can be determined from the root locus as $-55.9 \pm j18.0$.

14. The control ratio, using values of the roots obtained in steps 10 and 12, is

$$\frac{C(s)}{R(s)} = \frac{N_1 D_2}{\text{factors determined from root locus}}$$

$$= \frac{24{,}040(s + 25)}{(s + 6.6 + j11.4)(s + 6.6 - j11.4)(s + 55.9 + j18)(s + 55.9 - j18)}$$

$$= \frac{24{,}040(s + 25)}{(s^2 + 13.2s + 173.5)(s^2 + 111.8s + 3450)}$$

15. The response $c(t)$ for a unit step input is found from[3]

$$C(s) = \frac{24{,}040(s + 25)}{s(s^2 + 13.2s + 173.5)(s^2 + 111.8s + 3450)}$$

$$= \frac{A_0}{s} + \frac{A_1}{s + 6.6 - j11.4} + \frac{A_2}{s + 6.6 + j11.4}$$

$$+ \frac{A_3}{s + 55.9 - j18} + \frac{A_4}{s + 55.9 + j18}$$

The constants are

$$A_0 = 1.0 \qquad A_1 = 0.604\angle -201.7^\circ \qquad A_3 = 0.14\angle -63.9^\circ$$

Note that A_0 must be exactly 1 since $G(s)$ is Type 1 and the gain of $H(s)$ is unity. It can also be obtained from Eq. (7.71) for $C(s)/R(s)$ in step 8. Inserting $R(s) = 1/s$ and finding the final value gives $c(t)_{ss} = 1.0$. The response $c(t)$ is

$$c(t) = 1 + 1.21e^{-6.6t}\sin(11.4t - 111.7^\circ) + 0.28e^{-55.9t}\sin(18t + 26.1^\circ) \tag{7.72}$$

A plot of $c(t)$ is shown in Fig. 7.25.

Example 2. Many aircraft flight control systems use an adjustable gain in the feedback path. Also, the forward transfer function may be non-minimum phase and have a negative gain. This is illustrated by the control system shown in Fig. 7.18.

$$G(s) = \frac{-2(s + 6)(s - 6)}{s(s + 3)(s + 4 - j4)(s + 4 + j4)} \qquad H(s) = K_h > 0$$

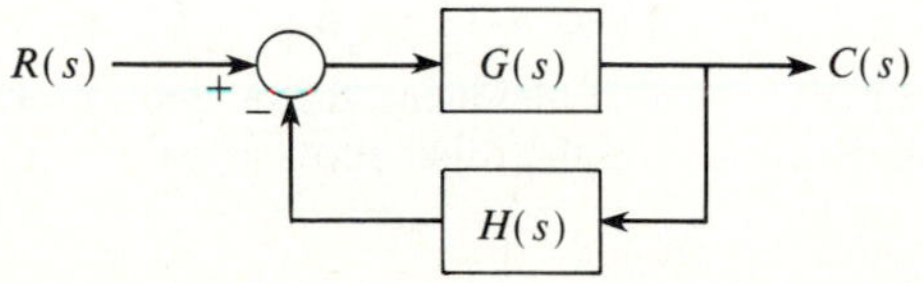

FIGURE 7.18
Nonunity-feedback control system.

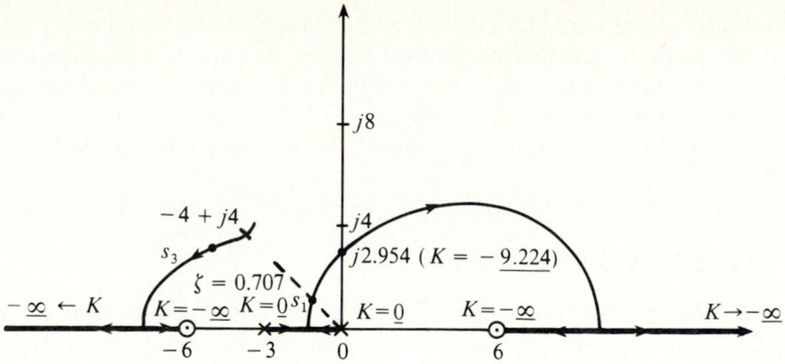

FIGURE 7.19
Root locus for Eq. (7.73) for negative values of K.

Thus the open-loop transfer function is

$$G(s)H(s) = \frac{K(s+6)(s-6)}{s(s+3)(s+4-j4)(s+4+j4)} \tag{7.73}$$

where $K = -2K_h$. The root-locus characteristics are summarized as follows:

1. $G(s)H(s)$ has zeros located at $s = -6, +6$. The zero at $s = 6$ means that this transfer function is non-minimum phase. The poles are located at $s = 0, -3, -4 + j4, -4 - j4$.
2. There are four branches of the root locus.
3. For positive values of K, there is a branch of the root locus located on the positive real axis from $s = 0$ to $s = 6$. Thus there is a positive real root and the system is unstable. For negative values of K (the values of K_h are positive), the root locus must satisfy the 360° angle condition given in Eq. (7.27). The root locus is shown in Fig. 7.19. This plot is readily obtained using a computer program (see App. B).
4. The closed-loop transfer function is

$$\frac{C(s)}{R(s)} = \frac{-2(s+6)(s-6)}{s^4 + 11s^3 + (56+K)s^2 + 96s - 36K} \tag{7.74}$$

   The Routhian array for the denominator polynomial shows that the imaginary axis crossing occurs at $s = \pm j2.954$ for $K = -9.224$. Thus the closed-loop system is stable for $-9.224 < K < 0$.
5. Using $\zeta = 0.707$ for selecting the dominant poles yields $s_{1,2} = -1.1806 \pm j1.1810$ and $s_{3,4} = -4.3194 \pm j3.4347$, with $K_h = 1.18$. The closed-loop transfer function is

$$\frac{C(s)}{R(s)} = \frac{-2(s+6)(s-6)}{(s+1.1806 \pm j1.1810)(s+4.3194 \pm j3.4347)} \tag{7.75}$$

6. The output response for a unit-step input is

$$c(t) = 0.84782 - 1.6978e^{-1.1806t}\sin(1.181t + 29.33°)$$
$$-0.20346e^{-4.3194t}\sin(3.4347t + 175.47°) \tag{7.76}$$

The figures of merit are $M_p = 0.8873$, $t_p = 2.89$ s, $t_s = 3.85$ s, and $c(\infty) = 0.8478$. It is important to note that the forward transfer function $G(s)$ is Type 1, but the steady-state value of the output is not equal to the input because of the nonunity feedback. Also note that the numerator of $C(s)/R(s)$ is the same as the numerator of $G(s)$; thus, it is not affected by the value selected for K.

7.10 FREQUENCY RESPONSE

Once the root locus has been determined and the system gain has been set for the desired performance, it is very easy to determine the steady-state frequency response.[3] The frequency response gives the ratio of phasor output to phase input for sinusoidal inputs over a band of frequencies. The plots of the magnitude M and the angle α of $\mathbf{C}(j\omega)/\mathbf{R}(j\omega)$ vs. the frequency ω define the frequency response of a control system. These curves are very useful in control-system design:

1. They enable a designer to minimize the effect of any undesirable noise within the system.
2. They present a qualitative picture of the system's transient response. The correlation between the transient and frequency responses is discussed in detail in Chap. 9.

For the problem of Sec. 7.9 the closed-loop control ratio is

$$\frac{C(s)}{R(s)} = \frac{24{,}040(s+25)}{(s+6.6-j11.4)(s+6.6+j11.4)(s+55.9-j18)(s+55.9+j18)} \tag{7.77}$$

To obtain the steady-state frequency response, let s assume values equal to $j\omega$. The control ratio as a function of frequency is

$$\frac{\mathbf{C}(j\omega)}{\mathbf{R}(j\omega)} = \frac{24{,}040(j\omega+25)}{(j\omega+6.6-j11.4)(j\omega+6.6+j11.4)(j\omega+55.9-j18)(j\omega+55.9+j18)} \tag{7.78}$$

For any frequency ω_1 each of the factors of Eq. (7.78) is a directed line segment and can be measured on the pole-zero plot, as shown in Fig. 7.20. When the magnitudes and angle for each term of Eq. (7.78) have been obtained, the value of $\mathbf{C}(j\omega)/\mathbf{R}(j\omega)$ can be determined. Note that the angles are measured as shown in Fig. 7.20, where clockwise angles are negative and counterclockwise angles are positive. Figure 7.21 shows a plot of the magnitude $|\mathbf{C}(j\omega)/\mathbf{R}(j\omega)|$ vs. ω obtained from the pole-zero plot of Fig. 7.20. The angle curve of $\mathbf{C}(j\omega)/\mathbf{R}(j\omega)$ can also be evaluated but is not shown here. The maximum value of the magnitude curve is labeled M_m, and the frequency at which this occurs is labeled ω_m. In Chap. 9 these quantities are related to the time response.

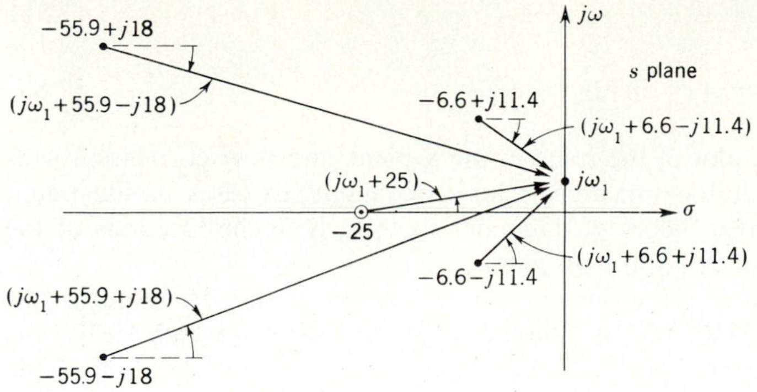

FIGURE 7.20
Frequency response from the plot in the s plane of the closed-loop poles and zeros.

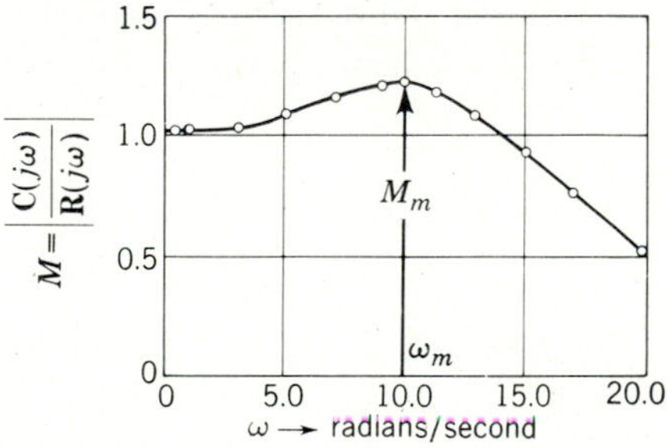

FIGURE 7.21
Closed-loop frequency response for the system of Fig. 7.20.

7.11 PERFORMANCE CHARACTERISTICS

As pointed out early in this chapter, the root-locus method incorporates the more desirable features of both the classical method and the steady-state sinusoidal phasor analysis. In the example of Sec. 7.9 a direct relationship is noted between the root locus and the time solution. This section is devoted to strengthening this relationship to enable the designer to synthesize and/or compensate a system.

General Introduction

In review, consider a simple second-order system whose control ratio is

$$\frac{C(s)}{R(s)} = \frac{K}{s^2 + 2\zeta\omega_n s + \omega_n^2} \tag{7.79}$$

and whose transient component of the response to a step input is

$$c(t)_t = C_1 e^{s_1 t} + C_2 e^{s_2 t} \tag{7.80}$$

where $$s_{1,2} = -\zeta\omega_n \pm j\omega_n\sqrt{1-\zeta^2} = \sigma \pm j\omega_d \qquad \zeta < 1 \tag{7.81}$$

Thus $$c(t)_t = Ae^{\sigma t}\sin(\omega_d t + \phi) \tag{7.82}$$

Consider now a plot of the roots in the s plane and their correlation with the transient solution in the time domain for a step input. Six cases are illustrated in Fig. 7.22 for different values of damping, showing both the locations of the roots and the corresponding transient plots.

The roots lie on the $\pm j\omega$ axis in the case of $\zeta = 0$ and the response has sustained oscillations (Fig. 7.22d). Those portions of the locus which yield roots in the right-half s plane result in unstable operation (Fig. 7.22f). Thus, the desirable roots are on that portion of the locus in the left-half s plane. (As shown in Chap. 9, the $\pm j\omega$ axis of the s plane corresponds to the $-1 + j0$ point of the Nyquist plot.)

Table 4.1 summarizes the information available from Fig. 7.22, i.e., the correlation between the location of the closed-loop poles and the corresponding transient component of the response. Thus the value of the root-locus method is that it is possible to determine all the forms of the transient component of the response that a control system may have.

Plot of Characteristic Roots for $0 < \zeta < 1$

The important desired roots lie in the region in which $0 < \zeta < 1$ (generally between 0.4 and 0.8). In Fig. 7.22a, the radius r from the origin to the root s_1 is

$$R = \sqrt{\omega_{d_1}^2 + \sigma_{1,2}^2} = \sqrt{\omega_n^2(1-\zeta^2) + \omega_n^2\zeta^2} = \omega_n \tag{7.83}$$

and $$\cos\eta = \left|\frac{-\sigma_{1,2}}{r}\right| = \frac{\zeta\omega_n}{\omega_n} = \zeta \tag{7.84}$$

or $$\eta = \cos^{-1}\zeta \tag{7.85}$$

From the above equations the constant-parameter loci are drawn in Fig. 7.23.

From Figs. 7.22 and 7.23 and Eqs. (7.83) to (7.85) the following conclusions are drawn pertaining to the s plane:

1. Horizontal lines represent lines of constant damped natural frequency ω_d. The closer these lines are to the real axis, the lower the value of ω_d. For roots lying on the real axis ($\omega_d = 0$) there is no oscillation.
2. Vertical lines represent lines of constant damping or constant rate of decay of the transient. The closer these lines (or the characteristic roots) are to the imaginary axis, the longer it takes for the transient response to die out.

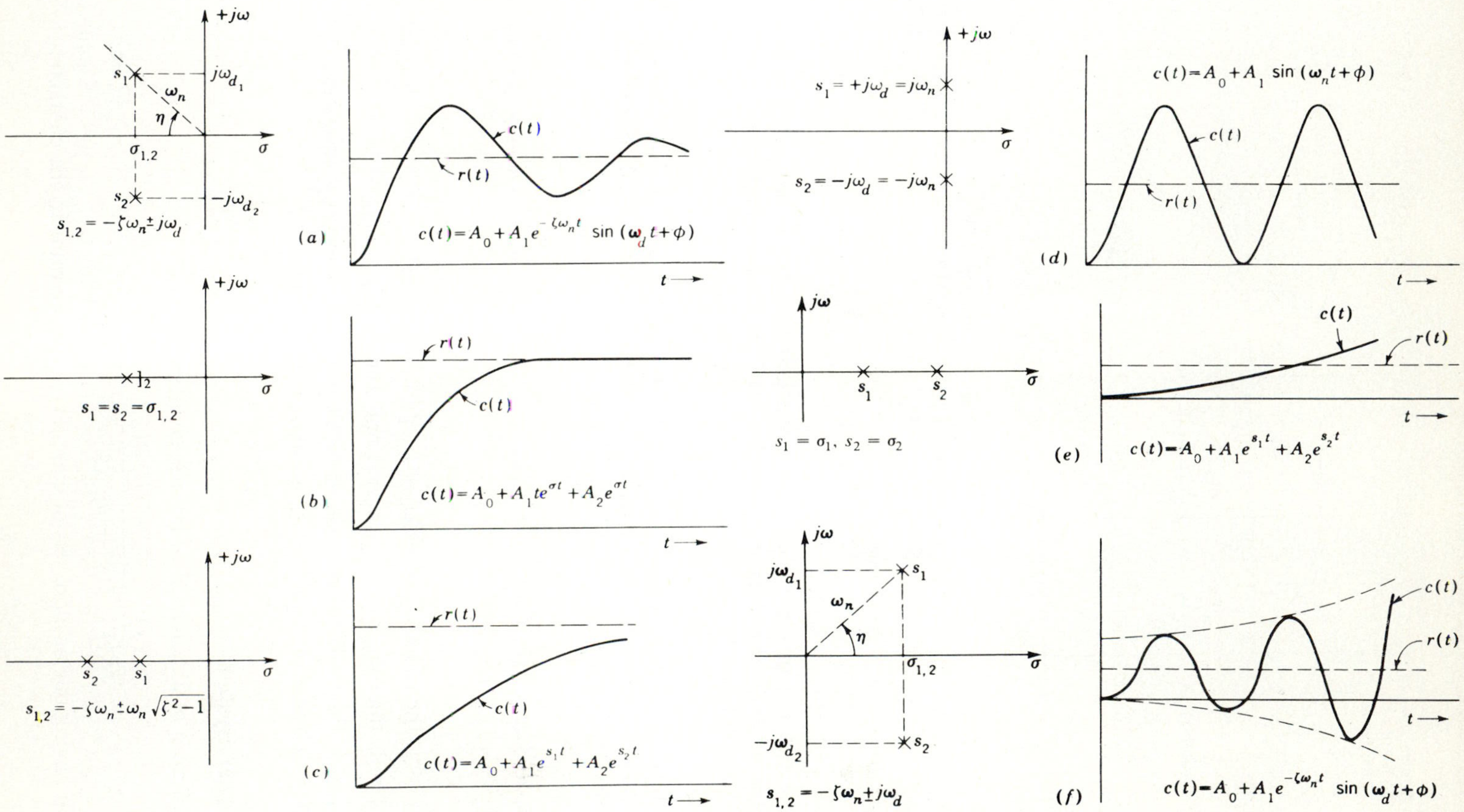

FIGURE 7.22
Plot of roots of a second-order characteristic equation in the s plane and their correlation to the transient solution in the time domain for a step input; (a) underdamped, stable, $0 < \zeta < 1$; (b) critically damped, stable, $\zeta = 1$; (c) real roots, overdamped, stable, $\zeta > 1$; (d) undamped, sustained oscillations, $\zeta = 0$; (e) real roots, unstable, $\zeta < -1$, (f) underdamped, unstable, $0 > \zeta > -1$.

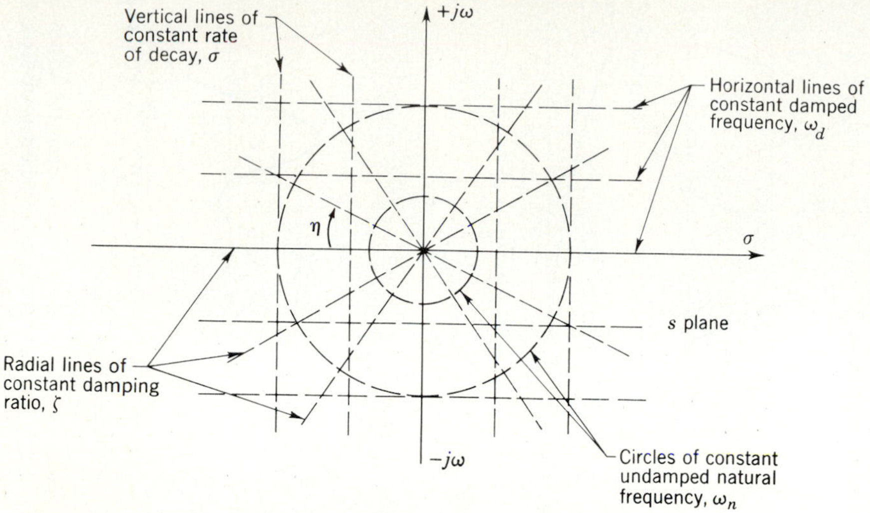

FIGURE 7.23
Constant-parameter curves on the *s* plane.

3. Circles about the origin are circles of constant undamped natural angular frequency ω_n. Since $\sigma^2 + \omega_d^2 = \omega_n^2$, the locus of the roots of constant ω_n is a circle in the s plane; that is, $|s_1| = |s_2| = \omega_n$. The smaller the circles, the lower ω_n.

4. Radial lines passing through the origin with the angle η are lines of constant damping ratio ζ. The angle η is measured clockwise from the negative real axis for positive ζ, as shown in Figs. 7.22*a* and 7.23.

Variation of Roots with ζ

Note in Fig. 7.24 the following:

For $\zeta > 1$, the roots $s_{1,2} = -\zeta\omega_n \pm \omega_n\sqrt{\zeta^2 - 1}$ are real.
For $\zeta = 1$, the roots $s_{1,2} = -\zeta\omega_n$ are real and equal.
For $\zeta < 1$, the roots $s_{1,2} = -\zeta\omega_n \pm j\omega_n\sqrt{1 - \zeta^2}$ are complex conjugates.
For $\zeta = 0$, the roots $s_{1,2} = \pm j\omega_n$ are imaginary.

Higher-Order Systems

The control ratio of a system of order n is given by

$$\frac{C(s)}{R(s)} = \frac{P(s)}{s^n + a_{n-1}s^{n-1} + \cdots + a_0} \tag{7.86}$$

In a system having one or more sets of complex-conjugate roots in the character-

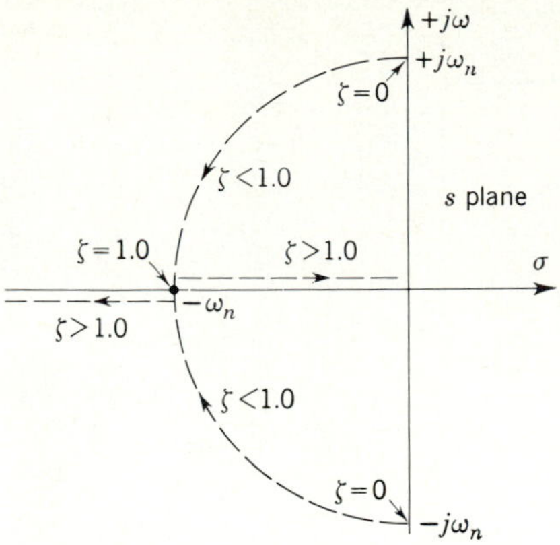

FIGURE 7.24
Variation of roots of a simple second-order system for a constant ω_n.

istic equation, each quadratic factor is of the form

$$s^2 + 2\zeta\omega_n s + \omega_n^2 \tag{7.87}$$

and the roots are

$$s_{1,2} = \sigma \pm j\omega_d \tag{7.88}$$

The relationships developed for ω_n and ζ earlier in this section apply equally well for each complex-conjugate pair of roots of an nth-order system. The distinction is that the dominant ζ and ω_n apply for that pair of complex-conjugate roots which lie closest to the imaginary axis. These values of ζ and ω_n are dominant because the corresponding transient term has the longest settling time and the largest magnitude. Thus the dominant values of ζ and ω_n are selected for the desired response. It must be remembered that in setting the values of the dominant ζ and ω_n the other roots are automatically set. Depending on the location of the other roots, they modify the solution obtained from the dominant roots.

In the example of Sec. 7.9 the response $c(t)$ with a step input given by Eq. (7.74) is

$$c(t) = 1 + 1.21e^{-6.6t}\sin(11.4t - 111.7°) + 0.28e^{-55.9t}\sin(18t + 26.1°) \tag{7.89}$$

Note that the transient term due to the roots of $s = -55.9 \pm j18$ dies out in approximately one-tenth the time of the transient term due to the dominant roots $s = -6.6 \pm j11.4$. Therefore, the solution can be approximated by the simplified equation

$$c(t) \approx 1 + 1.21e^{-6.6t}\sin(11.4t - 11.7°) \tag{7.90}$$

This expression is valid except for a short initial period of time while the other transient term dies out. Equation (7.90) suffices if an approximate solution is desired, i.e., for determining M_p, t_p, and T_s, since the neglected term does not appreciably affect these three quantities. In general, the designer can judge from the location of the roots which ones may be neglected. If an exact solution is desired, all roots must be considered. Plots of $c(t)$ corresponding to Eqs. (7.89)

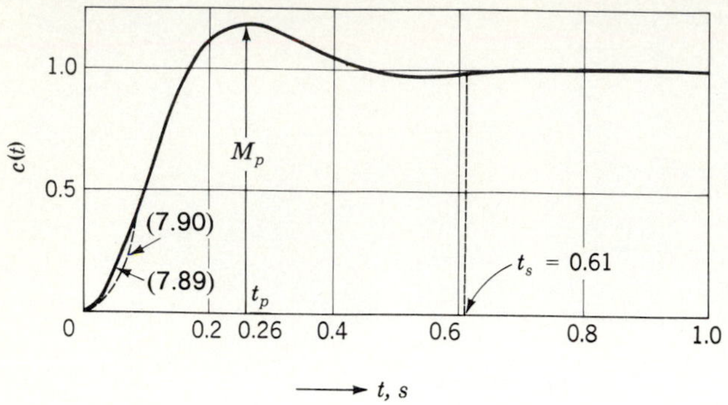

FIGURE 7.25
Plot of $c(t)$ vs. t for Eqs. (7.89) and (7.90).

and (7.90) are shown in Fig. 7.25. Since there is little difference between the two curves, the system can be approximated as a second-order system. The significant characteristics of the time response are $M_p = 1.19$, $t_p = 0.26$ s, $t_s = 0.61$ s, and $\omega_d \approx 11.4$ rad/s.

The transient response of any complex system, with the effect of all the roots taken into account, often can be considered to be the result of an *equivalent* second-order system. On this basis, it is possible to define effective (or equivalent) values of ζ and ω_n. Realize that in order to alter either effective quantity, the location of one or more of the roots must be altered.

7.12 TRANSPORT LAG[8]

Some elements in control systems are characterized by dead time or transport lag. This appears as a dead interval for which the output is delayed in response to an input. Figure 7.26a shows the block diagram of a transport-lag element. In Fig.

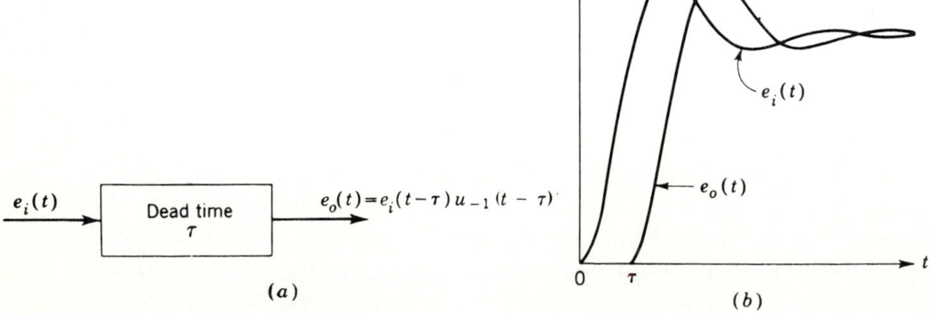

FIGURE 7.26
(a) Transport lag τ; (b) time characteristic.

7.26*b* a typical resultant output $e_o(t) = e_i(t - \tau)u_{-1}(t - \tau)$ is shown, which has the same form as the input but is delayed by a time interval τ. Dead time is a nonlinear characteristic which can be represented precisely in terms of the Laplace transform (see Sec. 4.4) as a transcendental function. Thus

$$E_o(s) = e^{-\tau s}E_i(s) \tag{7.91}$$

The transfer function of the transport-lag element, where $s = \sigma + j\omega$, is

$$G_\tau(s) = \frac{E_o(s)}{E_i(s)} = e^{-\tau s} = e^{-\sigma\tau}\angle{-\omega\tau} \tag{7.92}$$

It has a negative angle which is directly proportional to the frequency.

The root locus for a system containing a transport lag is illustrated by the following example:[8]

$$G(s)H(s) = \frac{Ke^{-\tau s}}{s(s - p_1)} = \frac{Ke^{-\sigma\tau}e^{-j\omega\tau}}{s(s - p_1)} \tag{7.93}$$

For $K > 0$ the angle condition is

$$\angle s + \angle s - p_1 + \angle \omega\tau = (1 + 2h)180° \tag{7.94}$$

The root locus for the system with transport lag is drawn in Fig. 7.27. The magnitude condition used to calibrate the root locus is

$$K = |s| \cdot |s - p_1|e^{\sigma\tau} \tag{7.95}$$

The same system without transport lag has only two branches (see Fig. 7.2) and is stable for all $K > 0$. The system with transport lag has an infinite number of branches and the asymptotes are all parallel to the real axis. The imaginary axis crossings of the asymptotes are

$$\omega = \frac{(1 + 2h)\pi}{\gamma} \tag{7.96}$$

where $h = 0, \pm 1, \pm 2, \ldots$. The values of K at the points where the branches cross the imaginary axis are

$$K = \omega\sqrt{p_1^2 + \omega^2} \tag{7.97}$$

The two branches closest to the origin thus have the largest influence on system stability and are therefore the principal branches. The maximum value of loop sensitivity K given by Eq. (7.97) is therefore determined by the frequency ω_1 (shown in Fig. 7.27) which satisfies the angle condition $\omega_1\tau = \tan^{-1}|p_1|/\omega_1$.

7.13 SYNTHESIS

The root-locus method lends itself very readily to the synthesis problem because of the direct relationship between the frequency and the time domains to the s domain. The desired response can be achieved by keeping in mind the following

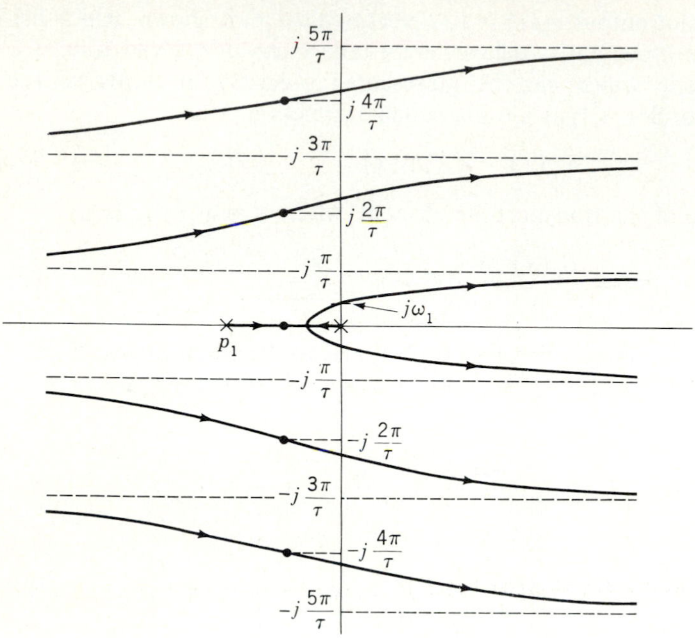

FIGURE 7.27
Root locus for

$$G(s)H(s) = \frac{Ke^{-\tau s}}{s(s - p_1)}$$

five points:

1. First plot the root locus; then the locations of the dominant closed-loop poles must be specified, based on the desired transient response.
2. Guidelines for selecting the roots:
   a. Since the desired time response is often specified to have a peak value M_p between 1.0 and 1.4, the locus has at least one set of complex-conjugate poles. Often there is a pair of complex poles near the imaginary axis which dominate the time response of the system; if so, these poles are referred to as the *dominant-pole pair*.
   b. To set the system gain, any of the following items must be specified for the dominant-pole pair: the damping ratio ζ, the settling time T_s (for 2 percent: $T_s = 4/\zeta\omega_n$) the undamped natural frequency ω_n, or the damped natural frequency ω_d. As previously determined, each of these factors corresponds to either a line or a circle in the s domain.
   c. When the line or circle corresponding to the given information has been drawn, as stated in point 2*b*, its intersection with the locus determines the dominant-pole pair and fixes the value of the system gain. By applying the

magnitude condition to this point the value of the gain required can be determined. Setting the value of the gain to this value results in the transient-response terms, corresponding to the dominant-pole pair, having the desired characterizing form.

3. When the gain corresponding to the dominant roots has been determined, the remaining roots can be found by applying the magnitude condition to each branch of the locus. Once the value of the gain on the dominant branch is fixed, thus fixing the dominant poles, the locations of all remaining roots on the other branches are also fixed.
4. If the root locus does not yield the desired response, the locus must be altered by compensation to achieve the desired results. The subject of compensation in the s domain is discussed in later chapters.

7.14 SUMMARY OF ROOT-LOCUS CONSTRUCTION RULES FOR NEGATIVE FEEDBACK

RULE 1. The number of branches of the root locus is equal to the number of poles of the open-loop transfer function.

RULE 2. For *positive* values of K, the root locus exists on those portions of the real axis for which the total number of real poles and zeros to the right is an odd number. For *negative* values of K, the root locus exists on those portions of the real axis for which the total number of real poles and zeros to the right is an even number (including zero).

RULE 3. The root locus starts ($K = 0$) at the open-loop poles and terminates ($K = \pm\infty$) at the open-loop zeros or at infinity.

RULE 4. The angles of the asymptotes of the root locus that end at infinity are determined by

$$\gamma = \frac{(1 + 2h)180^\circ \text{ for } K > 0 \qquad \text{or} \qquad h360^\circ \text{ for } K < 0}{[\text{number of poles of } G(s)H(s)] - [\text{number of zeros of } G(s)H(s)]}$$

RULE 5. The real-axis intercept of the asymptotes is

$$\sigma_o = \frac{\sum_{c=1}^{n} \operatorname{Re}(p_c) - \sum_{h=1}^{w} \operatorname{Re}(z_h)}{n - w}$$

RULE 6. The breakaway point for the locus between two poles on the real axis (or the break-in point for the locus between two zeros on the real axis) can be determined by taking the derivative of the loop sensitivity K with respect to s. Equate this derivative to zero and find the roots of the resulting equation. The root that occurs between the poles (or the zeros) is the breakaway (or break-in) point.

RULE 7. For $K > 0$ the angle of departure from a complex pole is equal to 180° minus the sum of the angles from the other poles plus the sum of the angles from the zeros. Any of these angles may be positive or negative. For $K < 0$ the departure angle is 180° from that obtained for $K > 0$.

For $K > 0$ the angle of approach to a complex zero is equal to the sum of the angles from the poles minus the sum of the angles from the other zeros minus 180°. For $K < 0$ the approach angle is 180° from that obtained from $K > 0$.

RULE 8. The imaginary-axis crossing of the root locus can be determined by setting up the Routhian array from the closed-loop characteristic equation. Equate the s^1 row to zero and form the auxiliary equation from the s^2 row. The roots of the auxiliary equation are the imaginary-axis crossover points.

RULE 9. The selection of the dominant roots of the characteristic equation is based on the specifications that give the required system performance; i.e., it is possible to evaluate σ, ω_d, and ζ from Eqs. (3.60), (3.61), and (3.64), which in turn determine the location of the desired dominant roots. The loop sensitivity for these roots is determined by means of the magnitude condition. The remaining roots are then determined to satisfy the same magnitude condition.

RULE 10. For those open-loop transfer functions for which $w \leq n - 2$, Grant's rule states that the sum of the closed-loop roots is equal to the sum of the open-loop poles. Thus, once the dominant roots have been located, Grant's rule

$$\sum_{j=1}^{n} p_j = \sum_{j=1}^{n} r_j$$

can be used to find one real or two complex roots. Factoring known roots from the characteristic equation can simplify the work of finding the remaining roots.

A root-locus digital-computer program[3] will produce an accurate calibrated root locus. This considerably simplifies the work required for the system design. By specifying ζ for the dominant roots or K_m, a computer program[3] can determine all the roots of the characteristic equation.

7.15 SUMMARY

In this chapter the root-locus method is developed to solve graphically for the roots of the characteristic equation. This method can be extended to solving for the roots of any polynomial. Any polynomial can be rearranged and put into the mathematical form of a ratio of factored polynomials which is equal to plus or minus unity, i.e.,

$$\frac{N(s)}{D(s)} = \pm 1$$

Once this form has been obtained, the procedures given in this chapter can be utilized to locate the roots of the polynomial.

The root locus permits the analysis of the performance of feedback systems and provides a basis for selecting the gain in order to best meet the performance specifications. Since the closed-loop poles are obtained explicitly, the form of the time response is directly available. A computer program is available for obtaining $c(t)$ vs. t. Computer-aided-design (CAD) packages are widely available (see Ref. 3, 10, and App. B) to assist in the analysis and design of control systems. These CAD programs are readily available and simplify considerably the analysis and design process. They provide information including root-locus plots, roots of interest for a given gain, closed-loop transfer function, and time response. If the performance specifications cannot be met, the root locus can be analyzed to determine the appropriate compensation to yield the desired results. This is covered in Chap. 10.

REFERENCES

1. Truxal, J. G.: *Automatic Feedback Control System Synthesis*, McGraw-Hill, New York, 1955.
2. Evans, W. R.: *Control-System Dynamics*, McGraw-Hill, New York, 1954.
3. Larimer, S. J.: "An Interactive Computer-Aided Design Program for Digital and Continuous System Analysis and Synthesis (TOTAL)," M.S. thesis, GE/GGC/EE/78-2, School of Engineering, Air Force Institute of Technology, Wright-Patterson Air Force Base, Ohio, 1978; available from Defense Documentation Denter (DDC), Cameron Station, Alexandria, Va. 22314.
4. Lorens, C. S., and R. C. Titsworth: "Properties of Root Locus Asymptotes," letter in *IRE Trans. Autom. Control*, vol. AC-5, pp. 71–72, January 1960.
5. Wilts, C. H.: *Principles of Feedback Control*, Addison-Wesley, Reading, Mass., 1960.
6. Grant, A. J. "The Conservation of the Sum of the System Roots as Applied to the Root Locus Method," unpublished paper, North American Aviation, Inc., Apr. 10, 1953.
7. Rao, S. N. "A Relation between Closed Loop Pole and Zeros Locations," *Int. J. Control*, vol. 24, p. 147, 1976.
8. Chang, C. S.: "Analytical Method for Obtaining the Root Locus with Positive and Negative Gain," *IEEE Trans. Autom. Control*, vol. AC-10, pp. 92–94, 1965.
9. Yeh, V. C. M.: "The Study of Transients in Linear Feedback Systems by Conformal Mapping and Root-Locus Method," *Trans. ASME*, vol. 76, pp. 349–361, 1954.
10. Thompson, P. M.: *USER's Guide to Program CC, Version 3*, Systems Technology, Inc., Hawthorne, Calif., March 1985.
11. Yeung, K. S.: "A Remark on the Use of Remec's Method of Finding Breakaway Points," *IEEE Trans. Autom. Control*, vol. AC-26, pp. 940–941, 1981.

CHAPTER 8

FREQUENCY RESPONSE

8.1 INTRODUCTION

In conventional control-system analysis there are two basic methods for predicting and adjusting a system's performance without resorting to the solution of the system's differential equation. One of these, the root-locus method, is discussed in detail in Chap. 7. The frequency-response method, to which the Nyquist stability criterion may be applied, is developed in this chapter.[1,2] For the comprehensive study of a system by conventional methods it is necessary to use both methods of analysis. The principal advantage of the root-locus method is that the actual time response is easily obtained by means of the inverse Laplace transform because the precise root locations are known. However, it is sometimes necessary to have performance requirements in terms of the frequency response. Also, the noise which is always present in any system can result in poor overall performance. The frequency response of a system permits analysis with respect to both these items. The design of a passband for the system response may exclude the noise and therefore improve the system performance as long as the dynamic performance specifications are met. The frequency response is also useful in situations for which the transfer functions of some or all of the blocks in a block diagram are unknown. The frequency response can be determined experimentally for these situations. Then an approximate expression for the transfer function can be obtained from the graphical plot of the experimental data. The frequency-response method is also a very powerful method for analyzing and designing a robust MIMO system with uncertain plant parameters (see Chap. 21). Thus no particular representation can be judged superior to the rest. Each has its particular use and advantage in a particular situation. As the reader becomes fully acquainted with all methods, the potentialities of each method become more apparent and the more appropriate method is selected.

Design of control systems by modern control-theory techniques is often based upon achieving an optimum performance according to a specified performance index PI; for example, minimizing the integral of squared error (ISE): $\mathrm{PI} = \int_0^\infty e(t)^2\,dt$, where $e \equiv r - c$. Both frequency-response and root-locus methods are valuable complementary tools for many of the techniques of modern control theory.

In this chapter two graphical representations of transfer functions are presented, the logarithmic and the polar plots. These plots are used to develop Nyquist's stability criterion and closed-loop design procedures.

8.2 CORRELATION OF THE SINUSOIDAL AND TIME RESPONSES[2]

As pointed out earlier in the text, solving for $c(t)$ by the classical method is laborious and impractical for synthesis purposes, especially when the input is not a simple analytical function. The use of Laplace transform theory lessens the work involved and permits the engineer to synthesize and improve a system. The root-locus method of Chap. 7 illustrates this fact. The advantages of the graphical representations in the frequency domain of the transfer functions are developed in the following pages.

Once the frequency response of a system has been determined, the time response can be determined by inverting the corresponding Fourier transform. The behavior in the frequency domain for a given driving function $r(t)$ can be determined by the Fourier transform as

$$\mathbf{R}(j\omega) = \int_{-\infty}^{\infty} r(t)e^{-j\omega t}\,dt \tag{8.1}$$

For a given control system the frequency response of the controlled variable is

$$\mathbf{C}(j\omega) = \frac{\mathbf{G}(j\omega)}{1 + \mathbf{G}(j\omega)\mathbf{H}(j\omega)}\mathbf{R}(j\omega) \tag{8.2}$$

By use of the inverse Fourier transform the controlled variable as a function of time is

$$c(t) = \frac{1}{2\pi}\int_{-\infty}^{\infty} \mathbf{C}(j\omega)e^{j\omega t}\,d\omega \tag{8.3}$$

This approach is much used in practice. If the design engineer cannot evaluate Eq. (8.3) by reference to a table of definite integrals, this equation can be evaluated by numerical or graphical integration. This is necessary if $\mathbf{C}(j\omega)$ is available only as a curve and cannot be simply expressed in analytical form, as is often the case. The procedure is described in several books.[10] In addition, methods have been developed based on the Fourier transform and a step input signal, relating $\mathbf{C}(j\omega)$ qualitatively to the time solution without actually taking the inverse Fourier transform. These methods permit the engineer to make an approximate determination of the response of his system through the interpretation of graphical plots in the frequency domain. This makes the design and improvement of feedback systems possible with a minimum effort and is emphasized in this chapter.

Sections 4.10 and 7.10 show that the frequency response is a function of the pole-zero pattern in the s plane. It is therefore related to the time response of the system. Two features of the frequency response are the maximum value M_m and the resonant frequency ω_m. Section 9.3 describes the qualitative relationship between the time response and the values M_m and ω_m. Since the location of the poles can be determined from the root locus, there is a direct relationship between the root-locus and frequency-response methods.

8.3 FREQUENCY-RESPONSE CURVES

The plots in the frequency domain that have found great use in graphical analysis in the design of feedback control systems belong to two categories. The first category is the plot of the magnitude of the output-input ratio vs. frequency in rectangular coordinates, as illustrated in Secs. 4.10 and 7.10. In logarithmic coordinates these are known as *Bode plots*. Associated with this plot is a second plot of the corresponding phase angle vs. frequency. In the second category the output-input ratio may be plotted in polar coordinates with frequency as a parameter. For this category there are two types of polar plots, direct and inverse. Polar plots are generally used only for the open-loop response and are commonly referred to as *Nyquist plots*.[3] The plots can readily be obtained experimentally or by computer.[4] The availability of programmable calculators with library programs capable of handling complex arithmetic facilitates the computation of frequency-response functions. When a computer or programmable calculator is not available, the Bode plots are easily obtained by a graphical procedure. The other plots can then be obtained from the Bode plots. Computer-aided-design (CAD) packages are readily available to assist in the analysis and design of control systems (see App. B). These programs assist in obtaining plots and the designs developed in this text.

For a given sinusoidal input signal, the input and steady-state output are of the following forms:

$$r(t) = R \sin \omega t \tag{8.4}$$

$$c(t) = C \sin(\omega t + \alpha) \tag{8.5}$$

The closed-loop frequency response is given by

$$\frac{\mathbf{C}(j\omega)}{\mathbf{R}(j\omega)} = \frac{\mathbf{G}(j\omega)}{1 + \mathbf{G}(j\omega)\mathbf{H}(j\omega)} = M(\omega)\underline{/\alpha(\omega)} \tag{8.6}$$

For each value of frequency, Eq. (8.6) yields a phasor quantity whose magnitude is M and whose phase angle α is the angle between $\mathbf{C}(j\omega)$ and $\mathbf{R}(j\omega)$.

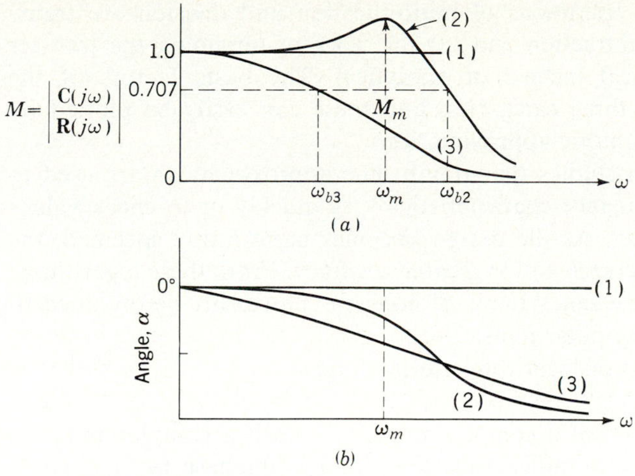

FIGURE 8.1
Frequency-response characteristics of $\mathbf{C}(j\omega)/\mathbf{R}(j\omega)$ in rectangular coordinates.

An ideal system may be defined as one where $\alpha = 0°$ and $\mathbf{R}(j\omega) = \mathbf{C}(j\omega)$ for $0 < \omega < \infty$. Curves 1 in Fig. 8.1 represent the characteristics of that ideal system. However, this definition implies an instantaneous transfer of energy from the input to the output. Such a transfer cannot be achieved in practice since any physical system has some energy dissipation and some energy-storage elements. Curves 2 and 3 in Fig. 8.1 represent the frequency responses of practical control systems. The passband, or bandwidth, of the frequency response is defined as the range of frequencies from 0 to the frequency ω_b, where $M = 0.707$ of the value at $\omega = 0$. The frequency ω_m is more easily obtained than ω_b. The values M_m and ω_m are often used as figures of merit.

In any system the input signal may contain spurious noise signals in addition to the true signal input, or there may be sources of noise within the closed-loop system. This noise is generally in a band of frequencies above the dominant frequency band of the true signal. Thus, in order to reproduce the true signal and attenuate the noise, feedback control systems are designed to have a definite passband. In certain cases the noise frequency may exist in the same frequency band as the true signal. When this occurs, the problem of eliminating it becomes complicated. Therefore, even if the ideal system were possible, it would not be desirable.

8.4 BODE PLOTS (LOGARITHMIC PLOTS)

The plotting of the frequency transfer function can be systematized and simplified by using logarithmic plots. The use of semilog paper eliminates the need to take logarithms of very many numbers and also expands the low-frequency range, which is of primary importance. The advantages of logarithmic plots are that (1) the mathematical operations of multiplication and division are transformed to addition and subtraction and (2) the work of obtaining the transfer function is largely graphical instead of analytical. The basic factors of the transfer function fall into three categories, and these can easily be plotted by means of straight-line asymptotic approximations.

In preliminary design studies the straight-line approximations are used to obtain approximate performance characteristics very quickly or to check values obtained from the computer. As the design becomes more firmly specified, the straight-line curves can be corrected for greater accuracy. From these logarithmic plots enough data in the frequency range of concern can readily be obtained to determine the corresponding polar plots.

Some basic definitions of logarithmic terms follow.

LOGARITHM. The logarithm of a complex number is itself a complex number. The abbreviation log is used to indicate the logarithm to the base 10:

$$\log |\mathbf{G}(j\omega)| e^{j\phi(\omega)} = \log |\mathbf{G}(j\omega)| + \log e^{j\phi(\omega)}$$

$$= \log |\mathbf{G}(j\omega)| + j0.434\phi(\omega) \tag{8.7}$$

The real part is equal to the logarithm of the magnitude, $\log |\mathbf{G}(j\omega)|$, and the imaginary part is proportional to the angle, $0.434\phi(\omega)$. In the rest of this book the factor 0.434 is omitted and only the angle $\phi(\omega)$ is used.

DECIBEL. In feedback-system work the unit commonly used for the logarithm of the magnitude is the *decibel*. Logarithms of transfer functions are used, where the transfer function is the ratio of output to input of a block. The variables are not necessarily in the same units; e.g., the output may be speed in radians per second, and the input may be voltage in volts.

LOG MAGNITUDE. The logarithm of the magnitude of a transfer function $\mathbf{G}(j\omega)$ expressed in decibels is

$$20 \log |\mathbf{G}(j\omega)| \qquad \text{dB}$$

This quantity is called the *log magnitude*, abbreviated Lm. Thus

$$\text{Lm}\, \mathbf{G}(j\omega) = 20 \log |\mathbf{G}(j\omega)| \qquad \text{dB} \tag{8.8}$$

Since the transfer function is a function of frequency, the log magnitude is also a function of frequency.

OCTAVE AND DECADE. Two units used to express frequency bands or frequency ratios are the octave and the decade. An octave is a frequency band from f_1 to f_2, where $f_2/f_1 = 2$. Thus, the frequency band from 1 to 2 Hz is 1 octave in width, and the frequency band from 17.4 to 34.8 Hz is also 1 octave in width. Note that 1 octave is not a fixed frequency bandwidth but depends on the frequency range being considered. The number of octaves in the frequency range

TABLE 8.1
Decibel values of some common numbers

Number	Decibels
0.01	−40
0.1	−20
0.5	−6
1.0	0
2.0	6
10.0	20
100.0	40
200.0	46

from f_1 to f_2 is

$$\frac{\log(f_2/f_1)}{\log 2} = 3.22 \log \frac{f_2}{f_1} \qquad \text{octaves} \tag{8.9}$$

There is an increase of 1 decade from f_1 to f_2 when $f_2/f_1 = 10$. The frequency band from 1 to 10 Hz or from 2.5 to 25 Hz is 1 decade in width. The number of decades from f_1 to f_2 is given by

$$\log \frac{f_2}{f_1} \qquad \text{decades} \tag{8.10}$$

The decibel values of some common numbers are given in Table 8.1. Note that the reciprocals of numbers differ only in sign. Thus, the decibel value of 2 is +6 dB and the decibel value of $\frac{1}{2}$ is −6 dB. Two properties are illustrated in Table 8.1.

Property 1. As a number doubles, the decibel value increases by 6 dB. The number 2.70 is twice as big as 1.35, and its decibel value is 6 dB more. The number 200 is twice as big as 100, and its decibel value is 6 dB greater

Property 2. As a number increases by a factor of 10, the decibel value increases by 20 dB. The number 100 is ten times as large as the number 10, and its decibel value is 20 dB more. The number 200 is one hundred times larger than the number 2, and its decibel value is 40 dB greater.

8.5 GENERAL FREQUENCY-TRANSFER-FUNCTION RELATIONSHIPS

The frequency transfer function can be written in generalized form as the ratio of polynomials:

$$\mathbf{G}(j\omega) = \frac{K_m(1 + j\omega T_1)(1 + j\omega T_2)^r \cdots}{(j\omega)^m(1 + j\omega T_a)\left[1 + (2\zeta/\omega_n)j\omega + (1/\omega_n^2)(j\omega)^2\right] \cdots} = K_m \mathbf{G}'(j\omega) \tag{8.11}$$

where K_m is the gain constant. The logarithm of the transfer function is a complex quantity; the real portion is proportional to the log of the magnitude, and the complex portion is proportional to the angle. Two separate equations are written, one for the log magnitude and one for the angle:

$$\text{Lm}\,\mathbf{G}(j\omega) = \text{Lm}\,K_m + \text{Lm}\,(1 + j\omega T_1) + r\,\text{Lm}\,(1 + j\omega T_2) + \cdots - m\,\text{Lm}\,j\omega$$
$$- \text{Lm}\,(1 + j\omega T_a) - \text{Lm}\left[1 + \frac{2\zeta}{\omega_n} j\omega + \frac{1}{\omega_n^2}(j\omega)^2\right] - \cdots \qquad (8.12)$$

$$\underline{/\mathbf{G}(j\omega)} = \underline{/K_m} + \underline{/1 + j\omega T_1} + r\underline{/1 + j\omega T_2} + \cdots - m\underline{/j\omega}$$
$$- \underline{/1 + j\omega T_a} - \underline{/1 + \frac{2\zeta}{\omega_n} j\omega + \frac{1}{\omega_n^2}(j\omega)^2} - \cdots \qquad (8.13)$$

The angle equation may be rewritten as

$$\underline{/G(j\omega)} = \underline{/K_m} + \tan^{-1}\omega T_1 + r\tan^{-1}\omega T_2 + \cdots - m90°$$
$$- \tan^{-1}\omega T_a - \tan^{-1}\frac{2\zeta\omega/\omega_n}{1 - \omega^2/\omega_n^2} - \cdots \qquad (8.14)$$

The gain K_m is a real number but may be positive or negative; therefore its angle is correspondingly 0° or 180°. Unless otherwise indicated, a positive value of gain is assumed in this book. Both the log magnitude and the angle given by these equations are functions of frequency. When the log magnitude and the angle are plotted as functions of the log of frequency, the resulting curves are referred to as the Bode plots or the *log magnitude diagram and the phase diagram*. Equations (8.12) and (8.13) show that the resultant curves are obtained by the addition and subtraction of the corresponding individual terms in the transfer-function equation. The two curves can be combined into a single curve of log magnitude vs. angle, with frequency ω as a parameter. This curve, called the Nichols or the *log magnitude–angle diagram*, corresponds to the Nyquist polar plot and is used for the quantitative design of the feedback system to meet specifications of required performance.

8.6 DRAWING THE BODE PLOTS

The properties of frequency-response plots are presented in this section, but the data for these plots usually are obtained from a computer program or a programmable calculator. The generalized form of the transfer function as given by Eq. (8.11) shows that the numerator and denominator have four basic types of factors:

$$K_m \qquad (8.15)$$

$$(j\omega)^{\pm m} \qquad (8.16)$$

$$(1 + j\omega T)^{\pm r} \qquad (8.17)$$

$$\left[1 + \frac{2\zeta}{\omega_n} j\omega + \frac{1}{\omega_n^2}(j\omega)^2\right]^{\pm p} \qquad (8.18)$$

Each of these terms except K_m may appear raised to an integral power other than 1. The curves of log magnitude and angle vs. the log frequency can easily be drawn for each factor. Then these curves for each factor can be added together graphically to get the curves for the complete transfer function. The procedure can be further simplified by using asymptotic approximations to these curves, as shown in the following pages.

Constants

Since the constant K_m is frequency-invariant, the plot of

$$\text{Lm}\, K_m = 20 \log K_m \qquad \text{dB}$$

is a horizontal straight line. The constant raises or lowers the log magnitude curve of the complete transfer function by a fixed amount. The angle, of course, is zero as long as K_m is positive.

$j\omega$ Factors

The factor $j\omega$ appearing in the denominator has a log magnitude

$$\text{Lm}\, \frac{1}{j\omega} = 20 \log \left| \frac{1}{j\omega} \right| = -20 \log \omega \tag{8.19}$$

When plotted against log ω, this curve is a straight line with a negative slope of 6 dB/octave or 20 dB/decade. Values of this function can be obtained from Table 8.1 for several values of ω. The angle is constant and equal to $-90°$.

When the factor $j\omega$ appears in the numerator, the log magnitude is

$$\text{Lm}\,(j\omega) = 20 \log |j\omega| = 20 \log \omega \tag{8.20}$$

This curve is a straight line with a positive slope of 6 dB/octave or 20 dB/decade. The angle is constant and equal to $+90°$. Notice that the only difference between the curves for $j\omega$ and for $1/j\omega$ is a change in the sign of the slope of the log magnitude and a change in the sign of the angle. Both curves go through the point 0 dB at $\omega = 1$.

For the factor $(j\omega)^{\pm m}$ the log magnitude curve has a slope of $\pm 6m$ dB/octave or $\pm 20m$ dB/decade, and the angle is constant and equal to $\pm m90°$.

$1 + j\omega T$ Factors

The factor $1 + j\omega T$ appearing in the denominator has a log magnitude

$$\text{Lm}\,(1 + j\omega T)^{-1} = 20 \log |1 + j\omega T|^{-1} = -20 \log \sqrt{1 + \omega^2 T^2} \tag{8.21}$$

For very small values of ω, that is, $\omega T \ll 1$,

$$\text{Lm}\,(1 + j\omega T)^{-1} \approx \log 1 = 0 \text{ dB} \tag{8.22}$$

The plot of the log magnitude at small frequencies is the 0-dB line. For very large

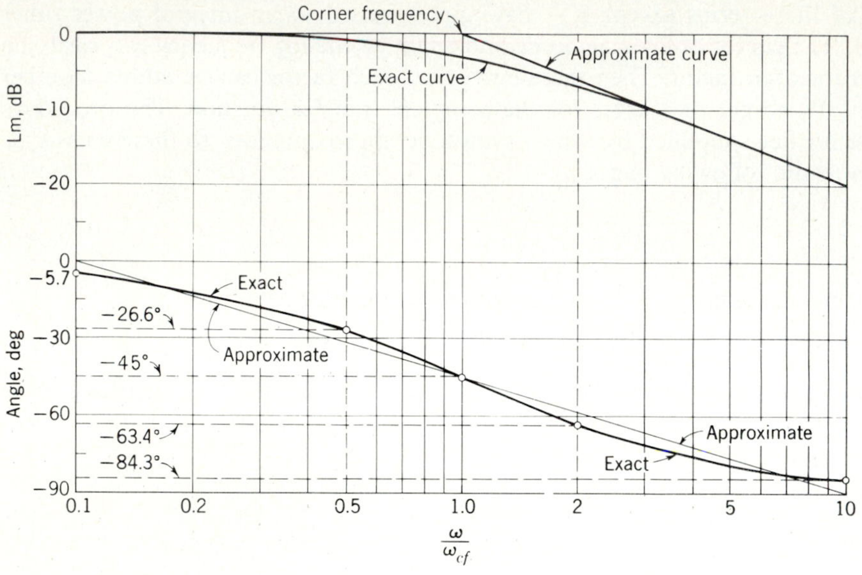

FIGURE 8.2
Log magnitude and phase diagram for

$$(1 + j\omega T)^{-1} = [1 + j(\omega/\omega_{cf})]^{-1}$$

values of ω, that is, $\omega T \gg 1$,

$$\text{Lm}\,(1 + j\omega T)^{-1} \approx 20 \log |j\omega T|^{-1} = -20 \log \omega T \tag{8.23}$$

The value of Eq. (8.23) at $\omega = 1/T$ is 0. For values of $\omega > 1/T$ this function is a straight line with a negative slope of 6 dB/octave. The asymptotes of the plot of $\text{Lm}\,(1 + j\omega T)^{-1}$ are two straight lines, one of zero slope below $\omega = 1/T$ and one of -6 dB/octave slope above $\omega = 1/T$. These asymptotes are drawn in Fig. 8.2.

The frequency at which the asymptotes to the log magnitude curve intersect is defined as the corner frequency ω_{cf}. The value $\omega_{cf} = 1/T$ is the corner frequency for the function $(1 + j\omega T)^{\pm r} = (1 + j\omega/\omega_{cf})^{\pm r}$.

The exact values of $\text{Lm}\,(1 + j\omega T)^{-1}$ are given in Table 8.2 for several frequencies in the range a decade above and below the corner frequency. The exact curve is also drawn in Fig. 8.2. The error, in decibels, between the exact curve and the asymptotes is approximately as follows:

1. At the corner frequency: 3 dB
2. One octave above and below the corner frequency: 1 dB
3. Two octaves from the corner frequency: 0.26 dB

Frequently the preliminary design studies are made by using the asymptotes only. The correction to the straight-line approximation to yield the true log

TABLE 8.2
Values of Lm $(1 + j\omega T)^{-1}$ for several frequencies

$\frac{\omega}{\omega_{cf}}$	Exact value, dB	Value of the asymptote, dB	Error, dB
0.1	−0.04	0	−0.04
0.25	−0.26	0	−0.26
0.5	−0.97	0	−0.97
0.76	−2.00	0	−2.00
1	−3.01	0	−3.01
1.31	−4.35	−2.35	−2.00
2	−6.99	−6.02	−0.97
4	−12.30	−12.04	−0.26
10	−20.04	−20.0	−0.04

magnitude curve is shown in Fig. 8.3. For more exact studies the corrections are put in at the corner frequency and at 1 octave above and below the corner frequency, and the new curve is drawn with a french curve. If a more accurate curve is desired, computer-calculated values should be used.

The phase curve for this function is plotted in Fig. 8.2. At zero frequency the angle is 0°; at the corner frequency $\omega = \omega_{cf}$ the angle is $-45°$; and at infinite frequency the angle is $-90°$. The angle curve is symmetrical about the corner-frequency value when plotted against $\log(\omega/\omega_{cf})$ or $\log \omega$. Since the abscissa of the curves in Fig. 8.2 is ω/ω_{cf}, the shapes of the angle and log magnitude curves are independent of the time constant T. Thus when the curves are plotted with the abscissa in terms of ω, changing T just "slides" the log magnitude and the angle curves left or right so that the -3 dB and the $-45°$ points occur at the frequency $\omega = \omega_{cf}$.

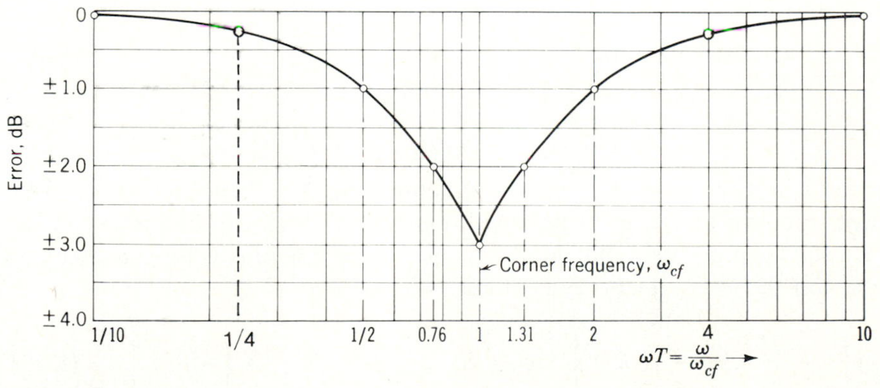

FIGURE 8.3
Log magnitude correction for $(1 + j\omega T)^{\pm 1}$.

TABLE 8.3
Angles of $(1 + j\omega/\omega_{cf})^{-1}$ for key frequency points

$\frac{\omega}{\omega_{cf}}$	**Angle, deg**
0.1	−5.7
0.5	−26.6
1.0	−45.0
2.0	−63.4
10.0	−84.3

For preliminary design studies, straight-line approximations of the phase curve can be used. The approximation is a straight line drawn through the following three points:

ω/ω_{cf}	0.1	1.0	10
Angle	0°	−45°	−90°

The maximum error resulting from this approximation is about $\pm 6°$. For greater accuracy, a smooth curve is drawn through the points given in Table 8.3.

The factor $1 + j\omega T$ appearing in the numerator has the log magnitude

$$\mathrm{Lm}\,(1 + j\omega T) = 20 \log \sqrt{1 + \omega^2 T^2}$$

This is the same function as its inverse $\mathrm{Lm}\,(1 + j\omega T)^{-1}$ except that it is positive. The corner frequency is the same, and the angle varies from 0 to 90° as the frequency increases from zero to infinity. The log magnitude and angle curves for the function $1 + j\omega T$ are symmetrical about the abscissa to the curves for $(1 + j\omega T)^{-1}$.

Quadratic Factors

Quadratic factors of the form

$$\left[1 + \frac{2\zeta}{\omega_n} j\omega + \frac{1}{\omega_n^2}(j\omega)^2\right]^{-1} \tag{8.24}$$

often occur in feedback-system transfer functions. For $\zeta > 1$ the quadratic can be factored into two first-order factors with real zeros which can be plotted in the manner shown previously. But for $\zeta < 1$ Eq. (8.24) contains conjugate-complex factors, and the entire quadratic is plotted without factoring:

$$\mathrm{Lm}\left[1 + \frac{2\zeta}{\omega_n} j\omega + \frac{1}{\omega_n^2}(j\omega)^2\right]^{-1} = -20 \log \left[\left(1 - \frac{\omega^2}{\omega_n^2}\right)^2 + \left(\frac{2\zeta\omega}{\omega_n}\right)^2\right]^{1/2} \tag{8.25}$$

$$\text{Angle}\left[1 + \frac{2\zeta}{\omega_n} j\omega + \frac{1}{\omega_n^2}(j\omega)^2\right]^{-1} = -\tan^{-1} \frac{2\zeta\omega/\omega_n}{1 - \omega^2/\omega_n^2} \tag{8.26}$$

From Eq. (8.25) it is seen that for very small values of ω the low-frequency asymptote is represented by log magnitude $= 0$ dB. For very high values of frequency, the log magnitude is approximately

$$-20 \log \frac{\omega^2}{\omega_n^2} = -40 \log \frac{\omega}{\omega_n}$$

and the high-frequency asymptote has a slope of -40 dB/decade. The asymptotes cross at the corner frequency $\omega_{cf} = \omega_n$.

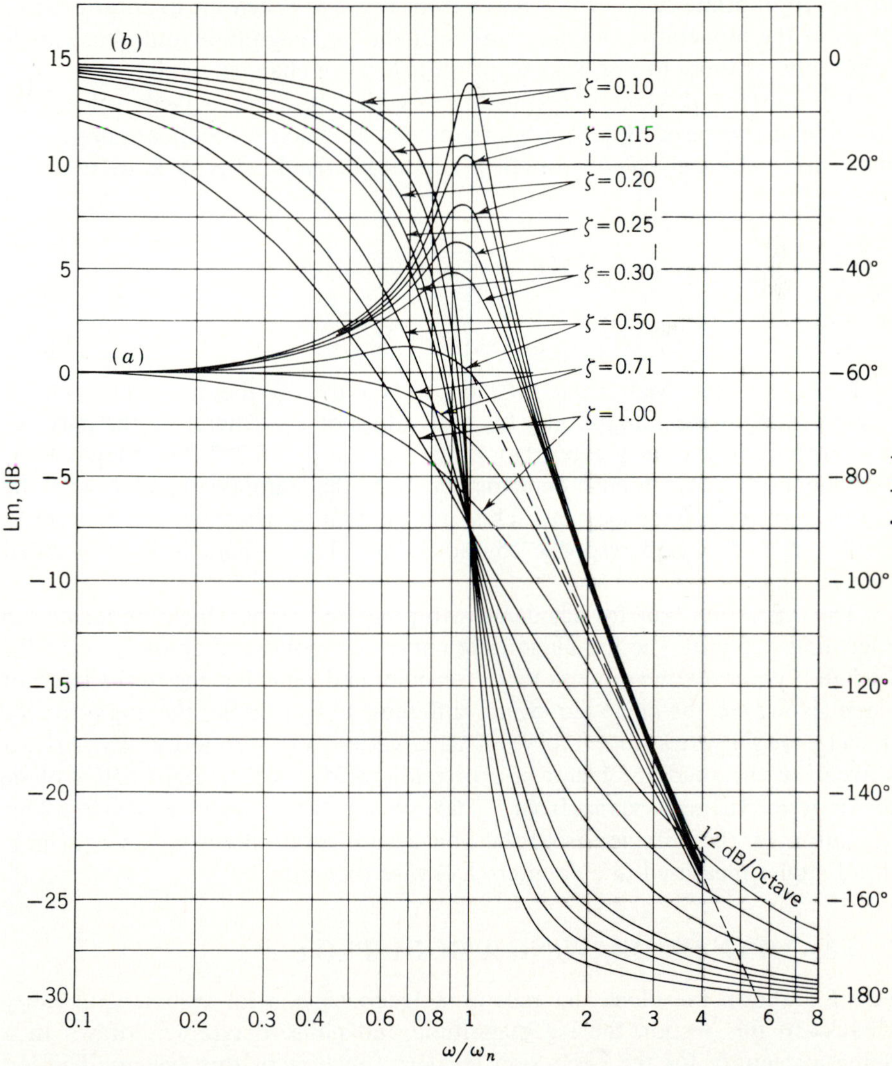

FIGURE 8.4
Log magnitude and phase diagram for $[1 + j2\zeta\omega/\omega_n + (j\omega/\omega_n)^2]^{-1}$.

From Eq. (8.25) it is seen that a resonant condition is exhibited in the vicinity of $\omega = \omega_n$, where the peak value of the Lm > 0 dB. Therefore there may be a substantial deviation of the log magnitude curve from the straight-line asymptotes, depending on the value of ζ. A family of curves of several values of $\zeta < 1$ is plotted in Fig. 8.4. For the appropriate ζ the curve can be traced directly on the graph being drawn or sufficient points can be picked from Fig. 8.4.

The phase-angle curve for this function also varies with ζ. At zero frequency the angle is 0°, at the corner frequency the angle is $-90°$, and at infinite frequency the angle is $-180°$. A family of curves for various values of $\zeta < 1$ is plotted in Fig. 8.4. Templates can be made of these curves, or enough values to draw the appropriate curve can be taken from Fig. 8.4. When the quadratic factor appears in the numerator, the magnitudes of the log magnitude and phase angle are the same as those in Fig. 8.4 except that they are changed in sign.

The $\text{Lm}[1 + j2\zeta\omega/\omega_n + (j\omega/\omega_n)^2]^{-1}$ with $\zeta < 1$ has a peak value. The magnitude of this peak value and the frequency at which it occurs are important terms. These values, given earlier in Sec. 4.10 and derived in Sec. 9.3, are repeated here:

$$M_m = \frac{1}{2\zeta\sqrt{1 - \zeta^2}} \tag{8.27}$$

$$\omega_m = \omega_m\sqrt{1 - 2\zeta^2} \tag{8.28}$$

Note that the peak value M_m depends only on the damping ratio ζ. Equation (8.28) is meaningful only for real values of ω_m. Therefore the curve of M vs. ω has a peak value greater than unity only for $\zeta < 0.707$. The frequency at which the peak value occurs depends on both the damping ratio ζ and the undamped natural frequency ω_n. This information is used when adjusting a control system for good response characteristics. These characteristics are discussed in Chap. 9.

The discussion thus far has dealt with poles and zeros which are located in the left-half s plane. The log magnitude curves for poles and zeros lying in the right-half s plane are the same as those for poles and zeros located in the left-half s plane. However, the angle curves are different. For example, the angle for the factor $(1 - j\omega T)$ varies from 0 to $-90°$ as ω varies from zero to infinity. Also, if ζ is negative, the quadratic factor of expression (8.24) contains right-half s plane poles or zeros. Its angle varies from $-360°$ at $\omega = 0$ to $-180°$ at $\omega = \infty$. This information can be obtained from the pole-zero diagram discussed in Sec. 4.10, with all angles measured in a counter-clockwise direction.

8.7 EXAMPLE OF DRAWING A BODE PLOT

Figure 8.5 shows the block diagram of a feedback control system with unity feedback. In this section the log magnitude and phase diagram is drawn in a systematic manner for the open-loop transfer function of this system. The log magnitude curve is drawn both for the straight-line approximation and for the

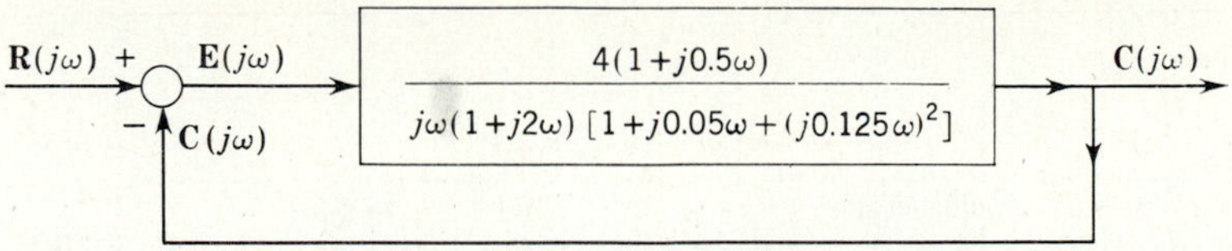

FIGURE 8.5
Block diagram of a control system with unity feedback.

exact curve. Table 8.4 lists the pertinent characteristics for each factor. The log magnitude asymptotes and angle curves for each factor are shown in Figs. 8.6 and 8.7, respectively. They are added algebraically to obtain the composite curve. The composite log magnitude curve using straight-line approximations is drawn directly, as outlined below.

Step 1. At frequencies less than ω_1, the first corner frequency, only the factors Lm 4 and $\text{Lm}(j\omega)^{-1}$ are effective. All the other factors have zero value. At ω_1, Lm 4 = 12 dB, and $\text{Lm}(j\omega_1)^{-1} = 6$ dB; thus at this frequency the composite curve has the value of 18 dB. Below ω_1 the composite curve has a slope of −20 dB/decade because of the $\text{Lm}(j\omega)^{-1}$ term.

Step 2. Above ω_1, the factor $\text{Lm}(1 + j2\omega)^{-1}$ has a slope of −20 dB/decade and must be added to the terms in step 1. When the slopes are added, the

TABLE 8.4
Characteristics of log magnitude and angle diagrams for various factors

Factor	Corner frequency ω_{cf}	Log magnitude	Angle characteristics
4	None	Constant magnitude of +12 dB	Constant 0°
$(j\omega)^{-1}$	None	Constant slope of −20 dB/decade	Constant −90°
$(1 + j2\omega)^{-1}$	$\omega_1 = 0.5$	0 slope below corner frequency; −20 dB/decade slope above corner frequency	Varies from 0 to −90°
$1 + j0.5\omega$	$\omega_2 = 2.0$	0 slope below corner frequency; +20 dB/decade slope above corner frequency	Varies from 0 to +90°
$[1 + j0.05\omega + (j0.125\omega)^2]^{-1}$ $\zeta = 0.2$ $\omega_n = 8$	$\omega_3 = 8.0$	0 slope below corner frequency; −40 dB/decade slope above corner frequency	Varies from 0 to −180°

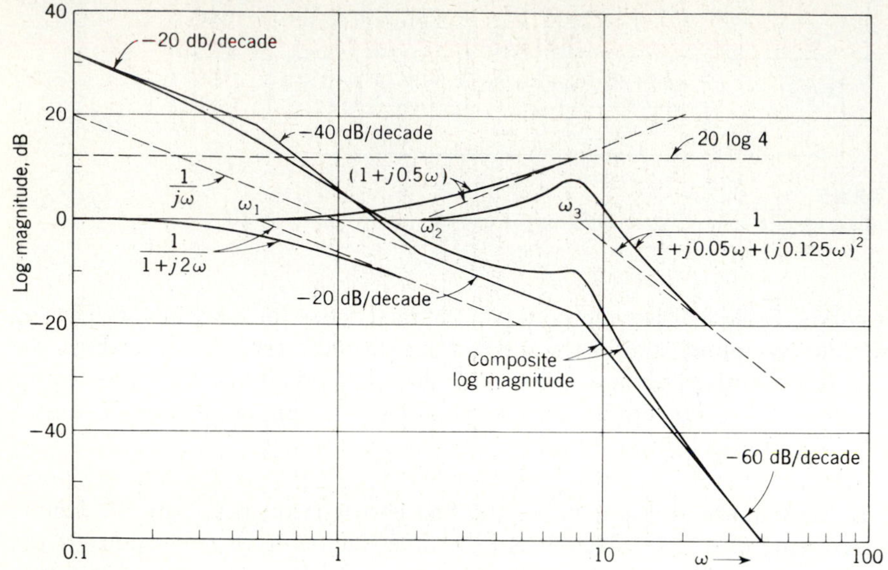

FIGURE 8.6
Log magnitude curve for

$$\mathbf{G}(j\omega) = \frac{4(1 + j0.5\omega)}{j\omega(1 + j2\omega)\left[1 + j0.05\omega + (j0.125\omega)^2\right]}$$

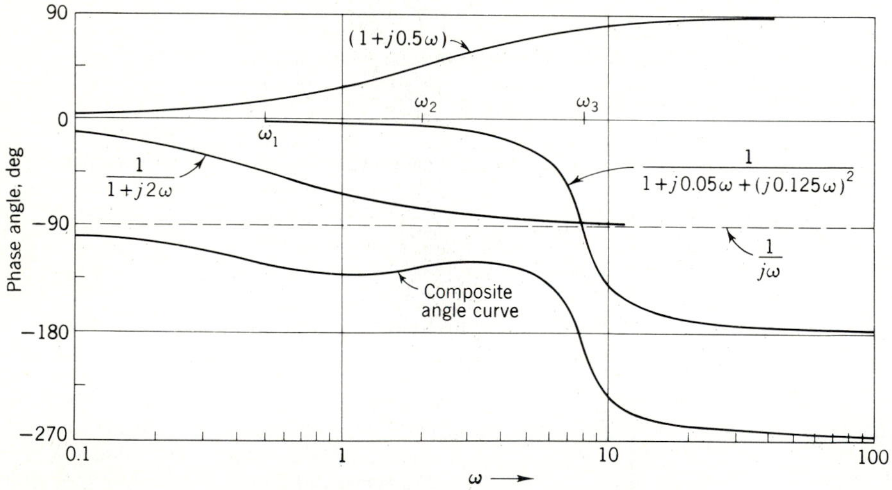

FIGURE 8.7
Phase-angle curve for

$$\mathbf{G}(j\omega) = \frac{4(1 + j0.5\omega)}{j\omega(1 + j2\omega)\left[1 + j0.05\omega + (j0.125\omega)^2\right]}$$

composite curve has a total slope of -40 dB/decade in the frequency band from ω_1 to ω_2. Since this bandwidth is 2 octaves, the value of the composite curve at ω_2 is -6 dB.

Step 3. Above ω_2, the factor $\text{Lm}(1 + j0.5\omega)$ is effective. This factor has a slope of $+20$ dB/decade above ω_2 and must be added to obtain the composite curve. The composite curve now has a total slope of -20 dB/decade in the frequency band from ω_2 to ω_3. The bandwidth from ω_2 to ω_3 is 2 octaves; therefore the value of the composite curve at ω_3 is -18 dB.

Step 4. Above ω_3 the last term $\text{Lm}[1 + j0.05\omega + (j0.125\omega)^2]^{-1}$ must be added. This factor has a slope of -40 dB/decade; therefore the total slope of the composite curve above ω_3 is -60 dB/decade.

Step 5. Once the asymptotic plot of the log magnitude of $\mathbf{G}(j\omega)$ has been drawn, the corrections can be added if desired. The corrections at each corner frequency and at an octave above and below the corner frequency are usually sufficient. For first-order terms the corrections are ± 3 dB at the corner frequencies and ± 1 dB at an octave above and below the corner frequency. The values for quadratic terms can be obtained from Fig. 8.4 since they are a function of the damping ratio. If the correction curve for the quadratic factor is not available, the correction at the frequencies $\omega = \omega_n$ and $\omega = 0.707\omega_n$ can easily be calculated from Eq. (8.25). The correction at ω_m can also be obtained using Eqs. (8.27) and (8.28).

The corrected log magnitude curves for each factor and for the composite curve are shown in Fig. 8.6.

Determination of the phase-angle curve of $\mathbf{G}(j\omega)$ can be simplified by using the following procedure.

1. For the $(j\omega)^{-m}$ term, draw a line at the angle of $(-m)90°$.
2. For each $(1 + j\omega T)^{\pm 1}$ term, locate the angles from Table 8.4 at the corner frequency, an octave above and an octave below the corner frequency, and a decade above and below the corner frequency. Then draw a curve through these points for each $(1 + j\omega T)^{\pm 1}$ term.
3. For each $1 + j2\zeta\omega/\omega_n + (j\omega/\omega_n)^2$ term:
   a. Locate the $\pm 90°$ point at the corner frequency.
   b. From Fig. 8.4, for the respective ζ, obtain a few points to draw the phase plot for each term with the aid of a french curve. The angle at $\omega = 0.707\omega_n$ may be sufficient and can be evaluated easily from Eq. (8.26) as $\tan^{-1} 2.828\zeta$.
4. Once the phase plot of each term of $\mathbf{G}(j\omega)$ has been drawn, the composite phase plot of $\mathbf{G}(j\omega)$ is determined by adding the individual phase curves.
   a. Use the line representing the angle equal to

$$\angle \lim_{\omega \to 0} \mathbf{G}(j\omega) = (-m)90°$$

as the base line, where m is the system type. Add or subtract the angles of each factor from this reference line.

b. At a particular frequency on the graph, measure the angle for each single-order and quadratic factor. Add and/or subtract them from the base line until all terms have been accounted for. The number of frequency points used is determined by the desired accuracy of the phase plots.

c. At $\omega = \infty$ the phase is 90° times the difference of the orders of the numerator and denominator $[-(n - w)90°]$.

8.8 SYSTEM TYPE AND GAIN AS RELATED TO LOG MAGNITUDE CURVES

The steady-state error of a closed-loop system depends on the system type and the gain. The system error coefficients are determined by these two characteristics, as noted in Chap. 6. For any given log magnitude curve the system type and gain can be determined. Also, with the transfer function given so that the system type and gain are known, they can expedite drawing the log magnitude curve. This is described for Type 0, 1, and 2 systems.

Type 0 System

A Type 0 system has a transfer function of the form

$$\mathbf{G}(j\omega) = \frac{K_0}{1 + j\omega T_a}$$

At low frequencies, $\omega < 1/T_a$, $\text{Lm}\,\mathbf{G}(j\omega) = 20 \log K_0$, which is a constant. The slope of the log magnitude curve is zero below the corner frequency $\omega_1 = 1/T_a$ and -20 dB/decade above the corner frequency. The log magnitude curve is shown in Fig. 8.8.

For a Type 0 system the characteristics are as follows:

1. The slope at low frequencies is zero.
2. The magnitude at low frequencies is $20 \log K_0$.
3. The gain K_0 is the steady-state step error coefficient.

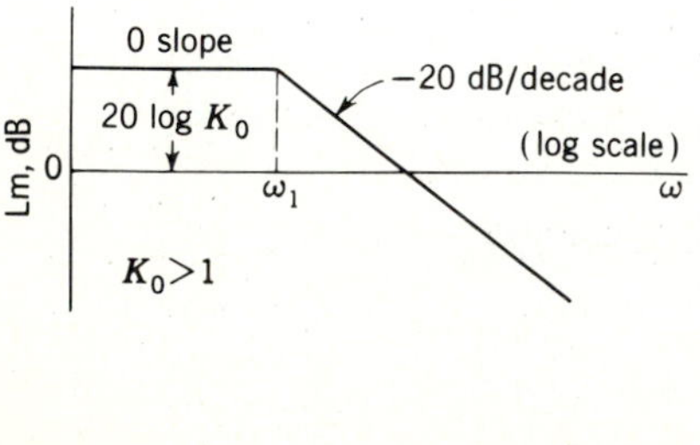

FIGURE 8.8
Log magnitude plot for $\mathbf{G}(j\omega) = K_0/(1 + j\omega T_a)$.

Type 1 System

A Type 1 system has a transfer function of the form

$$\mathbf{G}(j\omega) = \frac{K_1}{j\omega(1 + j\omega T_a)}$$

At low frequencies, $\omega < 1/T_a$, $\text{Lm}\,\mathbf{G}(j\omega) \approx \text{Lm}(K_1/j\omega) = \text{Lm}\,K_1 - \text{Lm}\,j\omega$, which has a slope of -20 dB/decade. At $\omega = K_1$, $\text{Lm}(K_1/j\omega) = 0$. If the corner frequency $\omega_1 = 1/T_a$ is greater than K_1, the low-frequency portion of the curve of slope -20 dB/decade crosses the 0-dB axis at a value of $\omega_x = K_1$, as shown in Fig. 8.9*a*. If the corner frequency is less than K_1, the low-frequency portion of the curve of slope -20 dB/decade may be extended until it does cross the 0-dB axis. The value of the frequency at which the extension crosses the 0-dB axis is $\omega_x = K_1$. In other words, the plot $\text{Lm}(K_1/j\omega)$ crosses the 0-dB value at $\omega_x = K_1$, as illustrated in Fig. 8.9*b*.

At $\omega = 1$, $\text{Lm}\,j\omega = 0$; therefore $\text{Lm}(K_1/j\omega)_{\omega=1} = 20 \log K_1$. For $T_a < 1$ this value is a point on the slope of -20 dB/decade. For $T_a > 1$ this value is a point on the extension of the initial slope, as shown in Fig. 8.9*b*. The frequency ω_x is smaller or larger than unity according as K_1 is smaller or larger than unity.

For a Type 1 system the characteristics are as follows:

1. The slope at low frequencies is -20 dB/decade.
2. The intercept of the low-frequency slope of -20 dB/decade (or its extension) with the 0-dB axis occurs at the frequency ω_x, where $\omega_x = K_1$.
3. The value of the low-frequency slope of -20dB/decade (or its extension) at the frequency $\omega = 1$ is equal to $20 \log K_1$.
4. The gain K_1 is the steady-state ramp error coefficient.

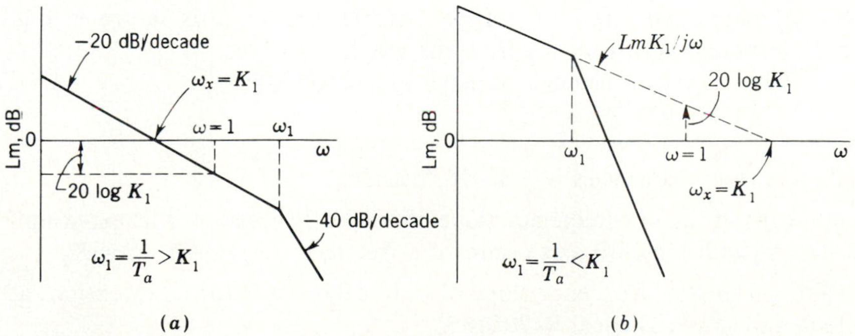

FIGURE 8.9
Log magnitude plot for $\mathbf{G}(j\omega) = K_1/j\omega(1 + j\omega T_a)$.

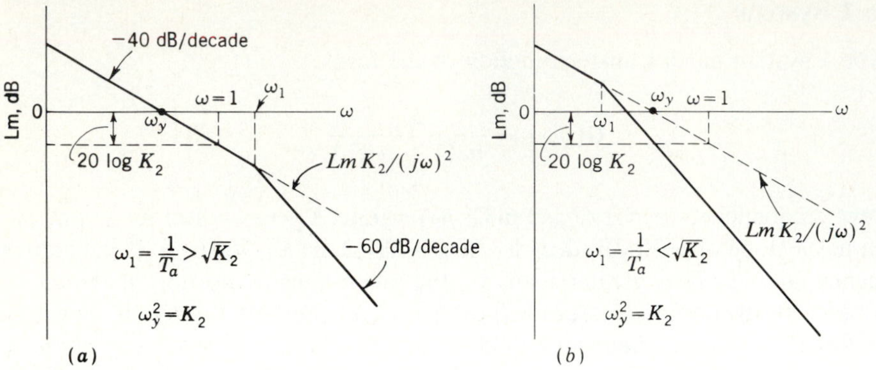

FIGURE 8.10
Log magnitude plot for $\mathbf{G}(j\omega) = K_2/(j\omega)^2(1 + j\omega T_a)$.

Type 2 System

A Type 2 system has a transfer function of the form

$$\mathbf{G}(j\omega) = \frac{K_2}{(j\omega)^2(1 + j\omega T_a)}$$

At low frequencies, $\omega < 1/T_a$, $\mathrm{Lm}\,\mathbf{G}(j\omega) = \mathrm{Lm}[K_2/(j\omega)^2] = \mathrm{Lm}\,K_2 - \mathrm{Lm}(j\omega)^2$, for which the slope is -40 dB/decade. At $\omega^2 = K_2$, $\mathrm{Lm}[K_2/(j\omega)^2] = 0$; therefore the intercept of the initial slope of -40 dB/decade (or its extension, if necessary) with the 0-dB axis occurs at the frequency ω_y, where $\omega_y^2 = K_2$.

At $\omega = 1$, $\mathrm{Lm}(j\omega)^2 = 0$; therefore, $\mathrm{Lm}[K_2/(j\omega)^2]_{\omega=1} = 20 \log K_2$. This point occurs on the initial slope or on its extension, according as $\omega_1 = 1/T_a$ is larger or smaller than $\sqrt{K_2}$. If $K_2 > 1$, the quantity $20 \log K_2$ is positive, and if $K_2 < 1$, the quantity $20 \log K_2$ is negative.

The log magnitude curve for a Type 2 transfer function is shown in Fig. 8.10. The determination of gain K_2 from the graph is shown.

For a Type 2 system the characteristics are as follows:

1. The slope at low frequencies is -40 dB/decade.
2. The intercept of the low-frequency slope of -40 dB/decade (or its extension, if necessary) with the 0-dB axis occurs at a frequency ω_y, where $\omega_y^2 = K_2$.
3. The value on the low-frequency slope of -40 dB/decade (or its extension) at the frequency $\omega = 1$ is equal to $20 \log K_2$.
4. The gain K_2 is the steady-state parabolic error coefficient.

8.9 EXPERIMENTAL DETERMINATION OF TRANSFER FUNCTIONS[5,9]

The log magnitude and phase-angle diagram is of great value when the mathematical expression for the transfer function of a given system is not known. The magnitude and angle of the ratio of the output to the input can be obtained experimentally for a steady-state sinusoidal input signal at a number of frequencies. These data are used to obtain the exact log magnitude and angle diagram. Asymptotes are drawn on the exact log magnitude curve, using the fact that their slopes must be multiples of ± 20 dB/decade. From these asymptotes the system type and the approximate time constants are determined. Thus, in this manner, the transfer function of the system can be synthesized.

Care must be exercised in determining whether any zeros of the transfer function are in the right-half s plane. A system that has no open-loop zeros in the right-half s plane is defined as a *minimum-phase* system,[6] and all factors have the form $1 + Ts$ and/or $1 + As + Bs^2$. A system that has open-loop zeros in the right-half s plane is a *non-minimum-phase* system. The stability is determined by the location of the poles and does not affect the designation of minimum or non-minimum phase.

The angular variation for poles or zeros in the right-half s plane is different from those in the left-half plane. For this situation, one or more terms in the transfer function have the form $1 - Ts$ and/or $1 \pm As \pm Bs^2$. As an example, consider the functions $1 + j\omega T$ and $1 - j\omega T$. The log magnitude plots of these functions are identical, but the angle diagram for the former goes from 0 to 90° whereas for the latter it goes from 0 to $-90°$. Therefore care must be exercised in interpreting the angle plot to determine whether any factors of the transfer function lie in the right-half s plane. Many practical systems are in the minimum-phase category.

8.10 DIRECT POLAR PLOTS

In the earlier sections of this chapter a simple graphical method is presented for plotting the characteristic curves of the transfer functions. These curves are the log magnitude and the angle of $\mathbf{G}(j\omega)$ vs. ω, plotted on semilog graph paper. The reason for presenting this method first is that these curves can be constructed easily and rapidly. Frequently, the polar plot of the transfer function is desired. The magnitude and angle of $\mathbf{G}(j\omega)$, for sufficient frequency points, are readily obtainable from the $\text{Lm}\,\mathbf{G}(j\omega)$ and $\angle \mathbf{G}(j\omega)$ vs. $\log \omega$ curves. This approach often takes less time than calculating $|\mathbf{G}(j\omega)|$ and $\phi(\omega)$ analytically for each frequency point desired, unless a digital computer or programmable calculator is available.[4]

The data for drawing the polar plot of the frequency response can also be obtained from the pole-zero diagram, as described in Sec. 4.10. It is possible to visualize the complete shape of the frequency-response curve from the pole-zero

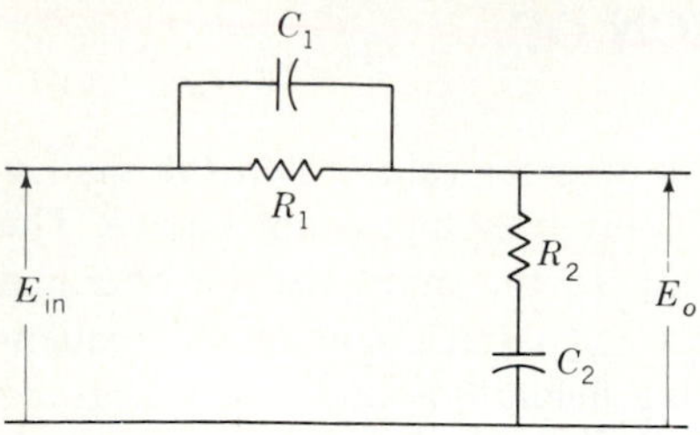

FIGURE 8.11
An *RC* circuit used as a lag-lead compensator.

diagram because the angular contribution of each pole and zero is readily apparent. The polar plot of $\mathbf{G}(j\omega)$ is called the *direct polar plot*, and the polar plot of $[\mathbf{G}(j\omega)]^{-1}$ is called the *inverse polar plot*. The shape of these curves is related to the system type.

Complex *RC* Network (Lag-Lead Compensator)

The circuit of Fig. 8.11 is used in later chapters as a compensator:

$$G(s) = \frac{E_o(s)}{E_{in}(s)} = \frac{1 + (T_1 + T_2)s + T_1 T_2 s^2}{1 + (T_1 + T_2 + T_{12})s + T_1 T_2 s^2} \tag{8.29}$$

where the time constants are

$$T_1 = R_1 C_1 \qquad T_2 = R_2 C_2 \qquad T_{12} = R_1 C_2$$

As a function of frequency, the transfer function is

$$\mathbf{G}(j\omega) = \frac{\mathbf{E}_o(j\omega)}{\mathbf{E}_{in}(j\omega)} = \frac{(1 - \omega^2 T_1 T_2) + j\omega(T_1 + T_2)}{(1 - \omega^2 T_1 T_2) + j\omega(T_1 + T_2 + T_{12})} \tag{8.30}$$

By the proper choice of the time constants, the circuit acts as a lag network in the lower-frequency range of 0 to ω_x and as a lead network in the higher-frequency range of ω_x to ∞. This means that the steady-state sinusoidal output E_o lags or leads the sinusoidal input E_{in}, according as ω is smaller than or larger than ω_x. The polar plot of this transfer function is a circle with its center on the real axis and lying in the first and fourth quadrants. Figure 8.12 illustrates the polar plot in nondimensionalized form for a typical circuit. When T_2 and T_{12} in Eq. (8.30) are expressed in terms of T_1, the expression of the form given in the title of Fig. 8.12 results. Its properties are:

1. $\lim_{\omega \to 0} \mathbf{G}(j\omega T_1) \to 1\angle 0°$.
2. $\lim_{\omega \to \infty} \mathbf{G}(j\omega T_1) \to 1\angle 0°$.
3. At the frequency $\omega = \omega_x$, for which $\omega_x^2 T_1 T_2 = 1$, Eq. (8.30) becomes

$$\mathbf{G}(j\omega_x T_1) = \frac{T_1 + T_2}{T_1 + T_2 + T_{12}} = |\mathbf{G}(j\omega_x T_1)|\angle 0° \tag{8.31}$$

Note that Eq. (8.31) represents the minimum value of the transfer function in the whole frequency spectrum. From Fig. 8.12 it is seen that for frequencies below ω_x

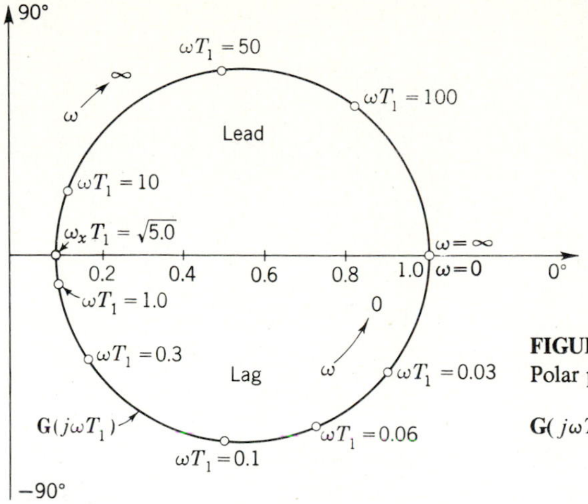

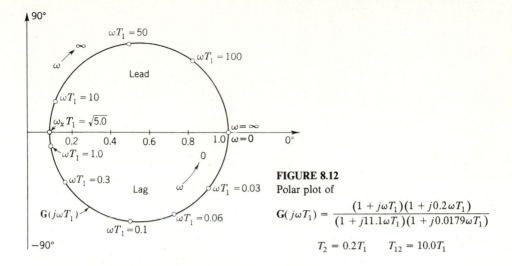

FIGURE 8.12
Polar plot of

$$G(j\omega T_1) = \frac{(1 + j\omega T_1)(1 + j0.2\omega T_1)}{(1 + j11.1\omega T_1)(1 + j0.0179\omega T_1)}$$

$T_2 = 0.2T_1 \qquad T_{12} = 10.0T_1$

the transfer function has a negative or lag angle. For frequencies above ω_x it has a positive or lead angle. The applications and advantages of this circuit are discussed in later chapters on compensation.

Type 0 Feedback Control System

The field-controlled servomotor described by Eq. (2.152) illustrates a typical Type 0 device. It has the transfer function

$$\mathbf{G}(j\omega) = \frac{\mathbf{C}(j\omega)}{\mathbf{E}(j\omega)} = \frac{K_0}{(1 + j\omega T_f)(1 + j\omega T_m)} \tag{8.32}$$

Note from Eq. (8.32) that

$$\mathbf{G}(j\omega) \to \begin{cases} K_0\angle 0^\circ & \omega \to 0^+ \\ 0\angle -180^\circ & \omega \to \infty \end{cases}$$

Also, for each term in the denominator the angular contribution to $\mathbf{G}(j\omega)$, as ω goes from 0 to ∞, goes from 0 to $-90°$. Thus the polar plot of this transfer function must start at $\mathbf{G}(j\omega) = K_0\angle 0^\circ$ for $\omega = 0$ and proceeds first through the fourth and then through the third quadrants to $\lim_{\omega \to \infty} \mathbf{G}(j\omega) = 0\angle -180^\circ$ as the frequency approaches infinity. In other words, the angular variation of $\mathbf{G}(j\omega)$ is continuously decreasing, going in a clockwise direction from 0° to $-180°$. The exact shape of this plot is determined by the particular values of the time constants T_f and T_m.

If Eq. (8.32) had another term of the form $1 + j\omega T$ in the denominator, the transfer function would be

$$\mathbf{G}(j\omega) = \frac{K_0}{(1 + j\omega T_f)(1 + j\omega T_m)(1 + j\omega T)} \tag{8.33}$$

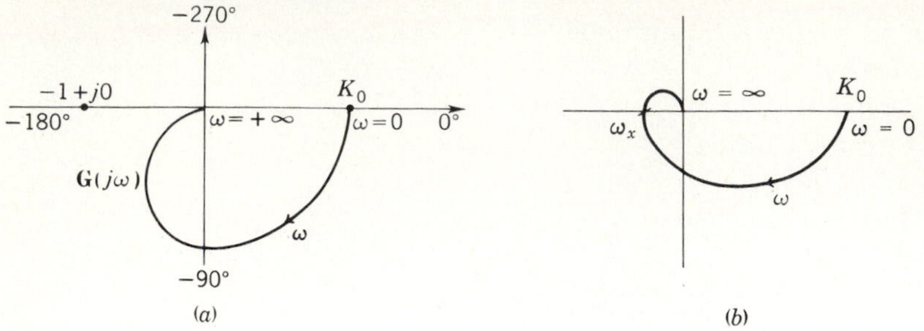

FIGURE 8.13
Polar plot for typical Type 0 transfer functions: (*a*) Eq. (8.32); (*b*) Eq. (8.33).

The point $\mathbf{G}(j\omega)|_{\omega=\infty}$ rotates clockwise by an additional 90°. In other words, when $\omega \to \infty$, $\mathbf{G}(j\omega) \to 0\angle -270°$. In this case the curve crosses the real axis at a frequency ω_x for which the imaginary part of the transfer function is zero.

When a term of the form $1 + j\omega T$ appears in the numerator, the transfer function experiences an angular variation of 0 to 90° (a counterclockwise rotation) as the frequency is varied from 0 to ∞. Thus the angle of $\mathbf{G}(j\omega)$ may not change continuously in one direction. Also, the resultant polar plot may not be as smooth as the one shown in Fig. 8.13. As an example, consider the transfer function

$$\mathbf{G}(j\omega) = \frac{K_0(1 + j\omega T_1)^2}{(1 + j\omega T_2)(1 + j\omega T_3)(1 + j\omega T_4)^2} \tag{8.34}$$

whose polar plot has a "dent" as shown in Fig. 8.14 when the time constants T_2 and T_3 are greater than T_1, and T_1 is greater than T_4. From the angular contribution of each factor and from the analysis above, it can be surmised that the polar plot has the general shape shown in the figure. In the event that T_1 is smaller than all the others, the polar plot is similar in shape to the one shown in Fig. 8.13*a*.

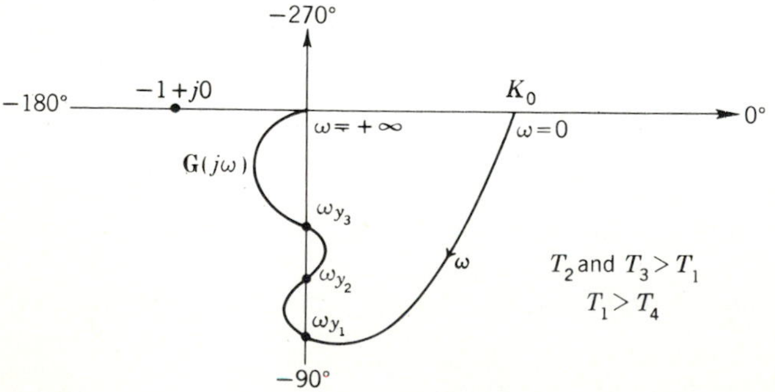

FIGURE 8.14
Polar plot for a more complex Type 0 transfer function [Eq. (8.34)].

In the same manner, a quadratic in either the numerator or the denominator of a transfer function results in an angular contribution of 0 to $\pm 180°$, respectively, and the polar plot of $\mathbf{G}(j\omega)$ is affected accordingly. It can be seen from the examples that the polar plot of a Type 0 system always starts at a value K_0 (step error coefficient) on the positive real axis for $\omega = 0$ and ends at zero magnitude (for $n > w$) and tangent to one of the major axes at $\omega = \infty$. The final angle is $-90°$ times the order of the denominator minus the order of the numerator of $\mathbf{G}(j\omega)$. The arrows on the plots of $\mathbf{G}(j\omega)$ in Figs. 8.13 and 8.14 indicate the direction of increasing frequency.

Type 1 Feedback Control System

A typical Type 1 system is

$$\mathbf{G}(j\omega) = \frac{\mathbf{C}(j\omega)}{\mathbf{E}(j\omega)} = \frac{K_1}{j\omega(1 + j\omega T_m)(1 + j\omega T_c)(1 + j\omega T_q)} \tag{8.35}$$

Note from Eq. (8.35) that

$$\mathbf{G}(j\omega) \to \begin{cases} \infty\angle -90° & \omega \to 0^+ \quad (8.36) \\ 0\angle -360° & \omega \to \infty \quad (8.37) \end{cases}$$

Note that the $j\omega$ term in the denominator contributes the angle $-90°$ to the total angle of $\mathbf{G}(j\omega)$ for all frequencies. Thus the basic difference between Eqs. (8.33) and (8.35) is the presence of the term $j\omega$ in the denominator of the latter equation. Since all the $1 + j\omega T$ terms of Eq. (8.35) appear in the denominator, its polar plot, as shown in Fig. 8.15, has no dents. From the remarks of this and previous sections, it can be seen that the angular variation of $\mathbf{G}(j\omega)$ decreases continuously in the same direction from -90 to $-360°$ as ω increases from 0 to ∞. The presence of any frequency-dependent factor in the numerator has the same general effect on the polar plot as that described previously for the Type 0 system.

It is seen from Eq. (8.36) that the magnitude of the function $\mathbf{G}(j\omega)$ approaches infinity as the value of ω approaches zero. There is a line parallel to

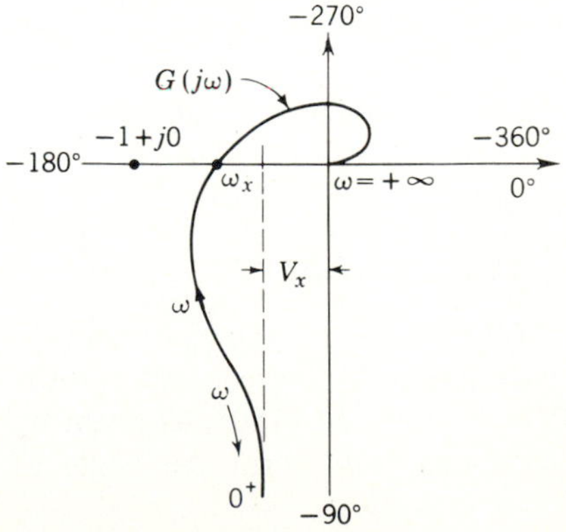

FIGURE 8.15
Polar plot for a typical Type 1 transfer function [Eq. (8.35)].

the $-90°$ axis but displaced to the left of the origin, to which $\mathbf{G}(j\omega)$ tends asymptotically as $\omega \to 0$. The true asymptote is determined by finding the value of the real part of $\mathbf{G}(j\omega)$ as ω approaches zero. Thus

$$V_x = \lim_{\omega \to 0} \operatorname{Re}\left[\mathbf{G}(j\omega)\right] \tag{8.38}$$

or, for this particular transfer function,

$$V_x = -K_1\left(T_q + T_c + T_m\right) \tag{8.39}$$

Equation (8.39) shows that the magnitude of $\mathbf{G}(j\omega)$ approaches infinity asymptotically to a vertical line whose real-axis intercept equals V_x, as illustrated in Fig. 8.15. Note that the value of V_x is a direct function of the ramp error coefficient.

The frequency of the crossing point on the negative real axis of the $\mathbf{G}(j\omega)$ function is that value of frequency ω_x for which the imaginary part of $\mathbf{G}(j\omega)$ is equal to zero. Thus

$$\operatorname{Im}\left[\mathbf{G}(j\omega_x)\right] = 0 \tag{8.40}$$

or, for this particular transfer function,

$$\omega_x = \left(T_cT_q + T_qT_m + T_mT_c\right)^{-1/2} \tag{8.41}$$

The significance of the real-axis crossing point is pointed out in later sections dealing with system stability.

Type 2 Feedback Control System

The transfer function of the Type 2 system illustrated in Sec. 6.6 and represented by Eq. (6.62) is

$$\mathbf{G}(j\omega) = \frac{\mathbf{C}(j\omega)}{\mathbf{E}(j\omega)} = \frac{K_2}{(j\omega)^2(1 + j\omega T_f)(1 + j\omega T_m)} \tag{8.42}$$

Its properties are

$$\mathbf{G}(j\omega) \to \begin{cases} \infty\angle -180° & \omega \to 0^+ \quad (8.43) \\ 0\angle -360° & \omega \to +\infty \quad (8.44) \end{cases}$$

The presence of the $(j\omega)^2$ term in the denominator contributes $-180°$ to the total angle of $\mathbf{G}(j\omega)$ for all frequencies. For the transfer function of Eq. (8.42) the polar plot (Fig. 8.16) is a smooth curve whose angle $\phi(\omega)$ decreases continuously from -180 to $-360°$.

The introduction of an additional pole and a zero can alter the shape of the polar plot. Consider the transfer function

$$\mathbf{G}_o(j\omega) = \frac{K_2'(1 + j\omega T_1)}{(j\omega)^2(1 + j\omega T_f)(1 + j\omega T_m)(1 + j\omega T_2)} \tag{8.45}$$

where $T_1 > T_2$. The polar plot can be obtained from the pole-zero diagram shown in Fig. 8.17. At $s = j\omega = j0^+$ the angle of each factor is zero except for the double pole at the origin. The angle at $\omega = 0^+$ is therefore $-180°$, as given by Eq. (8.43), which is still applicable. As ω increases from zero, the angle of

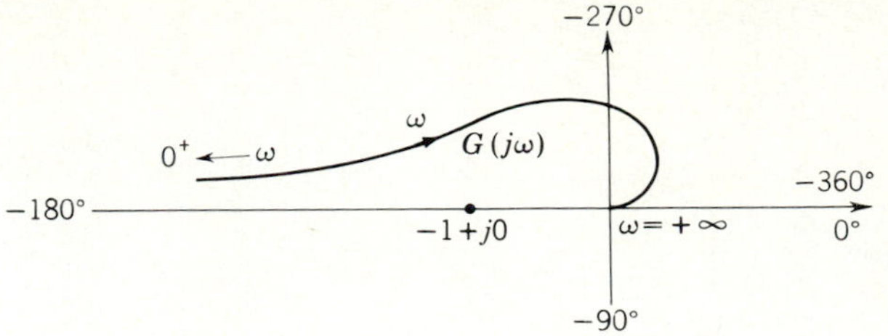

FIGURE 8.16
Polar plot for a typical Type 2 transfer function, resulting in an unstable feedback control system [Eq. (8.42)].

$j\omega + 1/T_1$ increases faster than the angles of the other poles. In fact, at low frequencies the angle due to the zero is larger than the sum of the angles due to the poles located to the left of the zero. This is shown qualitatively at the frequency ω_1 in Fig. 8.17. Therefore, the angle of $\mathbf{G}(j\omega)$ at low frequencies is greater than $-180°$. As the frequency increases to a value ω_x, the sum of the component angles of $\mathbf{G}(j\omega)$ is $-180°$ and the polar plot crosses the real axis, as shown in Fig. 8.18. As ω increases further, the angle of $j\omega + 1/T_1$ shows only a small increase, but the angles from the poles increase rapidly. In the limit, as $\omega \to \infty$, the angles of $j\omega + 1/T_1$ and $j\omega + 1/T_2$ are equal and opposite in sign, so the angle of $\mathbf{G}(j\omega)$ approaches $-360°$, as given by Eq. (8.44).

Fig. 8.18 shows the complete polar plot of $\mathbf{G}(j\omega)$. A comparison of Figs. 8.16 and 8.18 shows that both curves approach $-180°$ at $\omega = 0^+$, which is typical of Type 2 systems. As $\omega \to \infty$, the angle approaches $-360°$ since both $\mathbf{G}(j\omega)$ and $\mathbf{G}_o(j\omega)$ have the same degree, $n - w = 4$. When the Nyquist stability criterion described in Secs. 8.13 and 8.14 is used, the feedback system containing $\mathbf{G}(j\omega)$ can be shown to be unstable, while the system containing $\mathbf{G}_o(j\omega)$ is stable.

It can be shown that as $\omega \to 0^+$, the polar plot of a Type 2 system is below the real axis if

$$\sum(T_{\text{numerator}}) - \sum(T_{\text{denominator}})$$

is a positive value and is above the real axis if it is a negative value. Thus for this example the necessary condition is $T_1 > T_f + T_m + T_2$.

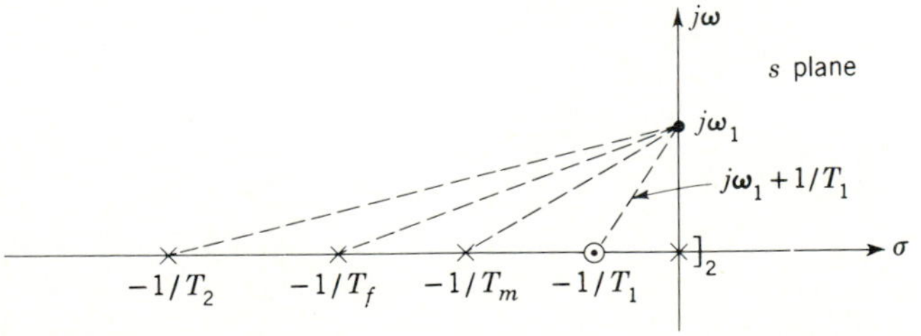

FIGURE 8.17
Pole-zero diagram for

$$G_o(s) = \frac{K_2' T_1}{T_f T_m T_2} \frac{s + 1/T_1}{s^2(s + 1/T_f)(s + 1/T_m)(s + 1/T_2)}$$

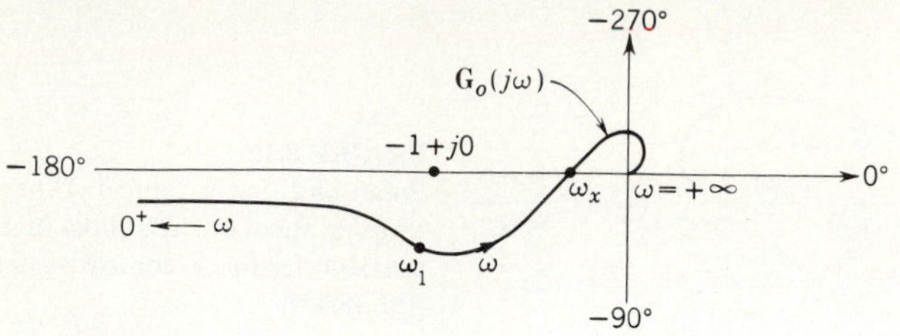

FIGURE 8.18
Polar plot for a typical Type 2 transfer function, resulting in a stable system [Eq. (8.45)].

8.11 SUMMARY: DIRECT POLAR PLOTS

To obtain the direct polar plot of a system's forward transfer function, the following criteria are used to determine the key parts of the curve.

Step 1. The forward transfer function has the general form

$$\mathbf{G}(j\omega) = \frac{K_m(1 + j\omega T_a)(1 + j\omega T_b) \cdots (1 + j\omega T_w)}{(j\omega)^m(1 + j\omega T_1)(1 + j\omega T_2) \cdots (1 + j\omega T_u)} \tag{8.46}$$

For this transfer function the system type is equal to the value of m and determines the portion of the polar plot representing the $\lim_{\omega \to 0}\mathbf{G}(j\omega)$. The low-frequency polar-plot characteristics (as $\omega \to 0$) of the different system types are summarized in Fig. 8.19. The angle at $\omega = 0$ is $m(-90°)$. The arrow on the polar plots indicates the direction of increasing frequency.

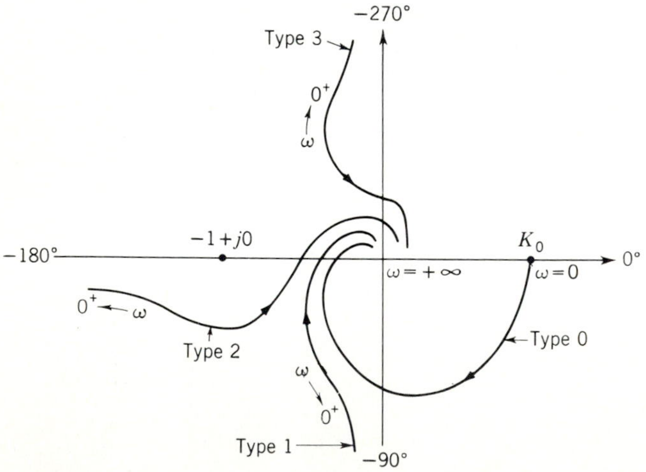

FIGURE 8.19
A summary of direct polar plots of different types of systems.

Step 2. The high-frequency end of the polar plot can be determined as follows:

$$\lim_{\omega \to +\infty} \mathbf{G}(j\omega) = 0\underline{/(w - m - u)90^\circ} \tag{8.47}$$

Note that since the degree of the denominator of Eq. (8.46) is always greater than the degree of the numerator, the high-frequency point ($\omega = \infty$) is approached (i.e., the angular condition) in the clockwise sense. The plot ends at the origin tangent to the axis determined by Eq. (8.47). Tangency may occur on either side of the axis.

Step 3. The asymptote that the low-frequency end approaches, for a Type 1 system, is determined by taking the limit as $\omega \to 0$ of the real part of the transfer function.

Step 4. The frequencies at the points of intersection of the polar plot with the negative real axis and the imaginary axis are determined, respectively, by setting

$$\text{Im}\,[\mathbf{G}(j\omega)] = 0 \tag{8.48}$$

$$\text{Re}\,[\mathbf{G}(j\omega)] = 0 \tag{8.49}$$

Step 5. If there are no frequency dependent terms in the numerator of the transfer function, the curve is a smooth one in which the angle of $\mathbf{G}(j\omega)$ continuously decreases as ω goes from 0 to ∞. With time constants in the numerator, and depending upon their values, the angle may not continuously vary in the same direction, thus creating "dents" in the polar plot.

Step 6. As is seen later in this chapter, it is important to know the exact shape of the polar plot of $\mathbf{G}(j\omega)$ in the vicinity of the $-1 + j0$ point and the crossing point on the negative real axis.

8.12 INVERSE POLAR PLOTS

The direct polar plots have certain drawbacks when they are used for systems that contain elements in the feedback path. In these cases it is found that the graphical analysis of the inverse polar plots is much simpler and therefore advantageous. The inverse polar plot is obtained by plotting the phasor quantity

$$\mathbf{G}^{-1}(j\omega) = \frac{1}{\mathbf{G}(j\omega)} = \frac{\mathbf{E}(j\omega)}{\mathbf{C}(j\omega)} \tag{8.50}$$

as a function of frequency.

Compensators

The inverse plot of the lag-lead compensator discussed in Sec. 8.10, Eq. (8.30) is illustrated in Fig. 8.20.

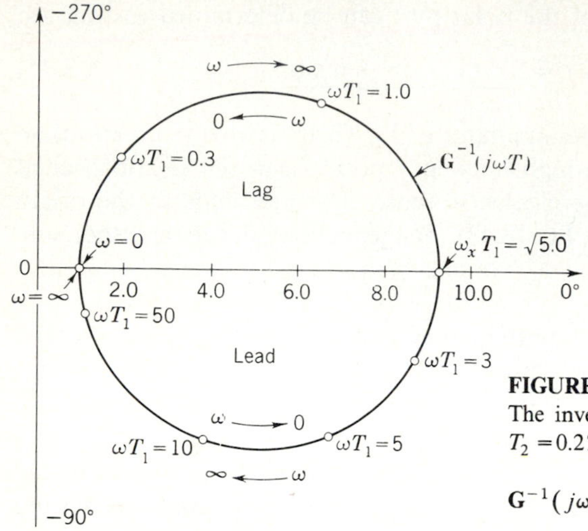

FIGURE 8.20
The inverse polar plot of a lag-lead network for $T_2 = 0.2T_1$, $T_{12} = 10.0T_1$:

$$\mathbf{G}^{-1}(j\omega T_1) = \frac{(1 + j11.1\omega T_1)(1 + j0.0179\omega T_1)}{(1 + j\omega T_1)(1 + j0.2\omega T_1)}$$

Type 0 Feedback Control System

The inverse of the transfer function used in Sec. 8.40 for the Type 0 system of Eq. (8.32) is

$$\mathbf{G}^{-1}(j\omega) = \frac{(1 + j\omega T_f)(1 + j\omega T_m)}{K_0} \tag{8.51}$$

The two limiting points are

$$\lim_{\omega \to 0^+} \mathbf{G}^{-1}(j\omega) = \frac{1}{K_0}$$

$$\lim_{\omega \to +\infty} \mathbf{G}^{-1}(j\omega) = \lim_{\omega \to +\infty} (j\omega)^2 T_f T_m / K_0 = \infty \angle 180^\circ$$

As ω increases from 0 to ∞, the angle of each term in the numerator of $\mathbf{G}^{-1}(j\omega)$ goes from 0 to 90°. Since there are no frequency-dependent terms in the denominator, the angle of $\mathbf{G}^{-1}(j\omega)$ continuously increases from 0° to 180° as ω goes from 0 to ∞. The inverse plot of this Type 0 system is shown in Fig. 8.21. Each additional $1 + j\omega T$ factor in the numerator of $\mathbf{G}^{-1}(j\omega)$ rotates the high-frequency portion of the plot counterclockwise by 90°.

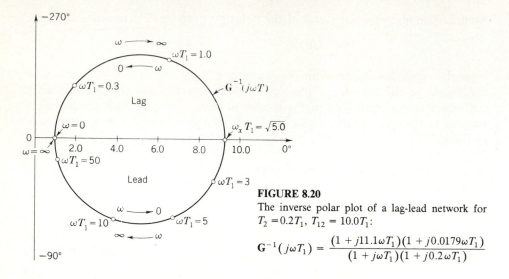

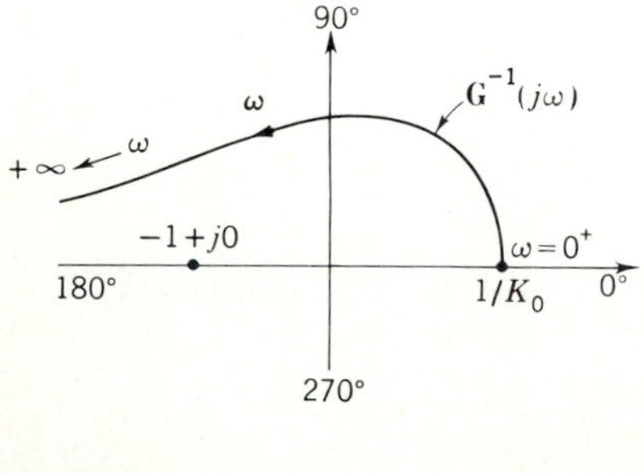

FIGURE 8.21
Inverse polar plot of a typical Type 0 transfer function [Eq. (8.51)].

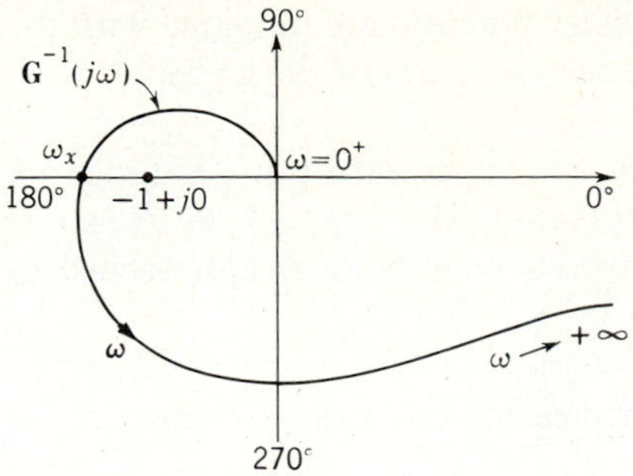

FIGURE 8.22
Inverse polar plot of a typical Type 1 transfer function [Eq. (8.52)].

Type 1 Feedback Control System

The inverse of the transfer function used in Sec. 8.10 for the Type 1 system of Eq. (8.35) is

$$\mathbf{G}^{-1}(j\omega) = \frac{j\omega(1 + j\omega T_m)(1 + j\omega T_c)(1 + j\omega T_q)}{K_1} \tag{8.52}$$

The two limiting points are

$$\lim_{\omega \to 0^+} \mathbf{G}^{-1}(j\omega) = 0\angle 90^\circ \qquad \lim_{\omega \to +\infty} \mathbf{G}^{-1}(j\omega) = \infty\angle 360^\circ$$

With these end points and previous analyses, the inverse polar plot of Eq. (8.52) is a smooth curve, as shown in Fig. 8.22. Additional terms in the numerator or the denominator alter the location of the end points and the shape of the plot accordingly.

Type 2 Feedback Control System

In the same manner, the inverse polar plot for the modified Type 2 system of Eq. (8.45) can be obtained. The inverse of the transfer function of Eq. (8.45) for this system is

$$\mathbf{G}_o^{-1}(j\omega) = \frac{(j\omega)^2(1 + j\omega T_f)(1 + j\omega T_m)(1 + j\omega T_2)}{K_2'(1 + j\omega T_1)} \tag{8.53}$$

and its polar plot is illustrated in Fig. 8.23.

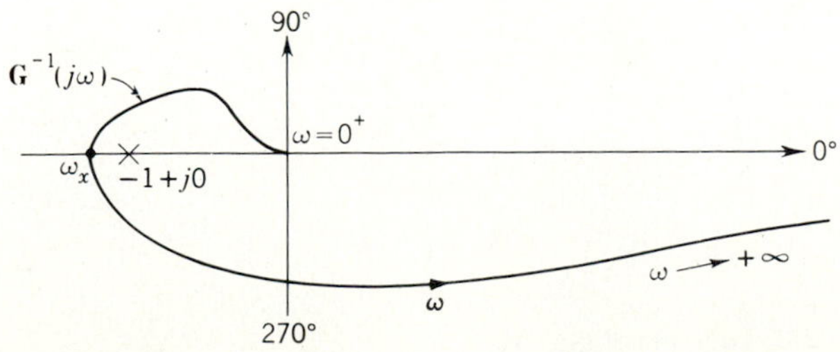

FIGURE 8.23
Inverse polar plot of a typical Type 2 transfer function [Eq. (8.53)].

The inverse polar plots of a forward transfer function are obtained with the aid of the following criteria, which determine the key parts of the curve.

Step 1. Once the system type is determined from the forward transfer function, the portion of the inverse polar plot representing the $\lim_{\omega \to 0^+} \mathbf{G}^{-1}(j\omega)$ is located. Note that in taking the limit the zero-frequency point is approached in the clockwise sense.

Step 2. By using the general form of the inverse forward transfer function as given by

$$G^{-1}(s) = \frac{s^m(1 + T_1 s)(1 + T_2 s) \cdots (1 + T_u s)}{K_m(1 + T_a s) \cdots (1 + T_w s)} \tag{8.54}$$

the high-frequency end of the polar plot is determined as follows:

$$\lim_{\omega \to +\infty} \mathbf{G}^{-1}(j\omega) = \infty \angle (m + u - w)90^\circ \tag{8.55}$$

Step 3. The frequencies at the points of intersection of the polar plot with the negative real axis and the imaginary axis are determined, respectively, by setting

$$\mathrm{Im}\left[\mathbf{G}^{-1}(j\omega)\right] = 0 \tag{8.56}$$

$$\mathrm{Re}\left[\mathbf{G}^{-1}(j\omega)\right] = 0 \tag{8.57}$$

Step 4. The exact shape of the polar plot in the vicinity of the negative real axis crossing is important, and points on $\mathbf{G}^{-1}(j\omega)$ should be accurately determined in this area with the aid of step 3 or a computer (see App. B).[4]

A summary of the inverse polar plots for the different types of systems is shown in Fig. 8.24. Just as for the direct polar plots, the low-frequency characteristic (as $\omega \to 0^+$) distinguishes the different system types.

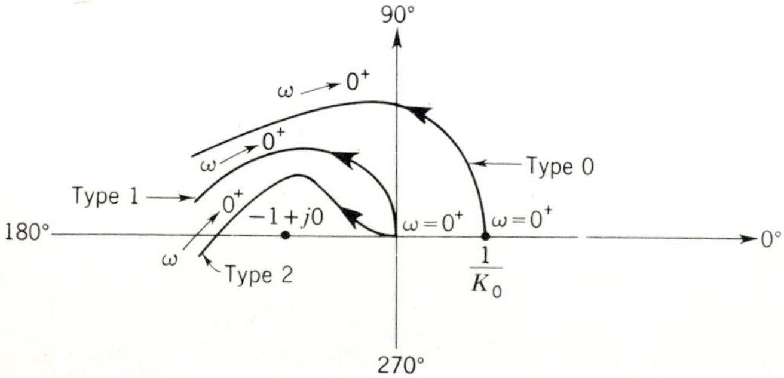

FIGURE 8.24
A summary of inverse polar plots of different types of systems.

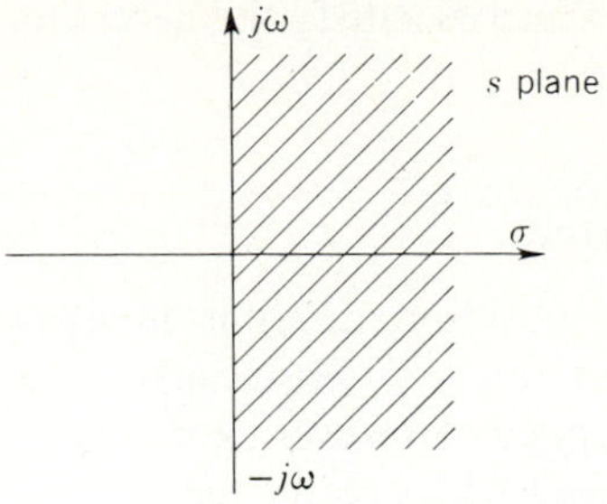

FIGURE 8.25
Prohibited region in the s plane.

8.13 NYQUIST'S STABILITY CRITERION

A system designer must be sure that the closed-loop system he designs is stable. The Nyquist stability criterion[2,3,7] provides a simple graphical procedure for determining closed-loop stability from the frequency-response curves of the open-loop transfer function $\mathbf{G}(j\omega)\mathbf{H}(j\omega)$. The application of this method in terms of the polar plot is covered in this section; application in terms of the log magnitude–angle (Nichols) diagram is covered in Sec. 8.19.

For a stable system the roots of the characteristic equation

$$B(s) = 1 + G(s)H(s) = 0 \tag{8.58}$$

cannot be permitted to lie in the right-half s plane or on the $j\omega$ axis, as shown in Fig. 8.25. In terms of $G = N_1/D_1$ and $H = N_2/D_2$, Eq. (8.58) becomes

$$B(s) = 1 + \frac{N_1 N_2}{D_1 D_2} = \frac{D_1 D_2 + N_1 N_2}{D_1 D_2} \tag{8.59}$$

Note that the numerator and denominator of $B(s)$ have the same degree. It is seen that the *poles of the open-loop transfer function* $G(s)H(s)$ *are the poles* of $B(s)$. The closed-loop transfer function of the system is

$$\frac{C(s)}{R(s)} = \frac{G(s)}{1 + G(s)H(s)} = \frac{N_1 D_2}{D_1 D_2 + N_1 N_2} \tag{8.60}$$

The denominator of $C(s)/R(s)$ is the same as the numerator of $B(s)$. The condition for stability may therefore be restated as: for a stable system none of the zeros of $B(s)$ can lie in the right-half s plane or on the imaginary axis. Nyquist's stability criterion relates the number of zeros and poles of $B(s)$ that lie in the right-half s plane to the polar plot of $G(s)H(s)$.

Limitations

In this analysis it is assumed that all the control systems are inherently linear or that their limits of operation are confined to give a linear operation. This yields a set of linear differential equations which describe the dynamic performance of the systems. Because of the physical nature of feedback control systems, the order of the denominator D_1D_2 is equal to or greater than the order of the numerator N_1N_2 of the open-loop transfer function $G(s)H(s)$. Mathematically, this means that $\lim_{s \to \infty} G(s)H(s) \to 0$ or a constant. These two factors satisfy the necessary limitations to the generalized Nyquist stability criterion.

Mathematical Basis for Nyquist's Stability Criterion

A rigorous mathematical derivation of Nyquist's stability criterion involves complex-variable theory and is available in the literature.[11] Fortunately, the result of the derivation is simple and is readily applied. However, a complete knowledge of its derivation ensures that it will be used with greater facility and sureness. A qualitative approach is now presented for the special case that $B(s)$ is a rational fraction. The characteristic function $B(s)$ given by Eq. (8.59) is rationalized, factored, and then written in the form

$$B(s) = \frac{(s - Z_1)(s - Z_2) \cdots (s - Z_n)}{(s - p_1)(s - p_2) \cdots (s - p_n)} \tag{8.61}$$

where $Z_1, Z_2, \ldots, Z_n$ are the zeros and $p_1, p_2, \ldots, p_n$ are the poles. (Remember that z_i is used to denote a zero of $G(s)H(s)$.) The poles p_i are the same as the poles of the open-loop transfer function $G(s)H(s)$ and include the s term for which $p = 0$, if it is present.

In Fig. 8.26 some of the poles and zeros of a generalized function $B(s)$ are arbitrarily drawn on the s plane. Also, an arbitrary *closed contour* Q' is drawn which encloses the zero Z_1. To the point O' on Q', whose coordinates are $s = \sigma + j\omega$, are drawn directed line segments from all the poles and zeros. The lengths of these directed line segments are given by $s - Z_1$, $s - Z_2$, $s - p_1$, $s - p_2$, etc. Not all the directed segments from the poles and zeros are indicated

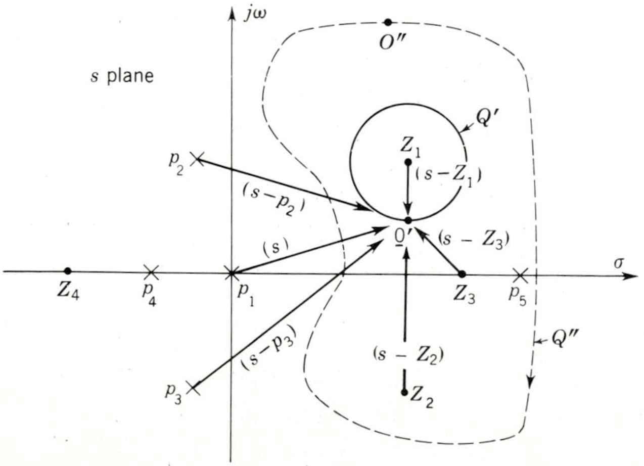

FIGURE 8.26
A plot of some poles and zeros of Eq. (8.61).

in the figure, as they are not necessary to proceed with this development. *As the point O' is rotated clockwise once around the closed contour Q', length $s - Z_1$ rotates through a net clockwise angle of* 360°. *All the other directed* segments have rotated through a *net angle of* 0°. Thus, by referring to Eq. (8.60) it is seen that the clockwise rotation of 360° for the length $s - Z_1$ *must simultaneously be realized* by the function $B(s)$ for the enclosure of the zero Z_1 by the path Q'.

Consider now a larger closed contour Q'' which includes the zeros Z_1, Z_2, Z_3 and the pole p_5. *As a point O'' is rotated clockwise once around the closed curve Q'', each of the directed line segments from the enclosed pole and zeros rotates through a net clockwise angle of* 360°. Since the angular rotation of the pole is experienced by the characteristic function $B(s)$ in its denominator, the net angular rotation realized by Eq. (8.61) must be equal to the net angular rotations due to the pole p_5 minus the net angular rotations due to the zeros Z_1, Z_2, and Z_3. In other words, the net angular rotation experienced by $1 + G(s)H(s)$ is $360° - (3)(360°) = -720°$. Therefore, for this case, the net number of rotations N experienced by $B(s) = 1 + G(s)H(s)$ for the clockwise movement of point O'' *once* about the closed contour Q'' is equal to -2; that is,

$$N = (\text{Number of poles enclosed}) - (\text{number of zeros enclosed}) = 1 - 3 = -2$$

where the minus sign denotes clockwise (cw) rotation. Note that if the contour Q'' includes only the pole p_5, $B(s)$ experiences one counterclockwise (ccw) rotation as the point O'' is moved clockwise around the contour. Also, for *any closed path* that may be chosen, all the poles and zeros that lie outside the closed path each contribute a net angular rotation of 0° to $B(s)$ as a point is moved once around this contour.

Generalizing Nyquist's Stability Criterion

Consider now a closed contour Q such that the whole right-half s plane is encircled (see Fig. 8.27), thus enclosing all zeros and poles of $B(s)$ that have positive real parts. As a consequence of the theory of complex variables used in

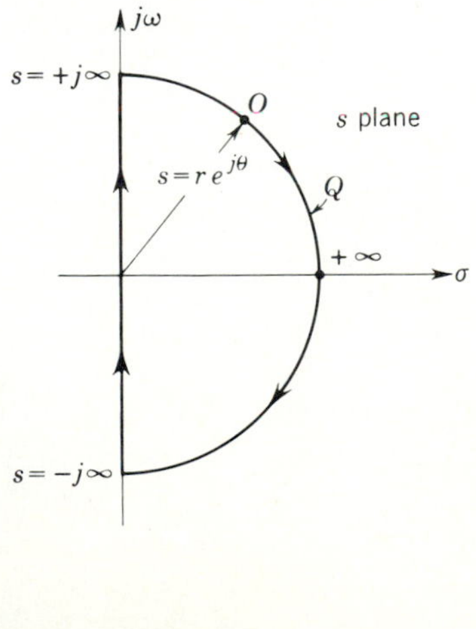

FIGURE 8.27
The contour that encloses the entire right-half s plane.

the derivation, the contour Q must not pass *through any poles or zeros of* $B(s)$. When the results of the preceding discussion are applied to the contour Q, the following properties are noted:

1. The total number of clockwise rotations of $B(s)$ due to its zeros is equal to the total number of zeros Z_R in the right-half s plane.
2. The total number of counterclockwise rotations of $B(s)$ due to its poles is equal to the total number of poles P_R in the right-half s plane.
3. The *net* number of rotations N of $B(s) = 1 + G(s)H(s)$ about the origin is equal to its total number of poles P_R minus its total number of zeros Z_R in the right-half s plane. N may be positive (ccw), negative (cw), or zero.

The essence of these three conclusions can be represented by the equation

$$N = \frac{\text{change in phase of } [1 + G(s)H(s)]}{2\pi} = P_R - Z_R \tag{8.62}$$

where counterclockwise rotation is defined as being positive and clockwise rotation is negative. In order for the characteristic function $B(s)$ to realize a net rotation N, the directed line segment representing $B(s)$ (see Fig. 8.28a) must rotate about the origin $360N$ degrees, or N complete revolutions. Solving for Z_R in Eq. (8.62) yields

$$Z_R = P_R - N \tag{8.63}$$

Since $B(s)$ can have no zeros Z_R in the right-half s plane for a stable system; it can therefore be concluded that, *for a stable system, the net number of rotations of* $B(s)$ *about the origin must be counterclockwise and equal to the number of poles* P_R *that lie in the right-half plane*. In other words, if $B(s)$ experiences a net clockwise rotation, this indicates that $Z_R > P_R$, where $P_R \geq 0$, and thus the closed-loop system is unstable. If there are zero net rotations, then $Z_R = P_R$ and the system may or may not be stable, according as $P_R = 0$ or $P_R > 0$.

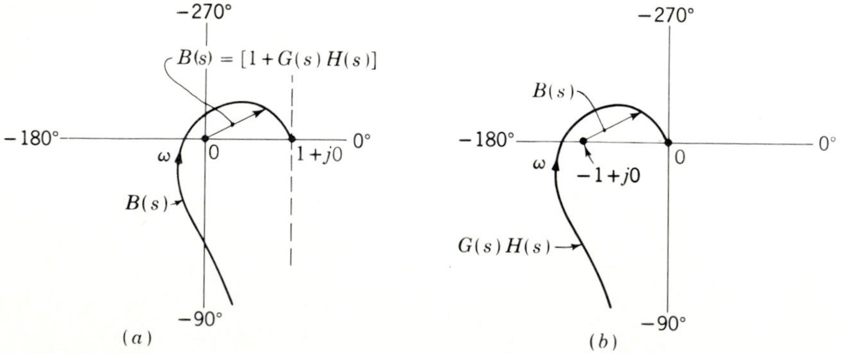

FIGURE 8.28
A change of reference for $B(s)$.

Obtaining a Plot of $B(s)$

Figure 8.28*a* and *b* shows a plot of $B(s)$ and a plot of $G(s)H(s)$, respectively. By moving the origin of Fig. 8.28*b* to the $-1 + j0$ point, the curve is now equal to $1 + G(s)H(s)$, which is $B(s)$. Since $G(s)H(s)$ is known, this function is plotted, and then the origin is moved to the -1 point to obtain $B(s)$.

In general, the open-loop transfer functions of many physical systems do not have any poles P_R in the right-half s plane. In this case, $Z_R = N$. *Thus for a stable system the net number of rotations about the* $-1 + j0$ *point must be zero when there are no poles of* $G(s)H(s)$ *in the right-half s plane.*

In the event that the function $G(s)H(s)$ has some poles in the right-half s plane and the denominator is not in factored form, the number P_R can be determined by applying Routh's criterion to D_1D_2. The Routhian array gives the number of roots in the right-half s plane by the number of sign changes in the first column.

Analysis of Path Q

In applying Nyquist's criterion, the whole right-half s plane must be encircled to ensure the inclusion of all poles or zeros in this portion of the plane. In Fig. 8.27 the entire right-half s plane is included by the closed path Q which is composed of the following two segments:

1. One segment is the imaginary axis from $-j\infty$ to $+j\infty$.
2. The other segment is a semicircle of infinite radius that encircles the entire right-half s plane.

The portion of the path along the imaginary axis is represented mathematically by $s = j\omega$. Thus, replacing s by $j\omega$ in Eq. (8.61) and letting ω take on all values from $-\infty$ to $+\infty$ gives the portion of the $B(s)$ plot corresponding to that portion of the closed contour Q which lies on the imaginary axis.

One of the requirements of the Nyquist criterion is that $\lim_{s\to\infty} G(s)H(s) \to 0$ or a constant. Thus $\lim_{s\to\infty} B(s) = \lim_{s\to\infty} [1 + G(s)H(s)] \to 1$ or 1 plus the constant. As a consequence, as the point O moves along the segment of the closed contour represented by the semicircle of infinite radius, the corresponding portion of the $B(s)$ plot is a fixed point. As a result, the movement of point O along only the imaginary axis from $-j\infty$ to $+j\infty$ results in the same net rotation of $B(s)$ as if the whole contour Q were considered. *In other words, all the rotation of* $B(s)$ *occurs while the point O goes from* $-j\infty$ *to* $+j\infty$ *along the imaginary axis.* More generally, this statement applies only to those transfer functions $G(s)H(s)$ that conform to the limitations stated earlier in this section.†

† A transfer function that does not conform to these limitations and to which the above italicized statement does not apply is $G(s)H(s) = s$.

Effect of Poles at the Origin on the Rotation of $B(s)$

Some transfer functions $G(s)H(s)$ have an s^m in the denominator. Since no poles or zeros can lie on the contour Q, the contour shown in Fig. 8.27 must be modified. For these cases the manner in which the $\omega = 0^-$ and $\omega = 0^+$ portions of the plot are joined is now investigated. This can best be done by taking an example. Consider the transfer function

$$G(s)H(s) = \frac{K_1}{s(1 + T_1 s)(1 + T_2 s)} \tag{8.64}$$

The direct polar plot of $\mathbf{G}(j\omega)\mathbf{H}(j\omega)$ of this function is obtained by substituting $s = j\omega$ into Eq. (8.64). Figure 8.29 represents the plot of $G(s)H(s)$ for values of s along the imaginary axis; that is, $j0^+ < j\omega < j\infty$ and $-j\infty < j\omega < j0^-$. The plot is drawn for both positive and negative frequency values. *The polar plot drawn for negative frequencies is the conjugate of the plot drawn for positive frequencies*. This means that the curve for negative frequencies is symmetrical to the curve for positive frequencies, with the real axis as the axis of symmetry.

The closed contour Q of Fig. 8.27, in the vicinity of $s = 0$, is modified to avoid passing through the origin, as shown in Fig. 8.30a in order to determine the system stability. In other words, the point O is moved along the negative imaginary axis from $s = -j\infty$ to a point where $s = -j\omega = 0^- \angle -\pi/2$ becomes very small; that is, $s = -j\varepsilon$. Then the point O moves along a semicircular path of radius $s = \varepsilon e^{j\theta}$ in the right-half s plane with a very small radius ε until it reaches the positive imaginary axis at $s = +j\omega = j0^+ = 0^+ \angle \pi/2$. From here the point O proceeds along the positive imaginary axis to $s = +j\infty$. Letting the radius approach zero, $\varepsilon \to 0$, for the semicircle around the origin ensures the inclusion of all poles and zeros in the right-half s plane. To complete the plot of $B(s)$ in Fig. 8.29, the effect of moving point O on this semicircle around the origin must be investigated.

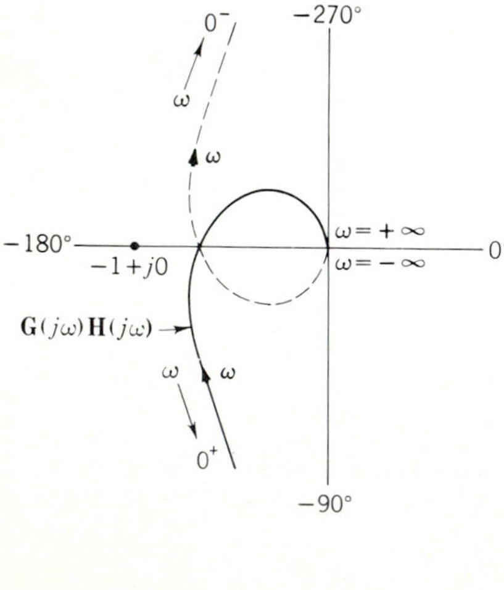

FIGURE 8.29
A plot of the transfer function $\mathbf{G}(j\omega)\mathbf{H}(j\omega)$ for Eq. (8.64).

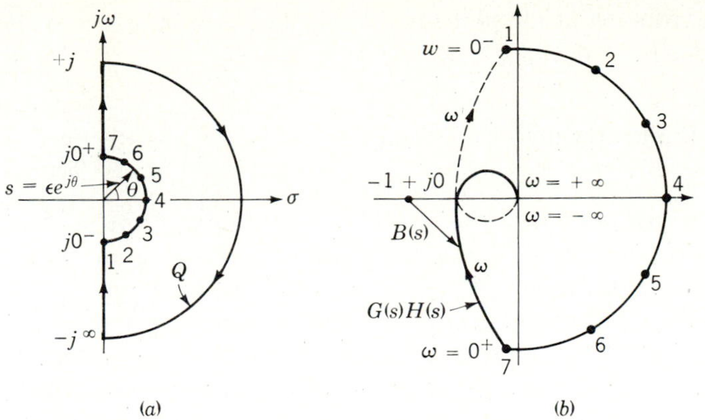

FIGURE 8.30
(a) The contour Q which encircles the right-half s plane. (b) Complete plot for Eq. (8.64).

For the semicircular portion of the path Q represented by $s = \varepsilon e^{j\theta}$, where $\varepsilon \to 0$ and $-\pi/2 \leq \theta \leq \pi/2$, Eq. (8.64) becomes

$$G(s)H(s) = \frac{K_1}{s} = \frac{K_1}{\varepsilon e^{j\theta}} = \frac{K_1}{\varepsilon}e^{-j\theta} = \frac{K_1}{\varepsilon}e^{j\psi} \tag{8.65}$$

where $K_1/\varepsilon \to \infty$ as $\varepsilon \to 0$, and $\psi = -\theta$ goes from $\pi/2$ to $-\pi/2$ as the directed segment s goes counterclockwise from $\varepsilon\underline{/-\pi/2}$ to $\varepsilon\underline{/+\pi/2}$. Thus, in Fig. 8.29, the end points from $\omega \to 0^-$ and $\omega \to 0^+$ are joined by a semicircle of infinite radius in the first and fourth quadrants. Figure 8.30 illustrates the above procedure and shows the completed contour of $G(s)H(s)$ as the point O moves along the modified contour Q in the s plane in the clockwise direction. When the origin is moved to the $-1 + j0$ point, the curve becomes $B(s)$.

The plot of $B(s)$ in Fig. 8.30b does not encircle the $-1 + j0$ point; therefore the encirclement N is zero. From Eq. (8.64) there are no poles within Q; that is, $P_R = 0$. Thus, when Eq. (8.63) is applied, $Z_R = 0$ and the system is stable.

Transfer functions that have the term s^m in the denominator have the general form, as $\varepsilon \to 0$,

$$G(s)H(s) = \frac{K_m}{s^m} = \frac{K_m}{(\varepsilon^m)e^{jm\theta}} = \frac{K_m}{\varepsilon^m}e^{-jm\theta} = \frac{K_m}{\varepsilon^m}e^{jm\psi} \tag{8.66}$$

where $m = 1, 2, 3, 4, \ldots$. With the reasoning used in the preceding example, it is seen from Eq. (8.66) that, as s moves from 0^- to 0^+, the plot of $G(s)H(s)$ traces m clockwise semicircles of infinite radius about the origin. If $m = 2$, then, as θ goes from $-\pi/2$ to $\pi/2$ in the s plane with radius ε, $G(s)H(s)$ experiences a net rotation of (2)(180°) = 360°.

Since the polar plots are symmetrical about the real axis, it is only necessary to determine the shape of the plot of $G(s)H(s)$ for a range of values of

$0 < \omega < +\infty$. The net rotation of the plot for the range of $-\infty < \omega < +\infty$ is twice that of the plot for the range of $0 < \omega < +\infty$.

When $\mathbf{G}(j\omega)\mathbf{H}(j\omega)$ Passes through the Point $-1 + j0$

When the curve of $\mathbf{G}(j\omega)\mathbf{H}(j\omega)$ passes through the $-1 + j0$ point, the number of encirclements N is indeterminate. This corresponds to the condition where $B(s)$ has zeros on the imaginary axis. A necessary condition for applying the Nyquist criterion is that the path encircling the specified area must not pass through any poles or zeros of $B(s)$. When this condition is violated, the value for N becomes indeterminate and the Nyquist stability criterion cannot be applied. Simple imaginary zeros of $B(s)$ mean that the closed-loop system will have a continuous steady-state sinusoidal component in its output which is independent of the form of the input. Unless otherwise stated, this condition is considered unstable.

8.14 EXAMPLES OF NYQUIST'S CRITERION USING DIRECT POLAR PLOT

Several polar plots are illustrated in this section. These plots can be obtained with the aid of pole-zero diagrams, from calculations for a few key frequencies, or from a computer program.[4] The angular variation of each term of a $G(s)H(s)$ function, as the contour Q is traversed, is readily determined from its pole-zero diagram. Both stable and unstable open-loop transfer functions $G(s)H(s)$ are illustrated in the following examples for which all time constants T_i are positive.

Example 1 Type 0. The direct polar plot is shown in Fig. 8.31 for the following Type 0 transfer function:

$$G(s)H(s) = \frac{K_0}{(1 + T_1 s)(1 + T_2 s)} \tag{8.67}$$

From this plot it is seen that $N = 0$, and Eq. (8.67) yields the value $P_R = 0$. Therefore, the value of Z_R is $Z_R = P_R - N = 0$ and the system represented by Eq.

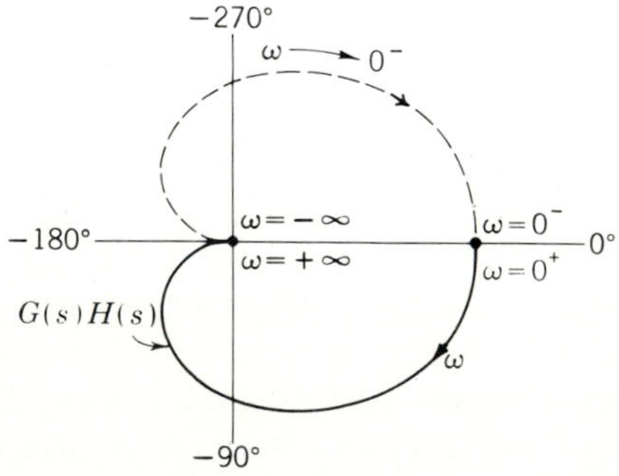

FIGURE 8.31
A plot of a typical $G(s)H(s)$ transfer function for $-\infty < \omega < \infty$.

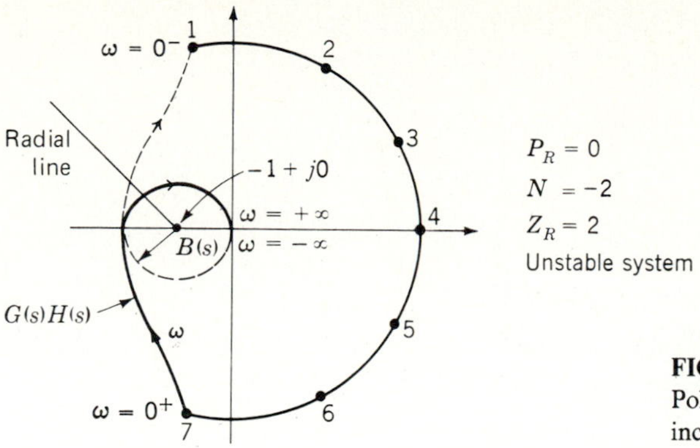

FIGURE 8.32
Polar plot of Fig. 8.30*b* with increased gain.

(8.67) is stable. Thus the Nyquist stability criterion indicates that, no matter how much the gain K_0 is increased, this system is always stable.

Example 2 Type 1. The example in Fig. 8.30 for the transfer function of the form

$$G(s)H(s) = \frac{K_1}{s(1 + T_1 s)(1 + T_2 s)} \tag{8.68}$$

is shown in the preceding section to be stable. However, if the gain K_1 is increased, the system is made unstable, as seen from Fig. 8.32. Note that the rotation of $B(s)$, in the direction of increasing frequency, produces a net angular rotation of $-720°$, or two complete clockwise rotations ($N = -2$) so that $Z_R = P_R - N = 2$.

The number of rotations N can be determined by drawing a radial line from the $-1 + j0$ point. Then, at each intersection of this radial line with the curve of $G(s)H(s)$, the direction of increasing frequency is noted. If the frequency along the curve of $G(s)H(s)$ is increasing as the curve crosses the radial line in a counterclockwise direction, this crossing is positive. Similarly, if the frequency along the curve of $G(s)H(s)$ is increasing as the curve crosses the radial line in a clockwise direction, this crossing is negative. The sum of the crossings, including the sign, is equal to N. In Fig. 8.32 both crossings of the radial line are cw and are negative. Thus the sum of the crossings is $N = -2$.

Example 3 Type 2

$$G(s)H(s) = \frac{K_2(1 + T_4 s)}{s^2(1 + T_1 s)(1 + T_2 s)(1 + T_3 s)} \tag{8.69}$$

where $T_4 > T_1 + T_2 + T_3$. Figure 8.33 shows the mapping of $G(s)H(s)$ for the contour Q of the s plane. The word *mapping*, as used here, means that for a given point in the s plane there corresponds a given value of $G(s)H(s)$ or $B(s)$.

As pointed out in the preceding section, the presence of the s^2 term in the denominator of Eq. (8.69) results in a net rotation of 360° in the vicinity of $\omega = 0$, as shown in Fig. 8.33. For the complete range of frequencies the net rotation is zero; thus, since $P_R = 0$, the system is stable. The value of N can be determined by drawing the line radiating from the $-1 + j0$ point (see Fig. 8.33) and noting one cw and one ccw crossing; thus $N = 0$. Like the previous example, this system can be

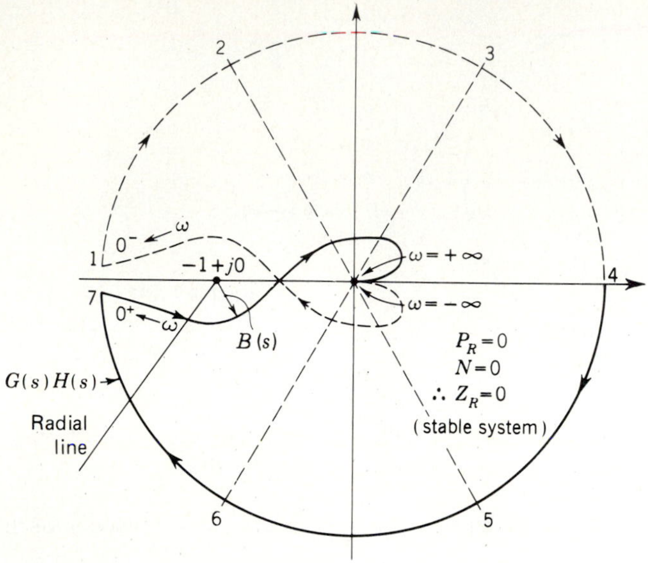

FIGURE 8.33
The complete polar plot of Eq. (8.69).

made unstable by increasing the gain sufficiently for the $G(s)H(s)$ plot to cross the negative real axis to the left of the $-1 + j0$ point.

Example 4 Conditionally stable system

$$G(s)H(s) = \frac{K_0(1 + T_1 s)^2}{(1 + T_2 s)(1 + T_3 s)(1 + T_4 s)(1 + T_5 s)^2} \tag{8.70}$$

where $T_5 < T_1 < T_2$, T_3, and T_4. The complete polar plot of Eq. (8.70) is illustrated in Fig. 8.34 for a particular value of gain. For the radial line drawn from the $-1 + j0$ point the number of rotations is readily determined as $N = 0$.

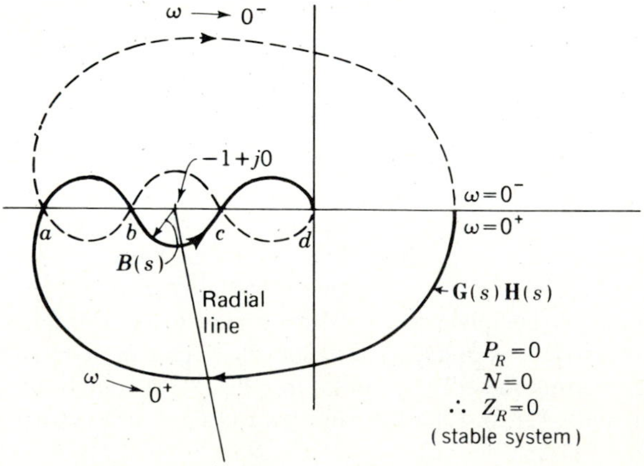

FIGURE 8.34
The complete polar plot of Eq. (8.70).

It is seen in this example that the system can be made unstable not only by increasing the gain but also by decreasing the gain. If the gain is increased sufficiently for the $-1 + j0$ point to lie between the points c and d of the polar plot, the net clockwise rotation is equal to 2. Therefore $Z_R = 2$ and the system is unstable. On the other hand, if the gain is decreased so that the $-1 + j0$ point lies between the points a and b of the polar plot, the net clockwise rotation is again 2 and the system is unstable. The gain can be further decreased so that the $-1 + j0$ point lies to the left of point a of the polar plot, resulting in a stable system. *This system is therefore conditionally stable*. A conditionally stable system is stable for a given range of values of gain but becomes unstable if the gain is either reduced or increased sufficiently. Such a system places a greater restriction on the stability and drift of amplifier gain and system characteristics. In addition, an effective gain reduction occurs in amplifiers that reach saturation with large input signals. This, in turn, may result in an unstable operation for the conditionally stable system.

Example 5 Open-loop unstable

$$G(s)H(s) = \frac{K_1(T_2 s + 1)}{s(T_1 s - 1)} \tag{8.71}$$

The complete polar plot for this equation is illustrated in Fig. 8.35 for a particular value of gain K_1. There is one cw intersection with the radial line drawn from the $-1 + j0$ point, and thus $N = -1$. Therefore, since $P_R = 1$, $Z_R = 2$ and the system is unstable. In this example it is seen that the system is unstable for low values of gain ($0 < K_1 < K_{1x}$) for which the real axis crossing b is to the right of the $-1 + j0$ point. For the range $0 < K_1 < K_{1x}$ the $-1 + j0$ point is located, as shown in Fig. 8.35, between the points a and b. For the range $K_{1x} < K_1 < \infty$ the $-1 + j0$ point lies between the points b and c, which yields $N = +1$, thus resulting in $Z_R = 0$ and a stable system.

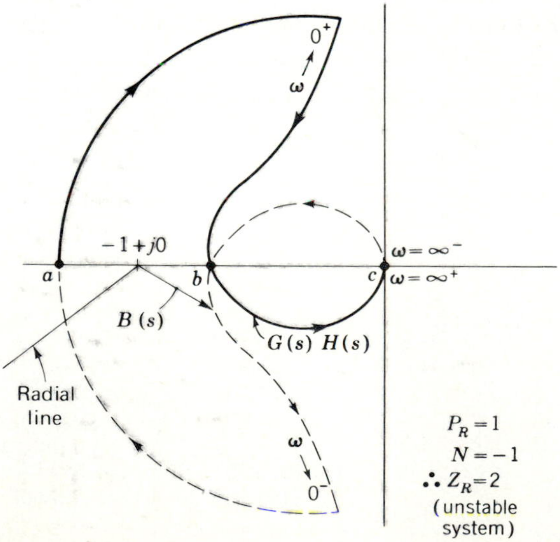

FIGURE 8.35
The complete polar plot of the open-loop unstable transfer function
$G(s)H(s) = \dfrac{K_1(T_2 s + 1)}{s(T_1 s - 1)}$

8.15 NYQUIST'S STABILITY CRITERION APPLIED TO SYSTEMS HAVING DEAD TIME

Transport lag, as described in Sec. 7.12, is represented by the transfer function $G_\tau(s) = e^{-\tau s}$. The frequency transfer function is

$$\mathbf{G}_\tau(j\omega) = e^{-j\omega\tau} = 1\angle -\omega\tau \tag{8.72}$$

It has a magnitude of unity and a negative angle whose magnitude increases directly in proportion to frequency. The polar plot of Eq. (8.72) is a unit circle which is traced indefinitely, as shown in Fig. 8.36. The log magnitude and phase-angle diagram shows a constant value of 0 dB and a phase angle which decreases with frequency.

An example can best illustrate the application of the Nyquist criterion to a control system having dead time, or transport lag. Figure 8.37*a* illustrates the complete polar plot of a stable system having a specified value of gain and the transfer function

$$G_x(s)H(s) = \frac{K_1}{s(1+T_1s)(1+T_2s)} \tag{8.73}$$

If dead time is added to this system, its transfer function becomes

$$G(s)H(s) = \frac{K_1e^{-\tau s}}{s(1+T_1s)(1+T_2s)} \tag{8.74}$$

and the resulting complete polar plot is shown in Fig. 8.37*b*. When the contour Q is traversed and the polar-plot characteristic of dead time, shown in Fig. 8.36, is included, the effects on the complete polar plot are as follows:

1. In traversing the imaginary axis of the contour Q between $0^+ < \omega < +\infty$, the polar plot of $\mathbf{G}(j\omega)\mathbf{H}(j\omega)$ in the third quadrant is shifted clockwise, closer to the $-1 + j0$ point. Thus, if the dead time is increased sufficiently, the $-1 + j0$ point will be enclosed by the polar plot and the system becomes unstable.

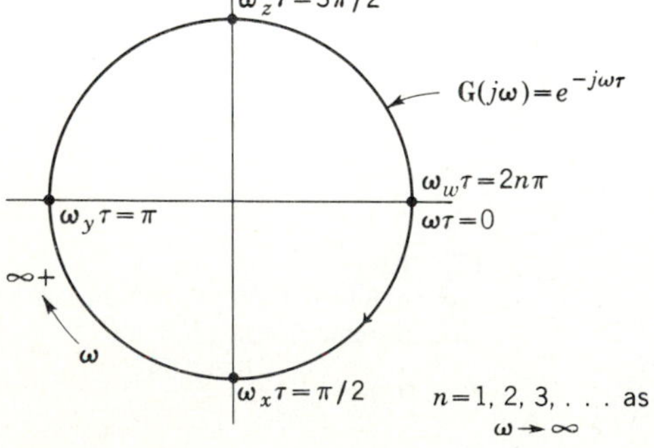

FIGURE 8.36
Polar-plot characteristic for transport lag.

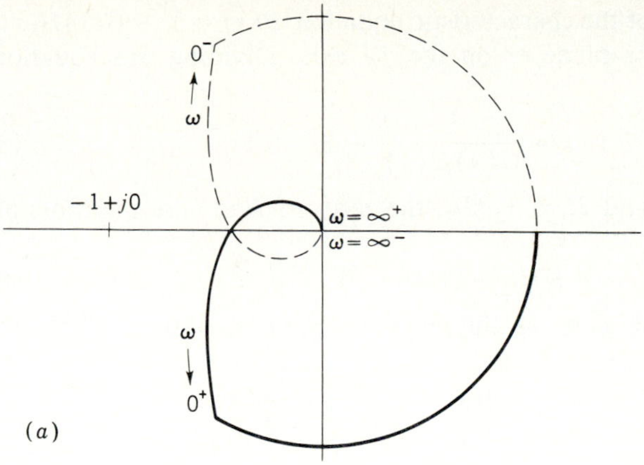

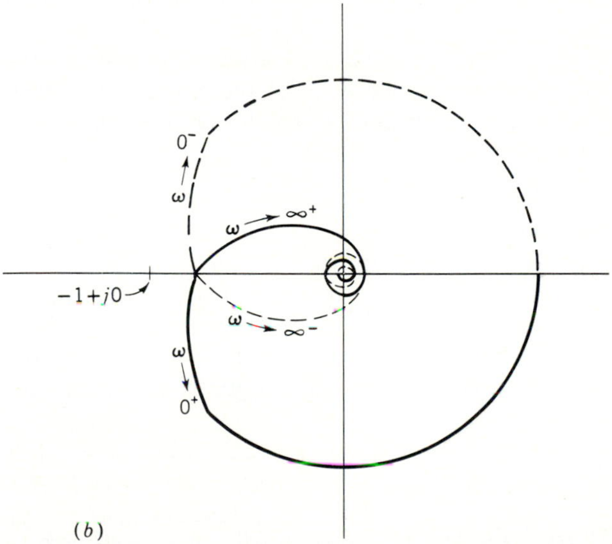

FIGURE 8.37
Complete polar plots of (*a*) Eq. (8.73) and (*b*) Eq. (8.74).

2. As $\omega \to +\infty$, the magnitude of the angle contributed by the transport lag increases indefinitely. This yields a spiraling curve as $|\mathbf{G}(j\omega)\mathbf{H}(j\omega)| \to 0$.

A transport lag therefore tends to make a system less stable.

8.16 NYQUIST'S STABILITY CRITERION APPLIED TO THE INVERSE POLAR PLOTS

The criterion applied to the inverse plots is the same as that for the direct plots except for one minor modification. As previously stipulated, for a system to be stable, none of the roots of the characteristic equation $B(s) = 1 + G(s)H(s) = 0$ can lie in the right-half s plane or on the $j\omega$ axis. Dividing this equation by $G(s)H(s)$ yields

$$B'(s) = \frac{1}{G(s)H(s)} + 1 = 0 \tag{8.75}$$

In terms of $G = N_1/D_1$ and $H = N_2/D_2$, the zeros of $B'(s)$ are the roots of the characteristic equation

$$D_1 D_2 + N_1 N_2 = 0 \tag{8.76}$$

which are seen to be the same as the zeros of $B(s)$. Equation (8.75) can be expressed as

$$B'(s) = \frac{(s - Z_1)(s - Z_2)\cdots(s - Z_n)}{(s - z_1)(s - z_2)\cdots(s - z_w)} \tag{8.77}$$

where $Z_1, Z_2, \ldots, Z_n$ are the zeros of the functions $B(s)$ and $B'(s)$, and $z_1, z_2, \ldots, z_w$ are the poles of $B'(s)$, which are the same as the zeros of $G(s)H(s)$.

The mathematical basis for Nyquist's stability criterion as applied to the inverse plots is the same as that stipulated in Sec. 8.13 for the direct plots. When $B'(s)$ takes on the values of s on the contour Q which encircles the entire right-half of the s plane, the resulting equation is

$$Z_R = P'_R - N' \tag{8.78}$$

where N' = number of net rotations of $B'(s)$
P'_R = number of zeros of $G(s)H(s)$ in right-half s plane
Z_R = number of roots of the characteristic equation in right-half plane

The plot of $B'(s) = 1/G(s)H(s) + 1$ is most easily obtained by first plotting the known function $1/G(s)H(s)$. Then the origin is moved to the $-1 + j0$ point to obtain $B'(s)$.

Analysis of Path Q

Replacing s by $j\omega$ in $[G(s)H(s)]^{-1}$ and letting ω take on all values from $-\infty$ to $+\infty$ gives the portion of the plot corresponding to that portion of the closed contour Q on the imaginary axis of Fig. 8.27. The value is a constant for the case of $\omega = 0$. Thus, as the point O on the closed contour Q passes through or around the point $0 + j0$, the corresponding portion of the $B'(s)$ plot crosses the real axis.

To complete the plot of $[G(s)H(s)]^{-1}$ the effect of moving point O on the semicircle of infinite radius on the contour, which ensures the inclusion of all poles and zeros in the right-half plane, must be investigated. This semicircle is represented by

$$s = re^{j\theta} \tag{8.79}$$

where $r \to \infty$ and $\pi/2 \geq \theta \geq -\pi/2$. The general open-loop transfer function is

$$G(s)H(s) = \frac{K_m(s - z_1)\cdots(s - z_w)}{s^m(s - p_1)(s - p_2)\cdots(s - p_u)} \tag{8.80}$$

The function $[G(s)H(s)]^{-1}$ for this portion of the contour Q has the value

$$\lim_{s\to\infty} \frac{1}{G(s)H(s)} = \frac{s^m s^u}{K_m s^w} = \frac{r^{(m+u-w)}}{K_m} e^{j(m+u-w)\theta} \tag{8.81}$$

From this equation it is easily seen that, as the point O, for $r = \infty$, goes clockwise from $\pi/2$ to $-\pi/2$ on the path Q in the s plane, the plot of $1/G(s)H(s)$ traces $m + u - w$ clockwise semicircles of infinite radius about the origin.

8.17 EXAMPLES OF NYQUIST'S CRITERION USING THE INVERSE POLAR PLOTS

Example 1 Type 0. The inverse polar plot of

$$[G(s)H(s)]^{-1} = \frac{(1 + T_1 s)(1 + T_2 s)}{K_0} \tag{8.82}$$

is illustrated in Fig. 8.38. For this example

$$\lim_{s\to\infty} [G(s)H(s)]^{-1} = \frac{T_1 T_2}{K_0} r^2 e^{j2\theta} \tag{8.83}$$

By referring to Fig. 8.38 and using Eq. (8.83), the complete polar plot of $[G(s)H(s)]^{-1}$ shown in Fig. 8.38a is obtained. By moving the origin to the $-1 + j0$ point, the curve becomes $B'(s)$. When Nyquist's criterion is applied to this plot, the net encirclement of the origin is zero. From Eq. (8.82) the value $P_R' = 0$;

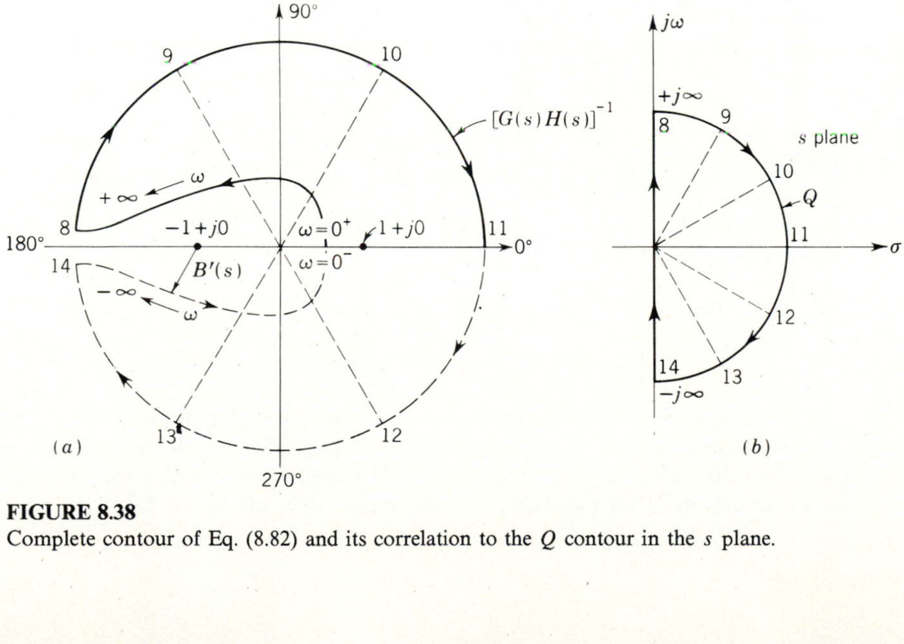

FIGURE 8.38
Complete contour of Eq. (8.82) and its correlation to the Q contour in the s plane.

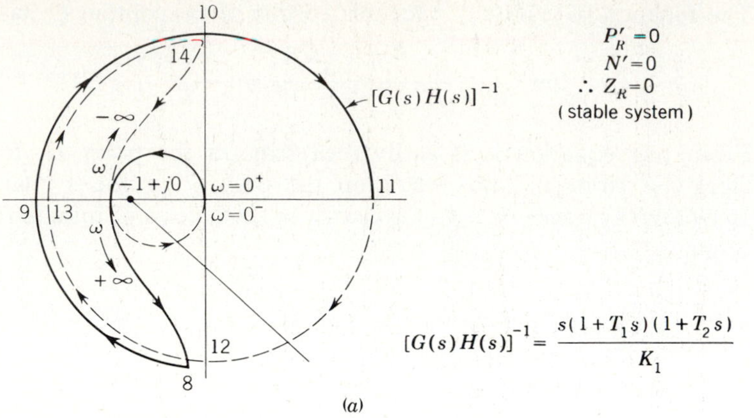

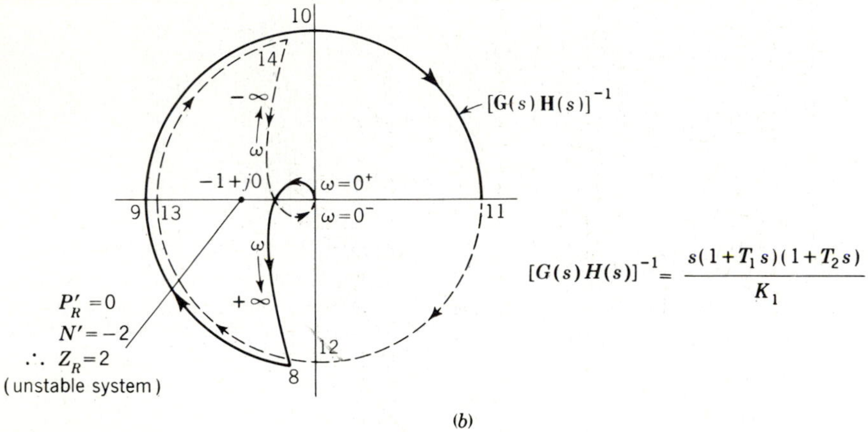

FIGURE 8.39
A complete inverse polar plot for Eq. (8.85), indicating correlation to Fig. 8.38*b*: (*a*) stable; (*b*) unstable.

therefore,

$$Z_R = P'_R - N' = 0 - 0 = 0 \tag{8.84}$$

In other words, the system is stable. No matter how much the gain K_0 is increased, this particular system is never unstable.

Example 2 Type 1. The inverse polar plot of

$$[G(s)H(s)]^{-1} = \frac{s(1 + T_1 s)(1 + T_2 s)}{K_1} \tag{8.85}$$

is given in Fig. 8.39*a*. The radial line drawn from the $-1 + j0$ point has one cw and one ccw crossing, so that $N' = 0$ and the system is therefore stable. When the gain is increased, the plot drawn in Fig. 8.39*b* shows that $N' = -2$ and therefore the system is unstable. Thus the stability of this system depends on the gain.

8.18 DEFINITIONS OF PHASE MARGIN AND GAIN MARGIN AND THEIR RELATION TO STABILITY[8]

The stability and approximate degree of stability can be determined from the log magnitude and phase diagram. The stability characteristic is specified in terms of the following quantities:

Gain crossover. This is the point on the plot of the transfer function at which the magnitude of $\mathbf{G}(j\omega)$ is unity [$\mathrm{Lm}\,\mathbf{G}(j\omega) = 0$ dB]. The frequency at gain crossover is called the *phase-margin frequency* ω_ϕ.

Phase margin. This is 180° plus the negative trigonometrically considered angle of the transfer function at the gain-crossover point. It is designated as the angle γ, which can be expressed as $\gamma = 180° + \phi$, where $\angle \mathbf{G}(j\omega_\phi) = \phi$ is negative.

Phase crossover. This is the point on the plot of the transfer function at which the phase angle is $-180°$. The frequency at which phase crossover occurs is called the *gain-margin frequency* ω_c.

Gain margin. The gain margin is the factor a by which the gain must be changed in order to produce instability. Expressed in terms of the transfer function at the frequency ω_c, it is

$$|\mathbf{G}(j\omega_c)|a = 1 \tag{8.86}$$

On the polar plot of $\mathbf{G}(j\omega)$ the value at ω_c is

$$|\mathbf{G}(j\omega_c)| = \frac{1}{a} \tag{8.87}$$

In terms of the log magnitude, in decibels, this is

$$\mathrm{Lm}\,a = -\mathrm{Lm}\,\mathbf{G}(j\omega_c) \tag{8.88}$$

which identifies the gain margin on the log magnitude diagram.

These quantities are illustrated in Fig. 8.40 on both the log and the polar curves. Note the algebraic sign associated with these two quantities as marked on the curves. Figure 8.40*a*, *b*, and *c* represents a stable system, and Fig. 8.40*d*, *e*, and *f* represents an unstable system.

The phase margin is the amount of phase shift at the frequency ω_ϕ that would just produce instability. The phase margin for minimum-phase systems must be positive for a stable system, whereas a negative phase margin means that the system is unstable.

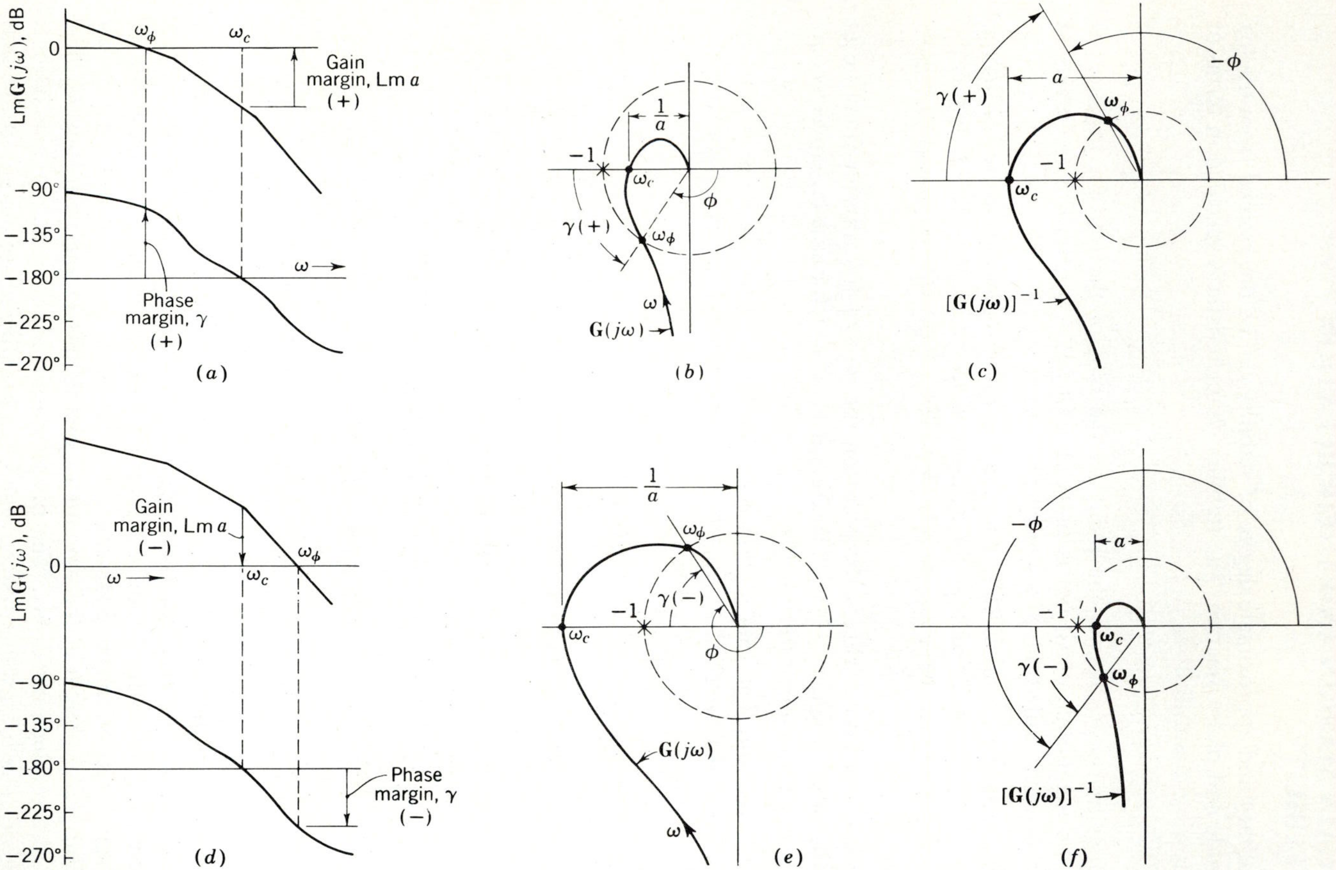

FIGURE 8.40
Log magnitude and phase diagram and polar plots of $\mathbf{G}(j\omega)$, showing gain margin and phase margin: (*a*) to (*c*) stable; (*d*) to (*f*) unstable.

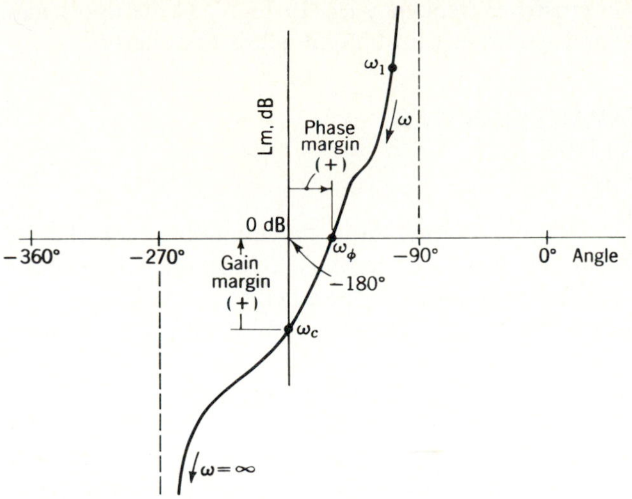

FIGURE 8.41
Log magnitude–angle diagram obtained from the curves of Figs. 8.6 and 8.7.

It can be shown that the phase margin is related to the effective damping ratio ζ of the system. This is discussed qualitatively in the next chapter. Satisfactory response is usually obtained with a phase margin of 45 to 60°. As an individual gains experience and develops his own particular technique, the value of γ to be used for a particular system becomes more evident. This guideline for system performance applies only to those systems where behavior is equivalent to that of a second-order system. The gain margin must be positive when expressed in decibels (greater than unity as a numeric) for a stable system. A negative gain margin means that the system is unstable. The damping ratio ζ of the system is also related to the gain margin. However, the phase margin gives a better estimate of damping ratio, and therefore of the transient overshoot of the system, than the gain margin.

The phase-margin frequency ω_ϕ, phase margin γ, crossover frequency ω_c, and the gain margin Lm a are also readily identified on the Nichols plot as shown in Fig. 8.41 and described in Sec. 8.20. Further information about the speed of response of the system can be obtained from the log magnitude–angle diagram, which defines the maximum value of the control ratio and the frequency at which this maximum occurs (see Chap. 9).

The relationship of stability and gain margin is modified for a conditionally stable system. This can be seen by reference to the polar plot of Fig. 8.34. The system for the particular gain represented in Fig. 8.34 is stable. Instability can occur with both an increase or a decrease in gain. Therefore, both "upper" and "lower" gain margins must be identified, corresponding to the upper crossover frequency ω_{cu} and the lower crossover frequency ω_{cl}. In Fig. 8.34 the upper gain margin is larger than unity and the lower gain margin is less than unity.

8.19 STABILITY CHARACTERISTICS OF THE LOG MAGNITUDE AND PHASE DIAGRAM

The earlier sections of this chapter discuss how the asymptotes of the log magnitude curve are related to each factor of the transfer function. For example, factors of the form $(1 + j\omega)^{-1}$ have a negative slope equal to -20 dB/decade at high frequencies. Also, the angle for this factor varies from 0° at low frequencies to $-90°$ at high frequencies.

The total phase angle of the transfer function at any frequency is closely related to the slope of the log magnitude curve at that frequency. A slope of -20 dB/decade is related to an angle of $-90°$; a slope of -40 dB/decade is related to an angle of $-180°$; a slope of -60 dB/decade is related to an angle of $-270°$; etc. Changes of slope at higher and lower corner frequencies, around the particular frequency being considered, contribute to the total angle at that frequency. The farther away the changes of slope are from the particular frequency, the less they contribute to the total angle at that frequency.

By observing the asymptotes of the log magnitude curve, it is possible to estimate the approximate value of the angle. With reference to the example of Sec. 8.7, the angle at $\omega = 4$ is now investigated. The slope of the curve at $\omega = 4$ is -20 dB/decade; therefore the angle is near $-90°$. The slope changes at $\omega = 2$ and $\omega = 8$, and so the slopes beyond these frequencies contribute to the total angle at $\omega = 4$. The actual angle, as read from the graph of Fig. 8.7, is $-122°$. The farther away the corner frequencies occur, the closer the angle is to $-90°$.

As has been seen, the stability of a minimum-phase system requires that the phase margin be positive. For this to be true, the angle at the gain crossover [Lm $\mathbf{G}(j\omega) = 0$ dB] must be greater than $-180°$. This places a limit on the slope of the log magnitude curve at the gain crossover. *The slope at the gain crossover should be more positive than* -40 *dB/decade if the adjacent corner frequencies are not close*. A slope of -20 dB/decade is preferable. This is derived from the consideration of a theorem by Bode. However, it can be seen qualitatively from the association of the slope of the log magnitude curve with the value of the phase angle. This guide should be used to assist in system design.

The log magnitude and phase diagram reveals some pertinent information, just as the polar plots do. For example, the gain can be adjusted (this raises or lowers the log magnitude curve) to produce a phase margin in the desirable range of 45 to 60°. The shape of the low-frequency portion of the curve determines system type and therefore the degree of steady-state accuracy. The system type and the gain determine the error coefficients and therefore the steady-state error. The phase-margin frequency ω_ϕ gives a qualitative indication of the speed of response of a system. However, this is only a qualitative relationship. A more detailed analysis of this relationship is made in the next chapter.

8.20 STABILITY FROM THE NICHOLS PLOT (LOG MAGNITUDE-ANGLE DIAGRAM)

The log magnitude–angle diagram is drawn by picking for each frequency the values of log magnitude and angle from the log magnitude and phase diagram. The resultant curve has frequency as a parameter. The curve for the example of Sec. 8.7, sketched in Fig. 8.41, shows a positive gain margin and phase margin; therefore this represents a stable system. Changing the gain raises or lowers the curve without changing the angle characteristics. Increasing the gain raises the curve, thereby decreasing the gain margin and phase margin, with the result that the stability is decreased. Increasing the gain so that the curve has a positive log magnitude at $-180°$ results in negative gain and phase margins; therefore an unstable system results. Decreasing the gain lowers the curve and increases stability. However, a large gain is desired to reduce steady-state errors, as shown in Chap. 6.

The log magnitude–angle diagram for $G(s)H(s)$ can be drawn for all values of s on the contour Q of Fig. 8.30a. The resultant curve for minimum-phase systems is a closed contour. Nyquist's criterion can be applied to this contour by determining the number of points (having the values 0 dB and odd multiples of 180°) enclosed by the curve of $G(s)H(s)$. This number is the value of N which is used in the equation $Z_R = N - P_R$ to determine the value of Z_R. As an example, consider a control system whose transfer function is given by

$$G(s) = \frac{K_1}{s(1 + Ts)} \tag{8.89}$$

Its log magnitude-angle diagram, for the contour Q, is shown in Fig. 8.42. From this figure it is seen that the value of N is zero and the system is stable. The log magnitude–angle contour for a non-minimum-phase system does not close; thus it is difficult to determine the value of N. For these cases the polar plot is easier to use to determine stability.

It is not necessary for minimum-phase systems to obtain the complete log magnitude–angle contour to determine stability. Only that portion of the contour is drawn representing $\mathbf{G}(j\omega)$ for the range of values $0^+ < \omega < \infty$. The stability is then determined from the position of the curve of $\mathbf{G}(j\omega)$ relative to the (0 dB, $-180°$) point. In other words, the curve is traced in the direction of increasing frequency. The system is stable if the (0 dB, $-180°$) point is to the right of the curve. This is a simplified rule of thumb which is based on Nyquist's stability criterion.

A conditionally stable system (as defined in Sec. 8.14, Example 4) is one in which the curve crosses the $-180°$ axis at more than one point. Figure 8.43 shows the transfer-function plot for such a system with two stable and two unstable regions. The gain determines whether the system is stable or unstable. The stable system represented by Fig. 8.43c has two values of gain margin, since instability can occur with both an increase or a decrease in gain. Corresponding

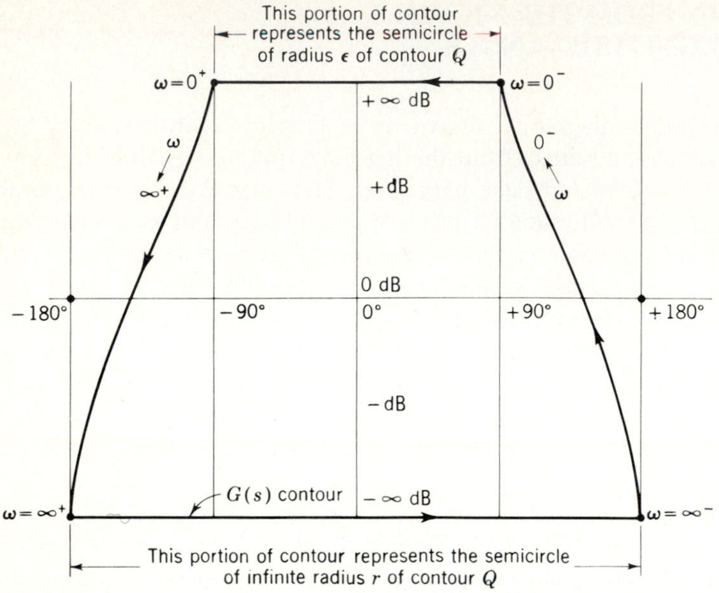

FIGURE 8.42
The log magnitude–angle contour for the minimum-phase system of Eq. (8.89).

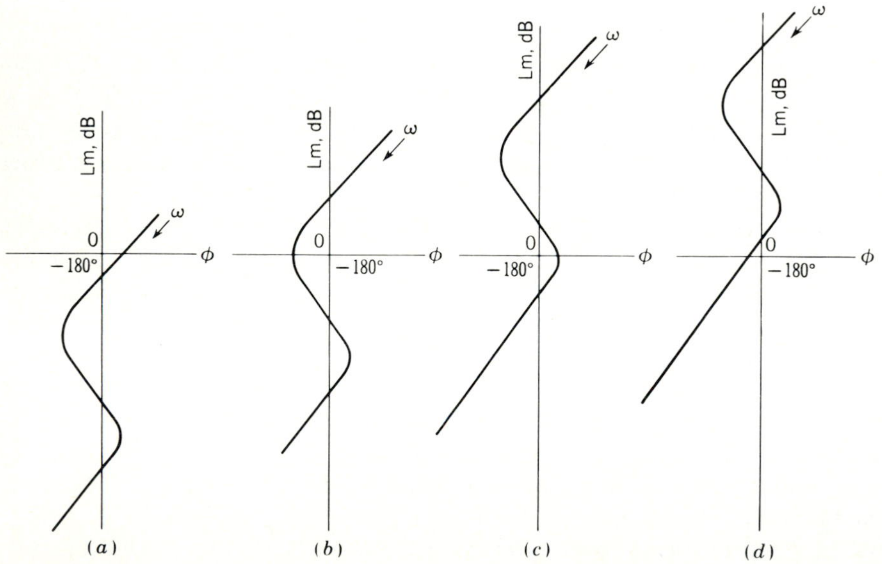

FIGURE 8.43
Log magnitude–angle diagram for a conditionally stable system: $(a),(c)$ stable; $(b),(d)$ unstable.

to the upper and lower crossover frequencies ω_{cu} and ω_{cl}, the gain margins Lm a_u and Lm a_l are positive and negative, respectively. Additional stability information can be obtained from the log magnitude–angle diagram. This is shown in the next chapter.

8.21 SUMMARY

In this chapter different types of frequency-response plots are introduced. From the log magnitude and phase-angle diagrams the polar plot can be obtained rapidly and with ease. All these plots indicate the type of system under consideration and the necessary gain adjustments that must be made to improve its response. How these adjustments are made is discussed in the following chapter.

The methods presented for obtaining the log magnitude frequency-response plots stress graphical techniques. For greater accuracy a programmable calculator or a digital-computer program can be used to calculate these data (see App. B).

The methods described in this chapter for obtaining frequency-response plots are based upon the condition that the transfer function of a given system is known. These plots can also be obtained from experimental data, which do not require the analytical expression for the transfer function of a system to be known. These experimental data can be used to synthesize an analytical expression for the transfer function using the log magnitude plot (see Sec. 8.9).

This chapter shows that the polar plot of the transfer function $G(s)H(s)$ or its inverse, in conjunction with Nyquist's stability criterion, gives a rapid means of determining whether a system is stable or unstable. The same information is obtained from the log magnitude and phase-angle diagrams and the log magnitude–angle diagram; the phase margin and gain margin are also used as a means of indicating stability. In the next chapter it is shown that other key information related to the time domain is readily obtained from any of these plots. Thus the designer can determine whether the given system is satisfactory.

Another useful application of the Nyquist criterion is the analysis of systems having the characteristic of dead time. The Nyquist criterion has one disadvantage in that poles or zeros on the $j\omega$ axis require special treatment. This treatment is not expounded since, in general, most problems are free of poles or zeros on the $j\omega$ axis.

REFERENCES

1. Maccoll, L. A.: *Fundamental Theory of Servomechanisms*, Van Nostrand, Princeton, N.J., 1945.
2. James, H. M., N. B. Nichols, and R. S. Phillips: *Theory of Servomechanisms*, McGraw-Hill, New York, 1947.
3. Nyquist, H.: "Regeneration Theory," *Bell Syst. Tech. J.*, vol. 11, pp. 126–147, 1932.
4. Larimer, S. J.: "An Interactive Computer-Aided Design Program for Digital and Continuous Control System Analysis and Synthesis (TOTAL)," M.S. thesis, GE/GGC/EE/78-2, School of Engineering, Air Force Institute of Technology, Wright-Patterson Air Force Base, Ohio, 1978; available from Defense Documentation Center (DDC), Cameron Station, Alexandria, Va. 22314.

5. Bruns, R. A., and R. M. Saunders: *Analysis of Feedback Control Systems*, McGraw-Hill, New York, 1955, chap. 14.
6. Balabanian, N., and W. R. LePage: "What Is a Minimum-Phase Network?," *Trans. AIEE*, vol. 74, pt. II, pp. 785–788, January 1956.
7. Bode, H. W.: *Network Analysis and Feedback Amplifier Design*, Van Nostrand, Princeton, N.J., 1945, chap. 8.
8. Chestnut, H., and R. W. Mayer: *Servomechanisms and Regulating System Design*, 2d ed., vol. 1, Wiley, New York, 1959.
9. Sanathanan, C. K., and H. Tsukui: "Synthesis of Transfer Function from Frequency Response Data," *Intern. J. Systems Science*, vol. 5, no. 1, pp. 41–54, January 1974.
10. Brown, G. S., and D. P. Campbell: *Principles of Servomechanisms*, Wiley, New York, 1948.
11. D'Azzo, J. J., and C. H. Houpis: *Feedback Control System Analysis and Synthesis*, 2d ed., McGraw-Hill, New York, 1966.

CHAPTER 9

CLOSED-LOOP TRACKING PERFORMANCE BASED ON THE FREQUENCY RESPONSE

9.1 INTRODUCTION

Chapter 8 is devoted to plotting the open-loop transfer function $\mathbf{G}(j\omega)\mathbf{H}(j\omega)$. The three types of plots that are useful are the direct polar plot (Nyquist plot), the inverse polar plot, and the log magnitude–angle plot (Nichols plot). Also included in Chap. 8 is the determination of closed-loop stability in terms of the open-loop transfer function by use of the Nyquist stability criterion, which is illustrated in terms of all three types of plot. The result of the stability study places bounds on the permitted range of values of gain.

This chapter develops a correlation between the frequency and the time responses of a system, leading to a method of gain setting in order to achieve a specified closed-loop frequency response.[1] The closed-loop frequency response is obtained as a function of the open-loop frequency response. Through the graphical representations in this chapter, the reader should obtain an appreciation of the responses which can be achieved by gain adjustment. The designer typically uses digital-computer programs[4,7] to perform a number of designs rapidly. In addition, although the design is performed in the frequency domain, the closed-loop responses in the time domain are also obtained. Then a "best"

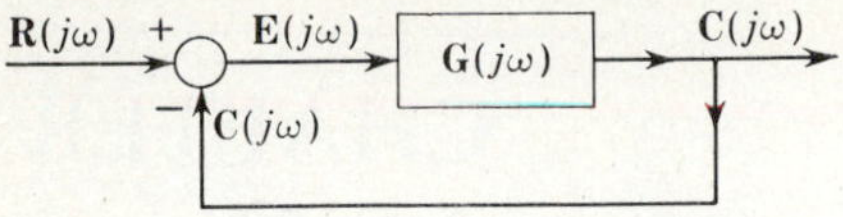

FIGURE 9.1
Block diagram of a simple control system.

design is selected by considering both the frequency and the time responses. Commercially available CAD programs (see App. B) may be used interactively to assist in achieving a satisfactory design. This considerably reduces the computational time of the designer in achieving a system that meets the design specifications.

9.2 DIRECT POLAR PLOT

Consider a simple control system with unity feedback, as shown in Fig. 9.1. The following equations describe the performance of this system with a sinusoidal input $\mathbf{R}(j\omega)$:

$$\mathbf{R}(j\omega) - \mathbf{C}(j\omega) = \mathbf{E}(j\omega) \tag{9.1}$$

$$\frac{\mathbf{C}(j\omega)}{\mathbf{E}(j\omega)} = \mathbf{G}(j\omega) = |\mathbf{G}(j\omega)|e^{j\phi} \tag{9.2}$$

$$\frac{\mathbf{C}(j\omega)}{\mathbf{R}(j\omega)} = \frac{\mathbf{G}(j\omega)}{1 + \mathbf{G}(j\omega)} = \mathbf{M}(j\omega) = M(\omega)e^{j\alpha(\omega)} \tag{9.3}$$

$$\frac{\mathbf{E}(j\omega)}{\mathbf{R}(j\omega)} = \frac{1}{1 + \mathbf{G}(j\omega)} \tag{9.4}$$

In Fig. 9.2 is shown the polar plot of $\mathbf{G}(j\omega)$ for this control system. As pointed out in Chap. 8, the directed line segment drawn from the $-1 + j0$ point to any point on the $\mathbf{G}(j\omega)\mathbf{H}(j\omega)$ curve represents the quantity $\mathbf{B}(j\omega) =$

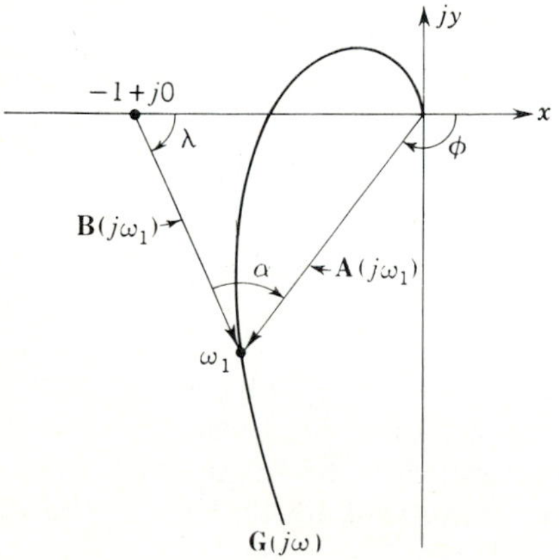

FIGURE 9.2
Polar plot of $\mathbf{G}(j\omega)$ for the control system of Fig. 9.1.

$1 + \mathbf{G}(j\omega)\mathbf{H}(j\omega)$. For the system under consideration $\mathbf{H}(j\omega)$ is unity; therefore

$$\mathbf{B}(j\omega) = |\mathbf{B}(j\omega)|e^{j\lambda(\omega)} = 1 + \mathbf{G}(j\omega) \tag{9.5}$$

The frequency control ratio $\mathbf{C}(j\omega)/\mathbf{R}(j\omega)$ is therefore equal to the ratio

$$\frac{\mathbf{C}(j\omega)}{\mathbf{R}(j\omega)} = \frac{\mathbf{A}(j\omega)}{\mathbf{B}(j\omega)} = \frac{|\mathbf{A}(j\omega)|e^{j\phi(\omega)}}{|\mathbf{B}(j\omega)|e^{j\lambda(\omega)}} = \frac{\mathbf{G}(j\omega)}{1 + \mathbf{G}(j\omega)} \tag{9.6}$$

$$\frac{\mathbf{C}(j\omega)}{\mathbf{R}(j\omega)} = \frac{|\mathbf{A}(j\omega)|}{|\mathbf{B}(j\omega)|}e^{j(\phi-\lambda)} = M(\omega)e^{j\alpha(\omega)} \tag{9.7}$$

Note that the angle $\alpha(\omega)$ can be determined directly from the construction shown in Fig. 9.2. Since the magnitude of the angle $\phi(\omega)$ is greater than the magnitude of the angle $\lambda(\omega)$, the value of the angle $\alpha(\omega)$ is negative. Remember that counterclockwise rotation is taken as positive.

Combining Eqs. (9.4) and (9.5) gives

$$\frac{\mathbf{E}(j\omega)}{\mathbf{R}(j\omega)} = \frac{1}{1 + \mathbf{G}(j\omega)} = \frac{1}{|\mathbf{B}(j\omega)|e^{j\lambda}} \tag{9.8}$$

From Eq. (9.8) it is seen that the greater the distance from the $-1 + j0$ point to a point on the $\mathbf{G}(j\omega)$ locus, for a given frequency, the smaller the steady-state sinusoidal error for a stated sinusoidal input. Thus the usefulness and importance of the polar plot of $\mathbf{G}(j\omega)$ have been enhanced.

9.3 DETERMINATION OF M_m AND ω_m FOR A SIMPLE SECOND-ORDER SYSTEM

The frequency at which the maximum value of $|\mathbf{C}(j\omega)/\mathbf{R}(j\omega)|$ occurs is referred to as the *resonant frequency* ω_m. The maximum value is labeled M_m. These two quantities are figures of merit of a system. The methods of compensation to improve system performance, using polar and log plots as shown in later chapters, are based upon a knowledge of these two factors.

Only for a *simple second-order system* can a direct and simple relationship be obtained for M_m and ω_m in terms of the system parameters. Consider the position-control system of Fig. 9.3, which utilizes the servomotor described by Eq. (2.149). The forward and closed-loop transfer functions are, respectively,

$$\frac{C(s)}{E(s)} = \frac{K_1}{s(T_m s + 1)} \tag{9.9}$$

$$\frac{C(s)}{R(s)} = \frac{1}{(T_m/K_1)s^2 + (1/K_1)s + 1} = \frac{1}{s^2/\omega_n^2 + (2\zeta/\omega_n)s + 1} \tag{9.10}$$

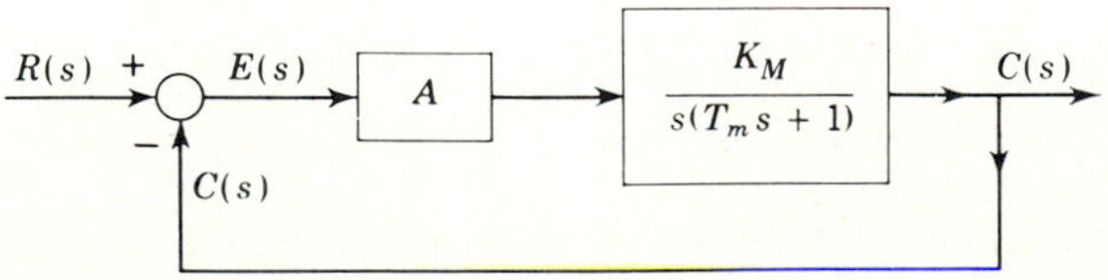

FIGURE 9.3
A position-control system.

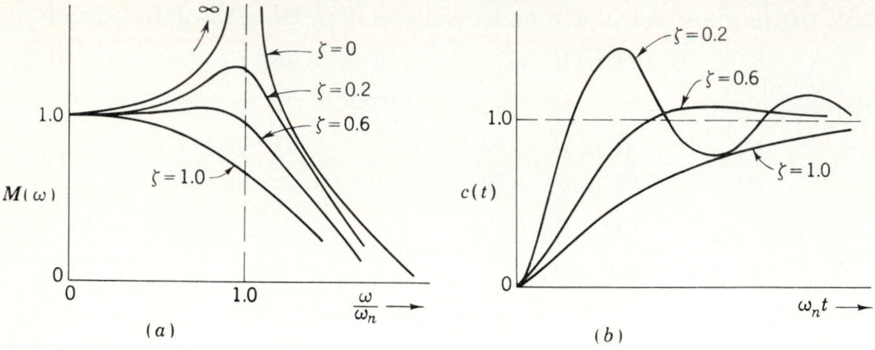

FIGURE 9.4
(a) Plots of M vs. ω/ω_n for a simple second-order system; (b) corresponding time plots for a step input.

where $K_1 = AK_M$. The damping ratio and the undamped natural frequency for this system, as determined from Eq. (9.10), are $\zeta = 1/(2\sqrt{K_1 T_m})$ and $\omega_n = \sqrt{K_1/T_m}$. The control ratio as a function of frequency is

$$\frac{\mathbf{C}(j\omega)}{\mathbf{R}(j\omega)} = \frac{1}{(1 - \omega^2/\omega_n^2) + j2\zeta(\omega/\omega_n)} = M(\omega)e^{j\alpha(\omega)} \tag{9.11}$$

Piots of $M(\omega)$ and $\alpha(\omega)$ vs. ω, for a particular value of ω_n, can be obtained for different values of ζ (see Fig. 8.4). For these plots to be applicable to all simple second-order systems with different values of ω_n, they are plotted vs. ω/ω_n. A sketch of $M(\omega)$ vs. ω/ω_n is shown in Fig. 9.4a.

The inverse Laplace transform of $C(s)$, for a unit step input, gives the time response, for $\zeta < 1$, as

$$c(t) = 1 - \frac{1}{\sqrt{1-\zeta^2}} e^{-\zeta\omega_n t} \sin\left(\omega_n\sqrt{1-\zeta^2}\,t + \cos^{-1}\zeta\right) \tag{9.12}$$

This time solution is derived in Sec. 3.9. The plot of $c(t)$ for several values of damping ratio ζ is drawn in Fig. 9.4b.

Next, consider the magnitude M^2, as derived from Eq. (9.11):

$$M^2 = \frac{1}{(1 - \omega^2/\omega_n^2)^2 + 4\zeta^2(\omega^2/\omega_n^2)} \tag{9.13}$$

To find the maximum value of M and the frequency at which it occurs, Eq. (9.13) is differentiated with respect to frequency and set equal to zero:

$$\frac{dM^2}{d\omega} = -\frac{-4(1 - \omega^2/\omega_n^2)(\omega/\omega_n^2) + 8\zeta^2(\omega/\omega_n^2)}{\left[(1 - \omega^2/\omega_n^2)^2 + 4\zeta^2(\omega^2/\omega_n^2)\right]^2} = 0 \tag{9.14}$$

The frequency ω_m at which the value M exhibits a peak (see Fig. 9.5), as found

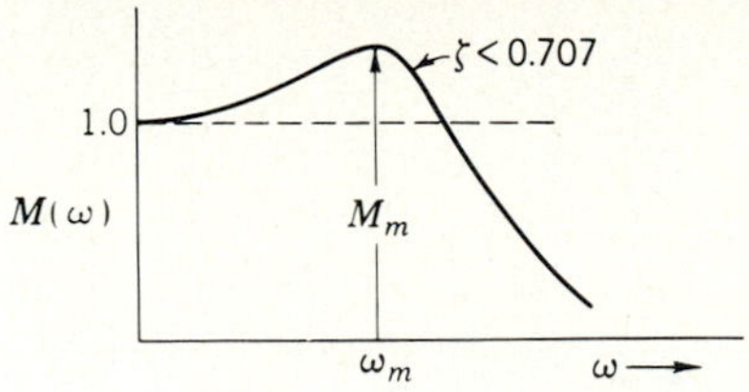

FIGURE 9.5
A closed-loop frequency-response curve indicating M_m and ω_m.

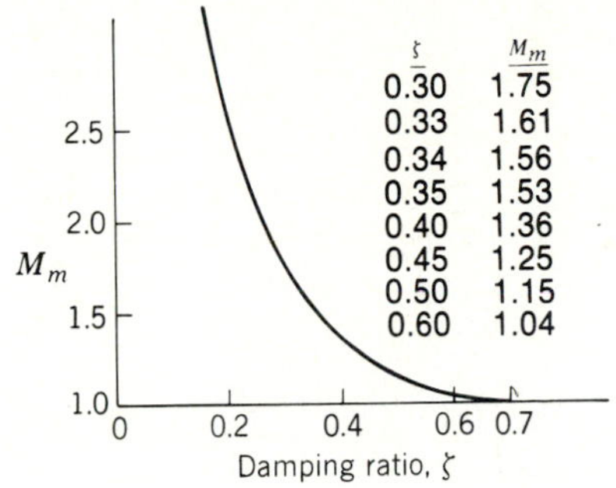

FIGURE 9.6
A plot of M_m vs. ζ for a simple second-order system.

from Eq. (9.14), is

$$\omega_m = \omega_n\sqrt{1 - 2\zeta^2} \tag{9.15}$$

This value of frequency is substituted into Eq. (9.13) to yield

$$M_m = \frac{1}{2\zeta\sqrt{1 - \zeta^2}} \tag{9.16}$$

From these equations it is seen that the curve of M vs. ω has a peak value, other than at $\omega = 0$, only for $\zeta < 0.707$. Figure 9.6 shows a plot of M_m vs. ζ for a simple second-order system. It is seen for values of $\zeta < 0.4$ that M_m increases very rapidly in magnitude; the transient oscillatory response is therefore excessively large and might damage the physical equipment.

In Secs. 3.5 and 7.11 it is determined that the damped natural frequency ω_d for the transient of the simple second-order system is

$$\omega_d = \omega_n\sqrt{1 - \zeta^2} \tag{9.17}$$

Also, in Sec. 3.9 the expression for the peak value M_p for a unit step input to this simple system is determined. It is repeated here:

$$M_p = \frac{c_p}{r} = 1 + \exp\left(-\frac{\zeta\pi}{\sqrt{1 - \zeta^2}}\right) \tag{9.18}$$

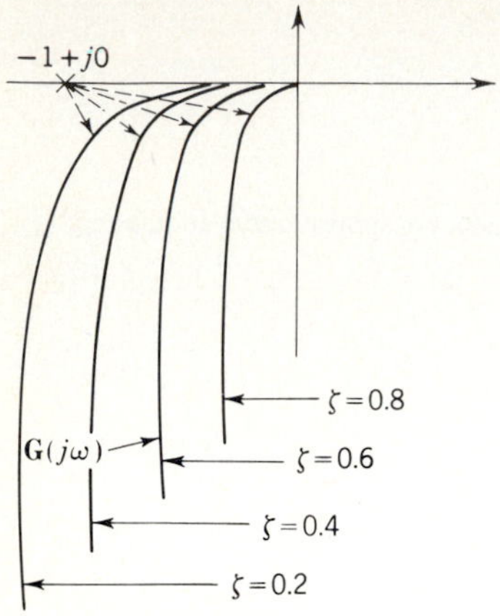

FIGURE 9.7
Polar plots of **G**($j\omega$) for different damping ratios for the system shown in Fig. 9.3.

Therefore, for this simple second-order system the following conclusions are obtained in correlating the frequency and time responses:

1. Inspection of Eq. (9.15) reveals that ω_m is a function of both ω_n and ζ. For a given ζ, the larger the value of ω_m, the larger ω_n, and the faster the transient time of response for this system.
2. Inspection of Eqs. (9.16) and (9.18) shows that both M_m and M_p are functions of ζ. The smaller ζ becomes, the larger in value M_m and M_p become. Thus, it is concluded that the larger the value of M_m, the larger the value of M_p. For values of $\zeta < 0.4$ the correspondence between M_m and M_p is only qualitative for this simple case. In other words, for $\zeta = 0$ the time domain yields $M_p = 2$, whereas the frequency domain yields $M_m = \infty$. When $\zeta > 0.4$, there is a close correspondence between M_m and M_p. As an example, for ζ equal to 0.6, $M_m = 1.04$ and $M_p = 1.09$.
3. In Fig. 9.7 are shown polar plots for different damping ratios for the simple second-order system. Note that the shorter the distance between the $-1 + j0$ point and a particular **G**($j\omega$) plot, as indicated by the dashed lines, the smaller the damping ratio. Thus M_m is larger and consequently M_p is also larger.

As a result of these characteristics, a designer can obtain a good approximation of the time response of a simple second-order system by knowing only the M_m and ω_m of its frequency response.

The procedure used above of setting to zero the derivative, with respect to ω, of **C**($j\omega$)/**R**($j\omega$) works very well with a simple system. But the differentiation of **C**($j\omega$)/**R**($j\omega$) and solution for ω_m and M_m become tedious for more complex systems. Although this analytical procedure can be simplified, a graphic procedure is generally used, as shown in the following sections.[2]

9.4 CORRELATION OF SINUSOIDAL AND TIME RESPONSES[3]

Although the correlation in the preceding section is for a simple second-order system, it has been found by experience that M_m is also a function of the *effective* ζ and ω_n for higher-order systems. The effective ζ and ω_n of a higher-order system (see Sec. 7.11) is dependent upon the ζ and ω_n of each second-order term and the values of the real roots in the characteristic equation of $C(s)/R(s)$. Thus, in order to alter the M_m, the location of some of the roots must be changed. Which ones should be altered depends on which are dominant in the time domain.

From the analysis for a simple second-order system, whenever the frequency response has the shape shown in Fig. 9.5, the following correlation exists between the frequency and time responses for systems of any order:

1. The larger ω_m is made, the faster the time of response for the system.
2. The value of M_m gives a good approximation of M_p within the acceptable range of the *effective* damping ratio $0.4 < \zeta < 0.707$. In terms of M_m, the acceptable range is $1 < M_m < 1.4$.
3. The closer the $\mathbf{G}(j\omega)$ curve comes to the $-1 + j0$ point, the larger the value of M_m.

The factor developed in Chap. 6 can be added to these three items, i.e., the larger K_p, K_v, or K_a is made, the greater the steady-state accuracy for a step, a ramp, and a parabolic input, respectively. In terms of the polar plot, the farther the point $\mathbf{G}(j\omega)|_{\omega=0} = K_0$ for a Type 0 system is from the origin, the more accurate is the steady-state time response for a step input. For a Type 1 system, the farther the low-frequency asymptote (as $\omega \to 0$) is from the imaginary axis, the more accurate is the steady-state time response for a ramp input.

All the factors mentioned above are merely *guideposts* in the *frequency domain* to assist the designer in obtaining an *approximate* idea of the time response of a system. They serve as "stop-and-go signals" to indicate if one is headed in the right direction in achieving the desired time response. If the desired performance specifications are not satisfactorily met, compensation techniques (see later chapters) must be used. After compensation the exact time response can be obtained by taking the inverse Laplace transform of $C(s)$. The approximate approach saves much valuable time. Exceptions to the above analysis do occur for higher-order systems. The digital computer is a valuable tool for obtaining the exact closed-loop frequency and time responses.[4] Thus a direct correlation between them can easily be made.

9.5 CONSTANT $M(\omega)$ AND $\alpha(\omega)$ CONTOURS OF $\mathbf{C}(j\omega)/\mathbf{R}(j\omega)$ ON THE COMPLEX PLANE (DIRECT PLOT)

The open-loop transfer function and its polar plot for a given feedback control system have provided the following information, so far:

1. The stability or instability of the system
2. If the system is stable, the degree of its stability
3. The system type
4. The degree of steady-state accuracy
5. A graphical method of determining $\mathbf{C}(j\omega)/\mathbf{R}(j\omega)$

These items provide a qualitative idea of the system's time response. The polar plot is useful since it permits the determination of M_m and ω_m. Also, it is used to determine the gain required to produce a specified value of M_m.

The contours of constant values of M drawn in the complex plane yield a rapid means of determining the values of M_m and ω_m and the value of gain required to achieve a desired value of M_m. In conjunction with the contours of constant values of $\alpha(\omega)$, also drawn in the complex plane, the plot of $\mathbf{C}(j\omega)/\mathbf{R}(j\omega)$ can be obtained rapidly. The M and α contours are developed only for unity-feedback systems. However, it is shown at the end of this chapter how these contours can be applied to nonunity-feedback systems.

Equation of a Circle

The equation of a circle with its center at the point (a, b) and having radius r is

$$(x-a)^2 + (y-b)^2 = r^2 \tag{9.19}$$

The circle is displaced to the left of the origin for a negative value of a, and for negative values of b the circle is displaced down below the origin. This equation of a circle in the xy plane is used later in this section to express contours of constant M and α.

$M(\omega)$ Contours

Figure 9.2 is the polar plot of a forward transfer function $\mathbf{G}(j\omega)$ of a *unity-feedback system*. From Eq. (9.3), the magnitude M of the control ratio is

$$M(\omega) = \frac{|\mathbf{A}(j\omega)|}{|\mathbf{B}(j\omega)|} = \frac{|\mathbf{G}(j\omega)|}{|1+\mathbf{G}(j\omega)|} \tag{9.20}$$

The question at hand is: How many points are there in the complex plane for which the ratios of the magnitudes of the phasors $\mathbf{A}(j\omega)$ and $\mathbf{B}(j\omega)$ have the same value of $M(\omega)$? For example, referring to Fig. 9.2, for the frequency

$\omega = \omega_1$, M has a value of M_a. It is desirable to determine all the other points in the complex plane for which

$$\frac{|\mathbf{A}(j\omega)|}{|\mathbf{B}(j\omega)|} = M_a$$

To derive the constant M locus, express the transfer function in the rectangular coordinates. That is,

$$\mathbf{G}(j\omega) = x + jy \tag{9.21}$$

Substituting this equation into Eq. (9.20) gives

$$M = \frac{|x + jy|}{|1 + x + jy|} = \left[\frac{x^2 + y^2}{(1 + x)^2 + y^2}\right]^{1/2} \qquad \text{or} \qquad M^2 = \frac{x^2 + y^2}{(1 + x)^2 + y^2}$$

Rearranging the terms of this equation yields

$$\left(x + \frac{M^2}{M^2 - 1}\right)^2 + y^2 = \frac{M^2}{(M^2 - 1)^2} \tag{9.22}$$

By comparison with Eq. (9.19) it is seen that Eq. (9.22) is the equation of a circle with its center on the real axis with M as a parameter. The center is located at

$$x_0 = -\frac{M^2}{M^2 - 1} \tag{9.23}$$

$$y_0 = 0 \tag{9.24}$$

and the radius is

$$r_0 = \left|\frac{M}{M^2 - 1}\right| \tag{9.25}$$

Inserting a given value of $M = M_a$ into Eq. (9.22) results in a circle in the complex plane having a radius r_0 and its center at (x_0, y_0). This circle is called a constant M contour for $M = M_a$.

The ratio of magnitudes of the phasors $\mathbf{A}(j\omega)$ and $\mathbf{B}(j\omega)$ drawn to any point on the M_a circle has the same value. As an example, refer to Fig. 9.8, in which the M_a circle and the $\mathbf{G}(j\omega)$ function are plotted. Since the circle intersects the $\mathbf{G}(j\omega)$ plot at the two frequencies ω_1 and ω_2, there are two frequencies for which

$$\frac{|\mathbf{A}(j\omega_1)|}{|\mathbf{B}(j\omega_1)|} = \frac{|\mathbf{A}(j\omega_2)|}{|\mathbf{B}(j\omega_2)|} = M_a$$

In other words, a given point (x_1, y_1) is simultaneously a point on a particular transfer function $\mathbf{G}(j\omega)$ and a point on the M circle passing through it.

The plot of Fig. 9.8 is redrawn in Fig. 9.9 with two more M circles added. In this figure the circle $M = M_b$ is just tangent to the $\mathbf{G}(j\omega)$ plot. There is only one point (x_3, y_3) for $\mathbf{G}(j\omega_3)$ in the complex plane for which the ratio $|\mathbf{A}(j\omega)/\mathbf{B}(j\omega)|$ is equal to M_b. Also, the M_c circle does not intersect and is not tangent to the $\mathbf{G}(j\omega)$ plot. This indicates that there are no points in the plane that can simultaneously satisfy Eqs. (9.20) and (9.22) for $M = M_c$.

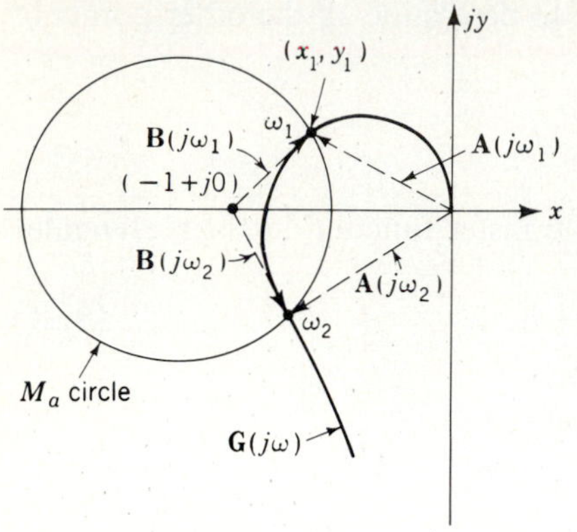

FIGURE 9.8
A plot of the transfer function $\mathbf{G}(j\omega)$ and the M_a circle.

Figure 9.10 shows a family of circles in the complex plane for different values of M. In this figure it is noticed that the larger the value M, the smaller its corresponding M circle. Thus, for the example shown in Fig. 9.9 the ratio $\mathbf{C}(j\omega)/\mathbf{R}(j\omega)$, for a unity-feedback control system, has a maximum value of M equal to $M_m = M_b$. A further inspection of Fig. 9.10 and Eq. (9.23) reveals the following:

1. For $M \to \infty$, which represents a condition of oscillation ($\zeta \to 0$), the center of the M circle $x_0 \to -1 + j0$ and the radius $r_0 \to 0$. This agrees with the statement made previously that as the $\mathbf{G}(j\omega)$ plot comes closer to the $-1 + j0$ point, the system's effective ζ becomes smaller and the degree of its stability decreases.
2. For $M(\omega) = 1$, which represents the condition where $\mathbf{C}(j\omega) = \mathbf{R}(j\omega)$, $r_0 \to \infty$ and the M contour becomes a straight line perpendicular to the real axis at $x = -\frac{1}{2}$.
3. For $M \to 0$, the center of the M circle $x_0 \to 0$ and the radius $r_0 \to 0$.

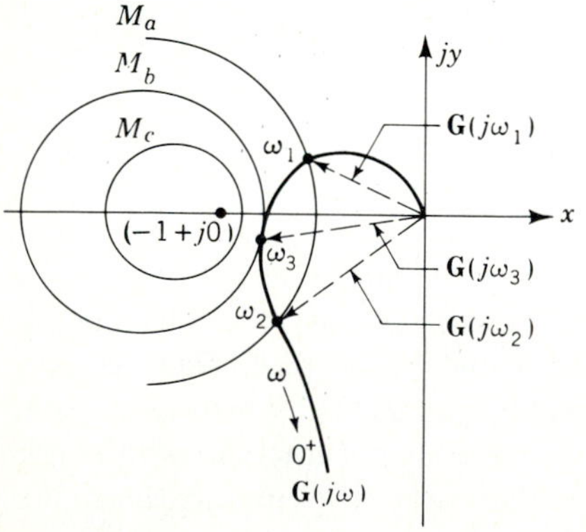

FIGURE 9.9
M contours and a $\mathbf{G}(j\omega)$ plot.

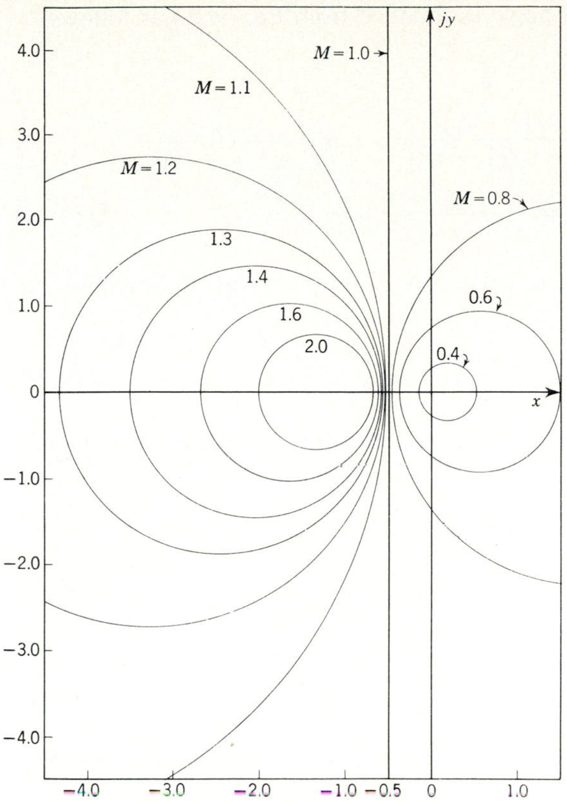

FIGURE 9.10
Constant M contours.

4. For $M > 1$ the centers of the circles lie to the left of $x = -1 + j0$, and for $M < 1$ the centers of the circles lie to the right of $x = 0$. All centers are on the real axis.

$\alpha(\omega)$ Contours

The $\alpha(\omega)$ contours, representing constant values of phase angle $\alpha(\omega)$ for $\mathbf{C}(j\omega)/\mathbf{R}(j\omega)$, can also be determined in the same manner as the M contours. Substituting Eq. (9.21) into Eq. (9.3) yields

$$M(\omega)e^{j\alpha(\omega)} = \frac{x + jy}{(1 + x) + jy} = \frac{\mathbf{A}(j\omega)}{\mathbf{B}(j\omega)} \tag{9.26}$$

The question at hand now is: How many points are there in the complex plane for which the ratio of the phasors $\mathbf{A}(j\omega)$ and $\mathbf{B}(j\omega)$ yields the same angle α? To answer this question, express the angle α obtained from Eq. (9.26) as follows:

$$\alpha = \tan^{-1}\frac{y}{x} - \tan^{-1}\frac{y}{1+x}$$

$$= \tan^{-1}\frac{y/x - y/(1+x)}{1 + (y/x)[y/(1+x)]} = \tan^{-1}\frac{y}{x^2 + x + y^2}$$

$$\tan\alpha = \frac{y}{x^2 + x + y^2} = N \tag{9.27}$$

For a constant value of the angle α, $\tan\alpha = N$ is also constant. Rearranging Eq. (9.27) results in

$$\left(x + \tfrac{1}{2}\right)^2 + \left(y - \frac{1}{2N}\right)^2 = \frac{1}{4}\frac{N^2 + 1}{N^2} \tag{9.28}$$

By comparing with Eq. (9.19) it is seen that Eq. (9.28) is the equation of a circle with N as a parameter. It has its center at

$$x_q = -\tfrac{1}{2} \tag{9.29}$$

$$y_q = \frac{1}{2N} \tag{9.30}$$

and has a radius of

$$r_q = \frac{1}{2}\left(\frac{N^2 + 1}{N^2}\right)^{1/2} \tag{9.31}$$

Inserting a given value of $N_q = \tan\alpha_q$ into Eq. (9.28) results in a circle (see Fig. 9.11) of radius r_q with its center at (x_q, y_q).

The tangent of angles in the first and third quadrant is positive. Therefore the y_q coordinate given by Eq. (9.30) is the same for an angle in the first quadrant and for the negative of its supplement, which is in the third quadrant. As a result,

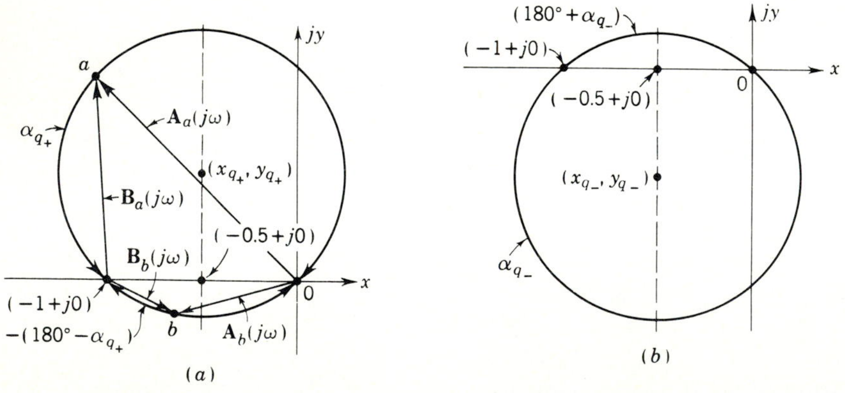

FIGURE 9.11
Arcs of constant α.

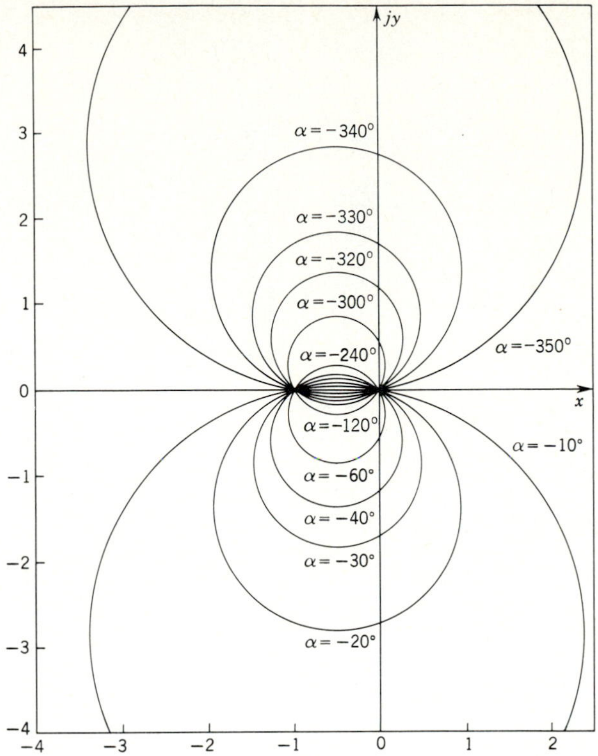

FIGURE 9.12
Constant α contours.

the constant α contour is only an arc of the circle. In other words, the $\alpha = -310°$ and $-130°$ arcs are part of the same circle, as shown in Fig. 9.11*a*. Similarly, angles α in the second and fourth quadrants have the same value y_q if they are negative supplements of each other. The constant α contours for these angles are shown in Fig. 9.11*b*.

The ratio of the complex quantities $\mathbf{A}(j\omega)$ and $\mathbf{B}(j\omega)$, for all points on the α_q arc, yields the same phase angle α_q, that is,

$$\angle \mathbf{A}(j\omega) - \angle \mathbf{B}(j\omega) = \alpha_q \tag{9.32}$$

A good procedure, in order to avoid ambiguity in determining the angle α, is to start at zero frequency, where $\alpha = 0$, and to proceed to higher frequencies, knowing that the angle is a continuous function.

There results, for different values of α, a family of circles in the complex plane with centers on the line represented by $(-\frac{1}{2}, y)$, as illustrated in Fig. 9.12.

Figure 9.13 shows a plot of $\mathbf{G}(j\omega)$ with constant M and α contours superimposed. From this figure, Table 9.1 gives the magnitude M and the angle α of the control ratio $\mathbf{C}(j\omega)/\mathbf{R}(j\omega)$ for a number of values of frequency. While a graphical interpretation of M and α as related to $\mathbf{G}(j\omega)$ is shown in Fig. 9.13, it should be realized that these data are easily obtainable by use of a digital-computer program.[4] The control-ratio characteristics are plotted in Fig. 9.14.

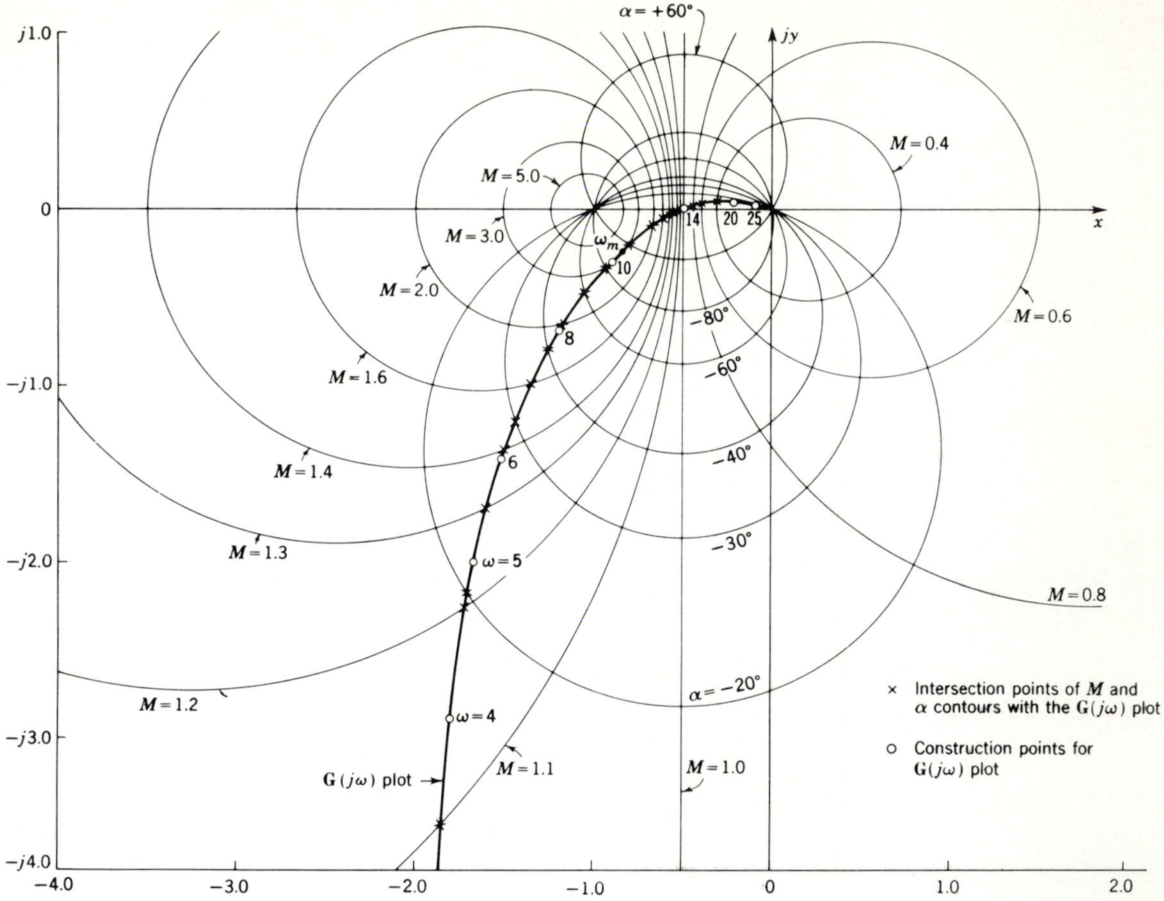

FIGURE 9.13
Determination of $|\mathbf{C}(j\omega)/\mathbf{R}(j\omega)|$ and α from the $\mathbf{G}(j\omega)$ plot with the aid of M-α contours.

TABLE 9.1
Values obtained from Fig. 9.13 to plot the curves in Fig. 9.14

Values for $|C(j\omega)/R(j\omega)|$ plot			
ω, rad / s	M	ω, rad / s	M
3.8	1.1	13.4	1.4
4.7	1.2	13.6	1.3
5.5	1.3	13.8	1.2
6.2	1.4	13.9	1.1
7.0	1.6	14.0	1.0
8.1	2.0	15.0	0.8
11.0	3.0	16.5	0.6
12.6	2.0	18.0	0.4
13.2	1.6		
Values for α plot			
ω, rad / s	α, deg	ω, rad / s	α, deg
3	−10	8.8	−60
4.8	−20	9.7	−80
6.5	−30	11.3	−120
7.7	−40	13.0	−180

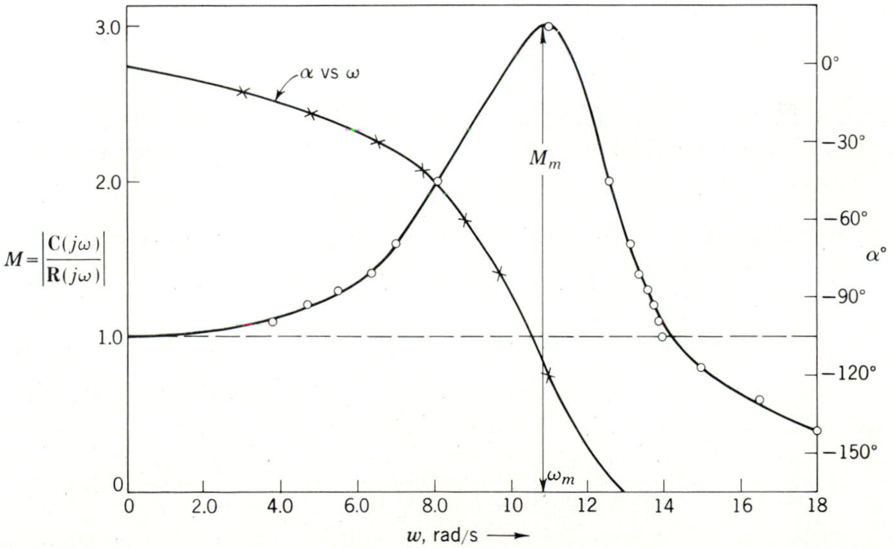

FIGURE 9.14
Resultant $|C(j\omega)/R(j\omega)|$ and α vs. ω, obtained from Fig. 9.13.

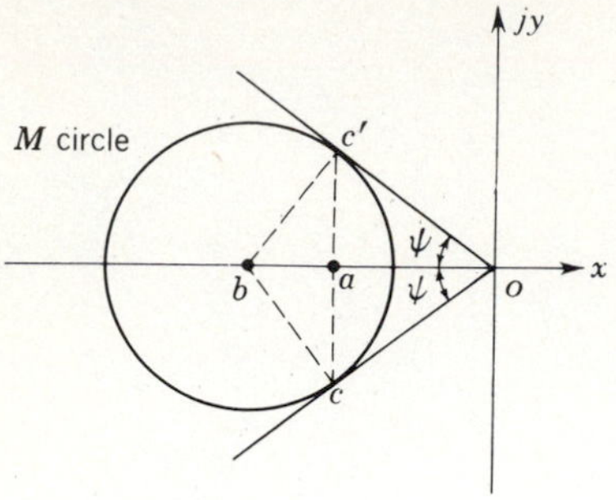

FIGURE 9.15
Determination of sin ψ.

Tangents to the *M* Circles

The line drawn through the origin of the complex plane tangent to a given M circle plays an important part in the gain setting of the $\mathbf{G}(j\omega)$ function. By referring to Fig. 9.15 and recognizing that $bc = r_0$ is the radius and $ob = x_0$ is the distance to the center of the particular M circle, then

$$\sin\psi = \frac{bc}{ob} = \frac{M/(M^2-1)}{M^2/(M^2-1)} = \frac{1}{M} \tag{9.33}$$

This relationship is utilized in the section on gain adjustment. Also, the point α in Fig. 9.15 is the $-1 + j0$ point. This is proved from

$$(oc)^2 = (ob)^2 - (bc)^2 \qquad ac = oc\sin\psi \qquad \text{and} \qquad (oa)^2 = (oc)^2 - (ac)^2$$

Combining these equations yields

$$(oa)^2 = \frac{M^2-1}{M^2}\left[(ob)^2 - (bc)^2\right] \tag{9.34}$$

The values of the distances to the center ob and of the radius bc of the M circle [see Eqs. (9.23) and (9.25)] are then substituted into Eq. (9.34), which results in

$$oa = 1 \tag{9.35}$$

The values necessary for constructing M and α contours, for typical values of M and α, are given in Tables 9.2 and 9.3, respectively.

Nonunity-Feedback Control System

The M and α contours are applied to a nonunity-feedback system by putting the control ratio in the form

$$\frac{\mathbf{C}(j\omega)}{\mathbf{R}(j\omega)} = \frac{\mathbf{G}(j\omega)}{1+\mathbf{G}(j\omega)\mathbf{H}(j\omega)} = \frac{1}{\mathbf{H}(j\omega)}\,\frac{\mathbf{G}(j\omega)\mathbf{H}(j\omega)}{1+\mathbf{G}(j\omega)\mathbf{H}(j\omega)} \tag{9.36}$$

or

$$\frac{\mathbf{C}(j\omega)}{\mathbf{R}(j\omega)} = \frac{1}{\mathbf{H}(j\omega)}\,\frac{\mathbf{G}_0(j\omega)}{1+\mathbf{G}_0(j\omega)} = \frac{1}{\mathbf{H}(j\omega)}\,\frac{\mathbf{C}_0(j\omega)}{\mathbf{R}_0(j\omega)} \tag{9.37}$$

TABLE 9.2
Values for constructing M circles

M	Center $x_0 = -\frac{M^2}{M^2-1}$	Radius $r_0 = \frac{M}{M^2-1}$	Angle, deg, $\psi = \sin^{-1}\frac{1}{M}$
0.5	0.333	0.67	
0.7	0.960	1.37	
0.9	4.26	4.74	
1.0	∞	∞	90
1.05	−10.74	10.24	72.3
1.1	−5.76	5.24	65.4
1.15	−4.1	3.57	60.3
1.2	−3.27	2.73	56.4
1.25	−2.78	2.22	53.2
1.3	−2.45	1.88	50.3
1.35	−2.215	1.64	47.7
1.4	−2.04	1.46	45.6
1.5	−1.8	1.20	41.8
1.6	−1.64	1.03	38.7
1.7	−1.53	0.90	36.0
1.8	−1.45	0.80	33.7
1.9	−1.38	0.729	31.7
2.0	−1.33	0.67	30.0

TABLE 9.3
Values for constructing α contours

$\alpha \pm 180° m$, deg	N	Radius $r_q = \frac{1}{2N}\sqrt{N^2+1}$	Center $y_q = \frac{1}{2N}$
−90	−∞	0.500	0
−80	−5.67	0.528	−0.0882
−70	−2.75	0.531	−0.182
−60	−1.73	0.577	−0.289
−50	−1.19	0.656	−0.420
−40	−0.838	0.775	−0.596
−30	−0.577	1.000	−0.866
−20	−0.364	1.460	−1.370
−10	−0.176	2.88	−2.84
0	0.0	∞	∞
10	0.176	2.88	2.84
30	0.577	1.000	0.866
50	1.19	0.656	0.42
70	2.75	0.531	0.182
90	∞	0.5	0

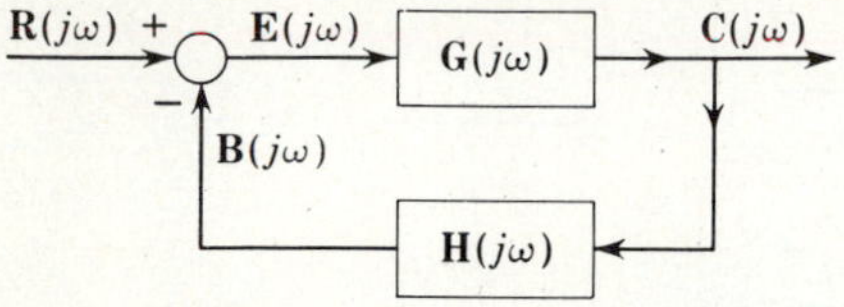

FIGURE 9.16
Block diagram of a control system with feedback.

where
$$\mathbf{G}_0(j\omega) = \mathbf{G}(j\omega)\mathbf{H}(j\omega) \tag{9.38}$$

The M and α contours can be applied to $\mathbf{G}_0(j\omega)$ to obtain $\mathbf{C}_0(j\omega)/\mathbf{R}_0(j\omega)$. Multiplying $\mathbf{C}_0(j\omega)/\mathbf{R}_0(j\omega)$ by $1/\mathbf{H}(j\omega)$ gives $\mathbf{C}(j\omega)/\mathbf{R}(j\omega)$.

9.6 CONTOURS IN THE INVERSE POLAR PLANE

The use of direct polar plots for feedback control systems that utilize nonunity feedback is rather tedious when compared with the use of inverse plots. Also, the effect of the $\mathbf{H}(j\omega)$ term is more evident in the inverse plot. Even for unity-feedback systems the value of required gain can be determined just as rapidly as on the direct plot, as described in Sec. 9.7.

Inverse Polar Plot

Figure 9.16 illustrates a control system with a feedback transfer function $\mathbf{H}(j\omega)$, whose performance is described by the following equations:

$$\frac{\mathbf{C}(j\omega)}{\mathbf{R}(j\omega)} = \frac{\mathbf{G}(j\omega)}{1 + \mathbf{G}(j\omega)\mathbf{H}(j\omega)} = Me^{j\alpha} \tag{9.39}$$

$$\frac{\mathbf{R}(j\omega)}{\mathbf{C}(j\omega)} = \frac{1}{\mathbf{G}(j\omega)} + \mathbf{H}(j\omega) = \frac{1}{M}e^{-j\alpha} \tag{9.40}$$

Note that Eq. (9.40) is composed of two complex quantities that can be readily plotted in the complex plane, as shown in Fig. 9.17. Thus, by plotting the complex quantities $1/\mathbf{G}(j\omega)$ and $\mathbf{H}(j\omega)$, the $\mathbf{R}(j\omega)/\mathbf{C}(j\omega)$ term can be obtained

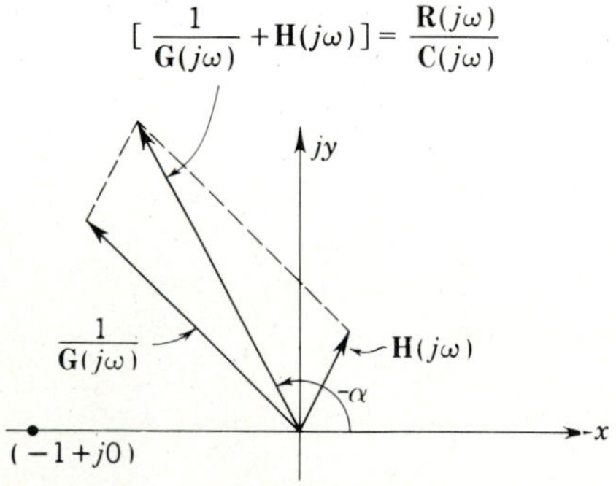

FIGURE 9.17
Complex representation of the quantities in Eq. (9.40).

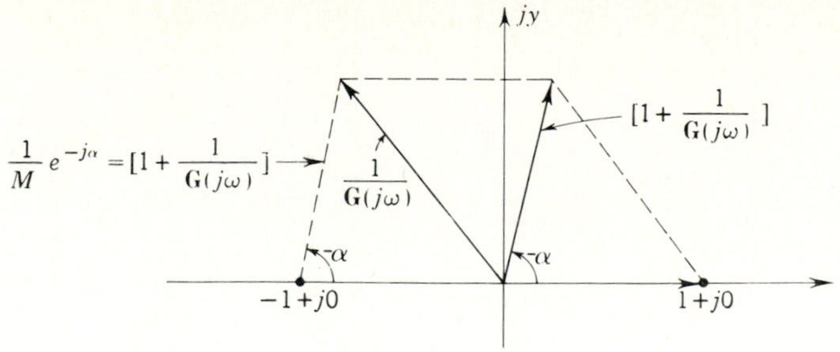

FIGURE 9.18
Representation of the quantities in Eq. (9.40) with **H**($j\omega$) = 1.

graphically. It is now seen that the effect on **R**($j\omega$)/**C**($j\omega$) of changing the **H**($j\omega$) term is more evident than it would be with direct plots.

Constant $1/M$ and α Contours (Unity Feedback)

Constant $1/M$ and α contours for the inverse plots are developed first for unity-feedback systems. Figure 9.18 illustrates the quantities in Eq. (9.40) for a unity-feedback system, with **H**($j\omega$) = 1. Note that, because of the geometry of construction, the directed line segment drawn from the $-1 + j0$ point is also the quantity $1 + 1/\mathbf{G}(j\omega)$ or $(1/M)e^{-j\alpha}$. Thus the inverse plot yields very readily the form and location of the contours of constant $1/M$ and α. In other words, in the inverse plane:

1. Contours of constant values of M are circles whose centers are at the $-1 + j0$ point, and the radii are equal to $1/M$.
2. Contours of constant values of $-\alpha$ are radial lines that pass through the $-1 + j0$ point.

Figure 9.19 indicates the contours for a particular magnitude of **R**($j\omega$)/**C**($j\omega$) and a tangent drawn to the M circle from the origin. In Fig. 9.19

$$\sin\psi = \left|\frac{ab}{ob}\right| = \frac{1/M}{1} = \frac{1}{M}$$

which is the same as for the contours in the direct plane. The directed line segment from $-1 + j0$ to the point (x_1, y_1) has a value of $1/M_q$ and an angle of $-\alpha_q$. If the polar plot of $1/\mathbf{G}(j\omega)$ passes through the (x_1, y_1) point, then $\mathbf{R}(j\omega_1)/\mathbf{C}(j\omega_1)$ has these values. The directed line segment from $-1 + j0$ to the point (x_2, y_2) has a magnitude of $1/M_q$ and an angle of $-\alpha_q - 180°$. If the polar plot of $1/\mathbf{G}(j\omega)$ passes through the (x_2, y_2) point, then $|\mathbf{R}(j\omega_2)/\mathbf{C}(j\omega_2)|$ has these values. Figure 9.20 illustrates families of typical M and α contours.

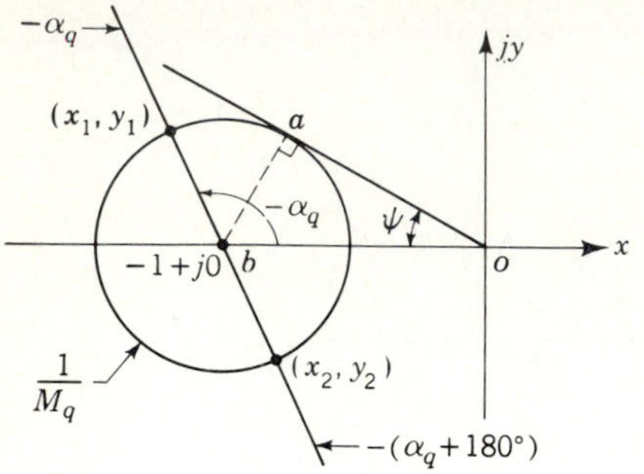

FIGURE 9.19
Contours for a particular magnitude $|\mathbf{R}(j\omega)/\mathbf{C}(j\omega)|$.

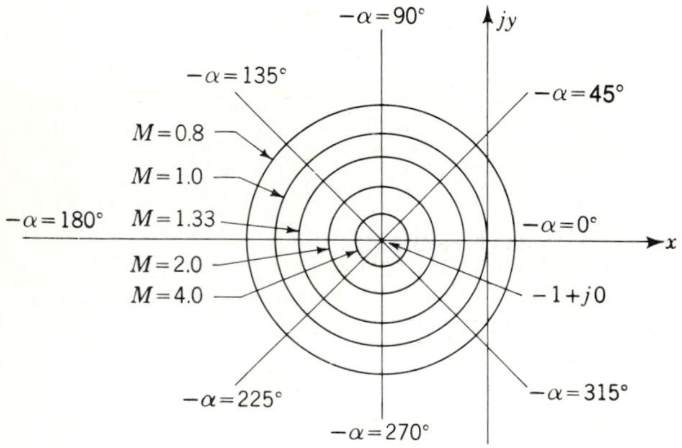

FIGURE 9.20
Typical M and α contours on the inverse polar plane for unity feedback.

The values of $\mathbf{C}(j\omega)/\mathbf{R}(j\omega)$ for different values of frequency can be determined in the manner given in the example in Sec. 9.5. By superimposing the M and α contours on the $1/\mathbf{G}(j\omega)$ plot, the points of intersection provide data for plotting M and α vs. ω.

Nonunity-Feedback Control System

The case of nonunity feedback is handled more simply on the inverse plot than on the direct plot since

$$\frac{\mathbf{R}(j\omega)}{\mathbf{C}(j\omega)} = \frac{1 + \mathbf{G}(j\omega)\mathbf{H}(j\omega)}{\mathbf{G}(j\omega)} = \frac{1}{\mathbf{G}(j\omega)} + \mathbf{H}(j\omega) \tag{9.41}$$

The curves $1/\mathbf{G}(j\omega)$ and $\mathbf{H}(j\omega)$ can be drawn separately and then added, as shown in Fig. 9.17. From the origin the quantity $\mathbf{R}(j\omega)/\mathbf{C}(j\omega)$ is measured directly in both magnitude and angle. The $1/M$ circles and α lines are the same as those for the unity-feedback case but are now drawn from the origin.

Example. Rate feedback

$$H(s) = K_1 s \tag{9.42}$$

The curves $1/\mathbf{G}(j\omega)$ and $\mathbf{H}(j\omega)$ are plotted in Fig. 9.21. Their sum, $1/\mathbf{G}(j\omega) + \mathbf{H}(j\omega) = \mathbf{R}(j\omega)/\mathbf{C}(j\omega)$, is also shown in this figure.

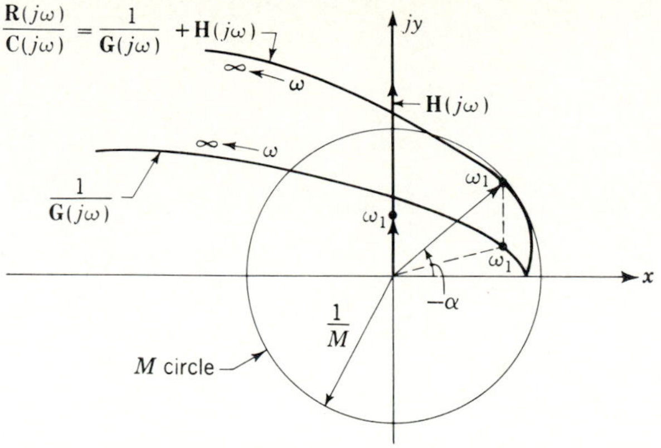

FIGURE 9.21
An example of M and α contours for nonunity feedback.

9.7 GAIN ADJUSTMENT FOR A DESIRED M_m OF A UNITY-FEEDBACK SYSTEM: DIRECT POLAR PLOT

Gain adjustment is the first step in adjusting the system for the desired performance. The procedure for adjusting the gain is outlined in this section. Figure 9.22*a* shows $\mathbf{G}_x(j\omega)$ with its respective M_m circle in the complex plane. Since

$$\mathbf{G}_x(j\omega) = x + jy = K_x\mathbf{G}'_x(j\omega) = K_x(x' + jy') \tag{9.43}$$

then

$$x' + jy' = \frac{x}{K_x} + j\frac{y}{K_x}$$

where $\mathbf{G}'_x(j\omega) = \mathbf{G}_x(j\omega)/K_x$ is defined as the frequency-sensitive portion of $\mathbf{G}_x(j\omega)$ with unity gain. Note that changing the gain merely changes the amplitude and not the angle of the locus of points of $\mathbf{G}_x(j\omega)$. Thus, if in Fig. 9.22*a* a change of scale is made by dividing the x, y coordinates by K_x so that the new coordinates are x', y', the following are true:

1. The $\mathbf{G}_x(j\omega)$ plot becomes the $\mathbf{G}'_x(j\omega)$ plot.
2. The M_m circle becomes *a circle* which is simultaneously tangent to $\mathbf{G}'_x(j\omega)$ and the line representing $\sin\psi = 1/M_m$.
3. The $-1 + j0$ point becomes the $-1/K_x + j0$ point.
4. The radius r_0 becomes $r'_0 = r_0/K_x$.

In other words, if $\mathbf{G}'_x(j\omega)$ is drawn on a separate graph sheet from that of $\mathbf{G}_x(j\omega)$ (see Fig. 9.22*b*), by superimposing the two graphs so that the axes coincide, the circles and the $\mathbf{G}_x(j\omega)$ and $\mathbf{G}'_x(j\omega)$ plots also coincide. Referring to Fig. 9.22*a* and *b*, note that $oa = -1$ and $oa' = -1/K_x$.

As a consequence, it is possible to determine the required gain to achieve a desired M_m for a given system by the following graphical procedure:

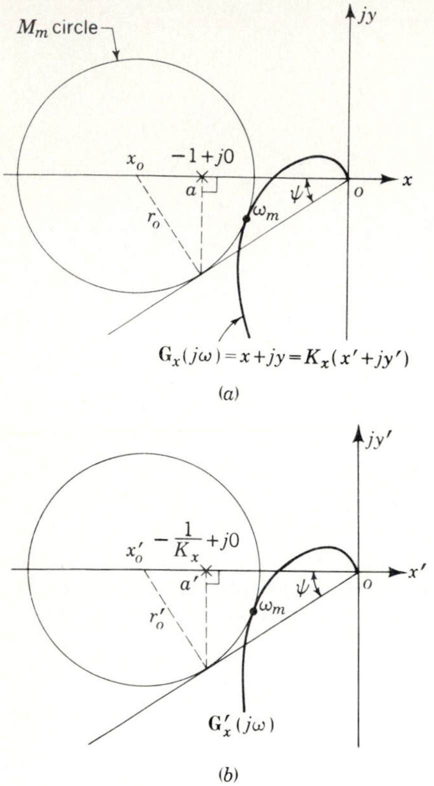

FIGURE 9.22
(a) Plot of $\mathbf{G}_x(j\omega)$ with the respective M_m circle. (b) Circle drawn tangent to both the plot of $\mathbf{G}'_x(j\omega)$ and to the line representing the angle $\psi = \sin^{-1}(1/M_m)$.

Step 1. If the original system has a transfer function

$$\mathbf{G}_x(j\omega) = \frac{K_x(1 + j\omega T_1)(1 + j\omega T_2)\cdots}{(j\omega)^m(1 + j\omega T_a)(1 + j\omega T_b)(1 + j\omega T_c)\cdots} = K_x\mathbf{G}'_x(j\omega) \quad (9.44)$$

with an original gain K_x, only the frequency-sensitive portion $\mathbf{G}'_x(j\omega)$ is plotted.

Step 2. Draw the straight line at the angle $\psi = \sin^{-1}(1/M_m)$, measured from the negative real axis.

Step 3. By trial and error, find a circle whose center lies on the negative real axis and is simultaneously tangent to the $\mathbf{G}'_x(j\omega)$ plot and the line representing the angle ψ.

Step 4. Having found this circle, locate the point of tangency on the ψ-angle line. Draw a line from the point of the tangency perpendicular to the real axis. Label the point where this line intersects the real axis as a'.

Step 5. For this circle to be an M circle representing M_m, the point a' must be the $-1 + j0$ point. Thus, the x', y' coordinates must be multiplied by a gain factor K_m in order to convert this plot into a plot of $\mathbf{G}(j\omega)$. From the graphical construction the value of K_m is

$$K_m = \frac{1}{oa'}$$

Step 6. The original gain must be changed by a factor

$$A = \frac{K_m}{K_x}$$

Note that if $\mathbf{G}_x(j\omega)$ which includes a gain K_x is already plotted, it is possible to work directly with the plot of the function $\mathbf{G}_x(j\omega)$. Following the procedure outlined above results in the determination of the *additional* gain required to produce the specified M_m; that is, the additional gain is

$$A = \frac{K_m}{K_x} = \frac{1}{oa''}$$

Example. It is desired that the closed-loop system which has the open-loop transfer function

$$\mathbf{G}_x(j\omega) = \frac{1.47}{j\omega(1 + j0.25\omega)(1 + j0.1\omega)}$$

have an $M_m = 1.3$. The problem is to determine the actual gain K_1 needed and the amount by which the original gain K_x must be changed to obtain this M_m. The procedure previously outlined is applied to this problem. $\mathbf{G}_x'(j\omega)$ is plotted in Fig. 9.23 and results in $\omega_m = 2.55$ and

$$K_1 = \frac{1}{oa'} \approx \frac{1}{0.34} = 2.94\,\text{s}^{-1}$$

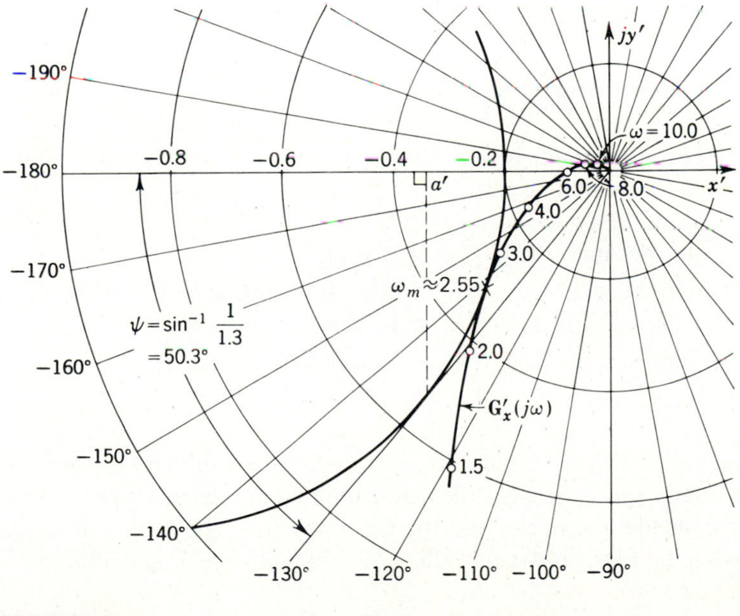

FIGURE 9.23
Gain adjustment using the procedure outlined in Sec. 9.7 for

$$\mathbf{G}_x'(j\omega) = 1/j\omega(1 + j0.25\omega)(1 + j0.1\omega).$$

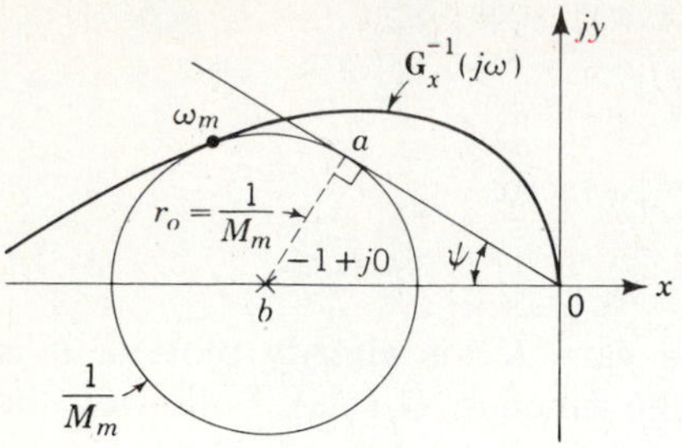

FIGURE 9.24
Plot of $\mathbf{G}_x^{-1}(j\omega)$ and the $1/M_m$ circle.

The additional gain required is

$$A = \frac{K_1}{K_x} = \frac{2.94}{1.47} = 2.0$$

In other words, the original gain must be doubled to obtain M_m of 1.3 for $\mathbf{C}(j\omega)/\mathbf{R}(j\omega)$.

9.8 GAIN ADJUSTMENT FOR A DESIRED M_m OF A UNITY-FEEDBACK SYSTEM: INVERSE POLAR PLOT

The approach is similar to that for the direct plots in establishing the procedure for gain setting for the inverse plot. From Fig. 9.24

$$\frac{1}{\mathbf{G}_x(j\omega)} = \frac{1}{K_x\mathbf{G}_x'(j\omega)} = \mathbf{G}_x^{-1}(j\omega) = x + jy \tag{9.45a}$$

Then

$$\frac{K_x}{\mathbf{G}_x(j\omega)} = \frac{1}{\mathbf{G}_x'(j\omega)} = K_x x + jK_x y = x' + jy' \tag{9.45b}$$

Thus, if a change of scale is made in Fig. 9.24 by multiplying the x, y coordinates by K_x so that the new coordinates are x', y', the following are true:

1. The $\mathbf{G}_x^{-1}(j\omega)$ plot becomes the $[\mathbf{G}_x'(j\omega)]^{-1}$ plot.
2. The $1/M_m$ circle becomes *a circle* which is simultaneously tangent to $[\mathbf{G}_x'(j\omega)]^{-1}$ and the line drawn at the angle $\psi = \sin^{-1}(1/M_m)$.
3. The $-1 + j0$ point becomes the $-K_x + j0$ point.
4. The radius r_0 becomes $r_0' = K_x r_0$.

In the same manner as with the direct plot, if $[\mathbf{G}_x'(j\omega)]^{-1}$ is drawn on a separate sheet from that of $\mathbf{G}_x^{-1}(j\omega)$ (see Fig. 9.25), and the two graphs are superimposed so that the axes coincide, the circles and the $\mathbf{G}_x^{-1}(j\omega)$ and $[\mathbf{G}_x'(j\omega)]^{-1}$ plots also coincide. Referring to Figs. 9.24 and 9.25, note that $ob = -1$ and that $ob' = -K_x$.

To determine the gain necessary for a given M_m from the inverse plots, the following graphical procedure can be used.

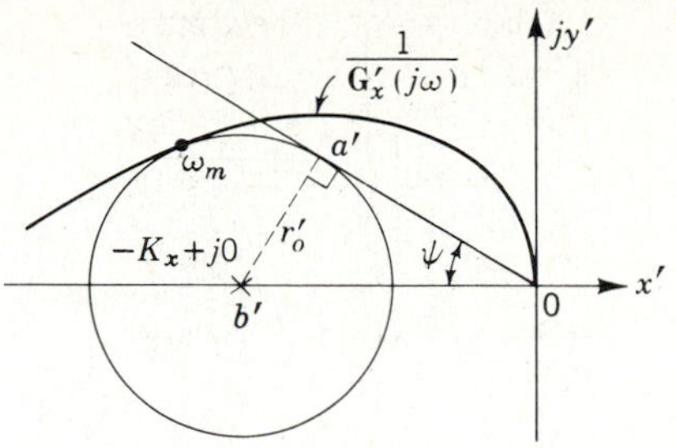

FIGURE 9.25
Plot of $[\mathbf{G}_x'(j\omega)]^{-1}$ with circle drawn tangent to it and to the line at the angle $\psi = \sin^{-1}(1/M_m)$.

Step 1. If the original system has an inverse transfer function

$$\mathbf{G}_x^{-1}(j\omega) = \frac{(j\omega)^m(1+j\omega T_a)(1+j\omega T_b)(1+j\omega T_c)\cdots}{K_x(1+j\omega T_1)(1+j\omega T_2)\cdots} \tag{9.46}$$

with an original gain K_x, only the frequency-sensitive portion $[\mathbf{G}_x'(j\omega)]^{-1}$ is plotted.

Step 2. Draw the line at the angle $\psi = \sin^{-1}(1/M_m)$.

Step 3. By trail and error, find a circle whose center lies on the negative real axis and is simultaneously tangent to both the $[\mathbf{G}_x'(j\omega)]^{-1}$ plot and the line representing the angle ψ.

Step 4. For this circle to be a $1/M_m$ circle representing M_m, the center of the circle b' must be the $-1 + j0$ point. Thus, the x', y' coordinates must be divided by the gain constant K_m in order to convert this plot into a plot of $\mathbf{G}^{-1}(j\omega)$. From the graphical construction the value of K_m is

$$K_m = ob'$$

Step 5. Thus the original gain must be changed by a factor

$$A = \frac{K_m}{K_x}$$

If $\mathbf{G}_x^{-1}(j\omega)$ is already plotted, this plot can become $[\mathbf{G}_x'(j\omega)]^{-1}$ by merely changing the scale, i.e., multiplying the x, y coordinates by K_x. However, the $\mathbf{G}_x^{-1}(j\omega)$ plot can be used to determine the *additional* gain required for the desired M_m. The value of the additional gain is equal to ob''.

Example. Assume that a system having the inverse transfer function

$$\mathbf{G}_x^{-1}(j\omega) = \frac{j\omega(1+j0.8\omega)(1+j0.25\omega)}{0.5}$$

must have an $M_m = 1.5$. In Fig. 9.26 the above procedure is followed in order to

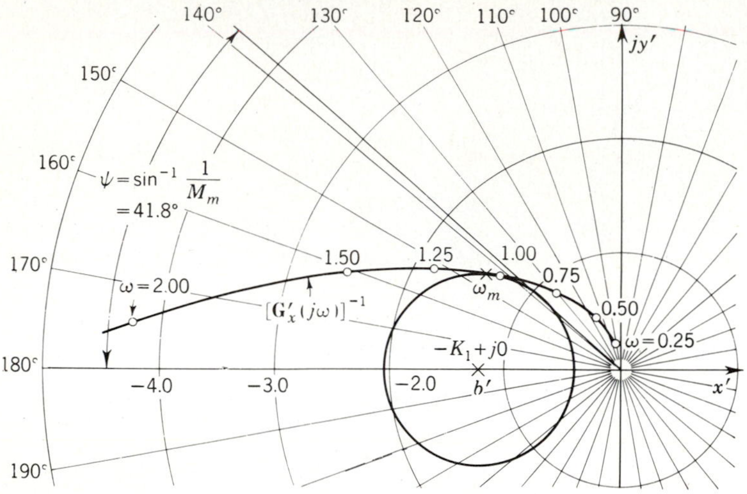

FIGURE 9.26
Gain adjustment using the procedure outlined in Sec. 9.8 for

$$[\mathbf{G}_x'(j\omega)]^{-1} = j\omega(1 + j0.8\omega)(1 + j0.25\omega).$$

determine the required gain K_1. Thus $\omega_m = 1.02$ and

$$K_1 = 1.23 \text{ s}^{-1}$$

Then the amount by which the original gain K_x must be changed is

$$A = \frac{K_1}{K_x} = \frac{1.23}{0.5} = 2.46$$

Therefore, increasing the original gain by a factor of 2.46 gives the desired $M_m = 1.5$ with $\omega_m = 1.02$ for $\mathbf{C}(j\omega)/\mathbf{R}(j\omega)$.

9.9 CONSTANT M AND α CURVES ON THE LOG MAGNITUDE–ANGLE DIAGRAM (NICHOLS CHART)[5]

As derived earlier in this chapter, the constant M curves on the direct and inverse polar plots are circles. The transformation of these curves to the log magnitude–angle diagram is done more easily by starting from the inverse polar plot since all the M circles have the same center. This requires a change of sign of the log magnitude and angle obtained, since the transformation is from the inverse transfer function on the polar plot to the direct transfer function on the log magnitude plot.

A constant M circle is shown in Fig. 9.27 on the inverse polar plot. The magnitude ρ and angle λ drawn to any point on this circle are shown.

The equation for this M circle is

$$y^2 + (1 + x)^2 = \frac{1}{M^2} \tag{9.47}$$

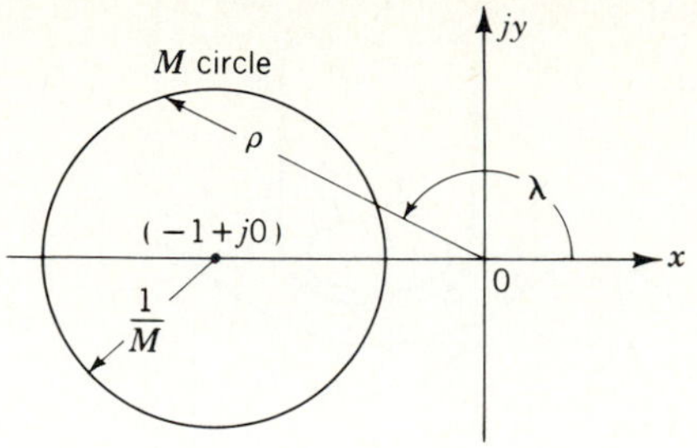

FIGURE 9.27
Constant M circle on the inverse polar plot.

where $$x = \rho \cos \lambda \qquad y = \rho \sin \lambda \tag{9.48}$$

Combining these equations produces

$$\rho^2 M^2 + 2\rho M^2 \cos \lambda + M^2 - 1 = 0 \tag{9.49}$$

Solving for ρ and λ yields

$$\rho = -\cos \lambda \pm \left(\cos^2 \lambda - \frac{M^2 - 1}{M^2} \right)^{1/2} \tag{9.50}$$

$$\lambda = \cos^{-1} \frac{1 - M^2 - \rho^2 M^2}{2\rho M^2} \tag{9.51}$$

These equations are derived from the inverse polar plot $1/\mathbf{G}(j\omega)$. Since the log magnitude–angle diagram is drawn for the transfer function $\mathbf{G}(j\omega)$ and not for its reciprocal, a change in the equations must be made by substituting

$$r = \frac{1}{\rho} \qquad \phi = -\lambda$$

Since $\text{Lm}\, r = -\text{Lm}\, \rho$, it is only necessary to change the sign of $\text{Lm}\, \rho$. For any value of M a series of values of angle λ can be inserted in Eq. (9.50) to solve for ρ. This magnitude must be changed to decibels. Alternatively, for any value of M a series of values of ρ can be inserted in Eq. (9.51) to solve for the corresponding angle λ. Therefore the constant M curve on the log magnitude–angle diagram can be plotted by using either of these two equations.

In a similar fashion the constant α curves can be drawn on the log magnitude–angle diagram. The α curves on the inverse polar plot are semi-infinite straight lines terminating on the $-1 + j0$ point and are given by

$$\tan \alpha + x \tan \alpha + y = 0 \tag{9.52}$$

Combining with Eq. (9.48) produces

$$\tan \alpha + \rho \cos \lambda \tan \alpha + \rho \sin \lambda = 0 \tag{9.53}$$

$$\tan \alpha = \frac{-y}{1 + x} = \frac{-\rho \sin \lambda}{1 + \rho \cos \lambda} \tag{9.54}$$

For constant values of α a series of values of λ can be inserted in this equation to solve for ρ. The constant α semi-infinite curves can then be plotted on the log magnitude–angle diagram.

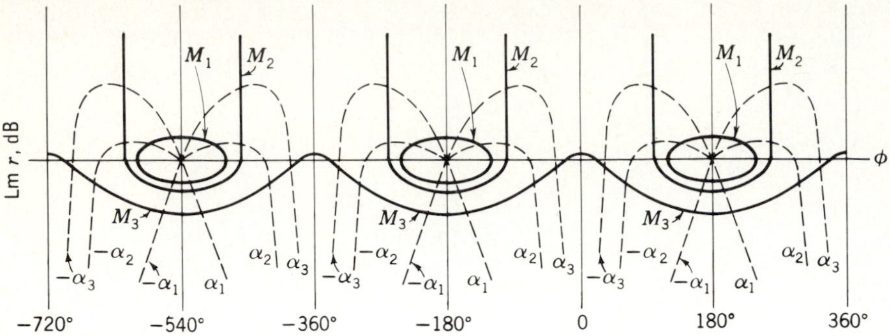

FIGURE 9.28
Constant M and α curves.

Constant M and α curves, as shown in Fig. 9.28, repeat for every 360°. Also, there is symmetry at every 180° interval. An expanded 300° section of the constant M and α graph is shown in Fig. 9.29. This graph is commonly referred to as the Nichols chart. Note that the $M = 1$ (0 dB) curve is asymptotic to $\phi = -90°$ and $-270°$ and the curves for $M < \frac{1}{2}$ (-6 dB) are always negative. $M = \infty$ is the point at 0 dB, $-180°$, and the curves for $M > 1$ are closed curves inside the limits $\phi = -90°$ and $\phi = -270°$.

It should be noted that the Nichols chart should be viewed as two charts. The first has the Cartesian coordinates of dB vs. phase angle ϕ which are used to plot the open-loop frequency response, with the frequencies noted along the plot. The second is the chart for loci of constant M and α for the closed-loop transfer function that is superimposed on the open-loop frequency-response plot. It should also be emphasized that the loci for constant M and α for the Nichols chart are applied only for stable unity-feedback systems.

9.10 ADJUSTMENT OF GAIN BY USE OF THE LOG MAGNITUDE–ANGLE DIAGRAM

The log magnitude–angle diagram for

$$\mathbf{G}(j\omega) = \frac{2.04(1 + j2\omega/3)}{j\omega(1 + j\omega)(1 + j0.2\omega)(1 + j0.2\omega/3)} \tag{9.55}$$

is drawn as the solid curve in Fig. 9.30 on graph paper which has the constant M and α contours. (The log magnitude and phase-angle diagram can be drawn first to get the data for this curve.) It is convenient to use the same log magnitude and angle scales on both the log magnitude and phase diagram and the log magnitude–angle diagram. A pair of dividers can then be used to move the transfer-function locus to the log magnitude–angle diagram. The $M = 1.12$ (1-dB) curve is tangent to the curve at $\omega_{m1} = 1.1$. These values are the maximum value of the control ratio M_m and the resonant frequency ω_m.

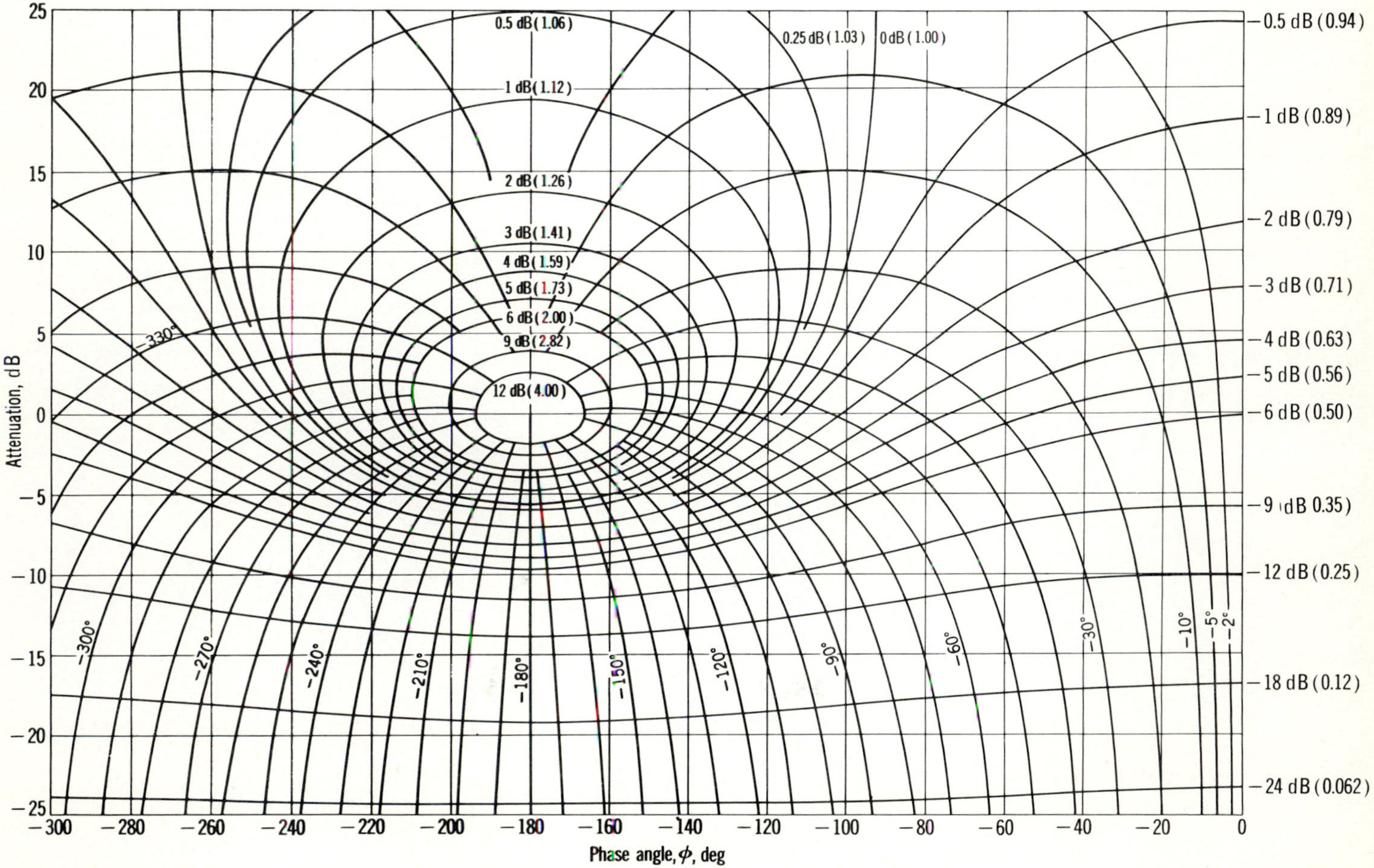

FIGURE 9.29
Constant M and α curves in the log magnitude–angle plane.

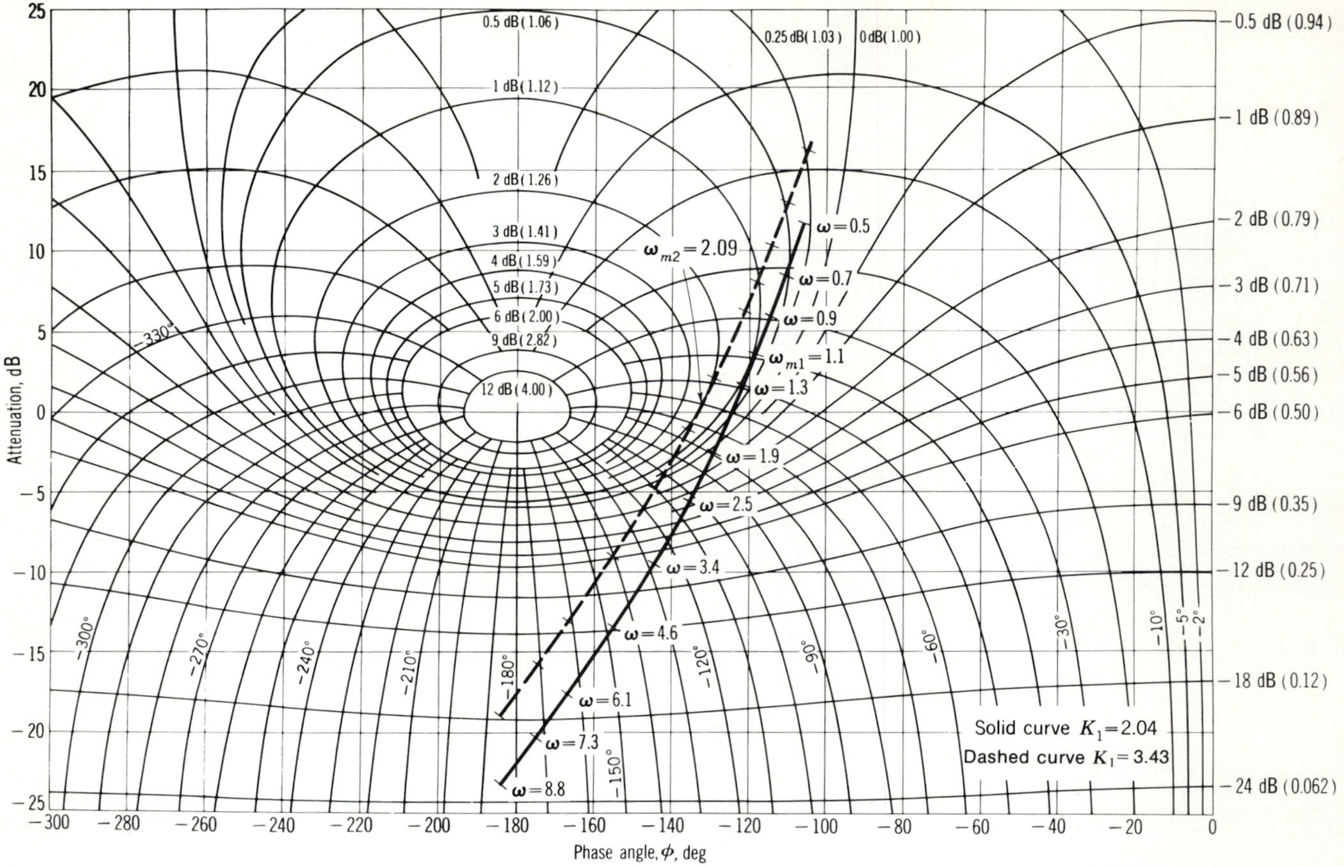

FIGURE 9.30
Log magnitude–angle diagram for Eq. (9.55).

It is specified that the closed-loop system be adjusted to produce an $M_m = 1.26$ (2 dB) by changing the gain. The dashed curve is obtained by raising the transfer-function curve until it is tangent to the $M_m = 1.26$ (2-dB) curve. The resonant frequency is now equal to $\omega_{m2} = 2.09$. The curve has been raised by the amount Lm $A = 4.5$ dB, meaning that an additional gain of $A = 1.679$ must be put into the system.

In practice it is easier to make a template of the desired M_m curve. Then the template can be moved up or down until it is tangent to the transfer-function curve. The amount the template is moved represents the gain change required. Once a graphical estimate of the gain is obtained by use of the template, a digital-computer program[4,7] can be used to refine the design. When the gain has been adjusted for the desired M_m, the closed-loop frequency response can be found either graphically or by use of the digital computer. Graph paper with the constant M and α curves superimposed simplifies system analysis and design. For any frequency point on the transfer-function curve the values of M and α can be read from the graph. For example, with M_m made equal to 1.26 in Fig. 9.30, the value of M at $\omega = 3.4$ is -0.5 dB and α is $-110°$. The closed-loop frequency response, both log magnitude and angle, obtained from Fig. 9.30 is plotted in Fig. 9.31 for both values of gain, $K_1 = 2.04$ and 3.43.

After the gain has been adjusted for the desired M_m, methods described in the literature can be used to obtain the approximate values of the closed-loop poles[6] from $\mathbf{M}(\omega)$. The feedback system in which the gain has been adjusted may be a portion of a larger system. Analysis and design of the larger system can proceed in a similar fashion.

The closed-loop response with the resultant gain adjustment may have too low a resonant frequency for the desired performance. The next design step

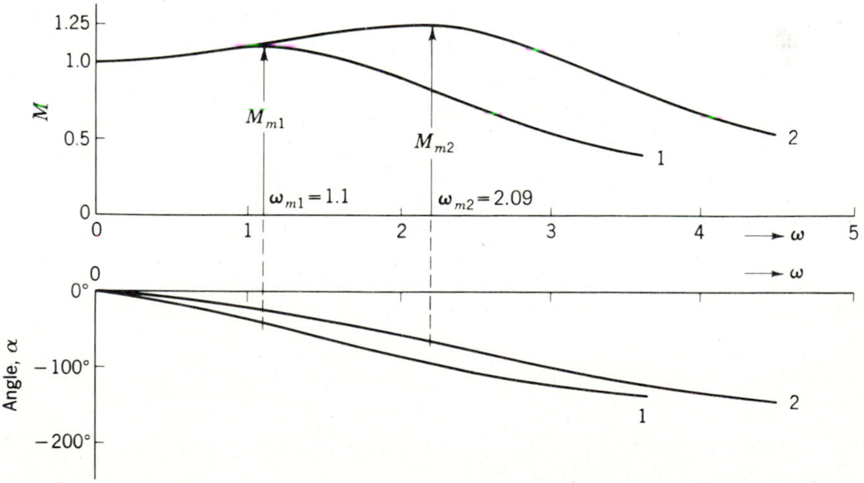

FIGURE 9.31
Control ratio vs. frequency obtained from the log magnitude–angle diagram of Fig. 9.30.

would be to compensate the system to improve the frequency response. Compensation is studied in detail in Chaps. 10, 11, 12, and 21.

9.11 CORRELATION OF POLE-ZERO DIAGRAM WITH FREQUENCY AND TIME RESPONSES

Whenever the closed-loop control ratio $\mathbf{M}(j\omega)$ has the characteristic form shown in Fig. 9.5, the system may be approximated as a simple second-order system. This usually implies that the poles, other than the dominant complex pair, are either far to the left of the dominant complex poles or are close to zeros. When these conditions are not satisfied, the frequency response may have other shapes. This can be illustrated by considering the following three control ratios:

$$\frac{C(s)}{R(s)} = \frac{1}{s^2 + s + 1} \tag{9.56}$$

$$\frac{C(s)}{R(s)} = \frac{0.313(s + 0.8)}{(s + 0.25)(s^2 + 0.3s + 1)} \tag{9.57}$$

$$\frac{C(s)}{R(s)} = \frac{4}{(s^2 + s + 1)(s^2 + 0.4s + 4)} \tag{9.58}$$

The pole-zero diagram, the frequency response, and the time response to a step input for each of these equations are shown in Fig 9.32.

The following characteristics are noted from Fig. 9.32*a*, which represents Eq. (9.56):

1. The control ratio has only two complex poles, which are therefore dominant, and no zeros.
2. The frequency-response curve has the following characteristics:
   a. A single peak $M_m = 1.157$ at $\omega_m = 0.7$.
   b. $1 < M < M_m$ in the frequency range $0 < \omega < 1$.
3. The time response has the typical waveform described in Chap. 3 for a simple second-order system. That is, the first maximum of $c(t)$ due to the oscillatory term is greater than $c(t)_{ss}$, and the $c(t)$ response after this maximum oscillates around the value of $c(t)_{ss}$.

The following characteristics are noted from Fig. 9.32*b*, for Eq. (9.57):

1. The control ratio has two complex poles and one real pole, all dominant, and one real zero.
2. The frequency-response curve has the following characteristics:
   a. A single peak, $M_m = 1.27$ at $\omega_m = 0.95$.
   b. $M < 1$ in the frequency range $0 < \omega < \omega_x$.
   c. The peak M_m occurs at $\omega_m = 0.95 > \omega_x$.

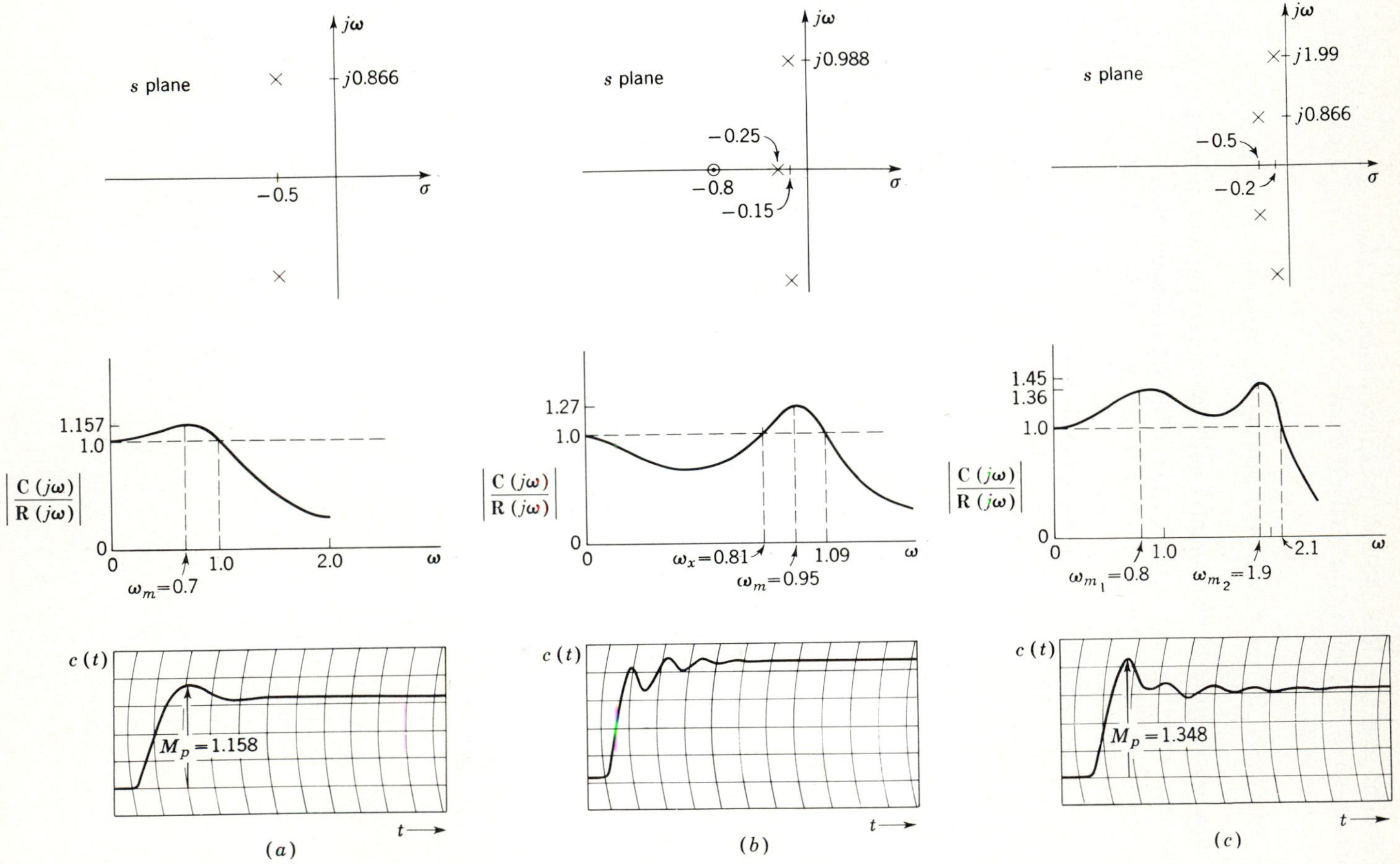

FIGURE 9.32
Comparison of frequency and time responses for three pole-zero patterns.

3. The time response does not have the conventional waveform. That is, the first maximum of $c(t)$ due to the oscillatory term is less than $c(t)_{ss}$ because of the transient term $A_3 e^{-0.25t}$.

From Fig. 9.32*c*, for Eq. (9.58), the following characteristics are noted:

1. The control ratio has four complex poles, all dominant, and no zeros.
2. The frequency-response curve has the following characteristics:
   a. There are two peaks, $M_{m1} = 1.36$ at $\omega_{m1} = 0.81$ and $M_{m2} = 1.45$ at $\omega_{m2} = 1.9$
   b. $1 < M < 1.45$ in the frequency range $0 < \omega < 2.1$.
   c. The time response does not have the simple second-order waveform. That is, the first maximum of $c(t)$ in the oscillation is greater than $c(t)_{ss}$, and the oscillatory portion of $c(t)$ does not oscillate about a value of $c(t)_{ss}$. This time response can be predicted from the pole locations in the s plane and from the plot of M vs. ω (the two peaks).

Another example is the system represented by

$$M(s) = \frac{K}{(s - p_1)(s^2 + 2\zeta\omega_n s + \omega_n^2)} \tag{9.59}$$

When the real pole p_1 and real part of the complex poles are equal as shown in Fig. 9.33*a*, the corresponding frequency response is as shown in Fig. 9.33*b*. The

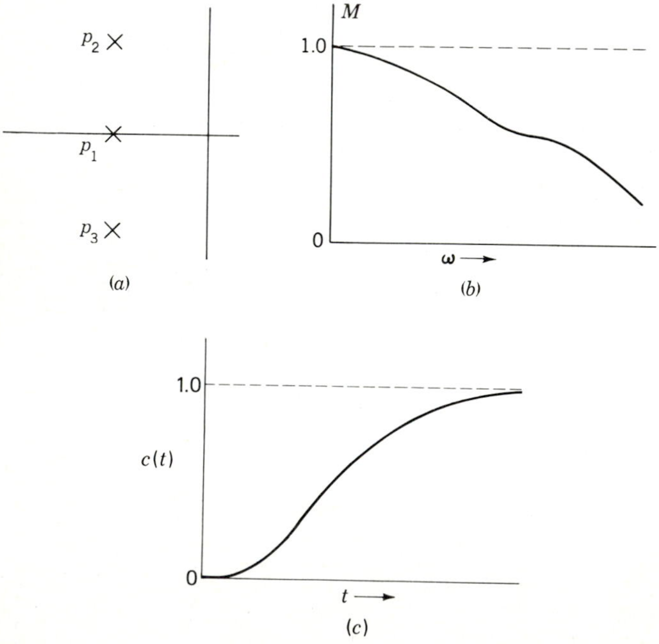

FIGURE 9.33
Form of the frequency and time responses for a particular pole pattern of Eq. (9.59).

magnitude at ω_m is less than unity. The corresponding time response to a step input is monotonic; i.e., there is no overshoot as shown in Fig. 9.33*c*. This may be considered as a critically damped response.

The examples discussed in this section show that the time-response waveform is closely related to the frequency response of the system. In other words, the time-response waveform of the system can be predicted from the shape of the frequency-response plot. This time-response correlation with the plot of $M = |\mathbf{C}(j\omega)/\mathbf{R}(j\omega)|$ vs. ω, when there is no distinct dominant complex pair of poles, is an important advantage of frequency-response analysis. Thus, as illustrated in this section, the frequency-response plot may be utilized as a guide in determining (or predicting) time-response characteristics.

9.12 SUMMARY

In summary, this chapter is devoted to indicating the correlation between the frequency and time responses. The figures of merit M_m and ω_m are established as guideposts for evaluating a system's tracking performance. The addition of a pole to an open-loop transfer function produces a clockwise shift of the direct polar plot, which results in a larger value of M_m. The time response also suffers because ω_m becomes smaller. The reverse is true if a zero is added to the open-loop transfer function. This agrees with the analysis of the root locus, which shows that the addition of a pole or zero results in a less stable or more stable system, respectively. Thus the qualitative correlation between the root locus and the frequency response is enhanced. The M and α contours are developed as a graphical aid in obtaining the closed-loop frequency response and in adjusting the gain to obtain a desired M_m. The methods described for setting the gain for a desired M_m are based on the fact that generally the desired values of M_m are greater than 1 in order to obtain an underdamped response. The procedure for gain adjustment may not yield a satisfactory value of ω_m. In this case the system must be compensated in order to increase ω_m without changing the value of M_m. Compensation procedures are covered in the following chapters.

The procedure used in the frequency-response method is summarized as follows:

Step 1. Derive the open-loop transfer function $G(s)H(s)$ of the system.

Step 2. Put the transfer function into the form $\mathbf{G}(j\omega)\mathbf{H}(j\omega)$.

Step 3. Arrange the various factors of the transfer function so that they are in the complex form $j\omega$, $1 + j\omega T$, and $1 + aj\omega + b(j\omega)^2$.

Step 4. Plot the log magnitude and phase-angle diagram for $\mathbf{G}(j\omega)\mathbf{H}(j\omega)$. The graphical methods of Chap. 8 may be used, but a digital computer or a calculator can provide more extensive data.[4,7]

Step 5. Transfer the data from the plots in step 4 to any of the following: (*a*) log magnitude–angle diagram, (*b*) inverse polar plot, or (*c*) direct polar plot.

Step 6. Apply the Nyquist stability criterion and adjust the gain for the desired degree of stability M_m of the system. Then check the correlation to the time response for a step input signal. This correlation reveals some qualitative information about the time response.

Step 7. If the qualitative response does not meet the desired specifications, determine the shape that the plot must have to meet these specifications.

Step 8. Synthesize the compensator that must be inserted into the system, if other than just gain adjustment is required, to make the necessary modification on the original plot. This procedure is developed in the following chapters.

A digital computer is frequently available and is used extensively in system design. A standard computer program can be used to obtain the frequency response of both the open-loop and closed-loop transfer functions.[4,7] The computer techniques are especially useful for systems containing frequency-sensitive feedback. The exact time response is then obtained to provide a correlation between the frequency and time responses. This chapter illustrates the use of the Nichols chart for analyzing the tracking performances of unity- and nonunity-feedback systems. The Nichols chart can also be used for analyzing the disturbance rejection performance of a nonunity-feedback system. This is discussed in detail in later chapters.

REFERENCES

1. Brown, G. S., and D. P. Campbell: *Principles of Servomechanisms*, Wiley, New York, 1948, chaps. 6 and 8.
2. Higgins, T. J., and C. M. Siegel: "Determination of the Maximum Modulus, or the Specified Gain of a Servomechanism by Complex Variable Differentiation," *Trans. AIEE*, vol. 72, pt. II, p. 467, January 1954.
3. Chu, Y.: "Correlation between Frequency and Transient Responses of Feedback Control Systems," *Trans. AIEE*, vol. 72, pt. II, pp. 81–92, May 1953.
4. Larimer, S. J.: "An Interactive Computer-Aided Design Program for Digital and Continuous System Analysis and Synthesis (TOTAL)," M.S. thesis, GE/GGC/EE/78-2, School of Engineering, Air Force Institute of Technology, Wright-Patterson Air Force Base, Ohio, 1978; available from Defense Documentation Center (DDC), Cameron Station, Alexandria, Va. 22314.
5. James, H. M., N. B. Nichols, and R. S. Phillips: *Theory of Servomechanisms*, McGraw-Hill, New York, 1947, chap. 4.
6. Chen, K.: "A Quick Method for Estimating Closed-Loop Poles of Control Systems," *Trans. AIEE*, vol. 76, pt. II, pp. 80–87, May 1957.
7. Thompson, P. M.: *USER's Guide to Program CC, Version 3*, Systems Technology, Inc., Hawthorne, Calif., March 1985.

CHAPTER 10

ROOT-LOCUS COMPENSATION

10.1 INTRODUCTION

The preceding chapters deal with basic feedback control systems composed of the minimum amount of equipment required to perform the control function and the necessary sensors and comparators to provide feedback. Refinements are not made until the designer has analyzed the performance of the basic system. This chapter presents additional topics which enhance the design of a control system. In order to increase the reader's understanding of how poles and zeros affect the time response, a graphical means is presented for calculating the figures of merit of the system and the effect of additional significant nondominant poles. Typical pole-zero patterns are employed to demonstrate the correlation between the pole-zero diagram and the frequency and time responses. As a result of this analysis, equipment may be added to the control system to achieve the desired time response. The next few chapters are devoted to achieving the necessary refinements.

Introducing additional equipment into a system to reshape its root locus in order to improve system performance is called *compensation* or stabilization. When the system is compensated, it is stable, has a satisfactory transient response, and has a large enough gain to ensure that the steady-state error does not exceed the specified maximum. Compensation devices may consist of electric networks or mechanical equipment containing levers, springs, dashpots, etc. The compensator (also called a filter) may be placed in cascade with the forward transfer function $G_x(s)$ (cascade or series compensation), as shown in Fig. 10.1*a*,

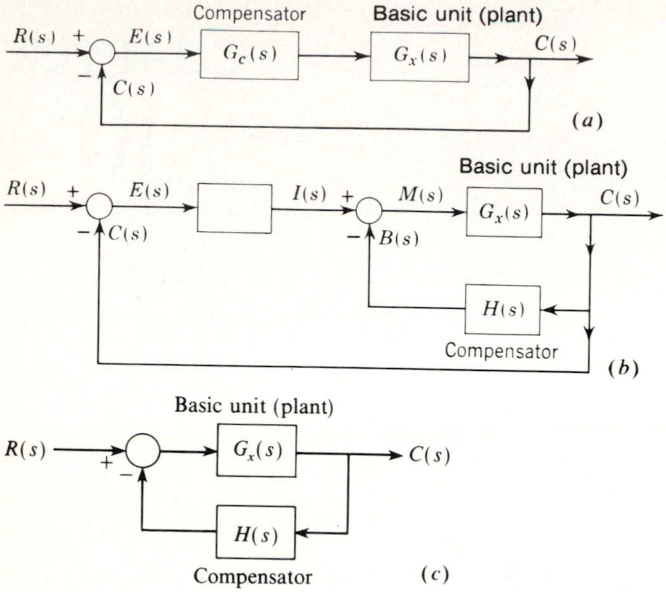

FIGURE 10.1
Block diagram showing the location of compensators: (a) cascade; (b) minor loop feedback; and (c) output feedback.

or in the feedback path (feedback or parallel compensation), as shown in Fig. 10.1b and c. In Figure 10.1b the compensation is introduced by means of a minor feedback loop. The selection of the location for inserting the compensator depends largely on the control system, the necessary physical modifications, and the results desired. The cascade compensator $G_c(s)$ is inserted at the low-energy point in the forward path so that power dissipation is very small. This also requires that the input impedance be high. Isolation amplifiers may be necessary to avoid loading of or by the compensating network. The networks used for compensation are generally called lag, lead, and lag-lead compensators. The examples used for each of these compensators are not the only ones available but are intended primarily to show the methods of applying compensation. Feedback compensation is used in later design examples to improve the system's tracking of a desired input $r(t)$ or to improve the system's rejection of a disturbance input $d(t)$.

The following factors must be considered when making a choice between cascade and feedback compensation:

1. The design procedures for a cascade compensator are more direct than those for a feedback compensator. The application of feedback compensators is sometimes more laborious, but it may be easier to implement.

2. Because of the physical form of the control system, i.e., whether it is electrical, hydraulic, mechanical, etc., a cascade or feedback compensator may not exist or be practical.
3. The type of signal input to the compensator must be considered. For example, if the system utilizes a 400-Hz carrier, the design of a feedback compensator may be more difficult than that of a cascade compensator.
4. The economics in the use of either technique for a given control system involves items such as the size, weight, and cost of components and amplifiers. In the forward path the signal goes from a low- to a high-energy level, whereas the reverse is true in the feedback loop. Thus, generally an amplifier may not be necessary in the feedback path. The cascade path generally requires an amplifier for gain and/or isolation. Also, the size and weight may be different for the cascade and the feedback compensators. These items are of great importance in aircraft, both commercial and military, and in spacecraft, where minimum size and weight of equipment are essential.
5. The environmental conditions in which the feedback control system is to be utilized affect the accuracy and stability of the controlled quantity. This is a serious problem in an airplane or space vehicle, which is subjected to rapid changes in altitude and temperature. It is shown in Chaps. 10 to 14 that control-system performance can be improved by the use of a feedback compensator or by state-variable feedback.
6. The problem of noise within a control system may determine the choice of compensator. The noise problem is accentuated in situations where a greater amplifier gain is required with a forward compensator than by the use of feedback networks. Also, the frequency characteristics of the compensator needed to give the desired system improvement may be such that it attenuates the high-frequency portion of the noise content.
7. The time of response desired for a control system is a determining factor. Often a faster time of response can be achieved by the use of feedback compensation.
8. Some systems require "tight-loop" stabilization to isolate the dynamics of one portion of a control system from other portions of the complete system. This can be accomplished by introducing an inner feedback loop around the portion to be isolated.
9. When $G_x(s)$ has a pair of dominant complex poles that yield the dominant poles of $C(s)/R(s)$, the simple first-order cascade compensators discussed in this chapter provide minimal improvement to the system's time-response characteristics. For this situation, as shown in Sec. 10.20, feedback compensation is more effective and is therefore more desirable.
10. Besides all these factors, the available components and the designer's experience and preferences influence the choice between a cascade and feedback compensator for achieving the desired tracking performance.

11. The feedback-compensator configuration of Fig. 10.1*c* is used in Chap. 12 for disturbance rejection to minimize the response $c(t)$ to a disturbance input $d(t)$.

Meeting the system-performance specification often requires the inclusion of a properly designed compensator. The design process is normally performed by successively changing the compensator parameter(s) and comparing the resulting system performance. This process can be expedited by the use of a CAD program. Some of the CAD packages which are available are listed in App. B. The availability of a graphics capability in a CAD program permits the designer to see the changes in system-performance in real time. An interactive capability allows the designer to quickly try a number of compensator designs in order to select a design that best meets the system performance requirements. The use of an interactive CAD package is standard practice for control system designers.

10.2 TRANSIENT RESPONSE: DOMINANT COMPLEX POLES[2]

The root-locus plot permits selection of the best poles for the control ratio. The criteria for determining which poles are best must come from the specifications of system performance and from practical considerations. For example, the presence of backlash, dead zone, and coulomb friction can produce a steady-state error with a step input, even though the linear portion of the system is Type 1 or higher. The response obtained from the pole-zero locations of the control ratio does not show the presence of a steady-state error because the nonlinearities have been neglected. The effect of some nonlinearities can be minimized by making the system underdamped.† Therefore, the system gain is adjusted so that there is a dominant pair of complex poles and the response to a step input has the form shown in Fig. 10.2*a*. In this case the transient contributions from the other poles must be small. From Eq. (4.68), the necessary conditions for the time response to be dominated by one pair of complex poles require the pole-zero pattern of Fig. 10.2*b* and have the following characteristics:

1. The other poles must be far to the left of the dominant poles, so that the transients due to these other poles are small in amplitude and die out rapidly.
2. Any other pole which is not far to the left of the dominant complex poles must be near a zero so that the magnitude of the transient term due to that pole is small.

With the system designed so that the response to a unit step input has the underdamped form shown in Fig. 10.2*a*, the following transient figures of merit

† For a detailed study of the effect of nonlinear characteristics on system performance, the reader is referred to the literature.[3,4]

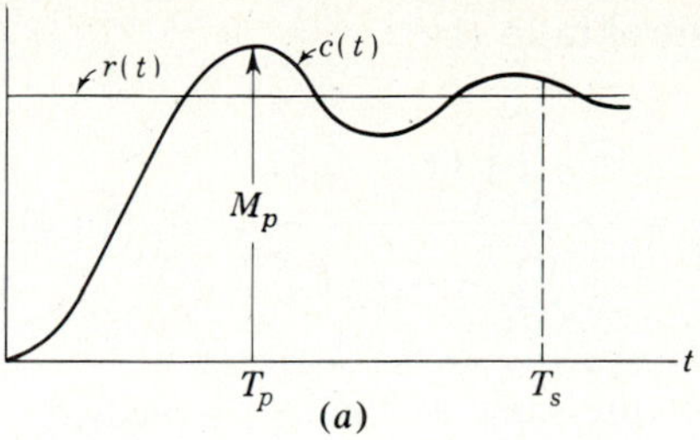

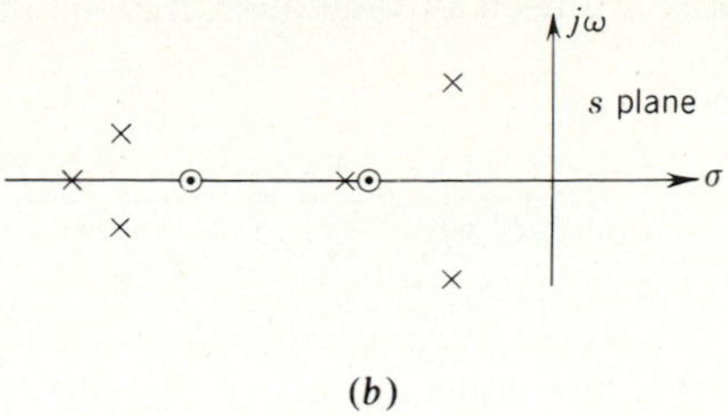

FIGURE 10.2
(*a*) Transient response to a step input with dominant complex poles. (*b*) Pole-zero pattern of $C(s)/R(s)$ for the desired response.

(described in Secs. 3.9 and 3.10) are used to judge its performance:

1. M_p, peak overshoot, is the amplitude of the first overshoot.
2. T_p, peak time, is the time to reach the peak overshoot.
3. T_s, settling time, is the time for the response envelope to first reach and thereafter remain within 2 percent of the final value.
4. N is the number of oscillations in the response up to the settling time.

Consider the nonunity-feedback system shown in Fig. 10.3, using

$$G(s) = \frac{N_1}{D_1} = \frac{K_G \prod (s - z_k)}{\prod (s - p_g)} \tag{10.1}$$

$$H(s) = \frac{N_2}{D_2} = \frac{K_H \prod (s - z_j)}{\prod (s - p_i)} \tag{10.2}$$

$$G(s)H(s) = \frac{N_1 N_2}{D_1 D_2} = \frac{K_G K_H \prod_{h=1}^{w} (s - z_h)}{\prod_{c=1}^{n} (s - p_c)} \tag{10.3}$$

The product $K_G K_H = K$ is defined as the loop sensitivity. For $G(s)H(s)$ the degree of the numerator is w and the degree of the denominator is n. These

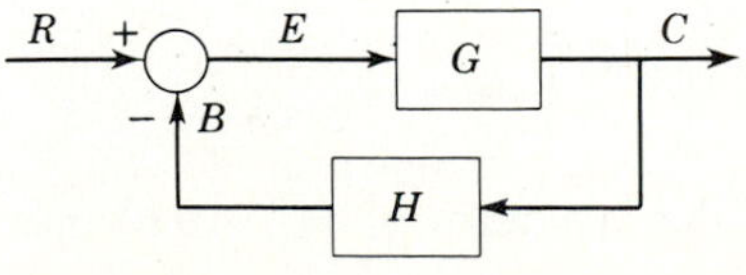

FIGURE 10.3
Feedback system.

symbols are used throughout Chap. 6. The control ratio is

$$\frac{C(s)}{R(s)} = \frac{P(s)}{Q(s)} = \frac{N_1 D_2}{D_1 D_2 + N_1 N_2} = \frac{K_G \prod_{m=1}^{w'} (s - z_m)}{\prod_{k=1}^{n} (s - p_k)} \tag{10.4}$$

Note that the constant K_G in Eq. (10.4) is not the same as the loop sensitivity unless the system has unity feedback. The degree n of the denominator of $C(s)/R(s)$ is the same as for $G(s)H(s)$, regardless of whether the system has unity or nonunity feedback. *The degree w' of the numerator of $C(s)/R(s)$ is equal to the sum of the degrees of N_1 and D_2.* For a unity-feedback system the degree of the numerator is $w' = w$. For a nonunity-feedback system the w' zeros of Eq. (10.4) include both the zeros of $G(s)$, i.e., N_1 and the poles of $H(s)$, i.e., D_2, as shown in Eq. (10.4).

For a unit step input the output of Fig. 10.3 is

$$C(s) = \frac{P(s)}{sQ(s)} = \frac{K_G \prod_{m=1}^{w'} (s - z_m)}{s \prod_{k=1}^{n} (s - p_k)}$$

$$= \frac{A_0}{s} + \frac{A_1}{s - p_1} + \cdots + \frac{A_k}{s - p_k} + \cdots + \frac{A_n}{s - p_n} \tag{10.5}$$

The pole $s = 0$ which comes from $R(s)$ must be included as a pole of $C(s)$, as shown in Eq. (10.5). The values of the coefficients can be obtained graphically from the pole-zero diagram by the method described in Sec. 4.9 and Eq. (4.68) or by the more practical use of a computer.[5] Assume that the system represented by Eq. (10.5) has a dominant complex pole $p_1 = \sigma \pm j\omega_d$. The complete time solution is

$$c(t) = \frac{P(0)}{Q(0)} + 2\left|\frac{K_G \prod_{m=1}^{w'} (p_1 - z_m)}{p_1 \prod_{k=2}^{n} (p_1 - p_k)}\right| e^{\sigma t} \cos\left[\omega_d t + \angle P(p_1) - \angle p_1 - \angle Q'(p_1)\right]$$

$$+ \sum_{k=3}^{n} \left[\frac{P(p_k)}{p_k Q'(p_k)}\right] e^{p_k t} \tag{10.6}$$

where $$Q'(p_k) = \left[\frac{dQ(s)}{ds}\right]_{s=p_k} = \left[\frac{Q(s)}{s - p_k}\right]_{s=p_k}$$

By assuming the pole-zero pattern of Fig. 10.2, the last term of Eq. (10.6) may often be neglected. The time response is therefore approximated by

$$c(t) \approx \frac{P(0)}{Q(0)} + 2\left|\frac{K_G \prod_{m=1}^{w'} (p_1 - z_m)}{p_1 \prod_{k=2}^{n} (p_1 - p_k)}\right| e^{\sigma t}\cos\left[\omega_d t + \angle P(p_1) - \angle p_1 - \angle Q'(p_1)\right] \tag{10.7}$$

Note that although the transient terms due to the other poles have been neglected, the effect of those poles on the amplitude and phase angle of the dominant transient has not been neglected.

The peak time T_p is obtained by setting the derivative with respect to time of Eq. (10.7) equal to zero. This gives

$$T_p = \frac{1}{\omega_d}\left[\frac{\pi}{2} - \angle P(p_1) + \angle Q'(p_1)\right] \tag{10.8}$$

which may be stated as follows:

$$T_p = \frac{1}{\omega_d}\left\{\frac{\pi}{2} - \left[\text{sum of angles from zeros of } \frac{C(s)}{R(s)} \text{ to dominant pole } p_1\right] + \left[\begin{array}{c}\text{sum of angles from all other poles of } \frac{C(s)}{R(s)} \text{ to} \\ \text{dominant pole } p_1 \text{, including conjugate pole}\end{array}\right]\right\} \tag{10.9}$$

From a physical consideration of Eq. (10.7) it can be seen that the phase angle $\phi = \angle P(p_1) - \angle p_1 - \angle Q'(p_1)$ cannot have a value greater than 2π. This same limitation must also be applied to the angles of Eq. (10.9), i.e., the value T_p must occur within one "cycle" of the transient. Inserting this value of T_p into Eq. (10.7) gives the peak overshoot M_p. By using the value $\cos(\pi/2 - \angle p_1) = \omega_d/\omega_n$, the value M_p can be expressed as

$$M_p = \frac{P(0)}{Q(0)} + \frac{2\omega_d}{\omega_n^2}\left|\frac{K_G \prod_{m=1}^{w'} (p_1 - z_m)}{\prod_{k=2}^{n} (p_1 - p_k)}\right| e^{\sigma T_p} \tag{10.10}$$

The first term in Eq. (10.10) represents the final value, and the second term represents the overshoot M_o. For a unity-feedback system which is Type 1 or higher, the equation for M_o can be put in an alternative form. For such a system there is zero steady-state error when the input is a step function. Therefore the first term on the right side of Eq. (10.7) is equal to unity. The same expression can also be obtained by applying the final-value theorem to $C(s)$ given by Eq.

(10.5). This equality is used to solve for K_G. Note that under these conditions $K_G = K$ and $w' = w$:

$$K_G = \frac{\prod_{k=1}^{n} (-p_k)}{\prod_{m=1}^{w} (-z_m)} \tag{10.11}$$

The value of the overshoot M_o can therefore be expressed as

$$M_o = \frac{2\omega_d}{\omega_n^2} \left| \frac{\prod_{k=1}^{n} (-p_k)}{\prod_{m=1}^{w} (-z_m)} \frac{\prod_{m=1}^{w} (p_1 - z_m)}{\prod_{k=2}^{n} (p_1 - p_k)} \right| e^{\sigma T_p} \tag{10.12}$$

Some terms inside the brackets in Eq. (10.12) cancel the terms in front so that M_o can be expressed in words as

$$M_o = \left[\frac{\begin{array}{l}\text{product of distances}\\ \text{from all poles of}\\ C(s)/R(s) \text{ to origin,}\\ \text{excluding distances}\\ \text{of two dominant}\\ \text{poles from origin}\end{array}}{\begin{array}{l}\text{product of distances}\\ \text{from all other poles}\\ \text{of } C(s)/R(s) \text{ to}\\ \text{dominant pole } p_1,\\ \text{excluding distance}\\ \text{between dominant}\\ \text{poles}\end{array}} \right] \left[\frac{\begin{array}{l}\text{product of distances}\\ \text{from all zeros of}\\ C(s)/R(s) \text{ to dominant}\\ \text{pole } p_1\end{array}}{\begin{array}{l}\text{product of distances}\\ \text{from all zeros of}\\ C(s)/R(s) \text{ to origin}\end{array}} \right] e^{\sigma T_p} \tag{10.13}$$

If there are no finite zeros of $C(s)/R(s)$, the factors in M_o involving zeros become unity. Equation (10.13) is valid only for a unity-feedback system that is Type 1 or higher. It may be sufficiently accurate for a Type 0 system that has a large value for K_0. The value of M_o can be calculated either from the right-hand term of Eq. (10.10) or from Eq. (10.13), whichever is more convenient. The effect on M_o of other poles, which cannot be neglected, is discussed in the next section.

The values of T_s and N can be approximated from the dominant roots. Section 3.9 shows that T_s is four time constants for 2 percent error:

$$T_s = \frac{4}{|\sigma|} = \frac{4}{\zeta\omega_n} \tag{10.14}$$

$$N = \frac{\text{settling time}}{\text{period}} = \frac{T_s}{2\pi/\omega_d} = \frac{2\omega_d}{\pi|\sigma|} = \frac{2}{\pi}\frac{\sqrt{1-\zeta^2}}{\zeta} \tag{10.15}$$

An examination of Eq. (10.8) or (10.9) reveals that zeros of the control ratio cause a decrease in the peak time T_p, whereas poles increase the peak time. Peak time can also be decreased by shifting zeros to the right or poles (other than the dominant poles) to the left. Equation (10.13) shows that the larger the value of T_p, the smaller the value of M_o because $e^{\sigma T_p}$ decreases. M_o can also be decreased by reducing the ratios $p_k/(p_k - p_1)$ and $(z_m - p_1)/z_m$. But this can have an adverse effect on T_p. The conditions on pole-zero locations that lead to a small M_o may therefore produce a large T_p. Conversely, the conditions that lead to a small T_p may be obtained at the expense of a large M_o. Thus a compromise is required in the peak time and peak overshoot that are attainable. Some improvements can be obtained by introducing additional poles and zeros into the system and locating them appropriately. This is covered in the following sections on compensation.

The approximate equations, given in this section, for the response of a system to a step-function input are based on the fundamental premise that there is a pair of dominant complex poles and that the effect of other poles is small. When this premise is satisfied, a higher-order system, $n > 2$, effectively acts like a simple second-order system. Although these approximate equations can be used, they are presented primarily to enhance the reader's understanding of how poles and zeros affect the time response. A computer program[5,14] can be used to calculate and plot $c(t)$ and to yield precise values of M_p, t_p, and t_s.

10.3 ADDITIONAL SIGNIFICANT POLES[6]

When there are two dominant complex poles, the approximations developed in Sec. 10.2 give accurate results. However, there are cases where an additional pole of $C(s)/R(s)$ is significant. Figure 10.4*a* shows a pole-zero diagram which contains dominant complex poles and an additional real pole p_3. The control ratio, with $K = -\omega_n^2 p_3$, is given by

$$\frac{C(s)}{R(s)} = \frac{K}{(s^2 + 2\zeta\omega_n s + \omega_n^2)(s - p_3)} \tag{10.16}$$

With a unit step input the time response is

$$c(t) = 1 + 2|A_1|e^{-\zeta\omega_n t}\sin\left(\omega_n\sqrt{1-\zeta^2}\,t + \phi\right) + A_3 e^{p_3 t} \tag{10.17}$$

The transient term due to the real pole p_3 has the form $A_3 e^{p_3 t}$, where A_3, based upon Eq. (4.68), is *always negative*. Thus the overshoot M_p is reduced, and settling time t_s may be increased or decreased. This is the typical effect of an additional real pole. The magnitude A_3 depends on the location of p_3 relative to the complex poles. The further to the left the pole p_3 is located, the smaller the magnitude of A_3, therefore the smaller its effect on the total response. A pole which is 6 times as far to the left as the complex poles has negligible effect on the time response. The typical time response is shown in Fig. 10.4*a*. As the pole p_3 moves to the right, the magnitude of A_3 increases and the overshoot becomes

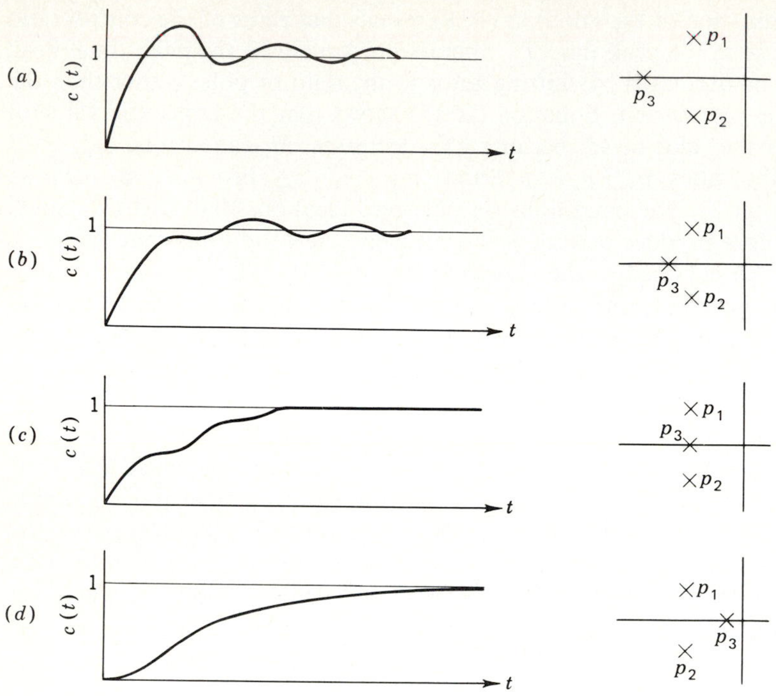

FIGURE 10.4
Typical time response as a function of the real-pole location.

smaller. As p_3 approaches but is still to the left of the complex poles, the first maximum in the time response is less than the final value. The largest overshoot can occur at the second or a later maximum as shown in Fig. 10.4*b* (see also Fig. 9.32*b*).

When p_3 is located at the real-axis projection of the complex poles, the response is monotonic; i.e., there is no overshoot. This represents the critically damped situation, as shown in Fig. 10.4*c*. The complex poles may contribute a "ripple" to the time response, as shown in the figure. When p_3 is located to the right of the complex poles, it is the dominant pole and the response is overdamped.

When the real pole p_3 is to the left of the complex poles, the peak time T_p is approximately given by Eq. (10.9). Although the effect of the *real* pole is to increase the peak time, this change is small if the real pole is fairly far to the left. A first-order correction can be made to the peak overshoot M_o given by Eq. (10.13) by adding the value of $A_3 e^{p_3 T_p}$ to the peak overshoot due to the complex poles. The effect of the real pole on the actual settling time t_s can be estimated by calculating $A_3 e^{p_3 T_s}$ and comparing it with the size and sign of the underdamped transient at time T_s obtained from Eq. (10.14). If both are negative at T_s, then the

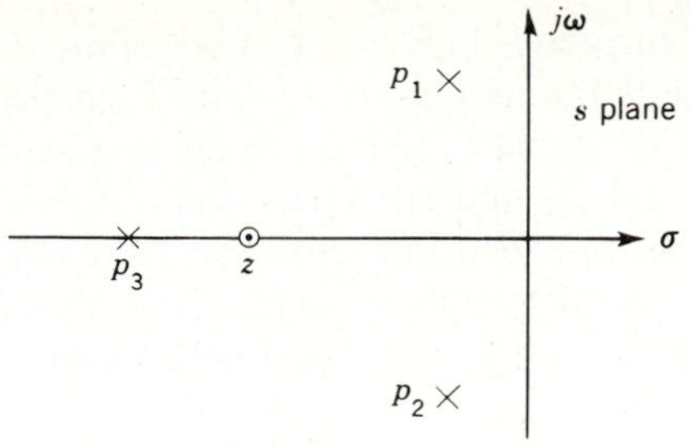

FIGURE 10.5
Pole-zero diagram of $C(s)/R(s)$ for Eq. (10.18).

true settling time t_s is increased. If they have opposite signs, then the value $T_s = 4T$ based on the complex roots is a good approximation for t_s.

The presence of a real zero in addition to the real pole further modifies the transient response. A possible pole-zero diagram for which the control ratio is

$$\frac{C(s)}{R(s)} = \frac{K(s - z)}{(s^2 + 2\zeta\omega_n s + \omega_n^2)(s - p_3)} \tag{10.18}$$

is shown in Fig. 10.5. The complete time response to a unit step-function input still has the form given in Eq. (10.17). However, the sign of A_3, based upon Eq. (4.68), depends on the relative location of the real pole and the real zero. A_3 is negative if the zero is to the left of p_3, and it is positive if the zero is to the right of p_3. Also, the magnitude of A_3 is proportional to the distance from p_3 to z [see Eq. (4.68)]. Therefore, if the zero is close to the pole, A_3 is small and the contribution of this transient term is correspondingly small. Compared with the response for Eq. (10.16), as shown in Fig. 10.4,

1. If the zero z is to the left of the real pole p_3, the response is qualitatively the same as that for a system with only complex poles but the peak overshoot is smaller.
2. If the zero z is to the right of the real pole p_3, the peak overshoot is greater than that for a system with only complex poles.

These characteristics are evident in Fig. 10.20.

When the real pole p_3 and the real zero z are close together, the value of peak time T_p obtained from Eq. (10.9) can be considered essentially correct. Actually T_p decreases if the zero is to the right of the real pole, and vice versa. A first-order correction to M_o from Eq. (10.13) can be obtained by adding the contribution of $A_3 e^{p_3 t}$ at the time T_p. The analysis of the effect on t_s is similar to that described above for the case of just an additional real pole.

Many control systems having an underdamped response can be approximated by one having the following characteristics: (1) two complex poles; (2) two complex poles and one real pole; and (3) two complex poles, one real pole, and one real zero. For case 1 the relations T_p, M_o, T_s, and N developed in Sec. 10.2

give an accurate representation of the time response. For case 2 these approximate values can be corrected if the real pole is far enough to the left. Then the contribution of the additional transient term is small, and the total response remains essentiallly the sum of a constant and an underdamped sinusoid. For case 3 the approximate values can be corrected provided the zero is near the real pole so that the amplitude of the additional transient term is small. More exact calculation of the figures of merit can be obtained by plotting the exact response as a function of time by use of a computer.[5,14]

10.4 ROOT-LOCUS DESIGN CONSIDERATIONS

The root-locus design technique presented in Chap. 7 stresses the desired transient figures of merit. Thus, the selection of the dominant complex root can be based on a desired damping ratio ζ_D which determines the peak overshoot, the desired damped natural frequency ω_d which is the frequency of oscillation of the transient, or the desired real part σ_D of the dominant root which determines the settling time. Two additional factors must be including in the design. First, the presence of additional poles will modify the peak overshoot, as described in Sec. 10.3. Second, it may be necessary to specify the value of the gain K_m. For example, in a Type 1 system it may be necessary to specify a minimum value of K_1 in order to limit the magnitude of the steady-state error e_{ss} with a ramp input, where $e(t)_{ss} = Dr(t)/K_1$. The following example incorporates these factors into a root-locus design.

Example. Consider a unity-feedback system where

$$G_x(s) = \frac{K_x}{s(s^2 + 4.2s + 14.4)}$$

$$= \frac{K_x}{s(s + 2.1 + j3.1607)(s + 2.1 - j3.1607)} \tag{10.19}$$

and $K_1 = K_x/14.4$. For this system it is specified that the following figures of merit must be satisfied: $1 < M_p \leq 1.123$, $t_s \leq 3$ s, $e(t)_{ss} = 0$, $t_p \leq 1.6$ s, $K_1 \geq 1.5\ \text{s}^{-1}$.

First Design

Assuming that the closed-loop system can be represented by an effective second-order model, Eqs. (3.60), (3.61), and (10.14) are used to determine the required values of ζ_D, ω_d, and ω_n:

$$M_p = 1.123 = 1 + \exp\frac{-\zeta\pi}{\sqrt{1-\zeta^2}} \rightarrow \zeta_D = 0.555$$

$$t_p = \frac{\pi}{\omega_d} \rightarrow \omega_d > 1.9635$$

$$T_s = \frac{4}{\zeta_D\omega_n} \rightarrow \omega_n > 2.4024$$

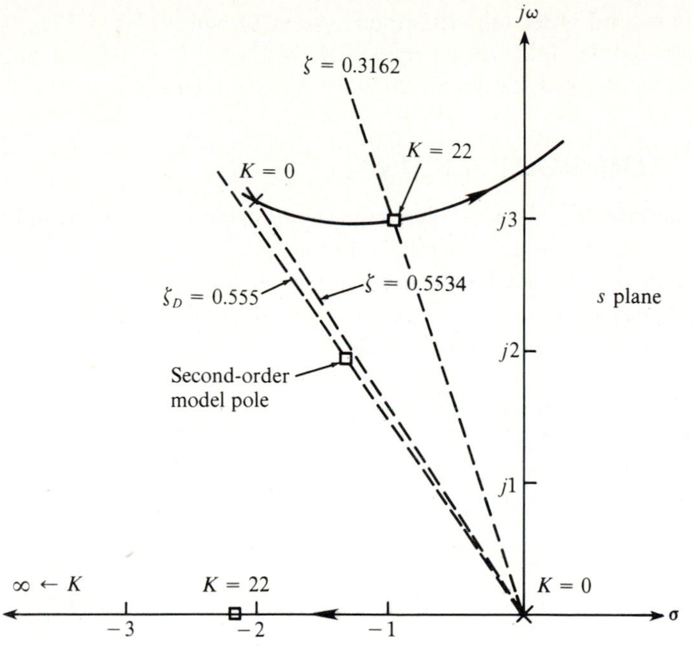

FIGURE 10.6
Root locus for Eq. (10.19).

Using the values $\zeta_D = 0.555$, $\omega_n = 2.4024$, and $\omega_d = 1.9984$, the required dominant complex roots are $s_{1,2} = -1.3333 \pm j1.9984$. Figure 10.6 contains the root locus for $G_x(s)$ and the line representing $\zeta_D = 0.555$. It is evident that the required dominant roots are not achievable.

Second Design

The first design does not consider the fact that there is a third root s_3. The associated transient $A_3 e^{s_3 t}$ is negative (see Sec. 10.3) and it will reduce the peak overshoot of the underdamped transient associated with the complex roots $s_{1,2}$. Therefore, it is possible to select a smaller value of ζ for the complex roots, and the effect of the third root can be used to keep M_p within the specified limits. In order to simultaneously satisfy the specification for $K_1 > 1.5$, the design is therefore based on obtaining the complex roots $s_{1,2} = -1 \pm j3$ on the root locus. The corresponding sensitivity is $K = 14.4K_1 = 22$ which yields $K_1 = 1.528$. The closed-loop transfer function is

$$\frac{C(s)}{R(s)} = \frac{22}{(s+1+j3)(s+1-j3)(s+2.2)} \tag{10.20}$$

The plot of $|\mathbf{C}(j\omega)/\mathbf{R}(j\omega)|$ vs. ω very closely approximates the frequency response for a simple second-order underdamped system shown in Fig. 9.32*a*. A unit-step input results in the following values: $M_p \approx 1.123$, $t_p \approx 1.51$ s, and $t_s \approx 2.95$ s. Therefore the desired specifications have been achieved.

10.5 RESHAPING THE ROOT LOCUS[5]

The root-locus plots described in Chap. 7 show the relationship between the gain of the system and the time response. Depending on the specifications established for the system, the gain that best achieves the desired performance is selected. The performance specifications may be based on the desired damping ratio, undamped natural frequency, time constant, or steady-state error. The root locus may show that the desired performance cannot be achieved just by adjustment of the gain. In fact, the system may be unstable for all values of gain. The control-systems engineer must then investigate the methods for reshaping the root locus to meet the performance specifications.

The purpose of reshaping the root locus generally falls into one of the following categories:

1. A given system is stable and its transient response is satisfactory, but its steady-state error is too large. Thus, the gain must be increased to reduce the steady-state error (see Chap. 6). This must be accomplished without appreciably reducing the system stability.
2. A given system is stable, but its transient response is unsatisfactory. Thus, the root locus must be reshaped so that it is moved farther to the left, away from the imaginary axis.
3. A given system is stable, but both its transient response and its steady-state response are unsatisfactory. Thus, the locus must be moved to the left and the gain must be increased.
4. A given system is unstable for all values of gain. Thus, the root locus must be reshaped so that part of each branch falls in the left-half s plane, thereby making the system stable.

Compensation of a system by the introduction of poles and zeros is used to improve the operating performance. However, each additional compensator pole increases the number of roots of the closed-loop characteristic equation. If an underdamped response of the form shown in Fig. 10.2*a* is desired, the system gain must be adjusted so that there is a pair of dominant complex poles. This requires that any other pole be far to the left or near a zero so that its transient has a small amplitude and therefore has a small effect on the total time response. The required pole-zero diagram is shown in Fig. 10.2*b*. The approximate values of peak overshoot M_o, peak time T_p, settling time T_s, and the number of oscillations up to settling time N can be obtained from the pole-zero pattern as described in Sec. 10.2. The effect of compensator poles and zeros on these quantities can be evaluated rapidly by use of a digital computer.[5,14]

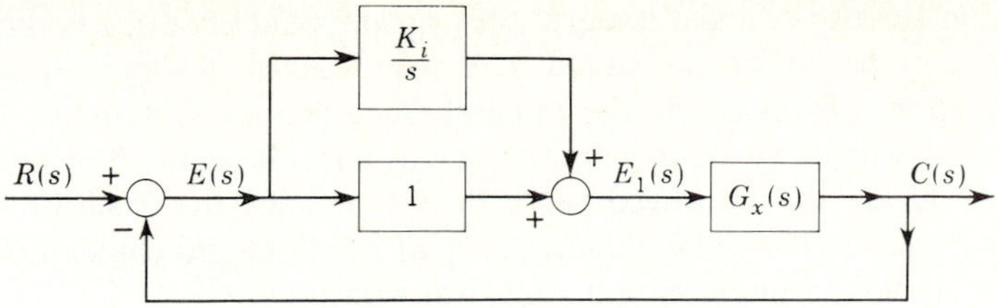

FIGURE 10.7
Ideal proportional plus integral (PI) control.

10.6 IDEAL INTEGRAL CASCADE COMPENSATION (PI CONTROLLER)

When the transient response of a feedback control system is considered satisfactory but the steady-state error is too large, it is possible to eliminate the error by increasing the system type. This must be accomplished *without appreciably changing the dominant roots of the characteristic equation*. The system type can be increased by operating on the actuating signal e to produce one that is proportional to both the magnitude and the integral of this signal. This proportional plus integral (PI) controller is shown in Fig. 10.7 where

$$E_1(s) = \left(1 + \frac{K_i}{s}\right) E(s) \tag{10.21}$$

$$G_c(s) = \frac{E_1(s)}{E(s)} = \frac{s + K_i}{s} \tag{10.22}$$

In this system the quantity $e_1(t)$ continues to increase as long as an error $e(t)$ is present. Eventually $e_1(t)$ becomes large enough to produce an output $c(t)$ equal to the input $r(t)$. The error $e(t)$ is then equal to zero. The constant K_i (generally very small) and the overall gain of the system must be selected to produce satisfactory roots of the characteristic equation. Since the system type has been increased, the corresponding error coefficient is equal to infinity. Provided that the new roots of the characteristic equation can be satisfactorily located, the transient response is still acceptable.

The locations of the pole and zero of this ideal proportional plus integral compensator are shown in Fig. 10.8. The pole alone would move the root locus to the right, thereby slowing down the time response. The zero must be near the origin in order to minimize the increase in response time of the complete system.

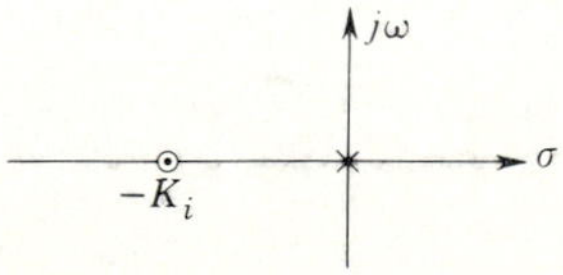

FIGURE 10.8
Location of the pole and zero of an ideal proportional plus integral compensator.

The main limitation on the use of ideal integral plus proportional control is the equipment required to obtain the integral signal. This may require an amplifier in a positive-feedback system. Mechanically the integral signal can be obtained by an integrating gyroscope, which is used in vehicles such as aircraft, space vehicles, ships, and submarines where the improved performance justifies the cost. Frequently, however, a passive electric network consisting of resistors and capacitors sufficiently approximates the proportional plus integral action.

10.7 CASCADE LAG COMPENSATION USING PASSIVE ELEMENTS

Figure 10.9*a* shows a network which approximates a proportional plus integral output and is used as a lag compensator. Putting an amplifier of gain A in series with this network yields the transfer function

$$G_c(s) = A\frac{1 + Ts}{1 + \alpha Ts} = \frac{A}{\alpha}\frac{s + 1/T}{s + 1/\alpha T} \tag{10.23}$$

where $\alpha = (R_1 + R_2)/R_2 > 1$ and $T = R_2 C$. The pole $s = -1/\alpha T$ is therefore to the right of the zero $s = -1/T$, as shown in Fig. 10.9*b*. The locations of the pole and zero of $G_c(s)$ on the s plane can be made close to those of the ideal compensator. Besides furnishing the necessary gain, the amplifier also acts as an isolating unit to prevent any loading effects between the compensator and the original system.

Assume that the original forward transfer function is

$$G_x(s) = \frac{K\prod_{h=1}^{w}(s - z_h)}{\prod_{c=1}^{n}(s - p_c)} \tag{10.24}$$

For the original system the loop sensitivity K for the selected closed-loop root s

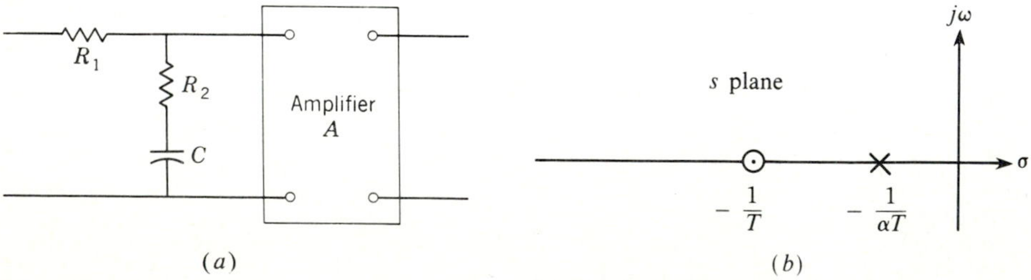

FIGURE 10.9
(*a*) Integral or lag compensator; (*b*) pole and zero location.

is

$$K = \frac{\prod_{c=1}^{n} |s - p_c|}{\prod_{h=1}^{w} |s - z_h|} \tag{10.25}$$

With the addition of the lag compensator in cascade, the new forward transfer function is

$$G(s) = G_c(s)G_x(s) = \frac{AK}{\alpha} \frac{\left(s + \dfrac{1}{T}\right)}{\left(s + \dfrac{1}{\alpha T}\right)} \frac{\prod_{h=1}^{w} (s - z_h)}{\prod_{c=1}^{n} (s - p_c)} \tag{10.26}$$

When the desired roots of the characteristic equation are located on the root locus, the magnitude of the loop sensitivity for the new root s' becomes

$$K' = \frac{AK}{\alpha} = \frac{\left|s' + \dfrac{1}{\alpha T}\right|}{\left|s' + \dfrac{1}{T}\right|} \frac{\prod_{c=1}^{n} |s' - p_c|}{\prod_{h=1}^{w} |s' - z_h|} \tag{10.27}$$

As an example, a lag compensator is applied to a Type 0 system. To improve the steady-state accuracy, the error coefficient must be increased. The value of K_0 before and after the addition of the compensator is calculated by using the definition $K_0 = \lim_{s \to 0} G(s)$ from Eqs. (10.24) and (10.26), respectively,

$$K_0 = \frac{\prod_{h=1}^{w} (-z_h)}{\prod_{c=1}^{n} (-p_c)} K \tag{10.28}$$

$$K_0' = \frac{\prod_{h=1}^{w} (-z_h)}{\prod_{c=1}^{n} (-p_c)} \alpha K' \tag{10.29}$$

The following procedure is used to design the passive lag cascade compensator. First, the pole $s = -1/\alpha T$ and the zero $s = -1/T$ of the compensator are placed very close together. This means that most of the original root locus remains practically unchanged. If the angle contributed by the compensator *at the original closed-loop dominant root* is less than 5°, the new locus is displaced only slightly. This 5° figure is only a guide and should not be applied arbitrarily. The new closed-loop dominant pole s' is therefore essentially unchanged from the uncompensated value. This satisfies the restriction that the transient response must not change appreciably. As a result, the values $s' + 1/\alpha T$ and $s' + 1/T$ are almost equal, and the values K and K' in Eqs. (10.25) and (10.27) are approximately equal. The values K_0 and K_0' in Eqs. (10.28) and (10.29) now differ only by the factor α so that $K_0' \approx \alpha K_0$. The gain required to produce the new root s' therefore increases approximately by the factor α, which is the ratio of the compensator zero and pole. Summarizing, the necessary conditions on the compensator are that (1) the pole and zero must be close together and (2) the ratio α of the zero and pole must approximately equal the desired increase in gain. These requirements can be achieved by placing the compensator pole and zero very close to the origin. The size of α is limited by the physical parameters required in the network. A value $\alpha = 10$ is often used.

Although the statements above are based on a Type 0 system, the same conditions apply equally well for a Type 1 or higher system.

Example of Lag Compensation Applied to a Type 1 System

A control system with unity feedback has a forward transfer function

$$G_x(s) = \frac{K_1}{s(1+s)(1+0.2s)} = \frac{K}{s(s+1)(s+5)} \tag{10.30}$$

which yields the root locus shown in Fig. 10.10. For the basic control system a

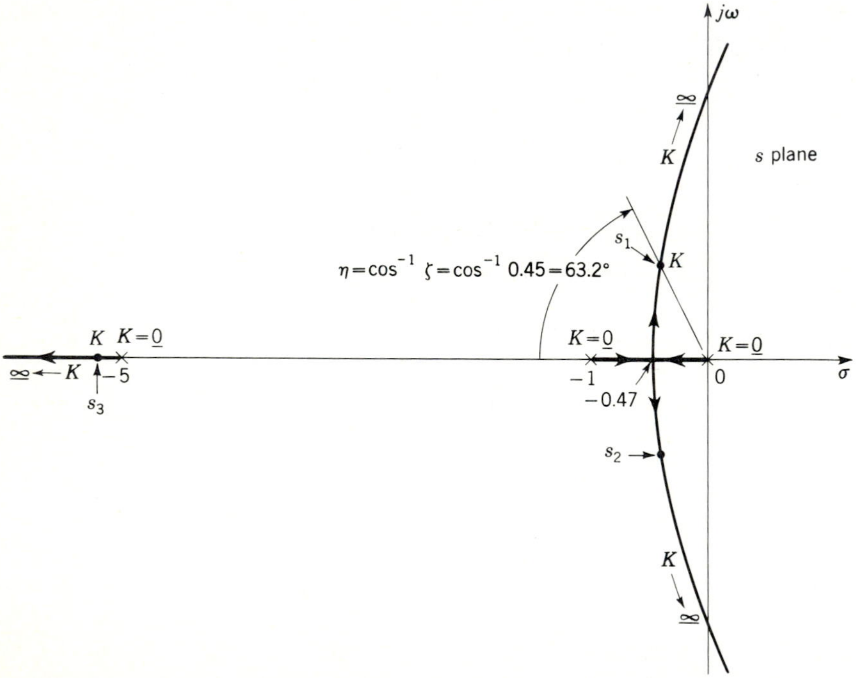

FIGURE 10.10
Root locus of

$$G_x(s) = \frac{K}{s(s+1)(s+5)}$$

damping ratio $\zeta = 0.45$ yields the following pertinent data:

Dominant roots: $s_{1,2} = -0.404 \pm j0.802$

Loop sensitivity:

$$K = |s_1| \cdot |s_1 + 1| \cdot |s_1 + 5|$$
$$= |-0.404 + j0.802| \cdot |0.596 + j0.802| \cdot |4.596 + j0.802| = 4.188$$

Ramp error coefficient: $K_1 = \lim_{s \to 0} sG(s) = \dfrac{K}{5} = \dfrac{4.188}{5} = 0.838\ \text{s}^{-1}$

Undamped natural frequency: $\omega_n = 0.898$ rad/s

Third root: $s_3 = -5.192$

The values of peak time T_p, peak overshoot M_o, and settling time T_s are obtained from Eqs. (10.9), (10.13), and (10.14) as $T_p = 4.12$ s, $M_o = 0.202$, and $T_s = 9.9$ s. These are approximate values whereas the computer program gives more accurate values (see Table 10.2) including $t_s = 9.48$ s.

To increase the gain, a lag compensator is put in cascade with the forward transfer function. With the criteria discussed in the preceding section and

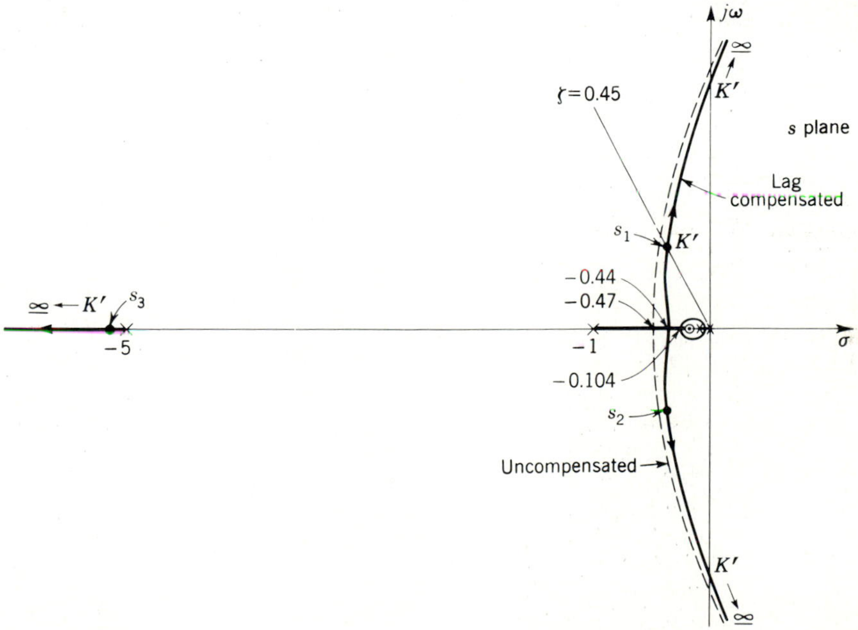

FIGURE 10.11
Root locus of

$$G(s) = \frac{K'(s + 0.05)}{s(s + 1)(s + 5)(s + 0.005)}$$

$\alpha = 10$, the compensator pole is located at $s = -0.005$ and the zero at $s = -0.05$. This selection is based on the requirements of a practical network. For the network of Fig. 10.9 the values $R_1 = 18$ MΩ, $R_2 = 2$ MΩ, and $C = 10$ μF produce the pole and zero required. The angle of the compensator at the original dominant roots is about 2.6° and is acceptable. Since the compensator pole and zero are very close together, they make only a small change in the new root locus in the vicinity of the original roots. The compensator transfer function is

$$G_c(s) = \frac{A}{\alpha}\frac{s + 1/T}{s + 1/\alpha T} = \frac{A}{10}\frac{s + 0.05}{s + 0.005} \tag{10.31}$$

Figure 10.11 (not to scale) shows the new root locus for $G(s) = G_x(s)G_c(s)$ as solid lines and the original locus as dashed lines. Note that for the damping ratio $\zeta = 0.45$ the new locus and the original locus are close together. The new dominant roots are $s_{1,2} = -0.384 \pm j0.763$; thus the roots are essentially unchanged.

For the compensated system adjusted to a damping ratio 0.45 the following results are obtained:

Dominant roots: $s_{1,2} = -0.384 \pm j0.763$

Loop sensitivity: $$K' = \frac{|s| \cdot |s+1| \cdot |s+5| \cdot |s+0.005|}{|s+0.05|} = 4.01$$

Ramp error coefficient: $$K_1' = \frac{K'\alpha}{5} = \frac{(4.01)(10)}{5} = 8.02 \text{ s}^{-1}$$

Increase in gain: $$A = \frac{K_1'}{K_1} = \frac{8.02}{0.838} = 9.57$$

Undamped natural frequency: $\omega_n = 0.854$ rad/s

Other roots: $s_3 = -5.183$ and $s_4 = -0.053$

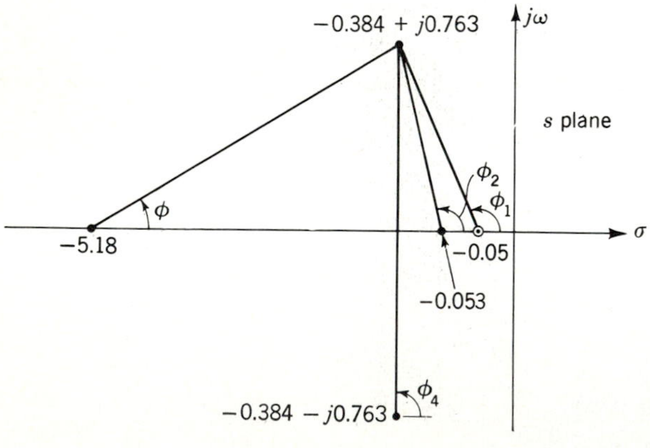

FIGURE 10.12
Angles used to evaluate T_p.

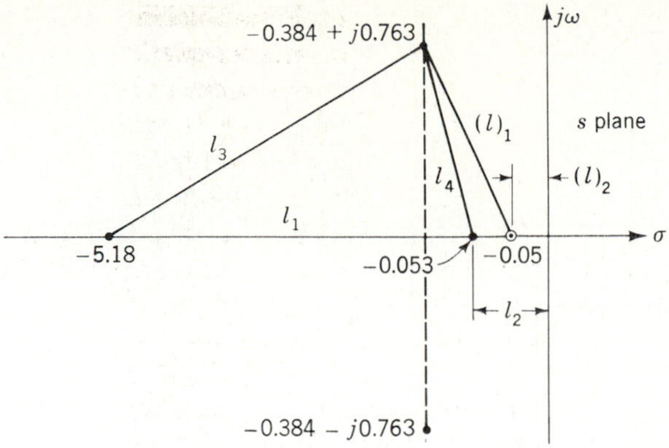

FIGURE 10.13
Lengths used to evaluate M_o.

The value of T_p for the compensated system is determined by use of Eq. (10.9), with the values shown in Fig. 10.12:

$$T_p = \frac{1}{\omega_d}\left[\frac{\pi}{2} - \phi_1 + (\phi_2 + \phi_3 + \phi_4)\right]$$

$$= \frac{1}{0.763}[90° - 113.6° + (113.4° + 9.04° + 90°)]\frac{\pi \text{ rad}}{180°} = 4.32 \text{ s}$$

The value of M_o for the compensated system is determined by use of Eq. (10.13), with the values shown in Fig. 10.13:

$$M_o = \frac{l_1 l_2}{l_3 l_4}\frac{(l)_1}{(l)_2}e^{\sigma T_p} = \frac{(5.18)(0.053)}{(4.76)(0.832)}\frac{0.834}{0.05}e^{(-0.384)(4.32)}$$

$$= 1.1563e^{-1.659} = 0.22$$

The values obtained from a computer[5] are $t_p = 4.30$ s and $M_o = 0.266$ and are listed in Table 10.2.

A comparison of the uncompensated system and the system with integral compensation shows that the ramp error coefficient K_1' has increased by a factor of 9.57 but the undamped natural frequency has been decreased slightly from 0.898 to 0.854 rad/s. This means that the steady-state error with a ramp input has decreased but the settling time has been increased. Provided that the increased settling time and peak overshoot are acceptable, the system has been improved. Although the complete root locus is drawn in Fig. 10.11, in practice this is not necessary. Only the general shape is sketched and the particular needed points are accurately determined.

10.8 IDEAL DERIVATIVE CASCADE COMPENSATION (PD CONTROLLER)

When the transient response of a feedback system must be improved, it is necessary to reshape the root locus so that it is moved farther to the left of the

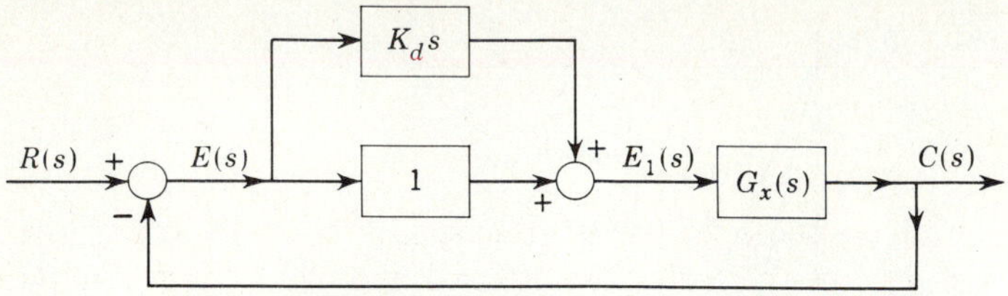

FIGURE 10.14
Ideal proportional plus derivative (PD) control.

imaginary axis. Section 7.3 shows that introducing an additional zero in the forward transfer function produces this effect. A zero can be produced by operating on the actuating signal to produce one that is proportional to both the magnitude and the derivative (rate of change) of this signal. This proportional plus derivative (PD) controller is shown in Fig. 10.14, where

$$E_1(s) = (1 + K_d s)E(s) \tag{10.32}$$

Physically the effect can be described as introducing anticipation into the system. The system reacts not only to the magnitude of the error but also to its rate of change. If the error is changing rapidly, then $e_1(t)$ is large and the system responds faster. The net result is to speed up the response of the system. On the root locus the introduction of a zero has the effect of shifting the curves to the left, as shown in the following example.

Example. The root locus for a Type 2 system which is unstable for all values of loop sensitivity K is shown in Fig. 10.15a. Adding a zero to the system by means of the compensator $G_c(s)$ (an ideal proportional plus derivative compensator) has the

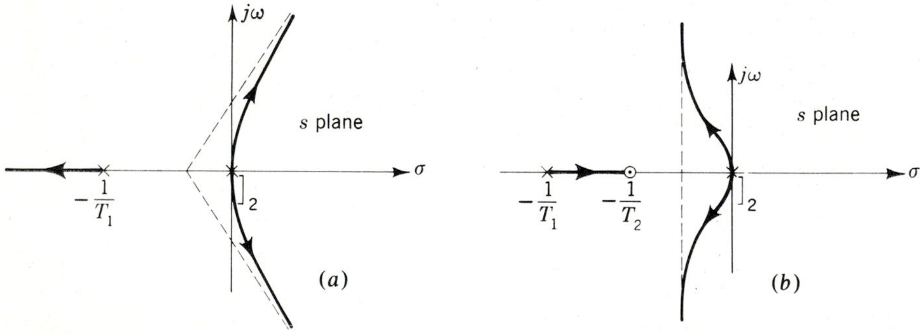

FIGURE 10.15
Root locus for

$$(a)\quad G_x(s) = \frac{K}{s^2(s + 1/T_1)}$$

$$(b)\quad G(s) = \frac{K(s + 1/T_2)}{s^2(s + 1/T_1)}$$

effect of moving the locus to the left, as shown in Fig. 10.15b. The addition of a zero $s = -1/T_2$ between the origin and the pole at $-1/T_1$ stabilizes the system for all positive values of gain.

This example shows how the introduction of a zero in the forward transfer function by a proportional plus derivative compensator has modified and improved the feedback control system. However, the synthesis of such derivative action requires the use of an active network. In addition, the derivative action amplifies any spurious signal or noise that may be present in the actuating signal. This noise amplification may saturate electronic amplifiers so that the system does not operate properly. A simple passive network is used to approximate proportional plus derivative action as shown in the next section.

10.9 LEAD COMPENSATION USING PASSIVE ELEMENTS

Figure 10.16a shows a derivative or lead compensator made up of electrical elements. To this network an amplifier of gain A is added in series. The transfer function is

$$G_c(s) = A\alpha\frac{1 + Ts}{1 + \alpha Ts} = A\frac{s + 1/T}{s + 1/\alpha T} = A\frac{s - z_c}{s - p_c} \tag{10.33}$$

where $\alpha = R_2/(R_1 + R_2) < 1$ and $T = R_1C$. This lead compensator introduces a zero at $s = -1/T$ and a pole at $s = -1/\alpha T$. By making α sufficiently small,

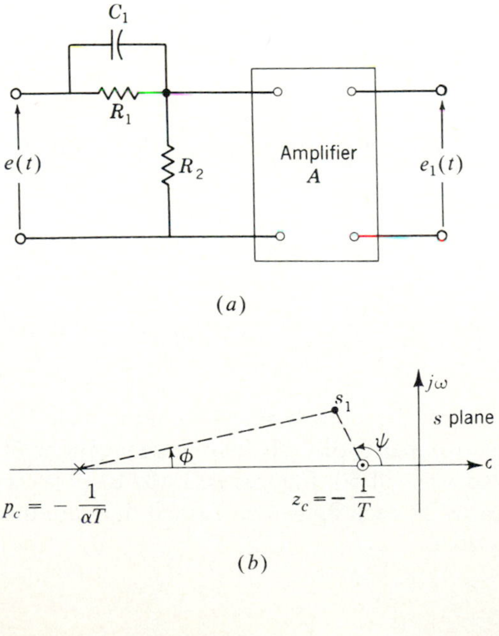

FIGURE 10.16
(a) Derivative or lead compensator. (b) Location of pole and zero of a lead compensator.

the location of the pole is far to the left and has small effect on the important part of the root locus. Near the zero the net angle of the compensator is due predominantly to the zero. Figure 10.16*b* shows the location of the lead-compensator pole and zero and the angles contributed by each at a point s_1. The best location of the zero must be determined by trial and error. It is also found that the gain of the compensated system is often increased. The maximum increases in gain and in the real part ($\zeta\omega_n$) of the dominant root of the characteristic equation do not coincide. The compensator zero location must then be determined by making several trials with the aid of a CAD program for the desired optimum performance.

The loop sensitivity is proportional to the ratio of $|s + 1/\alpha T|$ to $|s + 1/T|$. Therefore, as α decreases, the loop sensitivity increases. The minimum value of α is limited by the size of the parameters needed in the network to obtain the minimum input impedance required. Note also from Eq. (10.33) that a small α requires a large value of additional gain A from the amplifier. The value $\alpha = 0.1$ is a common choice.

Example of Lead Compensation Applied to a Type 1 System

The same system used in Sec. 10.7 is now used with lead compensation. The locations of the pole and zero of the compensator are first selected by trial. At the conclusion of this section some rules are given to show the best location.

The forward transfer function of the original system is

$$G_x(s) = \frac{K_1}{s(1+s)(1+0.2s)} = \frac{K}{s(s+1)(s+5)} \tag{10.34}$$

The transfer function of the compensator, using a value $\alpha = 0.1$, is

$$G_c(s) = 0.1A\frac{1+Ts}{1+0.1Ts} = A\frac{s+1/T}{s+1/0.1T} \tag{10.35}$$

The new forward transfer function is

$$G(s) = G_c(s)G_x(s) = \frac{AK(s+1/T)}{s(s+1)(s+5)(s+1/0.1T)} \tag{10.36}$$

Three selections for the position of the zero $s = -1/T$ are made. A comparison of the results shows the relative merits of each. The three locations of the zero are $s = -0.75$, $s = -1.00$, and $s = -1.50$. Figures 10.17 to 10.19 show the resultant root loci for these three cases. In each figure the dashed lines are the locus of the original uncompensated system.

TABLE 10.1
Results of lead compensation

Compensator	Dominant complex roots	ω_n	K_1	Additional roots	t_p, s	t_s, s	Additional Gain, A
Uncompensated	$-0.404 \pm j0.802$	0.898	0.84	-5.192	4.12	9.48	1.0
$\frac{s+0.75}{s+7.5}$	$-1.522 \pm j3.021$	3.38	2.05	-0.689; -9.77	1.18	2.78	24.4
$\frac{s+1.0}{s+10}$	$-1.588 \pm j3.152$	3.52	2.95	-11.83	1.09	2.45	35.1
$\frac{s+1.50}{s+15.0}$	$-1.531 \pm j3.039$	3.40	4.40	-1.76; -16.18	1.05	2.50	52.3

The loop sensitivity is evaluated in each case from the expression

$$K' = AK = \frac{|s| \cdot |s+1| \cdot |s+5| \cdot |s+1/\alpha T|}{|s+1/T|} \tag{10.37}$$

and the ramp error coefficient is evaluated from the equation

$$K_1' = \lim_{s \to 0} sG(s) = \frac{K'\alpha}{5} \tag{10.38}$$

The results, given in Table 10.1, show that the large increase in ω_n has been obtained by adding the lead compensators. Since the value of ζ is held constant for all three cases, the system has a much faster response time. A comparison of the individual lead compensators shows the following:

CASE a. A zero of the compensator to the right of the pole at $s = -1$ results in a root on the negative real axis close to the origin (see Fig. 10.17). From Eq. (4.68) it can be seen that the coefficient associated with this transient term is negative. Therefore it has the desirable effect of decreasing the peak overshoot and the undesirable characteristic of having a longer settling time. If the size of this transient term is small (the magnitude is proportional to the distance from the real root to the zero), the effect on the settling time is small.

The decrease in M_o can be determined by calculating the contribution of the transient term $A_3 e^{s_3 t}$ due to the real root at time t_p, as discussed in Sec. 10.3. At the time $t = T_s$, the sign of the underdamped transient is negative and the transient due to the real root is also negative; therefore the actual settling time is increased. The exact settling time obtained from a complete solution of the time response is $t_{sa} = 2.78$ s and is shown in Fig. 10.20.

CASE b. The best settling time occurs when the zero of the compensator cancels the pole $s = -1$ of the original transfer function (see Fig. 10.18). There is no real root near the imaginary axis in this case. The actual settling time is $t_{sb} = 2.45$ s. The order of the characteristic equation is unchanged.

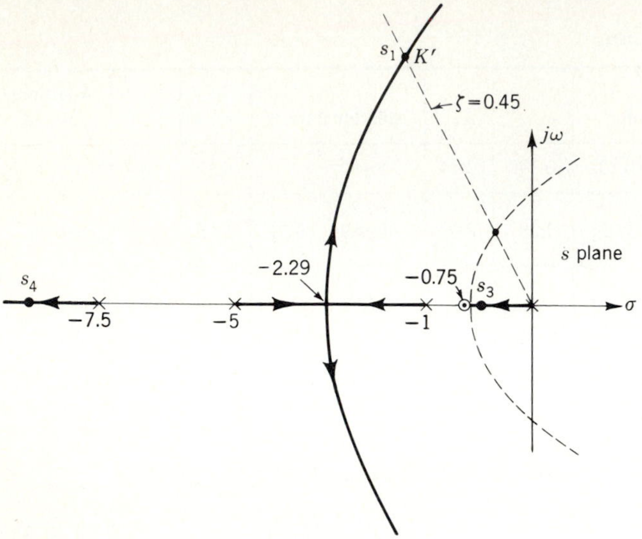

FIGURE 10.17
Root locus of

$$G(s) = \frac{AK(s + 0.75)}{s(s + 1)(s + 5)(s + 7.5)}$$

CASE c. The largest gain occurs when the zero of the compensator is located at the left of the pole $s = -1$ of the original transfer function (see Fig. 10.19). Since the transient due to the root $s_3 = -1.76$ is positive, it increases the peak overshoot of the response. The value of the transient due to the real root can be added to the peak value of the underdamped sinusoid to obtain the correct value of the peak overshoot. The real root may also affect the settling time. In this case

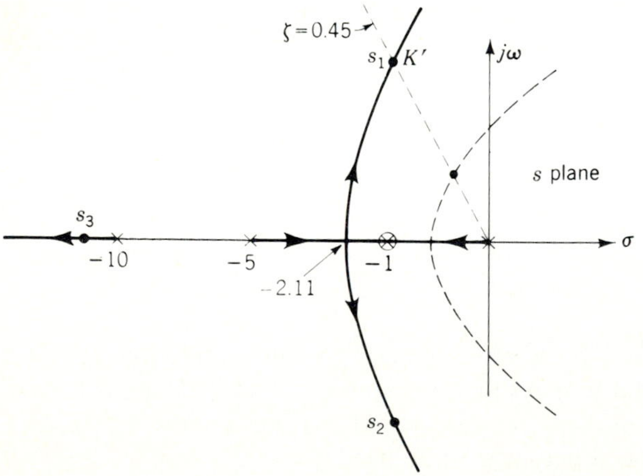

FIGURE 10.18
Root locus of

$$G(s) = \frac{AK(s + 1)}{s(s + 1)(s + 5)(s + 10)}$$

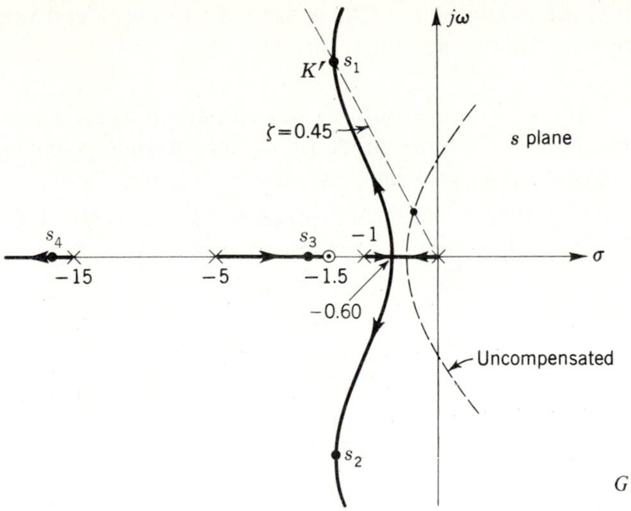

FIGURE 10.19
Root locus of

$$G(s) = \frac{AK(s+1.5)}{s(s+1)(s+5)(s+15)}$$

the real root decreases the peak time and increases the settling time. From Fig. 10.20 the actual values are $t_p = 1.05$ s and $t_{sc} = 2.50$ s.

From the root-locus plots it is evident that increasing the gain increases the peak overshoot and the settling time. The responses for each of the three lead compensators is shown in Fig. 10.20. The effect of the additional real root is to change the peak overshoot and the settling time. The designer must make a choice between the largest gain attainable, the smallest peak overshoot, and the shortest settling time.

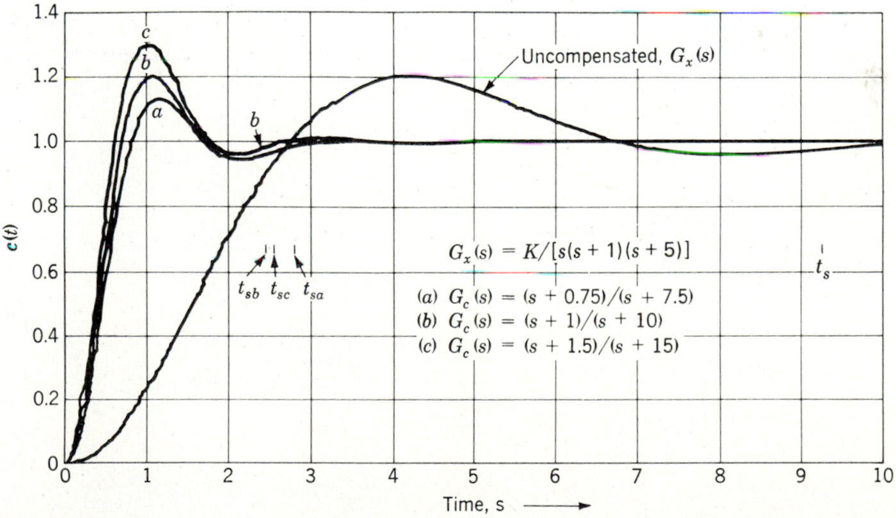

FIGURE 10.20
Time responses with a step input for the uncompensated and the compensated cases of Table 10.1.

Based on this example, the following guidelines are used to apply cascade lead compensators to a Type 1 or higher system:

1. If the zero $s = -1/T$ of the lead compensator is superimposed and cancels the largest real pole (excluding the pole at zero) of the original transfer function, a good improvement in the transient response is obtained.
2. If a larger gain is desired than that obtained by guideline 1, several trials should be made with the zero of the compensator moved to the left or right of the largest real pole of $G_x(s)$. The location of the zero that results in the desired gain and roots is selected.

For a Type 0 system it will often be found that a better time response and a larger gain can be obtained by placing the compensator zero so that it cancels or is close to the second largest real pole of the original transfer function.

10.10 GENERAL LEAD-COMPENSATOR DESIGN

Additional methods are available for the design of an appropriate lead compensator G_c which is placed in cascade with a basic transfer function G_x, as shown in Fig. 10.1. Figure 10.21 shows the original root locus of a control system. For a specified damping ratio ζ the dominant root of the uncompensated system is s_1. Also shown is s_2, which is the desired root of the system's characteristic equation. Selection of s_2 as a desired root is based on the performance required for the system. The design problem is to select a lead compensator that results in s_2 being a root. The first step is to find the sum of the angles at the point s_2 due to the poles and zeros of the original system. This angle is $180° + \phi$. For s_2 to be on the new root locus, it is necessary for the lead compensator to contribute an angle $\phi_c = -\phi$ at this point. The total angle at s_2 is then 180°, and it is a point

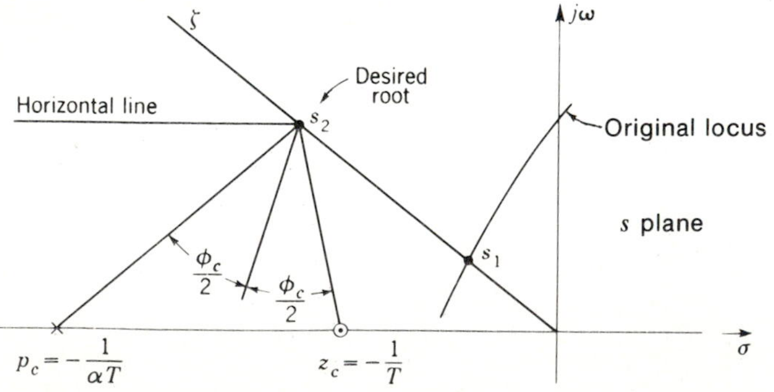

FIGURE 10.21
Graphical construction for locating the pole and zero of a simple lead compensator.

on the new root locus. A simple lead compensator represented by Eq. (10.33), with its zero to the right of its pole, can be used to provide the angle ϕ_c at the point s_2.

Method 1. Actually, there are many possible locations of the compensator pole and zero that will produce the necessary angle ϕ_c at the point s_2. Cancellation of poles of the original open-loop transfer function by means of compensator zeros may simplify the root locus and thereby reduce the complexity of the problem. The compensator zero is simply placed over a real pole. Then the compensator pole is placed further to the left at a location which makes s_2 a point on the new root locus. The pole to be canceled depends on the system type. For a Type 1 system the largest real pole (excluding the pole at zero) should be canceled. For a Type 0 system the second largest pole should be canceled.

Method 2. The following construction is based on obtaining the maximum value of α, which is the ratio of the compensator zero and pole. The steps in locating the lead-compensator pole and zero (see Fig. 10.21) are

1. Locate the desired root s_2. Draw a line from this root to the origin and a horizontal line to the left of s_2.
2. Bisect the angle between the two lines drawn in step 1.
3. Measure the angle $\phi_c/2$ on either side of the bisector drawn in step 2.
4. The intersections of these lines with the real axis locate the compensator pole p_c and zero z_c.

The construction outlined above is shown in Fig. 10.21. It is left for the reader to show that this construction results in the largest possible value of α. Since α is less than unity and appears as the gain of the compensator [see Eq. (10.33)], the largest value of α requires the smallest additional gain A.

Method 3. The procedures outlined above give the desired location of a root. However, no conditions have been placed on the desired system gain. A minimum value of gain may be specified to restrict the maximum steady-state error. Changing the locations of the compensator pole and zero can produce a range of values for the gain while maintaining the desired root s_2. A specific procedure for achieving both the desired root s_2 and the system gain K_m is given in Refs. 8 and 9.

10.11 LAG-LEAD CASCADE COMPENSATION

The preceding sections have shown that (1) the insertion of an integral or lag compensator in cascade results in a large increase in gain and a small reduction in the undamped natural frequency and (2) that the insertion of a derivative or lead compensator in cascade results in a small increase in gain and a large

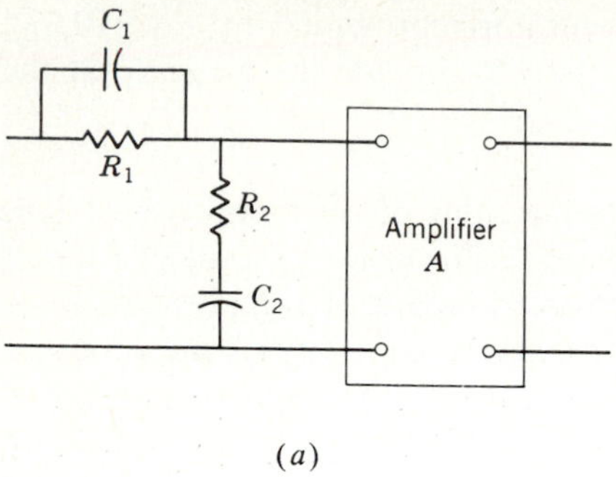

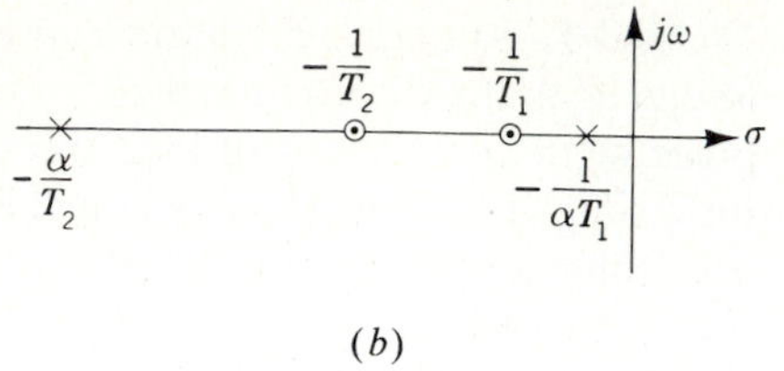

FIGURE 10.22
(a) Lag-lead compensator. (b) Pole and zero location.

increase in the undamped natural frequency. By inserting both the lag and the lead compensators in cascade with the original transfer function, the advantages of both can be realized simultaneously; i.e., a large increase in gain and a large increase in the undamped natural frequency can be obtained. Instead of using two separate networks, it is possible to use one network that acts as both a lead and a lag compensator. Such a network, shown in Fig. 10.22 with an amplifier added in cascade, is called a lag-lead compensator. The transfer function of this compensator (see Sec. 8.10) is

$$G_c(s) = A\frac{(1+T_1 s)(1+T_2 s)}{(1+\alpha T_1 s)[1+(T_2/\alpha)s]} = A\frac{(s+1/T_1)(s+1/T_2)}{(s+1/\alpha T_1)(s+\alpha/T_2)} \qquad (10.39)$$

where $\alpha > 1$ and $T_1 > T_2$.

The fraction $(1 + T_1 s)/(1 + \alpha T_1 s)$ represents the lag compensator, and the fraction $(1 + T_2 s)/[1 + (T_2/\alpha)s]$ represents the lead compensator. The values T_1, T_2, and α are selected to achieve the desired improvement in system performance. The specific values of the compensator components are obtained from the relationships $T_1 = R_1C_1$, $T_2 = R_2C_2$, $\alpha T_1 + T_2/\alpha = R_1C_1 + R_2C_2 + R_1C_2$. It may also be desirable to specify a minimum value of the input impedance over the passband frequency range.

The procedure for applying the lag-lead compensator is a combination of the procedures for the individual units.

1. For the integral or lag component:
   a. The zero $s = -1/T_1$ and the pole $s = -1/\alpha T_1$ are selected close together, with α set to a large value such as $\alpha = 10$.
   b. The pole and zero are located to the left of and close to the origin. This results in an increased gain.
2. For the derivative or lead component the zero $s = -1/T_2$ is superimposed on a pole of the original system. Since α is already selected in step 1, the new pole is located at $s = -\alpha/T_2$. This results in moving the root locus to the left and therefore increases the undamped natural frequency.

The relative positions of the poles and zeros of the lag-lead compensators are shown in Fig. 10.22*b*.

Example of Lag-Lead Compensation Applied to a Type 1 System

The system described in Secs. 10.7 and 10.9 is now used with a lag-lead compensator. The forward transfer function is

$$G_x(s) = \frac{K_1}{s(1+s)(1+0.2s)} = \frac{K}{s(s+1)(s+5)} \tag{10.40}$$

The lag-lead compensator used is shown in Fig. 10.22, and the transfer function $G_c(s)$ is given by Eq. (10.39). The new complete forward transfer function is

$$G(s) = G_c(s)G_x(s) = \frac{AK(s+1/T_1)(s+1/T_2)}{s(s+1)(s+5)(s+1/\alpha T_1)(s+\alpha/T_2)} \tag{10.41}$$

The poles and zeros of the compensator are selected in accordance with the principles outlined previously in this section. They coincide with the values used for the integral and derivative compensators when applied individually in Secs. 10.7 and 10.9. With an $\alpha = 10$, the compensator poles and zeros are

$$z_{\text{lag}} = \frac{1}{T_1} = 0.05 \qquad z_{\text{lead}} = \frac{1}{T_2} = 1 \qquad p_{\text{lag}} = \frac{1}{\alpha T_1} = 0.005 \qquad p_{\text{lead}} = \frac{\alpha}{T_2} = 10$$

The forward transfer function becomes

$$G(s) = G_c(s)G_x(s) = \frac{K'(s+1)(s+0.05)}{s(s+1)(s+0.005)(s+5)(s+10)} \tag{10.42}$$

The root locus (not to scale) for this system is shown in Fig. 10.23.

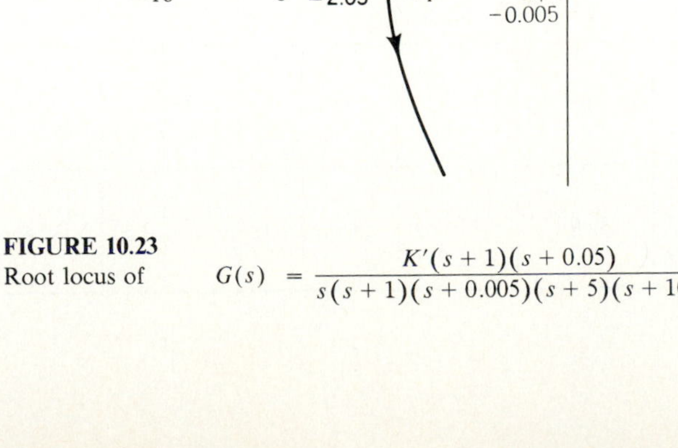

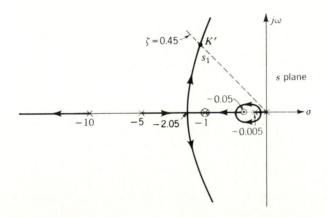

FIGURE 10.23
Root locus of $G(s) = \dfrac{K'(s+1)(s+0.05)}{s(s+1)(s+0.005)(s+5)(s+10)}$

For a damping ratio $\zeta = 0.45$ the following pertinent data are obtained for the compensated system:

Dominant roots: $s_{1,2} = -1.571 \pm j3.119$

Loop sensitivity (evaluated at $s = s_1$):

$$K' = \frac{|s| \cdot |s + 0.005| \cdot |s + 5| \cdot |s + 10|}{|s + 0.05|} = 146.3$$

Ramp error coefficient: $K_1' = \lim_{s \to 0} sG(s) = \dfrac{K'(0.05)}{(0.005)(5)(10)} = 29.3 \text{ s}^{-1}$

Undamped natural frequency: $\omega_n = 3.49$ rad/s

The additional roots are $s_3 = -11.81$ and $s_4 = -0.051$. These results, when compared with those of the uncompensated system, show a large increase in both the gain and the undamped natural frequency. They indicate that the advantages of the lag-lead compensator are equivalent to the combined improvement obtained by the lag and the lead compensators. Note the practicality in the choice of $\alpha = 10$, which is based on experience. The response characteristics are shown in Table 10.2.

TABLE 10.2
Comparison of performance of several cascade compensators for the system of Eq. (10.30) with $\zeta = 0.45$

Compensator	Dominant complex roots	Undamped natural frequency ω_n	Other roots	Ramp error coefficients K_1	t_p, s	M_o	t_s, s	Additional gain required A
Uncompensated	$-0.404 \pm j0.802$	0.898	-5.192	0.84	4.12	0.202	9.48	—
Lag: $\dfrac{s + 0.05}{s + 0.005}$	$-0.384 \pm j0.763$	0.854	-0.053, -5.183	8.02	4.30	0.266	22.0	9.57
Lead: $\dfrac{s + 1.5}{s + 15}$	$-1.531 \pm j3.039$	3.40	-1.76, -16.18	4.40	1.05	0.280	2.50	52.5
Lead: $\dfrac{s + 1}{s + 10}$	$-1.588 \pm j3.152$	3.53	-11.83	2.95	1.09	0.195	2.45	35.1
Lag-lead: $\dfrac{(s + 0.05)(s + 1)}{(s + 0.005)(s + 10)}$	$-1.571 \pm j3.119$	3.49	-0.051, -11.81	29.26	1.10	0.213	3.33	34.9

10.12 COMPARISON OF CASCADE COMPENSATORS

Table 10.2 shows a comparison of the system response obtained when a lag, a lead, and a lag-lead compensator are added in cascade with the same basic feedback system. These results are for the examples of Secs. 10.7, 10.9, and 10.11 and are obtained by use of a computer program (see App. B). These results are typical of the changes introduced by the use of cascade compensators, which can be summarized as follows:

1. Lag compensator:
   a. Results in a large increase in gain K_m (by a factor almost equal to α), which means a much smaller steady-state error.
   b. Decreases ω_n and therefore has the disadvantage of producing an increase in the settling time.

2. Lead compensator:
   a. Results in a moderate increase in gain K_m, thereby improving steady-state accuracy.
   b. Results in a large increase in ω_n and therefore significantly reduces the settling time.
   c. The transfer function of the lead compensator, using the passive network of Fig. 10.16*a*, contains the gain α, which is less than unity. Therefore the additional gain A, which must be added, is larger than the increase in K_m for the system.

3. Lag-lead compensator (essentially combines the desirable characteristics of the lag and the lead compensators):
   a. Results in a large increase in gain K_m, which improves the steady-state response.
   b. Results in a large increase in ω_n, which improves the transient-response settling time.

Figure 10.24 shows the frequency response for the original and compensated systems corresponding to Table 10.2. There is a close qualitative correlation between the time-response characteristics listed in Table 10.2 and the corresponding frequency-response curves shown in Fig. 10.24. The peak value M_p is directly proportional to the maximum value M_m, and the settling time t_s is a direct function of the passband frequency. Curves 3, 4 and 5 have a much wider passband than curves 1 and 2; therefore these systems have a faster settling time.

For systems other than the one used as an example, these simple compensators may not produce the desired changes. *However, the basic principles described here are applicable to any system for which the open-loop poles closest to the imaginary axis are on the negative real axis.* For these systems the conventional compensators may be used, and good results are obtained.

When the complex open-loop poles are dominant, the conventional compensators produce only small improvements. A pair of complex poles near the

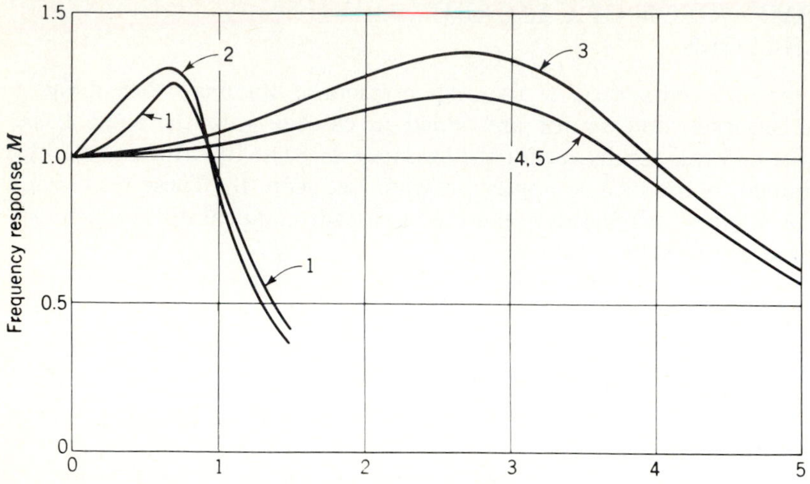

FIGURE 10.24
Frequency response for systems of Table 10.2: (1) original basic system; (2) lag-compensated; (3) lead-compensated, zero at -1.5; (4) lead-compensated, zero at -1.0; (5) lag-lead-compensated.

imaginary axis often occurs in aircraft and in missile transfer functions. Effective compensation could remove the open-loop dominant complex poles and replace them with poles located farther to the left. This means that the compensator should have a transfer function of the form

$$G_c(s) = \frac{1 + Bs + As^2}{1 + Ds + Cs^2} \tag{10.43}$$

Assuming exact cancellation, the zeros of $G_c(s)$ would coincide with the complex poles of the original transfer function. Consider the feedback system which has the forward transfer function

$$G_x(s) = \frac{K_0}{(1 + T_1 s)(1 + T_2 s)(1 + Bs + As^2)} \qquad T_1 < T_2 \tag{10.44}$$

The composite forward transfer function with the compensator would be

$$G(s) = G_c(s)G_x(s) = \frac{K_0'}{(1 + T_1 s)(1 + T_2 s)(1 + Ds + Cs^2)} \tag{10.45}$$

and the root locus, when the compensator has real poles, is shown in Fig. 10.25. The solid lines are the new locus, and the dashed lines are the original locus. Thus the transient response has been improved. The problem remaining to be solved is the synthesis of a network with the desired poles and zeros. The reader is referred to books that cover network synthesis.[10-12] An alternative design method for improving the performance of plants which have dominant complex poles is the use of rate feedback compensation, discussed in Sec. 10.19. This method does not involve the assumption of exact cancellation shown in Eq. (10.45).

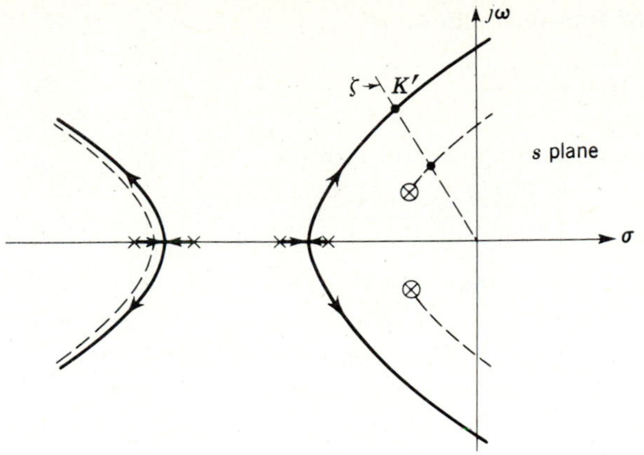

FIGURE 10.25
Root locus of

$$G_c(s)G_x(s) = \frac{K_0'}{(1 + T_1s)(1 + T_2s)(1 + T_3s)(1 + T_4s)}$$

10.13 PID CONTROLLER

The PI and PD controller of Secs. 10.6 and 10.8, respectively, can be combined into a single controller, i.e., a proportional plus integral plus derivative (PID) controller as shown in Fig. 10.26. The transfer function of this controller is

$$G_c(s) = \frac{E_1(s)}{E(s)} = K_d s + K_p + \frac{K_i}{s} = \frac{K_d(s - z_1)(s - z_2)}{s} \tag{10.46}$$

where $(z_1 + z_2) = -K_p/K_d$ and $z_1 z_2 = K_i/K_d$. The compensator of Eq. (10.46) is placed in cascade with the basic system of Eq. (10.30), with $z_1 = -1$ and

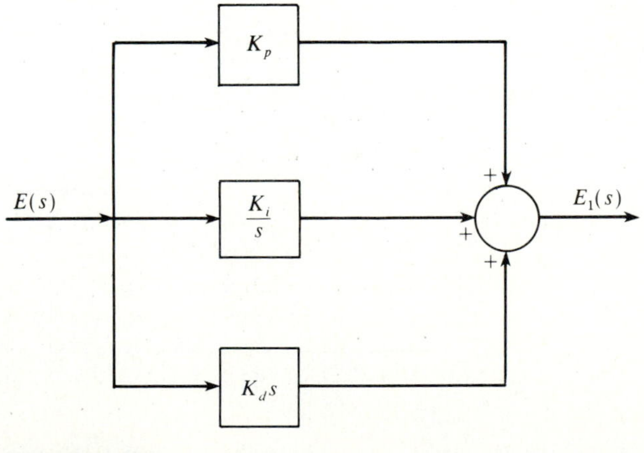

FIGURE 10.26
A PID controller.

$z_2 = -0.1$, so that the forward transfer function is

$$G_c(s)G_x(s) = \frac{KK_d(s + 0.1)}{s^2(s + 5)}$$

For a damping ratio $\zeta = 0.45$ the resulting control ratio is

$$\frac{C(s)}{R(s)} = \frac{30.13(s + 0.1)}{(s + 2.449 \pm j4.861)(s + 0.1017)} \tag{10.47}$$

The figures of merit for the PID compensated system are: $M_p = 1.225$, $t_p = 0.646$ s, $t_s = 2.10$ s, $K_v = \infty$, and $K_2 = 0.6026\ \text{s}^{-2}$. Comparing these results with those in Table 10.2 shows that the PID controller results in the best improvement in system performance. Also note that the use of a PID controller increases the system type from Type 1 to Type 2. With the advances in the state of the art of microelectronics, satisfactory derivative action may be achieved to produce the PID controller. However, to avoid saturation of the differentiated signal e due to noise which is present in the system, it is necessary to add a cascade filter to attenuate the high frequencies.

10.14 INTRODUCTION TO FEEDBACK COMPENSATION

System performance may be improved by using either cascade or feedback compensation. The factors that must be considered when making that decision are listed in Sec. 10.1. The effectiveness of cascade compensation is illustrated in the preceding sections of this chapter. The use of feedback compensation is demonstrated in the remainder of this chapter. The location of a feedback compensator depends on the complexity of the basic system, the accessibility of

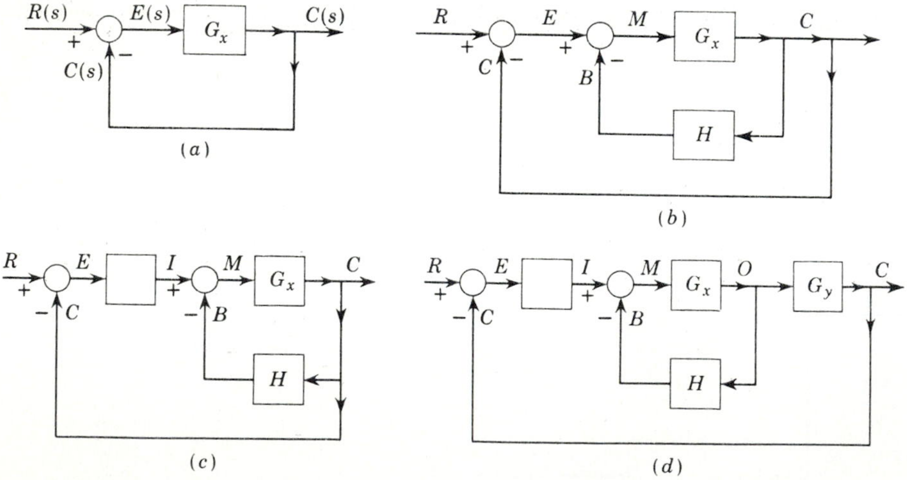

FIGURE 10.27
Various forms of feedback systems: (*a*) uncompensated system; (*b*) feedback-compensated system; (*c*) and (*d*) feedback-compensated systems with a cascade gain A or compensator G_c.

TABLE 10.3
Basic forms of $H(s)$ to be used to prevent change of system type

Basic form of $H(s)$	Type with which it can be used
K	0
Ks	0 and 1
Ks^2	0, 1, and 2
Ks^3	0, 1, 2, and 3

insertion points, the form of the signal being fed back, the signal with which it is being compared, and the desired improvement. Figure 10.27 shows various forms of feedback-compensated systems. For each case there is a unity-feedback loop (major loop) besides the compensation loop (minor loop) in order to maintain a direct correspondence (tracking) between the output and input.

It is often necessary to maintain the type of the basic system when applying feedback compensation. Care must therefore be exercised in determining the number of differentiating terms which are permitted in the numerator of $H(s)$. Table 10.3 presents the basic forms that $H(s)$ can have so that the system type remains the same after the minor feedback-compensation loop is added. $H(s)$ may have additional factors in the numerator and denominator. It can be shown by evaluating $G(s) = C(s)/E(s)$ that, to maintain the system type, the degree of the factor s in the numerator of $H(s)$ must be equal to or higher than the type of the forward transfer function $G_x(s)$ shown in Fig. 10.27.

By the use of a direct or unity-feedback loop on any non-Type 0 control element is converted into a Type 0 element. This effect is advantageous when going from a low-energy level input to a high-energy-level output for a given control element and also when the forms of the input and output energy are different. When $G_x(s)$ in Fig. 10.28 is Type 1 or higher, it may represent either a single control element or a group of control elements whose transfer function is of the general form

$$G_x(s) = \frac{K_m(1 + T_1 s)\cdots}{s^m(1 + T_a s)(1 + T_b s)\cdots} \tag{10.48}$$

where $m \neq 0$. The overall ratio of output to input is given by

$$G(s) = \frac{O(s)}{I(s)} = \frac{G_x(s)}{1 + G_x(s)} \tag{10.49}$$

or

$$G(s) = \frac{K_m(1 + T_1 s)\cdots}{s^m(1 + T_a s)(1 + T_b s)\cdots + K_m(1 + T_1 s)\cdots} \tag{10.50}$$

Equation (10.50) represents an equivalent forward transfer function between $O(s)$ and $I(s)$ of Fig. 10.28 and has the form of a Type 0 control element. Thus the transformation of type has been effected. Therefore, for a step input,

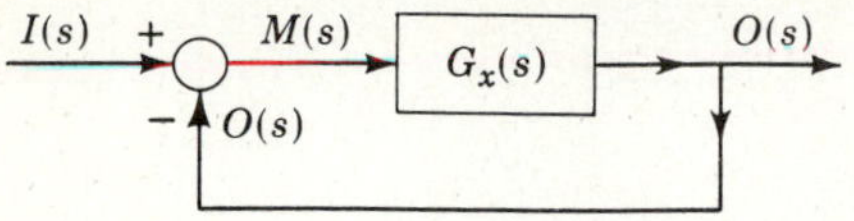

FIGURE 10.28
Transforming non-Type 0 control element to Type 0.

$O(t)_{ss} = \lim_{s \to 0}[sG(s)I(s)] = 1$. As a consequence, a given input signal tends to produce an output signal of equal magnitude but at a higher energy level and/or with a different form of energy. The polar plots of $\mathbf{G}_x(j\omega)$ and $\mathbf{O}(j\omega)/\mathbf{I}(j\omega)$ would show that for any given frequency the phase angle is less with feedback, thus indicating a more stable system (a smaller value of M_m).

10.15 FEEDBACK COMPENSATION: DESIGN PROCEDURES

Techniques are now described for applying feedback compensation to the root-locus method of system design.[7] The system to be investigated, see Fig. 10.29, has a minor feedback loop around the original transfer function $G_x(s)$. The minor-loop feedback transfer function $H(s)$ consists of elements and networks described in the following pages. Either the gain portion of $G_x(s)$ or the cascade amplifier A is adjusted to fix the system characteristics. For more complex systems one can develop a design approach based on the knowledge obtained for this simple system.

The method of attack is similar to that used for series compensation. The characteristic equation of the complete system may be obtained and the root locus plotted, by use of the *partition*[1] method, as a function of a gain that appears in the system. A more general approach is first to adjust the roots of the characteristic equation of the inner loop. These roots are then the poles of the forward transfer function and are used to draw the root locus for the overall system. Both methods, using various types of feedback compensators, are illustrated in the succeeding sections. The techniques developed with series compensation should be kept in mind. Basically, the addition of poles and zeros changes the root locus. To improve the system time response the locus must be moved to the left, away from the imaginary axis.

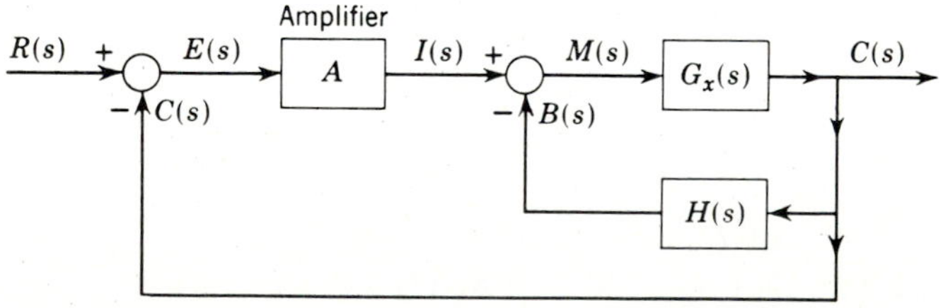

FIGURE 10.29
Block diagram for feedback compensation.

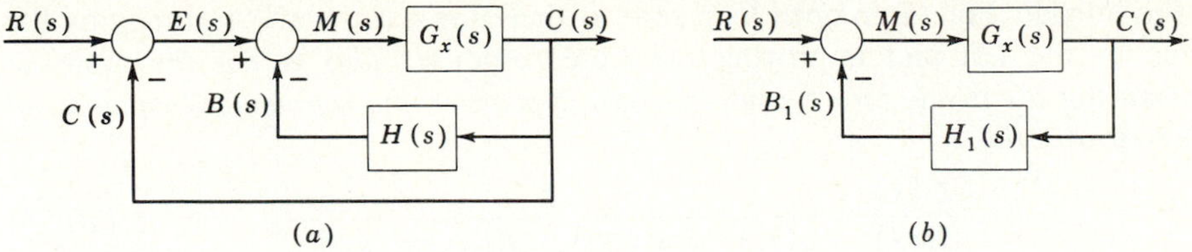

FIGURE 10.30
(a) Minor-loop compensation. (b) Simplified block diagram, $H_1(s) = 1 + H(s)$.

10.16 SIMPLIFIED RATE FEEDBACK COMPENSATION

The feedback system shown in Fig. 10.29 provides the opportunity for varying three functions, $G_x(s)$, $H(s)$, and A. The larger the number of variables inserted into a system, the better the opportunity for achieving a specified performance. However, the design problem becomes much more complex. In order to develop a feel for the changes that can be introduced by feedback compensation, a simplification is first made by letting $A = 1$. The system of Fig. 10.29 then reduces to that shown in Fig. 10.30a. It can be further simplified as shown in Fig. 10.30b.

The first and simplest case of feedback compensation to be considered is the use of a transducer which has the transfer function $H(s) = K_t s$. The original forward transfer function is

$$G_x(s) = \frac{K}{s(s+1)(s+5)} \tag{10.51}$$

The control ratio is

$$\frac{C(s)}{R(s)} = \frac{G_x(s)}{1 + G_x(s)H_1(s)} = \frac{G_x(s)}{1 + G_x(s)[1 + H(s)]} \tag{10.52}$$

The characteristic equation is

$$1 + G_x(s)[1 + H(s)] = 1 + \frac{K(1 + K_t s)}{s(s+1)(s+5)} = 0 \tag{10.53}$$

Equation (10.53) shows that the transducer plus the unity feedback have introduced the ideal proportional plus derivative control $(1 + K_t s)$ that can also be achieved by a cascade compensator. The results should therefore be qualitatively the same.

Partitioning the characteristic equation to obtain the general format $F(s) = -1$ which is required for obtaining a root locus yields

$$G_x(s)H_1(s) = \frac{KK_t(s + 1/K_t)}{s(s+1)(s+5)} = -1 \tag{10.54}$$

It is seen from Eq. (10.54) that introduction of minor loop rate feedback results in the introduction of an open-loop zero. It therefore has the effect of moving the locus to the left and improving the time response. *This is not the same as introducing a cascade zero because no zero appears in the rationalized equation of the control ratio*

$$\frac{C(s)}{R(s)} = \frac{K}{s(s+1)(s+5) + KK_t(s + 1/K_t)} \tag{10.55}$$

The rate constant K_t must be chosen before the root locus can be drawn for Eq. (10.54). *Care must be taken in selecting the value of K_t. For example*, if $K_t = 1$ is used, cancellation of $(s+1)$ in the numerator and denominator of Eq. (10.54) yields

$$\frac{K}{s(s+5)} = -1 \tag{10.56}$$

and it is possible to get the impression that $C(s)/R(s)$ has only two poles. *This is not the case*, as can be seen from Eq. (10.55). Letting $K_t = 1$ provides the common factor $s + 1$ in the characteristic equation so that it becomes

$$(s+1)[s(s+5) + K] = 0 \tag{10.57}$$

One closed-loop pole has therefore been made equal to $s = -1$, and the other two poles are obtained from the root locus of Eq. (10.56). If complex roots are selected from the root locus of Eq. (10.56), it is easily shown that the pole $s = -1$ is dominant and the response of the closed-loop system is overdamped (see Fig. 10.4*d*).

The proper design procedure is to select a number of trial locations for the zero $s = -1/K_t$, to tabulate or plot the system characteristics that result, and then to select the best combination. As an example, with $\zeta = 0.45$ as the criterion for the complex closed-loop poles and $K_t = 0.4$, the root locus is similar to that of Fig. 10.32 with $A = 1$. Table 10.4 gives a comparison of the original system with the rate-compensated system. From the root locus the roots are $s_{1,2} = -1 \pm j2$, and $s_3 = -4$, with $K = 20$. The gain K_1 must be obtained from the forward transfer function:

$$K_1 = \lim_{s\to 0} sG(s) = \lim_{s\to 0} \frac{sG_x(s)}{1 + G_x(s)H(s)} = \lim_{s\to 0} \frac{sK}{s[(s+1)(s+5) + KK_t]}$$

$$= \frac{K}{5 + KK_t} = 1.54 \tag{10.58}$$

TABLE 10.4
Comparison of performances

System	Dominant complex poles $\zeta = 0.45$	Other roots	Ramp error coefficient K_1, s^{-1}	t_p, s	M_o	t_s, s
Original	$-0.40 \pm j0.8$	-5.19	0.84	4.12	0.202	9.48
Rate feedback	$-1.0 \pm j2.0$	-4.0	1.54	1.87	0.17	3.95

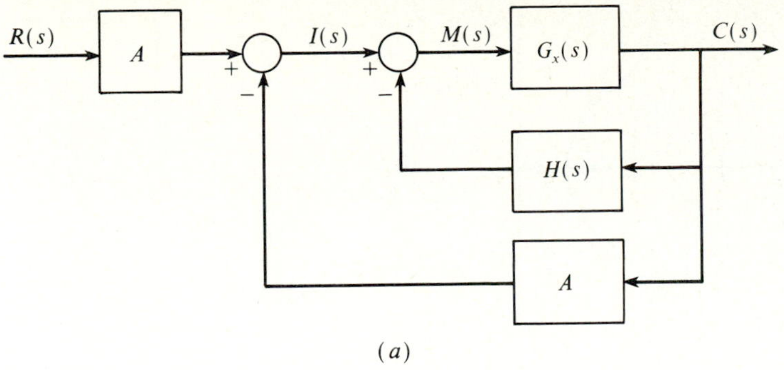

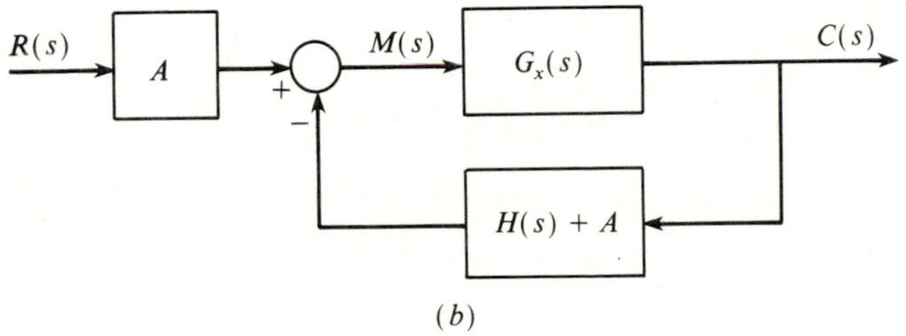

FIGURE 10.31
Block diagrams equivalent to Fig. 10.29.

An analysis of the results of Table 10.4 shows that the effects of rate feedback are similar to those of a cascade lead compensator, but the gain K_1 is not so large (see Table 10.2, Sec. 10.12). The control ratio of the system with rate feedback is

$$\frac{C(s)}{R(s)} = \frac{20}{(s+1-j2)(s+1+j2)(s+4)} \tag{10.59}$$

10.17 RATE FEEDBACK

The more general case of feedback compensation shown in Fig. 10.29 with $H(s) = K_t s$ and the amplifier gain A not equal to unity, is considered in this section. Two methods of attacking this problem are presented. The block diagrams of Fig. 10.31a and b are mathematically equivalent to the block diagram of Fig. 10.29 since they all yield the same closed-loop transfer function:

$$\frac{C(s)}{R(s)} = \frac{AG_x(s)}{1 + G_x(s)H(s) + AG_x(s)} = \frac{AG_x(s)}{1 + G_x(s)[H(s) + A]} \tag{10.60}$$

Method 1. The characteristic equation obtained from the denominator of Eq. (10.60) for the plant given by Eq. (10.51) is

$$1 + G_x(s)[H(s) + A] = 1 + \frac{K(K_t s + A)}{s(s+1)(s+5)} = 0 \tag{10.61}$$

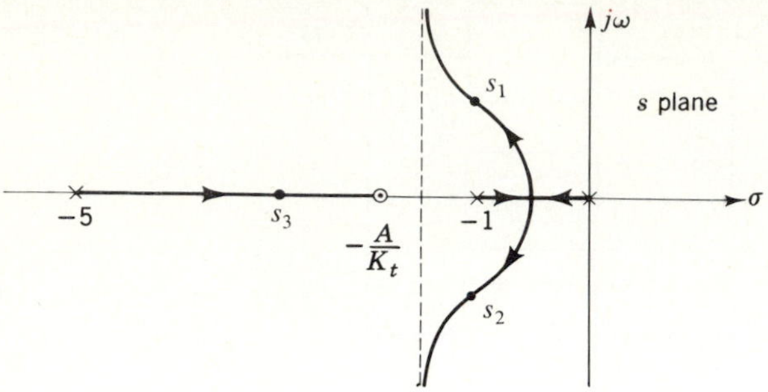

FIGURE 10.32
Root locus for Eq. (10.62).

Partitioning the characteristic equation to obtain the general format $F(s) = -1$ which is required for the root locus yields

$$\frac{KK_t(s + A/K_t)}{s(s + 1)(s + 5)} = -1 \tag{10.62}$$

Since the numerator and denominator are factored, the roots can be obtained by plotting a root locus. The angle condition of $(1 + 2h)180°$ must be satisfied for the root locus of Eq. (10.62). It should be kept in mind, however, that Eq. (10.62) is *not* the forward transfer function of this system; it is *just an equation whose roots are the poles of the control ratio.* The three roots of this characteristic equation vary as K, K_t, and A are varied. Obviously, it is necessary to fix two of these quantities and to determine the roots as a function of the third. A sketch of the root locus as a function of K is shown in Fig. 10.32 for arbitrary values of A and K_t. Note that rate feedback results in an equivalent open-loop zero, so the system is stable for all values of gain for the value of A/K_t chosen. This is typical of the stabilizing effect which can be achieved by derivative feedback and corresponds to the effect achieved with an ideal derivative plus proportional compensator in cascade.

The angles of the asymptotes of two of the branches of Eq. (10.62) are $\pm 90°$. For the selection A/K_t shown in Fig. 10.32, these asymptotes are in the left half of the s plane, and the system is stable for any value of K. However, if $A/K_t > 6$, the asymptotes are in the right half of the s plane and the system can become unstable if K is large enough. For the best transient response the root locus should be moved as far to the left as possible. This is determined by the location of the zero $s = -A/K_t$. To avoid having a dominant real root, the value of A/K_t is made greater than unity.

The forward transfer function obtained from Fig. 10.29 is

$$G(s) = \frac{C(s)}{E(s)} = \frac{AG_x(s)}{1 + G_x(s)H(s)} = \frac{AK}{s(s + 1)(s + 5) + KK_t s} \tag{10.63}$$

The ramp error coefficient obtained from Eq. (10.63) is

$$K_1' = \frac{AK}{5 + KK_t} \tag{10.64}$$

Selection of the zero ($s = -A/K_t$), the sensitivity KK_t for the root locus drawn for Eq. (10.62), and the ramp error coefficient of Eq. (10.64) are all interrelated. The designer therefore has some flexibility in assigning the values of A, K_t, and K.

The values $A = 2.0$ and $K_t = 0.8$ result in a zero at $s = -2.5$, the same as the example in Sec. 10.16. For $\zeta = 0.45$, the control ratio is given by Eq. (10.59) and the performance is the same as that listed in Table 10.4. The use of the additional amplifier A does not change the performance; instead, it permits greater flexibility in the allocation of gains. For example, K can be decreased and both A and K_t increased by a corresponding amount with no change in performance.

Method 2. A more general method is to start with the inner loop, adjusting the gain to select the poles. Then successive loops are adjusted, setting the poles for each loop to desired values. For the system being studied, the transfer function of the inner loop is

$$\frac{C(s)}{I(s)} = \frac{K}{s(s+1)(s+5) + KK_t s} = \frac{K}{s[(s+1)(s+5) + KK_t]}$$

$$= \frac{K}{s(s-p_1)(s-p_2)} \tag{10.65}$$

The equation for which the root locus is to be plotted to find the roots p_1 and p_2 is

$$(s+1)(s+5) + KK_t = 0$$

which can be rewritten as

$$\frac{KK_t}{(s+1)(s+5)} = -1 \tag{10.66}$$

The locus is plotted in Fig. 10.33, where several values of the inner-loop damping ratio ζ_i for the roots are shown. Selection of a ζ_i permits evaluation of the product KK_t by use of Eq. (10.66). Upon assignment of a value to one of these quantities, the other can then be determined. The poles of the inner loop are the

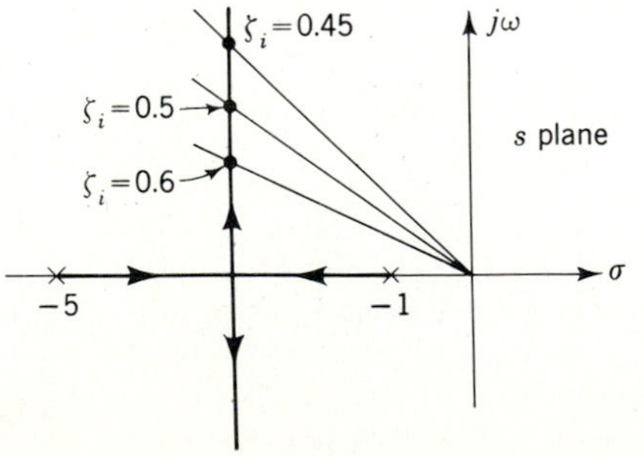

FIGURE 10.33
Poles of the inner loop for several values of ζ_i.

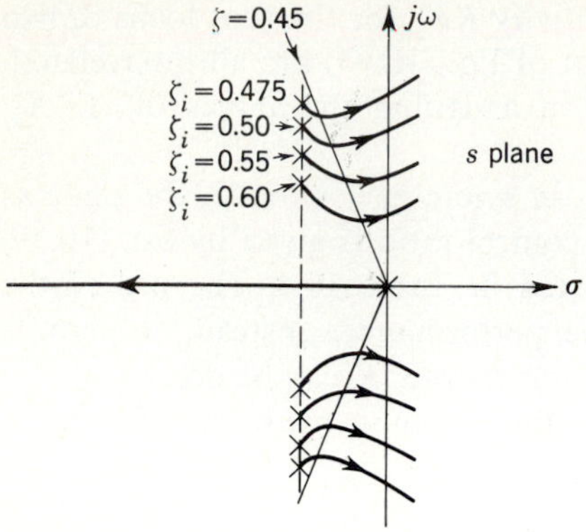

FIGURE 10.34
Root locus of complete system for several values of ζ_i.

points on the locus that are determined by the ζ_i selected. They also are the poles of the open-loop transfer function for the overall system as represented in Eq. (10.65). It is therefore necessary that the damping ratio ζ_i of the inner loop have a higher value than the damping ratio of the overall system. The forward transfer function of the complete system becomes

$$G(s) = \frac{C(s)}{E(s)} = \frac{AK}{s(s^2 + 2\zeta_i\omega_{n,i}s + \omega_{n,i}^2)} \tag{10.67}$$

The root locus of the complete system is shown in Fig. 10.34 for several values of the inner-loop damping ratio ζ_i. These loci are plotted as a function of AK.

The control ratio has the form

$$\frac{C(s)}{R(s)} = \frac{AK}{s(s^2 + 2\zeta_i\omega_{n,i}s + \omega_{n,i}^2) + AK}$$

$$= \frac{AK}{(s + \alpha)(s^2 + 2\zeta\omega_n s + \omega_n^2)} \tag{10.68}$$

In this problem the damping ratio ζ of the complex poles of the control ratio is used as the basis for design. Both the undamped natural frequency ω_n and the

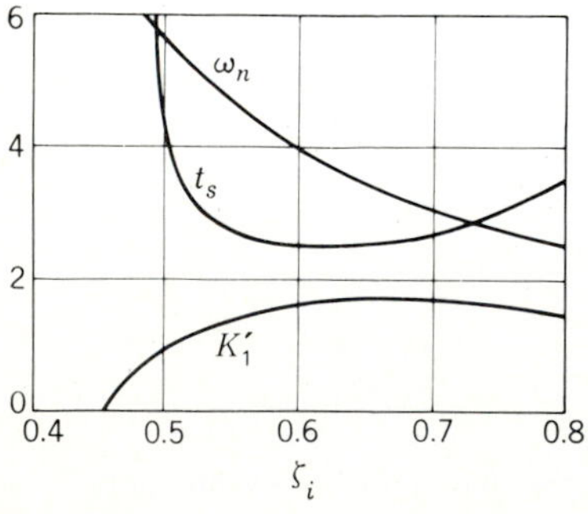

FIGURE 10.35
Variation of K_1', t_s, and ω_n with ζ_i.

TABLE 10.5
System response characteristics using method 2

ζ_i	KK_t	AK	Roots	t_p, s	M_o	t_s, s	K_1'
0.475	34.91	19.33	$-2.74 \pm j5.48$ -0.52	†	0	7.64	0.484
0.5	31.01	30.53	$-2.5 \pm j4.98$ -0.93	†	0	4.34	0.847
0.55	24.76	39.0	$-2.1 \pm j4.22$ -1.75	†	0	2.36	1.31
0.6	20.01	38.63	$-1.82 \pm j3.62$ -2.35	1.40	0.01	2.19	1.54
0.7	13.37	30.51	$-1.38 \pm j2.74$ -3.24	1.52	0.12	2.88	1.66
0.8	9.06	22.14	$-1.03 \pm j2.14$ -3.83	1.77	0.20	3.75	1.57

† Overdamped response.

ramp error coefficient $K_1' = AK/(5 + KK_t)$ depend on the selection of the inner-loop damping ratio ζ_i. Several trials are necessary to show how these quantities and the remaining figures of merit vary with the selection of ζ_i. This information is shown graphically in Fig. 10.35 and in Table 10.5. The system performance specifications must be used to determine a suitable design. The associated value of K_1' determines the required values of A, K, and K_t. The results in Table 10.5 show that $\zeta_i = 0.6$ produces the shortest settling time $t_s = 2.19$, an overshoot $M_o = 0.01$, and $K_1' = 1.54$. Also it shows for $\zeta < 0.6$ that a real root is most dominant, which yields an overdamped system with an increase in t_s.

10.18 FEEDBACK OF SECOND DERIVATIVE OF OUTPUT

To further improve the system performance, a signal proportional to the second derivative of the output may be fed back. For a position system an accelerometer will generate such a signal. An approximation to this desired signal may be generated by modifying the output of a rate transducer with a high-pass filter. An example of this technique for a position-control system is shown in Fig. 10.36.

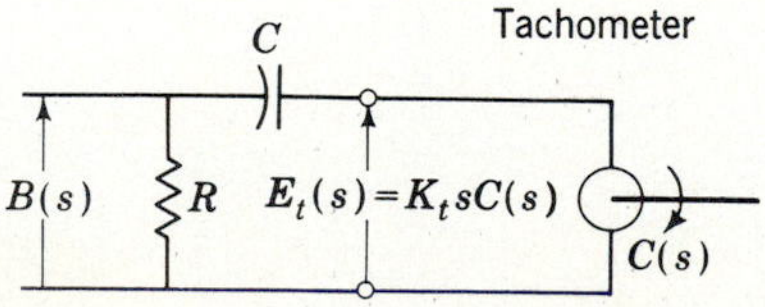

FIGURE 10.36
Tachometer and RC feedback network.

The transfer function for the inner feedback is

$$H(s) = \frac{B(s)}{C(s)} = K_t s\left[\frac{s}{s + 1/RC}\right] = \frac{K_t s^2}{s + 1/RC} \tag{10.69}$$

which consists of a tachometer in cascade with a lead compensator. Method 1 is used to show the application of this compensator. Using the denominator of Eq. (10.60), the root locus is drawn for $G_x(s)[H(s) + A] = -1$, which yields

$$\frac{KK_t(s^2 + As/K_t + A/K_tRC)}{s(s+1)(s+5)(s+1/RC)} = -1 \tag{10.70}$$

Since A/K_t and $1/RC$ must be selected in order to specify the zeros in Eq. (10.70) before the locus can be drawn, it is well to look at the expression for K_1' to determine the limits to be placed on these values. The forward transfer

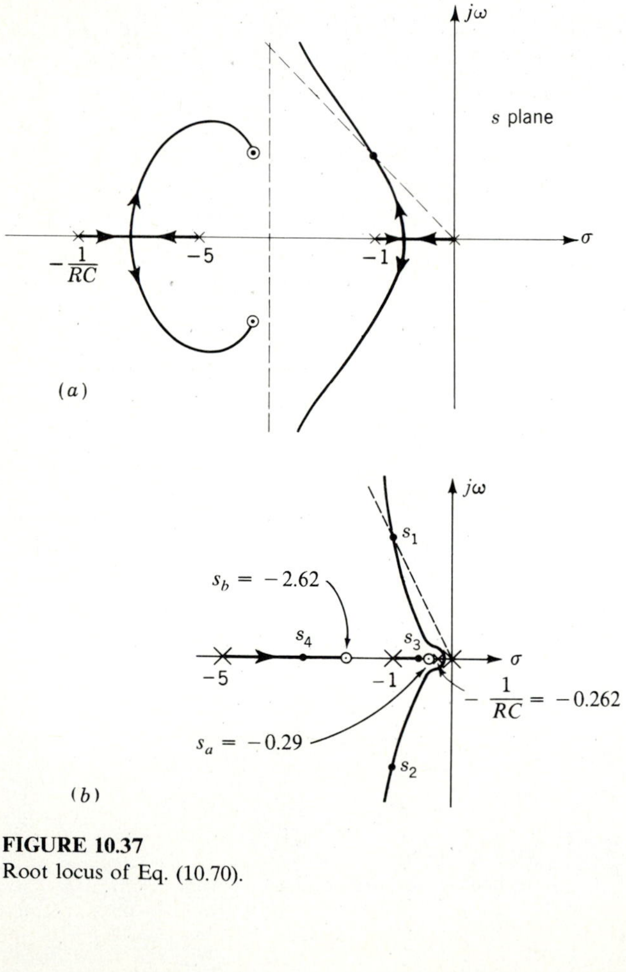

FIGURE 10.37
Root locus of Eq. (10.70).

function is

$$G(s) = \frac{AG_x(s)}{1 + G_x(s)H(s)} = \frac{AK(s + 1/RC)}{s(s+1)(s+5)(s+1/RC) + KK_t s^2} \tag{10.71}$$

Since it is a Type 1 transfer function, the gain K_1' is equal to

$$K_1' = \lim_{s \to 0} sG(s) = \frac{AK}{5} \tag{10.72}$$

Since K_1' is independent of RC, the value of RC is chosen to produce the most desirable root locus. The two zeros of $s^2 + (A/K_t)s + A/K_tRC$ may be either real or complex and are equal to

$$s_{a,b} = -\frac{A}{2K_t} \pm \sqrt{\left(\frac{A}{2K_t}\right)^2 - \frac{A}{K_tRC}} \tag{10.73}$$

Two possible sketches of the root locus are shown in Fig. 10.37 for an arbitrary selection of A, K_t, and RC.

For Fig. 10.37*b* the values $A = 1$, $K_t = 0.344$, and $1/RC = 0.262$ give zeros at $s_a = -0.29$ and $s_b = -2.62$. Note that the zero s_a and the pole $s = -1/RC$ have the properties of a cascade lag compensator; therefore an increase in gain can be expected. The zero s_b results in an improved time response. By specifying a damping ratio $\zeta = 0.35$ for the dominant roots, the result is $K = 53.3$, $K_1' = 10.6$, and $\omega_n = 3.79$, and the overall control ratio is

$$\frac{C(s)}{R(s)} = \frac{53.3(s + 0.262)}{(s + 1.32 + j3.55)(s + 1.32 - j3.55)(s + 3.34)(s + 0.291)} \tag{10.74}$$

Note that a passive lead network in the feedback loop acts like a passive lag network in cascade and results in a large increase in K_1'. Several trials may be necessary to determine the best selection of the closed-loop poles and ramp error coefficient. A $\zeta = 0.35$ is chosen instead of $\zeta = 0.45$ to obtain a larger value for K_1'. Since the third pole $s = -3.34$ reduces the peak overshoot, a smaller value of ζ is justifiable; however, the peak overshoot and settling time are larger than the best results obtained with just rate feedback (see Table 10.6).

TABLE 10.6
Comparison of cascade and feedback compensation using the root locus

System	K_1	ω_n	ζ	M_o	t_s, s
Uncompensated	0.84	0.89	0.45	0.202	9.48
Rate feedback	1.54	4.0	0.45	0.01	2.19
Second derivative of output	10.60	3.79	0.35	0.266	6.68
Cascade lag compensator	8.02	0.854	0.45	0.266	22.0
Cascade lead compensator	2.95	3.53	0.45	0.195	2.45
Cascade lag-lead compensator	29.26	3.49	0.45	0.213	3.33

The use of an *RC* network in the feedback path may not always produce a desired increase in gain. Also, selecting the proper location of the zeros and the corresponding values of A, K_t, and RC in Eq. (10.70) may require much work. Therefore, a possible alternative is to use only rate feedback to improve the time response, and then to add a cascade lag compensator to increase the gain. The corresponding design steps are essentially independent.

10.19 RESULTS OF FEEDBACK COMPENSATION

The preceding sections show the application of feedback compensation by the root-locus method. Basically, it has been shown that rate feedback is similar to ideal derivative control in cascade. The addition of a high-pass *RC* filter in the feedback path permits a large increase in the error coefficient. For good results the feedback compensator $H(s)$ should have a zero, $s = 0$, of higher order than the type of the original transfer function $G_x(s)$. The method of adjusting the poles of the inner loop and then adjusting successively larger loops is applicable to all multiloop systems. This method is used extensively in the design of aircraft flight control systems.[13] The results of feedback compensation applied in the preceding sections are summarized in Table 10.6. Results of cascade compensation are also listed for comparison. This tabulation shows that comparable results are obtained by cascade and feedback compensation.

10.20 RATE FEEDBACK: PLANTS WITH DOMINANT COMPLEX POLES

The simple cascade compensators of Secs. 10.7 to 10.11 yield very little system performance improvement for plants having the pole-zero pattern of Fig. 10.38*a* where the plant has dominant complex poles. The following transfer function, representing the longitudinal motion of an aircraft,[13] yields such a pattern:

$$G_x(s) = \frac{\theta(s)}{e_g(s)} = \frac{K_x(s - z_1)}{s(s - p_1)(s - p_2)(s - p_3)} \tag{10.75}$$

where θ is the pitch angle and e_g is the input to the elevator servo. With unity feedback the root locus of Fig. 10.38*a* shows, for $K_x > 0$, that only $\zeta < \zeta_x$ is achievable for the dominant roots. However, good improvement in system performance is obtained by using a minor loop with rate feedback $H(s) = K_t s$. Using method 1 of Sec. 10.17 for the system configuration of Fig. 10.29 yields

$$\frac{C(s)}{E(s)} = \frac{AK_x(s - z_1)}{s(s - p_1)(s - p_2)(s - p_3) + K_t K_x s(s - z_1)} \tag{10.76}$$

and

$$\frac{C(s)}{R(s)} = \frac{AK_x(s - z_1)}{s(s - p_1)(s - p_2)(s - p_3) + K_t K_x s(s - z_1) + AK_x(s - z_1)} \tag{10.77}$$

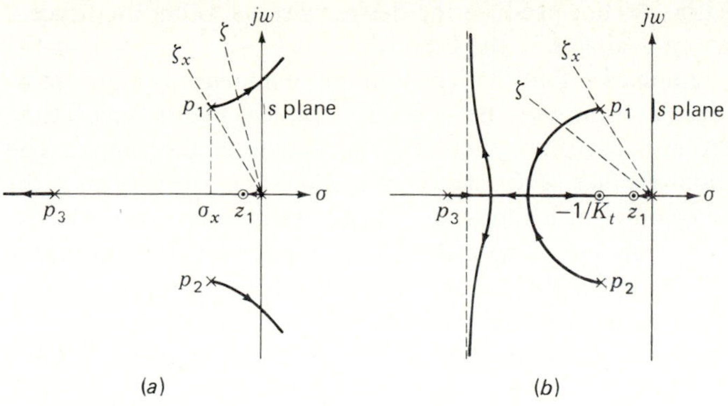

FIGURE 10.38
Root locus for the longitudinal motion of an aircraft: (a) uncompensated; (b) compensated (rate feedback).

The characteristic equation formed from the denominator of Eq. (10.60), or from the denominator of Eq. (10.77), is partitioned to yield

$$\frac{K_t K_x (s - z_1)(s + A/K_t)}{s(s - p_1)(s - p_2)(s - p_3)} = -1 \qquad (10.78)$$

A comparison of Eqs. (10.75) and (10.78) shows that rate feedback has added a zero to the root locus. One possible root locus, depending on the value of A/K_t, is shown in Fig. 10.38b. This illustrates that using rate feedback allows, for $K_t > 0$, a good improvement in system performance, since it is possible to achieve $\zeta > \zeta_x$ and $\sigma < \sigma_x$ for the complex roots. The real root near the origin may appear to be dominant, but it is close to the zero z_1. Therefore, it may not make a large contribution to the transient response.

Note that the cascade compensator of Eq. (10.43), as applied in Fig. 10.25, shows exact cancellation of the pair of complex poles by the compensator zeros. In practice, exact cancellation cannot be achieved because the aircraft poles vary with different flight conditions. Thus, as illustrated by this section, using simple rate feedback is a more practical approach for achieving improved system performance.

10.21 SUMMARY

This chapter shows the applications of cascade and feedback compensation. The procedures for applying the compensators are given so that they can be used with any system. They involve a certain amount of trial and error and the exercise of judgment based on past experience. The improvements produced by each compensator are shown by application to a specific system. In cases where the conventional compensators do not produce the desired results, other methods of applying compensators are available in the literature.

It is important to emphasize that the conventional compensators cannot be applied indiscriminately. For example, the necessary improvements may not fit the four categories listed in Sec. 10.6, or the use of the conventional compensator may cause a deterioration of other performance characteristics. In such cases the designer must use his ingenuity in locating the compensator poles and zeros to achieve the desired results. The use of several compensating networks in cascade and/or in feedback may be required to achieve the desired root locations.

Realization of compensator pole-zero locations by means of passive networks requires an investigation of network synthesis techniques. Detail design procedures are available for specialized networks which produce two poles and two zeros.[10-12]

It may occur to the reader that the methods of compensation in this and the following chapters can become laborious because there is some trial and error involved. The use of a computer-aided control-system design program[5,14] (see App. B) to solve design problems minimizes the work involved. Using a computer means that an acceptable design of the compensator can be determined in a shorter time. It is still useful to obtain by an analytical or graphical method a few values of the solution to ensure that the problem has been properly set up and that the computer program is operating correctly.

REFERENCES

1. D'Azzo, J. J., and C. H. Houpis: *Feedback Control System Analysis and Synthesis*, 2d ed., McGraw-Hill, New York, 1966.
2. Chu, Y.: "Synthesis of Feedback Control System by Phase-Angle Loci," *Trans. AIEE*, vol. 71, pt. II, pp. 330–339, November 1952.
3. Gibson, J. E.: *Nonlinear Automatic Control*, McGraw-Hill, New York, 1963.
4. Thaler, G. J., and M. P. Pastel: *Analysis and Design of Nonlinear Feedback Control Systems*, McGraw-Hill, New York, 1960.
5. Larimer, S. J.: "An Interactive Computer-Aided Design Program for Digital and Continuous System Analysis and Synthesis (TOTAL)," M.S. thesis, GE/GGC/EE/78-2, School of Engineering, Air Force Institute of Technology, Wright-Patterson Air Force Base, Ohio, 1978; available from Defense Documentation Center (DDC), Cameron Station, Alexandria, Va. 22314.
6. Elgerd, O. I., and W. C. Stephens: "Effect of Closed-Loop Transfer Function Pole and Zero Locations on the Transient Response of Linear Control Systems," *Trans. AIEE*, vol. 78, pt. II, pp. 121–127, May 1959.
7. Truxal, J. G.: *Automatic Feedback Control System Synthesis*, McGraw-Hill, New York, 1955.
8. Ross, E. R., T. C. Warren, and G. J. Thaler: "Design of Servo Compensation Based on the Root Locus Approach," *Trans. AIEE*, vol. 79, pt. II, pp. 272–277, September 1960.
9. Pena, L. Q.: "Designing Servocompensators Graphically," *Control Eng.*, vol. 82, pp. 79–81, January 1964.
10. Lazear, T. J., and A. B. Rosenstein: "On Pole-Zero Synthesis and the General Twin-T," *Trans. IEEE*, vol. 83, pt. II, pp. 389–393, November 1964.
11. Kuo, F. F.: *Network Analysis and Synthesis*, 2d ed., Wiley, New York, 1966.
12. Mitra, S. K.: *Analysis and Synthesis of Linear Active Networks*, Wiley, New York, 1969.
13. Blakelock, J. H.: *Automatic Control of Aircraft and Missiles*, Wiley, New York, 1965.
14. Thompson, P. M.: *USER's Guide to Program CC, Version 3*, Systems Technology, Inc., Hawthorne, Calif., March 1985.

CHAPTER 11

CASCADE AND FEEDBACK COMPENSATION: FREQUENCY-RESPONSE PLOTS

11.1 INTRODUCTION

General methods have been developed for improving the response of a feedback tracking control system and for disturbance rejection when the analysis is performed in the frequency domain. This chapter is devoted to frequency-domain tracking techniques and Chap. 12 is devoted to disturbance-rejection techniques. In the preceding chapter the determining factors for the selection of a lag, lead, or lag-lead network and their corresponding effects on the time response are presented. The choice of the poles and zeros of a cascade compensator by the use of the root-locus method is often readily determinable for any system. The designer can place the poles and zeros of the proposed compensator on the s plane and determine the poles of the closed-loop tracking system, which in turn permits evaluation of the closed-loop time response. Similarly, the new values of M_m, ω_m, and K_m can be determined by the frequency-response method, using either the polar plot or the log plot. However, the closed-loop poles are not explicitly determined. Furthermore, the correlation between the frequency-response parameters M_m and ω_m and the time response is only qualitative, as discussed in Secs. 9.3, 9.4, and 9.11. The presence of real roots near the complex dominant roots further changes the correlation between the frequency-response parameters and the time response, as discussed in Sec. 9.11.

This chapter is concerned with the changes that can be made in the frequency-response characteristics by use of the three types of cascade compensators (lag, lead, and lag-lead) and by feedback compensation. They are only representative of the compensators that can be used. A study is made of the effect of these compensators on the overall system. Design procedures to obtain improvement of the system performance are described. With practical experience behind him, a designer can extend or modify these design procedures, which are based on the presence of one pair of complex dominant roots, to those systems where there are dominant roots besides the main complex pair.[1] Both the log plots and the polar plots are utilized in this chapter in applying the frequency-response compensation criteria to show that either type of plot can be used; the choice depends on individual preference.

The performance of a closed-loop system can be described in terms of M_m, ω_m, and the system error coefficient K_m. The value of M_m essentially describes the damping ratio ζ and therefore the amount of overshoot in the transient response. For a specific value of M_m, the resonant frequency ω_m determines the undamped natural frequency ω_n, which in turn determines the response time of the system. The system error coefficient K_m is important because it determines the steady-state error with appropriate standard input. The design procedure is usually based on selecting a value of M_m and using the methods described in Chap. 9 to find the corresponding values of ω_m and the required gain K_m. Once this is accomplished, if the desired performance specifications are not met by the basic system, compensation must be used. This alters the shape of the frequency-response plot in order to try to meet the performance specifications. Also, for those systems that are unstable for all values of gain, it is mandatory that a stabilizing or compensating network be inserted in the system. The compensator may be placed in cascade or in a minor feedback loop.

The reasons for reshaping the frequency-response plot generally fall into the following categories:

1. A given system is stable, and its M_m and ω_m (and therefore the transient response) are satisfactory, but its steady-state error is too large. The gain K_m must therefore be increased to reduce the steady-state error (see Chap. 6) without appreciably altering the values of M_m and ω_m. It is shown later that in this case the high-frequency portion of the frequency-response plot is satisfactory but the low-frequency portion is not.
2. A given system is stable but its transient settling time is unsatisfactory; i.e., the M_m is set to a satisfactory value, but the ω_m is too low. The gain may be satisfactory, or a small increase may be desirable. The high-frequency portion of the frequency-response plot must be altered in order to increase the value of ω_m.

3. A given system is stable and has a desired M_m, but both its transient response and its steady-state response are unsatisfactory. Therefore the values of both ω_m and K_m must be increased. The portion of the frequency plot in the vicinity of the $-1 + j0$ or ($-180°$, 0-dB) point must be altered to yield the desired ω_m, and the low-frequency portion must be changed to obtain the increase in gain desired.
4. A given system is unstable for all values of gain. The frequency-response plot must be altered in the vicinity of the $-1 + j0$ or ($-180°$, 0-dB) point to produce a stable system with a desired M_m and ω_m.

Thus the objective of compensation is to reshape, by means of a compensator, the frequency-response plot of the basic system to try to achieve the performance specifications. Examples which demonstrate this objective are presented in the following sections.

The design methods in this chapter and the effects of compensation are illustrated by use of the frequency-response plots in order to provide a means for understanding the principles involved. It is expected, however, that a computer-aided design (CAD) program[4,5] (see App. B) will be used extensively in the design process. Such a program provides both the frequency response and the time response of the resultant system.

11.2 SELECTION OF A CASCADE COMPENSATOR

Consider the system shown in Fig. 11.1.

$\mathbf{G}_x(j\omega)$ = forward transfer function of a basic feedback control system

$\mathbf{G}(j\omega) = \mathbf{G}_c(j\omega)\mathbf{G}_x(j\omega)$ = forward transfer function which results in the required stability and steady-state accuracy

Dividing $\mathbf{G}(j\omega)$ by $\mathbf{G}_x(j\omega)$ gives the necessary cascade compensator

$$\mathbf{G}_c(j\omega) = \frac{\mathbf{G}(j\omega)}{\mathbf{G}_x(j\omega)} \tag{11.1}$$

A physical electrical network for the compensator described in Eq. (11.1) can be synthesized by the techniques of Foster, Cauer, Brune, and Guillemin. In the preceding and present chapters the compensators are limited to relatively simple RC networks. Based upon the design criteria presented, the parameters of

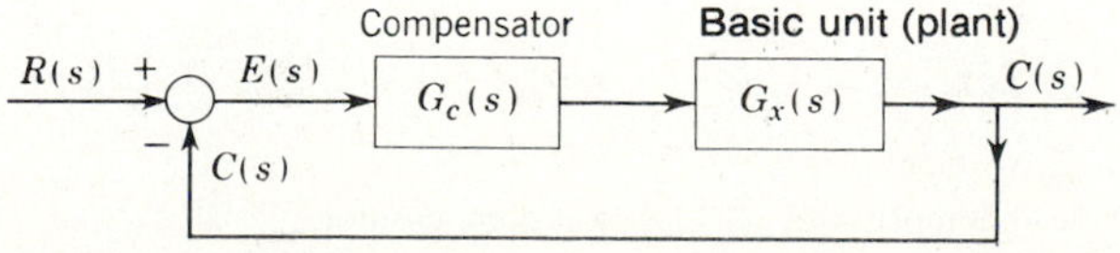

FIGURE 11.1
Cascade compensation.

these simple networks can be evaluated in a rather direct manner. The simple networks quite often provide the desired improvement in the control system's performance. When they do not, more complex networks must be synthesized.[2,3] Physically realizable networks are limited in their performance over the whole frequency spectrum; this makes it difficult to obtain the exact transfer-function characteristic given by Eq. (11.1). Actually this is not a serious limitation, since the desired performance specifications are interpreted in terms of a comparatively small bandwidth of the frequency spectrum. A compensator can be constructed to have the desired magnitude and angle characteristics over this bandwidth. Minor loop feedback compensation can also be used to achieve the desired $G(j\omega)$. This may yield a simpler network than the cascade compensator given by Eq. (11.1).

The characteristics of the lag and lead compensators and their effects upon a given polar plot are shown in Fig. 11.2a and b. The cascade lag compensator results in a clockwise rotation of $G(j\omega)$ and can produce a large increase in gain

Lag compensator

$$G_c(j\omega) = \frac{1 + j\omega T}{1 + j\omega \alpha T}$$

$$\alpha > 1$$

Lead compensator

$$G_c(j\omega) = \alpha \frac{1 + j\omega T}{1 + j\omega \alpha T}$$

$$\alpha < 1$$

Lag-lead compensator

$$G_c(j\omega) = \frac{1 + j\omega(T_1 + T_2) + (j\omega)^2 T_1 T_2}{1 + j\omega(T_1 + T_2 + T_{12}) + (j\omega)^2 T_1 T_2}$$

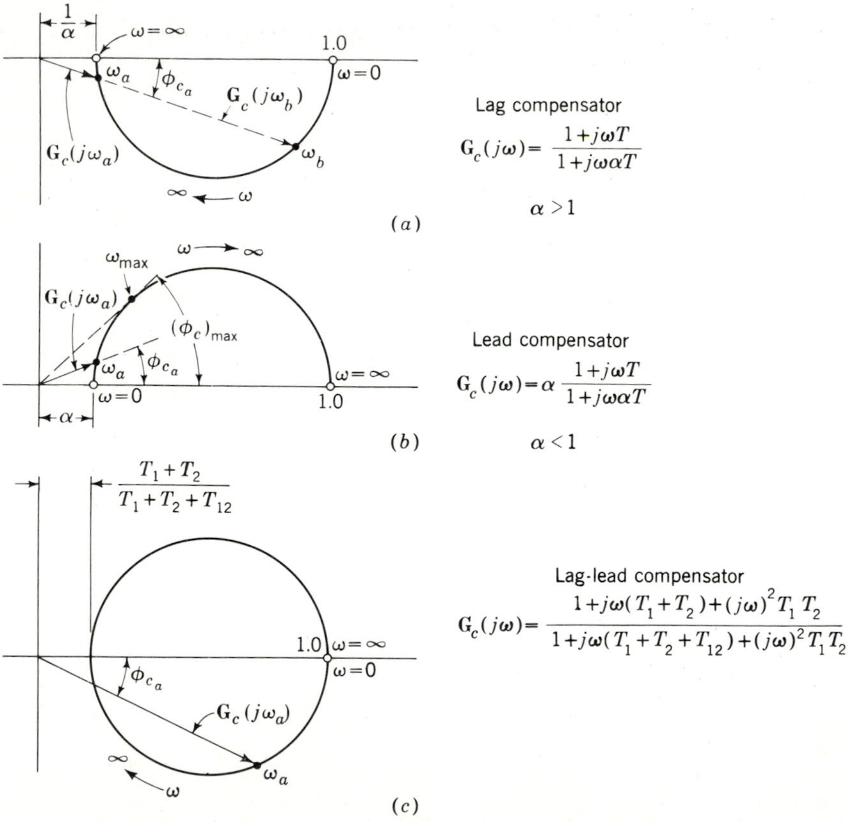

FIGURE 11.2
Polar plots: (a) lag compensator; (b) lead compensator; (c) lag-lead compensator.

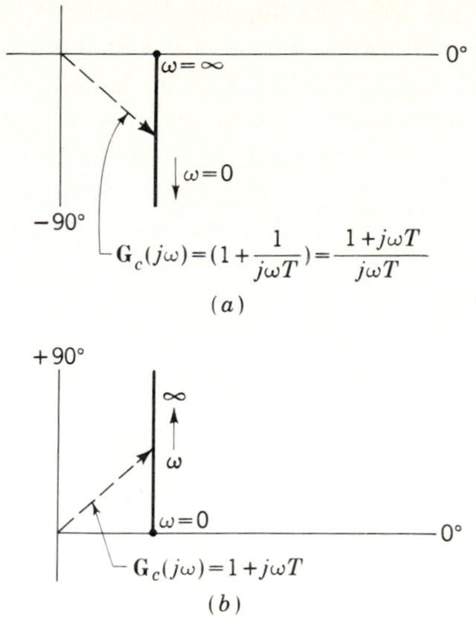

FIGURE 11.3
(*a*) Ideal integral plus proportional compensator.
(*b*) Ideal derivative plus proportional compensator.

K_m with a small decrease in the resonant frequency ω_m. The cascade lead compensator results in a counterclockwise rotation of $\mathbf{G}(j\omega)$ and can produce a large increase in ω_m with a small increase in K_m. When the characteristics of both the lag and the lead networks are necessary to achieve the desired specifications, a lag-lead compensator can be used. The transfer function and polar plot of this lag-lead compensator are shown in Fig. 11.2*c*. The corresponding log plots for each of these compensators are included with the design procedures later in the chapter.

The choice of a compensator depends on the characteristics of the given (or basic) feedback control system and the desired specifications. Since these networks are to be inserted in cascade, they must be of the (1) proportional plus integral and (2) proportional plus derivative types. The plot of the lag compensator shown in Fig. 11.2*a* approximates the ideal integral plus proportional characteristics shown in Fig. 11.3*a* at high frequencies. Also, the plot of the lead compensator shown in Fig. 11.2*b* approximates the ideal derivative plus proportional characteristics shown in Fig. 11.3*b* at low frequencies. These are the frequency ranges where the proportional plus integral and proportional plus derivative controls are needed to provide the necessary compensation.

The equations of $\mathbf{G}_c(j\omega)$ in Fig. 11.2*a* and 11.2*b*, which represent the lag and the lead compensators used in this chapter, can be expressed as

$$\mathbf{G}_c(j\omega) = |\mathbf{G}_c(j\omega)| \angle \phi_c \tag{11.2}$$

Solving for the angle ϕ_c yields

$$\phi_c = \tan^{-1}\omega T - \tan^{-1}\omega\alpha T = \tan^{-1}\frac{\omega T - \omega\alpha T}{1 + \omega^2\alpha T^2} \tag{11.3}$$

or

$$T^2 + \frac{\alpha - 1}{\omega\alpha\tan\phi_c}T + \frac{1}{\omega^2\alpha} = 0 \tag{11.4}$$

where $\alpha > 1$ for the lag compensator and $\alpha < 1$ for the lead compensator. It is shown in later sections that in the design procedure a frequency ω is specified at which a particular value of ϕ_c is desired. For the selected values of α, ϕ_c, and ω, Eq. (11.4) is used to determine the value T required for the compensator. For a lag compensator a small value of ϕ_c is required, whereas for a lead compensator a large value of ϕ_c is required. For a lead compensator the value ω_{max} at which the maximum phase shift occurs, for a given T and α, can be determined by setting the derivative of ϕ_c with respect to the frequency ω equal to zero. Therefore, from Eq. (11.3) the maximum angle occurs at the frequency

$$\omega_{max} = \frac{1}{T\sqrt{\alpha}} \tag{11.5}$$

Inserting this value of frequency into Eq. (11.3) gives the maximum phase shift

$$(\phi_c)_{max} = \sin^{-1}\frac{1-\alpha}{1+\alpha} \tag{11.6}$$

Typical values for a lead compensator are

α	0	0.1	0.2	0.3	0.4	0.5
$(\phi_c)_{max}$	90°	54.9°	41.8°	32.6°	25.4°	19.5°

A knowledge of the maximum phase shift is useful in the application of lead compensators.

The design procedures developed in the following sections utilize the characteristics of compensators discussed in this section.

11.3 CASCADE LAG COMPENSATOR

When the M_m and ω_m (that is, the transient response) of a feedback control system are considered satisfactory but the steady-state error is too large, it is necessary to increase the gain without appreciably altering that portion of the given log magnitude–angle plot in the vicinity of the (−180°, 0-dB) point (category 1). This can be done by introducing an integral or lag compensator in cascade with the original forward transfer function as shown in Fig. 11.1. Figure 10.9 shows a representative lag network having the transfer function

$$G_c(s) = A\frac{1+T_s}{1+\alpha Ts} \tag{11.7}$$

where $\alpha > 1$. As an example of its application, consider a basic Type 0 system. For an improvement in the steady-state accuracy the system step error coefficient K_p must be increased. Its values before and after the addition of the compensator are

$$K_p = \lim_{s\to 0} G_x(s) = K_0 \tag{11.8}$$

$$K_p' = \lim_{s\to 0} G_c(s)G_x(s) = AK_0 = K_0' \tag{11.9}$$

The compensator amplifier gain A must have a value which gives the desired increase in the value of the step error coefficient. Thus

$$A = \frac{K_0'}{K_0} \tag{11.10}$$

This increase in gain must be achieved while maintaining the same M_m and without appreciably changing the value of ω_m. However, the lag compensator has a negative angle which moves the original log magnitude–angle plot to the left, or closer to the $(-180°, 0\text{-dB})$ point. This has a destabilizing effect and reduces ω_m. To limit this decrease in ω_m, the lag compensator is designed so that it introduces a small angle, generally no more than $-5°$, at the original resonant frequency ω_{m1}. The value of $-5°$ at $\omega = \omega_{m1}$ is an empirical value, determined from practical experience in applying this type of network. A slight variation in method, which gives equally good results, is to select the lag compensator so that the magnitude of its angle is 5° or less at the original phase-margin frequency $\omega_{\phi 1}$.

The overall characteristics of lag compensators can best be visualized from the log plot. The log magnitude and phase-angle equations for the compensator of Eq. (11.7) are

$$\text{Lm}\,\mathbf{G}_c'(j\omega) = \text{Lm}\,(1 + j\omega T) - \text{Lm}\,(1 + j\omega\alpha T) \tag{11.11}$$

$$\angle \mathbf{G}_c(j\omega) = \angle 1 + j\omega T - \angle 1 + j\omega\alpha T \tag{11.12}$$

For various values of α there is a family of curves for the log magnitude and phase-angle diagram, as shown in Fig. 11.4. Engineers who use this compensator

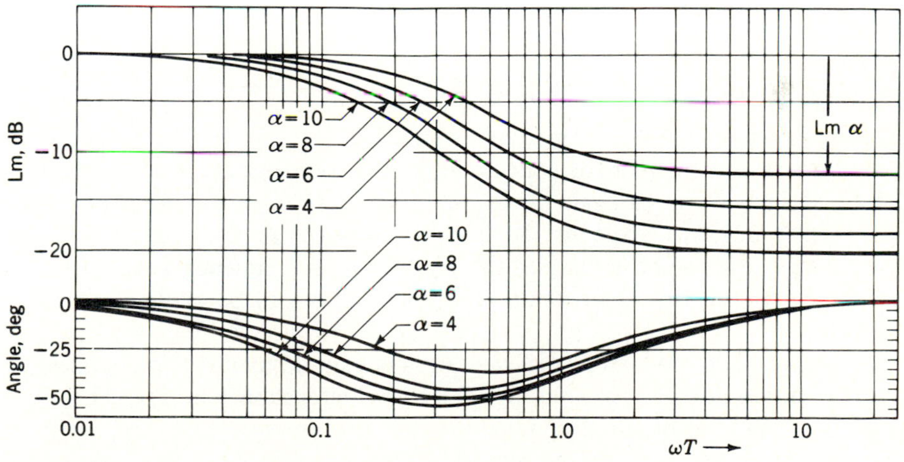

FIGURE 11.4
Log magnitude and phase-angle diagram for lag compensator

$$\mathbf{G}_c'(j\omega) = \frac{1 + j\omega T}{1 + j\omega\alpha T}$$

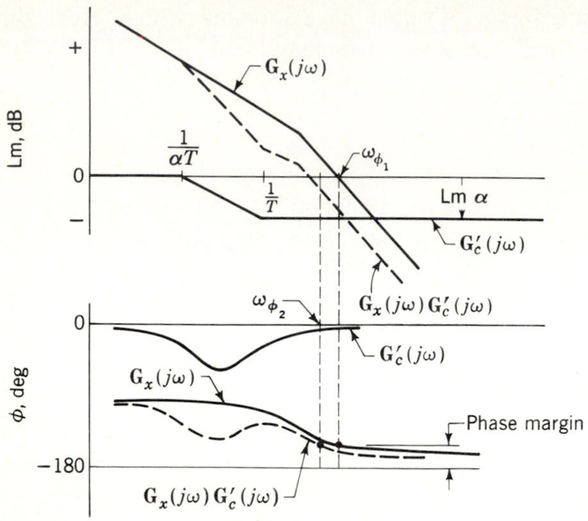

FIGURE 11.5
Original and lag-compensated log magnitude and phase-angle plots.

repeatedly may find it convenient to make templates of these curves. From the shape of these curves it is seen that an attenuation equal to Lm α is introduced above $\omega T = 1$. Assume that a system has the original forward transfer function $\mathbf{G}_x(j\omega)$ and that the gain has been adjusted for the desired phase margin. The log magnitude and phase-angle diagram of $\mathbf{G}_x(j\omega)$ is sketched in Fig. 11.5. The addition of the lag compensator $\mathbf{G}_c(j\omega)$ reduces the log magnitude by Lm α for the frequencies above $\omega = 1/T$. The value of T of the compensator is selected so that the attenuation Lm α occurs at the original phase-margin frequency $\omega_{\phi 1}$. Notice that the phase margin at $\omega_{\phi 1}$ has been reduced. To maintain the specified phase margin, the phase-margin frequency has been reduced to $\omega_{\phi 2}$. It is now possible to increase the gain of $\mathbf{G}_x(j\omega)\mathbf{G}_c'(j\omega)$ by selecting the value A so that the Lm curve will have the value 0 dB at the frequency $\omega_{\phi 2}$.

The effects of the lag compensator can now be analyzed. The reduction in log magnitude at the higher frequencies due to the compensator is desirable. This permits an increase in the gain to maintain the desired phase margin. Unfortunately, the negative angle of the compensator lowers the angle curve. To maintain the desired phase margin, the phase-margin frequency is reduced. To keep this reduction small, the value of $\omega_{cf} = 1/T$ should be made as small as possible.

The log magnitude–angle diagram permits gain adjustment for the desired M_m. The curves for $\mathbf{G}_x(j\omega)$ and $\mathbf{G}_x(j\omega)\mathbf{G}_c'(j\omega)$ are shown in Fig. 11.6. The increase in gain to obtain the same M_m is the amount that the curve of $\mathbf{G}_x(j\omega)\mathbf{G}_c'(j\omega)$ must be raised in order to be tangent to the M_m curve. Because of the negative angle introduced by the lag compensator, the new resonant frequency ω_{m2} is smaller than the original resonant frequency ω_{m1}. Provided that the decrease in ω_m is small and the increase in gain is sufficient to meet the specifications, the system can be considered satisfactorily compensated.

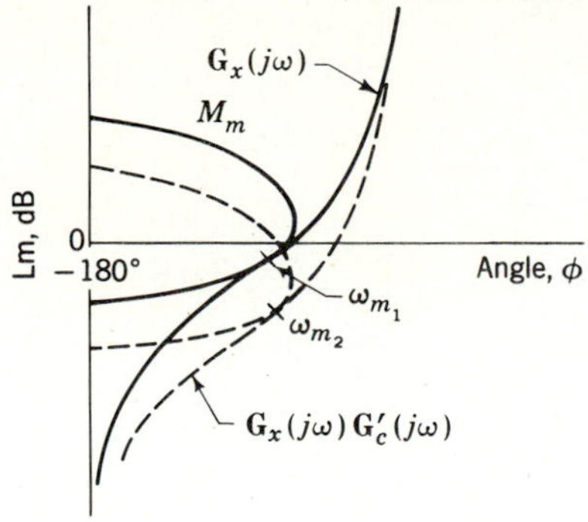

FIGURE 11.6
Original and lag-compensated Nichols plots.

In summary, the lag compensator is basically a low-pass filter: the low frequencies are passed and the higher frequencies are attenuated. The attenuation characteristic of the lag compensator is useful and permits an increase in the gain. The negative phase-shift characteristic is detrimental to system performance but must be tolerated. Because the predominant and useful feature of the compensator is attenuation, a more appropriate name for it is *high-frequency attenuation compensator*.

11.4 EXAMPLE: CASCADE LAG COMPENSATION

To give a basis for comparison of all methods of compensation, the examples of Chap. 10 are used in this chapter. For the unity-feedback Type 1 system of Sec. 10.7 the forward transfer function is

$$G_x(s) = \frac{K_1}{s(1+s)(1+0.2s)} \tag{11.13}$$

By the methods of Secs. 9.7 and 9.10 the gain required for this system to have an M_m of 1.26 (2 dB) is $K_1 = 0.872\ \text{s}^{-1}$, and the resulting resonant frequency is $\omega_{m1} = 0.715$ rad/s. With these values the transient response is considered to be satisfactory, but the steady-state error is too large. Putting a lag compensator in cascade with a basic system (category 1) permits an increase in gain and therefore reduces the steady-state error.

The cascade compensator to be used has the transfer function given by Eq. (11.7). Selection of an $\alpha = 10$ means that an increase in gain of almost 10 is desired. Based on the criterion discussed in the last section (a phase shift of $-5°$ at the original ω_m), the compensator time constant can be determined. Note that in Fig. 11.4, $\phi_c = -5°$ occurs for values of ωT approximately equal to 0.01 and 10. These two points correspond, respectively, to points ω_b and ω_a in Fig. 11.2*a*. The choice $\omega T = \omega_{m1}T = 10$ (or $\omega_a = \omega_{m1}$) is made since this provides the maximum attenuation at ω_{m1}. This ensures a maximum possible gain increase in $\mathbf{G}(j\omega)$ to achieve the desired M_m. Thus for $\omega_{m1} = 0.72$ the value $T \approx 13.9$ is

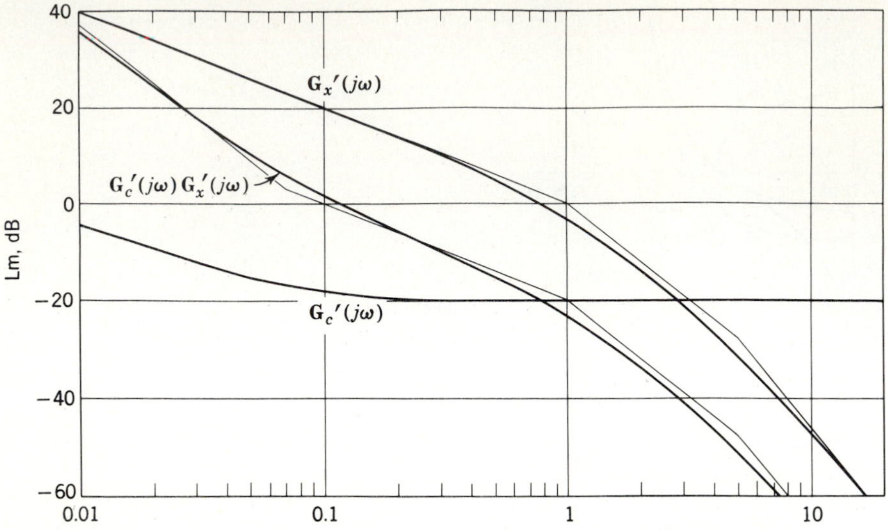

(a)

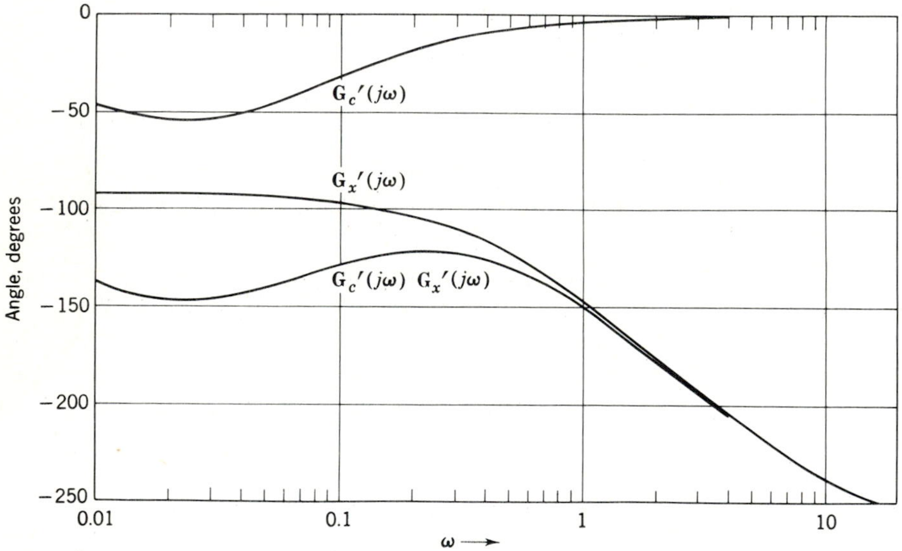

(b)

FIGURE 11.7
Log magnitude and phase-angle diagrams for

$$\mathbf{G}_x'(j\omega) = \frac{1}{j\omega(1 + j\omega)(1 + j0.2\omega)}$$

$$\mathbf{G}_c'(j\omega)\mathbf{G}_x'(j\omega) = \frac{1 + j14\omega}{j\omega(1 + j\omega)(1 + j0.2\omega)(1 + j140\omega)}$$

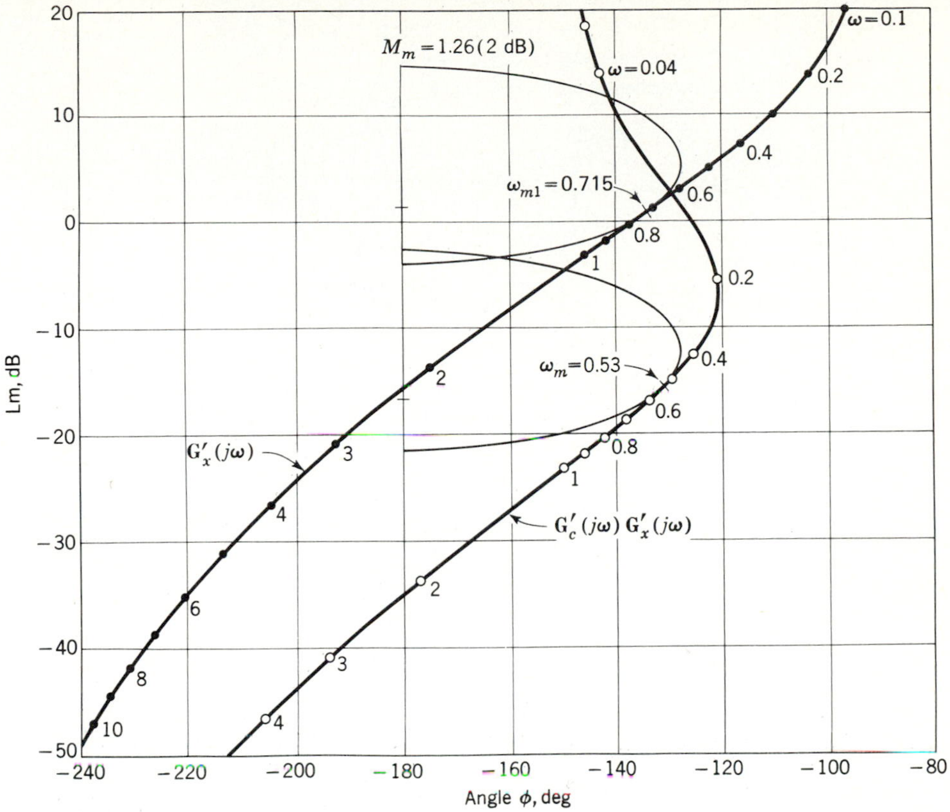

FIGURE 11.8
Log magnitude–angle diagram: original and lag-compensated systems.

obtained. Note that the value of T can also be obtained from Eq. (11.4). For ease of calculation, the value of the time constant is rounded off to 14.0, which does not appreciably affect the results.

In Figs. 11.7*a* and 11.7*b* the log magnitude and phase-angle diagrams for the compensator are shown, together with the curves for $\mathbf{G}_x'(j\omega)$. The sum of the curves is also shown and represents the function $\mathbf{G}_c'(j\omega)\mathbf{G}_x'(j\omega)$. The new curves of $\mathbf{G}_c'(j\omega)\mathbf{G}_x'(j\omega)$ from Fig. 11.7 are used to draw the log magnitude–angle diagram of Fig. 11.8. By using this curve, the gain is adjusted to obtain an $M_m = 1.26$. The results are $K_1' = 6.68$ and $\omega_m = 0.53$. Thus an improvement in K_1 has been obtained, but a penalty of a smaller ω_m results. An increase in gain of approximately $A = 7.7$ results from the use of a lag compensator with $\alpha = 10$, which is typical of the gain increase permitted by a given α. In actual practice this additional gain may be obtained from an amplifier present in the basic system if it is not already operating at maximum gain. Note that the value of gain is somewhat less than the chosen value of α. Provided that the decrease in ω_m and the increase in gain are acceptable, the system has been suitably compensated.

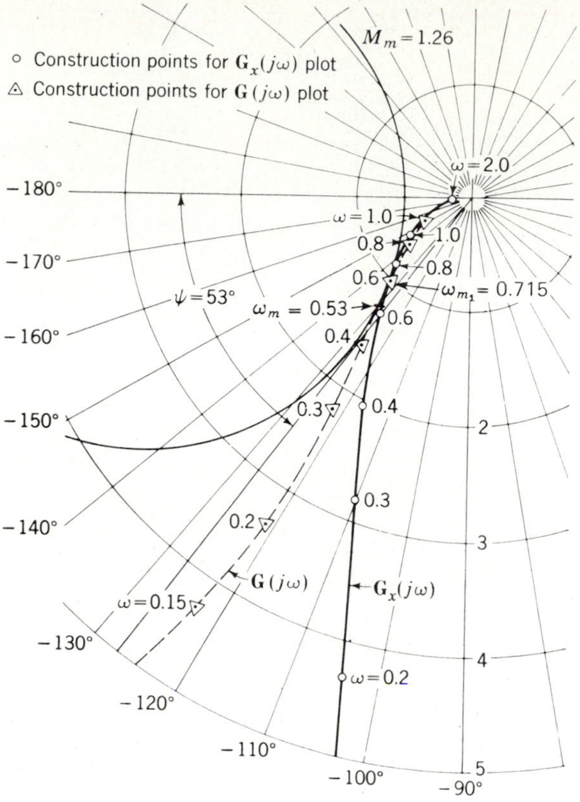

FIGURE 11.9
Polar plots of a Type 1 control system, with and without lag compensation:

$$\mathbf{G}_x(j\omega) = \frac{0.872}{j\omega(1 + j\omega)(1 + j0.2\omega)}$$

$$\mathbf{G}(j\omega) = \mathbf{G}_c(j\omega)\mathbf{G}_x(j\omega) = \frac{6.675(1 + j14\omega)}{j\omega(1 + j\omega)(1 + j0.2\omega)(1 + j140\omega)}$$

To further show the effects of compensation, the polar plots representing the basic and compensated systems are shown in Fig. 11.9. Both curves are tangent to the circle representing $M_m = 1.26$. Note that there is essentially no change of the high-frequency portion of $\mathbf{G}(j\omega)$ compared with $\mathbf{G}_x(j\omega)$. However, in the low-frequency region there is a clockwise rotation of $\mathbf{G}(j\omega)$ due to the lag compensator.

Table 11.1 presents a comprehensive comparison between the root-locus method and the frequency-plot method of applying a lag compensator to a basic control system. Note the similarity of results obtained by these methods. The selection of the method depends upon individual preference. The compensators developed in the root-locus and frequency-plot methods are different, but achieve comparable results. To show this explicitly, Table 11.1 includes the case of the compensator developed in the root-locus method and applied to the frequency-response method. In the frequency-response method, either the log or the polar

TABLE 11.1
Comparison between root-locus and frequency methods of applying a cascade lag compensator to a basic control system

<table>
<tr><th>$G_c'(s)$</th><th>ω_{m1}</th><th>ω_m</th><th>ω_{n1}</th><th>ω_n</th><th>K_1</th><th>K_1'</th><th>M_m</th><th>M_p</th><th>t_p, s</th><th>t_s, s</th></tr>
<tr><td colspan="11">**Root-locus method, ζ of dominant roots = 0.45**</td></tr>
<tr><td>$\frac{1 + 20s}{1 + 200s}$</td><td>—</td><td>0.66</td><td>0.898</td><td>0.854</td><td>0.840</td><td>8.02</td><td>1.31</td><td>1.266</td><td>4.30</td><td>22.0</td></tr>
<tr><td colspan="11">**Frequency-response method, M_m = 1.26 (effective ζ = 0.45)**</td></tr>
<tr><td>$\frac{1 + 14s}{1 + 140s}$</td><td rowspan="2">0.715</td><td>0.53</td><td rowspan="2">—</td><td rowspan="2">—</td><td rowspan="2">0.872</td><td>6.68</td><td rowspan="2">1.260</td><td>1.257</td><td>4.90</td><td>22.8</td></tr>
<tr><td>$\frac{1 + 20s}{1 + 200s}$</td><td>0.60</td><td>7.38</td><td>1.246</td><td>4.55</td><td>23.6</td></tr>
</table>

plot may be used. An analytical technique for designing a lag or PI compensator, using phase margin and a desired value of K_m, is presented in Ref. 6.

11.5 LEAD COMPENSATOR

Figure 10.6*a* shows a lead compensator made up of passive elements which has the transfer function

$$\mathbf{G}_c'(j\omega) = \alpha \frac{1 + j\omega T}{1 + j\omega\alpha T} \qquad \alpha < 1 \tag{11.14a}$$

This equation is marked with a prime since it does not contain the gain A. The log magnitude and phase-angle equations for this compensator are

$$\text{Lm}\,\mathbf{G}_c'(j\omega) = \text{Lm}\,\alpha + \text{Lm}\,(1 + j\omega T) - \text{Lm}\,(1 + j\omega\alpha T) \tag{11.14b}$$

$$\angle \mathbf{G}_c'(j\omega) = \angle 1 + j\omega T - \angle 1 + j\omega\alpha T \tag{11.14c}$$

A family of curves for various values of α is shown in Fig. 11.10. It may be convenient to make templates of these log magnitude and phase-angle curves for the lead compensator. It is seen from the shape of the log magnitude curves that an attenuation equal to $\text{Lm}\,\alpha$ is introduced at frequencies below $\omega T = 1$. Thus the lead network is basically a high-pass filter: the high frequencies are passed, and the low frequencies are attenuated. Also, an appreciable lead angle is introduced in the frequency range from $\omega = 1/T$ to $\omega = 1/\alpha T$ (from $\omega T = 1$ to $\omega T = 1/\alpha$). Because of its angle characteristic a lead network can be used to increase the bandwidth for a system that falls in category 2.

Application of the lead compensator can be based on adjusting the phase margin and the phase-margin frequency. Assume that a system has the original forward transfer function $\mathbf{G}_x(j\omega)$ and that the gain has been adjusted for the desired phase margin or for the desired M_m. The log magnitude and phase-angle diagram of $\mathbf{G}_x(j\omega)$ is sketched in Fig. 11.11. The purpose of the lead compensator is to increase the phase-margin frequency and therefore to increase ω_m. The lead compensator introduces a positive angle over a relatively narrow bandwidth.

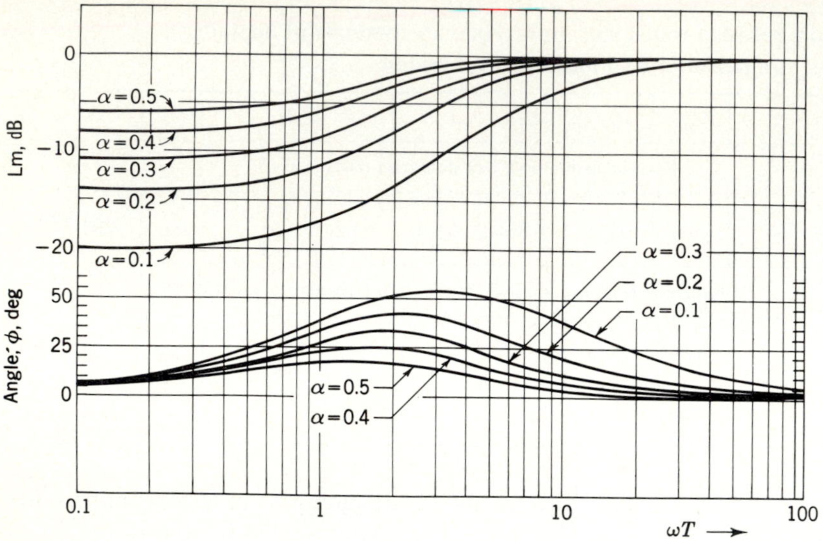

FIGURE 11.10
Log magnitude and phase-angle diagram for lead compensator

$$\mathbf{G}_c'(j\omega) = \alpha\frac{1 + j\omega T}{1 + j\omega\alpha T}$$

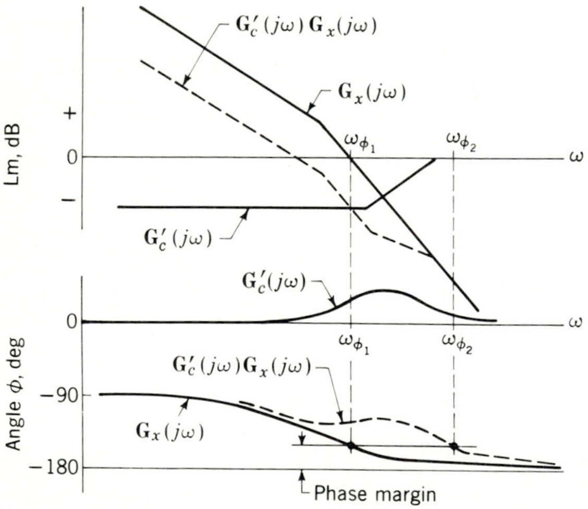

FIGURE 11.11
Original and lead-compensated log magnitude and phase-angle plots.

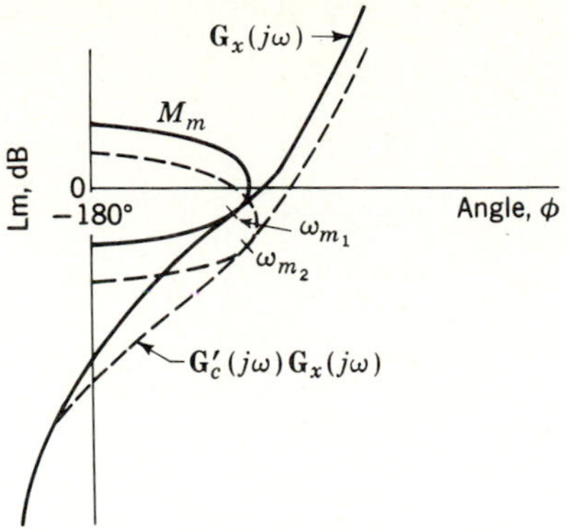

FIGURE 11.12
Original and lead-compensated log magnitude–angle curves

By properly selecting the value of T, the phase-margin frequency can be increased from $\omega_{\phi 1}$ to $\omega_{\phi 2}$. Selection of T can be accomplished by physically placing the angle curve for the compensator on the same graph with the angle curve of the original system. The location of the compensator angle curve must be such as to produce the specified phase margin at the highest possible frequency; this location determines the value of the time constant T. The gain of $\mathbf{G}_c'(j\omega)\mathbf{G}_x(j\omega)$ must be increased so that the log magnitude curve has the value 0 dB at the frequency $\omega_{\phi 2}$. For a given α, an analysis of the log magnitude curve of Fig. 11.10 shows that the farther to the right the compensator curves are placed, i.e., the smaller T is made, the larger the new gain of the system.

It is also possible to use the criterion derived in the previous chapter for the selection of T. This criterion is to select T, for Type 1 or higher systems, equal to or slightly smaller than the largest time constant of the original forward transfer function. This is the procedure used in the example of the next section. For a Type 0 system the compensator time constant T is made equal to or slightly smaller than the second largest time constant of the original system. Several locations of the compensator curves should be tested and the best results selected.

More accurate application of the lead compensator is based on adjusting the gain to obtain a desired M_m by use of the log magnitude–angle diagram. Figure 11.12 shows the original curve $\mathbf{G}_x(j\omega)$ and the new curve $\mathbf{G}_c'(j\omega)\mathbf{G}_x(j\omega)$. The increase in gain to obtain the same M_m is the amount that the curve $\mathbf{G}_c'(j\omega)\mathbf{G}_x(j\omega)$ must be raised to be tangent to the M_m curve. Because of the positive angle introduced by the lead compensator, the new resonant frequency ω_{m2} is larger than the original resonant frequency ω_{m1}. The increase in gain is not as large as that obtained by use of the lag compensator.

11.6 EXAMPLE: CASCADE LEAD COMPENSATION

For the unity-feedback Type 1 system of Sec. 10.7 the forward transfer function is

$$G_x(s) = \frac{K_1}{s(1+s)(1+0.2s)} \tag{11.15}$$

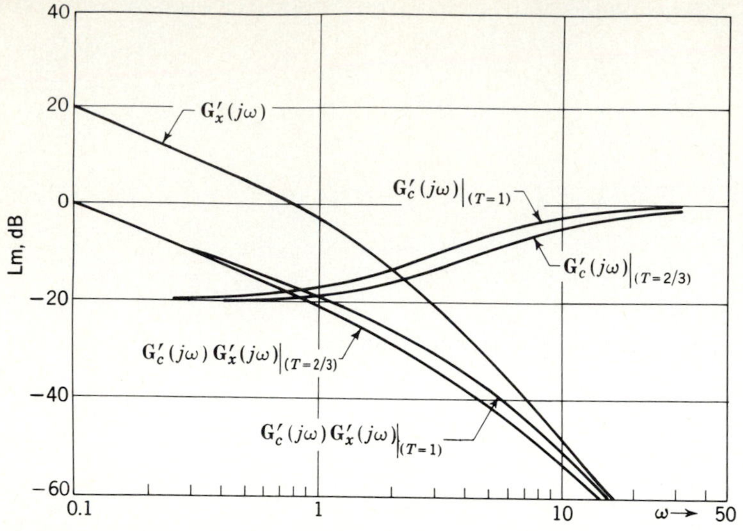

(a)

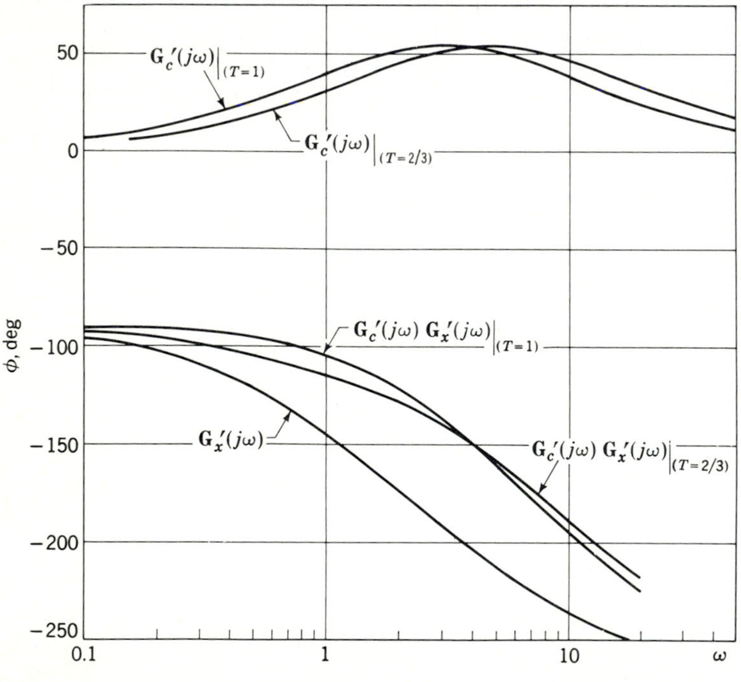

(b)

FIGURE 11.13
Log magnitude and phase-angle diagrams of

$$\mathbf{G}_x'(j\omega) = \frac{1}{j\omega(1 + j\omega)(1 + j0.2\omega)}$$

$$\mathbf{G}_c'(j\omega)\mathbf{G}_x'(j\omega) = \frac{0.1(1 + j\omega T)}{j\omega(1 + j\omega)(1 + j0.2\omega)(1 + j0.1\omega T)}$$

From Sec. 11.4 for $M_m = 1.26$ it is found that $K_1 = 0.872\ \text{s}^{-1}$ and $\omega_{m1} = 0.715$ rad/s. In this example the value of ω_m is considered to be too low. From previous considerations, putting a lead compensator in cascade with the basic system increases the value of ω_m and therefore improves the time response of the system. This situation falls in category 2.

The choice of values of α and T of the compensator must be such that the angle ϕ_c adds a sizeable positive phase shift at frequencies above ω_{m1} of the overall forward transfer function. The nominal value of $\alpha = 0.1$ is often used; thus only the determination of the value of T remains. Two values of T selected in Sec. 10.9 are used in this example so that a comparison can be made between the root-locus and frequency-response techniques. The results are used to establish the design criteria for lead compensators.

In Fig. 11.13a and 11.13b are drawn the log magnitude and phase-angle diagrams for the basic system $\mathbf{G}_x'(j\omega)$, the two compensators $\mathbf{G}_c'(j\omega)$, and the two composite transfer functions $\mathbf{G}_c(j\omega)\mathbf{G}_x'(j\omega)$. Using the log magnitude and phase-angle curves of $\mathbf{G}_c'(j\omega)\mathbf{G}_x'(j\omega)$ from Fig. 11.13, the log magnitude–angle diagrams are drawn in Fig. 11.14. The gain is again adjusted to obtain $M_m = 1.26$ (2 dB).

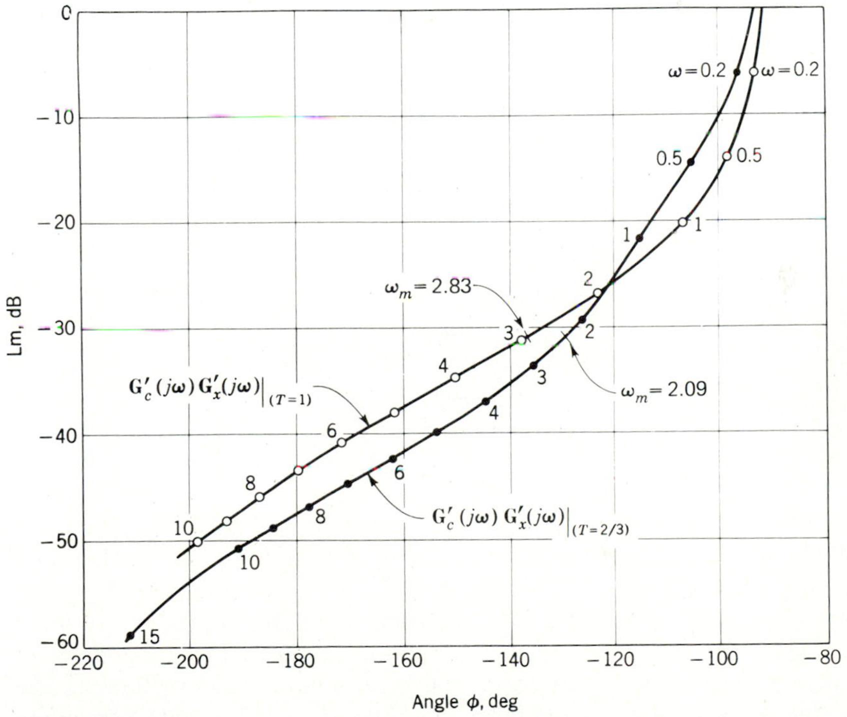

FIGURE 11.14
Log magnitude–angle diagrams with lead compensators from Fig. 11.13a and b.

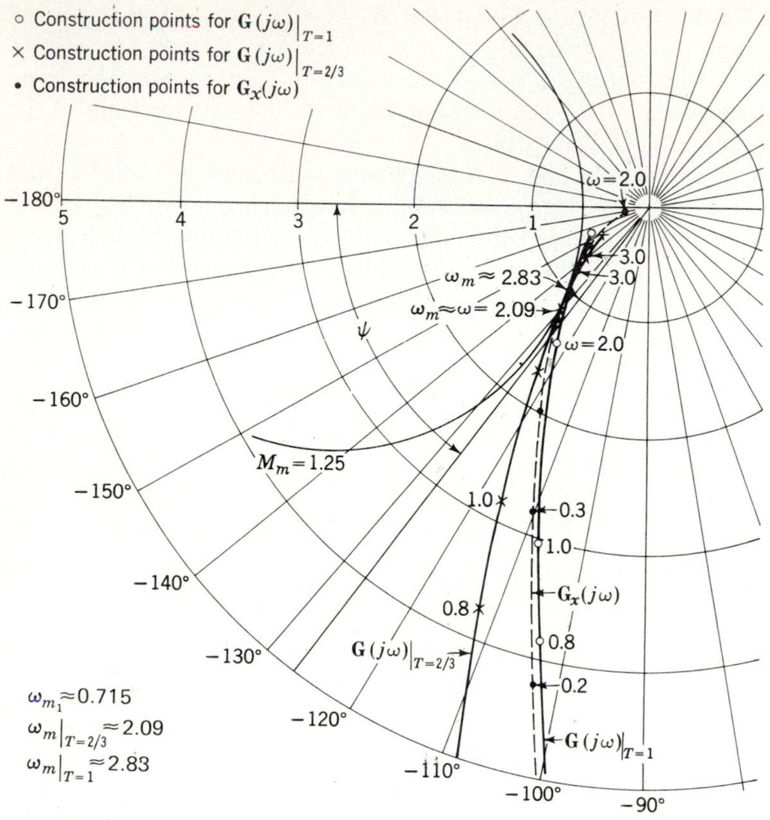

FIGURE 11.15
Polar plots of Type 1 control system with and without lead compensation:

$$\mathbf{G}_x(j\omega) = \frac{0.872}{j\omega(1 + j\omega)(1 + j0.2\omega)}$$

$$\mathbf{G}(j\omega)|_{T=2/3} = \frac{3.48(1 + j2\omega/3)}{j\omega(1 + j\omega)(1 + j0.2\omega)(1 + j0.2\omega/3)}$$

$$\mathbf{G}(j\omega)|_{T=1} = \frac{3.13}{j\omega(1 + j0.1\omega)(1 + j0.2\omega)}$$

The corresponding polar plots of the original system $\mathbf{G}_x(j\omega)$ and the compensated systems $\mathbf{G}(j\omega) = \mathbf{G}_c(j\omega)\mathbf{G}_x(j\omega)$ are shown in Fig. 11.15. Note that the value of ω_m is increased for both values of T.

Table 11.2 presents a comprehensive comparison between the root-locus method and the frequency-response method of applying a lead compensator to a basic control system. The results show that a lead compensator increases the resonant frequency ω_m. However, a range of values of ω_m is possible, depending on the compensator time constant used. The larger the value of ω_m, the smaller the value of K_1. The characteristics selected must be based on the system

TABLE 11.2
Comparison between root-locus and frequency-response methods of applying a lead compensator to a basic control system

$G_c'(s)$	ω_{m1}	ω_m	ω_{n1}	ω_n	K_1	K_1'	M_m	M_p	t_p, s	t_s, s
Root-locus method, ζ of dominant roots = 0.45										
$0.1\frac{1+s}{1+0.1s}$	—	2.66	0.89	3.52	0.84	2.95	1.215	1.196	1.09	2.45
$0.1\frac{1+0.667s}{1+0.0667s}$	—	2.67	0.89	3.40	0.84	4.40	1.38	1.283	1.05	2.50
Frequency-response method, M_m = 1.26 (effective ζ = 0.45)										
$0.1\frac{1+s}{1+0.1s}$	0.715	2.83	—	—	0.872	3.13	1.26	1.215	1.05	2.40
$0.1\frac{1+0.667s}{1+0.0667s}$	0.715	2.09	—	—	0.872	3.48	1.26	1.225	1.23	2.73

specifications. The increase in gain is not as large as that obtained with a lag compensator. Note in the table the similarity of results obtained by the root-locus and frequency-response methods. Because of this similarity, the design rules of Sec. 10.9 can be interpreted in terms of the frequency-response method as follows.

RULE 1. The value of the time constant T in the numerator of $\mathbf{G}_c(j\omega)$ given by Eq. (11.14a) is made equal to the value of the largest time constant in the denominator of the original transfer function $\mathbf{G}_x(j\omega)$ for a Type 1 or higher system. This usually results in the largest increase in ω_m (best time response), and there is an increase in system gain K_m.

RULE 2. The value of the time constant T in the numerator of $\mathbf{G}_c(j\omega)$ is made slightly smaller than the largest time constant in the denominator of the original transfer function for a Type 1 or higher system. This results in achieving the largest gain increase for the system with an appreciable increase in ω_m.

If the maximum improvement in the time response is desired with whatever gain increase is obtainable, Rule 1 is applied to the design of the lead compensator. Where maximum gain increase and a good improvement in the time of response are desired, Rule 2 is applicable.

For a Type 0 system, Rules 1 and 2 are modified so that the lead-compensator time constant is selected either equal to or slightly smaller than the second largest time constant of the original system.

Remember that the lag and lead networks are designed for the same basic system; thus from Tables 11.1 and 11.2 it is seen that both result in a gain increase. The distinction between the two types of compensation is that the lag network gives the largest increase in gain (with the best improvement in steady-state accuracy) at the expense of increasing the response time, whereas the lead network gives an appreciable improvement in the time of response and a small improvement in steady-state accuracy. The particular problem at hand dictates the type of compensation to be used. An analytical technique for designing a lead compensator is presented in Ref. 6, using phase margin as the basis of design.

11.7 LAG-LEAD COMPENSATOR

The previous sections demonstrate that the introduction of a lag compensator results in an increase in the gain, with a consequent reduction of the steady-state error in the system. Introducing a lead compensator results in an increase in the resonant frequency ω_m and a reduction in the system's settling time. If both the steady-state error and the settling time are to be reduced, a lag and a lead compensator must be utilized simultaneously. This improvement can be accomplished by inserting the individual lag and lead networks in cascade, but it is more economical in equipment to use a new network that has both the lag and the lead characteristics. Such a network, shown in Fig. 10.22*a*, is called a lag-lead compensator. Its transfer function is

$$\mathbf{G}_c(j\omega) = \frac{A(1 + j\omega T_1)(1 + j\omega T_2)}{(1 + j\omega\alpha T_1)(1 + j\omega T_2/\alpha)} \tag{11.16}$$

where $\alpha > 1$ and $T_1 > T_2$. The first half of this transfer function produces the lag effect, and the second half produces the lead effect. The log magnitude and phase-angle diagram for a representative lag-lead compensator is shown in Fig. 11.16, in which the selection is made $T_1 = 5T_2$. A simple design method is to make T_2 equal to a time constant in the denominator of the original system. An alternative procedure is to locate the compensator curves on the log plots of the original system to produce the highest possible phase-margin frequency. Figure 11.17 shows a sketch of the log magnitude and phase-angle diagram for the original forward transfer function $\mathbf{G}_x(j\omega)$, the compensator $\mathbf{G}_c'(j\omega)$, and the combination $\mathbf{G}_c(j\omega)\mathbf{G}_x'(j\omega)$. The phase-margin frequency is increased from $\omega_{\phi 1}$ to $\omega_{\phi 2}$, and the additional gain is labeled Lm A. The new value of M_m can be obtained from the log magnitude–angle diagram in the usual manner.

Another method is simply the combination of the design procedures given for the use of the lag and lead networks, respectively. Basically, the log magnitude and phase-angle curves of the lag compensator can be located as described in Sec. 11.3. The lag-compensator curves are given by Fig. 11.4 and are located so that the negative angle introduced at either the original phase-margin frequency $\omega_{\phi 1}$ or the original resonant frequency ω_{m1} is small—of the order of $-5°$. This permits determination of the time constant T_1. The log magnitude and phase-angle curves of the lead compensator can be located as described in Sec. 11.5. The lead-compensator curves are given by Fig. 11.10 except that the log magnitude curve should be raised to have a value of 0 dB at the low frequencies. This is

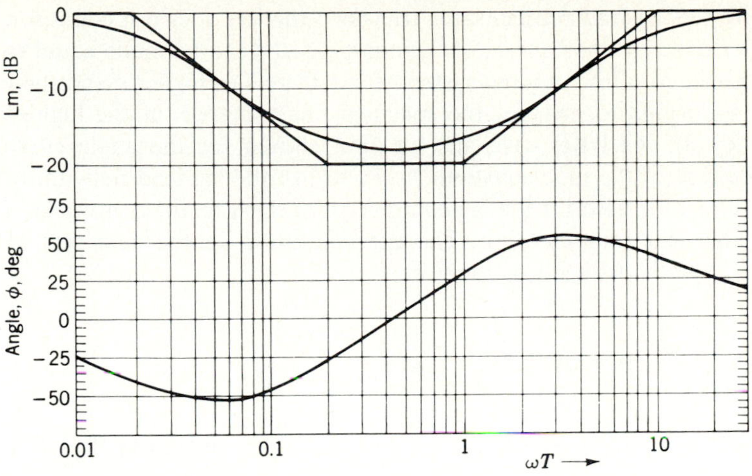

FIGURE 11.16
Log magnitude and phase-angle diagram for lag-lead compensator

$$\mathbf{G}'_c(j\omega) = \frac{(1 + j5\omega T_2)(1 + j\omega T_2)}{(1 + j50\omega T_2)(1 + j0.1\omega T_2)}$$

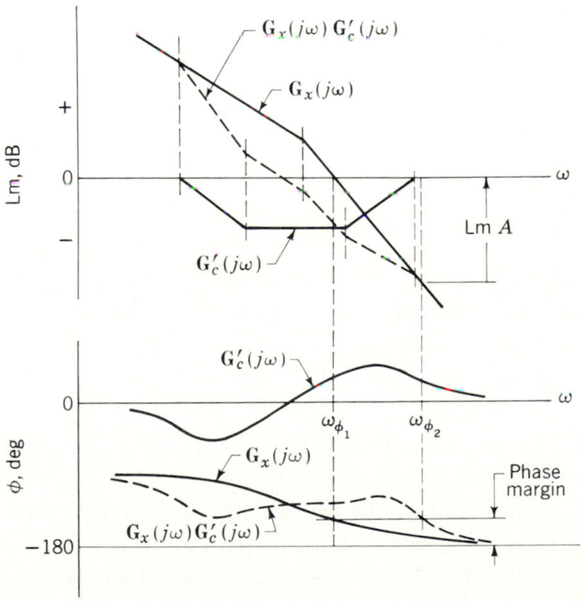

FIGURE 11.17
Original and lag-lead compensated log magnitude and phase angle plots.

necessary because the lag-lead compensator transfer function does not contain α as a factor. Either the time constant T_2 of the compensator can be made equal to the largest time constant of the original system (for a Type 1 or higher system), or the angle curve is located to produce the specified phase margin at the highest possible frequency. In the latter case, the location chosen for the angle curve permits determination of the time constant T_2. To utilize the lag-lead network of Eq. (11.16) the value of α used for the lag compensation must be the reciprocal of the α used for the lead compensation. The example given in the next section utilizes this second method since it is more flexible.

11.8 EXAMPLE: CASCADE LAG-LEAD COMPENSATION

For the basic system treated in Secs. 11.4 and 11.6, the forward transfer function is

$$\mathbf{G}_x(j\omega) = \frac{K_1}{j\omega(1 + j\omega)(1 + j0.2\omega)} \tag{11.17}$$

For an $M_m = 1.26$ the result is $K_1 = 0.872$ and $\omega_{m1} = 0.715$ rad/s. According to the second method described above, the lag-lead compensator is the combination of the individual lag and lead compensators designed in Secs. 11.4 and 11.6. Thus the compensator transfer function is

$$\mathbf{G}_c(j\omega) = A\frac{(1 + jT_1\omega)(1 + j\omega T_2)}{(1 + j\alpha T_1\omega)(1 + j0.1\omega T_2)} \qquad \text{for } \alpha = 10 \tag{11.18}$$

The new forward transfer function with $T_1 = 14$ and $T_2 = 1$ is

$$\mathbf{G}(j\omega) = \mathbf{G}_c(j\omega)\mathbf{G}_x(j\omega) = \frac{AK_1(1 + j14\omega)}{j\omega(1 + j0.2\omega)(1 + j140\omega)(1 + j0.1\omega)} \tag{11.19}$$

In Fig. 11.18*a* and 11.18*b* are drawn the log magnitude and phase-angle diagrams of the basic system $\mathbf{G}_x'(j\omega)$, the compensator $\mathbf{G}_c'(j\omega)$, and the composite transfer function $\mathbf{G}_c'(j\omega)\mathbf{G}_x'(j\omega)$. From these curves, the log magnitude–angle diagram $\mathbf{G}_a'(j\omega)$ is drawn in Fig. 11.19. Upon adjusting for an $M_m = 1.26$ the results are $K_1' = 29.93$ and $\omega_{ma} = 2.69$. When these values are compared with those of Tables 11.1 and 11.2, it is seen that lag-lead compensation results in a value of K_1 approximately equal to the product of the K_1's of the lag- and lead-compensated systems. Also, the ω_m is only slightly less than that obtained with the lead-compensated system. Thus a larger increase in gain is achieved than with lag compensation, and the value of ω_m is almost as large as the value obtained by using the lead-compensated system. One may therefore be inclined to use a lag-lead compensator exclusively. $\mathbf{G}_b$ in Fig. 11.19 is with $T_1 = 20$.

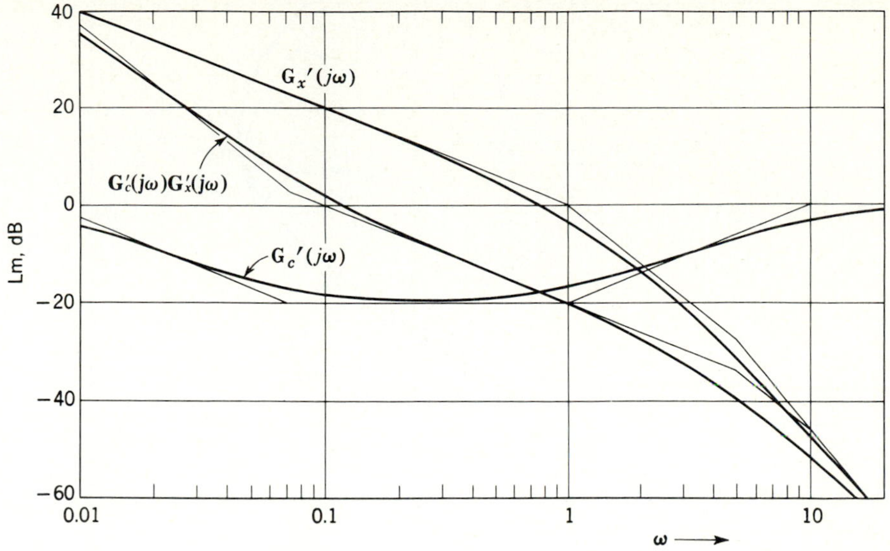

(a)

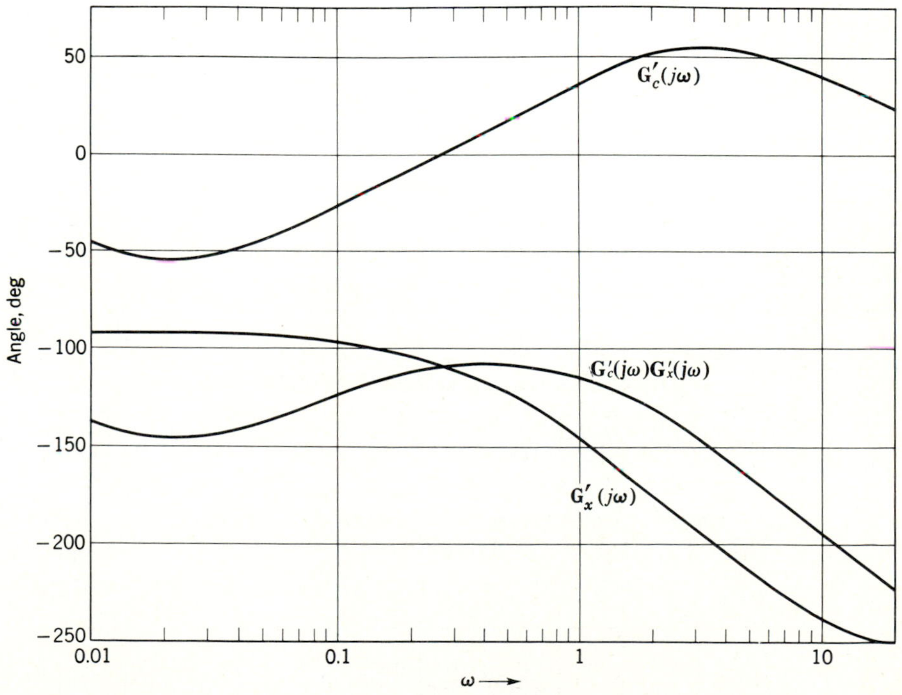

(b)

FIGURE 11.18
Log magnitude and phase-angle diagrams of Eqs. (11.17) to (11.19) with $T_1 = 14$.

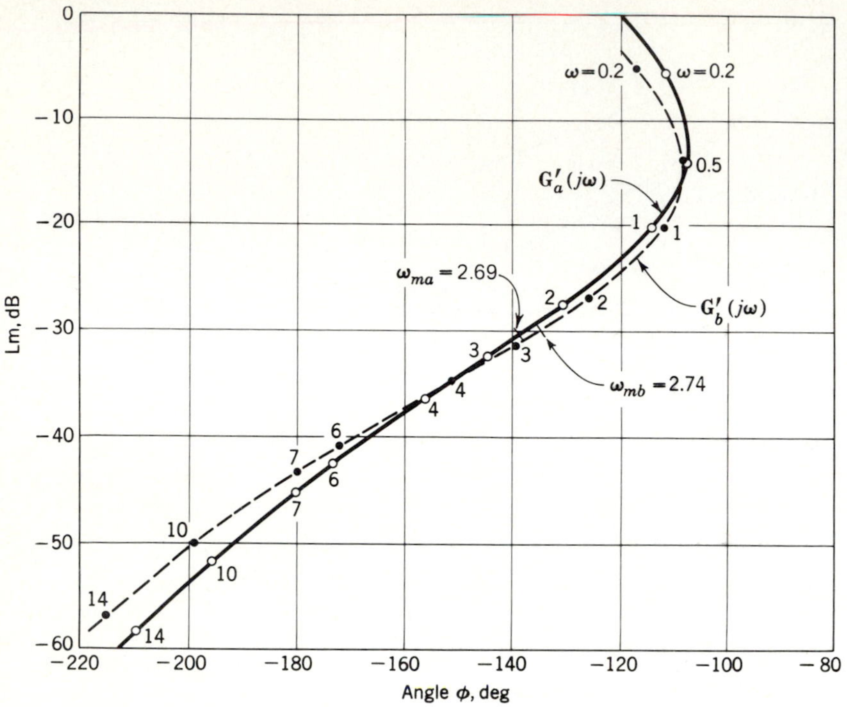

FIGURE 11.19
Log magnitude–angle diagram of $\mathbf{G}'(j\omega)$ from Eq. (11.19)—$\mathbf{G}'_a$ with $T_1 = 14$ and $\mathbf{G}'_b$ with $T_1 = 20$.

TABLE 11.3
Comparison between root-locus and frequency-response methods of applying a lag-lead compensator to a basic control system

$G'_c(s)$	ω_{m1}	ω_m	ω_{n1}	ω_n	K_1	K'_1	M_m	M_p	t_p, s	t_s, s
Root-locus method, ζ of dominant roots = 0.45										
$\dfrac{(1+s)(1+20s)}{(1+0.1s)(1+200s)}$	—	2.63	0.898	3.49	0.84	29.26	1.23	1.213	1.10	3.33
Frequency-response method, M_m = 1.26 (effective ζ = 0.45)										
$\dfrac{(1+s)(1+20s)}{(1+0.1s)(1+200s)}$	0.715	2.74	—	—	0.872	30.36	1.26	1.225	1.08	3.30
$\dfrac{(1+s)(1+14s)}{(1+0.1s)(1+140s)}$	0.715	2.69	—	—	0.872	29.93	1.26	1.229	1.09	3.55

The resulting error coefficient (gain) due to the insertion of a lag-lead compensator is approximately equal to the product of the following:

1. The original error coefficient K_m of the uncompensated system
2. The increase in K_m due to the insertion of a lag compensator
3. The increase in K_m due to the insertion of a lead compensator

Table 11.3 presents a comparison of the root-locus and log-plot methods of applying a lag-lead compensator to a basic control system. The compensator designed from the log plots is different from the one designed by the root-locus method. To show that the difference in performance is small, Table 11.3 includes the results obtained with both compensators when applying the frequency-response method. Both plots are shown in Fig. 11.19. The plot using Eq. (10.42) with $T_1 = 20$ yields $\omega_{mb} = 2.74$, as shown in this figure.

11.9 FEEDBACK COMPENSATION USING LOG PLOTS[1]

The previous sections demonstrate the procedures for applying cascade compensation. Those methods can be applied in a straightforward manner. When feedback compensation is applied to improve the tracking qualities of a control system, additional parameters must be determined. Besides the gain constant of the basic forward transfer function, the gain constant and frequency characteristics of the minor feedback must also be selected. Therefore, additional steps must be introduced into the design procedure. The general effects of feedback compensation are demonstrated by application to the specific system which is used throughout this text. The use of a minor loop for feedback compensation is shown in Fig. 11.20. The transfer function $\mathbf{G}_x(j\omega)$ represents the basic system, $\mathbf{H}(j\omega)$ is the feedback compensator forming a minor loop around $\mathbf{G}_x(j\omega)$, and A is an amplifier which is used to adjust the overall performance. The forward transfer function of this system is

$$\mathbf{G}_1(j\omega) = \frac{\mathbf{C}(j\omega)}{\mathbf{I}(j\omega)} = \frac{\mathbf{G}_x(j\omega)}{1 + \mathbf{G}_x(j\omega)\mathbf{H}(j\omega)} \tag{11.20}$$

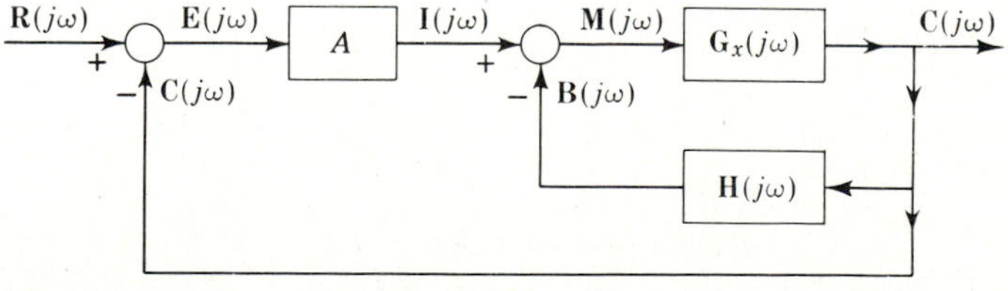

FIGURE 11.20
Block diagram for feedback compensation.

In order to apply feedback compensation new techniques must be developed. This is done by first using some approximations with the straight-line log magnitude curves and then developing an exact procedure. Consider the cases when

$$|\mathbf{G}_x(j\omega)\mathbf{H}(j\omega)| \ll 1 \qquad \text{and} \qquad |\mathbf{G}_x(j\omega)\mathbf{H}(j\omega)| \gg 1$$

The forward transfer function of Eq. (11.20) can be approximated by

$$\mathbf{G}_1(j\omega) \approx \mathbf{G}_x(j\omega) \qquad \text{for } |\mathbf{G}_x(j\omega)\mathbf{H}(j\omega)| \ll 1 \tag{11.21}$$

and

$$\mathbf{G}_1(j\omega) \approx \frac{1}{\mathbf{H}(j\omega)} \qquad \text{for } |\mathbf{G}_x(j\omega)\mathbf{H}(j\omega)| \gg 1 \tag{11.22}$$

Still undefined is the condition when $|\mathbf{G}_x(j\omega)\mathbf{H}(j\omega)| \approx 1$, in which case neither Eq. (11.21) nor Eq. (11.22) is applicable. In the approximate procedure this condition is neglected, and Eqs. (11.21) and (11.22) are used when $|\mathbf{G}_x(j\omega)\mathbf{H}(j\omega)| < 1$ and $|\mathbf{G}_x(j\omega)\mathbf{H}(j\omega)| > 1$, respectively. This approximation allows investigation of the qualitative results to be obtained. After these results are found to be satisfactory, the refinements for an exact solution are introduced.

An example illustrates the use of these approximations. Assume that $\mathbf{G}_x(j\omega)$ represents a motor having inertia and damping. The transfer function derived in Sec. 2.14 can be represented by

$$\mathbf{G}_x(j\omega) = \frac{K_M}{j\omega(1 + j\omega T_m)} \tag{11.23}$$

Let the feedback $\mathbf{H}(j\omega) = 1\angle 0°$. This problem is sufficiently simple to be solved exactly algebraically. However, use is made of the log magnitude curve and the approximate conditions. In Fig. 11.21 is sketched the log magnitude curve for $\mathbf{G}_x(j\omega)\mathbf{H}(j\omega)$.

From Fig. 11.21 it is seen that $|\mathbf{G}_x(j\omega)\mathbf{H}(j\omega)| > 1$ for all frequencies below ω_1. With the approximation of Eq. (11.22), $\mathbf{G}_1(j\omega)$ can be represented by $1/\mathbf{H}(j\omega)$ for frequencies below ω_1. Also, $|\mathbf{G}_x(j\omega)\mathbf{H}(j\omega)| < 1$ for all frequencies above ω_1. Therefore $\mathbf{G}_1(j\omega)$ can be represented, as shown, by the line of zero slope and 0 dB for frequencies up to ω_1 and the line of -12-dB slope above ω_1. The equation of $\mathbf{G}_1(j\omega)$ therefore has a quadratic in the denominator with $\omega_n = \omega_1$:

$$\mathbf{G}_1(j\omega) = \frac{1}{1 + 2\zeta j\omega/\omega_1 + (j\omega/\omega_1)^2} \tag{11.24}$$

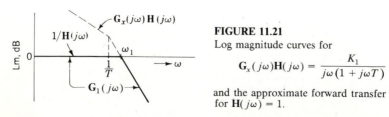

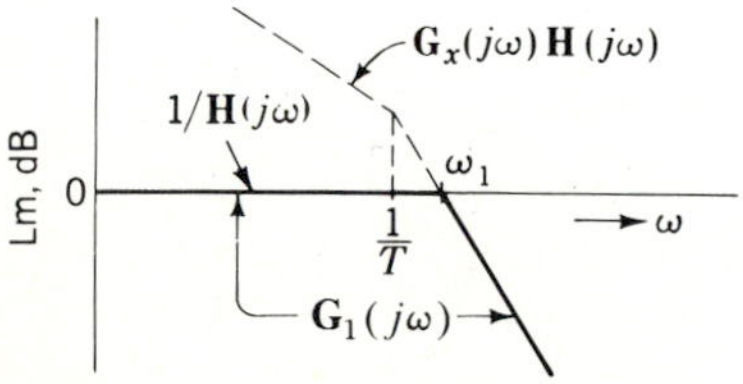

FIGURE 11.21
Log magnitude curves for

$$\mathbf{G}_x(j\omega)\mathbf{H}(j\omega) = \frac{K_1}{j\omega(1 + j\omega T)}$$

and the approximate forward transfer function $\mathbf{G}_1(j\omega)$ for $\mathbf{H}(j\omega) = 1$.

Of course $\mathbf{G}(j\omega)$ can be obtained algebraically for this simple case from Eq. (11.20), with the result given as

$$\mathbf{G}_1(j\omega) = \frac{1}{1 + (1/K_M)j\omega + (j\omega)^2 T/K_M} \tag{11.25}$$

where $\omega_1 = \omega_n = (K_M/T)^{1/2}$ and $\zeta = 1/[2(K_M T)^{1/2}]$. Note that the approximate result is basically correct but some detailed information, in this case the value of ζ, is missing. The approximate angle curve can be drawn to correspond to the approximate log magnitude curve of $\mathbf{G}_1(j\omega)$ or to Eq. (11.24).

11.10 EXAMPLE: FEEDBACK COMPENSATION (LOG PLOTS)

The system of Fig. 11.20 is investigated with the value of $\mathbf{G}_x(j\omega)$ given by

$$\mathbf{G}_x(j\omega) = \frac{K_x}{j\omega(1 + j\omega)(1 + j0.2\omega)} \tag{11.26}$$

The system having this transfer function has been used throughout this text for the various methods of compensation. The log magnitude and phase-angle diagram using the straight-line log magnitude curve is drawn in Fig. 11.22. A phase margin of 45° can be obtained and the phase-margin frequency is $\omega_{\phi 1} = 0.8$, provided the gain is changed by -2 dB.

The object in applying compensation is to increase both the phase-margin frequency and the gain. In order to select a feedback compensator $\mathbf{H}(j\omega)$,

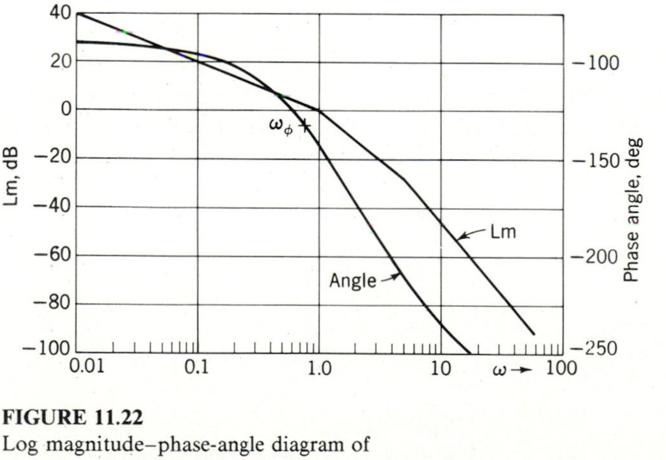

FIGURE 11.22
Log magnitude–phase-angle diagram of

$$\mathbf{G}'_x(j\omega) = \frac{1}{j\omega(1 + j\omega)(1 + j0.2\omega)}$$

consider the following facts:

1. The system type should be maintained. In accordance with the conditions outlined in Sec. 10.14 and the results given in Sec. 10.19 for the root-locus method, it is known that good improvement is achieved by using a feedback compensator $\mathbf{H}(j\omega)$ which has a zero, $s = 0$, of order equal to (or preferably higher than) the type of the original transfer function $\mathbf{G}_x(j\omega)$.
2. It is shown in Sec. 11.9 that the new forward transfer function can be approximated by

$$\mathbf{G}_1(j\omega) \approx \begin{cases} \mathbf{G}_x(j\omega) & \text{for } |\mathbf{G}_x(j\omega)\mathbf{H}(j\omega)| < 1 \\ \dfrac{1}{\mathbf{H}(j\omega)} & \text{for } |\mathbf{G}_x(j\omega)\mathbf{H}(j\omega)| > 1 \end{cases}$$

Now apply this information to a consideration of the proper selection of $\mathbf{H}(j\omega)$.

Consider the possibility of replacing a portion of the curves of $\mathbf{G}_x(j\omega)$ for a selected range of frequencies by the curves of $1/\mathbf{H}(j\omega)$ with the intent of increasing the values of ω_ϕ and the gain. This requires that the value of $\mathbf{H}(j\omega)$ be such that $|\mathbf{G}_x(j\omega)\mathbf{H}(j\omega)| > 1$ for that range of frequencies. A feedback unit using a tachometer and an *RC* derivative network is considered. The circuit is shown in Fig. 10.36, and the transfer function is

$$\mathbf{H}(j\omega) = \frac{(K_t/T)(j\omega T)^2}{1 + j\omega T} = K_h \mathbf{H}'(j\omega T) \tag{11.27}$$

where $K_h = K_t/T$. The log magnitude and phase-angle diagram for $1/\mathbf{H}'(j\omega)$ is plotted in Fig. 11.23 with a nondimensionalized frequency scale, ωT. When this scale is converted to a dimensionalized scale ω, the curves of Fig. 11.23 shift to the left or to the right according as T is increased or decreased. The angle curve $1/\mathbf{H}(j\omega)$ shows that the desired phase margin can be obtained at a frequency which is a function of the compensator time constant T. This shows promise for increasing the phase margin, provided that the magnitude $|\mathbf{G}_x(j\omega)\mathbf{H}(j\omega)|$ can be

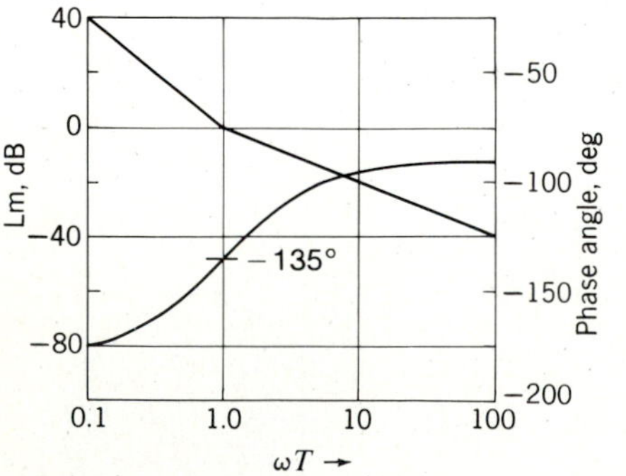

FIGURE 11.23
Log magnitude and phase-angle diagram of $\dfrac{1}{\mathbf{H}'(j\omega T)} = \dfrac{1 + j\omega T}{(j\omega T)^2}$

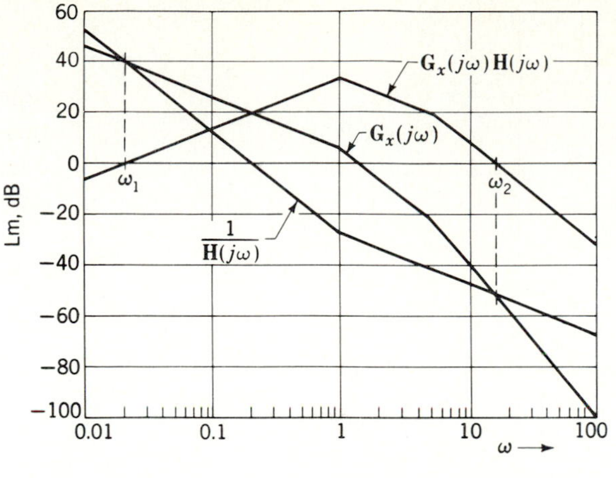

FIGURE 11.24
Log magnitude plots of $1/\mathbf{H}(j\omega)$, $\mathbf{G}_x(j\omega)$, and $\mathbf{G}_x(j\omega)\mathbf{H}(j\omega)$.

made greater than unity over the correct range of frequencies. Also the larger K_t/T is made, the smaller the magnitude of $1/\mathbf{H}(j\omega)$ at the phase-margin frequency. This permits a large value of A and therefore a large error coefficient.

The selection of T and K_t is based on a trial-and-error procedure. The object is to produce a section of the $\mathbf{G}_1(j\omega)$ log magnitude curve with a slope of -20 dB/decade and with a smaller magnitude than (i.e., it is below) the log magnitude curve of $\mathbf{G}_x(j\omega)$. This new section of the $\mathbf{G}_1(j\omega)$ curve must occur at a higher frequency than the original phase-margin frequency. Section 8.19 describes the desirability of a slope of -20 dB/decade to produce a large phase margin. To achieve compensation, the log magnitude curve $1/\mathbf{H}(j\omega)$ is placed over the $\mathbf{G}_x(j\omega)$ curve so that there are one or two points of intersection. There is no reason to restrict the value of K_x, but a practical procedure may be to use the value of K_x which produces the desired M_m without compensation. One such arrangement is shown in Fig. 11.24, with $K_x = 2$, $T = 1$, and $K_t = 25$. Intersections between the two curves occur at $\omega_1 = 0.021$ and $\omega_2 = 16$. The log magnitude plot of $\mathbf{G}_x(j\omega)\mathbf{H}(j\omega)$ is also drawn in Fig. 11.24 and shows that

$$|\mathbf{G}_x(j\omega)\mathbf{H}(j\omega)| \begin{cases} < 1 & \text{for } \omega_2 < \omega < \omega_1 \\ > 1 & \text{for } \omega_2 > \omega > \omega_1 \end{cases}$$

Therefore $\mathbf{G}_1(j\omega)$ can be represented approximately by

$$\mathbf{G}_1(j\omega) \approx \begin{cases} \mathbf{G}_x(j\omega) & \text{for } \omega_2 < \omega < \omega_1 \\ \dfrac{1}{\mathbf{H}(j\omega)} & \text{for } \omega_2 > \omega > \omega_1 \end{cases}$$

The composite curve has corner frequencies at $1/\omega_1$, $1/T$, and $1/\omega_2$ and can therefore be approximately represented, when $T = 1$, by

$$\mathbf{G}_1(j\omega) = \frac{2(1 + j\omega/T)}{j\omega(1 + j\omega/\omega_1)(1 + j\omega/\omega_2)^2} = \frac{2(1 + j\omega)}{j\omega(1 + j47.5\omega)(1 + j0.0625\omega)^2} \tag{11.28}$$

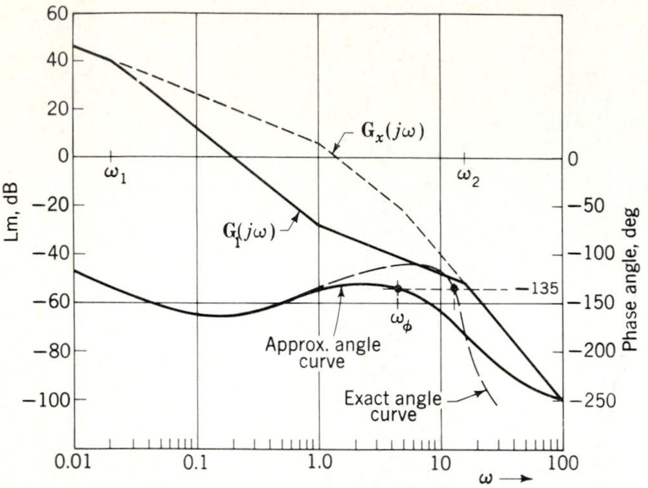

FIGURE 11.25
Log magnitude and phase-angle diagrams of

$$\mathbf{G}_1(j\omega) = \frac{2(1 + j\omega)}{j\omega(1 + j47.5\omega)(1 + j0.0625\omega)^2}$$

The log magnitude and phase-angle diagram for Eq. (11.28) is shown in Fig. 11.25. The new value of phase-margin frequency is $\omega_\phi = 4.5$, and the gain is Lm $A = 41$ dB. This shows a considerable improvement of the system performance.

The angle curve of $1/\mathbf{H}'(j\omega T)$ in Fig. 11.23 shows that a desired phase margin γ and phase-margin frequency ω_ϕ are a function of the value of T. For a given value of γ the angle $\underline{/1/\mathbf{H}'(j\omega T)} = -180° + \gamma$ is evaluated and the corresponding value of ωT is determined from Fig. 11.23 or from $1/\mathbf{H}'(j\omega T)$. Then the required value of T is

$$T = \frac{\omega T}{\omega_\phi} \tag{11.29}$$

Selecting $T = 1$, a phase margin $\gamma = 45°$ can occur at $\omega = 1$. However, the new approximate forward transfer function $\mathbf{G}_1(j\omega)$ has a second-order pole at $\omega = 1/\omega_2$ as shown in Eq. (11.28). Therefore, there is another higher frequency at which the phase margin of 45° can be achieved, as shown in Fig. 11.25. The higher frequency $\omega_\phi = 4.5$ is chosen in this example.

An exact curve of $\mathbf{G}_1(j\omega)$ can be obtained analytically. The functions $\mathbf{G}_x(j\omega)$ and $\mathbf{H}(j\omega)$ are

$$G_x(s) = \frac{2}{s(1 + s)(1 + 0.2s)} \qquad H(s) = \frac{25s^2}{1 + s}$$

The function $\mathbf{G}_1(j\omega)$ is obtained as follows:

$$\begin{aligned}\mathbf{G}_1(j\omega) &= \frac{\mathbf{G}_x(j\omega)}{1+\mathbf{G}_x(j\omega)\mathbf{H}(j\omega)}\\ &= \frac{2(1+j\omega)}{j\omega(1+j\omega)^2(1+j0.2\omega)+50(j\omega)^2}\\ &= \frac{2(1+j\omega)}{j\omega\left[0.2(j\omega)^3+1.4(j\omega)^2+52.2(j\omega)+1\right]}\\ &= \frac{2(1+j\omega)}{j\omega(1+j52\omega)\left[0.00385(j\omega)^2+0.027(j\omega)+1\right]}\end{aligned} \tag{11.30}$$

A comparison of Eqs. (11.28) and (11.30) shows that there is little difference between the approximate and the exact equations. The main difference is that the approximate equation has the term $(1 + j0.0625\omega)^2$, which assumes a damping ratio $\zeta = 1$ for this quadratic factor. The actual damping ratio is $\zeta = 0.217$ in the correct quadratic factor $[0.00385(j\omega)^2 + 0.027(j\omega) + 1]$. The corner frequency is the same for both cases, $\omega_n = 16$. As a result, the exact angle curve, given in Fig. 11.25, shows a higher phase-margin frequency ($\omega = 13$) than that obtained for the approximate curve ($\omega = 4.5$). If the basis of design is the value of M_m, the exact curve should be used. The data for the exact curve are readily available from a computer (see App. B).

The adjustment of the amplifier gain A is based on either the phase margin or M_m. An approximate value for A is the gain necessary to raise the curve of $\text{Lm}\,\mathbf{G}_1(j\omega)$ so that it crosses the 0-dB line at ω_ϕ. A more precise value for A is obtained by plotting $\mathbf{G}_1(j\omega)$ on the log magnitude–angle diagram and adjusting the system gain for a desired value of M_m. The exact $\text{Lm}\,\mathbf{G}_1(j\omega)$ curve should be used for best results.

The shape of the Nichols plot for $\mathbf{G}_1(j\omega)$ is shown in Fig. 11.26. When adjusting the gain for a desired M_m the objectives are to achieve a large resonant frequency ω_m and a large gain constant K_1. The tangency of the M_m curve at $\omega_c = 15.3$ requires a gain $A = 95.5$ and yields $K_1 = 191$. However, the rapid change in phase angle in the vicinity of ω_c can produce unexpected results. The M vs. ω characteristic curve c, shown in Fig. 11.27, has two peaks. This is not the form desired and must be rejected. This example serves to call attention to the fact that one point, the peak M value, is not sufficient to determine the suitability of the system response.

Tangency at $\omega_b = 1.2$ in Fig. 11.26 requires $A = 57.5$ and yields $K_1 = 115$. Curve b in Fig. 11.27 has the form desired and is acceptable. Compared with the original system, the minor-loop feedback compensation has produced a modest increase in ω_m from 0.715 to 1.2. There is a very large increase in K_1 from 0.872 to 115.

Tangency at ω_a is not considered, as it would represent a degradation of the uncompensated performance.

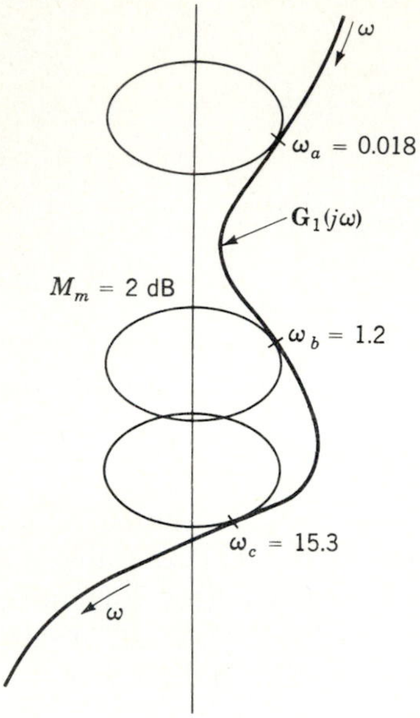

FIGURE 11.26
Nichols plot of $\mathbf{G}_1(j\omega)$.

This example demonstrates qualitatively that feedback compensation can be used to produce a section of the log magnitude curve of the forward transfer function $\mathbf{G}(j\omega)$ with a slope of -20 dB/decade. This section of the curve can be placed in the desired frequency range, with the result that the new phase-margin frequency is larger than that of the original system. However, to obtain the correct quantitative results the exact curves should be used instead of the straight-line approximations. Also, the use of the log magnitude–angle diagram as the last step in the design process, to adjust the gain for a specified M_m, gives a better indication of performance than the phase margin does.

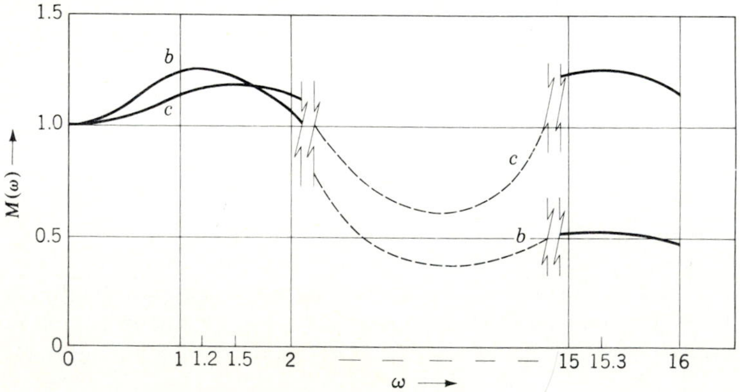

FIGURE 11.27
Closed-loop frequency response.

The reader may wish to try other values of K_t and T in the feedback compensator in order to modify the system performance. Also, it is worth considering whether the RC network should be used at all as part of the feedback compensator. The results using the root-locus method (see Secs. 10.16 to 10.19) show that rate feedback alone does improve system performance. The use of second-derivative feedback produces only limited improvements.

11.11 APPLICATION GUIDELINES: BASIC MINOR-LOOP FEEDBACK COMPENSATORS

Table 11.4 presents three basic feedback compensators and the corresponding angle characteristics of $1/\mathbf{H}(j\omega)$. One approach in designing these compensators is to determine the parameters of $H(s)$ which yield the desired value of the phase margin γ at a desired value of ω_ϕ.

For compensators 1 and 2 in Table 11.4 the desired values of γ ($< 90°$) and ω_ϕ should occur in the vicinity of the intersection of the $\mathrm{Lm}\,1/\mathbf{H}(j\omega)$ and the $\mathrm{Lm}\,\mathbf{G}_x(j\omega)$ plots, assuming that the angle of $\mathbf{G}_x(j\omega)$ varies from its initial value to $-90°i$ ($i = 2, 3, \ldots$) as the frequency increases from zero to infinity. Thus, to achieve an acceptable design of these compensators, an initial choice is made for the value of K_t or K_t and T. It may be necessary to lower the plot of $\mathrm{Lm}\,(1/\mathbf{H}(j\omega))$ to insure that the intersection value of ω is in the proper range, such that the desired value of ω_ϕ is attainable. The intersection of $\mathrm{Lm}\,[1/\mathbf{H}(j\omega)]$ with $\mathrm{Lm}\,\mathbf{G}_x(j\omega)$ must occur on that part of $\mathrm{Lm}\,\mathbf{G}(j\omega)$ that has a slope that will allow a $\gamma < 90°$ to be achievable in the vicinity of the intersection. By trial and error the parameters of $1/H(s)$ are varied to achieve the minimum acceptable value of ω_ϕ. For practical control systems there is an upper limit to an acceptable value of ω_ϕ, based upon the desire to attenuate high-frequency system noise.

For compensator 3, based upon Table 11.4 and Sec. 11.10, there are usually two intersections of $\mathrm{Lm}\,\mathbf{G}_x(j\omega)$ and $\mathrm{Lm}\,(1/\mathbf{H}(j\omega))$. The desired values of γ ($< 90°$) and ω_ϕ for $\mathrm{Lm}\,A\mathbf{G}_1$ of Fig. 11.20 can be made to occur on the

TABLE 11.4
Basic feedback compensators

$H(s)$	$1/H(s)$	Characteristic of $\angle 1/\mathbf{H}(j\omega)$ for $0 \le \omega \le \infty$
1. $K_t s$	$\dfrac{1}{K_t s}$	$-90°$
2. $\dfrac{K_t s}{Ts+1}$	$\dfrac{Ts+1}{K_t s}$	$-90°$ to $0°$
3. $\dfrac{K_t s^2}{Ts+1}$	$\dfrac{Ts+1}{K_t s^2}$	$-180°$ to $-90°$

-6-dB/octave portion of the plot of Lm $1/\mathbf{H}(j\omega)$. This requires that the intersections of the plots of Lm $\mathbf{G}_x(j\omega)$ and Lm $(1/\mathbf{H}(j\omega))$ occur at least one decade away from the value of ω that yields the desired values of ω_ϕ and γ on the -6-dB/octave portion of the Lm $(1/\mathbf{H})$ plot. With this separation, then the forward transfer function $\mathbf{G}_1(j\omega_\phi) \approx 1/\mathbf{H}(j\omega_\phi)$.

The guidelines presented in this section for the three compensators of Table 11.4 are intended to provide some fundamental insight to the reader in developing effective feedback compensator design procedures.

11.12 SUMMARY

A basic unity-feedback control system has been used throughout this and other chapters. It is therefore possible to compare the results obtained by adding each of the compensators to the system. The results obtained are consolidated in Table 11.5. The results achieved by using the frequency response are comparable with

TABLE 11.5
Summary of cascade and feedback compensation of a basic control system using the frequency-response method with $M_m = 1.26$ (effective $\zeta = 0.45$)

Compensator	ω_m	K_1	Additional gain required	M_p	t_p, s	t_s, s
Uncompensated	0.715	0.872	—	1.203	4.11	9.9
Lag: $\frac{1+14s}{1+140s}$	0.53	6.68	7.66	1.257	4.90	22.8
$\frac{1+20s}{1+200s}$	0.60	7.38	8.46	1.246	4.55	23.6
Lead: $0.1\frac{1+s}{1+0.1s}$	2.83	3.13	35.9	1.215	1.05	2.40
$0.1\frac{1+2s/3}{1+0.2s/3}$	2.09	3.48	39.9	1.225	1.23	2.73
Lag-lead: $\frac{(1+s)(1+14s)}{(1+0.1s)(1+140s)}$	2.69	29.93	34.3	1.229	1.088	3.55
$\frac{(1+s)(1+20s)}{(1+0.1s)(1+200s)}$	2.74	30.36	35.0	1.225	1.077	3.30
Feedback: $H(s) = \frac{25s^2}{1+s}$	1.20	115	57.5	1.203	1.49	3.27

those obtained by using the root locus (see Chap. 10). The design methods which have been demonstrated use some trial-and-error techniques. This design process can be expedited by use of appropriate CAD packages[4-6] (see App. B). The greater complexity of the feedback design procedures makes the reliance on these CAD packages more valuable.

The following properties can be attributed to each compensator:

Cascade lag compensator Results in an increase in the gain K_m and a small reduction in ω_m. This reduces steady-state error but increases transient settling time.

Cascade lead compensator Results in an increase in ω_m, thus reducing the settling time. There may be a small increase in gain.

Cascade lag-lead compensator Results in both an increase in gain K_m and in resonant frequency ω_m. This combines the improvements of the lag and the lead compensators.

Feedback compensators Results in an increase in both the gain K_m and resonant frequency ω_m. The basic improvement in ω_m is achieved by derivative feedback. This may be modified by adding a filter such as an *RC* network.

REFERENCES

1. Chestnut, H., and R. W. Mayer: *Servomechanisms and Regulating Systems Design*, 2d ed., vol. 1, Wiley, New York, 1959, chaps. 10 and 12.
2. Kuo, F. F.: *Network Analysis and Synthesis*, 2d ed., Wiley, New York, 1966.
3. Mitra, S. K.: *Analysis and Synthesis of Linear Active Networks*, Wiley, New York, 1969.
4. Larimer, S. J.: "An Interactive Computer-Aided Design Program for Digital and Continuous System Analysis and Synthesis (TOTAL)," M.S. thesis, GE/GGC/EE/78-2, School of Engineering, Air Force Institute of Technology, Wright-Patterson Air Force Base, Ohio, 1978; available from Defense Documentation Center (DDC), Cameron Station, Alexandria, Va. 22314.
5. Thompson, P. M.: *USER's Guide to Program CC, Version 3*, Systems Technology, Inc., Hawthorne, Calif., March, 1985.
6. Phillips, C. L.: "Analytical Bode Design of Controllers," *IEEE Trans. Educ.*, vol. E-28, no. 1, pp. 43–44, 1985.

CHAPTER 12

CONTROL-RATIO MODELING

12.1 INTRODUCTION

The conventional control-design techniques of the previous chapters determine closed-loop system performance based on the open-loop transfer function [$G(s)$ or $G(s)H(s)$] or the open-loop transfer function of an equivalent unity-feedback system. The frequency response or the root-locus technique yields a closed-loop response which is based on some of the closed-loop specifications. If necessary, compensation is then added in order to meet additional specifications. There are other techniques that first require the modeling of a desired control ratio which satisfies the desired figures of merit. The desired control ratio is used to determine the necessary compensation. Formulating the desired control ratio is the main objective of this chapter. Figure 12.1 represents a multiple-input single-output (MISO) control system in which $r(t)$ is the input, $y(t)$ is the output which is required to track the input, $d(t)$ is a disturbance input that must have minimal effect on the output, and $F(s)$ is a prefilter to assist in the tracking of $r(t)$ by $y(t)$. Thus, two types of control ratios need to be modeled: a tracking transfer function $M_T(s)$ and a disturbance transfer function $M_D(s)$. In Fig. 12.1 and the remainder of this text the output is now denoted by $y(t)$ or $Y(s)$, instead of $c(t)$ or $C(s)$. This conforms with the symbols typically found in the literature for this material. For single-input single-output (SISO) systems the symbol $c(t)$ is most often used to represent the output in most basic texts and agrees with the presentation in the earlier parts of this book.

The control systems considered in the previous chapters represent systems for which $F(s) = 1$ and $d(t) = 0$; thus they are single-input single-output (SISO)

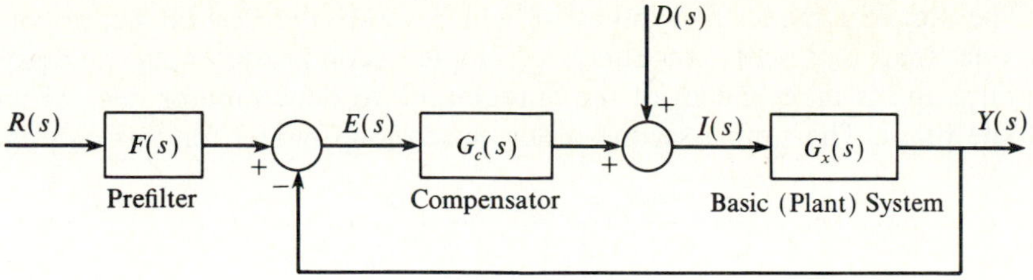

FIGURE 12.1
A MISO control system.

control systems. For these SISO systems the Guillemin-Truxal (GT) method, which requires the designer to specify a desired model $M_T(s)$, represents a method of designing the required compensator $G_c(s)$ of Fig. 12.1.

MISO control systems, in which both a desired input $r(t)$ and a disturbance input $d(t)$ are present, require the design of a cascade compensator $G_c(s)$ and of an input prefilter $F(s)$ as shown in Fig. 12.1. The *quantitative feedback theory* (QFT) technique presented in Chap. 21 addresses the design of such systems which have a desired and a disturbance input. This design technique requires both types of control-ratio models, $M_T(s)$ and $M_D(s)$, in order to effect a satisfactory design. Some state-variable design techniques (Chaps. 13 and 17) also require the specification of a model tracking control ratio. Disturbance rejection while tracking a command input is covered for multiple-input multiple-output systems in Chaps. 20 and 21. The use of proportional plus integral plus derivative (PID) controllers leads naturally to the rejection of disturbance and the tracking of the inputs.

This chapter is devoted to the discussion of tracking control-ratio models, the presentation of the Guillemin-Truxal design method, the presentation of a disturbance-rejection design method for a control system having $r(t) = 0$ and a disturbance input $d(t) \neq 0$, and the discussion of disturbance-rejection control-ratio models. CAD packages (see App. B) are available to assist the designer in synthesizing desired model control ratios.

12.2 MODELING A DESIRED TRACKING CONTROL RATIO

The pole-zero placement method of this section requires that a desired tracking control ratio $[Y(s)/R(s)]_T = M_T(s)$ be modeled in order to satisfy the desired tracking performance specifications for M_p, t_p, and T_s or t_s for a step input (transient characteristic) and the gain K_m (steady-state characteristic). The model must be consistent with the degrees of the numerator and denominator, w and n, respectively, of the basic plant $G_x(s)$. Before proceeding, it is suggested that the reader review Secs. 3.9, 3.10, 9.12, 10.2, and 10.3.

CASE 1. The simplest model to synthesize, which yields the desired figures of merit, is one that represents an effective simple second-order control ratio $[Y(s)/R(s)]_T$, and is independent of the numerator and denominator degrees w and n of the plant. The simple second-order closed-loop model for this case is

$$M_{T1}(s) = \frac{A_1}{s^2 + as + A_1} = \frac{\omega_n^2}{s^2 + 2\zeta\omega_n s + \omega_n^2}$$

$$= \frac{\omega_n^2}{(s - p_1)(s - p_2)} = \frac{G_{eq}(s)}{1 + G_{eq}(s)} \tag{12.1}$$

where $A_1 = \omega_n^2$ ensures that, with a step input of amplitude R_0, the output tracks the input so that $y(t)_{ss} = R_0$. The control ratio of Eq. (12.1) corresponds to a unity-feedback system with an equivalent forward transfer function

$$G_{eq}(s) = \frac{A_1}{s(s + a)} \tag{12.2}$$

Note that although there are four possible specifications, only two adjustable parameters, ζ and ω_n, are shown in Eq. (12.1). When Eqs. (3.60), (3.61), and (3.64) are analyzed for peak time t_p, peak overshoot M_p, and settling time T_s, respectively, and the appropriate value for K_m, it is evident that only two of these relationships can be used to obtain the two adjustable parameters which locate the two dominant poles $p_{1,2}$ of Eq. (12.1). Therefore, it may not be possible to satisfy all four specifications. For example, specifying the minimum value of the gain K_1 (where $K_1 = A_1/a$) yields the minimum acceptable value of ω_n. Also, from $M_p = 1 + \exp(-\pi\zeta/\sqrt{1 - \zeta^2})$, the required value of ζ is determined. Having determined ω_n and ζ, the values of $t_p = \pi/\omega_d$ and $T_s = 4/\zeta\omega_n$ can be evaluated. If these values of t_p and T_s are within the acceptable range of values, the desired model has been achieved. If not, an alternative procedure is to determine ζ and ω_n by using the desired M_p and T_s, or M_p and $t_p = \pi/\omega_n\sqrt{1 - \zeta^2}$.

In order to satisfy the requirement that the specifications $1 < M_p \le \alpha_p$, $t_p \le t_{\alpha_p}$, and $t_s \le t_{\alpha_s}$, two of these values α_p, t_{α_p} and t_{α_s} are used to determine the two unknowns ζ_α and ω_{n_α}. These values of ζ_α and ω_{n_α} yield the desired dominant poles $p_{1,2}$ shown in Fig. 12.2. Any dominant complex pole not lying in the shaded area satisfies these specifications. If an acceptable second-order $M_T(s)$, as given by Eq. (12.1), cannot be found that satisfies the desired performance specifications, a higher-order control ratio must be used.

CASE 2. It may be possible to satisfy the specifications by use of the following third-order model:

$$M_{T2}(s) = \frac{A}{s + d} M_{T1}(s) = \frac{A_2}{(s + d)(s^2 + 2\zeta\omega_n s + \omega_n^2)} \tag{12.3}$$

The dominant complex poles of Eq. (12.3) are determined in the same manner as for case 1; i.e., the values of ζ and ω_n are determined by using the specifications

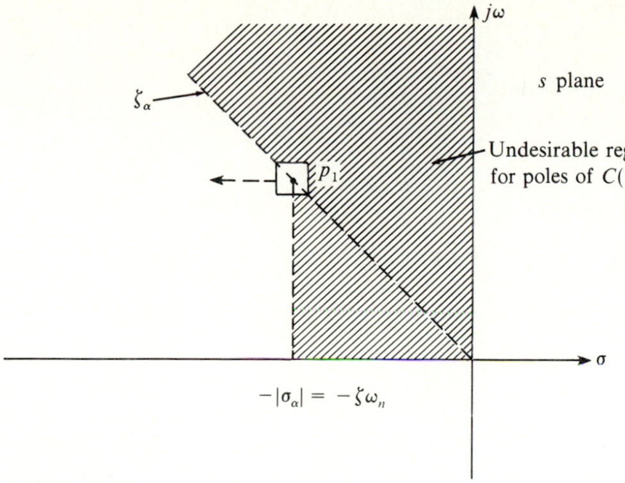

FIGURE 12.2
Location of the desired dominant pole.

for M_p, t_p, T_s, and K_m. Thus, since the model of Eq. (12.1) cannot satisfy all the specifications, the known effects of the real pole $s = -d$ in Eq. (12.3) can be used to try to achieve these specifications. With a digital-computer-aided control-system design program (see App. B), various trial values of d are used to check whether the desired specifications can be achieved. Note that for each value of d, the corresponding value of A_2 must be adjusted in order to satisfy the requirement that $e(t)_{ss} = 0$ for a step input. From Eq. (12.3), the required gain is $A_2 = d\omega_n^2$. As an example, the specifications $M_p = 1.125$, $t_p = 1.5$ s, and $T_s = 3.0$ s cannot be satisfied by the second-order control ratio of the form of Eq. (12.1), but they can be satisfied by

$$\frac{Y(s)}{R(s)} = \frac{22}{(s + 1 \pm j3)(s + 2.2)} \tag{12.4}$$

Note that if the model of Eq. (12.1) is satisfactory, then, for a third-order all-pole open-loop plant, the model of Eq. (12.3) must be used and the pole $-d$ is made a nondominant pole.

CASE 3. The third case deals with the third-order model [see Eq. (10.18)]

$$M_{T3}(s) = \frac{A(s + c)}{s + d} M_{T1}(s) = \frac{A_3(s + c)}{(s + d)(s^2 + 2\zeta\omega_n s + \omega_n^2)}$$

$$= \frac{A_3(s - z_1)}{(s - p_1)(s - p_2)(s - p_3)} \tag{12.5}$$

which has one zero. The zero can be a zero of the plant or a compensator zero

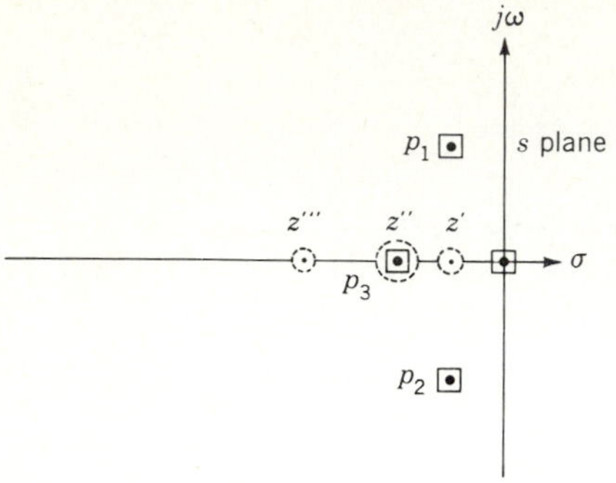

FIGURE 12.3
Pole-zero location of $Y(s)$ for Eq. (12.5), with a step input.

which is inserted into the system to be used in conjunction with the pole $-d$ in order to satisfy the specifications. The procedure used in synthesizing Eq. (12.5) is based upon the nature of the zero; i.e., the zero (1) is not close to the origin or (2) it is close to the origin and is therefore a troublesome zero. (A stable system with a zero close to the origin may have a large transient overshoot for a step input.)

Figure 12.3 represents the pole-zero diagram of $Y(s)$ of Eq. (12.5) with a step input. Consider three possible locations of the real zero: to the right of, canceling, and to the left of, the real pole $p_3 = -d$. When the zero cancels the pole p_3, the value of M_p is determined only by the complex-conjugate poles p_1 and p_2; when the zero is to the right of p_3, the value of M_p is larger (see Sec. 10.9); and when it is to the left, the value of M_p is smaller [see Eq. (4.68) and Sec. 10.9]. The amount of this increase or decrease is determined by how close the zero is to the pole p_3. Thus the optimum location of the real pole, to the right of the zero, can be determined which minimizes the effect of the troublesome zero, i.e., it reduces the peak of the transient overshoot.

In order to achieve M_T, it may be necessary to insert additional equipment in cascade and/or in feedback with $G_x(s)$. Also, in achieving the desired model for a plant containing a zero, the design method may use (1) the cancellation of the zero by a dominator pole, (2) the addition of a second zero, or (3) the replacement of the zero with one at a new location.

In synthesizing the desired control ratio, a system designer has a choice of one of the following approaches.

Method 1. Cancel some or all of the zeros of the control ratio by poles of the control ratio. The number of zeros α to be canceled must be equal to or less than the excess of the number of poles n minus the number of desired dominant poles β. If necessary, cascade compensators of the form $G_c(s) = 1/(s - P_i)$ can be inserted into the open-loop system in order to increase the number of poles of $Y(s)/R(s)$ so that cancellation of zeros can be accomplished. The number of compensator poles γ required must increase the order of the system sufficiently to ensure that $\alpha \leq n + \gamma - \beta$. Theoretically it is possible to insert a compensator of this form just preceding the block that contains the unwanted zero to be canceled by the pole of this compensator. However, in many systems it is not physically possible to locate the compensator in this manner. Also, setting P_i equal to the value of the unwanted zero z_h of the plant results in an unobservable state. All the states must be controllable and observable in order to achieve an *optimal response* (see Chap. 16). Thus, to achieve an optimal response and to cancel the unwanted zero, choose P_i such that $P_i \neq z_h$. It is possible to achieve the desired $Y(s)/R(s)$ by using a cascade compensator with $P_i = z_h$; however, an optimal response, as defined in Sec. 16.6 is not achieved.

Method 2. Relocate some or all of the zeros by using cascade compensators of the form $G_c(s) = (s - z_i)/(s - P_i)$. The number γ of these compensators to be inserted equals the number of zeros α to be relocated. The values of z_i represent the desired zeros. The values of P_i are selected to meet the condition that α *poles of the control ratio* are used to cancel the α unwanted zeros.

Method 3. A combination of the above two approaches can be used, i.e., canceling some zeros by closed-loop poles and replacing some of them by the zeros of the cascade compensators $G_c(s)$.

The modeling technique discussed in this section requires that the poles and zeros of $Y(s)/R(s)$ be located in a pattern which achieves the desired time response. This is called the *pole-zero placement technique*. The number of poles minus zeros of $(Y/R)_T$ must equal or exceed the number of poles minus zeros of the plant transfer function. For a plant having w zeros, w of the closed-loop poles are either located "very close" to these zeros or suitably placed for the desired time response to be obtained. If there are β dominant poles, the remaining $n - \beta - w$ poles are located in the nondominant region of the s plane. Another method of selecting the locations of the poles and zeros of $M_T(s)$ is given in Sec. 16.5; it is based on a performance index which optimizes the closed-loop performance. The specification of $Y(s)/R(s)$ is the basis for the Guillemin-Truxal method of Sec. 12.3 and for the state-feedback designs discussed in this and in the next chapter.

12.3 GUILLEMIN-TRUXAL DESIGN PROCEDURE[1]

In contrast to designing the cascade compensator $G_c(s)$ based upon the analysis of the open-loop transfer function, the Guillemin-Truxal method is based upon designing $G_c(s)$ to yield a specified or desired control ratio $Y(s)/R(s)$. The control ratio for the unity-feedback cascade-compensated control system of Fig. 12.4 is

$$M_T(s) = \frac{Y(s)}{R(s)} = \frac{N(s)}{D(s)} = \frac{G_c(s)G_x(s)}{1 + G_c(s)G_x(s)} \tag{12.6}$$

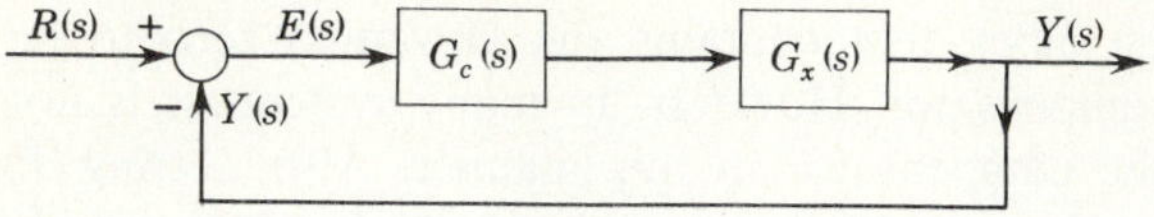

FIGURE 12.4
A cascade-compensated control system.

where $N(s)$ and $D(s)$ represent the specified zeros and poles, respectively, of the desired control ratio. Solving this equation for the required cascade compensator $G_c(s)$ yields

$$G_c(s) = \frac{M_T(s)}{[1 - M_T(s)]G_x(s)} = \frac{N(s)}{[D(s) - N(s)]G_x(s)} \tag{12.7}$$

In other words, when $N(s)$ and $D(s)$ have been specified, the required transfer function for $G_c(s)$ is obtained. Network-synthesis techniques are then utilized to synthesize a network having the required transfer function of Eq. (12.7). Usually a passive network containing only resistors and capacitors is desired.

Example. The desired control ratio of a unity-feedback control system is

$$\frac{Y(s)}{R(s)} = \frac{210(s + 1.5)}{(s + 1.75)(s + 16)(s + 1.5 \pm j3)} = \frac{N(s)}{D(s)} \tag{12.8}$$

The numerator constant has been selected to yield zero steady-state error with a step input and the desired figures of merit are $M_p = 1.28$, $t_p = 1.065$ s, and $t_s = 2.54$ s. The basic plant is

$$G_x(s) = \frac{4}{s(s + 1)(s + 5)} \tag{12.9}$$

The required cascade compensator, determined by substituting Eqs. (12.8) and (12.9) into Eq. (12.7), is

$$G_c(s) = \frac{52.5s(s + 1)(s + 1.5)(s + 5)}{s^4 + 20.75s^3 + 92.6s^2 + 73.69s}$$

$$= \frac{52.5s(s + 1)(s + 1.5)(s + 5)}{s(s + 1.02)(s + 4.88)(s + 14.86)} \approx \frac{52.5(s + 1.5)}{s + 14.86} \tag{12.10}$$

The approximation made in $G_c(s)$ yields a simple, physically realizable, minimum-phase lead network with an $\alpha = 0.101$. Based on a root-locus analysis of $G_c(s)G_x(s)$, the ratios of the terms $(s + 1)/(s + 1.02)$ and $(s + 5)/(s + 4.88)$s of $G_c(s)$ have a negligible effect on the location of the desired dominant poles of $Y(s)/R(s)$. With the simplified $G_c(s)$ the root-locus plot, using $G_c(s)G_x(s)$, is shown in Fig. 12.5, and the control ratio achieved is

$$\frac{Y(s)}{R(s)} = \frac{210(s + 1.5)}{(s + 1.755)(s + 16.0)(s + 1.52 \pm j3.07)}$$

Since this equation is very close to that of Eq. (12.8), it is satisfactory. Note that the root-locus plot of $G_c(s)G_x(s) = -1$ has asymptotes of 180° and ±60°; therefore the system becomes unstable for high gain.

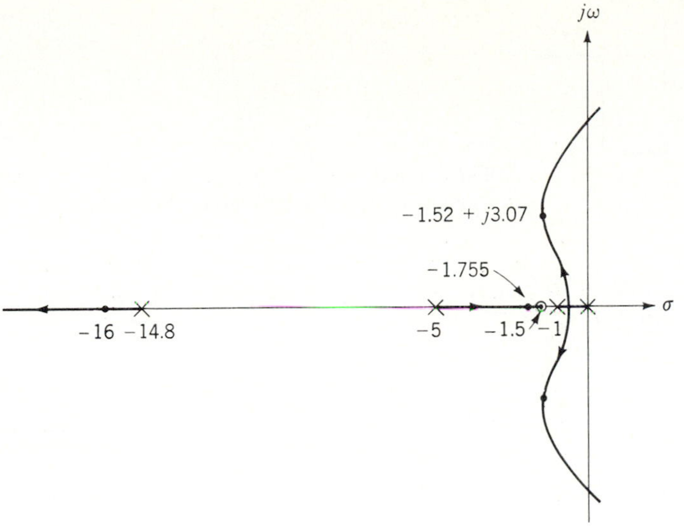

FIGURE 12.5
Root-locus plot for the example.

The Guillemin-Truxal design method, as illustrated by this example, involves three steps:

1. Specifying the desired zeros, poles, and numerator constant of the desired closed-loop function $Y(s)/R(s)$ in the manner discussed in Sec. 12.2.
2. Solving for the required cascade compensator transfer function $G_c(s)$ from Eq. (12.7).
3. Synthesizing a physically realizable compensator $G_c(s)$, preferably a passive network.

Using practical cascade compensators, the number of poles minus zeros of the closed-loop system must be equal to or greater than the number of poles minus zeros of the basic open-loop plant. This must include the poles and zeros introduced by the cascade compensator. The practical aspects of system synthesis using the Guillemin-Truxal method imposes the limitation that the poles of $G_x(s)$ lie in the left-half s plane. Failure to exactly cancel poles in the right-half s plane would result in an unstable closed-loop system. The state-variable-feedback method, described in the next chapter, is a compensation method that achieves the desired $Y(s)/R(s)$ without this limitation.

12.4 INTRODUCTION TO DISTURBANCE REJECTION[2,3]

The previous chapters and Sec. 12.3 deal with designing a control system whose output follows (tracks) the input signal. The design approach for the disturbance-rejection problem, by contrast, is based upon the performance

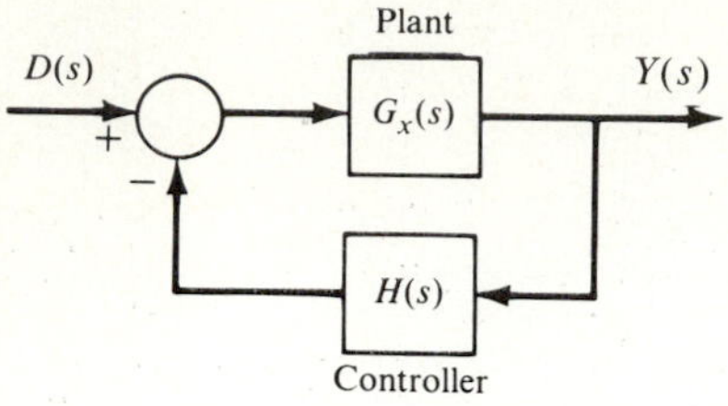

FIGURE 12.6
A simple disturbance-rejection control system.

specification that the output of the control system of Fig. 12.6 not be affected by a system disturbance input $d(t)$. In other words, the steady-state output $y(t)_{ss} = 0$ for the input $d(t) \neq 0$. Consider the control ratio

$$\frac{Y(s)}{D(s)} = \frac{K\left(s^w + c_{w-1}s^{w-1} + \cdots + c_1 s + c_0\right)}{s^n + q_{n-1}s^{n-1} + \cdots + q_1 s + q_0} \tag{12.11}$$

Assuming a stable system and a step disturbance $D(s) = D_0/s$, the output is given by

$$Y(s) = \frac{K\left(s^w + c_{w-1}s^{w-1} + \cdots + c_1 s + c_0\right)}{s\left(s^n + q_{n-1}s^{n-1} + \cdots + q_1 s + q_0\right)} \tag{12.12}$$

The desired output $y(t)_{ss} = 0$ is achieved only if $c_0 = 0$. *This requires that the numerator of $Y(s)/D(s)$ have at least one zero at the origin.*

It is desired for this type of control problem that the disturbance have no effect on the steady-state output; also, the resulting transient must die out as fast as possible with a limit α_p on the maximum magnitude of the output. This added constraint may require the system characteristic equation to have a pair of dominant complex-conjugate roots in order to achieve the smallest possible settling time t_s.

The next three sections present a frequency-domain design technique for minimizing the effect of a disturbance input on the output of a control system. This technique is enhanced by simultaneously performing a qualitative root-locus analysis for a proposed feedback-compensator structure.

12.5 A SECOND-ORDER DISTURBANCE-REJECTION MODEL[2]

A simple second-order continuous-time disturbance-rejection model transfer function for the system of Fig. 12.6 is

$$M_D(s) = \frac{Y(s)}{D(s)} = \frac{G_x}{1 + G_x H} = \frac{K_x s}{(s+a)^2 + b^2} = \frac{K_x s}{s^2 + 2\zeta\omega_n s + \omega_n^2} \tag{12.13}$$

where $K_x > 0$. For a unit-step disturbance $D(s) = 1/s$, Eq. (12.13) yields

$$y(t) = \mathscr{L}^{-1}\left[\frac{K_x}{(s+a)^2 + b^2}\right] = \frac{K_x}{b} e^{-at} \sin bt \tag{12.14}$$

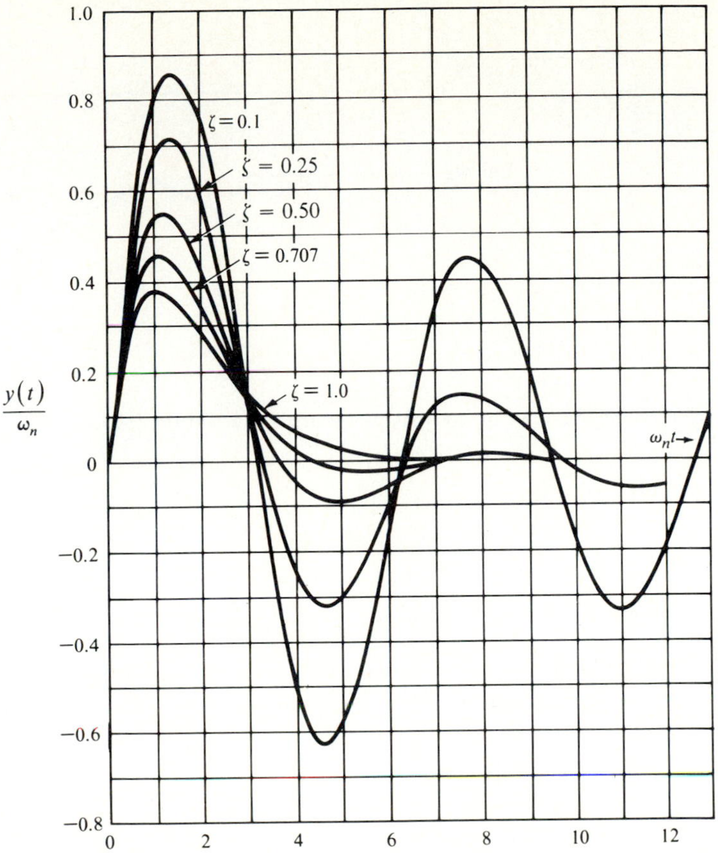

FIGURE 12.7
Time response with $K_x = \omega_n^2$ of

$$Y(s) = \frac{K_x}{s^2 + 2\zeta\omega_n s + K_x}$$

where $a^2 + b^2 = \omega_n^2$, $b = \omega_d$ and $a = |\sigma| = \zeta\omega_n$. The time—or frequency—approaches may be used to determine appropriate model values of ζ and ω_n, while K_x is assumed fixed.

Time domain

The time response of Eq. (12.14) is plotted vs. $\omega_n t$ in Fig. 12.7 for several values of $\zeta \leq 1$. By setting the derivative of $y(t)$ with respect to time equal to zero, it can be shown that the maximum overshoot occurs at the time

$$t_p = \frac{\cos^{-1}\zeta}{\omega_n\sqrt{1 - \zeta^2}} \tag{12.15}$$

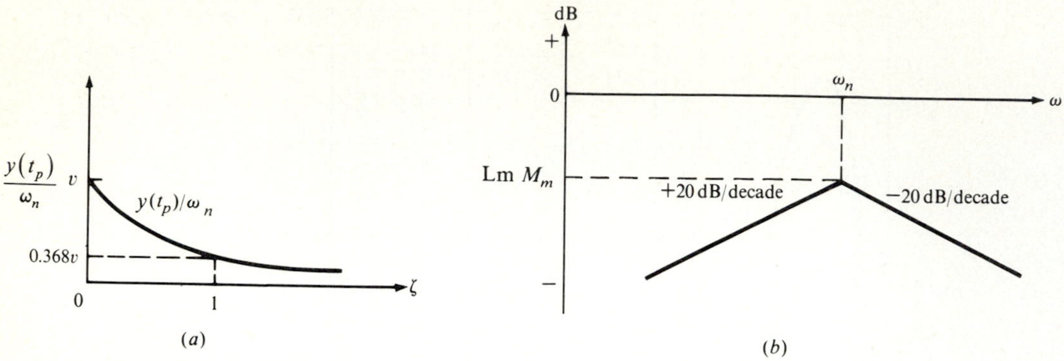

FIGURE 12.8
(a) Peak value of Eq. (12.14), using Eq. (12.17). (b) Bode plot characteristics of Eq. (12.18).

and the maximum value of $y(t)$ is

$$y(t_p) = \frac{K_x}{\omega_n} \exp - \frac{\zeta \cos^{-1} \zeta}{\sqrt{1 - \zeta^2}} \tag{12.16}$$

Letting $K_x = v\omega_n^2$ $(v > 0)$, Eq. (12.16) is rearranged to the form

$$\frac{y(t_p)}{\omega_n} = v \exp - \frac{\zeta \cos^{-1} \zeta}{\sqrt{1 - \zeta^2}} \tag{12.17}$$

A plot of Eq. (12.17) vs. ζ is shown in Fig. 12.8a.

Frequency domain

The frequency transient function of Eq. (12.13) is

$$\mathbf{M}_D(j\omega) = \frac{\mathbf{Y}(j\omega)}{\mathbf{D}(j\omega)} = \frac{(K_x/\omega_n^2)(j\omega)}{\left[1 - (\omega/\omega_n)^2\right] + j(2\zeta\omega/\omega_n)} = \frac{K_x}{\omega_n^2}\mathbf{M}'_D(j\omega) \tag{12.18}$$

A straight-line approximation of the Bode plot of Lm $\mathbf{M}_D(j\omega)$ vs. ω is shown in Fig. 12.8b for a selected value of ω_n. This plot reveals the following:

1. The closed-loop transfer function $\mathbf{M}_D(j\omega)$ must attenuate all frequencies present in the disturbance input $\mathbf{D}(j\omega)$. The plot of Lm $\mathbf{M}'_D(j\omega)$ vs. ω is raised or lowered by an amount equal to Lm(K_x/ω_n^2). The value of K_x/ω_n^2 therefore determines the maximum value M_m shown in Fig. 12.8b.
2. As ω_n increases, the peak value moves to the right to a higher frequency.

Selection of the desired disturbance model of Eq. (12.13) is based on both the time-response and the frequency-response characteristics. For example, selection of ω_n determines the bandwidth, as shown in Fig. 12.8b. Also, the maximum value of the permitted peak overshoot, as shown in Figs. 12.7 and 12.8a, is determined by the value of ζ. In addition, the peak time t_p and the settling time

$t_s = 4/\zeta\omega_n$ of the transient may be used to determine the parameters in the model of Eq. (12.13). Some trial and error may be necessary to insure that all specifications for the disturbance model are satisfied.

12.6 DISTURBANCE-REJECTION DESIGN PRINCIPLES FOR SISO SYSTEMS[2]

The desired disturbance model $M_D(s)$ of a system must be based on the specified time-domain and frequency-domain characteristics as described in Secs. 12.4 and 12.5. The next step is to design the required feedback controller $H(s)$ in Fig. 12.6 which will satisfy the characteristics of $M_D(s)$. A first trial may be $H(s) = K_H/s$. If this choice of $H(s)$ does not produce the desired characteristics, then a higher-order feedback transfer function must be used. The order of the transfer functions in the system may also require a disturbance model of higher than second order.

The approximations developed in Sec. 11.9 are now restated for the disturbance-rejection system.

$$\mathbf{M}_D(j\omega) = \frac{\mathbf{Y}(j\omega)}{\mathbf{D}(j\omega)} = \frac{\mathbf{G}_x(j\omega)}{1 + \mathbf{G}_x(j\omega)\mathbf{H}(j\omega)} \tag{12.19}$$

where $\mathbf{H}(j\omega) = \mathbf{H}_x(j\omega)\mathbf{H}_c(j\omega)$ (see Fig. 12.9). This control ratio can be approximated by

$$\mathbf{M}_D(j\omega) \approx \begin{cases} \mathbf{G}_x(j\omega) & \text{for } |\mathbf{G}_x(j\omega)\mathbf{H}(j\omega)| \ll 1 \quad (12.20) \\ \dfrac{1}{\mathbf{H}(j\omega)} & \text{for } |\mathbf{G}_x(j\omega)\mathbf{H}(j\omega)| \gg 1 \quad (12.21) \end{cases}$$

Still undefined is the condition when $|\mathbf{G}_x(j\omega)\mathbf{H}(j\omega)| \approx 1$, in which case neither of these conditions is applicable. In the approximate procedure the straight-line Bode plots are used and this condition is neglected. Therefore Eqs. (12.20) and (21.21) are used when $|\mathbf{G}_x(j\omega)\mathbf{H}(j\omega)| < 1$ and $|\mathbf{G}_x(j\omega)\mathbf{H}(j\omega)| > 1$, respectively. This approximation, along with a root-locus sketch of $G_x(s)H(s) = -1$, allows investigation of the qualitative results to be obtained. After these results are found to be satisfactory, the refinements for an exact solution are introduced.

The design procedure using the approximations of Eqs. (12.20) and (12.21) is as follows:

Step 1. Using the straight-line approximations, plot Lm $\mathbf{G}_x(j\omega)$ vs. ω.

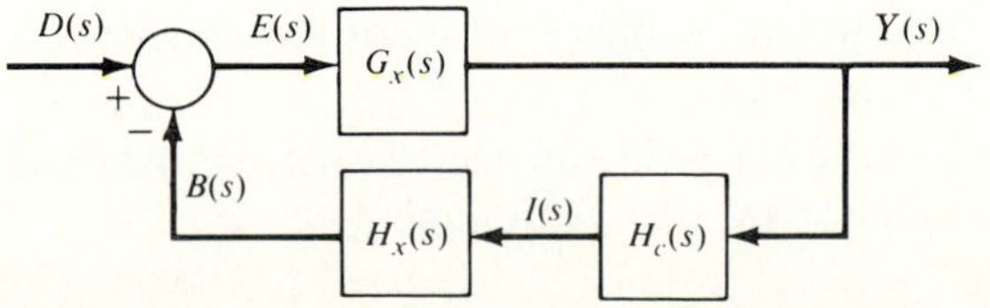

FIGURE 12.9
A control system with a disturbance input.

Step 2. Assume an $H_c(s)$ that may, when placed in cascade with $H_x(s)$, yield the desired system performance. For this $\mathbf{H}(j\omega) = \mathbf{H}_x(j\omega)\mathbf{H}_c(j\omega)$, using the straight-line approximations, plot $\text{Lm}\, 1/\mathbf{H}(j\omega)$ vs. ω.

Note:

1. The condition $\text{Lm}\, \mathbf{G}_x(j\omega)\mathbf{H}(j\omega) = 0$ may also be expressed as

$$\text{Lm}\, \mathbf{G}_x(j\omega) = -\text{Lm}\, \mathbf{H}(j\omega) = \text{Lm}\, \frac{1}{\mathbf{H}(j\omega)} \tag{12.22}$$

   This condition is satisfied at the intersection of the plots of $\mathbf{G}_x(j\omega)$ and $1/\mathbf{H}(j\omega)$. The frequency at the intersections represents the boundary between the two regions given by Eqs. (12.20) and (12.21). There must be at least one intersection between the plots in order for both approximations to be valid. Disturbance rejection over the frequency band of $0 \le \omega \le \omega_b$, the condition of Eq. (12.21), results in only one intersection.
2. For disturbance rejection it is desired to have $\text{Lm}\, \mathbf{Y}(j\omega)/\mathbf{D}(j\omega) < \text{Lm}\, \mathbf{M}_D$. Also, Eq. (12.21) must hold for $0 \le \omega \le \omega_b$ in order to yield the desired frequency-domain specifications. To satisfy these objectives it is usually necessary for the gain constant of $H(s)$ to be very high.

Step 3. Plot $\text{Lm}\, \mathbf{G}_x(j\omega)\mathbf{H}(j\omega)$ vs. ω by using the straight-line approximation and determine the slope of the plot at the 0-dB crossover point. Achieving a stable system requires (see Sec. 8.19) that this slope be greater than -40 dB/decade if the adjacent corner frequencies are not close. Also, the loop transmission frequency ω_ϕ (the 0-dB crossover or phase-margin frequency) of $\text{Lm}\, \mathbf{G}_x(j\omega)\mathbf{H}(j\omega)$ must be within the allowable specified value in order to attenuate high-frequency loop noise. Placing zeros of $H_c(s)$ "too close" to the imaginary axis may increase the value of ω_ϕ. It may be necessary at this stage of the design procedure to modify $H_c(s)$ in order to satisfy the stability requirement and the frequency-domain specifications.

Step 4. Once a trial $H(s)$ is found that satisfies the frequency-domain specifications, sketch the root locus $[G_x(s)H(s) = -1]$ to ascertain that (a) at least a conditionally stable system exists and (b) it is possible to satisfy the time-domain specifications. [Section 12.5 discusses the desired locations of the poles of $Y(s)$ which achieve the desired specifications (see also Prob. 12.7).] If the sketch "seems" reasonable (1) obtain the exact root-locus plots from which a value of the static loop sensitivity $K = K_G K_x K_c$ is chosen and (2) for this value of K, obtain a plot of $y(t)$ vs. t from which the figures of merit are obtained.

Step 5. If the results of step 4 do not yield satisfactory results, then use these results to select a new $H_c(s)$ and repeat steps 2 through 4.

Example 1. Consider the nonunity-feedback system of Fig. 12.9 where

$$G_x(s) = \frac{K_G}{s(s+1)} = \frac{1}{s(s+1)} \tag{12.23}$$

and

$$H_x(s) = \frac{K_x}{s+200} = \frac{200}{s+200} \tag{12.24}$$

Since a disturbance input is present, it is necessary to design the feedback compensator $H_c(s) = K_c H_c'(s)$ such that the control system of Fig. 12.9 satisfies the following specifications for a unit-step disturbance input $d(t)$: $|y(t_p)| \le 0.002$, $y(t)_{ss} = 0$, and $t_p \le 0.2$ s. Using the empirical estimate that $M_p = |y(t_p)/d(t)| \approx M_m$ gives the specification Lm $M_m \le$ Lm 0.002 = −54 dB. It is also estimated that the bandwidth of the disturbance transfer function model $\mathbf{Y}(j\omega)/\mathbf{D}(j\omega)$ is $0 \le \omega \le 10$ rad/s.

Solution

Step 1. The plot of Lm $\mathbf{G}_x(j\omega)$ vs. ω is shown in Fig. 12.10.

Step 2. Since $G_x(s)$ does not have a zero at the origin and $Y(s)/D(s)$ does require one for disturbance rejection, then $H(s)$ must have a pole at the origin. Thus the

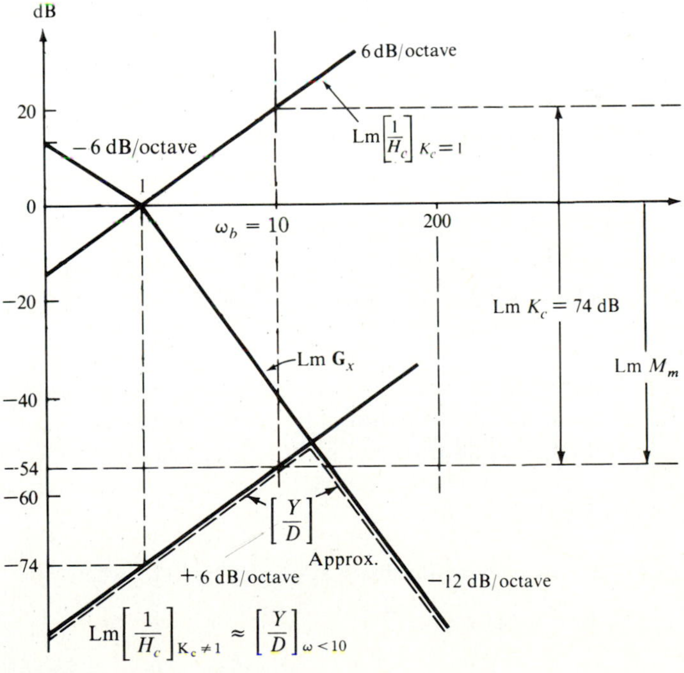

FIGURE 12.10
Log magnitude plots of Example 1.

feedback transfer function must have the form

$$H(s) = H_x(s)H_c(s) = \frac{K_H(s+a_1)\cdots}{s(s+200)(s+b_1)\cdots} \tag{12.25}$$

where the degree of the denominator is equal to or greater than the degree of the numerator. As a first trial, let

$$H_c(s) = \frac{K_c}{s} \tag{12.26}$$

Then $K_H = K_x K_c = 200K_c$. To determine the value of K_c required to satisfy $\text{Lm}\,\mathbf{Y}(j\omega)/\mathbf{D}(j\omega) \le -54$ dB for $0 \le \omega \le 10$ rad/s, initially plot the curve for $[\text{Lm}\,1/\mathbf{H}_c(j\omega)]_{K_c=1}$. The value of K_c is then selected so that the plot of $\text{Lm}\,1/\mathbf{H}_c(j\omega)$ is located such that it is on or below the -54 dB line for the frequency range $\omega \le \omega_b = 10$. Since the corner frequency of $\mathbf{H}(j\omega)$ at $\omega = 200$ lies outside the desired bandwidth, $\text{Lm}\,1/\mathbf{H}_c(j\omega) = \text{Lm}\,1/\mathbf{H}(j\omega)$ is the controlling factor in the frequency range $\omega \le 10$. Next, lower this plot, as shown in the figure, until the plot of $\text{Lm}\,1/\mathbf{H}(j\omega)$ yields the desired value of -54 dB at $\omega_b = 10$ rad/s. Equation (12.21), for this case, satisfies the requirement $|\mathbf{M}(j\omega)| \le M_m$ over the desired bandwidth and is below -54 dB. $\text{Lm}\,200K_c = 74$ dB is required to lower the plot of $\text{Lm}\,1/\mathbf{H}(j\omega)$ to the required location. With this value of K_c the resulting first-trial feedback transfer function is

$$H(s) = \frac{5000}{s(s+200)} \tag{12.27}$$

Step 3. An analysis of

$$\mathbf{G}_x(j\omega)\mathbf{H}(j\omega) = \frac{K_c}{(j\omega)^2(1+j\omega)\left(1+j\dfrac{\omega}{200}\right)} \tag{12.28}$$

reveals that the plot of $\text{Lm}[\mathbf{G}_x(j\omega)\mathbf{H}(j\omega)]$ vs. ω crosses the 0-dB axis with a minimum slope of -40 dB/decade regardless of the value of K_c. This crossover characteristic is indicative of an unstable system.

Step 4. The root-locus sketch for

$$G_x(s)H(s) = \frac{K_H}{s^2(s+1)(s+200)} = -1 \tag{12.29}$$

is shown in Fig. 12.11, which reveals a completely unstable system for the $H(s)$ of Eq. (12.27).

The next trial requires that at least one zero must be added to Eq. (12.27) in order to make the system at least conditionally stable while maintaining the desired bandwidth. A final acceptable design of $H_c(s)$ that results in meeting the system specifications requires a trial-and-error procedure. A solution to this design problem is given in Sec. 12.7.

Example 2. Consider the nonunity-feedback system of Fig. 12.9 where

$$G_x(s) = \frac{2}{s+2}$$

and

$$H(s) = H_x(s)H_c(s) = \frac{K_H}{s}$$

Determine the response $c(t)$ for a step disturbance input $d(t) = u_{-1}(t)$ as the gain K_H is increased.

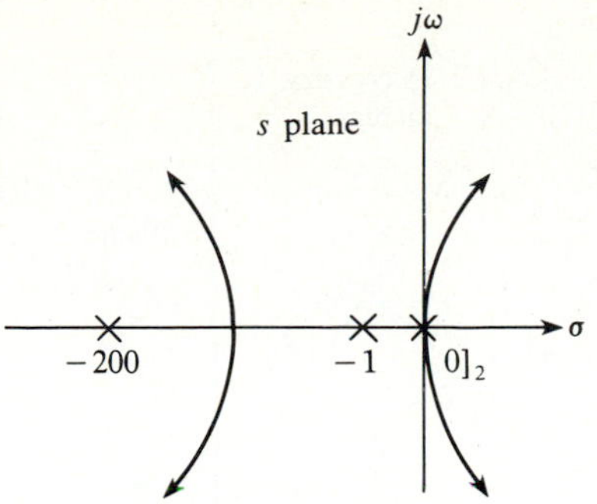

FIGURE 12.11
A root-locus sketch for Eq. (12.29).

The system output is

$$Y(s) = \frac{2}{s^2 + 2s + 2K_H} = \frac{2}{(s - p_1)(s - p_2)} = \frac{A_1}{(s - p_1)} + \frac{A_2}{(s - p_2)} \qquad (12.30)$$

Thus

$$y(t) = A_1 e^{p_1 t} + A_2 e^{p_2 t} \qquad (12.31)$$

where

$$p_{1,2} = -1 \pm j\sqrt{2K_H - 1} \qquad \text{for } K_H > 0.5 \qquad A_1 = \frac{0.1}{j\sqrt{2K_h - 1}} \qquad A_2 = -A_1$$

The coefficients of the output transient, as $K_H \to \infty$, are

$$\lim_{K_H \to \infty} A_1 = \lim_{K_H \to \infty} A_2 = 0$$

Therefore, for this example, as K_H is made larger, the coefficients A_1 and A_2 in Eq. (12.31) become smaller. Thus

$$\lim_{K_H \to \infty} y(t) = 0$$

which is the desired result for disturbance rejection. The output for a step disturbance input has two desired properties. First, the steady-state output is zero because the feedback contains an integrator. Second, the magnitude of the transient becomes smaller as the feedback gain K_H increases. (Note that the value of K_H is limited by practical considerations.) A similar analysis can be made for a more complicated system (see Probs. 12–8 and 12–9).

12.7 DISTURBANCE-REJECTION DESIGN EXAMPLE

In Example 1 the feedback compensator $H_c(s) = K_c/s$ results in an unstable system. An analysis of the root locus in Fig. 12.11 indicates that $H_c(s)$ must contain at least one zero near the origin in order to pull the root-locus branches into the left-half s plane. In order to minimize the effect of an unwanted disturbance on $y(t)$, it is necessary to minimize the magnitude of all the coefficients of

$$y(t) = A_1 e^{p_1 t} + \cdots + A_n e^{p_n t} \qquad (12.32)$$

Problem 4.18 shows that this requires the poles of $Y(s)/D(s)$ to be located as far to the left in the s plane as possible. Figure 12.12 graphically describes the result

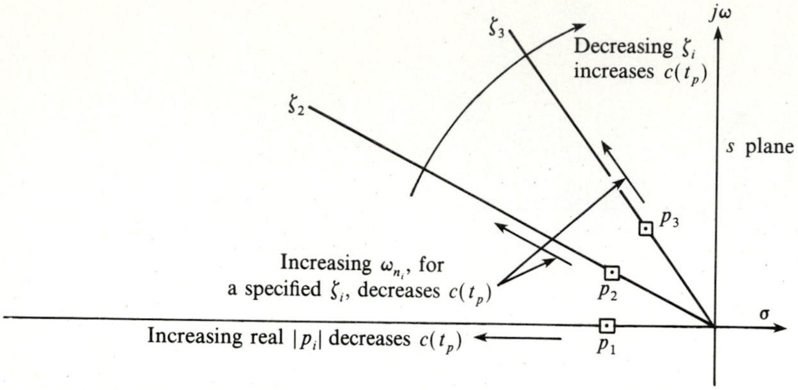

FIGURE 12.12
Effects on the peak overshoot to a disturbance input due to changes in the parameters ζ_i and ω_{n_i} of complex poles and the values of the real poles of the control ratio.

of the analysis made in Prob. 12.7 for the desired placement of the control-ratio poles. The root-locus branch which first crosses the imaginary axis determines the maximum value of gain for a stable system. The root selected on this branch should have $0.5 < \zeta < 0.7$, which has been determined empirically. When there is a choice of roots for a given ζ, the root selected should have the largest value of ω_n and gain. These properties are indicated in Fig. 12.12.

Example 3. The system of Example 1 is now redesigned in order to achieve a stable system. Also, the root-locus branches must be pulled as far to the left as possible. Therefore, assume the following structure which contains two zeros for the feedback compensator:

$$H_c(s) = \frac{K_c(s + a_1)(s + a_2)}{s(s + b_1)(s + b_2)} \tag{12.33}$$

As a trial, let $a_1 = 4$ and $a_2 = 8$. The poles $-b_1$ and $-b_2$ must be chosen as far left as possible and therefore values $b_1 = b_2 = 200$ are selected. The plot of $\text{Lm}\, 1/\mathbf{H}_x(j\omega)\mathbf{H}_c(j\omega) \le -54$ dB, where $|\mathbf{Y}(j\omega)/\mathbf{D}(j\omega)| \approx |1/\mathbf{H}_x(j\omega)\mathbf{H}_c(j\omega)|$ over the range $0 \le \omega \le \omega_b = 10$ rad/s, is shown in Fig. 12.13. This requires that $200K_c \ge 50 \times 10^7$. The open-loop transfer function of Fig. 12.9 is now

$$\begin{aligned} G_x(s)H_x(s)H_c(s) &= \frac{200K_c(s+4)(s+8)}{s^2(s+1)(s+200)^3} \\ &= \frac{(K_c/1250)(0.25s+1)(0.125s+1)}{s^2(s+1)(0.005s+1)^3} \end{aligned} \tag{12.34}$$

The poles and zeros of Eq. (12.34) are plotted in Fig. 12.14. The value of static-loop sensitivity $200K_c = 5 \times 10^8$ yields the set of roots

$$p_1 = -3.8469 \qquad p_2 = -10.079$$
$$p_{3,4} = -23.219 \pm j66.914 \qquad p_{5,6} = -270.32 \pm j95.873$$

and are shown in Fig. 12.14. This choice of gain results in a satisfactory design, that is, $M_p = 0.001076 \le 0.002$ (-54 dB), $t_p = 0.155$ s (< 0.2 s), $y(t)_{ss} = 0$, and

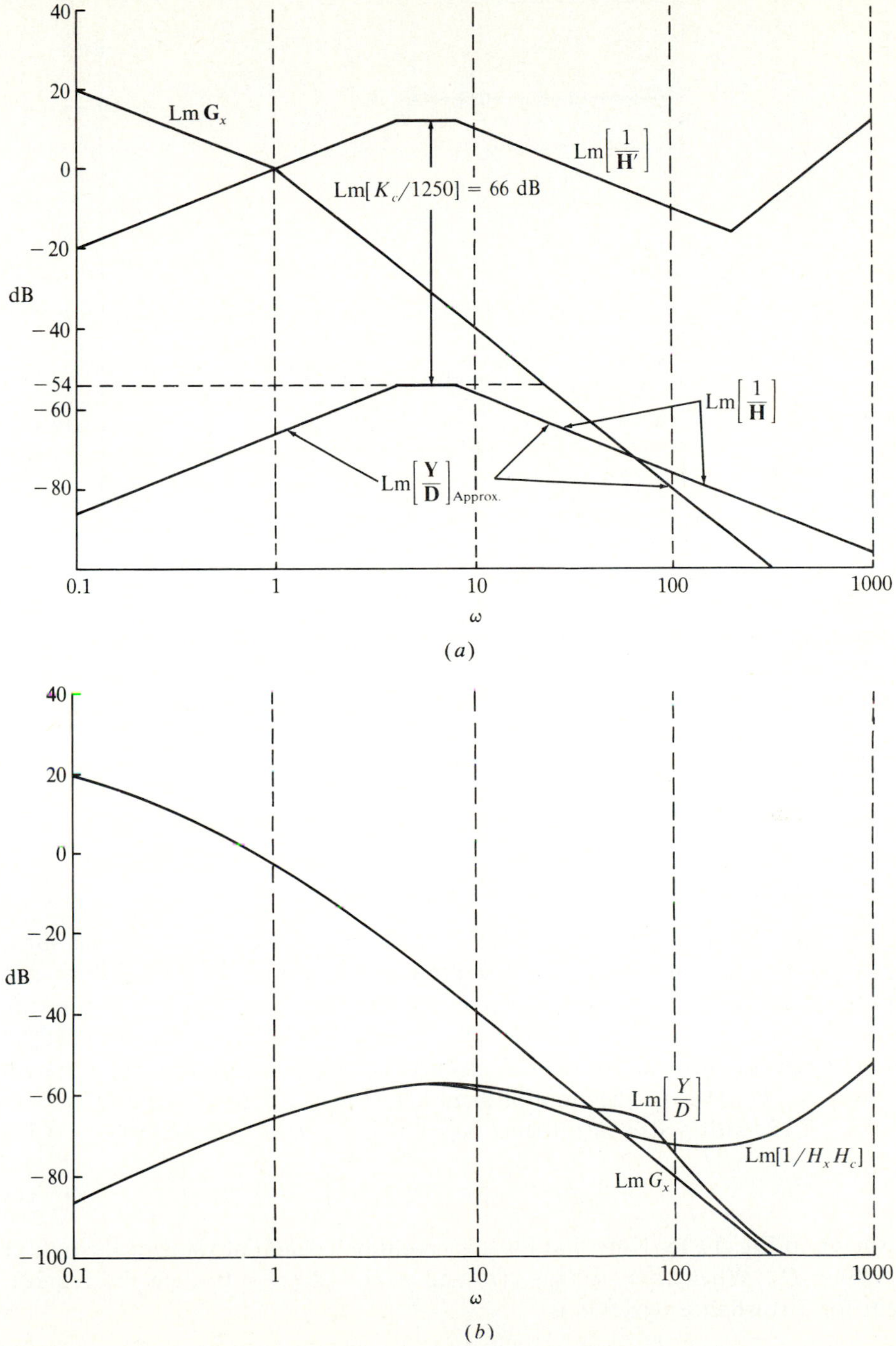

FIGURE 12.13
Bode plot for Example 3: (*a*) straight-line approximations; (*b*) exact poles.

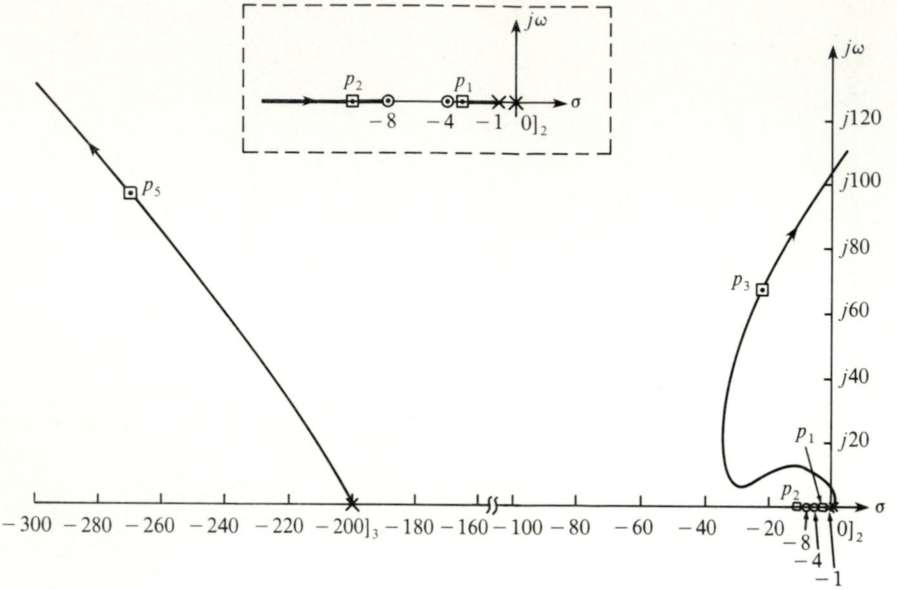

FIGURE 12.14
Sketch of root locus for Eq. (12.34). Not to scale.

$0 \le \omega \le \omega_b = 10$ rad/s. Note that the gain can be reduced until either $M_p = 0.002$ with $t_p \le 0.2$ s or $t_p = 0.2$ s with $M_p \le 0.002$ is achieved. Although not specified in this problem, the phase-margin frequency $\omega_\phi = 20.15$ rad/s must also be taken into account in the design of a practical system.

12.8 DISTURBANCE-REJECTION MODELS

The multiple-input single-output (MISO) control system of Fig. 12.15 has three inputs $r(t)$, $d_1(t)$, and $d_2(t)$ which represent a tracking and two disturbance inputs, respectively. The previous chapters are concerned with the tracking problem alone, with $d(t) = 0$. This section uses the pole-zero placement method for the development of a desired disturbance-rejection control-ratio model $(Y/R)_D = M_D$. The specification for disturbance rejection for the control system of Fig. 12.15, with a step disturbance input $d_2(t) = D_0 u_{-1}(t)$ and $r(t) = d_1(t) = 0$ is

$$|y(t)| \le \alpha_p \qquad \text{for } t \ge t_x \tag{12.35}$$

as shown in Fig. 12.16. Note that for this condition the initial value of the output is $y(0) = D_0$. When $d_1(t) = D_0 u_{-1}(t)$ and $r(t) = d_2(t) = 0$, then the requirement for disturbance rejection is

$$|y(t_p)| \le \alpha_p \tag{12.36}$$

as shown in Fig. 12.17, where $y(0) = 0$. The simplest control-ratio models for Figs. 12.16 and 12.17, respectively, are shown in Table 12.1.

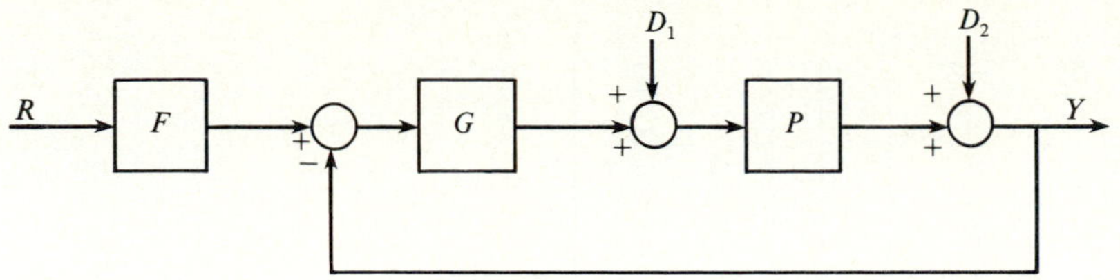

FIGURE 12.15
A multiple-input single-output (MISO) system.

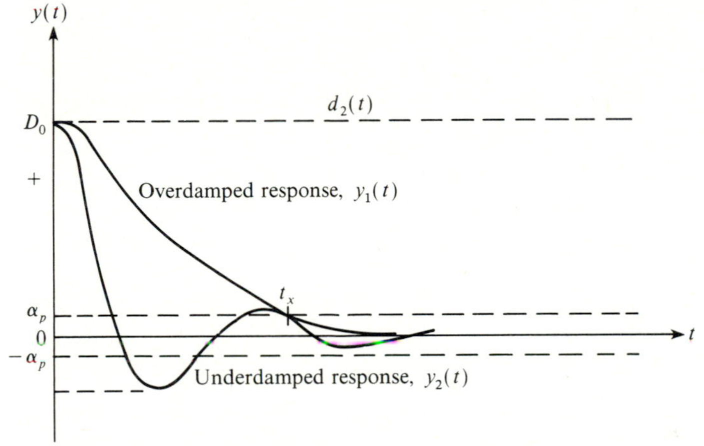

FIGURE 12.16
Disturbance-rejection time response for Fig. 12.15 with $r(t) = d_1(t) = 0$ and $d_2(t) = D_0 u_{-1}(t)$.

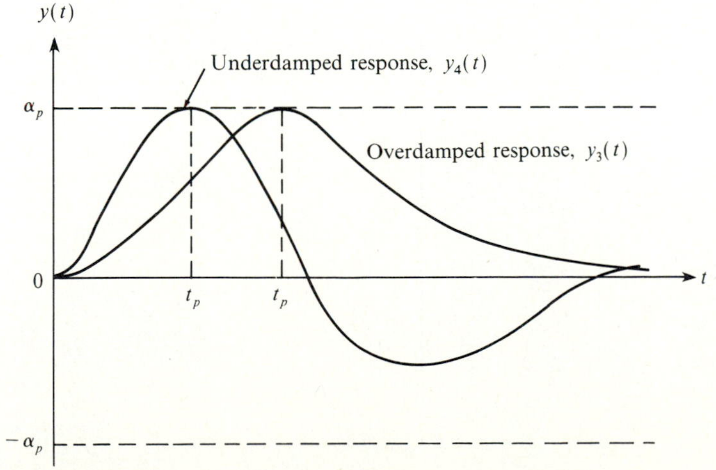

FIGURE 12.17
Disturbance-rejection time response for Fig. 12.15 with $r(t) = d_2(t) = 0$ and $d_1(t) = D_0 u_{-1}(t)$.

TABLE 12.1
Disturbance rejection control ratio models

$r(t) = d_1(t) = 0$ and $d_2(t) = D_0 u_{-1}(t)$ **(See Fig. 12.16)**					
$M_D = \dfrac{Y_1(s)}{D_2(s)} = \dfrac{s}{s+a}$	(12.37)	$Y_1(s) = \dfrac{D_0}{s+a}$		$y_1(t) = D_0 e^{-at}$	(12.39)
$M_D = \dfrac{Y_2(s)}{D_2(s)} = \dfrac{s(s+a)}{(s+a)^2+b^2}$	(12.38)	$Y_2(s) = \dfrac{D_0(s+a)}{(s+a)^2+b^2}$		$y_2(t) = D_0 e^{-at} \cos bt$	(12.40)
$r(t) = d_2(t) = 0$ and $d_1(t) = D_0 u_{-1}(t)$ **(See Fig. 12.17)**					
$M_D = \dfrac{Y_3(s)}{D(s)} = \dfrac{K_D s}{(s+a)(s+b)}$	(12.41)	$Y_3(s) = \dfrac{K_D D_0}{(s+a)(s+b)}$		$y_3(t) = \dfrac{K_D D_0}{b-a}(e^{-at} - e^{-bt})$	(12.43)
$M_D = \dfrac{Y_4(s)}{D_1(s)} = \dfrac{K_D s}{(s+a)^2+b^2}$	(12.42)	$Y_4(s) = \dfrac{K_D D_0}{(s+a)^2+b^2}$		$y_4(t) = \dfrac{K_D D_0}{b} e^{-at} \sin bt$	(12.44)

With the values of α_p, t_x, and D_0 specified, the value of a for the case represented by Eqs. (12.37) and (12.39) is given by

$$a = -\frac{1}{t_x}\ln\left(\frac{\alpha_p}{D_0}\right) \tag{12.45}$$

For the case represented by Eqs. (12.38) and (12.40), then

$$a = -\frac{1}{t_x}\ln\left(\frac{\alpha_p}{D_0\cos bt_x}\right) \tag{12.46}$$

In Eqs. (12.38) the two quantities a and b must be determined. Since Eq. (12.46) shows that a is a function of b, a trial-and-error procedure can be used to determine suitable values. Thus, assume a value of b and determine the corresponding value of a. If the time-response characteristics are satisfactory, the model is completely determined. If not, try another value for b.

Consider the case where the settling time specification is given by

$$|y(t_s)| = |D_0 e^{-at_s}\cos bt_s| = 0.02\alpha_p = \alpha_s \tag{12.47}$$

that is, the output reaches 2 percent of the maximum value at $t = t_s$. With t_s specified, then from Eq. (12.40) obtain the value of a, i.e.,

$$a = -\frac{1}{t_s}\ln\left(\frac{\alpha_s}{D_0\cos bt_s}\right) \tag{12.48}$$

If all four quantities (t_p, t_s, α_p, and D_0) are specified, then the intersection, if one exists, of the plots of Eqs. (12.46) and (12.48) yields the value of a that meets all the specifications.

For the case represented by Eqs. (12.41) and (12.43), assume that the values of α_p, α_s, t_p, t_s, and D_0 are specified. Then, from Eq. (12.43):

$$\alpha_p = A_\alpha(e^{-at_p} - e^{-bt_p}) > 0 \tag{12.49}$$

where $K_\alpha = K_D D_0/(b - a)$. Also from Eq. (12.43)

$$\left.\frac{dc(t)}{dt}\right|_{t=t_p} = K_\alpha(-ae^{-at_p} + be^{-bt_p}) = 0 \tag{12.50}$$

At $t = t_s$, Eq. (12.43) yields

$$\alpha_s = K_\alpha(e^{-at_s} - e^{-bt_s}) \tag{12.51}$$

From Eqs. (12.49) and (12.51) obtain, respectively,

$$e^{-at_p} = \frac{\alpha_p}{K_\alpha} + e^{-bt_p} \tag{12.52}$$

and

$$\frac{b}{a} = e^{(b-a)t_p} \tag{12.53}$$

Also, from Eq. (12.51) with $K_\alpha > 0$, obtain

$$e^{-at_s} > e^{-bt_s} \rightarrow 1 > \frac{e^{-(b/a)at_s}}{e^{-at_s}} = e^{(1-(b/a))at_s} \tag{12.54}$$

Equation (12.54) is valid only when $b/a > 1$, or

$$b > a \tag{12.55}$$

Thus, assume a value for b/a and for various values of b (or a) obtain values for a (or b) from Eqs. (12.51) and (12.53). Obtain a plot of a vs. b for each of these equations. If these plots intersect, then the values of a or b at the intersection result in a transfer function $M_D(s)$ that satisfies the desired performance specifications. If no intersection exists, assume another value for b/a and repeat the process.

If the desired performance specifications cannot be obtained by the simple models of Table 12.1, then poles and/or zeros must be added, by a trial-and-error procedure, until a model is synthesized that is satisfactory. It may be necessary, for either case, to add nondominant poles and/or zeros to $M_D(s)$ in order that the degrees of its numerator and denominator polynomials are consistent with the degrees of the numerator and denominator polynomials of the control ratio of the actual control system.

12.9 SUMMARY

This chapter presents the concepts of synthesizing a desired control-ratio model for a system's output to either track a desired input or to reject a disturbance input. These models are obtained based upon the known characteristics of a second-order system. Once these simple models are obtained, nondominant poles and/or zeros may be added to ensure that the final model is compatible with the degree of the numerator and denominator polynomials of the control ratio of the actual control system. The Guillemin-Truxal method is presented as one design method that requires the use of a tracking model. A disturbance-rejection frequency-domain design technique is presented. This technique is based upon the known correlation between the frequency domain, time domain, and the s-plane.

REFERENCES

1. Truxal, J. G.: *Automatic Feedback Control System Synthesis*, McGraw-Hill, New York, 1955.
2. Houpis, C. H., and G. B. Lamont: *Digital Control Systems: Theory, Hardware, Software*, McGraw-Hill, New York, 1985.
3. D'Azzo, J. J., and C. H. Houpis: *Feedback Control System Analysis and Synthesis*, 2nd ed., McGraw-Hill, New York, 1966.

CHAPTER 13

CLOSED-LOOP POLE-ZERO ASSIGNMENT (STATE-VARIABLE FEEDBACK)[9-14]

13.1 INTRODUCTION

Chapters 10, 11, and 12 deal with the improvement of system performance in the time and frequency domains, respectively, by conventional design procedures. Such procedures are based on an analysis of the open-loop transfer function and yield an improved closed-loop response as a consequence of properly designed cascade and/or feedback compensators. The design method presented in this chapter is based upon achieving a desired or model control ratio $M_T(s)$, i.e., this is a pole-zero placement method. Section 12.2 describes the modeling of a desired closed-loop transfer function and Sec. 12.3 presents the Guillemin-Truxal method of designing a cascade compensator to yield this closed-loop transfer function. Modern control theory introduces the concept of using all the system states to provide the desired improvement in system performance. This chapter illustrates how conventional control-theory design concepts are used to design state-variable feedback. The state-variable-feedback concept requires that all states be accessible in a physical system, but for most systems this requirement is not met; i.e., some of the states are inaccessible. Techniques for handling systems with inaccessible states are presented in Chaps. 14 and 18. The state-variable feedback design method presented in this chapter is based upon achieving a desired control ratio for a *single-input single-output* (SISO) system. Synthesis methods using

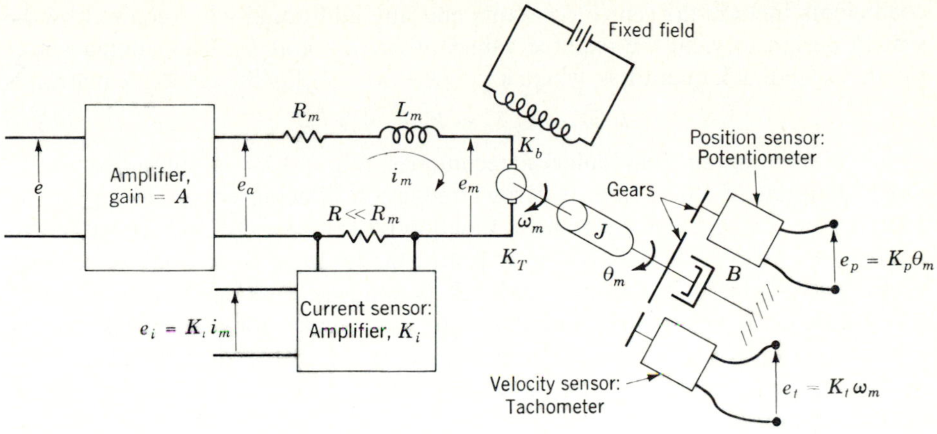

FIGURE 13.1
Open-loop position-control system and measurement of states.

matrix methods, for both SISO and *multiple-input multiple-output* (MIMO) *systems*, are presented in Chap. 18. Following the convention often used with the state-variable representation of systems, the output and actuating signals are represented by y and u instead of c and e, respectively, in the remainder of this text.

13.2 STATE-VARIABLE FEEDBACK[1]

The simple open-loop position-control system of Fig. 13.1 is used to illustrate the effects of state-variable feedback. The dynamic equations for this basic system are

$$e_a = Ae \qquad A > 0 \tag{13.1}$$

$$e_a - e_m = (R_m + L_m D)i_m \tag{13.2}$$

$$e_m = K_b\omega_m \tag{13.3}$$

$$T = K_T i_m = JD\omega_m + B\omega_m \tag{13.4}$$

$$\omega_m = D\theta_m \tag{13.5}$$

In order to obtain the maximum number of accessible (measurable) states, the *physical variables are utilized as state variables*. The two variables i_m and ω_m and the desired output quantity θ_m are identified as the three state variables: $x_1 = \theta_m = y$, $x_2 = \omega_m = \dot{x}_1$, and $x_3 = i_m$, and the input is $u = e$. Figure 13.1 shows that all three state variables are accessible for measurement. The sensors are selected to produce voltages which are proportional to the state variables.

Figure 13.2a is the block diagram of a closed-loop position-control system utilizing the system of Fig. 13.1. Each of the states is fed back through amplifiers whose gain values k_1, k_2, and k_3 are called *feedback coefficients*. These feedback coefficients include the sensor constants and any additional gain required by the system design to yield the required values of k_1, k_2, and k_3. The summation of the three feedback quantities, where $\mathbf{k}^T = [k_1 \quad k_2 \quad k_3]$ is the feedback matrix, is

$$k_1X_1 + k_2X_2 + k_3X_3 = \mathbf{k}^T\mathbf{X} \tag{13.6}$$

Using block-diagram manipulation techniques, Fig. 13.2a is simplified to the block diagram of Fig. 13.2b. A further reduction is achieved as shown in Fig. 13.2c. Note that the forward transfer function $G(s) = X_1(s)/U(s)$ shown in this figure is identical with that obtained from Fig. 13.2a with zero-state feedback coefficients. Thus, state-variable feedback is *equivalent* to inserting a feedback compensator $H_{eq}(s)$ in the feedback loop of a conventional feedback control system, as shown in Fig. 13.2d. The reduction of Fig. 13.2a to that of Fig. 13.2c or Fig. 13.2d is referred to as the *H-equivalent reduction*. The system control ratio† is therefore

$$\frac{Y(s)}{R(s)} = \frac{G(s)}{1 + G(s)H_{eq}(s)} \tag{13.7}$$

For Fig. 13.2c the following equations are obtained:

$$H_{eq}(s) = \frac{k_3s^2 + (2k_2 + k_3)s + 2k_1}{2} \tag{13.8}$$

$$G(s) = \frac{200A}{s(s^2 + 6s + 25)} \tag{13.9}$$

$$G(s)H_{eq}(s) = \frac{100A\left[k_3s^2 + (2k_2 + k_3)s + 2k_1\right]}{s(s^2 + 6s + 25)} \tag{13.10}$$

$$\frac{Y(s)}{R(s)} = \frac{200A}{s^3 + (6 + 100k_3A)s^2 + (25 + 200k_2A + 100k_3A)s + 200k_1A} \tag{13.11}$$

The equivalent feedback transfer function $H_{eq}(s)$ must be designed so that the system's desired figures of merit are satisfied. The following observations are noted from Eqs. (13.8) to (13.11) for this system in which $G(s)$ has no zeros; i.e., it is an all-pole plant:

1. In order to achieve zero steady-state error for a step input, $R(s) = R_0/s$, Eq. (13.11) must yield

$$y(t)_{ss} = \lim_{s \to 0} sY(s) = \frac{R_0}{k_1} = R_0$$

   This requires that $k_1 = 1$.
2. The numerator of $H_{eq}(s)$ is of second order; i.e., the denominator of $G(s)$ is of degree $n = 3$ and the numerator of $H_{eq}(s)$ is of degree $n - 1 = 2$.

† See also Prob. 13.1.

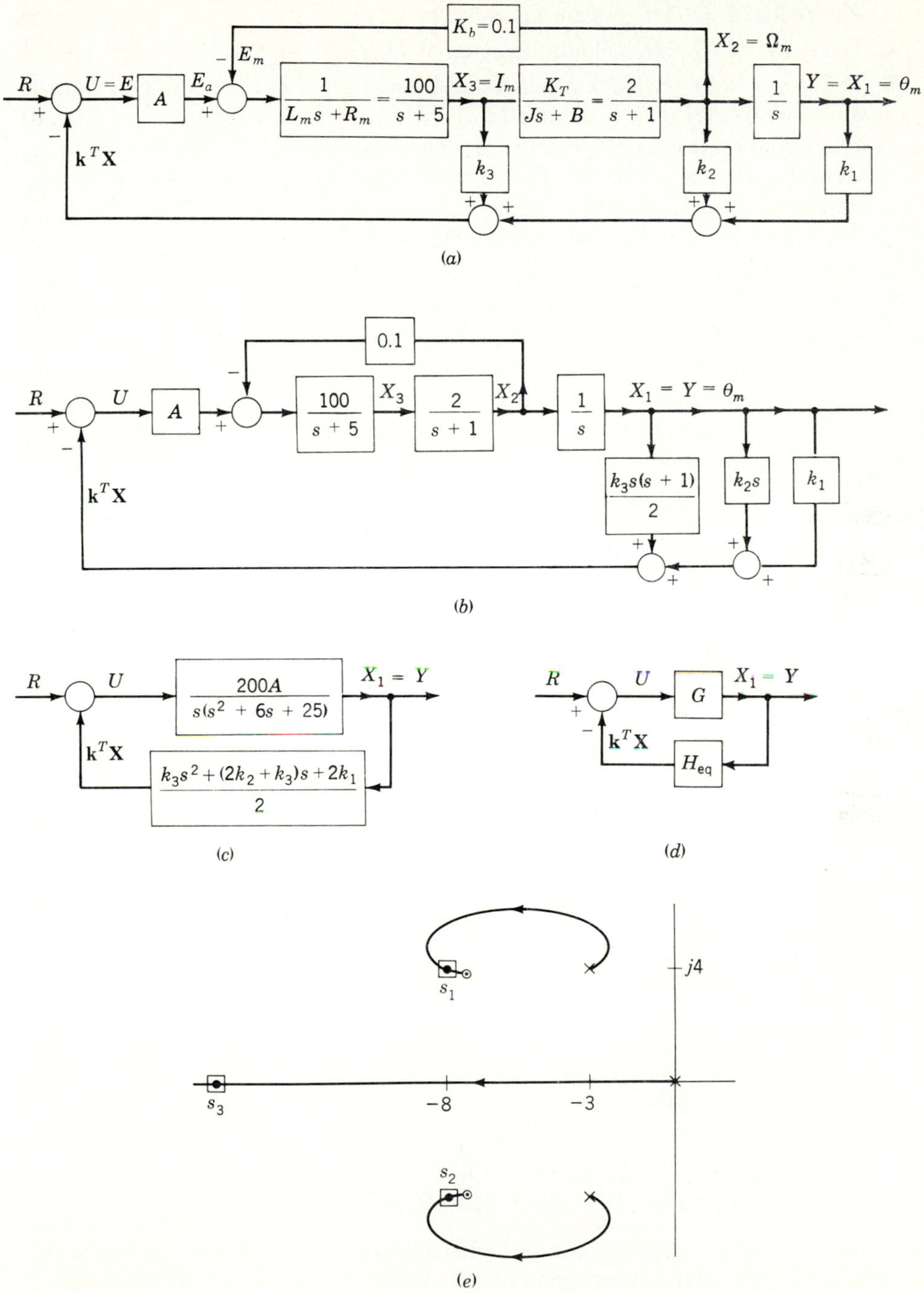

FIGURE 13.2
A position-control system with state feedback, its H_{eq} representation, and the corresponding root locus.

3. The poles of $G(s)H_{eq}(s)$ are the poles of $G(s)$.
4. The zeros of $G(s)H_{eq}(s)$ are the zeros of $H_{eq}(s)$.
5. The use of state feedback produces additional zeros in the open-loop transfer function $G(s)H_{eq}(s)$ without adding poles. This is in contrast to cascade compensation using passive networks. The zeros of $H_{eq}(s)$ are to be located to yield the desired response. It is known that the addition of zeros to the open-loop transfer function moves the root locus to the left; i.e., they make the system more stable and improve the time-response characteristics.
6. Since the root-locus plot of $G(s)H_{eq}(s) = -1$ has one asymptote, $\gamma = -180°$ for $k_3 > 0$, it is possible to pick the locations of the zeros of $H_{eq}(s)$ to ensure a completely stable system for all positive values of gain. Figure 13.2*e* shows a possible selection of the zeros and the resulting root locus. Note that when the conventional cascade compensators of Chap. 10 are used, the angles of the asymptotes yield branches that go, in general, into the right-half s plane. The maximum gain is therefore limited in order to maintain system stability. This restriction is removed with state feedback.
7. The desired figures of merit, for an underdamped response, are satisfied by selecting the desired locations of the poles of $Y(s)/R(s)$ of Eq. (13.11) at $s_{1,2} = -a \pm jb$ and $s_3 = -c$. For these specified pole locations the known characteristic equation has the form.

$$(s + a \pm jb)(s + c) = s^3 + d_2 s^2 + d_1 s + d_0 = 0 \tag{13.12}$$

Equating the denominator of Eq. (13.11) to Eq. (13.12) yields

$$d_2 = 6 + 100k_3 A \tag{13.13}$$

$$d_1 = 25 + 200k_2 A + 100k_3 A \tag{13.14}$$

$$d_0 = 200k_1 A = 200A \tag{13.15}$$

These three equations with three unknowns can be solved for the required values of A, k_2, and k_3.

It should be noted that if the basic system $G(s)$ has any zeros, they become poles of $H_{eq}(s)$. Thus, the zeros of $G(s)$ cancel the poles of $H_{eq}(s)$ in $G(s)H_{eq}(s)$. Also, the zeros of $G(s)$ become the zeros of the control ratio. The reader is able to verify this property by replacing $2/(s+1)$ in Fig. 13.2*a* by $2(s+\alpha)/(s+1)$ and performing an H-equivalent reduction. This property is also illustrated in the examples of Sec. 13.10

13.3 GENERAL PROPERTIES OF STATE FEEDBACK (USING PHASE VARIABLES)

In order to obtain general state-variable-feedback properties that are applicable to any $G(s)$ transfer function, consider the minimum-phase transfer function

$$G(s) = \frac{Y(s)}{U(s)} = \frac{K_G(s^w + c_{w-1}s^{w-1} + \cdots + c_1 s + c_0)}{s^n + a_{n-1}s^{n-1} + \cdots + a_1 s + a_0} \tag{13.16}$$

where $K_G > 0$, $w < n$, and the c_i elements are all positive quantities. The system characteristics and properties are most readily obtained by representing this system in terms of phase variables (see Secs. 5.6 and 5.8 with $c_w = 1$ and u replaced by $K_G u$):

$$\dot{\mathbf{x}} = \begin{bmatrix} 0 & 1 & 0 & \cdots & 0 \\ 0 & 0 & 1 & \cdots & 0 \\ \cdots & \cdots & \cdots & \cdots & 1 \\ -a_0 & -a_1 & -a_2 & \cdots & -a_{n-1} \end{bmatrix}\mathbf{x} + \begin{bmatrix} 0 \\ 0 \\ \cdot \\ K_G \end{bmatrix} u \qquad (13.17)$$

$$y = [c_0 \quad c_1 \quad \cdots \quad c_{w-1} \quad 1 \quad 0 \quad \cdots \quad 0]\mathbf{x} = \mathbf{c}_c^T\mathbf{x} \qquad (13.18)$$

From Fig. 13.2d and Eq. (13.18)

$$H_{eq}(s) = \frac{\mathbf{k}_c^T\mathbf{X}(s)}{Y(s)} = \frac{\mathbf{k}_c^T\mathbf{X}(s)}{\mathbf{c}_c^T\mathbf{X}(s)} = \frac{k_1X_1(s) + k_2X_2(s) + \cdots + k_nX_n(s)}{c_0X_1(s) + c_1X_2(s) + \cdots + X_{w+1}} \qquad (13.19)$$

The state-feedback matrix is $\mathbf{k}_c^T$ since the state variables are phase variables. Using the phase-variable relationship $X_j(s) = s^{j-1}X_1(s)$, Eq. (13.19) becomes

$$H_{eq}(s) = \frac{k_n s^{n-1} + k_{n-1}s^{n-2} + \cdots + k_2 s + k_1}{s^w + c_{w-1}s^{w-1} + \cdots + c_1 s + c_0} \qquad (13.20)$$

Thus, from Eqs. (13.16) and (13.20), after cancellation of common terms,

$$G(s)H_{eq}(s) = \frac{K_G(k_n s^{n-1} + k_{n-1}s^{n-2} + \cdots + k_2 s + k_1)}{s^n + a_{n-1}s^{n-1} + \cdots + a_1 s + a_0} \qquad (13.21a)$$

Factoring k_n from the numerator polynomial gives an alternate form of this equation

$$G(s)H_{eq}(s) = \frac{K_G k_n(s^{n-1} + \cdots + \alpha_1 s + \alpha_0)}{s^n + a_{n-1}s^{n-1} + \cdots + a_1 s + a_0} \qquad (13.21b)$$

where the loop sensitivity is $K_G k_n = K$. Substituting Eqs. (13.16) and (13.21a) into Eq. (13.7) yields

$$\frac{Y(s)}{R(s)} = \frac{K_G(s^w + c_{w-1}s^{w-1} + \cdots + c_0)}{s^n + (a_{n-1} + K_G k_n)s^{n-1} + \cdots + (a_0 + K_G k_1)} \qquad (13.22)$$

An analysis of Eqs. (13.17), (13.20), (13.21), and (13.22) yields the following conclusions:

1. The numerator of $H_{eq}(s)$ is a polynomial of degree $n - 1$ in s; that is, it has $n - 1$ zeros. The values of the coefficients k_i of this numerator polynomial, and thus the $n - 1$ zeros of $H_{eq}(s)$, can be selected by the system designer.
2. The numerator of $G(s)$, except for K_G, is equal to the denominator of $H_{eq}(s)$. Therefore, the open-loop transfer function $G(s)H_{eq}(s)$ has the same n poles as $G(s)$ and has the $n - 1$ zeros of $H_{eq}(s)$ which are determined in the design

process. [An exception to this occurs when physical variables are utilized and the transfer function $X_n(s)/U(s)$ contains a zero. In that case this zero does not appear as a pole of $H_{eq}(s)$, and the numerator and denominator of $G(s)H_{eq}(s)$ are both of degree n.]

3. The root-locus diagram, based on Eq. (13.21*b*) for $k_n > 0$, reveals that $n - 1$ branches terminate on the $n - 1$ arbitrary zeros which can be located anywhere in the s plane by properly selecting the values of the k_i. There is a single asymptote at $\gamma = -180°$, and therefore one branch terminates on the negative real axis at $s = -\infty$. System stability is ensured for high values of K_G if all the zeros of $H_{eq}(s)$ are located in the left-half s plane. For $k_n < 0$ a single asymptote exists at $\gamma = 0°$; thus the system becomes unstable for high values of K_G.
4. State-variable feedback provides the ability to select the locations of the zeros of $G(s)H_{eq}(s)$, as given by Eq. (13.21*b*), in order to locate closed-loop system poles which yield the desired system performance. From Eq. (13.22) it is noted that the result of feedback of the states through constant gains k_i is to change the closed-loop poles and to leave the system zeros, if any, the same as those of $G(s)$. (This requires the plant to be controllable, as described in Sec. 13.7.) It may be desirable to specify that one or more of the closed-loop poles cancel one or more unwanted zeros in $Y(s)/R(s)$.
5. In synthesizing the desired $Y(s)/R(s)$ the following limitation must be observed: the number of poles minus the number of zeros of $Y(s)/R(s)$ must equal the number of poles minus the number of zeros of $G(s)$.

A thorough steady-state error analysis of Eq. (13.22) is made in Secs. 13.4 to 13.6 for the standard step, ramp, and parabolic inputs. From the conclusions obtained in that section, some or all of the feedback gains k_1, k_2, and k_3 are specified.

Summarizing, the $H_{eq}(s)$ can be designed to yield the specified closed-loop transfer function in the manner described for the system of Fig. 13.2. A root-locus sketch of $G(s)H_{eq}(s) = \pm 1$ can be used as an aid in synthesizing the desired $Y(s)/R(s)$ function. From this sketch it is possible to determine the locations of the closed-loop poles, in relation to any system zeros, that will achieve the desired performance specifications. Note that although the zeros of $G(s)$ are *not* involved in obtaining the root locus for $G(s)H_{eq}(s)$, they do appear in $Y(s)/R(s)$ and thus affect the time function $y(t)$. The design procedure is summarized as follows:

Step 1. Draw the system state-variable block diagram when applicable.

Step 2. Assume that all state variables are accessible.

Step 3. Obtain $H_{eq}(s)$ and $Y(s)/R(s)$ in terms of the k_i's.

Step 4. Determine the value of k_1 required for zero steady-state error for a unit step input, $u_{-1}(t)$; that is $y(t)_{ss} = \lim_{s \to 0} sY(s) = 1$.

Step 5. Synthesize the desired $[Y(s)/R(s)]_D$ from the desired performance specifications for M_p, t_p, t_s, and steady-state error requirements (see Sec. 12.2). It may also be desirable to specify that one or more of the closed-loop poles should cancel unwanted zeros of $Y(s)/R(s)$.

Step 6. Set the closed-loop transfer functions obtained in steps 3 and 5 equal to each other. Equate the coefficients of like powers of s of the denominator polynomials and solve for the required values of the k_i's. If phase variables are used in the design, a linear transformation must be made to convert the phase-variable-feedback coefficients required for the physical variables present in the control system.

If some of the states are not accessible, suitable cascade or minor-loop compensators can be determined using the known values of the k_i's. This method is discussed in Sec. 14.5.

The procedure for designing *a state-variable-feedback system* is illustrated by means of examples using physical variables in Secs. 13.7 through 13.10. Before the examples are considered, however, the steady-state error properties are investigated in Secs. 13.4 through 13.6.

13.4 STATE VARIABLE FEEDBACK: STEADY-STATE ERROR ANALYSIS[2]

A steady-state error analysis of a stable control system is an important design consideration as pointed out in Chap. 6. The three standard inputs are utilized for a steady-state analysis of a state-variable-feedback control system. The phase-variable representation is used in the following developments, in which the necessary values of the feedback coefficients are determined. The results apply only when phase variables are used and do not apply when general state variables are used.

Step Input $r(t) = R_0 u_{-1}(t)$, $R(s) = R_0/s$

Solving Eq. (13.22), which is based upon the phase-variable representation, for $Y(s)$ and applying the final-value theorem yields, for a step input,

$$y(t)_{ss} = \frac{K_G c_0 R_0}{a_0 + K_G k_1} \tag{13.23}$$

The system error is defined by

$$e(t) \equiv r(t) - y(t) \tag{13.24}$$

For zero steady-state error, $e(t)_{ss} = 0$, the steady-state output must be equal to the input:

$$y(t)_{ss} = r(t) = R_0 = \frac{K_G c_0 R_0}{a_0 + K_G k_1} \tag{13.25}$$

where $c_0 \neq 0$. This requires that $K_G c_0 = a_0 + K_G k_1$, or

$$k_1 = c_0 - \frac{a_0}{K_G} \tag{13.26}$$

Thus, zero steady-state error with a step input can be achieved with state-variable feedback even if $G(s)$ is Type 0. For an all-pole plant $c_0 = 1$, and for a Type 1 or higher plant $a_0 = 0$. In that case the required value is $k_1 = 1$. *Therefore, the feedback coefficient k_1 is determined by the plant parameters once zero steady-state error for a step input is specified*. This specification is maintained for all state-variable-feedback control-system designs throughout the remainder of this text. From Eq. (13.22), the system characteristic equation is

$$\begin{aligned} s^n + (a_{n-1} + K_G k_n)s^{n-1} + \cdots + (a_2 + K_G k_3)s^2 \\ + (a_1 + K_G k_2)s + (a_0 + K_G k_1) = 0 \end{aligned} \tag{13.27}$$

A necessary condition for all closed-loop poles of $Y(s)/R(s)$ to be in the left-half s plane is that all coefficients in Eq. (13.27) must be positive. Therefore, the constant term of the polynomial $a_0 + K_G k_1 = K_G c_0$ must be positive.

Ramp Input $r(t) = R_1 u_{-2}(t) = R_1 t u_{-1}(t)$, $R(s) = R_1/s^2$

To perform the analysis for ramp and parabolic inputs the error function $E(s)$ must first be determined. Solving for $Y(s)$ from Eq. (13.22) and evaluating $E(s) = R(s) - Y(s)$ yields

$$E(s) = R(s)\frac{\begin{aligned} s^n + (a_{n-1} + K_G k_n)s^{n-1} + \cdots + (a_0 + K_G k_1) \\ - K_G\left(s^w + c_{w-1}s^{w-1} + \cdots + c_1 s + c_0\right) \end{aligned}}{s^n + (a_{n-1} + K_G k_n)s^{n-1} + \cdots + (a_1 + K_G k_2)s + (a_0 + K_G k_1)} \tag{13.28}$$

Substituting Eq. (13.26), the requirement for zero steady-state error for a step input, into Eq. (13.28) produces a cancellation of the constant terms in the numerator.

$$E(s) = R(s)\frac{\begin{aligned} s^n + (a_{n-1} + K_G k_n)s^{n-1} + \cdots + (a_1 + K_G k_2)s \\ - K_G\left(s^w + c_{w-1}s^{w-1} + \cdots + c_1 s\right) \end{aligned}}{s^n + (a_{n-1} + K_G k_n)s^{n-1} + \cdots + (a_1 + K_G K_2)s + K_G c_0} \tag{13.29}$$

For a ramp input the steady-state output is

$$e(t)_{ss} = \lim_{s \to 0} sE(s) = \frac{a_1 + K_G k_2 - K_G c_1}{K_G c_0} R_1 \tag{13.30}$$

In order to obtain $e(t)_{ss} = 0$ it is necessary that $a_1 + K_G k_2 - K_G c_1 = 0$. This is achieved when

$$k_2 = c_1 - \frac{a_1}{K_G} \tag{13.31}$$

For an all-pole plant $c_0 = 1$ and $c_1 = 0$; thus $k_2 = -a_1/K_G$. In that case the coefficient of s in the system characteristic equation (13.27) is zero, and the system is unstable. *Therefore, zero steady-state error for a ramp input cannot be achieved by state-variable feedback for an all-pole stable system.* If $a_1 + K_G k_2 > 0$, in order to maintain a positive coefficient of s in Eq. (13.27), the system may be stable but a finite steady-state error exists. The error in an all-pole plant is

$$e(t)_{ss} = \left(k_2 + \frac{a_1}{K_G}\right) R_1 \tag{13.32}$$

When a plant contains at least one zero so that $c_1 > 0$, it is possible to specify the value of k_2 given by Eq. (13.31) and to have a stable system. In that case the system has zero steady-state error with a ramp input. Substituting Eq. (13.31) into Eq. (13.27) results in the positive coefficient $K_G c_1$ for the first power of s. *Thus, a stable state-variable-feedback system having zero steady-state error with both a step and a ramp input can be achieved only when the control ratio has at least one zero.* Note that this occurs even though $G(s)$ may be Type 0 or 1.

Parabolic Input $r(t) = R_2 u_{-3}(t) = (R_2 t^2/2) u_{-1}(t)$, $R(s) = R_2/s^3$

The all-pole and pole-zero plants are considered separately for this type of input. For the all-pole plant with Eq. (13.26) satisfied, Eq. (13.28) reduces to

$$E(s) = \frac{R_2}{s^3} \frac{s^n + (a_{n-1} + K_G k_n)s^{n-1} + \cdots + (a_1 + K_G k_2)s}{s^n + (a_{n-1} + K_G k_n)s^{n-1} + \cdots + (a_1 + K_G k_2)s + K_G c_0} \tag{13.33}$$

Since $e(t)_{ss} = \lim_{s \to 0} sE(s) = \infty$, *it is concluded that an all-pole plant cannot follow a parabolic input*. This is true even if Eq. (13.26) is not satisfied.

Under the condition that $e(t)_{ss} = 0$ for both step and ramp inputs to a pole-zero stable system, Eqs. (13.26) and (13.31) are substituted into Eq. (13.29) to yield, for a plant containing at least one zero,

$$E(s) = \frac{R_2}{s^3} \frac{\begin{array}{r} s^n + (a_{n-1} + K_G k_n)s^{n-1} + \cdots + (a_2 + K_G k_3)s^2 \\ -K_G(c_w s^w + c_{w-1}s^{w-1} + \cdots + c_2 s^2) \end{array}}{s^n + (a_{n-1} + K_G k_n)s^{n-1} + \cdots + K_G c_1 s + K_G c_0} \tag{13.34}$$

Note that s^2 is the lowest power term in the numerator. Applying the final-value theorem to this equation yields

$$e(t)_{ss} = \lim_{s \to 0} sE(s) = \frac{a_2 + K_G k_3 - K_G c_2}{K_G c_0} R_2 \tag{13.35}$$

In order to achieve $e(t)_{ss} = 0$ it is necessary that $a_2 + K_G k_3 - K_G c_2 = 0$. Thus

$$k_3 = c_2 - \frac{a_2}{K_G} \tag{13.36}$$

With this value of k_3 the s^2 term in Eq. (13.27) has the positive coefficient $K_G c_2$ only when $w \geq 2$. *Thus, a stable system having zero steady-state error with a parabolic input can be achieved only for a pole-zero system with two or more zeros, even if $G(s)$ is Type 0.*

In a system having only one zero ($w = 1$), the value $c_2 = 0$ and $k_3 = -a_2/K_G$ produces a zero coefficient for the s^2 term in the characteristic equation (13.27), resulting in an unstable system. *Therefore, zero steady-state error with a parabolic input cannot be achieved by a state-variable-feedback pole-zero stable system with one zero.* If $k_3 \neq -a_2/K_G$ and $a_2 + K_G k_3 > 0$ are satisfied to maintain a positive coefficient of s^2 in Eq. (13.27), a finite error exists. The system can be stable, and the steady-state error is

$$e(t)_{ss} = \frac{R_2(a_2 + K_G k_3)}{K_G c_0} \tag{13.37}$$

13.5 USE OF STEADY-STATE ERROR COEFFICIENTS

For the example represented by Fig. 13.2*a* the feedback coefficient blocks are manipulated to yield the *H*-equivalent block diagram in Fig. 13.2*d*. Similarly a *G-equivalent* unity-feedback block diagram, as shown in Fig. 13.3, can be obtained. The overall transfer functions for Figs. 13.2*d* and 13.3 are, respectively,

$$M(s) = \frac{Y(s)}{R(s)} = \frac{G(s)}{1 + G(s)H_{eq}(s)} \tag{13.38}$$

$$M(s) = \frac{Y(s)}{R(s)} = \frac{G_{eq}(s)}{1 + G_{eq}(s)} \tag{13.39}$$

Solving for $G_{eq}(s)$ from Eq. (13.39) yields

$$G_{eq}(s) = \frac{M(s)}{1 - M(s)} \tag{13.40}$$

Using the control ratio $M(s)$ from Eq. (13.22) in Eq. (13.40) yields

$$G_{eq}(s) = \frac{K_G(s^w + \cdots + c_1 s + c_0)}{s^n + (a_{n-1} + K_G k_n)s^{n-1} + \cdots + (a_1 + K_G k_2)s + (a_0 + K_G k_1) - K_G(s^w + \cdots + c_1 s + c_0)} \tag{13.41}$$

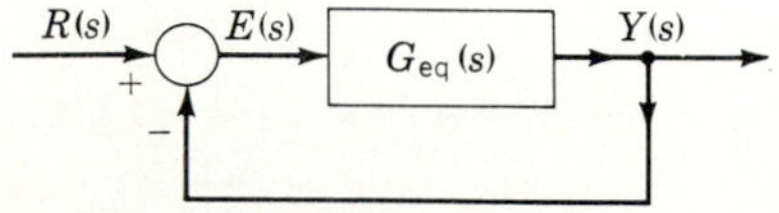

FIGURE 13.3
G-equivalent block diagram.

This is a Type 0 transfer function and has the step-error coefficient $K_p = K_G c_0/(a_0 + K_G k_1 - K_G c_0)$. When the condition $k_1 = c_0 - a_0/K_G$ is satisfied so that $e(t)_{ss} = 0$ with a step input, Eq. (13.41) reduces to

$$G_{eq}(s) = \frac{K_G(s^w + \cdots + c_1 s + c_0)}{s^n + (a_{n-1} + K_G k_n)s^{n-1} + \cdots + (a_1 + K_G k_2)s - K_G(s^w + \cdots + c_1 s)} \tag{13.42}$$

Since this is a Type 1 transfer function, the step error coefficient is $K_p = \lim_{s \to 0} [sG_{eq}(s)] = \infty$, and the ramp error coefficient is

$$K_v = K_G c_0/(a_1 + K_G k_2 - K_G c_1).$$

Satisfying the conditions for $e(t)_{ss} = 0$ for both step and ramp inputs, as given by Eqs. (13.26) and (13.31), Eq. (13.42) reduces to

$$G_{eq}(s) = \frac{K_G(s^w + \cdots + c_1 s + c_0)}{s^n + (a_{n-1} + K_G k_n)s^{n-1} + \cdots + (a_2 + K_G k_3)s^2 - K_G(s^w + \cdots + c_2 s^2)} \tag{13.43}$$

Since this is a Type 2 transfer function, there is zero steady-state error with a ramp input. It follows a parabolic input with a steady-state error. The error coefficients are $K_p = \infty$, $K_v = \infty$, and $K_a = K_G c_0/(a_2 + K_G k_3 - K_G c_2)$.

For the case when $w \geq 2$ and the conditions for $e(t)_{ss} = 0$ for all three standard inputs are satisfied [Eqs. (13.26), (13.31), and (13.42)], Eq. (13.43) reduces to

$$G_{eq}(s) = \frac{K_G(s^w + \cdots + c_1 s + c_0)}{s^n + (a_{n-1} + K_G k_n)s^{-1} + \cdots + (a_3 + K_G k_4)s^3 - K_G(s^w + \cdots + c_3 s^3)} \tag{13.44}$$

This is a Type 3 transfer function which has zero steady-state error with a parabolic input. The error coefficients are $K_p = K_v = K_a = \infty$. The results of the previous analysis are contained in Table 13.1. Table 13.2 summarizes the zero steady-state error requirements for a stable state-variable-feedback control system. As Tables 13.1 and 13.2 indicate, the specification on the type of input that

TABLE 13.1
State-variable-feedback system: steady-state error coefficients†

State-variable-feedback system	Number of zeros	K_p	K_v	K_a
All-pole plant	$w = 0$	∞	$\dfrac{K_G}{a_1 + K_G k_2}$	0
Pole-zero plant	$w = 1$	∞	∞	$\dfrac{K_G c_0}{a_2 + K_G k_3}$
	$w \geq 2$			∞

† For phase-variable representation.

TABLE 13.2
Stable state-variable-feedback system: requirements for zero steady-state error†

<table>
<tr><th rowspan="2">System</th><th rowspan="2">Number of zeros required</th><th colspan="3">Input</th><th rowspan="2">$G_{eq}(s)$ type</th></tr>
<tr><th>Step</th><th>Ramp</th><th>Parabola</th></tr>
<tr><td>All-pole plant</td><td>$w = 0$</td><td>$k_1 = 1 - \frac{a_0}{K_G}$</td><td>‡</td><td>§</td><td>1</td></tr>
<tr><td rowspan="2">Pole-zero plant</td><td>$w = 1$</td><td rowspan="2">$k_1 = c_0 - \frac{a_0}{K_G}$</td><td rowspan="2">$k_2 = c_1 - \frac{a_1}{K_G}$</td><td>‡</td><td>2</td></tr>
<tr><td>$w \geq 2$</td><td>$k_3 = c_2 - \frac{a_2}{K_G}$</td><td>3</td></tr>
</table>

† For phase-variable representation.
‡ A steady-state error exists.
§ Cannot follow.

the system must follow determines the number of zeros that must be included in $Y(s)/R(s)$. The number of zeros determines the ability of the system to follow a particular input with no steady-state error and to achieve a stable system.

With state-variable feedback any desired system type can be achieved for $G_{eq}(s)$, regardless of the type represented by $G(s)$. In order to achieve a Type m system the requirement, for $m > 0$, is that

$$a_{i-1} + K_G k_1 - K_G c_{i-1} = 0 \qquad \text{for all} \qquad i = 1, 2, 3, \ldots, m \tag{13.45}$$

Thus, a system in which the plant is not Type m can be made to behave as a Type m system by use of state feedback. This is achieved without adding pure integrators in cascade with the plant. *In order to achieve a Type m system it is seen from Table* 13.2 *that the control ratio must have a minimum of m* − 1 *zeros*. This is an important characteristic of state-variable-feedback systems.

The requirement for $e(t)_{ss} = 0$ for a ramp input can also be specified in terms of the locations of the zeros of the plant with respect to the desired locations of the poles of the control ratio (see Chap. 12). Expanding

$$\frac{E(s)}{R(s)} = \frac{1}{1 + G_{eq}(s)} \tag{13.46}$$

in a Maclaurin series in s yields

$$F(s) = \frac{E(s)}{R(s)} = e_0 + e_1 s + e_2 s^2 + \cdots + e_i s^i + \cdots \tag{13.47}$$

where the coefficients $e_0, e_1, e_2, \ldots$ are called the *generalized error coefficients*[3] and are given by

$$e_i = \frac{1}{i!}\left[\frac{d^i F(s)}{ds^i}\right]_{s=0} \tag{13.48}$$

From this equation, e_0 and e_1 are

$$e_0 = \frac{1}{1 + K_p} \tag{13.49}$$

$$e_1 = \frac{1}{K_v} \tag{13.50}$$

Note that e_0 and e_1 are expressed in terms of the system error coefficients defined in Chap. 6. From Eq. (13.42), in which the requirement of Eq. (13.26) is satisfied, $K_p = \infty$. Solving for $E(s)$ from Eq. (13.47) and substituting the result into $Y(s) = R(s) - E(s)$ yields

$$\frac{Y(s)}{R(s)} = 1 - e_0 - e_1 s - e_2 s^2 - \cdots \tag{13.51}$$

The relation between K_v and the poles and zeros of $Y(s)/R(s)$ is readily determined if the control ratio is written in factored form:

$$\frac{Y(s)}{R(s)} = \frac{K_G(s - z_1)(s - z_2)\cdots(s - z_w)}{(s - p_1)(s - p_2)\cdots(s - p_n)} \tag{13.52}$$

The first derivative of $Y(s)/R(s)$ in Eq. (13.51) with respect to s, evaluated at $s = 0$, yields $-e_1$ which is equal to $-1/K_v$.

When it is noted that $[Y(s)/R(s)]_{s=0} = 1$ for a Type 1 or higher system with unity feedback, the following equation can be written:

$$\frac{1}{K_v} = -\frac{\left[\dfrac{d}{ds}\left(\dfrac{Y}{R}\right)\right]_{s=0}}{\left[\dfrac{Y}{R}\right]_{s=0}} = -\left[\frac{d}{ds}\ln\frac{Y(s)}{R(s)}\right]_{s=0} \tag{13.53}$$

Substitution of Eq. (13.52) into Eq. (13.53) yields

$$\frac{1}{K_v} = -\left(\frac{1}{s - z_1} + \cdots + \frac{1}{s - z_w} - \frac{1}{s - p_1} - \cdots \frac{1}{s - p_n}\right)_{s=0}$$

$$= \sum_{j=1}^{n}\frac{1}{p_j} - \sum_{j=1}^{w}\frac{1}{z_j} \tag{13.54}$$

Thus, the condition for $e(t)_{ss} = 0$ for a ramp input, which occurs for $K_v = \infty$, that is, $1/K_v = 0$, is satisfied when

$$\sum_{j=1}^{n}\frac{1}{p_j} = \sum_{j=1}^{w}\frac{1}{z_j} \tag{13.55}$$

Therefore, in synthesizing the desired control ratio, its poles can be located not only to yield the desired transient response characteristics but also to yield the value $K_v = \infty$. In addition, it must contain at least one zero that is not canceled by a pole.

The values of k_1, k_2, and k_3 determined for $e(t)_{ss} = 0$ for the respective input from Eqs. (13.26), (13.31), and (13.36) are based upon the phase-variable representation of the system. These values may not be compatible with the desired roots of the characteristic equation. In that case it is necessary to leave at least one root unspecified so that these feedback gains may be used. If the resulting location of the unspecified root(s) is not satisfactory, other root locations may be tried in order to achieve a satisfactory performance. Once a system is designed which satisfies both the steady-state and the transient time responses, the feedback coefficients for the physical system can be calculated. This is accomplished by equating the corresponding coefficients of the characteristic equations representing, respectively, the physical- and the phase-variable systems. It is also possible to achieve the desired performance by working with the system represented entirely by physical variables, as illustrated by the examples in the following sections.

13.6 DEFINITION OF ZERO-ERROR SYSTEMS[4]

In general, the control ratio for a feedback control system has the form

$$\frac{Y(s)}{R(s)} = \frac{K_G\left(s^w + c_{w-1}s^{w-1} + \cdots + c_2s^2 + c_1s + c_0\right)}{s^n + q_{n-1}s^{n-1} + \cdots + q_2s^2 + q_1s + q_0} \tag{13.56}$$

The steady-state error for this system can be shown to be [see Eq. (13.47)]

$$e_{ss} = e_0r + e_1Dr + e_2D^2r + \cdots \tag{13.57}$$

The form of the input $r(t)$ determines the size of the steady-state error. Note that for the state-feedback systems of this chapter, a comparison of Eqs. (13.22) and (13.56) shows that the denominator coefficients are given by $q_{i-1} = a_{i-1} + K_Gk_i$, where $i = 1, 2, \ldots, n$. The values of $e_0, e_1, \ldots$ can be determined by putting Eq. (13.56) into the form of Eq. (13.51).

Since e_0 is a function of $q_0 - K_Gc_0$, the requirement for zero steady-state error with a step-function input is that $K_Gc_0 = q_0$. This also means that in a unity-feedback system the forward transfer function is Type 1 or higher. Since the degree of the numerator of $Y(s)/R(s)$ can be equal to or less than the degree of the denominator, there are many possible forms of $Y(s)/R(s)$ for which the steady-state error is zero with a step input. The system having the control ratio with only a constant in the numerator, that is, $w = 0$, is referred to as the *zero steady-state step-error system* and is given by

$$\frac{Y(s)}{R(s)} = \frac{q_0}{s^n + q_{n-1}s^{n-1} + \cdots + q_2s^2 + q_1s + q_0} \tag{13.58}$$

Since e_1, for $w = 1$, is a function of $q_0 - K_Gc_0$ and $q_1 - K_Gc_1$, zero steady-state error with a ramp-function input results when $K_Gc_0 = q_0$ and $K_Gc_1 = q_1$. Therefore the system is Type 2 or higher. *Zero steady-state ramp-error systems* are described by

$$\frac{Y(s)}{R(s)} = \frac{q_1s + q_0}{s^n + q_{n-1}s^{n-1} + \cdots + q_2s^2 + q_1s + q_0} \tag{13.59}$$

Since e_2, for $w = 2$, is a function of $q_0 - K_G c_0$, $q_1 - K_G c_1$, and $q_2 - K_G c_2$, zero steady-state error with a parabolic-function input results when $K_G c_0 = q_0$, $K_G c_1 = q_1$, and $K_G c_2 = q_2$. This means that the system is Type 3 or higher. Systems that have a quadratic numerator are referred to as *zero steady-state parabolic-error systems* and are described by

$$\frac{Y(s)}{R(s)} = \frac{q_2 s^2 + q_1 s + q_0}{s^n + q_{n-1}s^{n-1} + \cdots + q_2 s^2 + q_1 s + q_0} \tag{13.60}$$

These requirements for zero steady-state error, for all three standard type inputs, are also derived in Secs. 13.4 and 13.5.

13.7 CONTROLLABILITY AND OBSERVABILITY[5,6]

An important objective of modern control theory is the design of systems that have an optimum performance. The optimal control is based on the optimization of some specific performance criterion. The achievement of such optimal linear control systems is governed by the *controllability* and *observability* properties of the system. For example, in order to be able to relocate or reassign the open-loop plant poles to more desirable closed-loop locations in the s plane it is necessary that the plant satisfy the controllability property. Further, these properties establish the conditions for complete equivalence between the state-variable and transfer-function representations. A study of these properties, presented below, provides a basis for consideration of the optimal control problem, contained in Chap 16, and the assignment of the closed-loop eigenstructure (eigenvalue and eigenvector spectra) in Chaps. 18 through 20.

CONTROLLABILITY. A system is said to be completely *state-controllable* if, for any initial time t_0, each initial state $\mathbf{x}(t_0)$ can be transferred to any final state $\mathbf{x}(t_f)$ in a finite time, $t_f > t_0$, by means of an unconstrained control input vector $\mathbf{u}(t)$. An unconstrained control vector has no limit on the amplitudes of $\mathbf{u}(t)$. This definition implies that $\mathbf{u}(t)$ is able to affect each state variable in

$$\mathbf{x}(t) = \mathbf{\Phi}(t - t_0)\mathbf{x}(t_0) + \int_{t_0}^{t} \mathbf{\Phi}(t - \tau)\mathbf{B}\mathbf{u}(\tau)\, d\tau \tag{13.61}$$

OBSERVABILITY. A system is said to be completely *observable* if every initial state $\mathbf{x}(t_0)$ can be exactly determined from the measurements of the output $\mathbf{y}(t)$ over the finite interval of time $t_0 \le t \le t_f$. This definition implies that every state of $\mathbf{x}(t)$ affects the output $\mathbf{y}(t)$:

$$\mathbf{y}(t) = \mathbf{C}\mathbf{x}(t) = \mathbf{C}\mathbf{\Phi}(t - t_0)\mathbf{x}(t_0) + \mathbf{C}\int_{t_0}^{t} \mathbf{\Phi}(t - \tau)\mathbf{B}\mathbf{u}(\tau)\, d\tau \tag{13.62}$$

where the initial state $\mathbf{x}(t_0)$ is the result of control inputs prior to t_0.

The concepts of controllability and observability[7] can be illustrated graphically by the block diagram of Fig. 13.4. By the proper selection of the state

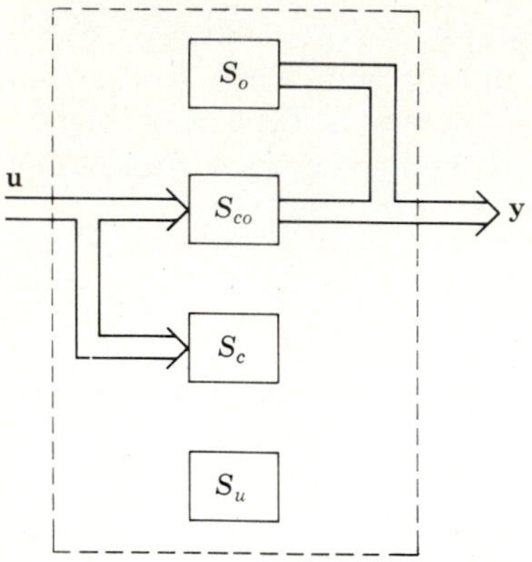

FIGURE 13.4
The four possible subdividions of a system.

variables it is possible to divide a system into the four subdivisions shown in Fig. 13.4.

S_{co} = completely controllable and completely observable subsystem
S_o = completely observable but uncontrollable subsystem
S_c = completely controllable but unobservable subsystem
S_u = uncontrollable and unobservable subsystem

An inspection of this figure readily reveals that only the controllable and observable subsystem S_{co} satisfies the definition of a transfer-function matrix, that is,

$$\mathbf{G}(s)\mathbf{U}(s) = \mathbf{Y}(s) \tag{13.63}$$

If the entire system is completely controllable and completely observable, then the state-variable and transfer-function matrix representations of a system are equivalent and accurately represent the system. In that case the resulting transfer function does carry all the information characterizing the dynamic performance of the system. As mentioned in Chap. 2, there is no unique method of selecting the state variables to be used in representing a system, but the transfer-function matrix is completely and uniquely specified once the state-variable representation of the system is known. In the transfer-function approach the state vector is suppressed, i.e., the transfer-function method is concerned only with the system's input-output characteristics. The state-variable method also includes a description of the system's internal behavior.

The determination of controllability and observability of a system by subdividing into its four possible subdivisions is not easy. A simple method for determining these system characteristics is developed by using Eq. (13.61) with $t_0 = 0$ and specifying that the final state vector $\mathbf{x}(t_f) = \mathbf{0}$. This yields

$$\mathbf{0} = e^{\mathbf{A}t_f}\mathbf{x}(0) + \int_0^{t_f} e^{\mathbf{A}(t_f-\tau)}\mathbf{B}\mathbf{u}(\tau)\, d\tau$$

or

$$\mathbf{x}(0) = -\int_0^{t_f} e^{-\mathbf{A}\tau}\mathbf{B}\mathbf{u}(\tau)\, d\tau \tag{13.64}$$

The Cayley-Hamilton method presented in Sec. 3.13 shows that $\exp(-\mathbf{A}\tau)$ can be expressed as a polynomial of order $n - 1$ as follows:

$$e^{-\mathbf{A}\tau} = \sum_{k=0}^{n-1} \alpha_k(\tau)\mathbf{A}^k \tag{13.65}$$

Inserting this equation into Eq. (13.64) yields

$$\mathbf{x}(0) = -\sum_{k=0}^{n-1} \mathbf{A}^k\mathbf{B}\int_0^{t_f}\alpha_k(\tau)\mathbf{u}(\tau)\,d\tau \tag{13.66}$$

With the input $\mathbf{u}(t)$ of dimension r, the integral in Eq. (13.66) can be evaluated, and the result is

$$\int_0^{t_f}\alpha_k(\tau)\mathbf{u}(\tau)\,d\tau = \boldsymbol{\beta}_k$$

Equation (13.66) can now be expressed as

$$\mathbf{x}(0) = -\sum_{k=0}^{n-1} \mathbf{A}^k\mathbf{B}\boldsymbol{\beta}_k = -\left[\mathbf{B} \mid \mathbf{AB} \mid \cdots \mid \mathbf{A}^{n-1}\mathbf{B}\right]\begin{bmatrix} \boldsymbol{\beta}_0 \\ \hline \boldsymbol{\beta}_1 \\ \hline \vdots \\ \hline \boldsymbol{\beta}_{n-1} \end{bmatrix} \tag{13.67}$$

According to the definition of complete controllability, each mode $\mathbf{v}_i e^{\lambda_i t}$ must be directly affected by the input $\mathbf{u}(t)$. This requires that the *controllability* matrix $\mathbf{M}_c$ have the property

$$\text{Rank}\,\mathbf{M}_c = \text{Rank}\left[\mathbf{B} \mid \mathbf{AB} \mid \cdots \mid \mathbf{A}^{n-1}\mathbf{B}\right] = n \tag{13.68}$$

For a single-input system the matrix $\mathbf{B}$ is the vector $\mathbf{b}$, and the matrix of Eq. (13.68) has dimension $n \times n$. When the matrix pair $(\mathbf{A}, \mathbf{B})$ is not completely controllable, the uncontrollable modes can be determined by transforming the state equation so that the plant matrix is in diagonal form (see Sec. 5.10). This approach is described below. An alternate method is given in Chap. 19.

When the state equation of a SISO system is in control canonical form, that is, the plant matrix $\mathbf{A}$ is in companion form and the control vector $\mathbf{b} = [0 \quad \cdots \quad 0 \quad K_G]^T$, then the controllability matrix $\mathbf{M}_c$ has full rank n (see Sec. 5.11). Thus, the system is completely controllable and it is not necessary to perform a controllability test.

A simple method for determining controllability when the system has nonrepeated eigenvalues is to use the transformation $\mathbf{x} = \mathbf{Tz}$ to convert the $\mathbf{A}$ matrix into diagonal (canonical) form, as described in Sec. 5.10, so that the modes are decoupled. The resulting state equation is

$$\dot{\mathbf{z}} = \mathbf{T}^{-1}\mathbf{ATz} + \mathbf{T}^{-1}\mathbf{Bu} = \boldsymbol{\Lambda}\mathbf{z} + \mathbf{B}'\mathbf{u} \tag{13.69}$$

Each transformed state z_i represents a mode and can be directly affected by the input $\mathbf{u}(t)$ only if $\mathbf{B}' = \mathbf{T}^{-1}\mathbf{B}$ has no zero row. Therefore, for nonrepeated eigenvalues, *a system is completely controllable if* $\mathbf{B}'$ has no zero row. For repeated eigenvalues see Refs. 5 and 7. $\mathbf{B}'$ is called the *mode-controllability* matrix. A zero row of $\mathbf{B}'$ indicates that the corresponding mode is uncontrollable. The eigenvalues associated with uncontrollable modes are called *input-decoupling zeros*.

In a similar manner the condition for observability is derived from the homogeneous system equations:

$$\dot{\mathbf{x}} = \mathbf{A}\mathbf{x} \tag{13.70}$$

$$\mathbf{y} = \mathbf{C}\mathbf{x} \tag{13.71}$$

When Eq. (13.65) is used, the output vector $\mathbf{y}(t)$ can be expressed as

$$\mathbf{y}(t) = \mathbf{C}e^{\mathbf{A}t}\mathbf{x}(0) + \sum_{k=0}^{n-1} \alpha_k(t)\mathbf{C}\mathbf{A}^k\mathbf{x}(0) \tag{13.72}$$

For observability the output $\mathbf{y}(t)$ must be influenced by each state $\mathbf{x}_i$. This imposes restrictions on $\mathbf{CA}^k$. It can be shown that the system is completely observable if the following *observability* matrix $\mathbf{M}_o$ has the property

$$\text{Rank}\,\mathbf{M}_o = \text{Rank}\left[\mathbf{C}^T \vdots \mathbf{A}^T\mathbf{C}^T \vdots (\mathbf{A}^T)^2\mathbf{C}^T \vdots \cdots \vdots (\mathbf{A}^T)^{(n-1)}\mathbf{C}^T\right] = n \tag{13.73}$$

For a single-output system the matrix $\mathbf{C}^T$ is a column matrix which can be represented by $\mathbf{c}$, and the matrix of Eq. (13.73) has dimension $n \times n$. When the matrix pair $(\mathbf{A}, \mathbf{C})$ is not completely observable, the unobservable modes can be determined by transforming the state and output equations so that the plant matrix is in diagonal form. This approach is described below. An alternate method is given in Chap. 19.

A simple method for determining observability that can be used when the system has nonrepeated eigenvalues is to put the equations in canonical form, that is, the plant matrix is in diagonal form Λ. The output equation is then

$$\mathbf{y} = \mathbf{C}\mathbf{x} = \mathbf{C}\mathbf{T}\mathbf{z} = \mathbf{C}'\mathbf{z} \tag{13.74}$$

If a column of $\mathbf{C}'$ has all zeros, then one mode is not coupled to *any* of the outputs and the system is unobservable. *The condition for observability, for the case of nonrepeated eigenvalues, is that* $\mathbf{C}'$ *have no zero columns.* $\mathbf{C}'$ is called the *mode-observability* matrix. The eigenvalues associated with an unobservable mode are called *output-decoupling zeros.*

Example 1. Figure 13.5 shows the block-diagram representation of a system whose state and output equations are

$$\dot{\mathbf{x}} = \begin{bmatrix} -2 & 0 \\ -1 & -1 \end{bmatrix}\mathbf{x} + \begin{bmatrix} 1 \\ 1 \end{bmatrix}u \tag{13.75}$$

$$y = [0 \quad 1]\mathbf{x} \tag{13.76}$$

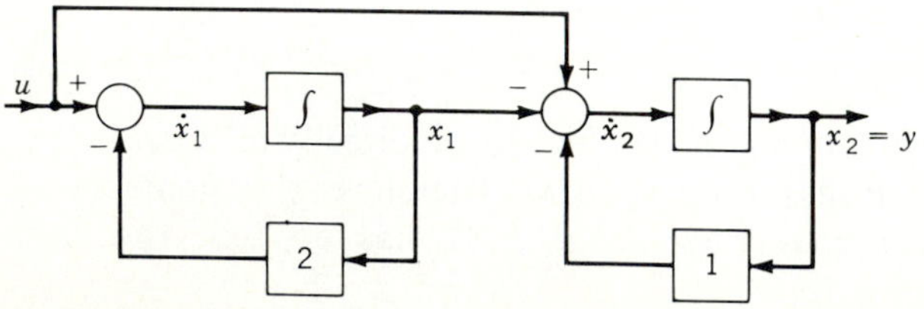

FIGURE 13.5
Simulation diagram for the system of Eq. (13.75).

Determine (a) the eigenvalues from the state equation, (b) the transfer function $Y(s)/U(s)$, (c) whether the system is controllable and/or observable, (d) the transformation matrix $\mathbf{T}$ that transforms the general state-variable representation of Eqs. (13.75) and (13.76) into the canonical form representation, and (e) draw a simulation diagram based on the canonical representation.

$$(a) \qquad |s\mathbf{I} - \mathbf{A}| = \begin{vmatrix} s+2 & 0 \\ 1 & s+1 \end{vmatrix} = (s+2)(s+1)$$

The eigenvalues are $\lambda_1 = -2$ and $\lambda_2 = -1$. Thus,

$$\boldsymbol{\Lambda} = \begin{bmatrix} -2 & 0 \\ 0 & -1 \end{bmatrix}$$

$$(b) \qquad \boldsymbol{\Phi}(s) = [s\mathbf{I} - \mathbf{A}]^{-1} = \frac{\begin{bmatrix} s+1 & 0 \\ -1 & s+2 \end{bmatrix}}{(s+1)(s+2)}$$

$$\mathbf{G}(s) = \mathbf{c}^T\boldsymbol{\Phi}(s)\mathbf{b} = \left([0 \quad 1]\begin{bmatrix} s+1 & 0 \\ -1 & s+2 \end{bmatrix}\begin{bmatrix} 1 \\ 1 \end{bmatrix}\right)\frac{1}{(s+1)(s+2)}$$

$$= \frac{s+1}{(s+1)(s+2)} = \frac{1}{s+2}$$

Note that there is a cancellation of the factor $s + 1$, and only the mode with $\lambda_1 = -2$ is both controllable and observable.

$$(c) \qquad \text{Rank}\,[\mathbf{b} \mid \mathbf{Ab}] = \text{Rank}\begin{bmatrix} 1 & -2 \\ 1 & -2 \end{bmatrix} = 1$$

The controllability matrix $\mathbf{M}_c = [\mathbf{b} \quad \mathbf{Ab}]$ is singular and has a rank deficiency of one; thus the system is *not* completely controllable and has one uncontrollable mode.

$$\text{Rank}\,[\mathbf{c} \mid \mathbf{A}^T\mathbf{c}] = \text{Rank}\begin{bmatrix} 0 & -1 \\ 1 & -1 \end{bmatrix} = 2$$

The observability matrix $[\mathbf{c} \quad \mathbf{A}^T\mathbf{c}]$ is nonsingular; thus the system is completely observable. Note that the mode with $\lambda_2 = -1$, which is canceled in the determination of the transfer function, is observable but not controllable.

(d) Using the third method of Sec. 5.10, with $\mathbf{x} = \mathbf{Tz}$, gives

$$\mathbf{AT} = \mathbf{T}\boldsymbol{\Lambda}$$

$$\begin{bmatrix} -2 & 0 \\ -1 & -1 \end{bmatrix}\begin{bmatrix} t_{11} & t_{12} \\ t_{21} & t_{22} \end{bmatrix} = \begin{bmatrix} t_{11} & t_{12} \\ t_{21} & t_{22} \end{bmatrix}\begin{bmatrix} -2 & 0 \\ 0 & -1 \end{bmatrix}$$

$$\begin{bmatrix} -2t_{11} & -2t_{12} \\ -t_{11} - t_{21} & -t_{12} - t_{22} \end{bmatrix} = \begin{bmatrix} -2t_{11} & -t_{12} \\ -2t_{21} & -t_{22} \end{bmatrix}$$

Equating elements on both sides of this equation yields

$$-2t_{11} = -2t_{11} \qquad -2t_{12} = -t_{12}$$

$$-t_{11} - t_{21} = -2t_{21} \qquad -t_{12} - t_{22} = -t_{22}$$

Since there are only two independent equations, assume that $t_{11} = t_{22} = 1$. Then $t_{21} = 1$ and $t_{12} = 0$.

$$\mathbf{b}' = \mathbf{T}^{-1}\mathbf{b} = \begin{bmatrix} 1 & 0 \\ -1 & 1 \end{bmatrix}\begin{bmatrix} 1 \\ 1 \end{bmatrix} = \begin{bmatrix} 1 \\ 0 \end{bmatrix}$$

$$\mathbf{c}'^T = \mathbf{c}^T\mathbf{T} = [0 \quad 1]\begin{bmatrix} 1 & 0 \\ 1 & 1 \end{bmatrix} = [1 \quad 1]$$

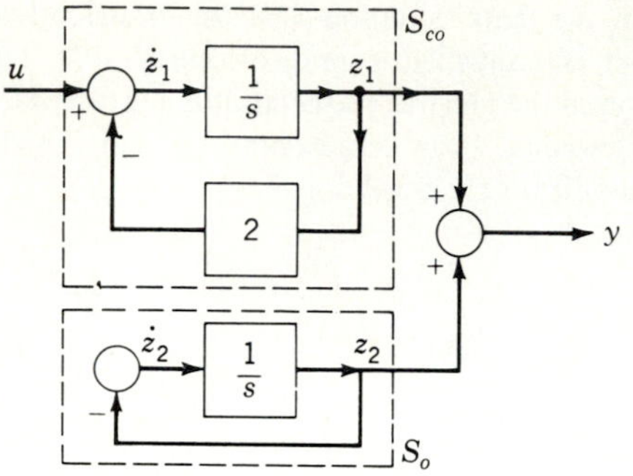

FIGURE 13.6
Simulation diagram for Eqs. (13.77) and (13.78).

The canonical equations for this system are

$$\dot{\mathbf{z}} = \begin{bmatrix} -2 & 0 \\ 0 & -1 \end{bmatrix}\mathbf{z} + \begin{bmatrix} 1 \\ 0 \end{bmatrix}u \tag{13.77}$$

$$\mathbf{y} = [1 \quad 1]\mathbf{z} \tag{13.78}$$

Since $\mathbf{b}'$ contains a zero row, the system is uncontrollable and $\lambda = -1$ is an input-decoupling zero. Since $\mathbf{c}'$ contains all nonzero elements, the system is completely observable. These results agree with the results obtained by the method used in part (*c*).

(*e*) The individual state and output equations are

$$\dot{z}_1 = -2z_1 + u \tag{13.79}$$

$$\dot{z}_2 = -z_2 \tag{13.80}$$

$$y = z_1 + z_2 \tag{13.81}$$

Equation (13.79) satisfies the requirement of controllability; i.e., for any t_0 the initial state $z_1(t_0)$ can be transferred to any final state $z_1(t_f)$ in a finite time $t_f \geq t_0$ by means of the unconstrained control vector u. Since u does not appear in Eq. (13.80), it is impossible to control the state z_2; thus $\lambda_2 = -1$ is an input-decoupling zero and the system is uncontrollable. Equation (13.81) satisfies the definition of observability, i.e., both the states appear in y and, therefore, $z_1(t_0)$ and $z_2(t_0)$ can be determined from the measurements of the output $y(t)$ over the finite interval of time $t_0 \leq t \leq t_f$.

From Eqs. (13.77) and (13.78) the simulation diagram of Fig. 13.6 is synthesized. The advantage of the canonical form is that the modes $\lambda_1 = -2$ and $\lambda_2 = -1$ are completely decoupled. This feature permits the establishment of the existence of some or all of the four subdivisions of Fig. 13.4 and the application of the definitions of controllability and observability.

The following example illustrates the requirement of controllability for closed-loop eigenvalue assignment by state feedback.

Example 2. A gain K is inserted in the plant of Example 1. Figure 13.7 shows a state-feedback system utilizing this plant whose control ratio is given by

$$\frac{Y(s)}{R(s)} = \frac{K(s+1)}{s^2 + (3 + Kk_1 + Kk_2)s + (2 + Kk_2 + Kk_1)} \tag{13.82}$$

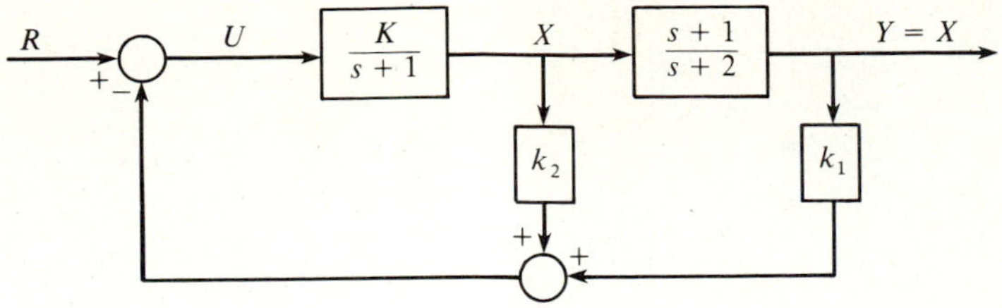

FIGURE 13.7
An uncontrollable state-feedback system.

In order for the system to follow a step input with no steady-state error the requirement is that $K = 2 + Kk_2 + Kk_1$, or

$$K = \frac{2}{1 - k_1 - k_2} \tag{13.83}$$

Since this system is uncontrollable, as determined in Example 1, the pole located at $s_2 = -1$ cannot be altered by state feedback. The characteristic equation of Eq. (13.82), which can be factored to the form

$$(s + 1)(s + 2 + Kk_2 + Kk_1) = 0 \tag{13.84}$$

illustrates this property. Therefore one eigenvalue, $s_1 = -2$, is altered by state feedback to the value $s_1 = -(2 + Kk_2 + Kk_1) = -K$, where K can be assigned as desired. Since the uncontrollable pole cancels the zero of Eq. (13.82), the control ratio becomes

$$\frac{Y(s)}{R(s)} = \frac{K(s + 1)}{(s + 1)(s + K)} = \frac{K}{s + K} \tag{13.85}$$

This example demonstrates that the uncontrollable pole $s_1 = -1$ cannot be changed by state feedback.

When deriving the transfer function from the system differential equations, cancellation of a pole by a zero is not permitted in developing the state representation. Cancellation of a pole by a zero in the transfer function may hide some of the dynamics of the entire system. This eliminates information pertaining to S_o, S_c, and S_u; that is, the observability and controllability of the system are determined from the state equations as illustrated in these examples.

13.8 STATE-VARIABLE FEEDBACK: ALL-POLE PLANT

The design procedure presented in Sec. 13.3 is now applied to an all-pole plant.

Example. For the basic plant having the transfer function shown in Fig. 13.8a, it is desired to use state-variable feedback in order to achieve a closed-loop response

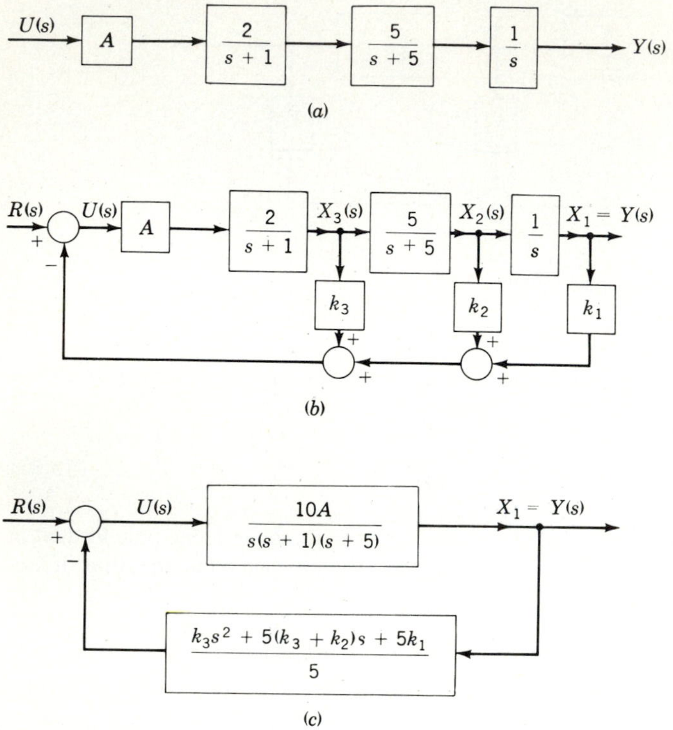

FIGURE 13.8
A state-variable-feedback system, all-pole plant: (*a*) plant; (*b*) state-variable feedback; (*c*) $H_{eq}(s)$ form.

with $M_p = 1.043$, $t_s = 5.65$ s, and zero steady-state error for a step input $r(t) = R_0 u_{-1}(t)$.

Steps 1 to 3. See Fig. 13.8*b* and *c*;

$$\frac{Y(s)}{R(s)} = \frac{10A}{s^3 + (6 + 2Ak_3)s^2 + [5 + 10A(k_3 + k_2)]s + 10Ak_1} \tag{13.86}$$

Step 4. Inserting Eq. (13.86) into $y(t)_{ss} = \lim_{s \to 0} sY(s) = R_0$, for a step input, yields the requirement that $k_1 = 1$.

Step 5. Assuming that Eq. (13.86) has complex poles which are truly dominant, Eq. (3.61) is used to determine the damping ratio:

$$M_p = 1.043 = 1 + \exp\left(-\frac{\zeta\pi}{\sqrt{1 - \zeta^2}}\right)$$

or

$$\zeta \approx 0.7079$$

From $T_s = 4/|\sigma|$, $|\sigma| = 4/5.65 = 0.708$. For these values of ζ and $|\sigma|$, the desired closed-loop complex-conjugate poles are $s_{1,2} = -0.708 \pm j0.7064$. Since the denominator of $G(s)$ is of third order, the control ratio also has a third-order denominator, as shown in Eq. (13.86). To ensure the dominancy of the complex poles the third pole is made significantly nondominant and is arbitrarily selected as $s_3 = -100$. Therefore, the desired closed-loop transfer function is

$$\frac{Y(s)}{R(s)} = \frac{10A}{(s+100)(s+0.708 \pm j0.7064)} = \frac{10A}{s^3 + 101.4s^2 + 142.6s + 100} \tag{13.87}$$

Step 6. In order to achieve zero steady-state error with a step input requires the selection $k_1 = 1$ (see Sec. 13.2). Then, equating corresponding terms in the denominators of Eqs. (13.86) and (13.87) yields

$$\begin{aligned} 10Ak_1 &= 100 & A &= 10 \\ 6 + 2Ak_3 &= 101.4 & k_3 &= 4.77 \\ 5 + 100(k_3 + k_2) &= 142.7 & k_2 &= -3.393 \end{aligned}$$

Inserting Eq. (13.87) into Eq. (13.40) to obtain the G-equivalent form (see Fig. 13.3) yields

$$G_{eq}(s) = \frac{100}{s(s^2 + 101.4s + 142.7)} \tag{13.88}$$

Thus, $G_{eq}(s)$ is Type 1 and the ramp error coefficient is $K_1 = 0.701$.

An analysis of the root locus for the unity-feedback and the state-feedback systems, shown in Fig. 13.9a and b, respectively, reveals several characteristics.

For large values of gain the unity-feedback system is unstable, even if cascade compensation is added. For any variation in gain the corresponding

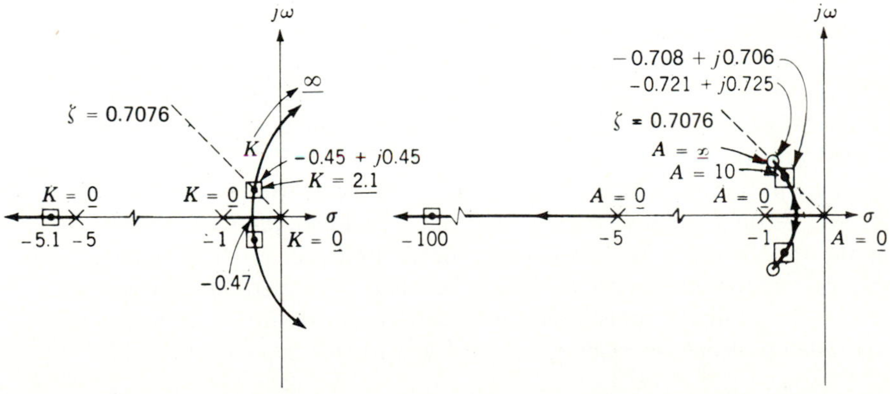

(a) $G(s) = \dfrac{K}{s(s+1)(s+5)}$

(b) $G(s)H_{eq}(s) = \dfrac{9.54A(s+0.721 \pm j0.726)}{s(s+1)(s+5)}$

FIGURE 13.9
Root-locus diagrams for (a) basic unity-feedback and (b) state-variable-feedback control systems.

change in the dominant roots changes the system response. Thus, the response is sensitive to changes of gain.

The state-variable-feedback system is stable for all $A > 0$. Since the roots are very close to the zeros of $G(s)H_{eq}(s)$, any increase in gain from the value $A = 10$ produces negligible changes in the location of the roots. In other words, for high-forward-gain (hfg) operation, $A > 10$, the system's response is essentially unaffected by variations in A, provided the feedback coefficients remain fixed. The value of the hfg $K = 9.54A \geq K_{min}$, for which the state-variable-feedback system's response is insensitive to gain variations, is obtained from the root-locus plot for $G(s)H_{eq}(s) = -1$. Thus, for the example represented by Fig. 13.9*b*, when $K = 9.54A \geq 95.4$, the change in the values of the dominant roots is small and there is little effect on the time response for a step input. In Chap. 16 another method is presented for determining K_{min}. The insensitivity of the system response to the variation of the gain A is related to the system's sensitivity function, which is discussed in the next chapter. *This insensitivity is due to the fact that the two desired dominant roots are close to the two zeros of $H_{eq}(s)$ and the third root is nondominant. To achieve this insensitivity to gain variation there must be β dominant roots which are located close to β zeros of $H_{eq}(s)$. The remaining $n - \beta$ roots must be nondominant.* In this example $\beta = n - 1 = 2$, but in general $\beta \leq n - 1$.

Both the conventional and state-feedback systems can follow a step input with zero steady-state error. Also, they both follow a ramp input with a steady-state error. The state-variable-feedback system cannot be designed to produce zero steady-state error for a ramp input because $G(s)$ does not contain a zero. In order to satisfy this condition a cascade compensator which adds a zero to $G(s)$ may be utilized in the state-variable-feedback system.

In general, as discussed in Sec. 12.2, when selecting the poles and zeros of $[Y(s)/R(s)]_T$ to satisfy the desired specifications, one pair of dominant complex-conjugate poles $s_{1,2}$ is often chosen. The other $n - 2$ poles are located far to the left of the dominant pair or close to zeros in order to ensure the dominancy of $s_{1,2}$. It is also possible to have more than just a pair of dominant complex-conjugate poles to satisfy the desired specifications. For example, an additional pole of $Y(s)/R(s)$ can be located to the left of the dominant complex pair. When properly located, it can reduce the peak overshoot and may also reduce the settling time. Choosing only one nondominant root, that is, $n - 1$ dominant roots, results in lower values of K and the k_i than when a larger number of nondominant roots is chosen. In Chap. 17 a method is presented that yields $n - 1$ dominant roots which may satisfy the desired specifications. *For state-variable feedback the value of the gain A is always taken as positive.*

13.9 PLANT COMPLEX POLES

This section illustrates how the state-variable representation of a plant containing complex poles is obtained. Since the system shown in Fig. 13.10*a* is of the fourth order, four state variables must be identified. The transfer function $O(s)/I(s)$ involves two states. If the detailed block-diagram representation of $O(s)/I(s)$

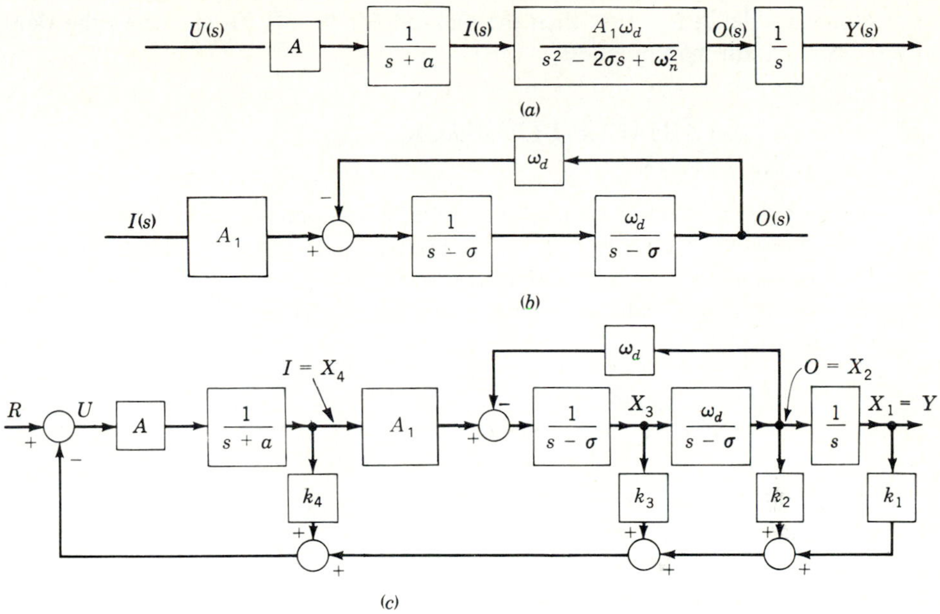

FIGURE 13.10
Representation of a plant containing complex poles.

contains two explicit states, it should not be combined into the quadratic form. By combining the blocks the identity of one of the states is lost. That situation is illustrated by the system of Fig. 13.1. As represented in Fig. 13.2a, the two states X_2 and X_3 are accessible, but in the combined mathematical form of Fig. 13.2c the physical variables are not identifiable. When one of the states is not initially identified, an equivalent model containing two states must be formed for the quadratic transfer function $O(s)/I(s)$ of Fig. 13.10a. This decomposition can be performed by the method of Sec. 5.12. The simulation diagram of Fig. 5.34 is shown in the block diagram of Fig. 13.10b. This decomposition is inserted into the complete control system utilizing state-variable feedback, as shown in Fig. 13.10c. If the state X_3 is inaccessible, then, once the values of k_i are determined, the physical realization of the state feedback can be achieved by the block-diagram manipulation discussed in previous sections. An alternate approach is to represent the transfer function $O(s)/I(s)$ in state-variable form, as shown in Fig. 13.11. This representation is particularly attractive when X_3 is recognizable as a

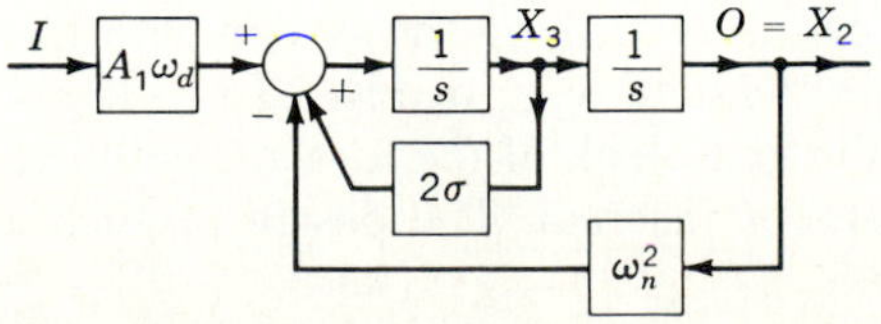

FIGURE 13.11
State-variable representation of $O(s)/I(s)$ in Fig. 13.10a.

physical variable. If it is not, then the pick-off for k_3 can be moved to the right by block-diagram manipulation.

13.10 STATE-VARIABLE FEEDBACK: POLE-ZERO PLANT

Section 13.8 deals with a system having an all-pole plant whose control ratio has no zeros. Locating the roots of the characteristic equation for such a system to yield the desired system response is relatively easy. This section deals with a system having a plant which contains both poles and zeros and has a control ratio of the form

$$\frac{Y(s)}{R(s)} = \frac{K_G(s-z_1)(s-z_2)\cdots(s-z_h)\cdots(s-z_w)}{(s-p_1)(s-p_2)\cdots(s-p_c)\cdots(s-p_n)} \tag{13.89}$$

which has the same zeros as the plant. When open-loop zeros are present, the synthesis of the overall characteristic equation becomes more complex than for the all-pole plant.

As discussed in Sec. 13.6 and summarized in Table 13.2, the presence of a zero in the control ratio is necessary to achieve $e(t)_{ss} = 0$ for a ramp input. In synthesizing the desired control ratio, Eq. (13.55) can also be used to assist in achieving this condition while simultaneously satisfying the other desired figures of merit.

If the locations of the zeros of $G_x(s)$ are satisfactory, the design procedure is the same as for the all-pole case of Sec 13.8. That is, the coefficients of the desired characteristic equation are equated to the corresponding coefficients of the numerator polynomial of $1 + G(s)H_{eq}(s)$ to determine the required K_G and k_i. It is possible to achieve low sensitivity to variation in gain by requiring that

1. $Y(s)/R(s)$ have β dominant poles and *at least one nondominant pole*.
2. The β dominant poles be close to β zeros of $H_{eq}(s)$.

The latter can be achieved by a judicious choice of the compensator pole(s) and/or the nondominant poles.

When a cascade compensator of the form $G_c(s) = (s - z_i)/(s - p_i)$ is required, it can be constructed in either of the forms shown in Fig. 13.12*a* or *b*. The input signal is shown as AU since, in general, this is the most accessible place to insert the compensator. The differential equations for the two forms are

For Fig. 13.12*a*: $$\dot{x}_{i+n} = P_i x_{i+n} - z_i A u + A\dot{u} \tag{13.90}$$

For Fig. 13.12*b*: $$\dot{x}'_{i+n} = P_i x'_{i+n} + (P_i - z_i) A u \tag{13.91}$$

The presence of $\dot{u}$ in Eq. (13.90) does not permit describing the system with this compensator by the state equation $\dot{\mathbf{x}} = \mathbf{A}\mathbf{x} + \mathbf{b}u$, and x_{i+n} in Fig. 13.12*a* is not a state variable. The presence of a zero in the first block of the forward path can also occur in a system before a compensator is inserted. The design of such a

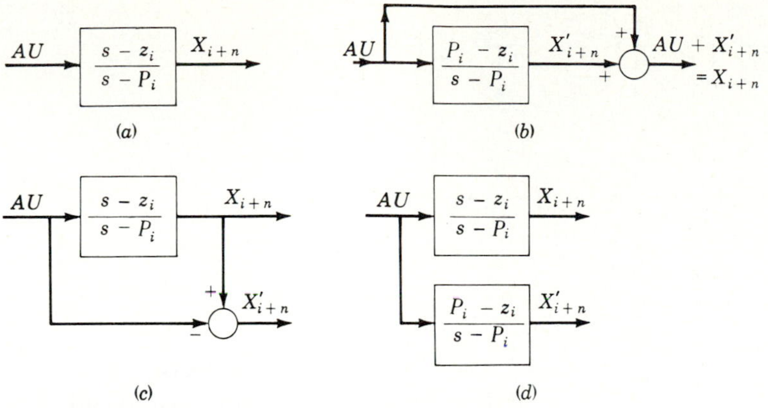

FIGURE 13.12
Cascade compensators: $G_c(s) = (s - z_i)/(s - P_i)$: (*a*) conventional; (*b*) feedforward form; (*c*) generating the state variable X'_{i+n}; (*d*) mathematical model for (*c*).

system can be treated like that of a system containing the compensator of Fig. 13.12*a*. The consequence of the $\dot{u}$ term in Eq. (13.90) is that the zero of $G_c(s)$ does not appear as a pole of $H_{eq}(s)$. Thus the root-locus characteristics discussed in Sec. 13.3 must be modified accordingly. There is no difficulty in expressing the system in state-variable form when the function of Fig. 13.12*a* is not the first frequency-sensitive term in the forward path. When the compensator of Fig. 13.12*a* is inserted at any place to the right of the G_1 block in the forward path shown in Fig. 13.13, the system can be described by a state equation. Thus, the root-locus characteristics discussed in Sec. 13.3 also apply for this case.

A system utilizing the feedforward compensator of Fig. 13.12*b* can be described by a state equation since $\dot{u}$ does not appear in Eq. (13.91). Thus, the poles of $H_{eq}(s)$ continue to be the zeros of $G(s)$. Therefore, the root-locus characteristics discussed in Sec. 13.3 also apply for this case. This is a suitable method of constructing the compensator $G_c(s) = (s - z_i)/(s - P_i)$ since it produces the accessible state variable X'_{i+n}. The implementation of $G_c(s)$ consists of the lag term $K_c/(s - P_i)$, where $K_c = P_i - z_i$, with the unity-gain feedforward path shown in Fig. 13.12*b*.

An additional possibility is to generate a new state variable X'_{i+n} from the physical arrangement shown in Fig. 13.12*c*. This is equivalent to the form shown in Fig. 13.12*d* and is mathematically equivalent to Fig. 13.12*b*. Thus, X'_{i+n} is a valid state variable and may be used with a feedback coefficient k_{i+n}.

These approaches are now illustrated by examples having one zero in the plant transfer function. The procedure is essentially the same for systems having

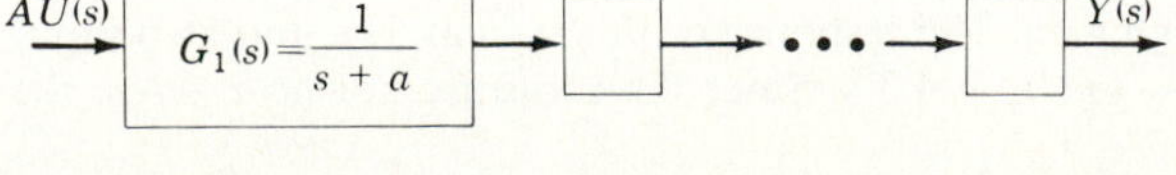

FIGURE 13.13
Forward path.

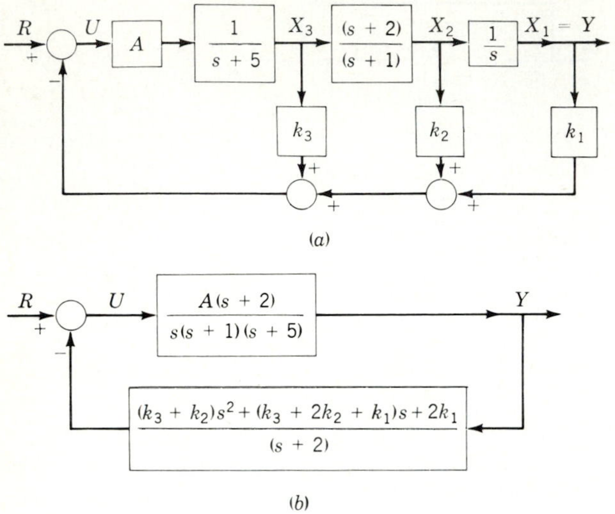

FIGURE 13.14
Closed-loop zero cancellation by state-variable feedback: (*a*) detailed block diagram; (*b*) *H*-equivalent reduction (where $K_G = A$).

more than one zero. A typical design technique is illustrated in each example and can be modified as desired.

Example 1. The control ratio for the system of Fig. 13.14 is

$$\frac{Y(s)}{R(s)} = \frac{A(s+2)}{s^3 + [6 + (k_3 + k_2)A]s^2 + [5 + (k_3 + 2k_2 + k_1)A]s + 2k_1A} \tag{13.92}$$

A trial function for the desired control ratio for this system is specified as

$$\left.\frac{Y(s)}{R(s)}\right|_{\text{desired}} = \frac{2}{s^2 + 2s + 2} \tag{13.93}$$

If Eq. (13.92) is to reduce to Eq. (13.93), one pole of Eq. (13.92) must have the value $s = -2$, in order to cancel the zero, and $A = 2$. Equation (13.93) is modified to reflect this requirement:

$$\left.\frac{Y(s)}{R(s)}\right|_{\text{desired}} = \frac{2(s+2)}{(s^2 + 2s + 2)(s+2)} = \frac{2(s+2)}{s^3 + 4s^2 + 6s + 4} \tag{13.94}$$

Equating the corresponding coefficients of the denominators of Eqs. (13.92) and (13.94) yields $k_1 = 1$, $k_2 = 1/2$, and $k_3 = -3/2$. Thus

$$G(s)H_{eq}(s) = \frac{-A(s^2 - s/2 - 2)}{s(s+1)(s+5)} = -1 \tag{13.95}$$

Canceling the minus sign shows that the 0° root locus is required. The zeros are $s_1 = -1.18$ and $s_2 = 1.68$, and the root locus is shown in Fig. 13.15. Note that since Eq. (13.93) has no other poles in addition to the dominant poles, the necessary condition of at least one nondominant pole is not satisfied. As a result, two branches of the root locus enter the right half of the s plane, yielding a system that can become unstable for high gain. The response of this system to a unit step input has $M_p = 1.043$, $t_p = 3.1$ s, and $t_s = 4.2$ s. Since the roots are not near zeros, the

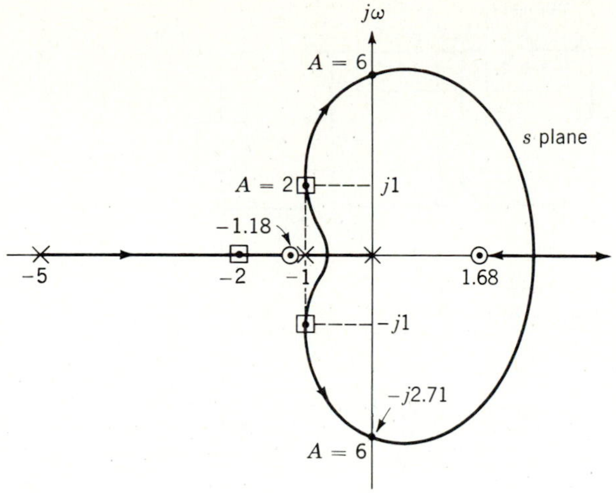

FIGURE 13.15
Root locus for Eq. (13.95).

system is highly sensitive to gain variation, as seen from the root locus in Fig. 13.15. Any increase in A results in a significant change in performance characteristics. If A exceeds the value of 6, the system becomes unstable. The ramp error coefficient obtained from $G_{eq}(s)$ is $K_1 = 1.0$

Example 1 illustrates that insensitivity to gain variation and stability for all positive values of gain is not achieved. In order to satisfy these conditions, $[Y(s)/R(s)]_D$ must have β dominant poles and at least one nondominant pole. The next example illustrates this design concept.

Example 2. For the third-order system of Fig. 13.14 it is desired to cancel the control ratio zero at $s = -2$ and to reduce the system output sensitivity to variations in the gain A, while effectively achieving the second-order response of Eq. (13.93). The time response is satisfied if $Y(s)/R(s)$ has three dominant poles $s_{1,2} = -1 + j1$ and $s_3 = -2$. In order to achieve insensitivity to gain changes, the transfer function $H_{eq}(s)$ must have three zeros. In accordance with Eq. (13.21a), the number of poles of $G(s)$ must be $n = 4$. Therefore, one pole must be added to $G(s)$ by using the cascade compensator $G_c(s) = 1/(s + a)$. Then $Y(s)/R(s)$ can have one nondominant pole p and the desired control ratio is given by

$$\left.\frac{Y(s)}{R(s)}\right|_{\text{desired}} = \frac{K_G(s+2)}{(s^2+2s+2)(s+2)(s-p)} \tag{13.96}$$

The selections of the values of $K_G = A$ and p are not independent since the condition of $e(t)_{ss} = 0$ for a step input $r(t) = R_0 u_{-1}(t)$ is required. Thus

$$y(t)_{ss} = \lim_{s\to 0} sY(s) = \frac{-K_G}{2p} R_0 = R_0 \tag{13.97}$$

or

$$K_G = -2p$$

As mentioned previously, the larger the value of the loop sensitivity for the root locus of $G(s)H_{eq}(s) = -1$, the closer the roots will be to the zeros of $H_{eq}(s)$.

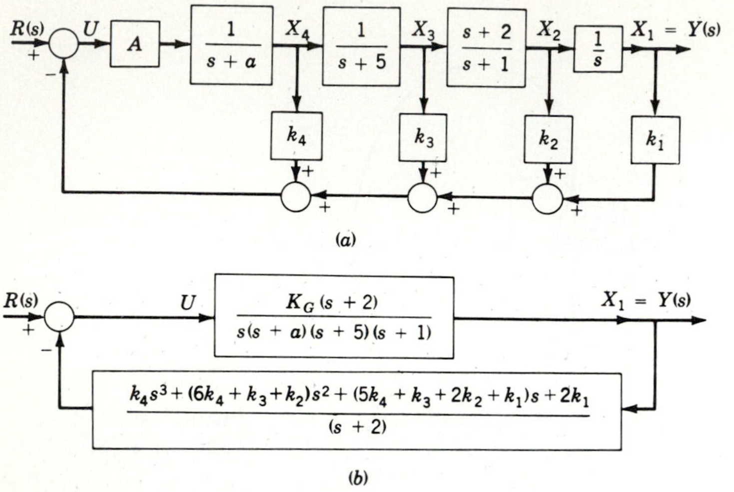

FIGURE 13.16
Closed-loop zero cancellation by state-variable feedback: (*a*) detailed block diagram; (*b*) *H*-equivalent reduction (where $K_G = A$).

It is this "closeness" that reduces the system's output sensitivity to gain variations. Therefore, the value of K_G should be chosen as large as possible consistent with the capabilities of the system components. This large value of K_G may have to be limited to a value that does not accentuate the system noise to an unacceptable level. For this example, selecting a value $K_G = 100$ ensures that p is a nondominant pole. This results in a value $p = -50$ so that Eq. (13.96) becomes

$$\left.\frac{Y(s)}{R(s)}\right|_{\text{desired}} = \frac{100(s+2)}{s^4 + 54s^3 + 206s^2 + 304s + 200} \tag{13.98}$$

The detailed state feedback and *H*-equivalent block diagrams for the modified control system are shown in Fig. 13.16*a* and *b*, respectively.

In order to maintain a controllable and observable system the value of a can be chosen to be any value *other than* the value of the zero term of $G(s)$. Thus, assume $a = 1$. With this value of a and with $K_G = 100$, the control ratio of the system of Fig. 13.16*b* is

$$\frac{Y(s)}{R(s)} = \frac{100(s+2)}{\begin{array}{c} s^4 + (7 + 100k_4)s^3 + [11 + 100(6k_4 + k_3 + k_2)]s^2 \\ + [5 + 100(5k_4 + k_3 + k_2 + k_1)]s + 200k_1 \end{array}} \tag{13.99}$$

For Eq. (13.99) to yield $e(t)_{ss} = 0$ with a step input requires $k_1 = 1$. Equating the corresponding coefficients of the denominators of Eqs. (13.98) and (13.99) and solving for the feedback coefficients yields $k_2 = 0.51$, $k_3 = -1.38$, and $k_4 = 0.47$. Thus, from Fig. 13.16*b*:

$$G(s)H_{eq}(s) = \frac{0.47K_G(s^3 + 4.149s^2 + 6.362s + 4.255)}{s^4 + 7s^3 + 11s^2 + 5s} \tag{13.100}$$

The root locus shown in Fig. 13.17 reveals that the three desired dominant roots are

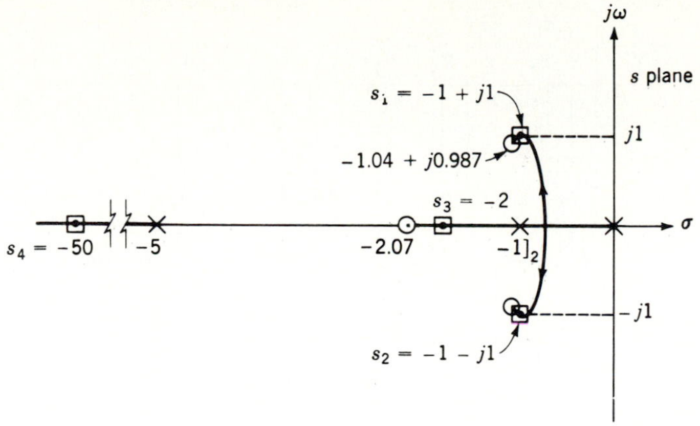

FIGURE 13.17
The root locus for Eq. (13.100).

close to the three zeros of $H_{eq}(s)$, the fourth root is nondominant, and the root locus lies entirely in the left-half s plane. Thus, the full benefits of state-variable feedback have been achieved. In other words, a completely stable system with low sensitivity to parameter variation has been realized. The figures of merit of this system are $M_p = 1.043$, $t_p = 3.2$ s, $t_s = 4.2$ s, and $K_1 = 0.98\ \text{s}^{-1}$.

Example 3. A further improvement in the time response can be obtained by adding a pole and zero to the overall system function of Example 2. The control ratio to be achieved is

$$\frac{Y(s)}{R(s)} = \frac{K_G(s+1.4)(s+2)}{(s+1)(s^2+2s+2)(s+2)(s-p)}$$
$$= \frac{K_G(s+1.4)(s+2)}{s^5+(5-p)s^4+(10-5p)s^3+(10-10p)s^2+(4-10p)s-4p} \quad (13.101)$$

Analysis of Eq. (13.101) shows the following features:

1. Since $Y(s)/R(s)$ has four dominant poles, the nondominant pole p yields the desired property of insensitivity to gain variation.
2. The pole-zero combination $(s + 1.4)/(s + 1)$ is added to reduce the peak overshoot in the transient response.
3. The factor $s + 2$ normally appears in the numerator because it is present in $G(s)$. Since it is not desired in $Y(s)/R(s)$, it is canceled by producing the same factor in the denominator.
4. In order to achieve zero steady-state error with a step input $r(t) = u_{-1}(t)$, the output is obtained by applying the final-value theorem to $Y(s)$. This yields $y(t)_{ss} = K_G \times 1.4 \times 2/(-4p) = 1$. Therefore, $p = -0.7\ K_G$. Assuming the large value of $K_G = 100$ to be satisfactory results in $p = -70$, which is nondominant, as required.

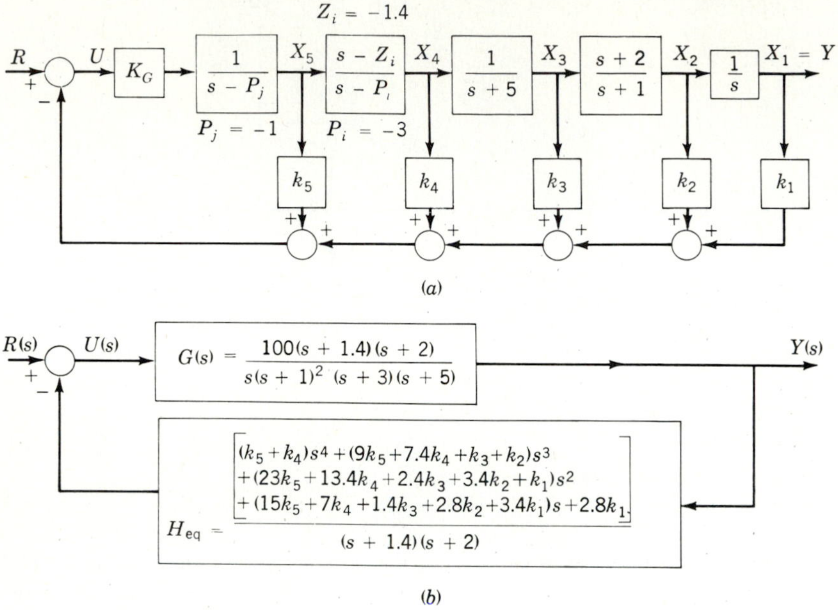

FIGURE 13.18
State feedback diagram for Example 3.

The required form of the modified system is shown in Fig. 13.18. The cascade compensator $1/(s+1)$ is added to the system in order to achieve the degree of the denominator of Eq. (13.101). The cascade compensator $(s+1.4)/(s+3)$ is included in order to produce the desired numerator factor $s+1.4$. The denominator is arbitrarily chosen as $s+3$. This increases the degree of the denominator in order to permit canceling the factor $s+2$ in $Y(s)/R(s)$.

The forward transfer function obtained from Fig. 13.18 is

$$G(s) = \frac{100(s+1.4)(s+2)}{s(s+1)^2(s+3)(s+5)} = \frac{100(s^2+3.4s+2.8)}{s^5+10s^4+32s^3+38s^2+15s} \tag{13.102}$$

The value of $H_{eq}(s)$ is evaluated from Fig. 13.18 and is

$$H_{eq}(s) = \frac{\begin{aligned}&(k_5+k_4)s^4+(9k_5+7.4k_4+k_3+k_2)s^3\\ &+(23k_5+13.4k_4+2.4k_3+3.4k_2+k_1)s^2\\ &+(15k_5+7k_4+1.4k_3+2.8k_2+3.4k_1)s+2.8k_1\end{aligned}}{(s+1.4)(s+2)} \tag{13.103}$$

The desired overall system function is obtained from Eq. (13.101), with the values $K_G = 100$ and $p = -70$, as follows:

$$\frac{Y(s)}{R(s)} = \frac{100(s+1.4)(s+2)}{s^5+75s^4+360s^3+710s^2+704s+280} \tag{13.104}$$

The system function obtained by using $G(s)$ and $H_{eq}(s)$ is

$$\frac{Y(s)}{R(s)} = \frac{100(s+1.4)(s+2)}{\begin{aligned} &s^5 + [10 + 100(k_5 + k_4)]s^4 \\ &+[32 + 100(9k_5 + 7.4k_4 + k_3 + k_2)]s^3 \\ &+[38 + 100(23k_5 + 13.4k_4 + 2.4k_3 + 3.4k_2 + k_1)]s^2 \\ &+[15 + 100(15k_5 + 7k_4 + 1.4k_3 + 2.8k_2 + 3.4k_1)]s + 100(2.8k_1)\end{aligned}} \tag{13.105}$$

Equating the coefficients in the denominators of Eqs. (13.101) and (13.102) yields $280k_1 = 280$, or $k_1 = 1$, and

$$\begin{aligned} 1500k_5 + 700k_4 + 140k_3 + 280k_2 &= 349 \\ 2300k_5 + 1340k_4 + 240k_3 + 340k_2 &= 572 \\ 900k_5 + 740k_4 + 100k_3 + 100k_2 &= 328 \\ 100k_5 + 100k_4 &= 65 \end{aligned} \tag{13.106}$$

Solving these four simultaneous equations gives $k_2 = 1.0$, $k_3 = -2.44167$, $k_4 = 0.70528$, and $k_5 = -0.05528$.

In order to demonstrate the desirable features of this system, the root locus is drawn in Fig. 13.19 for

$$G(s)H_{eq}(s) = \frac{100(0.65s^4 + 3.28s^3 + 6.72s^2 + 6.89s + 2.8)}{s(s+1)^2(s+3)(s+5)}$$

$$= \frac{65(s + 1.04532 \pm j1.05386)(s + 0.99972)(s + 1.95565)}{s(s+1)^2(s+3)(s+5)} \tag{13.107}$$

The root locus shows that the desired zero of $Y(s)/R(s)$ at $s = -2$ has been canceled by the pole at $s = -2$. Because the root $s_4 = -2$ is near a zero in Fig. 13.19, its value is insensitive to increases in gain. This ensures the desired cancellation. The dominant roots are all near zeros in Fig. 13.19, thus assuring that the desired control ratio $Y(s)/R(s)$ as given by Eq. (13.101) has been achieved. The system is stable for all values of gain K_G. Also, state-variable feedback has

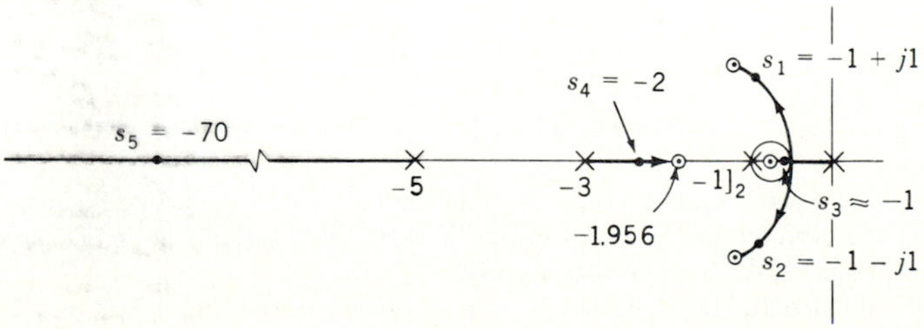

FIGURE 13.19
Root locus for Eq. (13.107).

produced a system in which $Y(s)/R(s)$ is unaffected by increases in A above 100. The time response $y(t)$ to a step input has a peak overshoot $M_p = 1.008$, a peak time $t_p = 3.92$ and a settling time $t_s = 2.94$. The ramp error coefficient evaluated for $G_{eq}(s)$ is $K_1 = 0.98$. These results show that the addition of the pole-zero combination $(s + 1.4)/(s + 1)$ to $Y(s)/R(s)$ has improved the performance. This system is insensitive to variation in gain.

The three examples in this section are intended to provide a firm foundation in the procedures for

1. Synthesizing the desired system performance and the overall control ratio.
2. Analyzing the root locus of $G(s)H_{eq}(s) = -1$ for the desired system performance and insensitivity to variations in gain.
3. Determining the values of k_i to yield the desired system performance.

These approaches are not intended to be all-inclusive, and variations may be developed.[1,8]

13.11 SUMMARY

A model control ratio incorporating the desired system performance specifications is used in this chapter to design a compensated state-variable-feedback control system. The root-locus properties and characterization, the steady-state error analysis, and the use of steady-state error coefficients, as they pertain to state-variable-feedback systems (all-pole and pole-zero plants) are discussed in detail. The presence of at least one nondominant closed-loop pole is required in order to minimize the sensitivity of the output response to gain variations. A comparison of the sensitivity to gain variation is made of unity-feedback and state-variable-feedback systems. System sensitivity for parameter variations is presented in more detail in Chap. 14. The use of computer programs can expedite the achievement of an acceptable design. These programs include the root-locus, time-response, and state-variable feedback-coefficient determination (see App. B).

REFERENCES

1. Schultz, D. G., and J. L. Melsa: *State Functions and Linear Control Systems*, McGraw-Hill, New York, 1967.
2. Houpis, C. H.: "The Relationship between the Conventional Control Theory Figures of Merit and the Performance Indices in Optimal Control Theory," Ph.D. dissertation, Department of Electrical Engineering, University of Wyoming, 1971.
3. James, H. M., N. B. Nichols, and R. S. Phillips: *Theory of Servomechanisms*, McGraw-Hill, New York, 1947.
4. King, L. H.: "Reduction of Forced Error in Closed-Loop Systems," *Proc. IRE*, vol. 41, pp. 1037–1042, August 1953.

5. Chen, C. T., and C. A. Desoer: "Proof of Controllability of Jordan Form State Equations," *Trans. IEEE*, vol. AC-13, pp. 195–196, 1968.
6. Athans, M., and P. L. Falb: *Optimal Control*, McGraw-Hill, New York, 1966.
7. Gilbert, E. G.: "Controllability and Observability in Multi-Variable Control Systems," *J. Soc. Ind. Appl. Math., Ser. A., Control*, vol. 1, no. 2, pp. 128–151, 1963.
8. Anderson, B. D. O., and J. B. Moore: *Linear Optimal Control*, Prentice-Hall, Englewood Cliffs, N.J., 1971.
9. Kailath, T.: *Linear Systems*, Prentice-Hall, Englewood Cliffs, N.J., 1980.
10. Friedland, B.: *Control System Design*, McGraw-Hill, New York, 1986.
11. Chen, C. T.: *Introduction to Linear System Theory*, Holt, Rinehart and Winston, New York, 1970.
12. Barnett, S.: *Polynomial and Linear Control Systems*, Marcel Dekker, New York, 1983.
13. Wolovich, W. A.: *Linear Multivariable Systems*, Springer-Verlag, New York, 1974.
14. Owens, D. H.: *Feedback and Multivariable Systems*, Peter Peregrinus Ltd. (on behalf of the Institution of Electrical Engineers), Stevenage, Herts, England, 1978.

CHAPTER 14

PARAMETER SENSITIVITY AND INACCESSIBLE STATES

14.1 INTRODUCTION

In Chap. 13 it is shown that an advantage of a state-feedback control system operating with high forward gain (hfg) is the insensitivity of the system output to gain variations of the amplifier A which has the actuating signal u as its input. The design method of Chap. 13 assumes that all states are accessible. This chapter investigates in depth the insensitive property of hfg operation of a state-feedback control system. This is followed by a treatment of inaccessible states.

14.2 SENSITIVITY

As stated in Sec. 10.1, the environmental conditions to which a control system is subjected affect the accuracy and stability of the system. The performance characteristics of most components are affected by their environment and by aging. Thus any change in the component characteristics causes a change in the transfer function and therefore in the controlled quantity. The effect of a parameter change on system performance can be expressed in terms of a *sensitivity function*. This sensitivity function S_δ^M is a measure of the sensitivity of the system's response to a system parameter variation and is given by

$$S_\delta^M = \frac{d(\ln M)}{d(\ln \delta)} = \frac{d(\ln M)}{d\delta}\frac{d\delta}{d(\ln \delta)} \tag{14.1}$$

where ln = logarithm to base e, M = system's output response (or its control ratio), and δ = system parameter that varies. Now

$$\frac{d(\ln M)}{d\delta} = \frac{1}{M}\frac{dM}{d\delta} \qquad \text{and} \qquad \frac{d(\ln \delta)}{d\delta} = \frac{1}{\delta} \tag{14.2}$$

Accordingly, Eq. (14.1) can be written

$$S_\delta^M\bigg|_{\substack{M=M_o\\ \delta=\delta_o}} = \frac{dM/M_o}{d\delta/\delta_o}$$

$$= \frac{\delta}{M}\left(\frac{dM}{d\delta}\right)\bigg|_{\substack{M=M_o\\ \delta=\delta_o}} = \frac{\text{fractional change in output}}{\text{fractional change in system parameter}} \tag{14.3}$$

where M_o and δ_o represent the nominal values of M and δ. When M is a function of more than one parameter, say $\delta_1, \delta_2, \ldots, \delta_k$, the corresponding formulas for the sensitivity entail partial derivatives. For a small change in δ from δ_o, M changes from M_o, and the sensitivity can be written as

$$S_\delta^M\bigg|_{\substack{M=M_o\\ \delta=\delta_o}} \approx \frac{\Delta M/M_o}{\Delta\delta/\delta_o} \tag{14.4}$$

To illustrate the effect of changes in the transfer function, four cases are considered for which the input signal $r(t)$ and its transform $R(s)$ are fixed. Although the response $Y(s)$ is used in these four cases, the results are the same when $M(s)$ is the control ratio.

Case 1: Open-Loop System of Fig. 14.1*a*

The effect of a change in $G(s)$, for a fixed $r(t)$ and thus a fixed $R(s)$, can be determined by differentiating, with respect to $G(s)$, the output expression

$$Y_o(s) = R(s)G(s) \tag{14.5}$$

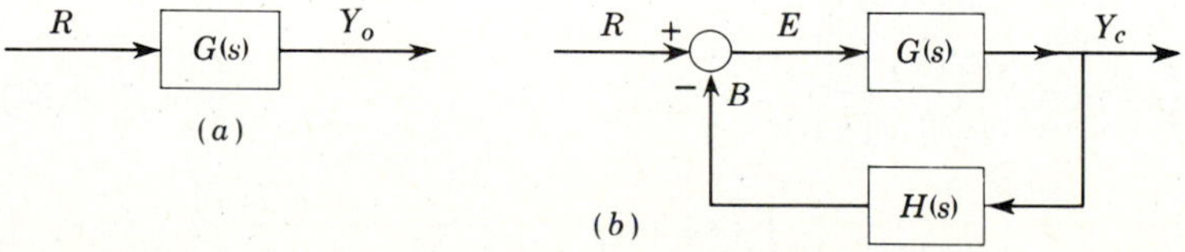

FIGURE 14.1
Control systems: (*a*) open loop; (*b*) closed loop.

giving

$$dY_o(s) = R(s)dG(s) \tag{14.6}$$

Combining these two equations gives

$$dY_o(s) = \frac{dG(s)}{G(s)} Y_o(s) \rightarrow S_{G(s)}^{Y(s)}(s) = \frac{dY_o(s)/Y_o(s)}{dG(s)/G(s)} = 1 \tag{14.7}$$

Therefore a change in the transfer function $G(s)$ causes a proportional change in the transform of the output $Y_o(s)$. This requires that the performance specifications of $G(s)$ be such that any variation still results in the degree of accuracy within the prescribed limits. In Eq. (14.7) the varying function in the system is the transfer function $G(s)$.

Case 2: Closed-Loop Unity-Feedback System of Fig. 14.1*b* [$H(s) = 1$]

Proceeding in the same manner as for case 1, for $G(s)$ varying, leads to

$$Y_c(s) = R(s)\frac{G(s)}{1 + G(s)} \tag{14.8}$$

$$dY_c(s) = R(s)\frac{dG(s)}{[1 + G(s)]^2} \tag{14.9}$$

$$dY_c(s) = \frac{dG(s)}{G(s)[1 + G(s)]} Y_c(s) = \frac{1}{1 + G(s)} \frac{dG(s)}{G(s)} Y_c(s)$$

$$S_{G(s)}^{Y(s)} = \frac{dY_c(s)/Y_c(s)}{dG(s)/G(s)} = \frac{1}{1 + G(s)} \tag{14.10}$$

Comparing Eq. (14.10) with Eq. (14.7) readily reveals that the effect of changes of $G(s)$ upon the transform of the output of the closed-loop control is reduced by the factor $1/|1 + G(s)|$ compared to the open-loop control. *This is an important reason why feedback systems are used.*

Case 3: Closed-Loop Nonunity-Feedback System of Fig. 14.1*b* [Feedback Function $H(s)$ Fixed and $G(s)$ Variable]

Proceeding in the same manner as for case 1, for $G(s)$ varying, leads to

$$Y_c(s) = R(s)\frac{G(s)}{1 + G(s)H(s)} \tag{14.11}$$

$$dY_c(s) = R(s)\frac{dG(s)}{[1 + G(s)H(s)]^2} \tag{14.12}$$

$$dY_c(s) = \frac{dG(s)}{G(s)[1 + G(s)H(s)]} Y_c(s) = \frac{1}{1 + G(s)H(s)} \frac{dG(s)}{G(s)} Y_c(s)$$

$$S_{G(s)}^{Y(s)} = \frac{dY_c(s)/Y_c(s)}{dG(s)/G(s)} = \frac{1}{1 + G(s)H(s)} \tag{14.13}$$

Comparing Eqs. (14.7) and (14.13) shows that the closed-loop variation is reduced by the factor $1/|1 + G(s)H(s)|$. In comparing Eqs. (14.10) and (14.13), if the term $|1 + G(s)H(s)|$ is larger than the term $|1 + G(s)|$, then there is an advantage to using a nonunity-feedback system. Further, $H(s)$ may be introduced both to provide an improvement in system performance and to reduce the effect of parameter variations within $G(s)$.

Case 4: Closed-Loop Nonunity-Feedback System of Fig. 14.1*b* [Feedback Function $H(s)$ Variable and $G(s)$ Fixed]

From Eq. (14.11),

$$dY_c(s) = R(s)\frac{-G(s)^2 dH(s)}{[1 + G(s)H(s)]^2} \tag{14.14}$$

Multiplying and dividing Eq. (14.14) by $H(s)$ and also dividing by Eq. (14.11) results in

$$dY_c(s) = -\left[\frac{G(s)H(s)}{1 + G(s)H(s)}\frac{dH(s)}{H(s)}\right]Y_c(s) \approx -\frac{dH(s)}{H(s)}Y_c(s)$$

$$S_{H(s)}^{Y(s)} = \frac{dY_c(s)/Y_c(s)}{dH(s)/H(s)} \approx -1 \tag{14.15}$$

The approximation applies for those cases where $|G(s)H(s)| \gg 1$. When Eq. (14.15) is compared with Eq. (14.7), it is seen that a variation in the feedback function has approximately a direct effect upon the output, the same as for the open-loop case. Thus, the components of $H(s)$ must be selected as precision fixed elements to maintain the desired degree of accuracy and stability in the transform $Y(s)$.

The two situations of cases 3 and 4 serve to point out the advantage of feedback compensation from the standpoint of parameter changes. Since the use of fixed feedback compensation minimizes the effect of variations in the components of $G(s)$, prime consideration can be given to obtaining the necessary power requirement in the forward loop rather than to accuracy and stability. $H(s)$ can be designed as a precision device so that the transform $Y(s)$ or the output $y(t)$ has the desired accuracy and stability. *In other words, by use of feedback compensation the performance of the system can be made to depend more on the feedback term than on the forward term.*

Applying the sensitivity equation (14.3) to each of the four cases, where $M(s) = Y(s)/R(s)$, where $\delta = G(s)$ (for cases 1, 2, and 3), and $\delta = H(s)$ (for case 4), yields the results shown in Table 14.1. This table reveals that S_δ^M *never exceeds a magnitude of* 1, and the smaller this value, the less sensitive the system is to a variation in the transfer function. For an increase in the variable function, a positive value of the sensitivity function means that the output increases from its nominal response. Similarly, a negative value of the sensitivity function means that the output decreases from its nominal response. It must be realized that the

TABLE 14.1
Sensitivity functions

Case	System variable parameter	S_δ^M
1	$G(s)$	$\dfrac{dY_o/Y_o}{dG/G} = 1$
2	$G(s)$	$\dfrac{dY_c/Y_c}{dG/G} = \dfrac{1}{1+G}$
3	$G(s)$	$\dfrac{dY_c/Y_c}{dG/G} = \dfrac{1}{1+GH}$
4	$H(s)$	$\dfrac{dY_c/Y_c}{dH/H} \approx -1$

results presented in this table are based upon a functional analysis, i.e, the " variations" considered are in $G(s)$ and $H(s)$. The results are easily interpreted when $G(s)$ and $H(s)$ are real numbers. When they are not real numbers, and where δ represents the parameter that varies within $G(s)$ or $H(s)$, the interpretation can be made as a function of frequency.

The analysis in this section so far has considered variations in the transfer functions $G(s)$ and $H(s)$. Take next the case when $r(t)$ is sinusoidal. Then the input can be represented by the phasor $\mathbf{R}(j\omega)$ and the output by the phasor $\mathbf{Y}(j\omega)$. The system is now represented by the frequency transfer functions $\mathbf{G}(j\omega)$ and $\mathbf{H}(j\omega)$. All the formulas developed earlier in this section are the same in form, but the arguments are $j\omega$ instead of s. As parameters vary within $\mathbf{G}(j\omega)$ and $\mathbf{H}(j\omega)$, the magnitude of the sensitivity function can be plotted as a function of frequency. The magnitude $|S_\delta^M(j\omega)|$ does not have the limits of 0 to 1 given in Table 14.1 but can vary from 0 to any large magnitude. An example of sensitivity to parameter variation is investigated in detail in the following section. That analysis shows that the sensitivity function can be considerably reduced with appropriate feedback included in the system structure.

14.3 SENSITIVITY ANALYSIS[1,2]

It is shown in this section how the use of complete state-variable or output feedback minimizes the sensitivity of the output response to a parameter variation. This feature of state feedback is in contrast to conventional control-system design and is illustrated in the example of Sec. 13.8 for the case of system gain variation. In order to illustrate the advantage of state-variable feedback over the conventional method of achieving the desired system performance, a system sensitivity analysis is made for both designs. The basic plant $G_x(s)$ of Fig. 13.8*a* is used for this comparison. The conventional control system is shown in Fig.

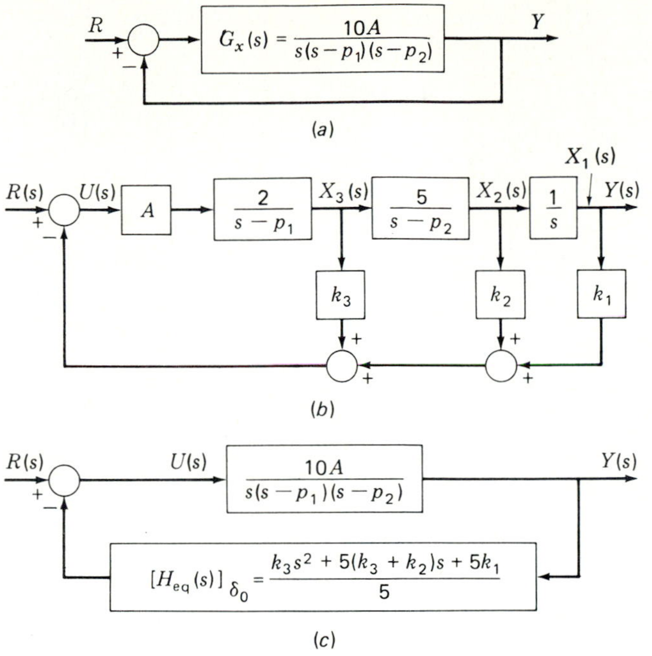

FIGURE 14.2
Control systems: (a) unity feedback; (b) state feedback; (c) state feedback, H-equivalent. The nominal values are $p_1 = -1$, $p_2 = -5$, and $A = 10$.

14.2a. The control-system design can be implemented by the complete state-feedback configuration of Fig. 14.2b or by the output feedback representation of Fig. 14.2c which uses H_{eq} as a fixed feedback compensator. The latter configuration is shown to be best for minimizing the sensitivity of the output response with a parameter variation which occurs between the state x_n and the output y. For both the state-feedback configuration and its output feedback equivalent, the coefficients k_i are determined for the nominal values of the parameters and are assumed to be invariant. The analysis is based upon determining the value of the passband frequency ω_b at $|\mathbf{M}_o(j\omega_b)| = M_o = 0.707$ for the nominal system values of the conventional and state-variable-feedback control systems. The system sensitivity function of Eq. (14.3), repeated here, is determined for each system of Fig. 14.2 and evaluated for the frequency range $0 \le \omega \le \omega_b$

$$S_\delta^M(s)\bigg|_{\substack{M=M_o\\ \delta=\delta_o}} = \left(\frac{\delta}{M}\frac{dM}{d\delta}\right)_{\substack{M=M_o\\ \delta=\delta_o}} \tag{14.16}$$

Note that M is the control ratio. The passband frequency is $\omega_b = 0.642$ for the system of Fig. 14.2a with nominal values, and $\omega_b = 1.0$ for the state-feedback system of Figs. 14.2b and 14.2c with nominal values.

The control ratios for each of the three cases to be analyzed are

Fig. 14.2*a*:

$$M(s) = \frac{10A}{s^3 - (p_1 + p_2)s^2 + p_1 p_2 s + 10A} \tag{14.17}$$

Fig. 14.2*b*:

$$M(s) = \frac{10A}{s^3 + (2Ak_3 - p_1 - p_2)s^2 + [p_1 p_2 + 2A(5k_2 - k_3 p_2)]s + 10Ak_1} \tag{14.18}$$

Fig. 14.2*c*:

$$M(s) = \frac{10A}{s^3 + (2Ak_3 - p_1 - p_2)s^2 + [p_1 p_2 + 10A(k_3 + k_2)]s + 10Ak_1} \tag{14.19}$$

The system sensitivity function, for any parameter variation, is readily obtained for each of the three configurations of Fig. 14.2 by using the respective control ratios of Eqs. (14.17) to (14.19). The plots of $|\mathbf{S}_{p_1}^M(j\omega)|$, $|\mathbf{S}_{p_2}^M(j\omega)|$, and $|\mathbf{S}_A^M(j\omega)|$ vs. ω, shown in Fig. 14.3, are drawn for the nominal values of $p_1 = -1$, $p_2 = -5$, and $A = 10$ for the state-variable-feedback system. For the conventional system the value of loop sensitivity equal to 2.1 corresponds to $\zeta = 0.7076$ as shown on the root locus in Fig. 13.9*a*. The same damping ratio is used in the state-variable-feedback system developed in the example in Sec. 13.8. Table 14.2 summarizes the values obtained.

The sensitivity function for each system, at $\omega = \omega_b$, is determined for a 100 percent variation of the plant poles $p_1 = -1$, $p_2 = -5$, and of the forward gain A, respectively. The term $\%|\Delta M|$ is defined as

$$\%|\Delta M| = \frac{|\mathbf{M}(j\omega_b)| - |\mathbf{M}_o(j\omega_b)|}{|\mathbf{M}_o(j\omega_b)|} 100 \tag{14.20}$$

where $|\mathbf{M}_o(j\omega_b)|$ is the value of the control ratio with nominal values of the plant parameters and $|\mathbf{M}(j\omega_b)|$ is the value with one parameter changed to the extreme value of its possible variation. The time-response data for systems 1 to 8 are given in Table 14.3 for a unit step input. Note that the response for the state-variable-feedback system (5 to 8, except 7*a*) is essentially unaffected by parameter variations.

As stated in Sec. 14.2, the sensitivity function S_δ^M represents a measure of the change to be expected in the system performance for a change in a system parameter. This measure is borne out by comparing Tables 14.2 and 14.3. That is, in the frequency-domain plots of Fig. 14.3, a value $|\mathbf{S}_A^M(j\omega)| = 1.29$ at $\omega = \omega_b = 0.642$ is indicated for the conventional system, compared with 0.051 for the state-variable-feedback system. Also, the percent change $\%|\Delta M(j\omega)|$ for a doubling of the forward gain is 74 percent for the conventional system and only 1.41 percent for the state-variable-feedback system. These are consistent with the changes in the peak overshoot in the time domain of $\%\Delta M_p(t) = 13.45$ and 0.192

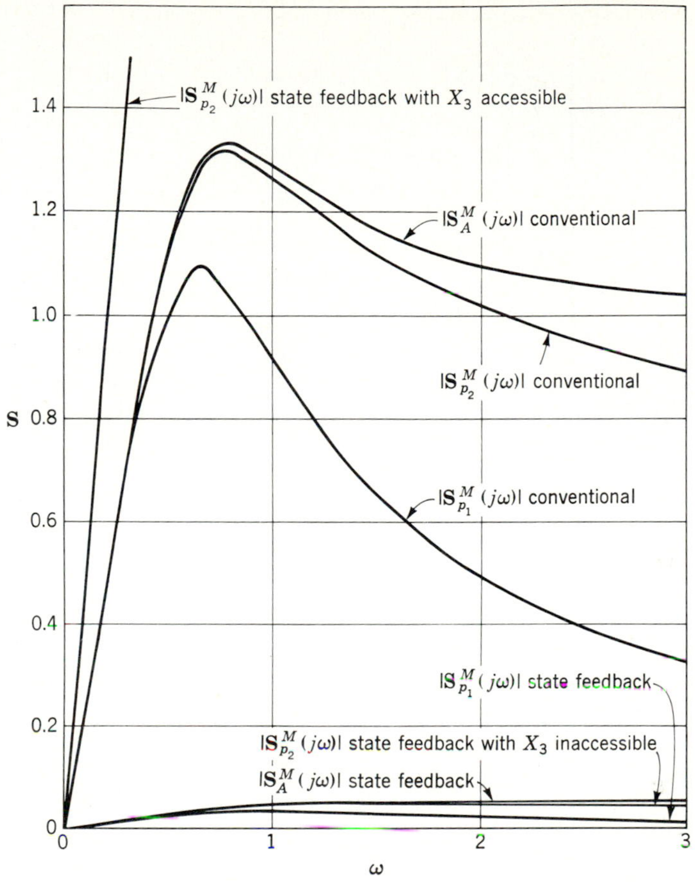

FIGURE 14.3
Sensitivity due to pole and gain variation.

percent, respectively, as shown in Table 14.3. Thus, the magnitudes of $\%\Delta M(j\omega)$ and $\%\Delta M_p(t)$ are consistent with the relative magnitudes of $\%|\mathbf{S}_A^M(j\omega_b)|$. Similar results are obtained for variations of the pole p_1.

An interesting result occurs when the pole $p_2 = -5$ is subject to variations. If the state X_3 (see Fig. 14.2b) is accessible and is fed back directly through k_3, the sensitivity $|\mathbf{S}_{p_2}^M(j\omega)|$ is much larger than for the conventional system, having a value of 3.40 at $\omega_b = 1.0$. However, if the equivalent feedback is obtained from X_1 through the fixed transfer function $[H_{\text{eq}}(s)]_{\delta_o = p_2}$, Fig. 14.2$c$, the sensitivity is considerably reduced to $|\mathbf{S}_{p_2}^M(j1)| = 0.050$, compared with 1.28 for the conventional system. If the components of the feedback unit are selected so that the transfer function $[H_{\text{eq}}(s)]$ is invariant, then the time-response characteristic for output feedback through $[H_{\text{eq}}(s)]$ is essentially unchanged when p_2 doubles in value (see Table 14.3) compared with the conventional system.

TABLE 14.2
System sensitivity analysis

Control system	Fig.	$\lvert \mathbf{M}(j\omega_b) \rvert$	$\%\lvert \Delta M(j\omega_b) \rvert$	$\lvert \mathbf{S}_{p_1}^M(j\omega_b) \rvert$	$\lvert \mathbf{S}_{p_2}^M(j\omega_b) \rvert$	$\lvert \mathbf{S}_A^M(j\omega_b) \rvert$
Conventional system design						
1. Nominal plant ($A = 0.21$, $p_1 = -1$, $p_2 = -5$): $G(s) = \dfrac{2.1}{s(s+1)(s+5)}$	14.2*a*	0.707 ($\omega_b = 0.642$)	—	1.09	1.28	1.29
2. $p_1 = -2$: $G(s) = \dfrac{2.1}{s(s+2)(s+5)}$	14.2*a*	0.338	52.4	—	—	—
3. $p_2 = -10$: $G(s) = \dfrac{2.1}{s(s+1)(s+10)}$	14.2*a*	0.658	6.9	—	—	—
4. $A = 0.42$: $G(s) = \dfrac{4.2}{s(s+1)(s+5)}$	14.2*a*	1.230	74.0	—	—	—

State-variable feedback system design [M is given in Eq. (13.83)]

5. Nominal plant: $G(s)H_{eq}(s) = \dfrac{95.4(s^2 + 1.443s + 1.0481)}{s(s+1)(s+5)}$	14.2b, c	0.707 ($\omega_b = 1.0$)	—	0.036	0.05 for X_3 inaccessible 3.40 for X_3 accessible	0.051
6. $p_1 = -2$: $G(s)H_{eq}(s) = \dfrac{95.4(s^2 + 1.443s + 1.0481)}{s(s+2)(s+5)}$	14.2b, c	0.683	3.39	—	—	—
7a. $p_2 = -10$, X_3 accessible: $G(s)H_{eq}(s) = \dfrac{95.4(s^2 + 6.443s + 1.0481)}{s(s+1)(s+10)}$	14.2b	0.160	77.4	—	—	—
7b. $p_2 = -10$, X_3 inaccessible: $G(s)H_{eq}(s) = \dfrac{95.4(s^2 + 1.443s + 1.0481)}{s(s+1)(s+10)}$	14.2c	0.682	3.54	—	—	—
8. $A = 20$: $G(s)H_{eq}(s) = \dfrac{190.8(s^2 + 1.443s + 1.0481)}{s(s+1)(s+5)}$	14.2b, c	0.717	1.41	—	—	—

TABLE 14.3
Time-response data

Control system	$M_p(t)$	t_p, s	t_s, s	$\%\Delta M_p(t) = \frac{\lvert M_p - M_{po}\rvert}{M_{po}} 100$
1	1.048	7.2	9.8	—
2	1.000	—	16.4	4.8
3	1.000	—	15.0	4.8
4	1.189	4.1	10.15	13.45
5	1.044	4.45	6	—
6	1.035	4.5	6^-	0.86
7*a*	1.000	—	24	4.4
7*b*	1.046	4.6	6.5	0.181
8	1.046	4.4	5.6^+	0.192

In analyzing Fig. 14.3, the values of the sensitivities must be considered over the entire passband ($0 \le \omega \le \omega_b$). In this passband the state-variable-feedback system has a much lower value of sensitivity than the conventional system. The performance of the state-variable-feedback systems is essentially unaffected by any variation in A, p_1, or p_2 (with restrictions) when the feedback coefficients remain fixed, provided that the forward gain satisfies the condition $K \ge K_{\min}$ (see Probs. 14.2 and 14.3). *The low-sensitivity state-variable feedback systems considered in this section satisfy the following conditions*: the system's characteristic equation must have (1) β dominant roots ($< n$) which are located close to zeros of $H_{eq}(s)$ and (2) at least one nondominant root.

A possible approach for determining the nominal value for the variable parameter δ for a state-variable-feedback design is to determine **k** for the minimum, midrange, and maximum value of δ. Then evaluate and plot $|\mathbf{S}_\delta^M(j\omega)|$ for the range of $0 \le \omega \le \omega_b$ for each of these values of δ. The value of δ that yields the minimum area under the curve of $|\mathbf{S}_\delta^M(j\omega)|$ vs. ω is selected as the nominal value δ_o.

The analysis of this section reveals that for the state-variable-feedback system any parameter variation between the control U and the state variable X_n has minimal effect on system performance. In order to minimize the effect on system performance for a parameter variation which occurs between the states X_n and X_1, the feedback signals should be moved to the right of the block containing the variable parameter (see Sec. 14.4) even if all the states are accessible. The output feedback system which incorporates an invariant $H_{eq}(s)$ has a response which is more invariant to parameter variations than a conventional system.

TABLE 14.4
Time-response and gain-sensitivity data

Example	$(Y/R)_D$	M_p	t_p, s	t_s, s	K_1, s^{-1}	$\lvert \mathbf{S}_{K_G}^M(j\omega_b) \rvert$ $\omega_b = 1.41$
1	$\dfrac{2(s+2)}{(s^2+2s+2)(s+2)}$	1.043	3.1	4.2	1.0	1.84
2	$\dfrac{100(s+2)}{(s^2+2s+2)(s+2)(s+50)}$	1.043	3.2	4.2	0.98	0.064
3	$\dfrac{100(s+1.4)(s+2)}{(s^2+2s+2)(s+1)(s+2)(s+70)}$	1.008	3.92	2.94	0.98	0.087

14.4 PARAMETER SENSITIVITY EXAMPLES

Example 1. The $(Y/R)_D$, Eq. (13.93), for Example 1 of Chap. 13 contains no nondominant pole. As seen from the root locus of Fig. 13.15 for this example, the system is highly sensitive to gain variations. Using Eq. (13.92) in Eq. (14.3) with $\delta = A$ yields $|\mathbf{S}_\delta^M(j\omega)| = 1.84$ at $\omega_b = 1.41$.

Example 2. The $(Y/R)_D$ of Example 1 is modified to contain one nondominant pole; see Eq. (13.96). Using Eq. (13.96), with $p = -50$ and $\delta = K_G$, in Eq. (14.3) yields $|\mathbf{S}_\delta^M(j\omega_b)| = 0.064$ at $\omega_b = 1.41$. Thus the presence of a nondominant pole drastically reduces $|\mathbf{S}_\delta^M(j\omega_b)|$ which is indicative of the reduction of the sensitivity magnitude for the frequency range $0 \le \omega \le \omega_b$.

Example 3. The $(Y/R)_D$ of Example 2 is modified to improve the value of t_s; see Eq. (13.101). Using Eq. (13.101), with $p = -70$ and $\delta = K_G$, in Eq. (14.3) yields $|\mathbf{S}_\delta^M(j\omega_b)| = 0.087$ at $\omega_b = 1.41$.

The data in Table 14.4 summarize the results of these three examples and show that the presence of at least one nondominant pole in $(Y/R)_D$ drastically reduces the system's sensitivity to gain variation while achieving the desired time-response characteristics.

14.5 INACCESSIBLE STATES

In order to achieve the full benefit of a state-variable-feedback system, all the states must be accessible. Although, in general, this requirement may not be satisfied, the design procedures presented in this chapter are still valid. That is, on the basis that all states are accessible, the required value of the state-feedback gain vector **k** which achieves the desired $Y(s)/R(s)$ is computed. Then, for states that are not accessible, the corresponding k_i blocks are moved by block-diagram manipulation techniques to states that are accessible. As a result of these manipulations the full benefits of state-variable feedback may not be achieved

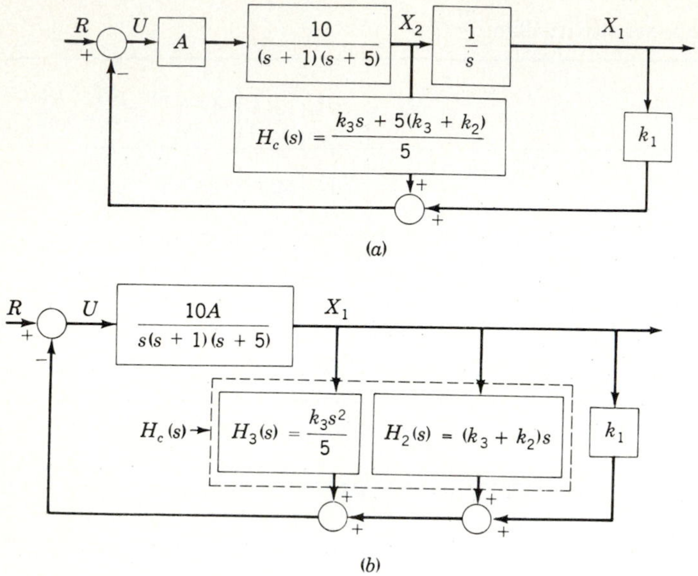

FIGURE 14.4
Minor-loop compensation (manipulation to the right) of the control system of Fig. 14.2*b* with $p_1 = -1$ and $p_2 = -2$.

(low sensitivity S_δ^M and a completely stable system), as described in Sec. 14.3. For the extreme case when the output $y = x_1$ is the only accessible state, the block-diagram manipulation to the *G*-equivalent reduces to the Guillemin-Truxal design. More sophisticated methods for reconstructing (estimating) the values of the missing states do exist (Luenberger observer theory) and are introduced in Chap. 18. They permit the accessible and the reconstructed (inaccessible) states all to be fed back through the feedback coefficients, thus eliminating the need for block-diagram manipulations and maintaining the full benefits of a state-variable feedback-designed system. Even when all states are theoretically accessible, design and/or cost of instrumentation may make it prohibitive to measure some of the states.

Assume that the X_3 state is inaccessible in the control system of Fig. 14.2*b*. Two possible block-diagram manipulations can be utilized: shift the k_3 block to the right (minor-loop compensation) or left (minor-loop or cascade compensation). Both approaches are now illustrated. Figure 14.4*a* represents the first case, where $H_c(s)$ is a proportional plus derivative device. This unit *requires the use of a transducer*; e.g., if X_2 represents velocity, an *accelerometer* can be used to obtain the derivative action of $H_c(s)$. In the event that both states X_2 and X_3 are inaccessible, the active minor-loop compensation of Fig. 14.4*b* can be used. This results in the requirement for first- and second-order derivative action in the minor loops; thus $H_c(s)$ is an improper function. It is possible to add poles to

$H_c(s)$ which are far to the left. Then a realizable $H_c(s)$ can be achieved, and the closed-loop response is not significantly affected. If X_1 represents position, an accelerometer and a tachometer can be used to realize $H_3(s)$ and $H_2(s)$, respectively. For the design requirement of the example of Sec. 13.8, the values of k_i have been determined. These values are used to design the feedback compensator $H_c(s)$. From this example the following conclusion is evident: shifting

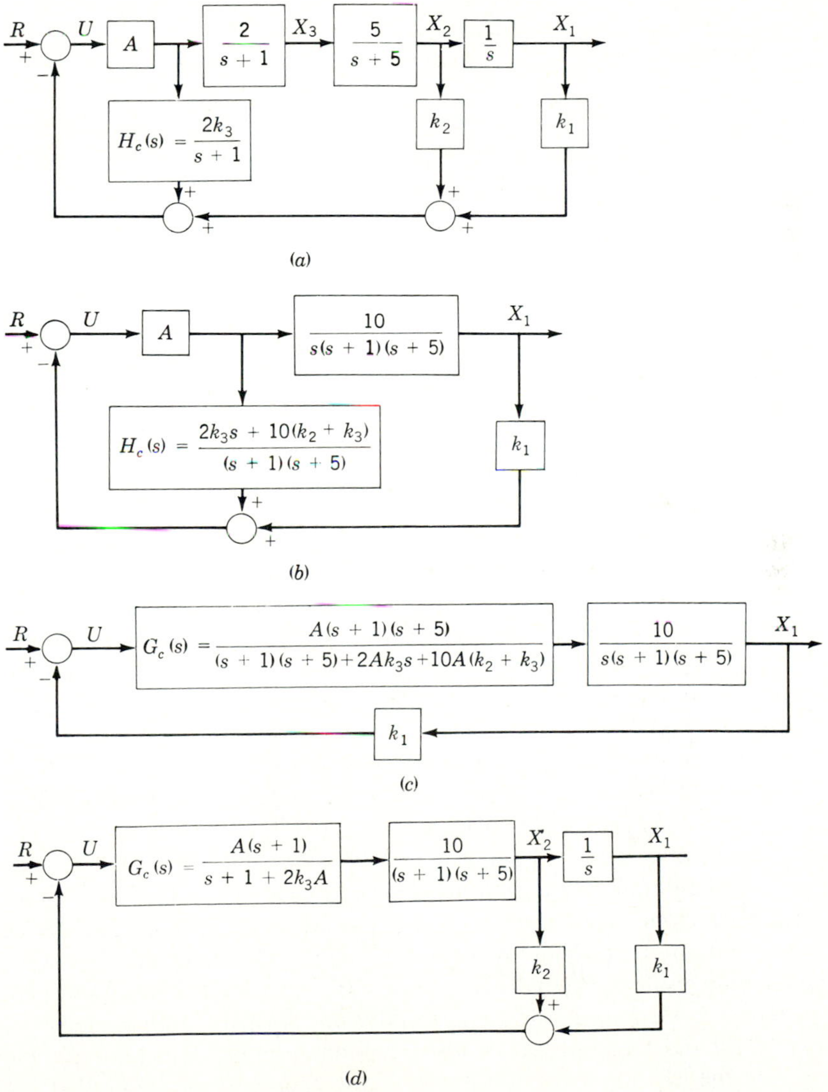

FIGURE 14.5
Manipulation to the left of the control system of Fig. 14.2*b*.

TABLE 14.5
Compensators and sensitivity functions for control systems of Figs. 14.4 and 14.5

Figure	Compensator	$\lvert \mathbf{S}_A^M(j\omega_b) \rvert$
14.4*a*	$H_c(s) = 0.954s + 1.377$	0.0508
14.4*b*	$H_c(s) = 0.954s^2 + 1.377s$	0.0508
14.5*a*	$H_c(s) = \dfrac{9.54}{s+1}$	0.0508
14.5*b*	$H_c(s) = \dfrac{9.54(s+1.443)}{(s+1)(s+5)}$	0.0508
14.5*c*	$G_c(s) = \dfrac{10(s+1)(s+5)}{(s+1.43)(s+99.98)}$	0.51
14.5*d*	$G_c(s) = \dfrac{10(s+1)}{s+96.4}$	0.51

k_i from the inaccessible state x_i to the right of the accessible state x_j requires $(i-j)$th-order derivative action in the minor-loop compensator $H_c(s)$, where $i > j$.

Figure 14.5 represents four possible cases of the manipulation to the left for the control system of Fig. 14.2*b*: Fig. 14.5*a* and *b* result in the use of minor-loop compensators; Fig. 14.5*c* results in the use of a cascade compensator with all feedback other than k_1 eliminated; and Fig. 14.5*d* results in the use of both a cascade compensator and state feedback. The first and fourth cases assume that only the state X_3 is inaccessible. The second and third cases assume that both the states X_2 and X_3 are inaccessible. The compensators in all four cases may be realized, using network-synthesis techniques, by passive elements in conjunction with a cascade amplifier.

The compensators required for each case represented in Figs. 14.4 and 14.5, using the values of $k_1 = 1$, $k_2 = -3.393$, $k_3 = 4.77$, and $A = 10$ determined in Sec. 13.8, are given in Table 14.5, which also gives the sensitivity function $\lvert \mathbf{S}_{K_G}^M(j\omega_b) \rvert$ for each case at $\omega_b = 1.0$ and $M = 0.707$. As noted in Table 14.5, the low sensitivity of the output response to gain variations is maintained as long as invariant feedback compensators $H_c(s)$ are used. In other words, the block-diagram manipulations that result in *only minor-loop* compensation, i.e., using only *feedback* compensators $H_c(s)$, have the same $H_{eq}(s)$ as before the manipulations are performed. Therefore, the same stability and sensitivity characteristics are maintained for the manipulated as for the nonmanipulated systems. When the shifting is to the left, to yield a resulting cascade compensator $G_c(s)$, the low sensitivity to parameter variation is not achieved. This sensitivity characteristic is

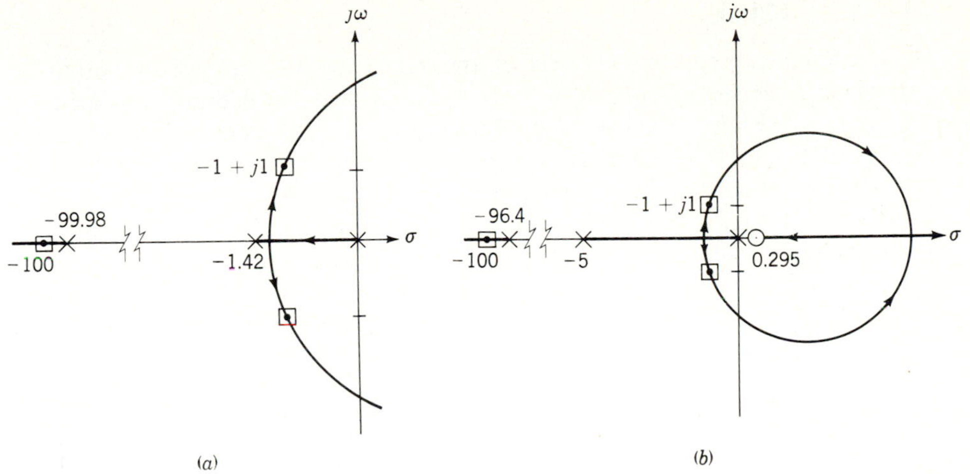

FIGURE 14.6
Root locus for (*a*) Fig. 14.5*c* and (*b*) Fig. 14.5*d*.

borne out by analyzing the root locus of Figs. 13.9*b* and 14.6. Figure 13.9*b* is the root locus representing the block-diagram manipulations using only $H_c(s)$ networks. Figure 14.6 shows the root locus representing the block-diagram manipulations using cascade $G_c(s)$ networks. In Fig. 13.9*b* an increase in the gain does not produce any appreciable change in the closed-loop system response. In contrast, any gain variation in the systems represented by Fig. 14.6 greatly affects the closed-loop system response.

Note that the manipulated system of Fig. 14.5*d*, using both cascade $G_c(s)$ and minor-loop $H_c(s) = k_2$ compensators, yields the same relative sensitivity as for the system of Fig. 14.5*c*, using only cascade compensation. In general, it may be possible to achieve a relative sensitivity for a system using both $G_c(s)$ and $H_c(s)$ compensators somewhere between that of a system using only minor-loop compensation and one using only cascade compensation. As pointed out in Sec. 14.3, in order to maintain low sensitivity for variations in parameters between the states X_1 and X_n, the feedback must be moved to the right.

The technique presented in this section for implementing a system having inaccessible states is rather simple and straightforward. It is very satisfactory for systems that are noise-free and time-invariant. For example, if in Fig. 14.2*b* the state X_3 is inaccessible and X_2 contains noise, it is best to shift the k_3 block to the left. This assumes that the amplifier output signal has minimal noise content. If all signals contain noise, the system performance is degraded from the desired performance. This technique is also satisfactory under parameter variations and/or if the poles and zeros of the transfer function $G_x(s)$ are not known exactly when operating under hfg.

14.6 SUMMARY

This chapter presents an analysis of the sensitivity of the system output to variations of system parameters. It is accomplished by defining a sensitivity function S_{δ}^{M}, where M is the control ratio and δ is the parameter that varies. A sensitivity analysis is performed for a specific example system, and it shows that a state-variable-feedback system has a much lower sensitivity than a conventional output feedback design. When there are inaccessible states, the states can be regenerated from the output provided that the system is completely observable. For SISO systems this requires derivative action, accentuating noise that may be present in the output. This can be avoided by designing an observer (see Chap. 18).

REFERENCES

1. "Sensitivity and Modal Response for Single-Loop and Multiloop Systems," *Flight Control Lab. ASD, AFSC Tech. Doc. Rep.* ASD-TDR-62-812, Wright-Patterson Air Force Base, Ohio, January 1963.
2. Cruz, J. B., Jr.: *Feedback Systems*, McGraw-Hill, New York, 1972.

CHAPTER 15

LIAPUNOV'S SECOND METHOD

15.1 INTRODUCTION

The definition of stability for a linear time-invariant system is an easy concept to understand. The complete response to any input contains a particular solution which has the same form as the input and a complementary solution containing terms of the form $Ae^{\lambda_i t}$. When the eigenvalues λ_i have negative real parts, the transients associated with the complementary solution decay with time, and the response is called stable. When the roots have positive real parts, the transients increase without bound and the system is called unstable. It is necessary to extend the concept of stability to nonlinear systems. Also, it is desired that stability be determined without explicitly solving for the eigenvalues. This is especially important for high-order systems. Liapunov attacked this problem through his *second, or direct, method*. The *sufficient but not necessary condition* of this method requires the determination of a scalar function V, called a *Liapunov function*. This function must approach an equilibrium point along a trajectory as time increases. A Liapunov function is readily determined for linear time-invariant systems; however, the determination is difficult for nonlinear and for time-variable systems. Despite these limitations, the Liapunov method provides a generalized approach to stability that is needed by the control engineer.

In order to develop Liapunov's second method properly, this chapter starts with an introduction to state-space trajectories. The kinds of singular points and the associated stability are categorized for both linear and nonlinear systems. The properties of quadratic functions and the definitions of definiteness are presented since the Liapunov function, which is a basis for ensuring an asymptotically stable response, may be a quadratic function for linear time-invariant systems. Some useful definitions of stability are given. The Liapunov method is presented and illustrated by examples. It is shown that the Liapunov function provides a measure of the system time response. This provides the basis for its use in Chap. 16 in defining optimum system response.

15.2 STATE-SPACE TRAJECTORIES[1-3]

The *state space* is defined as the n-dimensional space in which the components of the state vector represent its coordinate axes. The unforced response of a system having n state variables, released from any initial point $\mathbf{x}(t_0)$, traces a curve or trajectory in an n-dimensional state space. Time t is an implicit function along the trajectory. When the state variables are represented by phase variables, the state space may also be called a phase space. While it is impossible to visualize a space with more than three dimensions, the concept of a state space is nonetheless very useful in systems analysis. The behavior of second-order systems is conveniently viewed in the state space since it is two-dimensional and is in fact a state (or phase) plane. It is easy to obtain a graphical or geometrical representation and interpretation of second-order system behavior in the state plane. The concepts developed for the state plane can be extrapolated to the state space of higher-order systems. Examples of a number of second-order systems are presented in this section. These concepts are extended qualitatively to third-order systems. The symmetry that results in the state-plane trajectories is demonstrated for the cases when the **A** matrix is converted to normal (canonical) or modified normal (modified canonical) form. The second-order examples which are considered include the cases where the eigenvalues are (1) negative real and unequal (overdamped), (2) negative real and equal (critically damped), (3) imaginary (oscillatory), and (4) complex with negative real parts (underdamped). The system which is studied has no forcing function and is represented by the state equation

$$\dot{\mathbf{x}} = \begin{bmatrix} a_{11} & a_{12} \\ a_{21} & a_{22} \end{bmatrix} \mathbf{x} \tag{15.1}$$

Since there is no forcing function, the response results from energy stored in the system as represented by initial conditions.

Example 1 Overdamped response. Consider the following example with the given initial conditions:

$$\dot{\mathbf{x}} = \begin{bmatrix} 0 & 3 \\ -1 & -4 \end{bmatrix} \mathbf{x} \qquad \mathbf{x}(0) = \begin{bmatrix} 0 \\ 2 \end{bmatrix} \tag{15.2}$$

Since one state equation is $\dot{x}_1 = 3x_2$, then x_2 is a phase variable. Therefore, the resulting state plane is called a *phase plane*.

The state transition matrix $\boldsymbol{\Phi}(t)$ is obtained by using the Sylvester expansion theorem or any other convenient method. The system response is overdamped, with

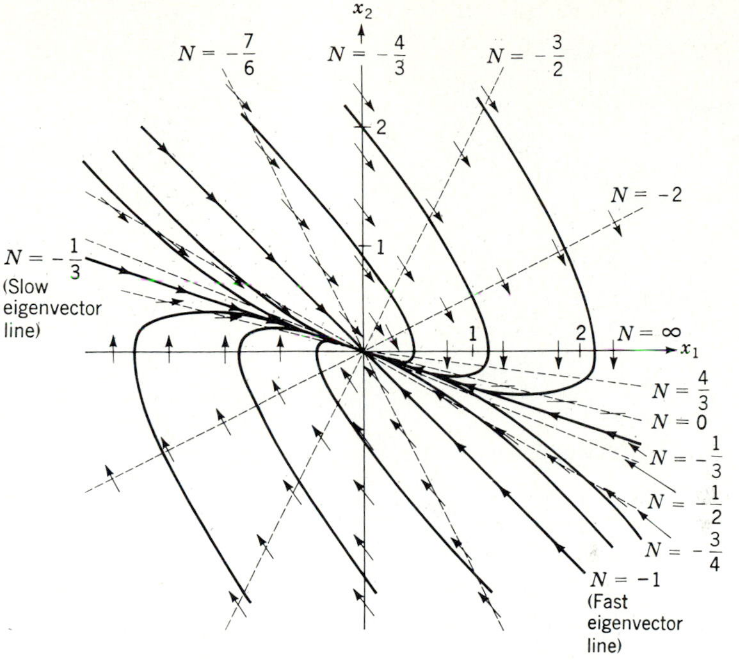

FIGURE 15.1
Phase-plane portrait for Eq. (15.2).

eigenvalues $\lambda_1 = -1$ and $\lambda_2 = -3$. For the given initial conditions the system response is

$$\mathbf{x}(t) = \mathbf{\Phi}(t)\mathbf{x}(0) = \begin{bmatrix} 3e^{-t} - 3e^{-3t} \\ -e^{-t} + 3e^{-3t} \end{bmatrix} \tag{15.3}$$

The trajectory represented by the equation can be plotted in the phase plane x_2 versus x_1. A family of trajectories in the phase plane for a number of initial conditions is drawn in Fig. 15.1. Such a family of trajectories is called a *phase portrait*. The arrows show the direction of the states along the trajectories for increasing time. Motion of the point on a trajectory is in the clockwise direction about the origin since $\dot{x}_1 = 3x_2$. Thus, when x_2 is positive, x_1 must be increasing in value, and when x_2 is negative, x_1 must be decreasing in value. The value of N in Fig. 15.2 is the slope of the trajectory at each point in the phase plane, i.e., $N = dx_2/dx_1$, as described in Eq. (15.4). The rate of change of x_1 is zero at the points where the trajectories cross the x_1 axis. Therefore, the trajectories always cross the x_1 axis in a perpendicular direction.

One method of drawing the phase trajectory is to insert values of time t into the solution $\mathbf{x}(t)$ and to plot the results. For the example above the solutions are $x_1 = 3e^{-t} - 3e^{-3t}$ and $x_2 = -e^{-t} + 3e^{-3t}$. Another approach is to eliminate t from these equations to obtain an analytical expression for the trajectory. This can

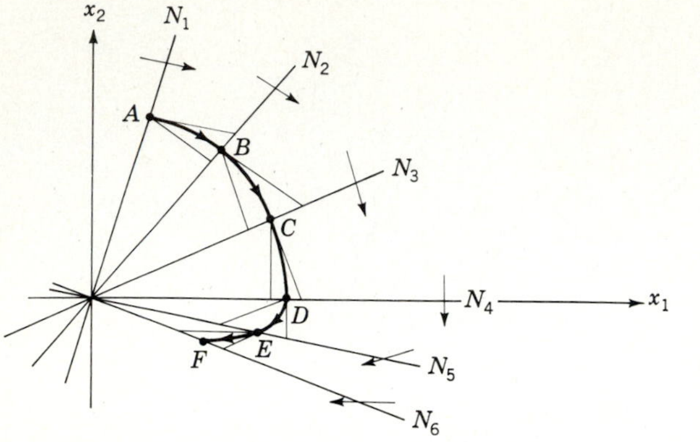

FIGURE 15.2
Construction of a trajectory using isoclines.

be done by first solving for e^{-t} and e^{-3t}.

$$e^{-t} = \frac{x_1 + x_2}{2}$$

$$e^{-3t} = \frac{x_1 + 3x_2}{6}$$

Raising e^{-t} to the third power and e^{-3t} to the first power and equating the results yields

$$\left(\frac{x_1 + x_2}{2}\right)^3 = \left(\frac{x_1 + 3x_2}{6}\right)^1$$

This equation is awkward to use and applies only for the specified initial conditions. Further, it is difficult to extend this procedure to nonlinear systems.

A practical graphical method for obtaining trajectories in the state plane is the *method of isoclines*. The previous example is used to illustrate this method. Taking the ratio of the state equations of this example yields

$$\frac{\dot{x}_2}{\dot{x}_1} = \frac{dx_2/dt}{dx_1/dt} = \frac{dx_2}{dx_1} = \frac{-x_1 - 4x_2}{3x_2} = N \tag{15.4}$$

This equation represents the slope of the trajectory passing through any point in the state plane. The integration of this equation traces the trajectory starting at any initial condition. The method of isoclines is a graphical technique for performing this integration. When the slope of the trajectory in Eq. (15.3) is fixed at any specific value N, the result is the equation of a straight line having a slope m and represented by

$$x_2 = -\frac{1}{4 + 3N}x_1 = mx_1 \tag{15.5}$$

This equation describes a family of curves which are called *isoclines*. They have the property that all trajectories crossing a particular isocline have the same slope at the crossing points. For linear time-invariant second-order systems the isoclines are all straight lines. A trajectory can be sketched by using the following procedure:

1. Draw a number of isoclines reasonably close together.
2. From any starting point A (see Fig. 15.2) draw two lines having slopes N_1 and N_2 and extending from the isocline for N_1 to the isocline for N_2. The trajectory must stay between these lines. Select point B midway in the segment of the isocline for N_2 between these lines.
3. Continue from point B using the procedure of step 2.
4. Connect the points with a smooth curve which crosses each isocline with the correct slope. An example is shown in Fig. 15.2.

The system represented by Example 1 is stable, as indicated by the negative eigenvalues. Since there is no forcing function, the response is due to the energy initially stored in the system. As this energy is dissipated, the response approaches equilibrium at $x_1 = x_2 = 0$. All the trajectories approach and terminate at this point. The slope of the trajectory going through this equilibrium point is therefore indeterminate, that is, $N = dx_2/dx_1 = 0/0$, representing a *singularity* point. This indeterminacy can be used to locate the equilibrium point by letting the slope of the trajectory in the state plane be $N = dx_2/dx_1 = 0/0$. Applying this condition to Eq. (15.4) yields the two equations $x_2 = 0$ and $-x_1 - 4x_2 = 0$. This confirms that the equilibrium point is $x_1 = x_2 = 0$. For the overdamped response this equilibrium point is called a *node*. The procedure can be extended to all systems of any order, even when they are nonlinear. This is done by solving $\dot{\mathbf{x}} = \mathbf{0}$ for the equilibrium point(s).

It is interesting to note the symmetry that results in the phase portrait when the states are transformed to canonical form in which the plant matrix $\mathbf{A}$ is diagonal [see Eq. (5.70)].[9] The methods for accomplishing this transformation are described in Sec. 5.10. The resulting state equations in canonical form for a second-order system are

$$\dot{z}_1 = \lambda_1 z_1 \tag{15.6}$$

$$\dot{z}_2 = \lambda_2 z_2 \tag{15.7}$$

When the eigenvalues $\lambda_1 \neq \lambda_2$, the ratio of these equations yields

$$\frac{dz_1}{dz_2} = \frac{\lambda_1 z_1}{\lambda_2 z_2} \tag{15.8}$$

The variables in this equation are separable, and the resulting equation can be integrated to obtain the equation of the state-plane trajectory

$$z_2 = C z_1^{\lambda_2/\lambda_1} \tag{15.9}$$

For a stable system with λ_1 and λ_2 both negative and real, where $\lambda_2 < \lambda_1 < 0$, the state-plane portraits are parabolic, as shown in Fig. 15.3*a* and *b*. All the trajectories are tangent to the z_1 axis. When returning to the x_2, x_1 plane, the trajectories are tangent to the line labeled z_1 in Fig. 15.3*c* and *d*. This line is

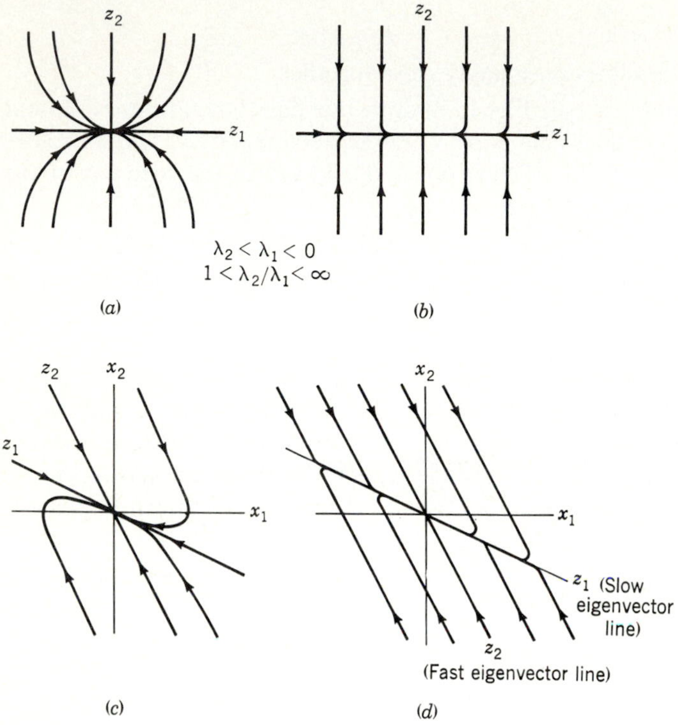

FIGURE 15.3
Stable node: (a) normal form: roots of similar magnitudes; (b) normal form: roots of widely different magnitudes; (c) x_2 vs. x_1: roots with similar magnitudes; (d) x_2 vs. x_1: roots with widely different magnitudes (see Ref. 9).

the transformation of the z_1 axis into the x_2, x_1 plane. The slope of this line can be obtained from the modal matrix $\mathbf{T}$ (see Sec. 5.10), which transforms the $\mathbf{z}$ plane into the $\mathbf{x}$ plane. Using $\mathbf{T} = [v_{ij}]$ and $\mathbf{z} = \mathbf{T}^{-1}\mathbf{x}$ yields

$$z_1 = \frac{1}{\Delta_T}(v_{22}x_1 - v_{12}x_2) \tag{15.10}$$

$$z_2 = \frac{1}{\Delta_T}(-v_{21}x_1 + v_{11}x_2) \tag{15.11}$$

The z_1 axis is defined by $z_2 = 0$. Therefore, from Eq. (15.11), the slope of the z_1 axis after transformation to the x_2, x_1 plane is

$$m_1 = \frac{x_2}{x_1} = \frac{v_{21}}{v_{11}} \tag{15.12}$$

Thus the slope is determined by that eigenvector $\mathbf{v}_1$ which is the first column of the modal matrix $\mathbf{T}$. The ratio v_{21}/v_{11} can be evaluated, using method 2 of Sec. 5.10, by equating the elements of $\mathbf{AT} = \mathbf{T\Lambda}$ [see Eq. (5.91)], which yields, for real eigenvalues,

$$m_1 = \frac{v_{21}}{v_{11}} = \frac{\lambda_1 - a_{11}}{a_{12}} = \frac{a_{21}}{\lambda_1 - a_{22}} \tag{15.13}$$

Since λ_1 yields the transient mode with the longest settling time, the line z_1 in Fig. 15.3*c* and *d* is labeled the "slow eigenvector line."

The trajectories in Fig. 15.3*a* and *b* are parallel to the z_2 axis at large distances from the origin. In the x_2, x_1 plane (see Fig. 15.3*c* and *d*) the trajectories retain this property except that they are parallel to the line labeled z_2. This line is the transformation of the z_2 axis into the x_2, x_1 plane. The z_2 axis is defined by the condition $z_1 = 0$. Using Eq. (15.10), the slope of this axis transformed into the $\mathbf{x}$ plane is

$$m_2 = \frac{x_2}{x_1} = \frac{v_{22}}{v_{12}} \tag{15.14}$$

This slope is therefore determined by the eigenvector, the second column of $\mathbf{T}$. From the coefficients of $\mathbf{AT} = \mathbf{T}\Lambda$ [see Eq. (5.91)] this slope is obtained as a function of the real eigenvalue λ_2:

$$m_2 = \frac{v_{22}}{v_{12}} = \frac{\lambda_2 - a_{11}}{a_{12}} = \frac{a_{21}}{\lambda_2 - a_{22}} \tag{15.15}$$

Since λ_2 yields the transient mode with the shortest settling time, the line z_2 in Fig. 15.3*c* and *d* is labeled the "fast eigenvector line."

For the system of Example 1 the eigenvectors are $\mathbf{v}_1 = [1 \quad -3]^T$ and $\mathbf{v}_2 = [1 \quad -1]^T$. Thus, the slopes are $m_1 = -1/3$ and $m_2 = -1$. The lines going through the equilibrium point with these slopes are drawn in Fig. 15.1. The transient term $e^{\lambda_2 t} = e^{-3t}$ decays faster than the term $e^{\lambda_1 t} = e^{-t}$. Therefore, near the origin the trajectories are tangent to the line having the slope $m_1 = -1/3$ determined by the slow eigenvalue $\lambda_1 = -1$.

When the real eigenvalues are equal and negative, the phase portrait is similar to that in Fig. 15.1 except that there is only one eigenvector line.

When any real eigenvalue is positive, the system response is unstable. For the second-order system with two positive eigenvalues the equilibrium point is an *unstable node*. In this case the trajectories depart from the equilibrium point and go to infinity.

Example 2 Underdamped response. The system considered here is

$$\dot{\mathbf{x}} = \begin{bmatrix} 0 & 1 \\ -4 & -2 \end{bmatrix} \mathbf{x} \tag{15.16}$$

The characteristic equation is $\lambda^2 + 2\lambda + 4 = 0$, and the eigenvalues are $\lambda_{1,2} = -1 \pm j\sqrt{3}$, representing a damping ratio $\zeta = 0.5$. The equilibrium point is determined from the equation of the slope of the trajectory

$$\frac{dx_2}{dx_1} = \frac{-4x_1 - 2x_2}{x_2} = N \tag{15.17}$$

The singularity condition $N = 0/0$ determines the equilibrium point $x_1 = x_2 = 0$.

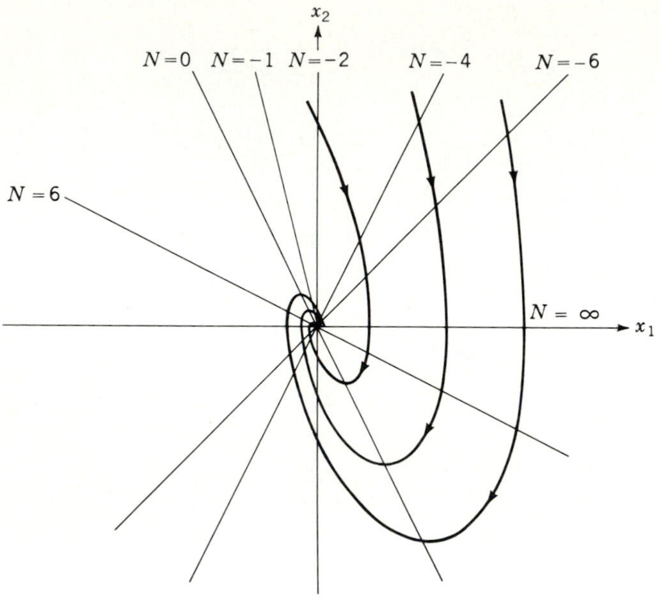

FIGURE 15.4
Underdamped phase-plane trajectories.

Using Eq. (15.17), the isoclines are the straight lines represented by

$$x_2 = \frac{-4}{2+N} x_1 \tag{15.18}$$

Since the matrix **A** in Eq. (15.15) is in companion form, the states are phase variables and the state plane is a phase plane. Several phase-plane trajectories are shown in Fig. 15.4. The trajectories are modified logarithmic spirals.

The equilibrium point for an underdamped response is called a *focus*. For stable responses the motion of the point on the spiral trajectories is clockwise toward the focus. Thus, when $x_2 = \dot{x}_1$ is positive, the trajectory moves to the right in the direction of increasing values of x_1. Also, when $x_2 = \dot{x}_1$ is negative, the trajectory moves to the left in the direction of decreasing values of x_1.

When the state variables are transformed to canonical form, the equation representing the trajectories is best represented in polar form, since the trajectories are known to be spiral. When the eigenvalues are complex conjugates, $\lambda_{1,2} = \sigma \pm j\omega_d$, a simplified form of the trajectory equation is obtained by assuming the canonical state variables are also complex conjugates:

$$z_1 = w_1 - jw_2 \qquad z_2 = w_1 + jw_2 \tag{15.19}$$

Inserting these values into the canonical state equations $\dot{z}_1 = \lambda_1 z_1$ and $\dot{z}_2 = \lambda_2 z_2$

yields

$$\dot{w}_1 - j\dot{w}_2 = (\sigma + j\omega_d)(w_1 - jw_2) \qquad \dot{w}_1 + j\dot{w}_2 = (\sigma - j\omega_d)(w_1 + jw_2) \tag{15.20}$$

Separating the real and imaginary components of these equations produces

$$\dot{w}_1 = \sigma w_1 + \omega_d w_2 \qquad \dot{w}_2 = -\omega_d w_1 + \sigma w_2 \tag{15.21}$$

These equations can be recognized as being in the modified normal form described in Sec. 5.12. The slope of the trajectories in the w_1, w_2 plane, obtained from Eq. (15.21), is

$$\frac{dw_2}{dw_1} = \frac{-\omega_d w_1 + \sigma w_2}{\sigma w_1 + \omega_d w_2} \tag{15.22}$$

This equation is converted from the rectangular coordinates w_1 and w_2 to the polar coordinates r and ϕ by using $w_1 = r\cos\phi$ and $w_2 = r\sin\phi$. Then,

$$dw_1 = \cos\phi\, dr - r\sin\phi\, d\phi \qquad \text{and} \qquad dw_2 = \sin\phi\, dr + r\cos\phi\, d\phi$$

The polar magnitude r and its angle ϕ are related to the rectangular coordinates w_1 and w_2 by

$$r(t) = \sqrt{w_1^2(t) + w_2^2(t)} \qquad \text{and} \qquad \phi(t) = \tan^{-1}[w_2(t)/w_1(t)] \tag{15.23}$$

These values apply for all values of time. After simplification of Eq. (15.22), the resulting equation representing the slope of the trajectory in the w_1, w_2 plane is

$$\frac{dr}{d\phi} = -\frac{\sigma}{\omega_d} r$$

The variables can be separated, and integration of the equation yields

$$r = Ce^{-(\sigma/\omega_d)\phi} = Ce^{\left(\zeta/\sqrt{1-\zeta^2}\right)\phi} \tag{15.24}$$

For any initial values, at time $t = 0$, of $w_1(0)$ and $w_2(0)$, see Fig. 15.5d, the corresponding values r_0 and ϕ_0 are obtained from Eq. (15.23). Then the value of C in Eq. (15.24) is given by

$$C = r_0 \Big/ e^{\left(\zeta/\sqrt{1-\zeta^2}\right)\phi_0}$$

The trajectories represented by this equation are logarithmic spirals. Motion along the trajectories is clockwise for increasing time. For positive values of damping ratio, $1 > \zeta > 0$, the system response is stable, and the trajectories approach the equilibrium point (*stable focus*). Figure 15.5a and b shows the shapes of the trajectories in the **w** and **x** planes for a stable response. For negative damping ratio, $-1 < \zeta < 0$, the response is unstable, and the trajectories diverge from the equilibrium point (*unstable focus*). Figure 15.5c shows the trajectories for an unstable response in the **x** plane.

For the state equation (15.16) the transformation matrix $\mathbf{T}_m$ from the **x** plane to the **w** plane [see Eq. (5.130)] is

$$\mathbf{T}_m = \begin{bmatrix} 1 & 0 \\ -1 & \sqrt{3} \end{bmatrix} \tag{15.25}$$

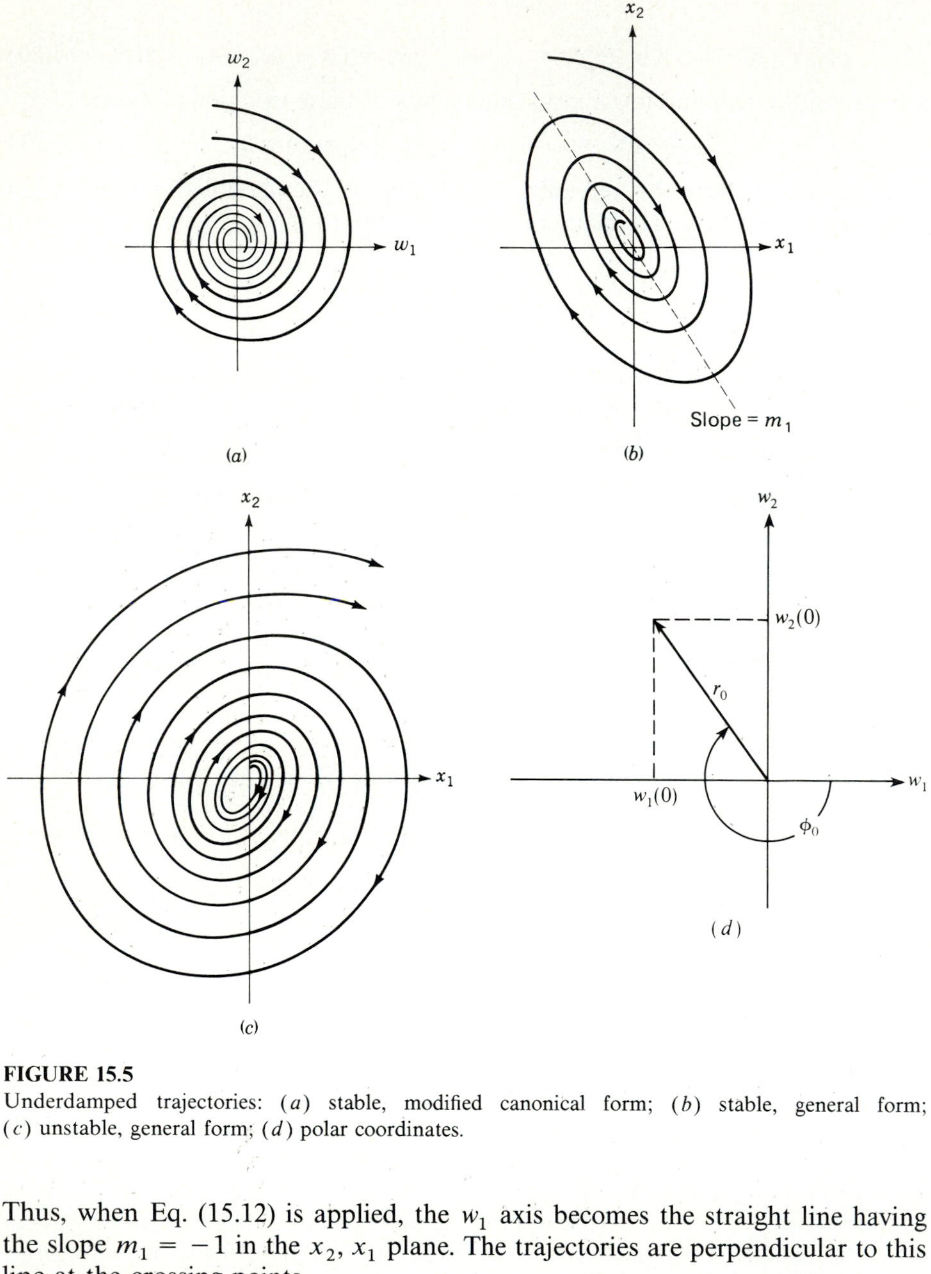

FIGURE 15.5
Underdamped trajectories: (a) stable, modified canonical form; (b) stable, general form; (c) unstable, general form; (d) polar coordinates.

Thus, when Eq. (15.12) is applied, the w_1 axis becomes the straight line having the slope $m_1 = -1$ in the x_2, x_1 plane. The trajectories are perpendicular to this line at the crossing points.

For the undamped case, $\zeta = 0$, the canonical and modified canonical coefficient matrices have the form

$$\Lambda = \begin{bmatrix} j\omega_d & 0 \\ 0 & -j\omega_d \end{bmatrix} \quad \text{and} \quad \Lambda_m = \begin{bmatrix} 0 & \omega_d \\ -\omega_d & 0 \end{bmatrix} \tag{15.26}$$

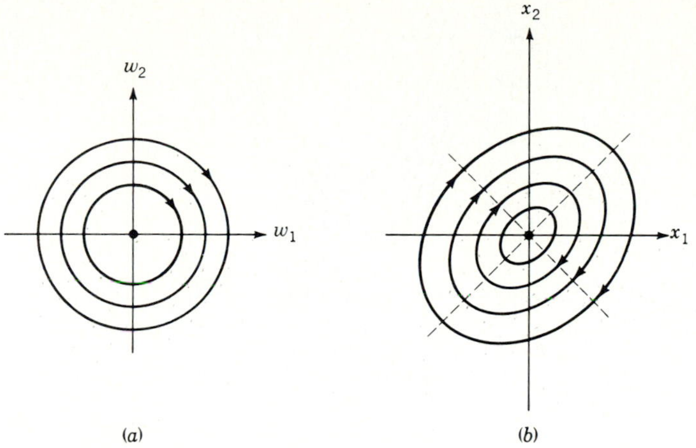

FIGURE 15.6
Center and trajectories with no damping in (*a*) normal form and (*b*) general form.

The trajectories, obtained from Eq. (15.24), are circles having constant radii in the **w** plane. In the **x** plane they become ellipses. The equilibrium point is called a *center* or *vortex*. The portraits for the **w** and **x** planes are illustrated in Fig. 15.6. The principal axis of the ellipses in the **x** plane is a function of the **A** matrix. As an example, the matrix

$$\mathbf{A} = \begin{bmatrix} 1 & 1 \\ -2 & -1 \end{bmatrix} \tag{15.27}$$

has the eigenvalues $\lambda_{1,2} = \pm j1$ and the transformation matrix $\mathbf{T}_m$ from the **x** plane to the **w** plane [see Eq. (5.130)] is

$$\mathbf{T}_m = \begin{bmatrix} 1 & 0 \\ -1 & 1 \end{bmatrix} \tag{15.28}$$

Thus, when Eq. (15.12) is applied, the w_1 axis becomes the straight line having the slope $m_1 = -1$ in the x_2, x_1 plane. The principal axes of the ellipses in the x_2, x_1 plane are shown in Fig. 15.6*b*.

Example 3 One positive and one negative real root. When the roots are real but of opposite sign, the equilibrium point is called a *saddle point* and represents an unstable equilibrium. The trajectories in canonical form are represented by Eq. (15.9), which can be rewritten as

$$z_2 = Cz_1^m \tag{15.29}$$

where $m = \lambda_2/\lambda_1$. The trajectories are hyperbolas which are asymptotic to the z_1 and z_2 axes, as shown in Fig. 15.7*a*. When transformed back to the **x** plane, the trajectories are rotated and modified in shape as shown in Fig. 15.7*b*. The transformation of the z axes to the **x** plane yields straight lines with the slopes given by Eqs. (15.12) to (15.15). These lines are called *separatrices* since they divide the state plane into separate regions.

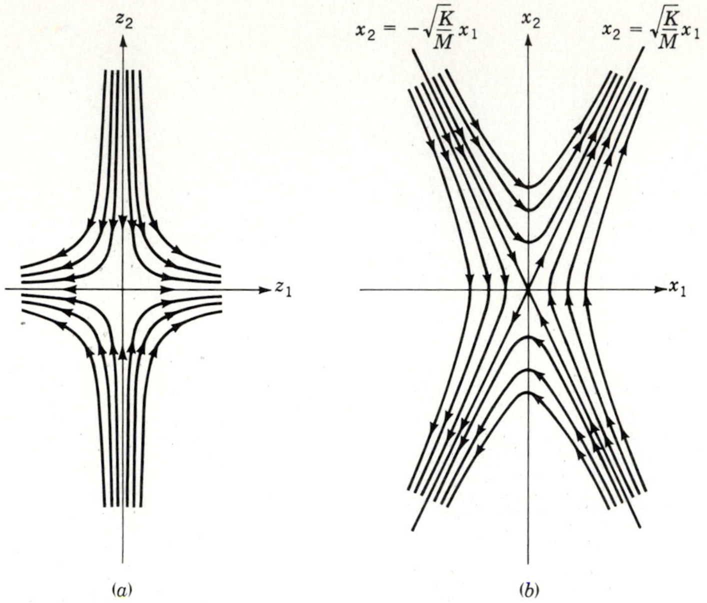

FIGURE 15.7
Saddle point and trajectories in (a) normal form and (b) general form.

The mechanical system containing mass, a "repulsive" spring force, and no damping is represented by the differential equation

$$M\ddot{x} - Kx = 0 \tag{15.30}$$

This equation describes the motion of a pendulum with no damping in the neighborhood of the upper unstable equilibrium position. In terms of phase variables the state equations are

$$\begin{bmatrix} \dot{x}_1 \\ \dot{x}_2 \end{bmatrix} = \begin{bmatrix} 0 & 1 \\ \dfrac{K}{M} & 0 \end{bmatrix} \begin{bmatrix} x_1 \\ x_2 \end{bmatrix} \tag{15.31}$$

The eigenvalues are $\lambda_{1,2} = \pm\sqrt{K/M}$. The slopes of the z_1 and z_2 axes transformed to the x_1, x_2 plane are obtained from Eqs. (15.13) and (15.15).

$$m_1 = \sqrt{\frac{K}{M}} = \lambda_1 \qquad m_2 = -\sqrt{\frac{K}{M}} = \lambda_2 \tag{15.32}$$

The trajectories are shown in Fig. 15.7b. If the initial value $\mathbf{x}(t_0)$ lies on the separatrix in the second or fourth quadrant, the trajectory terminates at the origin. That is, $\mathbf{x}(t_f) = \mathbf{0}$ satisfies the equation $M\dot{x}_2 = Kx_1$ and implies that the pendulum stops at the top equilibrium position where all the system energy is potential energy.

Third-order systems require a three-dimensional state space. At least one of the eigenvalues must be real. Figure 15.8 contains examples for third-order

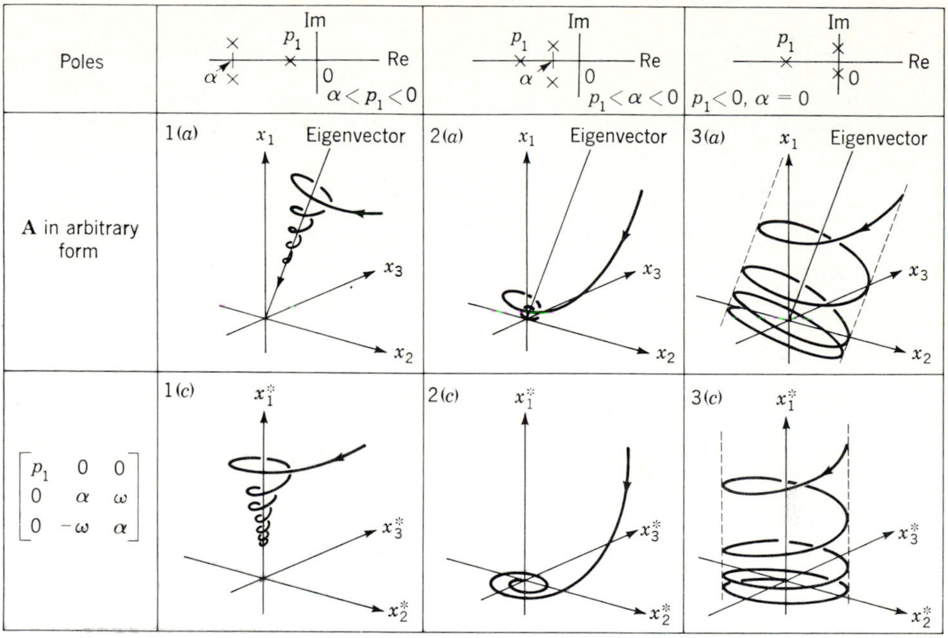

FIGURE 15.8
Trajectory patterns of third-order oscillatory systems. (*Reproduced by permission from Y. Takahashi, M. J. Rabins, and D. J. Anslander, Control and Dynamic Systems, Addison-Wesley Publishing Company, Reading, Mass.*, 1970.)

systems with a pair of complex-conjugate poles. It is observed that transforming the states to modified canonical form results in a symmetry of the trajectories with respect to the principal axes. The eigenvector for the real pole coincides with one of the axes.

The properties of trajectories in the state plane are developed in this section. The type of equilibrium points and trajectories can be extended conceptually to higher-order systems even when it is not possible to draw them in a higher-dimensional space. It is also possible to show the effect of nonlinearities on the shape of the trajectories, and an example is presented in Sec. 15.3. An important consideration in the stability analysis of such systems is that the system response in the neighborhood of the singularities can be determined from the linearized equations. Use is made of this principle in later sections of this chapter.

15.3 LINEARIZATION (JACOBIAN MATRIX)[1,8]

A linear system with no forcing function (an autonomous system) and with $|\mathbf{A}| \neq 0$ has only one equilibrium point $\mathbf{x}_0$. (See Prob. 15.17 for $|\mathbf{A}| = 0$.) All the examples in Sec. 15.2 have the equilibrium point at the origin $\mathbf{x}_0 = \mathbf{0}$. A

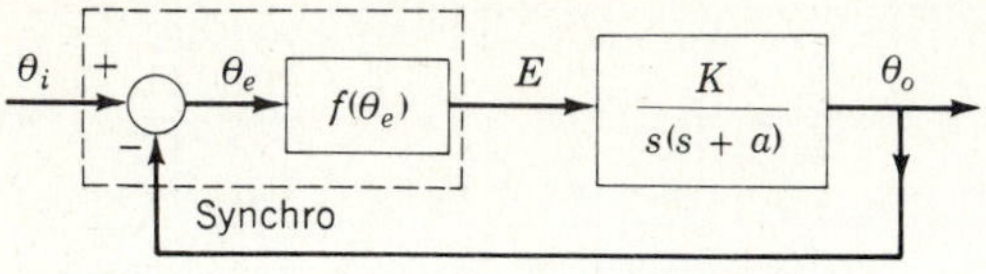

FIGURE 15.9
A nonlinear feedback control system.

nonlinear system, on the other hand, may have more than one equilibrium point. This is easily illustrated by considering the unity-feedback angular position-control system shown in Fig. 15.9. The feedback action is provided by synchros which generate the actuating signal $e = \sin(\theta_i - \theta_o)$. With no input, $\theta_i = 0$, the differential equation of the system is

$$\ddot{\theta}_o + a\dot{\theta}_o + K \sin\theta_o = 0 \tag{15.33}$$

This is obviously a nonlinear differential equation because of the term $\sin\theta_o$. With the phase variables $x_1 = \theta_o$ and $x_2 = \dot{x}_1 = \dot{\theta}_o$, the corresponding state equations are

$$\dot{x}_1 = x_2 \qquad \dot{x}_2 = -K\sin x_1 - ax_2 \tag{15.34}$$

The slope of the trajectories in the phase plane is obtained from

$$N = \frac{\dot{x}_2}{\dot{x}_1} = \frac{-K\sin x_1 - ax_2}{x_2} \tag{15.35}$$

In the general case the unforced nonlinear state equation is

$$\dot{\mathbf{x}} = \mathbf{f}(\mathbf{x}) \tag{15.36}$$

Since equilibrium points $\mathbf{x}_0$ exist at $\dot{\mathbf{x}} = \mathbf{f}(\mathbf{x}_0) = \mathbf{0}$, the singularities are $x_2 = 0$ and $x_1 = k\pi$, where k is an integer. The system therefore has multiple equilibrium points.

In a small neighborhood about each of the equilibrium points a nonlinear system behaves like a linear system. The states can therefore be written as $\mathbf{x} = \mathbf{x}_0 + \mathbf{x}^*$, where $\mathbf{x}^*$ represents the perturbation or *state deviation* from the equilibrium point $\mathbf{x}_0$. Each of the elements of $\mathbf{f}(\mathbf{x})$ can be expanded in a Taylor series about one of the equilibrium points $\mathbf{x}_0$. Assuming that $\mathbf{x}^*$ is restricted to a small neighborhood of the equilibrium point, the higher-order terms in the Taylor series may be neglected. Thus, the resulting linear variational state equation is

$$\dot{\mathbf{x}}^* = \begin{bmatrix} \frac{\partial f_1}{\partial x_1} & \frac{\partial f_1}{\partial x_2} & \cdots & \frac{\partial f_1}{\partial x_n} \\ \frac{\partial f_2}{\partial x_1} & \frac{\partial f_2}{\partial x_2} & \cdots & \frac{\partial f_2}{\partial x_n} \\ \cdots & \cdots & \cdots & \cdots \\ \frac{\partial f_n}{\partial x_1} & \frac{\partial f_n}{\partial x_2} & \cdots & \frac{\partial f_n}{\partial x_n} \end{bmatrix}_{\mathbf{x}=\mathbf{x}_0} \mathbf{x}^* = \mathbf{J}_x \mathbf{x}^* \tag{15.37}$$

where f_i is the ith row of $\mathbf{f}(\mathbf{x})$ and $\mathbf{J}_x = \partial\mathbf{f}/\partial\mathbf{x}^T$ is called the *Jacobian matrix and is evaluated at* $\mathbf{x}_0$.

For the system of Eq. (15.34) the motion about the equilibrium point $x_1 = x_2 = 0$ is represented by

$$\dot{\mathbf{x}}^* = \begin{bmatrix} \dot{x}_1^* \\ \dot{x}_2^* \end{bmatrix} = \begin{bmatrix} 0 & 1 \\ -K & -a \end{bmatrix} \begin{bmatrix} x_1^* \\ x_2^* \end{bmatrix} = \mathbf{J}_x \mathbf{x}^* \tag{15.38}$$

For these linearized equations the eigenvalues are $\lambda_{1,2} = -a/2 \pm \sqrt{(a/2)^2 - K}$. This equilibrium point is stable and is either a node or a focus, depending upon the magnitudes of a and K.

For motion about the equilibrium point $x_1 = \pi$, $x_2 = 0$ the state equations are

$$\dot{\mathbf{x}}^* = \begin{bmatrix} 0 & 1 \\ K & -a \end{bmatrix} \mathbf{x}^* \tag{15.39}$$

The eigenvalues of this $\mathbf{J}_x$ are $\lambda_{1,2} = -a/2 \pm \sqrt{(a/2)^2 + K}$. Thus, one eigenvalue is positive and the other is negative, and the equilibrium point represents an unstable saddle point. The motion around the saddle point is considered unstable since every point on all trajectories, except on the two separatrices, moves away from this equilibrium point. A phase-plane portrait for this system can be obtained by the method of isoclines (see Fig. 15.2 and Sec. 15.2). From Eq. (15.35) the isocline equation is

$$x_2 = \frac{-K \sin x_1}{N + a} \tag{15.40}$$

The phase portrait is shown in Fig. 15.10. Note that the linearized equations are applicable only in the neighborhood of the singular points. Thus, they describe stability *in the small*.

When an input $\mathbf{u}$ is present, the nonlinear state equation is

$$\dot{\mathbf{x}} = \mathbf{f}(\mathbf{x}, \mathbf{u}) \tag{15.41}$$

Its equilibrium point $\mathbf{x}_0$, when $\mathbf{u} = \mathbf{u}_0$ is a constant, is determined by letting

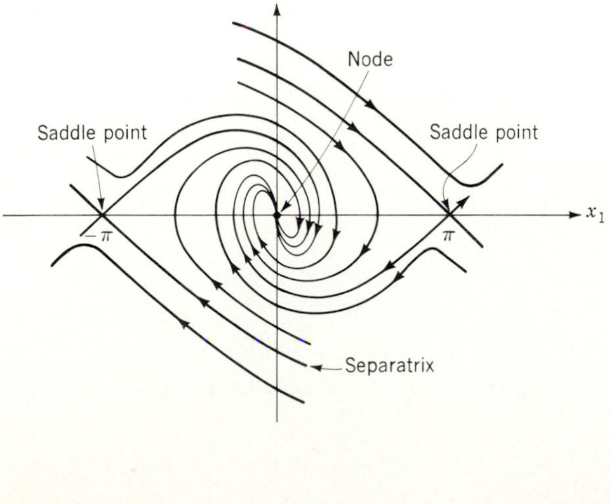

FIGURE 15.10
Phase portrait for Eq. (15.33).

$\dot{\mathbf{x}} = \mathbf{f}(\mathbf{x}_0, \mathbf{u}_0) = \mathbf{0}$. Then the linearized variational state equation describing the variation from the equilibrium point is obtained by using only the linear terms from the Taylor series expansion. For variations in $\mathbf{u}$, appearing as $\mathbf{u}^*$ in the input $\mathbf{u} = \mathbf{u}_0 + \mathbf{u}^*$, the linearized state equation is

$$\dot{\mathbf{x}}^* = \begin{bmatrix} \dfrac{\partial f_1(\mathbf{x},\mathbf{u})}{\partial x_1} & \dfrac{\partial f_1(\mathbf{x},\mathbf{u})}{\partial x_2} & \cdots & \dfrac{\partial f_1(\mathbf{x},\mathbf{u})}{\partial x_n} \\ \dfrac{\partial f_2(\mathbf{x},\mathbf{u})}{\partial x_1} & \dfrac{\partial f_2(\mathbf{x},\mathbf{u})}{\partial x_2} & \cdots & \dfrac{\partial f_2(\mathbf{x},\mathbf{u})}{\partial x_n} \\ \cdots & \cdots & \cdots & \cdots \\ \dfrac{\partial f_n(\mathbf{x},\mathbf{u})}{\partial x_1} & \dfrac{\partial f_n(\mathbf{x},\mathbf{u})}{\partial x_2} & \cdots & \dfrac{\partial f_n(\mathbf{x},\mathbf{u})}{\partial x_n} \end{bmatrix}_{\substack{\mathbf{x}=\mathbf{x}_0 \\ \mathbf{u}=\mathbf{u}_0}} \mathbf{x}^*$$

$$+ \begin{bmatrix} \dfrac{\partial f_1(\mathbf{x},\mathbf{u})}{\partial u_1} & \cdots & \dfrac{\partial f_1(\mathbf{x},\mathbf{u})}{\partial u_r} \\ \cdots & \cdots & \cdots \\ \dfrac{\partial f_n(\mathbf{x},\mathbf{u})}{\partial u_1} & \cdots & \dfrac{\partial f_n(\mathbf{x},\mathbf{u})}{\partial u_r} \end{bmatrix}_{\substack{\mathbf{x}=\mathbf{x}_0 \\ \mathbf{u}=\mathbf{u}_0}} \mathbf{u}^* = \mathbf{J}_x \mathbf{x}^* + \mathbf{J}_u \mathbf{u}^* \qquad (15.42)$$

In analyzing the system performance in the vicinity of equilibrium, it is usually convenient to translate the origin of the state space to that point. This is done by inserting $\mathbf{x} = \mathbf{x}_0 + \mathbf{x}^*$ into the original equations. With the origin at $\mathbf{x}_0$, the variations from this point are described by $\mathbf{x}^*$.

The characteristic equation describing the motion about the equilibrium point can be obtained from the linearized state equations. For a second-order system it has the form

$$s^2 + ps + q = (s - \lambda_1)(s - \lambda_2) = 0 \qquad (15.43)$$

The eigenvalues λ_1 and λ_2 determine whether the singular point is a node, a focus, or a saddle. For a node or focus the eigenvalues also determine whether the equilibrium point is stable or unstable. Figure 15.11 shows the type of equilibrium point, based on the coefficients p and q in Eq. (15.43).

Phase-plane portraits have been applied to provide a visual picture of the responses of many linear and nonlinear systems. They are described extensively in the literature. The purpose in presenting them here is to provide some insight

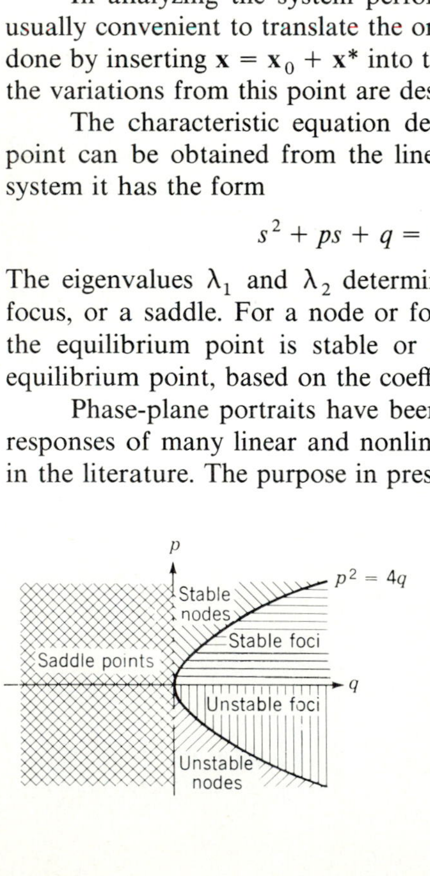

FIGURE 15.11
Type of equilibrium point for the characteristic equation $s^2 + ps + q = 0$.

into trajectories and their relationships to the singular points. This understanding of phase-plane trajectories should provide the basis for a clearer understanding of the stability analysis for higher-order systems. The definitions of system stability and the techniques for evaluating stability are covered in later sections of this chapter.

15.4 QUADRATIC FORMS[4,5]

Some of the techniques used in determining stability of control systems and for optimizing their response utilize scalar functions expressed in *quadratic form*. The necessary background for expressing functions in quadratic form is developed in this section. Then some important properties of quadratic forms are presented.

CONJUGATE MATRIX. The elements of a matrix may be complex quantities. For example, a matrix **A** may have the elements $a_{ij} = \alpha_{ij} + j\beta_{ij}$. A *conjugate matrix* **B** has elements with the same real component and with imaginary components of the opposite sign, that is, $b_{ij} = \alpha_{ij} - j\beta_{ij}$. This conjugate property is expressed by

$$\mathbf{B} = \mathbf{A}^* \tag{15.44}$$

INNER PRODUCT. The inner product is also called a *scalar* (or *dot*) *product* since it yields a scalar function. The scalar product of vectors **x** and **y** is defined by

$$\langle \mathbf{x}, \mathbf{y} \rangle = (\mathbf{x}^*)^T \mathbf{y} = \mathbf{y}^T \mathbf{x}^* = x_1^* y_1 + x_2^* y_2 + \cdots + x_n^* y_n = \sum_{i=1}^{n} x_i^* y_i \tag{15.45}$$

When **x** and **y** are real vectors, the inner product becomes

$$\langle \mathbf{x}, \mathbf{y} \rangle = \sum_{i=1}^{n} x_i y_i = x_1 y_1 + x_2 y_2 + \cdots + x_n y_n \tag{15.46}$$

It should be noted that when **x** and **y** are complex, $\langle \mathbf{x}, \mathbf{y} \rangle \neq \mathbf{x}^T \mathbf{y}^*$. However, when **x** and **y** are real,

$$\langle \mathbf{x}, \mathbf{y} \rangle = \mathbf{x}^T \mathbf{y} = \mathbf{y}^T \mathbf{x} = \langle \mathbf{y}, \mathbf{x} \rangle \tag{15.47}$$

BILINEAR FORM. A scalar homogeneous expression containing the product of the elements of vectors **x** and **y** is called a *bilinear form* in the variables x_i and y_i. When they are both of order n, the most general bilinear form in **x** and **y** is

$$\begin{aligned} f(\mathbf{x}, \mathbf{y}) = & a_{11} x_1 y_1 + a_{12} x_1 y_2 + \cdots + a_{1n} x_1 y_n \\ & + a_{21} x_2 y_1 + a_{22} x_2 y_2 + \cdots + a_{2n} x_2 y_n \\ & + \cdots\cdots\cdots\cdots\cdots\cdots\cdots\cdots \\ & + a_{n1} x_n y_1 + a_{n2} x_n y_2 + \cdots + a_{nn} x_n y_n \end{aligned} \tag{15.48}$$

This can be written more compactly as

$$f(\mathbf{x},\mathbf{y}) = \sum_{i=1}^{n} \sum_{j=1}^{n} a_{ij}x_i y_j$$

$$= [x_1 \quad x_2 \quad \cdots \quad x_n] \begin{bmatrix} a_{11} & a_{12} & \cdots & a_{1n} \\ a_{21} & a_{22} & \cdots & a_{2n} \\ \cdots & \cdots & \cdots & \cdots \\ a_{n1} & a_{n2} & \cdots & a_{nn} \end{bmatrix} \begin{bmatrix} y_1 \\ y_2 \\ \cdot \\ y_n \end{bmatrix}$$

$$= \mathbf{x}^T\mathbf{A}\mathbf{y} = \langle \mathbf{x}, \mathbf{A}\mathbf{y} \rangle \tag{15.49}$$

The matrix **A** is called the coefficient matrix of the bilinear form, and the rank of **A** is called the rank of the bilinear form. A bilinear form is called *symmetric* if the matrix **A** is symmetric.

QUADRATIC FORM. A quadratic form V is a real homogeneous polynomial in the real variables $x_1, x_2, \ldots, x_n$ of the form

$$V = \sum_{i=1}^{n} \sum_{j=1}^{n} a_{ij}x_i x_j \tag{15.50}$$

where all a_{ij} are real. This is the special case of Eq. (15.49) where $\mathbf{x} = \mathbf{y}$. The quadratic form V can be expressed as the inner product

$$V(\mathbf{x}) = \mathbf{x}^T\mathbf{A}\mathbf{x} = \langle \mathbf{x}, \mathbf{A}\mathbf{x} \rangle \tag{15.51}$$

A homogeneous polynomial can always be expressed in terms of a symmetric matrix **A**. In the expansion of Eq. (15.50) the cross-product terms $(i \neq j)$ all have the form $(a_{ij} + a_{ji})x_i x_j$. Choosing $a_{ij} = a_{ji}$ makes the matrix **A** for the quadratic form symmetric. This is illustrated in the following example.

Example 1

$$V(\mathbf{x}) = x_1^2 - 4x_2^2 + 5x_3^2 + 6x_1x_2 - 20x_2x_3$$

$$= \mathbf{x}^T \begin{bmatrix} 1 & 3 & 0 \\ 3 & -4 & -10 \\ 0 & -10 & 5 \end{bmatrix} \mathbf{x} = \mathbf{x}^T\mathbf{A}\mathbf{x} \tag{15.52}$$

The rank of the matrix **A** is called the rank of the quadratic form. If the rank of **A** is $r < n$, the quadratic form is singular. If the rank of **A** is n, the quadratic form is nonsingular.

A nonsymmetric matrix can be converted into an equivalent symmetric matrix by replacing all sets of elements a_{ij} and a_{ji} by the average value $(a_{ij} + a_{ji})/2$. As illustrated in the following example, it is obvious that the principal diagonal is preserved.

Example 2

$$\mathbf{A} = \begin{bmatrix} 1 & -1 & 2 \\ 5 & 3 & 7 \\ 0 & 1 & 2 \end{bmatrix} \qquad \mathbf{A}_{\text{sym}} = \begin{bmatrix} 1 & 2 & 1 \\ 2 & 3 & 4 \\ 1 & 4 & 2 \end{bmatrix}$$

CONGRUENT TRANSFORMATION. It is often desirable to introduce a transformation so that Eq. (15.51) can be transformed to the quadratic form $V = \mathbf{y}^T\mathbf{B}\mathbf{y}$, in which the matrix **B** contains only diagonal elements. This can be accomplished by means of a congruent transformation. Two square matrices **A** and **B** of order n are called congruent if there exists a nonsingular transformation matrix **P** such that

$$\mathbf{B} = \mathbf{P}^T\mathbf{A}\mathbf{P} \tag{15.53}$$

Then, with the linear transformation $\mathbf{x} = \mathbf{P}\mathbf{y}$, the quadratic form of Eq. (15.51) is transformed to

$$V = \mathbf{x}^T\mathbf{A}\mathbf{x} = \mathbf{y}^T(\mathbf{P}^T\mathbf{A}\mathbf{P})\mathbf{y} = \mathbf{y}^T\mathbf{B}\mathbf{y} \tag{15.54}$$

CANONICAL FORM. By means of a congruent transformation the canonical matrix **C**, Eq. (15.55), can be obtained from a matrix **A** of order n and rank r, so that only $+1$, -1, and 0 are present along the principal diagonal:

$$\mathbf{C} = \mathbf{P}^T\mathbf{A}\mathbf{P} = \left[\begin{array}{c|c|c} \mathbf{I}_p & 0 & 0 \\ \hline 0 & -\mathbf{I}_{r-p} & 0 \\ \hline 0 & 0 & 0 \end{array}\right] \tag{15.55}$$

The integer p is called the *index* of the matrix, and the integer $s = p - (r - p)$ is called the *signature* of the matrix. Congruent matrices are of the same order and have the same rank and index or the same rank and signature. The matrix **A** is transformed by the matrix **P** to obtain the matrix **C** from which p and s are determined.

The pairs of elementary transformations that are permitted by the transforming matrix **P** are

1. Interchanging the ith and jth rows and interchanging the ith and jth columns.
2. Multiplying the ith row and column by a nonzero scalar k.
3. Addition to the elements of the ith row of the corresponding elements of the jth row multiplied by a scalar k and addition to the ith column of the jth column multiplied by k.

Premultiplying and postmultiplying the matrix **A** by elementary matrices (see Sec. 4.14) performs these three operations. For example, for a third-order matrix **A**, premultiplying by the elementary matrices shown in Eq. (4.100) results in operations on the rows. Postmultiplying by similar elementary matrices performs the desired column operations.

Example 3. The matrix **A** of Eq. (15.52) is transformed to diagonal form using the following procedure.

Step 1. Form the augmented matrix [**A**¦**I**], where the unit matrix **I** has the same order as **A**:

$$[\mathbf{AI}] = \left[\begin{array}{ccc|ccc} 1 & 3 & 0 & 1 & 0 & 0 \\ 3 & -4 & -10 & 0 & 1 & 0 \\ 0 & -10 & 5 & 0 & 0 & 1 \end{array}\right]$$

Step 2. Subtract 3 times the first row from the second row; then subtract 3 times the first column from the second column. The resulting matrix is

$$\left[\begin{array}{ccc|ccc} 1 & 0 & 0 & 1 & 0 & 0 \\ 0 & -13 & -10 & -3 & 1 & 0 \\ 0 & -10 & 5 & 0 & 0 & 1 \end{array}\right]$$

Step 3. Add 2 times the third row to the second row; then add 2 times column 3 to column 2. This gives

$$\left[\begin{array}{ccc|ccc} 1 & 0 & 0 & 1 & 0 & 0 \\ 0 & -33 & 0 & -3 & 1 & 2 \\ 0 & 0 & 5 & 0 & 0 & 1 \end{array}\right]$$

Step 4. Interchange rows 2 and 3; then interchange columns 2 and 3. Multiply row 2 and column 2 by $1/\sqrt{5}$. Multiply row 3 and column 3 by $1/\sqrt{33}$. The resulting matrix is

$$\left[\begin{array}{ccc|ccc} 1 & 0 & 0 & 1 & 0 & 0 \\ 0 & 1 & 0 & 0 & 0 & \dfrac{1}{\sqrt{5}} \\ 0 & 0 & -1 & -\dfrac{3}{\sqrt{33}} & \dfrac{1}{\sqrt{33}} & \dfrac{2}{\sqrt{33}} \end{array}\right] = [\mathbf{CP}^T] \qquad (15.56)$$

The resulting canonical matrix **C** has rank $r = 3$, index $p = 2$, and signature $s = 1$. The required transformation $\mathbf{P}^T$ is identified directly in the final matrix. The original matrix **A** and the resulting canonical matrix **C** are symmetric. However, the matrix **P** is not symmetric.

When the matrix **A** is transformed to the canonical form, the quadratic form of Eq. (15.52) becomes

$$V = \mathbf{y}^T\mathbf{C}\mathbf{y} = y_1^2 + y_2^2 + \cdots + y_p^2 - y_{p+1}^2 - \cdots - y_r^2 \qquad (15.57)$$

where p is the index and r is the rank of the quadratic form. The rank and index remain the same in a transformation to the canonical form.

LENGTH OF A VECTOR. The length of a vector **x** is called the euclidean *norm* and is denoted by $\|\mathbf{x}\|$. It is defined as the square root of the inner product $\langle \mathbf{x}, \mathbf{x} \rangle$. For real vectors it is given by

$$\|\mathbf{x}\| = \sqrt{\langle \mathbf{x}, \mathbf{x} \rangle} = \sqrt{x_1^2 + x_2^2 + \cdots + x_n^2} \qquad (15.58)$$

A vector can be normalized so that its length is unity. In that case it is called a *unit vector*. The unit vector may be denoted by $\hat{\mathbf{x}}$ and is obtained by dividing each element of $\mathbf{x}$ by $\|\mathbf{x}\|$:

$$\hat{\mathbf{x}} = \frac{\mathbf{x}}{\|\mathbf{x}\|} \tag{15.59}$$

PRINCIPAL MINORS. A *principal minor* of a matrix $\mathbf{A}$ is obtained by deleting any row(s) and the same numbered column(s). The diagonal elements of a principal minor of $\mathbf{A}$ are therefore also diagonal elements of $\mathbf{A}$. The $|\mathbf{A}|$ is classified as a principal minor, with no rows and columns deleted. The number of principal minors can be determined from a Pascal triangle.[10] The number of principal minors for a matrix $\mathbf{A}$ of order n is: 1 for $n = 1$, 3 for $n = 2$, 7 for $n = 3$, 15 for $n = 4, \ldots$.

LEADING PRINCIPAL MINOR. There are n *leading principal minors*, formed by all the square arrays within $\mathbf{A}$ that contain a_{11}. Starting with $[a_{11}]$, the next square array is formed by including the *next* row and column. This process is continued until the matrix $\mathbf{A}$ is obtained. The leading principal minors are given by

$$\Delta_1 = |a_{11}| \quad \Delta_{12} = \begin{vmatrix} a_{11} & a_{12} \\ a_{21} & a_{22} \end{vmatrix} \quad \Delta_{123} = \begin{vmatrix} a_{11} & a_{12} & a_{13} \\ a_{21} & a_{22} & a_{23} \\ a_{31} & a_{32} & a_{33} \end{vmatrix} \quad \cdots \quad \Delta_{12\cdots n} = |\mathbf{A}| \tag{15.60}$$

The subscripts of Δ are the rows and columns of $\mathbf{A}$ used to form the minor. The minor contains the common elements of these rows and columns.

DEFINITENESS AND SEMIDEFINITENESS. The sign definiteness for a scalar function of a vector, such as $V(\mathbf{x})$, is defined for a spherical region S about the origin described by $\|\mathbf{x}\| \le K$ (a constant equal to the radius of S). The function of $V(\mathbf{x})$ and all $\partial V(\mathbf{x})/\partial x_i$, for $i = 1, \ldots, n$, must be continuous within S. *The definiteness of a quadratic form is determined by analyzing only the symmetric* $\mathbf{A}$ *matrix.* If $\mathbf{A}$ is given as a nonsymmetric matrix, it must first be converted to a symmetric matrix (see Example 2 of this section).

POSITIVE DEFINITE (PD). A scalar function, such as the quadratic form $V(\mathbf{x}) = \langle \mathbf{x}, \mathbf{A}\mathbf{x} \rangle$, is called positive definite when $V(\mathbf{x}) = 0$ for $\mathbf{x} = \mathbf{0}$ and $V(\mathbf{x}) > 0$ for all other $\|\mathbf{x}\| \le K$. The positive definite condition requires that $|\mathbf{A}| \ne 0$; thus, the rank of $\mathbf{A}$ is equal to $r = n$. By reference to Eq. (15.55) it is also seen that the index p must be equal to the rank.

When a real quadratic form $V(\mathbf{x}) = \mathbf{x}^T\mathbf{A}\mathbf{x}$ is positive definite, then the matrix $\mathbf{A}$ is also called positive definite. The matrix $\mathbf{A}$ has the property that it is positive definite iff there exists a nonsingular matrix $\mathbf{H}$ such that $\mathbf{A} = \mathbf{H}^T\mathbf{H}$.

The definiteness of the matrix $\mathbf{A}$ can be determined by reducing this matrix to diagonal or canonical form by means of a congruent transformation, as described earlier in this section. Then, if the index $p = n$, $\mathbf{A}$ is positive definite.

Another method of diagonalizing the **A** matrix is to determine a modal matrix **T** using the methods of Sec. 5.10. In that case the diagonalized matrix contains the eigenvalues obtained from $|\lambda\mathbf{I} - \mathbf{A}| = 0$. In order for the matrix **A** to be positive definite, all the eigenvalues must be positive.

An alternate method of determining positive definiteness is to calculate *all* the leading principal minors of the matrix **A**. If all the leading principal minors of **A** are positive, the real quadratic form is positive definite. This is sometimes called the Sylvester theorem. It should not be confused with the Sylvester expansion presented in Sec. 3.13. Note that when the leading principal minors are all positive, then all the other principal minors are also positive.

POSITIVE SEMIDEFINITE (PSD). The quadratic form $V(\mathbf{x})$ is called positive semidefinite when $V(\mathbf{x}) = 0$ for $\mathbf{x} = 0$ and $V(\mathbf{x}) \geq 0$ for all $\|\mathbf{x}\| \leq K$. The function $V(\mathbf{x})$ is permitted to equal zero at points in S other than the origin, i.e., for some $\mathbf{x} \neq \mathbf{0}$, but it may not be negative. In this case $|\mathbf{A}| = 0$, the rank $r < n$, and the rank is equal to the index, that is, $r = p < n$. The positive semidefinite quadratic form can therefore be reduced to the form $y_1^2 + y_2^2 + \cdots + y_r^2$, where $r < n$. When $V(\mathbf{x})$ is positive semidefinite, then the matrix **A** is also called positive semidefinite. The matrix **A** is positive semidefinite *iff* all its characteristic values are ≥ 0.

In order to be positive semidefinite, *all* the principal minors[6] must be nonnegative.

NEGATIVE DEFINITE (ND) AND NEGATIVE SEMIDEFINITE (NSD). The definitions of negative definite and negative semidefinite follow directly from the definitions above when the inequalities are applied to $-\mathbf{A}$. If a matrix **A** does not satisfy the conditions for positive definiteness or positive semidefiniteness, then $-\mathbf{A}$ is checked for these conditions. If $-\mathbf{A}$ satisfies the positive definite or positive semidefinite conditions, then **A** is said to be negative definite or negative semidefinite, respectively.

INDEFINITE. A scalar function $V(\mathbf{x})$ is *indefinite* if it assumes both positive and negative values within the region S described by $\|\mathbf{x}\| < K$.

If **A** is not positive definite, positive semidefinite, negative definite, or negative semidefinite, then it is said to be indefinite. Note that when **A**, in general form, has both positive and negative elements along the principal diagonal, it is indefinite. When **A** is transformed to a diagonal form, it is indefinite if some of the diagonal elements are positive and some are negative. This means that some of the eigenvalues are positive and some are negative.

Example 4. A quadratic form is given in Example 1. The matrix **A** is transformed in Example 3 to the canonical form **C** represented in Eq. (15.56). Since some of the diagonal elements of both **A** and **C** are positive and some are negative, the matrix is indefinite.

Example 5. Use the principal minors to determine the definiteness of

$$\mathbf{A} = \begin{bmatrix} 1 & 1 & 1 \\ 1 & 1 & 1 \\ 1 & 1 & 0 \end{bmatrix}$$

The leading principal minors are evaluated to check for positive definiteness:

$$\Delta_1 = 1 \qquad \Delta_{12} = \begin{vmatrix} 1 & 1 \\ 1 & 1 \end{vmatrix} = 0 \qquad \Delta_{123} = \Delta = 0$$

The conditions for positive definiteness are not satisfied. Next the remaining principal minors are evaluated to check for positive semidefiniteness

$$\Delta_{13} = \begin{vmatrix} 1 & 1 \\ 1 & 0 \end{vmatrix} = -1 \qquad \Delta_{23} = \begin{vmatrix} 1 & 1 \\ 1 & 0 \end{vmatrix} = -1 \qquad \Delta_2 = 1 \qquad \Delta_3 = 0$$

Since the principal minors are not all nonnegative, the matrix **A** is not positive semidefinite. Similarly, **A** is not negative definite or negative semidefinite; therefore **A** is indefinite.

Example 6. Transform **A** in Example 5 to the diagonal Jordan form Λ and check its definiteness. The characteristic equation is

$$|\lambda\mathbf{I} - \mathbf{A}| = \begin{vmatrix} \lambda - 1 & -1 & -1 \\ -1 & \lambda - 1 & -1 \\ -1 & -1 & \lambda \end{vmatrix} = \lambda^3 - 2\lambda^2 - 2\lambda = 0$$

The eigenvalues are

$$\lambda_1 = 1 + \sqrt{3} = 2.732 \qquad \lambda_2 = 1 - \sqrt{3} = -0.732 \qquad \lambda_3 = 0$$

Thus

$$\Lambda = \begin{bmatrix} 2.732 & 0 & 0 \\ 0 & -0.732 & 0 \\ 0 & 0 & 0 \end{bmatrix}$$

This confirms that the matrix is indefinite.

Example 7. Determine the definiteness of

$$\mathbf{A} = \begin{bmatrix} 2 & 1 & -1 \\ 1 & 2 & 0 \\ -1 & 0 & 2 \end{bmatrix}$$

The leading principal minors are

$$\Delta_1 = 2 \qquad \Delta_{12} = 3 \qquad \Delta = |\mathbf{A}| = 4$$

Therefore **A** is positive definite. Note that the other principal minors are also positive: $\Delta_2 = 2$, $\Delta_3 = 2$, $\Delta_{13} = 3$, $\Delta_{23} = 4$.

15.5 STABILITY[7,8]

The output stability of a linear time-invariant system is expressed simply by the condition that the output response must be bounded when the input is bounded. It follows directly that this requires all eigenvalues to have negative real parts.

This is a necessary and sufficient condition for stability. It tells one that the transient terms associated with those poles decrease with time and that the output response therefore approaches the particular solution which is determined by the input. An alternate way of expressing the necessary and sufficient condition for stability is that the integral of the system weighting function, or impulse response, $w(t)$ must be finite. This is expressed by

$$\int_0^\infty |w(t)|\,dt < \infty \tag{15.61}$$

A direct method of determining stability, when the characteristic equation is available, is to evaluate its roots. Forming the Routhian array or the Hurwitz determinants also permits the determination of stability without evaluating the roots.

The extension of stability evaluation to nonlinear systems is not so straightforward, since the concept of root location is no longer applicable. As a result, many different classes of stability have been described in the literature for such systems. In this chapter the intention is to present some new concepts for viewing system stability. Although the text is restricted primarily to linear time-invariant systems, these concepts can be extended to nonlinear and time-varying systems.

STABILITY IN THE SENSE OF LIAPUNOV. Consider a region ε in the state space enclosing an equilibrium point $\mathbf{x}_0$. This equilibrium point is stable provided that there is a region $\delta(\varepsilon)$, which is contained within ε, such that any trajectory starting in the region δ does not leave the region ε. Note that with this definition it is not necessary for the trajectory to approach the equilibrium point. It is necessary only for the trajectory to stay within the region ε. This permits the existence of a continuous oscillation about the equilibrium point. The state-space trajectory for such an oscillation is a closed path called a *limit cycle*. The performance specifications for a control system must be used to determine whether or not a limit cycle can be permitted. The amplitude and frequency of the oscillation may influence whether it represents acceptable performance. When limit cycles are not acceptable, more stringent restraints must be chosen to exclude their possible existence.

Limit cycles are usually characterized as stable or unstable. If a limit cycle is stable, it means that trajectories in the state space on either side will approach the limit cycle. An unstable limit cycle is one in which the trajectories on either side diverge from the limit cycle. These trajectories may approach other limit cycles or equilibrium points.

ASYMPTOTIC STABILITY. An equilibrium point is *asymptotically stable* if, in addition to being stable in the sense of Liapunov, all trajectories approach the equilibrium point. This means that the variational solution $\mathbf{x}^*(t)$ approaches $\mathbf{0}$ as time t approaches infinity. This is the stability definition usually used in control-system design.

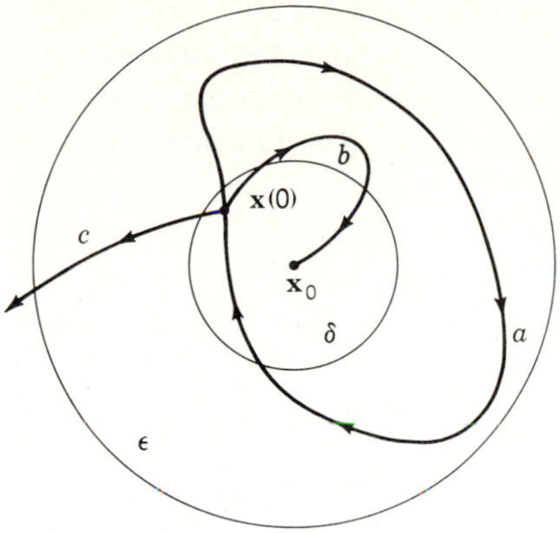

FIGURE 15.12
State plane trajectories indicating (*a*) Liapunov stability; (*b*) asymptotic stability; and (*c*) instability.

Figure 15.12 shows trajectories in the state plane illustrating the general principle of Liapunov stability, asymptotic stability, and instability. The trajectory *a* starting at the initial state point $\mathbf{x}(0)$ and remaining within the region ε meets the conditions for Liapunov stability. This trajectory is closed, indicating a continuous oscillation or limit cycle. Trajectory *b* terminates at the equilibrium point $\mathbf{x}_0$; thus it represents asymptotic stability. Trajectory *c* leaves the region ε and therefore indicates instability.

When the region δ includes the entire state space, the definitions of Liapunov and asymptotic stability are said to apply in a *global* sense. The stability or instability of a linear system is global because any initial state yields the same stability determination. A stable linear system is globally asymptotically stable.

In Sec. 15.4 the technique for linearizing nonlinear differential equations in the neighborhood of their singularities is presented. The validity of determining stability of the unperturbed solution near the singular points from the linearized equations was developed independently by Poincaré and Liapunov in 1892. Liapunov designated this as the *first method*. This stability determination is applicable only in a small region near the singularity and results in stability *in the small*. The next section considers Liapunov's *second method*, which is used to determine stability *in the large*. This larger region may include a finite portion or sometimes the whole region of the state space.

15.6 SECOND METHOD OF LIAPUNOV[7,8]

The second, or direct, method of Liapunov provides a means for determining the stability of a system without explicitly solving for the trajectories in the state space. This is in contrast to the first method of Liapunov, which requires the determination of the eigenvalues from the linearized equations about an equilibrium point. The second method is applicable for determining the behavior of higher-order systems which may be forced or unforced, linear or nonlinear, time-invariant or time-varying, and deterministic or stochastic. Solution of the differential equation is not required. The procedure requires the selection of a scalar function $V(\mathbf{x})$, which is tested for the conditions that indicate stability. When $V(\mathbf{x})$ successfully meets these conditions, it is called a *Liapunov function*. The principal difficulty in applying the method is in formulating a correct Liapunov function because the failure of one function to meet the stability conditions does not mean that a true Liapunov function does not exist. This difficulty is compounded by the fact that the Liapunov function is not unique. Nevertheless there is much interest in the second method. The further use of the Liapunov functions for evaluating quadratic performance criteria is covered in Chap. 16.

In order to show a simple example of a Liapunov function, consider the system of Sec. 15.3 represented by Eq. (15.33) which has multiple equilibrium points at $\dot{\theta}_o = 0$, $\theta_o = n\pi$, where n is an integer. The proposed function is the sum of the kinetic and potential stored energies, given by

$$V(\theta_o, \dot{\theta}_o) = \tfrac{1}{2}\dot{\theta}_o^2 + K(1 - \cos\theta_o) \qquad (15.62)$$

This function is positive for all values of θ_o and $\dot{\theta}_o$, except at the equilibrium points, where it is equal to zero. The rate of change of this energy function along any phase-plane trajectory is obtained by differentiating Eq. (15.62) and using Eq. (15.33):

$$\dot{V}(\theta_o, \dot{\theta}) = \dot{\theta}_o\ddot{\theta}_o + K(\sin\theta_o)\dot{\theta}_o = -a\dot{\theta}_o^2 \qquad (15.63)$$

The value of $\dot{V}$ is negative along any trajectory for all values of $\dot{\theta}_o$ except $\dot{\theta}_o = 0$. Note that the slope of $d\dot{\theta}_o/d\theta_o$ of the phase-plane trajectories [see Eq. (15.35)] is infinite along the line represented by $\dot{\theta}_o = 0$ except at the equilibrium points $\theta_o = n\pi$. Thus, the line $\dot{\theta}_o = 0$ does not represent an equilibrium, except at $\theta_o = n\pi$, and it is not a trajectory in the state plane. Since the energy stored in the system is continuously decreasing at all points except the equilibrium points, the equilibrium at the origin and at even multiples of $\theta_o = n\pi$ is asymptotically stable. This $\dot{V}$ is NSD.

This example demonstrates that the total system energy may be used as the Liapunov function. When the equations of a large system are given in mathematical form, it is usually difficult to define the energy of the system. Thus, alternate Liapunov functions must be obtained. For any system of order n a positive, constant value of the proper positive definite Liapunov function $V(\mathbf{x})$ represents a closed surface, a hyperellipsoid, in the state space with its center at the origin. The entire state space is filled with such nonintersecting closed surfaces, each representing a different positive value of $V(\mathbf{x})$. When $\dot{V}(\mathbf{x})$ is negative for all points in the state space, it means that all trajectories cross the closed surfaces from the outside to the inside and eventually converge at the equilibrium point at the origin.

A qualitatively correct Liapunov function is shown in Fig. 15.13 for a second-order system. The function $V(x_1, x_2) = x_1^2 + x_2^2$ is positive definite and is represented by the paraboloid surface shown. The value $V(x_1, x_2) = k_i$ (a constant) is represented by the intersection of the surface $V(x_1, x_2)$ and the plane $z = k_i$. The projection of this intersection on the x_1, x_2 plane is a closed curve,

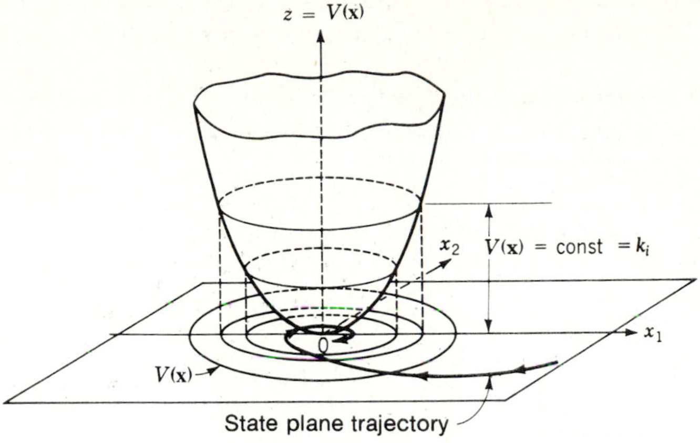

FIGURE 15.13
A positive definite function $V(x_1, x_2)$ and projections on the x_1, x_2 plane.

an oval, around the origin. There is a family of such closed curves in the x_1, x_2 plane for different values of k_i. The value $V(x_1, x_2) = 0$ is the point at the origin; it is the innermost curve of the family of curves representing different levels on the paraboloid for $V(x_1, x_2) = k_i$.

The gradient vector of $V(\mathbf{x})$ is defined by

$$\operatorname{grad} V(\mathbf{x}) = \nabla V(\mathbf{x}) = \begin{bmatrix} \dfrac{\partial V(\mathbf{x})}{\partial x_1} \\ \dfrac{\partial V(\mathbf{x})}{\partial x_2} \\ \vdots \\ \dfrac{\partial V(\mathbf{x})}{\partial x_n} \end{bmatrix} \tag{15.64}$$

The time derivative $\dot{V}(x)$ along any trajectory is

$$\begin{aligned} \frac{dV(\mathbf{x})}{dt} &= \dot{V}(\mathbf{x}) \\ &= \sum_{i=1}^{n} \frac{\partial V}{\partial x_i}\frac{dx_i}{dt} = \frac{\partial V(\mathbf{x})}{\partial x_1}\frac{dx_1}{dt} + \frac{\partial V(\mathbf{x})}{\partial x_2}\frac{dx_2}{dt} + \cdots + \frac{\partial V(\mathbf{x})}{\partial x_n}\frac{dx_n}{dt} \\ &= [\operatorname{grad} V(\mathbf{x})]^T \dot{\mathbf{x}} = [\nabla V(\mathbf{x})]^T \dot{\mathbf{x}} = \langle \nabla V, \dot{\mathbf{x}} \rangle \end{aligned} \tag{15.65}$$

It is important to note that $\dot{V}(\mathbf{x})$ can be evaluated without knowing the solution to the system state equation $\dot{\mathbf{x}} = f(\mathbf{x})$.

The gradient $\nabla V(x_1, x_2)$ describes the steepness between adjacent levels of $V(x_1, x_2)$. The function $V(x_1, x_2)$ is negative definite in Fig. 15.13, except at the origin where it is equal to zero. The state-plane trajectory shown crosses the ovals for successively smaller values of $V(\mathbf{x})$. Therefore, the system is asymptotically stable, and $V(\mathbf{x})$ is a proper Liapunov function.

The concepts described above may be summarized in the following theorem, which provides sufficient, but not necessary, conditions for stability.

Theorem 1 Liapunov asymptotic stability. A system is asymptotically stable in the vicinity of the equilibrium point at the origin if there exists a scalar function $V(\mathbf{x})$ such that:

1. $V(\mathbf{x})$ is continuous and has continuous first partial derivatives in a region S around the origin.
2. $V(\mathbf{x}) > 0$ for $\mathbf{x} \neq 0$.
3. $V(\mathbf{0}) = 0$.
4. $\dot{V}(\mathbf{x}) < 0$ for $\mathbf{x} \neq 0$.

Conditions 1 to 3 ensure that $V(\mathbf{x})$ is positive definite. Therefore, $V(\mathbf{x}) = k$ is a closed surface within the region S. Condition 4 means that $\dot{V}(\mathbf{x})$ is negative definite, and thus any trajectory in S crosses through the surface $V(\mathbf{x}) = k$ from the outside to the inside for all values of k. Therefore, the trajectory converges on the origin where $V(\mathbf{0}) = 0$.

The condition that $\dot{V}(\mathbf{x}) < 0$ for $\mathbf{x} \neq 0$ can be relaxed in Theorem 1 under the proper conditions. Condition 4 can be changed to $\dot{V}(\mathbf{x}) \leq 0$; that is, $\dot{V}(\mathbf{x})$ is negative semidefinite. This relaxed condition is sufficient provided that $\dot{V}(\mathbf{x})$ is not equal to zero at any solution of the original differential equation except at the equilibrium point at the origin. A test for this condition is to insert the solution of $\dot{V}(\mathbf{x}) = 0$ into the state equation $\dot{x} = f(\mathbf{x})$ to verify that it is satisfied only at the equilibrium point. Also, if it can be shown that no trajectory can stay forever at the points or on the line, other than the origin, at which $\dot{V} = 0$, then the origin is asymptotically stable. This is the case for the system of Sec. 15.3 as described at the beginning of this section. For linear systems there is only one equilibrium point which is at the origin; therefore it is sufficient for $\dot{V}(\mathbf{x})$ to be negative semidefinite. Theorem 1 may be extended so that it is applicable to the entire state space. In that case the system is said to have *global stability* or stability *in the large*. Including these conditions results in the following theorem.

Theorem 2 Liapunov global asymptotic stability. A system is globally asymptotically stable if there is only one stable equilibrium point and there exists a scalar function $V(\mathbf{x})$ such that:

1. $V(\mathbf{x})$ is continuous and has continuous first partial derivatives in the entire state space.
2. $V(\mathbf{x}) > 0$ for $\mathbf{x} \neq 0$.
3. $V(\mathbf{0}) = 0$.
4. $V(\mathbf{x}) \to \infty$ at $\|x\| \to \infty$.
5. $\dot{V}(\mathbf{x}) \leq 0$.
6. Either $\dot{V}(\mathbf{x}) \neq 0$ except at $\mathbf{x} = 0$ or any locus in the state space where $\dot{V}(\mathbf{x}) = 0$ is not a trajectory of the system.

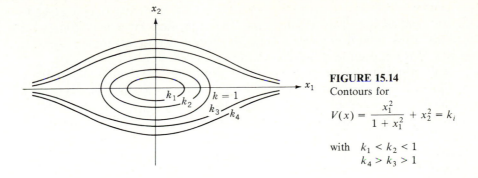

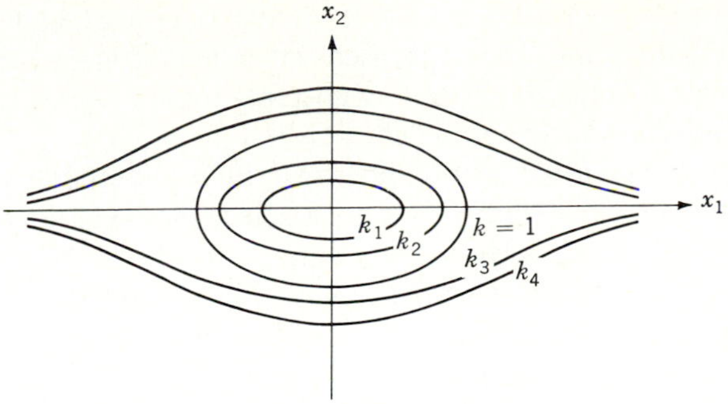

FIGURE 15.14
Contours for

$$V(x) = \frac{x_1^2}{1 + x_1^2} + x_2^2 = k_i$$

with $k_1 < k_2 < 1$
$k_4 > k_3 > 1$

Conditions 1 to 3 ensure that $V(\mathbf{x})$ is positive definite. Condition 4 is satisfied when $V(\mathbf{x})$ is positive definite, i.e., it is closed, in the *entire* state space. When $V(\mathbf{x})$ goes to infinity as any $x_i \to \infty$, then $V(\mathbf{x}) = k_i$ is a closed curve for any k_i. Conditions 5 and 6 mean that $V(\mathbf{x})$ is continuously decreasing along any trajectory in the entire plane and ensures that the system is asymptotically stable.

In order to check for global stability it is necessary to select $V(\mathbf{x})$ so that conditions 1 to 4 are all satisfied. For example, the following function, for a second-order system, does not satisfy condition 4:

$$V(\mathbf{x}) = \frac{x_1^2}{1 + x_1^2} + x_2^2 \tag{15.66}$$

With x_2 finite and $x_1 \to \infty$, then $V(\mathbf{x}) = 1 + x_2^2$, which is not infinite. Therefore, this $V(\mathbf{x})$ does not satisfy condition 4 and cannot be used to determine global stability. The curves for $V(\mathbf{x}) = k$ are closed for $k \leq 1$, and they are open if $k > 1$, as shown in Fig. 15.14. This limits the region for which this function can be used to determine stability.

Finding a proper Liapunov function $V(\mathbf{x})$ means that the system is stable, but this is just a sufficient and not a necessary condition for stability. Because a function $V(\mathbf{x})$ cannot be found does not mean that it does not exist. The Liapunov stability theorems are stated above as requiring $V(\mathbf{x})$ to be positive definite and $\dot{V}(\mathbf{x})$ to be negative definite or negative semidefinite. The stability conditions can also be specified in terms of a negative definite $V(\mathbf{x})$ and a positive definite or positive semidefinite $\dot{V}(\mathbf{x})$. The only requirement is that $V(\mathbf{x})$ and $\dot{V}(\mathbf{x})$ must be of opposite sign. It is more common to start with a $V(\mathbf{x})$ which is positive definite.

The instability theorem can be used as a *sufficient* condition to positively identify an unstable system.

Theorem 3 Liapunov instability theorem. A system $\dot{\mathbf{x}} = f(\mathbf{x})$ is unstable in a region Ω about the origin if there exists a scalar function $V(\mathbf{x})$ such that:

1. $V(\mathbf{x}) \geq 0$, $V(\mathbf{0}) = 0$, and $V(\mathbf{x})$ is continuous and has continuous partial derivatives in the region Ω.
2. $\dot{V}(\mathbf{x})$ is positive definite in the region Ω.

The response of such a system is unbounded as $t \to \infty$ unless $\dot{V}(\mathbf{x})$ is globally negative semidefinite. This theorem may be rephrased in terms of negative functions. Thus, if $\dot{V}$ is negative definite, the system is unstable in any region Ω in which V is not positive definite or positive semidefinite.

This instability theorem is potentially more powerful than the stability theorem because if the conditions for instability are not satisfied, the conditions for stability are automatically satisfied. This is in contrast to the stability theorems (Theorems 1 and 2), where failure to satisfy the stability criteria does not necessarily mean that the system is unstable. It is often possible to select a function V for which $\dot{V}$ is positive (or negative) definite. Then, if V is positive (or negative) definite, semidefinite, or indefinite, the response is unstable. On the other hand, if V is negative (or positive) semidefinite, the instability conditions are not satisfied. This means that the system is stable.

The Liapunov instability theorem has the disadvantage of requiring both V and $\dot{V}$ to be positive definite over the entire region Ω. This theorem can be modified so that instability can be determined in just a region, however small, of the state space. The modifications are presented in the Cetaev theorem.

Theorem 4 Cetaev instability theorem. The origin is unstable when there is a region Ω_1, however small, inside a region Ω and when the Liapunov function $V(\mathbf{x})$ for the differential equation $\dot{\mathbf{x}} = f(\mathbf{x})$ has the following properties:

1. $V(\mathbf{x})$ has continuous first partial derivatives in Ω.
2. $V(\mathbf{x})$ and $\dot{V}(\mathbf{x})$ are positive in Ω_1.
3. The value of the function is $V(\mathbf{x}) = 0$ at the boundary of Ω_1 inside of Ω.
4. The origin is a boundary point of Ω_1.

15.7 APPLICATION OF THE LIAPUNOV METHOD TO LINEAR SYSTEMS

The second method of Liapunov is applicable to time-varying and nonlinear systems. However, there is no simple general method of developing the Liapunov function $V(\mathbf{x})$. Methods have been developed for many such systems, and the literature in this area[8] is extensive. The remaining applications in this chapter are restricted to linear systems. Since linear systems may have only one equilibrium point, the stability or instability is necessarily global in nature. The Routh-Hurwitz stability criterion is available for determining the stability of linear systems. However, the necessity of obtaining the characteristic polynomial can be a disadvantage for higher-order systems. Thus, the following material is presented to develop familiarity with the more general second method of viewing stability. Confidence in the second method of Liapunov can be developed by showing that the results are identical to those obtained with the Routh-Hurwitz method.

The first approach presented is to select a $V(\mathbf{x})$ which is positive definite and to evaluate $\dot{V}(\mathbf{x})$. The coefficients of $V(\mathbf{x})$ and those restraints on the system parameters are then determined which make $\dot{V}(\mathbf{x})$ negative definite or negative

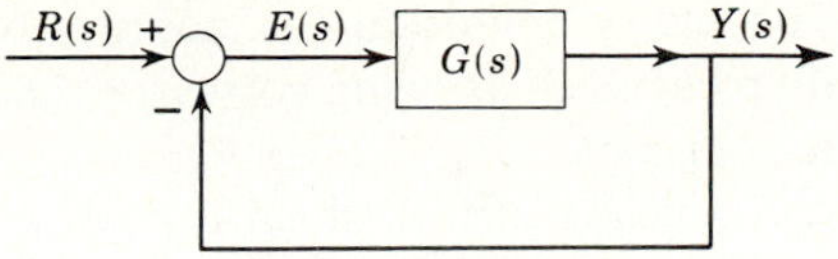

FIGURE 15.15
Unity-feedback system.

semidefinite. Consider the linear feedback system presented in Fig. 15.15 with $r(t) = 0$.

Example 1

$$G(s) = \frac{K}{s(s+a)} \qquad a > 0$$

The differential equation for the actuating signal, when $r = 0$, is

$$\ddot{e} + a\dot{e} + Ke = 0 \tag{15.67}$$

When phase variables with $x_1 = e$ are used, the state equations are

$$\dot{x}_1 = x_2 \qquad \dot{x}_2 = -Kx_1 - ax_2 \tag{15.68}$$

A simple Liapunov function which is positive definite is

$$V(\mathbf{x}) = \tfrac{1}{2}p_1 x_1^2 + \tfrac{1}{2}p_2 x_2^2 = \tfrac{1}{2}\mathbf{x}^T\mathbf{P}\mathbf{x} \tag{15.69}$$

where $p_1 > 0$ and $p_2 > 0$. Its derivative is

$$\dot{V}(\mathbf{x}) = p_1 x_1 \dot{x}_1 + p_2 x_2 \dot{x}_2 = p_1 x_1 x_2 - p_2 K x_1 x_2 - a p_2 x_2^2$$

$$= \mathbf{x}^T \begin{bmatrix} 0 & \dfrac{p_1 - p_2 K}{2} \\ \dfrac{p_1 - p_2 K}{2} & -ap_2 \end{bmatrix} \mathbf{x} = -\mathbf{x}^T\mathbf{N}\mathbf{x} \tag{15.70}$$

The function $\dot{V}(\mathbf{x})$ is always negative semidefinite if $-\mathbf{N}$ is negative semidefinite. The matrix $-\mathbf{N}$ is negative semidefinite if *all* the principal minors of $\mathbf{N}$ are positive or zero.

$$\Delta_2 = ap_2 \ge 0 \tag{15.71}$$

$$\Delta_{12} = \frac{-(p_1 - p_2 K)^2}{4} \ge 0 \tag{15.72}$$

Equation (15.71) is satisfied since it is required that both a and p_2 be greater than zero. Since the left side of Eq. (15.72) cannot be positive, only the equality condition $p_1 - p_2 K = 0$ can be satisfied, which yields $p_1 = p_2 K$. Since $p_1 > 0$ and $p_2 > 0$, this requires that $K > 0$. Then Eq. (15.70) yields $\dot{V}(\mathbf{x}) = -ap_2 x_2^2$ which is negative semidefinite. The Liapunov global asymptotic stability theorem (Theorem 2) is satisfied since, by condition 4, $V(\mathbf{x}) \to \infty$ as $\|\mathbf{x}\| \to \infty$. The condition $\dot{V}(\mathbf{x}) = 0$ exists along the x_1 axis where $x_2 = 0$ and x_1 has any value. A way of showing that $\dot{V}(\mathbf{x})$ being negative semidefinite is sufficient for global asymptotic stability is to show that the x_1 axis is not a trajectory of the system state equations (15.68). The first equation yields $\dot{x}_1 = 0$ or $x_1 = c$. The x_1 axis can be a trajectory only if $x_2 = 0$ and $\dot{x}_2 = 0$. But the second equation yields $\dot{x}_2 = -Kx_1 = -Kc = 0$. This is a contradiction since it requires $x_1 = c = 0$. Therefore, the x_1 axis is not a trajectory, and the system is asymptotically stable.

Using Liapunov's second method, the necessary condition for stability is shown above to be $K > 0$. This result is readily recognized as being correct, either from the Routh stability criterion or from the root locus.

The second approach is to apply Theorem 3, select a $\dot{V}(\mathbf{x})$ which is positive definite, and then to determine the conditions which make $V(\mathbf{x})$ positive definite.

Example 2

$$G(s) = \frac{K}{s(s-a)} \qquad a > 0$$

The phase-variable equations, using $x_1 = e$, are

$$\dot{x}_1 = x_2 \qquad \dot{x}_2 = -Kx_1 + ax_2 \tag{15.73}$$

A more general Liapunov function is chosen in this example

$$V(\mathbf{x}) = \tfrac{1}{2}p_{11}x_1^2 + p_{12}x_1x_2 + \tfrac{1}{2}p_{22}x_2^2$$

$$= \tfrac{1}{2}\mathbf{x}^T\begin{bmatrix} p_{11} & p_{12} \\ p_{12} & p_{22} \end{bmatrix}\mathbf{x} = \tfrac{1}{2}\mathbf{x}^T\mathbf{P}\mathbf{x} \tag{15.74}$$

The derivative $\dot{V}(\mathbf{x})$ is obtained from Eq. (15.74) with values of $\dot{x}_1$ and x_2 inserted from Eq. (15.73):

$$\dot{V}(\mathbf{x}) = -p_{12}Kx_1^2 + (p_{11} + p_{12}a - p_{22}K)x_1x_2 + (p_{12} + p_{22}a)x_2^2 \tag{15.75}$$

A positive definite form is selected for $\dot{V}(\mathbf{x})$. Then the coefficients p_{11}, p_{12}, and p_{22} are evaluated, and the defintieness of $V(\mathbf{x})$ is determined by use of Sylvester's theorem. Accordingly, select

$$\dot{V}(\mathbf{x}) = x_1^2 + x_2^2 \tag{15.76}$$

Equating the coefficients of Eqs. (15.75) and (15.76) yields

$$p_{12} = -\frac{1}{K} \qquad p_{22} = \frac{1+K}{Ka} \qquad p_{11} = \frac{a^2 + K(1+K)}{Ka}$$

The necessary conditions that $V(\mathbf{x})$ be positive definite, using Sylvester's theorem, are

$$\Delta_1 = p_{11} = \frac{a^2 + K(1+K)}{Ka} > 0 \tag{15.77}$$

$$\Delta_{12} = p_{11}p_{22} - p_{12}^2 = \frac{K\left[a^2 + (1+K)^2\right]}{K^2a^2} > 0 \tag{15.78}$$

Equation (15.78) is satisfied if $K > 0$. Since $a > 0$, this requirement also satisfies Eq. (15.77); therefore, $V(\mathbf{x})$ is positive definite. Since $\dot{V}(\mathbf{x})$ is also positive definite [see Eq. (15.76)], Theorem 3 is satisfied and the system is globally unstable. This result is in agreement with the results obtained using the Routh criterion. The reader may wish to investigate (1) the results of permitting $K < 0$ and (2) the use of the negative definite function $\dot{V}(\mathbf{x}) = -x_1^2 - x_2^2$.

Two approaches for applying the second method of Liapunov are demonstrated in these examples. In Example 1 the positive definite Liapunov function $V(\mathbf{x})$ is selected, and the definiteness of $\dot{V}(\mathbf{x})$ is determined. In Example 2 the definiteness of $\dot{V}(\mathbf{x})$ is selected, and then the definiteness of $V(\mathbf{x})$ is determined.

Another approach presented in this section is the use of a procedure developed by Lur'e. Consider the unexcited system represented by the state equation in which $\mathbf{A}$ is of order n:

$$\dot{\mathbf{x}} = \mathbf{A}\mathbf{x} \tag{15.79}$$

The quadratic Liapunov function $V(\mathbf{x})$ is expressed in terms of the symmetric matrix $\mathbf{P}$ by

$$V(\mathbf{x}) = \mathbf{x}^T\mathbf{P}\mathbf{x} \tag{15.80}$$

The time derivative of $V(\mathbf{x})$ is

$$\dot{V}(\mathbf{x}) = \dot{\mathbf{x}}^T\mathbf{P}\mathbf{x} + \mathbf{x}^T\mathbf{P}\dot{\mathbf{x}} = \mathbf{x}^T(\mathbf{A}^T\mathbf{P} + \mathbf{P}\mathbf{A})\mathbf{x} = -\mathbf{x}^T\mathbf{N}\mathbf{x} \tag{15.81}$$

where the symmetric matrix $\mathbf{N}$ is given by

$$\mathbf{N} = -(\mathbf{A}^T\mathbf{P} + \mathbf{P}\mathbf{A}) = -2(\mathbf{P}\mathbf{A})_{\text{sym}} \tag{15.82}$$

The following procedure is used.

Step 1. Select an *arbitrary* symmetric matrix $\mathbf{N}$ of order n which is either positive definite or positive semidefinite. For example, a simple diagonal matrix such as $\mathbf{N} = \mathbf{I}$ or $\mathbf{N} = 2\mathbf{I}$ is positive definite. A positive semidefinite matrix $\mathbf{N}$ can be chosen which contains all zero elements except one positive element in any position along the principal diagonal.

Step 2. Determine the elements of $\mathbf{P}$ by equating terms in Eq. (15.82). Since $\mathbf{P}$ is symmetric, this requires the solution of $n(n + 1)/2$ equations.

Step 3. Use the Sylvester theorem to determine the definiteness of $\mathbf{P}$.

Step 4. Since $\mathbf{N}$ is selected as positive definite (or positive semidefinite), the necessary and sufficient condition for asymptotic stability (instability) is that $\mathbf{P}$ be positive definite (negative definite).

The Sylvester conditions for the positive definiteness of $\mathbf{P}$ are the same as the Routh-Hurwitz stability conditions. The equivalence is derived in Ref. 11. Linear systems which are asymptotically stable are globally asymptotically stable.

Example 3. The reader may verify that Example 2 of this section is solved by using the Lur'e method with

$$\mathbf{N} = \begin{bmatrix} -2 & 0 \\ 0 & -2 \end{bmatrix}$$

Example 4. Examine the conditions of asymptotic stability for the second-order dynamic system

$$\ddot{x} + a_1\dot{x} + a_0x = 0$$

With phase variables the coefficient matrix is

$$\mathbf{A} = \begin{bmatrix} 0 & 1 \\ -a_0 & -a_1 \end{bmatrix}$$

The symmetric matrix $\mathbf{P}$ is

$$\mathbf{P} = \begin{bmatrix} p_{11} & p_{12} \\ p_{12} & p_{22} \end{bmatrix}$$

The matrix $\mathbf{N}$ is chosen as positive semidefinite:

$$\mathbf{N} = \begin{bmatrix} 2 & 0 \\ 0 & 0 \end{bmatrix}$$

The identity in Eq. (15.82) yields three simultaneous equations

$$n_{11} = 2 = 2a_0 p_{12}$$

$$n_{12} = 0 = -p_{11} + a_1 p_{12} + a_0 p_{22}$$

$$n_{22} = 0 = -2p_{12} + 2a_1 p_{22}$$

The solution of these equations yields

$$p_{12} = \frac{1}{a_0} \qquad p_{22} = \frac{1}{a_0 a_1} \qquad p_{11} = \frac{a_0 + a_1^2}{a_0 a_1}$$

The necessary conditions for $\mathbf{P}$ to be positive definite are obtained by applying the Sylvester theorem:

$$p_{11} = \frac{a_0 + a_1^2}{a_0 a_1} > 0 \qquad p_{11} p_{22} - p_{12}^2 = \frac{a_0}{a_0^2 a_1^2} > 0$$

The second equation requires that $a_0 > 0$. Using this condition in the first equation produces the necessary condition $a_1 > 0$. These are obviously the same conditions as are obtained from the Routh-Hurwitz conditions for stability.

Krasovskii has shown[8] that a similar approach may be used with a nonlinear system. However, only sufficient conditions for local asymptotic stability in the vicinity of an equilibrium point may be determined. For the Liapunov function $V(\mathbf{x}) = \mathbf{x}^T\mathbf{P}\mathbf{x}$, the time derivative is $\dot{V}(\mathbf{x}) = -\mathbf{x}^T\mathbf{N}\mathbf{x}$, where

$$-\mathbf{N} = \mathbf{J}_{\mathbf{x}}^T\mathbf{P} + \mathbf{P}\mathbf{J}_{\mathbf{x}} \tag{15.83}$$

The matrix $\mathbf{J}_{\mathbf{x}}$ is the Jacobian evaluated at the equilibrium point. Selecting $\mathbf{P} = \mathbf{I}$ may often lead to a successful determination of the conditions for asymptotic stability in the vicinity of the equilibrium point. This extension to nonlinear systems is presented to show the generality of the method.

15.8 ESTIMATION OF TRANSIENT SETTLING TIME[8]

When the Liapunov function $V(\mathbf{x}) = \mathbf{x}^T\mathbf{P}\mathbf{x}$ is positive definite and its derivative $\dot{V}(\mathbf{x}) = -\mathbf{x}^T\mathbf{N}\mathbf{x}$ is negative definite, it is known that the system is asymptotically stable. The functions $\mathbf{V}(x)$ and $\dot{\mathbf{V}}(x)$ may also be used to determine a measure of the settling time of the unforced transient response. The development is demonstrated by starting with the identity

$$\dot{V}(\mathbf{x}) = \frac{\dot{V}(\mathbf{x})}{V(\mathbf{x})} V(\mathbf{x}) \le -aV(\mathbf{x}) \tag{15.84}$$

where $-a$ is defined as the maximum value of the ratio $\dot{V}(\mathbf{x})/V(\mathbf{x})$ throughout the state space $\mathbf{x}$. This definition is represented symbolically by

$$-a \equiv \max_{\mathbf{x}} \frac{\dot{V}(\mathbf{x})}{V(\mathbf{x})} \tag{15.85}$$

where $a > 0$. Equation (15.85) can be rewritten as

$$\frac{\dot{V}(\mathbf{x})}{V(\mathbf{x})} \le -a \tag{15.86}$$

The integration of Eq. (15.86) from $t = 0$ to the settling time $t = T_s$ yields

$$V[\mathbf{x}(T_s)] \le V[\mathbf{x}(0)]\, e^{-aT_s} \tag{15.87}$$

Equation (15.87) gives the value of the upper bound of $V[\mathbf{x}(T_s)]$ at the settling time T_s. Thus, $-a$ may be considered proportional to the largest eigenvalue of the system, and $1/a$ is the upper bound on the largest time constant of the Liapunov function.

Equation (15.85) is now rewritten as

$$a = \min_{\mathbf{x}} \frac{\mathbf{x}^T\mathbf{N}\mathbf{x}}{\mathbf{x}^T\mathbf{P}\mathbf{x}} \tag{15.88}$$

Since $\mathbf{N}$ and $\mathbf{P}$ are positive definite, the denominator is not zero except when the numerator is also zero, at $\mathbf{x} = 0$. Therefore the division is allowable. Also, the relative values of $\dot{V}(\mathbf{x})$ and $V(\mathbf{x})$ are fixed throughout the whole state space. Therefore the evaluation of a in Eq. (15.88) may be constrained to the surface in the state space where

$$\mathbf{x}^T\mathbf{P}\mathbf{x} = 1 \tag{15.89}$$

Then Eq. (15.88) becomes

$$a = \min_{\mathbf{x}} \mathbf{x}^T\mathbf{N}\mathbf{x} \tag{15.90}$$

The Lagrange multiplier technique is used to evaluate the function of Eq. (15.90) with the constraint of Eq. (15.89). This is done by introducing the constant λ and forming the function

$$\mathbf{x}^T\mathbf{N}\mathbf{x} - \lambda\mathbf{I} = \mathbf{x}^T\mathbf{N}\mathbf{x} - \lambda\mathbf{x}^T\mathbf{P}\mathbf{x} = \mathbf{x}^T[\mathbf{N} - \lambda\mathbf{P}]\mathbf{x} \tag{15.91}$$

The minimization is performed by setting the derivative to zero

$$\frac{d}{d\mathbf{x}}\{\mathbf{x}^T[\mathbf{N} - \lambda\mathbf{P}]\mathbf{x}\} = 0 \tag{15.92}$$

The result (see Prob. 15.16) is

$$[\mathbf{N} - \lambda\mathbf{P}]\mathbf{x} = 0 \tag{15.93}$$

This confirms that $\mathbf{N}$ and $\mathbf{P}$ differ only by the proportionality constant λ.

Premultiplying by $\mathbf{x}^T$ and using Eq. (15.89) gives

$$\mathbf{x}^T\mathbf{N}\mathbf{x} = \lambda > 0 \tag{15.94}$$

From Eq. (15.93) it is seen that $|\mathbf{N}\mathbf{P}^{-1} - \lambda\mathbf{I}| = 0$. Thus, λ is the eigenvalue of $\mathbf{N}\mathbf{P}^{-1}$. Therefore, the right side of Eq. (15.90) is a minimum when λ is a minimum, giving

$$a = \lambda_{\text{min}} = \text{minimum eigenvalue of } \mathbf{N}\mathbf{P}^{-1} \tag{15.95}$$

When the eigenvalues of $\mathbf{N}\mathbf{P}^{-1}$ are complex, the real component is intended in Eq. (15.95).

Example. Apply the method of this section to the system of Example 1, Sec. 15.2, to determine an upper bound on the time for $x(t)$ to reach within the region

$$x_1^2 + x_2^2 = 0.02 \tag{15.96}$$

Using the Lur'e procedure of Sec. 15.7 with the selection $\mathbf{N} = 2\mathbf{I}$ permits the evaluation of the matrix $\mathbf{P}$ by use of Eq. (15.82) and yields

$$\mathbf{P} = \begin{bmatrix} 5/3 & 1 \\ 1 & 1 \end{bmatrix} \tag{15.97}$$

so that

$$V(\mathbf{x}) = (5/3)x_1^2 + 2x_1x_2 + x_2^2 \tag{15.98}$$

Since $\mathbf{P}$ is positive definite, the system is asymptotically stable. From Eq. (15.95), the value of a is the minimum eigenvalue of the matrix

$$\mathbf{N}\mathbf{P}^{-1} = \begin{bmatrix} 3 & -3 \\ -3 & 5 \end{bmatrix} \tag{15.99}$$

Thus, $a = 0.838$. Since the value of $V(\mathbf{x})$ varies along the boundary of Eq. (15.96), it is necessary to find the maximum value of $V(\mathbf{x})$ on this boundary. This is accomplished by solving for x_1 from Eq. (15.96) and substituting into Eq. (15.98). Then differentiate the resulting equation of $V(\mathbf{x})$ with respect to x_2 and equate to zero. The maximum value occurs at the point $x_1 = 0.0827$, $x_2 = -0.1147$. Since the trajectory crosses this boundary at $t = T_s$, $V[\mathbf{x}(T_s)] \le 0.0056$. Also, at the initial condition $\mathbf{x}(0) = [0 \quad 2]^T$, $V[\mathbf{x}(0)] = 4$. Solving Eq. (15.87) for T_s yields

$$T_s \le -\frac{1}{a}\ln\frac{V[\mathbf{x}(T_s)]}{V[\mathbf{x}(0)]} = -\frac{1}{0.838}\ln\frac{0.0056}{4} = 7.84 \text{ s} \tag{15.100}$$

Thus, any trajectory starting on the surface $V[\mathbf{x}(0)] = 4$ will cross the surface described by $V[\mathbf{x}(T_s)]$ within 7.84 s.

This example illustrates that the Liapunov function provides an estimate of the actual system time response.

15.9 SUMMARY

This chapter provides the foundation for the design of optimal feedback control systems in Chap. 16. The properties of state-space trajectories and singular points are introduced. Most of the examples used are second-order systems because they clearly and easily demonstrate the important characteristics, such as stability and the kinds of singular points. Both linear and nonlinear differential equations are considered. The Jacobian matrix is used to represent nonlinear systems by approximate linear equations in the vicinity of their singular points. The important property is demonstrated that a linear equation has only one singular point, whereas a nonlinear equation may have more than one singular point. The mathematical properties of quadratic forms and the definitions of definiteness are presented. The definitions of stability then lead to the Liapunov second method. Stability is evaluated in a linear system without determining the eigenvalues. The chapter concludes with a demonstration that the Liapunov function $V(\mathbf{x})$ provides a measure of system performance. This is the justification for its use in defining a performance index upon which the conditions of optimal performance are presented in Chap. 16.

REFERENCES

1. Andronow, A. A., et al.: *Theory of Oscillations*, Addison-Wesley, Reading, Mass., 1966.
2. Thaler, G. J., and M. P. Pastel: *Analysis and Design of Nonlinear Feedback Control Systems*, McGraw-Hill, New York, 1962.
3. Takahashi, Y., et al.: *Control and Dynamic Systems*, Addison-Wesley, Reading, Mass., 1970.
4. DeRusso, P. M., et al.: *State Variables for Engineers*, Wiley, New York, 1967.
5. Gantmacher, F. R.: *Applications of the Theory of Matrices*, Wiley-Interscience, New York, 1959.
6. Swamy, K. N.: "On Sylvester's Criterion for Positive-Semidefinite Matrices," *IEEE Trans. Autom. Control*, vol. AC-18, p. 306, June 1973.
7. Minorsky, N.: *Theory of Nonlinear Systems*, McGraw-Hill, New York, 1969.
8. Csaki, F.: *Modern Control Theories*, Akademiai, Kiado, Budapest, 1972.
9. Gibson, J. E.: *Nonlinear Automatic Control*, McGraw-Hill, New York, 1963.
10. Parzen, E.: *Modern Probability and Its Applications*, Wiley, New York, 1960.
11. Kalman, R. E., and J. E. Bertram: "Control System Analysis and Design via the Second Method of Liapunov," *J. Basic Eng.*, vol. 80, pp. 371–400, 1960.

CHAPTER 16

INTRODUCTION TO OPTIMAL CONTROL

16.1 INTRODUCTION

A control problem exists when a performance criterion or a set of performance specifications is stipulated for a system and these conditions are not met. Generally, in order to obtain the desired system performance additional equipment must be used in conjunction with the basic system. Either modern control or conventional design techniques are utilized to determine the configuration and parameters of the required additional equipment. By the nature of the modern control technique a desired performance criterion (for example, PI = $\int_0^\infty e^2\,dt$ to be minimum) is selected and a unique design is obtained. The performance of the system is then said to be *optimal* in terms of the defined performance criterion. In contrast, the conventional technique satisfies a required set of performance specifications. These requirements may be met by a number of different designs. In general, these designs do *not* simultaneously meet a defined optimal performance criterion. Therefore, these designs may be called *suboptimal*.

The synthesis and design of conventional feedback control systems are treated extensively in the literature, and the principal design methods for single-input single-output systems (SISO) are described in the earlier chapters of this book. The design process includes the use of compensation or stabilization to improve the transient and steady-state performance. Each of the methods described has its particular advantages and disadvantages. The frequency-response method suffers from difficulty in determining the exact transient response from the frequency-response characteristics. While a correlation is made between the frequency and time responses in Chaps. 7, 9, and 10, this is based on representing the system by an approximately equivalent second-order system. Effective damping ratio and effective undamped natural frequency are related to the maximum value of the frequency response M_m. This approximation may become less valid as the order of the system becomes higher and is of limited value for nonlinear systems. Precise methods for obtaining the time response from the frequency response involve the inverse Fourier transform and generally require the use of a computer. The root-locus method has the advantage that both the transient response and the frequency response can be obtained from the location of the closed-loop poles and zeros. Therefore, the improvements obtained by additional poles and zeros can be readily evaluated.

The principal difficulty remaining in feedback system design is the establishment of an optimum criterion for performance. In other words, what shape of the frequency response or what location of the poles and zeros gives the "best" system performance? This must be determined not only for linear time-invariant (LTI) systems, but also in systems with plants that have variable parameters.[26] The design process becomes more difficult when plant nonlinearities must be considered.

Evaluation of performance, given in Sec. 3.10, is based on the time response[25] to a step-function input. The characteristics used to judge performance are as follows:

1. Maximum overshoot M_p.
2. Time to reach the maximum overshoot t_p.
3. Time for the error to reach its first zero (duplicating time) t_0.
4. Settling time (also called solution time), which is the time for the response to settle within a given percentage of the final value t_s.
5. The time for the response, on its initial rise, to go from 0.1 to 0.9 of the steady-state value, commonly referred to as rise-time t_r.
6. Frequency of oscillation of the transient ω_d.
7. Steady-state error for a given input e_{ss}.

The conventional design methods of Chaps. 7 to 12 present methods of achieving a desired system performance based on the figures of merit such as the time response characteristics M_p, t_p, t_s, and K_m, or on the frequency response characteristics M_m, ω_m, and K_m. These design specifications are selected because of the convenience in graphical interpretation with respect to the root-locus, frequency plots, and/or the control-ratio pole-zero pattern. Thus, these design methods rely heavily on graphical analysis in achieving the desired system performance. Compensators or feedback gains are selected that yield, as closely as possible, the desired system performance. In general, it may not be possible to satisfy all the desired figures of merit. Then, through a trial-and-error procedure, assisted by a computer-aided control system program, an acceptable system performance is achieved, in which a trade-off in the values of the figures of merit is utilized. Generally, there are many system designs that can yield this acceptable system performance. In other words, there is, in general, no unique solution. The conventional design methods work very well for single-input single-output linear (LTI) systems and result in good control systems. The frequency-domain conventional design technique for SISO LTI control systems has been extended[26] to include the design of multiple-input multiple-output (MIMO) LTI systems and MIMO systems with plants that have variable parameters (see Chap. 21).

Optimal control theory, with the help of a digital computer, permits the achievement of an optimal system performance which meets some specified performance criterion. It involves minimizing (or maximizing) a performance index (PI). In contrast to the conventional design technique, an optimal control design method relies on the extensive use of mathematical analysis. The selection of a performance index is often based on mathematical convenience, i.e., the selected performance index permits the mathematical design of the system. While the method yields a unique mathematical solution, the actual system performance may not have all the desired performance characteristics. In other words, the resulting optimal control satisfies the mathematical performance index but may not yield desired values of M_p, t_p, t_s, etc. A compromise must be made between specifying a performance index which includes all the desired system characteristics and a performance index which can be achieved with a reasonable amount of computation.

It is desired to develop a single PI to judge the "goodness" of the time response of the system. Such a PI should have three basic properties: reliability, ease of application, and selectivity. It should be reliable for a given class of systems so that it can be applied with confidence. It should be easy to apply, and it should also be selective so that the resulting system is clearly optimal. The minimization of the PI should also yield a design which satisfies most of the conventional figures of merit. This chapter is devoted to presenting several PIs used for optimizing system response.

Many control-system designs are based upon conventional design techniques, but an increasing number of systems are designed using modern control techniques.[1] These have played a large role in the success of many aerospace control systems which require high precision. It is expected that modern control theory will play a larger role in the design of future control systems. Chapter 18 presents a design technique which assigns the entire closed-loop eigenstructure, i.e., both the eigenvalues and the associated eigenvectors. The multivariable tracker design technique which is presented in Chap. 20 is another MIMO system design method.

16.2 DEVELOPMENT OF THE SOLUTION-TIME CRITERION

To indicate the basis for establishing a performance criterion, the characteristics of a unity-feedback system are studied. Consider first a simple linear second-order

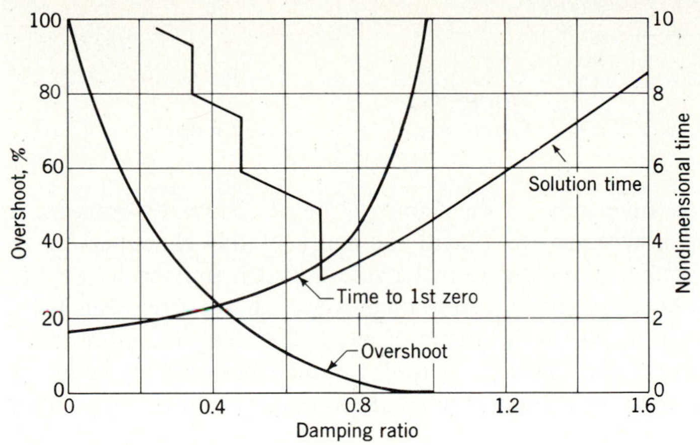

FIGURE 16.1
Transient performance of a simple second-order system with a step input.

system described by the control ratio

$$\frac{Y(s)}{R(s)} = \frac{\omega_n^2}{s^2 + 2\zeta\omega_n s + \omega_n^2} \tag{16.1}$$

This system has a zero steady-state error with a step input. Only such systems are considered. This means that the open-loop transfer function must be Type 1 for a unity-feedback system.

Figure 16.1 shows the following three quantities as a function of the damping ratio ζ for the system represented by Eq. (16.1):

1. The time for the error to reach its first zero.
2. The size of the first and largest overshoot, expressed as a percentage of the final value.
3. The solution time (settling time), which is the time to reach and thereafter remain within 5 percent of the final value.

A commonly desired response is one in which the output has a form identical to that of the input with no error at any time. Such a response is impractical, of course. Therefore the characteristics of Fig. 16.1 are studied to determine the best performance. It is seen that overshoot and time to first zero are conflicting characteristics; i.e., their minimum values occur at different damping ratios. Therefore, if these two characteristics are to be used as criteria, there must be a compromise between the overshoot and rapid rise time. The solution time appears to combine both properties and can be used as a criterion of performance. The optimum solution time for this second-order system specifies a damping ratio of 0.7 as optimum, which is a commonly accepted value.

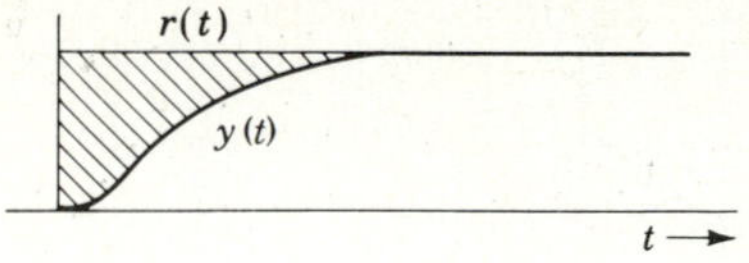

FIGURE 16.2
Control area for an overdamped system.

Therefore the solution time may be the figure of merit to use to optimize a system. Of course, it is necessary to extend the study of this characteristic to higher-order and to nonlinear systems to make sure that it is universally applicable. Some results for linear systems are presented in Sec. 16.5. There seems to be one possible weakness in the use of this figure of merit. The sharp minimum gives an exaggerated picture of the optimum damping ratio since slightly larger or slighlty smaller damping ratios will result in solution times not far removed from the optimum.

16.3 CONTROL-AREA CRITERION†

Another measure of the quality of the transient response of a control system to a step input is the *control area*,[3] which is shown as the shaded selection of Fig. 16.2. The "best" system is one that has the minimum control area since $c(t)$ is then very close to $r(t)$. The control area is given by the integral of the error, $e = r - c$, for some specified time. A typical performance index is

$$\text{PI}_1 = \int_0^{\infty} e\,dt \tag{16.2}$$

When this PI is a minimum for a specified input, the system performance is said to be optimal. The control ratio of the closed-loop system can be described by

$$\frac{Y(s)}{R(s)} = \frac{(1 + A_1 s)(1 + A_2 s)\cdots(1 + A_w s)}{(1 + a_1 s)(1 + a_2 s)\cdots(1 + a_n s)} \tag{16.3}$$

In this equation the values of a_i and A_j can be either real or complex. The value of the control area obtained from Eq. (16.2) is

$$\text{PI}_1 = (a_1 + a_2 + \cdots + a_n) - (A_1 + A_2 + \cdots + A_w) \tag{16.4}$$

If the values of the coefficients a_i in the denominator of $Y(s)/R(s)$ are complex the minimum value of control area PI_1 occurs when the damping ratio of the dominant poles is zero. This is also shown in Fig. 16.3, which contains a plot of PI_1 vs. ζ for the system represented by Eq. (16.1). The minimum value of PI_1 occurs at $\zeta = 0$, which is obviously not a practical control system. Since the effect of minimizing the control area without regard to the system damping results in highly oscillatory response, this PI does not serve the intended purpose and is rejected from further consideration.

---

† The material in Secs. 16.3 to 16.5 is based on Ref. 2.

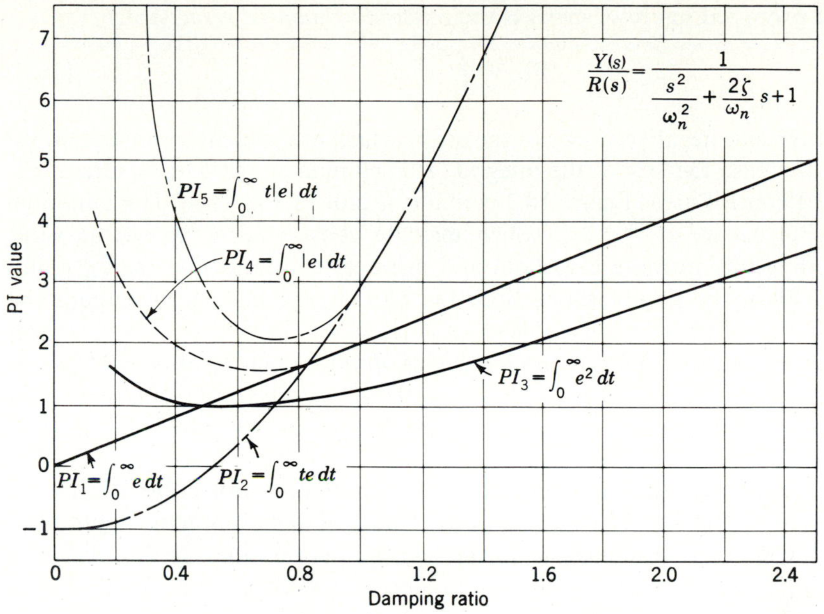

FIGURE 16.3
Criteria for second-order systems.

A modification of control area as a criterion, using time as a weighting factor, has also been proposed.[4] This modified criterion is defined by the integral

$$\mathrm{PI}_2 = \int_0^\infty te\,dt \tag{16.5}$$

The *time-weighted control area* adds a heavy penalty for errors that do not die out rapidly. It is intended that a minimum value of PI_2 should define an optimum system. Figure 16.3 contains a plot of PI_2 vs. ζ for the system represented by Eq. (16.1). Since the minimum value of PI_2 also occurs at $\zeta = 0$, this criterion fails in the same manner as the control-area criterion.

16.4 ADDITIONAL PERFORMANCE INDEXES

In this section additional PIs are investigated to demonstrate their suitability in defining optimum response. These PI are described below and are applied to the same second-order system used in Secs. 16.2 and 16.3:

$$\frac{Y(s)}{R(s)} = \frac{\omega_n^2}{s^2 + 2\zeta\omega_n s + \omega_n^2} \tag{16.6}$$

To make a comparison between them, these additional PIs are plotted vs. the damping ratio in Fig. 16.3.

One proposed figure of merit is the *integral of squared error* (ISE)

$$PI_3 = \int_0^\infty e^2 \, dt \tag{16.7}$$

Both positive and negative values of the error, which are present in underdamped response, increase the size of the integral. This eliminates the bad feature of the unmodified control area. Figure 16.3 contains a plot of PI_3 vs. ζ. The minimum value of PI_3 occurs at $\zeta = 0.5$, which may be considered a reasonable value although it results in more overshoot and a longer solution time than $\zeta = 0.7$. Also, this PI is not very selective since the plot of PI_3 has a broad minimum region.

Another figure of merit is the *integral of absolute value of error* (IAE)

$$PI_4 = \int_0^\infty |e| \, dt \tag{16.8}$$

By using the magnitude of the error, the integral increases for either positive or negative error. This should result in a good underdamped system. Figure 16.3 contains the plot of this function vs. ζ. The minimum value of PI_4 occurs at about $\zeta = 0.7$, which is known to give a good response (see Sec. 16.2). It also gives a slightly better selectivity than PI_3.

Time weighting the magnitude of error gives the PI defined by the integral[2]

$$PI_5 = \int_0^\infty t|e| \, dt \tag{16.9}$$

This function is known as the *integral of time multiplied by the absolute value of error* (ITAE) criterion and is plotted in Fig. 16.3. This curve has a minimum value at $\zeta = 0.7$ and is more selective than the others shown. This PI shows promise; it is used later in this chapter for more complex systems to check its suitability for determining their optimum response. This PI selects good Type 1 systems, but Type 2 systems have excessive overshoot.

Still other figures of merit can be formed with more complex combinations of error and time weighting. Three such PI are:

Integral of time multiplied by squared error (ITSE):

$$PI_6 = \int_0^\infty te^2 \, dt \tag{16.10}$$

Results indicate that this PI yields good performance.

Integral of squared time multiplied by squared error (ISTSE):

$$PI_7 = \int_0^\infty t^2e^2 \, dt \tag{16.11}$$

Integral of squared time multiplied by absolute value of error (ISTAE):

$$PI_8 = \int_0^\infty t^2|e| \, dt \tag{16.12}$$

The generalization of optimizing criteria has been expanded by Schultz and Rideout.[5] PI_5 to PI_8 are not mathematically attractive, but they are used because of the availability of tables which have been derived from extensive simulation studies.

16.5 ZERO STEADY-STATE STEP-ERROR SYSTEMS[6]

The zero steady-state step-error system (see Sec. 13.6) is now studied in terms of the various optimizing methods. The control ratio of the closed-loop system is given by

$$\frac{Y(s)}{R(s)} = \frac{q_0}{s^n + q_{n-1}s^{n-1} + \cdots + q_2 s^2 + q_1 s + q_0} \tag{16.13}$$

One criterion suggested as a standard for selecting the denominator coefficients is to use a pole-placement technique having the characteristic equation composed of equal critically damped modes.[7,8] Such a system has a stable output response. The coefficients are obtained from the binomial expansion for the control ratio given by

$$\frac{Y(s)}{R(s)} = \frac{\omega_0^n}{(s + \omega_0)^n} \tag{16.14}$$

Table 16.1 gives the standard binomial forms for the denominator of $Y(s)/R(s)$. This response is slow and becomes slower as the order of the system is increased. It therefore does not represent an optimum, but it can be used for comparison purposes.

The Butterworth[9] method locates the poles uniformly in the left-hand s plane on a circle of radius ω_0 with its center at the origin. Table 16.2 presents the standard Butterworth forms of the denominator of $Y(s)/R(s)$. The response to a step function using these standard forms is shown in Fig. 16.4 for systems of order 2 to 8.

With modern control, the objective is to apply an optimizing figure of merit to produce a table of standard forms in which the best value is specified for each

TABLE 16.1
Binomial standard forms

$s + \omega_0$
$s^2 + 2\omega_0 s + \omega_0^2$
$s^3 + 3\omega_0 s^2 + 3\omega_0^2 s + \omega_0^3$
$s^4 + 4\omega_0 s^3 + 6\omega_0^2 s^2 + 4\omega_0^3 s + \omega_0^4$
$s^5 + 5\omega_0 s^4 + 10\omega_0^2 s^3 + 10\omega_0^3 s^2 + 5\omega_0^4 s + \omega_0^5$
$s^6 + 6\omega_0 s^5 + 15\omega_0^2 s^4 + 20\omega_0^3 s^3 + 15\omega_0^4 s^2 + 6\omega_0^5 s + \omega_0^6$
$s^7 + 7\omega_0 s^6 + 21\omega_0^2 s^5 + 35\omega_0^3 s^4 + 35\omega_0^4 s^3 + 21\omega_0^5 s^2 + 7\omega_0^6 s + \omega_0^7$
$s^8 + 8\omega_0 s^7 + 28\omega_0^2 s^6 + 56\omega_0^3 s^5 + 70\omega_0^4 s^4 + 56\omega_0^5 s^3 + 28\omega_0^6 s^2 + 8\omega_0^7 s + \omega_0^8$

TABLE 16.2
Butterworth standard forms

$$s + \omega_0$$
$$s^2 + 1.41\omega_0 s + \omega_0^2$$
$$s^3 + 2.0\omega_0 s^2 + 2.0\omega_0^2 s + \omega_0^3$$
$$s^4 + 2.6\omega_0 s^3 + 3.4\omega_0^2 s^2 + 2.6\omega_0^3 s + \omega_0^4$$
$$s^5 + 3.24\omega_0 s^4 + 5.24\omega_0^2 s^3 + 5.24\omega_0^3 s^2 + 3.24\omega_0^4 s + \omega_0^5$$
$$s^6 + 3.86\omega_0 s^5 + 7.46\omega_0^2 s^4 + 9.14\omega_0^3 s^3 + 7.46\omega_0^4 s^2 + 3.86\omega_0^5 s + \omega_0^6$$
$$s^7 + 4.49\omega_0 s^6 + 10.1\omega_0^2 s^5 + 14.6\omega_0^3 s^4 + 14.6\omega_0^4 s^3 + 10.1\omega_0^5 s^2 + 4.49\omega_0^6 s + \omega_0^7$$
$$s^8 + 5.13\omega_0 s^7 + 13.14\omega_0^2 s^6 + 21.85\omega_0^3 s^5 + 25.69\omega_0^4 s^4 + 21.85\omega_0^5 s^3 + 13.14\omega_0^6 s^2 + 5.13\omega_0^7 s + \omega_0^8$$

coefficient of the denominator. Such a table of standard forms would assist the designer since he would try to duplicate these coefficients in his system. He would then be sure of having the "best" system possible.

The minimum ITAE criterion was applied to this system,[2] resulting in the standard forms given in Table 16.3. The procedure used was to vary each coefficient separately until the ITAE value became a minimum. Then the successive coefficients were varied in sequence to minimize the ITAE value. The response to a step function using these standard forms is shown in Fig. 16.5 for systems of order 2 to 8.

The next criterion used was to minimize the solution time, which is the time to reach and remain within 5 percent of the final value. Table 16.4 gives the standard forms using this criterion. The response to a step function using these standard forms is shown in Fig. 16.6 for systems of order 3 to 6. Results for higher orders can be obtained by further computation.

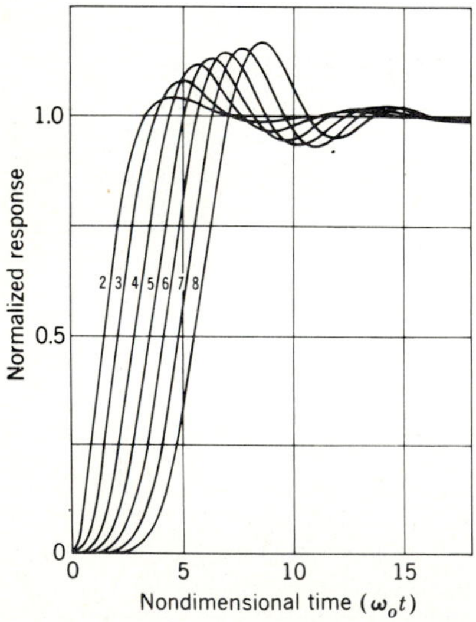

FIGURE 16.4
Response using Butterworth standard forms for zero steady-state step-error systems.

TABLE 16.3
ITAE standard forms for zero steady-state step-error systems

$s + \omega_0$
$s^2 + 1.41\omega_0 s + \omega_0^2$
$s^3 + 1.75\omega_0 s^2 + 2.15\omega_0^2 s + \omega_0^3$
$s^4 + 2.1\omega_0 s^3 + 3.4\omega_0^2 s^2 + 2.7\omega_0^3 s + \omega_0^4$
$s^5 + 2.8\omega_0 s^4 + 5.0\omega_0^2 s^3 + 5.5\omega_0^3 s^2 + 3.4\omega_0^4 s + \omega_0^5$
$s^6 + 3.25\omega_0 s^5 + 6.60\omega_0^2 s^4 + 8.60\omega_0^3 s^3 + 7.45\omega_0^4 s^2 + 3.95\omega_0^5 s + \omega_0^6$
$s^7 + 4.475\omega_0 s^6 + 10.42\omega_0^2 s^5 + 15.08\omega_0^3 s^4 + 15.54\omega_0^4 s^3 + 10.64\omega_0^5 s^2 + 4.58\omega_0^6 s + \omega_0^7$
$s^8 + 5.20\omega_0 s^7 + 12.80\omega_0^2 s^6 + 21.60\omega_0^3 s^5 + 25.75\omega_0^4 s^4 + 22.20\omega_0^5 s^3 + 13.30\omega_0^6 s^2 + 5.15\omega_0^7 s + \omega_0^8$

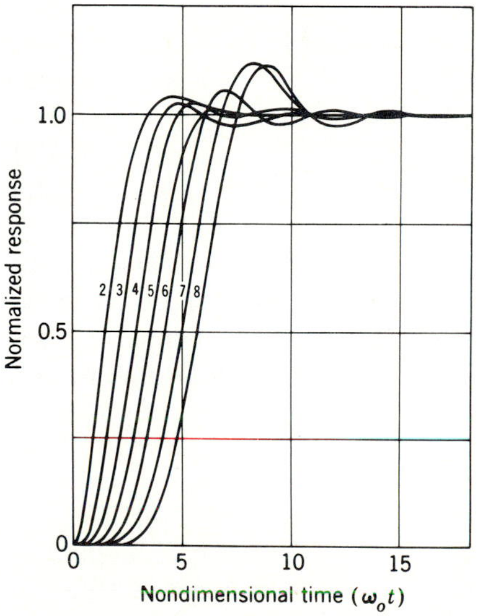

FIGURE 16.5
Response using ITAE standard forms for zero steady-state step-error systems.

TABLE 16.4
Solution-time standard forms for zero steady-state step-error systems

$s + \omega_0$
$s^2 + 1.4\omega_0 s + \omega_0^2$
$s^3 + 1.55\omega_0 s^2 + 2.10\omega_0^2 s + \omega_0^3$
$s^4 + 1.60\omega_0 s^3 + 3.15\omega_0^2 s^2 + 2.45\omega_0^3 s + \omega_0^4$
$s^5 + 1.575\omega_0 s^4 + 4.05\omega_0^2 s^3 + 4.10\omega_0^3 s^2 + 3.025\omega_0^4 s + \omega_0^5$
$s^6 + 1.45\omega_0 s^5 + 5.10\omega_0^2 s^4 + 5.30\omega_0^3 s^3 + 6.25\omega_0^4 s^2 + 3.425\omega_0^5 s + \omega_0^6$

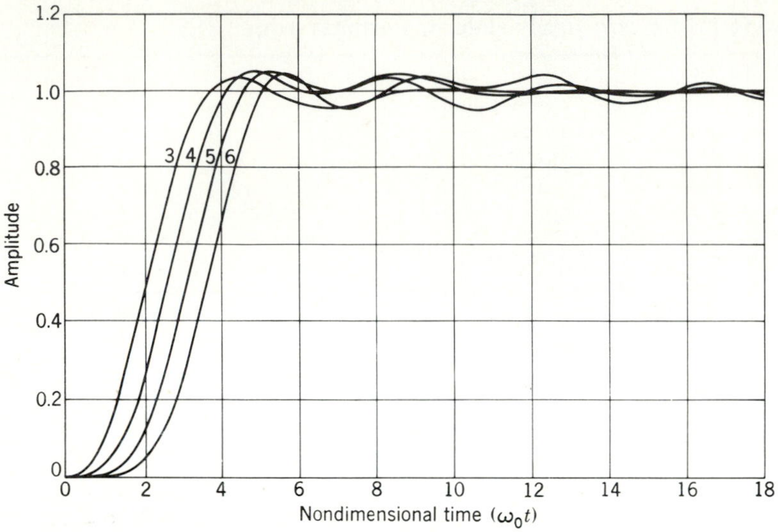

FIGURE 16.6
Response using the solution-time criterion for zero steady-state step-error systems.

A comparison of the response curves of Figs. 16.4 to 16.6 shows that all three criteria give good results. It must be realized that for a system of particular order, one of the criteria may be better in some respect. For example, for the fifth-order system the time to first zero is 4.25 s when the solution time is used and is 5.5 s for the ITAE criterion. But it must also be taken into account that the maximum overshoot is 5 percent for the solution time and only 2 percent for the ITAE criterion. Further, the solution-time response is more oscillatory than the ITAE response. The most suitable response must be determined in terms of which is most important—rise time, peak overshoot, or frequency of oscillation. Once the standard form has been chosen to yield the desired optimal performance, either the Guillemin-Truxal or the state-variable-feedback method described in Chap. 13 can be used to determine the required parameters of the cascade compensator $G_c(s)$ or the state-feedback vector **k**, respectively. A similar analysis can be made for zero steady-state ramp-error systems.[2]

16.6 MODERN CONTROL PERFORMANCE INDEX[10,24]

The PIs of the previous sections represent the first applications of a PI to achieve an optimal closed-loop performance based upon parameter optimization. That is, the plant parameters are fixed at values that satisfy a desired PI without constraining any variable within the system. These variables may have limits (constraints) on their maximum and/or minimum values, such as those on **u**. Because of the difficulty in constraining these variables in the parameter-optimi-

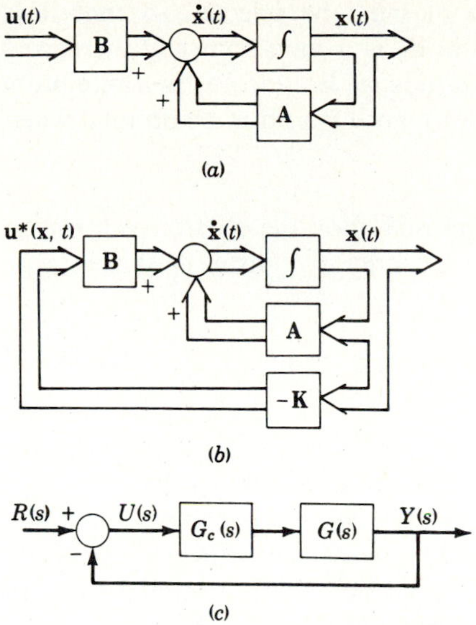

FIGURE 16.7
Various forms of control systems: (*a*) open-loop control system; (*b*) closed-loop optimal state-feedback control system $[\mathbf{r}(t) = \mathbf{0}]$; (*c*) conventional unity-feedback control system.

zation method another approach is necessary. Thus, to achieve an optimal performance when constraints are imposed upon the variables, modern control theory is used to design the system.

In modern control theory the PI is defined in terms of the state and control vectors, $\mathbf{x}(t)$ and $\mathbf{u}(t)$, respectively. The constraints are manifested by weighting the variables (signals) in the modern control PI. Whereas for the parameter-optimization approach the structure of the system is fixed, the modern control-theory approach requires that the structure of the system be selected. The modern optimal feedback-control problem considered in this text depends on the following conditions.

CONDITION 1. A system represented by the state equation $\dot{\mathbf{x}} = f(\mathbf{x}, \mathbf{u}, t)$ of order n is to be optimally controlled. (Figure 16.7*a* represents the open-loop linear control system, which is often called the plant.)

CONDITION 2. The initial (starting) time t_i and the initial state $\mathbf{x}(t_i)$ are specified. Design specifications are defined for either or both the final time t_f and the final state $\mathbf{x}(t_f)$. These quantities may be described implicitly, such as when minimum time is specified or when the final state must reach within a tolerance of the ideal end state.

CONDITION 3. An *integral performance index* is specified to yield the optimal control

$$\mathrm{PI} = \int_{t_i}^{t_f} L(\mathbf{x}, \mathbf{u}, t)\, dt \tag{16.15}$$

This results in selecting the control $\mathbf{u}$ to minimize the PI over the entire time interval of concern. The functional $L(\mathbf{x}, \mathbf{u}, t)$ must be selected so that it is positive. With $L(\mathbf{x}, \mathbf{u}, t) > 0$, the value of the PI is a monotonically increasing function of t within $t_i \le t \le t_f$. If the magnitude of Eq. (16.15) is a minimum over the time interval t_i to t_f, the system performance is said to be optimal when the state is transferred from $\mathbf{x}(t_i)$ to $\mathbf{x}(t_f)$.

CONDITION 4. Amplitude constraints are imposed upon the control vector $\mathbf{u}$. In physical systems there is a limit to the available energy to control the system; e.g., the amplitude constraints on $\mathbf{u}(t)$ are

$$\mathbf{u}_{\min} \le \mathbf{u}(t) \le \mathbf{u}_{\max} \tag{16.16}$$

CONDITION 5. When Eq. (16.15) is changed to the linear quadratic performance form [see Eq. (16.22)], the system must be controllable and observable.[11]

The object of the modern optimal feedback-control problem, under the above conditions, is to determine the *optimal control* law $\mathbf{u}^*(\mathbf{x}, t)$ which can transfer the system from its initial state to the final state while minimizing the PI of Eq. (16.15) (closed-loop operation). This optimal control design is based upon obtaining a system response to a set of initial conditions, $\mathbf{x}(t_i)$, with the input $\mathbf{r}(t)$ assumed to be zero. This is defined as the *optimal regulator problem*. When $\mathbf{r}(t) \neq \mathbf{0}$, it becomes the *optimal servo problem*. Thus, *the optimal control* $\mathbf{u}^*(\mathbf{x}, t)$ is a function of the states $\mathbf{x}(t)$ of the system. For example, for the linear-regulator problem to be described later, applying the quadratic PI yields the *optimal state feedback-control law* $\mathbf{u}^*(\mathbf{x}, t) = -\mathbf{K}(t)\mathbf{x}(t)$, where $\mathbf{K}(t)$ is the feedback matrix of order $m \times n$. For the linear system $\dot{\mathbf{x}} = \mathbf{A}\mathbf{x} + \mathbf{B}\mathbf{u}$, this optimal control law $\mathbf{u}^*(\mathbf{x}, t)$ is generated from the states $\mathbf{x}(t)$, assuming that all states are accessible, by the use of feedback as illustrated in Fig. 16.7*b*. This is in contrast to conventional control design, which is based upon obtaining a system response to a step or sinusoidal input function with all initial conditions assumed to be zero, as illustrated in Fig. 16.7*c*. As defined by Eq. (16.15) and in the previous sections, PIs are formulated in the time domain. Although there are other forms of modern control PIs, this text is restricted to the one expressed by Eq. (16.15). The other PIs are discussed in advanced texts on optimal control.[10,12,16,24]

The formulation of the particular PI to be used is based not only on mathematical convenience but also upon the particular control problem under consideration. For example, consider that the state of a system is to go from some initial value $\mathbf{x}(t_i)$ to some specified final value $\mathbf{x}(t_f)$ in minimum time, but t_f is unspecified. *This is called the minimum-time optimal control problem*. The quantity to be minimized is $t_f - t_i$, which is equal to $\int_{t_i}^{t_f} dt$. Therefore $L(\mathbf{x}, \mathbf{u}, t) = 1$ is used to evaluate the PI. Note that L is not a function of $\mathbf{x}$ or $\mathbf{u}$, and they are unconstrained.

$$\mathrm{PI} = \int_{t_i}^{t_f} 1\, dt = t_f - t_i \tag{16.17}$$

The minimum-time condition can be applied to the problem of a missile intercepting an aerospace vehicle. The time for the missile to travel from its launch point to the target-interception point depends principally on the thrust forces available. Clearly, this time can be made as small as desired if an unlimited thrust force **u** is available—obviously an impractical situation. Usually the control **u** must be constrained, as specified by Eq. (16.16), and this in turn affects the minimum time that is possible. The minimum value of the PI of Eq. (16.17) is zero. This requires the minimum time $t_f - t_i$ to be zero. The dynamic system equations would therefore be required to yield the change of state from $\mathbf{x}(t_i)$ to $\mathbf{x}(t_f)$ in zero time. This implies the use of infinite thrust, a clearly inadmissible solution.

In contrast to the minimum-time requirement of the previous problem, consider the requirement of minimum control effort **u** for a linear system. A suitable PI for this problem with $\mathbf{x}(t_i)$, $\mathbf{x}(t_f)$, t_i, and t_f all specified is

$$\text{PI} = \int_{t_i}^{t_f} \left(\sum_{i=1}^{r} u_i^2 \right) dt = \int_{t_i}^{t_f} (\mathbf{u}^T \mathbf{u})\, dt$$

This equation is usually generalized by introducing a proportionality factor for each control input. A more flexible performance index is achieved by introducing a weighting matrix **Z**; thus,

$$\text{PI} = \int_{t_i}^{t_f} (\mathbf{u}^T \mathbf{Z} \mathbf{u})\, dt \tag{16.18}$$

Thus, $L(\mathbf{x}, \mathbf{u}, t) = \mathbf{u}^T \mathbf{Z} \mathbf{u}$, where **Z** is a real time-varying or time-invariant weighting matrix. For analytical convenience, **Z** is often made symmetric and positive definite. Only the time-invariant case is considered in this text. When considering the scalar case where u may represent a variable such as a force, a current, or a voltage, this PI represents energy. Therefore, by association, problems using PIs of this form are termed *minimum-energy optimal control problems*. The PI of Eq. (16.18) is a minimum when $\mathbf{u}^*(\mathbf{x}, t) = \mathbf{0}$; that is, PI = 0. Solving dynamic system equations with the value $\mathbf{u}^*(\mathbf{x}, t) = \mathbf{0}$ leads to the analytical result for a linear system that the time to reach the desired final state $\mathbf{x}(t_f)$ must be $t_f = \infty$, which is inadmissible. This value of $\mathbf{u}(\mathbf{x}, t)$ violates condition 3, which requires $L(\mathbf{x}, \mathbf{u}, t) > 0$.

The ISE criterion is also applicable in minimizing the error states or state deviations. When the final state $\mathbf{x}(t_f)$ is specified to be the origin and the initial state $\mathbf{x}(t_i)$ describes the initial conditions, the state is then the error and the PI is

$$\text{PI} = \int_{t_i}^{t_f} \sum_{i=1}^{n} x_i^2\, dt = \int_{t_i}^{t_f} \mathbf{x}^T \mathbf{x}\, dt \tag{16.19}$$

where $L(\mathbf{x}, \mathbf{u}, t) = \mathbf{x}^T \mathbf{x}$ and the limits of the integral are specified. A more general performance index is obtained as in Eq. (16.18) by introducing a weighting matrix **Q**. The modified minimum ISE criterion yields

$$\text{PI} = \int_{t_i}^{t_f} \mathbf{x}^T \mathbf{Q} \mathbf{x}\, dt \tag{16.20}$$

where **Q** is a weighting matrix of order n and $L(\mathbf{x}, \mathbf{u}, t) = \mathbf{x}^T \mathbf{Q} \mathbf{x} > 0$. When this is applied to a specific problem, it is often necessary (see Sec. 16.8) or mathematically convenient to select **Q** as real, symmetric, and positive definite or positive semidefinite. The object of the design problem is to determine the optimal control $\mathbf{u}^*(\mathbf{x}, t)$ that transfers the system from its initial state $\mathbf{x}(t_i)$ to its final state $\mathbf{x}(t_f)$ while minimizing the PI.

As noted, impractical results are obtained for the minimum-time and the minimum-energy optimal control problems. Also, the PIs of Eqs. (16.17) to (16.20) can produce nonunique solutions. To overcome these difficulties, a more practical PI can be formulated by the combination of more than one of these simple indexes. These PIs lend flexibility to the optimal control design. The composite PI that is widely used in optimal control design is formed by the linear combination of Eqs. (16.18) and (16.20),

$$\mathrm{PI} = \int_{t_i}^{t_f} (\mathbf{x}^T\mathbf{Q}\mathbf{x} + \mathbf{u}^T\mathbf{Z}\mathbf{u})\,dt \tag{16.21}$$

This is referred to as the *quadratic performance index*.[11] Its application to a control-system design achieves an optimal system which represents a compromise between, or a blending of, the minimum-error and the minimum-energy criteria. For this combined PI the matrices **Q** and **Z** are restricted to being constant and real symmetric. As stated in condition 3, the functional $L(\mathbf{x}, \mathbf{u}, t) = \mathbf{x}^T\mathbf{Q}\mathbf{x} + \mathbf{u}^T\mathbf{Z}\mathbf{u}$ must be selected to be positive definite. As shown later, **Z** must be positive definite, and thus it has an inverse. The solution of the optimal control problem, Sec. 16.7, yields the requirement that **Q** be positive definite, but in some cases this condition may be relaxed to allow **Q** to be at least positive semidefinite. For a *linear* system represented by $\dot{\mathbf{x}} = \mathbf{A}\mathbf{x} + \mathbf{B}\mathbf{u}$, the determination of **u** which minimizes the quadratic performance index, with the limits of integration expanded to 0 and ∞,

$$\mathrm{PI} = \int_0^\infty L(\mathbf{x}, \mathbf{u}, t)\,dt = \int_0^\infty \left[\mathbf{x}^T(t)\mathbf{Q}\mathbf{x}(t) + \mathbf{u}^T(t)\mathbf{Z}\mathbf{u}(t)\right] dt \tag{16.22}$$

is often called the *linear quadratic control problem*. With these limits the resulting optimal control law $\mathbf{u}^*(\mathbf{x})$ is now an explicit function of only the state vector $\mathbf{x}(t)$ (see Sec. 16.8). This permits the implementation of the optimal control by means of a closed-loop state feedback control. Guidelines for selecting the elements of the weighting matrices **Q** and **Z** are discussed in Chap. 17. As a result of choosing the quadratic PI and an infinite range on time, the amplitudes u_i need not be explicitly constrained (the constraints on the amplitudes of u_i are embedded in the selection of the matrix **Z**). Hence, in this special case, the restraints of condition 4 may be relaxed. The latter portion of this text demonstrates that the linear quadratic control problem is amenable to the utilization of conventional control-theory techniques and figures of merit.

16.7 ALGEBRAIC RICCATI EQUATION[12]

In this section the results of the second method of Liapunov are utilized to derive a method for readily determining the optimal control $\mathbf{u}^*(\mathbf{x})$ for the linear control problem. This method is also applicable to multiple-input multiple-output systems. Assume[13,14] a Liapunov function of the quadratic form (see Sec. 15.6):

$$V[\mathbf{x}(t)] = \mathbf{x}(t)^T\mathbf{P}\mathbf{x}(t) \tag{16.23}$$

where **P** is a positive definite, symmetric, constant matrix with real elements, and the control vector has the form

$$\mathbf{u}^*[\mathbf{x}(t)] = -\mathbf{K}\mathbf{x}(t) \tag{16.24}$$

The feedback matrix **K** is assumed to be a constant matrix. Note that V and $\mathbf{u}^*$ are not explicit functions of time. Substitute Eq. (16.24) into Eq. (16.22) to obtain

$$\text{PI} = \int_{t_i=0}^{t_f=\infty} \left\{\mathbf{x}(\tau)^T[\mathbf{Q} + \mathbf{K}^T\mathbf{Z}\mathbf{K}]\mathbf{x}(\tau)\right\} d\tau \tag{16.25}$$

It is desired to determine an optimal control law that is independent of the value of the initial state $\mathbf{x}(t_i)$ and minimizes the PI of Eq. (16.25). Since the initial state can take on any value of $\mathbf{x}(t)$ for any value of t up to infinity, replacing the lower limit of the integral by t results in a generalization of Eq. (16.25). Thus

$$\text{PI} = \int_{t}^{\infty} \left\{\mathbf{x}(\tau)^T[\mathbf{Q} + \mathbf{K}^T\mathbf{Z}\mathbf{K}]\mathbf{x}(\tau)\right\} d\tau \tag{16.26}$$

where τ is a dummy variable. Note that Eq. (16.26) is a quadratic performance index. The quadratic Liapunov function of Eq. (16.23) may be selected as the PI to be used in Eq. (16.26). Thus

$$V[\mathbf{x}(t)] = \mathbf{x}(t)^T\mathbf{P}\mathbf{x}(t) = \int_{t}^{\infty} \left\{\mathbf{x}(\tau)^T[\mathbf{Q} + \mathbf{K}^T\mathbf{Z}\mathbf{K}]\mathbf{x}(\tau)\right\} d\tau \tag{16.27}$$

A unique solution of this equation for a positive definite matrix **P** exists if the matrix **Q** is positive definite.[12] In the linear-regulator problem, the final time is taken to be infinite. However, digital-computer programs available for solving this equation involve a finite time interval; that is, $\mathbf{t}_f < \infty$. Thus these programs yield a solution if **Q** is at least positive semidefinite. If **Q** is positive semidefinite, a minimum to Eq. (16.22) exists if and only if (iff) the following condition, established by Kalman,[13] is satisfied:

$$\text{Rank}\left[\mathbf{H}|\mathbf{A}^T\mathbf{H}|\ \cdots\ |(\mathbf{A}^T)^{n-1}\mathbf{H}\right] = n \tag{16.28}$$

where $\mathbf{H}\mathbf{H}^T = \mathbf{Q}$. It means that the system described by **A** and **H** is observable. This condition should be checked before proceeding with the solution of Eq. (16.27).

If the system input **r** is a constant vector other than zero and the feedback law $\mathbf{u}^* = -\mathbf{K}\mathbf{x}$ is used, the state equation for the closed-loop system becomes

$$\dot{\mathbf{x}} = \mathbf{A}\mathbf{x} + \mathbf{B}\mathbf{u} = [\mathbf{A} - \mathbf{B}\mathbf{K}]\mathbf{x} + \mathbf{B}\mathbf{r} \tag{16.29}$$

where $\mathbf{u} = \mathbf{r} - \mathbf{K}\mathbf{x}$. Therefore, the poles of the control ratio, for the output given by $\mathbf{y} = \mathbf{C}\mathbf{x}$, are the eigenvalues of $\mathbf{A} - \mathbf{B}\mathbf{K}$ iff the optimal closed-loop system is observable. The following two conditions ensure that **P** is positive definite: (1) the plant is controllable and (2) Eq. (16.28) is satisfied. These conditions also guarantee that the optimal control law yields a stable closed-loop system; i.e., all eigenvalues of $\mathbf{A} - \mathbf{B}\mathbf{K}$ have negative real parts.

If $(\mathbf{A}, \mathbf{C})$ is observable and $\mathbf{C} = \mathbf{H}$, it follows that $(\mathbf{A}, \mathbf{H})$ is observable. Letting $\mathbf{Q} = \mathbf{H}\mathbf{H}^T$ is then an appropriate method of ensuring that condition 5 of Sec. 16.6 will be satisfied, although it must be emphasized that the $\mathbf{Q}$ which results from this choice may be judged, from other consideration, to be unsuitable. Consequently the effectiveness of choosing $\mathbf{Q} = \mathbf{H}\mathbf{H}^T = \mathbf{C}\mathbf{C}^T$, in conjunction with the choice of weighting on the control, must be evaluated from the closed-loop time response.

Performing the integration in Eq. (16.27) yields a function of t since the upper limit of the integral is infinity. Also, $V[\mathbf{x}(\infty)] = 0$ since it must be positive definite. Therefore differentiating Eq. (16.27) with respect to its lower limit t yields†

$$\dot{V}[\mathbf{x}(t)] = \dot{\mathbf{x}}(t)^T\mathbf{P}\mathbf{x}(t) + \mathbf{x}(t)^T\mathbf{P}\dot{\mathbf{x}}(t) = -\mathbf{x}(t)^T[\mathbf{Q} + \mathbf{K}^T\mathbf{Z}\mathbf{K}]\mathbf{x}(t) \tag{16.30}$$

The minus sign is the result of the differentiation with respect to time, which is the lower limit of the integral. Since $\mathbf{Q}$ and $\mathbf{Z}$ are positive definite, $\dot{V}(\mathbf{x})$ is negative definite and the closed-loop system is asymptotically stable. Substituting $\dot{\mathbf{x}}$ from Eq. (16.29), with $\mathbf{r} = \mathbf{0}$, into Eq. (16.30) yields

$$\dot{V}(\mathbf{x}) = \mathbf{x}^T\left[(\mathbf{A} - \mathbf{B}\mathbf{K})^T\mathbf{P} + \mathbf{P}(\mathbf{A} - \mathbf{B}\mathbf{K})\right]\mathbf{x} = -\mathbf{x}^T[\mathbf{Q} + \mathbf{K}^T\mathbf{Z}\mathbf{K}]\mathbf{x} \tag{16.31}$$

This equation is satisfied when the bracketed terms on each side of the equation are equal, i.e.,

$$(\mathbf{A} - \mathbf{B}\mathbf{K})^T\mathbf{P} + \mathbf{P}(\mathbf{A} - \mathbf{B}\mathbf{K}) = -\mathbf{Q} - \mathbf{K}^T\mathbf{Z}\mathbf{K} \tag{16.32}$$

Since $\mathbf{A}$, $\mathbf{B}$, and $\mathbf{Q}$ are known constant matrices, Eq. (16.32) represents an equation with the unknown matrix $\mathbf{K}$ as a function of the variable $\mathbf{P}$. Substituting $\mathbf{Q} + \mathbf{K}^T\mathbf{Z}\mathbf{K}$ from Eq. (16.32) into Eq. (16.27) yields

$$V[\mathbf{x}(t)] = \int_t^\infty \left\{-\mathbf{x}^T\left[(\mathbf{A} - \mathbf{B}\mathbf{K})^T\mathbf{P} + \mathbf{P}(\mathbf{A} - \mathbf{B}\mathbf{K})\right]\mathbf{x}\right\} dt \tag{16.33}$$

Note that the function within the braces is equal to the functional $L(\mathbf{x}, \mathbf{u}, t)$ of Eq. (16.22), where $L(\mathbf{x}, \mathbf{u}, t) > 0$. The minimum value of this PI occurs when the integrand is a minimum. Therefore the problem now is to determine the values of the elements of $\mathbf{K}$ that make the integrand a minimum. This is accomplished by taking the derivative of Eq. (16.32) with respect to the matrix $\mathbf{K}$ and setting each $\partial p_{vu}/\partial k_{wx} = 0$, where p_{vu} is the vuth element of $\mathbf{P}$ and k_{wx} is the wxth element of $\mathbf{K}$. In doing this, a derivative of a matrix with respect to a matrix is required. The necessary additional matrix operations are now presented.

Consider an $n \times m$ $\mathbf{G}$ matrix which is a function of the elements of an $n \times r$ $\mathbf{M}$ matrix. These matrices are partitioned into column vectors

$$\mathbf{G} = [\mathbf{g}_1 \quad \mathbf{g}_2 \quad \cdots \quad \mathbf{g}_m] \qquad \mathbf{M} = [\mathbf{m}_1 \quad \mathbf{m}_2 \quad \cdots \quad \mathbf{m}_r]$$

† The Leibnitz rule for the differentiation of an integral may be used to obtain Eq. (16.30).

The derivative of a vector with respect to a vector yields the $n \times n$ *Jacobian matrix*

$$\frac{d\mathbf{g}_i}{d\mathbf{m}_j^T} = \begin{bmatrix} \dfrac{\partial g_{1i}}{\partial m_{1j}} & \dfrac{\partial g_{1i}}{\partial m_{2j}} & \cdots & \dfrac{\partial g_{1i}}{\partial m_{nj}} \\ \dfrac{\partial g_{2i}}{\partial m_{1j}} & \dfrac{\partial g_{2i}}{\partial m_{2j}} & \cdots & \dfrac{\partial g_{2i}}{\partial m_{nj}} \\ \cdots & \cdots & \cdots & \cdots \\ \dfrac{\partial g_{ni}}{\partial m_{1j}} & \dfrac{\partial g_{ni}}{\partial m_{2j}} & \cdots & \dfrac{\partial g_{ni}}{\partial m_{nj}} \end{bmatrix} \tag{16.34}$$

where $\mathbf{g}_i = \begin{bmatrix} g_{1i} \\ g_{2i} \\ \vdots \\ g_{ni} \end{bmatrix}$ and $\mathbf{m}_j = \begin{bmatrix} m_{1j} \\ m_{2j} \\ \vdots \\ m_{nj} \end{bmatrix}$

The derivative of a matrix with respect to a matrix yields the $m \times r$ *matrix of Jacobian matrices* (*or a matrix of Jacobians*)

$$\frac{d\mathbf{G}^T}{d\mathbf{M}} = \begin{bmatrix} \dfrac{\partial \mathbf{g}_1}{\partial \mathbf{m}_1} & \dfrac{\partial \mathbf{g}_1}{\partial \mathbf{m}_2} & \cdots & \dfrac{\partial \mathbf{g}_1}{\partial \mathbf{m}_r} \\ \dfrac{\partial \mathbf{g}_2}{\partial \mathbf{m}_1} & \dfrac{\partial \mathbf{g}_2}{\partial \mathbf{m}_2} & \cdots & \dfrac{\partial \mathbf{g}_2}{\partial \mathbf{m}_r} \\ \cdots & \cdots & \cdots & \cdots \\ \dfrac{\partial \mathbf{g}_m}{\partial \mathbf{m}_1} & \dfrac{\partial \mathbf{g}_m}{\partial \mathbf{m}_2} & \cdots & \dfrac{\partial \mathbf{g}_m}{\partial \mathbf{m}_r} \end{bmatrix} \tag{16.35}$$

The expanded matrix (with the Jacobian matrices inserted) is of order $mn \times rn$; thus, it contains mrn^2 elements.

Applying Eqs. (16.34) and (16.35) to the derivative with respect to $\mathbf{K}$ of Eq. (16.32) permits equating the elements of the resulting matrices on both sides of the equation, yielding mrn^2 equations. It is found that only nr equations are independent, and these nr independent equations yield

$$\mathbf{PB} = \mathbf{K}^T\mathbf{Z} \tag{16.36}$$

$\mathbf{PB}$ and $\mathbf{K}^T\mathbf{Z}$ are matrices of order $n \times r$. Since $\mathbf{Z}$ is restricted to being positive definite, the existence of $\mathbf{Z}^{-1}$ is assured. Thus

$$\mathbf{K}^T = \mathbf{PBZ}^{-1} \tag{16.37}$$

Since $\mathbf{P}$ and $\mathbf{Z}$ are symmetric,

$$\mathbf{K} = \mathbf{Z}^{-1}\mathbf{B}^T\mathbf{P} \tag{16.38}$$

Substituting these functions into Eq. (16.32) yields

$$[\mathbf{A} - \mathbf{BZ}^{-1}\mathbf{B}^T\mathbf{P}]^T\mathbf{P} + \mathbf{P}[\mathbf{A} - \mathbf{BZ}^{-1}\mathbf{B}^T\mathbf{P}] = -\mathbf{Q} - \mathbf{PBZ}^{-1}\mathbf{ZZ}^{-1}\mathbf{B}^T\mathbf{P} \tag{16.39}$$

To rearrange this equation it is necessary to use the matrix-transpose property, namely, $[\mathbf{G} + \mathbf{M}]^T = \mathbf{G}^T + \mathbf{M}^T$. When this matrix operation is used and the term $\mathbf{PBZ}^{-1}\mathbf{B}^T\mathbf{P}$ is canceled, Eq. (16.39) reduces to

$$\mathbf{A}^T\mathbf{P} - \mathbf{PBZ}^{-1}\mathbf{B}^T\mathbf{P} + \mathbf{PA} + \mathbf{Q} = \mathbf{O} \tag{16.40}$$

This is the *algebraic Riccati equation*, in which $\mathbf{P}$ is called the *Riccati matrix*. It is difficult to obtain the analytic solution of this equation for $\mathbf{P}$, except for low-order systems ($n \leq 3$). Fortunately, a numerical solution for $\mathbf{P}$ can be obtained on a digital computer.[15] Once $\mathbf{P}$ is known, the feedback matrix $\mathbf{K}$ is obtained from Eq. (16.38) and the optimal control law $\mathbf{u}^*[\mathbf{x}(t)] = -\mathbf{Kx}(t)$ is determined.

The solution of $\mathbf{P}$ from Eq. (15.40) minimizes the PI given by Eq. (16.22). The development of this result is based on Eq. (16.27), in which the lower limit of the integral is the initial time t so that $\mathbf{P}$ is independent of the initial conditions $\mathbf{x}(t_i)$. Thus, the minimum value of the PI given in Eq. (16.22), where $\mathbf{x}(\infty) = \mathbf{0}$, is evaluated from Eq. (16.27) at $t = t_i = 0$ as

$$\min \text{PI} = \mathbf{x}^T(0)\mathbf{Px}(0) \tag{16.41}$$

The optimal control law indicates that for an optimum system to exist when the quadratic PI is used, all state variables must be fed back as shown in Fig. 16.7*b*.

In practice all states may not be available for measurement. Thus, in order to synthesize this optimal control law, it is necessary to reconstruct the unavailable states. A method of doing this is the Luenberger observer theory,[11,16] presented in Chap. 18. For linear time-invariant systems with a single output another possibility is the technique of block-diagram manipulation, discussed in Chaps. 14 and 17.

The elements of the matrix $\mathbf{K}$ are the *feedback coefficients* and are the same as those described in Chap. 13. Phase-variable representation can be used in the study of optimal control systems because of the relative ease of the mathematical operations and analysis. In order to convert the feedback coefficients $\mathbf{K}_c$, determined from a phase-variable representation, into the feedback coefficients $\mathbf{K}_p$ for a physical-variable representation, the transformation matrix $\mathbf{T}$ (see Chap. 5) is used. Inserting $\mathbf{x} = \mathbf{T}^{-1}\mathbf{x}_p$ into the optimal control law yields

$$\mathbf{u}^* = -\mathbf{K}_c\mathbf{x} = -\mathbf{K}_c\mathbf{T}^{-1}\mathbf{x}_p = -\mathbf{K}_p\mathbf{x}_p \tag{16.42}$$

where

$$\mathbf{K}_p = \mathbf{K}_c\mathbf{T}^{-1} \tag{16.43}$$

The direct control problem starts with the specification that either (1) $\mathbf{Q}$ is positive definite or (2) $\mathbf{Q}$ is at least positive semidefinite but the rank condition given in Eq. (16.28) must be satisfied. This restriction on $\mathbf{Q}$ is necessary to guarantee an asymptotically stable optimal closed-loop system. In addition, $\mathbf{Z}$ must be positive definite in order to ensure finite control. The design leads to a feedback matrix $\mathbf{K}$ that yields roots of the closed-loop system in the left-half s plane.

The inverse control problem starts with the placement of the roots of the characteristic equation so that the system has the desired stable response characteristics. The methods presented in Chap. 13 can be used to obtain the feedback gains **K** which yield the desired root placement. It is now possible to evaluate, for a given value of **Z**, the matrix **Q** corresponding to the feedback matrix **K**. This may result in a **Q** for which the restrictions on sign definiteness are relaxed from those required in the direct control method.

When single-input time-invariant linear systems are considered, Eqs. (16.22), (16.24), (16.37), and (16.40) can be rewritten. Since **u** is now a scalar, the control weighting **Z** is also a scalar z. As a consequence **B** is a column matrix and **K** is a row matrix. Hence,

$$\mathrm{PI} = \int_0^\infty [\mathbf{x}^T\mathbf{Q}\mathbf{x} + zu^2]\,dt \tag{16.44}$$

$$u^*[\mathbf{x}(t)] = -\mathbf{k}^T\mathbf{x}(t) \tag{16.45}$$

$$\mathbf{k}^T = z^{-1}\mathbf{b}^T\mathbf{P} \tag{16.46}$$

$$\mathbf{A}^T\mathbf{P} - \mathbf{P}\mathbf{b}z^{-1}\mathbf{b}^T\mathbf{P} + \mathbf{P}\mathbf{A} + \mathbf{Q} = \mathbf{0} \tag{16.47}$$

Molinari[22,23] claims that if **KB** (or $\mathbf{k}^T\mathbf{b}$) is symmetric, a stable closed-loop system is optimal. Thus, a stable single-input state-variable-feedback system, with all states fed back, is an optimal system. Therefore, the state-variable-feedback design method of Chap. 13 represents a method for achieving an optimal design, satisfying the PI of Eq. (16.44). It results in the control u in the same form as given by Eq. (16.45) and does not explicitly utilize the Riccati matrix **P**.

In the literature, there are design methods which simultaneously result in a desired control-ratio transfer function and an optimal quadratic performance.[17,18] The next section presents two examples which illustrate the direct control method for achieving an optimal control system.

16.8 EXAMPLES

Example 1. It is desired to optimize the closed-loop response of the SISO system of Fig. 16.8*a* utilizing the Riccati equation of Eq. (16.47), which is based on the quadratic PI of Eq. (16.44). Because of the simplicity of the system the phase- and physical-variable representations of this system are identical; thus

$$\mathbf{A} = \begin{bmatrix} 0 & 1 \\ 0 & -1 \end{bmatrix} \qquad \mathbf{b} = \begin{bmatrix} 0 \\ 1 \end{bmatrix} \tag{16.48}$$

The weighting matrices to be used in the quadratic PI are selected as

$$\mathbf{Q} = \begin{bmatrix} 1 & 0 \\ 0 & 0.1 \end{bmatrix} \qquad z = 1 \tag{16.49}$$

A procedure for selecting these matrices is presented in Chap. 17. Substituting Eqs. (16.48) and (16.49) into Eq. (16.47) yields

$$\begin{bmatrix} 0 & 0 \\ p_{11} - p_{12} & p_{12} - p_{22} \end{bmatrix} - \begin{bmatrix} p_{12}^2 & p_{12}p_{22} \\ p_{12}p_{22} & p_{22}^2 \end{bmatrix} + \begin{bmatrix} 0 & p_{11} - p_{12} \\ 0 & p_{12} - p_{22} \end{bmatrix} + \begin{bmatrix} 1 & 0 \\ 0 & 0.1 \end{bmatrix} = \mathbf{0}$$

From this equation the following relationships are obtained by equating each

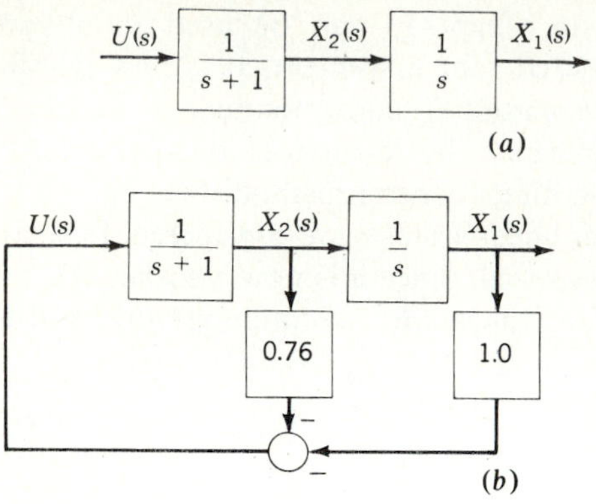

FIGURE 16.8
The (a) open-loop and (b) optimal closed-loop systems of Example 1.

element of the composite matrix to zero:

$$p_{12}^2 = 1 \qquad p_{11} - p_{12} - p_{12}p_{22} = 0 \qquad 2p_{12} - 2p_{22} - p_{22}^2 + 0.1 = 0$$

A choice must be made between using $p_{12} = -1$ and $p_{12} = 1$. Utilizing the value $p_{12} = -1$ yields complex-conjugate values for p_{22} which are not allowable. Using the value $p_{12} = 1$ gives

$$p_{11} = p_{22} + 1 \qquad p_{22}^2 + 2p_{22} - 2.1 = 0$$

Only the value $p_{22} = 0.76$ yields a positive definite $\mathbf{P}$ matrix. Thus

$$\mathbf{P} = \begin{bmatrix} 1.76 & 1 \\ 1 & 0.76 \end{bmatrix}$$

Therefore, from Eqs. (16.46) and (16.45), respectively, with $z = 1$,

$$\mathbf{k}^T = z^{-1}\mathbf{b}^T\mathbf{P} = [1 \quad 0.76]$$

$$u^*(\mathbf{x}) = -\mathbf{k}^T\mathbf{x} = -(x_1 + 0.76x_2)$$

If $y(t) = x_1(t)$, the damping ratio of the resulting optimal system is 0.88. Note that this is higher than for the ITAE and Butterworth forms but lower than the binomial form. The resulting optimal feedback control system, with $r(t) = 0$, is shown in Fig. 16.8b. From the analysis in Chap. 13, the value $k_1 = 1$ is required to produce $e(t)_{ss} = 0$ when the system has a step input. The same value is obtained in this problem because of the value assigned to q_{11}. (The criterion for this selection is described in Chap. 17.)

Example 2. The quadratic PI is used with the phase-variable representation of the open-loop transfer function

$$G(s) = \frac{100}{s(s+1)(s+5)}$$

Using the weighting matrices

$$\mathbf{Q} = \begin{bmatrix} 1 & 0 & 0 \\ 0 & 0.008 & 0 \\ 0 & 0 & 1.0 \end{bmatrix} \qquad z = 1$$

and the phase-variable representation, the feedback gain matrix required to yield an

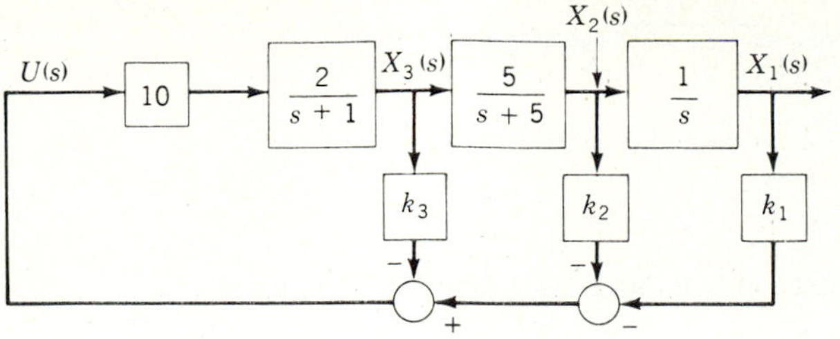

FIGURE 16.9
Optimal feedback control system for Example 2.

optimal closed-loop performance is $\mathbf{k}_c^T = [1 \quad 1.377 \quad 0.954]$. Determine $\mathbf{k}_p$ required for the physical-variable representation of this system as shown in Fig. 16.9. The plant matrices and control vectors for the phase- and physical-variable representations are, respectively,

Phase-variable representation:

$$\mathbf{A}_c = \begin{bmatrix} 0 & 1 & 0 \\ 0 & 0 & 1 \\ 0 & -5 & -6 \end{bmatrix} \qquad \mathbf{b}_c = \begin{bmatrix} 0 \\ 0 \\ 100 \end{bmatrix}$$

Physical-variable representation:

$$\mathbf{A}_p = \begin{bmatrix} 0 & 1 & 0 \\ 0 & -5 & 5 \\ 0 & 0 & -1 \end{bmatrix} \qquad \mathbf{b}_p = \begin{bmatrix} 0 \\ 0 \\ 20 \end{bmatrix}$$

The subscript p indicates the system representation in terms of the physical variables.

The change from phase variables to physical variables is accomplished by means of a similarity transformation $\mathbf{x} = \mathbf{T}^{-1}\mathbf{x}_p$. Using this transformation, the state equation in phase-variable form, $\dot{\mathbf{x}} = \mathbf{A}_c\mathbf{x} + \mathbf{B}_c\mathbf{u}$, becomes

$$\dot{\mathbf{x}}_p = \mathbf{T}\mathbf{A}_c\mathbf{T}^{-1}\mathbf{x}_p + \mathbf{T}\mathbf{b}_c\mathbf{u} = \mathbf{A}_p\mathbf{x}_p + \mathbf{b}_p\mathbf{u} \tag{16.50}$$

Using the method of Sec. 5.13 or simply equating the matrices of Eq. (16.50) yields the required conditions $\mathbf{A}_c\mathbf{T}^{-1} = \mathbf{T}^{-1}\mathbf{A}_p$ and $\mathbf{b}_c = \mathbf{T}^{-1}\mathbf{b}_p$. Equating the elements on both sides of these equalities gives the transformation matrix

$$\mathbf{T}^{-1} = \begin{bmatrix} 1 & 0 & 0 \\ 0 & 1 & 0 \\ 0 & -5 & 5 \end{bmatrix}$$

Thus, from Eq. (16.43),

$$\mathbf{k}_p^T = \mathbf{k}_c^T\mathbf{T}^{-1} = [1 \quad -3.393 \quad 4.770]$$

16.9 BODE DIAGRAM ANALYSIS OF OPTIMAL STATE-VARIABLE-FEEDBACK SYSTEMS[14,19]

Optimal state-variable-feedback systems can be designed by a frequency-domain method. The basis for this design method is the condition of optimality represented by the Kalman equation.[13] This equation represents the Riccati equation in which $\mathbf{k}$ appears explicitly and the Riccati matrix $\mathbf{P}$ is eliminated. The form of

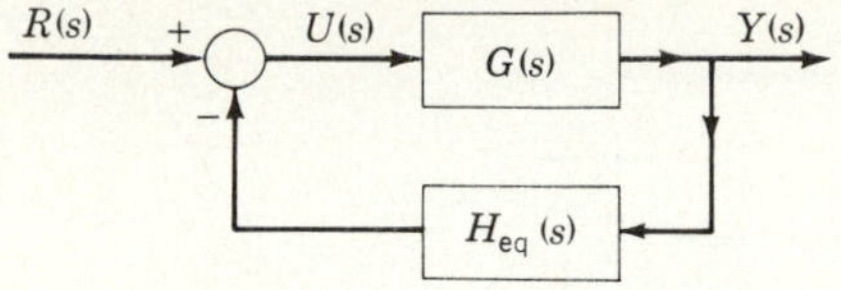

FIGURE 16.10
Equivalent feedback system.

the equation which is used in this approach for a SISO system is

$$|1 + \mathbf{k}^T\mathbf{\Phi}(s)\mathbf{b}|^2 = 1 + \frac{1}{z}|G_K(s)|^2 \tag{16.51}$$

where $s = j\omega$, $\mathbf{\Phi}(s) = [s\mathbf{I} - \mathbf{A}]^{-1}$, and

$$G_K(s) = \mathbf{h}^T\mathbf{\Phi}(s)\mathbf{b} \tag{16.52}$$

As indicated in Sec. 16.7, the **Q** matrix can be decomposed into the matrix product $\mathbf{Q} = \mathbf{H}\mathbf{H}^T$. The matrix $\mathbf{H}\mathbf{H}^T$ is limited to having a rank of 1. Hence, without loss of generality, $\mathbf{Q} = \mathbf{h}\mathbf{h}^T$, and the scalar $y_h = \mathbf{h}^T\mathbf{x}$ is an effective system output which appears in the minimization since†

$$\text{PI} = \int_0^\infty (\mathbf{x}^T\mathbf{h}\mathbf{h}^T\mathbf{x} + zu^2)\, dt = \int_0^\infty \left(y_h^2 + zu^2\right) dt \tag{16.53}$$

Once **Q** and z are specified, the only unknown quantity in Eq. (16.51) is **k**. Equation (16.51) is applicable to the linear system

$$\dot{\mathbf{x}} = \mathbf{A}\mathbf{x} + \mathbf{b}u \qquad \text{and} \qquad y = \mathbf{c}^T\mathbf{x} \tag{16.54}$$

Note that $G_K(s)$ represents a mathematical transfer function which is utilized in the graphical solution for the k_i.

The method presented in this section is an indirect graphical solution of the Riccati equation. When the $H_{eq}(s)$ representation of the state-variable feedback as developed in Chap. 13 is used, the system is represented in Fig. 16.10. The transfer functions are

$$G(s) = \frac{Y(s)}{U(s)} = \mathbf{c}^T\mathbf{\Phi}(s)\mathbf{b} \tag{16.55}$$

$$H_{eq}(s) = \frac{\mathbf{k}^T\mathbf{X}(s)}{Y(s)} = \frac{\mathbf{k}^T\mathbf{X}(s)}{\mathbf{c}^T\mathbf{X}(s)} = \frac{\mathbf{k}^T\mathbf{\Phi}(s)\mathbf{b}}{\mathbf{c}^T\mathbf{\Phi}(s)\mathbf{b}} \tag{16.56}$$

$$G(s)H_{eq}(s) = \mathbf{k}^T\mathbf{\Phi}(s)\mathbf{b} \tag{16.57}$$

When these relationships are used, Eq. (16.51) becomes

$$|1 + G(s)H_{eq}(s)|^2 = 1 + \frac{1}{z}|G_K(s)|^2 \tag{16.58}$$

---

† The selection $\mathbf{h} = \mathbf{c}$ restricts the design problem and results in $G_K(s) = \mathbf{c}^T\mathbf{\Phi}(s)\mathbf{b} = G(s)$. This is not a necessary condition for use of the Bode method.

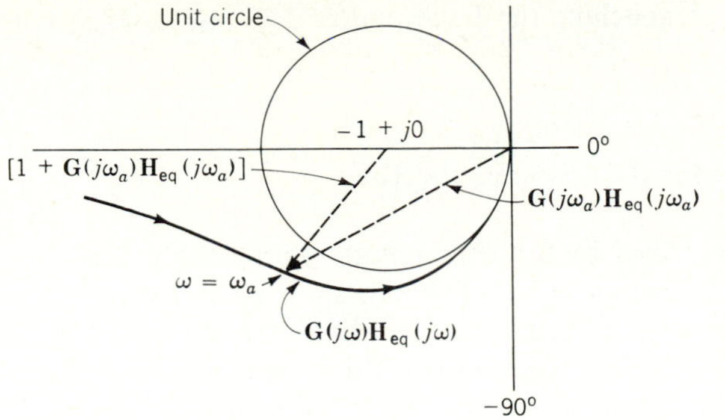

FIGURE 16.11
Polar plot of an optimal system.

For $s = j\omega$ and $0 \leq \omega < \infty$, the function $|\mathbf{G}_K(j\omega)|^2/z > 0$. Thus, from Eq. (16.58),

$$\left|1 + \mathbf{G}(j\omega)\mathbf{H}_{\text{eq}}(j\omega)\right| > 1 \tag{16.59}$$

Based on the condition of Eq. (16.59), the polar plot of $\mathbf{G}(j\omega)\mathbf{H}_{\text{eq}}(j\omega)$ must remain outside of a unit circle centered on the $-1 + j0$ point, for all values of frequency. A typical polar plot of $\mathbf{G}(j\omega)\mathbf{H}_{\text{eq}}(j\omega)$ is shown in Fig. 16.11. This leads to the interesting observation that many of the control systems designed by the conventional methods of Chap. 10 are not optimal according to the quadratic PI. In other words, the methods of Chap. 10 are satisfactory design methods which yield desired conventional figures of merit. However, when an optimal performance (quadratic PI) is desired, the methods presented in this and succeeding sections may be used. Although the design is designated as optimal, the time responses may not satisfy the system specifications. The method of combining these properties is presented in the following sections.

Properties of Bode Design

The Bode design method utilizes the following properties of Eq. (16.58), where properties 3 to 5 *are based upon the plant having at least one pure integration and a pole-zero excess of* 1 *or more*, i.e., $n \geq w + 1$.

1. The poles of $1 + G(s)H_{\text{eq}}(s)$ are known.
2. The zeros of $1 + G(s)H_{\text{eq}}(s)$ are the poles of the closed-loop system, which must be stable, and therefore they all lie in the left-half s plane.
3. For small ω, when $|\mathbf{G}_K(j\omega)/\sqrt{z}| \gg 1$, $|1 + \mathbf{G}(j\omega)\mathbf{H}_{\text{eq}}(j\omega)| \approx |\mathbf{G}_K(j\omega)/\sqrt{z}|$.
4. For large ω, when $|\mathbf{G}_K(j\omega)/\sqrt{z}| \ll 1$, $|1 + \mathbf{G}(j\omega)\mathbf{H}_{\text{eq}}(j\omega)| \approx 1$.
5. When $|\mathbf{G}_K(j\omega)/\sqrt{z}| = 1$, at $\omega = \omega_\phi$, $|1 + \mathbf{G}(j\omega_\phi)\mathbf{H}_{\text{eq}}(j\omega_\phi)| = \sqrt{2}$.

The design method consists of matching the Bode plot of $|1 + \mathbf{G}(j\omega)\mathbf{H}_{eq}(j\omega)|$ with:

1. The Bode plot of $|\mathbf{G}_K(j\omega)/\sqrt{z}|$ for $|\mathbf{G}_K(j\omega)/\sqrt{z}| > 1$.
2. Unity gain (the 0-dB line) for $|\mathbf{G}_K(j\omega)/\sqrt{z}| < 1$.

This matching can be accomplished by utilizing a Butterworth polynomial (see Table 16.2) with the corner frequency ω_ϕ at unity crossover for $|\mathbf{G}_K(j\omega)/\sqrt{z}|$ (see property 5).[20] A possible area of further research investigation would be to use another polynomial, such as one of the family of the Chebyshev[21] polynomials. The design procedure, which does not require the actual drawing of the Bode plots, is as follows:

Step 1. Determine the transfer function $G_K(s)$. If necessary, increase the gain so that the crossover frequency ω_ϕ is greater than all of the corner frequencies of $G_K(s)$ (see step 4).

Step 2. Synthesize the transfer function

$$T_s(s) = \left[\frac{G_K(s)}{\sqrt{z}}\right] B_b(s) \tag{16.60}$$

where $B_b(s)$ is a Butterworth polynomial having the order α, equal to the pole-zero excess of $G_K(s)$, and with $\omega_0 = \omega_\phi$. Some Butterworth polynomials are listed in Table 16.2. The term ω_0^α must be factored from these polynomials to obtain the standard form $B_b(s)$ with the constant term equal to unity. The following generalized equation can be employed to extend this table for values of $\alpha > 8$:

$$B_b(s) = \left(\frac{s}{\omega_0}\right)^\alpha + V_{\alpha-1}\left(\frac{s}{\omega_0}\right)^{\alpha-1} + \cdots + V_1\frac{s}{\omega_0} + 1 \tag{16.61}$$

where $$V_{\alpha-1} = \frac{1}{\sin(\pi/2\alpha)} \qquad V_{\alpha-2} = \frac{V_{\alpha-1}\cos(\pi/2\alpha)}{\sin(\pi/\alpha)}$$

$$V_{\alpha-\beta} = \frac{V_{\alpha-\beta+1}\cos[(\beta-1)(\pi/2\alpha)]}{\sin(\beta\pi/2\alpha)}$$

Choosing a Butterworth polynomial having $\omega_0 = \omega_\phi$ results in the polynomial having a value of $\sqrt{2}$ at unity crossover of $|G_K(j\omega)/\sqrt{z}|$, thus satisfying property 5. T_s is the *synthesized transfer function*.

Step 3. Determine $1 + G(s)H_{eq}(s)$ using Eq. (16.57). All the parameters of this expression except the feedback coefficients $\mathbf{k}$ are known.

Step 4. Set $T_s(s)$ from Eq. (16.60) equal to $1 + G(s)H_{eq}(s)$. Express the numerator and denominator polynomials of both sides of this equality so that the coefficient of the highest power of s in each polynomial is unity. In order to satisfy the equality, the gain constants on each side of the equation must be identical. Note that the gain constant of $1 + G(s)H_{eq}(s)$ is unity. If the plant gain is specified, the resultant gain constant of Eq. (16.60) contains only one unknown, that is, ω_ϕ, and thus ω_ϕ can be determined. If its value is greater than all the corner frequencies of $\mathbf{G}_K(j\omega)$, and if $G_K(s)$ is a Type 1 or higher function, properties 1 through 5 are satisfied. In the event the value of ω_ϕ is not large enough (or the plant gain is not specified) the plant gain must be increased (or made sufficiently large) to ensure that the resultant value of ω_ϕ will be greater than all the corner frequencies of $\mathbf{G}_K(j\omega)$. The result of this procedure is that the values of $\mathbf{G}_K(j\omega)$ at all the corner frequencies are above the 0-dB axis of the Bode plot. Thus

1. For $\omega < \omega_\phi$, the Bode diagram of $|1 + \mathbf{G}(j\omega)\mathbf{H}_{eq}(j\omega)|$ is identical with that of $|\mathbf{G}_K(j\omega)/\sqrt{z}\,|$ (in accordance with property 3).
2. For $\omega > \omega_\phi$, the Bode diagram of $|1 + \mathbf{G}(j\omega)\mathbf{H}_{eq}(j\omega)|$, by use of property 4, coincides with the 0-dB line.
3. The degree of the numerator and denominator polynomials of the synthesized transfer function must be equal to the degree of the numerator and denominator polynomials, respectively, of the exact transfer function.

Step 5. By equating the synthesized transfer function T_s of step 4 to the exact transfer function of step 3 the unknown values of k_i are determined.

Example. The plant of the example in Sec. 13.8 is used to illustrate the above procedure to solve for $\mathbf{k}_p$ of Fig. 13.7*b*.

Step 1. Assume that $z = 1$ and $\mathbf{h}_p^T = [1 \quad 1 \quad 1]$. Use the matrix $\mathbf{A}_p$ in Example 2, Sec. 16.8, to obtain $\Phi(s)$. Then

$$G_K(s) = \mathbf{h}_p^T\Phi(s)\mathbf{b} = [1 \quad 1 \quad 1]\begin{bmatrix} \frac{1}{s} & \frac{1}{s(s+5)} & \frac{5}{s(s+1)(s+5)} \\ 0 & \frac{1}{s+5} & \frac{5}{(s+1)(s+5)} \\ 0 & 0 & \frac{1}{s+1} \end{bmatrix}\begin{bmatrix} 0 \\ 0 \\ 2A \end{bmatrix}$$

$$= \frac{2A(s^2+10s+5)}{s(s+1)(s+5)} = \frac{2A(s/9.47+1)(s/0.53+1)}{s(s+1)(0.2s+1)} \tag{16.62}$$

Step 2. Since the pole-zero excess of $G_K(s)$ is 1, the synthesized T_s representing the approximate transfer function of $1 + G(s)H_{eq}(s)$ is obtained by multiplying $G_K(s)$ by a first-order polynomial $B_b(s)$, obtained from Table 16.2 after dividing by ω_0; thus

$$T_s(s) = 1 + G(s)H_{eq}(s) = G_K(s)\left(\frac{s}{\omega_\phi} + 1\right)$$

$$= \frac{(2A/\omega_\phi)\left[s^3 + (10+\omega_\phi)s^2 + (10\omega_\phi + 5)s + 5\omega_\phi\right]}{s(s+1)(s+5)} \tag{16.63}$$

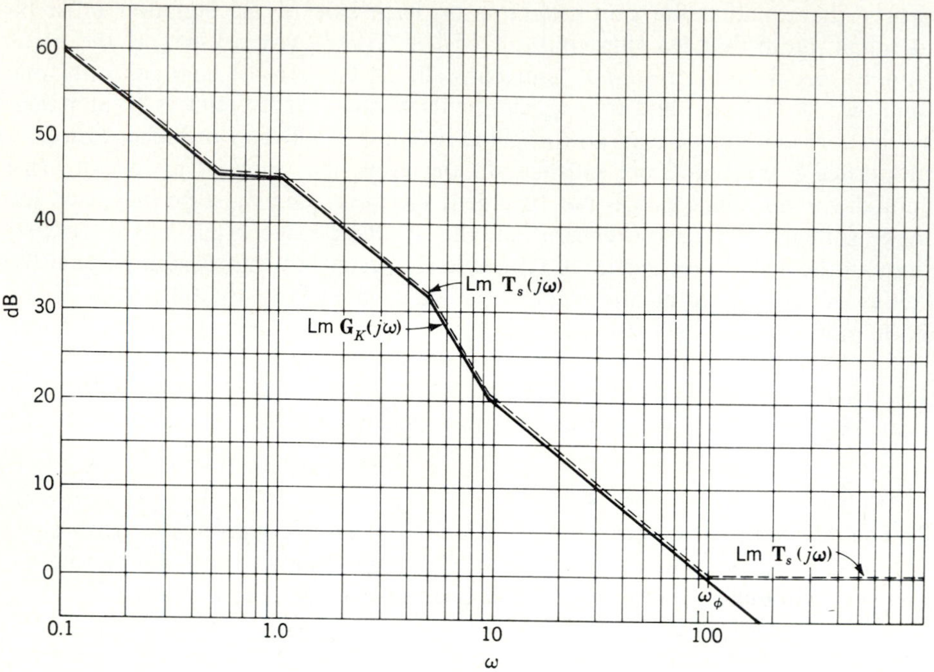

FIGURE 16.12
Plot of $\text{Lm}\,\mathbf{G}_K(j\omega)$ and the synthesized $\text{Lm}\,\mathbf{T}_s(j\omega)$.

Step 3. When Eq. (16.57) is used, the following exact expression of the characteristic polynomial is formed:

$$1 + G(s)H_{\text{eq}}(s) = \frac{s^3 + (6 + 2Ak_3)s^2 + [5 + 10A(k_3 + k_2)]s + 10Ak_1}{s(s+1)(s+5)} \tag{16.64}$$

Step 4. Equating the coefficient of the s^3 term of the numerators of Eqs. (16.63) and (16.64) yields $A = \omega_\phi/2$. Since the largest corner frequency of Eq. (16.62) is 9.47, selecting either $\omega_\phi > 9.47$ or $A > 9.47/2 = 4.735$ ensures that the values of $\mathbf{G}_K(j\omega)$ at all the corner frequencies lie above the 0-dB line. Using the value $A = 50$ yields $\omega_\phi = 100$. The Bode plot of $\mathbf{G}_K(j\omega)$ is drawn as the solid line in Fig. 16.12. The 0-dB crossover point is identified as ω_ϕ. A dashed line is drawn in Fig. 16.12 to match the curve of $\text{Lm}[1 + \mathbf{G}(j\omega)\mathbf{H}_{\text{eq}}(j\omega)]$ with that of $\text{Lm}\,\mathbf{G}_K(j\omega)$ for $\omega < \omega_\phi$. The dashed line is drawn along the 0-dB line for $\omega > \omega_\phi$, to represent the value of $\text{Lm}[1 + G(j\omega)H_{\text{eq}}(j\omega)]_{\omega > \omega_\phi} = 0$ dB. Since the break points are above the 0-dB line, two roots of the system characteristic equation are already known and are equal to the zeros of $G_K(s)$, namely, $s_1 = -9.47$ and $s_2 = -0.53$. The third root, $s_3 = -\omega_\phi = -100$, is given by the Butterworth polynomial, as seen from Eq. (16.64) (see Prob. 16.6).

Step 5. Equating the corresponding coefficients of Eq. (16.63) and (16.64), with $A = 50$ and $\omega_\phi = 100$, yields

$$6 + 100k_3 = 10 + \omega_\phi = 110 \qquad k_3 = 1.04$$

$$5 + 500(k_3 + k_2) = 10\omega_\phi + 5 = 1005 \qquad k_2 = 0.96$$

$$500k_1 = 500 \qquad k_1 = 1$$

For comparison, the $\mathbf{k}_p$ obtained by utilizing the Riccati computer program has the values $k_1 = 1$, $k_2 = 0.96263$, and $k_3 = 1.0371$. These values compare very favorably with those obtained by the graphical method for this example. These favorable results are assured by choosing a large value of A or ω_ϕ so that all the break points of $\mathbf{G}_K(j\omega)$ occur far enough above the 0-dB line. The closer the break points are to the 0-dB line, the less accurate the graphical solution for the k_i. An explanation of this requirement and the selection of ω_ϕ without drawing the Bode diagram is contained in Sec. 16.11.

16.10 ROOT-SQUARE-LOCUS[20] ANALYSIS OF OPTIMAL STATE-VARIABLE-FEEDBACK SYSTEMS[19]

In Secs. 12.3 and 13.2 the design procedures are based upon specifying the desired poles of the closed-loop transfer function (control ratio). This section presents a procedure whereby the poles of the state-variable-feedback system transfer function are determined from the synthetic transfer function $G_K(s)$ which comes from the Kalman equation (16.51). This analysis is based upon the root-square-locus (RSL) method.

When $s = j\omega$, Eq. (16.58) can be rewritten in the form

$$\left[1 + G(s)H_{\text{eq}}(s)\right]\left[1 + G(-s)H_{\text{eq}}(-s)\right] = 1 + \frac{G_K(s)G_K(-s)}{z} \tag{16.65}$$

In Sec. 13.2 it is shown that the poles and zeros of $1 + G(s)H_{\text{eq}}(s)$ for a state-feedback minimum-phase plane all lie in the left-half s plane (LHP) for positive gain. Also, its poles are the poles of $G(s)$. In a similar manner, the poles and zeros of $1 + G(-s)H_{\text{eq}}(-s)$ all lie in the right-half s plane (RHP). Further, the denominator polynomials on both sides of Eq. (16.65) must be identical, as seen in the example of the last section. As the function $1 + G_K(s)G_K(-s)/z$ is also a ratio of polynomials, the technique of *spectral factorization*[20] permits factoring it so that its poles and zeros can be separated into two groups, those lying in the LHP and those lying in the RHP. These two groups are denoted by

$$\left[1 + \frac{G_K(s)G_K(-s)}{z}\right]^L \qquad \text{and} \qquad \left[1 + \frac{G_K(s)G_K(-s)}{z}\right]^R$$

respectively. Correspondingly, utilizing the spectral factorization technique leads to

$$1 + G(s)H_{eq}(s) = \left[1 + \frac{G_K(s)G_K(-s)}{z}\right]^L \tag{16.66}$$

and

$$1 + G(-s)H_{eq}(-s) = \left[1 + \frac{G_K(s)G_K(-s)}{z}\right]^R \tag{16.67}$$

These equations are now valid in the entire s plane. The zeros of the term on the right side of Eq. (16.66) are the poles of the control ratio and are evaluated from the root square locus, which is presented in this section. Then the feedback coefficients k_i are determined by equating the coefficients of like powers of s in Eq. (16.66).

The zeros of $1 + G_K(s)G_K(-s)/z$ are determined by setting Eq. (16.65) equal to zero, which yields

$$\frac{G_K(s)G_K(-s)}{z} = -1 \tag{16.68}$$

This equation has the identical mathematical form as $G(s)H(s) = -1 = e^{-j(1+2h)\pi}$ used for the root-locus method of Chap. 7. Therefore, the root locus can be drawn for Eq. (16.68) to determine the zeros of the numerator polynomial of $1 + G_K(s)G_K(-s)/z$. The function $G_K(s)G_K(-s)/z$ is referred to as the associated or *squared transfer function*. Drawing the root locus for Eq. (16.68) is referred to as the *root-square locus* (RSL) method. Since $G_K(s)$ has n poles and w zeros, the magnitude and angle conditions for the RSL are, respectively,

$$\frac{|G_K(s)G_K(-s)|}{z} = 1 \tag{16.69}$$

$$\underline{/G_K(s)G_K(-s)} = \begin{cases} (1 + 2h)180^\circ & \text{for } n - w \text{ even} \\ (h)360^\circ & \text{for } n - w \text{ odd} \end{cases} \qquad h = 0, \pm 1, \pm 2, \ldots \tag{16.70}$$

Equation (16.69) is rewritten as

$$\mathbf{K}^2 = \frac{|s^m|^2 \cdot |s - p_1| \cdot |s + p_1| \cdot |s - p_2| \cdot |s + p_2| \cdots |s - p_u| \cdot |s + p_u|}{|s - z_1| \cdot |s + z_1| \cdots |s - z_w| \cdot |s + z_w|} \tag{16.71}$$

where $K > 0$ and K is the loop sensitivity. Note that the value of z is incorporated in K. The geometrical shortcuts for obtaining the RSL are summarized in Table 16.5. Only the branches in the second quadrant are drawn since the plot is symmetrical about both the real and imaginary axes. For large values of K the branches of the root square locus become straight lines approaching their asymptotes. *The value of K for which the dominant branch becomes essentially a straight line is defined as K_x.*

When $G_K(s)$ has poles only, the RSL for high forward gain (hfg) always has a pair of dominant complex-conjugate roots. When $G_K(s)$ has $w = n - 1$ zeros, the two asymptotes determined from Eq. (16.68) lie on the real axis. Thus, under hfg operation, the RSL of the system may have all real roots. A characteristic of

TABLE 16.5
Summary of RSL geometrical properties

Rule 1	Number of branches of the locus $= 2(m + u) = 2n$
Rule 2	Real-axis locus: if the total number of poles and zeros to the right of the s point on the real axis is odd, this point lies on the locus for $n - w =$ even number; if the total number of poles and zeros to the right of the s point on the real axis is even, this point lies on the locus for $n - w =$ odd number
Rule 3	Locus end points: locus starting points ($K = 0$) are the poles of $G_K(s)$; the locus ending points ($K = \infty$) are at zeros of $G_K(s)$ or at infinity
Rule 4	Asymptotes of the root square locus as s approaches infinity: $\gamma = \begin{cases} \dfrac{(1 + 2h)180°}{2(n - w)} & \text{for } n - w \text{ even} \\ \dfrac{(h)360°}{2(n - w)} & \text{for } n - w \text{ odd} \end{cases}$ where h $= 0, \pm 1, \pm 2$
Rule 5	Real-axis intercept: $\sigma_o = 0$
Rule 6	Breakaway point on the real axis: obtained from $dW/ds = 0$, where $W(s) = -K^2$ (or K^2)
Rule 7	Complex pole (or zero): angle of departure (approach angle): see rule 7 for root locus in Chap. 7; be sure to include the poles and zeros in the RHP when applying this rule
Rule 8	Imaginary-axis crossing point: none

the RSL plots is that as K becomes large, the poles and zeros of Eq. (16.68) or (16.69) can be represented by $2n - 2w$ poles located at the origin. Since the asymptotes emanate from the origin, the complex portions of the root locus, for $n - w > 1$, approach straight lines for $K > K_x$ and are essentially the same for any nth-order plant. The difference between the RSL of each nth-order plant is therefore the loop-sensitivity calibration. Also, for $K > K_x$, the $2(n - w)$ branches approach the Butterworth configuration for $2(n - w)$ poles.

Example. The example of Sec. 16.9 is reworked using the RSL method. Utilizing the transfer function in Eq. (16.62), for $z = 1$, yields

$$G_K(s)G_K(-s) = \frac{-10^4(s \pm 9.47)(s \pm 0.53)}{s^2(s \pm 1)(s \pm 5)} \tag{16.72}$$

Since $n - w$ is odd the constant in Eq. (16.72) is negative. Thus, the RSL satisfying the $(h)360°$ angle condition is obtained as shown in Fig. 16.13. For $K^2 = 10^4$ the roots of the characteristic equation are shown in the figure. Using these roots to form the characteristic polynomial yields $(s - s_1)(s - s_2)(s - s_3) = s^3 + 109.731s^2 + 1005.125s + 502.033$. The coefficients of this equation are equated to the corresponding coefficients of Eq. (16.64) to yield $k_1 = 1.0041$, $k_2 = 0.96294$, and $k_3 = 1.03731$. These values compare favorably with those obtained from the Bode and the Riccati computer solution methods. The advantages of the RSL are the same as for the root-locus method, i.e., there is a direct relationship between the s plane and the time domain. For the range of values of K^2 for which one dominant pair of roots

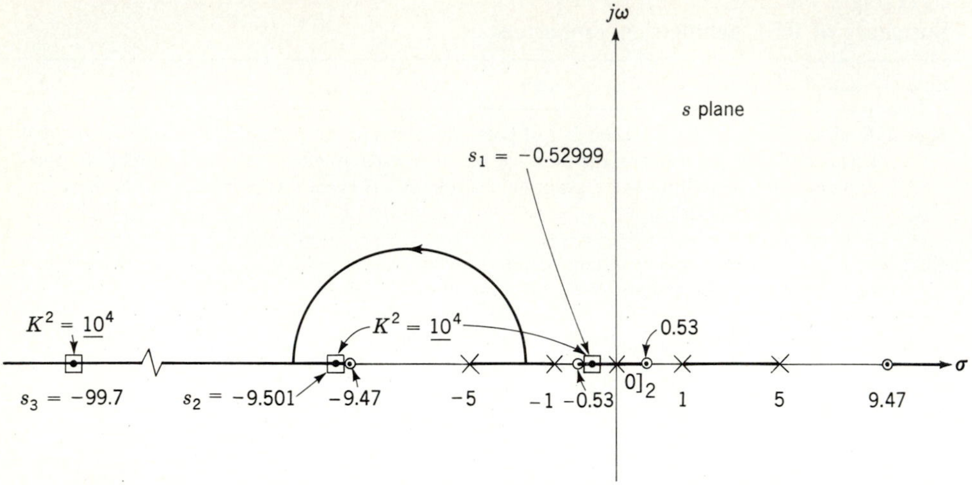

FIGURE 16.13
RSL for Eq. (16.72).

exists, the system loop sensitivity may be chosen to yield not only the optimal performance according to the quadratic PI but also satisfactory conventional figures of merit.

16.11 DETERMINATION OF HFG CONDITIONS

The solution for the feedback coefficients by the use of Eq. (16.51) is facilitated by using the phase-variable representation. Thus $G_K(s) = \mathbf{h}_c^T \mathbf{\Phi}(s)\mathbf{b}_c$ becomes

$$\frac{G_K(s)}{\sqrt{z}} = \frac{K_G}{\sqrt{z}} \frac{h_n s^{n-1} + \cdots + h_2 s + h_1}{s^n + a_{n-1}s^{n-1} + \cdots + a_0} \tag{16.73}$$

The pole-zero excess of $\mathbf{G}_K(j\omega)$ is exactly 1 for $h_n \neq 0$. Property 5 in Sec. 16.9 can be achieved by multiplying $\mathbf{G}_K(j\omega)$ by a first-order Butterworth polynomial, that is $T_s(s) = G_K(s)B_b(s)/\sqrt{z}$. Thus, using step 5 of Sec. 16.9

$$1 + G(s)H_{\text{eq}}(s) = \frac{G_K(s)B_b(s)}{\sqrt{z}} = T_s(s) \tag{16.74}$$

Using the form of $G(s)H_{\text{eq}}(s)$ given in Eq. (13.21a), this becomes

$$1 + K_G \frac{k_n s^{n-1} + \cdots + k_2 s + k_1}{s_n + a_{n-1}s^{n-1} + \cdots + a_1 s + a_0} = \frac{K_G}{\sqrt{z}}\left[\frac{h_n s^{n-1} + \cdots + h_1}{s^n + \cdots + a_0}\left(\frac{s}{\omega_\phi} + 1\right)\right] \tag{16.75}$$

The optimal closed-loop poles, represented by the zeros of the left-hand side of Eq. (16.75), asymptotically approach the w zeros of $G_K(s)$, while the remaining $n - w$ poles ($n - w = 1$ for $h_n \neq 0$) approach a Butterworth pattern. This is an important result due to Kalman.[13] Therefore Eq. (16.75) is a valid approximation for the case of hfg but is precisely true only in the limit. In other words,

$$|1 + G(s)H_{eq}(s)| \neq \frac{|G_K(s)(s/\omega_\phi + 1)|}{\sqrt{z}} \tag{16.76}$$

but for large gain,

$$\lim_{K_G \to \infty} |1 + G(s)H_{eq}(s)| = \frac{\lim_{K_G \to \infty} |G_K(s)(s/\omega_\phi + 1)|}{\sqrt{z}} \tag{16.77}$$

Equating the numerators of Eq. (16.75) gives

$$s^n + (a_{n-1} + K_G k_n)s^{n-1} + \cdots + (a_0 + K_G k_1)$$
$$= \frac{K_G}{\omega_\phi \sqrt{z}}\left[h_n s^n + (\omega_\phi h_n + h_{n-1})s^{n-1} + \cdots + (\omega_\phi h_2 + h_1)s + \omega_\phi h_1\right] \tag{16.78}$$

Equating coefficients in Eq. (16.78) yields

$$\omega_\phi = \frac{K_G h_n}{\sqrt{z}} \tag{16.79a}$$

$$k_1 = \frac{h_1}{\sqrt{z}} - \frac{a_0}{K_G} \tag{16.79b}$$

$$k_i = \frac{h_i}{\sqrt{z}} + \frac{h_{i-1}}{K_G h_n} - \frac{a_{i-1}}{K_G} \qquad i = 2, \ldots, n \tag{16.79c}$$

The first expression provides an opportunity to define a minimum gain for *reproducibility*. This is the gain above which the closed-loop poles become independent of the open-loop system parameters. This value of gain K_x can be determined from the RSL as discussed in Sec. 16.10 and in Chap. 17. In order to achieve zero steady-state error with a step input, k_1 must satisfy Eq. (13.26), $k_1 = c_0 - a_0/K_G$.

Consider the accuracy of the straight-line approximation of the Bode plot. Near the corner frequencies, significant corrections to the asymptotes are required. Thus, a rule of thumb for accuracy of the straight-line approximation to the Bode plot at crossover is

$$\omega_\phi \geq \xi \omega^* \tag{16.80}$$

where ω^* represents the highest corner frequency of $\mathbf{G}_K(j\omega)$ and $\xi > 1$ represents the value that yields the desired separation between ω^* and ω_ϕ. For example, with $\xi = 10$, all corner frequencies of $\mathbf{G}_K(j\omega)$ are at least 1 decade below the crossover frequency. This gives a minimum gain value for accurate pole placement and, equivalently, a minimum gain for use in Eqs. (16.79) for computing the feedback coefficients.

Substituting Eq. (16.79a) into Eq. (16.80) gives

$$K_G h_n \geq \xi \omega^* \sqrt{z} \tag{16.81}$$

First, consider a fixed value of K_G. Then Eq. (16.81) can be used to find the minimum value of h_n for which an exact pole placement is possible:

$$(h_n)_{\min} = \frac{\xi\omega^*\sqrt{z}}{K_G} \tag{16.82}$$

Alternatively, considering a given value of h_n, the minimum value of forward gain for exact pole placement is

$$(K_G)_{\min} = \frac{\xi\omega^*\sqrt{z}}{h_n} \tag{16.83}$$

It should be noted from Eq. (16.79) that under hfg operation

$$\mathbf{k}_c \approx \frac{\mathbf{h}_c}{\sqrt{z}} \tag{16.84}$$

Example. The example of Sec. 16.9 is utilized to illustrate this simpler approach of obtaining $\mathbf{k}_p$. This approach solves for $\mathbf{k}_c$, the phase-variable-feedback coefficients. In order to compare the results of the two methods, it is necessary to determine the $\mathbf{h}_c$ which is equivalent to $\mathbf{h}_p$. The transformation from physical to phase variable is performed by using $\mathbf{x}_p = \mathbf{T}\mathbf{x}$ and $\mathbf{b}_p = \mathbf{T}\mathbf{b}_c$. Using

$$\mathbf{b}_p = \begin{bmatrix} 0 \\ 0 \\ 100 \end{bmatrix} \qquad \mathbf{b}_c = \begin{bmatrix} 0 \\ 0 \\ 500 \end{bmatrix}$$

$$\mathbf{A}_p = \begin{bmatrix} 0 & 1 & 0 \\ 0 & -5 & 5 \\ 0 & 0 & -1 \end{bmatrix} \qquad \mathbf{A}_c = \begin{bmatrix} 0 & 1 & 0 \\ 0 & 0 & 1 \\ 0 & -5 & -6 \end{bmatrix}$$

yields

$$\mathbf{T} = \begin{bmatrix} 1 & 0 & 0 \\ 0 & 1 & 0 \\ 0 & 1 & 0.2 \end{bmatrix} \tag{16.85}$$

Substituting $\mathbf{x}_p = \mathbf{T}\mathbf{x}$ into $\mathbf{x}_p^T\mathbf{Q}_p\mathbf{x}_p = \mathbf{x}_p^T\mathbf{h}_p\mathbf{h}_p^T\mathbf{x}_p$ yields

$$\mathbf{x}^T\mathbf{T}^T\mathbf{h}_p\mathbf{h}_p^T\mathbf{T}\mathbf{x} = \mathbf{x}^T\mathbf{h}_c\mathbf{h}_c^T\mathbf{x}$$

Since $\mathbf{h}_p^T = [1 \quad 1 \quad 1]$,

$$\mathbf{h}_c^T = \mathbf{h}_p^T\mathbf{T} = [1 \quad 2 \quad 0.2] \tag{16.86}$$

For the transfer function

$$G(s) = \frac{10A}{s(s+1)(s+5)} = \frac{K_G}{s^3 + 6s^2 + 5s} \tag{16.87}$$

where $K_G = 500$, Eqs. (16.79*a*) to (16.79*c*) yield

$$\omega_\phi = 500 \times 0.2 = 100 \qquad k_1 = 1$$

$$k_2 = 2 + \frac{1}{(500)(0.2)} - \frac{5}{500} = 2 \qquad k_3 = 0.2 + \frac{2}{(500)(0.2)} - \frac{6}{500} = 0.208$$

The optimal control is $\mathbf{u}^* = -\mathbf{k}_c^T\mathbf{x} = -\mathbf{k}_c^T\mathbf{T}^{-1}\mathbf{x}_p = -\mathbf{k}_p^T\mathbf{x}_p$. Thus

$$\mathbf{k}_p^T = \mathbf{k}_c^T\mathbf{T}^{-1} = [1 \quad 2 \quad 0.208]\begin{bmatrix} 1 & 0 & 0 \\ 0 & 1 & 0 \\ 0 & -5 & 5 \end{bmatrix} = [1 \quad 0.96 \quad 1.04] \tag{16.88}$$

These agree with the values of the example in Sec. 16.9. Note that for $K_G = 500$, $\mathbf{k}_c \approx \mathbf{h}_c$, and $\mathbf{k}_p^T = \mathbf{k}_c^T\mathbf{T}^{-1} \approx \mathbf{h}_c^T\mathbf{T}^{-1} \approx \mathbf{h}_p^T\mathbf{T}\mathbf{T}^{-1} = \mathbf{h}_p^T$. For this example $\mathbf{k}_p^T \approx \mathbf{h}_p^T = [1 \quad 1 \quad 1]$.

16.12 SUMMARY

This chapter establishes the concept of using a performance index which yields an optimal system response. Typical PIs, in particular the quadratic PI, are discussed. The Riccati equation is developed, based on setting the quadratic PI equal to a Liapunov function $\mathbf{x}^T\mathbf{P}\mathbf{x}$. This equation is solved for the Riccati matrix $\mathbf{P}$, which in turn permits the determination of the optimal control law $\mathbf{u}^*(\mathbf{x}) = -\mathbf{K}\mathbf{x}$. Design methods are presented that satisfy the conditions for optimality and simultaneously yield satisfactory time-response characteristics. For single-input single-output (SISO) systems, solution of the optimal linear-regulator problem may also be effected by using the frequency-response and root-square-locus design methods. Chapter 17 presents a design method for selecting the matrix $\mathbf{Q}$ of the quadratic PI so that the conventional figures of merit can be achieved.

REFERENCES

1. Flügge-Lotz, I.: "The Importance of Optimal Control for the Practical Engineer," *Automatica*, pp. 749–753, November 1970.
2. Graham, D., and R. C. Lathrop: "The Synthesis of Optimum Response: Criteria and Standard Forms," *Trans. AIEE*, vol. 72, pt. II, pp. 273–288, November 1953.
3. Stout, T. M.: "A Note on Control Area," *J. Appl. Phys.*, vol. 21, pp. 1129–1131, November 1950.
4. Nims, P. T.: "Some Design Criteria for Automatic Controls," *Trans. AIEE*, vol. 70, pt. II, pp. 606–611, 1951.
5. Schultz, W. C., and V. C. Rideout: "The Selection and Use of Servo Performance Criteria," *Trans. AIEE*, vol. 76, pt. II, pp. 383–388, January 1958.
6. King, L. H.: "Reduction of Forced Error in Closed-Loop Systems," *Proc. IRE*, vol. 41, pp. 1037–1042, August 1953.
7. Oldenbourg, R. C., and H. Sartorius: *The Dynamics of Automatic Controls*, trans. and ed. by H. L. Mason, American Society of Mechanical Engineers, New York, 1948.
8. Whitely, A. L.: "The Theory of Servo Systems, with Particular Reference to Stabilization," *J. Inst. Elec. Eng.* (*Lond.*), vol. 93, pp. 353–377, 1946.
9. Butterworth, S.: "On the Theory of Filter Amplifiers," *Wireless Eng.*, vol. 7, pp. 536–541, 1930.
10. Kirk, D. E.: *Optimum Control Theory*, Prentice-Hall, Englewood Cliffs, N.J., 1970.
11. *IEEE Trans. Autom. Control*, vol. AC-16, Special Issue on Linear-Quadratic-Gaussian Problem, December 1971.
12. Athans, M., and P. L. Falb: *Optimal Control*, McGraw-Hill, New York, 1966.
13. Kalman, R. E.: "When is a Linear Control System Optimal?," *ASME J. Basic Eng.*, vol. 86, pp. 51–60, March 1964.
14. Schultz, D. G., and J. L. Melsa: *State Functions and Linear Control Systems*, McGraw-Hill, New York, 1967.
15. "Solution of Linear-Quadratic Optimal Control Problem (OPTCON)," Air Force Institute of Technology Computer Program Library, 1974.
16. Anderson, B. D. O., and J. B. Moore: *Linear Optimal Control*, Prentice-Hall, Englewood Cliffs, N.J., 1971.

17. Solheim, O. A.: "Design of Optimal Control Systems with Prescribed Eigenvalues," *Int. J. Control*, vol. 15, no. 1, pp. 143–160, 1972.
18. Woodhead, M. A., and B. Porter: "Optimal Modal Control," *Measurement Control*, vol. 6, pp. 301–303, July 1973.
19. Leake, R. J.: "Analytical Controller Design and the Servomechanism Problem," Syracuse Univ. Dept. Elec. Eng. Tech. Rep. 64-4, July 1964; "Return Difference Bode Diagram for Optimal System Design," *IEEE Trans. Autom. Control*, vol. AC-10, no. 3, pp. 342–344, July 1965.
20. Chang, S. S. L.: *Synthesis of Optimal Control Systems*, McGraw-Hill, New York, 1961.
21. Van Valkenburg, M. E.: *Introduction to Modern Network Synthesis*, Wiley, New York, 1960.
22. Willems, J. C.: "Least Squares Optimal Control and the Algebraic Riccati Equation," *IEEE Trans. Autom. Control*, vol. AC-16, December 1971.
23. Molinari, B. P.: "The Stable Regulator Problem and Its Inverse," *IEEE Trans. Autom. Control*, vol. AC-18, pp. 454–459, October 1973.
24. Bryson, A. E., and Y. C. Ho: *Applied Optimal Control*, Hemisphere Pub. Corp., Washington, 1975.
25. Thompson, P. M.: *User's Guide to Program CC, Version 3*, Systems Technology, Inc., Hawthorne, CA, 1985.
26. Horowitz, I. M.: "Improved Design Technique for Uncertain Multiple Input–Output Systems," *Int. J. Control*, vol. 36, pp. 977–988, 1982.

CHAPTER 17

OPTIMAL DESIGN BY USE OF QUADRATIC PERFORMANCE INDEX

17.1 INTRODUCTION[1,2]

The optimal control theory associated with the quadratic PI and the Riccati equation is developed in Chap. 16. In order to determine the state-feedback vector **k**, a weighting matrix **Q** and a weighting value z of the PI must be chosen. In that chapter these weighting quantities are assumed to be known in order to proceed with the presentation of the material. This chapter presents a method for choosing these weighting quantities for a linear time-invariant system in terms of *phase variables* for single-input single-output systems. Then the feedback coefficients for the physical-variable representation can be determined by means of a transformation matrix. Other approaches for determining these weighting quantities are available in the literature.[12]

17.2 BASIC SYSTEM

The controllable and observable plant having the transfer function

$$G(s) = \frac{K_G(s^w + c_{w-1}s^{w-1} + \cdots + c_0)}{s^n + a_{n-1}s^{n-1} + \cdots + a_0} \tag{17.1}$$

is represented in phase variables by

$$\dot{\mathbf{x}} = \mathbf{A}_c\mathbf{x} + \mathbf{b}_c\mathbf{u} \tag{17.2}$$

$$\mathbf{y} = \mathbf{c}_c^T\mathbf{x} \tag{17.3}$$

where
$$\mathbf{A}_c = \begin{bmatrix} 0 & 1 & 0 & \cdots & 0 \\ 0 & 0 & 1 & \cdots & 0 \\ \cdots & \cdots & \cdots & \cdots & \cdots \\ -a_0 & -a_1 & -a_2 & \cdots & -a_{n-1} \end{bmatrix} \tag{17.4}$$

$$\mathbf{b}_c^T = [0 \quad 0 \quad 0 \quad \cdots \quad 0 \quad K_G] \qquad \text{for} \qquad K_G > 0 \tag{17.5}$$

$$\mathbf{c}_c^T = [c_0 \quad c_1 \quad c_2 \quad \cdots \quad c_{w-1} \quad c_w \quad 0 \quad \cdots \quad 0] \tag{17.6}$$

where $c_w = 1$. For most practical systems $n > w$, thus the number of poles n of the plant is greater than the number of zeros w. The zeros of Eq. (17.1) affect only the output matrix $\mathbf{c}_c^T$ as shown in Eq. (17.6). The block diagram of the phase-variable representation is shown in Fig. 5.18, and the signal flow graph is shown in Fig. 5.30.

Note that for *an all-pole plant* $c_0 = 1$ and $c_1 = \cdots = c_w = 0$. In addition, for a Type 1 plant $a_0 = 0$, and for a Type 2 plant $a_1 = a_0 = 0$.

17.3 QUADRATIC PERFORMANCE INDEX

A suitable quadratic PI for a single-input single-output system is

$$\mathrm{PI} = \int_0^\infty (\mathbf{x}^T\mathbf{Q}\mathbf{x} + zu^2)\,dt \tag{17.7}$$

where z is a scalar and $\mathbf{Q}$ is at least a positive semidefinite symmetric matrix. The algebraic Riccati equation for this PI, from Eq. (16.47), is

$$\mathbf{PA} + \mathbf{A}^T\mathbf{P} - \frac{1}{z}[\mathbf{Pbb}^T\mathbf{P}] = -\mathbf{Q} \tag{17.8}$$

where $\mathbf{P}$ is the symmetric positive definite Riccati matrix. If $\mathbf{Q}$ is positive semidefinite, Eq. (16.28) must be satisfied in order for $\mathbf{P}$ to be positive definite. If $\mathbf{Q}$ is positive definite, then $\mathbf{P}$ is positive definite. The resulting optimal feedback

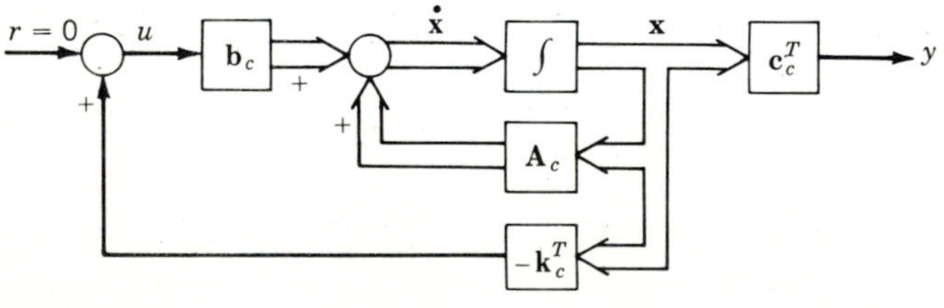

FIGURE 17.1
Optimal linear regulator system.

control law for the phase variable representation is

$$u^*(\mathbf{x}) = -z^{-1}\mathbf{b}^T\mathbf{P}\mathbf{x} = -\mathbf{k}_c^T\mathbf{x} \tag{17.9}$$

The elements of the vector $\mathbf{k}_c^T$ are the feedback coefficients (see Chaps. 13 and 16), and Eq. (17.9) indicates that all the state variables are to be fed back. This optimal system is represented by Fig. 17.1, with $r(t) = 0$, and is referred to as the optimal linear regulator system. In Eq. (17.8), for the phase-variable representation,

$$\mathbf{PA}_c = \begin{bmatrix} p_{11} & p_{12} & \cdots & p_{1n} \\ p_{12} & p_{22} & \cdots & p_{2n} \\ \cdots & \cdots & \cdots & \cdots \\ p_{1n} & p_{2n} & \cdots & p_{nn} \end{bmatrix} \begin{bmatrix} 0 & 1 & 0 & \cdots & 0 \\ 0 & 0 & 1 & \cdots & 0 \\ \cdots & \cdots & \cdots & \cdots & \cdots \\ -a_0 & -a_1 & -a_2 & \cdots & -a_{n-1} \end{bmatrix}$$

$$= \left[\begin{array}{c|c} -a_0p_{1n} & \\ -a_0p_{2n} & \hat{g}_{ij} \\ \cdots & \\ -a_0p_{nn} & \end{array}\right] \tag{17.10}$$

$$\mathbf{b}_c\mathbf{b}_c^T = \begin{bmatrix} 0 & \cdots & 0 \\ \cdots & \cdots & \cdots \\ 0 & \cdots & 0 \\ 0 & \cdots & K_G^2 \end{bmatrix} \tag{17.11}$$

This is a matrix of order n, with all zero elements except for the nn element which is K_G^2.

$$\mathbf{Pb}_c\mathbf{b}_c^T = \begin{bmatrix} 0 & \cdots & 0 & K_G^2p_{1n} \\ 0 & \cdots & 0 & K_G^2p_{2n} \\ \cdots & \cdots & \cdots & \cdots \\ 0 & \cdots & 0 & K_G^2p_{nn} \end{bmatrix} \tag{17.12}$$

$$\mathbf{Pb}_c\mathbf{b}_c^T\mathbf{P} = \left[\begin{array}{c|c} K_G^2p_{1n}^2 & \\ K_G^2p_{2n}p_{1n} & \\ \vdots & m_{ij} \\ K_G^2p_{nn}p_{1n} & \end{array}\right] \tag{17.13}$$

$$\mathbf{Q} = \begin{bmatrix} q_{11} & & & \mathbf{0} \\ & q_{22} & & \\ & & \ddots & \\ \mathbf{0} & & & q_{nn} \end{bmatrix} \tag{17.14}$$

It has been shown[3] that the matrix $\mathbf{Q}$ used to produce the optimal state-feedback matrix $\mathbf{k}_c^T$ is not unique; i.e., many $\mathbf{Q}$ matrices produce the same $\mathbf{k}_c^T$. Since an equivalent diagonal $\mathbf{Q}$ exists for each $\mathbf{k}_c^T$, $\mathbf{Q}$ is considered diagonal in the following analysis as shown in Eq. (17.14). Adding the corresponding elements in the 1, 1 position on the left-hand side of Eq. (17.8) and equating their sum to $-q_{11}$ yields the quadratic equation

$$p_{1n}^2 + \frac{2a_0 z}{K_G^2} p_{1n} - \frac{q_{11} z}{K_G^2} = 0 \tag{17.15}$$

Thus

$$p_{1n} = \frac{1}{K_G}\left[-\frac{a_0 z}{K_G} \pm \sqrt{\left(\frac{a_0 z}{K_G}\right)^2 + q_{11} z} \right] \tag{17.16}$$

For any plant higher than Type 0, $a_0 = 0$ and Eq. (17.16) reduces to

$$p_{1n} = \pm \frac{\sqrt{q_{11} z}}{K_G} \tag{17.17}$$

The feedback matrix [see Eq. (16.46)] is given by

$$\mathbf{k}_c^T = z^{-1}\mathbf{b}_c^T\mathbf{P} = \left[\frac{K_G p_{1n}}{z} \quad \frac{K_G p_{2n}}{z} \quad \cdots \quad \frac{K_G p_{nn}}{z} \right] = [k_1 \quad k_2 \quad \cdots \quad k_n] \tag{17.18}$$

Thus, for a Type 0 system, Eqs. (17.16) to (17.18) yield

$$k_1 = \frac{K_G p_{1n}}{z} = -\frac{a_0}{K_G} \pm \sqrt{\left(\frac{a_0}{K_G}\right)^2 + \frac{q_{11}}{z}} \tag{17.19}$$

and, for a Type 1 or higher-order plant,

$$k_1 = \pm \sqrt{\frac{q_{11}}{z}} \tag{17.20}$$

Equation (17.16) explicitly relates the Riccati matrix element p_{1n} to the fixed plant parameters and to the weighting matrix element q_{11}. This also applies for the feedback coefficient k_1, Eq. (17.19), for a Type 0 plant. For a Type 1 or higher plant, the coefficient k_1, Eq. (17.20), is a function of only q_{11}. Equations (17.16) and (17.19) serve *as checks on the accuracy of the resulting Riccati matrix and feedback coefficients obtained from a digital-computer program.*

17.4 STEADY-STATE ERROR REQUIREMENTS

The steady-state error characteristics for a stable unity-feedback control system obtained from the conventional control theory for the three standard inputs (step, ramp, and parabolic) are discussed in Chap. 6. A corresponding analysis is made in this section for an optimal control system which simultaneously satisfies the quadratic PI and one or more of the steady-state error requirements of Table 13.2. Thus, the overall PI is required to meet both the steady-state requirement and the optimal PI requirement. The analysis of the steady-state requirement enhances the correlation between conventional and modern control theory.

For a system having an input $r(t)$, the control signal obtained from Fig. 17.1 is

$$u = r - \mathbf{k}_c^T\mathbf{x} = r + u^* \tag{17.21}$$

where u^* is the optimal control and u is the control necessary to satisfy both the steady-state and optimal requirements. The feedback control structure ($r = 0$) is used to obtain the optimal control u^* as a function of a feedback vector $\mathbf{k}_c^T$, where $\mathbf{k}_c^T$ is a function of the unspecified q_{ii}'s (with z fixed). When this structure is used, the additional constraint of zero steady-state error for a step input ($r \neq 0$) is imposed to obtain the overall control u. This additional constraint allows the determination of q_{11} (which is a "free" choice in the optimal criterion). The overall relationship for Fig. 17.1, see Prob. 13.1, is

$$\frac{Y(s)}{R(s)} = \mathbf{c}_c^T\left[s\mathbf{I} - \left(\mathbf{A}_c - \mathbf{b}_c\mathbf{k}_c^T\right)\right]^{-1}\mathbf{b}_c \tag{17.22}$$

Substituting Eqs. (17.4) to (17.6) into Eq. (17.22) yields

$$\frac{Y(s)}{R(s)} = \frac{K_G(s+b_1)(s+b_2)\cdots(s+b_w)}{s^n + (a_{n-1} + K_Gk_n)s^{n-1} + \cdots + (a_0 + K_Gk_1)} \qquad n > w$$

$$= \frac{K_G\left(s^w + c_{w-1}s^{w-1} + \cdots + c_0\right)}{s^n + (a_{n-1} + K_Gk_n)s^{n-1} + \cdots + (a_0 + K_Gk_1)} \tag{17.23}$$

where $c_0 = b_1 \cdot b_2 \cdots b_w$. Equation (17.23) yields the system characteristic equation

$$s^n + (a_{n-1} + K_Gk_n)s^{n-1} + \cdots + (a_0 + K_Gk_1) = 0 \tag{17.24}$$

A necessary condition for stability is that all the coefficients in Eq. (17.24) be positive, i.e.,

$$a_{i-1} + K_Gk_i > 0 \tag{17.25}$$

Substituting Eq. (17.19) into $a_0 + K_Gk_1 > 0$ yields

$$\sqrt{(a_0)^2 + \frac{K_G^2 q_{11}}{z}} > 0 \tag{17.26}$$

This inequality is satisfied by using the positive value of the square root in Eqs. (17.19) and (17.20).

The system error is defined as $e = r - y$. From this definition the requirements on k_1, k_2, and k_3 for step, ramp, and parabolic inputs, respectively, are developed in Chap. 13 (see Table 13.2). Remember that in order to achieve the steady-state error characteristics of a Type m system, as discussed in Chap. 13, the control ratio *must have a minimum of* $w = m - 1$ zeros.

Step Input: $R(s) = R_0/s$ (for $w \geq 0$)

The requirement for zero steady-state error for a step input, as established by Eq. (13.26), is

$$k_1 = c_0 - \frac{a_0}{K_G} \tag{17.27}$$

Substituting this value into Eq. (17.19) and rearranging yields

$$q_{11} = \left[c_0^2 - \left(\frac{a_0}{K_G}\right)^2\right] z \geq 0 \tag{17.28}$$

For $c_0 \gg a_0/K_G$, hfg operation, Eqs. (17.27), (17.16), and (17.28) yield, respectively,

$$k_1 \approx c_0 \tag{17.29}$$

$$p_{1n} = \frac{k_1 z}{K_G} \approx \frac{c_0 z}{K_G} \tag{17.30}$$

$$q_{11} \approx c_0^2 z > 0 \tag{17.31}$$

In order to achieve the Butterworth characteristics of Table 16.2, the forward gain K_G must be sufficiently high to produce roots on the asymptotic portions of the root square locus. It is assumed in the remainder of this chapter that the hfg value of K_G is determined so that Eqs. (17.29) to (17.31) are valid. For Type 1 or higher plants, $a_0 = 0$, and Eqs. (17.29) to (17.31) become exact. Since $y(t)_{ss} = 0$ for $c_0 = 0$, zero steady-state error requires that none of the zeros of the system transfer function lie at the origin. Note that for an all-pole plant, $c_0 = 1$. Stability and optimality considerations are discussed at the end of this section.

Ramp Input: $R(s) = R_1/s^2$ (for $w \geq 1$)

Zero steady-state error, with a step and a ramp input, requires that Eq. (13.31) be satisfied; that is, $k_2 = c_1 - a_1/K_G$. Also, Eq. (17.18) specifies that $K_G p_{2n}/z = k_2$. For a Type 1 system under hfg operation with $a/K_G \ll c_1$, the results are

$$k_2 \approx c_1 \tag{17.32}$$

$$p_{2n} \approx \frac{c_1 z}{K_G} \tag{17.33}$$

For a Type 2 system in which $a_1 = 0$, the results are $k_2 = c_1$, and $p_{2n} = c_1 z/K_G$.

Parabolic Input: $R(s) = R_2/s^3$ (for $w \geq 2$)

Zero steady-state error with a parabolic input requires that Eq. (13.26) be satisfied; that is, $k_3 = c_2 - a_2/K_G$. Also, $K_G p_{3n}/z = k_3$. Under hfg operation

$$k_3 \approx c_2 \tag{17.34}$$

$$p_{3n} \approx \frac{c_2 z}{K_G} \tag{17.35}$$

Steady-State Error Coefficients

See Table 13.1.

Stability and Optimality Considerations

The optimality considerations which follow are based on the additional condition that zero steady-state error is required for step, ramp, and parabolic inputs.

STEP INPUT. Under the conditions of Eq. (13.26), $w \geq 0$ and $c_0 \neq 0$, the system characteristic equation given by Eq. (17.24), becomes

$$s^n + (a_{n-1} + K_G k_n)s^{n-1} + \cdots + (a_1 + K_G k_2)s + K_G c_0 = 0 \quad (17.36)$$

Thus, a necessary condition for stability is $c_0 > 0$. Further, with q_{11} satisfying Eq. (17.28) and with all remaining q_{ii}'s ≥ 0, the matrix $\mathbf{Q}$ is either positive definite or positive semidefinite. With the rank requirement of Eq. (16.28) satisfied, the condition $c_0 > 0$ ensures both the stability and the optimality of the system.

RAMP INPUT. When Eqs. (13.26) and (13.31) are satisfied, and when $w \geq 1$, the characteristic equation given by Eq. (17.24) becomes

$$s^n + (a_{n-1} + K_G k_n)s^{n-1} + \cdots + (a_2 + K_G k_3)s^2 + K_G c_1 s + K_G c_0 = 0 \quad (17.37)$$

Thus, necessary conditions for stability are $c_0 > 0$ and $c_1 > 0$. With these requirements and conditions satisfied, stability and optimality of the closed-loop system are assured when the $\mathbf{Q}$ matrix is at least positive semidefinite and Eq. (17.33) is satisfied.

PARABOLIC INPUT. When Eqs. (13.26), (13.31), and (13.36) are all satisfied and with $w \geq 2$, the system characteristic equation, from Eq. (17.24), is

$$s^n + (a_{n-1} + K_G k_n)s^{n-1} + \cdots + (a_3 + K_G k_4)s^3 + K_G c_2 s^2 + K_G c_1 s + K_G c_0 = 0 \quad (17.38)$$

Thus, necessary conditions for stability are $c_0 > 0$, $c_1 > 0$, and $c_2 > 0$. Optimality of the system is assured with the same restraints on $\mathbf{Q}$ as listed for a step input.

Example 1. Given an all-pole plant with Eq. (17.28) satisfied,

$$G(s) = \frac{100}{s(s+1)(s+5)} \qquad \mathbf{Q} = \begin{bmatrix} 1 & 0 & 0 \\ 0 & 0.008 & 0 \\ 0 & 0 & 0.06 \end{bmatrix} \qquad \text{and} \qquad z = 1$$

The solution of the Riccati equation, Eq. (17.8), yields the optimal feedback matrix

$$\mathbf{k}_c^T = [1 \quad 0.693 \quad 0.218]$$

The resulting control ratio, from Eq. (17.23), is

$$\frac{Y(s)}{R(s)} = \frac{100}{s^3 + 27.8s^2 + 74.3s + 100}$$

For an input $r(t) = 10tu_{-1}(t)$, the result obtained from Eq. (13.32) is that the system has a steady-state error $e(t)_{ss} = 7.43$. This is expected for an all-pole plant.

Example 2. Given

$$G(s) = \frac{100(s+2)}{s(s+1)(s+5)} \qquad \mathbf{Q} = \begin{bmatrix} 4 & 0 & 0 \\ 0 & 0.5396 & 0 \\ 0 & 0 & 4 \times 10^{-6} \end{bmatrix} \qquad \text{and} \qquad z = 1$$

The solution of Eq. (17.8) yields the optimal feedback matrix

$$\mathbf{k}_c^T = [2 \quad 0.95 \quad 0.090346]$$

The resulting control ratio, from Eq. (17.23), is

$$\frac{Y(s)}{R(s)} = \frac{100(s+2)}{s^3 + 15.03s^2 + 100s + 200}$$

This system follows a step and ramp input with no steady-state error. For an input $r(t) = 10t^2u_{-1}(t)$ the result obtained from Eq. (13.37) is that the system has a steady-state error $e(t)_{ss} = 1$. There is no steady-state error with a step or a ramp input.

Special Cases of Constant Nonzero Steady-State Error

In Chap. 13 it is pointed out that for a ramp input a stable all-pole system has a finite nonzero steady-state error [see Eq. (13.32)]. Defining the maximum acceptable error as E_e leads to

$$e(t)_{ss} = \left(\frac{a_1}{K_G} + k_2\right)R_1 \le E_e \tag{17.39}$$

Normalizing (dividing by R_1) yields

$$e_N(t)_{ss} = \frac{e(t)_{ss}}{R_1} = \frac{a_1}{K_G} + k_2 \le \frac{E_e}{R_1} = E_{eN} \tag{17.40}$$

or

$$k_2 \le E_{eN} - \frac{a_1}{K_G} \tag{17.41}$$

For a stable system response to exist, $E_{eN} > 0$ is required. Proof of this inequality is left to the reader. Using the relationship $K_G p_{2n}/z = k_2$ in Eq. (17.41) yields

$$p_{2n} \le \frac{zE_{eN}}{K_G} - \frac{a_1 z}{K_G^2} \tag{17.42}$$

For a Type 2 plant, since $a_1 = 0$, Eqs. (17.41) and (17.42) reduce to

$$k_2 \le E_{eN} \qquad \text{and} \qquad p_{2n} \le \frac{zE_{eN}}{K_G}$$

Also, in Chap. 13 it is pointed out that for a parabolic input a stable system containing a pole-zero plant ($w = 1$) has a finite nonzero steady-state error, given by Eq. (13.37). Let the maximum acceptable error be given by

$$e(t)_{ss} = \frac{R_2(a_2 + K_G k_3)}{K_G c_0} \leq E_e \tag{17.43}$$

Normalizing (dividing by R_2) and rearranging yields

$$k_3 \leq c_0 E_{eN} - \frac{a_2}{K_G} \tag{17.44}$$

TABLE 17.1
Summary of results of Sec. 17.4 ($z = 1$)†

Input $r(t)$	n	$e(t)_{ss}$	Fixed elements of P and Q
		All-pole plant	
$R_0 u_{-1}(t)$	≥ 1	0	$p_{1n} = \frac{1}{K_G} - \frac{a_0}{K_G^2}$ $q_{11} = 1 - \frac{a_0^2}{K_G^2}$
$R_1 t u_{-1}(t)$	≥ 2	$\left(\frac{a_1}{K_G} + k_2\right) R_1 \leq E_e$ or $\frac{a_1}{K_G} + k_2 \leq E_{eN}$	$p_{1n} = \frac{1}{K_G} - \frac{a_0}{K_G^2}$ $q_{11} = 1 - \frac{a_0^2}{K_G^2}$ $p_{2n} = \frac{E_{eN}}{K_G} - \frac{a_1}{K_G^2}$
$\frac{R_2 t^2 u_{-1}(t)}{2}$		∞	
		Pole-zero plant	
$R_0 u_{-1}(t)$ One zero	≥ 1	0	$p_{1n} = \frac{c_0}{K_G} - \frac{a_0}{K_G^2}$ $q_{11} = c_0^2 - \frac{a_0^2}{K_G^2}$
$R_1 t u_{-1}(t)$ One zero	≥ 2	0	$p_{1n} = \frac{c_0}{K_G} - \frac{a_0}{K_G^2}$ $q_{11} = c_0^2 - \frac{a_0^2}{K_G^2}$ $p_{2n} = \frac{c_1}{K_G} - \frac{a_1}{K_G^2}$
$\frac{R_2 t^2 u_{-1}(t)}{2}$ One zero	≥ 3	$\frac{a_2 + K_G k_3}{c_0 K_G} R_2 \leq E_e$ or $\frac{a_2 + K_G k_3}{c_0 K_G} \leq E_{eN}$	$p_{1n} = \frac{c_0}{K_G} - \frac{a_0}{K_G^2}$ $q_{11} = c_0^2 - \frac{a_0^2}{K_G^2}$ $p_{2n} = \frac{c_1}{K_G} - \frac{a_1}{K_G^2}$ $p_{3n} = \frac{c_0 E_{eN}}{K_G} - \frac{a_2}{K_G^2}$
More than one zero	≥ 3	0	$p_{1n} = \frac{c_0}{K_G} - \frac{a_0}{K_G^2}$ $q_{11} = c_0^2 - \frac{a_0^2}{K_G^2}$ $p_{2n} = \frac{c_1}{K_G} - \frac{a_1}{K_G^2}$ $p_{3n} = \frac{c_2}{K_G} - \frac{a_2}{K_G^2}$

† The requirement on q_{11} must satisfy the condition $q_{11} \geq 0$.

TABLE 17.2
Number of unknown quantities and equations available for solution when satisfying steady-state error requirements

Order of system n	$r(t)$	Total number of equations available from Eq. (17.8)	Unknowns		
			P	Q	Total
1	Step	0	0	0	0
2	Step	2	2	1	3
	Ramp		1		2
3	Step	5	5	2	7
	Ramp		4		6
	Parabolic†		3		5
4	Step	9	9	3	12
	Ramp		8		11
	Parabolic†		7		10

† For pole-zero plant only.

Using the relationship $K_G p_{3n}/z = k_3$ results in

$$p_{3n} \leq \frac{zc_0 E_{eN}}{K_G} - \frac{a_2 z}{K_G^2} \tag{17.45}$$

Summary

Steady-state error requirements for step, ramp, and parabolic inputs to linear optimal control systems, based on the quadratic performance index, result in the specification of some of the elements of the feedback matrix $\mathbf{k}_c$. In turn, this specifies some elements of the Riccati matrix $\mathbf{P}$. These elements are specified in terms of the parameters of the plant and/or the steady-state error. Table 17.1 summarizes the results of this section for Type 0, 1, and 2 plants with the weighting factor $z = 1$. As shown, two, four, or six elements of the $\mathbf{P}$ matrix and one element of the $\mathbf{Q}$ matrix are specified in order to achieve the desired steady-state error requirements. For an all-pole plant the optimal control system cannot follow a parabolic input. Further, for a ramp input $r(t) = R_1 t u_{-1}(t)$ the value p_{2n} is a function of the slope R_1. For a plant with one zero and a parabolic input $r(t) = (R_2 t^2/2)u_{-1}(t)$ the value of p_{3n} is a function of R_2. Table 17.2 lists, for various order systems and inputs, the number of unknown elements in the matrices $\mathbf{P}$ and $\mathbf{Q}$ which exist in the Riccati equation after satisfying the steady-state error requirements. Depending on the number of requirements and the order of the system, it may be necessary to assume values for the remaining elements. Equations (13.31) and (17.28) for the all-pole plant and Eqs. (13.31), (13.37), and (17.19) for the pole-zero plant serve as additional checks on the accuracy of the resulting values (computed either by hand or computer) of the elements of the Riccati matrix and of the feedback coefficients.

17.5 EXACT CORRELATION (CONVENTIONAL VS. MODERN)

The next few sections are devoted to describing the correlation between the conventional design figures of merit and the optimal control PI, based on the quadratic cost function. A correlation is presented for the simple second-order plant,[4] and from the knowledge gained for this simple plant, a correlation is presented for higher-order plants upon the condition that z is *assumed to be unity* and that *the off-diagonal elements of the* **Q** *matrix* are taken to be zero in Eq. (17.8). The correlation is made between M_p, t_p, t_s, and ω_d and the elements q_{ii}, for $i > 1$. As shown in Sec. 17.4, the element q_{11} is specified exactly in terms of the fixed parameters of the plant when zero steady-state error is required for a step input to the system. Consider the Type 1 plant transfer function

$$G(s) = \frac{Y(s)}{U(s)} = \frac{K_G}{s(s + a_1)} \tag{17.46}$$

Phase variables are utilized, and q_{11} and q_{22} are unspecified. The solution of the Riccati equation obtained by equating diagonal elements on both sides of Eq. (17.8) and using Eq. (17.20) yields

$$k_1 = \sqrt{q_{11}} \tag{17.47}$$

$$k_2 = -\frac{a_1}{K_G} + \frac{a_1}{K_G}\sqrt{1 + \frac{2K_G k_1}{a_1^2} + q_{22}\frac{K_G^2}{a_1^2}} \tag{17.48}$$

The characteristic equation of the closed-loop optimal system is

$$s^2 + (a_1 + K_G k_2)s + K_G k_1 = s^2 + 2\zeta\omega_n s + \omega_n^2 \tag{17.49}$$

From the RSL analysis it can be shown that a stable closed-loop performance may be achieved with an indefinite **Q** matrix. This is also shown by deriving ζ of Eq. (17.49) in terms of q_{11} and q_{22}. From Eq. (17.49)

$$\omega_n = \sqrt{K_G k_1} \text{ rad/s} \tag{17.50}$$

and

$$2\zeta\omega_n = a_1 + K_G k_2 \tag{17.51}$$

Hence,

$$\zeta = \frac{a_1 + K_G k_2}{2\omega_n} = \frac{a_1 + K_G k_2}{2\sqrt{K_G k_1}} \tag{17.52}$$

From Eqs. (17.47) to (17.52) the following are obtained:

$$\omega_n = (q_{11})^{1/4}\sqrt{K_G} \tag{17.53}$$

$$\zeta = \frac{a_1\sqrt{1 + \left[2K_G(q_{11})^{1/2}\right]/a_1^2 + q_{22}K_G^2/a_1^2}}{2\sqrt{K_G}(q_{11})^{1/4}} \tag{17.54}$$

Let $K_1 = a_1/\sqrt{K_G} \geq 0$; then Eq. (17.54) becomes

$$\zeta = \frac{K_1}{2}\sqrt{\frac{1}{(q_{11})^{1/2}} + \frac{2}{K_1^2} + \frac{K_G}{K_1^2}\frac{q_{22}}{(q_{11})^{1/2}}} \tag{17.55}$$

Also, let

$$K_2 = \frac{q_{22}}{\sqrt{q_{11}}} \tag{17.56}$$

so that Eq. (17.55) becomes

$$\zeta = \frac{1}{2}\sqrt{\frac{K_1^2}{(q_{11})^{1/2}} + 2 + K_G K_2} \tag{17.57}$$

Since Eq. (17.57) contains $\sqrt{q_{11}}$, then $q_{11} \geq 0$. If the restriction that $\mathbf{Q}$ be positive semidefinite is maintained, the minimum ζ that can be achieved is when $K_2 = 0$ (or $q_{22} = 0$) and $K_1^2/\sqrt{q_{11}} \ll 2$. This yields $\zeta_{\min} \approx 0.707$. Therefore, in order to achieve $0 < \zeta < 0.707$, Eqs. (17.56) and (17.57) require that $q_{22} < 0$. This in turn requires that $\mathbf{Q}$ be indefinite.

For a simple second-order plant the peak time can be derived in terms of q_{11} and q_{22} utilizing Eq. (3.60), namely,

$$t_p = \frac{\pi}{\omega_n\sqrt{1-\zeta^2}} = \frac{\pi}{\omega_d} \text{ s} \tag{17.58}$$

Substituting Eqs. (17.53) and (17.57) into Eq. (17.58) yields

$$t_p = \frac{\pi}{(q_{11})^{1/4}\sqrt{K_G}\sqrt{\frac{1}{2} - \frac{1}{4}\left(K_1^2/\sqrt{q_{11}} + K_G K_2\right)}} \tag{17.59}$$

When Eq. (17.50) is used, the damped natural frequency of the response is

$$\omega_d = \omega_n\sqrt{1-\zeta^2} = \sqrt{K_G k_1(1-\zeta^2)} \text{ rad/s} \tag{17.60}$$

The corresponding peak overshoot [Eq. (3.61)] for a unit step input is

$$M_p = y(t_p) = 1 + \exp\left(-\frac{\zeta\pi}{\sqrt{1-\zeta^2}}\right) \tag{17.61}$$

In Sec. 17.4 it is shown that, for an all-pole plant, $e(t)_{ss} = 0$ for a step input when $q_{11} = 1$ (Type 1 or higher system) or $q_{11} \approx 1$ (Type 0 plant under hfg operation). To achieve the Butterworth characteristic, namely, that $\zeta \approx 0.707$, it may be seen from the RSL[4] that an hfg condition must exist: thus $K_1 \approx 0$. Further, from Eq. (17.57), $K_G K_2 \approx 0$, which requires that $q_{22} \approx 0$. Thus, Eqs. (17.57) and (17.59) to (17.61), respectively, yield $\zeta \approx 0.707$ and

$$t_p \approx \pi \sqrt{\frac{2}{K_G}} \tag{17.62}$$

$$\omega_d = \frac{\pi}{t_p} \approx \sqrt{\frac{K_G k_1}{2}} = \sqrt{\frac{K_G}{2}} \tag{17.63}$$

$$M_p \simeq 1 + e^{-\pi} = 1.043 \tag{17.64}$$

The settling time is $T_s = 4/\zeta\omega_n$, and using Eq. (17.50) and $\zeta = 0.707$ yields

$$T_s \approx 4\sqrt{\frac{2}{K_G}} \tag{17.65}$$

Thus, curves of t_p, ω_d, and T_s vs. K_G are readily plotted for the simple second-order case under hfg operation.

For a ramp input, $R(s) = R_1/s^2$, and hfg operation, the steady-state error from Eq. (17.39) is

$$e(t)_{ss} = \left(\frac{a_1}{K_G} + k_2\right)R_1 \approx k_2 R_1 \le E_e \tag{17.66}$$

For the Type 1 plant of Eq. (17.46) the conventional system error coefficients are $K_p = \infty$ and $K_v = K_G/a_1$. For the optimal control system the coefficients, for hfg operation, from Table 13.1 are $K_p' = \infty$ and $K_v' \approx 1/k_2$. The step error coefficients for the conventional and optimal control systems are identical. To compare the two systems with a ramp input, let the acceptable error for the conventional system be E_e. Thus, for the conventional case

$$e(t)_{ss} = \frac{R_1}{K_v} = \frac{a_1}{K_G}R_1 = E_e \tag{17.67}$$

If this value of E_e is substituted into Eq. (17.41) for the optimal case, $k_2 = 0$. In the event $k_2 > 0$, Eq. (17.41) implies that $E_e'/R_1 > a_1/K_G$. The prime denotes the error obtained in the optimal case. Thus,

$$E_e' > \frac{a_1}{K_G}R_1 \tag{17.68}$$

Comparing Eqs. (17.67) and (17.68) shows that $E_e' > E_e$. Therefore, with a Type 1 plant for the condition $k_2 > 0$, the optimal control system does not perform as well in the steady state for ramp inputs as the conventional control system.

A similar analysis can be made for Type 0 and 2 plants with respect to ζ, ω_n, t_p, and t_s. This analysis is confined to the correlation of the steady-state errors achieved with the conventional and optimal control systems. Consider the Type 0 system

$$G(s) = \frac{K_G}{s^2 + a_1 s + a_0}$$

The system error coefficients are $K_p = K_G/a_0$ and $K_v = 0$. The optimal control system can be made to produce zero steady-state error with a step input by selecting $k_1 = 1 - a_0/K_G$. Then the system error coefficients obtained from Eq. (13.42) are $K_p' = \infty$ and $K_v' = K_G/(a_1 + K_G k_2)$. Thus, an optimal Type 0 control system can follow a ramp input but with a finite error. It is well known that a conventional Type 0 control system cannot follow a ramp input.

A similar analysis can be applied to the Type 2 plant described by

$$G(s) = \frac{K_G}{s^2}$$

The system error coefficients are $K_p = K_v = \infty$, $K_p' = \infty$, and $K_v' = 1/k_2$. Therefore, with a Type 2 plant, the optimal control system does not perform as well for ramp inputs as the conventional control system.

Under the hfg condition such that $a_1/K_G \approx 0$, with $q_{11} = 1$ ($k_1 = 1$) and with $z = 1$, Eq. (17.48) reduces to

$$k_2 \approx \frac{a_1}{K_G}\sqrt{1 + \frac{2K_G}{a_1^2} + q_{22}\left(\frac{K_G}{a_1}\right)^2} = \sqrt{\left(\frac{a_1}{K_G}\right)^2 + \frac{2}{K_G} + q_{22}} \approx \sqrt{q_{22}}$$

This implies that $q_{22} > 0$. This can be generalized to $k_n \approx \sqrt{q_{nn}}$. As is shown later in this chapter, this relationship applies to all nth-order systems under the conditions of hfg and reproducibility (see Sec. 16.11). These properties are defined in Sec. 17.6.

17.6 CORRELATION OF HIGHER-ORDER SYSTEMS WITH ALL-POLE PLANTS

Any linear system can be characterized in the frequency domain and has values of M_m and ω_m which are functions of the system parameters. These frequency-domain properties can be correlated directly with corresponding time-domain relationships M_p and ω_d only for a simple second-order system. This knowledge about the correlation between the frequency- and time-response specifications for a simple second-order system is often extrapolated qualitatively for higher-order systems. Experience has borne out the validity of this correlation for many higher-order systems. A similar approach can be used for an optimal control system utilizing the quadratic PI of Eq. (17.7).

For a simple second-order plant it is shown in Sec. 17.5 that the maximum overshoot attainable, with $\zeta = 0.707$, requires that $q_{22} = 0$, under the restriction that the **Q** matrix be positive semidefinite and $z = 1$. Also, direct relationships are obtained for ω_n, ω_d, ζ, and t_p in terms of the plant parameters. It is impossible to obtain similar simple relationships for higher-order plants. Thus, for higher-order plants with $z = 1$, the procedure is similar to that used for conventional control systems. In other words, the analysis technique for a simple second-order optimal control system is extrapolated qualitatively for higher-order plants such that correlation between the diagonal elements of the **Q** matrix and the time-response specifications is effected. This correlation is confirmed from results obtained by using a digital-computer simulation.

To illustrate how this correlation is made, consider a third-order all-pole plant with the **Q** matrix given by

$$\mathbf{Q} = \begin{bmatrix} q_1 & & \mathbf{0} \\ & q_2 & \\ \mathbf{0} & & q_3 \end{bmatrix} \tag{17.69}$$

and $z = 1$. For simplicity, *a single-subscript notation* for the matrix elements is utilized hereafter since the off-diagonal terms are always taken to be zero. Since the analysis of a simple second-order plant shows that $q_2 \approx 0$ yields the smallest possible ζ ($= 0.707$), the assumption is made that the value of q_2 in Eq. (17.69) has the greatest effect on M_p. Also, many trials with a computer simulation have shown that the value q_3 has the greatest effect on t_p, t_s, and ω_d. The values $q_2 = 0.008$ and $q_3 > 0$ are arbitrarily chosen to ensure that **Q** is positive definite. A very small value of q_2 is required in order to have an underdamped response. The correlation of the elements of **Q** with the conventional figures of merit for a step input is developed next for $n \geq 3$.

Constant M_p Criterion

Under the condition of an approximately constant M_p, the correlation requires the determination of the hfg values of $K_G \geq K_{\min}$. The value of $K_{\min}$ *for a given plant* can be determined from the RSL for

$$G(s)G(-s) = -1 \tag{17.70}$$

The dominant branch of the RSL for an all-pole plant becomes a straight line, approaching its asymptote as K_G increases. The value of K_G at which this branch becomes essentially a straight line is denoted K_x. It *should be noted that the* RSL of Eq. (17.70) is used in *this chapter solely for the purpose of determining* K_x, for $z = 1$, whereas the RSL of $G_K(s)G_K(-s) = -1$ (Sec. 16.10) is used to solve the Riccati equation. Note that this graphical procedure for determining K_x cannot be applied to a pole-zero plant having $w = n - 1$ zeros since its asymptotes lie on the real axis. Also note that the value of $K_{\min}$ must be greater than K_x.

A known characteristic of the RSL plots is that, as K_G becomes large, the poles and zeros of the plant can be approximated by $n - w$ poles at the origin. Since the asymptotes all emanate from the origin, the locus is essentially the same for any nth-order plant for $K_G \geq K_{\min}$. The difference between the loci for these nth-order plants is the gain calibration of the loci. This characteristic is the basis for the correlation, under a constant M_p criterion, between the conventional figures of merit and the **Q** matrix. This basis is independent of the plant transfer function. It should be noted that as the plant poles move further to the left in the s plane, $K_{\min}$ becomes larger.

The RSL for a third-order all-pole plant is shown in Fig. 17.2. From this figure is obtained $K_x \approx 262$. The following values, namely, $K_G = 500$, $q_1 = 1$

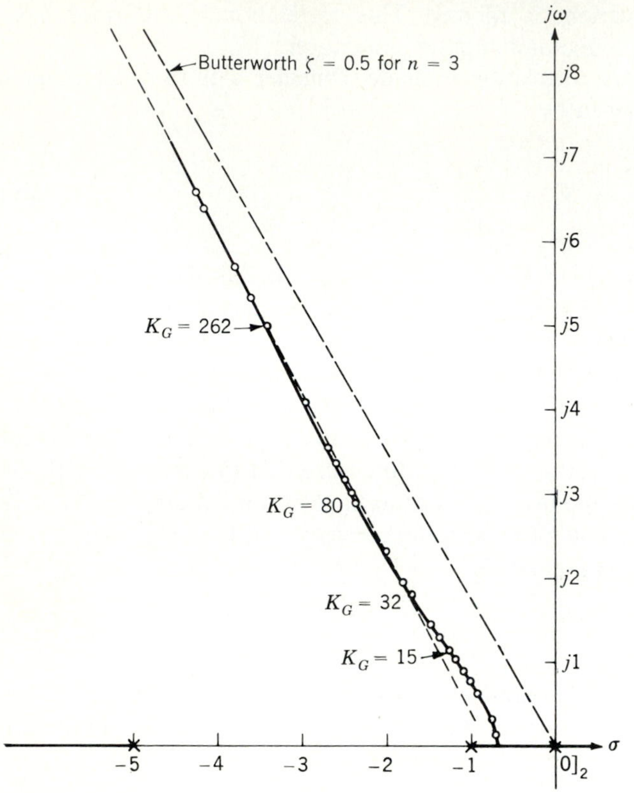

FIGURE 17.2
Third-order all-pole plant RSL for

$$G(s)G(-s) = \frac{-K_G^2}{s^2(s+1)(s+5)(s-1)(s-5)}$$

(required value for zero steady-state error with a step input), $q_2 = 0.008$, and $0.01 \le q_3 \le 10$, are utilized to solve for the values k_i and for the roots of the system characteristic equation. Once the feedback coefficients are determined for various values of q_3, the corresponding output time response $y(t)$ is obtained for a unit-step input. The data for M_p, t_p, and t_s obtained from such time responses are plotted in Fig. 17.3 as a function of q_3. The curve ω_d vs. q_3 represents the frequency of oscillation of the transient response as determined by the dominant roots. For comparison, similar data are obtained for $K_G = 100$ (see Table 17.3) and also plotted in Fig. 17.3. An analysis of the data and of Fig. 17.2 leads to the following conclusions.

For values of K_G equal to or greater than the value K_x at which the real part of the complex roots $s_{1,2}$ become truly dominant, the figures of merit become *reproducible*; i.e., *for a given* **Q**, *any value of* $K_G \ge K_{\min}$ *produces values*

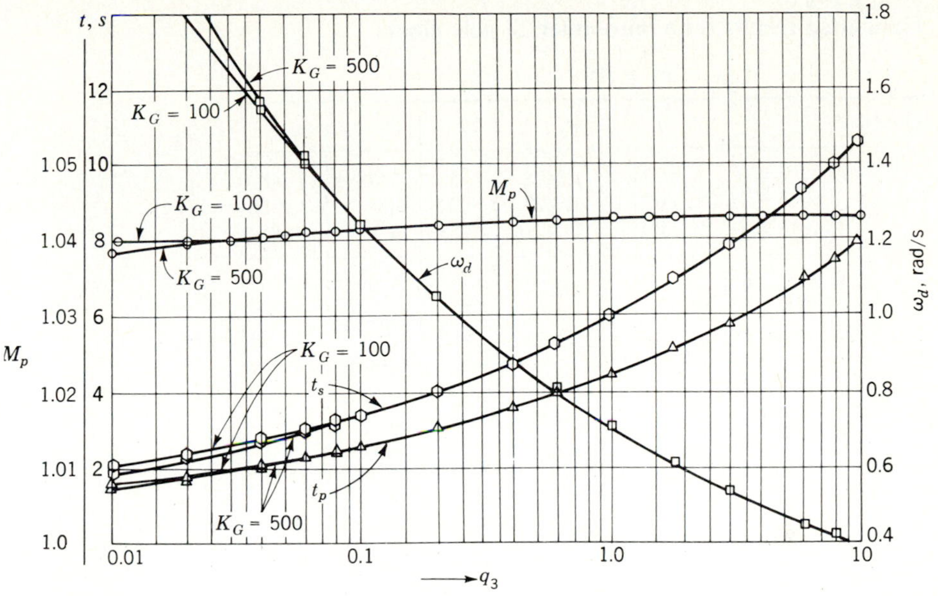

FIGURE 17.3
The parameters of the time-response characteristics for a third-order all-pole system with constant K_G and $z = 1$.

TABLE 17.3
Computer data for a third-order all-pole plant

$z = 1 \quad q_1 = 1 \quad q_2 = 0.008$

K_G	q_3	M_p	t_p	t_s	ω_d
100	0.01	1.0395	1.65	2.05^+	2.099
	0.04	1.0404	2.10	2.70^+	1.5415
	0.08	1.0410	2.45	3.20^+	1.3086
	0.1	1.0413	2.55	3.40	1.2410
150	0.01	1.0391	1.55	1.95^+	2.150
	0.04	1.0405	2.05	2.70^+	1.554
	0.08	1.0412	2.40	3.20^-	1.315
	0.1	1.0415	2.55	3.40	1.2458
200	0.01	1.0387	1.50	1.95^+	2.168
	0.04	1.0405	2.05	2.65^+	1.5591
	0.08	1.0412	2.40	3.20^-	1.317
300	0.01	1.0382	1.50	1.90^+	2.181
	0.04	1.0405	2.05	2.65^+	1.563
400	0.01	1.0382	1.45	1.90^+	2.185
	0.04	1.0405	2.00	2.65^+	1.564
500	0.01	1.03813	1.45	1.85^+	2.187
	0.04	1.0405	2.00	2.65^+	1.564

TABLE 17.4
Computed results for a third-order all-pole plant

$K_G = 100 \qquad z = 1$

Q matrix	q_3	M_p	t_p	t_s	ω_d
$\mathbf{Q}_1 = \begin{bmatrix} 1 & & \mathbf{0} \\ & 0 & \\ \mathbf{0} & & q_3 \end{bmatrix}$	0.1†	1.0428	2.55	3.40^+	1.248
	0.2†	1.0429	3.00	4.00^+	1.0528
	0.4†	1.04296	3.55	4.75^+	0.8869
$\mathbf{Q}_2 = \begin{bmatrix} 1 & & \mathbf{0} \\ & 0.8 & \\ \mathbf{0} & & q_3 \end{bmatrix}$	0.1	1.0000	—	3.70^+	—
	0.2	1.0000	—	3.70^+	0.3469
	0.4	1.0013	5.85	3.75^+	0.5383
$\mathbf{Q}_3 = \begin{bmatrix} 1 & & \mathbf{0} \\ & 1.274 & \\ \mathbf{0} & & q_3 \end{bmatrix}$	0.1	1.00000	—	4.60^+	—
	0.2	1.00000	—	4.65^+	—
	0.4	1.00000	—	4.65^+	—
	q_2				
$\mathbf{Q}_4 = \begin{bmatrix} 1 & & \mathbf{0} \\ & q_2 & \\ \mathbf{0} & & 0.2 \end{bmatrix}$	0†	1.0429	3.00	4.00^+	1.053
	0.008	1.0417	3.00	4.00^+	1.0481
	0.08	1.032	3.15	3.90^+	1.005
	0.6	1.00086	5.20	3.25^+	0.6064
	0.8	1.0000	—	3.70^+	0.3469
	0.9	1.0000	—	3.90^+	—
	1.274	1.0000	—	4.65^+	—

† **P** is positive definite.

of M_p, t_p, t_s, and ω_d which remain essentially constant. (Another method for determining K_x is presented in Sec. 16.11.) Reproducibility becomes evident at about $K_G = 500$ for $q_3 \geq 0.01$ and at about $K_G = 100$ for $q_3 \geq 0.1$. Figure 17.3 shows the variation in M_p, t_p, t_s, and ω_d with values of K_G in the range 100 to 500. For the plant of Fig. 17.2 it may be inferred that reproducibility of results occurs for $K_G \geq 500$. The variation in M_p is about 0.5 percent, and its average value is 1.0418 for $0.01 \leq q_3 \leq 10$ with $K_G = 500$. For $q_3 \geq 0.1$ and $K_G \geq 100$, M_p has a variation of about 0.16 percent. The identification of $K_G \geq 500$ for reproducibility is compatible with $K_{min} \approx 262$, which is determined from Fig. 17.2. Associated with these characteristics is the fact that the root s_3 moves closer to the imaginary axis as the value of q_3 gets closer to q_2. This results in s_3 exerting a greater effect on the time-response characteristics and increasing the variation in the value of M_p.

Table 17.4 presents data showing the effect of the value of q_2 on the time response of the system when K_G is not restricted to $K_G \geq K_{min}$. This table shows that increasing q_2 decreases the value of M_p and, when it is increased sufficiently, results in an overdamped system response. Note for some of the values of q_2 that **Q** is positive semidefinite. Although confirmatory data are not presented in this text, the value of q_2 affects the minimum value of K_G which yields reproducibility of the results on the lower bound of q_3. Thus, the minimum value of q_3 to be chosen is associated with reproducibility, approximately constant M_p, and the value of q_2 to be selected under hfg conditions ($K_G \geq K_{\min}$). It should also be noted that matrix $\mathbf{Q}_1$ in Table 17.4, for $q_3 \geq 0.1$ and $K_G \geq 100$, yields essentially the same results as those in Fig. 17.3. (For $q_3 < 0.1$ the values of M_p and ω_d are slightly higher.) This is to be expected since there is little difference between the values $q_2 = 0$ and $q_2 = 0.008$. Also, it should be noted from matrices $\mathbf{Q}_2$ and $\mathbf{Q}_3$ that, for a desired value of $M_p = 1.0$, the value of q_2 should be kept as small as possible because the value of t_s should be kept as low as possible. Matrix $\mathbf{Q}_4$ illustrates the variation of q_2 for a fixed value of q_3. The data corresponding to this matrix support the comments made about the other matrices.

The conclusions (for a $\mathbf{Q}$ matrix that is positive definite) drawn from the analysis above are (1) that M_p is determined primarily by q_2 and (2) that t_p, t_s, and ω_d are determined primarily by q_3. However, it should be noted that q_2 does have a secondary effect on t_p, t_s, and ω_d.

Under the condition $q_1 = 1$ for zero steady-state error for a step input, the following design procedure, based upon the data in Tables 17.3 and 17.4 and Fig. 17.3, may be used for a third-order all-pole system:

1. Adjust q_2 to achieve the desired approximate M_p value (from 1.0 up to about 1.043). Select $q_3 > q_2$.
2. Determine $K_{\min}$ from the RSL.
3. If an overshoot of approximately 4 percent is acceptable, go to Fig. 17.3 and select q_3 to yield acceptable values of t_p, t_s, and ω_d. If a smaller overshoot is desired, a set of curves similar to Fig. 17.3 must be obtained using a larger value of q_2.

Once the values of q_2 and q_3 have been selected to satisfy the desired conventional figures of merit, the feedback gains k_i can be determined to satisfy the optimal performance index of Eq. (17.7) with $z = 1$ by solving the Riccati equation and then using Eq. (17.19).

Other all-pole third-order plants and their respective $K_G \geq K_{\min}$ for the other corresponding $q_3|_{\min} \leq q_3$ (in this example $q_3|_{\min} = 0.01$) yield curves identical to those in Fig. 17.3. Therefore, the concept of reproducibility of results is established; i.e., Fig. 17.3 is applicable to any third-order all-pole plant where $z = 1$, $K_G \geq K_{\min}$, and $q_3 \geq q_3|_{\min}$ is observed for each respective plant. In other words, if $M_p \approx 1.04$ is acceptable, the value of q_3 can be selected to yield the acceptable values of t_p, t_s, and ω_d under the optimal control criteria of Eq. (17.7), $q_1 = 1$ (for zero steady-state error with a step input), and $q_2 = 0.008$.

Variable M_p Criteria

Under the constant M_p criterion, a desired constant M_p value for a constant value of K_G is achieved by adjusting q_2. The value of M_p can also be varied by

TABLE 17.5
Data for a third-order all-pole plant

P is positive definite and $z = 1$†

K_G	M_p	t_p	t_s	ω_d
$\mathbf{q} = \begin{bmatrix} 1 & & \mathbf{0} \\ & 0.008 & \\ \mathbf{0} & & 0 \end{bmatrix}$				
80	1.046	1.35	1.75^+	2.891
100	1.047	1.25	1.60	3.221
150	1.047	1.05	1.35^+	3.883
$\mathbf{Q} = \begin{bmatrix} 1 & & \mathbf{0} \\ & 0 & \\ \mathbf{0} & & 0 \end{bmatrix}$				
80	1.053	1.35	1.75^+	2.919
100	1.056	1.20	1.60^+	3.251
150	1.060	1.05	1.35^+	3.915
200	1.063	0.90	1.20^+	4.434
400	1.069	0.70	0.95^-	5.875
600	1.071	0.60	0.80	6.861
850	1.073	0.55	0.70^+	7.809
1000	1.074	0.50	0.70^-	8.286

† For $t_p < 1$ s the data are approximate since the calculation interval of the computer programs was 0.05 s.

varying K_G with either $q_3 = 0$ and q_2 held constant or $q_2 = q_3 = 0$. The necessary requirement for tracking is $q_1 = 1$. Table 17.5 presents the computed data for both conditions for the transfer function of $G(s)$ in Fig. 17.2. The results for the condition $q_2 = q_3 = 0$ are plotted in Fig. 17.4. For different third-order all-pole plants, under the conditions of $q_3 = 0$ and/or $q_2 = 0$, the reproducibility criterion is not satisfied. That is, Fig. 17.4, unlike Fig. 17.3 which is for the condition $q_3 > q_2 > 0$, cannot be used for all third-order all-pole plants. Table 17.5 indicates that

1. Values of $M_p > 1.043$ can be achieved.
2. With $q_2 = q_3 = 0$, the third-order Butterworth characteristic,[5] $M_p \approx 1.085$ is approached as K_G is increased.
3. With variable K_G and $q_2 = q_3 = 0$, smaller values of t_p and t_s can be achieved at the expense of a larger M_p compared with the case of a positive definite matrix **Q**.

Curves representing the parameters of the time response for the constant M_p criterion, with $r(t) = u_{-1}(t)$, can be obtained for higher-order systems[1,12]

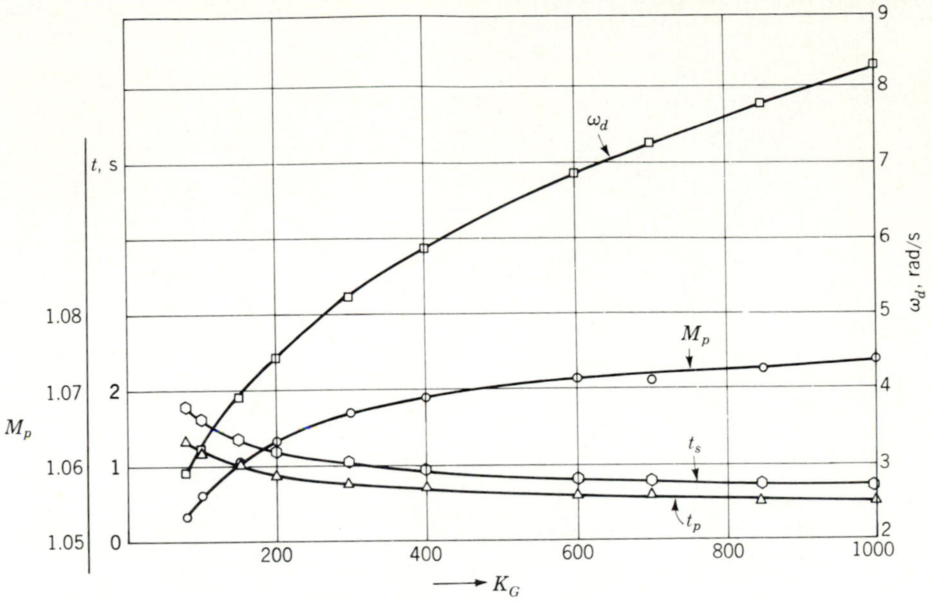

FIGURE 17.4
The parameters of the time-response characteristics for a third-order all-pole system with $z = 1$, $q_2 = q_3 = 0$.

under the conditions $K_G \geq K_{\min}$, $q_1 = 1$, and $q_n > q_i = 0.01$ ($i = 2, \ldots, n - 1$). An analysis of these curves and their associated data permits conclusions similar to those obtained for the all-pole third-order plant to be made for the $n > 3$ all-pole plant for both constant and variable M_p criteria. The complete results for $n \geq 3$ hfg all-pole plant systems, *for the constant M_p and reproducibility criteria, reveals that there are $n - 1$ dominant roots and that the nondominant root is given by $s_n \approx -K_G\sqrt{q_n}$*. A general design procedure for satisfying the conventional control figures of merit and the quadratic cost is presented in Sec. 17.9.

17.7 HIGHER-ORDER SYSTEM CORRELATION (POLE-ZERO PLANT)

The control ratio, Eq. (17.23), is for a pole-zero plant whose transfer function is

$$G(s) = \frac{K_G\left(s^w + c_{w-1}s^{w-1} + \cdots + c_0\right)}{s^n + a_{n-1}s^{n-1} + \cdots + a_0}$$

The approach to the pole-zero correlation is based upon knowledge of the all-pole plant. Again, the hfg condition, $K_G \geq K_{\min}$, is observed in the same manner as for the all-pole plant and for the condition $w \leq n - 2$. Also, to achieve zero steady-state error for a step input for a pole-zero plant, the value of q_1 in the **Q** matrix is specified by Eq. (17.28). The correlation analysis is

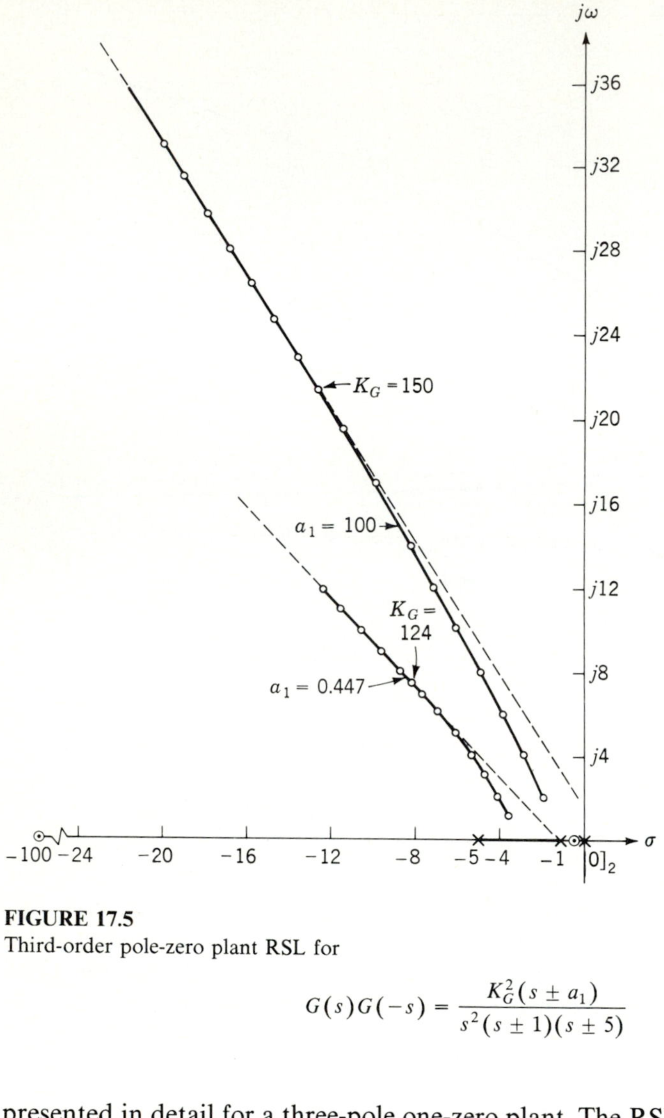

FIGURE 17.5
Third-order pole-zero plant RSL for

$$G(s)G(-s) = \frac{K_G^2(s \pm a_1)}{s^2(s \pm 1)(s \pm 5)}$$

presented in detail for a three-pole one-zero plant. The RSL is shown in Fig. 17.5. From this figure $K_x \approx 124$ for $a_1 = 0.4472$ and $K_x \approx 150$ for $a_1 = 100$. Table 17.6 contains representative data (M_p, t_p, t_s, ω_{d_e}, and $\mathbf{k}_c$) obtained by means of a digital computer, with $r(t) = u_{-1}(t)$ and $z = 1$, for $a_1 = \sqrt{q_1} = 0.4472$ and also for $a_1 = \sqrt{q_1} = 10$. For each value of q_1 data are given for two values of gain, $K_G = 150$ and $K_G = 300$. The value of the damped natural frequency of the system can be estimated from the response of $y(t)$. From Eq. (17.63), the effective damped natural frequency is approximately

$$\omega_{d_e} \approx \frac{\pi}{t_p} \tag{17.71}$$

The data for $K_G = 150$ and for $K_G = 300$ are plotted in Fig. 17.6*a* and *b*, respectively. Note that for $q_3 < 0.06$ the reproducibility characteristic for $K_G \geq K_{min}$ is not achieved. Analysis of Fig. 17.6, Table 17.6, the roots of the characteristic equation of the system, and additional time-response data for higher values of K_G (not included in this text) results in the following conclusion.

For values of K_G equal to or greater than the value for which the roots $s_{1,2}$ become definitely dominant, the figures of merit become reproducible. Reproducibility is achieved at approximately $K_G = 400$ for $q_1 = 0.2$ and $q_3 \geq 0.01$ and at $K_G = 500$ for $q_1 = 100$ and $q_3 \geq 0.01$. For $q_3 \geq 0.06$ reproducibility occurs for $K_G \geq 150$ for both values of a_1. These values of K_G for reproducibility are compatible with the values of K_x determined from the RSL of Fig. 17.5. Associated with these characteristics is the fact that as the value of q_3 becomes close to q_2, the s_3 root moves closer to the origin. The correlation data indicate that for a given value of a_1 the dominant roots remain essentially unchanged for given minimum values of K_G and q_3 or for higher values of K_G.

As in the all-pole case, the value of q_2 affects the value of K_{min} used for reproducibility of the results on the lower bounds of q_3 for each transfer function, that is, $q_3 \geq q_2$.

For $q_1 = 0.2$ and $K_G = 300$, Table 17.6 indicates that M_p varies from 2.4814 to 1.0816. For $q_1 = 100$, M_p varies from a value of 1.1007 to 1.0440. This shows that as a_1 is increased, a constant M_p condition can be achieved: the value of M_p approaches the average value, 1.0418, of the third-order all-pole case. Comparing Figs. 17.3 and 17.6 ($q_1 = 100$ and $q_3 \geq 0.1$) shows that the pole-zero case has improved the values of t_s and t_p for approximately the same value of M_p. This effect is identical to that achieved in conventional control-theory design when a zero is added to the original plant. For the entire range of $0.447 \leq a_1 \leq 10$ and $0.01 \leq q_3 \leq 10$, the value of t_p has been decreased by the addition of a zero to the plant. Although the data are not shown, for values of $a_1 \geq \sqrt{2}$ (approximately) the values of both t_p and t_s are less for the pole-zero plant than for the all-pole plant.

The choice of a minimum value of q_3 is associated with the reproducibility ($K_G \geq K_{min}$), the value of a_1, and the value of q_2 to be selected. Table 17.7 presents additional data showing the effect of the value of q_2 on the time-response parameters of the system. Using a value of q_2 larger than 0.008 decreases the value of M_p that is obtainable. Thus, for the desired value of q_2, under reproducibility conditions, the minimum value of q_3 to be selected is based upon the maximum acceptable forward gain of the system.

The design procedures for the third-order plant with one zero can be essentially the same as those for the all-pole plant. Once the values of a_1, q_2, and q_3 have been selected to satisfy the desired conventional figures of merit, the feedback gains k_i can be determined to satisfy, for $z = 1$, the optimal performance index of Eq. (17.7). For a desired overdamped response the value of q_2 should be kept as small as possible.

Larger values of M_p are obtained for a pole-zero plant than with an all-pole plant. Because the value of M_p varies for a given value of a_1 and small values of

TABLE 17.6
Experimental data: third-order plant, one-zero case, $K_G = 150$ and 300, $z = 1$

$$G(s) = \frac{K_G\left(s + \sqrt{q_1}\right)}{s(s+1)(s+5)} \qquad \mathbf{A} = \begin{bmatrix} 0 & 1 & 0 \\ 0 & 0 & 1 \\ 0 & -5 & -6 \end{bmatrix} \qquad \mathbf{Q} = \begin{bmatrix} q_1 & & \mathbf{0} \\ & 0.008 & \\ \mathbf{0} & & q_3 \end{bmatrix}$$

K_G	q_3	M_p	t_p	t_s	ω_{d_e}	k_1	k_2	k_3
				$q_1 = 0.2$				
150	0.01	2.4322	0.72	3.76–3.78	4.36	0.4472	0.3153	0.0857
	0.02	2.1617	0.86	4.30–4.32	3.65	0.4472	0.3596	0.1225
	0.04	1.9182	1.05	4.90–4.95	2.99	0.4472	0.4176	0.1772
	0.06	1.792	1.16	5.32–5.34	2.71	0.4472	0.4585	0.2202
	0.1	1.6498	1.38	5.84–5.86	2.275	0.4472	0.5178	0.2894
	0.2	1.4868	1.72	4.76–4.78	1.827	0.4472	0.6138	0.4180
	0.6	1.2939	2.62–2.64	6.36–6.38	1.192	0.4472	0.8087	0.7426
	1	1.2292	3.20	7.30	0.982	0.4472	0.9268	0.9774
	2	1.1625	4.25–4.30	8.80–8.85	0.735	0.4472	1.0977	1.3799
	5.5	1.1023	6.45	11.65^+	0.487	0.4472	1.4195	2.3096
	10	1.0817	8.15	13.75^+	0.386	0.4472	1.6522	3.1260
300	0.01	2.4814	0.68	3.68–3.70	4.62	0.4472	0.3213	0.0917
	0.02	2.1839	0.86	4.24–4.26	3.83	0.4472	0.3620	0.1310
	0.04	1.9286	1.00	4.90–4.95	3.14	0.4472	0.4241	0.1879
	0.06	1.7984	1.15	5.30–5.35	2.73	0.4472	0.4668	0.2320
	0.1	1.6535	1.36	5.84–5.86	2.31	0.4472	0.5280	0.3024
	0.2	1.4886	1.74	4.74–4.76	1.805	0.4472	0.6259	0.4323
	0.6	1.2945	2.62	6.34–6.36	1.198	0.4472	0.8227	0.7584
	1	1.2292	3.20	7.25–7.30	0.982	0.4472	0.9350	0.983
	2	1.1626	4.25	8.80–8.85	0.735	0.4472	1.1128	1.3970
	10	1.0816	8.15	13.75^+	0.386	0.4472	1.6680	3.1441

$q_1 = 100$

150	0.01	1.1064	0.36	0.60–0.62	8.72	10.0000	1.9358	0.1534
	0.02	1.0865	0.42	0.66–0.68	7.48	10.0000	2.0759	0.1820
	0.04	1.0688	0.50	0.78–0.80	6.28	10.0000	2.2858	0.2285
	0.06	1.0621	0.58	0.86–0.88	5.42	10.0000	2.4462	0.2670
	0.1	1.0563	0.67	0.98–0.99	4.68	10.0000	2.6915	0.3308
	0.2	1.0513	0.82	1.16–1.17	3.83	10.0000	3.1082	0.4530
	0.6	1.0472	1.12	1.56–1.57	2.8	10.0000	3.9907	0.7692
	1	1.0461	1.30	1.75–1.80	2.415	10.0000	4.5072	0.99039
	2	1.0452	1.55	2.10–2.15	2.03	10.0000	5.3335	1.3997
	5.5	1.0443	2.05	2.75–2.80	1.53	10.0000	6.8447	2.3249
	10	1.0440	2.40	3.25^+	1.31	10.0000	7.9412	3.1392
300	0.01	1.1007	0.32	0.54–0.56	9.81	10.0000	1.7033	0.1275
	0.02	1.0834	0.39	0.64–0.65	8.05	10.0000	1.8918	0.1617
	0.04	1.0673	0.49	0.76–0.77	6.4	10.0000	2.1482	0.2140
	0.06	1.0612	0.56	0.84–0.85	5.62	10.0000	2.3328	0.2556
	0.1	1.0558	0.65	0.97–0.98	4.83	10.0000	2.6047	0.3232
	0.2	1.0512	0.81	1.16–1.17	3.88	10.0000	3.0500	0.4498
	0.6	1.0472	1.11	1.56–1.57	2.83	10.0000	3.9636	0.7717
	1	1.0461	1.30	1.75–1.80	2.42	10.0000	4.4899	0.9951
	2	1.0452	1.55	2.10–2.15	2.028	10.0000	5.3261	1.4069
	10	1.0440	2.40	3.25^+	1.31	10.0000	7.9471	3.1507

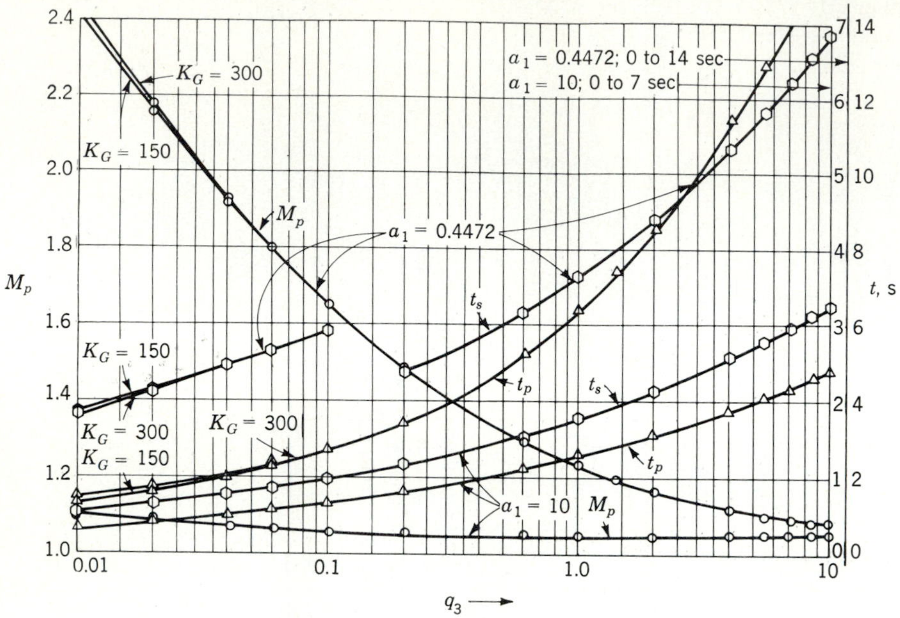

(a)

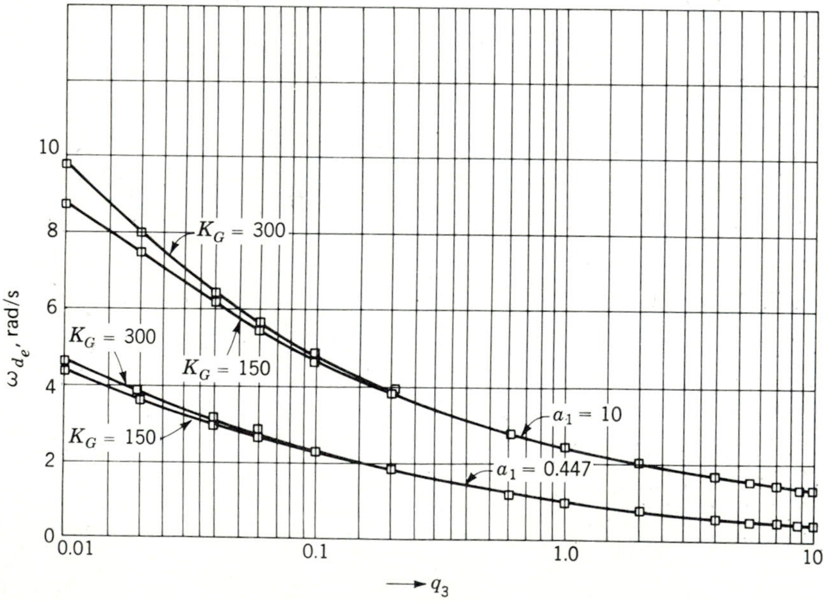

(b)

FIGURE 17.6
Time-response parameters of third-order system with one zero: (a) M_p, t_p, and t_s vs. q_3; (b) ω_{d_e} vs. q_3.

TABLE 17.7
Time-response data for a third-order plant with one zero

$$K_G = 150 \qquad z = 1 \qquad \mathbf{Q} = \begin{bmatrix} q_1 & & 0 \\ & 2 & \\ 0 & & q_3 \end{bmatrix}$$

q_1	q_3	M_p	t_p	t_s	ω_{d_e}
0.2	0.01	1.00000	—	8.60^+	—
	1.0	1.00000	—	8.85	—
	10	1.00203	15.15	9.05^+	0.207
100	0.01	1.00000	—	0.40^+	—
	1.0	1.03322	1.35	1.70^+	2.328
	10	1.03977	2.45	3.20^+	1.282

q_i, the plots of t_s and/or t_p vs. q_n can experience a discontinuity over the range of q_n. The effect is shown in Fig. 17.6a for $a_1 = 0.447$ and $q_2 = 0.008$.

Other one-zero third-order plants and their respective $K_G \geq K_{\min}$ for the corresponding values of a_1 and $q_3|_{\min} \leq q_3$ (in this example $q_3|_{\min} = 0.01$) yield curves identical to Fig. 17.7. Therefore, the concept of reproducibility of results is established; i.e., Fig. 17.6 is applicable to any one-zero third-order plant where $z = 1$, $K_G \geq K_{\min}$, and $q_3 \geq q_3|_{\min}$ is observed for each respective plant. In other words, from a reproducibility curve having an acceptable range of M_p values, the value of q_3 can be selected which yields acceptable values of t_p, t_s, and ω_{d_e} under the optimal control criterion of Eq. (17.7) with $z = 1$ and $q_1 \approx c_0^2$ (for zero steady-state error for a step input).

Curves representing the parameters of the time response with $r(t) = u_{-1}(t)$ can be obtained for higher-order plants with a single zero under the conditions of $K_G \geq K_{\min}$, Eq. (17.29), $z = 1$, and $q_n > q_i$, and for specified values of q_i. An analysis of these curves[1,2] and their associated data, for systems of order greater than 3 and with a single zero, yield conclusions similar to those made for the one-zero third-order plant. *The entire data for higher-order systems with hfg operation reveal that, under the reproducibility condition, there are $n - 1$ dominant roots and the nondominant root is given by the expression $s_n \approx -K_G\sqrt{q_n}$, just as for the all-pole plant.* (See Sec. 17.9 for a general design procedure.)

Extending the correlation to a plant having two or more zeros in order to derive straightforward conclusions becomes more difficult. Consider the two-zero case for which the numerator of the plant transfer function is $s^2 + c_1 s + c_0$. Under the reproducibility condition (hfg operation, where $K_G \geq K_{\min}$) the following general conclusions can be drawn:

1. A large value of c_0 together with a low value of c_1 provides the best overall results; i.e., acceptable values of M_p are achieved with a decrease in values of t_p and t_s.
2. Increasing the value of q_i for $i > 1$ decreases the overshoot.

3. A two-zero plant may yield better performance than an all-pole or a one-zero plant.

17.8 ANALYSIS OF CORRELATION RESULTS

The $a_{i-1} + K_G k_i$ terms of the system characteristic equation, Eq. (17.24), under the high-gain condition ($K_G \geq K_{\min}$) can be approximated by $K_G k_i$, that is, $a_{i-1} \ll K_G k_i$. Thus, Eq. (17.24) reduces to

$$s^n + K_G k_n s^{n-1} + K_G k_{n-1} s^{n-2} + \cdots + K_G k_1 = 0 \tag{17.72}$$

An analysis of the correlation data of Secs. 17.6 and 17.7, for $z = 1$, reveals that when the reproducibility condition is achieved, the actual value of the nondominant root s_n is

$$s_n \approx -K_G \sqrt{q_n} \tag{17.73}$$

At a value of K_G sufficiently large to achieve reproducibility, the actual root is very close (within approximately 0.5 percent) to the value given by Eq. (17.73). When reproducibility has been achieved, it is noted from the data that

$$k_n \approx \sqrt{q_n} \tag{17.74}$$

As the values of q_n and n become larger, this approximation improves. *Equation* (17.73) *serves as a necessary and sufficient condition for satisfying the reproducibility criterion.*

The analysis of the correlation data also shows that for a given nth-order system, as long as Eq. (17.73) is satisfied, the dominant roots remain unchanged and are independent of $G(s)$. The only difference is the value of the nth root, given by Eq. (17.73), since each nth-order $G(s)$ function has a different value of $K_G \geq K_{\min}$. Thus, for all nth-order plants, for any given $\mathbf{Q}$ the time-response characteristics are identical as long as $K_G \geq K_{\min}$ for the respective plant. This is to be expected since, if reproducibility is to occur, the dominant roots of an nth-order system utilizing any nth-order transfer function must not change. This is compatible with the development of Eq. (17.72). The $n - 1$ dominant roots, which do not change for $K_G \geq K_{\min}$ and are independent of the plant, yield the polynomial

$$s^{n-1} + \varepsilon_{n-1} s^{n-2} + \cdots + \varepsilon_1 \tag{17.75}$$

Multiplying Eq. (17.75) by $s + K_G\sqrt{q_n}$ yields the system characteristic equation

$$s^n + K_G k_n s^{n-1} + \cdots + K_G k_1$$
$$= s^n + \left(\varepsilon_{n-1} + K_G\sqrt{q_n}\right) s^{n-1} + \cdots + \varepsilon_1 K_G \sqrt{q_n} = 0 \tag{17.76}$$

Thus

$$k_n = \frac{\varepsilon_{n-1} + K_G\sqrt{q_n}}{K_G}, \quad k_{n-1} = \frac{\varepsilon_{n-2} + \varepsilon_{n-1} K_G \sqrt{q_n}}{K_G}, \ldots, \quad k_1 = \varepsilon_1 \sqrt{q_n} \tag{17.77}$$

17.9 A GENERAL DESIGN PROCEDURE

In the preceding sections a correlation between the conventional figures of merit and the q_i elements of the **Q** matrix has been presented. For zero steady-state error with a step input to the system the q_1 element is specified in terms of the plant parameters and the weighting factor z. The correlation shows that the elements q_i $(i = 2, \ldots, n-1)$ have a primary effect on the value of M_p and that the element q_n has the primary effect on the values of t_p, t_s, and ω_d (or ω_{d_e}). The values of q_i chosen to yield a desired value (or range of values) of M_p have the secondary effect of determining the lower bound of the values of t_p and t_s and the upper bound of ω_d (or ω_{d_e}) that can be achieved by adjusting the value of q_n. From this correlation, a suitable design procedure under hfg operation is developed as follows.

Step 1. Set $q_1 = c_0^2 z$ for zero steady-state error for a step input.

Step 2. For the all-pole or the pole-zero system adjust q_i $(i = 2, \ldots, n-1)$ to achieve the desired M_p value.

Step 3. Make certain that the value of K_G is high enough at the lower bound of q_n to ensure reproducibility. This is determined by use of the reproducibility criterion, which states that the actual value of the nondominant root must be within, approximately, 0.5 percent of the value given by $-K_G\sqrt{q_n}$.

Step 4. Once steps 1 and 3 have been achieved, obtain the reproducibility curves.

Step 5. Select the value of q_n from the reproducibility curves to yield acceptable values of M_p, t_p, t_s, and ω_d (or ω_{d_e}) for the all-pole or pole-zero case.

Step 6. When the desired values of the elements of the **Q** matrix have been established, a computer program can be used to determine the feedback gains k_i to satisfy the optimal performance index of Eq. (17.7). Where applicable, the method outlined in Sec. 17.8 can be used to determine the values of k_i.

Step 7. For the overdamped case the values of q_i should be kept as low as possible in order to maintain $M_p = 1.0$ and to achieve the lowest value of t_s.

The all-pole system with the fastest response is achieved with the elements q_i and q_n equal to zero and with the conditions for observability and controllability satisfied.[1,5] The condition for reproducibility is not achieved with these values of the elements of **Q**. The conventional figures of merit vary with the value of K_G. A one-zero plant can respond faster than an all-pole system for a suitable value of c_0. The same situation is possible for the two-zero case for reasonable values of M_p.

17.10 EFFECT OF CONTROL WEIGHTING

The preceding sections have presented a method for choosing the matrix **Q** for a control weighting $z = 1$ to achieve desired conventional figures of merit. The effect of varying the value of z is now illustrated for the third-order all-pole and one-zero plants of Secs. 17.6 and 17.7. In both cases q_1 is selected to satisfy Eq. (17.28), and the **Q** matrix utilized is

$$\mathbf{Q} = \begin{bmatrix} q_1 & & \mathbf{0} \\ & 0.008 & \\ \mathbf{0} & & q_3 \end{bmatrix} \tag{17.78}$$

For the all-pole plant a value of $K_G = 500$ is used. For the one-zero plant, $c_0 = a_1 = 10$, a value of $K_G = 300$ is used.

In order to illustrate the effect of z on system performance, three values of z are selected for each plant, and the corresponding values of k_i are obtained by solving the algebraic Riccati equation. The resulting data of the associated time response are plotted in Figs. 17.7 to 17.9. Analyzing Fig. 17.7 for the all-pole third-order plant reveals that increasing the value of z:

1. Improves the constant M_p characteristic under hfg operation.
2. Improves the time response, i.e., decreases t_p and t_s.
3. Requires smaller values of feedback gains except for $k_1 = 1$, which remains fixed.

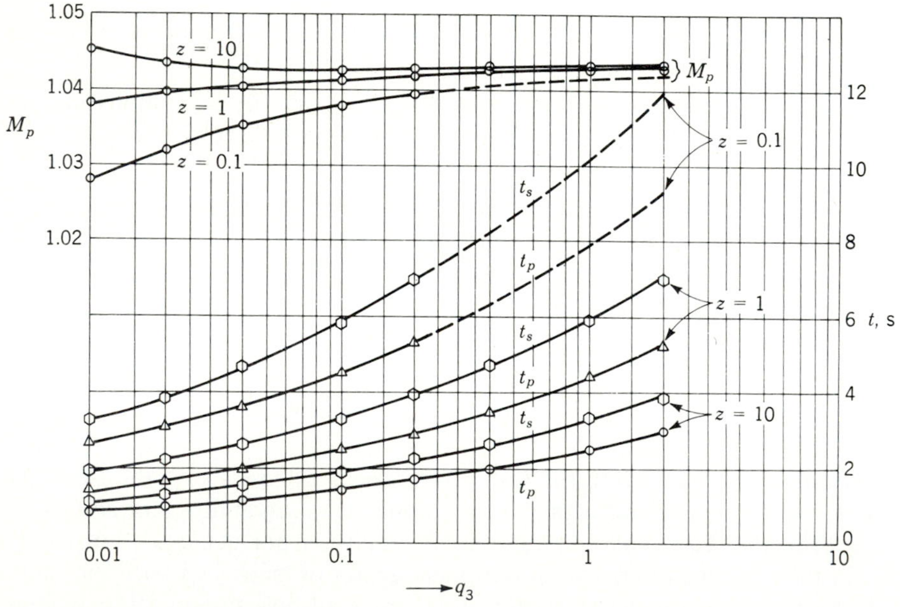

FIGURE 17.7
Time-response parameters for a third-order all-pole system for $K_G = 500$ and several values of z.

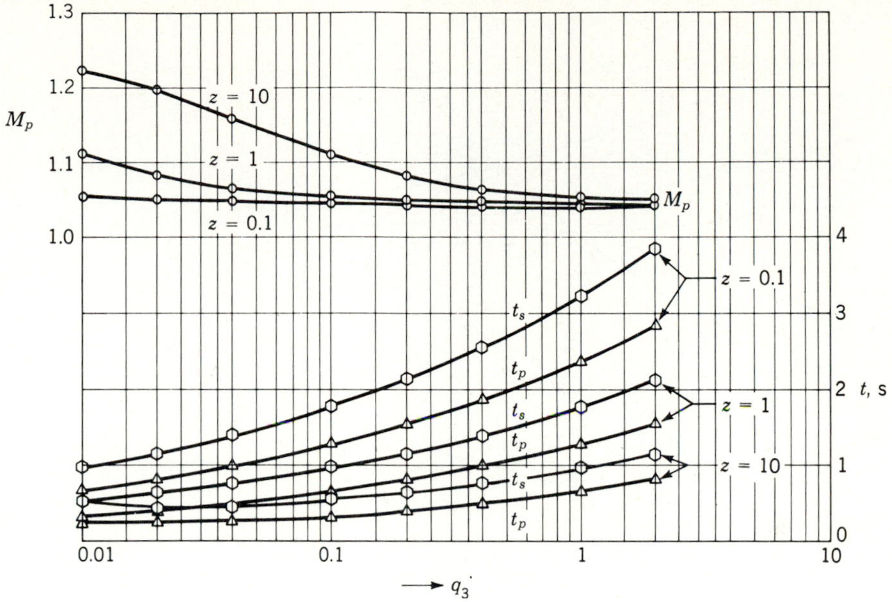

FIGURE 17.8
Time-response parameters for a third-order one-zero system for $K_G = 300$, $c_0 = 10$, and several values of z.

For the one-zero plant, Figs. 17.8 and 17.9 reveal that increasing the value of z:

1. Increases the value of M_p significantly for low values of q_3.
2. Decreases the values of t_p and t_s.
3. Requires smaller values of feedback gains except for $k_1 = c_0$, which remains fixed.

Therefore, as noted in Sec. 17.7, one-zero plants with a small value for a_1 have large overshoots. These large overshoots, for a given value of a_1, can be decreased by decreasing the value of z. In other words, decreasing the value of z, for a pole-zero plant, has the same effect on the value of M_p as increasing the value of q_i ($i = 2, 3, \ldots, n-1$). Similar results occur for higher-order plants. The value of K_G required to maintain reproducibility, from Eq. (16.83), is

$$K_{G_{\min}} = \xi \omega^* \frac{\sqrt{z}}{h_n} \tag{17.79}$$

where $\xi \geq 1$. Since ω^* and h_n are independent of K_G and ξ for hfg, then, for a given ξ,

$$\left[K_{G_{\min}}\right]_{z \neq 1} = \left[K_{G_{\min}}\right]_{z=1} \sqrt{z} \tag{17.80}$$

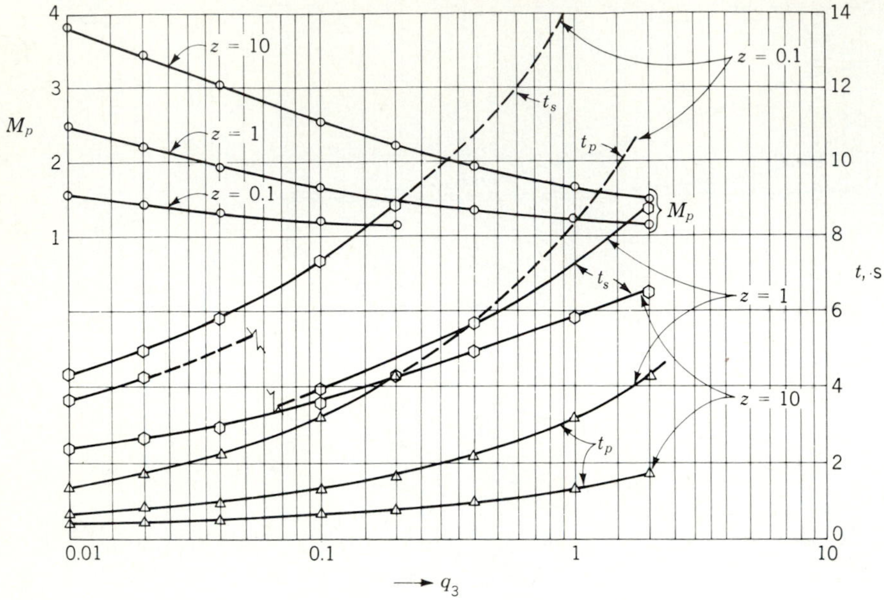

FIGURE 17.9
Time-response parameters for a third-order one-zero system for $K_G = 300$, $c_0 = \sqrt{0.2}$, and several values of z.

The value of $[K_{G_{\min}}]_{z=1}$ can be determined from the RSL of Eq. (17.70), and then $[K_{G_{\min}}]_{z\neq 1}$ can be obtained from Eq. (17.80).

17.11 ALTERNATE DESIGN METHOD

The previous sections present a method of selecting the diagonal elements of the matrix **Q** with zero off-diagonal terms to minimize the performance index

$$\text{PI} = \int_0^\infty (\mathbf{x}^T\mathbf{Q}\mathbf{x} + zu^2)\, dt \tag{17.81}$$

This section presents another method of obtaining the matrix **Q** by determining the elements of the vector **h**, so that $\mathbf{Q} = \mathbf{h}\mathbf{h}^T$. This matrix is not diagonal; however, an equivalent diagonal matrix can be obtained.[6,10] Using the phase-variable representation, this method is based upon selecting a *defined dynamic equation* (DDE) for the desired optimal performance. This approach is used to simplify the pole-placement method of Chaps. 13 and 14. Also, it is used to design a feedback system to meet desired performance specifications even though some of the parameters of the plant may vary.

If it were permissible to set $z = 0$ in Eq. (17.81), the PI would attain an absolute minimum of zero[14] for $\mathbf{h}^T\mathbf{x}(t) = 0$. When phase-variable representation of the plant dynamics is used, this minimum can be expressed as

$$h_1 X_1(s) + h_2 X_2(s) + \cdots + h_n X_n(s) = \left(h_1 + h_2 s + \cdots + h_n s^{n-1}\right) X_1(s) = 0 \tag{17.82}$$

because $sX_{i-1}(s) = X_i(s)$. For a solution to the optimal control problem,

$$h_1 + h_2 s + \cdots + h_n s^{n-1} = 0 \tag{17.83}$$

may be taken as the DDE of the system. The limiting form of the closed-loop system as $z \to 0$ is the model described by Eq. (17.83). For other values of z there is a mismatch. For the system design presented in this section, $z = 1$ is assumed. The necessary constraint on h_1, for zero steady-state error with a step input, using Eqs. (16.84), (17.20), and (17.28), is

$$q_{11} = h_1^2 = c_0^2 - \left(\frac{a_0}{K_G}\right)^2 \approx c_0^2 \tag{17.84}$$

This constraint, under hfg operation, is observed in this section.

Equation (17.83) is one degree less than the characteristic equation of the system. The coefficients of this equation are obtained in the following manner: (1) Take all denominator factors of the desired $Y(s)/R(s)$ except the factor containing one nondominant pole. (2) Multiply these factors and form a polynomial equation. (3) Multiply this polynomial equation by a constant so that the coefficient of the s^0 term has the required value of h_1 obtained from Eq. (17.84). This is now the DDE whose coefficients are the h_i given in Eq. (17.83) and whose corresponding $\mathbf{h}_c$ must satisfy Eq. (17.28).

Consider as a first example a third-order all-pole plant which is Type 1 ($a_0 = 0$). The quadratic equation formed from steps 1 and 2 above is

$$\omega_N^2 + 2\zeta_N \omega_N s + s^2 = 0 \tag{17.85}$$

where the values of ζ_N and ω_N are specified. For an all-pole plant $h_1 = c_0 = 1$. When Eq. (17.85) is compared with Eq. (17.83), the value $h_1 = 1$ can be obtained by dividing Eq. (17.85) by ω_N^2. This yields

$$1 + \frac{2\zeta_N}{\omega_N} s + \left(\frac{1}{\omega_N}\right)^2 s^2 = 1 + d_1 s + d_2 s^2 \tag{17.86}$$

Thus

$$\mathbf{h}_c^T = [1 \quad d_1 \quad d_2] \tag{17.87}$$

The second example is the three-pole one-zero case. As shown in Sec. 17.7, large overshoots may exist due to the presence of the zero. Thus, it is necessary to choose $\mathbf{h}_c$ to minimize the effect of that zero. The obvious way to accomplish this is to place a root of the characteristic equation near the plant zero $-\alpha$. Thus, the quadratic equation formed from steps 1 and 2 above is

$$(s + \alpha)(s + \beta) = \alpha\beta + (\alpha + \beta)s + s^2 = 0 \tag{17.88}$$

where the value of β is selected to yield the desired time response. In order for

TABLE 17.8
$\mathbf{h}_c^T = [1 \quad d_1 \quad d_2 \quad \cdots \quad d_n]$ for all-pole plants

n	Defined dynamic equation
3	$1 + \frac{2\zeta_N}{\omega_N}s + \left(\frac{1}{\omega_N}\right)^2 s^2 = 1 + d_1 s + d_2 s^2 = 0$
4	$(s+\alpha)(s^2 + 2\zeta_N\omega_N s + \omega_N^2) = 0$ or $1 + \frac{\omega_N^2 + 2\zeta_N\omega_N\alpha}{\alpha\omega_N^2}s + \frac{2\zeta_N\omega_N + \alpha}{\alpha\omega_N^2}s^2 + \frac{1}{\alpha\omega_N^2}s^3 = 1 + d_1 s + d_2 s^2 + d_3 s^3 = 0$
5	$(s+\alpha)(s+\beta)(s^2 + 2\zeta_N\omega_N s + \omega_N^2) = 0$ or $1 + d_1 s + d_2 s^2 + d_3 s^3 + d_4 s^4 = 0$

TABLE 17.9
$\mathbf{h}_c^T = [\alpha \quad d_1 \quad d_2 \quad \cdots \quad d_n]$ for one-zero plants

n	Defined dynamic equation
3	$(s+\alpha)(s+\beta) = 0$ or $\alpha + d_1 s + d_2 s^2 = 0$
4	$(s+\alpha)(s^2 + 2\zeta_N\omega_N s + \omega_N^2) = 0$ or $\alpha + d_1 s + d_2 s^2 + d_3 s^3 = 0$
5	$(s+\alpha)(s+\beta)(s^2 + 2\zeta_N\omega_N s + \omega_N^2) = 0$ or $\alpha + d_1 s + d_2 s^2 + d_3 s^3 + d_4 s^4 = 0$

$h_1 = \alpha = c_0$, Eq. (17.88) is divided by β to obtain the DDE

$$\alpha + \frac{\alpha + \beta}{\beta}s + \frac{1}{\beta}s^2 = \alpha + d_1 s + d_2 s^2 \tag{17.89}$$

Comparing Eq. (17.89) with Eq. (17.83) yields

$$\mathbf{h}_c^T = [\alpha \quad d_1 \quad d_2] \tag{17.90}$$

The vector $\mathbf{h}_c$ for higher-order systems can be determined in a similar manner. For typical desired DDEs, Table 17.8 lists the vector $\mathbf{h}_c$ for $n = 3, 4, 5$ all-pole plants. Table 17.9 applies for the one-zero plants for which a root of the closed-loop characteristic equation is made equal to the zero.† These tables can readily be extended for higher-order systems. The desired DDE can also be obtained by using the approach of Sec. 12.3.

Once $\mathbf{h}_c$ is determined, the method of Sec. 16.11 may be applicable for obtaining a vector $\mathbf{k}_c$ which ensures an optimal performance. The necessary condition for this method is that the minimum value of the forward gain satisfy

† If cancellation of the zero of the control ratio is not desired, then the value of α used in Table 17.9 is any specified pole.

Eq. (16.83); i.e.,

$$K_x = \frac{\xi\omega^*}{h_n} \qquad \text{for } \xi > 1 \tag{17.91}$$

With this value of hfg and the matrix $\mathbf{Q} = \mathbf{h}_c\mathbf{h}_c^T$, the feedback coefficients $\mathbf{k}$ can also be determined by solving the Riccati equation by use of a computer program.[11]

Thus the value of gain used in the plant for which the DDE is to be determined must be large enough to satisfy the condition $\xi > 1$ in Eq. (17.91). An alternate approach is to obtain the Bode plot of the $G_K(s)$ and select the gain so that $\omega_\phi \geq \xi\omega^*$, where $\xi > 1$. Once $\mathbf{h}$ and K_x are determined, the value of $\mathbf{k}$ can be calculated without a computer by utilizing Eq. (16.79).

The following two examples illustrate the method, based upon the selection of a DDE, for achieving an optimal performance.

Example 1. In the example in Sec. 13.8, the plant and desired control ratio are

$$G(s) = \frac{100}{s(s+1)(s+5)} = \frac{100}{s^3 + 6s^2 + 5s} \tag{17.92}$$

$$\frac{Y(s)}{R(s)} = \frac{100}{(s+100)(s^2 + 1.417s + 1)} \tag{17.93}$$

For this third-order system the dominant poles of Eq. (17.93) are used to obtain the desired second-order DDE. Thus, since $h_1 = 1$, from Table 17.8

$$\mathbf{h}_c^T = [1 \quad 1.417 \quad 1] \tag{17.94}$$

Use the $\mathbf{A}_c$ and $\mathbf{b}_c$ in Example 2, Sec. 16.8, to obtain $G_K(s) = \mathbf{h}_c^T\mathbf{\Phi}(s)\mathbf{b}_c$. An analysis of $\mathbf{G}_K(j\omega)$ reveals that $\omega^* = 5$, and Eq. (17.91) yields the value $\xi = K_x h_3/\omega^* = 20$. Optimal performance is achieved for this system with the value of $\mathbf{k}_c$ determined from Eq. (16.79). Therefore,

$$\mathbf{k}_c^T = [1 \quad 1.377 \quad 0.95417] \tag{17.95}$$

The feedback vector $\mathbf{k}_c$ could also be obtained by using the method of Chap. 13. This is accomplished by equating the coefficients in the denominator polynomial of Eq. (17.93) to the numerator coefficients of $1 + G(s)H_{eq}(s)$. However, the resulting system is not guaranteed to be optimal. When the transformation matrix $\mathbf{T}^{-1}$ in Example 2 of Sec. 16.8 is used, the feedback coefficients for the physical-variable representation of Fig. 13.7*b* are

$$\mathbf{k}_p^T = \mathbf{k}_c^T\mathbf{T}^{-1} = [1 \quad -3.39385 \quad 4.77085] \tag{17.96}$$

These values agree very closely with those obtained by the method of Chap. 13.

Example 2. Use of Table 17.9, with $n = 3$, for the plant

$$G(s) = \frac{100(s+2)}{s(s+1)(s+5)} \tag{17.97}$$

yields the DDE

$$\alpha + d_1 s + d_2 s^2 = 0 \tag{17.98}$$

The value $-\alpha$ is chosen to coincide with the value of the zero, $s = -2$, in Eq. (17.97). Assuming that an overdamped system response is acceptable, the particular quadratic equation is chosen as

$$(2 + s)(4 + s) = 8 + 6s + s^2 = 0$$

Dividing this equation by 4 yields the DDE

$$2 + 1.5s + 0.25s^2 = 0 \tag{17.99}$$

and

$$\mathbf{h}_c^T = [2 \quad 1.5 \quad 0.25] \tag{17.100}$$

Equation (16.79) yields

$$\mathbf{k}_c^T = [2 \quad 1.53 \quad 0.25] \tag{17.101}$$

With this value of $\mathbf{k}_c$ the actual characteristic equation of the system, in phase-variable notation, is

$$\begin{aligned} s^3 + (a_2 + K_G k_3)s^2 + (a_1 + K_G k_2)s + K_G k_1 &= s^3 + 31s^2 + 158s + 200 \\ &= (s + 2)(s + 4)(s + 25) \end{aligned} \tag{17.102}$$

Thus, with this design concept, a plant zero is canceled by a pole of the control ratio. Using the matrices $\mathbf{A}_c$ and $\mathbf{b}_c$ in Example 2, Sec. 16.8, the transfer function $G_K(s)$ for the plant of Eq. (17.97) is obtained from Eq. (16.51) [see Sec. 16.9]. For this $G_K(s)$, $\omega_\phi = 25$ and $\omega^* = 5$ [see Eq. (16.80)] which satisfy the condition $\xi = \omega_\phi/\omega^* = 5 > 1$ required to achieve an optimal performance.

17.12 PLANT WITH VARIABLE PARAMETERS[6-8]

A design method is presented which permits a system with *variable* plant parameters to satisfy the desired performance specifications. The method uses state-variable feedback with constant feedback gains and satisfies the quadratic performance index of Eq. (17.7). The design application is an aircraft pitch-rate control system. A block diagram of the basic plant is shown in Fig. 17.10. The control system must be insensitive to parameter variations which occur over a range of flight conditions. The values of these parameters for six flight conditions (FC) and speeds (M_δ, Mach number) are listed in Table 17.10. In order to design a control system with fixed feedback gains for such a plant it is necessary to determine the parameter which has the most influence on the time response. The closed-loop system performance specifications are as follows: the response to a

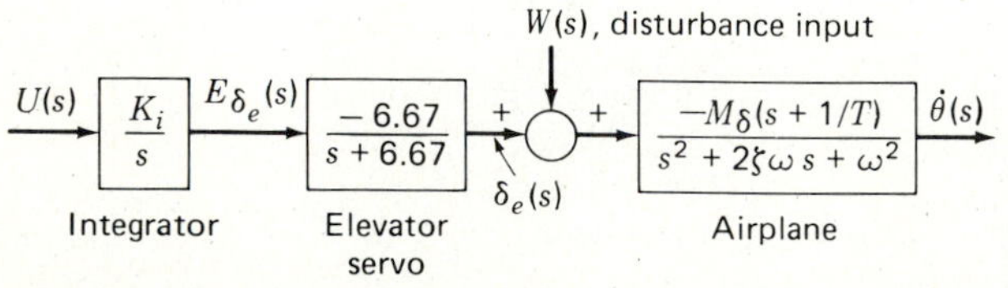

FIGURE 17.10
Block diagram of basic plant.

TABLE 17.10
System parameters

FC	M_δ	$1/T$	$2\zeta\omega$	ω^2
1	0.22	0.0356	0.302	2.28
2	16.29	1.163	2.226	6.51
3	52.95	2.070	4.980	56.10
4	20.86	0.325	0.652	18.71
5	2.24	0.0366	0.0792	3.68
6	0.70	0.00794	0.0165	0.65

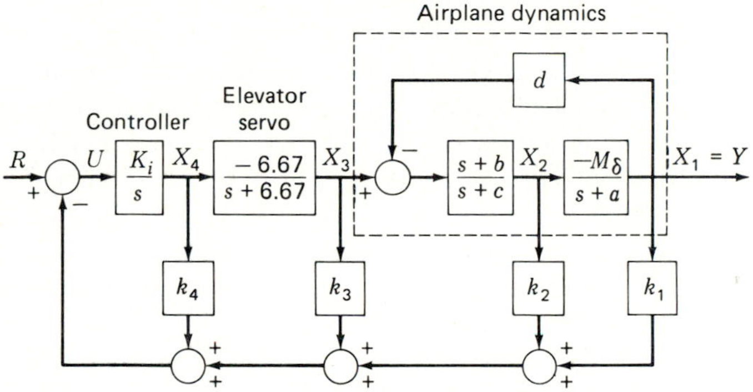

FIGURE 17.11
State-variable-feedback control system.

step command must have less than 25 percent overshoot and must damp to at least one-eighth amplitude within one cycle; the response must rise to within 90 percent of the command value in 3 s or less; and the system must damp the response to a step disturbance input to less than one-fourth amplitude in one cycle. If first-order effects dominate, 90 percent or more of the response to the disturbance should be eliminated within 3 s.

The configuration of the closed-loop system with the disturbance $W(s) = 0$ is shown in Fig. 17.11. The parameters shown are $a = \zeta\omega$, $b = 1/T$, $c = (\omega^2 - ab)/(a - b)$, and $d = (\omega^2 - a^2)/M_\delta(a - b)$, and the state x_2 is inaccessible. The basic plant transfer function is

$$G(s) = \frac{\dot{\theta}(s)}{U(s)} = \frac{6.67M_\delta K_i(s + b)}{s(s + 6.77)(s^2 + 2\zeta\omega s + \omega^2)} = \frac{K_G N(s)}{D(s)} \tag{17.103}$$

The system is to be designed by minimizing the PI

$$\text{PI} = \int_0^\infty \left(\mathbf{x}^T \mathbf{h}_c \mathbf{h}_c^T \mathbf{x} + u^2\right) dt \tag{17.104}$$

The equivalent feedback transfer function is

$$H_{eq}(s) = \frac{k_4(s^3 + p_3 s^2 + p_2 s + p_1)}{6.67 M_\delta (s + b)} = \frac{k_4 N_H(s)}{D_H(s)} \tag{17.105}$$

where the coefficients p_1, p_2, and p_3 are functions of the plant parameters and the feedback coefficients. For each FC a root locus can be obtained for this system, using $G(s)H_{eq}(s) = -1$. To minimize the effect of the variable coefficients of the plant upon the performance of the system, it is desired to design the system to minimize the movement of the zeros of $N_H(s)$. This procedure should yield the dominant roots of the system characteristic equation. Then, for $K_x > K_{x_{min}}$ the location of the zeros of Eq. (17.105) should not change appreciably in the s plane. To accomplish this minimum movement of zeros in the implementation of the system design, move all k_i to the x_1 state. This will require differentiation for $H_{eq}(s)$, which is, in general, not practical. An alternative choice is to move the maximum number feasible. Under this configuration choose the nominal $(\mathrm{FC})_N$ for which the values of K_i and the fixed-gain feedback coefficients are to be determined. The choice of $(\mathrm{FC})_N$ is based upon which coefficient has the principal influence on $y(t)$. For this system M_δ exerts the greatest influence since it changes by a ratio of 240:1. The FC having the minimum value of M_δ is FC 1. Hence, this flight condition is denoted as $(\mathrm{FC})_N$. For this nominal flight condition the pertinent equations are

$$\frac{Y(s)}{R(s)} = \frac{6.67 M_\delta^* K_i (s + 1/T^*)}{s^4 + (6.972 + K_i k_4^*)s^3 + (4.29 + K_i k_4^* p_3)s^2 + (15.21 + K_i k_4^* p_2)s + K_i k_4^* p_1} \tag{17.106}$$

where

$$p_1 = 6.67\left[\omega^2\left(1 - \frac{k_3^*}{k_4^*}\right) + \frac{M_\delta^*}{k_4^*}\beta\left(\frac{1}{T^*}\right)\right] \tag{17.107}$$

$$p_2 = 6.67\left[\left(1 - \frac{k_3^*}{k_4^*}\right)(2\zeta\omega)^* + \frac{M_\delta^*}{k_4^*}\left(\beta + \frac{A}{T^*}\right)\right] \tag{17.108}$$

$$p_3 = 6.67\left(1 + \frac{M_\delta^* A - k_3^*}{k_4^*}\right) + (2\zeta\omega)^* \tag{17.109}$$

$$A = \frac{-k_2^*}{M_\delta^*} \tag{17.110}$$

$$\beta = k_1^* + \left(\frac{\zeta\omega}{M_\delta}\right)^* k_2^* \tag{17.111}$$

Nominal values of parameters for $(\mathrm{FC})_N$ are denoted by an asterisk. For the aircraft under consideration, it is desirable to place a closed-loop system pole α coincident with the plant zero to reduce dependence of the system time response on the zero. If the two remaining poles are a complex-conjugate pair, the *nominal system characteristic equation* (DDE) is

$$(s + \alpha)(s^2 + 2\zeta_N \omega_N s + \omega_N^2) = 0 \tag{17.112}$$

For zero position error with a step input, $h_1 = \alpha$. Thus, the weighting vector for the phase-variable representation of a fourth-order plant with one zero (see Table 17.9) is

$$\mathbf{h}_c^T = \left[a \quad 1 + \frac{2\zeta_N}{\omega_N}\alpha \quad \frac{\alpha}{\omega_N^2} + \frac{2\zeta_N}{\omega_N} \quad \frac{1}{\omega_N^2} \right] \tag{17.113}$$

An analysis of the control ratio, Eq. (17.106), reveals that zero steady-state error occurs with a step input for all flight conditions when the following conditions are satisfied:

$$k_3^* = k_4^* \qquad \text{and} \qquad \beta^* = k_1^* - \left(\frac{a^*}{M_\delta^*}\right) k_2^* = 1 \tag{17.114}$$

The k_i or Eq. (17.114) are the values determined by use of the physical-variable representation of Fig. 17.11. Based upon the criterion of Eq. (16.80), the minimum value of K_i required for an optimal solution is

$$(K_i)_{\min} = \frac{\xi\omega^*\omega_N^2}{6.67M_\delta^*} \tag{17.115}$$

where ω^* represents the highest corner frequency of $\mathbf{G}_k(j\omega) = \mathbf{h}^T\mathbf{\Phi}(j\omega)\mathbf{b}$ [see Eq. (16.52)] and its value is chosen to be the highest corner frequency for all FCs.

A digital-computer control-system design program can be used to solve this problem. This program permits a trial-and-error approach with different values of ζ_N and ω_N for Eq. (17.112) in order to determine a satisfactory DDE (see Sec. 17.11) that satisfies the desired performance specifications for

$$y(t)\Big|_{\substack{w(t)=0 \\ r(t)=u_{-1}(t)}} \qquad \text{and} \qquad y(t)\Big|_{\substack{w(t)=u_{-1}(t) \\ r(t)=0}}$$

Based upon this trial-and-error procedure the values $K_i = 800$, $\xi = 2.8$, $\omega_N = 7.5$, $\zeta_N = 0.7$, $\omega^* = 7.5$, $\alpha^* = 0.0356$, and $M_\delta^* = 0.22$ yield satisfactory system performance. These values result in

$$\mathbf{h}_c^T = [0.0356 \quad 1.00665 \quad 0.18730 \quad 0.01778] \tag{17.116}$$

The matrix $\mathbf{h}_c$ is transformed to $\mathbf{h}_p$ which in turn results in

$$(\mathbf{k}_p^*)^T = [0.90526 \quad -0.04868 \quad 0.03102 \quad 0.03123] \tag{17.117}$$

Inserting the values from Eq. (17.117) into Eq. (17.111) yields $\beta = 0.93867$. Thus, in order to achieve the normalized value of $\beta^* = 1$, the k_i^* of Eq. (17.117) are divided by this value of β. Also, in order to achieve the condition of $k_3^* = k_4^*$, their normalized values of 0.0330466 and 0.0332704, respectively, are approximated by $k_{3N}^* = 0.033$. Thus the normalized $\mathbf{k}_N^*$ is

$$\mathbf{k}_N^* = [0.9644 \quad -0.05186 \quad 0.033 \quad 0.033] \tag{17.118}$$

A computer program is used to obtain the time responses for all FCs. An analysis of Eqs. (17.107) to (17.111) shows that the values of p_1, p_2, and p_3 remain unchanged by the normalization. Since $(k_{4N}^* - k_4^*)/k_4^*$ represents approxi-

TABLE 17.11
Roots of the characteristic equation

FC	Dominant pole(s)	Nondominant poles	Zero $s = -1/T$
Nominal			
1	$-5.125 + j5.0228$	$-23.0865, -0.03515$	-0.0356
2	-4.2550	$-14.939 \pm j142.14, -1.16306$	-1.163
3	-4.2386	$-15.871 \pm j257.76, -2.069$	-2.070
4	-4.2574	$-14.57 \pm j161.127, -0.32464$	-0.325
5	-4.4382	$-14.337 \pm j49.93, -0.03652$	-0.0366
6	-5.0468	$-14.016 \pm j23.335, -0.007931$	-0.00794

mately a 6.5 percent change in k_4^*, using k_{4N}^* is not detrimental to achieving the desired performance specifications.

Table 17.11 lists the poles and zero of the system control ratio for all six flight conditions. In all cases the closed-loop transfer function contains two real poles and a complex-conjugate pair, and one of the real poles essentially coincides with the plant zero. This was a desired result of the design: to minimize the effect of the plant zero on $y(t)$. For all cases, except for $(\text{FC})_N$, the remaining real pole is the dominant pole, whereas for $(\text{FC})_N$ the complex-conjugate poles are the dominant poles. Note that the value of all real dominant poles and the real part of the dominant poles for FC 1 do not vary significantly (-4.23 to -5.125). This is a desired result of the design; i.e., the system meets the desired performance specifications under fixed-gain feedback for all flight conditions.

The implementation of the final design values is shown in Fig. 17.12, where states x_3 and x_4 are fed back through a common feedback unit to ensure the condition of $k_{3N}^* = k_{4N}^*$ and the feedback signal $(x_2 k_{2N}^*)$ is obtained from the output state x_1. This requires a proportional plus rate feedback unit. The system

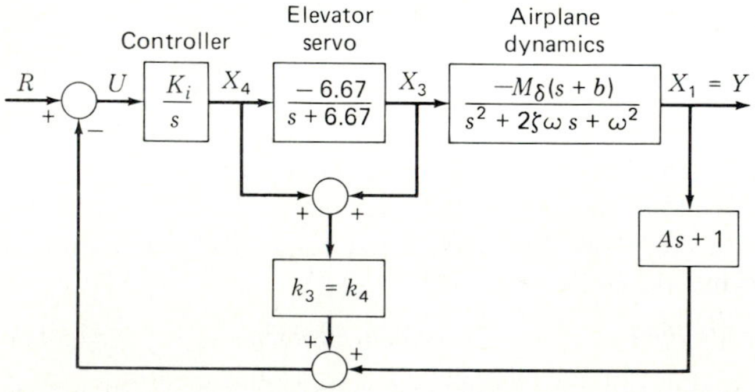

FIGURE 17.12
Final design configuration.

of Fig. 17.12 yields a suboptimal performance for all FCs. The settling time t_s [2 percent of the steady-state value of $y(t)$] for FC 1 is in excess of 15 s; for all other FCs $t_s < 15$ s. The design technique presented in this section, using the DDE method, illustrates a method of obtaining a fixed-gain feedback system that satisfies the desired performance specifications over the entire range of plant parameter variations.

17.13 SUMMARY

This chapter presents methods for selecting the elements of the matrix **Q** or the vector **h** under hfg operation in order to achieve desired conventional figures of merit in addition to satisfying the quadratic PI. Once the **Q** or **h** is specified, the corresponding **k** is determined in terms of the phase-variable representation. The feedback gains k_{p_i} for the physical-variable system are then solved by use of a linear transformation. When some states are not accessible, they can be generated from states that are accessible (see Sec. 18.3). The state-estimation techniques of Kalman and Luenberger may be used. For a linear time-invariant system the block-manipulation method for generating states still results in an optimal system. An alternate design method is to correlate with the conventional figures of merit[6–8] the elements h_i of the vector **h**, where $\mathbf{Q} = \mathbf{h}\mathbf{h}^T$. The hfg design method presented in this chapter can be used for systems which have plants with varying parameters.[8] The objective is to determine a feedback vector **k** that maintains satisfactory characteristics for the time response of the system, even though the plant parameters may vary. By this method the feedback vector is determined for an appropriate choice of nominal values of the plant parameters. Optimal performance exists when the system is operating with its nominal plant parameters; otherwise the performance is suboptimal.

Chapters 13 to 17 present methods for designing state-variable-feedback systems. These methods are not all-inclusive but are intended to illustrate the concepts of state-variable feedback and optimal control. Other complementary design methods, such as modal control theory,[9] are available in the literature. Two other methods for designing feedback control systems using the entire eigenstructure and the quantitative feedback theory techniques are presented in Chaps. 18 to 20, and 21, respectively. These methods are well suited to the synthesis of multiple-input multiple-output systems.

REFERENCES

1. Houpis, C. H.: "The Relationship between the Conventional Control Theory Figures of Merit and the Performance Indices in Optimal Control Theory," Ph.D. dissertation, Department of Electrical Engineering, University of Wyoming, 1971.
2. Houpis, C. H., and C. T. Constantinides: "Functional Relationship between the Conventional Steady-State Characteristics and the Weighting Matrices in the Quadratic Performance Index," *Int. J. Control*, vol. 15, no. 6, pp. 1147–1156, 1972; "Correlation between Conventional Figures of Merit and the **Q** Matrix on the Quadratic Cost Function: Third-Order Plant," *Int. J. Control*, vol. 16, no. 4, pp. 695–704, 1972; "Relationship between Conventional Control Theory Figures of

Merit and Quadratic Performance Index in Optimal Control Theory for a Single-Input/Single-Output System," *Proc. IEE*, vol. 120, no. 1, pp. 138–142, 1973.

3. Bullock, T. E., and J. M. Elder: "Quadratic Performance Index Generation for Optimal Regulator Design," *1971 IEEE Conf. Decision Control*, pp. 123–124.
4. Rynaski, E. G., and R. F. Whitbeck: "The Theory and Application of Linear Optimal Control," Air Force Flight Dynam. Lab. Tech. Rep. AFFDL-TR-65-28, Wright-Patterson Air Force Base, Ohio, 1966.
5. Kalman, R. E.: "When Is a Linear Control System Optimal?," *ASME J. Basic Eng.*, pp. 51–60, March 1964.
6. Mirmak, E. V.: "Some Techniques for Optimal Linear Regulator Design to Satisfy Conventional Figures of Merit," M.Sc. thesis, GA/EE/73A-4, Air Force Institute of Technology, Wright-Patterson Air Force Base, Ohio, 1974.
7. Wilt, D. C.: "Design of an Optimal Linear Regulator to Satisfy Conventional Figures of Merit by Utilizing $\mathbf{Q} = \mathbf{hh}^T$ and the Defined Dynamic Equation Approach," GE/EE/74-66, Air Force Institute of Technology, Wright-Patterson Air Force Base, Ohio, 1974.
8. Ray, R. A.: "A State-Variable Design Approach for a High-Performance Aerospace Vehicle Pitch Orientation System with Variable Coefficients," M.Sc. thesis, GGC/EE/73-15, Air Force Institute of Technology, Wright-Patterson Air Force Base, Ohio, 1973.
9. Porter, B., and T. R. Crossley: *Modal Control Theory and Applications*, Barnes and Noble, New York, 1972.
10. Kriendler, E.: "Synthesis of Flight Control Systems Subject to Vehicle Parameter Variations," Air Force Flight Dynam. Lab. Tech. Rep. TR-66-209, Wright-Patterson Air Force Base, Ohio, April 1967.
11. *Solution on Linear-Quadratic Optimal Control Problem (OPTCON)*, Air Force Institute of Technology Computer Program Library, 1974.
12. Sage, A. P., and C. C. White, III: *Optimum Systems Control*, 2d ed., Prentice-Hall, Englewood Cliffs, N.J., 1977.
13. Schultz, D. G., and J. L. Melsa: *State Functions and Linear Control Systems*, McGraw-Hill, New York, 1967.

CHAPTER 18

STRUCTURAL PROPERTIES OF LINEAR MULTIVARIABLE CONTROL SYSTEMS

18.1 INTRODUCTION

The foundations for control-system synthesis presented in this text apply to multivariable systems, i.e., systems with multiple inputs and multiple outputs (MIMO). However, the emphasis of the root-locus, frequency-response, and state-variable methods in the preceding chapters is upon applications to single-input single-output (SISO) systems. The synthesis methods developed in Chaps. 18 to 20 are based upon the time-domain representation of the basic plant in terms of vector state and output equations. Therefore, the matrix properties presented in the earlier chapters are used extensively. The principal objective of the state-feedback method of Chap. 13 is the assignment of the closed-loop eigenvalues for SISO systems. Similar results are obtained in Sec. 18.2 directly from the state equations. This design process can be simplified for higher-order systems by first transforming the state equation into control canonical form $\dot{\mathbf{x}} = \mathbf{A}_c\mathbf{x} + \mathbf{b}_c\mathbf{u}$. For this representation of the state equation the plant matrix $\mathbf{A}_c$ is in companion form, and the control matrix $\mathbf{b}_c$ contains all zeros except for the unit element in the last row. This form leads directly to the determination of the required state-feedback matrix. The assignment of the closed-loop eigenvalue spectrum for MIMO systems is covered in Sec. 18.6, where the state equation is first transformed into generalized control canonical form. In this representation the plant matrix is block diagonal, where each block is in companion form, and readily leads to the determination of a state-feedback matrix which assigns the desired closed-loop eigenvalues.

In the SISO case a unique feedback matrix is required to assign a desired eigenvalue spectrum. With each eigenvalue there is an associated eigenvector which is also unique. However, in the MIMO case there are an infinite number of feedback matrices which can assign a specified eigenvalue spectrum. Also, with each feedback matrix there results a new set of associated eigenvectors. Since the eigenvectors also affect the time response of the system (see Sec. 18.8), it is important to assign both the eigenvalues and the associated eigenvectors. Thus, the principles which are the basis of *entire eigenstructure assignment*, consisting of both the eigenvalues and the associated eigenvectors, are presented in the remainder of this chapter. The synthesis of regulators and trackers by the method of entire eigenstructure assignment, using state feedback, is illustrated in Chap. 19. The design of output feedback systems is covered in Chap. 20.

18.2 STATE FEEDBACK FOR SISO SYSTEMS

This section covers the design of a single-input single-output (SISO) system in which the set of closed-loop eigenvalues is assigned by state feedback. The plant state and output equations have the form

$$\dot{\mathbf{x}} = \mathbf{A}\mathbf{x} + \mathbf{B}\mathbf{u} \tag{18.1}$$

$$\mathbf{y} = \mathbf{C}\mathbf{x} \tag{18.2}$$

The state-feedback control law (see Fig. 18.1) is

$$\mathbf{u} = \mathbf{r} + \mathbf{K}\mathbf{x} \tag{18.3}$$

where $\mathbf{r}$ is the input vector and $\mathbf{K}$ is the state-feedback matrix having dimension $m \times n$. The closed-loop state equation obtained by combining Eqs. (18.1) and

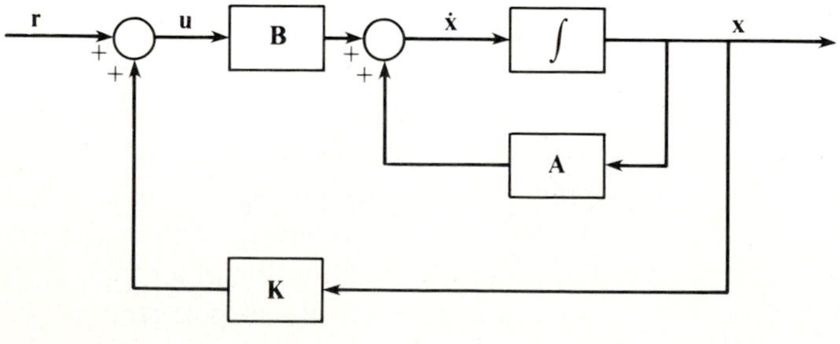

FIGURE 18.1
State feedback control system.

(18.3) yields

$$\dot{\mathbf{x}} = (\mathbf{A} + \mathbf{BK})\mathbf{x} + \mathbf{Br} \tag{18.4}$$

Thus the closed-loop system matrix produced by state feedback is

$$\mathbf{A}_{\text{cl}} = \mathbf{A} + \mathbf{BK} \tag{18.5}$$

which results in the closed-loop characteristic equation

$$Q(\lambda) = |\lambda\mathbf{I} - \mathbf{A}_{\text{cl}}| = |\lambda\mathbf{I} - \mathbf{A} - \mathbf{BK}| = 0 \tag{18.6}$$

Equation (18.6) reveals that the closed-loop eigenvalues are therefore assigned by the proper selection of the state-feedback matrix **K**. A necessary and sufficient condition for the selection of **K** is that the plant be completely controllable[16] (see Sec. 13.7).

Example. A second-order system is represented by the state and output equations

$$\dot{\mathbf{x}} = \begin{bmatrix} -1 & 3 \\ 0 & -2 \end{bmatrix}\mathbf{x} + \begin{bmatrix} 1 \\ 1 \end{bmatrix}\mathbf{u} \tag{18.7}$$

$$\mathbf{y} = [1 \quad 0]\mathbf{x} = x_1 \tag{18.8}$$

The open-loop eigenvalue spectrum is $\sigma(\mathbf{A}) = \{-1, -2\}$. The system is completely controllable since, from Eq. (13.68),

$$\text{Rank}\,\mathbf{M}_c = \text{Rank}\begin{bmatrix} 1 & 2 \\ 1 & -2 \end{bmatrix} = 2 \tag{18.9}$$

The closed-loop eigenvalues are determined from Eq. (18.6):

$$|\lambda\mathbf{I} - \mathbf{A} - \mathbf{BK}| = \left|\begin{bmatrix} \lambda & 0 \\ 0 & \lambda \end{bmatrix} - \begin{bmatrix} -1 & 3 \\ 0 & -2 \end{bmatrix} - \begin{bmatrix} 1 \\ 1 \end{bmatrix}[k_1 \quad k_2]\right|$$

$$= \lambda^2 + (3 - k_1 - k_2)\lambda + (2 - 5k_1 - k_2) \tag{18.10}$$

Assigning the closed-loop eigenvalues $\lambda_1 = -5$ and $\lambda_2 = -6$ requires the characteristic polynomial to be $\lambda^2 + 11\lambda + 30$. Equating the coefficients of this desired characteristic equation with those of Eq. (18.10) results in the values $k_1 = -5$ and $k_2 = -3$, or $\mathbf{K} = [-5 \quad -3]$. The resulting closed-loop state equation is

$$\dot{\mathbf{x}} = \begin{bmatrix} -6 & 0 \\ -5 & -5 \end{bmatrix}\mathbf{x} + \begin{bmatrix} 1 \\ 1 \end{bmatrix}\mathbf{r} \tag{18.11}$$

This state equation has the required eigenvalue spectrum $\sigma(\mathbf{A}_{\text{cl}}) = \{-5, -6\}$.

18.3 OBSERVERS[17-20]

The synthesis of state-feedback control presented in Chap. 13 and in Sec. 18.2 assumes that all the states **x** are measurable or that they can be generated from the output (see Sec. 14.5). In many practical control systems it is physically or economically impractical to install all the transducers which would be necessary to measure all of the states. The ability to reconstruct the plant states from the output requires that all the states be observable. The necessary condition for complete observability is given by Eq. (13.73) and is repeated here.

$$\text{Rank}\,\mathbf{M}_o = \text{Rank}\left[\mathbf{C}^T \quad \mathbf{A}^T\mathbf{C}^T \quad (\mathbf{A}^T)^2\mathbf{C}^T \cdots (\mathbf{A}^T)^{n-\ell}\mathbf{C}^T\right] = n \tag{18.12}$$

where n is the dimension of **A** and ℓ is the dimension of y.

The purpose of this section is to present methods of reconstructing the states from the measured outputs by a dynamical system, which is called an *observer*. The reconstructed state vector $\hat{\mathbf{x}}$ may then be used to implement a state-feedback control law $\mathbf{u} = \mathbf{K}\hat{\mathbf{x}}$. The basic method of reconstructing the states is to simulate the state and output equations of the plant on a computer. Thus, consider a system represented by the state and output equations of Eqs. (18.1) and (18.2). These equations are simulated on a computer with the same input **u** that is applied to the actual physical system. The states of the simulated system and of the actual system will then be identical only if the initial conditions of the simulation and of the physical plant are the same. However, the physical plant may be subjected to unmeasurable disturbances which cannot be applied to the simulation. Therefore, the difference between the actual plant output **y** and the simulation output $\hat{\mathbf{y}}$ is used as another input in the simulation equation. Thus, the observer state and output equations become

$$\dot{\hat{\mathbf{x}}} = \mathbf{A}\hat{\mathbf{x}} + \mathbf{B}\mathbf{u} + \mathbf{L}(\mathbf{y} - \hat{\mathbf{y}}) \tag{18.13}$$

and

$$\hat{\mathbf{y}} = \mathbf{C}\hat{\mathbf{x}} \tag{18.14}$$

where **L** is the $n \times \ell$ *observer matrix*.

A method for synthesizing the matrix **L** uses the observer *reconstruction error* defined by

$$\mathbf{e} \equiv \mathbf{x} - \hat{\mathbf{x}} \tag{18.15}$$

Subtracting Eq. (18.13) from Eq. (18.1) and using Eqs. (18.2), (18.14), and (18.15) yields the observer-error state equation

$$\dot{\mathbf{e}} = (\mathbf{A} - \mathbf{L}\mathbf{C})\mathbf{e} \tag{18.16}$$

By an appropriate choice of the observer matrix **L**, all the eigenvalues of $\mathbf{A} - \mathbf{LC}$ can be assigned to the left-half plane, so that the steady-state value of $\mathbf{e}(t)$ for *any* initial condition is zero:

$$\lim_{t \to \infty} \mathbf{e}(t) = \mathbf{0} \tag{18.17}$$

Equation (18.16) indicates that the error equation has no input and is excited only by the initial condition, and thus the observer error is not determined by the system input. The steady-state value of the error is therefore equal to zero. The significance of this is that the observer states approach the plant states in the steady-state for *any* input to the plant. The eigenvalues of $\mathbf{A} - \mathbf{LC}$ are usually selected so that they are to the left of the eigenvalues of **A**. Thus, the observer states have an initial "start up" error when the observer is turned on, but they rapidly approach the plant states.

The selection of the observer matrix **L** can be used to assign both the eigenvalues and the eigenvectors of the matrix $\mathbf{A} - \mathbf{LC}$. Only eigenvalue assignment is considered in this section. The method of Sec. 18.9 is used to assign the eigenvectors also.

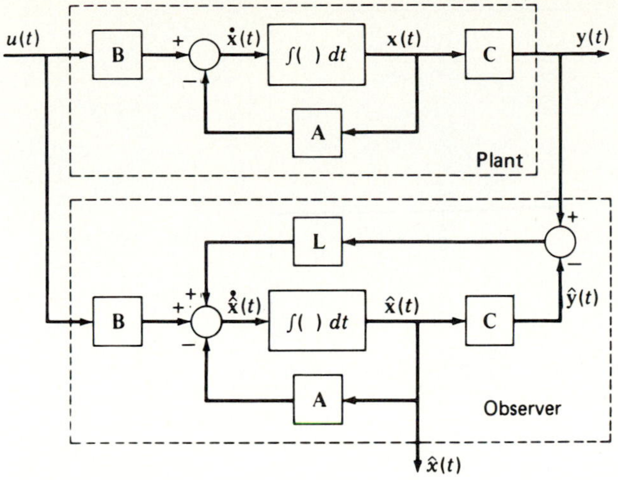

FIGURE 18.2
Plant with inaccessible states and an observer which reconstructs the states.

The representation of the physical plant represented by Eqs. (18.1) and (18.2) and the observer represented by Eqs. (18.13) and (18.14) are shown in Fig. 18.2.

Example. In the example of Sec. 18.2, the output $y = x_1$ is used to implement a state feedback control law. An observer is to be synthesized to reconstruct the inaccessible state x_2.

The system is completely observable since, from Eq. (18.12),

$$\text{Rank}\,\mathbf{M}_o = \text{Rank}\begin{bmatrix} 1 & -1 \\ 0 & 3 \end{bmatrix} = 2 = n \tag{18.18}$$

The eigenvalues of $\mathbf{A} - \mathbf{LC}$ are determined by the characteristic polynomial

$$|\lambda\mathbf{I} - (\mathbf{A} - \mathbf{LC})| = \left| \begin{bmatrix} \lambda & 0 \\ 0 & \lambda \end{bmatrix} - \begin{bmatrix} -1 & 3 \\ 0 & -2 \end{bmatrix} + \begin{bmatrix} l_1 \\ l_2 \end{bmatrix} [1 \quad 0] \right|$$

$$= \lambda^2 + (3 + l_1)\lambda + (2 + 2l_1 + 3l_2) = (\lambda + 9)(\lambda + 10) \tag{18.19}$$

Assigning the eigenvalues as $\lambda_1 = -9$ and $\lambda_2 = -10$ requires that $l_1 = 16$ and $l_2 = 56/3$. The effectiveness of the observer in reconstructing the state x_2 is shown by comparing the responses of the plant and the observer. For $u(t) = u_{-1}(t)$, $x_1(0) = 0$, $x_2(0) = 1$, $\hat{x}_1(0) = 0$, and $\hat{x}_2(0) = 0$ the initial state reconstruction errors are $e_1(0) = 0$ and $e_2(0) = 1$. The responses obtained from Eqs. (18.1), (18.16), and (18.13), respectively, are

$$x_2(t) = 0.5 + 0.5e^{-2t} \tag{18.20}$$

$$e_2(t) = 8e^{-9t} - 7e^{-10t} \tag{18.21}$$

$$\hat{x}_2(t) = 0.5 + 0.5e^{-2t} - 8e^{-9t} + 7e^{-10t} \tag{18.22}$$

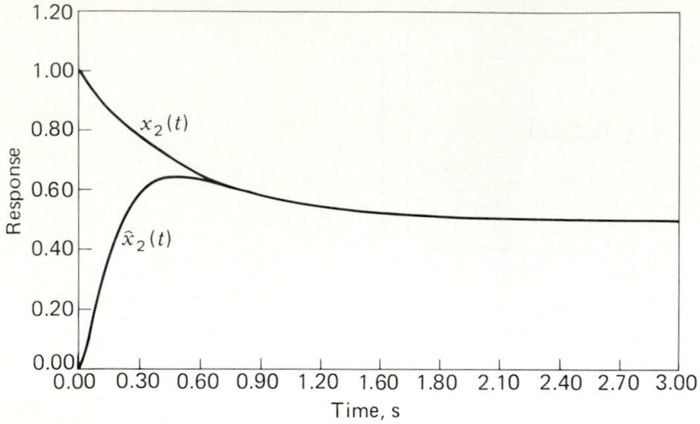

FIGURE 18.3
Plant and observer states, Eqs. (18.20) and (18.22).

These quantities are plotted in Fig. 18.3 and show that $\hat{x}_2(t)$ converges rapidly to the actual state $x_2(t)$.

The observer described by Eqs. (18.13) and (18.14) is called a *full-order observer* since it generates estimates of all the states. If some of the states are accessible, as in the example above in which the state $x_1 = y$ is measurable, the order of the observer may be reduced. The description of such *reduced-order observers* is contained in the literature.[4]

18.4 CONTROL SYSTEMS CONTAINING OBSERVERS

This section covers the synthesis of control systems for plants with inaccessible states. In such cases an observer can be used to generate estimates of the states, as described in Sec. 18.3. Consider the case of a controllable and observable system governed by the state and output equations of Eqs. (18.1) and (18.2). The observer governed by Eq. (18.13) produces the states $\hat{\mathbf{x}}$ which are estimates of the plant states $\mathbf{x}$. The reconstruction error between the plant states $\mathbf{x}$ and the observer states $\hat{\mathbf{x}}$ is given by Eq. (18.16). The matrix $\mathbf{L}$ is selected, as shown in Sec. 18.3, so that the eigenvalues of $\mathbf{A} - \mathbf{LC}$ are far to the left of the eigenvalues of $\mathbf{A}$. The control law used to modify the performance of the plant is

$$\mathbf{u} = \mathbf{r} + \mathbf{K}\hat{\mathbf{x}} = \mathbf{r} + \mathbf{K}(\mathbf{x} - \mathbf{e}) \tag{18.23}$$

where $\mathbf{r}$ is the input. Then Eq. (18.1) becomes

$$\dot{\mathbf{x}} = (\mathbf{A} + \mathbf{BK})\mathbf{x} - \mathbf{BKe} + \mathbf{Br} \tag{18.24}$$

The composite closed-loop system consists of the interconnected plant and observer (see Fig. 18.2) and the controller of Eq. (18.23). This closed-loop system is represented by Eqs. (18.24) and (18.16), which are written in the composite matrix form

$$\begin{bmatrix} \dot{\mathbf{x}} \\ \dot{\mathbf{e}} \end{bmatrix} = \begin{bmatrix} \mathbf{A} + \mathbf{BK} & -\mathbf{BK} \\ \mathbf{0} & \mathbf{A} - \mathbf{LC} \end{bmatrix} \begin{bmatrix} \mathbf{x} \\ \mathbf{e} \end{bmatrix} + \begin{bmatrix} \mathbf{B} \\ \mathbf{0} \end{bmatrix} \mathbf{r} \tag{18.25}$$

The characteristic polynomial for the matrix of Eq. (18.25) is

$$Q(\lambda) = |\lambda \mathbf{I} - (\mathbf{A} + \mathbf{BK})| \cdot |\lambda \mathbf{I} - (\mathbf{A} - \mathbf{LC})| \tag{18.26}$$

Therefore, the eigenvalues of $\mathbf{A} + \mathbf{BK}$ and $\mathbf{A} - \mathbf{LC}$ can be assigned independently by the selection of appropriate matrices $\mathbf{K}$ and $\mathbf{L}$. This is an important property which permits the observer and the controller to be designed separately.

Example. For the system in the example of Sec. 18.2, the eigenvalues of $\mathbf{A} + \mathbf{BK}$ are assigned as $\lambda_1 = -5$ and $\lambda_2 = -6$ by selecting $k_1 = -5$ and $k_2 = -3$. For $u(t) = u_{-1}(t)$, $x_1(0) = 0$, and $x_2(0) = 1$ the state response of the system of Eq. (18.11) when the states are accessible is

$$\begin{bmatrix} \mathbf{x}_1(t) \\ \mathbf{x}_2(t) \end{bmatrix} = \begin{bmatrix} 1/6 - (1/6)e^{-6t} \\ 1/30 + (9/5)e^{-5t} - (5/6)e^{-6t} \end{bmatrix} \tag{18.27}$$

It is noted, since $y = x_1$, that the output does not track the input. The method presented in Chap. 19 causes the output to track the input in the steady state.

When the states are not accessible and the observer of the example in Sec. 18.3 is used, the composite representation of Eq. (18.25) is

$$\begin{bmatrix} \dot{\mathbf{x}} \\ \dot{\mathbf{e}} \end{bmatrix} = \begin{bmatrix} -6 & 0 & 5 & 3 \\ -5 & -5 & 5 & 3 \\ 0 & 0 & -17 & 3 \\ 0 & 0 & -56/3 & -2 \end{bmatrix} \begin{bmatrix} \mathbf{x} \\ \mathbf{e} \end{bmatrix} + \begin{bmatrix} 1 \\ 1 \\ 0 \\ 0 \end{bmatrix} \mathbf{r} \tag{18.28}$$

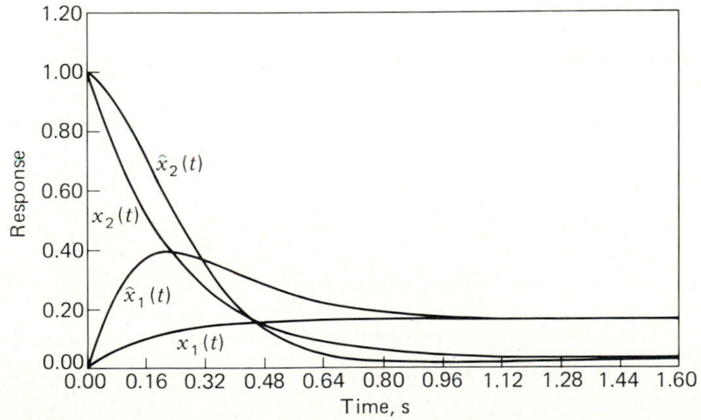

FIGURE 18.4
Plant and observer states, Eqs. (18.27) and (18.29).

The solution of the plant states from Eq. (18.28) with the input $r(t) = u_{-1}(t)$, $x_1(0) = 0$, $x_2(0) = 1$, $e_1(0) = 0$, and $e_2(0) = 1$ is

$$\begin{bmatrix} \mathbf{x}_1(t) \\ \mathbf{x}_2(t) \end{bmatrix} = \begin{bmatrix} 1/6 + (23/6)e^{-6t} - 13e^{-9t} + 9e^{-10t} \\ 1/30 - (42/5)e^{-5t} + (115/6)e^{-6t} - 26e^{-9t} + (81/5)e^{-10t} \end{bmatrix} \tag{18.29}$$

The difference between Eqs. (18.27) and (18.29) is due to the presence of the observer in the latter case. The responses are plotted in Fig. 18.4 and show that the states reach the same steady-state values.

The contribution of the observer to the plant-state response can be changed by assigning the eigenvectors as well as the eigenvalues. This can be accomplished by the method of Chaps. 18 and 19.

18.5 STATE-FEEDBACK DESIGN FOR SISO SYSTEMS USING THE CONTROL CANONICAL FORM

The example of Sec. 18.2 is of second order, and therefore the algebraic solution for the **K** matrix is computationally simple and direct. For higher-order systems the computations are simplified by first transforming the state equation to control canonical form, so that the plant matrix is in companion form. That procedure is presented in this section.

The model of an open-loop SISO system can be represented by a physical state variable and an output equation

$$\dot{\mathbf{x}}_p = \mathbf{A}_p\mathbf{x}_p + \mathbf{b}_p\mathbf{u} \tag{18.30}$$

$$\mathbf{y} = \mathbf{c}_p^T\mathbf{x}_p \tag{18.31}$$

The actuating signal **u** is formed in accordance with the feedback control law

$$\mathbf{u} = \mathbf{r} + \mathbf{k}_p^T\mathbf{x}_p \tag{18.32}$$

where **r** is the reference input and

$$\mathbf{k}_p^T = \begin{bmatrix} k_{p1} & k_{p2} & \cdots & k_{pn} \end{bmatrix} \tag{18.33}$$

The closed-loop state equation obtained from Eqs. (18.30) and (18.32) is

$$\dot{\mathbf{x}}_p = \left[\mathbf{A}_p + \mathbf{b}_p\mathbf{k}_p^T\right]\mathbf{x}_p + \mathbf{b}_p\mathbf{r} = \mathbf{A}_{p_{cl}}\mathbf{x}_p + \mathbf{b}_p\mathbf{r} \tag{18.34}$$

The design procedure which assigns the desired set of closed-loop eigenvalues, $\sigma(\mathbf{A}_{p_{cl}}) = \{\lambda_i\}$, $i = 1, 2, \ldots, n$, is accomplished as follows:

1. Use the method of Sec. 5.13 to obtain the matrix **T**, which transforms the state equation (18.30), by the change of state variables $\mathbf{x}_p = \mathbf{T}\mathbf{x}$ into the *control canonical form* $\dot{\mathbf{x}} = \mathbf{A}_c\mathbf{x} + \mathbf{b}_c\mathbf{u}$, where $\mathbf{A}_c$ is in companion form and $\mathbf{b}_c = [0 \quad \cdots \quad 0 \quad 1]^T$.

2. Select the set of desired closed-loop eigenvalues and obtain the corresponding closed-loop characteristic polynomial

$$\begin{aligned} Q_{\text{cl}}(\lambda) &= (\lambda - \lambda_1)(\lambda - \lambda_2)\cdots(\lambda - \lambda_n) \\ &= \lambda^n + \alpha_{n-1}\lambda^{n-1} + \cdots + \alpha_1\lambda + \alpha_0 \end{aligned} \tag{18.35}$$

3. Write the companion form of the closed-loop plant matrix

$$A_{c_{\text{cl}}} = \begin{bmatrix} 0 & 1 & & & & \\ & 0 & 1 & & & \\ & & 0 & \ddots & & \\ & & & \ddots & 1 & \\ & & & & 0 & 1 \\ -\alpha_0 & -\alpha_1 & & \cdots & -\alpha_{n-2} & -\alpha_{n-1} \end{bmatrix} \tag{18.36}$$

4. Use the feedback matrix

$$\mathbf{k}_c^T = [k_{c1} \quad k_{c2} \quad \cdots \quad k_{cn}] \tag{18.37}$$

to form the closed-loop matrix

$$\mathbf{A}_{c_{\text{cl}}} = \mathbf{A}_c + \mathbf{b}_c\mathbf{k}_c^T$$

$$= \begin{bmatrix} 0 & 1 & \cdots & 0 & 0 \\ 0 & 0 & \cdots & 0 & 0 \\ \cdots & \cdots & \cdots & \cdots & \cdots \\ 0 & 0 & \cdots & 1 & 0 \\ 0 & 0 & \cdots & 0 & 1 \\ -a_0 + k_{c1} & -a_1 + k_{c2} & \cdots & -a_{n-2} + k_{c(n-1)} & -a_{n-1} + k_{cn} \end{bmatrix} \tag{18.38}$$

5. Equate the matrices of Eqs. (18.36) and (18.38) and solve for the elements of the matrix $\mathbf{k}_c^T$.
6. Compute the required physical feedback matrix using

$$\mathbf{k}_p^T = \mathbf{k}_c^T\mathbf{T}^{-1} \tag{18.39}$$

Example. Using the physical system shown, assign the set of eigenvalues $\{-1 + j1, -1 - j1, -4\}$.

$$\mathbf{A}_p = \begin{bmatrix} 1 & 6 & -3 \\ -1 & -1 & 1 \\ -2 & 2 & 0 \end{bmatrix} \qquad \mathbf{b}_p = \begin{bmatrix} 1 \\ 1 \\ 1 \end{bmatrix} \tag{18.40}$$

This open-loop system is unstable since $\sigma(\mathbf{A}) = \{1, 1, -2\}$.

Step 1. As derived in the example of Sec. 5.13,

$$\mathbf{A}_c = \begin{bmatrix} 0 & 1 & 0 \\ 0 & 0 & 1 \\ -2 & 3 & 0 \end{bmatrix} \qquad \mathbf{b}_c = \begin{bmatrix} 0 \\ 0 \\ 1 \end{bmatrix} \tag{18.41}$$

and

$$\mathbf{T} = \begin{bmatrix} -5 & 4 & 1 \\ -6 & -1 & 1 \\ -13 & 0 & 1 \end{bmatrix} \tag{18.42}$$

Step 2

$$Q_{\text{cl}}(\lambda) = (\lambda + 1 - j1)(\lambda + 1 + j1)(\lambda + 4) = \lambda^3 + 6\lambda^2 + 10\lambda + 8 \tag{18.43}$$

Step 3

$$A_{c_{\text{cl}}} = \begin{bmatrix} 0 & 1 & 0 \\ 0 & 0 & 1 \\ -8 & -10 & -6 \end{bmatrix} \tag{18.44}$$

Step 4

$$\mathbf{A}_c + \mathbf{b}_c\mathbf{k}_c^T = \begin{bmatrix} 0 & 1 & 0 \\ 0 & 0 & 1 \\ -2 + k_{c1} & 3 + k_{c2} & k_{c3} \end{bmatrix} \tag{18.45}$$

Step 5. Equating the coefficients of $\mathbf{A}_{c_{\text{cl}}}$ and $\mathbf{A}_c + \mathbf{b}_c\mathbf{k}_c^T$ yields

$$\mathbf{k}_c^T = [-6 \quad -13 \quad -6] \tag{18.46}$$

Step 6. The feedback matrix required is

$$\mathbf{k}_p^T = \mathbf{k}_c^T\mathbf{T}^{-1} = \tfrac{1}{36}[-175 \quad -232 \quad 191] \tag{18.47}$$

The reader should verify that the physical closed-loop matrix

$$\mathbf{A}_{p_{\text{cl}}} = \mathbf{A}_p + \mathbf{b}_p\mathbf{k}_p^T = \frac{1}{36}\begin{bmatrix} -139 & -16 & 83 \\ -211 & -268 & 227 \\ -247 & -160 & 191 \end{bmatrix} \tag{18.48}$$

has the required set of eigenvalues $\sigma(\mathbf{A}_{p_{cl}}) = \{-1 + j1, -1 - j1, -4\}$.

18.6 GENERALIZED CONTROL CANONICAL FORM (GCCF)[1-5] FOR MIMO SYSTEMS

The basic mathematical structure of the matrix state-equation representation of a system is incorporated in the *generalized control canonical form*. This canonical form exhibits inherent properties of the system which are extremely useful in state-feedback system design. This form is readily applied to closed-loop system design which achieves a desired placement of the eigenvalues. While it is not the only form which can be used for closed-loop eigenvalue assignment, it has definite advantages for higher-order systems. The design objective in using the

GCCF is as follows:

1. Obtain the state equations in terms of physical state variables.
2. Transform to a new set of state variables for which the plant and control matrices are in a mathematically more tractable form; i.e., the design is more readily performed in this new state space.
3. Perform the design in the new state space.
4. Return to the original state space to produce the required state-feedback matrix.

The state equation for the open-loop system is

$$\dot{\mathbf{x}} = \mathbf{A}\mathbf{x} + \mathbf{B}\mathbf{u} \tag{18.49}$$

where $\mathbf{x}$ is $n \times 1$, $\mathbf{A}$ is $n \times n$, $\mathbf{B}$ is $n \times m$, $\mathbf{u}$ is $m \times 1$, and the rank of $\mathbf{B}$ is m.† By means of similarity transformations on $\mathbf{x}$ and $\mathbf{u}$ and by the addition of state feedback, this equation can be converted into the generalized canonical form[1-3]

$$\dot{\mathbf{z}} = \mathbf{A}_G\mathbf{z} + \mathbf{B}_G\mathbf{v} \tag{18.50}$$

where
$$\mathbf{A}_G = \text{block diagonal}\left(\mathbf{A}_{\gamma_1} \quad \mathbf{A}_{\gamma_2} \quad \cdots \quad \mathbf{A}_{\gamma_m}\right) \tag{18.51}$$

and
$$\mathbf{B}_G = \text{block diagonal}\left(\mathbf{b}_{\gamma_1} \quad \mathbf{b}_{\gamma_2} \quad \cdots \quad \mathbf{b}_{\gamma_m}\right) \tag{18.52}$$

All the eigenvalues are zero in the generalized canonical form. Additional state feedback is used to assign the desired eigenvalues, as described in Sec. 18.7. The γ_i in Eqs. (18.51) and (18.52) are an ordered set of integers whose sum is n; they are known as the *control invariants*[2] and are uniquely associated with the matrix pair $(\mathbf{A}, \mathbf{B})$. The method of evaluating the γ_i is illustrated later in this section. They are m in number, i.e., the number of control invariants is equal to the dimension of the input $\mathbf{u}$. They are ordered such that

$$\gamma_1 \geq \gamma_2 \geq \cdots \geq \gamma_m \geq 0 \tag{18.53}$$

$$\sum_{i=1}^{m} \gamma_i = \gamma_1 + \gamma_2 + \cdots + \gamma_m = n \tag{18.54}$$

The matrices $\mathbf{A}_{\gamma_i}$ and $\mathbf{b}_{\gamma_i}$ are of order $\gamma_i \times \gamma_i$ and $\gamma_i \times 1$, respectively, and have the forms

$$\mathbf{A}_{\gamma_i} = \begin{bmatrix} 0 & 1 & 0 & \cdots & 0 & 0 \\ 0 & 0 & 1 & \cdots & 0 & 0 \\ \cdot & \cdot & \cdot & \cdots & \cdot & \cdot \\ 0 & 0 & 0 & \cdots & 1 & 0 \\ 0 & 0 & 0 & \cdots & 0 & 1 \\ 0 & 0 & 0 & \cdots & 0 & 0 \end{bmatrix} \qquad \mathbf{b}_{\gamma_i} = \begin{bmatrix} 0 \\ 0 \\ \cdot \\ 0 \\ 0 \\ 1 \end{bmatrix} \tag{18.55}$$

† If $\mathbf{B}$ does not have full rank m, the inputs u_i are not independent and can be reduced in number.

As shown, the submatrices $\mathbf{A}_{\gamma_i}$ are companion matrices containing unit elements on the superdiagonal and the remaining elements are zero, and the vectors $\mathbf{b}_{\gamma_i}$ are null except for a unit element in the γ_i row (which is the last row).

An example of the canonical matrices for a system which has $\gamma_1 = 3$ and $\gamma_2 = 2$ is

$$\mathbf{A}_G = \begin{bmatrix} \mathbf{A}_1 & \mathbf{0} \\ \mathbf{0} & \mathbf{A}_2 \end{bmatrix} = \left[\begin{array}{ccc|cc} 0 & 1 & 0 & 0 & 0 \\ 0 & 0 & 1 & 0 & 0 \\ 0 & 0 & 0 & 0 & 0 \\ \hline 0 & 0 & 0 & 0 & 1 \\ 0 & 0 & 0 & 0 & 0 \end{array}\right] \tag{18.56}$$

$$\mathbf{B}_G = \begin{bmatrix} \mathbf{b}_1 & \mathbf{0} \\ \mathbf{0} & \mathbf{b}_2 \end{bmatrix} = \left[\begin{array}{c|c} 0 & 0 \\ 0 & 0 \\ 1 & 0 \\ \hline 0 & 0 \\ 0 & 1 \end{array}\right] \tag{18.57}$$

An important property of $\mathbf{A}_G$ is that $\mathbf{A}_G^3 = \mathbf{0}$ and therefore it has an *index of nilpotency* $\nu = 3$, which corresponds to the largest value of γ_i; that is $\nu = \gamma_1 = 3$.

The three transformations[2] which can be applied to the state equation of Eq. (18.49) in order to achieve the generalized control canonical form of Eq. (18.50) are

1. A change of basis in the state variable $\mathbf{x}$, that is, $\mathbf{x} = \mathbf{T}\mathbf{z}$, where $\det \mathbf{T} \neq 0$. Equation (18.49) then assumes the form

$$\dot{\mathbf{z}} = \mathbf{T}^{-1}\mathbf{A}\mathbf{T}\mathbf{z} + \mathbf{T}^{-1}\mathbf{B}\mathbf{u} = \overline{\mathbf{A}}_G\mathbf{z} + \overline{\mathbf{B}}_G\mathbf{u} \tag{18.58}$$

2. A change of basis in the control variable $\mathbf{u}$, that is,

$$\mathbf{u} = \mathbf{F}\mathbf{w} \tag{18.59}$$

   where $\mathbf{F}$ is $m \times m$ and $\det \mathbf{F} \neq 0$. When this transformation is applied, Eq. (18.58) assumes the form

$$\dot{\mathbf{z}} = \mathbf{T}^{-1}\mathbf{A}\mathbf{T}\mathbf{z} + \mathbf{T}^{-1}\mathbf{B}\mathbf{F}\mathbf{w} = \overline{\mathbf{A}}_G\mathbf{z} + \mathbf{B}_G\mathbf{w} \tag{18.60}$$

3. The introduction of state feedback

$$\mathbf{w} = \mathbf{v} - \mathbf{H}\mathbf{z} = \mathbf{v} - \mathbf{H}\mathbf{T}^{-1}\mathbf{x} \tag{18.61}$$

   where $\mathbf{H}$ is an $m \times n$ matrix and $\mathbf{v}$ is an $m \times 1$ input vector. Substituting $\mathbf{w}$ from Eq. (18.61) into Eq. (18.60) yields

$$\dot{\mathbf{z}} = [\mathbf{T}^{-1}\mathbf{A}\mathbf{T} - \mathbf{T}^{-1}\mathbf{B}\mathbf{F}\mathbf{H}]\mathbf{z} + [\mathbf{T}^{-1}\mathbf{B}\mathbf{F}]\mathbf{v} = [\overline{\mathbf{A}}_G - \mathbf{B}_G\mathbf{H}]\mathbf{z} + \mathbf{B}_G\mathbf{v} = \mathbf{A}_G\mathbf{z} + \mathbf{B}_G\mathbf{v} \tag{18.62}$$

The block-diagram representation of the open-loop system of Eq. (18.49) is shown in Fig. 18.5. Although a single-line drawing is used, all signals are vectors. The transformations and the addition of feedback are represented in Fig. 18.6. By

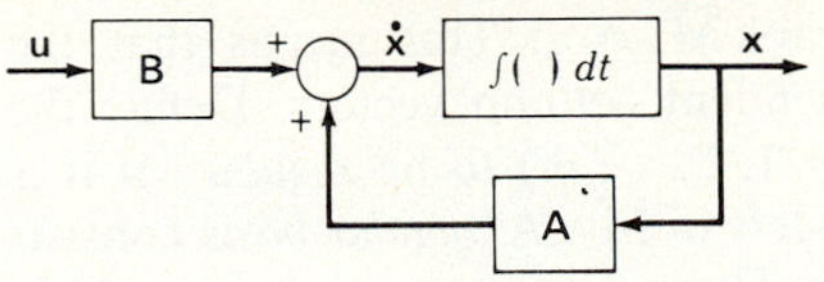

FIGURE 18.5
Open-loop system.

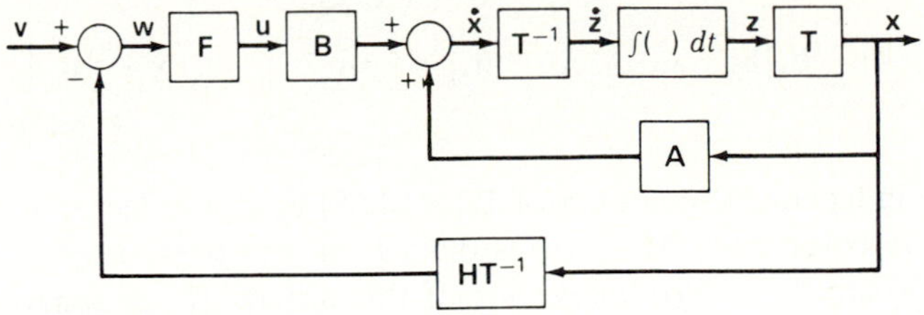

FIGURE 18.6
Original system with transformation of the variables and the addition of feedback to achieve the generalized control form.

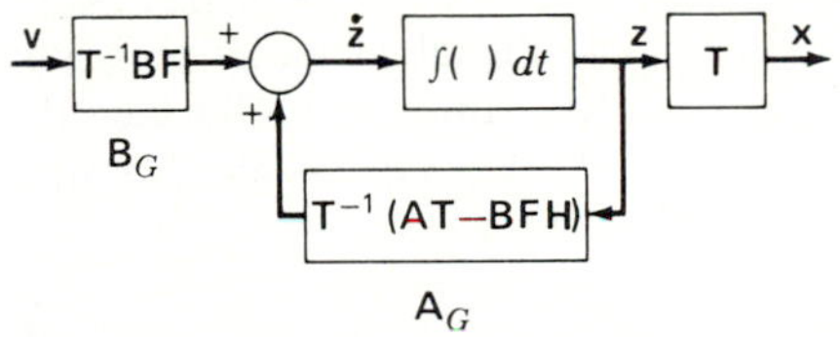

FIGURE 18.7
Transformed system represented by Eq. (18.62).

means of block-diagram manipulations, Fig. 18.6 can be converted into the system shown in Fig. 18.7 and represented by Eq. (18.62).

It may not be necessary to apply all three transformations in every case; only the necessary transformations are used to achieve the canonical form of the state equation. It should be noted that transformations 1 and 2 produce equivalent matrices; i.e., the control invariants and the eigenvalues do not change.

Control Invariants[2,4,5]

The control invariants can be calculated from the $n \times nm$ controllability matrix (see Sec. 12.7)

$$\mathbf{M}_c = [\mathbf{B} \quad \mathbf{AB} \quad \mathbf{A}^2\mathbf{B} \quad \cdots \quad \mathbf{A}^{n-1}\mathbf{B}] \tag{18.63}$$

For a multi-input system, controllability can also be evaluated from the modified controllability matrix $\overline{\mathbf{M}}_c$

$$\overline{\mathbf{M}}_c = [\mathbf{B} \quad \mathbf{AB} \quad \mathbf{A}^2\mathbf{B} \quad \cdots \quad \mathbf{A}^{n-m}\mathbf{B}] \tag{18.64}$$

The columns of this matrix are

$$\overline{\mathbf{M}}_c = \big[\mathbf{b}_1 \quad \mathbf{b}_2 \quad \cdots \quad \mathbf{b}_m \quad \mathbf{Ab}_1 \quad \mathbf{Ab}_2 \quad \cdots \quad \mathbf{Ab}_m \quad \mathbf{A}^2\mathbf{b}_1 \quad \cdots \quad \mathbf{A}^2\mathbf{b}_m$$
$$\cdots \mathbf{A}^3\mathbf{b}_1 \quad \cdots \quad \mathbf{A}^3\mathbf{b}_m \quad \cdots \quad \mathbf{A}^{n-m}\mathbf{b}_m\big] \tag{18.65}$$

For a controllable system the rank $\mathbf{M}_c$ = rank $\overline{\mathbf{M}}_c = n$. This means that the matrices $\mathbf{M}_c$ and $\overline{\mathbf{M}}_c$ have n linearly independent column vectors. Define the column vector $\mathbf{A}^j\mathbf{b}_i$ ($j = 1, 2, \ldots, n-m$; $i = 1, 2, \ldots, m$) to be *regular*[5] if it is linearly independent of all the columns to its left in $\overline{\mathbf{M}}_c$. A *regular basis* consists of the first n regular vectors contained in $\overline{\mathbf{M}}_c$. These n vectors are rearranged in the form

$$\hat{\mathbf{M}}_c = \begin{bmatrix}\mathbf{b}_1 & \mathbf{A}\mathbf{b}_1 & \cdots & \mathbf{A}^{\gamma_1-1}\mathbf{b}_1 | \mathbf{b}_2 & \cdots & \mathbf{A}^{\gamma_2-1}\mathbf{b}_2 | \cdots | \mathbf{b}_m & \cdots & \mathbf{A}^{\gamma_m-1}\mathbf{b}_m\end{bmatrix} \tag{18.66}$$

Although any n linearly independent columns of Eq. (18.65) can be selected to form $\hat{\mathbf{M}}_c$, using the first m columns of $\overline{\mathbf{M}}_c$ ensures that all m inputs are used to control the system. The γ_i are the *control invariants* of the system. If necessary, the columns of $\mathbf{B}$ are permuted so that the control invariants are in *decreasing order*, as given by Eq. (18.53). For this case the permuted $\mathbf{B}$ matrix is designated as $\hat{\mathbf{B}}$ and is used in Eqs. (18.58) to (18.62).

Example Transformation to Canonical Form. The transformation of the pair $(\mathbf{A}, \mathbf{B})$ into the canonical pair $(\mathbf{A}_G, \mathbf{B}_G)$ can be obtained from digital-computer programs using a coding based on methods like those of Aplevich.[6] In order to illustrate the transformation, the method of Luenberger[7] (based on his second canonical form) is illustrated below for a system with two inputs:

$$\dot{\mathbf{x}} = \begin{bmatrix} 3 & 5 & 1 \\ 0 & 0 & 1 \\ 0 & -2 & 3 \end{bmatrix}\mathbf{x} + \begin{bmatrix} 0 & 1 \\ 0 & 0 \\ 1 & 3 \end{bmatrix}\mathbf{u} + \mathbf{A}\mathbf{x} + \mathbf{B}\mathbf{u} \tag{18.67}$$

$$Q(\lambda) = (\lambda - 1)(\lambda - 2)(\lambda - 3) \qquad \sigma(\mathbf{A}) = \{1, 2, 3\}$$

$$\lambda_1 = 1 \qquad \lambda_2 = 2 \qquad \lambda_3 = 3$$

$$\text{Rank}\,\overline{\mathbf{M}}_c = \text{Rank}\,[\mathbf{B} \quad \mathbf{AB}] = \text{Rank}\left[\begin{array}{cc|cc} 0 & 1 & 1 & 6 \\ 0 & 0 & 1 & 3 \\ 1 & 3 & 3 & 9 \end{array}\right] = 3 \tag{18.68}$$

Thus, the system is controllable. The first three columns of $\overline{\mathbf{M}}_c$ are linearly independent, and the new matrix $\hat{\mathbf{M}}_c$ is formed as

$$\hat{\mathbf{M}}_c = [\mathbf{b}_1 \quad \mathbf{A}\mathbf{b}_1 \,\vdots\, \mathbf{b}_2] = \left[\begin{array}{cc|c} 0 & 1 & 1 \\ 0 & 1 & 0 \\ 1 & 3 & 3 \end{array}\right] \tag{18.69}$$

for which the control invariants are $\gamma_1 = 2$ and $\gamma_2 = 1$. Forming the inverse of this matrix and partitioning in conformity with Eq. (18.69) yields

$$\hat{\mathbf{M}}_c^{-1} = \left[\begin{array}{ccc} -3 & 0 & 1 \\ 0 & 1 & 0 \\ \hline 1 & -1 & 0 \end{array}\right] = \left[\begin{array}{c} \mathbf{e}_{11}^T \\ \mathbf{e}_{12}^T \\ \hline \mathbf{e}_{21}^T \end{array}\right] = \begin{bmatrix} \vdots \\ \mathbf{e}_{ij}^T \\ \vdots \end{bmatrix} \tag{18.70}$$

where $j = 1, 2, \ldots, \gamma_i$ and $i = 1, 2, \ldots, m$. The bottom row of each partitioned submatrix of $\hat{\mathbf{M}}_c^{-1}$ is used to form the transformation matrix $\mathbf{T}^{-1}$, using the format[7]

$$\mathbf{T}^{-1} = \begin{bmatrix} \mathbf{e}_{12}^T \\ \mathbf{e}_{12}^T\mathbf{A} \\ \hline \mathbf{e}_{21}^T \end{bmatrix} = \begin{bmatrix} 0 & 1 & 0 \\ 0 & 0 & 1 \\ \hline 1 & -1 & 0 \end{bmatrix} \tag{18.71}$$

When the transformation

$$\mathbf{x} = \mathbf{T}\mathbf{z} = \begin{bmatrix} 1 & 0 & 1 \\ 1 & 0 & 0 \\ 0 & 1 & 0 \end{bmatrix}\mathbf{z} \tag{18.72}$$

is used, the state equation becomes

$$\dot{\mathbf{z}} = [\mathbf{T}^{-1}\mathbf{A}\mathbf{T}]\mathbf{z} + [\mathbf{T}^{-1}\mathbf{B}]\mathbf{u} = \overline{\mathbf{A}}_G\mathbf{z} + \overline{\mathbf{B}}_G\mathbf{u} \tag{18.73}$$

where $$\overline{\mathbf{A}}_G = \mathbf{T}^{-1}\mathbf{A}\mathbf{T} = \begin{bmatrix} 0 & 1 & 0 \\ -2 & 3 & 0 \\ 8 & 0 & 3 \end{bmatrix} \qquad \overline{\mathbf{B}}_G = \mathbf{T}^{-1}\mathbf{B} = \begin{bmatrix} 0 & 0 \\ 1 & 3 \\ 0 & 1 \end{bmatrix} \tag{18.74}$$

(The matrices $\mathbf{A}_G$ and $\mathbf{B}_G$ can also be determined by using the method of Hickin and Sinha.[14]) In order to complete the transformation to canonical form, it is necessary to use the additional transformations $\mathbf{u} = \mathbf{F}\mathbf{w}$ and $\mathbf{w} = \mathbf{v} - \mathbf{H}\mathbf{z}$. The $\mathbf{F}$ matrix is 2×2 and hence has the form

$$\mathbf{F} = \begin{bmatrix} f_{11} & f_{12} \\ f_{21} & f_{22} \end{bmatrix} \tag{18.75}$$

Using Eqs. (18.52), (18.55), (18.60), and (18.74) yields

$$\mathbf{B}_G = \overline{\mathbf{B}}_G\mathbf{F} = \left[\begin{array}{c|c} 0 & 0 \\ f_{11} + 3f_{21} & f_{12} + 3f_{22} \\ \hline f_{21} & f_{22} \end{array}\right] = \left[\begin{array}{c|c} 0 & 0 \\ 1 & 0 \\ \hline 0 & 1 \end{array}\right] \tag{18.76}$$

The required $\mathbf{F}$ matrix is therefore

$$\mathbf{F} = \begin{bmatrix} 1 & -3 \\ 0 & 1 \end{bmatrix} \tag{18.77}$$

The matrix $\mathbf{H}$ has the form

$$\mathbf{H} = \begin{bmatrix} h_{11} & h_{12} & h_{13} \\ h_{21} & h_{22} & h_{23} \end{bmatrix} \tag{18.78}$$

Thus, using Eqs. (18.51), (18.55), and (18.62) yields

$$\mathbf{A}_G = \overline{\mathbf{A}}_G - \mathbf{B}_G\mathbf{H} = \begin{bmatrix} 0 & 1 & 0 \\ -2 - h_{11} & 3 - h_{12} & -h_{13} \\ 8 - h_{21} & -h_{22} & 3 - h_{23} \end{bmatrix} = \left[\begin{array}{cc|c} 0 & 1 & 0 \\ 0 & 0 & 0 \\ \hline 0 & 0 & 0 \end{array}\right] \tag{18.79}$$

Equating the elements in Eq. (18.79) gives

$$\mathbf{H} = \begin{bmatrix} -2 & 3 & 0 \\ 8 & 0 & 3 \end{bmatrix} \tag{18.78a}$$

The three transformations $\mathbf{x} = \mathbf{T}\mathbf{z}$, $\mathbf{u} = \mathbf{F}\mathbf{w}$, and $\mathbf{w} = \mathbf{v} - \mathbf{H}\mathbf{z}$ have transformed the original system representation into the generalized control canonical form of Eq. (18.50):

$$\dot{\mathbf{z}} = \left[\begin{array}{cc|c} 0 & 1 & 0 \\ 0 & 0 & 0 \\ \hline 0 & 0 & 0 \end{array}\right]\mathbf{z} + \left[\begin{array}{c|c} 0 & 0 \\ 1 & 0 \\ \hline 0 & 1 \end{array}\right]\mathbf{v} = \mathbf{A}_G\mathbf{z} + \mathbf{B}_G\mathbf{v} \tag{18.80}$$

When the values of the control invariants γ_i are larger than 2, a more general form for obtaining the transformation matrix $\mathbf{T}^{-1}$ given in Eq. (18.71) is[7]

$$\mathbf{T}^{-1} = \begin{bmatrix} \mathbf{e}_{1\gamma_1}^T \\ \mathbf{e}_{1\gamma_1}^T\mathbf{A} \\ \mathbf{e}_{1\gamma_1}^T\mathbf{A}^2 \\ \vdots \\ \mathbf{e}_{1\gamma_1}^T\mathbf{A}^{\gamma_1-1} \\ \hline \vdots \\ \hline \mathbf{e}_{m\gamma_m}^T \\ \vdots \\ \mathbf{e}_{m\gamma_m}^T\mathbf{A}^{\gamma_m-1} \end{bmatrix} \tag{18.81}$$

18.7 POLE PLACEMENT BY STATE FEEDBACK: MIMO SYSTEMS[3]

Given an open-loop plant represented by

$$\dot{\mathbf{x}} = \mathbf{A}\mathbf{x} + \mathbf{B}\mathbf{u} \tag{18.82}$$

the object is to determine a state-feedback control law

$$\mathbf{u} = \mathbf{K}\mathbf{x} \tag{18.83}$$

as shown in Fig. 18.8, which produces specified eigenvalues for the closed-loop system; i.e., they are the eigenvalues of

$$\dot{\mathbf{x}} = [\mathbf{A} + \mathbf{B}\mathbf{K}]\mathbf{x}$$

The design procedure is not to solve for $\mathbf{K}$ directly but to make use of the generalized control canonical form, Eq. (18.50). In Fig. 18.9 the feedback law represented by

$$\mathbf{v} = \boldsymbol{\Gamma}\mathbf{z} \tag{18.84}$$

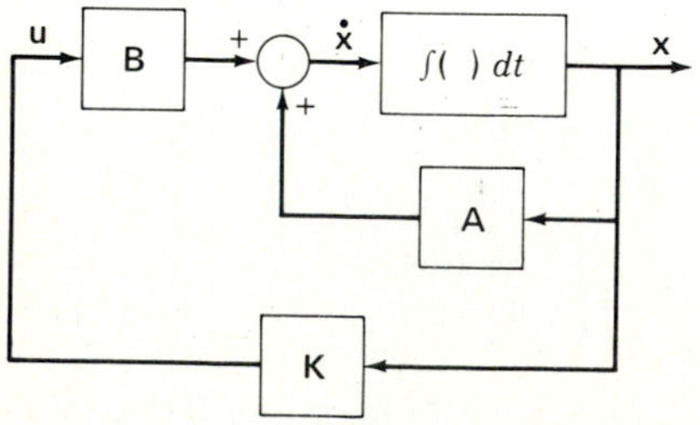

FIGURE 18.8
Plant with state feedback.

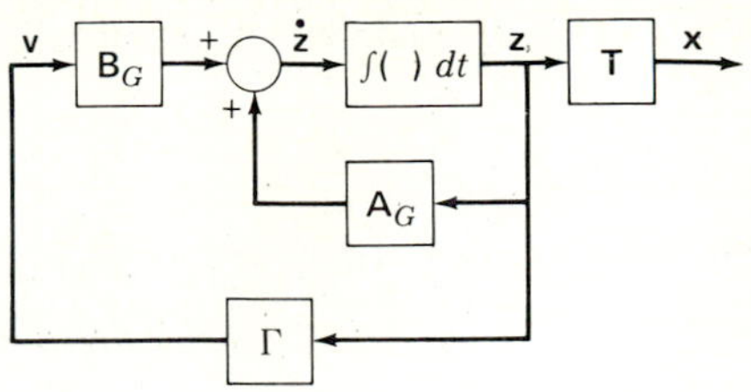

FIGURE 18.9
Representation of the generalized control canonical form with additional state feedback.

is applied to the control canonical state equation and produces

$$\dot{\mathbf{z}} = \mathbf{A}_G\mathbf{z} + \mathbf{B}_G\mathbf{v} = [\mathbf{A}_G + \mathbf{B}_G\boldsymbol{\Gamma}]\mathbf{z} = \mathbf{A}_d\mathbf{z} \tag{18.85}$$

where $\mathbf{A}_d$ is the desired closed-loop plant matrix. It is selected to have the form

$$\mathbf{A}_d = \text{block diagonal}(\mathbf{A}_{d_1} \quad \mathbf{A}_{d_2} \quad \cdots \quad \mathbf{A}_{d_m}) \tag{18.86}$$

where each $\mathbf{A}_{d_i}$ is a matrix block which has the same size as the blocks $\mathbf{A}_{\gamma_i}$ in $\mathbf{A}_G$ and is in companion form. In Eq. (18.85) the matrices $\mathbf{A}_G$, $\mathbf{B}_G$, and $\mathbf{A}_d$ are now known, and $\boldsymbol{\Gamma}$ can be evaluated by equating the elements of

$$\mathbf{A}_G + \mathbf{B}_G\boldsymbol{\Gamma} = \mathbf{A}_d \tag{18.87}$$

A formal expression for $\boldsymbol{\Gamma}$ can be obtained since

$$\mathbf{B}_G^T\mathbf{B}_G = \mathbf{I}_m \tag{18.88}$$

This leads to

$$\boldsymbol{\Gamma} = \mathbf{B}_G^T[\mathbf{A}_d - \mathbf{A}_G] \tag{18.89}$$

In order to return to the original state space, it is observed from Eqs. (18.59) and (18.61) that

$$\mathbf{u} = \mathbf{F}\mathbf{w} = \mathbf{F}\mathbf{v} - \mathbf{F}\mathbf{H}\mathbf{z} = \mathbf{F}\boldsymbol{\Gamma}z - \mathbf{F}\mathbf{H}\mathbf{z} = \mathbf{F}[\boldsymbol{\Gamma} - \mathbf{H}]\mathbf{z} = \mathbf{F}[\boldsymbol{\Gamma} - \mathbf{H}]\mathbf{T}^{-1}\mathbf{x} = \mathbf{K}\mathbf{x} \tag{18.90}$$

where

$$\mathbf{K} = \mathbf{F}[\boldsymbol{\Gamma} - \mathbf{H}]\mathbf{T}^{-1} \tag{18.91}$$

Thus, the feedback gain matrix $\mathbf{K}$ of Eq. (18.83) is determined through use of the generalized control canonical form. The advantages of this method are illustrated by the following examples.

Example 1. For the example of Sec. 18.6 the desired closed-loop eigenvalue spectrum is $\sigma(\mathbf{A} + \mathbf{BK}) = \{-1 + j1, -1 - j1, -4\}$. Since the subblocks of $\mathbf{A}_G$ in the generalized canonical state equation, Eq. (18.80), have dimensions 2 and 1, the corresponding desired characteristic polynomials are selected so that $Q(\lambda) = Q_1(\lambda)Q_2(\lambda)$:

$$Q_1(\lambda) = (\lambda + 1 - j1)(\lambda + 1 + j1) = \lambda^2 + 2\lambda + 2 \qquad Q_2(\lambda) = \lambda + 4$$

Therefore, the matrix $\mathbf{A}_d$ containing subblocks in companion form having these characteristic polynomials is

$$\mathbf{A}_d = \begin{bmatrix} \mathbf{A}_{d_1} & \mathbf{0} \\ \mathbf{0} & \mathbf{A}_{d_2} \end{bmatrix} = \left[\begin{array}{cc|c} 0 & 1 & 0 \\ -2 & -2 & 0 \\ \hline 0 & 0 & -4 \end{array}\right] \tag{18.92}$$

Using $\mathbf{A}_d$ and the matrices $\mathbf{A}_G$ and $\mathbf{B}_G$ in Eq. (18.80), the matrix $\mathbf{\Gamma}$ is determined from Eq. (18.89) as

$$\mathbf{\Gamma} = \begin{bmatrix} -2 & -2 & 0 \\ 0 & 0 & -4 \end{bmatrix} \tag{18.93}$$

Then the state-feedback matrix for the physical system is computed from Eq. (18.91) as

$$\mathbf{K} = \begin{bmatrix} 21 & 3 & -5 \\ -7 & -1 & 0 \end{bmatrix} \tag{18.94}$$

The closed-loop matrix is therefore

$$\mathbf{A}_{cl} = \mathbf{A} + \mathbf{BK} = \begin{bmatrix} -4 & 4 & 1 \\ 0 & 0 & 1 \\ 0 & -2 & -2 \end{bmatrix} \tag{18.95}$$

It can readily be determined that this closed-loop matrix has the specified eigenvalue spectrum.

It should be evident that the matrix $\mathbf{A}_d$ in Eq. (18.92) is not the only matrix with the desired set of closed-loop eigenvalues. For example, the zero-valued elements in the third row of $\mathbf{A}_d$ could be changed to any arbitrary values without changing its eigenvalues. Thus, the feedback matrix **K** which assigns the desired eigenvalues is not unique. In Sec. 18.9 the closed-loop eigenvalues and eigenvectors are both specified, leading to the selection of a unique feedback matrix **K**.

Example 2. Suppose the five desired closed-loop eigenvalues are $\lambda_{1,2} = -2 \pm j3$, $\lambda_3 = -6$, $\lambda_4 = -7$, $\lambda_5 = -8$, and the control invariants of $(\mathbf{A}, \mathbf{B})$ are $\gamma_1 = 3$ and $\gamma_2 = 2$. The factors of the characteristic polynomial of order 3 and 2, respectively, are selected so that $Q(\lambda) = Q_1(\lambda)Q_2(\lambda)$:

$$Q_1(\lambda) = (\lambda + 6)(\lambda + 7)(\lambda + 8) = \lambda^3 + 21\lambda^2 + 146\lambda + 336$$

$$Q_2(\lambda) = (\lambda + 2 - j3)(\lambda + 2 + j3) = \lambda^2 + 4\lambda + 13$$

The corresponding matrix blocks, in companion form, of the desired closed-loop matrix $\mathbf{A}_d$ are

$$\mathbf{A}_{d_1} = \begin{bmatrix} 0 & 1 & 0 \\ 0 & 0 & 1 \\ -336 & -146 & -21 \end{bmatrix} \qquad \mathbf{A}_{d_2} = \begin{bmatrix} 0 & 1 \\ -13 & -4 \end{bmatrix}$$

Since $\mathbf{A}_G$ and $\mathbf{B}_G$ are given in Eqs. (18.56) and (18.57), the feedback matrix $\mathbf{\Gamma}$ of order 2×5 is found directly from Eq. (18.89) as

$$\mathbf{\Gamma} = \begin{bmatrix} -336 & -146 & -21 & 0 & 0 \\ 0 & 0 & 0 & -13 & -4 \end{bmatrix}$$

The system feedback matrix **K** is then determined by using Eq. (18.91).

18.8 EFFECT OF EIGENSTRUCTURE ON TIME RESPONSE[9]

The previous sections illustrate methods of assigning the set of closed-loop eigenvalues $\sigma(\mathbf{A} + \mathbf{BK})$ by selecting an appropriate feedback matrix **K** which is used in the state-feedback control law

$$\mathbf{u} = \mathbf{Kx} \tag{18.96}$$

The feedback matrix **K** is unique for SISO systems but not for MIMO systems as shown in Sec. 18.7. In this section it is shown that the feedback matrix **K** determines the eigenvectors as well as the eigenvalues of the closed-loop plant matrix $\mathbf{A}_{cl} = \mathbf{A} + \mathbf{BK}$, and both of these quantities determine the time response.

For the open-loop system represented by the state and output equations

$$\dot{\mathbf{x}} = \mathbf{Ax} + \mathbf{Bu} \qquad \mathbf{y} = \mathbf{Cx} \tag{18.97}$$

the eigenvalue spectrum $\sigma(\mathbf{A})$ is the set of roots of the characteristic equation which is formed from

$$Q(\lambda) = |\lambda \mathbf{I} - \mathbf{A}| = \lambda^n + a_{n-1}\lambda^{n-1} + \cdots + a_1\lambda + a_0 = 0 \tag{18.98}$$

In Sec. 5.10 it is shown that when all the eigenvalues of **A** are distinct, the *modal matrix* **T** (see Sec. 5.10) can be determined such that

$$\mathbf{T}^{-1}\mathbf{AT} = \Lambda \tag{18.99}$$

The matrix Λ is a diagonal matrix in which the eigenvalues appear on the diagonal [see Eq. (5.82)]. The eigenvectors $\mathbf{v}_i$ are the columns of **T** and satisfy the equation

$$[\lambda_i \mathbf{I} - \mathbf{A}]\mathbf{v}_i = \mathbf{0} \tag{18.100}$$

The rows of $\mathbf{T}^{-1}$ are the row vectors $\mathbf{w}_i^T$, which are called the reciprocal or left eigenvectors and satisfy the equation

$$\mathbf{w}_i^T[\lambda_i \mathbf{I} - \mathbf{A}] = \mathbf{0} \tag{18.101}$$

Thus

$$\mathbf{T} = [\mathbf{v}_1 \quad \mathbf{v}_2 \quad \cdots \quad \mathbf{v}_n] \qquad \mathbf{T}^{-1} = \begin{bmatrix} \mathbf{w}_1^T \\ \mathbf{w}_2^T \\ \vdots \\ \mathbf{w}_n^T \end{bmatrix} \tag{18.102}$$

It is evident from Eqs. (18.102), since $\mathbf{TT}^{-1} = \mathbf{I}$, that the sets of eigenvectors $\mathbf{v}_i$ and reciprocal eigenvectors $\mathbf{w}_i^T$ are orthogonal, i.e.,

$$\mathbf{w}_i^T\mathbf{v}_j = \begin{cases} 1 & \text{for } i = j \\ 0 & \text{for } i \neq j \end{cases} \tag{18.103}$$

Solving for **A** in Eq. (18.99) yields $\mathbf{A} = \mathbf{T}\Lambda\mathbf{T}^{-1}$, which can be substituted into the solution of the state equation (see Sec. 3.14)

$$\mathbf{x}(t) = e^{\mathbf{A}t}\mathbf{x}(0) + \int_0^t e^{\mathbf{A}\tau}\mathbf{Bu}(t-\tau)\,d\tau \tag{18.104}$$

Thus, it is apparent that the state transition matrix $e^{\mathbf{A}t}$ can be expressed in terms of the eigenvectors and reciprocal eigenvectors. Using the representation of $e^{\mathbf{A}t}$ given in Eq. (5.115) yields

$$e^{\mathbf{A}t} = \mathbf{T}e^{\Lambda t}\mathbf{T}^{-1} \tag{18.105}$$

The matrix $e^{\Lambda t}$ has the diagonal form shown in Eq. (5.116). Therefore, the state transition matrix can be written as

$$e^{\mathbf{A}t} = [\mathbf{v}_1 \quad \mathbf{v}_2 \quad \cdots \quad \mathbf{v}_n] \begin{bmatrix} e^{\lambda_1 t} & & & 0 \\ & e^{\lambda_2 t} & & \\ & & \ddots & \\ 0 & & & e^{\lambda_2 t} \end{bmatrix} \begin{bmatrix} \mathbf{w}_1^T \\ \mathbf{w}_2^T \\ \vdots \\ \mathbf{w}_n^T \end{bmatrix}$$

$$= \mathbf{v}_1 e^{\lambda_1 t}\mathbf{w}_1^T + \mathbf{v}_2 e^{\lambda_2 t}\mathbf{w}_2^T + \cdots + \mathbf{v}_n e^{\lambda_n t}\mathbf{w}_n^T = \sum_{i=1}^{n} \mathbf{v}_i e^{\lambda_i t}\mathbf{w}_i^T \qquad (18.106)$$

When Eqs. (18.97), (18.104), and (18.106) are used, the output equation, when the dimension of the input $\mathbf{u}$ is m, is given by

$$\mathbf{y}(t) = \sum_{i=1}^{n} (\mathbf{C}\mathbf{v}_i) e^{\lambda_i t}\mathbf{w}_i^T \mathbf{x}(0) + \sum_{j=1}^{m} \sum_{i=1}^{n} (\mathbf{C}\mathbf{v}_i)(\mathbf{w}_i^T \mathbf{b}_j) \int_0^t e^{\lambda_i \tau} u_j(t-\tau)\, d\tau \qquad (18.107)$$

The transient response of the system is therefore a linear combination of n functions of the form

$$\mathbf{v}_i e^{\lambda_i t} \qquad i = 1, 2, \ldots, n \qquad (18.108)$$

which describe the *dynamical modes*[9] of the system. (This is an extension of the description in Sec. 3.5 which defines a mode as $e^{\lambda_i t}$.) The *shape* of a mode is determined by the eigenvector $\mathbf{v}_i$, and its time-domain characteristic is determined by the eigenvalue λ_i. It is evident from Eq. (18.107) that the entire eigenstructure determines the time response of the system; i.e., the eigenvalues λ_i, the associated eigenvectors $\mathbf{v}_i$, and the reciprocal eigenvectors $\mathbf{w}_i$ all contribute to the time response. *Therefore, the design of a state-feedback control law should be based upon the entire eigenstructure in order to be most effective in determining the time response*. The terms $\mathbf{c}_k^T \mathbf{v}_i$, $\mathbf{w}_i^T \mathbf{x}(0)$, and $\mathbf{w}_i^T \mathbf{b}_j$ are scalars and determine the magnitude of the modal responses $e^{\lambda_i t}$. The ability to select $\mathbf{v}_i$ and $\mathbf{w}_i^T$ provides the potential for adjusting the magnitude of each mode which appears in each of the outputs.[15] The selection of the eigenvectors as well as the eigenvalues is a logical extension of the methods previously available, which assigned only the eigenvalues (or poles). The constraints on the possible closed-loop eigenvectors are described in the next section.

18.9 ENTIRE EIGENSTRUCTURE ASSIGNMENT[8,10-13]

A linear time-invariant multiple-input open-loop system is represented in the continuous-time domain by the matrix state equation

$$\dot{\mathbf{x}} = \mathbf{A}\mathbf{x} + \mathbf{B}\mathbf{u} \qquad (18.109)$$

The control law for applying state-variable feedback is

$$\mathbf{u} = \mathbf{r} + \mathbf{K}\mathbf{x} \qquad (18.110)$$

and leads to the closed-loop equation

$$\dot{\mathbf{x}} = [\mathbf{A} + \mathbf{BK}]\mathbf{x} + \mathbf{Br} = \mathbf{A}_{\text{cl}}\mathbf{x} + \mathbf{Br} \tag{18.111}$$

The purpose in applying state feedback is to assign both a closed-loop eigenvalue spectrum

$$\sigma(\mathbf{A} + \mathbf{BK}) = \{\lambda_1, \lambda_2, \ldots, \lambda_n\} \tag{18.112}$$

and an associated set of eigenvectors

$$\mathbf{v}(\mathbf{A} + \mathbf{BK}) = \{\mathbf{v}_1, \mathbf{v}_2, \ldots, \mathbf{v}_n\} \tag{18.113}$$

which are selected in order to achieve the desired time-response characteristics. The closed-loop eigenvalues and eigenvectors are related by the equation

$$[\mathbf{A} + \mathbf{BK}]\mathbf{v}_i = \lambda_i \mathbf{v}_i \tag{18.114}$$

This equation can be put in the form

$$[\mathbf{A} - \lambda_i \mathbf{I} \quad \mathbf{B}]\begin{bmatrix} \mathbf{v}_i \\ \mathbf{q}_i \end{bmatrix} = \mathbf{0} \qquad \text{for } 1, 2, \ldots, n \tag{18.115}$$

where $\mathbf{v}_i$ is the eigenvector and

$$\mathbf{q}_i = \mathbf{K}\mathbf{v}_i \tag{18.116}$$

In order to satisfy Eq. (18.115), the vector $[\mathbf{v}_i^T \quad \mathbf{q}_i^T]^T$ must lie in the *kernel* or *null space* of the matrix

$$\mathbf{S}(\lambda_i) = [\mathbf{A} - \lambda_i \mathbf{I} \quad \mathbf{B}] \qquad \text{for } i = 1, 2, \ldots, n \tag{18.117}$$

The notation $\ker \mathbf{S}(\lambda_i)$ is used to define the null space which contains all the vectors $[\mathbf{v}_i^T \quad \mathbf{q}_i^T]^T$ for which Eq. (18.115) is satisfied. Equation (18.116) can be used to form the matrix equality

$$[\mathbf{q}_1 \quad \mathbf{q}_2 \quad \cdots \quad \mathbf{q}_n] = [\mathbf{K}\mathbf{v}_1 \quad \mathbf{K}\mathbf{v}_2 \quad \cdots \quad \mathbf{K}\mathbf{v}_n] \tag{18.118}$$

Since $\mathbf{K}$ can be factored from the right-hand matrix of Eq. (18.118),

$$\mathbf{K} = [\mathbf{q}_1 \quad \mathbf{q}_2 \quad \cdots \quad \mathbf{q}_n][\mathbf{v}_1 \quad \mathbf{v}_2 \quad \cdots \quad \mathbf{v}_n]^{-1} = \mathbf{Q}\mathbf{V}^{-1} \tag{18.119}$$

Therefore, if the desired eigenvalue spectrum of Eq. (18.112) is specified and the associated eigenvectors are selected to satisfy Eq. (18.115), then Eq. (18.119) specifies the required state-feedback matrix $\mathbf{K}$. The eigenvalues λ_i are real or appear in complex-conjugate sets, and the associated vectors $\mathbf{v}_i$ and $\mathbf{q}_i$ are also real or appear in complex-conjugate sets. The sets of eigenvalues and eigenvectors are therefore described as being *self-conjugate*. Thus, $\mathbf{K}$ is a real matrix.

The $\ker \mathbf{S}(\lambda_i)$ imposes constraints on the eigenvector $\mathbf{v}_i$ that may be associated with the assigned eigenvalue λ_i. The $\ker \mathbf{S}(\lambda_i)$ identifies a specific subspace, and the selected eigenvector $\mathbf{v}_i$ must be located within this subspace. In addition the selected eigenvectors must be linearly independent (see Sec. 4.14) so that the inverse matrix $\mathbf{V}^{-1}$ in Eq. (18.119) exists. Also, in order that $\mathbf{K}$ be a real matrix, the selected set of eigenvectors must be self-conjugate; i.e., they must be complex conjugates for pairs of eigenvalues which are complex conjugates. The procedure is illustrated by the examples of Chap. 19.

18.10 EIGENSTRUCTURE ASSIGNMENT FOR OBSERVERS

When the plant states are inaccessible for implementing a state-feedback control law, an observer (see Sec. 18.3) can be used to generate estimates of the plant states. With these estimates, the control law becomes

$$\mathbf{u} = \mathbf{r} + \mathbf{K}\hat{\mathbf{x}} = \mathbf{r} + \mathbf{K}(\mathbf{x} - \mathbf{e}) \tag{18.120}$$

where $\mathbf{r}$ = input
$\hat{\mathbf{x}}$ = state estimate produced by observer
$\mathbf{x}$ = actual plant state vector
$\mathbf{e}$ = observer reconstruction error

The development in Sec. 18.3 shows that when the states are observable in the output, the observer performance is determined by Eq. (18.16):

$$\dot{\mathbf{e}} = [\mathbf{A} - \mathbf{LC}]\mathbf{e} \tag{18.121}$$

where $\mathbf{L}$ is the observer matrix. This requires that $(\mathbf{A}, \mathbf{C})$ be an observable pair. The observer may be designed by synthesizing a matrix $\mathbf{L}$ which assigns the eigenvalues of the matrix $\mathbf{A} - \mathbf{LC}$, as illustrated in Sec. 18.3 for a SISO plant. For MIMO plants the assignments of *only* the observer eigenvalues may be accomplished by dualizing the development of Sec. 18.6. This requires that the matrix $\mathbf{A}^T - \mathbf{C}^T\mathbf{L}^T$, which is the transpose of the matrix in Eq. (18.121), be used.

The method of entire eigenstructure assignment can also be applied to the observer such that there is assigned to the matrix $\mathbf{A} - \mathbf{LC}$ a self-conjugate eigenvalue spectrum

$$\sigma(\mathbf{A} - \mathbf{LC}) = \{\lambda_1, \lambda_2, \ldots, \lambda_n\} \tag{18.122}$$

and an associated set of reciprocal eigenvectors

$$\mathbf{w}^T(\mathbf{A} - \mathbf{LC}) = \{\mathbf{w}_1^T, \mathbf{w}_2^T, \ldots, \mathbf{w}_n^T\} \tag{18.123}$$

The eigenvalues and the reciprocal eigenvectors are related by the equation

$$\mathbf{w}_i^T[\mathbf{A} - \mathbf{LC}] = \lambda_i \mathbf{w}_i^T \tag{18.124}$$

The transpose of this equation is

$$[\mathbf{A}^T - \mathbf{C}^T\mathbf{L}^T]\mathbf{w}_i = \lambda_i \mathbf{w}_i \tag{18.125}$$

so that

$$\begin{bmatrix}\mathbf{A}^T - \lambda_i\mathbf{I} & \mathbf{C}^T\end{bmatrix}\begin{bmatrix}\mathbf{w}_i \\ \boldsymbol{\zeta}_i\end{bmatrix} = \mathbf{0} \tag{18.126}$$

where $\mathbf{w}_i$ is the transpose of the reciprocal eigenvector and

$$\boldsymbol{\zeta}_i = -\mathbf{L}^T\mathbf{w}_i \tag{18.127}$$

In order to satisfy Eq. (18.126), the vector $[\mathbf{w}_i^T \quad \boldsymbol{\zeta}_i^T]^T$ must lie in the kernel or null space of the matrix

$$\mathbf{S}_o(\lambda_i) = \begin{bmatrix}\mathbf{A}^T - \lambda_i\mathbf{I} & \mathbf{C}^T\end{bmatrix} \tag{18.128}$$

When the transposes of the reciprocal eigenvectors have been selected from $\ker \mathbf{S}(\lambda_i)$, the transpose of the observer matrix $\mathbf{L}$ can be determined by the use of Eq. (18.127) as

$$\mathbf{L}^T = -[\boldsymbol{\zeta}_1 \quad \boldsymbol{\zeta}_2 \quad \cdots \quad \boldsymbol{\zeta}_n][\mathbf{w}_1 \quad \mathbf{w}_2 \quad \cdots \quad \mathbf{w}_n]^{-1} = -\mathbf{Z}\mathbf{W}^{-1} \qquad (18.129)$$

This procedure is illustrated in Chap. 19.

18.11 SUMMARY

The structural properties of linear multivariable control systems are presented in this chapter. First, the assignment of only the closed-loop eigenvalue spectrum is presented by introducing the generalized control canonical form of the state equation. This state equation has a mathematical structure which is characterized by the control invariants. The control invariants determine the sizes of the subblocks in the generalized control canonical form. Since each subblock is in companion form, the selection of a state-feedback matrix which assigns the desired eigenvalue spectrum is easily accomplished.

The use of a state-feedback control law permits the assignment of the entire eigenstructure, consisting of the eigenvalues and the associated eigenvectors. The closed-loop eigenvalues can be assigned arbitrarily, provided that the matrix pair $(\mathbf{A}, \mathbf{B})$ is controllable. The associated eigenvectors must be located within specific subspaces which are determined by $\ker[\mathbf{A} - \lambda\mathbf{I} \quad \mathbf{B}]$. In Chap. 19 there is presented a simple algorithm that is readily implemented on a computer for obtaining these subspaces. The use of this algorithm, the selection of the eigenvectors, and the synthesis of the feedback matrix is illustrated by means of examples in Chap. 19. These examples include regulators, trackers, and observers.

REFERENCES

1. Rosenbrock, H. H.: *State Space and Multivariable Theory*, Nelson, London, 1970.
2. Kalman, R. E.: "Kronecker Invariants and Feedback," *Proc. Conf. on Ordinary Differential Equations, Nav. Res. Lab., Washington, 1971*, Academic, New York, 1972, pp. 459–471.
3. Porter, B.: "Eigenvalue Assignment by State Feedback in Multi-Input Linear Systems Using the Generalized Canonical Form, in F. Fallside (ed.), *Control System Design by Pole-Zero Assignment*, Academic, London, 1977.
4. Morse, A. S.: "Structural Invariants of Linear Multivariable Systems," *SIAM J. Control*, vol. 11, pp. 446–465, 1973.
5. Popov, V. M.: "Invariant Description of Linear Time Invariant Control Systems," *SIAM J. Control*, vol. 10, pp. 252–264, 1972.
6. Aplevich, J. D.: "Direct Computation of Canonical Forms for Linear Systems by Elementary Matrix Operations," *IEEE Trans. Autom. Control*, vol. AC-19, pp. 124–126, 1974.
7. Luenberger, D. G.: "Canonical Forms for Linear Multivariable Systems," *IEEE Trans. Autom. Control*, vol. AC-12, pp. 290–293, 1967.
8. Porter, B., and J. J. D'Azzo: "Algorithm for the Synthesis of State-Feedback Regulators by Entire Eigenstructure Assignment," *Electron. Lett.*, vol. 13, pp. 230–231, 1977.
9. Porter, B.: *Synthesis of Dynamical Systems*, Nelson, London, 1969.
10. D'Azzo, J. J.: "Synthesis of Linear Multivariable Sampled-Data Feedback Control Systems by Entire Eigenstructure Assignment," Ph.D. thesis, University of Salford, England, 1978.

11. Moore, B. C.: "On the Flexibility Offered by State Feedback in Multivariable Systems Beyond Closed-Loop Eigenvalue Assignment," *IEEE Trans. Autom. Control*, vol. AC-21, pp. 689–692, 1976.
12. Porter, B., and J. J. D'Azzo: "Closed-Loop Eigenstructure Assignment by State Feedback in Multivariable Linear Systems," *Int. J. Control*, vol. 27, pp. 487–492, 1978.
13. Porter, B., and J. J. D'Azzo: "Algorithm for Closed-Loop Eigenstructure Assignment by State Feedback in Multivariable Linear Systems," *Int. J. Control*, vol. 27, pp. 943–947, 1978.
14. Hickin, J., and N. K. Sinha: "On the Transformation of Linear Multivariable Systems to Canonical Forms," *Int. J. Syst. Sci.*, vol. 10, pp. 783–796, 1979.
15. Houpis, C. H.: "Extension of the Synthesis Techniques for Eigenvector Selection in MIMO Systems," Air Force Wright Aeronautical Laboratories, AFWAL-TM-80-67-FIGL, Wright-Patterson Air Force Base, Ohio, 1980.
16. Kalman, R. E.: " When Is a Linear Control System Optimal?," *ASME J. Basic Eng.*, vol. 86, pp. 51–60, March 1964.
17. Luenberger, D. G.: "Observing the State of a Linear System," *IEEE Trans. Mil. Electron.*, vol. 8, pp. 74–80, 1964.
18. Luenberger, D. G.: "Observers for Multivariable Systems," *IEEE Trans. Autom. Control*, vol. AC-11, pp. 190–197, 1966.
19. Luenberger, D. G.: "An Introduction to Observer," *IEEE Trans. Autom. Control*, vol. AC-16, pp. 569–602, 1971.
20. Kwakernaak, H., and R. Sivan: *Linear Optimal Control Systems*, Wiley-Interscience, New York, 1972.

CHAPTER 19

SYNTHESIS OF LINEAR MULTIVARIABLE REGULATORS, OBSERVERS, AND TRACKERS USING ENTIRE EIGENSTRUCTURE ASSIGNMENT

19.1 INTRODUCTION

The power and versatility of entire eigenstructure assignment by state feedback is demonstrated in this chapter by the synthesis of regulators and tracking systems. Regulators are intended to restore the system output to equilibrium in the presence of initial conditions and disturbance inputs. Tracking systems are required to track a command input vector containing polynomial signals. In these tracking systems, specified outputs y_i are required to follow corresponding inputs r_i. This objective is achieved in a straightforward manner for SISO systems by adding integrators to increase the system type. The problem is more involved for MIMO systems since each input is required to control only a corresponding output, without affecting the other system outputs. The method of entire eigen-structure assignment is illustrated by the design of a tracker which follows piecewise-constant input commands. In the case of systems with uncontrollable modes, the eigenvalues associated with such modes cannot be changed by state feedback. However, the eigenvectors associated with such uncontrollable modes can be changed. This property is demonstrated by an illustrative example.

19.2 EXAMPLES OF ENTIRE EIGENSTRUCTURE ASSIGNMENT FOR REGULATORS

This section contains two examples illustrating the synthesis of regulators by the method of entire eigenstructure assignment. In Example 1 the closed-loop eigenvalues are real, while in Example 2 some of the eigenvalues are complex. The eigenvector subspace corresponding to a real eigenvalue can be obtained by either of the methods illustrated in these examples. For a complex eigenvalue the corresponding subspace is more readily obtained by the method of Example 2.

In these examples, $\ker \mathbf{S}(\lambda_i)$ is evaluated for each λ_i. However, it may be simpler to obtain $\ker \mathbf{S}(\lambda)$ without inserting the specific values of λ. This minimizes the work since the column operations need be performed only once. The desired values of λ to be assigned can then be inserted in the expression for $\ker \mathbf{S}(\lambda)$. The answers to selected problems are given in this form.

Example 1 Real closed-loop eigenvalues. A system whose plant, control, and output matrices are

$$\mathbf{A} = \begin{bmatrix} 0 & 1 & 0 \\ 0 & 0 & 1 \\ 4 & 4 & -1 \end{bmatrix} \qquad \mathbf{B} = \begin{bmatrix} 1 & 0 \\ 0 & 0 \\ 0 & 1 \end{bmatrix} \qquad \mathbf{C} = \begin{bmatrix} 1 & 0 & 0 \\ 0 & 1 & 1 \end{bmatrix}$$

has the open-loop eigenvalue spectrum

$$\sigma(\mathbf{A}) = \{-1, -2, 2\} \tag{19.1}$$

Since the open-loop system is unstable, it is desired to use state feedback, $\mathbf{u} = r + \mathbf{Kx}$, to assign the closed-loop eigenvalue spectrum

$$\sigma(\mathbf{A} + \mathbf{BK}) = \{-2, -3, -4\} \tag{19.2}$$

This is possible since $(\mathbf{A}, \mathbf{B})$ is a controllable pair. When the system is a regulator, the command input vector is $\mathbf{r} = \mathbf{0}$. The required steps for determining the spaces in which the eigenvectors must be located are as follows:

1. The matrix $\mathbf{S}(\lambda_i)$ is formed in accordance with Eq. (18.117)

$$\mathbf{S}(\lambda_i) = [\mathbf{A} - \lambda_i \mathbf{I} \quad \mathbf{B}] = \begin{bmatrix} -\lambda_i & 1 & 0 & 1 & 0 \\ 0 & -\lambda_i & 1 & 0 & 0 \\ 4 & 4 & -\lambda_i - 1 & 0 & 1 \end{bmatrix} \tag{19.3}$$

2. This matrix is augmented with sufficient rows of zeros to form a square matrix.
3. The augmented matrix is put into HNF by means of elementary row operations (see Sec. 4.14).
4. The rows are interchanged, if necessary, so that the leading 1 s in each row are on the principal diagonal.

5. Wherever a 0 appears on the principal diagonal,[13] it is replaced with -1 and the matrix is designated as $\hat{\mathbf{S}}(\lambda_i)$.
6. The columns of this matrix $\mathbf{S}(\lambda_i)$ which contain -1 on the principal diagonal are vectors which span the null space of $\mathbf{S}(\lambda_i)$. They can be multiplied by arbitrary constants. There are m such columns.
7. As verification that the correct $\ker \mathbf{S}(\lambda_i)$ has been achieved, check to see that

$$[\mathbf{A} - \lambda_i \mathbf{I} \quad \mathbf{B}] \cdot \ker \mathbf{S}(\lambda_i) = \mathbf{0}.$$

8. Select the desired vector $[\mathbf{v}_i^T \quad \mathbf{q}_i^T]^T$ from $\ker \mathbf{S}(\lambda_i)$.

Inserting $\lambda_1 = -2$ in Eq. (19.3) and following these steps yields

$$\hat{\mathbf{S}}(-2) = \begin{bmatrix} 1 & 0 & -\frac{1}{4} & 0 & \frac{1}{4} \\ 0 & 1 & \frac{1}{2} & 0 & 0 \\ 0 & 0 & -1 & 0 & 0 \\ 0 & 0 & 0 & 1 & -\frac{1}{2} \\ 0 & 0 & 0 & 0 & -1 \end{bmatrix} \tag{19.4}$$

Since there are two columns in this matrix with -1 on the principal diagonal, the null space consists of all vectors formed from combinations of these columns. The two columns are multiplied by -4 and 4, respectively, in order to eliminate fractions. Thus, the null space is indicated by the notation

$$\ker \mathbf{S}(-2) = \operatorname{span}\left\{ \begin{bmatrix} 1 \\ -2 \\ 4 \\ 0 \\ 0 \end{bmatrix}, \begin{bmatrix} 1 \\ 0 \\ 0 \\ -2 \\ -4 \end{bmatrix} \right\} = \operatorname{span}\{\mathbf{s}_1(-2), \mathbf{s}_2(-2)\} \tag{19.5}$$

Similarly, for $\lambda_2 = -3$

$$\hat{\mathbf{S}}(-3) = \begin{bmatrix} 1 & 0 & 0 & \frac{1}{5} & \frac{1}{10} \\ 0 & 1 & 0 & \frac{2}{5} & -\frac{3}{10} \\ 0 & 0 & 1 & -\frac{6}{5} & \frac{9}{10} \\ 0 & 0 & 0 & -1 & 0 \\ 0 & 0 & 0 & 0 & -1 \end{bmatrix} \tag{19.6}$$

$$\ker \mathbf{S}(-3) = \operatorname{span}\left\{ \begin{bmatrix} 1 \\ 2 \\ -6 \\ -5 \\ 0 \end{bmatrix}, \begin{bmatrix} 1 \\ -3 \\ 9 \\ 0 \\ -10 \end{bmatrix} \right\} = \operatorname{span}\{\mathbf{s}_1(-3), \mathbf{s}_2(-3)\} \tag{19.7}$$

Also, for $\lambda_3 = -4$

$$\hat{\mathbf{S}}(-4) = \begin{bmatrix} 1 & 0 & 0 & \frac{2}{9} & \frac{1}{36} \\ 0 & 1 & 0 & \frac{1}{9} & -\frac{4}{36} \\ 0 & 0 & 1 & -\frac{4}{9} & \frac{16}{36} \\ 0 & 0 & 0 & -1 & 0 \\ 0 & 0 & 0 & 0 & -1 \end{bmatrix} \tag{19.8}$$

$$\ker \mathbf{S}(-4) = \operatorname{span}\left\{ \begin{bmatrix} 2 \\ 1 \\ -4 \\ -9 \\ 0 \end{bmatrix}, \begin{bmatrix} 1 \\ -4 \\ 16 \\ 0 \\ -36 \end{bmatrix} \right\} = \operatorname{span}\{\mathbf{s}_1(-4), \mathbf{s}_2(-4)\} \tag{19.9}$$

The null-space vectors are selected by taking any desired combination of the two vectors in $\ker \mathbf{S}(\lambda_i)$, using the form

$$\begin{bmatrix} \mathbf{v}_i \\ \mathbf{q}_i \end{bmatrix} = \alpha_1 s_1(\lambda_i) + \alpha_2 \mathbf{s}_2(\lambda_i) \tag{19.10}$$

where α_1 and α_2 are arbitrary constants. For this example, the selected null-space vectors are

$$\lambda_1 = -2: \qquad \begin{bmatrix} \mathbf{v}_1 \\ \hline \mathbf{q}_1 \end{bmatrix} = \begin{bmatrix} 0 \\ 1 \\ -2 \\ \hline -1 \\ -2 \end{bmatrix} \qquad \alpha_1 = -\tfrac{1}{2}, \quad \alpha_2 = \tfrac{1}{2}$$

$$\lambda_2 = -3: \qquad \begin{bmatrix} \mathbf{v}_2 \\ \hline \mathbf{q}_2 \end{bmatrix} = \begin{bmatrix} 1 \\ 0 \\ 0 \\ \hline -3 \\ -4 \end{bmatrix} \qquad \alpha_1 = \tfrac{3}{5}, \quad \alpha_2 = \tfrac{2}{5} \tag{19.11}$$

$$\lambda_3 = -4: \qquad \begin{bmatrix} \mathbf{v}_3 \\ \hline \mathbf{q}_3 \end{bmatrix} = \begin{bmatrix} 0 \\ 1 \\ -4 \\ \hline -1 \\ -8 \end{bmatrix} \qquad \alpha_1 = \tfrac{1}{9}, \quad \alpha_2 = -\tfrac{2}{9}$$

The values of α_1 and α_2 used to form the vectors selected in Eq. (19.11) are shown in this equation. The vectors are also partitioned into the vector segments corresponding to $\mathbf{v}_i$ and $\mathbf{q}_i$, respectively. This selection of eigenvectors produces state responses for which $x_1(t)$ contains only the mode e^{-3t} and $x_2(t)$ and $x_3(t)$ contain only the modes e^{-2t} and e^{-4t}. The state-feedback matrix required to assign the closed-loop eigenvalue spectrum of Eq. (19.2) and the associated eigenvectors of Eq. (19.11) is obtained from Eq. (18.119) as

$$\begin{aligned} \mathbf{K} &= \begin{bmatrix} -1 & -3 & -1 \\ -2 & -4 & 8 \end{bmatrix} \begin{bmatrix} 0 & 1 & 0 \\ 1 & 0 & 1 \\ -2 & 0 & -4 \end{bmatrix}^{-1} \\ &= \begin{bmatrix} -1 & -3 & -1 \\ -2 & -4 & 8 \end{bmatrix} \begin{bmatrix} 0 & 2 & \frac{1}{2} \\ 1 & 0 & 0 \\ 0 & -1 & -\frac{1}{2} \end{bmatrix} \\ &= \begin{bmatrix} -3 & -1 & 0 \\ -4 & -12 & -5 \end{bmatrix} \end{aligned} \tag{19.12}$$

The closed-loop matrix is

$$\mathbf{A}_{\text{cl}} = \mathbf{A} + \mathbf{BK} = \begin{bmatrix} -3 & 0 & 0 \\ 0 & 0 & 1 \\ 0 & -8 & -6 \end{bmatrix} \tag{19.13}$$

This closed-loop matrix has the assigned eigenvalue spectrum of Eq. (19.2) and the eigenvectors of Eq. (19.11), as required. It should be noted that one open-loop eigenvalue $\lambda_1 = -2$ has been retained in the closed-loop plant matrix but its eigenvector has been changed from $[1 \quad -2 \quad 4]^T$ to $[0 \quad 1 \quad -2]^T$.

The state response of the closed-loop system computed for the initial condition $\mathbf{x}(0) = [1 \quad 1 \quad 2]^T$ and $\mathbf{r}(t) = \mathbf{0}$ is $\mathbf{x}(t) = \mathscr{L}^{-1}[s\mathbf{I} - \mathbf{A}_{cl}]^{-1}\mathbf{x}(0)$:

$$\begin{bmatrix} x_1(t) \\ x_2(t) \\ x_3(t) \end{bmatrix} = \begin{bmatrix} e^{-3t} & 0 & 0 \\ 0 & 2e^{-2t} - e^{-4t} & 0.5e^{-2t} - 0.5e^{-4t} \\ 0 & -4e^{-2t} + 4e^{-4t} & -e^{-2t} + 2e^{-4t} \end{bmatrix} \begin{bmatrix} 1 \\ 1 \\ 2 \end{bmatrix}$$

$$= \begin{bmatrix} e^{-3t} \\ 3e^{-2t} - 2e^{-4t} \\ -6e^{-2t} + 8e^{-4t} \end{bmatrix} \tag{19.14}$$

Note that the selection of the eigenvectors in Eq. (19.11) produces the result that only the mode e^{-3t} appears in $x_1(t)$. Also, only the modes e^{-2t} and e^{-4t} appear in $x_2(t)$ and $x_3(t)$. The same separation of the modes appears in the output response for this example obtained from $\mathbf{y} = \mathbf{Cx}$.

$$\begin{bmatrix} y_1(t) \\ y_2(t) \end{bmatrix} = \begin{bmatrix} e^{-3t} \\ -3e^{-2t} + 6e^{-4t} \end{bmatrix} \tag{19.15}$$

Example 2 Complex closed-loop eigenvalues. For the system of Example 1 the desired closed-loop eigenvalue spectrum is changed to

$$\sigma(\mathbf{A}_{cl}) = \sigma(\mathbf{A} + \mathbf{BK}) = \{-2, -1 + j1, -1, -j1\} \tag{19.16}$$

The purpose of this example is to determine the feedback matrix $\mathbf{K}$, the state response $\mathbf{x}(t)$ when $\sigma(\mathbf{A}_{cl})$ has complex eigenvalues, and an alternate method of obtaining $\ker \mathbf{S}(\lambda_i)$. The rules[15,16] for obtaining $\ker \mathbf{S}(\lambda_i)$ are

1. Form the augmented matrix

$$\left[\begin{array}{cc} \mathbf{A} - \lambda_i \mathbf{I}_n & \mathbf{B} \\ \hline \multicolumn{2}{c}{\mathbf{I}_{n+m}} \end{array}\right] \tag{19.17}$$

2. Use elementary column operations on this matrix to obtain m zero columns of $[\mathbf{A} - \lambda_i \mathbf{I} \quad \mathbf{B}]$.

3. The columns below these zero columns span $\ker \mathbf{S}(\lambda_i)$. This can be verified by checking that $[\mathbf{A} - \lambda_i \mathbf{I} \quad \mathbf{B}] \cdot \ker \mathbf{S}(\lambda_i) = \mathbf{0}$.

Forming the matrix of Eq. (19.17) gives

$$\left[\begin{array}{ccccc} -\lambda_i & 1 & 0 & 1 & 0 \\ 0 & -\lambda_i & 1 & 0 & 0 \\ 4 & 4 & -1-\lambda_i & 0 & 1 \\ \hline 1 & 0 & 0 & 0 & 0 \\ 0 & 1 & 0 & 0 & 0 \\ 0 & 0 & 1 & 0 & 0 \\ 0 & 0 & 0 & 1 & 0 \\ 0 & 0 & 0 & 0 & 1 \end{array}\right] \tag{19.18}$$

Ker $\mathbf{S}(\lambda_1)$ for $\lambda_1 = -2$ is given in Example 1. Thus, to obtain $\ker \mathbf{S}(\lambda_2)$, insert $\lambda_2 = -1 + j1$ in the matrix of (19.18) and use column operations as specified by rule 2; this yields

$$\left[\begin{array}{ccccc} 0 & 0 & 0 & 1 & 0 \\ 0 & 0 & 1 & 0 & 0 \\ 0 & 0 & -j1 & 0 & 1 \\ \hline 1 & 0 & 0 & 0 & 0 \\ 0 & 1 & 0 & 0 & 0 \\ 0 & -1+j1 & 1 & 0 & 0 \\ -1+j1 & -1 & 0 & 1 & 0 \\ -4 & -5-j1 & 0 & 0 & 1 \end{array}\right] \tag{19.19}$$

In accordance with rule 3, since the top portion of the first two columns is $[0 \quad 0 \quad 0]^T$, $\ker \mathbf{S}(\lambda_2)$ is

$$\ker \mathbf{S}(-1+j1) = \operatorname{span}\left\{\begin{bmatrix} 1 \\ 0 \\ 0 \\ -1+j1 \\ -4 \end{bmatrix}\begin{bmatrix} 0 \\ 1 \\ -1+j1 \\ -1 \\ -5-j1 \end{bmatrix}\right\}$$

$$= \operatorname{span}\{s_1(-1+j1), s_2(-1+j1)\} \tag{19.20}$$

It is not necessary to compute $\ker \mathbf{S}(\lambda_3) = \ker \mathbf{S}(-1-j1)$ since its spanning vectors are the complex conjugates of the spanning vectors of $\ker \mathbf{S}(-1+j1)$. From Eqs. (19.5), (19.10), and (19.20) the selected null-space vectors are

$$\begin{bmatrix} \mathbf{v}_1 \\ \hline \mathbf{q}_1 \end{bmatrix} = \left[\begin{array}{c} 1 \\ -2 \\ 4 \\ \hline 0 \\ 0 \end{array}\right] \qquad \begin{bmatrix} \mathbf{v}_2 \\ \hline \mathbf{q}_2 \end{bmatrix} = \left[\begin{array}{c} 0 \\ 1 \\ -1+j1 \\ \hline -1 \\ -5-j1 \end{array}\right] \qquad \begin{bmatrix} \mathbf{v}_3 \\ \hline \mathbf{q}_3 \end{bmatrix} = \left[\begin{array}{c} 0 \\ 1 \\ -1-j1 \\ \hline -1 \\ -5+j1 \end{array}\right] \tag{19.21}$$

Equation (18.119), which is used to compute the state-feedback matrix $\mathbf{K}$, can be modified[8] when the eigenvalues are complex conjugates. The following form for $\mathbf{K}$, where $n = 3$, uses only real vectors and therefore avoids operations with complex matrices[8]

$$\mathbf{K} = [\mathbf{q}_1 \quad \operatorname{Re}(\mathbf{q}_2) \quad \operatorname{Im}(\mathbf{q}_2)][\mathbf{v}_1 \quad \operatorname{Re}(\mathbf{v}_2) \quad \operatorname{Im}(\mathbf{v}_2)]^{-1} \tag{19.22}$$

This format for computing $\mathbf{K}$ can be extended for the case when there is more than one set of complex eigenvalues and eigenvectors. Applying Eq. (19.22) with the values of Eq. (19.21) yields

$$\mathbf{K} = \begin{bmatrix} 0 & -1 & 0 \\ 0 & -5 & -1 \end{bmatrix}\begin{bmatrix} 1 & 0 & 0 \\ -2 & 1 & 0 \\ 4 & -1 & 1 \end{bmatrix}^{-1}$$

$$= \begin{bmatrix} 0 & -1 & 0 \\ 0 & -5 & -1 \end{bmatrix}\begin{bmatrix} 1 & 0 & 0 \\ 2 & 1 & 0 \\ -2 & 1 & 1 \end{bmatrix} = \begin{bmatrix} -2 & -1 & 0 \\ -8 & -6 & -1 \end{bmatrix} \tag{19.23}$$

The closed-loop matrix is

$$\mathbf{A}_{cl} = \mathbf{A} + \mathbf{BK} = \begin{bmatrix} -2 & 0 & 0 \\ 0 & 0 & 1 \\ -4 & -2 & -2 \end{bmatrix} \tag{19.24}$$

This matrix has the assigned eigenvalue spectrum of Eq. (19.16) and the eigenvectors of Eq. (19.21). It should be noted that one open-loop eigenvalue, $\lambda_1 = -2$, and its associated eigenvector have been retained in the closed-loop matrix $\mathbf{A}_{cl}$.

The state response of the closed-loop system can be obtained in several ways for the initial condition $\mathbf{x}(0) = [1 \quad 1 \quad 2]$ and $\mathbf{r}(t) = \mathbf{0}$. Using the Laplace method yields

$$\mathbf{x}(t) = \mathscr{L}^{-1}[s\mathbf{I} - \mathbf{A}_{cl}]^{-1}\mathbf{x}(0) = \mathscr{L}^{-1}\begin{bmatrix} \dfrac{1}{s+2} \\ \dfrac{s^2+5s+2}{(s+2)(s^2+2s+2)} \\ \dfrac{s^2-4s-4}{(s+2)(s^2+2s+2)} \end{bmatrix}$$

$$= \begin{bmatrix} e^{-2t} \\ -2e^{-2t} + 3e^{-t}\cos t \\ 4e^{-2t} - 3e^{-t}\cos t \end{bmatrix} \tag{19.25}$$

Note that the selection of the eigenvectors in Eq. (19.21) produces the result that only e^{-2t} appears in $x_1(t)$ and all the modes appear in $x_2(t)$ and $x_3(t)$. (A suggested exercise for the reader is to assign the eigenvector $[1 \quad 0 \quad 0]^T$ for $\lambda_1 = -2$ and to compare the results with those shown above.)

19.3 UNCONTROLLABLE SYSTEMS

A system is completely controllable (see Secs. 12.7 and 18.6) when

$$\text{Rank}\,\mathbf{M}_c = \text{Rank}\,[\mathbf{B} \quad \mathbf{AB} \quad \mathbf{A}^2\mathbf{B} \quad \cdots \quad \mathbf{A}^{n-m}\mathbf{B}] = n \tag{19.26}$$

Complete controllability implies that all eigenvalues can be assigned by the use of state feedback. When Eq. (19.26) is not satisfied, the rank deficiency of $\mathbf{M}_c$ is equal to the number of modes for which the eigenvalues cannot be changed by the use of state feedback; i.e., these modes are uncontrollable. For distinct eigenvalues these uncontrollable modes have the property[14]

$$\text{Rank}\,\mathbf{S}(\lambda_u) = \text{Rank}\,[\mathbf{A} - \lambda_u\mathbf{I} \quad \mathbf{B}] = g < n \tag{19.27}$$

The eigenvalues λ_u associated with uncontrollable modes are defined[17] as *input-decoupling zeros*. While the eigenvalue λ_u associated with an uncontrollable mode cannot be changed, it is possible to change the associated eigenvector by the use of state feedback. For controllable modes the null space of $\mathbf{S}(\lambda_i)$ has dimension m, which is equal to the dimension of $\mathbf{B}$. For uncontrollable modes the dimension of $\ker \mathbf{S}(\lambda_u)$ is $n + m - g > m$. The larger dimension of this null space provides greater flexibility[8] in the selection permitted for the associated eigenvector. This is demonstrated in the following example.

Example. A system is represented by a state equation having the plant and control matrices

$$\mathbf{A} = \begin{bmatrix} -1 & 0 & 0 \\ -2 & -2 & -2 \\ -1 & 0 & -3 \end{bmatrix} \qquad \mathbf{B} = \begin{bmatrix} 1 & 0 \\ 0 & 2 \\ 0 & 1 \end{bmatrix} \tag{19.28}$$

and the open-loop eigenvalue spectrum is

$$\sigma(\mathbf{A}) = \{-1, -2, -3\} \tag{19.29}$$

Thus, for this system

$$\text{Rank}\,\mathbf{M}_c = \text{Rank}\,[\mathbf{B} \quad \mathbf{AB}] = \text{Rank}\begin{bmatrix} 1 & 0 & -1 & 0 \\ 0 & 2 & -2 & -6 \\ 0 & 1 & -1 & -3 \end{bmatrix} = 2 \tag{19.30}$$

Since rank $\mathbf{M}_c < n$, the system is not completely controllable. The mode associated with $\lambda = -2$ is uncontrollable since

$$\text{Rank}\,\mathbf{S}(-2) = \text{Rank}\begin{bmatrix} 1 & 0 & 0 & 1 & 0 \\ -2 & 0 & -2 & 0 & 2 \\ -1 & 0 & -1 & 0 & 1 \end{bmatrix} = 2 < n \tag{19.31}$$

The closed-loop eigenvalue spectrum must include $\lambda = -2$ and is selected as

$$\sigma(\mathbf{A} + \mathbf{BK}) = \{-2, -4, -5\} \tag{19.32}$$

Using the method of Example 2 in Sec. 19.2 with $\lambda_1 = -2$, $\lambda_2 = -4$, and $\lambda_3 = -5$ yields

$$\ker \mathbf{S}(\lambda_1) = \text{span}\left\{\begin{bmatrix} 1 \\ 0 \\ 0 \\ -1 \\ 1 \end{bmatrix}, \begin{bmatrix} 0 \\ 1 \\ 0 \\ 0 \\ 0 \end{bmatrix}, \begin{bmatrix} 0 \\ 0 \\ 1 \\ 0 \\ 1 \end{bmatrix}\right\} = \text{span}\{\mathbf{s}_1(\lambda_1), \mathbf{s}_2(\lambda_1), \mathbf{s}_3(\lambda_1)\} \tag{19.33}$$

$$\ker \mathbf{S}(\lambda_2) = \text{span}\left\{\begin{bmatrix} 1 \\ 0 \\ 0 \\ -3 \\ 1 \end{bmatrix}, \begin{bmatrix} 0 \\ 2 \\ 1 \\ 0 \\ -1 \end{bmatrix}\right\} = \text{span}\{\mathbf{s}_1(\lambda_2), \mathbf{s}_2(\lambda_2)\} \tag{19.34}$$

$$\ker \mathbf{S}(\lambda_3) = \text{span}\left\{\begin{bmatrix} 1 \\ 0 \\ 0 \\ -4 \\ 1 \end{bmatrix}, \begin{bmatrix} 0 \\ 2 \\ 1 \\ 0 \\ -2 \end{bmatrix}\right\} = \text{span}\{\mathbf{s}_1(\lambda_3), \mathbf{s}_2(\lambda_3)\} \tag{19.35}$$

The selected eigenvectors must be linearly independent. It is also noted from Eq. (19.33) that any desired eigenvector can be associated with $\lambda_1 = -2$. The linearly independent vectors selected from these subspaces are

$$\begin{bmatrix} \mathbf{v}_1 \\ \mathbf{q}_1 \end{bmatrix} = \begin{bmatrix} 0 \\ 0 \\ 1 \\ \hline 0 \\ 1 \end{bmatrix} \qquad \begin{bmatrix} \mathbf{v}_2 \\ \mathbf{q}_2 \end{bmatrix} = \begin{bmatrix} 1 \\ 0 \\ 0 \\ \hline -3 \\ 1 \end{bmatrix} \qquad \begin{bmatrix} \mathbf{v}_3 \\ \mathbf{q}_3 \end{bmatrix} = \begin{bmatrix} 0 \\ 2 \\ 1 \\ \hline 0 \\ -2 \end{bmatrix} \tag{19.36}$$

The required feedback matrix is computed using Eq. (18.118) as

$$\mathbf{K} = \begin{bmatrix} 0 & -3 & 0 \\ 1 & 1 & -2 \end{bmatrix} \begin{bmatrix} 0 & 1 & 0 \\ 0 & 0 & 2 \\ 1 & 0 & 1 \end{bmatrix}^{-1}$$

$$= \begin{bmatrix} 0 & -3 & 0 \\ 1 & 1 & -2 \end{bmatrix} \begin{bmatrix} 0 & -0.5 & 1 \\ 1 & 0 & 0 \\ 0 & 0.5 & 0 \end{bmatrix} = \begin{bmatrix} -3 & 0 & 0 \\ 1 & -1.5 & 1 \end{bmatrix} \quad (19.37)$$

The closed-loop matrix is

$$\mathbf{A}_{cl} = \mathbf{A} + \mathbf{BK} = \begin{bmatrix} -4 & 0 & 0 \\ 0 & -5 & 0 \\ 0 & -1.5 & -2 \end{bmatrix} \quad (19.38)$$

This closed-loop matrix has the required eigenvalue spectrum of Eq. (19.32) and the eigenvectors of Eq. (19.36). For the initial condition $\mathbf{x}(0) = [1 \quad 1 \quad 2]^T$ and $\mathbf{r}(t) = 0$, the state response is

$$\begin{bmatrix} x_1(t) \\ x_2(t) \\ x_3(t) \end{bmatrix} = \begin{bmatrix} e^{-4t} & 0 & 0 \\ 0 & e^{-5t} & 0 \\ 0 & 0.5e^{-2t} - 0.5e^{-5t} & e^{-2t} \end{bmatrix} \mathbf{x}(0) = \begin{bmatrix} e^{-4t} \\ e^{-5t} \\ 2.5e^{-2t} - 0.5e^{-5t} \end{bmatrix} \quad (19.39)$$

The selection of the eigenvectors of Eq. (19.36) results in the presence of only the mode e^{-4t} in $x_1(t)$ and only the mode e^{-5t} in $x_2(t)$. Also, the uncontrollable mode e^{-2t} appears only in $x_3(t)$. This is verified by the response given in Eq. (19.39).

19.4 TRACKING SYSTEMS

A controllable open-loop system is represented by the nth-order state and ℓth-order output equations of the form

$$\dot{\mathbf{x}} = \mathbf{Ax} + \mathbf{Bu} \quad (19.40)$$

$$\mathbf{y} = \mathbf{Cx} = \begin{bmatrix} \mathbf{E} \\ \mathbf{F} \end{bmatrix} \mathbf{x} \quad (19.41)$$

where $\mathbf{y}$ is an $\ell \times 1$ vector and $\mathbf{w} = \mathbf{Ex}$ is a $p \times 1$ vector representing the outputs which are required to follow a $p \times 1$ input vector $\mathbf{r}$. In order to maintain controllability, the number of outputs p which can be tracked cannot exceed the number of control inputs m as described below ($\ell \geq p$).

The feedback controller is required to cause the output vector $\mathbf{w}$ to track the command input $\mathbf{r}$ in the sense that the steady-state response

$$\lim_{t \to \infty} \mathbf{w}(t) = \mathbf{r}(t) \quad (19.42)$$

when $\mathbf{r}$ consists of piecewise-constant command inputs. The design method[1,8] consists of the addition of a vector comparator and integrator which satisfies the equation

$$\dot{\mathbf{z}} = \mathbf{r} - \mathbf{w} = \mathbf{r} - \mathbf{Ex} \quad (19.43)$$

The composite open-loop system is therefore governed by the augmented state

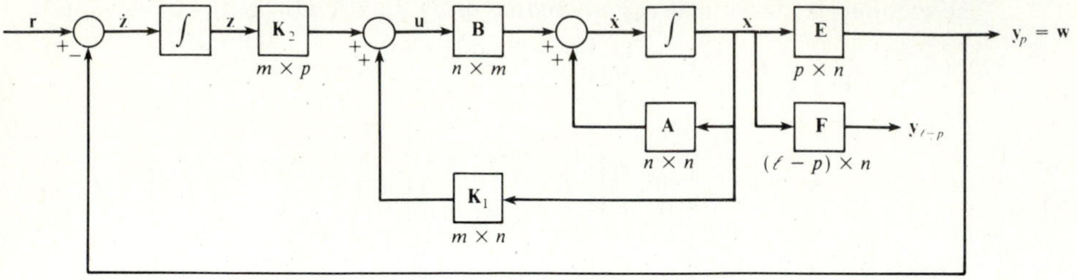

FIGURE 19.1
Tracking system.

and output equations formed from Eqs. (19.40) to (19.43):

$$\begin{bmatrix} \dot{\mathbf{x}} \\ \dot{\mathbf{z}} \end{bmatrix} = \begin{bmatrix} \mathbf{A} & 0 \\ -\mathbf{E} & 0 \end{bmatrix} \begin{bmatrix} \mathbf{x} \\ \mathbf{z} \end{bmatrix} + \begin{bmatrix} \mathbf{B} \\ \mathbf{0} \end{bmatrix} \mathbf{u} + \begin{bmatrix} \mathbf{0} \\ \mathbf{I} \end{bmatrix} \mathbf{r} \tag{19.44}$$

and

$$\mathbf{y} = [\mathbf{C} \quad \mathbf{0}] \begin{bmatrix} \mathbf{x} \\ \mathbf{z} \end{bmatrix} \tag{19.45}$$

where

$$\overline{\mathbf{A}} = \begin{bmatrix} \mathbf{A} & \mathbf{0} \\ -\mathbf{E} & \mathbf{0} \end{bmatrix} \qquad \overline{\mathbf{B}} = \begin{bmatrix} \mathbf{B} \\ \mathbf{0} \end{bmatrix} \qquad \overline{\mathbf{C}} = [\mathbf{C} \quad \mathbf{0}] \tag{19.46}$$

The state-feedback control law to be used is

$$\mathbf{u} = \mathbf{K}_1\mathbf{x} + \mathbf{K}_2\mathbf{z} = [\mathbf{K}_1 \quad \mathbf{K}_2] \begin{bmatrix} \mathbf{x} \\ \mathbf{z} \end{bmatrix} \tag{19.47}$$

where

$$\overline{\mathbf{K}} = [\mathbf{K}_1 \quad \mathbf{K}_2] \tag{19.48}$$

A block diagram representing the feedback control system, consisting of the plant state and output equations given by Eqs. (19.40) and (19.41) and the control law given by Eq. (19.47), is shown in Fig. 19.1. The dimensions of all matrices are shown in the figure. This control law assigns the desired closed-loop eigenvalue spectrum if and only if the augmented plant and control matrix pair $(\overline{\mathbf{A}}, \overline{\mathbf{B}})$ is controllable. It has been shown[1,18] that this condition is satisfied if $(\mathbf{A}, \mathbf{B})$ is a controllable pair and

$$\text{Rank} \begin{bmatrix} \mathbf{B} & \mathbf{A} \\ \mathbf{0} & -\mathbf{E} \end{bmatrix} = n + p \tag{19.49}$$

This condition is much easier to use than the determination of the rank of the controllability matrix for $(\overline{\mathbf{A}}, \overline{\mathbf{B}})$. *Equation* (19.49) *can be satisfied only if the number of outputs p which are required to track the input* $\mathbf{r}$ *is less than or equal to the number of control inputs m.* Satisfaction of the condition of Eq. (19.49) guarantees that a control law of the form of Eq. (19.47) can be synthesized such that the closed-loop output tracks the command input. In that case the closed-loop state equation is

$$\begin{bmatrix} \dot{\mathbf{x}} \\ \dot{\mathbf{z}} \end{bmatrix} = \begin{bmatrix} \mathbf{A} + \mathbf{B}\mathbf{K}_1 & \mathbf{B}\mathbf{K}_2 \\ -\mathbf{E} & \mathbf{0} \end{bmatrix} \begin{bmatrix} \mathbf{x} \\ \mathbf{z} \end{bmatrix} + \begin{bmatrix} \mathbf{0} \\ \mathbf{I} \end{bmatrix} \mathbf{r} \tag{19.50}$$

The feedback matrix of Eq. (19.48) can be selected so that the eigenvalues are in the left-half plane for the closed-loop plant matrix of Eq. (19.50). Then, since the inputs are constants, the steady-state values of the states $\mathbf{x}$ and $\mathbf{z}$ are also constants. Therefore, $\dot{\mathbf{z}} = \mathbf{0}$ in the steady state and Eq. (19.43) indicates that Eq.

(19.42) is satisfied. Thus, the outputs $\mathbf{w}(t)$ track the piecewise constant command vector $\mathbf{r}(t)$ in the steady state.

19.5 TRACKING-SYSTEM DESIGN EXAMPLE

Let the control system represented by the state and output equations of Eqs. (19.40) and (19.41) have the following plant, control, and output matrices

$$\mathbf{A} = \begin{bmatrix} 0 & 1 & 0 \\ 0 & 0 & 1 \\ 4 & 4 & -1 \end{bmatrix} \qquad \mathbf{B} = \begin{bmatrix} 1 & 0 \\ 0 & 0 \\ 0 & 1 \end{bmatrix} \qquad \mathbf{C} = \begin{bmatrix} 1 & 0 & 0 \\ 0 & 1 & 1 \end{bmatrix} \tag{19.51}$$

Since $\sigma(\mathbf{A}) = \{-1, -2, 2\}$, the open-loop system is unstable. The purpose of the design is to achieve a stable system and tracking of the command $\mathbf{r}(t)$ by the output vector $\mathbf{y}(t)$, where

$$\mathbf{r}(t) = \begin{bmatrix} 1 \\ -2 \end{bmatrix} \tag{19.52}$$

The system is augmented by the comparator and integrator given in Eq. (19.43). Since there are two outputs ($\ell = 2$) and two control inputs ($m = 2$), tracking of the inputs $\mathbf{r}$ by the outputs $\mathbf{y}$ is possible. Thus, $\mathbf{E} = \mathbf{C}$. Then the composite open-loop state and output equations are represented by Eqs. (19.44) and (19.45). The composite system matrices using Eqs. (19.46) are

$$\overline{\mathbf{A}} = \begin{bmatrix} 0 & 1 & 0 & 0 & 0 \\ 0 & 0 & 1 & 0 & 0 \\ 4 & 4 & -1 & 0 & 0 \\ -1 & 0 & 0 & 0 & 0 \\ 0 & -1 & -1 & 0 & 0 \end{bmatrix} \tag{19.53}$$

$$\overline{\mathbf{B}} = \begin{bmatrix} 1 & 0 \\ 0 & 0 \\ 0 & 1 \\ 0 & 0 \\ 0 & 0 \end{bmatrix} \tag{19.54}$$

$$\overline{\mathbf{C}} = \begin{bmatrix} 1 & 0 & 0 & 0 & 0 \\ 0 & 1 & 1 & 0 & 0 \end{bmatrix} \tag{19.55}$$

The augmented system is controllable since $(\mathbf{A}, \mathbf{B})$ is a controllable pair and the necessary condition of Eq. (19.49) is satisfied, that is, $n + p = 5$ and

$$\operatorname{Rank} \begin{bmatrix} \mathbf{B} & \mathbf{A} \\ \mathbf{0} & -\mathbf{C} \end{bmatrix} = \operatorname{Rank} \begin{bmatrix} 1 & 0 & 0 & 1 & 0 \\ 0 & 0 & 0 & 0 & 1 \\ 0 & 1 & 4 & 4 & -1 \\ 0 & 0 & -1 & 0 & 0 \\ 0 & 0 & 0 & -1 & -1 \end{bmatrix} = 5 \tag{19.56}$$

The eigenvalue spectrum of the closed-loop plant matrix in Eq. (19.50) is selected as

$$\sigma(\overline{\mathbf{A}} + \overline{\mathbf{B}}\mathbf{K}) = \{-2, -3, -4, -5, -6\} \tag{19.57}$$

The associated eigenvectors must lie in the null spaces of

$$\bar{\mathbf{S}}(\lambda_i) = [\bar{\mathbf{A}} - \lambda_i \mathbf{I} \quad \bar{\mathbf{B}}]$$
$$= \begin{bmatrix} -\lambda_i & 1 & 0 & 0 & 0 & 1 & 0 \\ 0 & -\lambda_i & 1 & 0 & 0 & 0 & 0 \\ 4 & 4 & -1-\lambda_i & 0 & 0 & 0 & 1 \\ -1 & 0 & 0 & -\lambda_i & 0 & 0 & 0 \\ 0 & -1 & -1 & 0 & -\lambda_i & 0 & 0 \end{bmatrix} \tag{19.58}$$

Using the method[15,16] shown in Sec. 19.2, Example 2, these null spaces are

$$\ker \bar{\mathbf{S}}(-2) = \text{span}\left\{ \begin{bmatrix} 2 \\ 0 \\ 0 \\ 1 \\ 0 \\ -4 \\ -8 \end{bmatrix}, \begin{bmatrix} 0 \\ 2 \\ -4 \\ 0 \\ 1 \\ -2 \\ -4 \end{bmatrix} \right\} \tag{19.59a}$$

$$\ker \bar{\mathbf{S}}(-3) = \text{span}\left\{ \begin{bmatrix} 3 \\ 0 \\ 0 \\ 1 \\ 0 \\ -9 \\ -12 \end{bmatrix}, \begin{bmatrix} 0 \\ 3 \\ -9 \\ 0 \\ -2 \\ -3 \\ 6 \end{bmatrix} \right\} \tag{19.59b}$$

$$\ker \bar{\mathbf{S}}(-4) = \text{span}\left\{ \begin{bmatrix} 4 \\ 0 \\ 0 \\ 1 \\ 0 \\ -16 \\ -16 \end{bmatrix}, \begin{bmatrix} 0 \\ 4 \\ -16 \\ 0 \\ -3 \\ -4 \\ 32 \end{bmatrix} \right\} \tag{19.59c}$$

$$\ker \bar{\mathbf{S}}(-5) = \text{span}\left\{ \begin{bmatrix} 5 \\ 0 \\ 0 \\ 1 \\ 0 \\ -25 \\ -20 \end{bmatrix}, \begin{bmatrix} 0 \\ 5 \\ -25 \\ 0 \\ -4 \\ -5 \\ 80 \end{bmatrix} \right\} \tag{19.59d}$$

$$\ker \bar{\mathbf{S}}(-6) = \text{span}\left\{ \begin{bmatrix} 6 \\ 0 \\ 0 \\ 1 \\ 0 \\ -36 \\ -24 \end{bmatrix}, \begin{bmatrix} 0 \\ 6 \\ -36 \\ 0 \\ -5 \\ -6 \\ 156 \end{bmatrix} \right\} \tag{19.59e}$$

The following vectors are selected from $\ker \mathbf{S}(\lambda_i)$ in Eqs. (19.59):

$$\begin{bmatrix} \mathbf{v}_1 \\ \mathbf{q}_1 \end{bmatrix} = \begin{bmatrix} 2 \\ 0 \\ 0 \\ 1 \\ 0 \\ \hline -4 \\ -8 \end{bmatrix} \quad \begin{bmatrix} \mathbf{v}_2 \\ \mathbf{q}_2 \end{bmatrix} = \begin{bmatrix} 0 \\ 3 \\ -9 \\ 0 \\ -2 \\ \hline -3 \\ 6 \end{bmatrix} \quad \begin{bmatrix} \mathbf{v}_3 \\ \mathbf{q}_3 \end{bmatrix} = \begin{bmatrix} 0 \\ 4 \\ -16 \\ 0 \\ -3 \\ \hline -4 \\ 32 \end{bmatrix} \tag{19.60}$$

$$\begin{bmatrix} \mathbf{v}_4 \\ \mathbf{q}_4 \end{bmatrix} = \begin{bmatrix} 0 \\ 5 \\ -25 \\ 0 \\ -4 \\ \hline -5 \\ 80 \end{bmatrix} \quad \begin{bmatrix} \mathbf{v}_5 \\ \mathbf{q}_5 \end{bmatrix} = \begin{bmatrix} 6 \\ 0 \\ 0 \\ 1 \\ 0 \\ \hline -36 \\ -24 \end{bmatrix}$$

The eigenvectors $\mathbf{v}_i$ $(i = 1, \ldots, 5)$ in Eqs. (19.60) are linearly independent, and therefore the feedback matrix $\overline{\mathbf{K}}$ can be computed by use of Eq. (18.118) as

$$\overline{\mathbf{K}} = \begin{bmatrix} -4 & -3 & -4 & -5 & -36 \\ -8 & 6 & 32 & 80 & -24 \end{bmatrix} \begin{bmatrix} 2 & 0 & 0 & 0 & 6 \\ 0 & 3 & 4 & 5 & 0 \\ 0 & -9 & -16 & -25 & 0 \\ 1 & 0 & 0 & 0 & 1 \\ 0 & -2 & -3 & -4 & 0 \end{bmatrix}^{-1}$$

$$= \begin{bmatrix} -4 & -3 & -4 & -5 & -36 \\ -8 & 6 & 32 & 80 & -24 \end{bmatrix} \left(\frac{1}{4}\right) \begin{bmatrix} -1 & 0 & 0 & 6 & 0 \\ 0 & 22 & -2 & 0 & 40 \\ 0 & -28 & 4 & 0 & -60 \\ 0 & 10 & -2 & 0 & 24 \\ 1 & 0 & 0 & -2 & 0 \end{bmatrix}$$

$$= \begin{bmatrix} -8 & -1 & 0 & 12 & 0 \\ -4 & 9 & -11 & 0 & 60 \end{bmatrix} \tag{19.61}$$

The closed-loop composite plant matrix is

$$\overline{\mathbf{A}} + \overline{\mathbf{B}}\overline{\mathbf{K}} = \begin{bmatrix} -8 & 0 & 0 & 12 & 0 \\ 0 & 0 & 1 & 0 & 0 \\ 0 & 13 & -12 & 0 & 60 \\ -1 & 0 & 0 & 0 & 0 \\ 0 & -1 & -1 & 0 & 0 \end{bmatrix} \tag{19.62}$$

This closed-loop plant matrix has the assigned eigenvalue spectrum of Eq. (19.57) and the eigenvectors of Eq. (19.60), as required. It should be noted that the selection of the eigenvectors in Eq. (19.60) produces the result that only the modes e^{-2t} and e^{-6t} appear in the states $x_1(t)$ and $z_1(t)$. Also, only the modes e^{-3t}, e^{-4t}, and e^{-5t} appear in the states $x_2(t)$, $x_3(t)$, and $z_2(t)$. This is verified by computing the state response from Eq. (19.11) for an initially quiescent system $[\mathbf{x}(0) = \mathbf{0}, \mathbf{z}(0) = \mathbf{0}]$ and the input of Eq. (19.52). The Laplace transform of Eqs.

(19.50) and (19.45) yields

$$\begin{bmatrix} \mathbf{X}(s) \\ \mathbf{Z}(s) \end{bmatrix} = \begin{bmatrix} \dfrac{12}{s(s+2)(s+6)} \\ \dfrac{-120}{s(s+3)(s+4)(s+5)} \\ \dfrac{-120}{(s+3)(s+4)(s+5)} \\ \dfrac{s+8}{s(s+2)(s+6)} \\ \dfrac{-2(s^2+12s-13)}{s(s+3)(s+4)(s+5)} \end{bmatrix} \tag{19.63}$$

$$\mathbf{Y}(s) = \begin{bmatrix} \dfrac{12}{s(s+2)(s+6)} \\ \dfrac{-120(s+1)}{s(s+3)(s+4)(s+5)} \end{bmatrix} \tag{19.64}$$

Thus, the output time response is

$$\mathbf{y}(t) = \begin{bmatrix} 1 - 1.5e^{-2t} + 0.5e^{-6t} \\ -2 - 40e^{-3t} + 90e^{-4t} - 48e^{-5t} \end{bmatrix} \tag{19.65}$$

and its final value is

$$\mathbf{y}(t)_{ss} = \begin{bmatrix} 1 \\ -2 \end{bmatrix} \tag{19.66}$$

Therefore $\mathbf{y}(t)_{ss}$ is equal to the command input given in Eq. (19.52), as required. This example demonstrates that a tracker can be synthesized which results in each output following a specified constant input command, that is, $y_1(t)$ tracks $r_1(t)$ and $y_2(t)$ tracks $r_2(t)$. Also, the use of the method of entire eigenstructure assignment permits the selection of the eigenvectors associated with each mode. The result of this selection may produce designs such that only certain modes appear in each state and in each output. This feature is a function of the spaces [see Eqs. (19.59)] from which the closed-loop eigenvectors must be selected.

19.6 EXAMPLE OF OBSERVER SYNTHESIS

An observer is to be synthesized for the plant of Example 1, Sec. 19.2. The closed-loop system eigenvalue spectrum is assigned as $\{-2, -3, -4\}$. In order to achieve rapid convergence of the observer states to the plant states, the observer eigenvalues are placed farther to the left of the system eigenvalues and are selected as

$$\sigma(\mathbf{A} - \mathbf{LC}) = \{-10, -11, -12\} \tag{19.67}$$

Therefore, using Eq. (18.128),

$$\mathbf{S}_o(\lambda) = [\mathbf{A}^T - \lambda\mathbf{I} \quad \mathbf{C}^T] = \begin{bmatrix} -\lambda & 0 & 4 & 1 & 0 \\ 1 & -\lambda & 4 & 0 & 1 \\ 0 & 1 & -\lambda - 1 & 0 & 1 \end{bmatrix} \tag{19.68}$$

Using the method shown in Example 2, Sec. 19.2, yields

$$\ker \mathbf{S}_o(\lambda) = \text{span}\left\{ \begin{bmatrix} \lambda + 1 \\ 1 \\ 0 \\ \lambda^2 + \lambda \\ -1 \end{bmatrix}, \begin{bmatrix} \lambda + 5 \\ 0 \\ -1 \\ \lambda^2 + 5\lambda + 4 \\ -\lambda - 1 \end{bmatrix} \right\} \tag{19.69}$$

The selected vectors from the subspaces of Eq. (19.69) for the eigenvalues of Eq. (19.67) are

$$\begin{bmatrix} \mathbf{w}_1 \\ \zeta_1 \end{bmatrix} = \begin{bmatrix} -9 \\ 1 \\ 0 \\ \hline 90 \\ -1 \end{bmatrix} \qquad \begin{bmatrix} \mathbf{w}_2 \\ \zeta_2 \end{bmatrix} = \begin{bmatrix} -10 \\ 1 \\ 0 \\ \hline 110 \\ -1 \end{bmatrix} \qquad \begin{bmatrix} \mathbf{w}_3 \\ \zeta_3 \end{bmatrix} = \begin{bmatrix} -7 \\ 0 \\ -1 \\ \hline 88 \\ 11 \end{bmatrix} \tag{19.70}$$

Using Eq. (18.129) to compute the observer matrix yields

$$\mathbf{L}^T = -\mathbf{Z}\mathbf{W}^{-1} = -\begin{bmatrix} 90 & 110 & 88 \\ -1 & -1 & 11 \end{bmatrix} \begin{bmatrix} -9 & -10 & -7 \\ 1 & 1 & 0 \\ 0 & 0 & -1 \end{bmatrix}^{-1}$$

$$= -\begin{bmatrix} 90 & 110 & 88 \\ -1 & -1 & 11 \end{bmatrix} \begin{bmatrix} 1 & 10 & -7 \\ -1 & -9 & 7 \\ 0 & 0 & -1 \end{bmatrix} = \begin{bmatrix} 20 & 90 & -52 \\ 0 & 1 & 11 \end{bmatrix} \tag{19.71}$$

The resulting observer plant matrix is

$$\mathbf{A} - \mathbf{LC} = \begin{bmatrix} -20 & 1 & 0 \\ -90 & -1 & 0 \\ 56 & -7 & -12 \end{bmatrix} \tag{19.72}$$

This matrix has the eigenvalue spectrum of Eq. (19.67) and the associated set of eigenvectors

$$\{\mathbf{v}_1, \ \mathbf{v}_2, \ \mathbf{v}_3\} = \left\{ \begin{bmatrix} 1 \\ 10 \\ -7 \end{bmatrix}, \begin{bmatrix} -1 \\ -9 \\ 7 \end{bmatrix}, \begin{bmatrix} 0 \\ 0 \\ -1 \end{bmatrix} \right\} \tag{19.73}$$

These eigenvectors of $\mathbf{A} - \mathbf{LC}$ are obtained from the rows of the matrix $\mathbf{W}^{-1}$ in Eq. (19.71).

19.7 SUMMARY

This chapter demonstrates the successful application of entire eigenstructure assignment to the synthesis of MIMO regulators and tracking systems. Such tracking systems achieve the objective of causing all output components to track specified inputs. Additional techniques need to be developed on the proper guidelines for selection of eigenvectors to satisfy specifications on output response. The selection of eigenvectors is dependent upon the particular application and the characteristics of the plant. In Ref. 12, guidelines are presented for the selection of eigenvectors for the longitudinal and lateral directional modes of aircraft flight control systems.

The assignment of the eigenvalue spectrum alone does not lead to a unique feedback matrix because the feedback matrix also assigns the eigenvectors associated with the eigenvalues. The assignment of the entire eigenstructure, consisting of both the eigenvalues and the eigenvectors, leads to a unique feedback matrix. The method of entire eigenstructure assignment using state feedback presented in this chapter identifies the subspaces in which the eigenvectors associated with each assigned eigenvalue must be located. The ability to select from many possible eigenvectors for a given eigenvalue provides the means for adjusting the magnitude of each mode which appears in the output. This method of shaping the output provides the potential for achieving the "best possible" response characteristics. Following the selection of the eigenvectors, a simple formula is used to compute the required feedback gain matrix. These methods are easily implemented on a digital computer.[16]

The method of entire eigenstructure assignment is useful in the case of systems which are not completely controllable. In such systems some of the eigenvalues cannot be changed but the associated eigenvectors can be changed. The eigenvector determines the magnitude of the uncontrollable mode in each of the states and is adjustable. The method of entire eigenstructure assignment is also applicable to the synthesis of observers. This is based on the dual of the state-feedback method and assigns the observer eigenvalue and reciprocal eigenvector spectra.

The designer must be aware of the fact that speeding up of the output response by use of feedback requires an increase in plant effort. This means that the signal $u(t)$ must increase in order to increase the speed of response of $y(t)$. The input $u(t)$ to the plant may have maximum limits which cannot be exceeded. An example of this is the aircraft in which the control surfaces have maximum displacements. Increasing the signal inputs to the servo actuators which drive the control surfaces may cause these surfaces to reach their mechanical maximum limits. Further increases in the inputs causes a saturation of the control surface deflections, and the system becomes nonlinear. It is therefore necessary in the design process to obtain a simulation of the complete system performance. Bounds on the maximum inputs for which the system remains linear can be determined from the simulation. Stability of the system in the presence of saturated control signals must be carefully investigated.

The material in Chaps. 18 and 19 provides the preparation for further study of modern control-system synthesis methods. Since the states are not usually all accessible for feedback, the use of output feedback may be more appropriate. Design of output feedback systems is covered in Chap. 20. Additional important topics from the technical literature include output feedback regulators,[2,5] dynamic compensators,[3] output feedback trackers,[4,3] and high-gain output feedback regulators.[6]

An important area of control theory is the use of digital controllers to achieve the desired system response. In such cases a discrete domain model representation of the plant may be necessary.[7] The method of entire eigenstructure assignment using state feedback is readily applied to such systems.[8-11] Chapter 20 provides an introduction to the design of output feedback control systems. The use of output feedback eliminates the requirement to either measure all of the states or to use an observer to generate an estimate of the state vector.

REFERENCES

1. Porter, B., and A. Bradshaw: "Design of Linear Multivariable Continuous-Time Tracking Systems," *Int. J. Syst. Sci.*, vol. 5, pp. 1155–1164, 1974.
2. Porter, B., and A. Bradshaw: "Design of Linear Multivariable Continuous-Time Output-Feedback Regulators," *Int. J. Syst. Sci.*, vol. 9, pp. 445–450, 1978.
3. Brasch, F. M., and J. B. Pearson: "Pole Placement Using Dynamic Compensators," *IEEE Trans. Autom. Control*, vol. AC-15, pp. 34–43, 1970.
4. Porter, B., and A. Bradshaw: "Design of Linear Multivariable Continuous-Time Tracking Systems Incorporating Error-Actuated Dynamic Controllers," *Int. J. Syst. Sci.*, vol. 9, pp. 627–637, 1978.
5. Bradshaw, A., L. R. Fletcher, and B. Porter: "Synthesis of Output-Feedback Control Laws for Linear Multivariable Continuous-Time Systems," *Int. J. Syst. Sci.*, vol. 9, pp. 1331–1340, 1978.
6. Porter, B., and A. Bradshaw: "Design of Linear Multivariable Continuous Time High-Gain Output-Feedback Regulators," *Int. J. Syst. Sci.*, vol. 10, pp. 113–121, 1979.
7. Kwakernaak, H., and R. Sivan: *Linear Optimal Control Systems*, Wiley-Interscience, New York, 1972.
8. D'Azzo, J. J.: "Synthesis of Linear Multivariable Sampled-Data Feedback Control Systems by Entire Eigenstructure Assignment," Ph.D. thesis, University of Salford, England, 1978.
9. D'Azzo, J. J., and B. Porter: "Synthesis of Digital Flight Control Systems by the Method of Entire Eigenstructure Assignment," *Proc. IEEE Natl. Aerosp. Electron. Conf. (NAECON)*, Dayton, 1978, pp. 91–99.
10. D'Azzo, J. J., and T. A. Kennedy: "Synthesis of Digital Flight Control Tracking Systems by the Method of Entire Eigenstructure Assignment," *Proc. IEEE Natl. Aerosp. Electron. Conf. (NAECON)*, Dayton, 1979, pp. 343–347.
11. D'Azzo, J. J.: "Design of Disturbance-Rejection Controllers for Linear Multivariable Discrete-Time Systems Using Entire Eigenstructure Assignment," *Proc. 18th Annu. Conf. Decision Control (CDC)*, Ft. Lauderdale, 1979, pp. 315–321.
12. Shapiro, E. Y., A. N. Andry, and J. C. Chung: "On Eigenstructure Assignment for Linear Systems," *IEEE Trans. Aerosp. Electron. Syst.*, vol. 19, pp. 711–729, 1983.
13. Reid, J. G.: *Linear System Fundamentals: Continuous and Discrete, Classic and Modern*, McGraw-Hill, New York, 1983.
14. Hautus, M. L. J.: "Controllability and Observability Conditions of Linear Autonomous Systems," *Proc. Ned. Acad.*, ser. A, vol. 72, pp. 443–448, 1969.
15. Porter, B., and J. J. D'Azzo: "Closed-Loop Eigenstructure Assignment by State Feedback in Multivariable Linear Systems," *Int. J. Control*, vol. 27, pp. 487–492, 1978.
16. Porter, B., and J. J. D'Azzo: "Algorithm for Closed-Loop Eigenstructure Assignment by State Feedback in Multivariable Linear Systems," *Int. J. Control.*, vol. 27, pp. 943–947, 1978.
17. Rosenbrock, H. H.: *State Space and Multivariable Theory*, Nelson, London, 1970.
18. Davison, E. J., and H. W. Smith: "Pole Assignment in Linear Time-Invariant Multivariable Systems with Constant Disturbances," *Automatica*, vol. 7, pp. 489–498, 1971.

CHAPTER 20

DESIGN OF TRACKING SYSTEMS USING OUTPUT FEEDBACK

20.1 INTRODUCTION

In the MIMO tracking systems described in Chap. 19 the system outputs $\mathbf{y}(t)$ effectively follow constant input commands $\mathbf{r}(t)$ in the steady state. The outputs are also decoupled and noninteracting so that each output follows only its respective input, that is, $y_1(t)$ tracks $r_1(t)$, $y_2(t)$ tracks $r_2(t)$, etc. There is only a transient interaction from other inputs and the amplitudes of those transients are very small. The design method of Chap. 19 requires that all the states must be fed back as shown in Fig. 19.1. Depending upon the dimension of the state vector, it may not be feasible to physically include sensors which will measure all of the states. Therefore, it is often necessary to use a state estimator or observer to reconstruct the states, and then these estimated states are fed back.

The control system configuration of Fig. 20.1 uses only output feedback to generate an error vector $\mathbf{e}(t) = \mathbf{r}(t) - \mathbf{y}(t)$. Therefore, the direct use of the output measurements to produce the input signal to the controller avoids the requirement for measuring or reconstructing the entire state vector. Further, based upon a high-gain approach, the closed-loop system exhibits a distinctive asymptotic structure in which there are *slow* and *fast* modes. The slow modes are

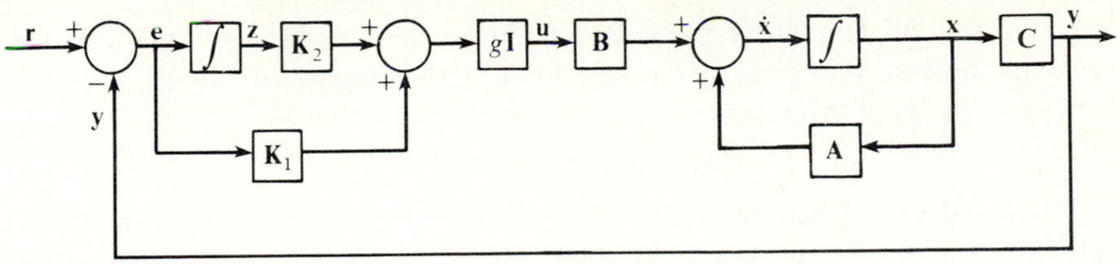

FIGURE 20.1
Output feedback tracking system containing a proportional plus integral (P1) controller.

asymptotically uncontrollable or unobservable, and thus the output response is dominated by the fast modes. This leads to very fast tracking of the command input by the output. This chapter presents a method for designing robust control systems, as shown in Fig. 20.1.

The design is dependent upon the first Markov parameter which is equal to the matrix product **CB**. When **CB** has maximal rank the plant is described as *regular*. When **CB** does not have maximal rank, then the plant is described as *irregular*. In the case of regular plants the controller implements a proportional plus integral control law as shown in Fig. 20.1. In the case of irregular plants, the proportional plus integral control is augmented with an inner-loop which provides extra measurements for control purposes. (See Sec. 20.7.)

20.2 OUTPUT FEEDBACK TRACKING SYSTEM

The multi-input multi-output (MIMO) plant is represented by the state and output equations of the respective forms

$$\dot{\mathbf{x}}(t) = \mathbf{A}\mathbf{x}(t) + \mathbf{B}\mathbf{u}(t) \tag{20.1}$$

$$\mathbf{y}(t) = \mathbf{C}\mathbf{x}(t) \tag{20.2}$$

where the dimension of **A** is $n \times n$, **B** is $n \times m$, **C** is $\ell \times n$, and the rank of **B** is m. These equations are often available or may be transformed to the form

$$\begin{bmatrix} \dot{\mathbf{x}}_1(t) \\ \dot{\mathbf{x}}_2(t) \end{bmatrix} = \begin{bmatrix} \mathbf{A}_{11} & \mathbf{A}_{12} \\ \mathbf{A}_{21} & \mathbf{A}_{22} \end{bmatrix} \begin{bmatrix} \mathbf{x}_1(t) \\ \mathbf{x}_2(t) \end{bmatrix} + \begin{bmatrix} \mathbf{0} \\ \mathbf{B}_2 \end{bmatrix} \mathbf{u}(t) \tag{20.3}$$

$$\mathbf{y}(t) = [\mathbf{C}_1 \quad \mathbf{C}_2] \begin{bmatrix} \mathbf{x}_1(t) \\ \mathbf{x}_2(t) \end{bmatrix} \tag{20.4}$$

where $\mathbf{x}_2$ is $m \times 1$, $\mathbf{B}_2$ is $m \times m$ and has rank m, $\mathbf{C}_2$ is $m \times m$ and has rank m, and the remaining elements in these equations have appropriate dimensions. The design method for the tracker presented in this chapter requires that the number of controlled outputs $\mathbf{y}(t)$ must be equal to the number of controls $\mathbf{u}(t)$. Thus, $\ell = m$. A high-gain controller implements a proportional plus integral (PI) control law represented by

$$\mathbf{u}(t) = g\{\mathbf{K}_1\mathbf{e}(t) + \mathbf{K}_2\mathbf{z}(t)\} \tag{20.5}$$

where g is a scalar gain. The error vector between the constant command input $\mathbf{r}(t)$ and the output $\mathbf{y}(t)$ is $\mathbf{e}(t) = \mathbf{r}(t) - \mathbf{y}(t)$. The integral of the error is the vector $\mathbf{z}(t) = \int_0^t \mathbf{e}(t)\,dt$ which therefore satisfies the equation

$$\dot{\mathbf{z}}(t) = \mathbf{r}(t) - \mathbf{y}(t) \tag{20.6}$$

The closed-loop tracking system of Fig. 20.1 is represented by Eqs. (20.3) through (20.6). Combining these equations, the composite closed-loop state and output equations have the respective forms

$$\begin{bmatrix} \dot{\mathbf{z}}(t) \\ \dot{\mathbf{x}}_1(t) \\ \dot{\mathbf{x}}_2(t) \end{bmatrix} = \left[\begin{array}{cc|c} \mathbf{0} & -\mathbf{C}_1 & -\mathbf{C}_2 \\ \mathbf{0} & \mathbf{A}_{11} & \mathbf{A}_{12} \\ \hline g\mathbf{B}_2\mathbf{K}_2 & \mathbf{A}_{21} - g\mathbf{B}_2\mathbf{K}_1\mathbf{C}_1 & \mathbf{A}_{22} - g\mathbf{B}_2\mathbf{K}_1\mathbf{C}_2 \end{array}\right] \begin{bmatrix} \mathbf{z}(t) \\ \mathbf{x}_1(t) \\ \mathbf{x}_2(t) \end{bmatrix} + \begin{bmatrix} \mathbf{I}_\ell \\ \mathbf{0} \\ g\mathbf{B}_2\mathbf{K}_1 \end{bmatrix} \mathbf{r}(t) \tag{20.7}$$

and

$$\mathbf{y}(t) = [\mathbf{0} \quad \mathbf{C}_1 \mid \mathbf{C}_2] \begin{bmatrix} \mathbf{z}(t) \\ \mathbf{x}_1(t) \\ \mathbf{x}_2(t) \end{bmatrix} \tag{20.8}$$

The partitioning of the matrices in these equations is used in the following section. These closed-loop equations may be represented in the forms, respectively,

$$\dot{\bar{\mathbf{x}}}(t) = \begin{bmatrix} \bar{\mathbf{A}}_1 & \bar{\mathbf{A}}_2 \\ \bar{\mathbf{A}}_3 & \bar{\mathbf{A}}_4 \end{bmatrix} \bar{\mathbf{x}}(t) + \begin{bmatrix} \bar{\mathbf{B}}_1 \\ \bar{\mathbf{B}}_2 \end{bmatrix} \mathbf{r}(t) = \bar{\mathbf{A}}\bar{\mathbf{x}}(t) + \bar{\mathbf{B}}\mathbf{r}(t) \tag{20.9}$$

$$\mathbf{y}(t) = [\bar{\mathbf{C}}_1 \quad \bar{\mathbf{C}}_2]\bar{\mathbf{x}}(t) = \bar{\mathbf{C}}\bar{\mathbf{x}}(t) \tag{20.10}$$

where the dimension of $\bar{\mathbf{A}}_1$ is $n \times n$, $\bar{\mathbf{A}}_4$ is $m \times m$, $\mathbf{B}_2$ is $m \times m$ and has rank m, and the remaining elements have corresponding dimensions. The closed-loop transfer function for Eqs. (20.7) and (20.8) is

$$\mathbf{G}(\lambda) = [\mathbf{0} \quad \mathbf{C}_1 \quad \mathbf{C}_2] \begin{bmatrix} \lambda\mathbf{I}_\ell & \mathbf{C}_1 & \mathbf{C}_2 \\ \mathbf{0} & \lambda\mathbf{I}_{n-\lambda} - \mathbf{A}_{11} & -\mathbf{A}_{12} \\ -g\mathbf{B}_2\mathbf{K}_2 & -\mathbf{A}_{21} + g\mathbf{B}_2\mathbf{K}_1\mathbf{C}_1 & \lambda\mathbf{I}_\ell - \mathbf{A}_{22} + g\mathbf{B}_2\mathbf{K}_1\mathbf{C}_2 \end{bmatrix}^{-1} \times \begin{bmatrix} \mathbf{I}_\ell \\ \mathbf{0} \\ g\mathbf{B}_2\mathbf{K}_1 \end{bmatrix} \tag{20.11}$$

The form of the $\mathbf{B}$ matrix in Eq. (20.3) and the corresponding partitioning in Eqs. (20.3) and (20.4) are introduced in order to develop the properties of the closed-loop system under high-gain operation. This is achieved for the control law of Eq. (20.5) when g becomes large and, in the limit, asymptotically approaches infinity, $g \to \infty$. These properties are derived by first using singular perturbation methods in Sec. 20.3 to block diagonalize the closed-loop plant matrix in Eq. (20.7). The resulting block diagonalization is presented in Sec. 20.3, and the analysis in Sec. 20.4 leads to definitions of the asymptotic slow and fast modes of the closed-loop system as $g \to \infty$. This further leads to a design procedure which is outlined in Sec. 20.5. Finally, a design example is presented in Sec. 20.6.

20.3 BLOCK DIAGONALIZATION

In this section the closed-loop system properties are determined when the scalar gain g in the control law of Eq. (20.5) has large values approaching infinity. In order to obtain these properties the state and output equations given in Eqs. (20.7) and (20.8) are transformed to the block diagonal form[1-6] given by

$$\begin{bmatrix} \dot{\mathbf{x}}_s(t) \\ \dot{\mathbf{x}}_f(t) \end{bmatrix} = \begin{bmatrix} \mathbf{A}_s & \mathbf{0} \\ \mathbf{0} & \mathbf{A}_f \end{bmatrix} \begin{bmatrix} \mathbf{x}_s(t) \\ \mathbf{x}_f(t) \end{bmatrix} + \begin{bmatrix} \mathbf{B}_s \\ \mathbf{B}_f \end{bmatrix} \mathbf{r}(t) \tag{20.12}$$

$$\mathbf{y}(t) = \begin{bmatrix} \mathbf{C}_s & \mathbf{C}_f \end{bmatrix} \begin{bmatrix} \mathbf{x}_s(t) \\ \mathbf{x}_f(t) \end{bmatrix} \tag{20.13}$$

The transformation is achieved by using

$$\bar{\mathbf{x}}(t) = \mathbf{T} \begin{bmatrix} \mathbf{x}_s(t) \\ \mathbf{x}_f(t) \end{bmatrix} \tag{20.14}$$

The transformation matrix $\mathbf{T}$ is selected[2-5] as

$$\mathbf{T} = \begin{bmatrix} \mathbf{I}_1 & \mathbf{M} \\ -\mathbf{L} & \mathbf{I}_2 - \mathbf{LM} \end{bmatrix} \tag{20.15}$$

where $\mathbf{I}_1$ and $\mathbf{I}_2$ are identity matrices of appropriate dimensions. The inverse of this matrix is equal to

$$\mathbf{T}^{-1} = \begin{bmatrix} \mathbf{I}_1 - \mathbf{ML} & -\mathbf{M} \\ \mathbf{L} & \mathbf{I}_2 \end{bmatrix} \tag{20.16}$$

Inserting the transformation of Eqs. (20.14) and (20.15) into Eq. (20.7) yields the new plant matrix

$$\mathbf{A}_T = \mathbf{T}^{-1}\bar{\mathbf{A}}\mathbf{T} = \begin{bmatrix} \mathbf{I}_1 - \mathbf{ML} & -\mathbf{M} \\ \mathbf{L} & \mathbf{I}_2 \end{bmatrix} \begin{bmatrix} \bar{\mathbf{A}}_1 & \bar{\mathbf{A}}_2 \\ \bar{\mathbf{A}}_3 & \bar{\mathbf{A}}_4 \end{bmatrix} \begin{bmatrix} \mathbf{I}_1 & \mathbf{M} \\ -\mathbf{L} & \mathbf{I}_2 - \mathbf{LM} \end{bmatrix} = \begin{bmatrix} \mathbf{A}_s & \mathbf{A}_x \\ \mathbf{A}_y & \mathbf{A}_f \end{bmatrix} \tag{20.17}$$

where

$$\mathbf{A}_s = \bar{\mathbf{A}}_1 - \mathbf{ML}\bar{\mathbf{A}}_1 - \mathbf{M}\bar{\mathbf{A}}_3 - \bar{\mathbf{A}}_2\mathbf{L} + \mathbf{ML}\bar{\mathbf{A}}_2\mathbf{L} + \mathbf{M}\bar{\mathbf{A}}_4\mathbf{L} \tag{20.18}$$

$$\begin{aligned} \mathbf{A}_x = {} & \bar{\mathbf{A}}_1\mathbf{M} - \mathbf{ML}\bar{\mathbf{A}}_1\mathbf{M} - \mathbf{M}\bar{\mathbf{A}}_3\mathbf{M} + \bar{\mathbf{A}}_2 - \mathbf{ML}\bar{\mathbf{A}}_2 \\ & - \mathbf{M}\bar{\mathbf{A}}_4 - \bar{\mathbf{A}}_2\mathbf{LM} + \mathbf{ML}\bar{\mathbf{A}}_2\mathbf{LM} + \mathbf{M}\bar{\mathbf{A}}_4\mathbf{LM} \end{aligned} \tag{20.19}$$

$$\mathbf{A}_y = \mathbf{L}\bar{\mathbf{A}}_1 + \bar{\mathbf{A}}_3 - \mathbf{L}\bar{\mathbf{A}}_2\mathbf{L} - \bar{\mathbf{A}}_4\mathbf{L} \tag{20.20}$$

$$\mathbf{A}_f = \mathbf{L}\bar{\mathbf{A}}_1\mathbf{M} + \bar{\mathbf{A}}_3\mathbf{M} + \mathbf{L}\bar{\mathbf{A}}_2 + \bar{\mathbf{A}}_4 - \mathbf{L}\bar{\mathbf{A}}_2\mathbf{LM} - \bar{\mathbf{A}}_4\mathbf{LM} \tag{20.21}$$

In order for the resulting matrix of Eq. (20.17) to have the block diagonal form shown in Eq. (20.12), it is necessary that $\mathbf{A}_x = \mathbf{A}_y = \mathbf{0}$. Inserting these conditions into Eqs. (20.18) and (20.21) yields

$$\mathbf{A}_s = \overline{\mathbf{A}}_1 - \overline{\mathbf{A}}_2\mathbf{L} \tag{20.22}$$

$$\mathbf{A}_f = \mathbf{L}\overline{\mathbf{A}}_2 + \overline{\mathbf{A}}_4 \tag{20.23}$$

The value of the matrix $\mathbf{L}$ is determined by using the series form

$$\mathbf{L} = \mathbf{L}_0 + g^{-1}\mathbf{L}_1 + g^{-2}\mathbf{L}_2 + \cdots \tag{20.24}$$

Inserting Eq. (20.24) into the equation $\mathbf{A}_y = \mathbf{0}$ and taking the limit as $g \to \infty$ yields (see Prob. 20.7)

$$\mathbf{L}_0 = \left[-\mathbf{C}_2^{-1}\mathbf{K}_1^{-1}\mathbf{K}_2 \quad \mathbf{C}_2^{-1}\mathbf{C}_1\right] \tag{20.25}$$

The first-order approximation of $\mathbf{L}$ as $g \to \infty$ is, therefore

$$\mathbf{L} = \mathbf{L}_0 = \left[-\mathbf{C}_2^{-1}\mathbf{K}_1^{-1}\mathbf{K}_2 \quad \mathbf{C}_2^{-1}\mathbf{C}_1\right] \tag{20.26}$$

The evaluation of $\mathbf{M}$ is determined by using the series form

$$\mathbf{M} = \mathbf{M}_0 + g^{-1}\mathbf{M}_1 + \cdots \tag{20.27}$$

The necessary conditions for obtaining the block diagonal plant matrix of Eq. (20.12) require that $\mathbf{A}_x = \mathbf{0}$ and $\mathbf{A}_y = \mathbf{0}$. Combining these equations for the values of $\mathbf{A}_x$ and $\mathbf{A}_y$ given in Eqs. (20.19) and (20.20) yields

$$\mathbf{A}_x = \left(\overline{\mathbf{A}}_1 - \overline{\mathbf{A}}_2\mathbf{L}\right)\mathbf{M} - \mathbf{M}\left(\mathbf{L}\overline{\mathbf{A}}_2 + \overline{\mathbf{A}}_4\right) + \overline{\mathbf{A}}_2 = \mathbf{0} \tag{20.28}$$

The expansion of the left side of Eq. (20.28), in terms of $\mathbf{L}$ and $\mathbf{M}$ given by Eqs. (20.24) and (20.27), produces the term containing the highest power of g as $g\mathbf{M}_0\mathbf{B}_2\mathbf{K}_1\mathbf{C}_2 = \mathbf{0}$. Then, as $g \to \infty$, since $\mathbf{B}_2\mathbf{K}_1\mathbf{C}_2$ is nonsingular, the necessary condition is that $\mathbf{M}_0 = \mathbf{0}$. Therefore, a first-order approximation of $\mathbf{M}$ is $\mathbf{M} = g^{-1}\mathbf{M}_1$. Inserting this value into Eq. (20.28) and taking the limit as $g \to \infty$, the nonzero terms remaining are

$$\mathbf{M}_1\mathbf{B}_2\mathbf{K}_1\mathbf{C}_2 + \begin{bmatrix} -\mathbf{C}_2 \\ \mathbf{A}_{12} \end{bmatrix} = \mathbf{0} \tag{20.29}$$

The first-order approximation of $\mathbf{M}$ is therefore

$$\mathbf{M} = g^{-1}\mathbf{M}_1 = \begin{bmatrix} g^{-1}\mathbf{K}_1^{-1}\mathbf{B}_2^{-1} \\ g^{-1}\mathbf{A}_{12}\mathbf{C}_2^{-1}\mathbf{K}_1^{-1}\mathbf{B}_2^{-1} \end{bmatrix} \tag{20.30}$$

With the values of $\mathbf{L}$ and $\mathbf{M}$ determined by Eqs. (20.26) and (20.30), respectively, the transformation matrix $\mathbf{T}$ of Eq. (20.15) can now be obtained.

As $g \to \infty$, using the value of $\mathbf{L}$ in Eq. (20.26), the matrix $\mathbf{A}_s$ of Eq. (20.22) therefore becomes

$$\mathbf{A}_s = \begin{bmatrix} -\mathbf{K}_1^{-1}\mathbf{K}_2 & \mathbf{0} \\ \mathbf{A}_{12}\mathbf{C}_2^{-1}\mathbf{K}_1^{-1}\mathbf{K}_2 & \mathbf{A}_{11} - \mathbf{A}_{12}\mathbf{C}_2^{-1}\mathbf{C}_1 \end{bmatrix} \tag{20.31}$$

Similarly, as $g \to \infty$, Eq. (20.21) yields

$$\mathbf{A}_f = \mathbf{L}\overline{\mathbf{A}}_2 + \overline{\mathbf{A}}_4 = -g\mathbf{B}_2\mathbf{K}_1\mathbf{C}_2 \tag{20.32}$$

Inserting the transformation of Eqs. (20.14) into Eq. (20.7) yields the new control matrix

$$\begin{bmatrix} \mathbf{B}_s \\ \mathbf{B}_f \end{bmatrix} = \mathbf{T}^{-1}\overline{\mathbf{B}} = \begin{bmatrix} \mathbf{I}_1 - \mathbf{ML} & -\mathbf{M} \\ \mathbf{L} & \mathbf{I}_2 \end{bmatrix} \begin{bmatrix} \overline{\mathbf{B}}_1 \\ \overline{\mathbf{B}}_2 \end{bmatrix} = \begin{bmatrix} \mathbf{I}_1 & -g^{-1}\mathbf{M}_1 \\ \mathbf{L}_0 & \mathbf{I}_2 \end{bmatrix} \begin{bmatrix} \mathbf{I}_\ell \\ \mathbf{0} \\ \hline g\mathbf{B}_2\mathbf{K}_1 \end{bmatrix} \tag{20.33}$$

Thus,

$$\mathbf{B}_s = \begin{bmatrix} \mathbf{0} \\ \mathbf{A}_{12}\mathbf{C}_2^{-1} \end{bmatrix} \tag{20.34}$$

and, as $g \to \infty$,

$$\mathbf{B}_f = g\mathbf{B}_2\mathbf{K}_1 \tag{20.35}$$

The output equation is also transformed to

$$\mathbf{y}(t) = \overline{\mathbf{C}}\overline{\mathbf{x}}(t) = \overline{\mathbf{C}}\mathbf{T}\begin{bmatrix} \mathbf{x}_s(t) \\ \mathbf{x}_f(t) \end{bmatrix} = [\overline{\mathbf{C}}_1 \quad \overline{\mathbf{C}}_2] \begin{bmatrix} \mathbf{I}_1 & g^{-1}\mathbf{M}_1 \\ -\mathbf{L}_0 & \mathbf{I}_2 - \mathbf{L}_0\mathbf{M}_1 \end{bmatrix} \begin{bmatrix} \mathbf{x}_s(t) \\ \mathbf{x}_f(t) \end{bmatrix}$$

$$= [\mathbf{C}_s \quad \mathbf{C}_f] \begin{bmatrix} \mathbf{x}_s(t) \\ \mathbf{x}_f(t) \end{bmatrix} \tag{20.36}$$

Using the values obtained from Eqs. (20.8), (20.25), and (20.30), the elements of the output matrix, as $g \to \infty$, are

$$\mathbf{C}_s = [\mathbf{K}_1^{-1}\mathbf{K}_2 \quad \mathbf{0}] \tag{20.37}$$

$$\mathbf{C}_f = \mathbf{C}_2 \tag{20.38}$$

20.4 ANALYSIS OF CLOSED-LOOP SYSTEM PERFORMANCE

The system of Fig. 20.1 contains a proportional plus integral controller that generates the input vector to the plant. The signal $\mathbf{u}(t)$ is formed by the control law of Eq. (20.5). The composite closed-loop system is represented by the state and output equations (20.7) and (20.8), and the exact overall transfer function is given by Eq. (20.11).

In order to analyze the closed-loop system performance, the transformation of Eq. (20.14) is introduced in Sec. 20.3. By properly selecting the transformation matrix $\mathbf{T}$ given by Eq. (20.15), the closed-loop state and output equations are then given by Eqs. (20.12) and (20.13) for the case when the scalar gain $g \to \infty$. In Eq. (20.12) the plant matrix is in block diagonal form, and the subblocks $\mathbf{A}_s$ and $\mathbf{A}_f$ are represented by Eqs. (20.31) and (20.32), respectively. These submatrices are first analyzed in order to determine the nature of the closed-loop eigenvalues. They are then used to synthesize the matrices $\mathbf{K}_1$ and $\mathbf{K}_2$ in order to implement the control law of Eq. (20.5).

The closed-loop characteristic equation when $g \to \infty$, obtained from the plant matrix of Eq. (20.12), is

$$\begin{vmatrix} \lambda\mathbf{I}_n - \mathbf{A}_s & \mathbf{0} \\ \mathbf{0} & \lambda\mathbf{I}_m - \mathbf{A}_f \end{vmatrix} = |\lambda\mathbf{I}_n - \mathbf{A}_s| \cdot |\lambda\mathbf{I}_m - \mathbf{A}_f| = 0 \tag{20.39}$$

Using the matrices for $\mathbf{A}_s$ and $\mathbf{A}_f$ given in Eqs. (20.31) and (20.32), the closed-loop characteristic equation is

$$\left|\lambda\mathbf{I}_m + \mathbf{K}_1^{-1}\mathbf{K}_2\right| \cdot \left|\lambda\mathbf{I}_{n-m} - \left(\mathbf{A}_{11} - \mathbf{A}_{12}\mathbf{C}_2^{-1}\mathbf{C}_1\right)\right| \cdot |\lambda\mathbf{I}_m + g\mathbf{B}_2\mathbf{K}_1\mathbf{C}_2| = 0 \tag{20.40}$$

The components of Eq. (20.40) lead to three classes of eigenvalues.[1]

Eigenvalue set S_1**.** The first class of m eigenvalues is designated by the set S_1 and is determined by

$$\left|\lambda\mathbf{I}_m + \mathbf{K}_1^{-1}\mathbf{K}_2\right| = 0 \tag{20.41}$$

These eigenvalues lead to *slow* modes in the output. The designation of *slow* is based on the fact that, although $g \to \infty$, these eigenvalues have finite values that are determined by the values of $\mathbf{K}_1$ and $\mathbf{K}_2$. These control law matrices are selected by the designer, and therefore the set of eigenvalues S_1 can be assigned as desired. Since $\mathbf{A}_s$ is lower triangular and $\mathbf{B}_s$ has an upper zero block [see Eqs. (20.31) and (20.34)], the modes associated with the set of eigenvalues S_1 are uncontrollable.

Eigenvalue set S_2**.** The second class of $n - m$ eigenvalues is designated by the set S_2 and is determined by

$$\left|\lambda\mathbf{I}_{n-m} - \left[\mathbf{A}_{11} - \mathbf{A}_{12}\mathbf{C}_2^{-1}\mathbf{C}_1\right]\right| = 0 \tag{20.42}$$

Since this equation is a function only of the plant [see Eqs. (20.3) and (20.4)], the set of eigenvalues S_2 cannot be changed by selection of the control law matrices $\mathbf{K}_1$ and $\mathbf{K}_2$. These eigenvalues have finite values, and the associated modes are therefore also designated as *slow* modes. Since these eigenvalues cannot be changed by the control law, it is essential that they be located in the stable region. These eigenvalues are also identified in the technical literature as *transmission zeros*.[1,7,8,10] Since $\mathbf{A}_s$ is lower triangular and $\mathbf{C}_s$ has a right-hand zero block [see Eqs. (20.31) and (20.37)], the modes associated with the set of eigenvalues S_2 are unobservable.

Eigenvalue set S_3**.** The third class of m eigenvalues is designated by the set S_3 and is determined by

$$|\lambda\mathbf{I}_m + g\mathbf{B}_2\mathbf{K}_1\mathbf{C}_2| = 0 \tag{20.43}$$

Since the matrix $\mathbf{C}_2$ is a square matrix having full rank, Eq. (20.43) can be shown to be equivalent to (see Prob. 20.8)

$$|\lambda\mathbf{I}_m + g\mathbf{C}_2\mathbf{B}_2\mathbf{K}_1| = 0 \tag{20.44}$$

The set S_3 therefore consists of m eigenvalues which are all located at infinity because $g \to \infty$. Consequently there are m *fast* modes in the output. The designation of *fast* is based on the fact that they can be located in the stable region at infinity by the selection of the proportional matrix $\mathbf{K}_1$.

Using Eqs. (20.12) and (20.13), the asymptotic transfer function as $g \to \infty$ is obtained as

$$\Gamma(\lambda) = \Gamma_s(\lambda) + \Gamma_f(\lambda) = \begin{bmatrix} \mathbf{C}_s & \mathbf{C}_f \end{bmatrix} \begin{bmatrix} \lambda\mathbf{I}_n - \mathbf{A}_s & \mathbf{0} \\ \mathbf{0} & \lambda\mathbf{I}_m - \mathbf{A}_f \end{bmatrix}^{-1} \begin{bmatrix} \mathbf{B}_s \\ \mathbf{B}_f \end{bmatrix} \tag{20.45}$$

where $\Gamma_s(\lambda)$ is the slow transfer function and $\Gamma_f(\lambda)$ is the fast transfer function. The slow transfer function is computed from Eq. (20.45). Using the matrices for $\mathbf{A}_s$, $\mathbf{B}_s$, and $\mathbf{C}_s$ given in Eqs. (20.31), (20.34), and (20.37) yields

$$\Gamma_s(\lambda) = \mathbf{C}_s[\lambda\mathbf{I}_n - \mathbf{A}_s]^{-1}\mathbf{B}_s = \mathbf{0} \tag{20.46}$$

This is a consequence of the uncontrollability of the modes associated with the set of eigenvalues S_1 and the unobservability of the modes associated with the set of eigenvalues S_2. Therefore, the output $\mathbf{y}(t)$ contains no slow modes.

The fast transfer function is computed from Eq. (20.45). Using the matrices for $\mathbf{A}_f$, $\mathbf{B}_f$, and $\mathbf{C}_f$ given in Eqs. (20.32), (20.35), and (20.38) yields

$$\Gamma_f(\lambda) = \mathbf{C}_f[\lambda\mathbf{I}_m - \mathbf{A}_f]^{-1}\mathbf{B}_f = \mathbf{C}_2[\lambda\mathbf{I}_m + g\mathbf{B}_2\mathbf{K}_1\mathbf{C}_2]^{-1}g\mathbf{B}_2\mathbf{K}_1 \tag{20.47}$$

By matrix manipulation (see Prob. 20.8), Eq. (20.47) may be rewritten as

$$\Gamma_f(\lambda) = [\lambda\mathbf{I}_m + g\mathbf{C}_2\mathbf{B}_2\mathbf{K}_1]^{-1}g\mathbf{C}_2\mathbf{B}_2\mathbf{K}_1 \tag{20.48}$$

In view of Eqs. (20.46) and (20.48), the exact transfer function $\mathbf{G}(\lambda)$ represented by Eq. (20.11) has the asymptotic form, as $g \to \infty$, given by

$$\Gamma(\lambda) = [\lambda\mathbf{I}_m + g\mathbf{C}_2\mathbf{B}_2\mathbf{K}_1]^{-1}g\mathbf{C}_2\mathbf{B}_2\mathbf{K}_1 \tag{20.49}$$

The output therefore contains only the fast modes corresponding to the set of eigenvalues S_3. In the case where a finite value is assigned to the scalar gain g, then the output will also contain slow modes. The dominancy of the fast modes is determined by the magnitude of the scalar gain g.

20.5 DESIGN PROCEDURE FOR REGULAR PLANTS

The design procedure is based on the requirement that all of the closed-loop eigenvalues must be located in the left-half plane so that the system is stable. In addition, noninteracting or decoupled control is desirable. This means that each output tracks its respective command input, that is, $y_1(t)$ follows $r_1(t)$, $y_2(t)$ follows $r_2(t)$, etc. Also, the amplitudes of any transients due to other inputs must be very small.

The design process consists of the following steps:

1. Check the controllability and observability of the plant given by the state and output equations given by Eqs. (20.1) and (20.2), respectively.

2. Check the controllability of the augmented plant containing the proportional plus integral control. This requires, in addition to controllability of the plant (step 1), that the following matrix have full rank.[9]

$$\text{Rank}\begin{bmatrix} \mathbf{B} & \mathbf{A} \\ \mathbf{0} & \mathbf{C} \end{bmatrix} = n + \ell \tag{20.50}$$

3. Determine the transmission zeros[10] which comprise set S_2 of the closed-loop eigenvalues. These are determined from

$$\left| \lambda \mathbf{I}_{n-m} - \mathbf{A}_{11} + \mathbf{A}_{12}\mathbf{C}_2^{-1}\mathbf{C}_1 \right| = 0 \tag{20.51}$$

The transmission zeros must be located in the left-half plane or the design cannot proceed. In the case of a controllable and observable plant, the transmission zeros can also be computed from the determinant of the system matrix which is defined[7] by

$$\begin{vmatrix} s\mathbf{I}_n - \mathbf{A} & \mathbf{B} \\ -\mathbf{C} & \mathbf{0} \end{vmatrix} = 0 \tag{20.52}$$

The transmission zeros are obtained from Eq. (20.52) even when the **B** matrix of the state equation is not in the partitioned form shown in Eq. (20.3). The number of transmission zeros[8] for a system having the same number of outputs and inputs, i.e., for $\ell = m$, is equal to $n - m - d$, where d is the rank deficiency of the matrix product **CB**.

4. Assign the fast eigenvalue set S_3 by use of Eq. (20.44). Since decoupling of the outputs is desired, an arbitrary diagonal matrix $\mathbf{\Sigma}$ is selected and used in

$$\mathbf{C}_2\mathbf{B}_2\mathbf{K}_1 = \mathbf{\Sigma} = \begin{bmatrix} \sigma_1 & & & \mathbf{0} \\ & \sigma_2 & & \\ & & \ddots & \\ \mathbf{0} & & & \sigma_m \end{bmatrix} \tag{20.53}$$

The values $-g\sigma_1, g\sigma_2, \ldots, g\sigma_m$ comprise the fast eigenvalue set S_3. From Eq. (20.53), the value of the proportional matrix $\mathbf{K}_1$ is computed from

$$\mathbf{K}_1 = (\mathbf{C}_2\mathbf{B}_2)^{-1}\mathbf{\Sigma} \tag{20.54a}$$

It is evident from Eq. (20.54) that $\mathbf{C}_2\mathbf{B}_2$ must have full rank in order for the inverse to exist. When this is true, the plant is described as regular. (Irregular plants for which $\mathbf{C}_2\mathbf{B}_2$ does not have full rank are covered in Sec. 20.7).

The development of Eq. (20.54*a*) is based on the state and output equations given in Eqs. (20.3) and (20.4), in which the **B** matrix has the form $\mathbf{B} = [\mathbf{0} \quad \mathbf{B}_2^T]^T$. When the **B** matrix is not in this form, it can be achieved by introducing the transformation $\mathbf{x} = \mathbf{T}\mathbf{x}'$ so that the state and output equations take the form

$$\dot{\mathbf{x}}' = [\mathbf{T}^{-1}\mathbf{A}\mathbf{T}]\mathbf{x}' + [\mathbf{T}^{-1}\mathbf{B}]\mathbf{u}$$

$$\mathbf{y} = [\mathbf{C}\mathbf{T}]\mathbf{x}'$$

where the new control matrix has the form $\mathbf{T}^{-1}\mathbf{B} = [\mathbf{0} \quad \mathbf{B}_2^T]^T$ and the new output matrix is $\mathbf{CT} = [\mathbf{C}_1 \quad \mathbf{C}_2]$. It is now evident that the product of these matrices yields

$$[\mathbf{CT}][\mathbf{T}^{-1}\mathbf{B}] = \mathbf{CB} = [\mathbf{C}_1 \quad \mathbf{C}_2]\begin{bmatrix}\mathbf{0}\\ \mathbf{B}_2\end{bmatrix} = \mathbf{C}_2\mathbf{B}_2$$

Therefore, the matrix product **CB**, known as the first *Markov parameter*, is invariant with change of the state space. It is a measure only of the input-output characteristics of the system. Equation (20.54*a*) can therefore be written as

$$\mathbf{K}_1 = [\mathbf{CB}]^{-1}\boldsymbol{\Sigma} \tag{20.54b}$$

5. Assign the slow eigenvalue set S_1 and determine the integral matrix $\mathbf{K}_2$ by use of Eq. (20.41). A simple and often effective selection of these eigenvalues is to assign them all to have the same value λ_o. This is achieved by using $\mathbf{K}_1^{-1}\mathbf{K}_2 = -\lambda_o\mathbf{I}$ which requires that

$$\mathbf{K}_2 = -\lambda_o\mathbf{K}_1 \tag{20.55}$$

6. Completion of the design process produces a diagonal asymptotic closed-loop transfer function given by Eq. (20.49) as

$$\Gamma(\lambda) = \text{diag}\left\{\frac{g\sigma_1}{\lambda + g\sigma_1} \quad \frac{g\sigma_2}{\lambda + g\sigma_2} \cdots \frac{g\sigma_m}{\lambda + g\sigma_m}\right\} \tag{20.56}$$

The value of g which produces satisfactory performance is determined from a simulation of the system response.

20.6 REGULAR SYSTEM DESIGN EXAMPLE

The design procedure outlined in Sec. 20.5 is applied to the development of a controller for an unstable chemical reactor.[1,11] The state and output equations are given by

$$\begin{bmatrix}\dot{x}(t)\\ \dot{x}_2(t)\\ \dot{x}_3(t)\\ \dot{x}_4(t)\end{bmatrix} = \begin{bmatrix}1.38 & -0.2077 & 6.715 & -5.676\\ -0.5814 & -4.29 & 0 & 0.675\\ 1.067 & 4.273 & -6.654 & 5.893\\ 0.048 & 4.273 & 1.343 & -2.104\end{bmatrix}\begin{bmatrix}x_1(t)\\ x_2(t)\\ x_3(t)\\ x_4(t)\end{bmatrix}$$
$$+\begin{bmatrix}0 & 0\\ 5.679 & 0\\ 1.136 & -3.146\\ 1.136 & 0\end{bmatrix}\begin{bmatrix}u_1(t)\\ u_2(t)\end{bmatrix} \tag{20.57}$$

and

$$\begin{bmatrix}y_1(t)\\ y_2(t)\end{bmatrix} = \begin{bmatrix}1 & 0 & 1 & -1\\ 0 & 1 & 0 & 0\end{bmatrix}\begin{bmatrix}x_1(t)\\ x_2(t)\\ x_3(t)\\ x_4(t)\end{bmatrix} \tag{20.58}$$

The plant is unstable as shown by the eigenvalue spectrum

$$\sigma(\mathbf{A}) = \{0.06351, 1.991, -5.057, -8.666\} \tag{20.59}$$

The state equation given in Eq. (20.57) is not in the partitioned form shown in Eq. (20.3). Therefore, the procedure uses Eq. (20.54*b*) to take this into account.

Step 1. The open-loop plant is controllable since the controllability matrix M_c, given in Eq. (13.68), has full rank $n = 4$. The open-loop plant is observable since the observability matrix $\mathbf{M}_o$, given in Eq. (13.73), has full rank $n = 4$.

Step 2. The augmented system with the proportional plus integral controller is controllable since the rank of the matrix formed in accordance with Eq. (20.50) has full rank $n + m = 6$.

Step 3. The transmission zeros cannot be obtained from Eq. (20.51) since the **B** matrix in Eq. (20.57) is not in the partitioned form of Eq. (20.3). Therefore, Eq. (20.52) is used and yields the transmission zeros

$$S_2 = \{-1.192, -5.039\} \tag{20.60}$$

Since these transmission zeros are in the stable region, the design can proceed.

Step 4. The fast eigenvalues in the set S_3 are all assigned the values of $-g$, thus $\mathbf{\Sigma} = \mathbf{I}$ is used in Eq. (20.54) to compute the proportional matrix $\mathbf{K}_1$. Since the state equation of Eq. (20.57) does not have the **B** matrix in the partitioned form of Eq. (20.3), the computation of $\mathbf{K}_1$ is obtained by using Eq. (20.54*b*).

$$\mathbf{K}_1 = [\mathbf{CB}]^{-1}\mathbf{\Sigma} = \left\{\begin{bmatrix} 1 & 0 & 1 & -1 \\ 0 & 1 & 0 & 0 \end{bmatrix}\begin{bmatrix} 0 & 0 \\ 5.679 & 0 \\ 1.136 & -3.146 \\ 1.136 & 0 \end{bmatrix}\right\}^{-1}\begin{bmatrix} 1 & 0 \\ 0 & 1 \end{bmatrix}$$

$$= \begin{bmatrix} 0 & 0.1761 \\ -0.3179 & 0 \end{bmatrix} \tag{20.61}$$

Step 5. The two slow eigenvalues contained in the set S_1 are assigned the values $\lambda_o = -2$. Thus, from Eq. (20.55)

$$\mathbf{K}_2 = 2\mathbf{K}_1 = \begin{bmatrix} 0 & 0.3522 \\ -0.6358 & 0 \end{bmatrix} \tag{20.62}$$

Step 6. The closed-loop eigenvalues, as $g \to \infty$, are asymptotic to the values $S_1 = \{-2, -2\}$, $S_2 = \{-1.192, -5.039\}$, $S_3 = \{-g, -g\}$. Only the fast eigenvalues from the set S_3 appear in the asymptotic transfer function given by Eq. (20.56):

$$\Gamma(\lambda) = \begin{bmatrix} \dfrac{g}{\lambda + g} & 0 \\ 0 & \dfrac{g}{\lambda + g} \end{bmatrix} \tag{20.63}$$

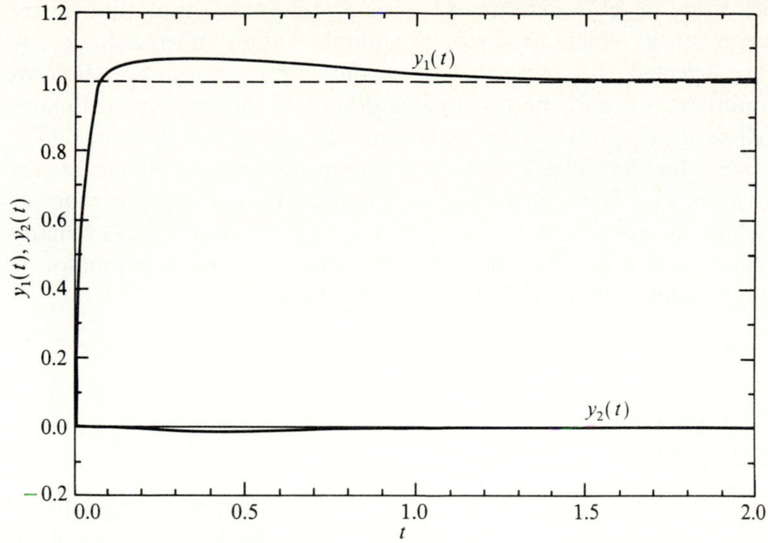

FIGURE 20.2
Step response $\mathbf{y}(t) = \{y_1(t), y_2(t)\}$ for $\mathbf{r}(t) = [1 \quad 0]^T$.

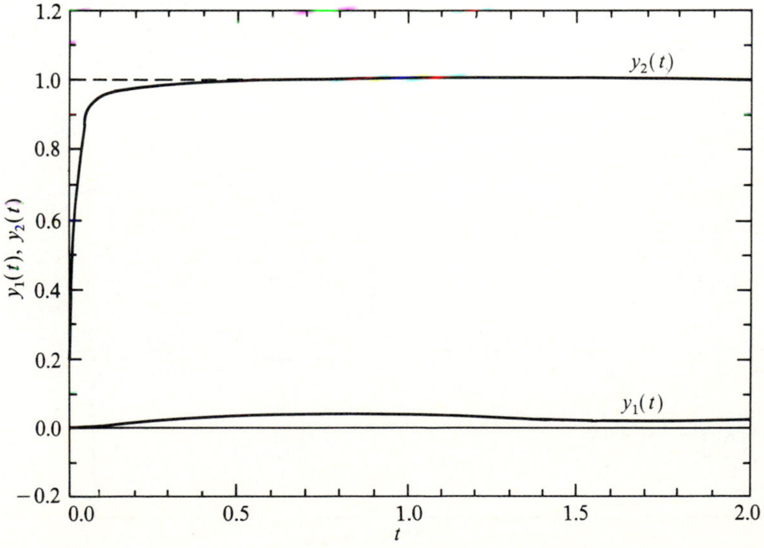

FIGURE 20.3
Step response $\mathbf{y}(t) = \{y_1(t), y_2(t)\}$ for $\mathbf{r}(t) = [0 \quad 1]^T$.

A suitable value of g is determined from a computer simulation of the system. The value of g which produces acceptable values of overshoot and settling time are selected. As the value of g increases, the outputs become increasingly noninteractive and the fast modes dominate the responses. Figures 20.2 and 20.3 show the responses with the commands $\mathbf{r}(t) = [1 \quad 0]^T$ and $\mathbf{r}(t) = [0 \quad 1]^T$, respectively, for the value $g = 50$. It is further necessary to determine the values of $u_1(t)$ and $u_2(t)$. The system must be capable of providing this required control effort while the linear system model of Eqs. (20.57) and (20.58) remains valid. Plots of $u_1(t)$ and $u_2(t)$ must be obtained from a computer simulation to be sure that they remain within the linear operating range.

20.7 IRREGULAR PLANT CHARACTERISTICS

The fast eigenvalues for the output feedback high-gain system with proportional plus integral control are determined from Eq. (20.54) which states that $\mathbf{K}_1 = (\mathbf{CB})^{-1}\mathbf{\Sigma}$. In the case where $\mathbf{CB}$ does not have full rank, the inverse $(\mathbf{CB})^{-1}$ does not exist and $\mathbf{K}_1$ cannot be determined. When $\mathbf{CB}$ does not have full rank the system is described as irregular. Such systems can be controlled by introducing[12] the minor loop feedback shown in Fig. 20.4. In this figure there is shown a new output described by

$$\mathbf{w}(t) = \mathbf{y}(t) + \mathbf{M}\dot{\mathbf{x}}_1 \tag{20.64}$$

When the command input to the system $\mathbf{r}(t)$ consists of step functions, steady-state values of the states for a stable system also are step functions. Since the state equations for $\dot{\mathbf{x}}_1(t)$ in Eq. (20.3) have zero forcing function, the steady-state value of $\mathbf{x}_1(t)$ is therefore

$$\lim_{t\to\infty} \dot{\mathbf{x}}_1(t) = 0 \tag{20.65}$$

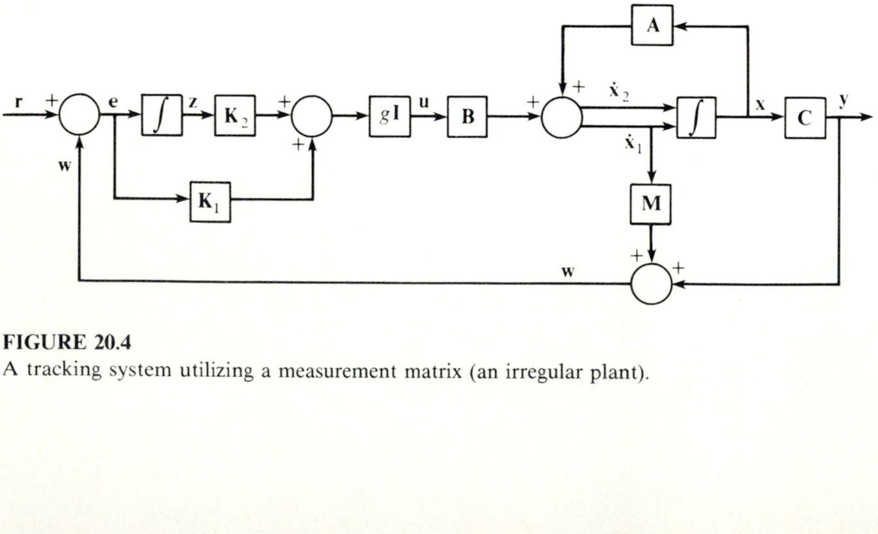

FIGURE 20.4
A tracking system utilizing a measurement matrix (an irregular plant).

Thus Eq. (20.64) shows that $\mathbf{w}(t)$ equals $\mathbf{y}(t)$ in the steady state:

$$\mathbf{w}(t)_{ss} = \mathbf{y}(t)_{ss} \tag{20.66}$$

Inserting the values obtained from Eqs. (20.3) and (20.4) into Eq. (20.64) yields the new output equation

$$\begin{aligned} \mathbf{w}(t) &= [\mathbf{C}_1 \quad \mathbf{C}_2]\begin{bmatrix} \mathbf{x}_1 \\ \mathbf{x}_2 \end{bmatrix} + \mathbf{M}[\mathbf{A}_{11} \quad \mathbf{A}_{12}]\begin{bmatrix} \mathbf{x}_1 \\ \mathbf{x}_2 \end{bmatrix} \\ &= [(\mathbf{C}_1 + \mathbf{M}\mathbf{A}_{11}) \qquad (\mathbf{C}_2 + \mathbf{M}\mathbf{A}_{12})]\begin{bmatrix} \mathbf{x}_1 \\ \mathbf{x}_2 \end{bmatrix} \\ &= [\mathbf{F}_1 \quad \mathbf{F}_2]\begin{bmatrix} \mathbf{x}_1 \\ \mathbf{x}_2 \end{bmatrix} \end{aligned} \tag{20.67}$$

By the proper selection[13] of the measurement matrix $\mathbf{M}$, the matrix $\mathbf{F}_2$ in Eq. (20.67) will have full rank m.

The composite continuous-time tracking system of Fig. 20.4, consisting of the plant of Eq. (20.3), the control law of Eq. (20.5), and the output of Eq. (20.67), has the state and output equations given by

$$\begin{bmatrix} \dot{\mathbf{z}}(t) \\ \dot{\mathbf{x}}_1(t) \\ \dot{\mathbf{x}}_2(t) \end{bmatrix} = \left[\begin{array}{cc|c} \mathbf{0} & -\mathbf{F}_1 & -\mathbf{F}_2 \\ \mathbf{0} & \mathbf{A}_{11} & \mathbf{A}_{12} \\ \hline g\mathbf{B}_2\mathbf{K}_2 & \mathbf{A}_{21} - g\mathbf{B}_2\mathbf{K}_1\mathbf{F}_1 & \mathbf{A}_{22} - g\mathbf{B}_2\mathbf{K}_1\mathbf{F}_2 \end{array}\right]\begin{bmatrix} \mathbf{z}(t) \\ \mathbf{x}_1(t) \\ \mathbf{x}_2(t) \end{bmatrix} + \begin{bmatrix} \mathbf{I}_m \\ \mathbf{0} \\ g\mathbf{B}_2\mathbf{K}_1 \end{bmatrix}\mathbf{r}(t) \tag{20.68}$$

and

$$\mathbf{y}(t) = \left[\begin{array}{cc|c} \mathbf{0} & \mathbf{C}_1 & \mathbf{C}_2 \end{array}\right]\begin{bmatrix} \mathbf{z}(t) \\ \mathbf{x}_1(t) \\ \mathbf{x}_2(t) \end{bmatrix} \tag{20.69}$$

The closed-loop transfer function for Eqs. (20.68) and (20.69) is

$$\mathbf{G}(\lambda) = [\mathbf{0} \quad \mathbf{C}_1 \quad \mathbf{C}_2]\begin{bmatrix} \lambda\mathbf{I}_l & \mathbf{F}_1 & \mathbf{F}_2 \\ \mathbf{0} & \lambda\mathbf{I}_{n-\ell} - \mathbf{A}_{11} & -\mathbf{A}_{12} \\ -g\mathbf{B}_2\mathbf{K}_2 & -\mathbf{A}_{21} + g\mathbf{B}_2\mathbf{K}_1\mathbf{F}_1 & \lambda\mathbf{I}_\ell - \mathbf{A}_{22} + g\mathbf{B}_2\mathbf{K}_1\mathbf{F}_2 \end{bmatrix}^{-1} \times \begin{bmatrix} \mathbf{I}_\ell \\ \mathbf{0} \\ g\mathbf{B}_2\mathbf{K}_1 \end{bmatrix} \tag{20.70}$$

Using the procedure illustrated in Sec. 20.3, the state and output equations given in Eqs. (20.3) and (20.68) are transformed to the forms of Eqs. (20.12) and (20.13) where the composite plant matrix is block diagonal. Taking the limit as

$g \to \infty$ yields the components

$$\mathbf{A}_s = \begin{bmatrix} -\mathbf{K}_1^{-1}\mathbf{K}_2 & \mathbf{0} \\ \mathbf{A}_{12}\mathbf{F}_2^{-1}\mathbf{K}_1^{-1}\mathbf{K}_2 & \mathbf{A}_{11} - \mathbf{A}_{12}\mathbf{F}_2^{-1}\mathbf{F}_1 \end{bmatrix} \tag{20.71}$$

$$\mathbf{B}_s = \begin{bmatrix} \mathbf{0} \\ \mathbf{A}_{12}\mathbf{F}_2^{-1} \end{bmatrix} \tag{20.72}$$

$$\mathbf{C}_s = \begin{bmatrix} \mathbf{C}_2\mathbf{F}_2^{-1}\mathbf{K}_1^{-1}\mathbf{K}_2 & \mathbf{C}_1 - \mathbf{C}_2\mathbf{F}_2^{-1}\mathbf{F}_1 \end{bmatrix} \tag{20.73}$$

$$\mathbf{A}_f = -g\mathbf{B}_2\mathbf{K}_1\mathbf{F}_2 \tag{20.74}$$

$$\mathbf{B}_f = g\mathbf{B}_2\mathbf{K}_1 \tag{20.75}$$

$$\mathbf{C}_f = \mathbf{C}_2 \tag{20.76}$$

20.8 IRREGULAR SYSTEM PERFORMANCE

The system of Fig. 20.4 shows a tracker for a plant which is irregular, that is, **CB** does not have full rank. A new output $\mathbf{w}(t)$ is formed by

$$\mathbf{w}(t) = \mathbf{y}(t) + \mathbf{M}\dot{\mathbf{x}}_1(t) = [\mathbf{F}_1 \quad \mathbf{F}_2]\begin{bmatrix} \mathbf{x}_1 \\ \mathbf{x}_2 \end{bmatrix} \tag{20.77}$$

where

$$\mathbf{F}_1 = \mathbf{C}_1 + \mathbf{M}\mathbf{A}_{11} \tag{20.78}$$

$$\mathbf{F}_2 = \mathbf{C}_2 + \mathbf{M}\mathbf{A}_{12} \tag{20.79}$$

The measurement matrix **M** in Eqs. (20.78) and (20.79) is selected[13] to be sparse, containing only enough elements so that $\mathbf{F}_2$ has full rank. This is essential to insure that the matrix product $\mathbf{F}_2\mathbf{B}_2$ has rank m.

Eigenvalue set S_1. An evaluation similar to that presented in Sec. 20.4 leads to the determination of the closed-loop eigenvalues. The slow eigenvalue set S_1 is determined by

$$|\lambda\mathbf{I}_m + \mathbf{K}_1^{-1}\mathbf{K}_2| = 0 \tag{20.80}$$

These finite eigenvalues are associated with uncontrollable modes because of the form of $\mathbf{B}_s$.

Eigenvalue set S_2. The slow eigenvalue set S_2 is determined by

$$|\lambda\mathbf{I}_{n-m} - [\mathbf{A}_{11} - \mathbf{A}_{12}\mathbf{F}_2^{-1}\mathbf{F}_1]| = 0 \tag{20.81}$$

These finite eigenvalues are the transmission zeros of the plant with the addition of minor-loop feedback. This adds transmission zeros to those of the plant. Because of the form of the matrix $\mathbf{C}_s$, the associated modes are observable in the output. This observability characteristic of the set S_2 is a major difference from the case of a regular plant. It is a consequence of the minor loop containing the measurement matrix **M** which has been added to the system.

Eigenvalue set S_3. The fast eigenvalue set S_3 is determined by

$$|\lambda\mathbf{I}_m + g\mathbf{B}_2\mathbf{K}_1\mathbf{F}_2| = 0 \tag{20.82}$$

This characteristic equation yields the same eigenvalues as

$$|\lambda \mathbf{I}_m + g\mathbf{F}_2\mathbf{B}_2\mathbf{K}_1| = 0 \tag{20.83}$$

The modes associated with the fast eigenvalues of the set S_3 are observable in the output.

The slow transfer function determined from Eqs. (20.71), (20.72), and (20.73) is

$$\Gamma_s(\lambda) = \left[\mathbf{C}_1 - \mathbf{C}_2\mathbf{F}_2^{-1}\mathbf{F}_1\right]\left[\lambda \mathbf{I}_{n-\ell} - \mathbf{A}_{11} + \mathbf{A}_{12}\mathbf{F}_2^{-1}\mathbf{F}_1\right]^{-1}\mathbf{A}_{12}\mathbf{F}_2^{-1} \tag{20.84}$$

This transfer function contains only the transmission zeros.

The fast transfer function determined from Eqs. (20.74), (20.75), and (20.76) is

$$\Gamma_f(\lambda) = \mathbf{C}_2\mathbf{F}_2^{-1}[\lambda \mathbf{I}_\ell + g\mathbf{F}_2\mathbf{B}_2\mathbf{K}_1]^{-1} g\mathbf{F}_2\mathbf{B}_2\mathbf{K}_1 \tag{20.85}$$

The total asymptotic transfer function as $g \to \infty$, obtained from Eq. (20.45), is the sum of the slow and fast transfer functions given by Eqs. (20.84) and (20.85):

$$\begin{aligned}\Gamma(\lambda) = \Gamma_s(\lambda) + \Gamma_f(\lambda) &= \left[\mathbf{C}_1 - \mathbf{C}_2\mathbf{F}_2^{-1}\mathbf{F}_1\right]\left[\lambda \mathbf{I}_{n-\ell} - \mathbf{A}_{11} + \mathbf{A}_{12}\mathbf{F}_2^{-1}\mathbf{F}_1\right]^{-1}\mathbf{A}_{12}\mathbf{F}_2^{-1} \\ &\quad + \mathbf{C}_2\mathbf{F}_2^{-1}[\lambda \mathbf{I}_\ell + g\mathbf{F}_2\mathbf{B}_2\mathbf{K}_1]^{-1} g\mathbf{F}_2\mathbf{B}_2\mathbf{K}_1\end{aligned} \tag{20.86}$$

This transfer function contains the slow eigenvalues (transmission zeros) of the set S_2 and the fast eigenvalues of the set S_3. Since the eigenvalue set S_1 is associated with asymptotically uncontrollable modes as $g \to \infty$, these eigenvalues do not appear in the transfer function of Eq. (20.86).

It is important to note the difference between the characteristics of the system containing a regular plant and the system containing an irregular plant. The transfer function of the tracking system containing the regular plant, given in Eq. (20.49), contains only fast eigenvalues. The transfer function of the tracking system containing the irregular plant, given in Eq. (20.86), contains both the transmission zeros and the fast eigenvalues. The transmission zeros include those of the plant and those determined by the measurement matrix **M**. Therefore, the presence of the associated slow modes in the output limits the speed of response which is achievable. This is shown in the example in Sec. 20.10.

20.9 DESIGN OF THE MEASUREMENT MATRIX M

The design of the control law for the tracking system may include the desirable requirement that the outputs be decoupled. This minimizes the interaction between the outputs. In order to achieve this decoupling, the component transfer functions $\Gamma_s(\lambda)$ and $\Gamma_f(\lambda)$ must each be diagonal. It may be possible to make these transfer functions diagonal by selecting the measurement matrix **M**, but there is no guarantee that this is achievable. The procedure presented in Ref. 13 provides a method for accomplishing the selection of the measurement matrix.

The design procedure is modified from that used for regular plants because it is necessary to select the measurement matrix **M** and to achieve, if possible, a diagonal transfer function $\Gamma(\lambda)$. The design process consists of the following steps:

1. Check the controllability and observability of the plant given by the state and output equations of Eqs. (20.1) and (20.2).
2. Check that the rank of the following matrix is satisfied:

$$\text{Rank}\begin{bmatrix} \mathbf{B} & \mathbf{A} \\ \mathbf{0} & \mathbf{C} \end{bmatrix} = n + \ell \tag{20.87}$$

3. Determine the rank of $\mathbf{C}_2\mathbf{B}_2$. If this matrix is rank deficient, select the most sparse matrix **M** which produces a matrix $\mathbf{F}_2 = \mathbf{C}_2 + \mathbf{M}\mathbf{A}_{12}$ which has full rank. Simultaneously the matrix product $\mathbf{C}_2\mathbf{F}_2^{-1}$ must be made diagonal, if possible. This is accomplished[13] by the following procedure:
   (*a*) First form the matrix

$$\mathbf{B}^* = \begin{bmatrix} \mathbf{c}_1^T\mathbf{A}_{11}^{d_1}\mathbf{A}_{12} \\ \vdots \\ \mathbf{c}_m^T\mathbf{A}_{11}^{d_m}\mathbf{A}_{12} \end{bmatrix} \tag{20.88}$$

   where m is the number of control inputs, $\mathbf{c}_i^T$ is the ith row of $\mathbf{C}_1$, and

$$d_i = \min\left\{ j\colon \mathbf{c}_i^T\mathbf{A}_{11}^j\mathbf{A}_{12} \neq \mathbf{0},\ j = 0,1,\ldots,n-1 \right\} \tag{20.89}$$

   Equation (20.89) specifies that d_i is the smallest value of j for which $\mathbf{c}_i^T\mathbf{A}_{11}^j\mathbf{A}_{12} \neq \mathbf{0}$. The permissible values of j are $0, 1, \ldots, n-1$. If all of these values of j result in $\mathbf{c}_i^T\mathbf{A}_{11}^j\mathbf{A}_{11} = \mathbf{0}$, then use $d_i = n - 1$.
   (*b*) Form $\mathbf{F}_2 = \mathbf{C}_2 + \mathbf{M}\mathbf{A}_{12}$ using a general form of the matrix $\mathbf{M} = [m_{ij}]$. The elements m_{ij} appearing in $\mathbf{F}_2$ are permitted nonzero values only if $\mathbf{B}^*$ has a nonzero element in a corresponding position. Set all other elements of **M** to zero.
   (*c*) Form the matrix $\mathbf{C}_2\mathbf{F}_2^{-1}$, where $\mathbf{F}_2$ is obtained in step *b*. Make the matrix $\mathbf{C}_2\mathbf{F}_2^{-1}$ diagonal, if possible, by setting off-diagonal elements to zero. This is accomplished by setting selected values of m_{ij} to zero.
   (*d*) Make sure that $\mathbf{F}_2$ has enough elements m_{ij} so that it has full rank. The nonzero elements of **M** are now all identified.
4. Determine the set of eigenvalues S_2 (the transmission zeros) using Eq. (20.81) to assure that they are in the left-half plane (stable).
5. Select the matrix $\mathbf{K}_1$ which makes the fast transfer function $\Gamma_f(\lambda)$ of Eq. (20.85) diagonal. This is accomplished by selecting a diagonal matrix $\boldsymbol{\Sigma}$ such that

$$\mathbf{F}_2\mathbf{B}_2\mathbf{K}_1 = \boldsymbol{\Sigma} = \begin{bmatrix} \sigma_1 & & & \mathbf{0} \\ & \sigma_2 & & \\ & & \ddots & \\ \mathbf{0} & & & \sigma_\ell \end{bmatrix} \tag{20.90}$$

The proportional matrix is computed as

$$\mathbf{K}_1 = (\mathbf{F}_2\mathbf{B}_2)^{-1}\Sigma \tag{20.91}$$

Then the fast eigenvalues are

$$S_3 = \{-\sigma_i g, -\sigma_2 g, \ldots, -\sigma_\ell g\} \tag{20.92}$$

6. Assign the integral matrix using

$$\mathbf{K}_2 = -\lambda_o \mathbf{K}_1 \tag{20.93}$$

where λ_o is adjusted to give good transient performance in the simulation.
7. Determine the asymptotic transfer function $\Gamma(\lambda)$.

20.10 IRREGULAR SYSTEM DESIGN EXAMPLE

The design procedure of Sec. 20.9 is applied to the development of a controller for an aircraft flight control system.[13,14] The model represents the longitudinal dynamics of an aircraft and the variables are perturbations from a steady-state straight and level trim condition. The cartesian coordinate system shown in Fig. 20.5 has three perpendicular axes labeled X, Y, and Z, and the rotation rates around these axes are labeled p, q, and r, respectively. Motion which is confined to the X-Z plane is described as longitudinal motion. The longitudinal dynamics are represented by the state and output equations which are partitioned as shown in Eqs. (20.3) and (20.4).

$$\begin{bmatrix} \dot\theta \\ \dot u \\ \dot w \\ \dot q \end{bmatrix} = \left[\begin{array}{c|ccc} 0 & 0 & 0 & 1.0 \\ \hline -32.1 & -0.0822 & 0.0472 & -19.7 \\ -0.705 & -0.0558 & -1.68 & 898 \\ 0 & -0.317\times 10^{-2} & 0.0303 & -0.253 \end{array}\right] \begin{bmatrix} \theta \\ u \\ w \\ q \end{bmatrix} + \left[\begin{array}{ccc} 0 & 0 & 0 \\ \hline 0.359 & -0.0479 & 0.218\times 10^{-2} \\ -0.634 & -2.007 & -0.744\times 10^{-4} \\ 0.527 & -0.065 & -0.484\times 10^{-5} \end{array}\right] \begin{bmatrix} \delta_c \\ \delta_f \\ \delta_t \end{bmatrix} \tag{20.94}$$

$$\begin{bmatrix} \theta \\ \gamma \\ v \end{bmatrix} = \left[\begin{array}{c|ccc} 1 & 0 & 0 & 0 \\ 1 & 2.45\times 10^{-5} & -1.12\times 10^{-3} & 0 \\ 0 & 1 & 0.0219 & 0 \end{array}\right] \begin{bmatrix} \theta \\ u \\ w \\ q \end{bmatrix} \tag{20.95}$$

In the state equations $\theta(t)$ is the pitch angle around the Y axis, $q(t)$ is the pitch rate around the Y axis, $u(t)$ is the perturbation velocity in the X direction, and $w(t)$ is the velocity in the Z direction. The outputs to be controlled are the pitch angle $\theta(t)$, the flight path angle $\gamma(t)$, and the total aircraft velocity $v(t)$. This unconventional aircraft is intended to perform uncommon maneuvers. It has the three longitudinal inputs described by δ_c (canards), δ_f (flaperons), and δ_t (thrust).

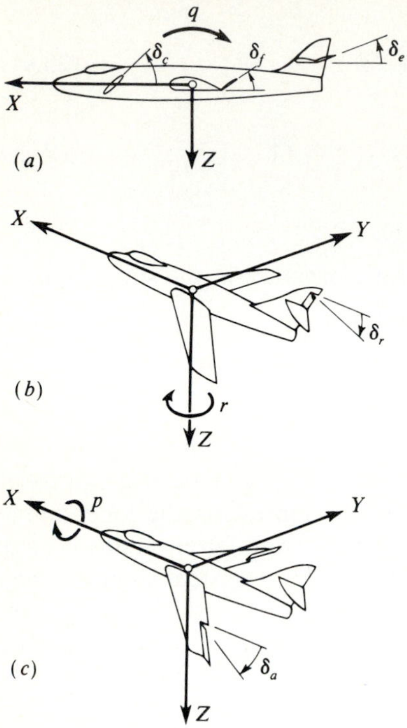

FIGURE 20.5
Airplane control surfaces. (*a*) Elevator, flaperon, and canard deflections produce pitching velocity q. (*b*) Rudder deflection produces yawing velocity r. (*c*) Aileron deflection produces rolling velocity p.

The equations are partitioned in accordance with Eqs. (20.3) and (20.4). The design steps follow:

Step 1. The controllability matrix $\mathbf{M}_c$ and the observability matrix $\mathbf{M}_o$ each have full rank equal to 4, and thus the plant is completely controllable and observable.

Step 2. The matrix of Eq. (20.87) has rank $n + m = 7$, and thus the complete system containing the proportional plus integral control is controllable.

Step 3. The matrix product **CB** is

$$\mathbf{CB} = \begin{bmatrix} 0 & 0 & 0 \\ 8.0 \times 10^{-4} & 2.25 \times 10^{-3} & 1.37 \times 10^{-7} \\ 3.45 & -0.092 & 0.0022 \end{bmatrix} \tag{20.96}$$

This matrix is rank deficient, having only rank 2, and thus the plant is irregular.

(*a*) Since $\mathbf{A}_{11} = [0]$, Eq. (20.89) yields the values $d_1 = d_2 = d_3 = 0$. Then $\mathbf{B}^*$ is formed in accordance with Eq. (20.88):

$$\mathbf{B}^* = \begin{bmatrix} 0 & 0 & 1 \\ 0 & 0 & 1 \\ 0 & 0 & 0 \end{bmatrix} \tag{20.97}$$

(*b*) In the matrix $\mathbf{F}_2 = \mathbf{C}_2 + \mathbf{MA}_{12}$, the dimension of $\mathbf{C}_2$ is 3×3. Since the dimension of $\mathbf{A}_{12}$ is 1×3, the necessary dimension of $\mathbf{M}$ is 3×1. Thus, using

$$\mathbf{M} = \begin{bmatrix} m_1 \\ m_2 \\ m_3 \end{bmatrix} \tag{20.98}$$

the matrix $\mathbf{F}_2$ is

$$\mathbf{F}_2 = \begin{bmatrix} 0 & 0 & m_1 \\ 2.45 \times 10^{-5} & 1.12 \times 10^{-3} & m_2 \\ 1 & 0.0219 & m_3 \end{bmatrix} \tag{20.99}$$

The assignable elements of $\mathbf{F}_2$ are permitted nonzero values only if $\mathbf{B}^*$ has a nonzero element in the same position. Thus it is required that $m_3 = 0$.

(*c*) The matrix $\mathbf{C}_2\mathbf{F}_2^{-1}$ is now obtained.

$$\mathbf{C}_2\mathbf{F}_2^{-1} = \begin{bmatrix} 0 & 0 & 0 \\ -m_2/m_1 & 1 & 0 \\ 0 & 0 & 1 \end{bmatrix} \tag{20.100}$$

Since $\mathbf{C}_2\mathbf{F}_2^{-1}$ must be diagonal, assign $m_2 = 0$.

(*d*) The resulting matrices $\mathbf{F}_1 = \mathbf{C}_1 + \mathbf{MA}_{11}$ and $\mathbf{F}_2 = \mathbf{C}_2 + \mathbf{MA}_{12}$ are therefore

$$\mathbf{F}_1 = \begin{bmatrix} 1 \\ 1 \\ 0 \end{bmatrix} \qquad \mathbf{F}_2 = \begin{bmatrix} 0 & 0 & m_1 \\ 2.45 \times 10^{-5} & 1.12 \times 10^{-3} & 0 \\ 1 & 0.0219 & 0 \end{bmatrix} \tag{20.101}$$

The determination of the value of $m_1 = 0.25$ is discussed in step 4 of this design.

Step 4. Using Eq. (20.81), there is one transmission zero $S_2 = \{-1/m_1\}$. A stable system is therefore achieved for a positive value of m_1.

Step 5. The proportional matrix, in accordance with Eq. (20.91), is

$$\mathbf{K}_1 = (\mathbf{F}_2\mathbf{B}_2)^{-1}\boldsymbol{\Sigma} = (\mathbf{F}_2\mathbf{B}_2)^{-1}\begin{bmatrix} \sigma_1 & & \\ & \sigma_2 & \\ & & \sigma_3 \end{bmatrix}$$

At this point in the design a complete simulation of the system is obtained, starting with assumed nominal initial values for m_1, g and $\boldsymbol{\Sigma} = \operatorname{diag}\{\sigma_1 \quad \sigma_2 \quad \sigma_3\}$. By trial and error suitable values of m_1, g, and σ_i are determined which yield acceptable performance. Based on the simulation, the values $m_1 = 0.25$, $\boldsymbol{\Sigma} = \operatorname{diag}\{1, 0.05, 0.2\}$, and $g = 75$ are selected. This yields

$$\mathbf{K}_1 = \begin{bmatrix} 7.2999 & 2.6441 & 0.0007 \\ -2.2589 & 21.4008 & -0.0015 \\ -1269.2 & 490.24 & 92.923 \end{bmatrix} \tag{20.102}$$

Step 6. A simulation of the complete system is now obtained. The value $\mathbf{K}_2 = \mathbf{K}_1$ yields good responses.

Step 7. The asymptotic closed-loop transfer function is computed using Eq. (20.86).

$$\Gamma(\lambda) = \Gamma_s(\lambda) + \Gamma_f(\lambda) = \begin{bmatrix} \dfrac{1/m_1}{\lambda + 1/m_1} & 0 & 0 \\ 0 & 0 & 0 \\ 0 & 0 & 0 \end{bmatrix} + \begin{bmatrix} 0 & 0 & 0 \\ 0 & \dfrac{g\sigma_2}{\lambda + g\sigma_2} & 0 \\ 0 & 0 & \dfrac{g\sigma_3}{\lambda + g\sigma_3} \end{bmatrix}$$

$$= \begin{bmatrix} \dfrac{1/m_1}{\lambda + 1/m_1} & 0 & 0 \\ 0 & \dfrac{g\sigma_2}{\lambda + g\sigma_2} & 0 \\ 0 & 0 & \dfrac{g\sigma_3}{\lambda + g\sigma_3} \end{bmatrix} \tag{20.103}$$

20.11 TRACKER SIMULATION

The performance of the tracking system must be verified from a simulation of the complete system. The simulation is also used to adjust the individual parameters until the responses are satisfactory and meet the performance specifications as mentioned in step 4 of the design of Sec. 20.10. The control inputs are also obtained so that their magnitudes and rates of change are within acceptable values. For this design the aircraft is required to perform three maneuvers which are identified as pitch pointing, vertical translation, and straight climb.

In a pitch pointing maneuver the aircraft is flying straight and level at a given velocity. Then a pitch up or pitch down of the angle θ is commanded, while the flight-path angle γ and velocity v are simultaneously commanded to remain unchanged. For the command $\mathbf{r}(t) = [\theta \quad \gamma \quad v]^T = [1 \quad 0 \quad 0]^T$ the responses are shown in Fig. 20.6 and the required controls are shown in Fig. 20.7. There is a maximum flight-path angle perturbation γ of only 0.23° and the maximum velocity perturbation v is only 1.1 ft/s from a nominal velocity of 881 ft/s.

In a vertical translation command the pitch angle and forward velocity remain constant, while the aircraft rises vertically. The flight-path angle shows a climb or a descent without any pitch change of the aircraft. For the command $\mathbf{r}(t) = [\theta \quad \gamma \quad v]^T = [0 \quad 1 \quad 0]^T$ the responses are shown in Fig. 20.8 and the controls are shown in Fig. 20.9. The response is excellent as the maximum pitch angle and velocity perturbations are only 0.05° and 0.4 ft/s, respectively.

The straight climb maneuver consists of simultaneous pitch angle and flight-path angle commands while holding the velocity constant. For the command $\mathbf{r}(t) = [\theta \quad \gamma \quad v]^T = [1 \quad 1 \quad 0]^T$ the responses are shown in Fig. 20.10 and the controls are shown in Fig. 20.11. Excellent responses are indicated.

The aircraft is able to perform all three maneuvers with a fixed control law. Effective decoupling is achieved for all commands. The control surface and thrust

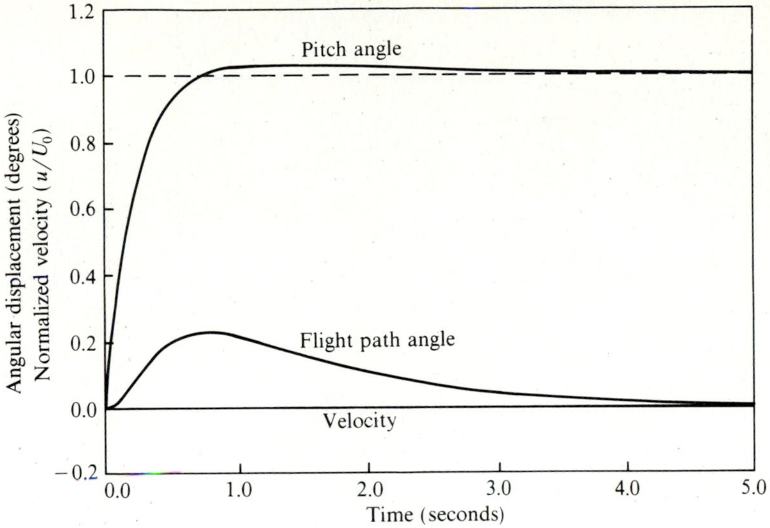

FIGURE 20.6
Pitch pointing maneuver.

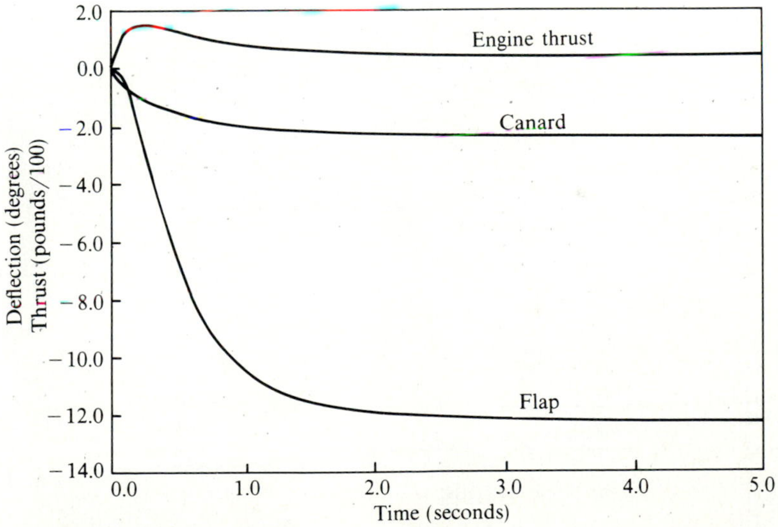

FIGURE 20.7
Controls for pitch pointing.

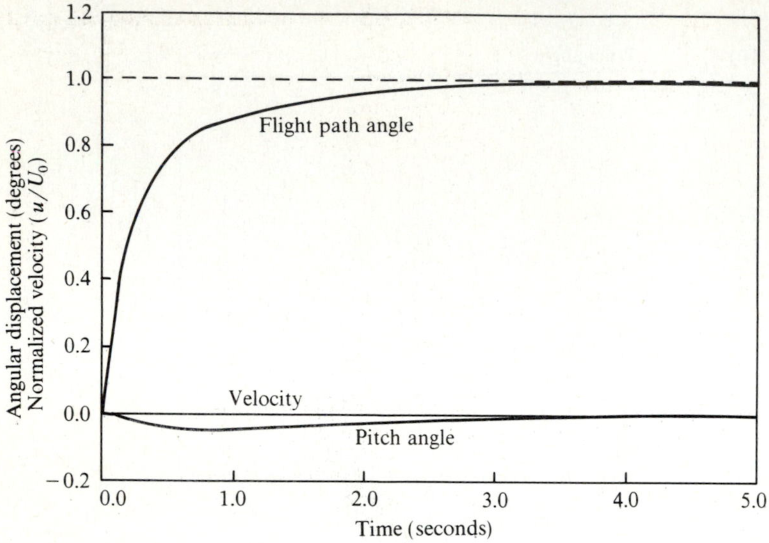

FIGURE 20.8
Vertical translation maneuver.

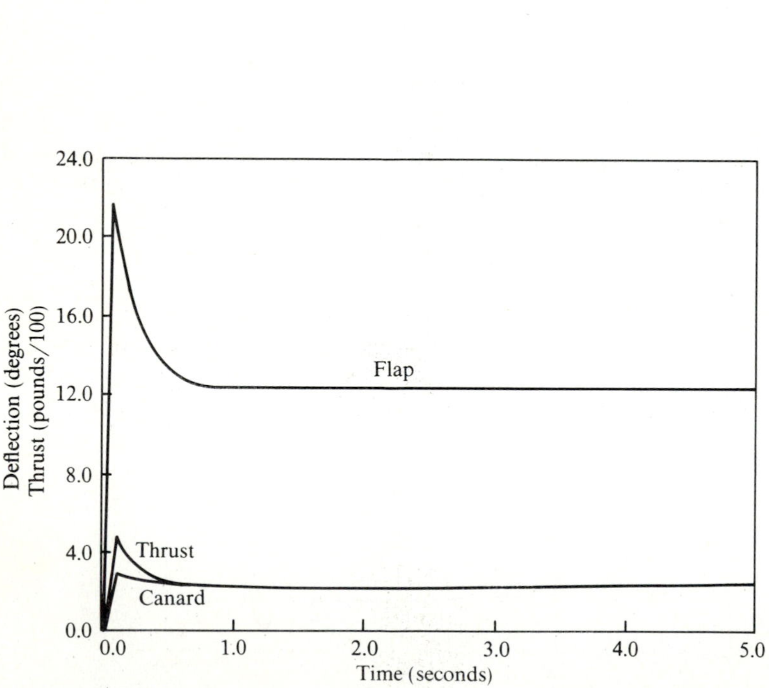

FIGURE 20.9
Controls for vertical translation.

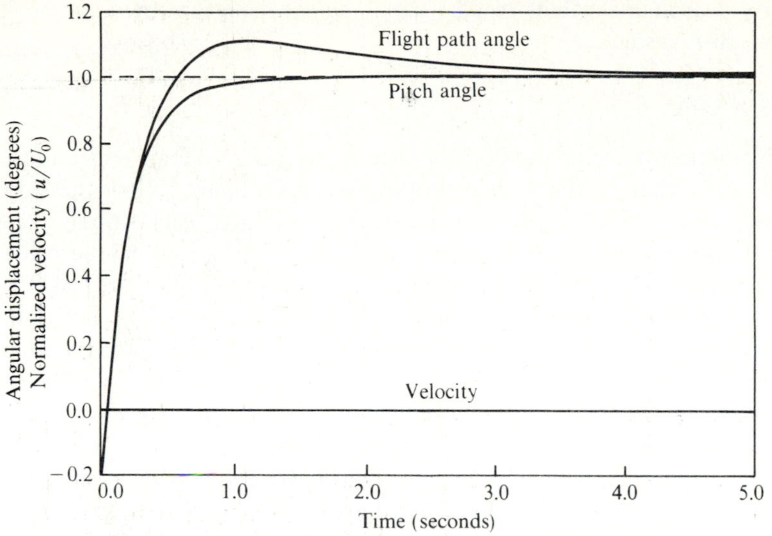

FIGURE 20.10
Straight climb maneuver.

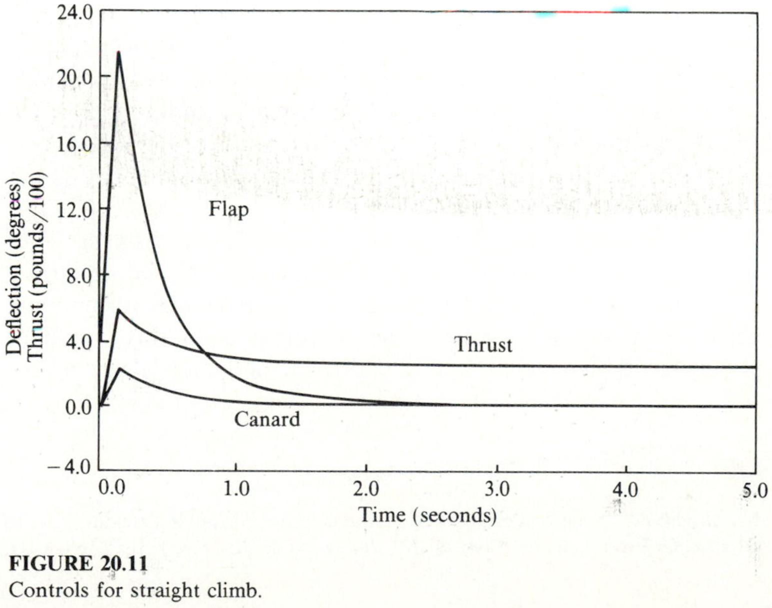

FIGURE 20.11
Controls for straight climb.

commands have reasonable values in all cases. Thus this design yields a satisfactory flight control system.

20.12 SUMMARY

This chapter introduces a design method for output feedback systems which must function as trackers, that is, the system outputs $\mathbf{y}(t)$ effectively follow the step command inputs $\mathbf{r}(t)$. The outputs are decoupled, so that each input $r_i(t)$ commands only one output $y_i(t)$, and any transient perturbations in the other outputs are very small. The basis for the high-gain proportional plus integral (PI) control law is presented and the asymptotic transfer function, as the gain $g \to \infty$, is developed. The asymptotic closed-loop eigenvalues approach three sets. The set S_1 consists of finite slow eigenvalues, the set S_2 consists of transmission zeros that are determined only by the plant and not by the control law, and the set S_3 consists of eigenvalues of the form $-g\sigma_i$ which approach infinity as $g \to \infty$. The designer selects a diagonal matrix $\mathbf{\Sigma}$ which assigns the fast eigenvalue set S_3, a location λ_o for the members of the slow eigenvalue set S_1, and a value of the scalar gain g.

In the case of regular plants, for which the matrix product **CB** has full rank, the set S_1 becomes uncontrollable and the set S_2 becomes unobservable; thus these eigenvalues do not appear in the asymptotic transfer functions. Therefore, rapid response is achieved since only the fast eigenvalues affect the output response.

In the case of irregular plants, for which the matrix product **CB** does not have full rank, it is necessary to add additional measurements through a measurement matrix **M**. A procedure for selecting a sparse matrix **M** is illustrated. When the measurement matrix **M** is added to the system the eigenvalue set S_2 is observable and the associated modes do appear in the output.

Excellent closed-loop system performance is demonstrated by means of two examples. One example is an unstable chemical plant which has a regular plant model. The second example is a flight control system for an aircraft which is required to perform a number of maneuvers. The Markov parameter **CB** is rank deficient, and therefore a minor loop feedback containing a measurement matrix **M** is required. The design is achieved with relative ease.

The methods presented in this chapter apply to continuous systems. They have the desirable features that they are extendable to a direct digital design[15,16] for the controller. A dual of the analog design leads to a very similar formulation for the digital design. By designing directly in the discrete domain, this avoids the common process of first designing an analog controller and then digitizing the resulting design.

REFERENCES

1. Porter, B., and A. Bradshaw: "Singular Perturbation Methods in the Design of Tracking Systems Incorporating High-Gain Error-Actuated Controllers," *Int. J. Syst. Sci.*, vol. 1, pp. 1169–1179, 1981.

2. Porter, B., and A. T. Shenton: "Singular Perturbation Analysis of the Transfer Function Matrices of a Class of Multivariable Linear Systems," *Int. J. Control*, vol. 21, pp. 655–660, 1975.
3. Phillips, R. G.: "A Two-State Design of Linear Feedback Controls," *IEEE Trans. Autom. Control*, vol. AC-25, pp. 1220–1223, 1980.
4. Tran, M. T., and M. E. Sawan: "Reduced Order Discrete-Time Models," *Int. J. Syst. Sci.*, vol. 14, pp. 745–752, 1983.
5. Mahmond, M. S., Y. Chen, and M. G. Singh: "On Eigenvalue Assignment in Discrete Systems with Fast and Slow Modes," *Int. J. Syst. Sci.*, vol. 16, pp. 61–70, 1985.
6. Kokotovic, P. V., and A. H. Haddad: "Controllability and Time-Optimal Control of Systems with Slow and Fast Modes," *IEEE Trans Autom. Control*, vol. AC-20, pp. 111–113, 1975.
7. Rosenbrock, H. H.: "The Zeros of a System," *Int. J. Control*, vol. 18, pp. 297–299, 1973 and "Correction to 'The Zeros of a System'," *Int. J. Control*, vol. 20, pp. 525–527, 1974.
8. MacFarlane, A. G. J., and N. Karcanias: "Poles and Zeros of Linear Multivariable Systems: A Survey of the Algebraic, Geometric and Complex-Variable Theory," *Int. J. Control*, vol. 24, pp. 33–74, 1976.
9. Porter, B., and A. Bradshaw: "Design of Linear Multivariable Continuous-Time Tracking Systems," *Int. J. Syst. Sci.*, vol. 5, pp. 1155–1164, 1974.
10. Porter, B., and J. J. D'Azzo, "Transmission Zeros of Linear Multivariable Continuous-Time Systems," *Electron. Lett.*, vol. 13, pp. 753–754, 1977.
11. MacFarlane, A. G. J., and B. Kouvaritakis: "Geometric Approach to Analysis and Synthesis of Systems Zeros, Part 1. Square System," *Int. J. Control*, vol. 23, pp. 149–166, 1976.
12. Porter, B., and A. Bradshaw: "Singular Perturbation Methods in the Design of Tracking Systems Incorporating Inner-Loop Compensators and High-Gain Error-Actuated Controllers," *Int. J. Syst. Sci.*, vol. 12, pp. 1193–1205, 1981.
13. Ridgely, D. B., S. S. Banda, and J. J. D'Azzo: "Decoupling of High Gain Multivariable Tracking Systems," *AIAA J. Guidance, Control, Dynamics*, vol. 8, pp. 44–49, 1985.
14. Smyth, J. S.: "Digital Flight Control System Design Using Singular Perturbation Methods," M.S. thesis, AFIT/EE/GE/81D-55, Air Force Institute of Technology, Wright-Patterson Air Force Base, Ohio, 1981.
15. Bradshaw, A., and B. Porter: "Singular perturbation methods in the design of tracking systems incorporating fast-sampling error-actuated controllers," *Int. J. Syst. Sci.*, vol. 12, pp. 1181–1191, 1981.
16. Bradshaw, A., and B. Porter: "Singular perturbation methods in the design of tracking systems incorporating inner-loop compensators and fast-sampling error-actuated controllers," *Int. J. Syst. Sci.*, vol. 12, pp. 1207–1220, 1981.

CHAPTER 21

QUANTITATIVE FEEDBACK THEORY (QFT) TECHNIQUE

21.1 INTRODUCTION[1-3]

Quantitative feedback theory (QFT) is a unified theory that emphasizes the use of feedback for achieving the desired system performance tolerances despite plant uncertainty and plant disturbances. QFT *quantitatively* formulates these two factors in the form of (a) sets $\mathscr{T}_R = \{T_R\}$ of acceptable command or tracking input-output relations and $\mathscr{T}_D = \{T_D\}$ of acceptable disturbance input-output relations, and (b) a set $\mathscr{P} = \{P\}$ of possible plants. The object is to guarantee that the control ratio $T_R = Y/R$ is a member of $\mathscr{T}_R$ and $T_D = Y/D$ is a member of $\mathscr{T}_D$, for all P in $\mathscr{P}$. QFT has been developed for both linear and nonlinear, time-invariant and time-varying, continuous and sampled data, uncertain MISO and MIMO plants, and for both output and internal variable feedback. It has also been extended to some classes of uncertain distributed systems (in which the plant is described by partial differential equations) and the feedback and specifications are also distributed.

The representation of a MIMO plant with m inputs and ℓ outputs is shown in Fig. 21.1. The QFT synthesis technique for highly uncertain linear time-invariant MIMO plants has the following features[13]:

1. The MIMO synthesis problem is converted into a number of single-loop feedback problems in which parameter uncertainty, external disturbances, and performance tolerances are derived from the original MIMO problem. The

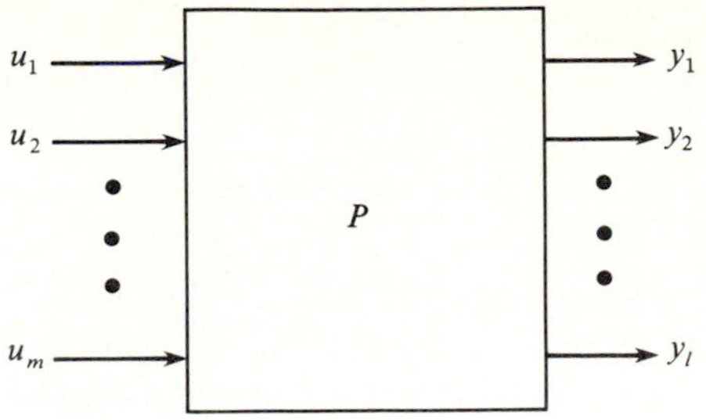

FIGURE 21.1
A MIMO uncertain plant.

solutions to these single-loop problems are guaranteed to work for the MIMO plant. It is not necessary to consider the complete system characteristic equation.

2. The design is tuned to the extent of the uncertainty and the performance tolerances.

This design technique is applicable to the following problem classes:

1. SISO linear time-invariant (LTI) systems
2. SISO nonlinear systems. These are rigorously converted to equivalent class 1 systems whose solutions are guaranteed to work for a large problem class.
3. MIMO LTI systems. The performance specifications on each individual closed-loop system transfer function and on all the closed-loop disturbance response functions must be specified.
4. MIMO nonlinear systems. They are rigorously converted to equivalent class 3 systems whose solutions are guaranteed to work for a large class.
5. Distributed systems.
6. Sampled-data systems as well as continuous systems for all of the preceding.

QFT is essentially a frequency-domain technique that is applied in this text to multiple-input single-output (MISO) systems. Problem classes 3 and 4 are converted into equivalent sets of MISO systems to which the QFT design technique is applied. The objective to is to solve the MISO problems, i.e., to find compensation functions which guarantee that the performance tolerances for each MISO problem are satisfied for all P in $\mathscr{P}$. The amount of feedback designed into the system is then tuned to the desired performance sets $\mathscr{T}_R$ and $\mathscr{T}_D$ and the given plant uncertainty set $\mathscr{P}$. Also, time-varying and nonlinear uncertain plant sets can be converted into equivalent MISO LTI plant problems to which the MISO frequency-domain technique can be readily applied and where the fundamental tradeoffs are highly visible.[11]

This chapter is devoted to presenting a basic understanding and appreciation of the power of the QFT technique. This is accomplished by presenting first an introduction of the QFT technique using the frequency-response method as applied to a single-loop system. This is followed by a discussion of minimum-phase

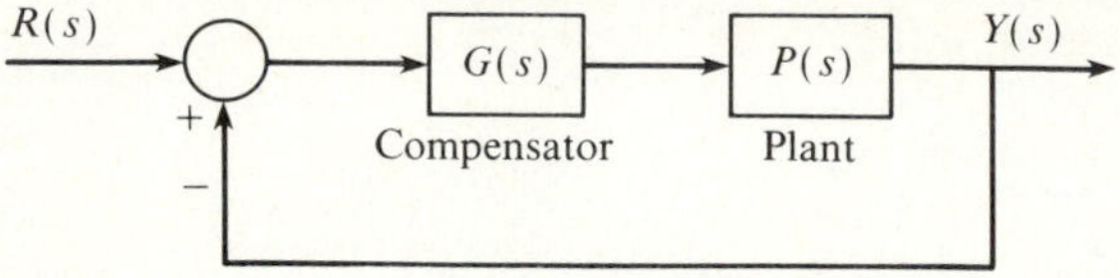

FIGURE 21.2
Cascade compensation.

performance specifications. The next few sections are devoted to the MIMO system and the development of a suitable mapping that permits the analysis and synthesis of a MIMO control system by a set of equivalent single-loop control systems. The remaining portion of the chapter is devoted to an in-depth development of the QFT technique as applied to the design of robust single-loop control systems having two inputs, a tracking and a disturbance input, respectively, and a single output (a MISO system). For a more complete and thorough presentation of this technique as applied to MIMO control systems, especially to the design of an aircraft flight control system with control surface failures, for both minimum and nonminimum phase plants, the reader is referred to the literature.[1-45]

21.2 FREQUENCY RESPONSES WITH PARAMETER VARIATIONS

Insight into the QFT technique is provided by considering the unity-feedback cascade-compensated control system of Fig. 21.2, where G is a compensator and P is the plant in which the plant parameters vary over some known range or there is plant parameter uncertainty. The loop transmission L is defined as

$$L = GP \tag{21.1}$$

and the control ratio of the unity-feedback system of Fig. 21.2 is

$$T = \frac{Y}{R} = \frac{L}{1 + L} \tag{21.2}$$

The plant with nominal plant parameters is denoted as P_o, thus $L_o = GP_o$. For a given $\mathbf{G}(j\omega)$ and $\mathbf{P}_o(j\omega)$ a plot of $\text{Lm}\,\mathbf{L}_o(j\omega)$ vs. $\angle \mathbf{L}_o$ on the Nichols chart (NC) is shown in Fig. 21.3. From this plot the closed-loop frequency-response data can be obtained for plotting M_o and α vs. ω, as shown in Fig. 21.4, where

$$M_o(j\omega)\angle \alpha(j\omega) = \frac{\mathbf{Y}(j\omega)}{\mathbf{R}(j\omega)} = \frac{\mathbf{L}_o(j\omega)}{1 + \mathbf{L}_o(j\omega)} \tag{21.3}$$

Note that for the nominal plant $\mathbf{P}_o(j\omega)$, the nominal loop transmission is

$$\text{Lm}\,\mathbf{L}_o = \text{Lm}\,\mathbf{GP}_o = \text{Lm}\,\mathbf{G} + \text{Lm}\,\mathbf{P}_o \tag{21.4}$$

whereas for the nonnominal plant $\mathbf{P}(j\omega)$

$$\text{Lm}\,\mathbf{L} = \text{Lm}\,\mathbf{GP} = \text{Lm}\,\mathbf{G} + \text{Lm}\,\mathbf{P} \tag{21.5}$$

Thus, for $\omega = \omega_i$, the variation $\delta_p(j\omega_i)$ in $\text{Lm}\,\mathbf{L}(j\omega_i)$ is given by

$$\delta_p(j\omega_i) = \text{Lm}\,\mathbf{L}(j\omega_i) - \text{Lm}\,\mathbf{L}_o(j\omega_i) = \text{Lm}\,\mathbf{P}(j\omega_i) - \text{Lm}\,\mathbf{P}_o(j\omega_i) \tag{21.6}$$

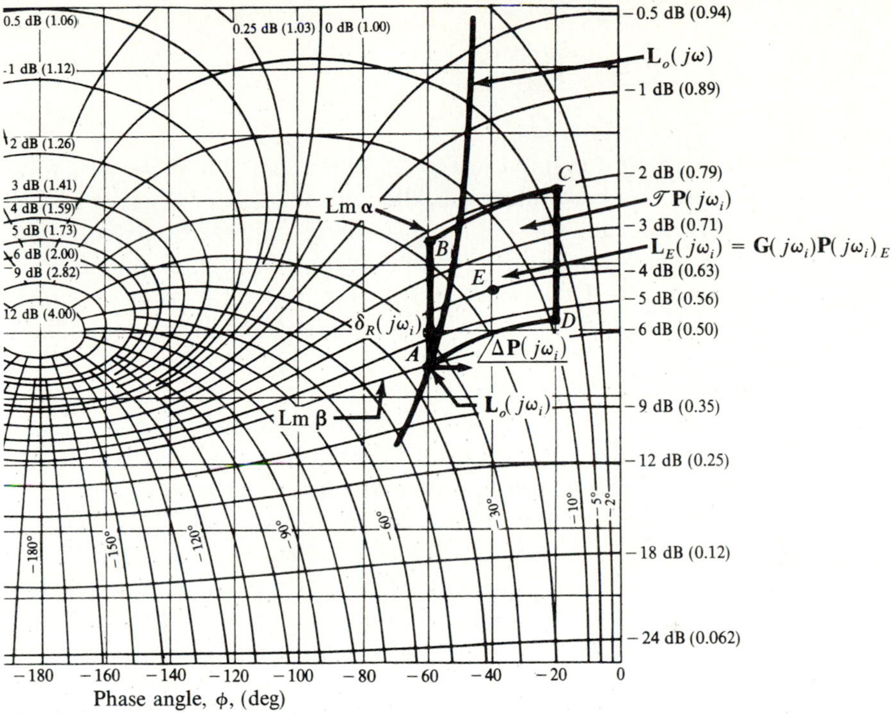

FIGURE 21.3
Nominal loop transmission plot with plant parameter area of uncertainty.

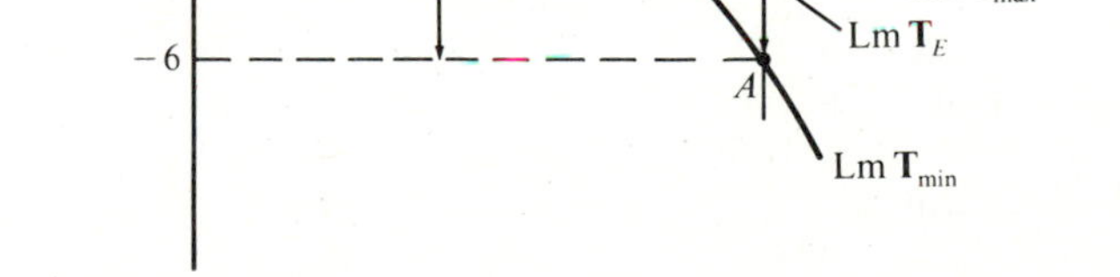

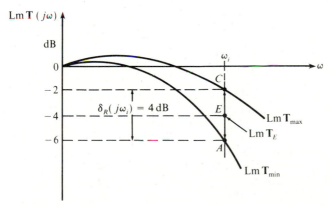

FIGURE 21.4
Closed-loop responses obtained from Fig. 21.3.

and

$$\underline{/\Delta \mathbf{P}(j\omega_i)} = \underline{/\mathbf{L}} - \underline{/\mathbf{L}_o} = (\underline{/\mathbf{G}} + \underline{/\mathbf{P}}) - (\underline{/\mathbf{G}} + \underline{/\mathbf{P}_o}) = \underline{/\mathbf{P}} - \underline{/\mathbf{P}_o} \tag{21.7}$$

Thus, assuming $\mathbf{G} = 1\underline{/0°}$, if point A in Fig. 21.3 represents $\text{Lm}\,\mathbf{P}_o$ vs. $\underline{/\mathbf{P}_o}$, then a variation in $\mathbf{P}$ results in a horizontal translation in the angle of $\mathbf{P}$, given by Eq. (21.7), and a vertical translation in the log magnitude value of $\mathbf{P}$, given by Eq. (21.6). These translations are shown in Fig. 21.3 at the points B, C, and D.

The variation $\delta_p(j\omega_i)$ of the plant, and in turn $\mathbf{L}(j\omega_i)$, from the nominal value for a given frequency ω_i, over the range of plant parameter variation is described by the template $\mathcal{T}\mathbf{P}(j\omega_i)$ shown in Fig. 21.3 (see Sec. 21.10 for a specific numerical example). Assume that point A on the template represents the nominal plant $\mathbf{P} = \mathbf{P}_o(j\omega_i)$ and that its corresponding closed-loop frequency response is $\text{Lm}\,M_A = \text{Lm}\,\beta = -6$ dB. When $\mathbf{P} = \mathbf{P}_C(j\omega)$, represented by point C in Fig. 21.3, the corresponding closed-loop frequency response is $\text{Lm}\,M_C = \text{Lm}\,\alpha = -2$ dB. These values are plotted in Fig. 21.4. Note that point A represents the minimum value of $\text{Lm}\,M(j\omega_i)$ at $\omega = \omega_i$, i.e., $\text{Lm}\,\mathbf{T}_{\min} = [\text{Lm}\,M(j\omega_i)]_{\min} = \text{Lm}\,\beta$. Point C represents the largest value of $\text{Lm}\,M(j\omega_i)$ at $\omega = \omega_i$, i.e., $\text{Lm}\,\mathbf{T}_{\max} = [\text{Lm}\,M(j\omega_i)]_{\max} = \text{Lm}\,\alpha$, with the range of plant parameter variation described by the template. Thus, the maximum variation in $\text{Lm}\,M$, denoted by $\delta_R(j\omega_i)$, for this example is

$$\delta_R(j\omega_i) = \text{Lm}\,\alpha - \text{Lm}\,\beta = -2 - (-6) = 4\text{ dB}$$

Also shown in Fig. 21.4 is point E which is midway between points A and C; it corresponds to point E in Fig. 21.3 which is within the variation template $\mathcal{T}\mathbf{P}(j\omega_i)$. For different frequencies the procedure is repeated to obtain the maximum variation $\delta_R(j\omega)$ at each frequency, as shown in Fig. 21.4. From this figure it is possible to determine the variation in the control system's figures of merit due to the plant uncertainty. This graphical description of the effect of plant uncertainty is the basis of the QFT technique.

21.3 INTRODUCTION TO THE QFT METHOD (SINGLE-LOOP SYSTEM)

Figure 21.5 represents the feedback control system in which P is the transfer function of the plant which may contain parameter uncertainty, G is the cascade compensator, and F is an input filter transfer function. The output $y(t)$ is required to track the command input $r(t)$ and to reject the disturbances $d_1(t)$ and $d_2(t)$. The compensator G in Fig. 21.5 is to be designed so that the variation of $y(t)$ to the uncertainty in the plant P is within tolerances and the effects of the disturbances $d_1(t)$ and $d_2(s)$ on $y(t)$ are acceptably small. Also, the filter properties of $F(s)$ must be designed to achieve the desired tracking by the output $y(t)$ of the input $r(t)$. Since the control system in Fig. 21.5 has two measurable quantities, $r(t)$ and $y(t)$, it is referred to as a two degree-of-freedom (DOF)

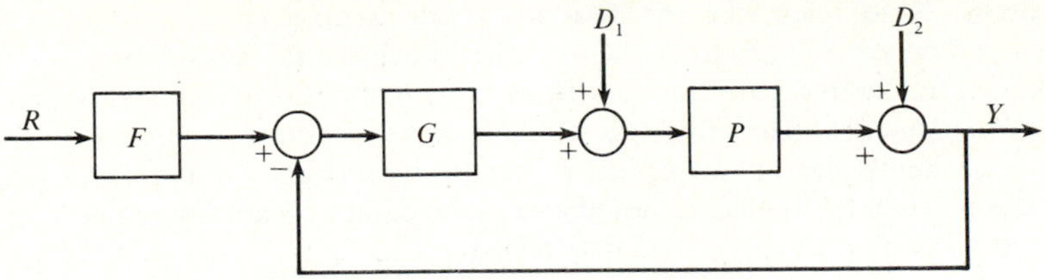

FIGURE 21.5
A feedback structure.

feedback structure. If the two disturbance inputs are measurable, then it represents a four DOF structure.

The actual design is closely related to the extent of the uncertainty and to the narrowness of the performance tolerances. The uncertainty of the plant transfer function is denoted by the set

$$\mathscr{P} \equiv \{\mathbf{P}\} \tag{21.8}$$

and is illustrated by Example 1.

Example 1. The plant transfer function is

$$P(s) = \frac{K}{s(s+a)} \tag{21.9}$$

where the values of K and a are fixed, but the value of K is in the range [1,10] and a is in the range [−2,2]. For example, an aircraft operates over a wide range of speed, altitude, and weight and it has a different transfer function for each flight condition. The region of parameter uncertainty of the plant is illustrated by Fig. 21.6.

The uncertainty of the disturbance is denoted by

$$\mathscr{D} \equiv \{\mathbf{D}\} \tag{21.10}$$

and the acceptable closed-loop transmittances (control ratios) are denoted by

$$\mathscr{T} \equiv \{\mathbf{T}\} \tag{21.11}$$

The design objective is to guarantee that $T_R(s) = C(s)/R(s)$ and $T_D(s) = C(s)/D(s)$ are members of the sets of acceptable $\mathscr{T}_R$ and $\mathscr{T}_D$, respectively, for all $\mathbf{P} \in \mathscr{P}$ and all $\mathbf{D} \in D$. In a feedback system, the principal challenge in the control

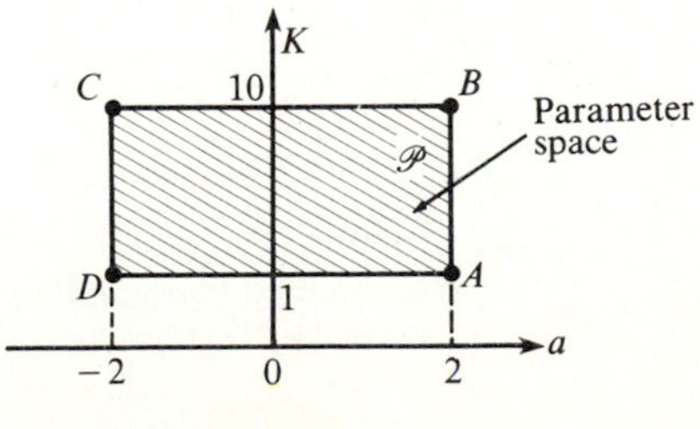

FIGURE 21.6
Plant uncertainty.

system design is to relate the specifications to the requirements on the loop transmission function $L(s) = G(s)P(s)$ in order to achieve the desired benefits of feedback, i.e., the desired reduction in sensitivity to plant uncertainty and desired disturbance attenuation. The advantage of the frequency domain is that $L(s) = G(s)P(s)$ is simply the multiplication of complex numbers. In the frequency domain it is possible to evaluate $\mathbf{L}(j\omega)$ at every ω_i separately, and thus, at each ω_i, the optimal bounds on $\mathbf{L}(j\omega_i)$ can be determined.

21.4 MINIMUM-PHASE SYSTEM PERFORMANCE SPECIFICATIONS[10,11,45]

In order to apply the QFT technique it is necessary to synthesize the desired or model control ratio, based upon the system's performance specifications in the time domain. For the structure of Fig. 21.5 the control ratios are

$$\frac{\text{Tracking}}{(D_1 = D_2 = 0)} : T_R(s) = \frac{F(s)G(s)P(s)}{1 + G(s)P(s)} = \frac{FL}{1+L} \tag{21.12}$$

$$\frac{\text{Disturbance}}{(R = 0)} : T_{D_1}(s) = \frac{P(s)}{1 + G(s)P(s)} = \frac{P}{1+L} \qquad \text{with } D_2 = 0 \tag{21.13}$$

$$T_{D_2}(s) = \frac{1}{1 + G(s)P(s)} = \frac{1}{1+L} \qquad \text{with } D_1 = 0 \tag{21.14}$$

Tracking Models

The tracking thumbprint specifications, based upon satisfying some or all of the step forcing function figures of merit for underdamped (M_p, t_p, t_s, t_r, K_m) and overdamped (t_s, t_r, K_m) responses, respectively, for a simple second-order system, are depicted in Fig. 21.7. The time responses $y(t)_U$ and $y(t)_L$ in this figure represent the upper and lower bounds, respectively, of the thumb print specifications; that is, an acceptable response $y(t)$ must lie between these bounds. The Bode plots of the upper bound B_U and lower bound B_L for Lm $\mathbf{T}_R(j\omega)$ vs. ω are shown in Fig. 21.8. Note that for the minimum-phase plants, only the tolerance on $|\mathbf{T}_R(j\omega_i)|$ need be satisfied for a satisfactory design. For *nonminimum-phase* plants, tolerances on $\angle \mathbf{T}_R(j\omega_i)$ must also be specified and satisfied in the design process. The modeling of a desired transmittance $T(s)$ is discussed in Chap. 12. It

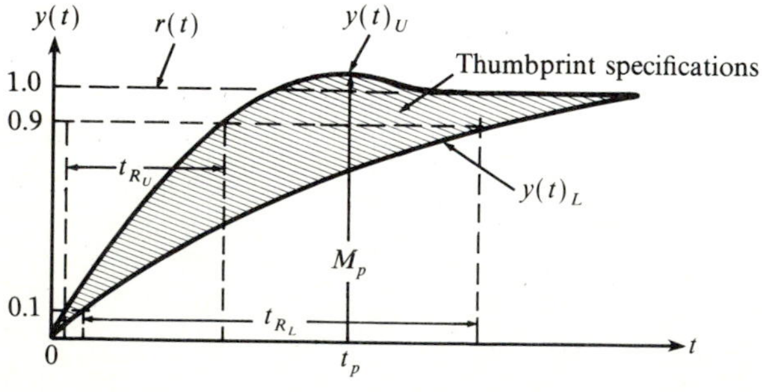

FIGURE 21.7
The tracking thumbprint specifications.

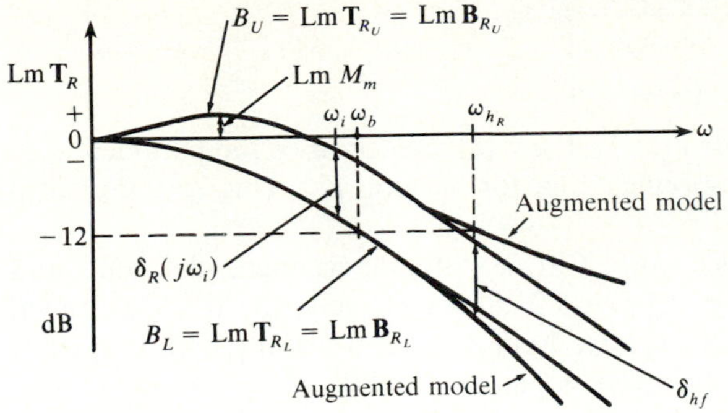

FIGURE 21.8
Bode plots of $\mathbf{T}_R$.

is desirable to synthesize the control ratios corresponding to the upper and lower bounds T_{R_U} and T_{R_L}, respectively, so that $\delta_R(j\omega_i)$ increases as ω_i increases above the 0-dB crossing frequency of T_{R_U}. This characteristic of $\delta_R(j\omega_i)$ simplifies the process of synthesizing a loop transmission $L_o(s) = G(s)P_o(s)$ as discussed in Sec. 21.16. To synthesize $L_o(s)$ it is necessary to determine the tracking bounds $\mathbf{B}_R(j\omega_i)$ (see Sec. 21.13) which are obtained based upon $\delta_R(j\omega_i)$. This characteristic of $\delta_R(j\omega_i)$ ensures that the tracking bounds $\mathbf{B}_R(j\omega)$ decrease as ω_i increases (see Sec. 2.1.12).

The QFT technique requires that the desired control ratios be modeled in the frequency domain to satisfy the required gain K_m and the desired time-domain performance specifications for a step input. Thus, M_p, t_p, t_s, and t_r must be satisfied for T_{R_U} and T_s (or t_s) and t_r must be satisfied for T_{R_L}. The modeling of the desired control ratio T_{R_U} is described in Sec. 12.2. It may have the form

$$T_{R_U}(s) = \frac{\omega_n^2}{s^2 + 2\zeta\omega_n s + \omega_n^2} = \frac{\omega_n^2}{(s - p_1)(s - p_2)} \tag{21.15}$$

where $\omega_n^2 = p_1 p_2$. The control ratio $T_{R_U}(s)$ of Eq. (21.15) can be represented by an equivalent unity-feedback system so that

$$T_{R_U}(s) = \frac{G_{eq}(s)}{1 + G_{eq}(s)} \tag{21.16}$$

where
$$G_{eq}(s) = \frac{\omega_n^2}{s(s + 2\zeta\omega_n)} \tag{21.17}$$

The gain constant of this equivalent Type 1 transfer function $G_{eq}(s)$ is $K_1 = \lim_{s\to 0}[sG_{eq}(s)] = \omega_n/2\zeta$. Equation (21.15) satisfies the requirement that $y(t)_{ss} = R_0 u_{-1}(t)$ for $r(t) = R_0 u_{-1}(t)$. The frequency ω_b for which $|\mathbf{T}_{R_U}(j\omega_b)| = 0.7071$ is defined as the system bandwidth frequency ω_{b_U}.

The simplest overdamped model for $T_{R_L}(s)$ is of the form

$$T_{R_L}(s) = \frac{Y(s)}{R(s)} = \frac{K}{(s - \sigma_1)(s - \sigma_2)} = \frac{G_{eq}(s)}{1 + G_{eq}(s)} \tag{21.18}$$

where $$G_{eq}(s) = \frac{\sigma_1\sigma_2}{s[s-(\sigma_1+\sigma_2)]}$$

and $$K_1 = -\sigma_1\sigma_2/(\sigma_1+\sigma_2)$$

For this system $y(t)_{ss} = R_o$ for $r(t) = R_o u_{-1}(t)$. Selection of the parameters σ_1 and σ_2 is used to meet the specifications for t_s and K_1. This selection also determines the bandwidth ω_{b_L}.

Once the models $\mathbf{T}_{R_b}(j\omega)$ and $\mathbf{T}_{R_L}(j\omega)$ are determined, the time and frequency response plots of Figs. 21.7 and 21.8, respectively, can then be drawn. Because the models for T_{R_U} and T_{R_L} are both second order, the high-frequency asymptotes in Fig. 21.8 have the same slope. The high-frequency range in Fig. 21.8 is defined as $\omega \geq \omega_b$ where ω_b is the model bandwidth frequency of B_U. In the high-frequency (hf) range an increasing spread between B_U and B_L is required, that is,

$$\delta_{hf} = B_U - B_L \tag{21.19}$$

must increase with increasing frequency. This is achieved by changing B_U and B_L, without violating the desired time-response characteristics, by augmenting T_{R_L} with a negative real pole which is as close to the origin as possible but far enough away not to significantly affect the time response. This additional pole lowers B_L in the high-frequency range. If necessary, the spread can be further increased by augmenting T_{R_U} with a zero as close to the origin as possible without significantly affecting the time response. This additional zero raises the curve B_U in the high-frequency range. At high frequencies δ_{hf} must be larger than the actual variation in the plant, ΔP. This comes about because of the Bode derived theorem which states that $\int_0^\infty \mathrm{Lm}\, \mathbf{S}_P^T\, d\omega = 0$, i.e., the reduction in sensitivity $\mathbf{S}_P^T$ at the lower frequencies must be compensated by an increase in sensitivity at the higher frequencies. At some high frequency $\omega_i \geq \omega_h$ (see Fig. 21.8), since $|\mathbf{T}_{R_L}| \approx 0$, then

$$\lim_{\omega\to\infty} \delta_R(j\omega_i) \approx B_U - (-\infty) = \infty \text{ dB} \tag{21.20}$$

For the case where $y(t)$, corresponding to T_{R_U}, is to have an allowable "large" overshoot followed by a small tolerable undershoot, a dominant complex-pole pair is not suitable for T_{R_U}. An acceptable overshoot with no undershoot for T_{R_U} can be achieved by T_{R_U} having two real dominant poles $p_1 > p_2$, a dominant real zero ($z_1 > p_1$) "close" to p_1, and a far off pole $p_4 \ll p_2$. The closeness of the zero dictates the value of M_p. Thus, a designer selects a pole-zero combination to yield the form of the desired time-domain response. This also applies for synthesizing a desired T_D.

Disturbance Rejection Models[6]

A detailed discussion on synthesizing disturbance-rejection models, based upon desired performance specifications, is given in Secs. 12.5 and 12.8. The reader is urged to review these sections before proceeding in this chapter. The Bode plots of Eqs. (12.37), (12.38), (12.41), and (12.42) are shown in Fig. 21.9. As seen from this figure, there is only an upper boundary for $\mathrm{Lm}\,\mathbf{T}_D$.

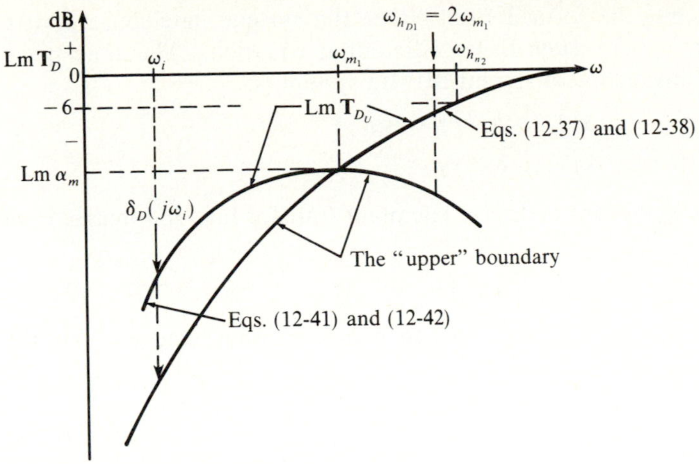

FIGURE 21.9
Bode plots of disturbance models for $\mathbf{T}_D(j\omega)$.

21.5 MULTIPLE-INPUT MULTIPLE-OUTPUT (MIMO) UNCERTAIN PLANTS

The design for a MIMO system involves the design of an equivalent set of single-loop MISO systems. In general, an $\ell \times m$ open-loop MIMO plant can be represented in matrix notation as

$$\mathbf{Y}(s) = \mathbf{PU}(s) \tag{21.21}$$

where $\mathbf{Y}(s)$ = ℓ-dimensional plant output vector
$\mathbf{U}(s)$ = m-dimensional plant input vector
$\mathbf{P}$ = $\ell \times m$ plant transfer-function matrix relating $\mathbf{U}(s)$ to $\mathbf{Y}(s)$

Consider the MIMO plant of Fig. 21.1 for which $\mathbf{P}$ is a member of $\mathcal{P}$ (i.e., $\mathbf{P} \in \mathcal{P}$), the set of all possible plant transfer functions relating each input to each output. When a system has variable parameters, known or unknown, the transfer-function matrix of the plant may be represented by an associated set of transfer-function matrices $\{\mathbf{P}_1, \mathbf{P}_2, \dots\}$ which are contained in $\mathcal{P}$. The QFT design method requires that the uncertainty in $\mathbf{P}$ be known or is at least bounded. In any MIMO system with m inputs there are at most m outputs which can be independently controlled[3]; therefore, the dimensions of both the input and output vectors are equal, i.e., $\ell = m$. If the model defines an unequal number of inputs and outputs, the first step is to modify the model so that the dimensions of the input and output are the same. This requires a judicious selection of inputs if $m > \ell$. The system is then defined as being of order $m \times m$. Using fixed point theory[14,20] (see Sec. 21.7), the MIMO problem for an $m \times m$ system can be separated into m equivalent single-loop MISO systems and m^2 prefilter disturbance problems. Each MISO system design is based upon the specifications relating its output and all of its inputs.

The **P** matrix may be formed from either the system state-space matrix representation or from the system linear differential equations. The state-space representation for a linear time-invariant MIMO system is:

$$\begin{aligned} \dot{\mathbf{x}}(t) &= \mathbf{A}\mathbf{x}(t) + \mathbf{B}\mathbf{u}(t) \\ \mathbf{y}(t) &= \mathbf{C}\mathbf{x}(t) \end{aligned} \tag{21.22}$$

where **A**, **B**, and **C** are constant matrices. The plant transfer-function matrix $\mathbf{P}(s)$ is evaluated as

$$\mathbf{P}(s) = \mathbf{C}[s\mathbf{I} - \mathbf{A}]^{-1}\mathbf{B} \tag{21.23}$$

The general plant model for a MIMO system with two inputs and two outputs has the form

$$\begin{aligned} a(s)y_1(s) + b(s)y_2(s) &= fu_1(s) + gu_2(s) \\ c(s)y_1(s) + d(s)y_2(s) &= hu_1(s) + iu_2(s) \end{aligned} \tag{21.24}$$

where $a(s)$ through $d(s)$ are polynomials in s, f through i are constant coefficients, $y_1(s)$ and $y_2(s)$ are the outputs, and $u_1(s)$ and $u_2(s)$ are the inputs. In matrix notation the system is represented by

$$\begin{bmatrix} a(s) & b(s) \\ c(s) & d(s) \end{bmatrix} \mathbf{Y}(s) = \begin{bmatrix} f & g \\ h & i \end{bmatrix} \mathbf{U}(s) \tag{21.25}$$

This is defined as a 2×2 system. In the general case with m inputs and m outputs, the system is defined as $m \times m$. Let the matrix premultiplying the output vector $\mathbf{Y}(s)$ be $\mathbf{M}(s)$ and the matrix premultiplying the input vector $\mathbf{U}(s)$ be **N**. Equation (20.25) may then be written as

$$\mathbf{M}(s)\mathbf{Y}(s) = \mathbf{N}\mathbf{U}(s) \tag{21.26}$$

The solution of Eq. (20.26) for the output $\mathbf{Y}(s)$ yields

$$\mathbf{Y}(s) = \mathbf{M}^{-1}(s)\mathbf{N}\mathbf{U}(s) = \mathbf{P}(s)\mathbf{U}(s) \tag{21.27}$$

Thus, the $m \times m$ plant transfer function matrix $\mathbf{P}(s)$ is

$$\mathbf{P}(s) = \mathbf{M}^{-1}(s)\mathbf{N} \tag{21.28}$$

This plant matrix $\mathbf{P}(s) = [p_{ij}(s)]$ is a member of the set $\mathscr{P} = \{\mathbf{P}(s)\}$ of possible plant matrices which are functions of the uncertainty in the plant parameters. In practice, only a finite set of **P** matrices is formed, representing the extreme boundaries of the plant uncertainty under varying conditions.

21.6 INTRODUCTION TO MIMO COMPENSATION

Figure 21.10 represents an $m \times m$ MIMO closed-loop system in which **F**, **G**, and **P** are each $m \times m$ matrices, and $\mathscr{P} = \{\mathbf{P}\}$ is a set of matrices due to plant uncertainty. There are m^2 closed-loop system transfer functions (transmissions) $t_{ij}(s)$ contained within its system transmission matrix, i.e., $\mathbf{T}(s) = \{t_{ij}(s)\}$,

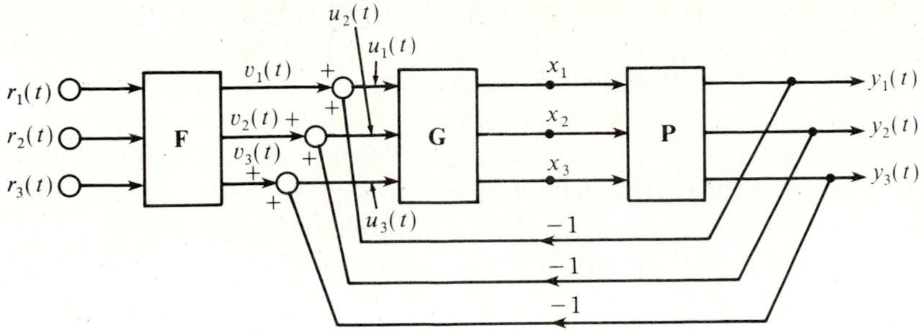

FIGURE 21.10
Three-input three-output MIMO feedback structure.

relating the outputs $y_i(s)$ to the inputs $r_j(s)$, e.g., $y_i(s) = t_{ij}(s)r_j(s)$. In a quantitative problem statement there are tolerance bounds on each $t_{ij}(s)$, giving a set of m^2 acceptable regions $\tau_{ij}(s)$ which are to be specified in the design, thus $t_{ij}(s) \in \tau_{ij}(s)$ and $\mathscr{T}(s) = \{\tau_{ij}(s)\}$. The QFT design technique for 2 × 2 and 3 × 3 systems has been developed and applied to the design of flight control systems.[12, 16, 38, 40, 41, 44]

From Fig. 21.10 the following equations can be written:

$$\mathbf{y} = \mathbf{Px} \qquad \mathbf{x} = \mathbf{Gu} \qquad \mathbf{u} = \mathbf{v} - \mathbf{y} \qquad \mathbf{v} = \mathbf{Fr}$$

In these equations $\mathbf{P}(s) = [p_{ij}(s)]$ is the matrix of plant transfer functions. $\mathbf{G}(s)$ is the matrix of compensator transfer functions and is often simplified so that it is diagonal, that is, $\mathbf{G}(s) = \text{diag}\{g_i(s)\}$. $\mathbf{F}(s) = [f_{ij}(s)]$ is the matrix of prefilter transfer functions which may also be a diagonal matrix. The combination of these equations yields

$$\mathbf{y} = [\mathbf{I} + \mathbf{PG}]^{-1}\mathbf{PGFr} \tag{21.29}$$

where the system control ratio relating **r** to **y** is

$$\mathbf{T} = [\mathbf{I} + \mathbf{PG}]^{-1}\mathbf{PGF} \tag{21.30}$$

To appreciate the difficulty of the design problem, note the very complex expression of Eq. (21.31) for t_{11} for the case $m = 3$ with a diagonal **G** matrix.

$$\begin{aligned} t_{11} = \{&[p_{11}f_{11}g_1 + p_{12}f_{21}g_2 + p_{13}f_{31}g_3][(1 + p_{22}g_2)(1 + p_{33}g_3) - p_{23}p_{32}g_2g_3] \\ &-[p_{21}f_{11}g_1 + p_{22}f_{21}g_2 + p_{23}f_{31}g_3][p_{12}g_2(1 + p_{33}g_3) - p_{32}p_{13}g_2g_3] \\ &+[p_{31}f_{11}g_1 + p_{32}f_{21}g_2 + p_{33}f_{31}g_3][p_{23}p_{12}g_2g_3 - (1 + p_{22}g_2)p_{13}g_3]\}/ \\ &\{(1 + p_{11}g_1)[(1 + p_{22}g_2)(1 + p_{33}g_3) - p_{23}p_{32}g_2g_3] \\ &-p_{21}g_1[p_{12}g_2(1 + p_{33}g_3) - p_{32}p_{13}g_2g_3] \\ &+p_{31}g_1[p_{12}p_{23}g_2g_3 - p_{13}g_3(1 + p_{22}g_2)]\} \end{aligned} \tag{21.31}$$

For the 3 × 3 plant there are $m^2 = 9$ such $t_{ij}(s)$ expressions (all have the same denominator), and there may be considerable uncertainty in the nine plant transfer functions $p_{ij}(s)$.

The QFT design procedure systematizes and simplifies the manner of achieving a satisfactory system design. The objective is to design a system which behaves as desired for the entire range of plant uncertainty. This requires finding nine $f_{ij}(s)$ and three $g_i(s)$ such that each $t_{ij}(s)$ stays within its acceptable region $\tau_{ij}(s)$, no matter how the $p_{ij}(s)$ may vary. Clearly, this is a very difficult problem. Therefore, another mathematical system description is presented in Sec. 21.7 that overcomes this problem. The MIMO system is converted into an equivalent set of single-loop systems, and then the MISO design method of Sec. 21.9 is used.

21.7 MIMO COMPENSATION

The basic MIMO compensation structure for a 2 × 2 MIMO system is shown in Fig. 21.11. The structure for a 3 × 3 MIMO system is shown in Fig. 21.10. They consist of the uncertain plant matrix **P**, the diagonal compensation matrix **G**, and prefilter **F**. These matrices are defined as follows:

$$\mathbf{G} = \left[\begin{array}{cc|ccc} g_1 & 0 & \cdots & 0 \\ 0 & g_2 & \cdots & 0 \\ \hline \vdots & \vdots & & \vdots \\ 0 & 0 & \cdots & g_m \end{array}\right] \qquad \mathbf{F} = \left[\begin{array}{cc|cc} f_{11} & f_{12} & \cdots & f_{1m} \\ f_{21} & f_{22} & \cdots & f_{2m} \\ \hline \vdots & \vdots & & \vdots \\ f_{m1} & f_{m2} & \cdots & f_{mm} \end{array}\right]$$

$$\mathbf{P} = \left[\begin{array}{cc|cc} p_{11} & p_{12} & \cdots & p_{1m} \\ p_{21} & p_{22} & \cdots & p_{2m} \\ \hline \vdots & \vdots & & \vdots \\ p_{m1} & p_{m2} & \cdots & p_{mm} \end{array}\right] \tag{21.32}$$

Although only a diagonal **G** matrix is considered in this chapter, the use of a nondiagonal **G** matrix may allow the designer more design flexibility.[19] The partitions in Eq. (21.32) denote the **G**, **F**, and **P** matrices for a 2 × 2 system. Substituting these matrices into Eq. (21.30) yields the control ratios $t_{ij}(s)$ relating the ith output to the jth input.

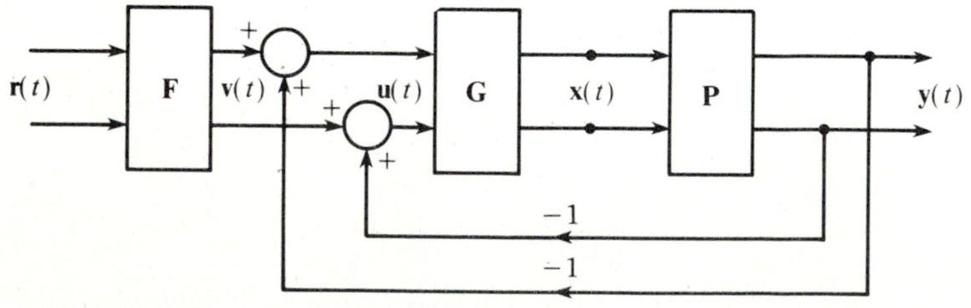

FIGURE 21.11
The MIMO control structure (2 × 2 system).

21.8 JUSTIFICATION OF THE MISO EQUIVALENT METHOD[13]

This section develops a mapping that permits the analysis and synthesis of a MIMO control system by a set of equivalent MISO control systems. This mapping results in m^2 equivalent systems, each with two inputs and one output. One input is designated as a "desired" input and the other as a "disturbance" input. The inverse of the plant matrix is represented by

$$\mathbf{P}^{-1} = \begin{bmatrix} p_{11}^* & p_{12}^* & \cdots & p_{1m}^* \\ p_{21}^* & p_{22}^* & \cdots & p_{2m}^* \\ \vdots & \vdots & & \vdots \\ p_{m1}^* & p_{m2}^* & \cdots & p_{mm}^* \end{bmatrix} \tag{21.33}$$

The m^2 effective plant transfer functions are formed as

$$q_{ij} = 1/p_{ij}^* = \frac{\det \mathbf{P}}{\operatorname{adj} \mathbf{P}_{ij}} \tag{21.34}$$

There is a requirement that $\det \mathbf{P}$ be minimum phase. The $\mathbf{Q}$ matrix is then formed as

$$\mathbf{Q} = \begin{bmatrix} q_{11} & q_{12} & \cdots & q_{1m} \\ q_{21} & q_{22} & \cdots & q_{2m} \\ \vdots & \vdots & & \vdots \\ q_{m1} & q_{m2} & \cdots & q_{mm} \end{bmatrix} = \begin{bmatrix} 1/p_{11}^* & 1/p_{12}^* & \cdots & 1/p_{1m}^* \\ 1/p_{21}^* & 1/p_{22}^* & \cdots & 1/p_{2m}^* \\ \vdots & \vdots & & \vdots \\ 1/p_{m1}^* & 1/p_{m2}^* & \cdots & 1/p_{mm}^* \end{bmatrix} \tag{21.35}$$

where $\mathbf{P} = [p_{ij}]$, $\mathbf{P}^{-1} = [p_{ij}^*] = [1/q_{ij}]$, and $\mathbf{Q} = [q_{ij}] = [1/p_{ij}^*]$. The matrix $\mathbf{P}^{-1}$ is partitioned to the form

$$\mathbf{P}^{-1} = [p_{ij}^*] = [1/q_{ij}] = \mathbf{\Lambda} + \mathbf{B} \tag{21.36}$$

where $\mathbf{\Lambda}$ is the diagonal part and $\mathbf{B}$ is the balance of $\mathbf{P}^{-1}$; thus, $\lambda_{ii} = 1/q_{ii} = p_{ii}^*$, $b_{ii} = 0$, and $b_{ij} = 1/q_{ij} = p_{ij}^*$ for $i \neq j$.

Premultiplying Eq. (21.30) by $[\mathbf{I} + \mathbf{PG}]$ yields

$$[\mathbf{I} + \mathbf{PG}]\mathbf{T} = \mathbf{PGF} \tag{21.37}$$

When $\mathbf{P}$ is nonsingular, premultiplying both sides of Eq. (21.37) by $\mathbf{P}^{-1}$ yields

$$[\mathbf{P}^{-1} + \mathbf{G}]\mathbf{T} = \mathbf{GF} \tag{21.38}$$

Using Eq. (21.36) and with $\mathbf{G}$ diagonal, Eq. (21.38) can be rearranged to the form

$$\mathbf{T} = [\mathbf{\Lambda} + \mathbf{G}]^{-1}[\mathbf{GF} - \mathbf{BT}] \tag{21.39}$$

This is used to define the desired fixed point mapping where each of the m^2 matrix elements on the right side of Eq. (21.39) can be interpreted as a MISO problem. Proof of the fact that design of each MISO system yields a satisfactory MIMO design is based on the Schauder fixed point theorem.[14] This theorem is described by defining a mapping $\mathbf{Y}(\mathbf{T})$ by

$$\mathbf{Y}(\mathbf{T}) \equiv [\mathbf{\Lambda} + \mathbf{G}]^{-1}[\mathbf{GF} - \mathbf{BT}] \tag{21.40}$$

where each member of $\mathbf{T}$ is from the acceptable set $\mathscr{T}$. If this mapping has a

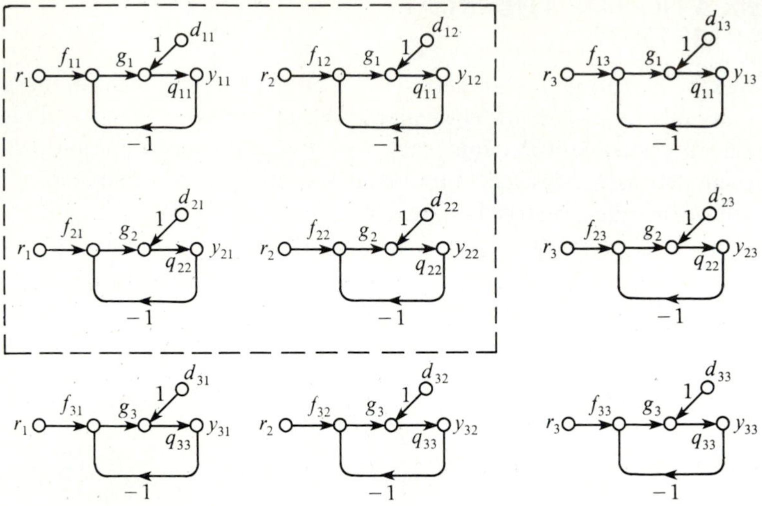

FIGURE 21.12
Effective MISO loops 2 × 2 (boxed-in loops) and 3 × 3 (all nine loops).

fixed point, i.e., $\mathbf{T} \in \mathscr{T}$ such that $\mathbf{Y(T)} = \mathbf{T}$, then this $\mathbf{T}$ is a solution of Eq. (21.39).

Figure 21.12 shows the four effective MISO loops (in the boxed area) resulting from a 2 × 2 system and the nine effective MISO loops resulting from a 3 × 3 system.[13] Since Λ and $\mathbf{G}$ in Eq. (21.36) are both diagonal, the (1, 1) element on the right side of Eq. (21.40) for the 3 × 3 case, for a *unit impulse* input, yields the output

$$y_{11} = \frac{q_{11}}{1 + g_1 q_{11}} \left[g_1 f_{11} - \left(\frac{t_{21}}{q_{12}} + \frac{t_{31}}{q_{13}} \right) \right] \tag{21.41}$$

This corresponds precisely to the first structure in Fig. 21.12. Similarly, each of the nine structures in Fig. 21.12 corresponds to one of the elements of $\mathbf{Y(T)}$ of Eq. (21.40). The control ratios for the desired tracking of the inputs r_i by the corresponding outputs y_i for each feedback loop of Eq. (21.40) have the form

$$y_{ii} = w_{ij}(v_{ij} + d_{ij}) \tag{21.42}$$

where $w_{ii} = q_{ii}/(1 + g_i q_{ii})$ and $v_{ij} = g_i f_{ij}$. The interaction between the loops has the form

$$d_{ij} = -\sum_{k \neq i} \left[\frac{t_{kj}}{q_{ik}} \right] \qquad k = 1, 2, \ldots, m \tag{21.43}$$

and appears as a "disturbance" input in each of the feedback loops. Thus Eq. (21.42) represents the control ratio of the ith MISO system. The transfer function $w_{ii}v_{ij}$ relates the "desired" ith output to the jth input r_j and the transfer function $w_{ii}d_{ij}$ relates the ith output to the jth "disturbance" input d_{ij}. The outputs given in Eq. (20.42) can thus be expressed as

$$y_{ij} = (y_{ij})_{r_i} + (y_{ij})_{d_{ij}} = y_{r_i} + y_{d_{ij}} \tag{21.44}$$

or, based on a unit impulse input,

$$t_{ij} = t_{r_{ij}} + t_{d_{ij}} \tag{21.45}$$

where

$$t_{r_{ij}} = y_{r_i} = w_{ii}v_{ij} \tag{21.46}$$

$$t_{d_{ij}} = y_{d_{ij}} = w_{ii}d_{ij} \tag{21.47}$$

and where now the upper bound, in the low-frequency range, is expressed as b'_{ij}. Thus,

$$\tau_{d_{ij}} = b_{ij} - b'_{ij} \tag{21.48}$$

represents the maximum portion of b_{ij} allocated toward disturbance rejection and b'_{ij} represents the upper bound for the tracking portion, respectively, of t_{ij}. For each MISO system there is a disturbance input which is a function of all the other loop outputs. The object of the design is to have each loop track its desired input while minimizing the outputs due to the disturbance inputs.

In each of the nine structures of Fig. 21.12 it is necessary that the control ratio $t_{ij}(s)$ must be a member of the acceptable $t_{ij} \in \tau_{ij}(s)$. All the $g_i(s)$ and $f_{ij}(s)$ must be chosen to ensure that this condition is satisfied, thus constituting nine MISO design problems. If all of these MISO problems are solved, there exists a fixed point, and then $y_{ij}(s)$ on the left side of Eq. (21.40) may be replaced by t_{ij} and all the elements of **T** on the right side by t_{kj}. This means that there exist nine t_{ij} and t_{kj}, each in its acceptable set, which is a solution to Fig. 21.11. If each element is 1 : 1, then this solution must be unique. A more formal and detailed treatment is given in Ref. 14.

21.9 EFFECTIVE MISO LOOPS OF THE MIMO SYSTEM

There are two design methods for designing MIMO systems. In the first method each MISO system in Fig. 21.12 is treated as an individual MISO design problem, which is solved using the procedures explained in the remainder of this chapter.

The disturbances $d_{ij}(s)$ expressed by Eq. (21.43) represent the interaction between the loops where now the value of t_{kj} is the upper bound b_{kj} (T_{R_U} or T_D in Figs. 21.8 and 21.9, respectively) for the respective input/output relationship. These are obtained from the design specifications. The first subscript k refers to the output variable, and the second subscript j refers to the input variable. Therefore, b_{kj} is a function of the response requirements on the output y_k due to the input r_j. The lower bound a_{kj} needs defining only when there is a command input.

For the disturbance-rejection problem, the responses must be less than some bound [i.e., $(y_{ij})_{d_{ij}} \le b_{ij}$]. Thus the loop equations become

$$b_{ij} \ge (y_{ij})_{d_{ij}} = \frac{d_{ij}|\mathbf{q}_{ii}|}{|1 + \mathbf{g}_i\mathbf{q}_{ii}|} \tag{21.49}$$

where $\mathbf{g}_i \mathbf{q}_{ii} = \mathbf{L}_i$. Reorganizing yields

$$|1 + \mathbf{L}_i| \geq \frac{d_{ij}|\mathbf{q}_{ii}|}{b_{ij}} \tag{21.50}$$

Using Eq. (21.43), and rewriting Eq. (21.50):

$$|1 + \mathbf{L}_i| \geq \frac{\left[\left|-\sum_{k \neq i} \frac{b_{kj}}{\mathbf{q}_{ik}}\right|\right]|\mathbf{q}_{ii}|}{b_{ij}} \tag{21.51}$$

For example, in a 3 × 3 system, for the first loop L_1, $i = 1$, $j = 2$, in the first term $k = 2$, and in the second term $k = 3$. Therefore, Eq. (21.51) becomes

$$|1 + \mathbf{L}_1| \geq \frac{\left|\frac{b_{22}}{|\mathbf{q}_{12}|} + \frac{b_{32}}{|\mathbf{q}_{13}|}\right| |\mathbf{q}_{11}|}{\tau_{d_{12}}}$$

Only the magnitude is used in the disturbance calculations. This assumes the worst case.

The second design is an improvement over the first method. It involves using the information on the $g_i(s)$ and $f_i(s)$ from the loops previously designed. This reduces the overdesign inherent in the early part of the design process. It may involve a trade-off in the design parameters. The final loop is designed using the exact equation.[17] The exact equation represents the last loop and the interactions caused by all the other loops. The order in which the loops are designed may be significant. Any order can be used, but some orders produce less overdesign (less bandwidth) than others. Other factors that enter into the selection of which loop to design first are: the sizes of the templates and making sure that det **P** in Eq. (21.34) is minimum phase. The last loop designed has the least amount of overdesign, therefore the most constrained loop is done first by the first method. Then the design is continued through all the loops by the second method. For more details on both methods the reader is referred to the literature.[12, 8, 9, 10, 14, 17, 45]

Only the design of a MISO system is covered in the remainder of this chapter. The reader is referred to the literature listed previously for design examples of MIMO systems.

21.10 PLANT TEMPLATES OF $P(s)$, $\mathscr{T}\mathbf{P}(j\omega_i)$

The tracking control ratio for Fig. 21.5 with $D_1 = D_2 = 0$ is

$$T_R(s) = F\left[\frac{L}{1 + L}\right] \tag{21.52}$$

where $L = GP$. From Eq. (21.52)

$$\text{Lm}\,\mathbf{T}_R = \text{Lm}\,\mathbf{F} + \text{Lm}\left[\frac{\mathbf{L}}{1 + \mathbf{L}}\right] \tag{21.53}$$

The change in T_R due to the uncertainty in P is

$$\Delta(\text{Lm}\,\mathbf{T}_R) = \text{Lm}\,\mathbf{T}_R - \text{Lm}\,\mathbf{F} = \text{Lm}\left[\frac{\mathbf{L}}{1+\mathbf{L}}\right] \tag{21.54}$$

By the proper design of $L = L_o$ and F, this change in T_R is restricted so that the actual value of $\text{Lm}\,\mathbf{T}_R$ always lies between B_U and B_L of Fig. 21.8. The first step in synthesizing an L_o is to make templates which characterize the variation of the plant uncertainty for various values of ω_i over a frequency range $\omega_x \le \omega_i \le \omega_h$, where ω_h is the largest value obtained from Figs. 21.8 (ω_{h_R}) and 21.9 (ω_{h_D}). The guidelines for selecting the frequency range for the templates are to select three frequency values, no less than an octave apart, up to approximately the -12 dB value of the $\mathbf{B}_U$ plot in Fig. 21.8. In addition, for a Type 0 plant select $\omega_x = 0$ and for Type 1 or higher-order plants select $\omega_x \neq 0$.

The simple plant

$$P(s) = \frac{Ka}{s(s+a)} \tag{21.55}$$

where

$$K \in \{1, 10\} \qquad a \in \{1, 10\} \tag{21.56}$$

is used to illustrate how the templates are obtained for a plant with variable parameters. The region of plant uncertainty for the plant of Eq. (21.56) is depicted in Fig. 21.13. The boundary of the plant template can be obtained by mapping the boundary of the plant parameter uncertainty region. A number of points on the perimeter of $ABCD$ of Fig. 21.13 are selected, and values of $\text{Lm}\,\mathbf{P}(j\omega_i)$ and $\angle\mathbf{P}(j\omega_i)$ are obtained at each point. These data, for each frequency $\omega = \omega_i$, are plotted on a Nichols chart (NC). A curve is drawn through these points and becomes the template $\mathscr{T}\mathbf{P}(j\omega_i)$ at the frequency ω_i. A sufficient number of points must be selected so that the contour of $\mathscr{T}\mathbf{P}(j\omega_i)$ accurately reflects the region of plant uncertainty. In addition to the points A, B, C, and D marked in Fig. 21.13, it may also be necessary to include additional points on the perimeter. The data for the points A, B, C, and D at the frequency $\omega = 1$, from Eq. (21.55), is $\mathbf{P}_A(j1) = \sqrt{2}/2\angle{-135°}$, $\mathbf{P}_B(j1) = 5\sqrt{2}\angle{-135°}$, $\mathbf{P}_C(j1) = 100/\sqrt{101}\angle{-95.7°}$, and $\mathbf{P}_D(j1) = 10/\sqrt{101}\angle{-95.7°}$.

These data are plotted on the Nichols chart shown in Fig. 21.14. A curve is drawn through the points A, B, C, and D and the shaded area is labeled $\mathscr{T}\mathbf{P}(j1)$. The contour A, B, C, and D in Fig. 21.14 may be drawn on a plastic sheet (preferably colored) so that a plastic template for $\mathscr{T}\mathbf{P}(j1)$ can be cut and labeled. The templates for other values of ω_i are obtained in a similar manner.

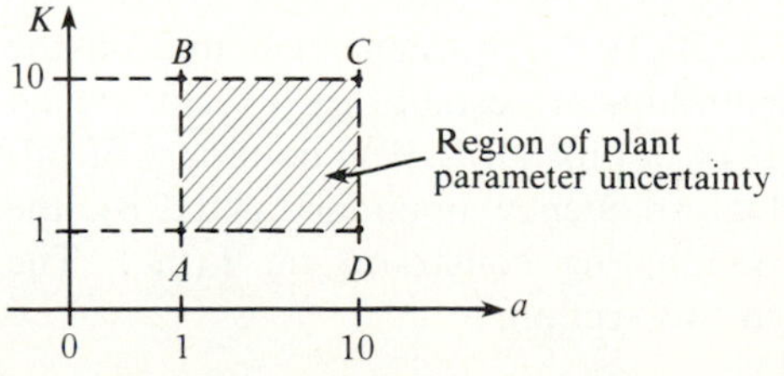

FIGURE 21.13
Region of plant uncertainty for Eqs. (21.55) and (21.56).

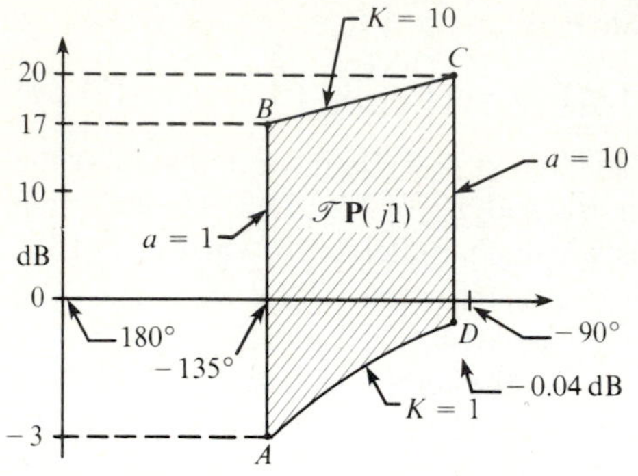

FIGURE 21.14
The template $\mathscr{T}\mathbf{P}(j1)$ on the Nichols chart characterizing Eq. (21.55) over the region of uncertainty.

For an aircraft, each point A, B, C, D, etc. in Fig. 21.14 represents a given flight condition (FC) at $\omega = \omega_i$, i.e., $[\mathbf{P}_{A/C}(j\omega_i)]_{\text{FC}_j}$. One of the flight conditions may be identified as the nominal plant P_o. For the plant of Eq. (21.55) the values $K = a = 1$ represent the lowest point of each of the templates $\mathscr{T}\mathbf{P}(j\omega_i)$ and may be selected as the nominal plant $\mathbf{P}_o$ for all frequencies. However, any plant in $\mathscr{P}$ can be chosen as the nominal plant.[13] With $\mathbf{L} = \mathbf{GP}$ and $\mathbf{L}_o = \mathbf{GP}_o$, as given in Eqs. (21.6) and (21.7):

$$\delta_P(j\omega_i) = \text{Lm}\,\mathbf{L} - \text{Lm}\,\mathbf{L}_o = (\text{Lm}\,\mathbf{G} + \text{Lm}\,\mathbf{P}) - (\text{Lm}\,\mathbf{G} + \text{Lm}\,\mathbf{P}_o)$$
$$= (\text{Lm}\,\mathbf{P} - \text{Lm}\,\mathbf{P}_o)\ \text{dB}$$

and $$\angle\Delta\mathbf{P} = \angle\mathbf{P} - \angle\mathbf{P}_o$$

Thus, if point A in Fig. 21.16 represents $\text{Lm}\,\mathbf{P}_o$, a variation in $\mathbf{P}$ results in a horizontal translation in the angle of $\mathbf{P}$ and a vertical translation in the log magnitude value of $\mathbf{P}$. When $\mathbf{G}(j\omega)$ represents a specific transfer function, the template of Fig. 21.14 can be converted into a template of $\mathbf{L}(j\omega_i)$ by translating it vertically by $\text{Lm}\,\mathbf{G}(j\omega_i)$ and horizontally by $\angle\mathbf{G}(j\omega_i)$. For the template of $\mathbf{L}(j\omega_i)$, the values of the M-contours at the intersections with the template are the values of the control ratio $\text{Lm}\,\mathbf{T}(j\omega_i) = \text{Lm}\,[\mathbf{L}(j\omega_i)/(1 + \mathbf{L}(j\omega_i)]$. The range of values of $\mathbf{T}(j\omega_i)$ for the entire range of parameter variation given in Eq. (2.56) can therefore be determined.

21.11 *U*-CONTOUR[2,14]

The specifications on system performance in the time domain (see Fig. 21.7) identify a minimum damping ratio ζ for the dominant roots of the closed-loop system. In the frequency domain (see Fig. 21.8) this becomes a bound on the value of M_m. On the Nichols chart this bound on M_m establishes a region which must not be penetrated by the template of $\mathbf{L}(j\omega)$ for all ω. The boundary of this region is referred to as the universal high-frequency boundary (UHFB), the U-contour, because this becomes the dominating constraint on $\mathbf{L}(j\omega)$. The formation of the U-contour is described in this section.

For the two cases of disturbance rejection depicted in Fig. 21.5 the control ratios are, respectively, as given in Eqs. (21.13) and (21.14):

$$T_{D_2} = \frac{1}{1+L} \qquad \text{and} \qquad T_{D_1} = \frac{P}{1+L}$$

Thus, it is necessary to synthesize an $L_o(s)$ so that the disturbances are properly attenuated. For the present, only one aspect of this disturbance-response problem is considered, namely a constraint is placed on the damping ratio ζ of the dominant complex-pole pair of T_D nearest the $j\omega$-axis.[6] This damping ratio is related to the peak value of

$$|\mathbf{T}(j\omega)| = \left| \frac{\mathbf{L}(j\omega)}{1 + \mathbf{L}(j\omega)} \right| \tag{21.57}$$

For example, consider that the dominant complex-pole pair of Eq. (21.57) results in a peak of Lm $\mathbf{T}$ = 8 dB for $\zeta = 0.2$, and 2.7 dB for $\zeta = 0.4$, etc. Although this large peak does not appear in T_R due to the design of the filter $F(s)$, it does affect the response for T_D. If $d(t)$ is very small, a peak of T_D, due to $\zeta = 0.2$, can be "very large" and be such that the restriction on the peak overshoot α_p of the time response [see Eqs. (12.35) and (12.36)]

$$|c(t)| \le \alpha_p$$

may be difficult to achieve. Therefore, it is reasonable to add the requirement

$$|\mathbf{T}| = \left| \frac{\mathbf{L}}{1 + \mathbf{L}} \right| \le M_L \tag{21.58}$$

where M_L is a constant for all ω and over the whole range of P parameter values. This results in a constraint on ζ of the dominant complex-pole pair of T_D. This constraint can therefore be translated into a constraint on the maximum value T_m of Eq. (21.57). For example, for Lm M_m = 2 dB, the oval, *agefa*, in Fig. 21.15 corresponds to the 2-dB M-contour on the NC. This results in limiting the peak of the disturbance response. A value of M_L can be selected to correspond to the maximum value of T_R. Therefore, the top portion, *efa*, of the M-contour on the Nichols chart, which corresponds to the value of the selected value of M_L, becomes part of the U-contour.

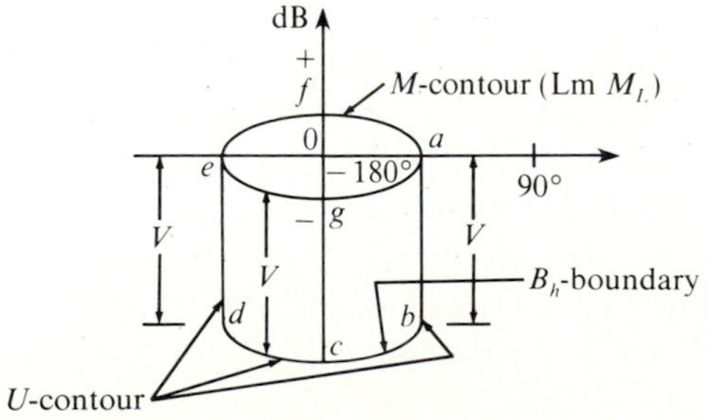

FIGURE 21.15
U-contour construction.

For a large problem class, as $\omega \to \infty$, the limiting value of the plant transfer function approaches

$$\lim_{\omega \to \infty} [P(j\omega)] = \frac{K}{\omega^{\lambda}}$$

where λ represents the excess of poles over zeros of $P(s)$. The plant template, for this problem class, approaches a vertical line of length equal to

$$\Delta \operatorname{Lm} \mathbf{P} \equiv \lim_{\omega \to \infty} [\operatorname{Lm} \mathbf{P}_{\max} - \operatorname{Lm} \mathbf{P}_{\min}] = \operatorname{Lm} K_{\max} - \operatorname{Lm} K_{\min} = V \text{ dB} \quad (21.59)$$

If the nominal plant is chosen at $K = K_{\min}$, then the constraint M_L gives a boundary which approaches the U-contour *abcdefa* of Fig. 20.15. (*Note*: For a MIMO plant, as $\omega \to \infty$, the templates may not approach a vertical line if the λ_{ij} are not the same for all p_{ij} elements of the plant matrix **P**. When the λ_{ij} are different, then the widths of the templates are a multiple of 90°.)

For a plant of Eqs. (21.55) and (21.56), applying the limiting condition, $\omega \to \infty$, to Eq. (21.59) yields

$$\begin{aligned} V &= \Delta \operatorname{Lm} \mathbf{P} \\ &= \lim_{\omega \to \infty} \left[\left\{ \operatorname{Lm} (Ka)_{\max} - \operatorname{Lm} (j\omega)^2 \right\} - \left\{ \operatorname{Lm} (Ka)_{\min} - \operatorname{Lm} (j\omega)^2 \right\} \right] \\ &= \operatorname{Lm} (Ka)_{\max} - \operatorname{Lm} (Ka)_{\min} = \operatorname{Lm} 100 - \operatorname{Lm} 1 = 40 \text{ dB} \end{aligned} \quad (21.60)$$

For the minimum-phase plant of Eq. (21.55), where the poles are real, the plant templates have the typical shape of Fig. 21.14.

The high-frequency boundary B_h, the *bcd* portion of the U-contour in Fig. 21.15, is obtained by measuring down V dB from the *ega* portion of the M-contour as illustrated in this figure. V is determined by Eq. (21.60) which, for this example, is 40 dB. The remaining portions of the U-contour, portions *ab* and *de*, not necessarily straight lines, are determined by satisfying the requirement of Eq. (21.58) and $\delta_R(j\omega_1)$. The $\mathscr{T}\mathbf{P}(j\omega_1)$ is used to determine the corresponding tracking bounds $\mathbf{B}_R(j\omega_i)$ on the NC in the manner described in Sec. 21.12.

21.12 TRACKING BOUNDS $\operatorname{Lm} \mathbf{B}_R(j\omega)$ ON THE NC

As an introduction to this section, the procedure for adjusting the gain of a unity-feedback system to achieve a desired value of M_m by use of the Nichols chart is reviewed. Consider the plot of $\operatorname{Lm} \mathbf{P}(j\omega)$ vs. $\angle \mathbf{P}(j\omega)$ for a plant shown in Fig. 21.16 (the solid curve). With $G(s) = A = 1$ and $F(s) = 1$ in Fig. 21.5, $\mathbf{L} = \mathbf{P}$. The plot of $\operatorname{Lm} \mathbf{L}(j\omega)$ vs. $\angle \mathbf{L}(j\omega)$ is tangent to the $M = 1$ dB curve with a resonant frequency $\omega_{m_1} = 1.1$. When Lm $M_m = 2$ dB is specified for $\operatorname{Lm} \mathbf{T}_R$, the gain A is increased, raising $\operatorname{Lm} \mathbf{L}(j\omega)$, until it is tangent to the 2-dB M-curve. For this example the curve is raised by Lm $A = 4.5$ dB ($G = A = 1.679$) and the resonant frequency is $\omega_m = 2.09$.

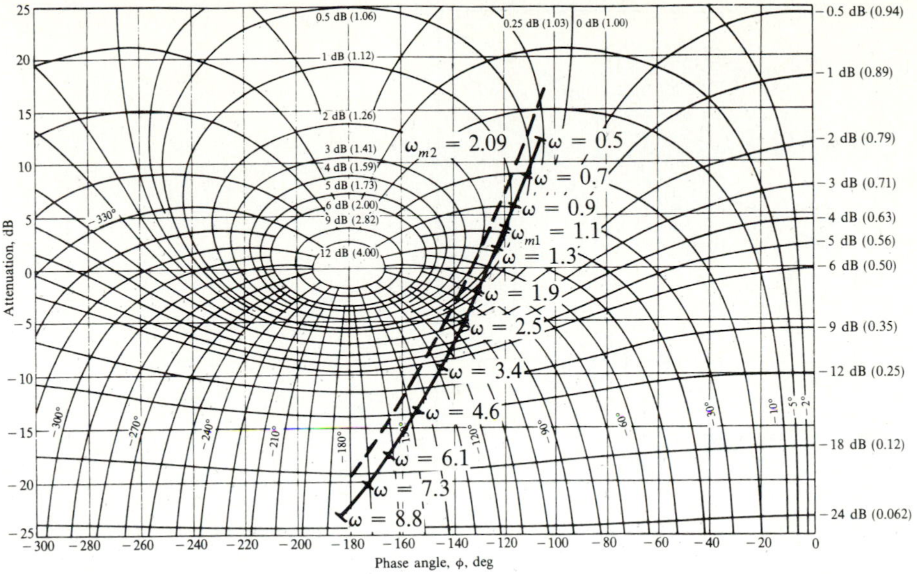

FIGURE 21.16
Log magnitude–angle diagram.

Now consider that the plant uncertainty involves only the variation in gain A between the values of 1 and 1.679. It is desired to find a cascade compensator $\mathbf{G}(j\omega)$, in Fig. 21.5, such that the specification 1 dB $\leq$ Lm $M_m \leq$ 2 dB is always maintained for this plant gain variation while the resonant frequency ω_m remains constant. This requires that the loop transmission $\mathbf{L}(j\omega) = \mathbf{G}(j\omega)\mathbf{P}(j\omega)$ must be tangent to an M-contour in the range of $1 \leq$ Lm $M \leq 2$ dB for the entire range of $1 \leq A \leq 1.679$, at the same resonant frequency $\omega_m = 2.09$. The manner of achieving this and other time-response specifications is the subject of the remaining portion of this chapter.

It is assumed for Eq. (21.54) that the compensators F and G are fixed, that is, they have negligible uncertainty. Thus, only the uncertainty in P contributes to the change in T_R given by Eq. (21.54). The solution requires that the actual Δ Lm $T_R(j\omega_i) \leq \delta_R(j\omega_i)$ dB in Fig. 21.8. Thus it is necessary to determine the resulting constraint, or bound $\mathbf{B}_R(j\omega_i)$, on $\mathbf{L}(j\omega_i)$. The procedure is to pick a nominal plant $P_o(s)$ and to derive the bounds on the resulting nominal loop transfer function $L_o(s) = G(s)P_o(s)$.

As an illustration, consider the plot of Lm $\mathbf{P}(j2)$ vs. $\angle \mathbf{P}(j2)$ for the plant of Eqs. (21.55) and (21.56). As shown in Fig. 21.17, the plant's region of uncertainty $\mathscr{T}\mathbf{P}(j2)$ is given by the contour $ABCD$, i.e., Lm $\mathbf{P}(j2)$ lies on or within the boundary of this contour. The nominal plant transfer function, with $K_o = 1$ and $a_o = 1$, is

$$P_o = \frac{1}{s(s+1)} \tag{21.61}$$

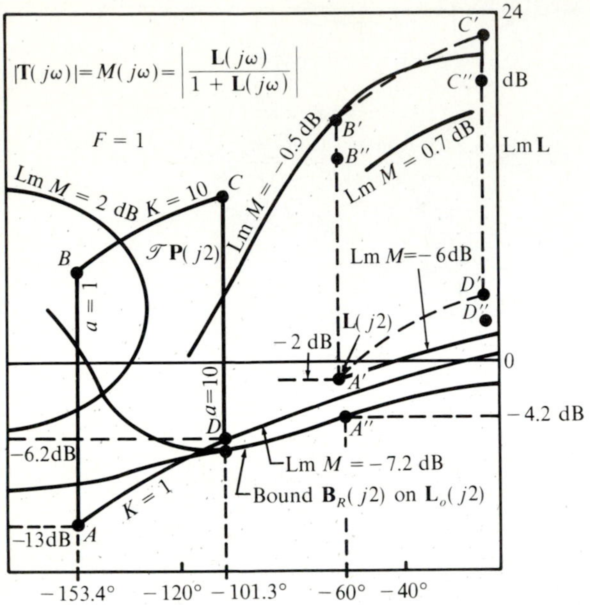

FIGURE 21.17
Derivation of bounds $\mathbf{B}_R(j\omega)$ on $\mathbf{L}_o(j\omega)$ for $\omega = 2$.

and is represented in Fig. 21.17 by point A for $\omega = 2$ [-13.0 dB, $-153.4°$]. Note, once a nominal plant is chosen, it must be used for determining all the bounds $\mathbf{B}_R(j\omega_i)$. Since $\text{Lm}\,\mathbf{L}(j2) = \text{Lm}\,\mathbf{G}(j2) + \text{Lm}\,\mathbf{P}(j2)$, then $\mathcal{T}\mathbf{P}(j2)$ is translated on the Nichols chart vertically by the value of $\text{Lm}\,\mathbf{G}(j2)$ and horizontally by the angle $\angle \mathbf{G}(j2)$. The templates $\mathcal{T}\mathbf{P}(j\omega_1)$ are relocated to find the position of $\mathbf{L}_o(j\omega)$ which satisfies the specifications in Fig. 21.8 of $\delta_R(j\omega_1)$. For example, if a trial design of $\mathbf{L}(j2)$ requires sliding $\mathcal{T}\mathbf{P}(j2)$ to the position $A'B'C'D'$ in Fig. 21.17, then

$$|\text{Lm}\,\mathbf{G}(j2)| = \big||\text{Lm}\,\mathbf{L}(j2)|_{A'} - |\text{Lm}\,\mathbf{P}(j2)|_{A}\big| = ||-2| - |-13|| = 11 \text{ dB} \tag{21.62}$$

$$\angle \mathbf{G}(j2) = \angle \mathbf{L}(j2)_{A'} - \angle \mathbf{P}(j2)_{A} = -60° - (-153.4°) = 93.4° \tag{21.63}$$

Using the contours of constant $\text{Lm}\,M = \text{Lm}[\mathbf{L}/(1+\mathbf{L})]$ on the NC in Fig. 21.17, the maximum occurs at point C' ($M = -0.49$ dB) and the minimum at point A' ($M = -5.7$ dB) so that the maximum change in $\text{Lm}\,\mathbf{T}$ is, in this case, $(-0.49) - (-5.7) = 5.2$ dB. If the specifications tolerate a change of 6.5 dB at $\omega = 2$, the above trial position of $\text{Lm}\,\mathbf{L}_o(j2)$ is well within the permissible tolerance. Lowering the template on the NC to $A''B''C''D''$, where the extreme values of $\text{Lm}[\mathbf{L}/(1+\mathbf{L})]$ are at C'' (-0.7 dB) and A'' (-7.2 dB), yields $\text{Lm}\,\mathbf{L}(j2)_{C''} - \text{Lm}\,\mathbf{L}(j2)_{A''} = -0.7 - (-7.2) = 6.5\text{ dB} = \delta_R(j2)$. Thus, if $\angle \mathbf{L}_o(j2) = -60°$, then -4.2 dB is the smallest or minimum value of $\text{Lm}\,\mathbf{L}_o(j2)$ which satisfies the 6.5 dB specifications for $\delta_R(j\omega_i)$. Any smaller magnitude is satisfactory but represents overdesign at that frequency. The manipulation of the

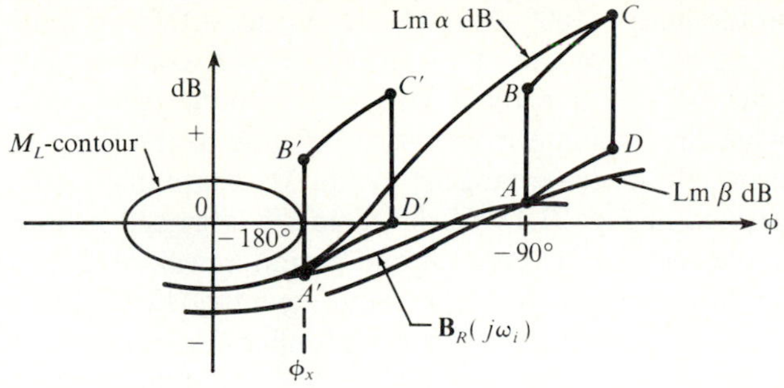

FIGURE 21.18
Graphical determination of $\mathbf{B}_R(j\omega_i)$.

$\omega = 2$ template, for ease of the design process, is repeated along a new angle (vertical) line, and a corresponding new minimum of $\mathbf{L}_o(j2)$ is found. Sufficient points are obtained in this manner to permit drawing a continuous curve of the bound $\mathbf{B}_R(j2)$ on $\mathbf{L}_o(j2)$, as shown in Fig. 21.17. The above procedure is repeated at other frequencies, resulting in a family of boundaries $\mathbf{B}_R(j\omega_i)$ of permissible $\mathbf{L}_o(j\omega)$. The procedure for determining the boundaries $\mathbf{B}_R(j\omega_i)$ is summarized as follows:

1. From Fig. 21.8 obtain values of $\delta_R(j\omega_i)$ for a range of values of $\omega_i(\omega_1, \omega_2, \ldots, \omega_i, \ldots, \omega_h)$, preferably an octave apart, over the specified bandwidth. [The selection of ω_h results in the bound $\mathbf{B}_R(j\omega_h)$ passing under the *U*-contour.]

2. Place $\mathscr{T}\mathbf{P}(j\omega_i)$ on the Nichols chart containing the *U*-contour to determine the bound $\mathbf{B}_R(j\omega_i)$ as follows:

(*a*) Use major angle divisions of the Nichols chart for lining up the $\mathscr{T}\mathbf{P}(j\omega_i)$.

(*b*) Select P_o to represent, in general, the lowest point of $\mathscr{T}\mathbf{P}(j\omega_i)$. For the design example shown in this chapter:

(1) Select point *A* in Fig. 21.18 as P_o.

(2) Line up side *A*-*B* of $\mathscr{T}\mathbf{P}(j\omega_i)$ on the $-90°$ line, as shown in Fig. 21.18. Move the template up or down until the difference $\Delta\text{Lm}\, T_R(j\omega_i)$ between the values of two adjacent *M*-contours (Lm α and Lm β, respectively) is equal to $\delta_R(j\omega_i)$ of Fig. 21.8. Thus, in Fig. 21.18, determine the locations of the template on the $-90°$ line where

$$\Delta\,\text{Lm}\,\mathbf{T}_R(j\omega_i) = \text{Lm}\,\alpha - \text{Lm}\,\beta = \delta_R(j\omega_i) \tag{21.64}$$

When Eq. (21.64) is satisfied, then point *A*, on the *M*-contour = Lm β, lies on the bound $\mathbf{B}_R(j\omega_i)$. Mark this point on the Nichols chart. *For other plants the shape of the template may be such that if point A represents* $\mathbf{P}_o$, *another point of the template may be the lowest point that satisfies Eq.* (21.64). *When this equation is satisfied, point A still yields points* $\mathbf{B}_R(j\omega_i)$.

(3) Repeat step 2 on the lines $-100°$, $-110°$, etc., up to $-180°$ or until a point of the template becomes tangent to the M_L-contour. No intersection of the M_L-contour by a template is permissible. For example, in moving the template from the $-90°$ line to the left, the template may eventually become tangent to the M_L-contour at some angle ϕ_x as illustrated in Fig. 21.18. If the template is moved further to the left it will intersect the M_L contour and permit a peak of $T(j\omega)$ greater than M_L. In order to satisfy the requirement of Eq. (21.58), point A' on the $\phi = \phi_x$ line becomes the left boundary or terminating point for the $\mathbf{B}_R(j\omega_i)$ contour and a point on the U-contour (a point on ab of Fig. 21.15).[2] Draw a curve through all the points to obtain the contour for $\mathbf{B}_R(j\omega_i)$. For the plant of this example the U-contour is symmetrical about the $-180°$ axis. Note that obtaining the bounds $\mathbf{B}_R(j\omega_i)$ as described in step 2(b)(2) only guarantees that the difference between the upper bound $\text{Lm}\,\mathbf{T}_U$ and the lower bound $\text{Lm}\,\mathbf{T}_L$ for $\text{Lm}\,\mathbf{T} = \text{Lm}[\mathbf{L}/(1 + \mathbf{L})]$ will be

$$\begin{aligned}\delta_R(j\omega_i) &= \text{Lm}\,\mathbf{T}_{R_U}(j\omega_i) - \text{Lm}\,\mathbf{T}_{R_L}(j\omega_i)\\ &= \text{Lm}\,\mathbf{T}_U(j\omega_i) - \text{Lm}\,\mathbf{T}_L(j\omega_i) \qquad (21.65)\end{aligned}$$

(4) Repeat steps 2 and 3, generally octaves apart, over the range of $\omega_x \le \omega_i \le \omega_h$, until the highest bound $\mathbf{B}_R(j\omega_i)$ and lowest bound $\mathbf{B}_R(j\omega_h)$ on the Nichols chart clear the U-contour. For reasonably damped plants ($\zeta > 0.6$), over the entire region of plant uncertainty, the magnitudes of the bounds $\mathbf{B}_R(j\omega_i)$ usually decrease as ω increases. Thus, for this type of plant, it is desirable to have $\delta_R(j\omega_i)$ increasing as ω_i increases, as discussed in Sec. 21.4. When this characteristic of $\delta_R(j\omega_i)$ is not observed, it is possible to have $|\mathbf{B}_R(j\omega_j)| > |\mathbf{B}_R(j\omega_i)|$ for $\omega_j > \omega_i$. For a plant that is highly underdamped ($\zeta \le 0.6$), over some portion of the region of plant uncertainty, avoid using an underdamped nominal plant $P_o(s)$. For this latter situation it is desirable, but not necessary, to synthesize T_{R_U} and T_{R_L} based upon an overdamped $P_o(s)$ model.

21.13 DISTURBANCE BOUNDS $\mathbf{B}_D(j\omega_i)$: CASE 1 [$d_2(t) = D_o u_{-1}(t)$, $d_1(t) = 0$]

Two disturbance inputs are shown in Fig. 21.5. It is assumed that only one disturbance input exists at a time. Both cases are analyzed.

CONTROL RATIO. From Fig. 21.5 the disturbance control ratio for input $d_2(t)$ is

$$T_D(s) = \frac{1}{1 + L} \qquad (21.66)$$

Substituting $L = 1/\ell$ into Eq. (21.66) yields

$$T_D(s) = \frac{\ell}{1 + \ell} \qquad (21.67)$$

which has the mathematical format required to use the NC. Over the specified bandwidth it is desired that $|\mathbf{T}_D(j\omega)| \ll 1$ which results in the requirement, from Eq. (20.67), that $|\mathbf{L}(j\omega)| \gg 1$ (or $|\ell(j\omega)| \ll 1$), i.e.,

$$|\mathbf{T}_D(j\omega)| \approx \frac{1}{|\mathbf{L}(j\omega)|} = |\ell(j\omega)| \tag{21.68}$$

DISTURBANCE RESPONSE CHARACTERISTIC. A time-domain tracking response characteristic based upon $r(t) = u_{-1}(t)$ often specifies a maximum allowable peak overshoot M_p. In the frequency domain this specification may be approximated by

$$|\mathbf{M}_R(j\omega)| = |\mathbf{T}_R(j\omega)| = \left|\frac{\mathbf{C}(j\omega)}{\mathbf{R}(j\omega)}\right| \le M_m \approx M_P \tag{21.69}$$

The corresponding time- and frequency-domain response characteristics, based upon a step disturbance forcing function are, respectively,

$$|M_D(t)| = \left|\frac{c(t)}{d(t)}\right| \le \alpha_p \qquad \text{for } t \ge t_x \tag{21.70}$$

and

$$|\mathbf{M}_D(j\omega)| = |\mathbf{T}_D(j\omega)| = \left|\frac{\mathbf{C}(j\omega)}{\mathbf{D}(j\omega)}\right| \le \alpha_m \approx \alpha_p \tag{21.71}$$

APPLICATION. Let $L = KL'$ in the tracking ratio $T = L/(1 + L)$, where K is an unspecified gain, and the specification on system performance is $M_m = 1.12$ (1 dB). By means of a NC determine the value of K required and obtain the data to plot $|\mathbf{M}(j\omega)|$ vs. ω. The plot of Lm $\mathbf{L}'$ vs. $\angle \mathbf{L}'$ on the NC is tangent to the $M = 1$ dB contour, resulting in the plot of Lm $\mathbf{L}(j\omega)$ vs. $\angle \mathbf{L}(j\omega)$ in Fig. 21.19. Intersections of Lm $\mathbf{L}(j\omega)$ with the M-contours provide the data to plot the tracking control ratio $|\mathbf{M}(j\omega)|$ vs. ω.

Now consider the corresponding disturbance control ratio for the same control system. The disturbance transfer function is $T_D = 1/(1 + L) = \ell/(1 + \ell)$, having the desired BW $0 \le \omega \le \omega_2$ for which $|\mathbf{L}| \gg 1$ and $|\ell| \ll 1$. Thus Eq. (21.69) applies within the BW region. Table 21.1 contains data for two points on the Nichols plot of Fig. 21.19. The plot of Lm ℓ vs. $\angle \ell$ for these two points is also shown in this figure. The NC of Fig. 21.19 can be rotated 180° and is redrawn in Fig. 21.20. Since Lm $\ell(j\omega)$ = Lm $(1/\mathbf{L}(j\omega))$ = $-$Lm $\mathbf{L}(j\omega)$, then a negative value of Lm ℓ yields a positive value for Lm $\mathbf{L}$ as shown in Fig. 21.20. Since $L = KL' = 1/\ell$, then

$$\ell(j\omega) = K^{-1}\ell'(j\omega) \tag{21.72}$$

If $\ell'(j\omega)$ is given and it is required to determine K^{-1} to satisfy Eq. (21.66), then the plot Lm $\ell'(j\omega)$ vs. $\angle \ell'(j\omega)$ must be raised or lowered until it is tangent to the Lm α_p-contour ($|\mathbf{T}_D|_{\max} = \alpha_m$). The amount Δ by which this plot is raised or lowered yields the value of K, i.e., Lm $K^{-1} = \Delta$. Note that this is the same procedure used for the tracking example of Fig. 21.19, except that the adjustment in Lm $\ell(j\omega)$ is K^{-1}.

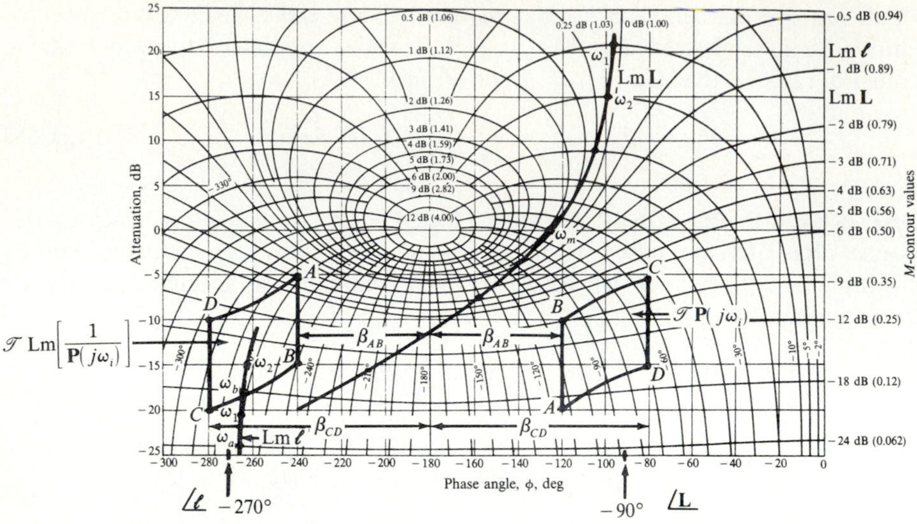

FIGURE 21.19
Regular Nichols chart.

TABLE 21.1
Data points for a Nichols plot

ω	Lm **L**	$\angle$**L**	Lm ℓ	$\angle \ell$
ω_1	21	$-96°$	-21	$96°$ (or $-264°$)
ω_2	15	$-98°$	-15	$98°$ (or $-262°$)

TEMPLATES. For a given plant **P** having uncertain parameters, consider that its template $\mathscr{T}\mathbf{P}(j\omega_i)$ for a given ω_i has equal dB differences along its *A-B* and *C-D* boundaries, i.e., for a given $\angle \mathbf{P}(j\omega_i)$,

$$\Delta(\text{Lm}\,\mathbf{P}_B - \text{Lm}\,\mathbf{P}_A) = \Delta(\text{Lm}\,\mathbf{P}_C - \text{Lm}\,\mathbf{P}_D) = 10 \text{ dB}$$

This template is arbitrarily set on the NC as shown in Fig. 21.19. Data corresponding to the template location shown in Fig. 21.19 are given in Table 21.2.

The template of the reciprocal, Lm $[1/\mathbf{P}(j\omega_i)]$, is arbitrarily set on the NC as shown in Fig. 21.19 for the same frequency as for the template of Lm $\mathbf{P}(j\omega_i)$ and for the angles of Table 21.2. Note that the template of Lm $[1/\mathbf{P}(j\omega_i)]$ is the same as the template of Lm $\mathbf{P}(j\omega_i)$ but is rotated by 180°. Thus it is located by first reflecting the template of Lm $\mathbf{P}(j\omega_i)$ about the $-180°$ axis, "flipping it over" vertically, and then moving it up or down so that it lies between -5 and

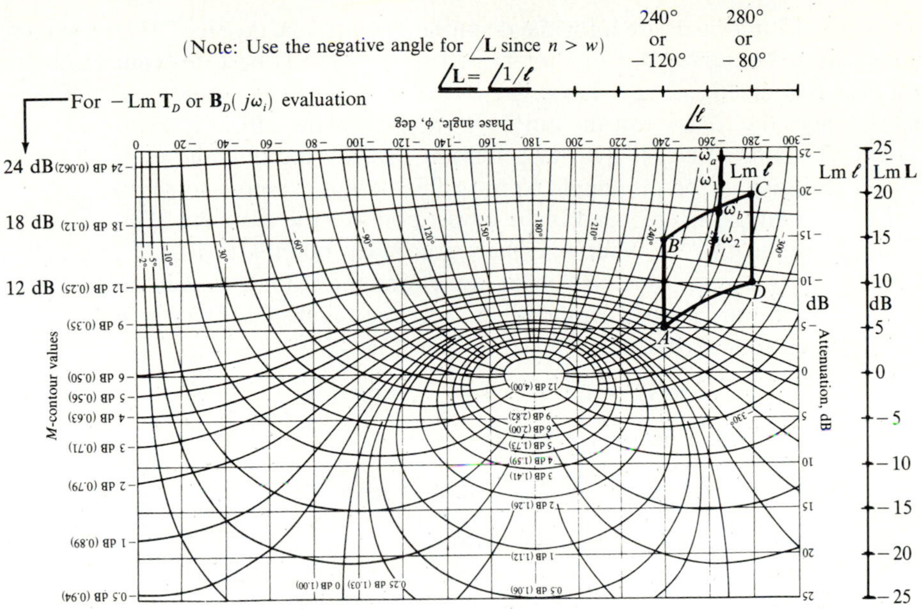

FIGURE 21.20
Rotated Nichols chart.

TABLE 21.2
Data points for the templates of Fig. 21.19

Points	$\angle \mathbf{P}$	$\angle 1/\mathbf{P}$
A, B	$-120°$	$120°$ (or $-240°$)
C, D	$-80°$	$80°$ (or $-280°$)

-20 dB. For the *arbitrary* location of the template of $\text{Lm}[1/\mathbf{P}(j\omega_i)]$, note that:

(*a*)
$$\beta_{AB} = 180° + \angle \mathbf{P} = 180° - 120° = 60°$$
$$\beta_{CD} = 180° + \angle \mathbf{P} = 180° - 80° = 100°$$

For $1/\mathbf{P}$, the corresponding angles are

$$\angle 1/\mathbf{P}_{AB} = -180° - \beta_{AB} = -180° - 60° = -240°$$
$$\angle 1/\mathbf{P}_{CD} = -180° - \beta_{CD} = -180° - 100° = -280°$$

(*b*) The templates of $\text{Lm}\,\mathbf{P}(j\omega_i)$ are used for the tracker case $T = L/(1 + L)$ and the templates of $\text{Lm}[1/\mathbf{P}(j\omega_i)]$ are used for the disturbance case of $T_D = 1/(1 + L) = \ell/(1 + \ell)$.

Since Lm **L** is desired, for the disturbance-rejection problem, Eq. (21.6), the disturbance boundary $\mathbf{B}_D(j\omega_i)$ for $\mathbf{L}(j\omega_i) = 1/\ell(j\omega_i)$ is best determined on the rotated NC of Fig. 21.20. Thus, the NC of Fig. 21.19 is rotated clockwise by 180°, where the rotation of the $\text{Lm}[1/\mathbf{P}(j\omega_i)]$ template $ABCD$ is reflected in Fig. 21.20. The rotated NC is used to determine directly the boundaries $\mathbf{B}_D(j\omega_i)$ for $\mathbf{L}_D(j\omega_i)$. Point A for the simple plant of this design example corresponds to the nominal plant parameters and is the lowest point of the template $\mathscr{T}\mathbf{P}(j\omega_i)$. This point is *again used* to determine the disturbance bounds $\mathbf{B}_D(j\omega_i)$. The lowest point of the template must be used to determine the bounds and, in general, may or may not be the point corresponding to the nominal plant parameters.

Based upon Eqs. (21.66) and (21.67) in Fig. 21.9

$$-\text{Lm}\,\mathbf{T}_D = \text{Lm}[1 + \mathbf{L}] \geq -\text{Lm}\,\alpha(j\omega_i) > 0 \text{ dB} \tag{21.73}$$

where $\alpha(j\omega_i) < 0$. Since in the BW $|\mathbf{L}| \gg 1$, then

$$-\text{Lm}\,\mathbf{T}_D \cong \text{Lm}\,\mathbf{L} \geq -\text{Lm}\,\alpha(j\omega_i) = -\delta_D(j\omega_i) \tag{21.74}$$

In terms of $\mathbf{L}(j\omega_i)$, the constant M-contours of the NC can be used to obtain the disturbance performance T_D. This requires the change of sign of the vertical axis in dB and the M-contours, as shown in Fig. 21.20.

BOUNDS $\mathbf{B}_D(j\omega_i)$. The procedure for determining the boundaries $\mathbf{B}_D(j\omega_i)$ is:

1. From Fig. 21.9 obtain values of $\delta_D(j\omega_i)$ from the curve representing Eqs. (12.37) and (12.38) for the same values of frequency as for the tracking boundary $\mathbf{B}_R(j\omega_i)$.
2. Select the lowest point of $\mathscr{T}\mathbf{P}(j\omega_i)$ to represent the nominal plant P_o in Fig. 21.20. For the design example used in this chapter select point A in Fig. 21.20 as $\mathbf{P}_o$. The same nominal point must be used in obtaining the tracking and disturbance bounds.
3. Use major angle divisions of the NC for lining up the $\mathscr{T}\mathbf{P}(j\omega_i)$. Line up side A-B of $\mathscr{T}\mathbf{P}(j\omega_i)$, for example, on the $-280°$ line for ℓ (or the $-80°$ line for **L**). Move the template up or down until point A lies on the M-contour that represents $\delta_D(j\omega_i)$. Mark this point on the NC.
4. Repeat step 3 on the vertical lines for $-100°$, $-120°$, etc., up to the $-180°$ line or the U-contour. Draw a curve through all the points to obtain the contour for $\mathbf{B}_D(j\omega_i)$.
5. Repeat steps 3 and 4 over the desired frequency range $\omega_x \leq \omega_i \leq \omega_h$.
6. Transcribe these $\mathbf{B}_D(j\omega_i)$ onto the NC that contains the bounds $\mathbf{B}_R(j\omega_i)$.

Note that when $|\mathbf{L}| \gg 1$, then, from Eq. (21.73), $|\mathbf{L}(j\omega_i)| \gg \alpha(j\omega_i)$. Thus the M-contour corresponding to $\mathbf{T}_D(j\omega_i)$ becomes the boundary $\mathbf{B}_D(j\omega_i) = -\text{Lm}\,M$ for $\mathbf{L}(j\omega_i)$. For example, if $\alpha(j\omega_i) = 0.12$, then $|\mathbf{L}(j\omega_i)| = 8.33$ and thus $\text{Lm}\,\alpha(j\omega_i) = -18.4$ dB and $\text{Lm}\,\mathbf{L}(j\omega_i) = 18.4$ dB.

21.14 DISTURBANCE BOUNDS $\mathbf{B}_D(j\omega_i)$: CASE 2 $[d_1(t) = D_o u_{-1}(t),\ d_2(t) = 0]$

CONTROL RATIO. From Fig. 21.5 the disturbance control ratio for the input $d_1(t)$ is

$$\mathbf{T}_D = \frac{\mathbf{P}}{1 + \mathbf{GP}} \tag{21.75}$$

Assuming point A of the template represents the nominal plant P_o, Eq. (21.76) is rearranged to

$$\mathbf{T}_D = \frac{\mathbf{P}_o}{\mathbf{P}_o}\left[\frac{1}{\dfrac{1}{\mathbf{P}} + \mathbf{G}}\right] = \frac{\mathbf{P}_o}{\dfrac{\mathbf{P}_o}{\mathbf{P}} + \mathbf{GP}_o} = \frac{\mathbf{P}_o}{\dfrac{\mathbf{P}_o}{\mathbf{P}} + \mathbf{L}} = \frac{\mathbf{P}_o}{\mathbf{W}} \tag{21.76}$$

where $\qquad \mathbf{W} = (\mathbf{P}_o/\mathbf{P}) + \mathbf{L} \qquad$ (21.77)

Thus, Eq. (21.76) with $\text{Lm}\,\mathbf{T}_D = \delta_D$ yields

$$\text{Lm}\,\mathbf{W} = \text{Lm}\,\mathbf{P}_o - \delta_D \tag{21.78}$$

DISTURBANCE RESPONSE CHARACTERISTICS. Based on Eq. (21.69), the time- and frequency-domain response characteristics, for a unit-step disturbance forcing function, are given, respectively, by

$$|M_D(t)| = \left|\frac{c(t_p)}{d(t)}\right| = |c(t_p)| \le \alpha_p \tag{21.79}$$

and

$$|\mathbf{M}_D(j\omega)| = |\mathbf{T}_D(j\omega)| = \left|\frac{\mathbf{C}(j\omega)}{\mathbf{D}(j\omega)}\right| \le \alpha_m \equiv \alpha_p \tag{21.80}$$

where t_p is the peak time.

BOUNDS $\mathbf{B}_D(j\omega_i)$. The procedure for determining the boundaries $\mathbf{B}_D(j\omega_i)$ is:

1. From Fig. 21.9 obtain values of $\delta_D(j\omega_i)$ from the plot representing Eqs. (21.40) and (21.41) for the same values of frequency as for the tracker boundaries $\mathbf{B}_R(j\omega_i)$.
2. *Evaluate* in tabular form for each value of ω_i the following items in the order given:

$$\text{Lm}\,\mathbf{P}_o(j\omega_i) \qquad \delta_D(j\omega_i) \qquad \text{Lm}\,\mathbf{W}(j\omega_i) \qquad |\mathbf{W}(j\omega_i)| \qquad \frac{\mathbf{P}_o(j\omega_i)}{\mathbf{P}(j\omega_i)}$$

   The ratio $\mathbf{P}_o/\mathbf{P}$ is evaluated at each of the four points of Fig. 21.13 for each value of ω_i. It may be necessary to evaluate this ratio at additional points around the perimeter of the $ABCD$ contour.
3. Before presenting the procedure for the graphical determination of $\mathbf{B}_D(j\omega_i)$ it is necessary to first review, graphically, the phasor relationship between $\mathbf{B}_D$,

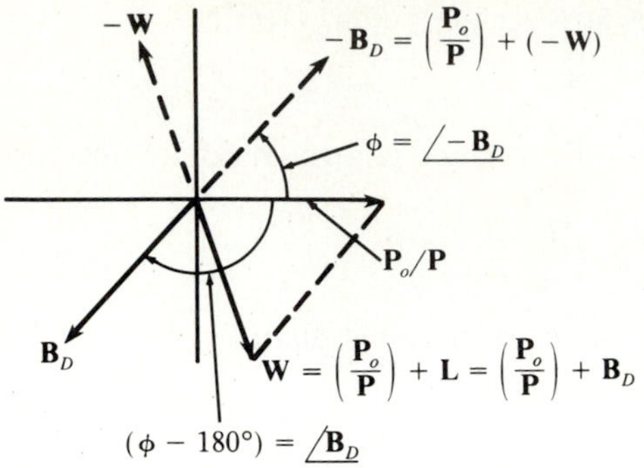

FIGURE 21.21
Phasor relationship of Eq. (21.77) with $\mathbf{L} = \mathbf{B}_D$.

$\mathbf{P}_o/\mathbf{P}$, and $\mathbf{W}$ for $\omega = \omega_i$. Equation (21.77), with $\mathbf{L}$ replaced by its bound $\mathbf{B}_D$, is rearranged to the form

$$-\mathbf{B}_D = \frac{\mathbf{P}_o}{\mathbf{P}} + (-\mathbf{W}) \tag{21.81}$$

For arbitrary values of $\mathbf{P}_o(j\omega_i)/\mathbf{P}(j\omega_i)$ and $\mathbf{W}(j\omega_i)$, Fig. 21.21 presents the phasor relationship of Eq. (21.81). Since the values of $\mathbf{P}_o(j\omega_i)/\mathbf{P}(j\omega_i)$ for $\mathbf{P} \in \mathscr{P}$ and $|\mathbf{W}(j\omega_i)|$ are known, the following procedure can be used to evaluate $\mathbf{B}_D(j\omega_i)$:

(*a*) On polar or rectangular graph paper draw $\mathscr{T}[\mathbf{P}_o/\mathbf{P}]$ for each ω_i as shown in Fig. 21.22, where point A is the nominal point.

(*b*) Based upon Fig. 21.21 and the location of the phasor $-\mathbf{W}(j\omega)$, the solution for $-\mathbf{B}_D$ is obtained from

$$-\mathbf{B}_D = \mathscr{T}[\mathbf{P}_o/\mathbf{P}] - \mathbf{W} \tag{21.82}$$

For one value of $\mathbf{P}_o/\mathbf{P}$ shown in Fig. 21.21, the value of $-\mathbf{W}$ is plotted and $-\mathbf{B}_D = |-\mathbf{B}_D|\angle\phi$ is obtained. This graphical evaluation of $\mathbf{B}_D = |\mathbf{B}_D|\angle\phi - 180°$ is performed for various points around the perimeter of $\mathscr{T}[\mathbf{P}_o/\mathbf{P}]$ in Fig. 21.22. A simple graphical evaluation yields a more restrictive bound (the worst case). By use of a compass mark off arcs with a radius equal to the distance $|\mathbf{W}(j\omega_i)|$ at a number of points on the perimeter of $\mathscr{T}[(\mathbf{P}_o(j\omega_i))/(\mathbf{P}(j\omega_i))]$. Draw a curve which is tangent to these arcs to form the first quadrant portion of the Q-contour shown in Fig. 21.23. Depending on the plant type desired for $\mathbf{L}$ it may be necessary to extend this contour into the second and fourth quadrants.

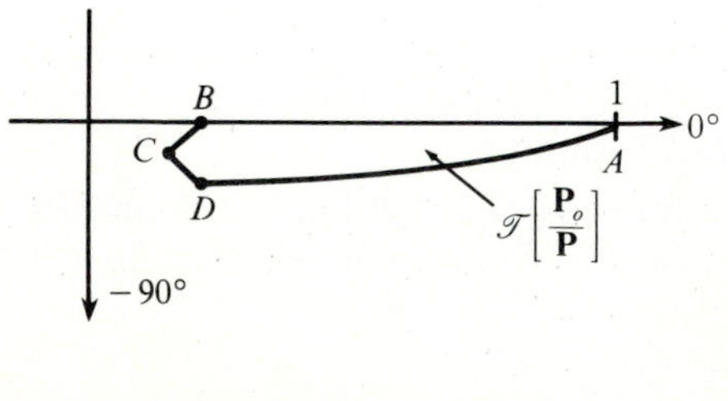

FIGURE 21.22
Template in polar coordinates.

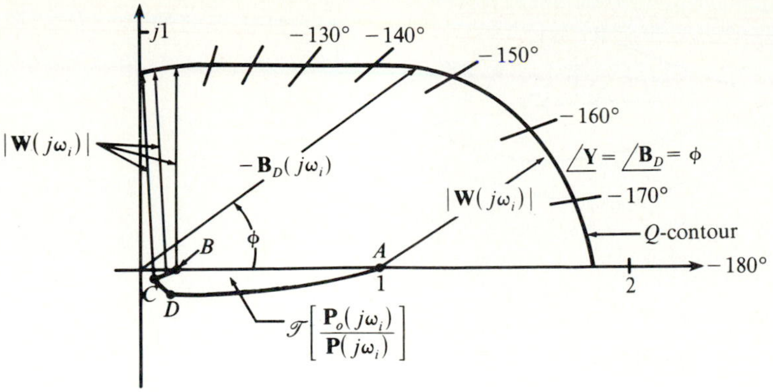

FIGURE 21.23
Graphical evaluation of $\mathbf{B}_D(j\omega_i)$.

4. Based upon Eq. (21.82) and Fig. 21.22, the phasor from the origin of Fig. 21.23 to the Q-contour represents $-\mathbf{B}_D(j\omega_i)$. This contour includes the plant uncertainty as represented by $\mathscr{T}[\mathbf{P}_o(j\omega_i)/\mathbf{P}(j\omega_i)]$. In the frequency range $\omega_x \le \omega_i \le \omega_h$, if $|\mathbf{W}(j\omega_i)| \gg |\mathbf{P}_o(j\omega_i)/\mathbf{P}(j\omega_i)|$, then the Q-contour is essentially a circle about the origin with radius $|\mathbf{W}(j\omega_i)| \equiv |\mathbf{B}_D(j\omega_i)|$.
5. Assuming the partial Q-contour of Fig. 21.23 is sufficient, measure from the graph the length $\mathbf{B}_D(j\omega_i)$ for every 10° of $\mathbf{B}_D(j\omega_i)$ and create Table 21.3 for each value of ω_i:
6. Plot the values from Table 21.3 for each value of ω_i, on the same NC as $\mathbf{B}_R(j\omega_i)$.

21.15 THE COMPOSITE BOUNDARY $\mathbf{B}_o(j\omega_i)$

The composite bound $\mathbf{B}_o(j\omega_i)$ that is used to synthesize the desired loop transmission transfer function $L_o(s)$ is obtained in the manner shown in Fig. 21.24. The composite bound $\mathbf{B}_o(j\omega_i)$, for each value of ω_i, is composed of those portions of each respective bound $\mathbf{B}_R(j\omega_i)$ and $\mathbf{B}_D(j\omega_i)$ that are the most

TABLE 21.3
Data points for the boundary $\mathrm{B}_D(j\omega_i)$

$\angle\mathbf{B}_D(j\omega_i)$	$\lvert\mathbf{B}_D(j\omega_i)\rvert$	Lm $\mathbf{B}_D(j\omega_i)$
−180°		
−170°		
−160°		
⋮		

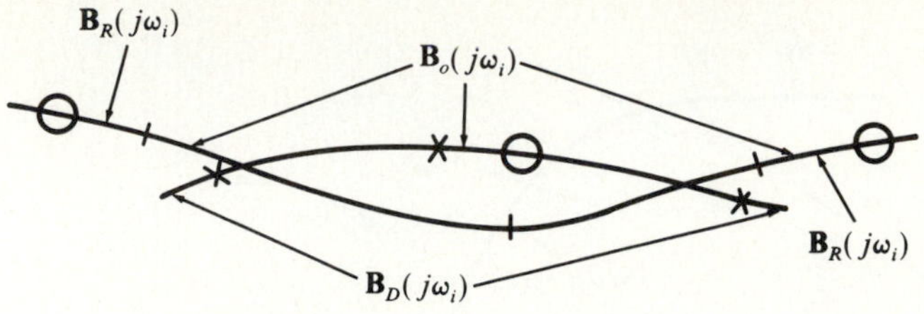

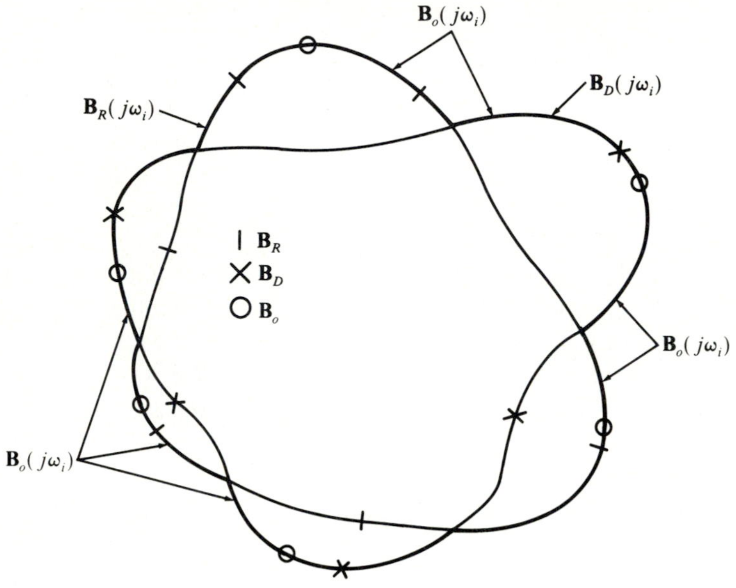

FIGURE 21.24
Composite $\mathbf{B}_o(j\omega_i)$.

restrictive. For the case shown in Fig. 21.24(a) the bound $\mathbf{B}_o(j\omega_i)$ is composed of those portions of each respective bound $\mathbf{B}_R(j\omega_i)$ and $\mathbf{B}_D(j\omega_i)$ that have the largest values. For the situation of Fig. 21.24(b), the outermost of the two boundaries $\mathbf{B}_R(j\omega_i)$ and $\mathbf{B}_D(j\omega_i)$ becomes the perimeter of $\mathbf{B}_o(j\omega_i)$. The situations of Fig. 21.24 occur when the two bounds have one or more intersections. If there are no intersections, then the bound with the largest value or with the outermost boundary dominates. The synthesized $\mathbf{L}_o(j\omega_i)$, for the situation of Fig. 21.24(a), must lie on or just above the bound $\mathbf{B}_o(j\omega_i)$. For the situation of Fig. 21.24(b) the synthesized $\mathbf{L}_o(j\omega_i)$ *must not lie in the interior* of the $\mathbf{B}_o(j\omega_i)$ contour.

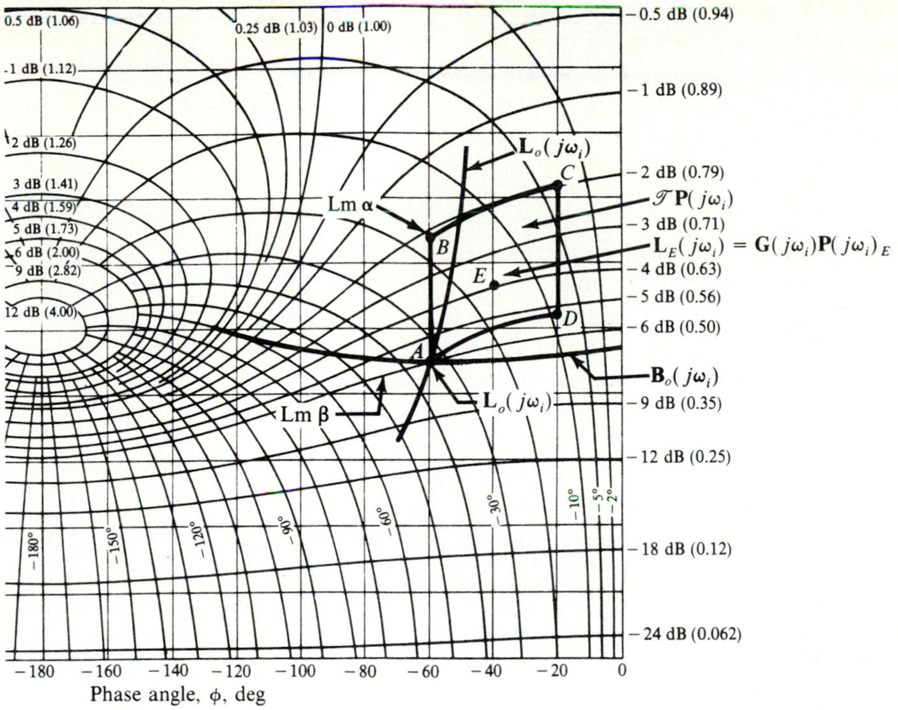

FIGURE 21.25
Graphical determination of Lm $\mathbf{T}(j\omega_i)$ for $\mathbf{P} \in \mathscr{P}$.

21.16 SHAPING OF $\mathbf{L}_o(j\omega)$[1,2,45]

A realistic definition of optimum[9,10] in an LTI system is the minimization of the high-frequency loop gain K. This gain affects the high-frequency response since $\lim_{\omega \to \infty} [\mathbf{L}(j\omega)] = K(j\omega)^{-\lambda}$, where λ is the excess of poles over zeros assigned to $\mathbf{L}(j\omega)$. Thus, only the gain K has a significant effect on the high-frequency response, and the effect of the other parameter uncertainty is negligible. It has been shown that the optimum $\mathbf{L}(j\omega)$ exists; it lies on the boundary $\mathbf{B}_o(j\omega_i)$ at all ω_i, and it is unique.[9,10]

Previous sections describe how tolerances on the closed-loop system frequency response, in combination with plant uncertainty templates, are translated into bounds on a nominal loop transmission function $\mathbf{L}(j\omega)$. In Fig. 21.25 the template $\mathscr{T}\mathbf{P}(j\omega_i)$ is located on the NC so that point A is on the constant M-curve Lm β, and point C is on the constant M-curve Lm α, such that

$$\delta_R(j\omega_i) = \text{Lm}\,\alpha - \text{Lm}\,\beta = 4 \text{ dB} \tag{21.83}$$

where $\alpha = T_{\max}$, $\beta = T_{\min}$, and $\delta_R(j\omega_i) = 4$ dB are obtained from Fig. 21.4.

Repeat this procedure for the required frequency range ω to determine the entire curve for $\mathbf{L}_o(j\omega_i)$ shown in Fig. 21.25.

For the set of plant parameters that correspond to point E within $\mathscr{T}\mathbf{P}(j\omega_i)$, as shown in Fig. 21.25, the open-loop transfer function is $\mathbf{L}_E = \mathbf{G}\mathbf{P}_E$. The control ratio

$$\mathbf{T}_E = \frac{\mathbf{L}_E}{1 + \mathbf{L}_E} \tag{21.84}$$

obtained from the constant M-contours on the NC has the value $\mathrm{Lm}\,\mathbf{T}_E = -4$ dB. Thus, any value of $\mathbf{P}$ within $\mathscr{T}\mathbf{P}(j\omega_i)$ yields a value of $\mathrm{Lm}\,\mathbf{T}$ between points A and C in Fig. 21.25. Therefore, any value of $\mathbf{P}$ that lies within the region of uncertainty (see Fig. 21.13 for the example of this chapter) yields a maximum variation in $\mathbf{T}(j\omega_i)$ that satisfies the requirement

$$\mathrm{Lm}\,\mathbf{T}_{\max} - \mathrm{Lm}\,\mathbf{T}_{\min} \le \delta_R(j\omega_i) \tag{21.85}$$

Thus, proper design of the prefilter F (see Fig. 21.5) yields a tracking control ratio $\mathbf{T}_R$ that lies between $\mathrm{Lm}\,\mathbf{T}_U$ and $\mathrm{Lm}\,\mathbf{T}_L$ in Fig. 21.8.

For the plant of Eqs. (21.55) and (21.56), the shaping of $\mathbf{L}_o(j\omega)$ is shown by the dashed curve in Fig. 21.26. A point such as $\mathrm{Lm}\,\mathbf{L}_o(j2)$ must be on or above the curve labeled $\mathbf{B}_o(j2)$. Further, in order to satisfy the specifications, $\mathbf{L}_o(j\omega)$ cannot violate the U-contour. In this example a reasonable $\mathbf{L}_o(j\omega)$ closely follows the U-contour up to $\omega = 40$ and must then stay below it as shown in Fig. 21.26. Additional specifications are $\lambda = 4$ and $L_o(s)$ must be Type 1 (one pole at the origin).

A representative procedure for choosing a rational function $L_o(s)$ which satisfies the above specifications is now described. It involves building up the function

$$\mathbf{L}_o(j\omega) = \mathbf{L}_{ok}(j\omega) = \mathbf{P}_o(j\omega)\prod_{k=0}^{w}\left[K_k\mathbf{G}_k(j\omega)\right]$$

where for $k = 0$, $\mathbf{G}_0 = 1\angle 0^\circ$, and $K = \prod_{k=0}^{w} K_k$. $\mathbf{L}_o(j\omega)$ is built up term-by-term, in order to stay just outside the U-contour in the NC of Fig. 21.26. The first step is to find the $\mathbf{B}_o(j\omega)$ which "dominates" $\mathbf{L}_o(j\omega)$. For example, suppose $L_{o0}(j4) = K_o\mathbf{P}_o'(j4) = 0$ dB $\angle -135^\circ$ (point A in Fig. 21.26) and that at $\omega = 1$ the required $\mathrm{Lm}\,\mathbf{L}_{o0}(j1)$ is approximately 27 dB. In order for $\mathrm{Lm}\,\mathbf{L}_o(j\omega)$ to decrease from 27 to about 0 dB in two octaves, the slope of $\mathbf{L}_o(j\omega)$ must be about -14 dB/octave, with $\angle\mathbf{L}_o < -180^\circ$ (see Fig. 21.27). Since absolute stability is required, $\mathbf{L}_o(j\omega)$ must have a phase margin of 40° over the entire frequency range for which $\mathbf{L}_o(j\omega)$ follows the vertical right-hand side of the U-contour and not just at the 0 dB crossover. Hence $\mathbf{B}_o(j1)$ dominates $\mathbf{L}_o(j\omega)$ more than does $\mathbf{B}_o(j4)$. In the same way it is seen that $\mathbf{B}_o(j1)$ dominates all other $\mathbf{B}_o(j\omega)$ in Fig. 21.26.

The $\mathbf{B}_o(j\omega)$ curves for $\omega < 1$ are not shown in Fig. 21.26 because it is assumed that a slope of -6 dB/octave for $\omega < 1$ suffices (additional values are 33 dB at $\omega = 0.5$, 39 dB at $\omega = 0.25$, etc.). To maintain -140° for $\omega \ge 1$, the first denominator factor of $\mathbf{L}(j\omega)$ has a corner frequency at $\omega = 1$ (i.e., a pole at -1 is chosen), and the value $\mathrm{Lm}\,\mathbf{L}_o(j1)$ on the straight line approximation is

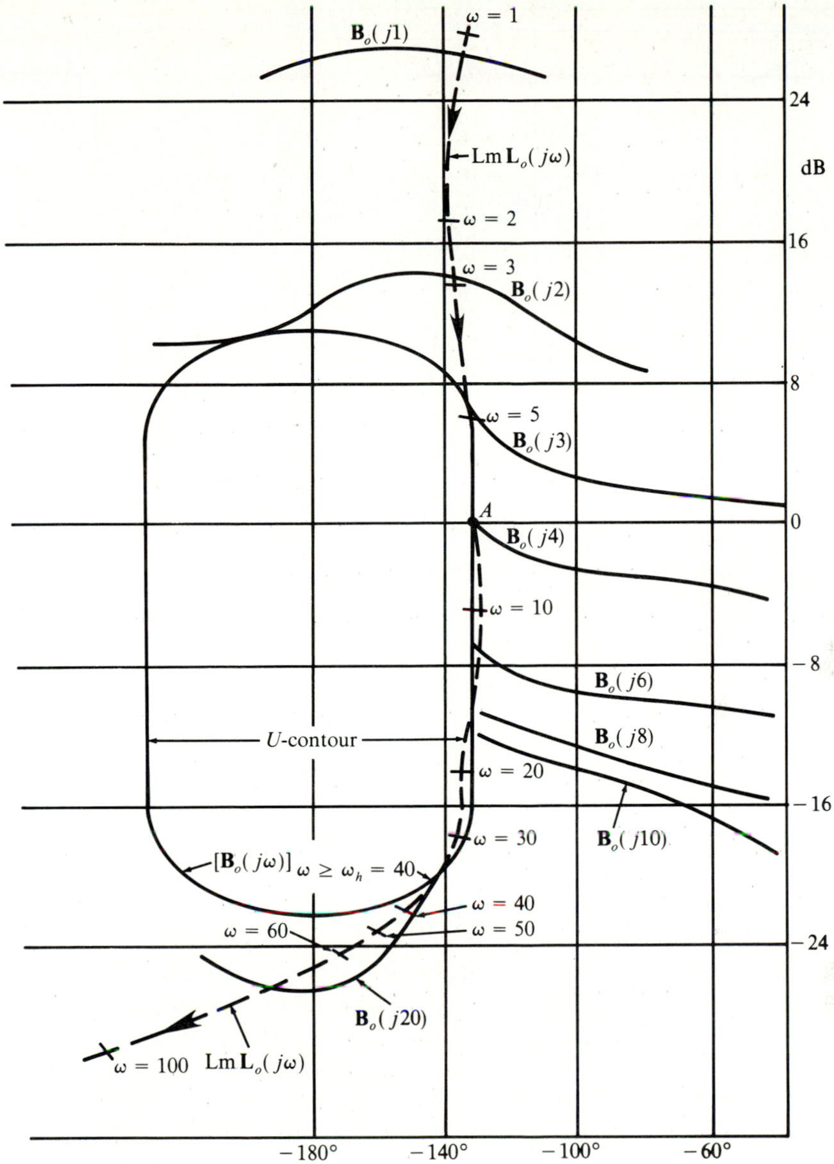

FIGURE 21.26
Shaping of $\mathbf{L}_o(j\omega)$ on the Nichols chart for the plant of Eqs. (21.55) and (21.56).

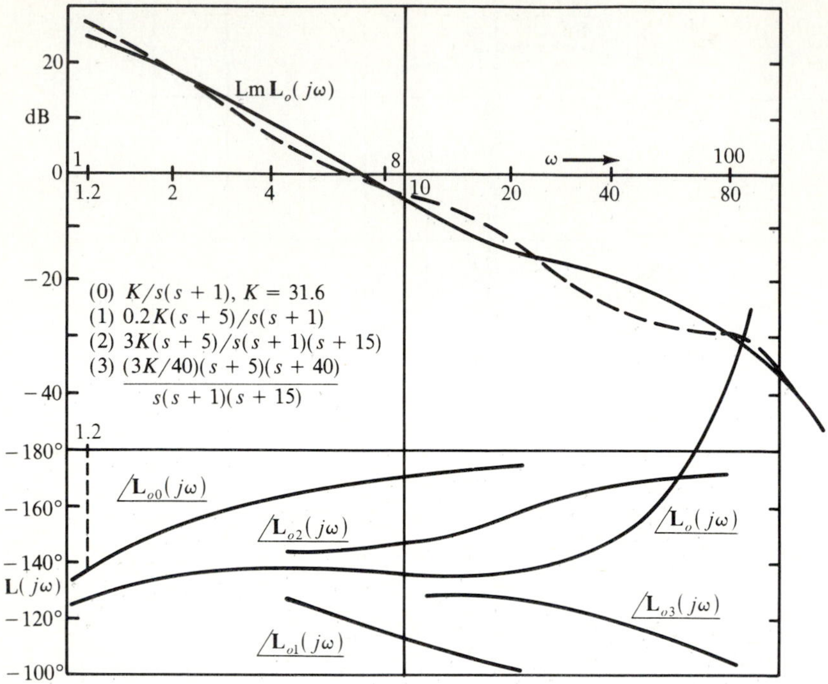

FIGURE 21.27
Shaping of $\mathbf{L}_o(j\omega)$ on the Bode plot.

selected at 30 dB (to allow for the -3 dB correction at the corner frequency). The function $L_o(s)$ determined so far is

$$L_{o0}(s) = 31.6/s(s+1)$$

whose angle $\angle \mathbf{L}_{o0}(j\omega)$ is sketched in Fig. 21.27.

$\mathbf{L}_{o0}(j\omega)$ violates the $-140°$ bound at $\omega \geq 1.2$, so a numerator term $(1 + j\omega T_2)$ must be added. At $\omega = 5$, $\angle \mathbf{L}_{o0}(j5) = -169°$ (see Fig. 21.27), and therefore a lead angle of 29° is needed at this frequency. Since a second denominator term $(1 + j\omega T_3)$ will be needed, allow an additional 15° for this factor, giving a total of $15° + 29° = 45°$ lead angle required at $\omega = 5$. This is achieved by selecting $T_2 = 0.2$, i.e., a zero at -5. This results in the composite value

$$L_{o1}(s) = \frac{31.6(1 + s/5)}{s(1+s)}$$

whose phase angle $\angle \mathbf{L}_{o1}(j\omega)$ is sketched in Fig. 21.27. In the NC of Fig. 21.26, $\omega \geq 5$ is the region where the maximum phase *lag* allowed is $-135°$ (i.e., $\angle \mathbf{L}_o(j\omega)$ must be $\geq -135°$). At $\omega = 10$, since $\angle \mathbf{L}_{o1}(j10) = -112°$, then $135° - 112° = 23°$ more lag angle is needed, and is provided with an additional denominator term $(1 + j\omega T_3)$. However, this is to be followed by an additional numerator term $(1 + j\omega T_4)$, so allow about 10° for it, giving 23° + 10° = 33° more lag allowable from $(1 + j\omega T_3)$. This requires selecting the corner frequency at $\omega = 15$ (or $T_3 = 1/15$), giving

$$L_{o2}(s) = \frac{31.6(1 + s/5)}{s(1 + s)(1 + s/15)}$$

A sketch of $\angle \mathbf{L}_{o2}(j\omega)$ is shown in Fig. 21.27.

Looking ahead at $\omega = 40$, $\text{Lm}\,\mathbf{L}_{o2}(j40) = -20$ dB, thus $\mathbf{L}_o(j\omega)$ can make its asymptotic left turn *under* the *U*-contour. The plan is to add two more numerator factors, $(1 + j\omega T_4)$ and $(1 + j\omega T_5)$, and finally two complex-pole pairs, in order to have an excess $\lambda = 4$ of poles over zeros and to minimize the *BW*. A zero is assigned at $\omega = 40$ with $T_4 = 1/40$, giving

$$L_{o3}(s) = \frac{31.6(1 + s/5)(1 + s/40)}{s(1 + s)(1 + s/15)}$$

A sketch of $\angle \mathbf{L}_{o3}(j\omega)$ is shown in Fig. 21.27.

In order to achieve (an asymptotic) horizontal segment for $\text{Lm}\,\mathbf{L}_o(j\omega)$, before the final -24 dB/octave slope is achieved, a final zero obtained from $(1 + j\omega T_5)$ is needed. Since the bottom of the *U*-contour (see Fig. 21.26) is at -22.5 dB, allow a 2 dB safety margin, a 3 dB correction due to $(1 + j\omega T_5)$, and 1.5 dB for the effect of $(1 + j\omega T_4)$, giving a total of $-(22.5 + 2 + 3 + 1.5) = -29$ dB. A damping ratio of $\zeta = 0.5$ is selected for the two complex-pole pairs, so no correction is needed for them. Thus, the final corner frequency for the straight-line curve of $\text{Lm}\,\mathbf{L}_{o3}(j\omega)$ is at -29 dB. Since $\text{Lm}\,\mathbf{L}_{o3}(j\omega)$ achieves this at $\omega = 60$, the last corner frequency is at $\omega = 60$, i.e., $T_5 = 1/60$. The resulting phase angle, due to L_{o3} and $(1 + j\omega/60)$, is $-66°$. An angle of $-180°$ could be selected at this point, but an additional 15° margin is allowed (this is a matter of judgment which depends on the problem, which may include the presence of higher-order modes, etc.). This means that 100° phase lag is permitted: 50° due to each complex-pole pair (180° − 66° − 15° ≈ 100°). Thus, a different value of damping ratio, with the appropriate dB correction applied to the log magnitude plot, is chosen i.e., $\zeta = 0.6$. This locates the corner frequency at $\omega = 100$. Thus

$$L_o(s) = \frac{31.6\left(1 + \frac{s}{5}\right)\left(1 + \frac{s}{40}\right)\left(1 + \frac{s}{60}\right)}{s(1 + s)\left(1 + \frac{s}{15}\right)\left[1 + \frac{(1.2)}{100}s + \frac{s^2}{10^4}\right]^2} = KP_o(s)\prod_{k=1}^{4} G_k(s) \tag{21.86}$$

The optimal loop transfer function $\mathbf{L}_o(j\omega)$ is sketched in Fig. 21.26 and is shown by the solid curves in Fig. 21.27. A well designed, i.e., an "economical" $\mathbf{L}_o(j\omega)$ is close to the $\mathbf{B}_o(j\omega)$ boundary at each ω_i. The vertical line at $-140°$ in Fig. 21.26 is the dominating vertical boundary for $\mathbf{L}_o(j\omega)$ for $\omega < 5 = \omega_x$, and the right side of the *U*-contour line at $-135°$ is the vertical boundary effectively for

$\omega_x \approx 5 < \omega < 30 \approx \omega_y$. The final $\mathbf{L}_o(j\omega)$ is good in this respect since it is close to these boundaries.

There is a trade-off between complexity of $L_o(s)$ (the number of its poles and zeros) and its final cut-off corner frequency which is $\omega = 100$ and the phase margin frequency $\omega_\phi = 7$. There is some phase to spare between $\mathbf{L}_o(j\omega)$ and the boundaries. Use of more poles and zeros in $L_o(s)$ permits this cut-off frequency to be reduced a bit below 100, but not by much. On the other hand, if it is desired to reduce the number of poles and zeros of $L_o(s)$, then the price in achieving this is a larger cut-off frequency. It is possible to economize significantly by allowing more phase lag in the low-frequency range. If $-180°$ is permitted at $\omega = 1$, then a decrease of Lm $\mathbf{L}_o(j\omega)$ at a rate of 12 dB/octave can be achieved, e.g., with Lm $\mathbf{L}_o(j1) = 25$ dB, then it will be 13 dB at $\omega = 2$ (instead of the present 18 dB). Even with no more saving, this 5-dB difference allows a cut-off frequency of about $\omega = 70$ instead of 100.

Figure 21.26 reveals immediately, without any shaping of $\mathbf{L}_o(j\omega)$ required, that reduction (i.e., easing) of the specifications at $\omega = 1$ to about 21 dB (instead of about 26 dB) has the same effect as the above. How badly the specifications are compromised by such easing can easily be checked. The design technique is thus highly "transparent" in revealing the trade-offs between performance tolerances, complexity of the compensation, stability margins, and the "cost of feedback" in bandwidth.

21.17 GUIDELINES FOR SHAPING $\mathbf{L}_o(j\omega)$

Some general guidelines for the shaping of $\mathbf{L}_o(j\omega)$ are:

1. Do not use a CAD program at the onset of your design problem. Use the straight-line approximations on the Bode diagram for the log magnitude.
2. On the graph paper for the Bode diagram plot the points representing Lm $\mathbf{B}_o(j\omega_i)$ and the angles corresponding to the right side of the *U*-contour (the desired phase margin angle γ) for the frequency range of $\omega_x \le \omega \le \omega_y$. For the example illustrated in Fig. 21.26 the phase margin angle $\gamma = -45°$ must be maintained for the frequency range $\omega_x = 5 \le \omega \le \omega_y \approx 30$.
3. Do the shaping of Lm $\mathbf{L}_o(j\omega)$ on the Bode plot using straight-line approximations for Lm $\mathbf{L}_o(j\omega)$, with the plotted information of step 2, and employ the shaping discussion just below Eq. (21.86) for the frequency range $\omega < \omega_x$ as guidelines in achieving Eq. (21.86).
4. Use frequencies an octave above and below and a decade above and below a corner frequency for both first- and second-order terms,[3,6] while maintaining the phase-margin angle corresponding to the right side of the *U*-contour in shaping $\mathbf{L}_o(j\omega)$.
5. The last two poles that are added to $\mathbf{L}_o(j\omega)$ are generally a complex pair (the nominal range is $0.5 < \zeta < 0.7$) which tends to minimize the *BW*.
6. Once $\mathbf{L}_o(j\omega)$ has been shaped, determine $F(s)$ (see Sec. 21.18) and then verify that $\mathbf{L}_o(j\omega)$ does meet the design objectives by use of a CAD program. If the

synthesized $\mathbf{L}_o(j\omega)$ yields the desired performance, then the required compensator is given by

$$G(s) = \frac{L_o(s)}{P_o(s)}$$

Specific guidelines for shaping $\mathbf{L}_o(j\omega)$ are:

1. An optimum design of $\mathbf{L}_o(j\omega)$ requires that $\mathbf{L}_o(j\omega_i)$ be on the corresponding bound. In practice, place $\mathbf{L}_o(j\omega_i)$ as close as possible to the bound $\mathbf{B}_o(j\omega_i)$, but above it, in order to keep the BW of $\mathbf{L}_o/[1 + \mathbf{L}_o]$ to a minimum.
2. Since exact cancellation of a pole by a zero is rarely possible, any right-hand plane poles and/or zeros of $P_o(s)$ should be included in $L_o(s)$. A good starting $L_o(s)$ is $L_{o0}(s) = K_0 P_o(s)$. If it is desired that $y(\infty) = 0$ for $d(t) = u_{-1}(t)$, it is necessary to insure that $T_D(s)$ has a zero at the origin. For this situation a possible starting point is $L_{o0}(s) = K_0 P_o(s)/s$.
3. If $P(s)$ has an excess of poles over zeros, $\#p$'s $- \#z$'s $= \lambda$, which is denoted by e^{λ}, then, in general, the *final form* of $L_o(s)$ must have an excess of poles over zeros of at least $e^{\lambda+i}$ where $i \geq 1$. If the BW is too large, then increase the value of i. Experience shows that a value of $\lambda + i$ of 3 or more for $L(s)$ yields satisfactory results.[2,10]
4. Generally the BW of $L_o(s)/[1 + L_o(s)]$ is larger than required for an acceptable rise time t_R for the tracking of $r(t)$ by $y(t)$. An acceptable rise time can be achieved by the proper design of the prefilter $F(s)$ (see Fig. 21.18).
5. Generally it is desirable to first find the bounds $\mathbf{B}_D(j\omega_i)$ and then the bounds $\mathbf{B}_R(j\omega_i)$. It may be evident, after finding the first $\mathbf{B}_R(j\omega_i)$, that all or some of the $\mathbf{B}_D(j\omega_i)$ are completely dominant compared to $\mathbf{B}_R(j\omega_i)$. In that case the $\mathbf{B}_D(j\omega_i)$ boundaries are the optimal boundaries, i.e., $\mathbf{B}_D(j\omega_i) = \mathbf{B}_o(j\omega_i)$.
6. If $\delta_R(j\omega_i)$ is not continuously increasing as ω_i increases, then it will be necessary to utilize complex poles and/or zeros in $G(s)$ in order to achieve an optimal $L_o(s)$. This assumes that the tracking model is not designed to yield this increasing characteristic for $\delta_R(j\omega_i)$.
7. The ability to shape the nominal loop transmission $L_o(s)$ is an art developed by the designer only after much practice and patience.

21.18 DESIGN OF THE PREFILTER $F(s)$[1,12]

Design of a proper $L_o(s)$ guarantees only that the variation in $|\mathbf{T}_R(j\omega)|$ is less than or equal to that allowed. The purpose of the prefilter is to position $\mathrm{Lm}\,[\mathbf{T}(j\omega)]$ within the frequency domain specifications. For the example of this chapter the magnitude of the frequency response must lie within the bounds B_U and B_L shown in Fig. 21.8, which are redrawn in Fig. 21.28. One method for determining the bounds on the prefilter $F(s)$ is as follows: Place the nominal point of the ω_i plant template on the $\mathbf{L}_o(j\omega_i)$ point on the $\mathbf{L}_o(j\omega)$ curve on the NC (see Fig. 21.29). Traversing the template, determine the maximum $\mathrm{Lm}\,\mathbf{T}_{max}$

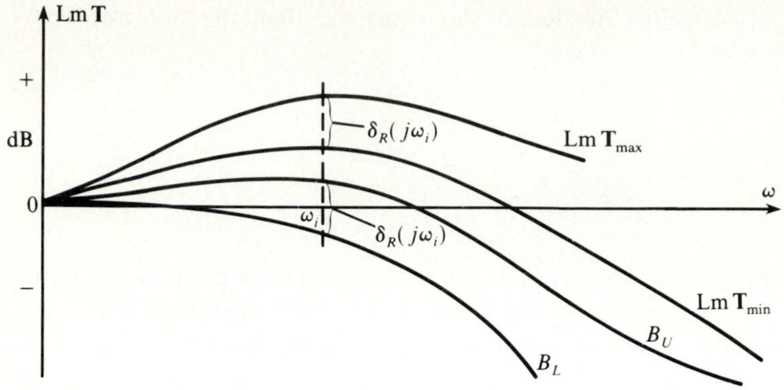

FIGURE 21.28
Requirements on $F(s)$.

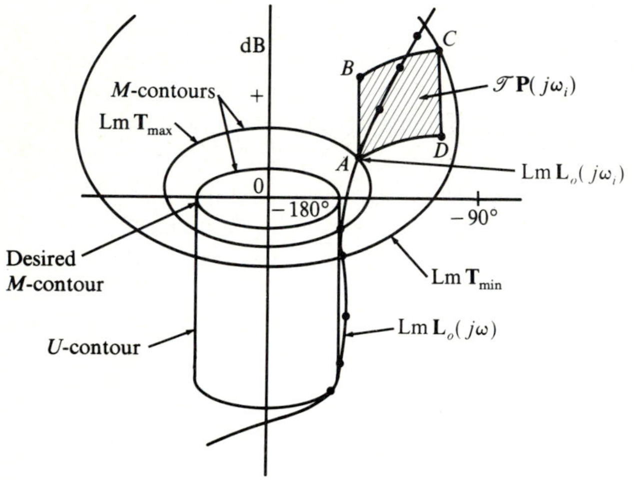

FIGURE 21.29
Prefilter determination.

and minimum Lm $\mathbf{T}_{min}$ values of

$$\text{Lm}\,\mathbf{T}(j\omega_i) = \frac{\mathbf{L}(j\omega_i)}{1 + \mathbf{L}(j\omega_i)} \tag{21.87}$$

obtained from the M-contours. These values are plotted as shown in Fig. 21.28. The tracking control ratio is $T_R = FL/[1 + L]$ and

$$\text{Lm}\,\mathbf{T}_R(j\omega_i) = \text{Lm}\,\mathbf{F}(j\omega_i) + \text{Lm}\,\mathbf{T}(j\omega_i) \tag{21.88}$$

The variations in Eqs. (21.87) and (21.88) are both due to the variation in P; thus

$$\delta_R(j\omega_i) = \text{Lm}\,\mathbf{T}_{max} - \text{Lm}\,\mathbf{T}_{min} = B_U - B_L \tag{21.89}$$

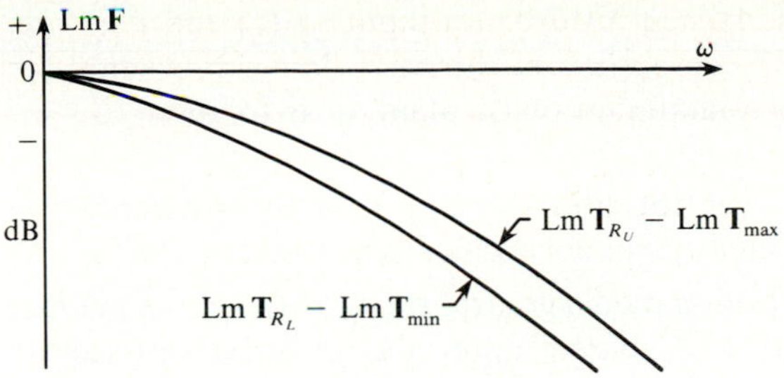

FIGURE 21.30
Frequency bounds on the prefilter $F(s)$.

Therefore, based upon Eq. (21.88), it is necessary to determine the range in dB that $\text{Lm}\,\mathbf{T}(j\omega_i)$ must be raised or lowered to fit within the bounds of the specifications by use of the prefilter $\mathbf{F}(j\omega_i)$. The process is repeated for each frequency corresponding to the templates used in the design of $\mathbf{L}_o(j\omega)$. Therefore, in Fig. 21.30 the difference between the $\text{Lm}\,\mathbf{T}_R - \text{Lm}\,\mathbf{T}_{\max}$ and the $\text{Lm}\,\mathbf{T}_R - \text{Lm}\,\mathbf{T}_{\min}$ curves yields the requirement for $\text{Lm}\,\mathbf{F}(j\omega)$, i.e., from Eq. (21.88)

$$\text{Lm}\,\mathbf{F}(j\omega) = \text{Lm}\,\mathbf{T}_R(j\omega) - \text{Lm}\,\mathbf{T}(j\omega) \qquad (21.90)$$

The procedure for designing $F(s)$ is summarized as follows:

1. Use templates in conjunction with the $\mathbf{L}_o(j\omega)$ plot on the NC to determine $T_{\max}$ and $T_{\min}$ for each ω_i. This is done by placing $\mathscr{T}\mathbf{P}(j\omega_i)$ with its nominal point on the point $\text{Lm}\,\mathbf{L}_o(j\omega_i)$. Then use the M-contours to determine $\mathbf{T}_{\max}(j\omega_i)$ and $\mathbf{T}_{\min}(j\omega_i)$ (see Fig. 21.29).
2. Obtain the values of $\text{Lm}\,\mathbf{T}_{R_U}$ and $\text{Lm}\,\mathbf{T}_{R_L}$ for various values of ω_i from Fig. 21.8.
3. From the values obtained in steps 1 and 2, plot

$$\left[\text{Lm}\,\mathbf{T}_{R_U} - \text{Lm}\,\mathbf{T}_{\max}\right] \text{ and } \left[\text{Lm}\,\mathbf{T}_{R_L} - \text{Lm}\,\mathbf{T}_{\min}\right] \text{ vs. } \omega$$

   as shown in Fig. 21.30.
4. Use straight-line approximations to synthesize an $F(s)$ so that $\text{Lm}\,\mathbf{F}(j\omega)$ lies within the plots of step 3. For step forcing functions the resulting $F(s)$ must satisfy

$$\lim_{s\to 0}\,[F(s)] = 1 \qquad (21.91)$$

21.19 BASIC DESIGN PROCEDURE FOR A MISO SYSTEM

The basic concepts of the QFT technique are explained by means of a design example. The system configuration shown in Fig. 21.5 contains three inputs. Parameter uncertainty for the plant of Eq. (21.9) is shown in Fig. 21.6. The first objectives are to track a step input $r(t) = u_{-1}(t)$ with no steady-state error and to satisfy the thumb print specifications of Fig. 21.7. An additional objective is to attenuate the system response due to step disturbance inputs $d_1(t)$ and $d_2(t)$, as described in Figs. 12.16 and 12.17. An outline of the basic design procedure for the QFT technique, as applied to a minimum phase plant, is as follows:

1. Synthesize the tracking model control ratio $T_R(s)$ in the manner described in Sec. 21.4, based upon the desired tracking specifications (see Fig. 21.8).
2. Synthesize the disturbance-rejection model control ratios $T_D(s)$ in the manner described in Secs. 12.5 and 12.8, based upon the disturbance-rejection specifications.
3. Obtain templates of $\mathbf{P}(j\omega_i)$ that pictorially describe the plant uncertainty on the Nichols chart (NC) for the desired passband frequency range.
4. Select a nominal plant from the set of Eq. (21.8) and denote it as $P_o(s)$.
5. Determine the U-contour based upon the specified values of $\delta_R(j\omega_i)$ for tracking, M_L for disturbance rejection, and V for the universal high-frequency boundary (UHFB) B_h in conjunction with steps 6 through 8.
6. Use the data of steps 2 and 3 and the values of $\delta_D(j\omega_i)$ (see Fig. 21.9) to determine the disturbance bound $\mathbf{B}_D(j\omega_i)$ on the loop transmission $\mathbf{L}_D(j\omega_i) = \mathbf{G}(j\omega_i)\mathbf{P}(j\omega_i)$. For minimum-phase systems this requires that the synthesized loop transmission $\text{Lm}\,\mathbf{L}_D(j\omega_i)$ must be on or above the curve for $\text{Lm}\,\mathbf{B}_D(j\omega_i)$ on the Nichols diagram (see Fig. 21.26 assuming $\mathbf{B}_D = \mathbf{B}_o$).
7. Determine the tracking bound $\mathbf{B}_R(j\omega_i)$ on the nominal transmission $\mathbf{L}_o(j\omega_i) = \mathbf{G}(j\omega_i)\mathbf{P}(j\omega_i)$, using the tracking model (step 1), the templates $\mathbf{P}(j\omega_i)$ (step 3), the values of $\delta_R(j\omega_i)$ (see Fig. 21.8), and M_L [see Eq. (21.60)]. For minimum-phase systems this requires that the synthesized loop transmission must satisfy the requirement that $\text{Lm}\,\mathbf{L}_R(j\omega_i)$ is on or above the curve for $\text{Lm}\,\mathbf{B}_R(j\omega_i)$ on the Nichols diagram.
8. Plot curves of $\text{Lm}\,\mathbf{B}_R(j\omega_i)$ vs. $\phi_R = \angle\mathbf{B}_R(j\omega_i)$ and $\text{Lm}\,\mathbf{B}_D(j\omega_i)$ versus $\phi_D = \angle\mathbf{B}_D(j\omega_i)$ on the same NC. For a given value of ω_i at various values of the angle ϕ, select the value of $\text{Lm}\,\mathbf{B}_D(j\omega_i)$ or $\text{Lm}\,\mathbf{B}_R(j\omega_i)$, whichever is the largest value (termed the "worst" or "most severe" boundary). Draw a curve through these points. The resulting plot defines the overall boundary $\text{Lm}\,\mathbf{B}_o(j\omega_i)$ vs. ϕ. Repeat this procedure for sufficient values of ω_i.
9. Design $\mathbf{L}_o(j\omega_i)$ to be as close as possible to the boundary value $\mathbf{B}_o(j\omega_i)$ by selecting an appropriate compensator transfer function $\mathbf{G}(j\omega)$. Synthesize an $\mathbf{L}_o(j\omega) = \mathbf{G}(j\omega)\mathbf{P}(j\omega)$ using the $\text{Lm}\,\mathbf{B}_o(j\omega_i)$ boundaries and U-contour so that $\text{Lm}\,\mathbf{L}_o(j\omega_i)$ is on or above the curve for $\text{Lm}\,\mathbf{B}_o(j\omega_i)$ on the Nicholas diagram. This procedure achieves the lowest possible value of the loop transmission frequency (phase margin frequency). Note that $\mathbf{L}_o(j\omega_i) \geq \mathbf{B}_o(j\omega_i)$ represents the loop transfer function that satisfies the most severe boundary $\mathbf{B}_R$ and $\mathbf{B}_D$.
10. Based upon the information available from steps 1 and 9, synthesize an $F(s)$ that results in a $\text{Lm}\,\mathbf{T}_R$ [Eq. (21.12)] vs. ω that lies between B_U and B_L of Fig. 21.8.

11. Obtain the time-response data for $y(t)$: (a) with $d(t) = u_{-1}(t)$ and $r(t) = 0$ and (b) with $r(t) = u_{-1}(t)$ and $d(t) = 0$ for sufficient points around the parameter space describing the plant uncertainty (see Fig. 21.15).

For the $\mathbf{L}_o(j\omega)$ obtained, the plot of $\text{Lm}\,(\mathbf{L}(j\omega)/[1 + \mathbf{L}(j\omega)]$ may be larger or smaller than $\text{Lm}\,\mathbf{T}_{R_U}$ or $\text{Lm}\,\mathbf{T}_{R_L}$ of Fig. 21.8, but $\delta_R(j\omega_i)$ is satisfied. By the proper design of the input filter $F(s)$, $\text{Lm}\,\mathbf{T}_R = \text{Lm}\,\mathbf{FL}/[1 + \mathbf{L}]$ will lie within the bounds of $\text{Lm}\,\mathbf{T}_{R_U}$ and $\text{Lm}\,\mathbf{T}_{R_L}$. In problems with very large uncertainty and in disturbance rejection requiring a very large $|\mathbf{L}(j\omega)|$ over a "large" BW, then

$$\frac{\mathbf{L}}{1 + \mathbf{L}} \approx 1 \tag{21.92}$$

and $T \approx F$. For these situations design $F(s)$ in the same manner as for the tracking models of Secs. 12.2 and 21.4.

The simple plant of Eq. (21.9) is used in the following sections to illustrate the details in applying this QFT design procedure.

21.20 DESIGN EXAMPLE

This design example is for the control system of Fig. 21.5 with $r(t) = d_2(t) = u_{-1}(t)$ and $d_1(t) = 0$. The plant transfer function

$$P(s) = \frac{Ka}{s(s + a)} \qquad 1 \le K \le 10 \qquad 1 \le a \le 10 \tag{21.93}$$

has the nominal values: $a = K = 1$.

Step 1. Modeling the tracking control ratio $T_R(s) = Y(s)/R(s)$:

$$T_{R_U}\text{:} \quad M_p = 1.2,\ t_s \le 2\text{ s}$$

$$T_{R_L}\text{:} \quad \text{Overdamped},\ t_s \le 2\text{ s}$$

(a) A tracking model for the upper bound, based upon the given desired performance specifications, is tentatively identified by

$$T_{R_U}(s) = \frac{19.753}{(s + 2 \pm j3.969)} \tag{21.94}$$

A zero is inserted in Eq. (21.94) which does not affect the desired performance specifications but widens δ_{hf} between T_{R_U} and T_{R_L} in the high-frequency range. Thus

$$T_{R_U}(s) = \frac{0.6584(s + 30)}{(s + 2 \pm j3.969)} \tag{21.95}$$

The figures of merit for this transfer function with a step input are $M_p = 1.2078$, $t_R = 0.342$ s, $t_p = 0.766$ s, $t_s = 1.84$ s, $\text{Lm}\, M_m = 1.95$ dB, and $\omega_m = 3.4$ r/s.

(b) A tracking model (see Chap. 12) for the lower bound, based upon the desired performance specifications, is tentatively identified by

$$T_{R_L}(s) = \frac{120}{(s + 3)(s + 4)(s + 10)} \tag{21.96}$$

and is modified with the addition of a pole to yield

$$T_{R_L}(s) = \frac{8400}{(s+3)(s+4)(s+10)(s+70)} \tag{21.97}$$

The pole at $s = -70$ is inserted in Eq. (21.97) to further widen δ_{hf} between T_{R_U} and T_{R_L} at high frequencies. The figures of merit for Eq. (21.97) are: $M_p = 1$, $t_R = 1.02$ s, and $t_s = 1.849$ s.

(*c*) Determination of $\delta_R(j\omega_i)$. From the data for the log magnitude plots of Eqs. (21.95) and (21.97), the following values are obtained for $\delta_R(j\omega_i)$

ω	$\delta_R(j\omega_i)$, dB
0.5	0.25
1	1.06
2	3.728
5	10.57
10	9.851
15	11.522
100	40

Due to the disturbance-rejection specifications (see step 6) the above values are sufficient. Normally, values of ω an octave apart up to ω_h should be obtained.

Step 2. Modeling the disturbance control ratio $T_{D_U}(s) = Y(s)/D_2(s)$. Since $y(0) = 1$, it is desired that $y(t)$ decay "as fast as possible," i.e., $|c(t)| \leq 0.01$ for $t \geq t_x = 60$ ms. Let the model disturbance control ratio be of the form

$$T_{D_U}(s) = \frac{Y(s)}{D_2(s)} = \frac{s(s+g)}{(s+g)^2 + h^2} \tag{21.98}$$

For $d_2(t) = u_{-1}(t)$ the output response has the form $y(t) = e^{-gt}\cos ht$. The desired specifications (see Fig. 21.9) are satisfied by choosing

$$T_{D_U}(s) = \frac{s(s+70)}{s+70 \pm j18} = \frac{1}{1+L_D} \tag{21.99}$$

For Eq. (21.99) the output decays rapidly so that $|y(t)| \leq 0.01$ at $t_x \approx 0.0565$ s.

Steps 3 and 4. Forming the templates of $P(j\omega_i)$. Analysis of $P(s)$ is based upon the nominal values $K = a = 1$ (point A of Fig. 21.31). The template for each frequency is based on the data listed in Table 21.4 which is computed using Eq. (21.93). The values in Table 21.4 are used to draw the templates $\mathcal{T}\mathbf{P}(j\omega_i)$ shown in Fig. 21.32.

Step 5. Determination of the *U*-contour.

(*a*) Determination of V [see Eq. (21.60)] for the plant of Eq. (21.93) yields

$$\lim_{s\to\infty} [\mathbf{P}(s)] = \left[\frac{Ka}{s^2}\right] \qquad \text{and} \qquad \Delta\,\text{Lm}\,[\mathbf{P}]_{\max} = \text{Lm}\,[Ka]_{\max} = 40 \text{ dB} \tag{21.100}$$

To determine the value of ω_h, where $\Delta\,\text{Lm}\,\mathbf{P} \approx V = 40$ dB, select various values

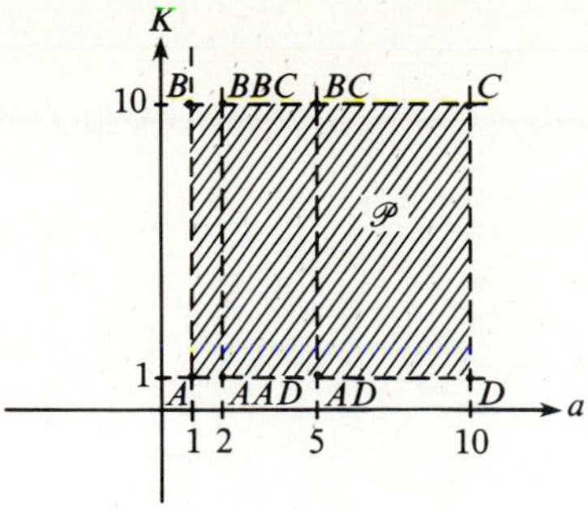

FIGURE 21.31
Template points.

TABLE 21.4
Data points for $\mathscr{T}\mathbf{P}(j\omega_i)$

Point ↓	ω_i →	0.5	1	2	5	10	15
A	Lm **P**	5	−3	−13	−28.1	−40	−47
	∠**P**	−116.6°	−135°	−153.5°	−168.7°	−174.3°	−176.2°
B	Lm **P**	25	17	7	−8.1	−20	−27
	∠**P**	−116.6°	−135°	−153.5°	−168.7°	−174.3°	−176.2°
BBC	Lm **P**	25.8	19	11	−2.6	−14.2	−21.1
	∠**P**	−104.1°	−116.6°	−135°	−158.2°	−168.7°	−171.4°
BC	Lm **P**	26	19.8	13.3	3	−7	−13.5
	∠**P**	−95.7°	−101.3°	−111.8°	−135°	−153.5°	−161.6°
C	Lm **P**	26	20	13.8	5	−3	−8.64
	∠**P**	−93°	−95.7°	−101.3°	−116.6°	−13.5°	−146.4°
D	Lm **P**	6	−0.04	−6.2	−14.95	−23	−28.6
	∠**P**	−93°	−95.7°	−101.3°	−116.6°	−135°	−146.4°
AD	Lm **P**	6	−0.17	6.67	−17	−27	−33.5
	∠**P**	−95.7°	−101.3°	−111.8°	−135°	−153.5°	−116.6°
ADD	Lm **P**	5.65	0.97	−9	−22.6	−34.2	−41.1
	∠**P**	−104.1°	−116.6°	−135°	−158.2°	−168.7°	−172.4°

Note: Values of Lm are in dB.

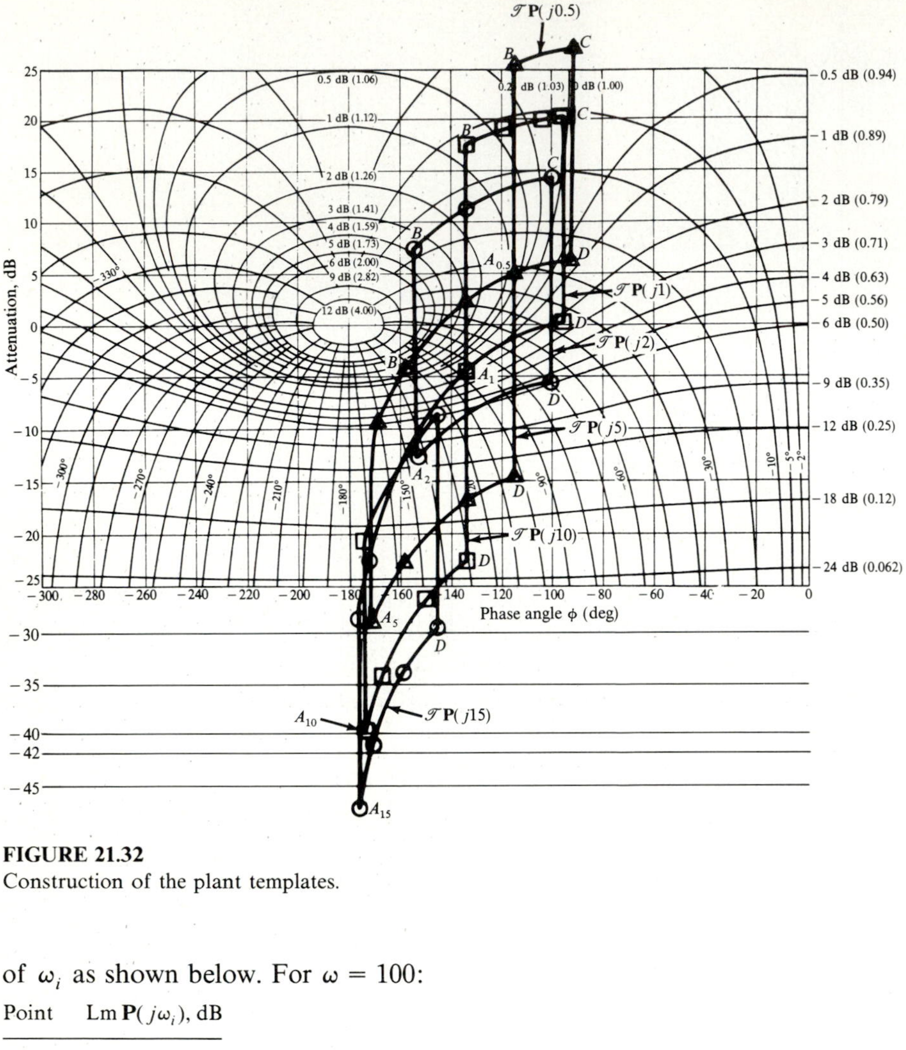

FIGURE 21.32
Construction of the plant templates.

of ω_i as shown below. For $\omega = 100$:

Point	Lm $\mathbf{P}(j\omega_i)$, dB
A	−80
B	−60
C	−40
D	−60

From this data it is seen that the maximum value of Δ Lm $\mathbf{P}(j\omega_i)$ occurs between points A and C. For other values of ω_i the maximum values of Δ Lm $\mathbf{P}(j\omega_i)$ are:

Frequency ω_i	max Δ Lm $\mathbf{P}(j\omega_i)$ dB
15	38.4
16	38.6
20	39.0
40	39.65

Therefore it can be seen, for this example, that V is achieved essentially at $\omega_h \approx 40$.

(b) Determination of the B_h boundary. Select the M-contour that represents the desired value of M_L for T (see Sec. 21.12). At various points around the lower half of this contour measure down 40 dB and draw the B_h boundary in the manner shown in Fig. 21.21.

Step 6. Determination of bounds $\mathbf{B}_D(j\omega_i)$ for

$$T_{D_U} = \frac{Y}{D} = \frac{1}{1+L} = \frac{\ell}{1+\ell} \tag{21.101}$$

where $\ell = 1/L = 1/PG$ and $Y = DT_{D_U}$. For the disturbance-rejection case:

$$\Delta\,\mathrm{Lm}\,\mathbf{Y} = \Delta\,\mathrm{Lm}\left[\frac{\ell}{1+\ell}\right] = \Delta\,\mathrm{Lm}\,\mathbf{T}_{D_U} \tag{21.102}$$

Note that there are no lower bounds on the response $y(t)$. Thus, choose

$$\mathbf{L} \geq \mathbf{B}_D(j\omega_i)$$

to satisfy

$$\mathrm{Lm}\left[\frac{\mathbf{Y}}{\mathbf{D}}\right] = \mathrm{Lm}\left[\frac{\ell}{1+\ell}\right] \leq \mathrm{Lm}\left[\mathbf{T}_{D_U}\right] \tag{21.103}$$

for each frequency ω_i of $\mathbf{T}_D(j\omega_i)$ over the frequency range $0 \leq \omega \leq 15$ r/s. From Eq. (21.99) the data of Table 21.5 are obtained. Point A of the templates of step 3 yields the maximum value for $\ell = 1/\mathbf{L}$. For $0 \leq \omega \leq 15$ the $\mathrm{Lm}\,\mathbf{T}_D(j\omega_i)$ contours (see Fig. 21.33) become the contours for $\mathbf{B}_D$, i.e., $\mathrm{Lm}\,\mathbf{T}_D = \mathrm{Lm}\,\mathbf{B}_D$. For

TABLE 21.5
Frequency data for Eq. (21.99)

ω	$\mathrm{Lm}\,\mathbf{T}_D$, dB	$\angle\mathbf{T}_D$, degrees	$\mathrm{Lm}\,\mathbf{L}_D$
0.5	−43.5	89.65	43.5
1	−37.5	89.3	37.5
2	−31.44	88.6	31.4
5	−23.5	86.4	23.5
10	−17.52	82.85	17.5
15	−14.1	79.31	14.1
20	−11.67	75.81	11.4
50	−4.84	56.8	4.46
100	−1.67	36.17	−2.91
200	−0.453	19.55	−9.05
500	−0.0737	7.99	−17.1
1000	−0.01845	4.007	−21.6
2000	−0.002954	2.005	−31.0
3000	−0.002052	1.337	−32.6
4000	−0.001154	1.003	−35.2
5000	−0.0007386	0.802	−37.1

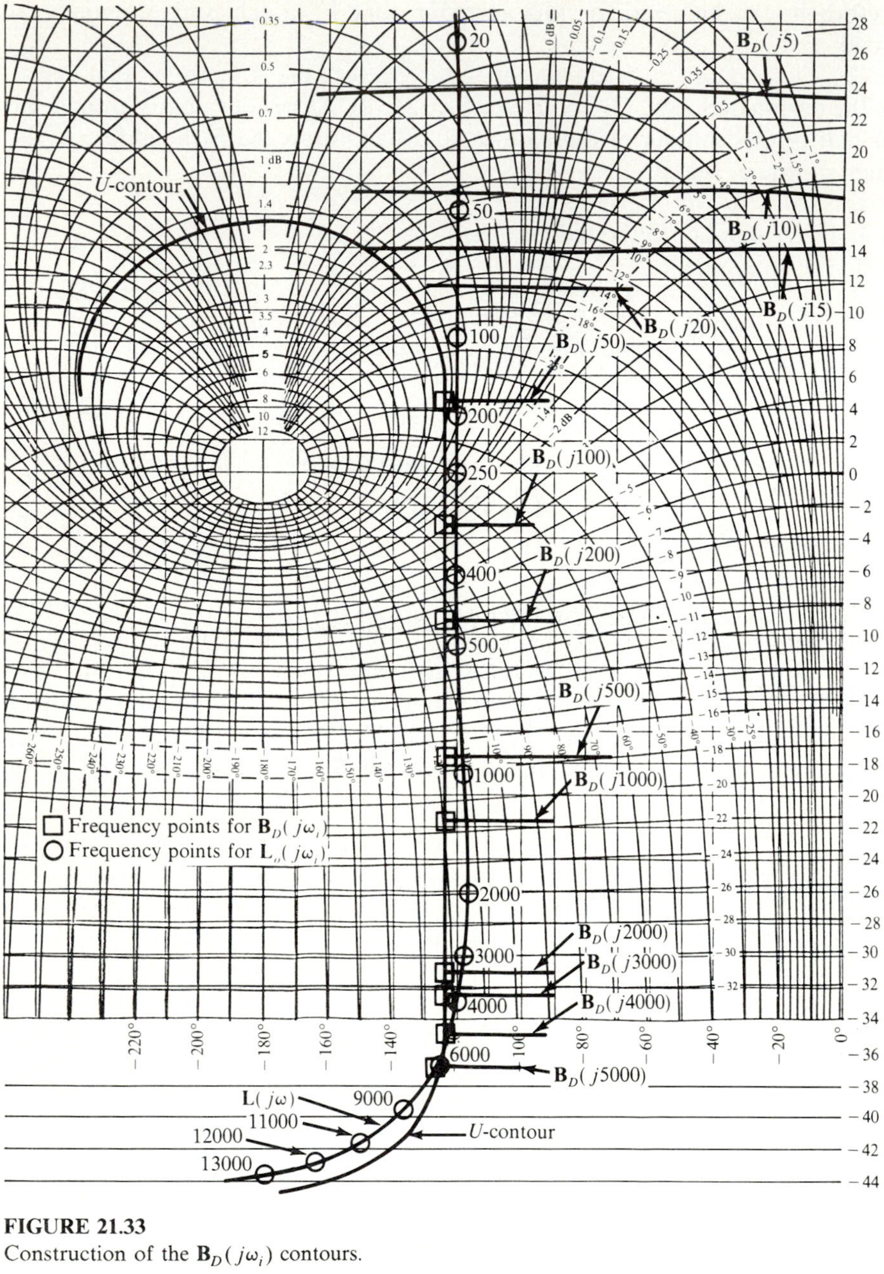

FIGURE 21.33
Construction of the $\mathbf{B}_D(j\omega_i)$ contours.

$\omega \geq 20$, the values of $\text{Lm}\,\mathbf{L}_D$ in Table 21.5, obtained by use of Eq. (21.99), are points of $\mathbf{B}_D(j\omega_i)$ on and in the vicinity of the U-contour.

Step 7. Determination of the bounds $\mathbf{B}_R(j\omega_i)$. By use of the templates, the values of $\delta_R(j\omega_i)$ given in step 2, the M_L contour, and the B_h boundary, the bounds $\mathbf{B}_R(j\omega_i)$ and the U-contour are determined and are drawn in Fig. 21.34. From this figure it is seen that $\gamma \approx 58°$.

Step 8. Determination of the composite bounds $\mathbf{B}_o(j\omega_i)$. An analysis of Figs. 21.33 and 21.34 reveals that the bounds $\mathbf{B}_D(j\omega_i)$ for this example all lie above the tracking contours $\mathbf{B}_R(j\omega_i)$. Thus the $\mathbf{B}_D$ contours of Fig. 21.33 are more severe and become the $\mathbf{B}_o(j\omega_i)$ contours for the overall system of Fig. 21.35.

Step 9. Synthesizing or shaping of $L_o(s)$.

(*a*) Minimum structure. Since $P(s)$ is Type 1, then $L_o(s)$ must be at least Type 1 in order to maintain the Type 1 tracking characteristic for $L/(1 + L)$. Thus, the initial $L_o(s)$ has the form

$$L_{o0}(s) = \frac{K_{o1}}{s^m} \tag{21.104}$$

where $m = 1$ for this example. For the case where $T_D = P/(1 + L)$, it is desired that T_D have a zero at the origin so that $y(\infty) = 0$ for $d_1(t) = u_{-1}(t)$. Therefore $G(s)$ must have a pole at the origin. This requirement may place a severe restriction on trying to synthesize an L_o.

(*b*) Synthesis of $L_o(s)$. Using four cycle semilog graph paper and the data in Table 21.6, construct or shape $L_o(s)$ so that it comes as close as possible to the U-contour. $\mathbf{L}_o(j\omega_i)$ must be as close as possible to $\mathbf{B}_o(j\omega_i)$ but never below it. For the approximate range $\omega_x \approx 15 \leq \omega_i \leq \omega_y$, the angle $\angle \mathbf{L}_o(j\omega_i) \geq -123°$ must be satisfied. For this example the frequency ω_y is the value of $\mathbf{L}_o(j\omega_y)$ which results in $\text{Lm}\,\mathbf{L}_o(j\omega_i) \approx -36$ dB. These restrictions can be satisfied by assuming the format of the optimal transfer function as

$$\mathbf{L}_o(j\omega) = \mathbf{L}_{o0}(j\omega)\left[\frac{(j\omega - z_1)(j\omega - z_2)\cdots}{(j\omega - p_3)(j\omega - p_4)\cdots\left[\left(1 - \frac{\omega^2}{\omega_n^2}\right) + j2\zeta\frac{\omega}{\omega_n}\right]}\right] \tag{21.105}$$

which has w real zeros, $n = w + \lambda - 2$ real poles, and a pair of complex-conjugate poles. For this example a table having the format shown in Table 21.6 is used to assist in obtaining the desired $L_o(s)$.

The resulting loop transfer function is

$$L_o(s) = \frac{19 \times 10^9(s + 18)(s + 150)(s + 750)(s + 4000)}{s(s + 7)(s + 80)(s + 400)(s + 2900)(s^2 + 2\zeta\omega_n s + \omega_n^2)} \tag{21.106}$$

where $\zeta = 0.5$, $\omega_n = 16{,}000$ r/s, and $\omega_\phi \approx 200$ r/s. $\mathbf{L}_o(j\omega)$ is drawn on the Bode plot of Fig. 21.36. It is also plotted on the NC of Fig. 21.33. Since L_o is Type 1, then $T_D = 1/(1 + L)$ has a zero at the origin.

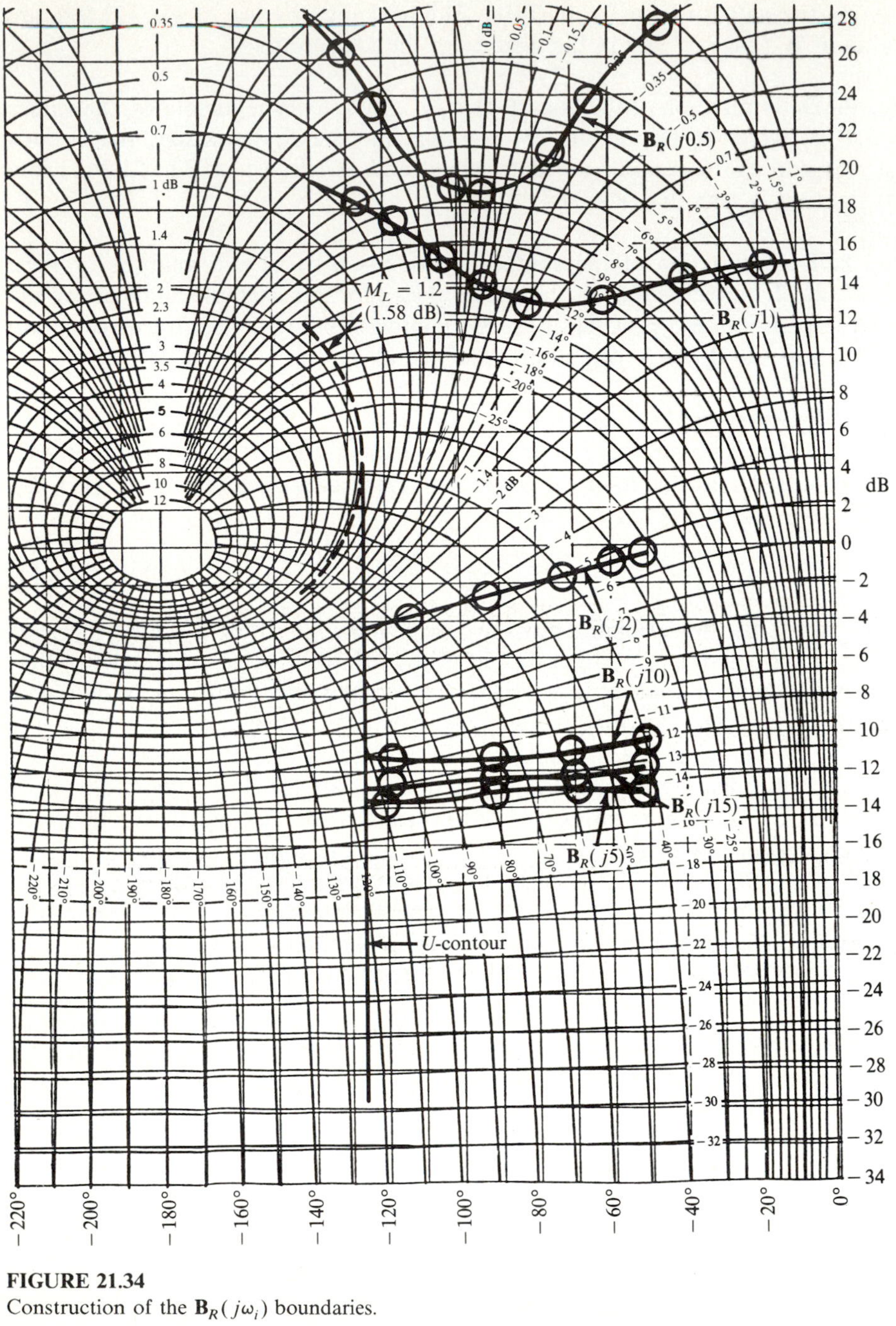

FIGURE 21.34
Construction of the $\mathbf{B}_R(j\omega_i)$ boundaries.

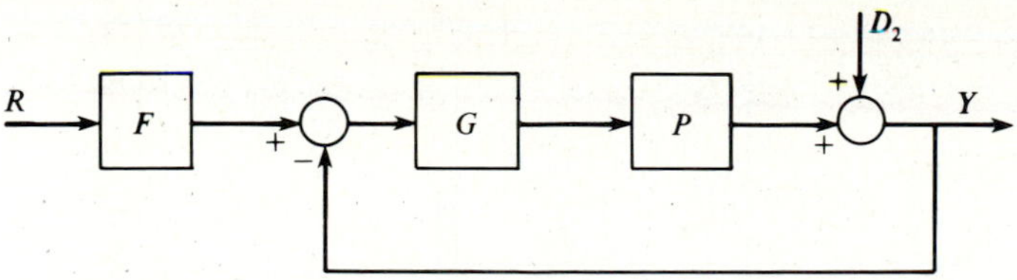

FIGURE 21.35
The MISO system.

TABLE 21.6
Data for loop shaping

ω	$\angle \mathbf{L}_{o1}$	**Angle contribution of pole (p) and zero (z) in degrees** p_2 -7	z_1 -18	p_3 -80	z_2 -150	p_4 -400	z_3 -750	p_5 -2900	z_4 -4000	$p_{6,7}$ $-8000 \pm j13856$	$\angle \mathbf{L}_o$
1	−90	−8.1	3.2	—	—	—	—	—	—	—	−94.9
5	−90	−35.5	15.5	−3.5	—	—	—	—	—	—	−113.5
10	−90	−55	29	−7.1	3.8	—	—	—	—	—	−119.3
15	−90	−65	40	−10.6	5.7	−2.2	1	—	—	—	−121.1
50	−90	−82	70.2	−32	18.5	−7.1	3.8	−1	—	—	−119.6
100	−90	−86	79.8	−51.5	33.5	−14	7.6	−2	1.5	—	−121.1
200	−90	−88	84.9	−68.6	53.2	−26.6	14.9	−3.9	3	—	−121.1
1000	−90	−89.6	88.9	−84.5	81.5	−68.2	53.2	−19.1	14	−4	−117.8
2000	−90	−90	89.48	−86.9	85.7	−78.6	69.5	−34.5	27	−7	−115.2

(c) Determination of $G(s)$. Use $L_o(s)$ and $P_o(s)$, where $P_o(s) = 1/s(s+1)$ is the nominal plant, to obtain the transfer function $G(s) = L_o(s)/P_o(s)$.

Step 10. The input filter $F(s)$ now needs to be synthesized to yield the desired tracking of the input by the output $y(t)$ in the manner described in Sec. 21.18. This is left as an exercise for the reader (see Prob. 21.9).

Step 11. Time responses for a disturbance input: $d(t) = u_{-1}(t)$. For each point of Fig. 21.31 substitute the corresponding data into

$$Y_D(s) = \frac{1}{1 + P(s)G(s)} D(s) \tag{21.107}$$

and determine $y_D(t)$ for each point. As indicated in Table 21.7, the disturbance time response characteristics for points A, B, C, and D satisfy the specification that $|y(t_x)| \le 0.01$ for $t_x \ge 60$ ms.

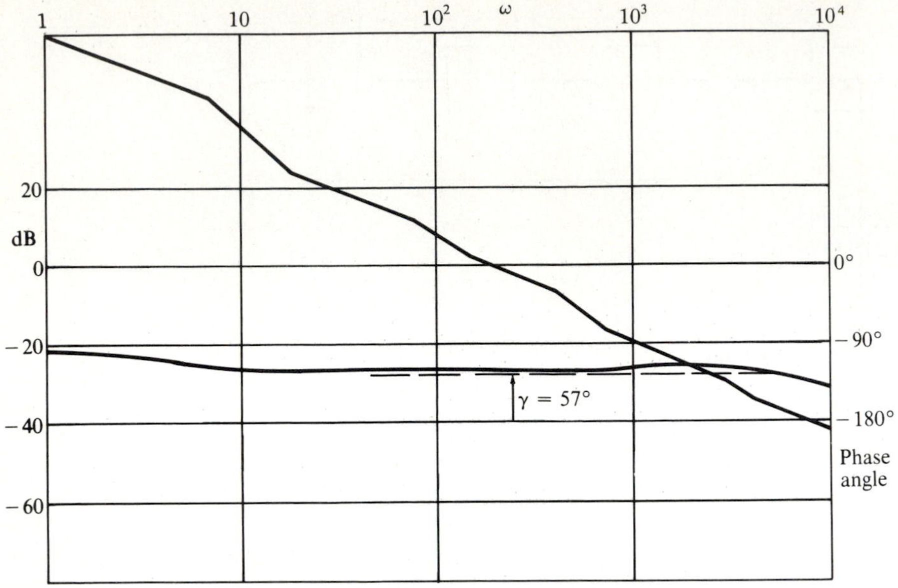

FIGURE 21.36
Bode plot of $\mathbf{L}_o(j\omega)$ of Eq. (21.106).

TABLE 21.7
Time response characteristics for Eq. (21.107)

Point	$y_D(t_x)$	Specified t_x	Actual t_x for $\lvert y_D(t)\rvert = 0.01$	$y(\infty)$
A	0.00998	0.06	0.06	0
B	0.0090	0.06	0.0074	0
C	0.00004	0.06	0.00226	0
D	0.000399	0.06	0.007	0

21.21 UNSTABLE PLANT TEMPLATE GENERATION

In the generation of templates for unstable plants, proper care must be exercised in analyzing the angular variation of $\mathbf{P}(j\omega)$ as the frequency is varied from zero to infinity. As an example, consider the plant

$$\mathbf{P}(j\omega) = \frac{K(j\omega - z_1)}{j\omega(j\omega - p_1)(j\omega - p_2)(j\omega - p_3)} \tag{21.108}$$

where p_1 and p_2 are a complex-conjugate pair and

$$\angle \lim_{\omega \to \infty} \mathbf{P}(j\omega) = \angle \frac{K}{(j\omega)^3} = -270° \tag{21.109}$$

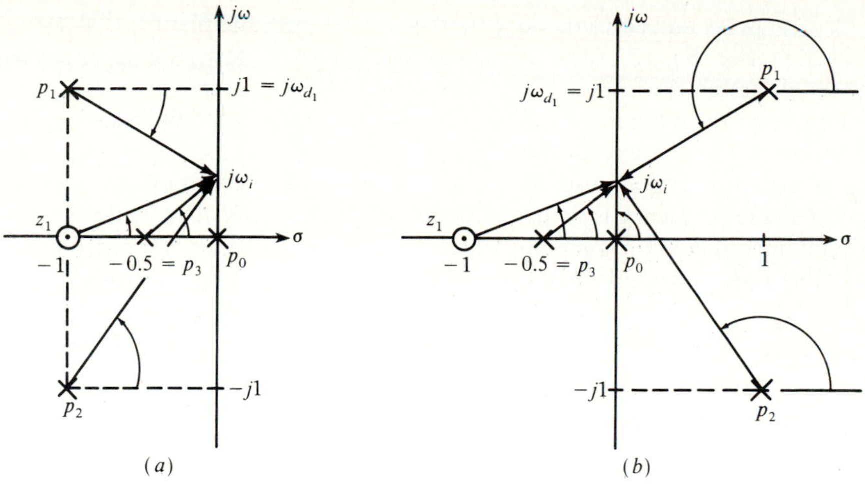

FIGURE 21.37
Poles and zeros in the s plane for (a) a stable plant and (b) an unstable plant.

In Eq. (21.108) the convention is assumed that, in the limit as $\omega \to \infty$, each $j\omega$ term contributes an angle of $+90°$. Two cases are analyzed for Eq. (21.108) as follows:

Case 1. The poles and zeros of Eq. (21.108) for a stable plant are plotted in Fig. 21.37a. In Eq. (21.108) the angle of $(j\omega_i - p_1)$ is negative (clockwise) for $\omega_i < \omega_d$ and is positive (counterclockwise) for $\omega_i > \omega_{d_1}$. All other angles are positive (counterclockwise) for all $0 \le \omega \le \infty$ as illustrated in Fig. 21.37a, and in Table 21.8.

TABLE 21.8
Angular variations of the minimum phase stable and unstable plants P($j\omega$) [see Eq. (21.108)] shown in Fig. 21.37

Case	ω_i	Angle					
		$\angle p_0$	$\angle p_1$	$\angle p_2$	$\angle p_3$	$\angle z_1$	$\angle P(j\omega_i)$
1	0	+90°	−45°	+45°	0°	0°	−90°
	1	+90°	0°	+63.5°	+63.5°	+45°	−172°
	∞	+90°	+90°	+90°	+90°	+90°	−270°
2	0	+90°	+225°	+135°	0°	0°	−450°
	1	+90°	+180°	+116.5°	+63.5°	+45°	−405°
	∞	+90°	+90°	+90°	+90°	+90°	−270°

Case 2. The poles and zeros of Eq. (21.108), for an unstable plant are plotted in Fig. 21.37*b*. For Eq. (21.108), all angles of first-order factors $(j\omega_i - p_i)$ are positive (counterclockwise), as illustrated in Fig. 21.37*b* and in Table 21.8.

Note that the difference in the angle of $\mathbf{P}(j0)$ between the two cases is 360° and is necessary to account for the right-half plane poles. Thus, because all angles are measured counterclockwise, the angle of $\mathbf{P}(j\omega)$ varies continuously in a given direction as the frequency is varied between zero and infinity. This feature is very important in obtaining a template where $\mathscr{P}$ contains both stable and unstable plants. The following guidelines should be used in the angular determination of $\angle\mathbf{P}(j\omega_i)$:

1. For stable plants, all angular directions are taken so that their values always lie within the range of $-90°$ to $+90°$.
2. For plants with right-half plane poles and zeros: (a) the angular directions for left-half plane poles and zeros of $P(s)$ are taken in the same manner as for the stable plant; (b) the angular directions of all right-half plane poles and zeros are all taken counterclockwise.

21.22 SUMMARY

A general introduction to the QFT technique is presented in this chapter. It is essentially a frequency-response design method applied to the design of a MIMO control system with an uncertain MIMO plant **P**. Guidelines for finding the **P** matrix are given. The method of representing a MIMO system by an equivalent set of MISO systems is presented using $\mathbf{P}^{-1}$. Two design approaches are available in which the equivalent MISO loops are designed according to the MISO design method presented in this chapter. This design is based upon

1. specifying the tolerance in the ω-domain by means of the sets of plant transfer functions and closed-loop control ratios, $\mathscr{P} = \{\mathbf{P}(j\omega)\}$ and $\mathscr{T}(j\omega)\} = \{\mathbf{T}(j\omega)\}$, respectively.
2. finding the resulting bounds on the loop transfer functions L_i and input filter transfer functions F_i of Figs. 21.2 and 21.17.

The robust design technique of this chapter permits an analog control system to satisfy the desired performance specifications within the specified range of plant parameter variation. Since the QFT technique is based upon the design of the loop transmission function $L(s)$ of the multiple-input single-output (MISO) control system of Fig. 21.2, it is referred to as a SISO design technique. Although this chapter deals with minimum phase plants, the QFT technique can be applied to nonminimum phase plants and to digital control systems.[4,13,45]

REFERENCES

1. Horowitz, I. M.: "Advanced Control Theory and Applications," unpublished notes, The Weizmann Institute of Science, Rehovot, Israel, 1982.
2. Horowitz, I. M., and M. Sidi: "Synthesis of Feedback Systems with Large Plant Ignorance for Prescribed Time Domain Tolerances," *Int. J. Control*, vol. 16, pp. 287–309, 1972.
3. Horowitz, I. M.: *Synthesis of Feedback Systems*, Academic, New York, 1963.
4. Schneider, D. L.: "QFT Digital Flight Control Design as Applied to the AFTI/F-16," M.S. Thesis, AFIT/GE/EE(86D-4), School of Engineering, Air Force Institute of Technology, Wright-Patterson AFB, Ohio, December 1986.
5. Bode, H. W.: *Network Analysis and Feedback Amplifier Design*, Van Nostrand, New York, 1945.
6. Houpis, C. H., and G. B. Lamont: *Digital Control Systems, Theory, Hardware, Software*, McGraw-Hill, New York, 1985.
7. Horowitz, I. M., and U. Shaked: "Superiority of Transfer Function over State-Variable Methods in Linear, Time Invariant Feedback System Design," *IEEE Trans. Autom. Control*, vol. AC-20, pp. 84–97, 1975.
8. Horowitz, I. M., et al.: "Research in Advanced Flight Control Design," Rep. AFFDL-TR-79-3120, Air Force Wright Aeronautical Laboratories, Wright-Patterson AFB, Ohio, 1979.
9. Horowitz, I. M., and M. Sidi: "Optimum Synthesis of Nonminimum-Phase Feedback Systems with Parameter Uncertainty," *Int. J. Control*, vol. 27, pp. 361–386, 1978.
10. Horowitz, I. M.: "Optimum Loop Transfer Function in Single-Loop Minimum Phase Feedback Systems," *Int. J. Control*, vol. 22, pp. 97–113, 1973.
11. Horowitz, I. M.: "Synthesis of Feedback Systems with Non-Linear Time Uncertain Plants to Satisfy Quantitative Performance Specifications," *IEEE Proc.*, vol. 64, pp. 123–130, 1976.
12. Betzold, R. W.: "Multiple Input-Multiple Output Flight Control Design with Highly Uncertain Parameters, Application to the C-135 Aircraft," M.S. Thesis, AFIT/GE/EE(83D-11), School of Engineering, Air Force Institute of Technology, Wright-Patterson AFB, Ohio, December 1983.
13. Horowitz, I. M., and C. Loecher: "Design of a 3 × 3 Multivariable Feedback System with Large Plant Uncertainty," *Int. J. Control*, vol. 33, pp. 677–699, 1981.
14. Horowitz, I. M.: "Quantitative Synthesis of Uncertain Multiple Input-Output Feedback Systems," *Int. J. Control*, vol. 30, pp. 81–106, 1979.
15. Rosenbrock, H. H.: *Computer-Aided Control System Design*, Academic, New York, 1974.
16. Horowitz, I. M., et al.: "Multivariable Flight Control Design with Uncertain Parameters (YF16CCV)," AFWAL-TR-83-3036, Air Force Wright Aeronautical Laboratories, Wright-Patterson AFB, Ohio, 1982.
17. Horowitz, I. M.: "Improved Design Technique for Uncertain Multiple Input-Output Feedback Systems," *Int. J. Control*, vol. 36, pp. 977–988, 1982.
18. Zames, G., and D. Bensoussan: "Multivariable Feedback Sensitivity and Optimal Robustness," *IEEE Trans. Autom. Control*, vol. AC-28, pp. 1030–1035, 1983.
19. Horowitz, I. M., and T. Kopelman: "Multivariable Flight Control Design with Uncertain Parameters," Final Report, The Weizmann Institute of Science, Rehovot, Israel, October 1981.
20. Horowitz, I. M.: "A Synthesis Theory for Linear Time-Varying Feedback Systems with Plant Uncertainty," *IEEE Trans. Autom. Control*, vol. AC-20, pp. 454–463, 1975.
21. Krishnan, K., and A. Cruickshanks: "Frequency Domain Design Feedback Systems for Specified Insensitivity of Time-Domain Response to Parameter Variation," *Int. J. Control*, vol. 25, pp. 609–620, 1977.
22. Sidi, M.: "Synthesis of Feedback Systems with Large Plant Uncertainty," Ph.D. thesis, The Weizmann Institute of Science, Rehovot, Israel, 1972.
23. East, D. J.: "A New Approach to Optimum Loop Synthesis," *Int. J. Control*, vol. 34, pp. 731–748, 1981.
24. Horowitz, I. M., and P. Rosenbaum: "Nonlinear Design for Cost of Feedback Reduction in Systems with Large Plant Uncertainty," *Int. J. Control*, vol. 21, pp. 977–1001, 1975.

25. Horowitz, I. M., and M. Sidi: "Synthesis of Cascaded Multiple-Loop Feedback Systems with Large Plant Parameter Ignorance," *Automatica*, vol. 9, pp. 589–600, 1973.
26. Horowitz, I. M., and T. S. Wang: "Quantitative Synthesis of Multiple-Loop Feedback Systems with Large Uncertainty," *Int. J. Syst. Sci.*, vol. 10, pp. 1235–1268, 1979.
27. Horowitz, I. M., and T. S. Wang: "Synthesis of a Class of Uncertain Multiple-Loop Feedback Systems," *Int. J. Control*, vol. 29, pp. 645–668, 1979.
28. Horowitz, I. M., and B. C. Wang: "Quantitative Synthesis of Uncertain Cascade Feedback System with Plant Modification," *Int. J. Control*, vol. 30, pp. 837–862, 1979.
29. Rosenbaum, P.: "Reduction in the Cost of Feedback in Systems with Large Parameter Uncertainties," D.E.E. Thesis, The Weizmann Institute of Science, Rehovot, Israel, 1977.
30. Horowitz, I. M., and D. Shur: "Control of a Highly Uncertain Van der Pol Plant," *Int. J. Control*, vol. 32, pp. 199–219, 1980.
31. Horowitz, I. M., and M. Breiner: "Quantitative Synthesis of Feedback Systems with Uncertain Nonlinear Multivariable Plants," *Int. J. Syst. Sci.*, vol. 12, pp. 593–563, 1981.
32. Horowitz, I. M.: "Improvement in Quantitative Nonlinear Feedback Design by Cancellation," *Int. J. Control*, vol. 34, pp. 547–560, 1981.
33. Horowitz, I. M.: "Design of Feedback Systems with Nonminimum-Phase Unstable Plants," *Int. J. Syst. Sci.*, vol. 10, pp. 1025–1040, 1979.
34. Horowitz, I. M.: "Quantitative Synthesis of Uncertain Nonlinear Feedback Systems with Nonminimum-Phase Inputs," *Int. J. Syst. Sci.*, vol. 12, pp. 55–76, 1981.
35. Horowitz, I. M.: "Nonlinear Uncertain Feedback Systems with Initial State Values," *Int. J. Control*, vol. 34, pp. 749–764, 1981.
36. Laban, M.: "Feedback System with Uncertain Nonlinear Plant," M.Sc. Thesis, The Weizmann Institute of Science, Rehovot, Israel, 1981.
37. Zeevi, G.: "Design and Simulation of a Nonlinear Feedback System," M.Sc. Thesis, The Weizmann Institute of Science, Rehovot, Israel, 1981.
38. Horowitz, I. M., B. Golubev, and T. Kopelman: "Flight Control Design Based on Nonlinear Model with Uncertain Parameters," *AIAA J. Guidance Control*, vol. 3, pp. 113–118, 1980.
39. Horowitz, I. M.: "Feedback Systems with Nonlinear Uncertain Plants," *Int. J. Control*, vol. 36, pp. 155–172, 1982.
40. Horowitz, I. M., et al.: "A Synthesis Technique for Highly Uncertain and Interacting Multivariable Flight Control Systems," *Proc. Natl. Aerosp. Electron. Conf. (NAECON), Dayton*, pp. 1276–1283, 1981.
41. Abrams, C. R.: "A Design Criterion for Highly Augmented Fly-By-Wire Aircraft," U.S. Naval Air. Dev. Center, Warminister, Pa., 1982.
42. Ramage, J. K., C. R. Abrams, and J. H. Watson: "AFTI/F-16 Digital Flight Control System Development Status," presented at 4th AIAA/IEEE Digital Avionics Systems Conf., St. Louis, Mo., Nov. 17–19, 1981.
43. Kochenburger, R. J.: "Limiting in Feedback Control Systems," *Trans. AIEE*, pt. II, *Appl. Ind.*, p. 180, July 1953.
44. Horowitz, I. M.: "A Quantitative Inherent Reconfiguration Theory for a Class of Systems," *Int. J. Syst. Sci.*, vol. 16, pp. 1377–1390, 1985.
45. Houpis, C. H.: "Quantitative Feedback Theory (QFT): Technique for Designing Multivariable Control Systems," AFWAL-TR-86-3107, Air Force Wright Aeronautical Laboratories, Wright-Patterson AFB, Ohio, January 1987. (Available from Defense Technical Information Center, Cameron Station, Alexandria, VA 22314, document number AD-A176883.)

CHAPTER 22

DIGITAL CONTROL SYSTEMS

22.1 INTRODUCTION

The methods presented in the previous chapters of this book deal with the fundamental properties of continuous systems. However, digital computers are available not only for the design of control systems but also to perform the control function. Recent advances in digital computers and microprocessors have made their use very attractive as components in control systems. The size and cost advantages associated with the microcomputer make its use as a compensator or controller economically and physically feasible. Using such a digital processor with a plant which operates in the continuous-time domain requires that its input signal be discrete, which in turn requires the sampling of the signals used by the controller. Such sampling may be an inherent characteristic of the system. For example, a radar tracking system supplies information on an airplane's position and motion to a digital processor at discrete periods of time. This information is therefore available as a succession of data points. When there is no inherent sampling, an analog-to-digital (A/D) converter must be incorporated in a digital or sampled-data (SD) control system. The sampling process can be performed at a constant rate or at a variable rate, or it may be random. The discussion in this book is based on a constant-rate sampling. The output of the controller must be converted from discrete form into an analog signal by a digital-to-analog (D/A) converter. A functional block diagram of such a system is shown in Fig. 22.1, in which the signals e^* and u^* are in discrete form, u is piecewise-continuous, and the remaining signals are continuous. The notation e^*

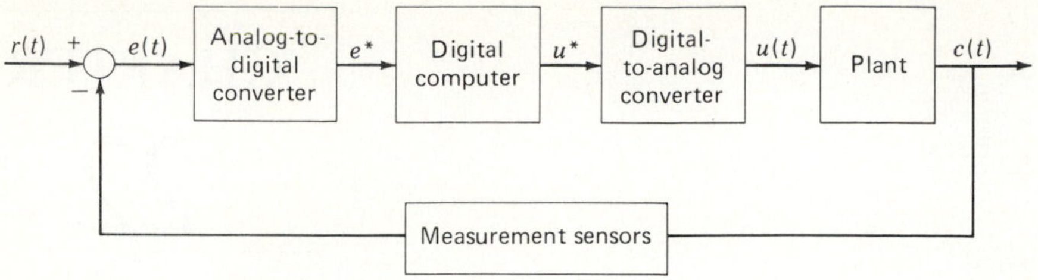

FIGURE 22.1
Control system incorporating a digital computer.

and u^* denotes that these signals are sampled at specified time intervals and are therefore in discrete form. Systems which include a digital computer are known as *digital control systems*. The synthesis techniques for such systems are based on representing the entire system as an equivalent sampled or discrete or an equivalent *pseudo-continuous-time* (PCT) system.

There are various approaches that may be used in analyzing the stability and time-response characteristics of sampled-data systems. These approaches may be divided into two distinct categories: (1) *direct* (DIR) and (2) *digitization* (DIG) or *discrete* digital control analysis techniques. The first category involves carrying out the analysis entirely in the discrete domain (z plane). The second category permits the analysis and synthesis of the sampled-data system to be carried out entirely by transformation to the w' plane or, by use of the Padé approximation (see Ref. 1, App. C), entirely in the s plane. The use of the Padé approximation, along with the information provided by a Fourier analysis of the sampled signal, results in the modeling of a sampled-data control system by a PCT control system. The w'-plane analysis is not covered in this text but the reader is referred to Ref. 1 where an overview of the DIR and DIG methods is presented.

22.2 SAMPLING

Sampling may occur at one or more places in a system. The sampling operation is represented in a block diagram by the symbol for a switch. Figure 22.2 shows a system with sampling of the actuating signal. Note that the output $c(t)$ is a continuous function of time.

The sampling process can be considered a modulation process in which a pulse train $p(t)$ with magnitude $1/\gamma$ and period T multiplies a continuous-time

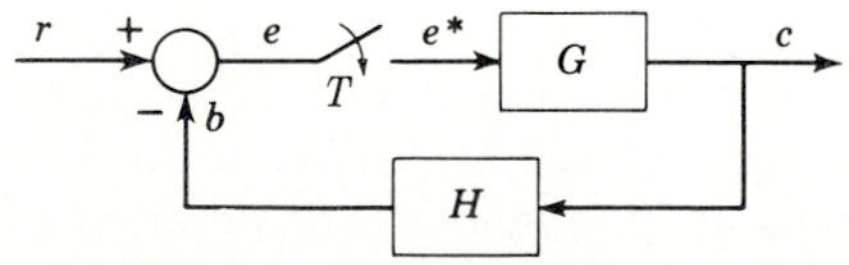

FIGURE 22.2
Block diagram of a system containing sampling of the actuating signal.

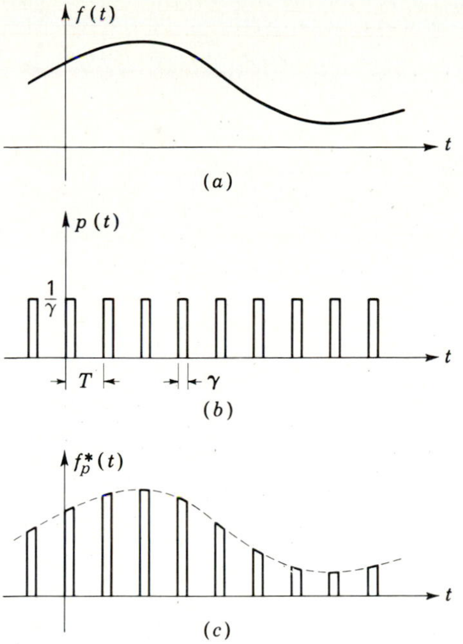

FIGURE 22.3
(a) Continuous function $f(t)$; (b) sampling pulse train $p(t)$; (c) sampled function $f_p^*(t)$.

function $f(t)$ and produces the sampled function $f_p^*(t)$. This is represented by

$$f_p^*(t) = p(t)f(t) \tag{22.1}$$

These quantities are shown in Fig. 22.3. A Fourier-series expansion of $p(t)$ is

$$p(t) = \frac{1}{\gamma} \sum_{n=-\infty}^{+\infty} C_n e^{jn\omega_s t} \tag{22.2}$$

where the sampling frequency is $\omega_s = 2\pi/T$ and the Fourier coefficients C_n are given by

$$C_n = \frac{1}{T} \int_0^T p(t) e^{-jn\omega_s t}\, dt = \frac{1}{T} \frac{\sin(n\omega_s\gamma/2)}{n\omega_s\gamma/2} e^{-jn\omega_s\gamma/2} \tag{22.3}$$

The sampled function is therefore

$$f_p^*(t) = \sum_{n=-\infty}^{+\infty} C_n f(t) e^{jn\omega_s t} \tag{22.4}$$

If the Fourier transform of the continuous function $f(t)$ is $\mathbf{F}(j\omega)$, the Fourier transform of the sampled function $f_p^*(t)$ is[2,3]

$$\mathbf{F}_p^*(j\omega) = \sum_{n=-\infty}^{+\infty} C_n \mathbf{F}(j\omega + jn\omega_s) \tag{22.5}$$

A comparison of the Fourier spectra of the continuous and sampled functions is shown in Fig. 22.4. It is seen that the sampling process produces a fundamental

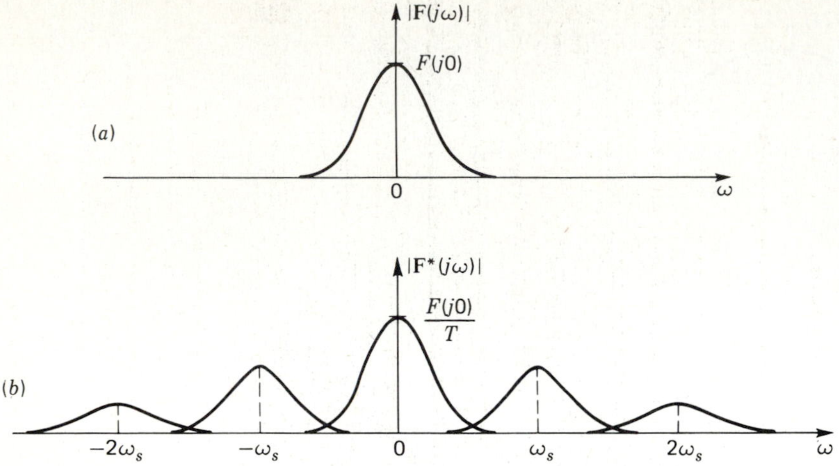

FIGURE 22.4
Frequency spectra for (a) a continuous function $f(t)$ and (b) a pulse-sampled function $f_p^*(t)$.

spectrum similar in shape to that of the continuous function. It also produces a succession of spurious complementary spectra which are shifted periodically by a frequency separation $n\omega_s$.

If the sampling frequency is sufficiently high, there is very little overlap between the fundamental and complementary frequency spectra. In that case a low-pass filter could extract the spectrum of the continuous input signal by attenuating the spurious higher-frequency spectra. The forward transfer function of a control system generally has a low-pass characteristic, so the system responds with more or less accuracy to the continuous signal.

If the sampling pulse has a duration γ which is small compared with the sampling time T, the pulse train $p(t)$ can be represented by an ideal sampler. Since the area of each pulse of $p(t)$ has unit value, the impulse train also has a magnitude of unity. The ideal sampler produces the impulse train $\delta_T(t)$, which is shown in Fig. 22.5 and is represented by

$$\delta_T(t) = \sum_{n=-\infty}^{+\infty} \delta(t - nT) \tag{22.6}$$

where $\delta(t - nT)$ is the unit impulse which occurs at $t = nT$. The frequency

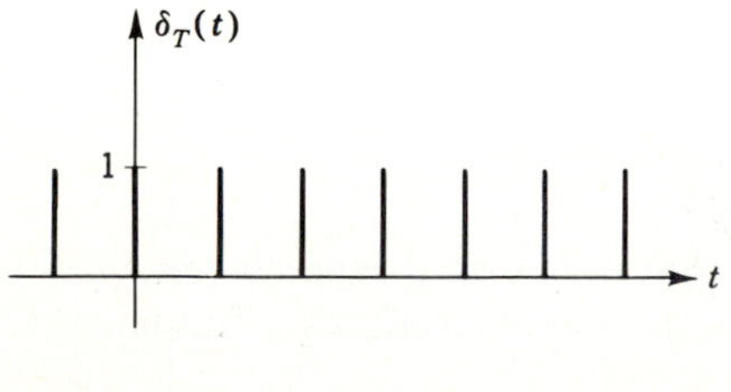

FIGURE 22.5
Impulse train.

spectrum of the function which is sampled by the ideal impulse train is similar to that shown in Fig. 22.4, except that the complementary spectra have the same amplitude as the fundamental spectrum. Since the forward transfer function of a control system attenuates the higher frequencies, the overall system response is essentially the same with the idealized impulse sampling as with the actual pulse sampling. The use of impulse sampling simplifies the mathematical analysis of sampled systems and is therefore used extensively to represent the sampling process.

When $f(t) = 0$ for $t < 0$, the Laplace transform of the impulse sequence

$$f^*(t) = f(t)\delta_T(t) = \sum_{n=0}^{+\infty} f(t)\delta_T(t - nT) \tag{22.7}$$

is given by the infinite series

$$F^*(s) = \sum_{n=0}^{+\infty} f(nT)e^{-nTs} \tag{22.8}$$

where $f(nT)$ represents the function $f(t)$ at the sampling times nT. The expression $F^*(s)$ contains the term e^{Ts}, which means that it is not an algebraic expression but a transcendental one. Therefore a change of variable is made:

$$z = e^{Ts} \tag{22.9}$$

Equation (22.8) can now be written as

$$[F^*(s)]_{s=(1/T)\ln z} = F(z) = \sum_{n=0}^{+\infty} f(nT)z^{-n} \tag{22.10}$$

For functions $f(t)$ which have zero value for $t < 0$, the one-sided z transform is the infinite series

$$F(z) = f(0) + f(T)z^{-1} + f(2T)z^{-2} + \cdots \tag{22.11}$$

If the Laplace transform of $f(t)$ is a rational function, it is possible to write $F^*(s)$ in closed form. When the degree of the denominator of $F(s)$ is at least 2 higher than the degree of the numerator, the closed form can be obtained from

$$F^*(s) = \sum_{\substack{\text{at poles}\\ \text{of } F(p)}} \text{residues of} \left[F(p)\frac{1}{1 - e^{-(s-p)T}} \right] \tag{22.12}$$

where $F(p)$ is the Laplace transform of $f(t)$ with s replaced by p. The z transform in closed form may be obtained from Eq. (22.12) and is written as[4]

$$F(z) = \hat{F}(z) + \beta = \sum_{\substack{\text{at poles}\\ \text{of } F(p)}} \text{residues of} \left[F(p)\frac{1}{1 - e^{pT}z^{-1}} \right] + \beta \tag{22.13}$$

where

$$\beta = \lim_{s\to\infty} sF(s) - \lim_{z\to\infty} \hat{F}(z) \tag{22.14}$$

The value of β given by Eq. (22.14) ensures that the initial value $f(0)$ represented by $F(s)$ and $F(z)$ are identical. *$F(z)$ is called the z transform of $f^*(t)$.* The starred and z forms of the *impulse response transfer function* are easily obtained for the ordinary Laplace transfer function.

Example 1. Consider the transfer function

$$G(s) = \frac{K}{s(s+a)} \tag{22.15}$$

The corresponding impulse transfer functions are

$$G^*(s) = \frac{Ke^{-sT}(1 - e^{-aT})}{a(1 - e^{-sT})(1 - e^{-(s+a)T})} \tag{22.16}$$

in the s domain and

$$G(z) = \frac{Kz^{-1}(1 - e^{-aT})}{a(1 - z^{-1})(1 - e^{-aT}z^{-1})} \tag{22.17}$$

in the z domain. The poles of $G^*(s)$ in Eq. (22.16) are infinite in number. These poles exist at $s = jn\omega_s$ and $s = -a + jn\omega_s$ for all values of $-\infty < n < +\infty$. Note that they are uniformly spaced with respect to their imaginary parts with a separation $j\omega_s$. By contrast, there are just two poles of Eq. (22.17), located at $z = 1$ and $z = e^{-aT}$. The root-locus method can therefore be applied in the z plane, whereas for sampled functions it is not very convenient to use in the s plane because of the infinite number of poles.

The transformation of the s plane into the z plane can be investigated by inserting $s = \sigma + j\omega$ into

$$z = e^{Ts} = e^{\sigma T}e^{j\omega T} = e^{\sigma T}e^{j2\pi\omega/\omega_s} \tag{22.18}$$

1. Lines of constant σ in the s plane map into circles of radius equal to $e^{\sigma T}$ in the z plane. Specifically, the segment of the imaginary axis in the s plane of width ω_s maps into the circle of unit radius in the z plane; successive segments map into overlapping circles. This fact shows that the *proper* consideration of sampled-data systems in the z plane requires the use of a multiple-sheeted surface, i.e., a Riemann surface. But by virtue of the uniform repetition of the roots of the characteristic equation, the condition for stability is that all roots of the characteristic equation contained in the principal branch lie within the unit circle in the principal sheet of the z plane.
2. Lines of constant ω in the s plane map into radial rays drawn at the angle ωT in the z plane. The portion of the constant ω line in the left half of the s plane becomes the radial ray within the unit circle in the z plane. The negative part of the real axis $-\infty < \sigma \leq 0$ in the s plane is mapped on the segment of the real axis defined by $0 < z \leq 1$.
3. The constant-damping-ratio ray in the s plane is defined by the equation

$$s = \sigma + j\omega = -\omega \cot \eta + j\omega$$

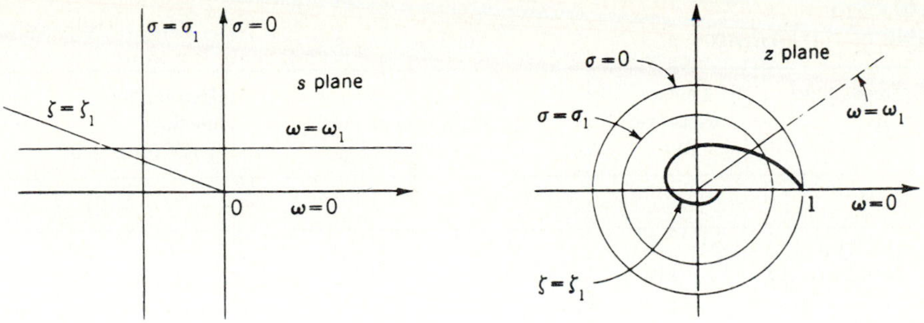

FIGURE 22.6
Transformation from the s plane to the z plane.

where $\eta = \cos^{-1}\zeta$. Therefore

$$z = e^{sT} = e^{-\omega T \cot \eta + j\omega T} = e^{-\omega T \cot \eta} \angle \omega T \tag{22.19}$$

The corresponding map describes a logarithmic spiral in the z plane.

The corresponding paths, as discussed above, are shown in Fig. 22.6.

22.3 z-TRANSFORM THEOREMS

A short list of z transforms is given in Table 22.1. Several simple properties of the one-sided z transform permit the extension of the table and facilitate its use in solving difference equations. These properties are presented as theorems. They apply when the z transform of $f^*(t)$, denoted by $\mathscr{Z}[f^*(t)]$, is $F(z)$ and the sampling time is T.

Theorem 1 Translation in time (time shift). (a) Shifting to the right (delay) yields

$$\mathscr{Z}[f^*(t-pT)] = z^{-p}F(z) \tag{22.20}$$

(b) Shifting to the left (advance) yields

$$\mathscr{Z}[f^*(t+pT)] = z^{p}F(z) - \sum_{i=0}^{p-1} f(iT)z^{p-i} \tag{22.21}$$

Theorem 2 Final value. If $F(z)$ converges for $|z| > 1$ and all poles of $(1-z)F(z)$ are inside the unit circle, then

$$\lim_{n\to\infty} f(nT) = \lim_{z\to 1}\left[(1-z^{-1})F(z)\right] \tag{22.22}$$

Theorem 3 Initial value. If $\lim_{z\to\infty} F(z)$ exists, then

$$\lim_{n\to 0} f(nT) = \lim_{z\to\infty} F(z) \tag{22.23}$$

TABLE 22.1
Table of z transforms

Continuous time function $f(t)$	Laplace transform $F(s)$	z transform of $f^*(t)$ $F(z)$	Discrete time function $f(nT)$ for $n \geq 0$
1. $\delta_0(t)$	1	1	$\delta_0(0)$
2. $\delta(t-kT)$ k is any integer	e^{-kTs}	z^{-1}	$\delta(t-kT)$
3. $u_{-1}(t)$	$\dfrac{1}{s}$	$\dfrac{z}{z-1}$	$\delta(t-nT)$
4. $tu_{-1}(t)$	$\dfrac{1}{s^2}$	$\dfrac{Tz}{(z-1)^2}$	$nT\delta(t-nT)$
5. $\dfrac{t^2}{2}u_{-1}(t)$	$\dfrac{1}{s^3}$	$\dfrac{T^2z(z+1)}{2(z-1)^3}$	$(nT)^2\delta(t-nT)$
6. e^{-at}	$\dfrac{1}{s+a}$	$\dfrac{z}{z-e^{-aT}}$	e^{-anT}
7. $\dfrac{e^{-bt}-e^{-at}}{a-b}$	$\dfrac{1}{(s+a)(s+b)}$	$\dfrac{1}{a-b}\left[\dfrac{z}{z-e^{-bt}}-\dfrac{z}{z-e^{-aT}}\right]$	$\dfrac{1}{a-b}(e^{-bnT}-e^{-anT})$
8. $u_{-1}(t)-e^{-at}$	$\dfrac{a}{s(s+a)}$	$a\dfrac{(1-e^{-aT})z}{(z-1)(z-e^{-aT})}$	$\delta(t-nT)e^{-anT}$
9. $t-\dfrac{1-e^{-aT}}{a}$	$\dfrac{a}{s^2(s+a)}$	$\dfrac{Tz}{(z-1)^2}-\dfrac{(1-e^{-aT})z}{a(z-1)(z-e^{-aT})}$	$nT-\dfrac{1-e^{-anT}}{a}$
10. $\sin at$	$\dfrac{a}{s^2+a^2}$	$\dfrac{z\sin aT}{z^2-2z\cos aT+1}$	$\sin anT$
11. $\cos at$	$\dfrac{s}{s^2+a^2}$	$\dfrac{z(z-\cos aT)}{z^2-2z\cos aT+1}$	$\cos ant$
12. $e^{-at}\sin bt$	$\dfrac{b}{(s+a)^2+b^2}$	$\dfrac{ze^{-aT}\sin bT}{z^2-2ze^{-aT}\cos bT+e^{-2aT}}$	$e^{-anT}\sin bnT$
13. $e^{-at}\cos bt$	$\dfrac{s+a}{(s+a)^2+b^2}$	$\dfrac{z^2-ze^{-aT}\cos bT}{z^2-2ze^{-aT}\cos bT+e^{-2aT}}$	$e^{-anT}\cos bnT$
14. te^{-at}	$\dfrac{1}{(s+a)^2}$	$\dfrac{Tze^{-aT}}{(z-e^{-aT})^2}$	nTe^{-anT}

22.4 SYNTHESIS IN THE z DOMAIN (DIRECT METHOD)

For the block diagram of Fig. 22.2, the system equations are

$$C(s) = G(s)E^*(s) \tag{22.24}$$

$$E(s) = R(s) - B(s) = R(s) - G(s)H(s)E^*(s) \tag{22.25}$$

The starred transform of Eq. (22.25) is

$$E^*(s) = R^*(s) - GH^*(s)E^*(s) \tag{22.26}$$

where $GH^*(s) = [G(s)H(s)]^*$ and, in general, $GH^*(s) \neq G^*(s)H^*(s)$.

Solving for $C(s)$ from Eqs. (22.24) and (22.26) gives

$$C(s) = \frac{G(s)R^*(s)}{1 + GH^*(s)} \tag{22.27}$$

The starred transform $C^*(s)$ obtained from Eq. (22.27) is

$$C^*(s) = \frac{G^*(s)R^*(s)}{1 + GH^*(s)} \tag{22.28}$$

The z transform is obtained by replacing each starred transform by the corresponding function of z:

$$C(z) = \frac{G(z)R(z)}{1 + GH(z)} \tag{22.29}$$

Although the output $c(t)$ is continuous, the inverse of $C(z)$ in Eq. (22.29) yields only the set of values $c(nT)$, $n = 0, 1, 2, \ldots,$ corresponding to the values at the sampling instants. Thus, $c(nT)$ is the set of impulses from an ideal sampler located at the output which operates in synchronism with the sampler of the actuating signal.

Analysis of the performance of the sampled system, corresponding to the response given by Eq. (22.29), can be performed by the frequency-response or root-locus method. It should be kept in mind that the geometric boundary for stability in the z plane is the unit circle. As an example of the root-locus method, consider the sampled feedback system represented by Fig. 22.2 with $H(s) = 1$. Using $G(s)$ given by Eq. (22.15) with $a = 1$ and a sampling time $T = 1$, the equation of $G(z)$ is

$$G(z) = GH(z) = \frac{0.632Kz}{(z-1)(z-0.368)} \tag{22.30}$$

The characteristic equation is $1 + GH(z) = 0$ or

$$GH(z) = -1 \tag{22.31}$$

The usual root-locus techniques can be used to obtain a plot of the roots of Eq. (22.31) as a function of the sensitivity K. The root locus is drawn in Fig. 22.7. The maximum value of K for stability is obtained from the magnitude condition which yields $K_{max} = 2.73$. This occurs at the crossing of the root locus and the unit circle. The selection of the desired roots can be based on the damping ratio ζ

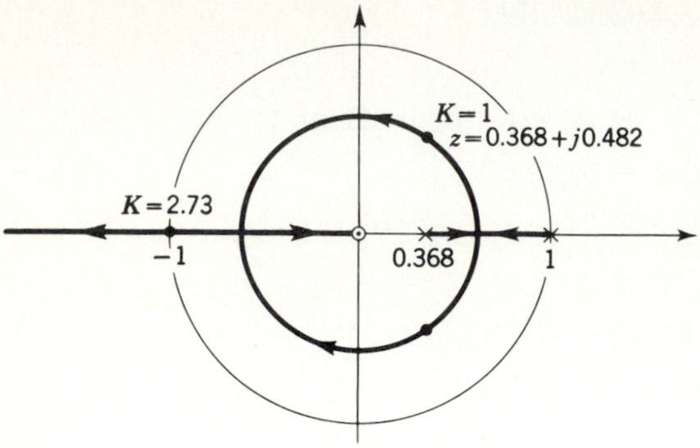

FIGURE 22.7
Root locus for Eqs. (22.30) and (22.31).

desired. For the specified value of ζ the spiral given by Eq. (22.19) and shown in Fig. 22.6 must be drawn. The intersection of this curve with the root locus determines the roots. Alternatively, it is possible to specify the settling time. This determines the value of σ in the s plane, so the circle of radius $e^{\sigma T}$ can be drawn in the z plane. The intersection of this circle and the root locus determines the roots. For the value $\zeta = 0.48$ the roots are $z = 0.368 \pm j0.482$, shown on the root locus in Fig. 22.7, and the value of $K = 1$. The control ratio is therefore

$$\frac{C(z)}{R(z)} = \frac{0.632z}{z - 0.368 \pm j0.482} = \frac{0.632z}{z^2 - 0.736z + 0.368} \tag{22.32}$$

For a unit-step input the value of $R(z)$ is

$$R(z) = \frac{z}{z-1} \tag{22.33}$$

so that

$$C(z) = \frac{0.632z^2}{(z-1)(z^2 - 0.736z + 0.368)} \tag{22.34}$$

The expression for $C(z)$ can be expanded by dividing its denominator into its numerator to get a power series in z^{-1}:

$$C(z) = 0.632z^{-1} + 1.096z^{-2} + 1.205z^{-3} + 1.120z^{-4} + 1.014z^{-5} + 0.98z^{-6} + \cdots \tag{22.35}$$

The inverse transform of $C(z)$ is

$$c(nT) = 0\delta(t) + 0.632\delta(t - T) + 1.096\delta(t - 2T) \cdots \tag{22.36}$$

Hence, comparing Eq. (22.35) with Eq. (22.11), the values of $c(nT)$ at the sampling instants are the coefficients of the terms in the series of Eq. (22.36) at the corresponding sampling instant. A plot of the values of $c(nT)$ is shown in Fig. 22.8. The curve of $c(t)$ is drawn as a smooth curve through these plotted points.

This section presents several basic points in the analysis of sampled-data systems. Additional topics of importance include the reconstruction of the continuous signal from the sampled signal by use of *hold* circuits and the problem of compensation to improve performance. These topics are presented in the following sections.

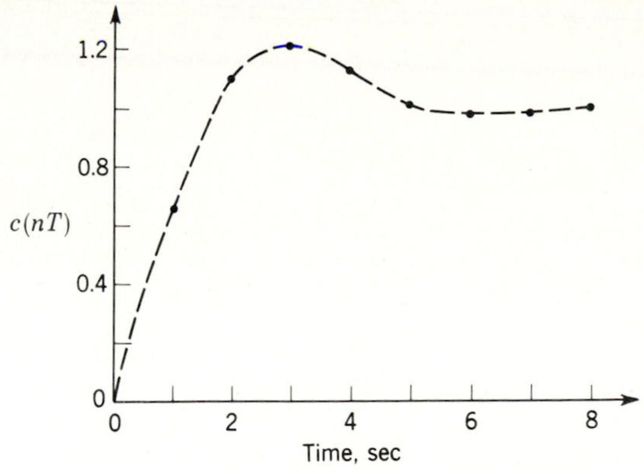

FIGURE 22.8
Plot of $c(nT)$ for Eq. (22.36).

22.5 THE INVERSE z TRANSFORM

In the example of Sec. 22.4 the inverse of $C(z)$ is obtained by expanding this function into an infinite series in terms of z^{-n}. This is referred to as the *power-series method*. Then the coefficients of z^{-n} are the values of $c(nT)$. The division required to obtain $C(z)$ as an infinite series (open form) is easily performed on a digital computer. However, if an analytical expression in closed form for $c^*(t)$ is desired, $C(z)$ can be expanded into partial fractions which appear in Table 22.1. Since the z transforms in Table 22.1 contain z in the numerator, the Heaviside partial fraction is first performed on $C(z)/z$ which must be in proper form.

Example 2. For $C(z)$ given by Eq. (22.34) with $T = 1$

$$\frac{C(z)}{z} = \frac{A}{z-1} + \frac{Bz + C}{z^2 - 0.736z + 0.368} \tag{22.37}$$

The coefficients are evaluated by the usual Heaviside partial-fraction methods and yield $A = 1$, $B = -1$, and $C = 0.368$. Inserting these values into Eq. (22.37) and using the form of entries 3 and 13 of Table 22.1, the response transform $C(z)$ is

$$C(z) = \frac{z}{z-1} - \frac{z(z - e^{-a}\cos b)}{z^2 - 2ze^{-a}\cos b + e^{-2a}} \tag{22.38}$$

where $a = 0.5$ and $b = 1$. Thus, from Table 22.1 the system output of Eq. (22.38) in closed form is

$$c(nT) = 1 - e^{-0.5n}\cos n \tag{22.39}$$

The open form introduces an error that is propagated in the division process due to roundoff. This does not occur with the closed form.

22.6 ZERO-ORDER HOLD

The function of a zero-order hold (ZOH) is to reconstruct a piecewise-continuous signal from the sampled function $f^*(t)$. It holds the output amplitude constant at the value of the impulse for the duration of the sampling period T. Thus, it has

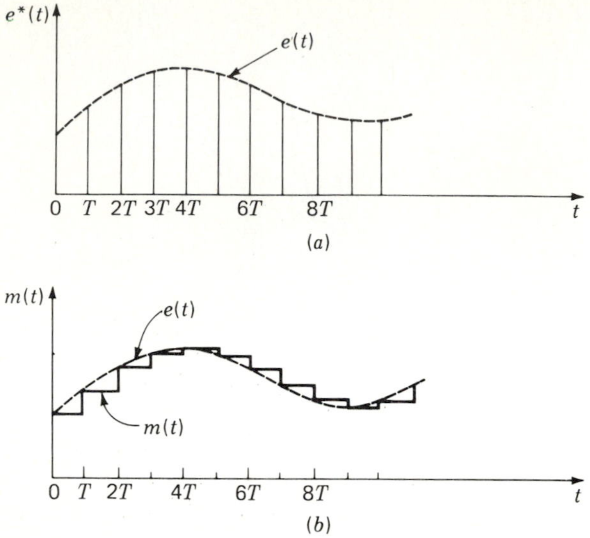

FIGURE 22.9
Input and output signals for a zero-order hold: (a) continuous input signal $e(t)$ and the sampled signal $e^*(t)$; (b) continuous signal $e(t)$ and the piecewise-constant output $m(t)$ of the zero-order hold.

the transfer function

$$G_{\text{ZOH}}(s) = \frac{1 - e^{-Ts}}{s} \tag{22.40}$$

The action of the ZOH is shown in Fig. 22.9. When the sampling time T is small or the signal is slowly varying, the output of the ZOH is frequently used in digital control systems. It is used to convert the discrete signal obtained from a digital computer into a piecewise-continuous signal which is the input to the plant. The ZOH is a low-pass filter which, together with the basic plant, attenuates the complementary frequency spectra introduced by the sampling process. The ZOH does introduce a lag angle into the system, and therefore it can affect the stability and time-response characteristics of the system.

Example 3. Figure 22.10 shows a unity-feedback system in which a zero-order hold is placed after the sampler. The z transform of the forward transfer function is

$$G(z) = \frac{C(z)}{E(z)} = \mathscr{Z}\left[\frac{1 - e^{-Ts}}{s}\,\frac{K}{s(s+1)}\right] \tag{22.41}$$

Since $\mathscr{Z}(1 - e^{-Ts}) = 1 - z^{-1}$, this term can be factored from Eq. (22.41) to give, for $T = 1$,

$$G(z) = (1 - z^{-1})\mathscr{Z}\left[\frac{K}{s^2(s+1)}\right] = \frac{K(ze^{-1} + 1 - 2e^{-1})}{(z-1)(z-e^{-1})} = \frac{0.368K(z + 0.717)}{(z-1)(z-0.368)} \tag{22.42}$$

A comparison with the z transform without the ZOH as given by Eq. (22.30) shows that the zero is moved from the origin to $z = -0.717$.

The root locus for the system of Fig. 22.10 is shown in Fig. 22.11 and may be compared with the root locus without the ZOH in Fig. 22.7. The maximum value of

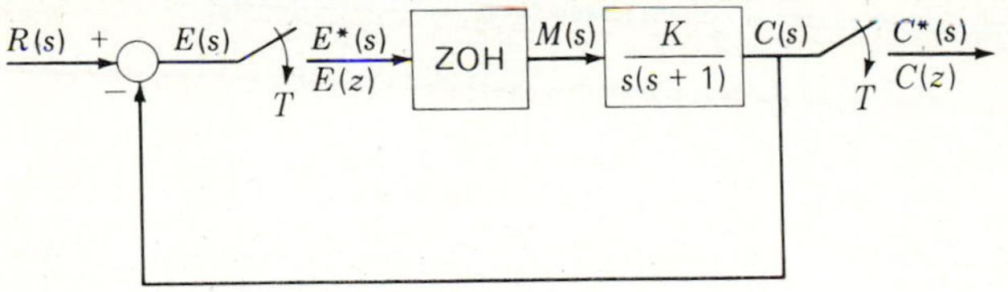

FIGURE 22.10
Sampled system containing a ZOH.

the gain K for stability is $K_{max} = 0.8824$. With $\zeta = 0.707$, the closed-loop control ratio is

$$\frac{C(z)}{R(z)} = \frac{0.118(z + 0.717)}{(z - 0.625 + j0.249)(z - 0.625 - j0.249)} = \frac{0.118(z + 0.717)}{z^2 - 1.25z + 0.453} \tag{22.43}$$

For this control ratio, $e^{\sigma T} = 0.673$ so that the real part of the closed-loop poles is $\sigma = -0.396$ and the settling time is 10.1 s.

This example illustrates that a second-order sampled system becomes unstable for large values of gain. This is in contrast to a continuous second-order system, which is stable for all positive values of gain. The advantage of a discrete control system is the greater flexibility of compensation that can be achieved with a digital compensator.

22.7 DIGITAL-COMPUTER COMPENSATION

A damping ratio of $\zeta = 0.707$ and a settling time of $T_s \approx 4$ s are desired for the system of Fig. 22.10. The settling time requires that the roots be located inside the circle of radius $e^{\sigma T}$. Since $T_s = 4/|\sigma|$, the requirement is that $\sigma = -1$. The circle of radius $e^{-1} = 0.368$ is drawn in Fig. 22.11. It is evident that the root locus does not cross this circle and therefore the settling time cannot be achieved.

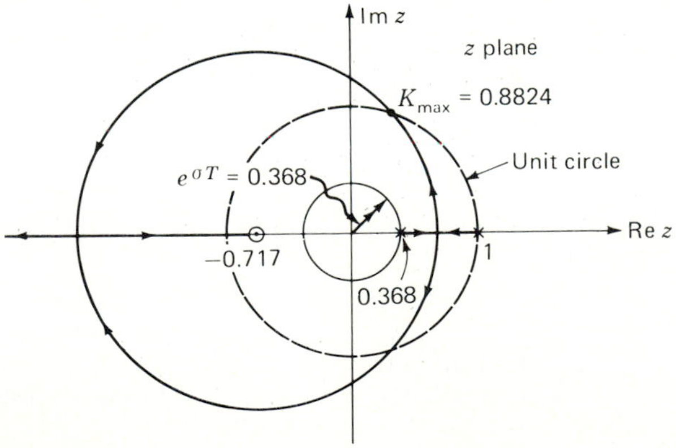

FIGURE 22.11
Root locus for the system of Fig. 22.10.

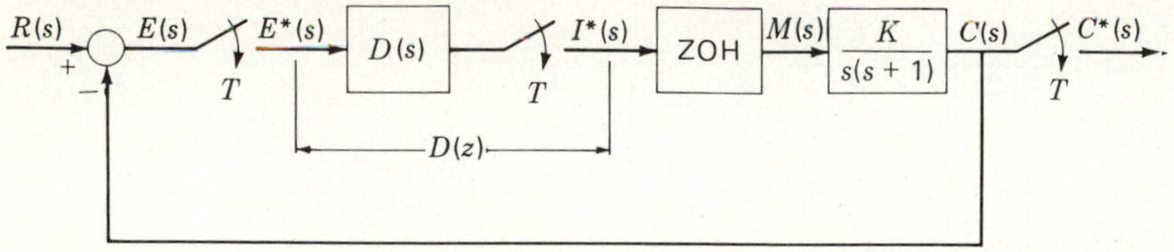

FIGURE 22.12
A sampled control system containing a digital compensator $D(z)$.

A digital cascade compensator $D(z)$ can be added to the system to improve the performance and is shown in Fig. 22.12. The digital compensator must be selected to bring the root locus within the circle of radius $e^{\sigma T} = 0.368$. A possible compensator is

$$D(z) = \frac{I(z)}{E(z)} = \frac{z - 0.368}{z + 0.230} \tag{22.44}$$

The zero of this compensator cancels the pole of $G(z)$ at $z = 0.368$ [see Eq. (22.42)] so that

$$D(z)G(z) = \frac{0.368K(z + 0.717)}{(z - 1)(z + 0.230)} \tag{22.45}$$

The new root locus is shown in Fig. 22.13. Since a portion of the root locus is within the circle of radius equal to 0.368, achieving the desired settling time depends on the damping ratio. For $\zeta = 0.707$ the closed-loop sampled control ratio is

$$\frac{C(z)}{R(z)} = \frac{0.473(z + 0.717)}{(z - 0.149 + j0.295)(z - 0.149 - j0.295)}$$

$$= \frac{0.473(z + 0.717)}{z^2 + 0.298z + 0.109} \tag{22.46}$$

which yields $\sigma = -1.1$. Based on continuous-time criteria, the transient settling time is $T_s = 4/|\sigma| = 3.64$ s. The use of the cascade digital compensator has reduced the settling time from 10.1 to 3.64 s and meets the specifications. The

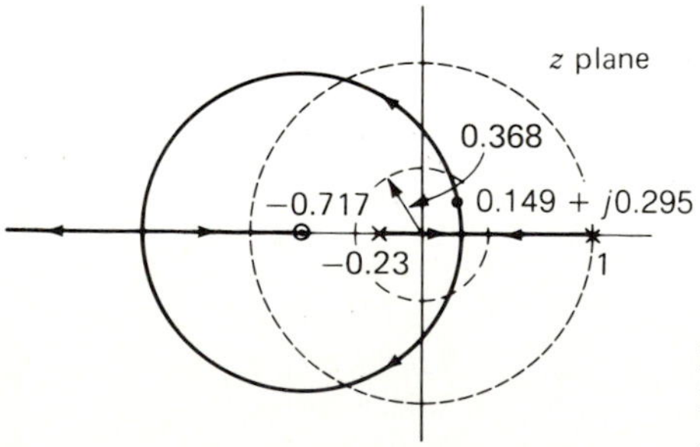

FIGURE 22.13
Root locus for the system of Fig. 22.12, where $D(z)G(z)$ is given by Eq. (22.45).

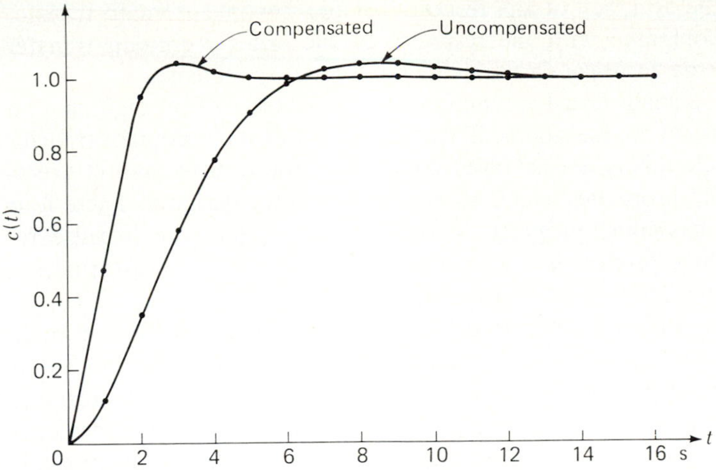

FIGURE 22.14
Step input responses for the uncompensated system of Eq. (22.43) and the compensated system of Eq. (22.45).

responses of both the compensated and uncompensated systems with a unit-step input are shown in Fig. 22.14. The characteristic values for the basic system are $M_p = 1.038$, $t_p = 8.5$ s, and $t_s = 10.75$ s. For the compensated systems the values are $M_p = 1.043$, $t_p = 3$ s, and $t_s = 3.9$ s. Thus, the digital compensator has reduced the settling time, as desired.

The digital compensator given by Eq. (22.44) is expressed in the discrete-time domain by the difference equation

$$i(nT) = e(nT) - 0.368e[(n-1)T] - 0.230i[(n-1)T] \qquad (22.47)$$

This equation is very easily implemented on a digital computer.

22.8 *s*- TO *z*-PLANE TRANSFORMATION METHODS

Several methods have been developed for obtaining the approximate z-domain transfer function from the s-domain transfer function representation, i.e., an approximation of $z = e^{sT}$. The transformations from the s to the z domain are accomplished by substituting functions of z for the s^q terms in the s-domain transfer function. One of the most popular of these approximation methods is the *Tustin algorithm*.[5,6] The function of z that is substituted for s^q in implementing the Tustin transformation is

$$s^q \approx \left(\frac{2}{T}\frac{1-z^{-1}}{1+z^{-1}}\right)^q \qquad (22.48)$$

One advantage of the Tustin algorithm is that it is comparatively easy to implement. Also, the accuracy of the response of the Tustin z-domain transfer function is good compared with the response of the exact z-domain transfer function; however, the accuracy decreases as the frequency increases.

Ideally, the z-domain transfer function, which is obtained by digitizing an s-domain transfer function, maintains all the properties of the s-domain transfer function. Thus, while this is not achieved completely, the transformation exemplifies four particular properties which are of interest to the designer. These four properties are the cascading property, stability, dc gain, and the impulse response. The first three properties are maintained by the Tustin transformation method, i.e., the three properties are invariant.

The Tustin algorithm maintains the cascading property of the s-domain transfer function. Thus, cascading two z-domain Tustin transfer functions yields the same result as cascading the s-domain counterparts and Tustin-transforming the result to the z domain. This cascading property implies that no ZOH device is involved in the cascading of the s-domain transfer functions. The Tustin transformation of a stable s-domain transfer function is stable. The dc gains for the s-domain and the z-domain Tustin transfer functions are identical; that is, $F(s)|_{s=0} = F(z)|_{z=1}$. Note that for the exact $\mathscr{Z}$ transfer functions the dc gains are not identical.

Example 4. This example illustrates the properties of the Tustin transformation of a function $e(t)$ that is sampled. Given $E(s) = E_x(s) = 1/[(s+1)(s+2)]$, *with no ZOH involved*, and $T = 0.1$ s, the exact $\mathscr{Z}$ transform yields

$$E(s) = \mathscr{Z}[e^*(t)] = \frac{0.086107z}{(z-0.81873)(z-0.90484)}$$

whereas the Tustin transformation, indicated by the subscript TU, yields

$$[E(z)]_{\mathrm{TU}} = \frac{0.0021645(z+1)^2}{(z-0.81818)(z-0.90476)}$$

The initial- and final-value theorems yield, respectively, for each transform, the results shown in the following table.

	$E(s)$	$E(z)$	$[E(z)]_{\mathrm{TU}}$	$\frac{1}{T}[E(z)]_{\mathrm{TU}}$
Initial value	0	0	0.0021645	0.021646
Final value	0	0	0	0
dc gain	0.5	4.9918	0.499985	4.99985
n	2	2	2	2
w	0	1	2	2

This example demonstrates that the Tustin transformation results in a $\mathscr{Z}$ transform for which, in general, the degrees of the numerator and denominator are the same, i.e., $n = w$. Thus, since $n = w$, the initial value of $[E(z)]_{\mathrm{TU}}$ is always different from zero, that is,

$$\lim_{z \to \infty} E(z) \neq \lim_{z \to \infty} [E(z)]_{TU}$$

Based upon the dc gain, it is evident that the values of $E(z)$ and $[E(z)]_{TU}$ differ by a factor of $1/T$; that is,

$$E(z) = \mathscr{Z}[e^*(t)] \approx \frac{1}{T}[E(z)]_{TU} \tag{22.49}$$

The reason for this difference is that the Tustin transformation does not take into account the attenuation factor $1/T$ due to the sampling process, whereas $E(z) = \mathscr{Z}[e^*(t)]$ does take this attenuation factor into account. This factor must be taken into account when the DIG technique is used for system analysis and design.

Tustin Transformation

The Tustin transformation for $q = 1$, as defined by Eq. (22.48), becomes

$$s \equiv \frac{2}{T}\frac{1 - z^{-1}}{1 + z^{-1}} \tag{22.50}$$

which is a bilinear transformation. Equation (22.50) can be derived by approximating $z = e^{Ts}$ as a finite series. The operator equation (22.50) is applicable to matrix equations, just as it is to scalar differential equations. The following discussion is useful in understanding the mapping result for the vector model also. Solving Eq. (22.50) for z yields

$$z = \frac{1 + sT/2}{1 - sT/2} \tag{22.51}$$

Letting $s = j\hat{\omega}_{sp}$ gives

$$z = \frac{1 + j\hat{\omega}_{sp}T/2}{1 - j\hat{\omega}_{sp}T/2} \tag{22.52}$$

The exact z transform yields $z = e^{j\omega_{sp}T}$, where ω_{sp} is an equivalent s-plane frequency. Equation (22.51) can be manipulated to yield

$$e^{j\omega_{sp}T} = \exp\left(j2\tan^{-1}\frac{\hat{\omega}_{sp}T}{2}\right) \tag{22.53}$$

Thus $\omega_{sp}T/2 = \tan^{-1}(\hat{\omega}_{sp}T/2)$, or

$$\tan\frac{\omega_{sp}T}{2} = \frac{\hat{\omega}_{sp}T}{2} \tag{22.54}$$

When $\omega_{sp}T/2 < 17°$, or about 0.30 rad, then

$$\omega_{sp} \approx \hat{\omega}_{sp} \tag{22.55}$$

This means that in the frequency domain the Tustin approximation is good for small values of $\omega_{sp}T/2$.

Using Eq. (22.53), it is easy to realize that the imaginary axis of the s plane is mapped as the unit-circle in the z plane as shown in Fig. 22.15. The left-half s plane is mapped into the unit circle. The same stability regions exist for the exact $\mathscr{Z}$ transform and the Tustin approximation.

Also, in this approximation, the entire imaginary axis s plane is mapped once, and only once, onto the unit circle. The Tustin approximation prevents pole

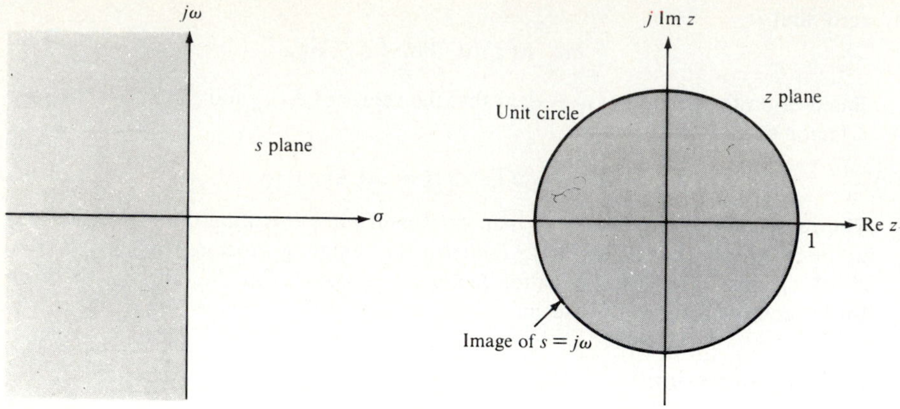

FIGURE 22.15
Tustin approximation s- to z-plane mapping.

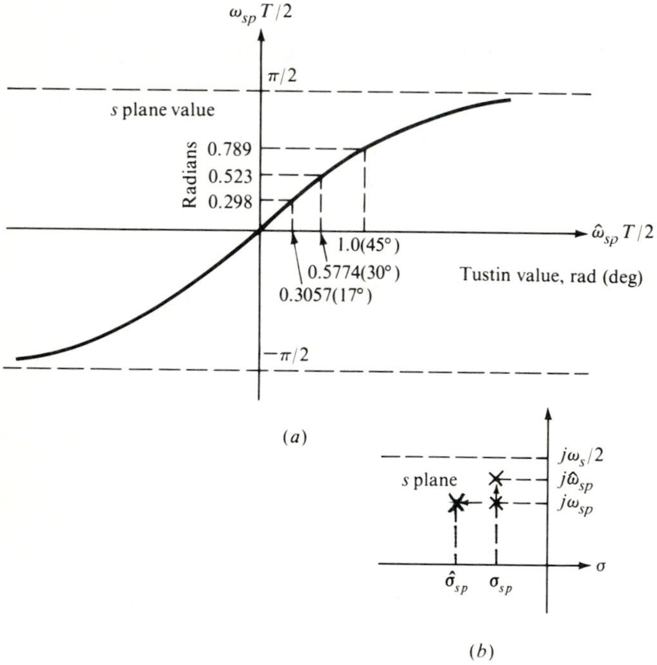

FIGURE 22.16
Map of $\hat{\omega}_{sp} = 2(\tan \omega_{sp}T/2)/T$. ($a$) Plot of Eq. (22.53) and (b) warping effect.

and zero *aliasing* since the folding phenomenon does not occur with this method. However, the warping of Eq. (22.54) is depicted by the arrow in Fig. 22.16b. Compensation can be accomplished by prewarping of ω_{sp} by using Eq. (22.54) to generate $\hat{\omega}_{sp}$. The continuous controller is mapped into the z plane by means of Eq. (22.51), using the prewarped frequency $\hat{\omega}_{sp}$. The digital compensator (controller) must be tuned (i.e., its numerical coefficients adjusted) to finalize the design since approximations have been employed. As seen from Fig. 22.16a, $\omega_{sp} \approx \hat{\omega}_{sp}$ is a good approximation when $\omega_{sp}T/2$ and $\hat{\omega}_{sp}T/2$ are both less than 0.3 rad.

The prewarping approach for the Tustin approximation takes the imaginary axis s plane and folds it back to $\pi/2$ to $-\pi/2$ as seen from Fig. 22.16a. The spectrum of the input must also be taken into consideration when selecting an approximation procedure with or without prewarping. In the previous discussion only the frequency has been prewarped due to the interest in the controller frequency response. The real part of the s-plane pole influences such parameters as rise time, overshoot, and settling time. Consideration of the warping of the real pole component is now analyzed. Substituting $z = e^{\sigma_{sp}T}$ and $s = \hat{\sigma}_{sp}$ into Eq. (22.51) yields

$$e^{\sigma_{sp}T} = \frac{1 + \hat{\sigma}_{sp}T/2}{1 - \hat{\sigma}_{sp}T/2} \tag{22.56}$$

Replacing $e^{\sigma_{sp}T}$ by its exponential series and dividing the numerator by the denominator in Eq. (22.56) results in an expression which can be put in closed form by forming the sum of the infinite series:

$$1 + \sigma_{sp}T + \frac{(\sigma_{sp}T)^2}{2} + \cdots = 1 + \frac{\hat{\sigma}_{sp}T}{1 - \hat{\sigma}_{sp}T/2} \tag{22.57}$$

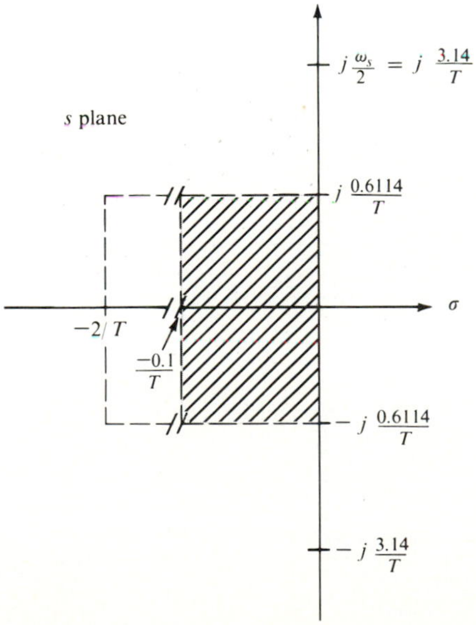

FIGURE 22.17
Allowable location (cross-hatched area) of dominant poles and zeros in s plane for a good Tustin approximation.

If $|\sigma_{sp}T| \gg (\sigma_{sp}T)^2/2$ (or $1 \gg |\sigma_{sp}T/2|$) and $1 \gg |\hat{\sigma}_{sp}T/2|$, then

$$|\hat{\sigma}_{sp}| \approx |\sigma_{sp}| \ll \frac{2}{T} \tag{22.58}$$

Thus with Eqs. (22.55) and (22.58) satisfied, the Tustin approximation in the s domain is good for small magnitudes of the real and imaginary components of the variable s. The cross-hatched area in Fig. 22.17 represents the allowable location of the poles and zeros in the s plane for a good Tustin approximation. Because of the mapping properties and its ease of use, the Tustin transformation is employed for the DIG technique in the next section.

22.9 PSEUDO-CONTINUOUS-TIME (PCT) CONTROL SYSTEM (DIG METHOD)

The DIG method of designing a sampled-data system, in the complex-frequency s plane, requires a satisfactory pseudo-continuous-time (PCT) model of the sampled-data system. In other words, for the sampled-data system of Fig. 22.18, the sampler and the ZOH units must be approximated by a linear continuous-time unit $G_A(s)$, as shown in Fig. 22.19c. The DIG method requires that the dominant poles and zeros of the PCT model should lie in the shaded area of Fig. 22.17 for a high level of correlation with the sampled-data system. To determine $G_A(s)$, first consider the frequency component of $\mathbf{E}^*(j\omega)$ representing the continuous-time signal $\mathbf{E}(j\omega)$, where all its sidebands are multiplied by $1/T$ (see Sec. 22.2).[1] Because of the low-pass filtering characteristics of a sampled-data system, only the primary component is considered in the analysis of the system. Therefore, the PCT approximation of the sampler of Fig. 22.18 is shown in Fig. 22.19b.

Using the first-order Padé approximation,[1] the transfer function of the ZOH, when the value of T is small enough, is approximated as follows:

$$G_{zo}(s) = \frac{1 - e^{-Ts}}{s} \approx \frac{2T}{Ts + 2} = G_{pa}(s) \tag{22.59}$$

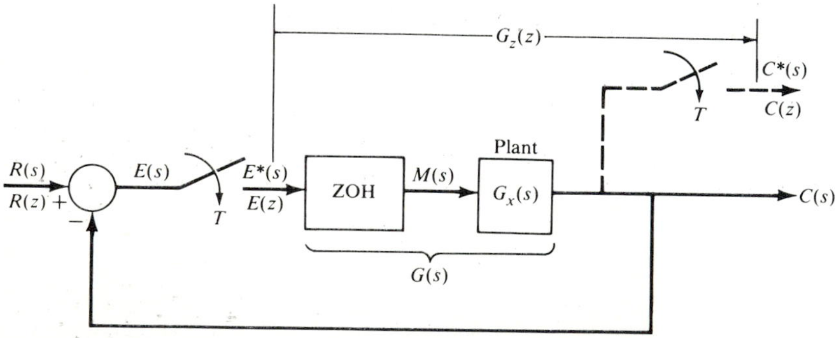

FIGURE 22.18
The uncompensated sampled-data control system.

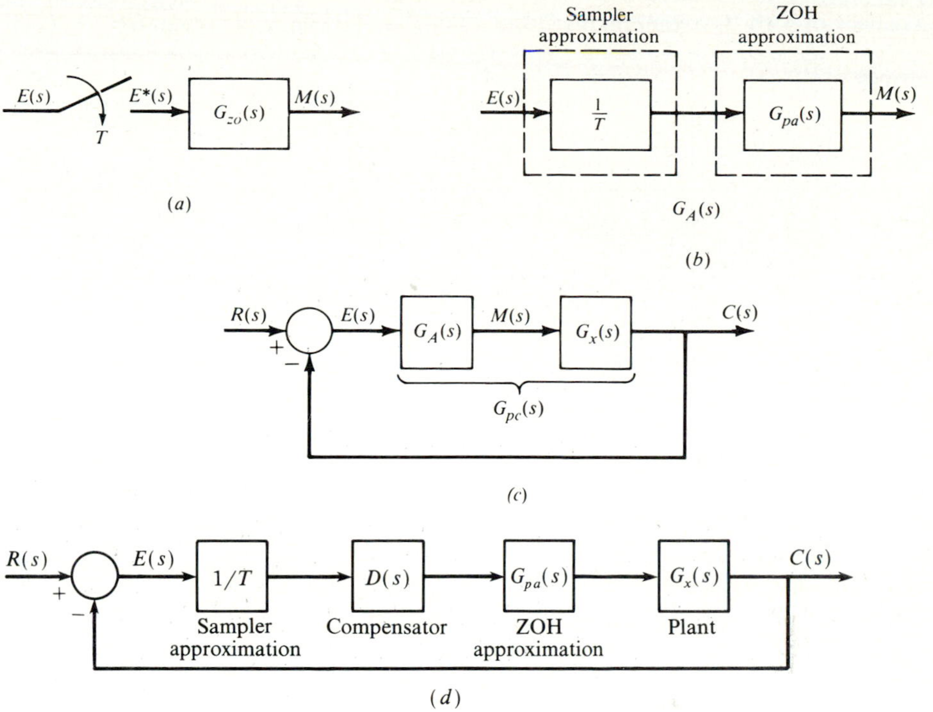

FIGURE 22.19
(*a*) Sampler and ZOH. (*b*) Approximations of the sampler and ZOH. (*c*) and (*d*) The approximate continuous-time control system equivalent of Figs. 22.18 and 22.19, respectively.

Thus the Padé approximation $G_{pa}(s)$ is used to replace $G_{zo}(s)$ as shown in Fig. 22.19*a* and *b*. This approximation is good for $\omega_c \le \omega_s/10$, whereas the second-order approximation is good for $\omega_c \le \omega_s/3$.[1] Therefore the sampler and ZOH of a sampled-data system are approximated in the PCT system of Fig. 22.19*c* by the transfer function

$$G_A(s) = \frac{1}{T}G_{pa}(s) = \frac{2}{Ts + 2} \tag{22.60}$$

Since $\lim_{T\to 0} G_A(s) = 1$, Eq. (22.60) is an accurate PCT representation of the sampler and ZOH units, because it satisfies the requirement that as $T \to 0$ the output of $G_A(s)$ must equal its input. Further note that in the frequency domain as $\omega_s \to \infty$ ($T \to 0$), the primary strip becomes the entire frequency-spectrum domain which is the representation for the continuous-time system.

Note that in obtaining PCT *systems for the sampled-data systems of Fig.* 22.12 *the factor* $1/T$ *replaces only the sampler that is sampling the continuous-time signal.* This multiplier of $1/T$ is reflected in the Fourier series analysis of $e^*(t)$ for ideal impulse sampling, which reveals that the fundamental, the frequency of the sampled signal, and all its harmonics are attenuated by $1/T$. *The sampler on the*

TABLE 22.2
Analysis of a PCT system representing a sampled-data control system for $\zeta = 0.45$

Method	T, s	Domain	K_x	M_p	t_p, s	t_s, s
DIR	0.01	z	4.147	1.202	4.16	9.53
DIG		s	4.215	1.206	4.11	9.478
DIR	0.1	z	3.892	1.202	4.2–4.3	9.8–9.9
DIG		s	3.906	1.203	4.33^-	9.90
DIR	1	z	2.4393	1.199	6	13–14
DIG		s	2.496	1.200	6.18	13.76

output of the digital controller is replaced by a factor of 1. To illustrate the effect of the value of T on the validity of the results obtained by the DIG method, consider the sampled-data closed-loop control system of Fig. 22.18 for the $G_x(s)$ of Sec. 10.6 with $K_x = 4.2$. The closed-loop system performance for three values of T and $\zeta = 0.45$ are determined in both the s and z domains, i.e., both the DIG and DIR methods, respectively. Table 22.2 presents the required value of K_x and time-response characteristics for each value of T. Note that for $T \leq 0.1$ s there is a high level of correlation between the DIG and DIR models. For $T \leq$ 1 s there is still a relatively good correlation. (The designer must specify, for a given application, what is considered to be "good correlation.")

22.10 ANALYSIS OF A BASIC (UNCOMPENSATED) SYSTEM

Figure 22.18 represents a basic or uncompensated sampled-data control system where

$$G_x(s) = \frac{K_x}{s(s+1)} \qquad \textbf{Case 1} \qquad (22.61)$$

is used to illustrate the approaches for improving the performance of a basic system. As mentioned in Sec. 22.6 the lag characteristic of $G_{zo}(s)$ reduces the degree of system stability. This degradation is illustrated in some of the examples of this chapter.

PCT Control-System Model

One approach for designing a sampled-data unity-feedback control system is to first obtain a suitable closed-loop model $[C(s)/R(s)]_M$ for the PCT unity-feedback control system of Fig. 22.19, utilizing the plant of the sampled-data control system. This model is then used as a guide for selecting an acceptable $C(z)/R(z)$.

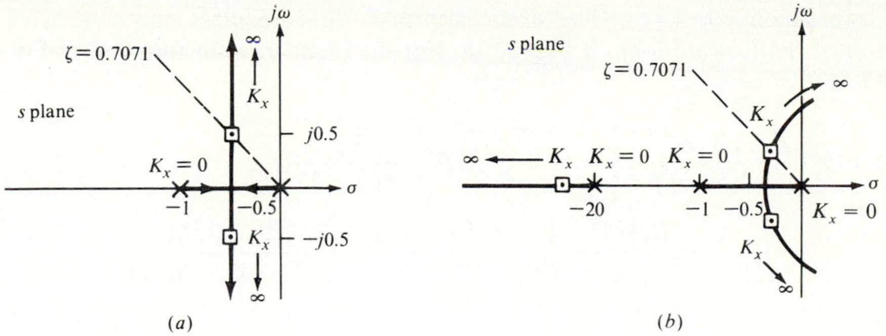

FIGURE 22.20
Root locus for (*a*) Case 1, Eq. (22.61); (*b*) Case 2, Eq. (22.62). Not to scale.

Thus, for the plant of Eq. (22.61),

$$G_{PC}(s) = G_A(s)G_x(s) = \frac{2K_x/T}{s(s+1)(s+2/T)} \tag{22.62}$$

or $T = 0.1$ s.

$$\left[\frac{C(s)}{R(s)}\right]_M = \frac{G_{PC}(s)}{1+G_{PC}(s)} = \frac{20K_x}{s^3+21s^2+20s+20K_x} \tag{22.63}$$

The root locus for $G_{PC}(s) = -1$ is shown in Fig. 22.20*b*. For comparison purposes the root-locus plot for $G_x(s) = -1$ is shown in Fig. 22.20*a*. These figures illustrate the effect of inserting a lag network in cascade in a feedback control system; i.e., it transforms a completely stable system into a conditionally stable system. Thus for a given value of ζ the values of t_p and t_s (and T_s) are increased. Therefore, the ZOH degrades the degree of system stability.

For the model it is assumed that the desired value of the damping ratio ζ for the dominant roots is 0.7071. Thus for a unit-step input,

$$[C(s)]_M = \frac{9.534}{s(s^3+21s^2+20s+9.534)}$$

$$= \frac{9.534}{s(s+0.4875 \pm j0.4883)(s+20.03)} \qquad \textbf{Case 2} \tag{22.64}$$

where $K_x = 0.4767$. The real and imaginary parts of the desired roots of Eq. (22.64) for $T = 0.1$ s, lie in the acceptable region of Fig. 22.17 for a good Tustin approximation.

Sampled-Data Control System

The determination of the time-domain performance of the sampled-data system of Fig. 22.18 may be achieved either by obtaining the exact expression for $C(z)$ or by applying the Tustin transformation to Eq. (22.63) to obtain the approximate expression for $C(z)$. The exact approach first requires the z-transfer function of the forward loop of Fig. 22.18. For the plant transfer function of Eq. (22.61),

$$G_z(z) = \mathscr{Z}\left[\frac{K_x(1 - e^{-sT})}{s^2(s+1)}\right] = (1 - z^{-1})\mathscr{Z}\left[\frac{K_x}{s^2(s+1)}\right]$$

$$= \frac{K_x[(T - 1 + e^{-T})z + (1 - Te^{-T} - e^{-T})]}{z^2 - (1 + e^{-T} + K_x - TK_x - K_x e^{-T})z + e^{-T} + K_x - K_x(T+1)e^{-T}} \tag{22.65}$$

Thus for $T = 0.1$ s and $K_x = 0.4767$,

$$G_z(z) = \frac{0.002306(z + 0.9672)}{(z-1)(z-0.9048)} \tag{22.66}$$

or

$$\frac{C(z)}{R(z)} = \frac{0.002306(z + 0.9672)}{(z - 0.9513 \pm j0.04649)} \qquad \textbf{Case 3} \tag{22.67}$$

The DIG technique requires that the s-domain model control ratio be transformed into a z-domain model. Applying the Tustin transformation to

$$\left[\frac{C(s)}{R(s)}\right]_M = \frac{G_{PC}(s)}{1 + G_{PC}(s)} = F_M(s) \tag{22.68}$$

yields

$$\frac{[C(z)]_{TU}}{[R(z)]_{TU}} = [F(z)]_{TU} \tag{22.69}$$

This equation is rearranged to

$$[C(z)]_{TU} = [F(z)]_{TU}[R(z)]_{TU} \tag{22.70}$$

As stated in Sec. 22.8,

$$R(z) = \mathscr{Z}[r^*(t)] = \frac{1}{T}[R(z)]_{TU} \tag{22.71}$$

$$C(z) = \mathscr{Z}[c^*(t)] = \frac{1}{T}[C(z)]_{TU} \tag{22.72}$$

Substituting from Eqs. (22.71) and (22.72) into Eq. (22.69) yields

$$\frac{C(z)}{R(z)} = [F(z)]_{TU} \tag{22.73}$$

Substituting from Eq. (22.72) into Eq. (22.70) and rearranging yields

$$C(z) = \frac{1}{T}[F(z)]_{TU}[R(z)]_{TU} = \frac{1}{T}[\text{Tustin of } F_M(s)R(s)] \tag{22.74}$$

TABLE 22.3
Comparison of time responses between $C(z)$ and $[C(z)]_{TU}$ for a unit-step input and $T = 0.1$ s

	$c(kT)$			$c(kT)$	
k	**Case 3 (exact),** $C(z)$	**Case 4 (Tustin),** $[C(z)]_{TU}$	k	**Case 3 (exact),** $C(z)$	**Case 4 (Tustin),** $[C(z)]_{TU}$
0	0.	0.5672E − 03	42	0.9460	0.9452
2	0.8924E − 02	0.9823E − 02	44	0.9668	0.9660
4	0.3340E − 01	0.3403E − 01	46	0.9844	0.9836
6	0.7024E − 01	0.7064E − 01	48	0.9992	0.9984
8	0.1166	0.1168	50	1.011	1.011
10	0.1701	0.1701	52	1.021	1.020
12	0.2284	0.2283	54	1.029	1.028
14	0.2897	0.2894	56	1.035	1.034
16	0.3525	0.3521	58	1.039	1.038
18	0.4153	0.4148	60	1.042	1.041
20	0.4771	0.4766	61	1.042	1.042
22	0.5370	0.5364	62	1.043	1.043
24	0.5943	0.5937	63	1.043	1.043
26	0.6485	0.6478	64	1.043 } M_p	1.043 } M_p
28	0.6992	0.6984	65	1.043 }	1.043 }
30	0.7461	0.7453	66	1.043	1.043
32	0.7890	0.7883	67	1.043	1.043
34	0.8281	0.8273	68	1.042	1.042
36	0.8632	0.8624	85	1.022	1.022
38	0.8944	0.8936	86	1.021 } t_s	1.021 } t_s
40	0.9220	0.9212	87	1.019 }	1.019 }

Thus based upon Eq. (22.73) the Tustin transformation to Eq. (22.63), with $K_x = 0.4767$, results in a Tustin model of the control ratio

$$\frac{C(z)}{R(z)} = \left[\frac{C(z)}{R(z)}\right]_{TU}$$

$$= \frac{5.672 \times 10^{-4}(z+1)^3}{(z - 0.9513 \pm j0.04651)(z + 6.252 \times 10^{-4})} \qquad \textbf{Case 4} \quad (22.75)$$

Note that the dominant poles of Eq. (22.75) are essentially the same as those of Eq. (22.67) due to the value of T used which resulted in the dominant roots lying in the good Tustin region of Fig. 22.17. In using the exact $\mathscr{Z}$ transformation the order of the numerator polynomial of $C(z)/R(z)$, Eq. (22.67), is one less than the order of its denominator polynomial. When using the Tustin transformation the order of the numerator polynomial of the resulting $[C(z)/R(z)]_{TU}$ is in general equal to the order of its corresponding denominator polynomial [see Eq. (22.75)]. Thus $[C(z)]_{TU}$ results in a value of $c^*(t) \neq 0$ at $t = 0$ which is in error

TABLE 22.4
Time-response characteristics of the closed-loop system

	M_p	t_p, s	t_s, s	K_1	Case
		Continuous-time system			
$G(s) = G_x(s)$	1.04821	6.3	8.40 to 8.45	0.4767	1
$G_M(s) = G_A(s)G_x(s)$	1.04342	6.48	8.69	0.4767	2
		Sampled-data system			
$C(z)$	1.043	6.45	8.65	0.4765	3
$[C(z)]_{TU}$	1.043	6.45	8.6	0.4765	4

based upon zero initial conditions. Table 22.3 illustrates the effect of this characteristic of the Tustin transformation on the time response due to a unit-step forcing function. The difference in the time response by use of the Tustin transformation is minimal; i.e., the figures of merit are in close agreement with those obtained by the use of the exact $\mathscr{Z}$ transformation. Therefore, the Tustin transformation is a valid design tool when the dominant zeros and poles of $[C(s)/R(s)]_M$ lie in the acceptable region of Fig. 22.17.

Table 22.4 summarizes the time-response characteristics for a unit-step forcing function of (1) the continuous-time system of Fig. 22.19 for the two cases of $G(s) = G_x(s)$ and $G(s) = G_A(s)G_x(s)$ with the continuous-time representation of the sampler and ZOH included, and (2) the sampled-data system of Fig. 22.18 based upon the exact and Tustin expressions for $C(z)$.

Steady-State Error Coefficients

The steady-state error coefficients have the same meaning and importance for sampled-data systems as for continuous-time systems; i.e., how well can the system output follow a given type of input forcing function in the steady state. *The following definitions, given without proof,*[1] *of the error coefficients are independent of the system type. They apply to any system type and are defined for specific forms of the input. These error coefficients are useful only for stable unity-feedback systems.*

STEP INPUT, $R(z) = R_0 z/(z-1)$. The step error coefficient K_p is defined as

$$K_p \equiv \frac{c^*(\infty)}{e^*(\infty)} = \lim_{z \to 1} G(z) \tag{22.76}$$

which applies only for a step input, $r(t) = R_0 u_{-1}(t)$.

RAMP INPUT, $R(z) = R_1Tz/(z-1)^2$. The ramp error coefficient K_v is defined as

$$K_v \equiv \frac{\text{steady-state value of derivative of output}}{e^*(\infty)}$$

$$= \frac{1}{T} \lim_{z\to 1}\left[\frac{z-1}{z}G(z)\right] \tag{22.77}$$

which applies only for a ramp input, $r(t) = R_1tu_{-1}(t)$.

PARABOLIC INPUT, $R(z) = R_2T^2z(z+1)/2(z-1)^3$. The parabolic error coefficient K_a is defined as

$$K_a \equiv \frac{\text{steady-state value of second derivative of output}}{e^*(\infty)}$$

$$= \frac{1}{T^2} \lim_{z\to 1}\left[\frac{(z-1)^2}{z^2}G(z)\right] \tag{22.78}$$

which applies only for a parabolic input, $r(t) = R_2t^2u_{-1}(t)/2$.

The ramp error coefficients for Cases 2 to 4 are obtained by applying Eq. (22.77). Table 22.4 reveals that:

1. In converting a continuous-time system into a sampled-data system the time-response characteristics are degraded.
2. The time-response characteristics of the sampled-data system, using the value of gain obtained from the continuous-time model, agrees favorably with those of the continuous-time model. As may be expected, there is some variation in the values obtained when utilizing the exact $C(z)$ and $[C(z)]_{TU}$.

The results of this section demonstrate that when the Tustin approximation is valid, the PCT approximation of the sampled-data system is a practical design method. When a cascade compensator $D(s)$ (see Fig. 22.19*d*) is designed to yield the desired performance specifications in the pseudo-continuous-time domain, the discrete cascade compensator $D(z)$ (see Fig. 22.12 where the controller does not include a ZOH) is accurately obtained by using the Tustin transformation.

22.11 SUMMARY

The availability of inexpensive microprocessors has led to their extensive use in digital control systems. Although there is a degradation of performance by the introduction of sampling, this is more than overcome by the greater range of compensators that can be achieved digitally. The flexibility which is provided by microprocessors in synthesizing digital compensators yields greatly improved closed-loop performance.

Two approaches for analyzing a sampled-data control system are presented: the DIG (digitization) and DIR (direct) techniques. The former requires the use of the Padé approximation and the Tustin transformation for an initial design in the s plane (PCT control-system configuration). If the Tustin approximation criteria of Sec. 22.8 are satisfied, the correlation between the two planes is very good. If the approximations are not valid, the analysis and design of the sampled-data control system should be done by the DIR technique.

REFERENCES

1. Houpis, C. H., and G. B. Lamont: *Digital Control Systems, Theory, Hardware, Software*, McGraw-Hill, New York, 1985.
2. Franklin, G. F., and J. D. Powell: *Digital Control of Dynamic Systems*, Addison-Wesley, Reading, Mass., 1980.
3. Kuo, B. C.: *Discrete-Data Control Systems*, Prentice-Hall, Englewood Cliffs, N.J., 1970.
4. Cadzow, J. A., and H. R. Martens, *Discrete-Time and Computer Control Systems*, Prentice-Hall, Englewood Cliffs, N.J., 1970.
5. Franklin, G. F., and J. David Powell: *Digital Control of Dynamic Systems*, Addison-Wesley, Reading, Mass., 1980.
6. Tustin, A.: "A Method of Analyzing the Behavior of Linear Systems in Terms of Time Series," *JIEE* (*Lond.*), vol. 94, pt. IIA, 1947.

APPENDIX A

TABLE OF LAPLACE TRANSFORM PAIRS

$F(s)$	$f(t) \qquad 0 \le t$
1. 1	$u_0(t)$ unit impulse at $t = 0$
2. $\frac{1}{s}$	1 or $u_{-1}(t)$ unit step starting at $t = 0$
3. $\frac{1}{s^2}$	$tu_{-1}(t)$ ramp function
4. $\frac{1}{s^n}$	$\frac{1}{(n-1)!}t^{n-1}$ n = positive integer
5. $\frac{1}{s}e^{-as}$	$u_{-1}(t-a)$ unit step starting at $t = a$
6. $\frac{1}{s}(1 - e^{-as})$	$u_{-1}(t) - u_{-1}(t-a)$ rectangular pulse
7. $\frac{1}{s+a}$	e^{-at} exponential decay
8. $\frac{1}{(s+a)^n}$	$\frac{1}{(n-1)!}t^{n-1}e^{-at}$ n = positive integer
9. $\frac{1}{s(s+a)}$	$\frac{1}{a}(1 - e^{-at})$
10. $\frac{1}{s(s+a)(s+b)}$	$\frac{1}{ab}\left(1 - \frac{b}{b-a}e^{-at} + \frac{a}{b-a}e^{-bt}\right)$

$F(s)$	$f(t) \quad 0 \le t$
11. $\dfrac{s+\alpha}{s(s+a)(s+b)}$	$\dfrac{1}{ab}\left[\alpha - \dfrac{b(\alpha-a)}{b-a}e^{-at} + \dfrac{a(\alpha-b)}{b-a}e^{-bt}\right]$
12. $\dfrac{1}{(s+a)(s+b)}$	$\dfrac{1}{b-a}(e^{-at} - e^{-bt})$
13. $\dfrac{s}{(s+a)(s+b)}$	$\dfrac{1}{a-b}(ae^{-at} - be^{-bt})$
14. $\dfrac{s+\alpha}{(s+a)(s+b)}$	$\dfrac{1}{b-a}[(\alpha-a)e^{-at} - (\alpha-b)e^{-bt}$
15. $\dfrac{1}{(s+a)(s+b)(s+c)}$	$\dfrac{e^{-at}}{(b-a)(c-a)} + \dfrac{e^{-bt}}{(c-b)(a-b)} + \dfrac{e^{-ct}}{(a-c)(b-c)}$
16. $\dfrac{s+\alpha}{(s+a)(s+b)(s+c)}$	$\dfrac{(\alpha-a)e^{-at}}{(b-a)(c-a)} + \dfrac{(\alpha-b)e^{-bt}}{(c-b)(a-b)} + \dfrac{(\alpha-c)e^{-ct}}{(a-c)(b-c)}$
17. $\dfrac{\omega}{s^2+\omega^2}$	$\sin \omega t$
18. $\dfrac{s}{s^2+\omega^2}$	$\cos \omega t$
19. $\dfrac{s+\alpha}{s^2+\omega^2}$	$\dfrac{\sqrt{\alpha^2+\omega^2}}{\omega}\sin(\omega t+\phi) \qquad \phi = \tan^{-1}\dfrac{\omega}{\alpha}$
20. $\dfrac{s\sin\theta + \omega\cos\theta}{s^2+\omega^2}$	$\sin(\omega t+\theta)$
21. $\dfrac{1}{s(s^2+\omega^2)}$	$\dfrac{1}{\omega^2}(1-\cos\omega t)$
22. $\dfrac{s+\alpha}{s(s^2+\omega^2)}$	$\dfrac{\alpha}{\omega^2} - \dfrac{\sqrt{\alpha^2+\omega^2}}{\omega^2}\cos(\omega t+\phi) \qquad \phi = \tan^{-1}\dfrac{\omega}{\alpha}$
23. $\dfrac{1}{(s+a)(s^2+\omega^2)}$	$\dfrac{e^{-at}}{a^2+\omega^2} + \dfrac{1}{\omega\sqrt{a^2+\omega^2}}\sin(\omega t-\phi) \qquad \phi = \tan^{-1}\dfrac{\omega}{a}$
24. $\dfrac{1}{(s+a)^2+b^2}$	$\dfrac{1}{b}e^{-at}\sin bt$
24*a*. $\dfrac{1}{s^2+2\zeta\omega_n s+\omega_n^2}$	$\dfrac{1}{\omega_n\sqrt{1-\zeta^2}}e^{-\zeta\omega_n t}\sin\omega_n\sqrt{1-\zeta^2}\,t$
25. $\dfrac{s+a}{(s+a)^2+b^2}$	$e^{-at}\cos bt$

APPENDIX B

INTERACTIVE COMPUTER-AIDED DESIGN (CAD) PROGRAMS FOR DIGITAL AND CONTINUOUS CONTROL-SYSTEM ANALYSIS AND SYNTHESIS

B.1 INTRODUCTION

Various CAD software packages[B1–B15] are available for the development of continuous- and discrete-time control systems. These packages can assist the control engineer in performing time-consuming calculations for both scalar and multivariable models. Some of the available CAD packages provide the user with an integrated set of design tools and perform a broad range of calculations applicable to control-system analysis and synthesis. The computer program called TOTAL[B9] (TOTAL-I) is presented in this appendix as an example of a specific

CAD package. Note that this CAD package is the one employed in the text examples. Extension[B5, B10] to TOTAL is also summarized.

B.2 OVERVIEW OF TOTAL

TOTAL is designed as a tool to be used with as much speed, agility, and confidence as a familiar hand calculator. To assist the user in obtaining an overview of TOTAL, the following information is provided. A complete description of this FORTRAN program, with examples of input and output data, is given in Ref. B9, which can be obtained from the Defense Documentation Center (DDC), Cameron Station, Alexandria, Va. 22314.

1. TOTAL is built around twelve general-purpose polynomials and seven general-purpose matrices.
2. Eight of the polynomials may be paired to form the numerators and denominators of four general-purpose transfer functions.
3. With just these polynomials, matrices, and transfer functions the user is able to use the entire spectrum of TOTAL's capabilities, which include:
   a. Adding, subtracting, multiplying, dividing, factoring, expanding, and copying polynomials.
   b. Adding, subtracting, multiplying, inverting, transposing, obtaining eigenvalues, presetting, and copying matrices.
   c. Obtaining root locus, frequency response, and time response of open- and closed-loop transfer functions in both the continuous and discrete domains.
   d. Performing block-diagram reduction.
   e. Computing transfer functions from the state-space equations.
   f. Transforming between continuous and discrete domains using a variety of methods.
4. A built-in 20-register scientific calculator with a four-register stack is available to the user at all times.
5. The user can quickly list, transfer, or modify any variable, at any time, anywhere in the program.
6. Complete error detection, diagnostics, and abnormal-termination protection are provided. Specific help is available at any time by simply typing a question mark.

The TOTAL package currently executes on the CDC Cyber, SEL 32, and the DEC VAX-11/780.

B.3 INTRODUCTION TO TOTAL

TOTAL contains over 100 options and a few special commands with which the user can manipulate a data base of 12 polynomials, 7 matrices, and roughly 60 scalar variables. Each option uses information stored in these variables to perform a particular function and saves the results for output or future use. During execution of TOTAL, the user has complete freedom to select options, modify variables, and give commands at will. The following sections summarize the information needed to make use of the full power of this program.

B.4 TOTAL'S INPUT MODES

TOTAL has three modes in which it pauses for the user to input information: OPTION, DATA, and CALCULATOR. Each mode has its own vocabulary of allowable inputs and its own characteristic prompt to the user. The OPTION mode is the primary command mode of TOTAL. When TOTAL pauses in this mode, the user is free to type any command, select any option, modify any variable, or list any data. It is from this mode that the user controls the program. When an option has been selected and data are needed, TOTAL will ask for them. This is called the DATA mode. The CALCULATOR mode can be entered at any time by typing C. The calculator operates like an HP-45 calculator (using reverse Polish notation). The CALCULATOR mode is terminated by typing another C.

B.5 TOTAL'S OPTIONS

TOTAL presently contains over 100 options which are divided into groups of 10 according to general function. Table B.1 lists some of the options currently available.

B.6 HELP

Help is available to the user at all times and can be requested in two ways. In option mode, typing the command "HELP, option number" gives the user a short explanation of the option number specified. In all modes, the user may type "?" for assistance.

B.7 TOTAL ENHANCEMENTS

The TOTAL CAD package was developed as an interactive tool based upon a line terminal. Because of the increasing use of CRT terminals, TOTAL has been rehosted on a DEC VAX-11/780 with a screen-oriented interface permitting easier and faster utilization.[B5, B10] This new package is called ICECAP and is written in the FORTRAN and PASCAL languages. This package also includes the quantitative feedback theory (QFT) CAD routines that are a helpful design tool for Chap. 21. The referenced documents are also available from the Defense Documentation Center (DDC) Cameron Station, Alexandria, Va. 22314. Currently TOTAL is being modified and expanded and is designated as TOTAL-II.

TABLE B.1
Some TOTAL options

No.	Option
	Transfer-function input options
0	LIST OPTIONS
1	RECOVER ALL DATA FROM FILE MEMORY
2	POLYNOMIAL FORM—GTF (forward transfer function)
3	POLYNOMIAL FORM—HTF (feedback transfer function)
4	POLYNOMIAL FORM—OLTF (open-loop transfer function)
5	POLYNOMIAL FORM—CLTF (closed-loop transfer function)
6	FACTORED FORM—GTF
7	FACTORED FORM—HTF
8	FACTORED FORM—OLTF
9	FACTORED FORM—CLTF
	Matrix input options for state equations
10	LIST OPTIONS
11	AMAT—CONTINUOUS PLANT MATRIX
12	BMAT—CONTINUOUS INPUT MATRIX
13	CMAT—OUTPUT MATRIX
14	DMAT—DIRECT TRANSMISSION MATRIX
15	KMAT—STATE VARIABLE FEEDBACK MATRIX
16	FMAT—DISCRETE PLANT MATRIX
17	GMAT—DISCRETE INPUT MATRIX
18	SET UP STATE SPACE MODEL OF SYSTEM
19	EXPLAIN USE OF ABOVE MATRICES
	Block diagram manipulation and state-space options
20	LIST OPTIONS
21	FORM OLTF = GTF * HTF (IN CASCADE)
22	FORM CLTF = (GAIN * GTF)/(1 + GAIN * GTF * HTF)
23	FORM CLTF = (GAIN * OLTF)/(1 + GAIN * OLTF)
24	FORM CLTF = GTF + HTF (IN PARALLEL)
25	GTF(s) AND HTF(s) FROM CONTINUOUS STATE SPACE MODEL
26	GTF(z) AND HTF(z) FROM DISCRETE STATE SPACE MODEL
27	WRITE ADJOINT ($s\mathbf{I}$ − AMAT) TO FILE ANSWER
28	FIND HTF FROM CLTF & GTF FOR CLTF = GTF * HTF/(1 + GTF * HTF)
29	FIND HTF FROM CLTF & GTF FOR CLTF = GTF/(1 + GTF * HTF)
	Time-response options, continuous $F(t)$ and discrete $F(kT)$†
30	LIST OPTIONS
31	TABULAR LISTING OF $F(t)$ OR $F(kT)$
32	PLOT $F(t)$ OR $F(kT)$ AT USER'S TERMINAL
33	PRINTER PLOT (WRITTEN TO FILE ANSWER)
34	CALCOMP PLOT (WRITTEN TO FILE PLOT)
35	PRINT TIME OR DIFFERENCE EQUATION [$F(t)$ OR $F(kT)$]
36	PARTIAL FRACTION EXPANSION OF CLTF (OR OLTF)
37	LIST T-PEAK, T-RISE, T-SETTLING, T-DUP, M-PEAK, FINAL VALUE
38	QUICK SKETCH AT USER'S TERMINAL
39	SELECT INPUT: STEP, RAMP, PULSE, IMPULSE, SIN ωT

† Option 37 is not available for the discrete domain.

TABLE B.1
Some TOTAL options—continued

No.	Option
	Root-locus options
40	LIST OPTIONS
41	GENERAL ROOT LOCUS
42	ROOT LOCUS WITH A GAIN OF INTEREST
43	ROOT LOCUS WITH A ZETA (DAMPING RATIO) OF INTEREST
44	LIST N POINTS ON A BRANCH OF INTEREST
45	LIST ALL POINTS ON A BRANCH OF INTEREST
46	LIST LOCUS ROOTS AT A GAIN OF INTEREST
47	LIST LOCUS ROOTS AT A ZETA OF INTEREST
48	PLOT ROOT LOCUS AT USER'S TERMINAL
49	LIST CURRENT VALUES OF ALL ROOT LOCUS VARIABLES
	Frequency-response options
50	LIST OPTIONS
51	TABULAR LISTING
52	TWO CYCLE SCAN OF MAGNITUDE (OR DB)
53	TWO CYCLE SCAN OF PHASE (DEGREES OR RADIANS)
54	PLOT $F(j\omega)$ AT USER'S TERMINAL
55	CALCOMP PLOT—LINEAR FREQUENCY AXIS
56	CALCOMP PLOT—LOG FREQUENCY AXIS
57	TABULATE POINTS OF INTEREST: PEAKS, BREAKS, ETC.
58	CALCOMP PLOT—NICHOLS LOG-MAG/ANGLE PLOT
	Polynomial operations
60	LIST OPTIONS
61	FACTOR POLYNOMIAL (POLYA)
62	ADD POLYNOMIALS (POLYC = POLYA + POLYB)
63	SUBTRACT POLYNOMIALS (POLYC = POLYA − POLYB)
64	MULTIPLY POLYS (POLYC = POLYA * POLYB)
65	DIVIDE POLYS (POLYC + REM = POLYA/POLYB)
66	STORE (POLY_) INTO POLYD
67	EXPAND ROOTS INTO A POLYNOMIAL
68	$(s + a)^n$ EXPANSION INTO A POLYNOMIAL
69	ACTIVATE POLYNOMIAL CALCULATOR
	Matrix operations
70	LIST OPTIONS
71	ROOTA = EIGENVALUES OF AMAT
72	CMAT = AMAT + BMAT
73	CMAT = AMAT − BMAT
74	CMAT = AMAT * BMAT
75	CMAT = AMAT INVERSE
76	CMAT = AMAT TRANSPOSED
77	CMAT = IDENTITY MATRIX I
78	DMAT = ZERO MATRIX 0
79	COPY ONE MATRIX TO ANOTHER

TABLE B.1
Some TOTAL options—continued

No.	Option
	Digitization options
80	LIST OPTIONS
81	CLTF(s) TO CLTF(z) BY IMPULSE INVARIANCE
82	CLTF(s) TO CLTF(z) BY FIRST DIFFERENCE APPROXIMATION
83	CLTF(s) to CLTF(z) BY TUSTIN TRANSFORMATION
84	CLTF(z) TO CLTF(s) BY IMPULSE INVARIANCE
85	CLTF(z) TO CLTF(s) BY INVERSE FIRST DIFFERENCE
86	CLTF(z) TO CLTF(s) BY INVERSE TUSTIN
87	FIND FMAT AND GMAT FROM AMAT AND BMAT
89	CLTF(X) TO CLTF(Y) BY X = ALPHA*(Y + A)/(Y + B)
	Miscellaneous options
90	LIST OPTIONS
91	REWIND AND UPDATE MEMORY FILE WITH CURRENT DATA
92	ERASE SCREEN (TEKTRONIX GRAPHICS TERMINALS ONLY)
93	LIST CURRENT SWITCH SETTINGS (ECHO, ANSWER, ETC.)
94	HARD COPY (TEK TERMINAL WITH HARD COPY UNIT ONLY)
95	STAR TREK GAME
96	LIST SPECIAL COMMANDS ALLOWED IN OPTION MODE
97	LIST VARIABLE NAME DIRECTORY
98	LIST MAIN OPTIONS OF TOTAL
99	PRINT NEW FEATURES BULLETIN
199	AUGMENT AMAT:[AMAT] = [AMAT] − GAIN*[BMAT]*[KMAT]
131	(INTEGRAL OF (CLTF SQUARED))/2PI =
	Double-precision discrete transform options
140	LIST OPTIONS
141	CLTF(s) TO CLTF(z) (IMPULSE VARIANCE)
142	CLTF(s) TO CLTF(w) W = (Z − 1)/(Z + 1)
143	CLTF(s) TO CLTF(w') W′ = (2/T)(Z − 1)/(Z + 1)
144	HI-RATE CLTF(z) TO LO-RATE CLTF(z)
145	HI-RATE CLTF(w) TO LO-RATE CLTF(w)
146	HI-RATE CLTF(w') TO LO-RATE CLTF(w')
147	OPTION 144 (AVOIDING INTERNAL FACTORING)
148	OPTION 144 (ALL CALCULATIONS IN Z PLANE)

B.8 OTHER COMPUTER-AIDED-DESIGN PACKAGES

There are many other computer-aided-design packages available to assist a control engineer in the analysis, design, and simulation of a control system design. Some of these packages are: CTRL-C program by Systems Control Technology, Inc., Program CC by Systems Technology, Inc., MATRIX_x program by Integrated Systems, Inc., and MACSYMA program by Math Lab Group, Laboratory for Computer Science, M.I.T. The latter two programs are very powerful programs: the MATRIX_x program is a data analysis, system identification, control design, and simulation package, and the MACSYMA program is a large computer program that performs symbolic mathematical computations. Other programs available are MATLB, L-A-S, LOGALPHA to mention just a few.[B17]

The program MULTI[B16] assists in the design of high-gain multiple-input multiple-output digital tracking systems similar to those described in Chap. 21. This is an output feedback system with a proportional plus integral controller. The controller gains are computed and a simulation of the closed-loop system provides state and output responses.

REFERENCES

B1. "A Computer Graphics Program for the Analysis and Design of Digital Filters," *Proc. Int. Conf. Cybernetics Soc.*, October 1980, pp. 1053–1057.

B2. Brubaker, A.: "Development of Improved Design Methods for Digital Filtering Systems," AFAL-TR-77-207, Colorado State University, Fort Collins, Colo., 1977.

B3. Colgate, A.: "INTERAC—An Interactive Software Package for Direct Digital Control Design," M.S. thesis, GGC/EE/77-1, School of Engineering, Air Force Institute of Technology, Wright-Patterson AFB, Ohio, 1977.

B4. Emami-Naeini, A., and G. F. Franklin: "Interactive Computer-Aided Design of Control Systems," *IEEE Control Syst. Mag.*, vol. 10, pp. 31–35, December 1981.

B5. Gembrowski, C. J.: "Continued Development of an Interactive Computer-Aided Design (CAD) Package for Discrete and Continuous Control System Analysis and Synthesis," M.S. thesis, GE/EE/82D, School of Engineering, Air Force Institute of Technology, Wright-Patterson AFB, Ohio, 1982.

B6. Herget, C. J., and P. Weis: "Linear Systems Analysis Program User's Manual," UCID-30184, Lawrence Livermore Laboratory, Livermore, Calif., 1980.

B7. Hernandez, G.: "An Interactive Computational Aerodynamics Analysis Program," M.S. thesis, GAE/AA/80D-9, School of Engineering, Air Force Institute of Technology, Wright-Patterson AFB, Ohio, 1980.

B8. Kennedy, A.: "The Design of Digital Controllers for the C-141 Aircraft Using Entire Eigenstructure Assignment and the Development of an Inter-Active Computer Design Program," M.S. thesis, GGC/EE/79-1, School of Engineering, Air Force Institute of Technology, Wright-Patterson AFB, Ohio, 1979.

B9. Larimer, S. J.: "An Interactive Computer-Aided Design Program for Digital and Continuous System Analysis and Synthesis (TOTAL)," M.S. thesis, GE/GGC/EE/78-2, School of Engineering, Air Force Institute of Technology, Wright-Patterson AFB, Ohio, 1978.

B10. Logan, G. T.: "Development of an Interactive Computer-Aided Design Program for Digital and Continuous Control System Analysis and Synthesis," M.S. thesis, GE/EE/82M-2, School of Engineering, Air Force Institute of Technology, Wright-Patterson AFB, Ohio, 1982.

B11. Manchini, J.: "Computer Aided Control System Design Using Frequency Domain Specifications," Naval Postgraduate School, Monterey, Calif., 1976.

B12. O'Brien, L.: "A Consolidated Computer Program for Control System Design," M.S. thesis, GE/EE/76-38, School of Engineering, Air Force Institute of Technology, Wright-Patterson AFB, Ohio, 1976.

B13. Spielman, C.: "Frequency Methods in Computer Aided Design of Control Systems," M.S. thesis, University of Illinois, Urbana, 1976.

B14. Vines, P.: "Computer Aided Design of Systems," Naval Postgraduate School, Monterey, Calif., 1975.
B15. "Multivariable Analysis Package (MAP)," Lockheed-Georgia Company, Marietta, Ga., 1983.
B16. Porter, D. S.: "Design and Analysis of a Multivariable, Digital Controller for the A-7D Digitac II Aircraft and the Development of an Interactive Computer Design Program," M.S. thesis, AFIT/GE/EE/81D-48, School of Engineering, Air Force Institute of Technology, Wright-Patterson AFB, Ohio, 1981.
B17. Putz, P., and M. J. Wozny: "A New Computer Graphics Approach to Parameter Space Design of Control Systems," *IEEE Trans. AC*, vol. AC-32, pp. 294–302, April 1987.

PROBLEMS

Chapter 2

2.1. Write the (a) loop, (b) node, and (c) state equations for the circuit shown after the switch is closed. Let $y = v_b$. (d) Determine the transfer function v_b/e.

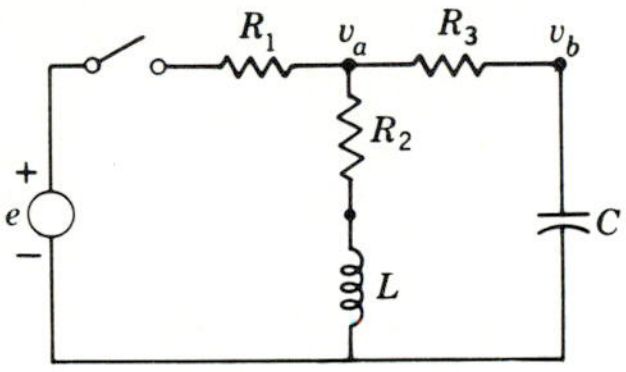

2.2. Write the (a) loop, (b) node, and (c) state equations for the circuit shown after the switch S is closed.

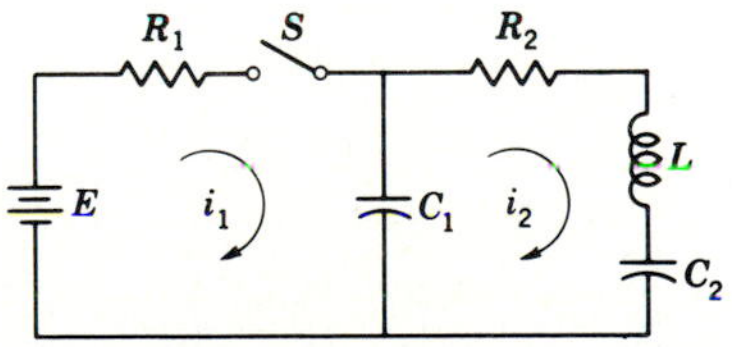

2.3. The circuits shown are in the steady state with the switch S closed. At time $t = 0$, S is opened. (a) Write the necessary differential equations for determining $i_1(t)$. (b) Write the state equations.

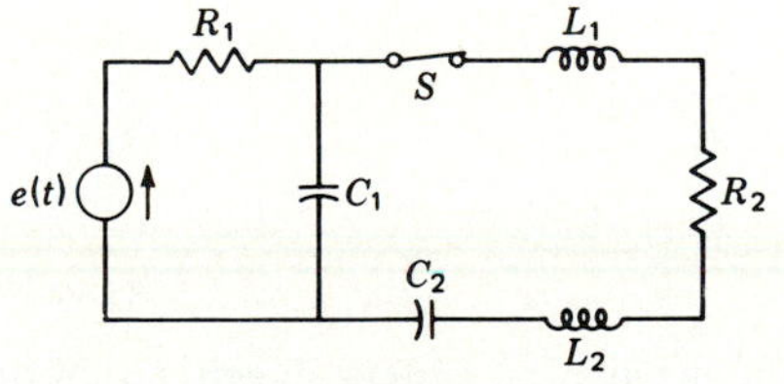

Circuit 1

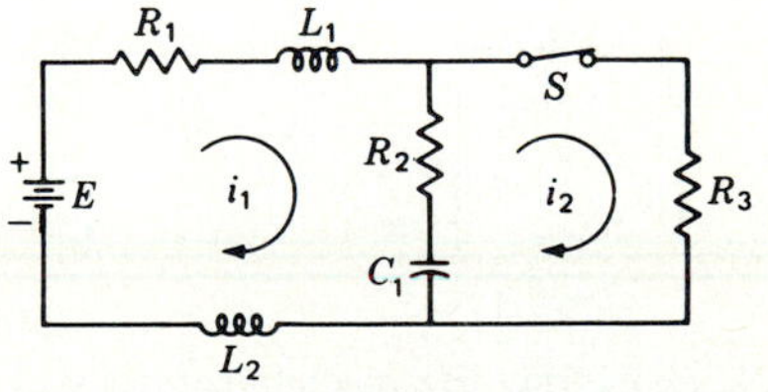

Circuit 2

2.4. Write all the necessary equations to determine v_o. (a) Use nodal equations. (b) Use loop equations. (c) Write the state and output equations.

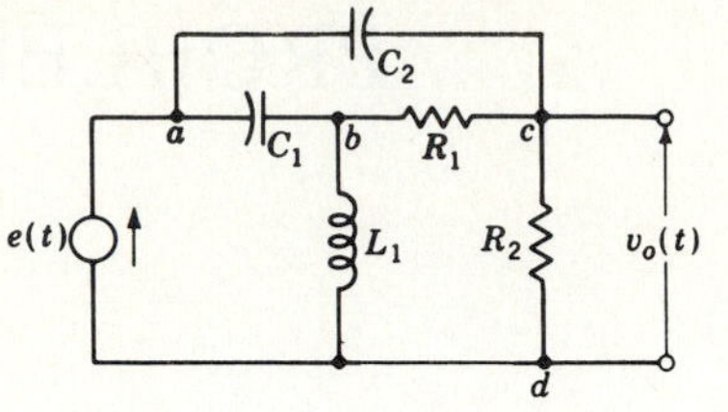

2.5. Derive the state equations. Note that there are only three independent state variables.

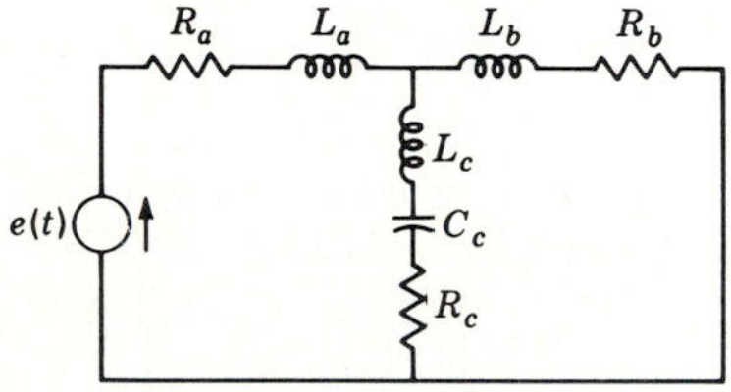

2.6. Write the state equations for the circuit of Fig. 2.3.

2.7. For the spring, damper, and mass system shown: (a) draw the mechanical network; (b) from the mechanical network write the differential equations of performance; (c) write the state equations; (d) determine the transfer function x_1/f.

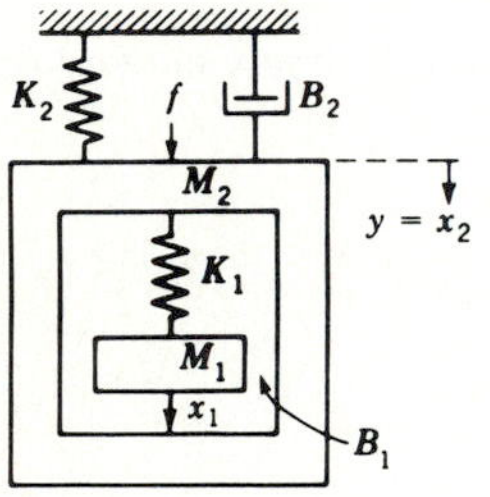

2.8. (a) Derive the differential equation relating the position $y(t)$ and the force $f(t)$. (b) Draw an analogous electric circuit. List all the analogous quantities. (c) Determine the transfer function $G(D) = y/f$. (d) Identify a suitable set of independent state variables. Write the state equation in matrix form.

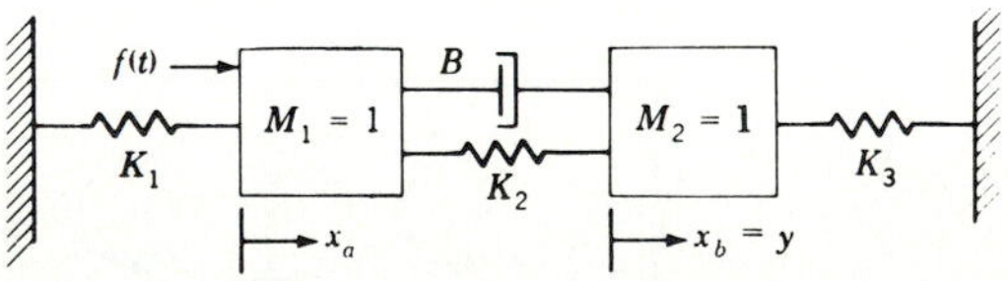

2.9. (a) Draw the mechanical network for the mechanical system shown. (b) Write the differential equations of performance. (c) Draw the analogous electric circuit in which force is analogous to current. (d) Write the state equations.

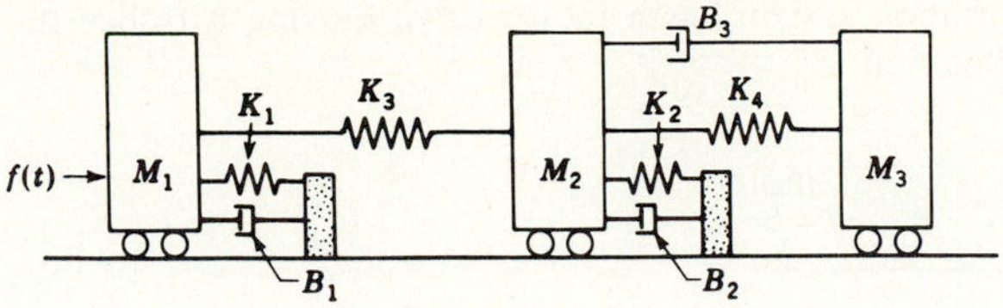

2.10. Write the differential equations describing the motion of the following system, assuming small displacements. (*b*) Write the state equations.

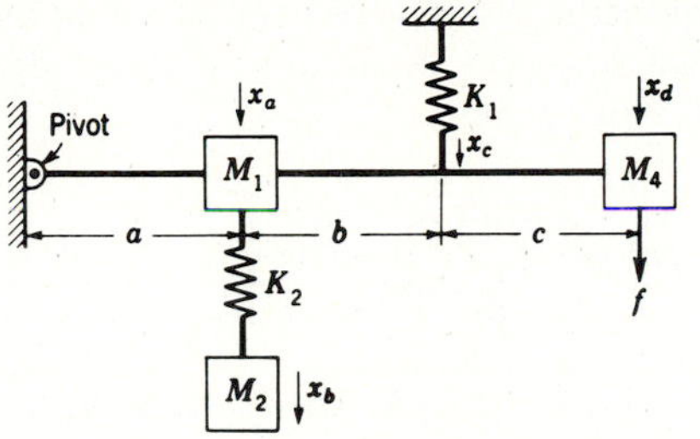

2.11. A model for the vertical suspension of an automobile is shown in the figure. (*a*) Draw the mechanical network; (*b*) write the differential equations of performance; (*c*) derive the state equations; (*d*) determine the transfer function $x_1/\hat{u}_2$, where $\hat{u}_2 = (B_2 D + K_2)u$ is the force exerted by the road.

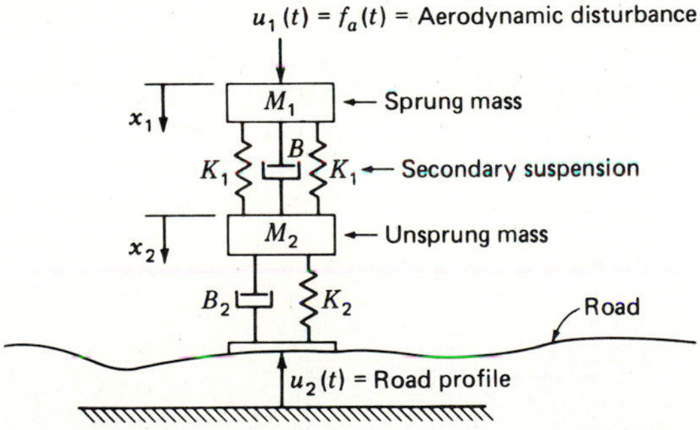

2.12. An electromagnetic actuator contains a solenoid which produces a magnetic force proportional to the current in the coil, $f = K_i i$. The coil has resistance and inductance. (*a*) Write the differential equations of performance. (*b*) Write the state equations.

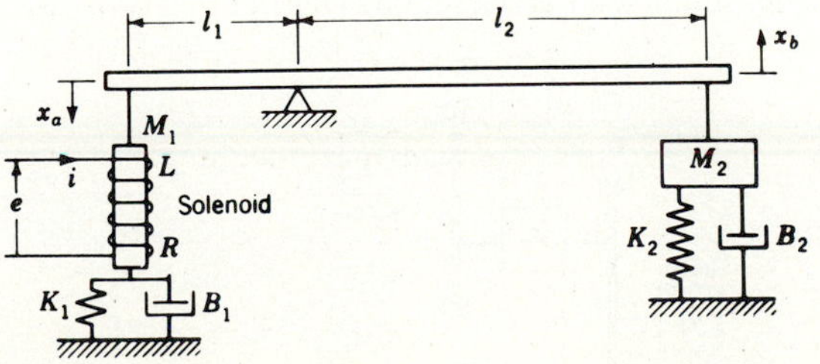

2.13. A warehouse transportation system has a motor drive moving a trolley on a rail. Write the equation of motion.

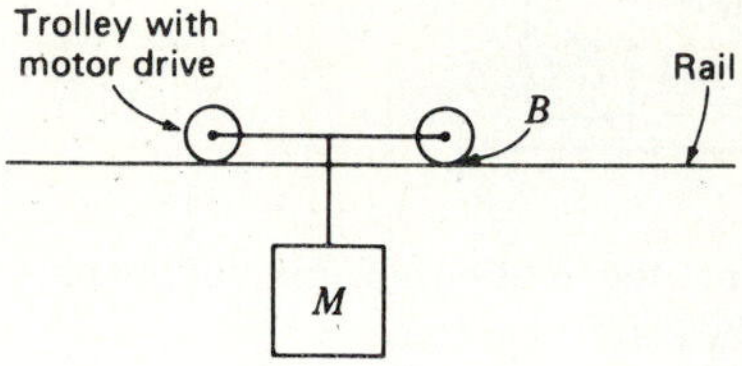

2.14. For the spring, damper, and moment-of-inertia system shown: (*a*) draw the mechanical network; (*b*) write the differential equations of performance; (*c*) draw the analogous electric circuit; (*d*) write the state equations.

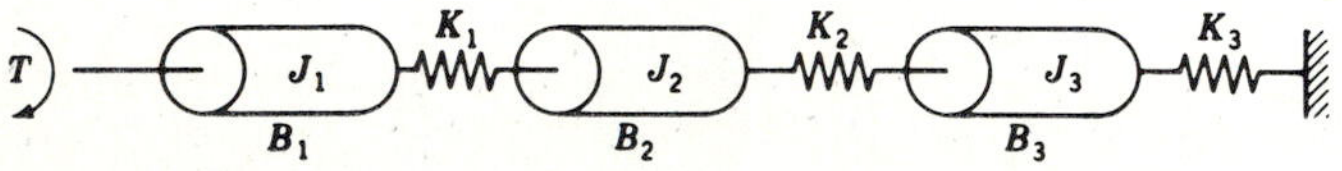

2.15. (*a*) Write the equations of motion for this system. (*b*) Using the physical energy variables, write the matrix state equation.

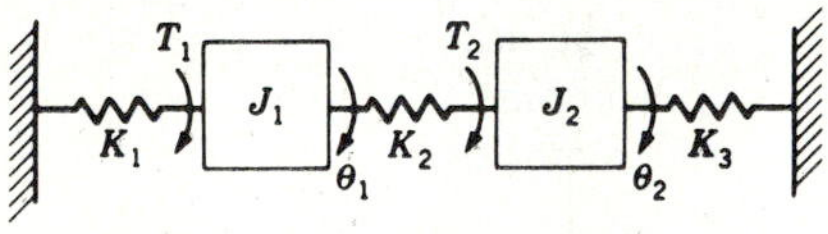

2.16. The figure represents a cylinder of inertia J_1 inside a sleeve of inertia J_2. There is viscous damping B_1 between the cylinder and the sleeve. The springs K_1 and K_2 are fastened to the inner cylinder. (*a*) Draw the mechanical network. (*b*) Write the system equations. (*c*) Draw the analogous electric circuit. (*d*) Write the state and output equations. The outputs are θ_1 and θ_2. (*e*) Determine the transfer function $G = \theta_2/T$.

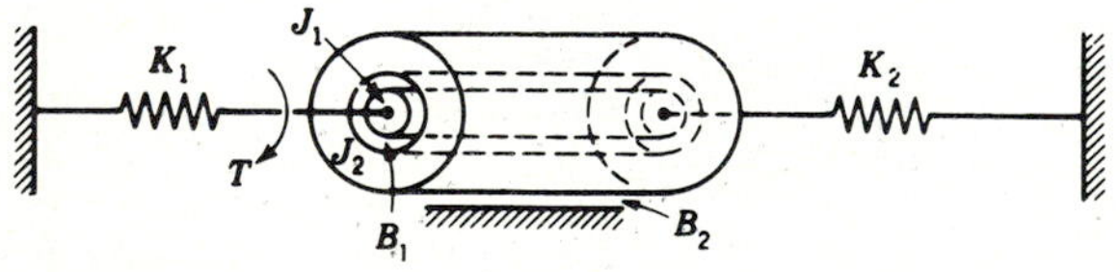

2.17. The two gear trains have an identical net reduction, have identical inertias at each stage, and are driven by the same motor. The number of teeth on each gear is indicated on the figures. At the instant of starting, the motor develops a torque T. Which system has the higher initial load acceleration? (*a*) Let $J_1 = J_2 = J_3$; (*b*) let $J_1 = 40J_2 = 4J_3$.

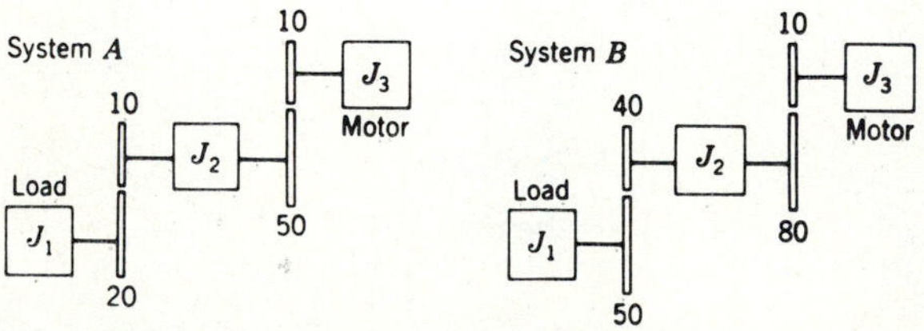

2.18. In the mechanical system shown, r_2 is the radius of the drum. (a) Write the necessary differential equations of performance for this system. (b) Obtain a differential equation expressing the relationship of the output x_3 in terms of the input θ_1 and the corresponding transfer function. (c) Write the state equations with θ_1 as the input.

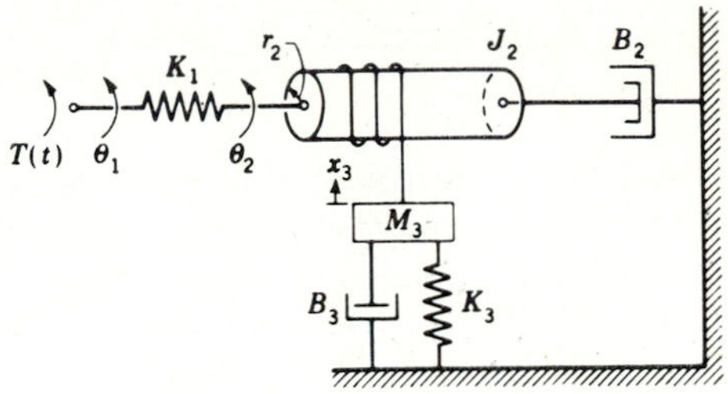

2.19. Write the state and system output equations for the rotational mechanical system of Fig. 2.16. (a) With T as the input, use $x_1 = \theta_3$, $x_2 = D\theta_3$, $x_3 = D\theta_2$ and (b) with θ_1 as the input, use $x_1 = \theta_3$, $x_2 = D\theta_3$, $x_3 = \theta_2$, $x_4 = D\theta_2$.

2.20. For the hydraulic preamplifier write the differential equations of performance relating x_1 to y_1. (a) Neglect the load reaction. (b) Do not neglect load reaction. There is only one fixed pivot, as shown in the figure.

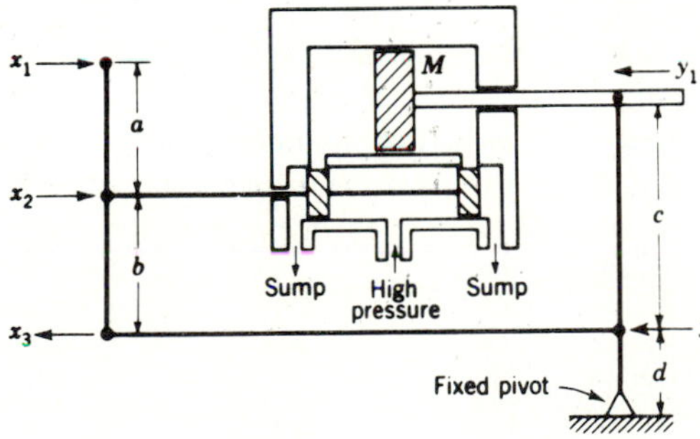

2.21. The mechanical load on the hydraulic translational actuator in Sec. 2.10 is changed to that below. Determine the new state equations and compare with Eq. (2.116).

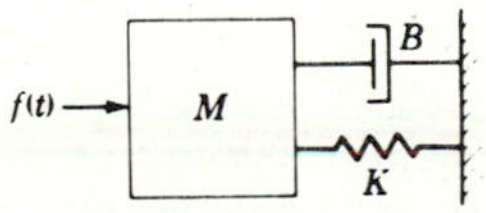

2.22. The sewage system leading to a treatment plant is shown. The variables q_A and q_B are input flow rates into tanks 1 and 2, respectively. Pipes 1, 2, and 3 have resistances as shown. Derive the state equations.

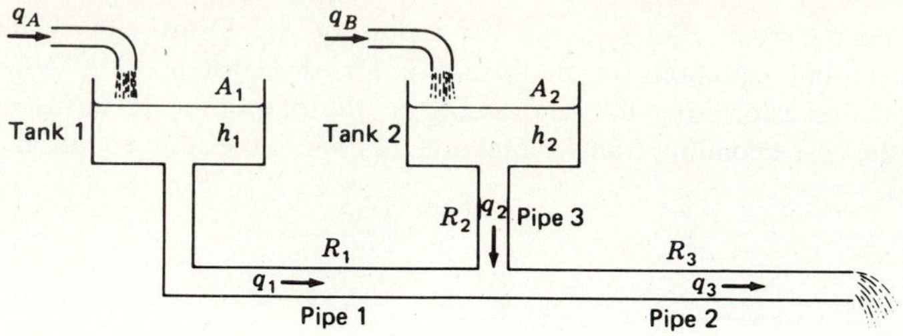

2.23. Most control systems require some type of motive power. One of the most commonly used units is the electric motor. Write the differential equation for the angular displacement of a moment of inertia, with damping, connected directly to a dc motor shaft when a voltage is suddenly applied to the armature terminals with the field separately energized.

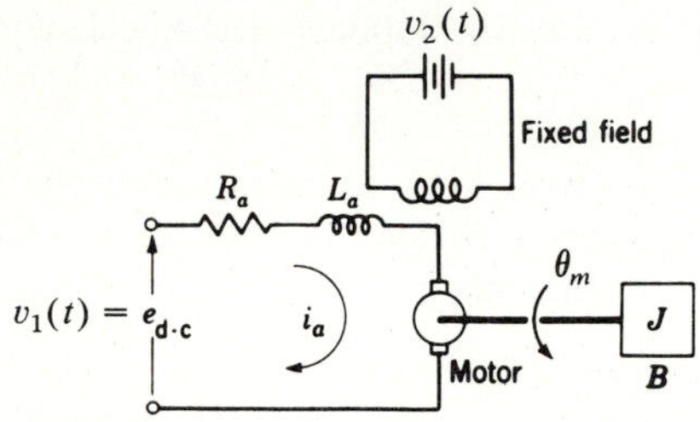

2.24. A sketch of a moving-coil microphone is shown. The diaphragm has the spring elastance K, mass M, and damping B. Fastened to the diaphragm is a coil which moves in the magnetic field produced by the permanent magnets. (a) Derive the differential equations of the system considering changes from the equilibrium conditions. (b) Draw an analogous electric circuit.

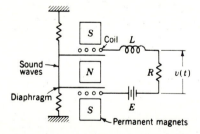

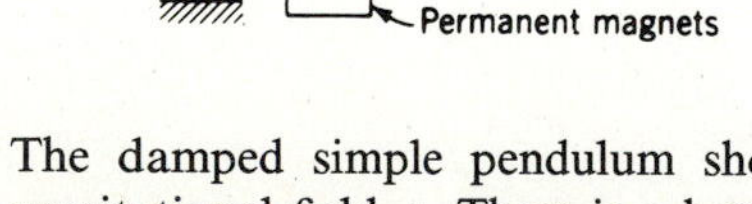

2.25. The damped simple pendulum shown in the figure is suspended in a uniform gravitational field g. There is a horizontal force f and a rotational viscous friction B. (a) Determine the equation of motion (note: $J = M\ell^2$). (b) Linearize the equation of part (a) by assuming $\cos\theta \approx 1$ and $\sin\theta \approx \theta$ for small values of θ. (c) Write the state and output equations of the system using $x_1 = \theta$, $x_2 = \dot{\theta}$, and $y = x_1$.

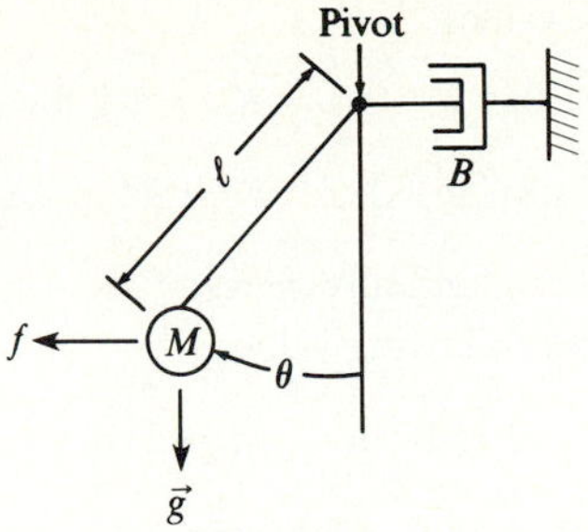

2.26. Given the mechanical system shown, the load J_2 is coupled to the rotor J_1 of a motor through a reduction gear and springs K_1 and K_2. The gear teeth ratio is $1:N$. The torque T is the mechanical torque generated by the motor. B_1 and B_2 are damping coefficients, and θ_1, θ_2, θ_3, and θ_4 are angular displacements. (a) Draw a mechanical network for this system. (b) Write the necessary differential equations in order to solve for θ_4.

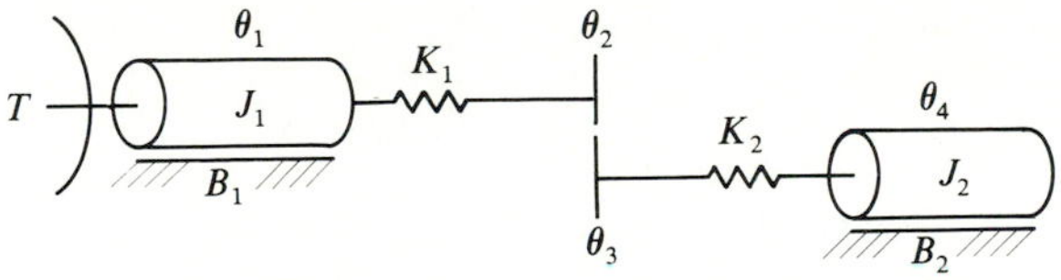

2.27. In some applications the motor shaft of Prob. 2.23 is of sufficient length so that its elastance K must be taken into account. Consider that $v_2(t)$ is constant and that the velocity $\omega(t) = D\theta_m(t)$ is being controlled. (a) Write the necessary differential equation for determining $\omega(t)$. (b) Determine the transfer function $\omega(t)/v_1(t)$.

Chapter 3

3.1. (a) With $v_c(0-) = 0$, what are the initial values of current in all elements when the switch is closed? (b) Write the state equations. (c) Solve for the voltage across C as a function of time from the state equations. (d) Find M_p, t_p, and t_s.

$$R_1 = R_2 = 1\ \text{k}\Omega \qquad C = 50\ \mu\text{F} \qquad L = 1\ \text{H} \qquad E = 10\ \text{V}$$

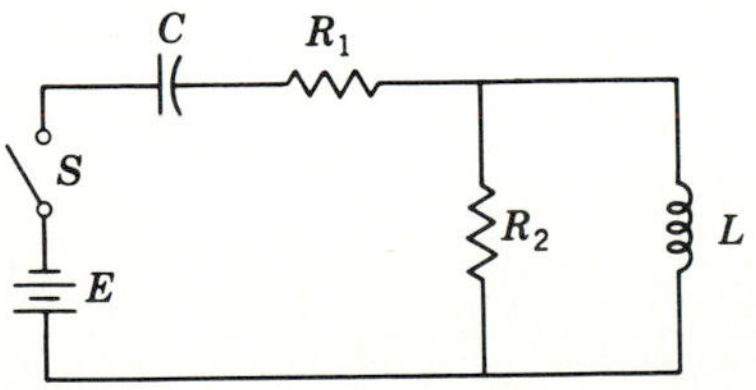

3.2. In Prob. 2.14, the parameters have the following values:

$J_1 = 74{,}150\ \text{oz}\cdot\text{in}^2$ $\quad B_1 = 0.5\ \text{lb}\cdot\text{ft}/(\text{rad/s})$ $\quad K_1 = 0.5\ \text{lb}\cdot\text{ft/rad}$

$J_2 = 1.0\ \text{lb}\cdot\text{ft}\cdot\text{s}^2$ $\quad B_2 = 12.8\ \text{oz}\cdot\text{ft}/(\text{rad/s})$ $\quad K_2 = 8.0\ \text{oz}\cdot\text{ft/rad}$

$J_3 = 1.0\ \text{slug}\cdot\text{ft}^2$ $\quad B_3 = 3.35\ \text{oz}\cdot\text{in}/(\text{deg/s})$ $\quad K_3 = 96.0\ \text{oz}\cdot\text{in/rad}$

Solve for $\theta_3(t)$ if $T(t) = tu_{-1}(t)$.

3.3. In Prob. 2.12, the parameters have the following values:

$$M_1 = M_2 = 0.05 \text{ slug} \qquad L = 1 \text{ H} \qquad l_1 = 10 \text{ in} \qquad K_1 = K_2 = 1.2 \text{ lb/in}$$

$$R = 10\ \Omega \qquad l_2 = 20 \text{ in} \qquad B_1 = B_2 = 15 \text{ oz/(in/s)} \qquad K_i = 24 \text{ oz/A}$$

Solve for $x_b(t)$ with $e(t) = u_{-1}(t)$ and the system is initially at rest.

3.4. In part (a) of Prob. 2.20, the parameters have the following values:

$$a = b = 6 \text{ in} \qquad C_1 = 6.0 \text{ (in/s)/in} \qquad c = 10 \text{ in} \qquad d = 2 \text{ in}$$

Solve for $y_1(t)$ with $x_1(t) = 0.1u_{-1}(t)$ in and zero initial conditions.

3.5. (a) $r(t) = (D^3 + 6D^2 + 11D + 6)c(t)$. With $r(t) = \sin t$ and zero initial conditions, determine $c(t)$. (b) $r(t) = [(D+1)(D^2 + 2D + 2)]c(t)$. With $r(t) = tu_{-1}(t)$ and all initial conditions zero, determine the complete solution with all constants evaluated. (c) $D^3c + 16D^2c + 85Dc + 150c = 37.5r$. Find $c(t)$ for $r(t) = u_{-1}(t)$, $D^2c(0) = Dc(0) = 0$, and $c(0) = -2$. One of the eigenvalues is $\lambda = -6$. (d) For each equation determine $\mathbf{C}(j\omega)/\mathbf{R}(j\omega)$ and $c(t)_{ss}$ for $r(t) = 10 \sin 5t$.

3.6. The rotational hydraulic transmission described in Sec. 2.11 has an inertial load coupled through a spring. The system equation is

$$4(D^2 + 1)x = (3D^2 + 3D + 1)\, D\theta_m$$

The system is originally stationary. With $x(t) = 5 \cos 5t$, find the motor velocity $\omega_m(t)$. *Note:* $\theta_m(0) = \theta_L(0) = \omega_L(0) = 0$, but $\omega_m(0) \neq 0$. Its value must be determined by using Eqs. (2.119) and (2.123).

3.7. A hydraulic motor with inertia load is driven by a variable-displacement pump. Use Eq. (2.122) to determine (a) the undamped natural frequency ω_n; (b) the damping ratio ζ; (c) the damped natural frequency of the system ω_d. (d) Determine the speed of response $D\theta_m(t)$ with $x(t) = 0$, $D\theta_m(0) = 0$, and $D^2\theta_m(0) = 1$. The parameters have the following values:

$$S_p = 100 \text{ in}^3\text{/s} = d_p\omega_p \qquad L = 0.01 \text{ (in}^3\text{/s)/(lb/in}^2\text{)} \qquad d_m = 1.0 \text{ in}^3\text{/rad}$$

$$C = 1 \text{ in}^3 \qquad V = 20 \text{ in}^3 \qquad K_B = 2.5 \times 10^5 \text{ lb/in}^2 \quad J = 200 \text{ lb} \cdot \text{in}^2$$

3.8. The system shown is initially at rest. At time $t = 0$ the string connecting M to W is severed at X. Find $x(t)$.

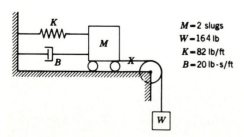

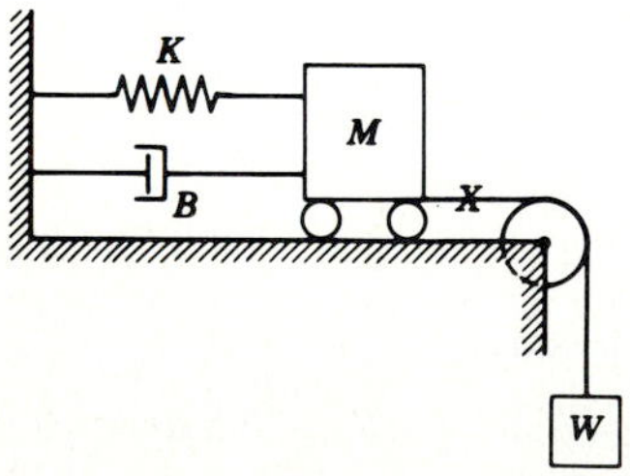

$M = 2$ slugs
$W = 164$ lb
$K = 82$ lb/ft
$B = 20$ lb-s/ft

3.9. Switch S_1 is open, and there is no energy stored in the circuit. (a) Write the state equations. (b) Find $v_o(t)$ after switch S_1 is closed.

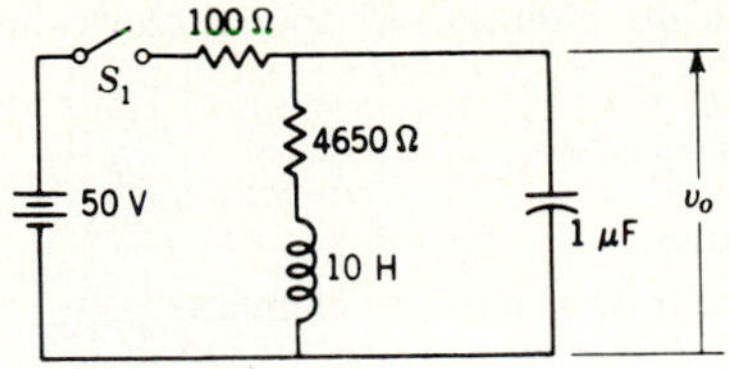

3.10. Solve the following differential equations. Assume zero initial conditions. Sketch the solutions.

(a) $D^2x + 16x = 1$ (b) $D^2x + 4Dx + 3x = 9$

(c) $D^2x + Dx + 4.25x = t + 1$ (d) $D^3x + 3D^2x + 4Dx + 2x = 10 \sin 10t$

3.11. For Prob. 2.2, solve for $i_2(t)$, where

$$E = 50 \text{ V} \qquad L = 1 \text{ H} \qquad C_1 = C_2 = 0.001 \text{ F} \qquad R_1 = 100\ \Omega \qquad R_2 = 100\ \Omega$$

3.12. For the circuit of Prob. 2.2: (a) obtain the differential equation relating $i_2(t)$ to the input E; (b) write the state equations using the physical energy variables; (c) solve the state and output equations and compare with the solution of Prob. 3.11; (d) write the state and output equations using the phase variables, starting with the differential equation from part (a).

3.13. In Prob. 2.3, the parameters have the values $R_1 = R_2 = R_3 = 10\ \Omega$, $C_1 = C_2 = 1\ \mu\text{F}$, and $L_1 = L_2 = 50$ H. If $E = 100$ V, (a) find $i_1(t)$; (b) sketch $i_1(t)$ and compute M_o, t_p, and t_s; (c) determine the value of T_s (± 2 percent of $i(0)$); (d) solve for $i_1(t)$ from the state equation.

3.14. For $L = C = 1$ and (1) $R = 1$, (2) $R = 0.4$: (a) show that the differential equation is $(RCD^2 + D + R/L)v_0 = De$; ($b$) find $v_o(t)_{ss}$ for $e(t) = u_{-1}(t)$; (c) find the roots m_1 and m_2; (d) find $v_o(0+)$; (e) find $Dv_o(0+)$ [use $i_c(0+)$]; (f) determine the complete solution $v_o(t)$; (g) sketch and dimension the plot of $v_o(t)$.

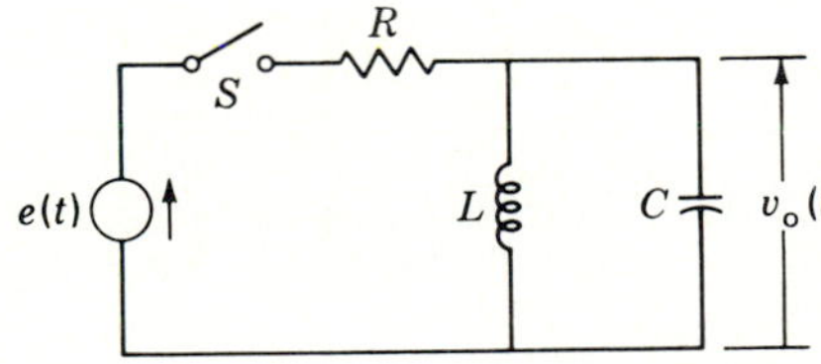

3.15. Solve the following equations. Show explicitly $\mathbf{\Phi}(t)$, $\mathbf{x}(t)$, and $\mathbf{y}(t)$.

$$\dot{\mathbf{x}} = \begin{bmatrix} -6 & 4 \\ -2 & 0 \end{bmatrix}\mathbf{x} + \begin{bmatrix} 0 \\ 1 \end{bmatrix}u \qquad \mathbf{y} = [1 \quad 0]\mathbf{x}$$

$$x_1(0) = 2 \qquad x_2(0) = 0 \qquad u(t) = u_{-1}(t)$$

3.16. A single-degree-of-freedom representation of the rolling dynamics of an aircraft, together with a first-order representation of the aileron servomotor, is given by

$$p(t) = \dot{\phi}(t) \qquad J_x\dot{p}(t) = L_{\delta_A}\delta_A(t) + L_p p(t)$$

$$T\dot{\delta}_A(t) = (\delta_A)_{\text{comm}}(t) - \delta_A(t)$$

(a) Derive a state-variable representation of the aircraft and draw a block diagram of the control system with

$$(\delta_A)_{\text{comm}}(t) = AV_{\text{ref}} - \mathbf{P}\mathbf{x}(t)$$

where $\mathbf{x}(t)$ = state vector of the aircraft system = $[\phi \quad p \quad \delta_A]^T$
$\mathbf{P}$ = matrix of correct dimensions and with constant nonzero elements p_{ij}
A = scalar gain
V_{ref} is the reference input and $(\delta_A)_{\text{comm}}$ is the input to the servo.
(b) Determine the aircraft state equations in response to the reference input V_{ref}.

3.17. Use the Sylvester expansion of Sec. 3.13 to obtain $\mathbf{A}^{10}$ for

$$\mathbf{A} = \begin{bmatrix} 0 & 1 & 0 \\ 0 & 0 & 1 \\ -6 & -11 & -6 \end{bmatrix}$$

3.18. For the mechanical system of Fig. 2.11a the state equation for Example 2, Sec. 2.6, is given in *phase-variable form*. With the input as $u = x_a$, the state variables are $x_1 = x_b$ and $x_2 = \dot{x}_b$. Use $M = 1$, $K = 12$, and $B = 7$. The initial conditions are $x_b(0) = 1$ and $\dot{x}_b(0) = -2$. (a) Find the homogeneous solution for $\mathbf{x}(t)$. (b) Find the complete solution with $u(t) = u_{-1}(t)$.

3.19. For the autonomous system

(1) $\dot{\mathbf{x}} = \begin{bmatrix} 0 & 1 & 0 \\ 0 & 0 & 1 \\ -2 & -4 & -3 \end{bmatrix}\mathbf{x} \qquad \mathbf{y} = [1 \quad 0 \quad 0]\mathbf{x}$

(2) $\dot{\mathbf{x}} = \begin{bmatrix} 0 & 1 & 0 \\ 0 & 0 & 1 \\ -12 & -13 & -8 \end{bmatrix}\mathbf{x} \qquad \mathbf{y} = [1 \quad 0 \quad 0]\mathbf{x}$

(a) Find the system eigenvalues. (b) Evaluate the state transmission matrix $\mathbf{\Phi}(t)$ by means of Sylvester's expansion theorem.

Chapter 4

4.1. Take the inverse Laplace transform of

(a) $F(s) = \dfrac{10}{(s+1)^2(s^2+2s+2)}$ (b) $F(s) = \dfrac{39}{s(s^2+4s+13)(s+3)}$

(c) $F(s) = \dfrac{10(s+5)}{(s+1)^2(s^2+4s+5)}$

4.2. Find $x(t)$ for

(a) $X(s) = \dfrac{1}{s^3+7s^2+20s+24}$ (b) $X(s) = \dfrac{20}{s^4+9s^3+30s^2+42s+20}$

(c) $X(s) = \dfrac{0.9524(s^2+2s+2.1)}{s(s+1)(s^2+2s+2)}$ (d) $X(s) = \dfrac{1.82(s+1.1)}{s(s+1)(s^2+2s+2)}$

(e) $X(s) = \dfrac{16(s+1)(s+10)}{s(s+2)^3(s^2+6s+10)}$ (f) Sketch $x(t)$ for parts (a) to (e).

4.3. Given an ac servomotor with inertia load, find $\omega(t)$ by (*a*) the classical method; (*b*) the Laplace-transform method.

$$\omega(0) = 10^4 \text{ rad/s} \qquad K_\omega = -3 \times 10^{-3} \text{ oz} \cdot \text{in/(rad/s)} \qquad K_c = 1.2 \text{ oz} \cdot \text{in/V}$$

$$e(t) = 5u_{-1}(t) \text{ V} \qquad J = 11.59 \text{ oz} \cdot \text{in}^2$$

4.4. Repeat the following problems using the Laplace transform:
(*a*) Prob. 3.1 (*b*) Prob. 3.3 (*c*) Prob. 3.5
(*d*) Prob. 3.8 (*e*) Prob. 3.12 (*f*) Prob. 3.13(*d*)
(*g*) Prob. 3.15 (*h*) Prob. 3.16

4.5. Find the partial-fraction expansions of the following:

(*a*) $F(s) = \dfrac{5}{(s+1)(s+5)}$ (*b*) $F(s) = \dfrac{10}{s(s+2)(s+5)}$

(*c*) $F(s) = \dfrac{10}{s(s^2+2s+10)}$ (*d*) $F(s) = \dfrac{2(s+2)}{s^2(s+1)(s+4)}$

(*e*) $F(s) = \dfrac{10(s+5)}{s^2(s^2+4s+5)}$ (*f*) $F(s) = \dfrac{4(s+3)}{(s+1)(s+2)(s+6)}$

(*g*) $F(s) = \dfrac{30}{s(s+3)(s^2+6s+10)}$ (*h*) $F(s) = \dfrac{30(s+1)}{s(s+3)(s^2+6s+10)}$

(*i*) $F(s) = \dfrac{10(s+1.01)}{s(s+1)(s^2+2s+10)}$ (*j*) $F(s) = \dfrac{10(s^2+2.2s+10.21)}{s(s+1)(s^2+2s+10)}$

(*k*) $F(s) = \dfrac{20}{s(s^2+2s+2)(s^2+6s+10)}$

(*l*) $F(s) = \dfrac{52}{s(s^2+2s+2)(s^2+10s+26)}$

(*m*) $F(s) = \dfrac{10(s^2+2s+2.21)}{s(s^2+6s+10)(s^2+2s+2)}$

(*n*) $F(s) = \dfrac{100}{s^2(s+10)(s^2+6s+10)}$

4.6. Solve the differential equations of Prob. 3.10 by means of the Laplace transform.

4.7. Write the Laplace transforms of the following equations and solve for $x(t)$; the initial conditions are given to the right.
(*a*) $Dx + 4x = 0$ $\qquad x(0) = 2$
(*b*) $D^2x + 2Dx + 5x = 10$ $\qquad x(0) = 2,\ Dx(0) = -4$
(*c*) $D^2x + 3Dx + 2x = t$ $\qquad x(0) = 0,\ Dx(0) = -2$
(*d*) $D^3x + 4D^2x + 14Dx + 20x = \sin 5t$ $\qquad x(0) = -4,\ Dx(0) = 1,\ D^2x(0) = 0$

4.8. Determine the final value for:
(*a*) Prob. 4.1 (*b*) Prob. 4.2 (*c*) Prob. 4.5

4.9. Determine the initial value for:
(*a*) Prob. 4.1 (*b*) Prob. 4.2 (*c*) Prob. 4.5

4.10. For the functions of Prob. 4.5, plot M vs. ω and α vs. ω.

4.11. Find the complete solution of $x(t)$ with zero initial conditions for

(1) $(D^2 + 2D + 2)(D + 4)x = (D + 2)f(t)$

(2) $(D^2 + 3D + 2)(D + 3)x = (D + 4)f(t)$

The forcing function $f(t)$ is

(*a*) $u_0(t)$ (*b*) $10u_{-1}(t)$ (*c*) $tu_{-1}(t)$ (*d*) $t^2u_{-1}(t)$

4.12. A linear system is described by

(1) $\dot{\mathbf{x}} = \begin{bmatrix} -2 & 1 \\ 3 & -4 \end{bmatrix}\mathbf{x} + \begin{bmatrix} 0 \\ 1 \end{bmatrix}\mathbf{u} \qquad \mathbf{y} = [1 \quad 0]\mathbf{x}$

(2) $\dot{\mathbf{x}} = \begin{bmatrix} 0 & 3 \\ -1 & -4 \end{bmatrix}\mathbf{x} + \begin{bmatrix} 1 \\ 0 \end{bmatrix}\mathbf{u} \qquad \mathbf{y} = [1 \quad 1]\mathbf{x}$

where $u(t) = u_{-1}(t)$ and the initial conditions are $x_1(0) = 0$, $x_2(0) = 1$. (*a*) Using Laplace transforms, find $\mathbf{X}(s)$. Put the elements of this vector over a common denominator. (*b*) Find the transfer function $G(s)$.

4.13. A linear system is represented by

(1) $\dot{\mathbf{x}} = \begin{bmatrix} -6 & 4 \\ -2 & 0 \end{bmatrix}\mathbf{x} + \begin{bmatrix} 1 \\ 1 \end{bmatrix}\mathbf{u} \qquad \mathbf{y} = \begin{bmatrix} 1 & 0 \\ 1 & 1 \end{bmatrix}\mathbf{x} \qquad u = u_{-1}(t)$

(2) $\dot{\mathbf{x}} = \begin{bmatrix} 0 & 3 \\ -1 & -4 \end{bmatrix}\mathbf{x} + \begin{bmatrix} 1 & 1 \\ 0 & -1 \end{bmatrix}\mathbf{u} \qquad \mathbf{y} = \begin{bmatrix} 1 & 0 \\ 1 & 2 \end{bmatrix}\mathbf{x} \qquad \mathbf{u} = \begin{bmatrix} 1 \\ 1 \end{bmatrix}u_{-1}(t)$

(3) $\dot{\mathbf{x}} = \begin{bmatrix} 0 & 5 \\ 0 & -4 \end{bmatrix}\mathbf{x} + \begin{bmatrix} 0 \\ 1 \end{bmatrix}\mathbf{u} \qquad \mathbf{y} = \begin{bmatrix} 0 & 1 \\ 1 & 1 \end{bmatrix}\mathbf{x} \qquad u = u_{-1}(t)$

(*a*) Find the complete solution for $\mathbf{y}(t)$ when $x_1(0) = 1$ and $x_2(0) = 0$. (*b*) Determine the transfer functions. (*c*) Draw a block diagram representing the system.

4.14. A system is described by

(1) $\dot{\mathbf{x}} = \begin{bmatrix} -3 & -1 \\ 2 & 0 \end{bmatrix}\mathbf{x} + \begin{bmatrix} 1 \\ 0 \end{bmatrix}\mathbf{u} \qquad y = x_1$

(2) $\dot{\mathbf{x}} = \begin{bmatrix} 0 & 1 \\ -3 & -4 \end{bmatrix}\mathbf{x} + \begin{bmatrix} 1 \\ 1 \end{bmatrix}\mathbf{u} \qquad y = x_1$

(*a*) Find $\mathbf{x}(t)$ with $\mathbf{x}(0) = \mathbf{0}$ and $u = u_{-1}(t)$. (*b*) Determine the transfer function $G(s) = Y(s)/U(s)$.

4.15. Repeat Prob. 3.19 using the Laplace transform.

4.16. Show that Eqs. (4.52) and (4.53) are equivalent expressions.

4.17. Given

$$Y(s) = \frac{K(s + a_1)\cdots(s + a_w)}{s(s + \zeta_1\omega_{n_1} \pm j\omega_{d_1})(s + \zeta_3\omega_{n_3} \pm j\omega_{d_3})(s + b_5)\cdots(s + b_n)}$$

all the a_i's and b_k's are positive and real and $\zeta_1\omega_{n_1} = \zeta_3\omega_{n_3}$ with $\omega_{n_3} > \omega_{n_1}$. Determine the effect on the Heaviside partial-fraction coefficients associated with the pole $p_3 = -\zeta_3\omega_{n_3} + j\omega_{d_3}$ when ω_{n_3} is allowed to approach infinity. *Hint:* Analyze

$$\lim_{\omega\to\infty} A_3 = \lim_{\omega\to\infty} [(s - p_k)Y(s)]_{s=p_3=-\zeta_3\omega_{n_3}+j\omega_{d_3}}$$

K has the value required for $y(t)_{ss} = 1$ for a unit step input. Consider the cases (1) $n = w$, (2) $n > w$.

4.18. Repeat Prob. 4.17 with $a_1 = 0$, $w = 2$, $n = 6$, and K is a fixed value.

4.19. Use the method of Sec. 4.14 to determine (*a*) the characteristic polynomial (*Hint:* For (1) use $\mathbf{b} = [1 \quad 0 \quad 1]^T$ and for (2) use $\mathbf{b} = [1 \quad 0 \quad 0]^T$); (*b*) the matrix inverse $\mathbf{A}^{-1}$.

$$(1)\ \mathbf{A} = \begin{bmatrix} -1 & 0 & 0 \\ -2 & -2 & -2 \\ -1 & 0 & -3 \end{bmatrix} \qquad (2)\ \mathbf{A} = \begin{bmatrix} 0 & 2 & -4 \\ 1 & 3 & 5 \\ 2 & 0 & 4 \end{bmatrix}$$

4.20. Repeat Prob. 4.19 with

$$\mathbf{A} = \begin{bmatrix} 0 & 1 & 0 & 0 \\ 1 & 0 & 1 & -1 \\ 0 & 0 & 0 & 1 \\ 1 & 1 & 0 & 0 \end{bmatrix}$$

Hint: use $\mathbf{b} = [0 \quad 1 \quad 0 \quad 1]^T$.

4.21. For the linearized system of Prob. 2.25 determine (*a*) the transfer function and (*b*) the value of B that makes this system critically damped.

4.22. Determine the transfer function and draw a block diagram.

$$\dot{\mathbf{x}} = \begin{bmatrix} 2 & 0 \\ -3 & 1 \end{bmatrix}\mathbf{x} + \begin{bmatrix} 1 & 0 \\ 1 & 2 \end{bmatrix}\mathbf{u} \qquad \mathbf{y} = \begin{bmatrix} 0 & 1 \\ 2 & 0 \end{bmatrix}\mathbf{x}$$

Chapter 5

5.1. For the temperature-control system of Fig. 5.1, some of the pertinent equations are $b = K_b\theta$, $f_s = K_s i_s$, $q = K_q x$, $\theta = K_c q/(D + a)$. The solenoid coil has resistance R and inductance L. The solenoid armature and valve have mass M, damping B, and a restraining spring K. (*a*) Determine the transfer function of each block in Fig. 5.1*b*. (*b*) Determine the forward transfer function $G(s)$. (*c*) Write the state and output equations in matrix form. Use $x_1 = \theta$, $x_2 = x$, $x_3 = \dot{x}$, $x_4 = i_s$.

5.2. Find an example of a practical closed-loop control system not covered in this book. Briefly describe the system and show a block diagram.

5.3. A satellite-tracking system is shown in the schematic diagram. The transfer function of the tracking receiver is

$$\frac{V_1}{\Theta_{\text{comm}} - \Theta_L} = \frac{5}{1 + s/40\pi}$$

The following parameters apply:

K_a = gain of servoamplifier = 60
K_t = tachometer constant = 0.04 V · s
K_T = motor torque constant = 0.5 N · m/A
K_b = motor back-emf constant = 0.75 V · s
J_L = antenna inertia = 2,880 kg · m²
J_m = motor inertia = 7×10^{-3} kg · m²
1 : N = gearbox stepdown ratio = 1 : 12,000

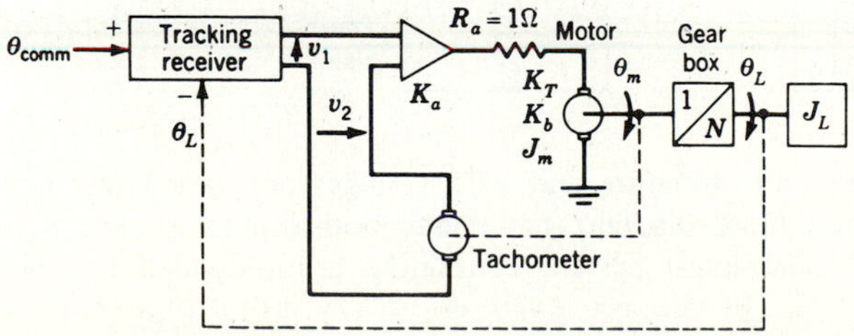

(a) Draw a detailed block diagram showing all the variables. (b) Derive the transfer function $\Theta_L(s)/\Theta_{comm}(s)$.

5.4. A photographic control system is shown in simplified form in the diagram. The aperture slide moves to admit the light from the high-intensity lamp to the sensitive plate, the illumination of which is a linear function of the exposed area of the aperture. The maximum area of the aperture is 4 m^2. The plate is illuminated by 1 candela (cd) for every square meter of aperture area. This luminosity is detected by the photocell, which provides an output voltage of 1 V/cd. The motor torque constant is 1 N · m/A, the motor inertia is 0.1 kg · m^2, and the motor back-emf constant is 1 V · s. The viscous friction is 0.2 N · m · s, and the amplifier gain is 50 V/V. (a) Draw the block diagram of the system with the appropriate transfer function inserted in each block. (b) Derive the overall transfer function.

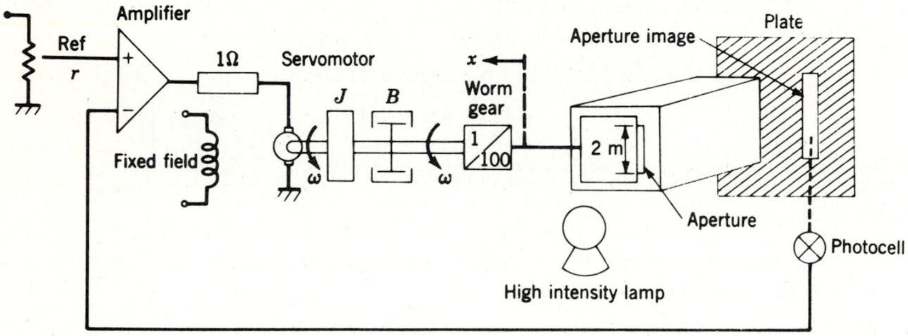

5.5. The longitudinal motion of an aircraft is represented by the vector differential equation

$$\dot{\mathbf{x}} = \begin{bmatrix} -0.08 & 1 & -0.01 \\ -8.0 & -0.05 & -6.0 \\ 0 & 0 & -10 \end{bmatrix} \mathbf{x} + \begin{bmatrix} 0 \\ 0 \\ 10 \end{bmatrix} \delta_E$$

where x_1 = angle of attack
x_2 = rate of change of pitch angle
x_3 = incremental elevator angle
δ_E = control input into the elevator actuator

Derive the transfer function relating the system output, rate of change of pitch angle, to the control input into the elevator actuator, x_2/δ_E.

5.6. (a) Use the hydraulic valve and power piston in a closed-loop position control system. Derive the transfer function of the system. (b) Draw a diagram of a control system for the elevators on an airplane. Use a hydraulic actuator.

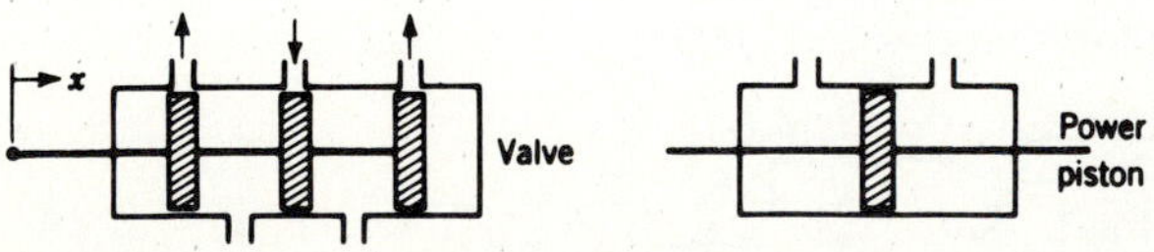

5.7. In the circuit shown, both armature and field voltages are varied to control the output θ_m. (a) Draw a block diagram that relates both inputs, v_1 and v_2, to the output θ_m. This is a nonlinear system containing a multiplication. The block diagram should be labeled in the *D*-operator form. (*b*) Linearize the system and use Laplace transforms. Draw the new block diagram. Multiplication is replaced by addition for small excursions of the variables by using

$$df(x, y) = \frac{\partial f}{\partial x}\,dx + \frac{\partial f}{\partial y}\,dy = K_x\,dx + K_y\,dy$$

(c) Write the state equation for the linearized system.

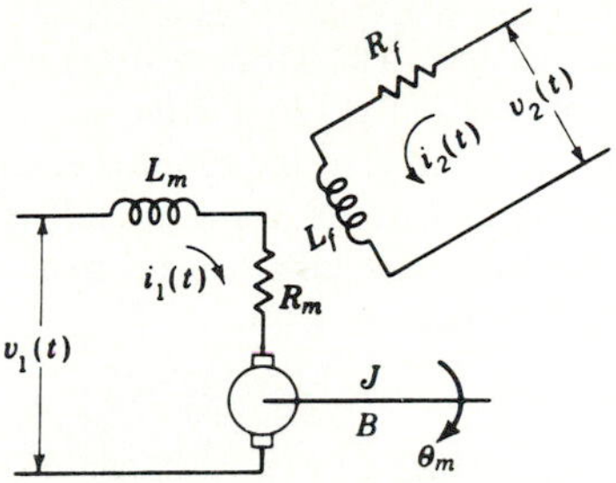

5.8. In the diagram $[q_c]$ represents compressibility flow, and the pressure p is the same in both cylinders. Assume there is no leakage flow around the pistons. Draw a block diagram that relates the output $Y(s)$ to the input $X(s)$.

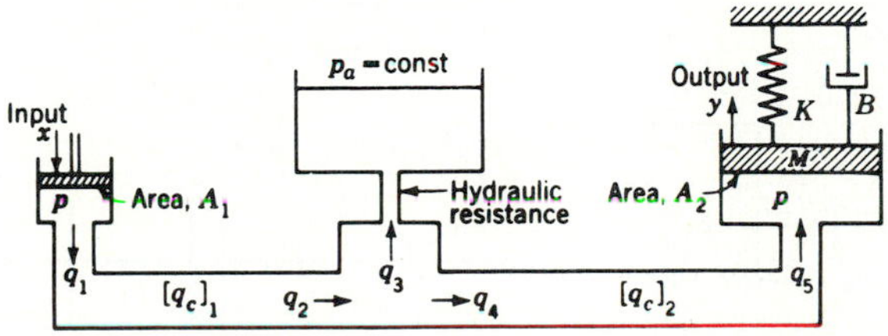

5.9. The diagram represents a gyroscope that is often used in autopilots, automatic gunsights, etc. Assume that the speed of the rotor is constant and that the total developed torque about the output axis is

$$T_0 = H\frac{d\theta_i}{dt}$$

where H is a constant. The inner gimbal has a moment of inertia J about the output axis. Draw a block diagram that relates the output $\Theta_o(s)$ to the input $\Theta_i(s)$.

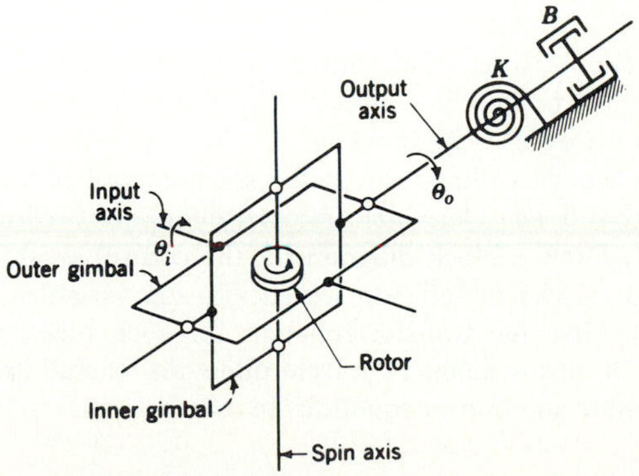

5.10. The figure shows a system for controlling the output movement $y(t)$ of a hydraulic linear actuator. The load reaction on the actuator can be considered negligible. The solenoid constant is K_c (lb/A), the output potentiometer constant is K_p (V/in), and the output of the tachometer is given by

$$e_t = K_t \frac{dy}{dt} \quad \text{V}$$

where k_t has the units of volts per (inch per second). (a) Draw a detailed block diagram that shows all variables of the system explicitly. (b) Derive the transfer functions in terms of the Laplace operator for each block in the block diagram. (c) Draw the SFG and use the Mason gain formula to determine the overall transfer function. *Note*: The lever at the input to the hydraulic unit has no fixed pivot. Consider that x, x_1, and y have only horizontal motion.

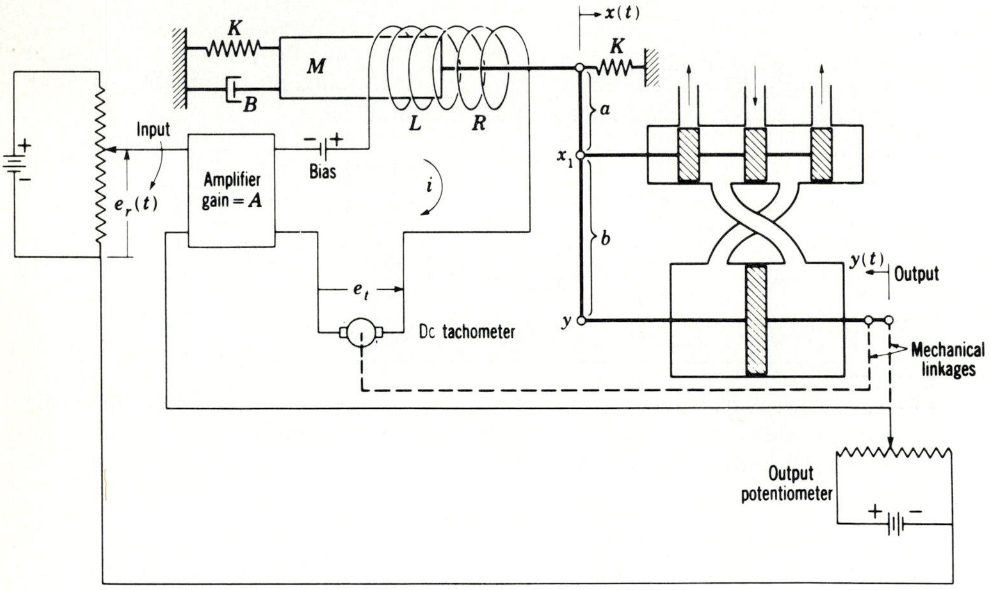

5.11. For the control system shown: (1) The force of attraction on the solenoid is given by $f_c = K_c i_c$, where K_c has the units pounds per ampere. (2) The voltage appearing across the generator field is given by $e_f = K_x x$, where K_x has the units volts per inch and x is in inches. (3) When the voltage across the solenoid coil is zero, the spring K_s is unstretched and $x = 0$. (a) Derive all necessary equations relating all the variables in the system. (b) Draw a block diagram for the control system. The diagram should include enough blocks to indicate specifically the variables $I_c(s)$, $X(s)$, $I_f(s)$, $E_g(s)$, and $T(s)$. Give the transfer function for each block in the diagram. (c) Draw an SFG for this system. (d) Determine the overall transfer function. (e) Write the system state and output equations in matrix form.

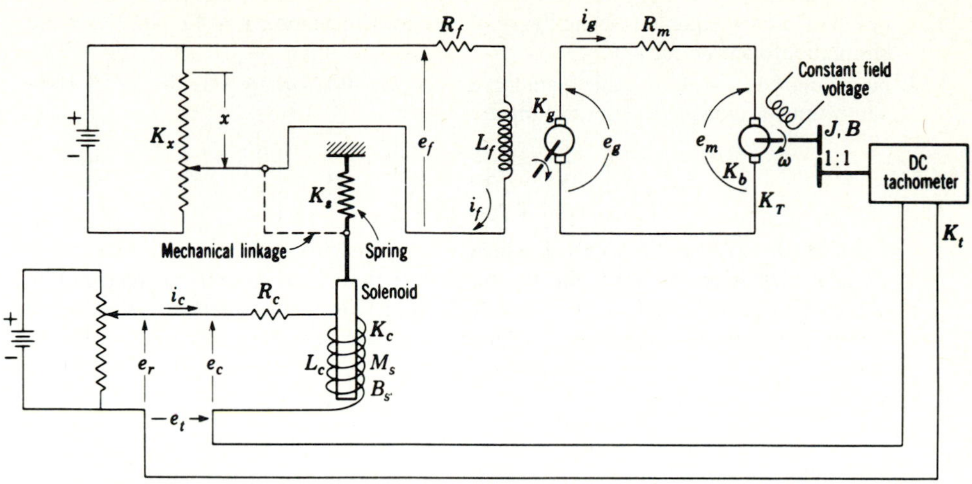

5.12. Given

(1) $D^3y + 5D^2y + 17Dy + 13y = D^3u + 7D^2u + 14Du + 8u$

(2) $D^4y + 10D^3y + 35D^2y + 50Dy + 24y = 5D^2u + 10u$

(3) $D^3y - 9D^2y + 23Dy - 15y = 2D^2u + 4Du + u$

Obtain state and output equations using (*a*) phase variables, (*b*) canonical variables.

5.13. Given

$$G(s) = \frac{b_1 s + b_0}{(s - \sigma)^2 + \omega_d^2}$$

write the state equation with the **A** matrix in modified canonical form so that all elements are real. Draw the SFG.

5.14. The state and output equations of a system are

(1) $\dot{\mathbf{x}} = \begin{bmatrix} 0 & 1 & 0 \\ 0 & 0 & 1 \\ -6 & -11 & -6 \end{bmatrix}\mathbf{x} + \begin{bmatrix} 1 \\ 1 \\ 0 \end{bmatrix}\mathbf{u} \qquad \mathbf{y} = [1 \quad 0 \quad 1]\mathbf{x}$

(2) $\dot{\mathbf{x}} = \begin{bmatrix} -6 & 1 & 0 \\ -11 & 0 & 1 \\ -6 & 0 & 0 \end{bmatrix}\mathbf{x} + \begin{bmatrix} 0 \\ 3 \\ 6 \end{bmatrix}\mathbf{u} \qquad \mathbf{y} = [1 \quad 0 \quad 0]\mathbf{x}$

(*a*) Draw a state transition SFG for these equations. (*b*) Determine the eigenvalues. (*c*) Determine the modal matrix **T** which uncouples the states. (*d*) Write the new state and output equations in matrix form. (*e*) Draw a simulation diagram for the new equations, and determine the uncontrollable mode.

5.15. For the system of Prob. 4.14: (*a*) draw an SFG; (*b*) from the transfer function determine a new state equation and draw the SFG using phase variables; (*c*) change the state equation to normal form and draw the SFG.

5.16. Use the state and output equations of Prob. 3.13. (*a*) Draw a simulation diagram for the given equations. (*b*) Determine the modal matrix **T**. Show that $\mathbf{T}^{-1}\mathbf{A}\mathbf{T} = \mathbf{\Lambda}$.

(c) Convert the equations using the similarity transformation $\mathbf{x} = \mathbf{Tz}$. (d) Draw the simulation diagram for (c).

5.17. A system has a state-variable equation $\dot{\mathbf{x}} = \mathbf{A}_p\mathbf{x} + \mathbf{B}_p u$, where $\mathbf{x}$ is the 3×1 state vector and u is the scalar control input.

$$\mathbf{A}_p = \begin{bmatrix} -1 & 10 & 0 \\ 0 & 0 & 1 \\ 0 & -20 & -10 \end{bmatrix} \qquad \mathbf{B}_p = \begin{bmatrix} 0 \\ 0 \\ 5 \end{bmatrix}$$

(a) Find a nonsingular matrix $\mathbf{T}$ which transforms the state variables into phase variables. Be sure to satisfy the $\mathbf{B}_0$ matrix for the phase variable representation. (This will lead to a unique matrix $\mathbf{T}$.) (b) Write the new state equation of the system in terms of the phase variables. (c) Draw a state-variable diagram representing the system.

5.18. Given

$$(1)\ \dot{\mathbf{x}} = \begin{bmatrix} 0 & 1 & 0 \\ 0 & 0 & 1 \\ -10 & -16 & -7 \end{bmatrix}\mathbf{x} + \begin{bmatrix} 0 \\ 0 \\ 1 \end{bmatrix}\mathbf{u}$$

$$(2)\ \dot{\mathbf{x}} = \begin{bmatrix} -1 & 2 & -1 \\ 0 & -2 & 0 \\ 1 & 0 & -2 \end{bmatrix}\mathbf{x} + \begin{bmatrix} 0 \\ 0 \\ 1 \end{bmatrix}\mathbf{u}$$

(a) Find a modified matrix Λ_m so that its elements are all real. (b) Determine the modified modal matrix $\mathbf{T}_m$. (c) Draw a simulation diagram for the transformed equations.

5.19. For the system described by the state equation

$$(1)\ \dot{\mathbf{x}} = \begin{bmatrix} 0 & 5 \\ 0 & -4 \end{bmatrix}\mathbf{x} + \begin{bmatrix} 0 \\ 2 \end{bmatrix}\mathbf{u}$$

$$(2)\ \dot{\mathbf{x}} = \begin{bmatrix} -3 & -1 \\ 2 & 0 \end{bmatrix}\mathbf{x} + \begin{bmatrix} 1 \\ 0 \end{bmatrix}\mathbf{u}$$

(a) Derive the system transfer function if $y = x_1$. (b) Draw an appropriate state-variable diagram. (c) For zero initial state and a unit step input evaluate $x_1(t)$ and $x_2(t)$.

5.20. For $\dot{\mathbf{x}} = \mathbf{Ax} + \mathbf{Bu}$

$$(1)\ \mathbf{A} = \begin{bmatrix} -1 & 0 & 0 \\ 1 & -2 & 0 \\ -1 & -2 & -4 \end{bmatrix} \qquad \mathbf{B} = \begin{bmatrix} 1 \\ 0 \\ 1 \end{bmatrix}$$

$$(2)\ \mathbf{A} = \begin{bmatrix} 0 & 1 & 0 \\ -3 & -4 & 0 \\ 0 & -2 & -2 \end{bmatrix} \qquad \mathbf{B} = \begin{bmatrix} 0 \\ 1 \\ 1 \end{bmatrix}$$

Determine the matrix $\mathbf{T}$ so that the transformation $\mathbf{x} = \mathbf{Tz}$ produces (a) a state equation in control canonical form and (b) a state equation with a diagonalized plant matrix Λ.

5.21. Repeat Prob. 5.20 for

$$\mathbf{A} = \begin{bmatrix} -1 & 0 & 0 \\ -1 & -1 & 0 \\ -2 & 2 & 0 \end{bmatrix} \qquad \mathbf{B} = \begin{bmatrix} 1 \\ 1 \\ 1 \end{bmatrix}$$

5.22. Given the real symmetric matrix

$$\mathbf{A} = \begin{bmatrix} 7 & 4 & -4 \\ 4 & -8 & -1 \\ -4 & -1 & -8 \end{bmatrix}$$

Determine (a) the eigenvalues, (b) the eigenvectors, and (c) the modal matrix **T**. Show that $\mathbf{T}^{-1}\mathbf{AT} = \mathbf{\Lambda}$. Note that this matrix can be diagonalized even though there are repeated eigenvalues.

5.23. Repeat Prob. 5.22 for

$$(1)\ \mathbf{A} = \begin{bmatrix} 0 & 1 & 0 \\ 0 & 0 & 1 \\ 3 & 1 & -3 \end{bmatrix} \qquad (2)\ \mathbf{A} = \begin{bmatrix} 7 & 4 & -1 \\ 4 & 7 & -1 \\ -4 & -4 & 4 \end{bmatrix}$$

5.24. Determine the modal matrix **T** which will uncouple the states

$$(1)\ \dot{\mathbf{x}} = \begin{bmatrix} -2 & -2 & 0 \\ 0 & 0 & 1 \\ 0 & -3 & -4 \end{bmatrix}\mathbf{x} + \begin{bmatrix} 1 \\ 0 \\ 1 \end{bmatrix}\mathbf{u} \qquad (2)\ \dot{\mathbf{x}} = \begin{bmatrix} 0 & 1 & 0 \\ -3 & -4 & 0 \\ 0 & -2 & -5 \end{bmatrix}\mathbf{x} + \begin{bmatrix} 1 \\ 0 \\ 1 \end{bmatrix}\mathbf{u}$$

Find $\mathbf{x}(t)$ for $u(t) = u_{-1}(t)$. Use the method of Sec. 5.11 where $\mathbf{x}(0) = \mathbf{0}$.

5.25. Draw a simulation diagram for the state equations of Prob. 3.9 using physical variables.

5.26. Given the following system:

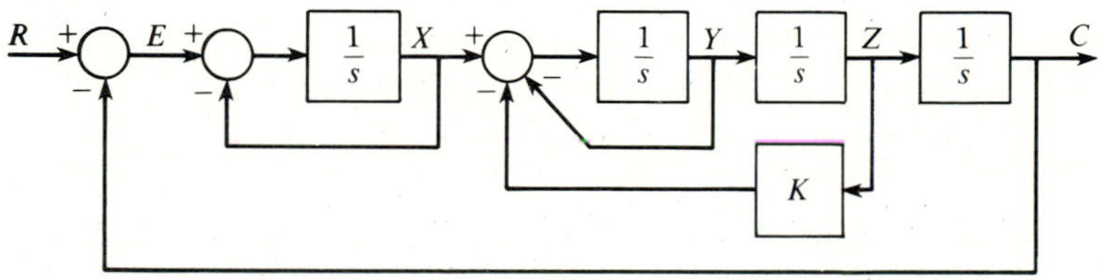

(a) Draw an equivalent signal flow graph. (b) Apply Mason's gain rule to find $C(s)/R(s)$. (c) Write the four differential and three algebraic equations which form the math model depicted by the block diagram above. Use only the signals $r(t)$, $e(t)$, $x(t)$, $y(t)$, $z(t)$, and $c(t)$ in these equations.

5.27. For the following system:

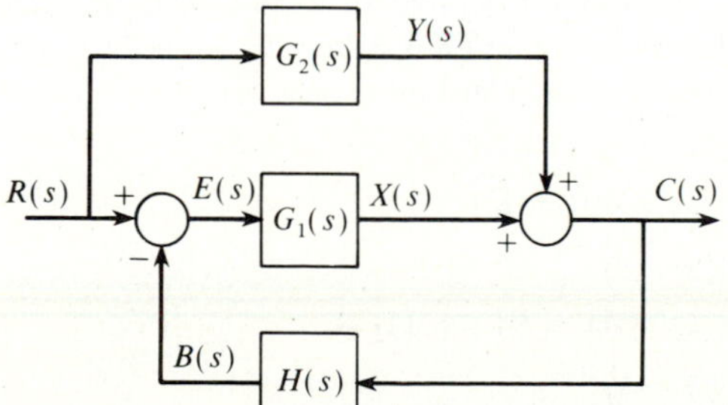

(a) Draw a signal flow graph. (b) Derive transfer functions for $E(s)/R(s)$, $C(s)/R(s)$, $X(s)/R(s)$, $B(s)/R(s)$, and $Y(s)/R(s)$.

5.28. For Prob. 5.27, given $G_1 = 1/(s+1)$, $G_2 = 1/s$, $G_2 = 1/s$, and $H = s/(s^2 + 2s + 2)$, write the differential equations (*a*) relating x to e, (*b*) relating y to r, and (*c*) relating b to c. Then write the algebraic equations (*d*) relating e to r and b and (*e*) relating c to x and y.

5.29. The pump in the system shown below supplies the pressure $p_1(t) = 4u_{-1}(t)$. Calculate the resulting velocity $v_2(t)$ of the mass M_2. Include the effect of the mass of the power piston. (*a*) Neglect load reaction. (*b*) Include load reaction.

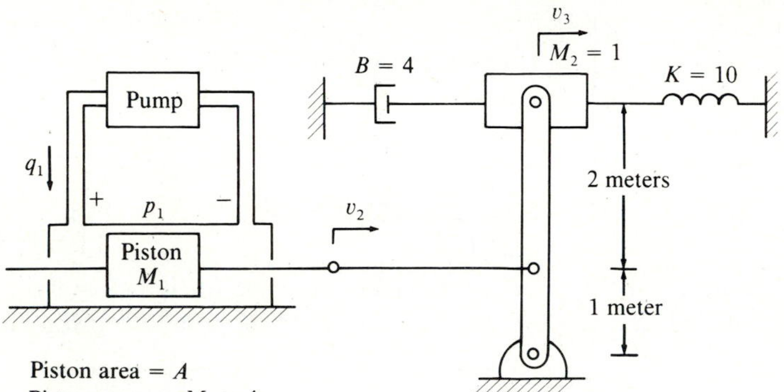

5.30. The state and output equations of a system are:

$$\dot{\mathbf{x}} = \begin{bmatrix} 0 & 1 & 0 \\ 0 & 0 & 1 \\ -6 & -11 & -6 \end{bmatrix}\mathbf{x} + \begin{bmatrix} 1 \\ 1 \\ 0 \end{bmatrix}\mathbf{u} \qquad \mathbf{y} = [1 \quad 0 \quad 1]\mathbf{x}$$

(*a*) Determine the eigenvalues of the system. (*b*) Determine the modal matrix **T** which uncouples the states. (*c*) Draw the simulation diagram for the new equations.

Chapter 6

6.1. For Prob. 5.3, (*a*) use the Routh criterion to check the stability of the system. (*b*) Compute the steady-state tracking error of the system in response to a command input which is a ramp of 0.04°/s. (*c*) Repeat (*b*) with the gain of the tracking receiver increased from 5 to 50.

6.2. For each of the following cases, determine the range of values of K for which the response $c(t)$ is stable, where the driving function is a step function. Determine the roots that lie on the imaginary axis that yield sustained oscillations.

$$(a)\quad C(s) = \frac{K}{s[s(s+2)(s^2+2s+10)+K]}$$

$$(b)\quad C(s) = \frac{K(s+1)}{s[s^2(s^3+3s^2+2s+4)+K(s+1)]}$$

$$(c)\quad C(s) = \frac{K(s+5)}{s[(s+5)(s^2+8s+20)+K(s+5)]}$$

$$(d)\quad C(s) = \frac{K}{s[(s+1)(s+3)(s+4)+K]}$$

(*e*) $C(s) = \dfrac{K}{s[s^3 + 6s^2 + 11s + (6 + K)]}$

(*f*) $C(s) = \dfrac{K(s + 4)}{s[s(s + 1)(s + 2)(s + 5) + K(s + 4)]}$

(*g*) $C(s) = \dfrac{K(s + 1)}{s[s(s^3 + 5s^2 + s + 1) + K(s + 1)]}$

(*h*) For Prob. 5.26.

6.3. Factor the following equations:

(*a*) $s^3 + 6s^2 + 12s + 9 = 0$
(*b*) $s^3 + 4s^2 + 21s + 34 = 0$
(*c*) $s^4 + 2s^3 + 2s^2 + 3s + 6 = 0$
(*d*) $s^4 + 6s^3 + 26s^2 + 56s + 80 = 0$
(*e*) $s^3 + 7s^2 + 14s + 8 = 0$
(*f*) $s^3 + 7s^2 + 16s + 10 = 0$
(*g*) $s^4 + 7s^3 + 9s^2 - 7s - 10 = 0$
(*h*) $s^4 + 7s^3 + 20s^2 + 50s = 0$
(*i*) $s^5 + 3s^4 + 28s^3 + 226s^2 + 600s + 400 = 0$
(*j*) $s^5 + 3s^4 + 4s^3 + 4s^2 = 0$

6.4. Use Routh's criterion to determine the number of roots in the right-half s plane for the equations of Prob. 6.3.

6.5. A unity-feedback system has the forward transfer function

$$G(s) = \frac{K(0.1s + 1)}{s(s + 2)(s + 3)}$$

To obtain the best possible ramp error coefficient, the highest possible gain K is desirable. Do stability requirements limit this choice of K?

6.6. The equation relating the output $y(t)$ of a control system to its input is

(*a*) $[s^5 + s^4 + 3s^3 + 2s^2 + (K + 1)s + (K + 2)]\,Y(s) = 5X(s)$

(*b*) $[s^3 - 7s^2 + 16s + 5 + K(s + 2)]\,Y(s) = X(s)$

Determine the range of K for stable operation of the system. Consider both positive and negative values of K.

6.7. The output of a control system is related to its input $r(t)$ by

$$[s^4 + 2s^3 + 2s^2 + (2 + K)s + K]\,C(s) = K(s + 1)\,R(s)$$

Determine the range of K for stable operation of the system. Consider both positive and negative values of K. Find K which yields imaginary roots.

6.8.

$$F(s) = \frac{1}{s[(s^4 + 2s^3 + 3s^2 + s + 1) + K(s + 1)]}$$

K is real but may be positive or negative. (*a*) Find the range of values of K for which the time response is stable. (*b*) Select a value of K which will produce imaginary poles for $F(s)$. Find these poles. What is the physical significance of imaginary poles as far as the time response is concerned?

6.9. For the system shown, (*a*) find $C(s)/R(s)$. (*b*) What type of system does $C(s)/E(s)$ represent? (*c*) Find the step-, ramp-, and parabolic-error coefficients. (*d*) Find the steady-state value of $c(t)$ if $r(t) = 10u_{-1}(t)$ rad. (*e*) Is the closed-loop system stable?

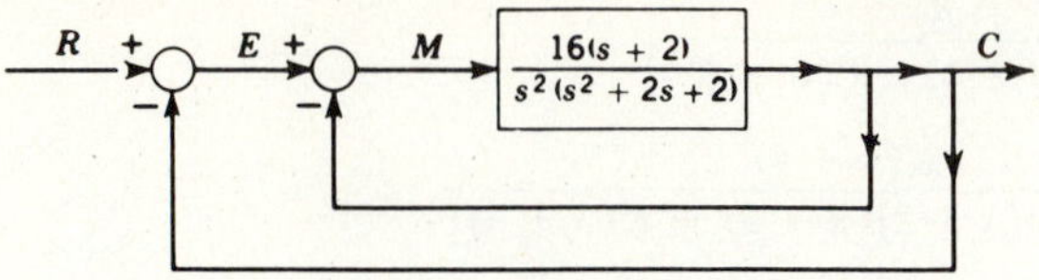

6.10. For a unity feedback control system:

$$(1)\ G(s) = \frac{12(s+4)}{s(s+1)(s+3)(s^2+2s+10)} \qquad (2)\ G(s) = \frac{10(s+8)}{s^2(s+5)(s^2+2s+2)}$$

(a) Determine the step-, ramp-, and parabolic-error coefficients (K_p, K_v, and K_a) for this system. (b) Using the appropriate error coefficients from part (a), determine the steady-state actuating signal $e(t)_{ss}$ for $r_1(t) = (16 + 2t)u_{-1}(t)$ and for $r_2(t) = 5t^2 u_{-1}(t)$. (c) Is the closed-loop system stable? (d) Use the results of part (b) to determine $c(t)_{ss}$.

6.11. (a) Determine the open-loop transfer function $G(s)H(s)$ of Fig. (a). (b) Determine the overall transfer function. (c) Figure (b) is an equivalent block diagram of Fig. (a). What must the transfer function $H_x(s)$ be in order for Fig. (b) to be equivalent to Fig. (a)? (d) Figure (b) represents what type of system? (e) Determine the system error coefficients of Fig. (b). What is the significance of the minus sign for the ramp-error coefficient? (f) If $r(t) = u_{-1}(t)$, determine the final value of $c(t)$. (g) What are the values of $e_x(t)_{ss}$ and $e(t)_{ss}$?

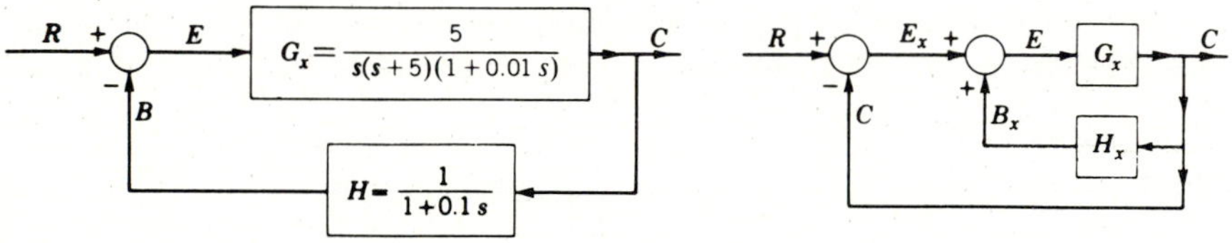

(a) (b)

6.12. Repeat Prob. 6.11 with $H(s)$ replaced in Fig. (a) by

$$H(s) = \frac{s+3}{s+1}$$

6.13. Find the step-, ramp-, and parabolic-error coefficients for unity-feedback systems which have the following forward transfer functions:

$$(a)\ G(s) = \frac{10}{(0.4s+1)(0.5s+1)}$$

$$(b)\ G(s) = \frac{108}{s^2(s^2+4s+4)(s^2+3s+12)}$$

$$(c)\ G(s) = \frac{10(s+5)}{s(s^2+2s+2)}$$

$$(d)\ G(s) = \frac{14(s+3)}{s(s+6)(s^2+2s+2)}$$

$$(e)\ G(s) = \frac{20(s+3)}{(s+2)(s^2+2s+2)}$$

$$(f)\ G(s) = \frac{11(s+30)}{s^3(s+1)(0.2s+1)(s^2+5s+15)}$$

$$(g)\ G(s) = \frac{-6(1+0.04s)}{s(1+0.1s)(1+0.1s+0.01s^2)}$$

$$(h)\ G(s) = \frac{4(s^2+10s+50)}{s^2(s+5)(s^2+6s+10)}$$

$$(i)\ G(s) = \frac{100(s-1)}{s(s+2)(1+0.25s)}$$

$$(j)\ G(s) = \frac{39}{s^2(s-1)(s^2+6s+13)}$$

6.14. For Prob. 6.13, find $e(\infty)$ by use of the error coefficients, with the following inputs: $(a)\ r(t) = 5$ $(b)\ r(t) = 2t$ $(c)\ r(t) = t^2$

6.15. A unity-feedback control system has

$$(1)\ G(s) = \frac{K_1}{s(s+1)(0.5s+1)} \qquad (2)\ G(s) = \frac{K(s+1.5)}{s(s+1)(s+2)(s+3)}$$

where $r(t) = 3t$. (a) If $K_1 = 3$, determine $e(t)_{ss}$. (b) It is desired that for a ramp input $e(t)_{ss} \leq 0.1$. What minimum value must K_1 have for this condition to be satisfied? (c) For the value of K_1 determined in part (b), is the system stable?

6.16. (a) Derive the ratio $G(s) = C(s)/E(s)$. (b) Based upon $G(s)$, the figure has the characteristic of what type of system? (c) Derive the control ratio for this control system. (d) Repeat with the minor loop feedback replaced by K_h. (e) What conclusion can be reached about the effect of minor loop feedback on system type?

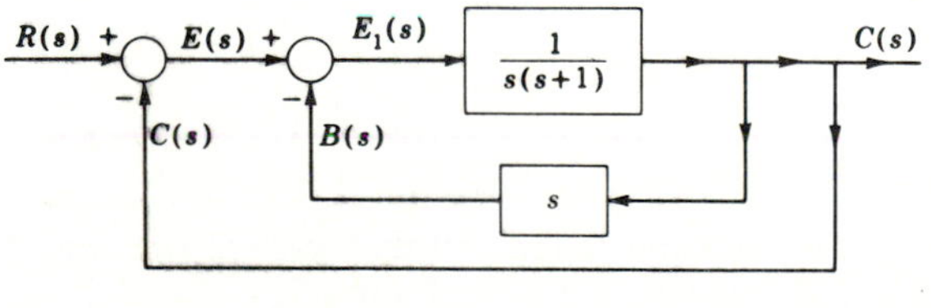

6.17. A unity-feedback system has the forward transfer function

$$G(s) = \frac{K_1(2s+1)}{s(4s+1)(s+1)^2}$$

The input $r(t) = 1 + t$ is applied to the system. (a) It is desired that the steady-state value of the error be equal to or less than 0.1 for the given input function. Determine the minimum value that K_1 must have to satisfy this requirement. (b) By use of Routh's stability criterion, determine whether the system is stable for the minimum value of K_1 determined in part (a).

6.18. The angular position θ of a radar antenna is required to follow a command signal θ_c. The command signal e_c is proportional to θ_c with a proportionality constant K_θ V/deg. The positioning torque is applied by an ac motor which is activated by a position error signal e. The shaft of the electric motor is geared to the antenna with a gear ratio n. A potentiometer having a proportionality constant K_θ V/deg produces a feedback signal e_θ proportional to θ. The actuating signal $e = e_c - e_\theta$ is amplified with gain K_α V/V to produce the voltage e_α which drives the motor. The amplifier output e_a is an ac voltage. With the motor shaft clamped, the motor stall torque measured at the motor shaft is proportional to e_a with a proportionality constant K_c in lb/V. The motor torque decreases linearly with increasing motor speed $\dot{\theta}_m$ and is zero at the no-load speed $\dot{\theta}_0$. The no-load speed is proportional to e_a with proportionality constant K_0. The moment of inertia of the motor, gearing, and antenna, referred to the output θ, is J and the viscous friction B is negligible. The motor torque is therefore $T = J\ddot{\theta} + T_L$, where T_L is a wind torque. (a) Draw a completely labeled detailed block diagram for this system. (b) Neglecting T_L, find the open-loop transfer function $\theta(s)/E(s)$ and the closed-loop transfer function $\theta(s)/\theta_c(s)$. (c) For a step input θ_c, describe the response characteristics as the gain K_a is increased. (d) For a constant wind torque T_L and $\theta_c(t) = 0$ find the steady-state position error as a function of K_a.

6.19. The components of a single axis stable platform with direct drive can be described by the following equations:

Controlled platform: $T = JD\omega + B\omega + T_L$

Integrating gyro: $De_{ig} = K_{ig}(\omega_c - \omega + \omega_d + \omega_b)$

Servomotor: $T = (K_a/K_{ig})e_{ig}$

where J = moment of platform and servomotor
B = damping
ω = angular velocity of platform with respect to inertial frame of reference
T = servomotor torque
T_L = interfering torque
e_{ig} = gyro output voltage
ω_c = command angular velocity
ω_d = drift rate of gyro
ω_b = angular velocity of the base
K_{ig} = gyro gain
K_a/K_{ig} = amplifier gain

Consider that the base has a given orientation and that $\omega_b = 0$. (a) Show that the overall equation of performance for the stable platform is

$$(JD^2 + BD + K_a)\omega = K_a(\omega_c + \omega_d) - DT_L$$

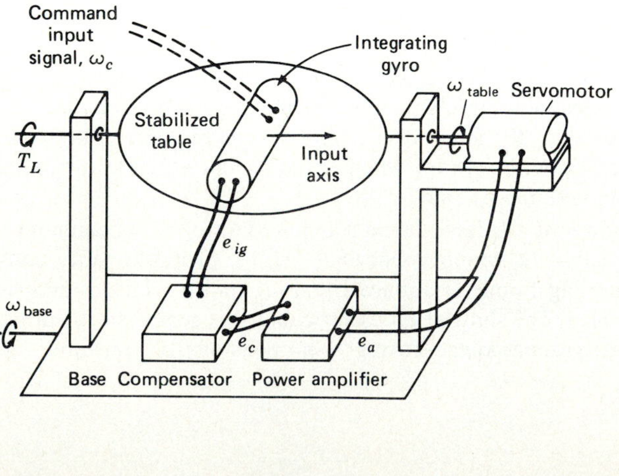

(b) Draw a detailed block diagram representing the system. (c) Using e_{ig} and ω as state variables, write the state equations of the system. (d) With $\omega_c = 0$ and $T_L = 0$ determine the effect of a constant drift rate ω_d. (e) With $\omega_c = 0$ and $\omega_d = 0$, determine the effect of a constant interfering torque T_L on ω.

6.20. In the feedback system shown, K_h is adjustable and

$$G(s) = \frac{K_G(s^w + \cdots + c_0)}{s^2 + \cdots + a_1 s + a_0}$$

The output is required to follow a step input with no steady-state error. Determine the necessary conditions on K_h for a stable system.

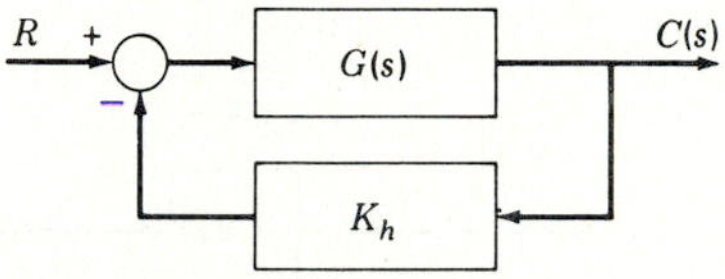

6.21. For Prob. 2.27: (a) derive the transfer function $G_x(s) = \Omega(s)/V_1(s)$. Figure ($a$) shows a velocity control system, where K_x is the velocity sensor coefficient. The equivalent unity-feedback system is shown in Fig. (b), where $\Omega(s)/R(s) = K_x G_x(s)/[1 + K_x G_x(s)]$. ($b$) What system type does $G(s) = K_x G_x(s)$ represent? Determine $\omega(t)_{ss}$, when $r(t) = R_0 u_{-1}(t)$. Assume the system is stable.

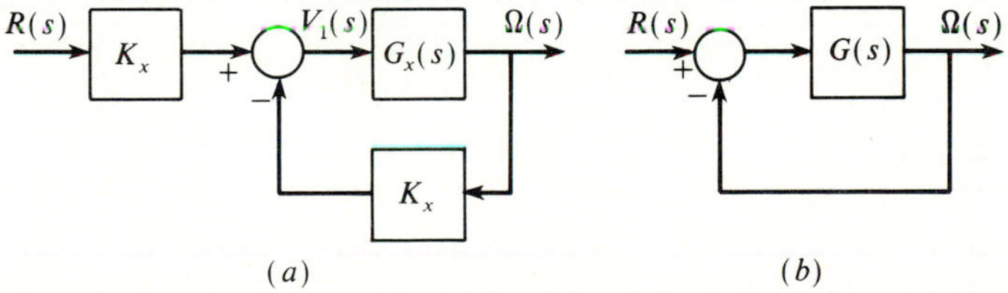

6.22. Given the following system, find (a) $C(s)/R(s)$, (b) the range of values of K for which the system is stable, (c) determine $G_{eq}(s)$, and (d) the equivalent system type.

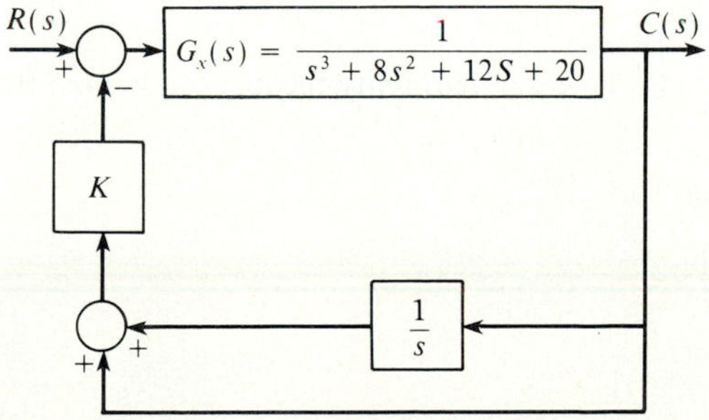

6.23. Given the system below find $C(s)/R(s)$. (a) Find the range of values of K for which the system is stable. (b) Find the *positive* value of K which yields pure oscillations in the homogeneous system and find the frequency associated with these oscillations.

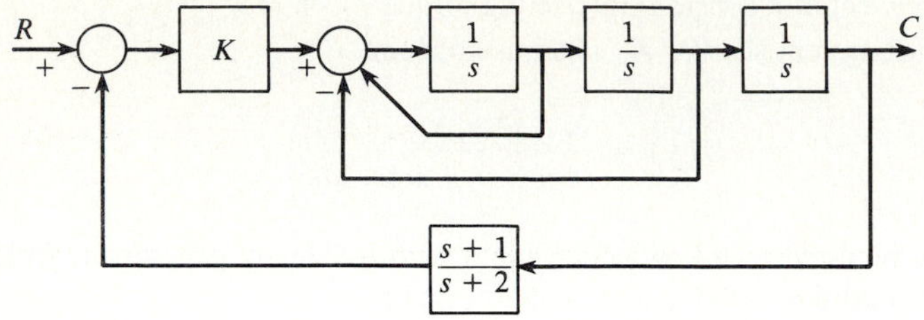

6.24. Given

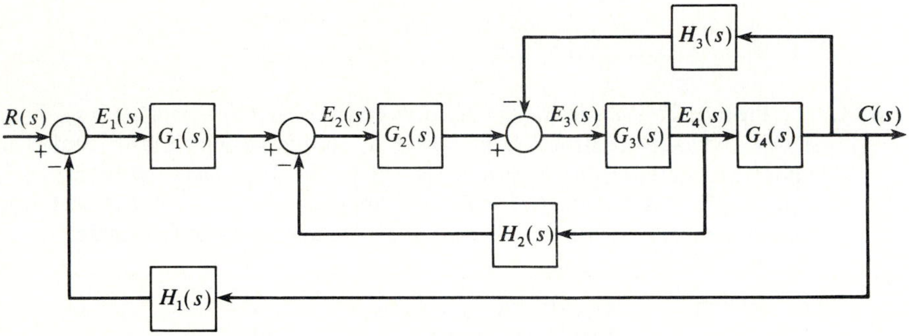

(a) By block-diagram reduction find $G(s)$ and $H(s)$ such that the system can be represented in the form:

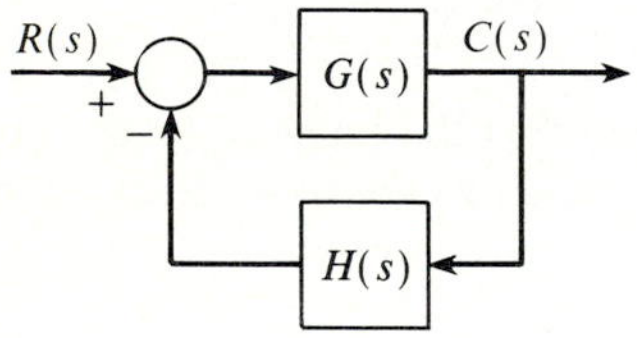

(b) For this system find $G_{eq}(s)$ for an equivalent unity-feedback system of the form:

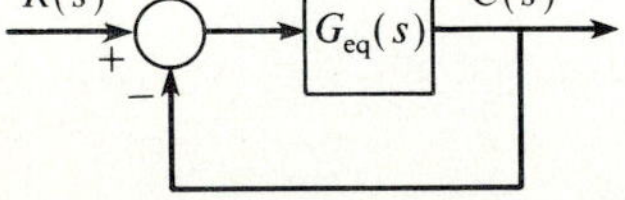

6.25. By block-diagram manipulations or the use of SFG simplify the following systems to the form of Fig. 6.1.

(1)

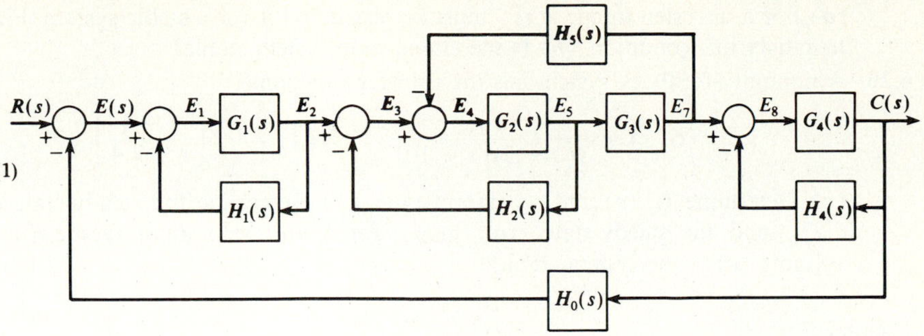

(2)

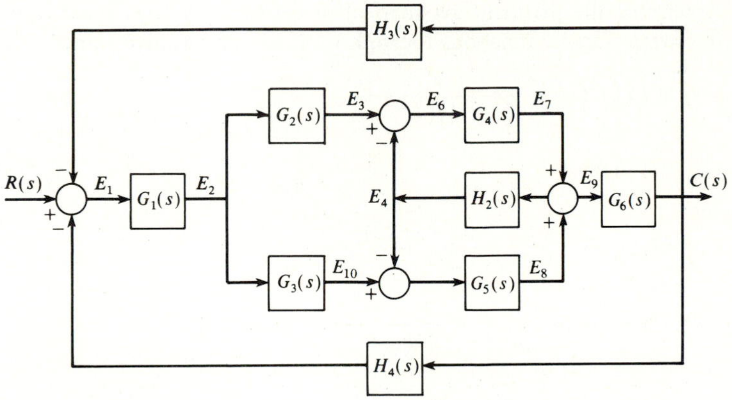

(3)

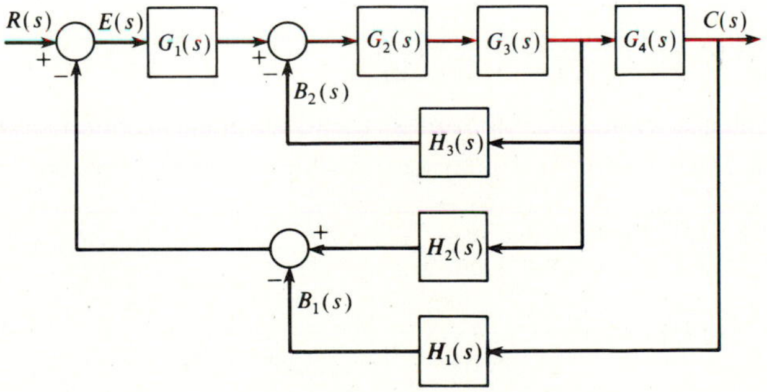

6.26. A unity-feedback system has the forward transfer function

$$G(s) = \frac{K}{s(s+1)^2(0.5s+1)}$$

(a) What is the maximum value of K for stable operation of the system? (b) For K equal to one half of the maximum value, what is the $e(t)_{ss}$ for the ramp input $r(t) = 10tu_{-1}(t)$?

6.27. A unity-feedback system has the forward transfer function

$$G(s) = \frac{K}{(s+1)(s+2)(s+3)}$$

(a) For a unit step input, $e(t)_{ss}$ must be less than 0.1 for a stable system. Find K_m to satisfy this condition. (b) Is the closed-loop system stable?

6.28. A nonunity-feedback system has the transfer functions

$$G(s) = \frac{10}{s(s+5)(s+10)} \qquad H(s) = 0.6\left[\frac{s+4}{s+2.4}\right]$$

(a) Determine $G_{eq}(s)$ and its system type. (b) Determine the steady-state output $c(t)_{ss}$ and the steady-state error $e(t)_{ss}$ for a unit step input (assume a stable system). (c) Is the system stable?

Chapter 7

7.1. Determine the pertinent geometrical properties and *sketch* the root locus for the following transfer functions for both positive and negative values of K:

$$(a)\quad G(s)H(s) = \frac{K}{(s+1)(s+5)}$$

$$(b)\quad G(s)H(s) = \frac{K}{(s+1)(s^2+8s+20)}$$

$$(c)\quad G(s)H(s) = \frac{K(s+12)}{s(s^2+16s+100)}$$

$$(d)\quad G(s)H(s) = \frac{K}{s^2(s+1)(s+2)}$$

$$(e)\quad G(s)H(s) = \frac{K_0}{(1+0.5s)(1+0.2s)(1+s)^2}$$

$$(f)\quad G(s)H(s) = \frac{K(s+3)^2}{s(s^2+2s+2)}$$

Determine the range of values of K for which the closed-loop system is stable.

7.2. Repeat Prob. 7.1 for the following open-loop transfer functions:

$$(a)\quad G(s)H(s) = \frac{Ks}{(s+1)^2}$$

$$(b)\quad G(s)H(s) = \frac{K}{s(s^2+2s+2)(s^2+6s+10)}$$

$$(c)\quad G(s)H(s) = \frac{K(s-1)}{s(s^2+4s+4)}$$

$$(d)\quad G(s)H(s) = \frac{K(s^2+8s+20)}{(s+2)(s^2+2s+2)}$$

7.3. Draw the root locus of the following for a unity-feedback system with $K \geq 0$:

$$(a)\quad G(s) = \frac{K_0}{(1+0.5s)(1+0.2s)(1+s)^2}$$

$$(b)\quad G(s) = \frac{K_1}{s(20s^2+10s+1)(s+2)}$$

(c) $G(s) = \dfrac{K}{(s^2 + 1)(s^2 + 4s + 5)}$

(d) $G(s) = \dfrac{K(s + 1)}{s^2(s^2 + 10s + 29)}$

(e) $G(s) = \dfrac{K}{s(s^2 + 2s + 2)(s^2 + 6s + 10)}$

(f) $G(s) = \dfrac{K}{(s + 1)(s^2 + 6s + 25)(s + 5)}$

(g) $G(s) = \dfrac{K}{s(s^2 + 8s + 28)}$

(h) $G(s) = \dfrac{K}{s(s^2 + 4s + 8)(s + 4)}$

Does the root locus cross any of the asymptotes?

7.4. A system has the following transfer functions:

$$G(s) = \frac{K}{s(s^2 + 8s + 20)} \qquad H(s) = 1 \qquad K > 0$$

(a) Plot the root locus. (b) A damping ratio of 0.58 is required for the dominant roots. Find $C(s)/R(s)$. The denominator should be in factored form. (c) With a unit step input, find $c(t)$. (d) Evaluate graphically $\mathbf{C}(j\omega)/\mathbf{R}(j\omega)$ vs. ω. Plot the magnitude and angle of $\mathbf{C}(j\omega)/\mathbf{R}(j\omega)$ vs. ω.

7.5. A feedback control system with unity feedback has a transfer function

(1) $G(s) = \dfrac{K_1}{s(1 + 0.02s)(1 + 0.01s)}$ (2) $G(s) = \dfrac{100K_x}{s(s^2 + 12s + 52)}$

(a) Plot the locus of the roots of $1 + G(s) = 0$ as the loop sensitivity K (> 0) is varied. (b) Determine the value of K_1 that just makes the system unstable. (c) From the root-locus plot determine the value of K_1 for a $\zeta = 0.5$. (d) For the value of K_1 found in part (c), determine $e(t)$ for $r(t) = u_{-1}(t)$. (e) Plot $M(j\omega)$ vs. ω for the closed-loop system. Determine the data graphically as indicated in Sec. 7.10.

7.6. (a) Sketch the root locus for the control system having the following open-loop transfer function. (b) Calculate the value of K_1 that causes instability. (c) Determine $C(s)/R(s)$ for $\zeta = 0.3$. (d) Evaluate $c(t)$ for $r(t) = u_{-1}(t)$.

(1) $G(s) = \dfrac{K_1(1 + 0.04s)}{s(1 + 0.1s)(1 + 0.1s + 0.0125s^2)}$ (2) $G(s) = \dfrac{K(s - 5)(s + 5)}{s(s^2 + 4s + 13)}$

7.7. (a) Sketch the root locus for a control system having the following forward and feedback transfer functions:

(1) $G(s) = \dfrac{K_2(1 + s/5)}{s^2(1 + s/12)}$ $H(s) = 1 + \dfrac{s}{12}$

(2) $G(s) = \dfrac{K}{s^2 + 2s + 2}$ $H(s) = \dfrac{1}{s(s + 1)}$

(b) For system (1), choose closed-loop pole locations which produce a time constant $T = 1/3$ s for the complex roots and indicate these locations on the root locus.

Using these locations, write the factored form of the closed-loop transfer function. For system (2), what is the maximum value of K if the system is to be stable? For this value of K determine the poles and zeros of $C(s)/R(s)$. Determine $c(\infty)$ for both systems assuming a stable value of K and a unit step forcing function.

7.8. For

$$G(s)H(s) = \frac{K_x}{(s+5)(0.4s+1.2)(0.5s^2+2s+4)}$$

determine the value of K from the root locus that makes the closed-loop system a perfect oscillator.

7.9. A unity-feedback control system has the transfer function

$$G(s) = \frac{K}{(s+1)(1+0.2s)(s^2+6s+13)}$$

(a) Sketch the root locus for positive and negative values of K. (b) For what values of K does the system become unstable? (c) Determine the value of K for which all the roots are equal.

7.10. A unity-feedback control system has the transfer function

$$(1)\quad G(s) = \frac{K_1(1-s)}{s(1+s)(1+0.5s)(1+0.25s)}$$

$$(2)\quad G(s) = \frac{K_1}{s(0.1s+1)(0.0016s^2+0.048s+1)}$$

(a) Sketch the root locus for positive and negative values of K_1. (b) What range of values of K_1 makes the system unstable?

7.11. For positive values of gain, sketch the root locus for unity-feedback control systems having the following open-loop transfer functions. For what value or values of gain does the system become unstable in each case?

$$(a)\quad G(s) = \frac{K(s^2+6s+13)}{s(s+1)(s+5)^2}$$

$$(b)\quad G(s) = \frac{K(1+0.04s)}{s(1+0.1s)(0.0125s^2+0.1s+1)}$$

$$(c)\quad G(s) = \frac{K_0(1-100s)}{(1+10s)(1+0.001s)(1+s+s^2)}$$

$$(d)\quad G(s) = \frac{K(s^2+4s+5)}{s^2(s+1)(s+3)}$$

$$(e)\quad G(s) = \frac{K(s+2)^2}{s(s^2-2s+2)}$$

$$(f)\quad G(s) = \frac{K(s+2)}{(s+18)(s^2+2.4s+12.33)}$$

7.12. For the system of part (f) of Prob. 7.11 determine $C(s)/R(s)$ for $\zeta = 0.707$. Select the *best* roots.

7.13. A nonunity-feedback control system has the transfer functions

$$G(s) = \frac{K_G(1 + s/3.9)}{(1 + s/10)\left[1 + 2(0.7)s/23 + (s/23)^2\right]\left[1 + 2(0.49)s/7.6 + (s/7.6)^2\right]}$$

$$H(s) = \frac{K_h(1 + s/10)}{1 + 2(0.89)s/42.7 + (s/42.7)^2}$$

(a) Sketch the root locus for the system, using as few trial points as possible. Determine the angles and locations of the asymptotes and the angles of departure of the branches from the open-loop poles. (b) Determine the gain $K_G K_H$ at which the system becomes unstable, and determine the approximate locations of all the closed-loop poles for this value of gain. (c) Using the closed-loop configuration determined in part (b), write the expression for the closed-loop transfer function of the system.

7.14. A nonunity-feedback control system has the transfer functions

$$(1)\quad G(s) = \frac{10A(s^2 + 8s + 20)}{s(s + 4)} \qquad H(s) = \frac{0.2}{s + 2}$$

$$(2)\quad G(s) = \frac{K(s^2 + 8s + 20)}{(s + 1)(s + 2)} \qquad H(s) = \frac{1}{s + 4}$$

For system (1): (a) determine the value of the amplifier gain A that will produce complex roots having the *minimum* possible value of ζ. (b) Express $C(s)/R(s)$ in terms of its poles, zeros, and constant term. For system (2): (a) determine the value of K that will produce complex roots with $\zeta = 0.707$ and $T_s \le 2$. (b) Express $C(s)/R(s)$ in terms of its poles, zeros, and constant term.

7.15. For the figure shown, only the feedback gain K_h is adjustable. Determine (a) $C(s)/R(s)$ for $\zeta = 0.6$ and (b) $c(t)$ for a unit step input. Note that K_h does not appear in the numerator of $C(s)/R(s)$.

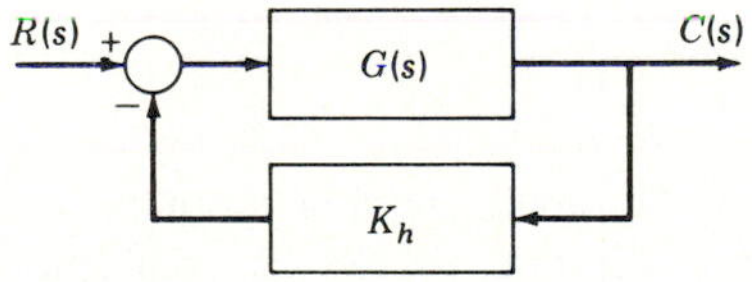

$$G(s) = \frac{10(s + 2)}{s(s + 1)(s + 5)}$$

7.16. Yaw rate feedback and sideslip feedback have been added to an aircraft for turn coordination. The transfer functions of the resulting aircraft and the aileron servo are, respectively,

$$\frac{\dot{\phi}(s)}{\delta_a(s)} = \frac{20}{s + 2} \qquad \text{and} \qquad \frac{\delta_a(s)}{e_{\delta_a}(s)} = \frac{10}{s + 10}$$

For the bank-angle control system shown in the figure, design a suitable system by drawing the root-locus plots on which your design is based. (a) Use $\zeta = 0.8$ for the inner loop and determine K_{rrg}. (b) Use $\zeta = 0.7$ for the outer loop and determine K_{vg}. (c) Determine $\phi(s)/\phi_{\text{comm}}(s)$.

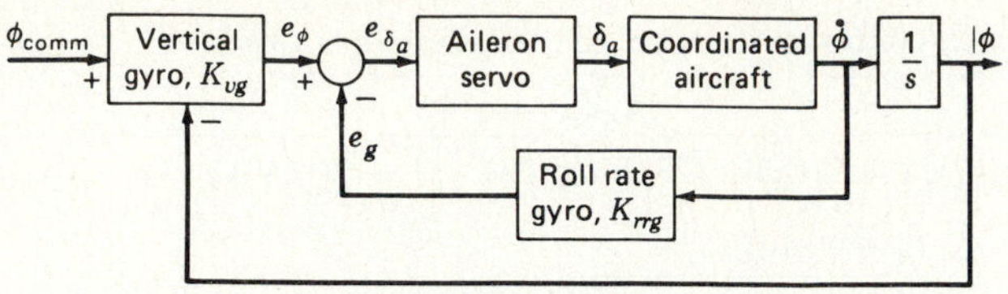

7.17. A heading control system using the bank-angle control system is shown in the following figure, where g is the acceleration of gravity and V_T is the true airspeed. Use the design of Prob. 7.16 for $\phi(s)/\phi_{comm}(s)$. (*a*) Draw a root locus for this system and use $\zeta = 0.6$ to determine K_{dg}. (*b*) Determine $\psi(s)/\psi_{comm}(s)$ and the time response with a step input for $V_T = 500$ ft/s. (The design can be repeated using $\zeta = 1.0$ for obtaining K_{rrg}.)

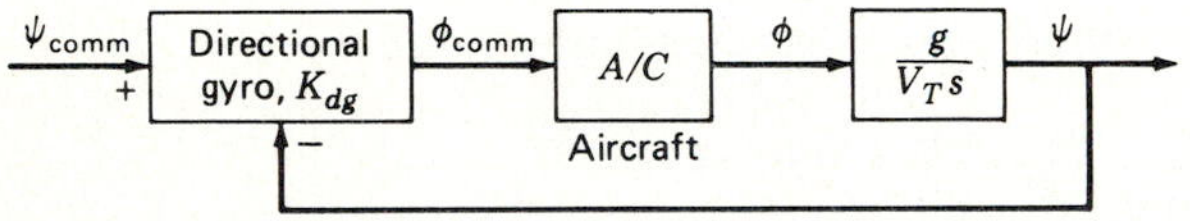

7.18. A nonunity-feedback control system has the transfer functions

$$G(s) = \frac{K(s^2 + 6s + 13)}{s(s+3)} \qquad H(s) = \frac{1}{s+1}$$

(*a*) Draw the root locus for this system. $\zeta = 0.56$ is required for the dominant roots. Select the dominant roots which will provide the *best* performance. (*b*) Find $C(s)/R(s)$ for the dominant roots found in (*a*). The denominator should be in factored form and the value of K specified. (*c*) Indicate why your selection provides the best performance.

7.19. The simplified open-loop transfer function for a T-38 aircraft at 35,000 feet and 0.9 Mach is

$$G(s) = \frac{\theta(s)}{\delta_e(s)} = \frac{22.3K_x(0.995s + 1)}{(s+50)(s^2 + 1.76s + 7.93)}$$

The open-loop transfer function contains the simplified longitudinal aircraft dynamics (a second-order system relating pitch rate to elevator deflection). The $K_x/(s+50)$ term models the hydraulic system between the pilot's control and the elevator. Use a CAD package to plot the root locus for this system. Select characteristic roots that have a damping ratio of 0.4. Find the gain value K_x that yields these roots. Determine the system response to an impulse. What is the minimum value of damping ratio available at these flight conditions?

Chapter 8

8.1. For each of the transfer functions

(1) $G(s) = \dfrac{10}{s(1 + 0.2s)(1 + 0.4s)}$

(2) $G(s) = \dfrac{1}{s(1 + 0.04s)(1 + s/50 + s^2/2500)}$

(3) $G(s) = \dfrac{20(s+2)}{s(s^2+6s+25)}$

(4) $G(s) = \dfrac{40(1-0.5s)}{s(1+s)(1+0.1s)}$

(a) draw the log magnitude (exact and asymptotic) and phase diagrams; (b) from the curves of part (a) obtain the data for plotting the direct polar plots; (c) from the curves of part (a) obtain the data for plotting the inverse polar plots. Determine ω_ϕ and ω_c for each transfer function.

8.2. For each of the transfer functions

(1) $G(s) = \dfrac{20}{(1+0.2s)(1+0.4s)(1+s)}$ (2) $G(s) = \dfrac{52}{s(s^2+8s+52)}$

(3) $G(s) = \dfrac{2(1+0.4s)}{s^2(1+0.1s)(1+0.05s)}$ (4) $G(s) = \dfrac{10}{(1+s)(1+0.2s)}$

(5) $G(s) = \dfrac{2}{s^2(1+0.1s)(1+0.4s)}$ (6) $G(s) = \dfrac{2(1+0.5s)}{s(1+0.1s)(1+0.4s)}$

(a) draw the log magnitude (exact and asymptotic) and phase diagrams; (b) from the curves of part (a) obtain the data for plotting the direct polar plots; (c) from the curves of part (a) obtain the data for plotting the inverse polar plots.

8.3. Plot to scale the log magnitude and angle vs. log ω curves for $\mathbf{G}(j\omega)$. Is it an integral or a derivative compensating network?

(a) $\mathbf{G}(j\omega) = \dfrac{10(1+j2\omega)}{1+j6\omega}$ (b) $\mathbf{G}(j\omega) = \dfrac{10(1+j10\omega)}{1+j\omega}$

8.4. A control system with unity feedback has the forward transfer function, with $K_m = 1$,

(1) $G(s) = \dfrac{K_1}{s(1+0.1s)(1+30s/625+s^2/625)}$

(2) $G(s) = \dfrac{K_0(1+0.5s)}{(1+s)(1+0.25s)(1+0.1s)^2}$

(a) Draw the log magnitude and angle diagrams. (b) Draw the polar plot of $\mathbf{G}'(j\omega)$. Determine all key points of the curve. (c) Plot the polar plot of $\mathbf{G}'(j\omega)^{-1}$. Determine all key points of the curve.

8.5. A system has

$$G(s) = \frac{2(1+T_2 s)(1+T_3 s)}{s^2(1+T_1 s)^2(1+T_4 s)}$$

where $T_1 = 4$, $T_2 = 1$, $T_3 = 1/2$, $T_4 = 1/4$. (a) Draw the asymptotes of $\mathbf{G}(j\omega)$ on a decibel vs. log ω plot. Label the corner frequencies on the graph. (b) What is the *total* correction from the asymptotes at $\omega = 2$?

8.6. (a) What characteristic must the plot of magnitude in decibels vs. log ω possess if a velocity servo system (ramp input) is to have no *steady-state velocity error* for a constant velocity input $dr(t)/dt$? (b) What is true of the corresponding phase-angle characteristic?

8.7. Explain why the phase-angle curve cannot be calculated from the plot of $|\mathbf{G}(j\omega)|$ in decibels vs. log ω if some of the factors are not minimum phase.

8.8. The asymptotic gain vs. frequency curve of the open-loop minimum-phase transfer function is shown for a unity-feedback control system. (*a*) Evaluate the open-loop transfer function. (*b*) What is the frequency at which $|\mathbf{G}(j\omega)|$ is unity? What is the phase angle at this frequency? (*c*) Draw the polar diagram of the open-loop control system.

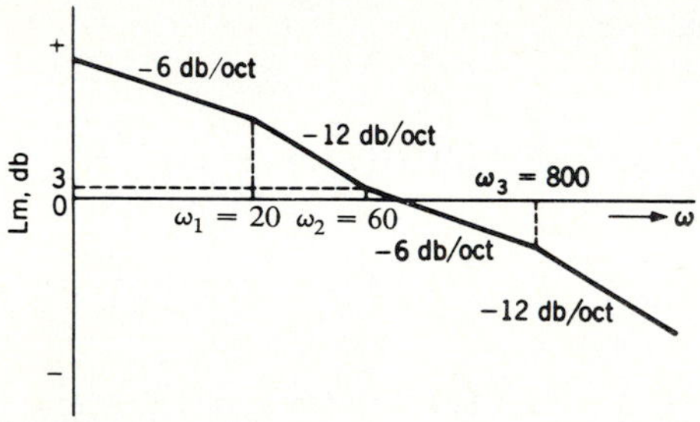

8.9. For each plot shown, (*a*) Evaluate the transfer function; (*b*) find the correction that should be applied to the straight-line curve at $\omega = 4$; (*c*) determine K_m for Cases I and II.

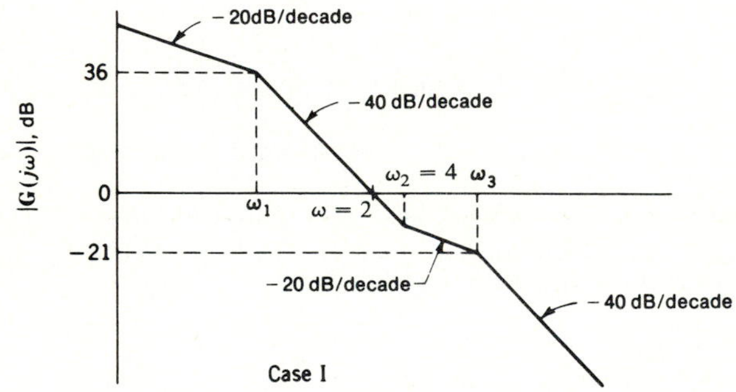

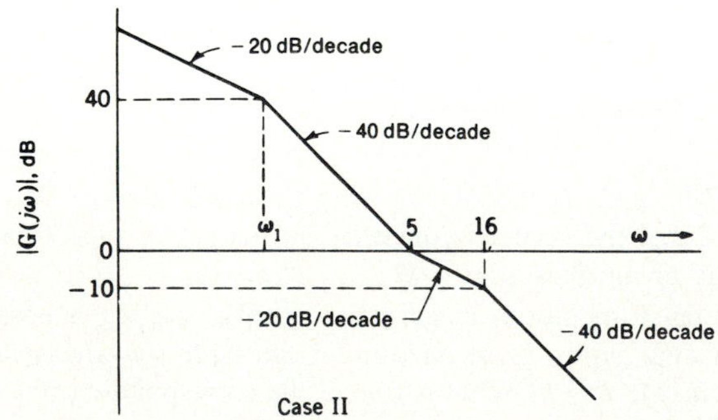

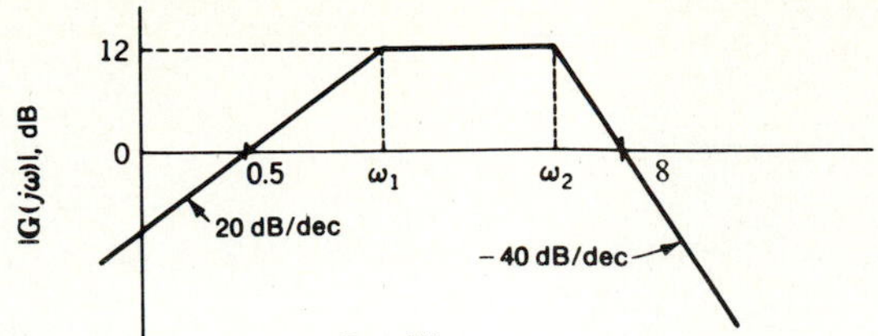

Case III

8.10. Determine the value of the error coefficient from parts (*a*) and (*b*) of the figure.

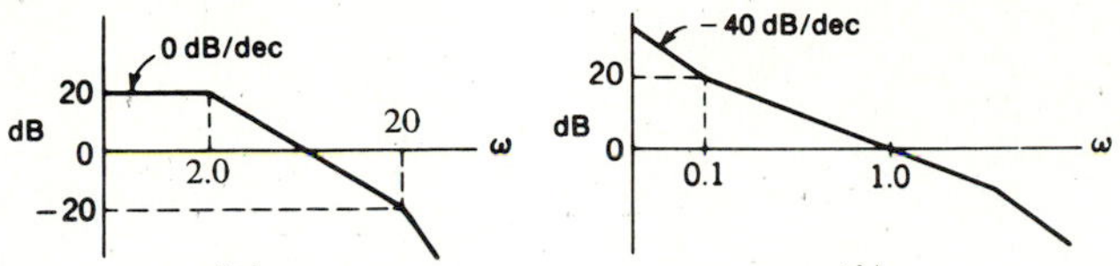

(*a*) (*b*)

8.11. An experimental transfer function gave the following results:

ω	Lm G(*jω*), dB	Angle G(*jω*), deg	ω	Lm G(*jω*), dB	Angle G(*jω*), deg
0.10	46.02	−179.1	14.0	−33.56	−192.8
0.50	18.14	−175.8	20.0	−40.34	−207.9
1.0	6.34	−171.8	40.0	−55.74	−234.2
2.0	−4.90	−166.1	80.0	−72.92	−251.2
4.0	−15.0	−162.9	140.0	−87.29	−259.1
8.0	−24.60	−173.1	200.0	−96.53	−262.4
10.0	−27.96	−180.0	240.0	−101.3	−236.6

(*a*) Determine the transfer function represented by the above data. (*b*) What type of system does it represent?

8.12. Repeat Prob. 8.11 for each of the following:

(*a*)

ω	|G(*jω*)|	Angle, deg
0.1	9.95	−96.9
0.3	3.19	−110.1
0.6	1.42	−127.8
0.8	0.963	−137.8
1.0	0.693	−146.3
2.0	0.21	−175.2
3.0	0.092	−192.5
5.0	0.028	−213.7
8.0	0.0082	−230.9
12.0	0.0027	−242.6
16.0	0.0012	−249.0
20.0	0.0006	−253.1

(*b*)

ω	Lm G(*jω*), dB	Angle G(*jω*), deg
1	−0.08	−92.9
2	−5.71	−95.9
4	−10.77	−103.4
6	−12.55	−115.1
8	−12.68	−138.0
10	−13.98	−180.0
15	−26.80	−239.0
20	−36.00	−251.6
30	−47.75	−259.4
40	−55.64	−262.4
50	−61.63	−264.1
55	−64.17	−264.6

8.13. Determine the transfer function by using the straight-line asymptotic log plot shown and the fact that the correct angle is $-129.7°$ at $\omega = 0.8$. Assume a minimum-phase system.

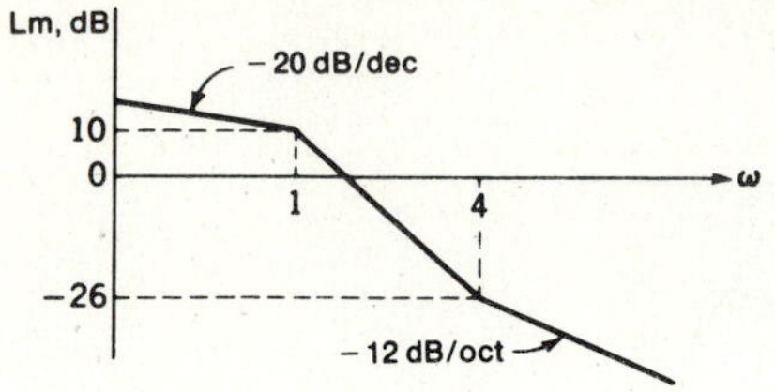

8.14. Determine whether each system shown is stable or unstable in the absolute sense by sketching the *complete* Nyquist diagrams. $H(s) = 1$.

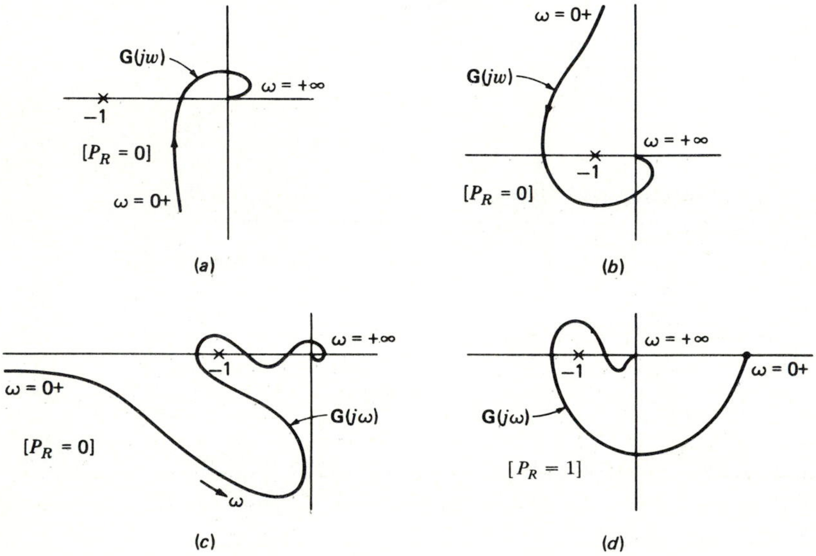

8.15. Use the Nyquist stability criterion and the polar plot to determine the range of values K (positive or negative) for which the closed-loop system is stable.

$$(a)\quad G(s)H(s) = \frac{K(1+s)^2}{s^3}$$

$$(b)\quad G(s)H(s) = \frac{K}{s^2(-1+5s)(1+s)}$$

$$(c)\quad G(s)H(s) = \frac{K}{s^2(1-0.5s)}$$

$$(d)\quad G(s)H(s) = \frac{K}{s^2(s+15)(s^2+6s+10)}$$

$$(e)\quad G(s)H(s) = \frac{K}{s(s^2+4s+5)}$$

$$(f)\quad G(s)H(s) = \frac{K}{s^2(s+9)}$$

8.16. Shown are plots of the open-loop transfer function for a number of control systems. Only the curves for positive frequencies and $K > 0$ are given. *Using Nyquist's criterion*, determine the closed-loop system stability. $G(s)$ has no poles in the right-half plane.

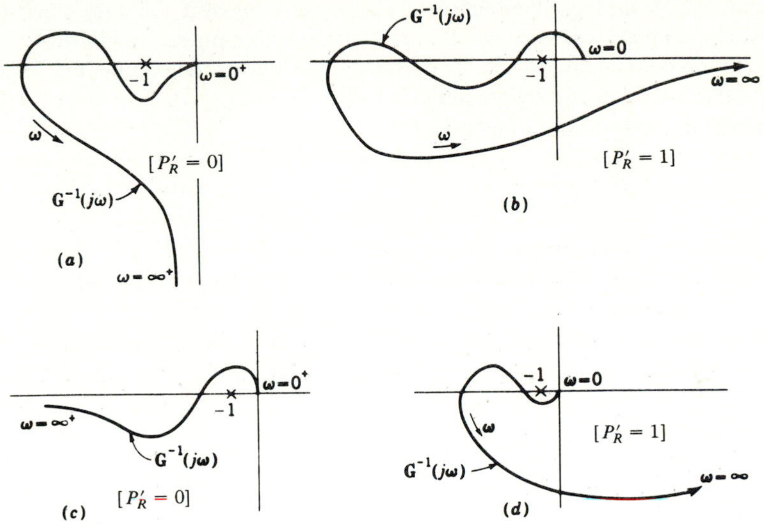

8.17. For the following transfer functions, sketch a direct and an inverse Nyquist locus to determine the closed-loop stability. Determine the range of values of K that produce stable closed-loop operation and those which produce unstable closed-loop operation. $H(s) = 1$.

(a) $G(s) = \dfrac{K}{1 + s}$ (b) $G(s) = K\dfrac{2 + s}{s(1 - s)}$

(c) $G(s) = \dfrac{K(1 + 0.5s)}{s^2(4 + s)}$ (d) $G(s) = \dfrac{K}{s(7 + s)(2 + s)}$

(e) $G(s) = \dfrac{K(s + 2)}{(s + 3)(s - 1)}$ (f) $G(s) = \dfrac{K}{(s - 1)(s + 2)(s + 4)}$

(g) $G(s) = \dfrac{K_1(1 + 2s)}{s(1 + s)(1 + s + s^2)}$ (h) $G(s) = \dfrac{K(s^2 + 2s + 2)}{s^3(1 + 0.2s)}$

(i) $G(s) = \dfrac{K(1 - s)}{(1 + 2s/3)(1 + s)}$ (j) $G(s) = \dfrac{K(s + 0.5)}{s^2(s + 1)(s + 4)}$

8.18. For the control systems having the transfer functions

(1) $G(s) = \dfrac{K_1}{s(1 + 0.01s)(1 + 0.025s)(1 + 0.10s)}$

(2) $G(s) = \dfrac{K(1 + 0.2s)}{(1 + 0.1s)(2 + 3s + s^2)}$

$$(3)\ G(s) = \frac{K_0(1 + 0.2s)}{(1 + s)(1 + 0.36s + 0.04s^2)}$$

$$(4)\ G(s) = \frac{K(s + 1)}{s(s + 2)(s^2 + 4s + 5)}$$

determine from the logarithmic curves the required value of K_m and the phase-margin frequency so that each system will have (a) a positive phase margin of 45°; (b) a positive phase margin of 60°. (c) From these curves determine the maximum permissible value of K_m for stability.

8.19. A system has the transfer functions

$$G(s) = \frac{K}{s(s + 5)(s^2 + 32s + 1552)} \qquad H(s) = 1$$

(a) Draw the log magnitude and phase diagram of $\mathbf{G}'(j\omega)$. Draw both the straight-line and the corrected log magnitude curves. (b) Draw the log magnitude–angle diagram. (c) Determine the maximum value of K_1 for stability.

8.20. What gain values would just make the systems in Prob. 8.2 unstable?

8.21. By use of the Nyquist stability criterion, determine whether the system having the following inverse polar plot is stable or unstable.

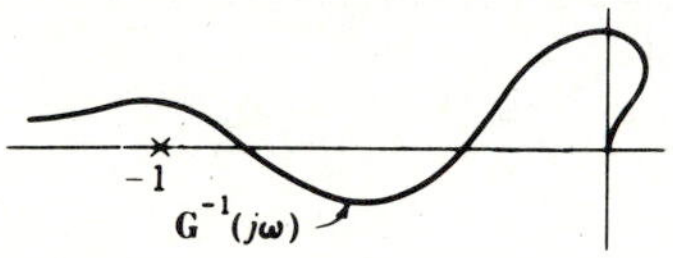

8.22. A system has the transfer function

$$G(s) = \frac{K_0}{(1 + s)^3}$$

(a) Draw the log magnitude vs. phase-angle diagram for $G'(s)$. (b) Determine the phase margin γ for $K_0 = 4$. (c) What is the maximum value of K_0 for stability?

8.23. For

$$G(s)^{-1} = \frac{s^2(0.5s + 1)(3s + 2)}{K_x(s - 1)}$$

draw the inverse Nyquist locus and determine the closed-loop stability for (a) $K_x = 2$; (b) $K_x = -2$.

8.24. Use the Nyquist criterion to determine the maximum value of K_m for stability of the closed-loop systems having the following transfer functions:

$$(a)\ G(s)H(s) = \frac{K_1\varepsilon^{-0.5s}}{s(1 + s)(1 + 0.5s)} \qquad (b)\ G(s)H(s) = \frac{K\varepsilon^{-2s}}{s^2 + 2s + 2}$$

(c) The transfer functions of Prob. 8.18 with the transport lag $\varepsilon^{-1.5s}$ included in the numerator.

8.25. Use the Nyquist criterion to determine the maximum value of T for stability of the closed-loop system which has the open-loop transfer function

$$G(s)H(s) = \frac{2\varepsilon^{-Ts}}{s(s^2 + 4s + 13)}$$

Hint: Determine the frequency ω_x for which $|\mathbf{G}(j\omega)\mathbf{H}(j\omega)| = 1$. What additional angle ωT, due to the transport lag, will make the angle of $\mathbf{G}(j\omega_x)\mathbf{H}(j\omega_x)$ equal to $-180°$?

8.26. A satellite control system controlled from the earth has the transfer function

$$G(s) = \frac{K_1}{s(1+s)(1+0.2s)}$$

There is a time T for the actuating signal e to reach the satellite from the earth and for the feedback signal to reach the earth from the satellite. (a) Draw a block diagram showing the presence of the time delays. (b) When $T = 1$ draw a Nyquist plot and show the effect of the time delays on stability.

8.27. In Stark and Young's classic paper, The Transfer Function of a Photoreceptor Organ, the open-loop transfer function of the photoreceptor organ of a crawfish is derived. The transfer function is

$$G(s) = \frac{55e^{-0.025s}}{(s+10)^{1/2}}$$

The main contribution of this paper was the novel approach developed for "opening the loop" of the crawfish's photoreceptor. Derive Bode construction rules for the nonstandard entries in the transfer function. Sketch the resulting Bode plot.

Chapter 9

9.1. For Prob. 8.1, find K_1, M_m, ω_c, ω_ϕ, and ω_m for a $\gamma = 45°$ by (a) the direct-polar-plot method; (b) the inverse-polar-plot method; (c) the log magnitude–angle diagram; (d) a computer program. (e) Repeat (a) through (c) by determining the value of K_1 for an $M_m = 1.32$. (f) Compare the values of ω_m, ω_c, ω_ϕ, and K_1 obtained in (e) with those obtained in (a) through (c).

9.2. For the feedback control systems of Prob. 7.5, (a) determine, by use of the polar-plot method, the value of K_1 that just makes the system unstable; (b) determine the value of K_1 that makes $M_m = 1.16$; (c) for the value of K_1 found in part (b), find $c(t)$ for $r(t) = u_{-1}(t)$; (d) determine graphically from the polar plot of $\mathbf{G}(j\omega)$ the data for the curve of M vs. ω for the closed-loop system and plot this curve; (e) compare the results of parts (a) to (d) with the results obtained in Prob. 7.5.

9.3. Determine the value K_1 must have for an $M_m = 1.3$.

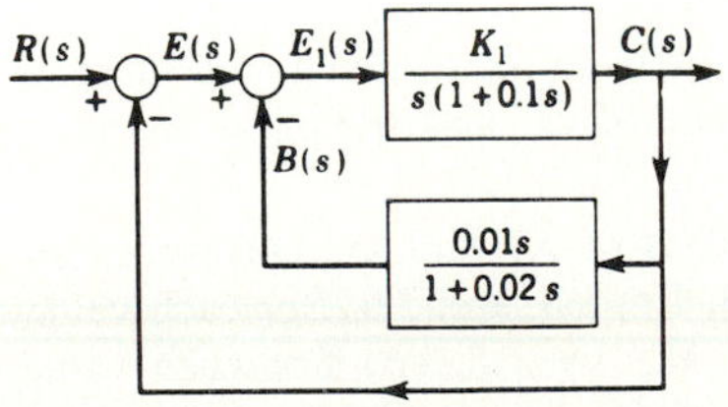

9.4. For the $1/\mathbf{G}'(j\omega)$ curve shown it is found that two values of gain, K_a and K_b, will produce a desired M_m. The corresponding resonant frequencies are ω_a and ω_b. Which value of gain gives the better performance for the system? Give the reasons for your choice.

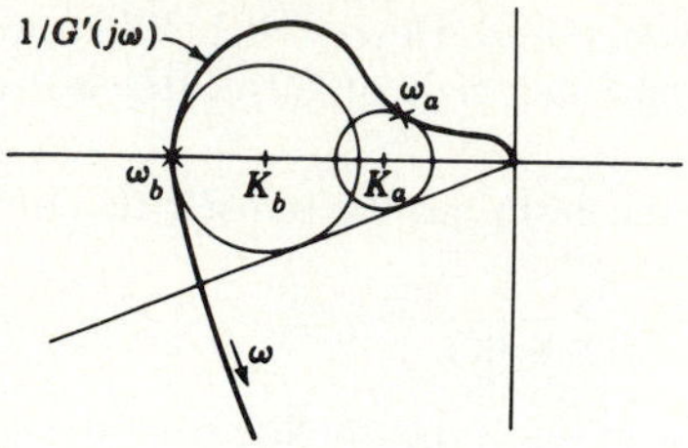

9.5. Using the plot of $\mathbf{G}'(j\omega)$ shown, determine the number of values of gain K_m which produce the same value of M_m. Which of these values yields the best system performance? Give the reasons for your answer.

$$G(s) = \frac{K_1(1 + 0.125s)^2}{s(1 + s)^2(1 + 0.005s)^3}$$

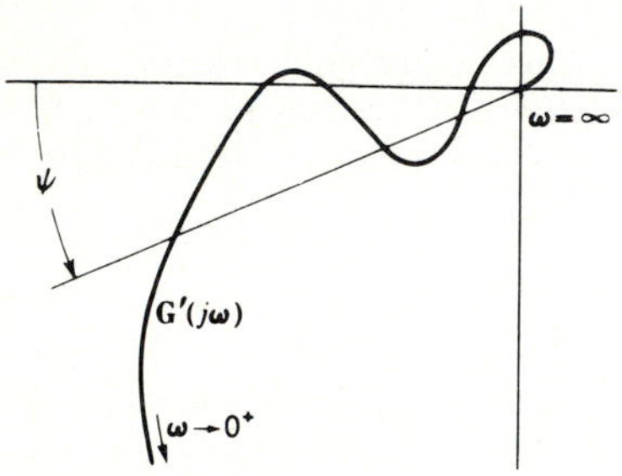

9.6. (a) Determine the values of M_m and ω_m for the transfer functions of Prob. 8.18 with $K_m = 2$. (b) Repeat with the gain-constant values obtained in part (a) of Prob. 8.18; (c) part (b) of Prob. 8.18. For part (c) plot M vs. ω and obtain $c(t)$ for a step input.

9.7. (a) In Prob. 8.19, adjust the gain for Lm $M_m = 2$ dB and determine ω_m. For this value of gain find the phase margin, and plot M vs. ω and α vs. ω. (b) Repeat for Prob. 8.4.

9.8. For Prob. 8.15, part (a), with $H(s) = 1$ determine how much gain must be added to achieve an $M_m = 1.26$. What is the value of ω_m for this value of M_m? (a) Use log plots. (b) Use polar plots. (c) Use a computer.

9.9. For Prob. 8.2, determine the values of M_m and ω_m. How much gain must be added to achieve an $M_m = 1.12$? What is the value of ω_m and the phase margin for this value of M_m for each case?

9.10. Refer to Prob. 8.1. (a) Determine the values of K_1 and ω_m corresponding to the following values of M_m: 1.05, 1.1, 1.2, 1.4, 1.6, 1.8, and 2.0. For each value of M_m determine ζ_{eff} from Eq. (9.16) and plot M_m vs. ζ_{eff}. (b) For each value of K_1 determined in part (a), calculate the value of M_p. Use available computer programs. For each value of M_p determine ζ_{eff} from Fig. 3.7 and plot M_p vs. ζ_{eff}. (See the last paragraph of Sec. 7.11.) (c) What is the correlation between M_m vs. ζ_{eff} and M_p vs. ζ_{eff}? What is the effect of a third real root which is also dominant?

9.11. For Prob. 8.24, determine the value of K_m for Lm $M_m = 3$ dB with $H(s) = 1$.

9.12. For the given transfer function: (a) specify the value of K_2 that will make the peak M_m in the frequency response as small as possible. (b) At what frequency does this peak occur? (c) What value does $\mathbf{C}(j\omega)/\mathbf{R}(j\omega)$ have at the peak?

(1) $$G(s) = \frac{K_2(1+s)}{s^2(1+0.1s)(1+0.05s)}$$

(2) $$G(s) = \frac{K(s+0.0465)}{s^2(s+1.46)}$$

9.13. The closed-loop frequency response of three simple second-order systems are sketched below. Sketch the time response to a step input for each of the three systems on the same time scale. Compare M_p, T_p, and T_s.

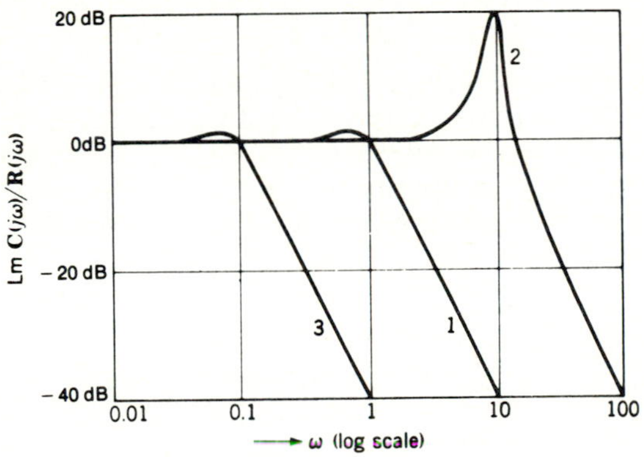

9.14. Evaluate the transfer function for the frequency-response plot shown below. The plot is not drawn to scale.

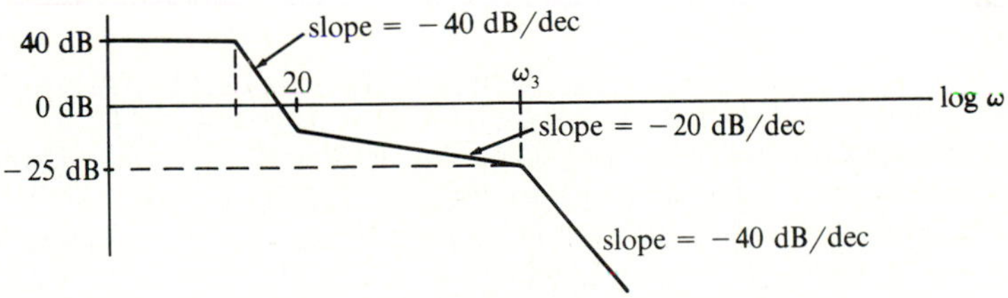

9.15. For a unity-feedback system, let $G(s)$ be

$$G(s) = \frac{K}{s(1+0.1s)^2(s^2+121)}$$

(a) Sketch the normalized Nyquist polar plot and determine the range of K for stability by using Nyquist's stability criterion. (b) Sketch the normalized Nyquist inverse polar plot and determine the range of K for stability for Nyquist's stability criterion. (c) Assume $K = 40$. Plot the log magnitude and phase-angle curves (Bode diagrams). Then find the gain margin and phase margin. Is the system stable for this gain?

Chapter 10

10.1. It is desired that a control system have a damping ratio of 0.5 for the dominant complex roots. Using the root-locus method, (*a*) add a lag compensator, with $\alpha = 10$, so that this value of ζ can be obtained; (*b*) add a lead compensator with $\alpha = 0.1$; (*c*) add a lag-lead compensator with $\alpha = 10$. Indicate the time constants of the compensator in each case. Compare the results obtained by the use of each type of compensator with respect to the error coefficient K_m, ω_d, T_s, and M_o.

$$(1)\ G(s) = \frac{K_x}{(s+1)(s+3)(s+5)} \qquad (2)\ \frac{K_x}{s(s+2)(s+5)(s+7)}$$

10.2. A control system has the forward transfer function

$$G_x(s) = \frac{K_x}{s^2(1+0.2s)}$$

The closed-loop system is to be made stable by adding a compensator $G_c(s)$ and an amplifier A in cascade with $G_x(s)$. A ζ of 0.5 is desired with a value of $\omega_n \approx 1.0$ rad/s. By use of the root-locus method, determine the following: (*a*) What kind of compensator is needed? (*b*) Select an appropriate α and T for the compensator. (*c*) Determine the value of the error coefficient K_2. (*d*) Plot the compensated locus. (*e*) Plot M vs. ω for the compensated system. (*f*) From the plot of part (*e*) determine the values of M_m and ω_m. With this value of M_m determine the effective ζ of the system by the use of $M_m = (2\zeta\sqrt{1-\zeta^2})^{-1}$. Compare the effective ζ and ω_m with the values obtained from the dominant pair of complex roots. *Note*: The effective $\omega_m = \omega_n\sqrt{1-2\zeta^2}$. (*g*) Obtain $c(t)$ for a unit step input.

10.3. With

$$(1)\ G_x(s) = \frac{K_x}{s(s+10)(s^2+30s+625)} \qquad H(s) = 1$$

$$(2)\ G_x(s) = \frac{K_x(s+25)}{s(s+10)(s^2+30s+625)} \qquad H(s) = 1$$

find K_1, ω_n, M_o, T_p, T_s, N for each of the following cases: (*a*) original system; (*b*) lag compensator added, $\alpha = 10$; (*c*) lead compensator added, $\alpha = 0.1$; (*d*) lag-lead compensator added, $\alpha = 10$. Use $\zeta = 0.5$ for the dominant roots. *Note*: Except for the original system it is not necessary to obtain the complete root locus for each type of compensation. When the lead and lag-lead compensators are added to the original system, there may be other dominant roots in addition to the complex pair. When a real root is also dominant, a $\zeta = 0.3$ may produce the desired improvement (see Sec. 10.3).

10.4. A unity-feedback system has the forward transfer function

$$G_x(s) = \frac{K_2}{s^2}$$

(*a*) Design a cascade compensator which will produce a stable system and which meets the following requirements without reducing the system type: (1) The dominant poles of the closed-loop control ratio are to have a damping ratio $\zeta = 0.5$. (2) The settling time is to be $T_s = 5$ s. Is it a lag or lead compensator? (*b*) Using the cascade compensator, determine the control ratio $C(s)/R(s)$. (*c*) Find

$c(t)$ for a step input. What is the effect of the real pole of $C(s)/R(s)$ on the transient response?

10.5. Using the root-locus plot of Prob. 7.9, adjust the damping ratio to $\zeta = 0.5$ for the dominant roots of the system. Find K_0, ω_n, M_o, T_p, T_s, N, $C(s)/R(s)$ for (a) the original system and (b) the original system with a cascade lag compensator using $\alpha = 10$; (c) design a cascade compensator that will improve the response time, i.e., will move the dominant branch to the left in the s plane.

10.6. A control system has

$$G_x(s)H(s) = \frac{K}{(s+1)(s+2)(s+5)(s+6)} \qquad H(s) = 1$$

(a) Determine the roots of the characteristic equation of this system for $\zeta = 0.5$. (b) Determine the ratio $\sigma_{3,4}/\sigma_{1,2}$. (c) It is desired that this ratio be increased to at least 10 with no increase in T_s for $\zeta = 0.5$. A compensator that may accomplish this has the transfer function

$$G_c = A\frac{s+a}{s+b}$$

Determine appropriate positive values of a and b that can be achieved with a practical value of α.

10.7. A unity-feedback system has the transfer function $G_x(s)$. The closed-loop roots must satisfy the specifications $\zeta = 0.707$ and $T_s = 2$ s. A suggested compensator $G_c(s)$ must maintain the same degree as the characteristic equation for the basic system:

$$G_x(s) = \frac{K_x(s+3)}{s(s+2)} \qquad G_c(s) = \frac{A(s+a)}{s+b}$$

(a) Determine the values of a and b. (b) Determine the value of α. (c) Is this a lag or a lead compensator?

10.8. A unity-feedback control system contains the forward transfer function shown below. The system specifications are $\zeta = 0.6$ and $T_s \le 1.2$ s. Without compensation the value of K_x is 57.33, and the control ratio has an undesirable dominant real pole. The proposed cascade compensator has the form indicated.

$$G_x(s) = \frac{K_x(s+2)}{s(s+5)(s+7)} \qquad G_c(s) = A\frac{s+a}{s+b}$$

(a) Determine the values of a and b such that the desired complex-conjugate poles of $C(s)/R(s)$ are truly dominant. (b) Determine the values of A and α for this compensator. (c) Is this $G_c(s)$ a lag or lead compensator?

10.9. A unity-feedback system has

$$G_x(s) = \frac{K_x}{(s+1)(s+2)(s+6)}$$

(a) For $\zeta = 0.5$, determine the roots and the value of K_x. (b) Add a lead compensator which cancels the pole at $s = -1$. (c) Add a lead compensator which cancels the pole at $s = -2$. (d) Compare the results of parts (b) and (c). Establish a "rule" for adding a lead compensator to a Type 0 system.

10.10. For the system of Prob. 7.5(1), the desired dominant roots are $-25 \pm j35$. (a) Design a compensator which will achieve these characteristics. (b) Determine $c(t)$

with a unit step input. (*c*) Are the desired complex roots dominant? (*d*) Repeat this problem with the roots at $-20 \pm j35$.

10.11. A unity-feedback system has the transfer function

$$G_x(s) = \frac{K_x}{s(s+1)(s+10)}$$

The poles of the closed-loop system must be $s = -1 \pm j2$. (*a*) Design a lead compensator with the maximum possible value of α which will produce these roots. (*b*) Determine the control ratio for the compensated system. (*c*) Add a compensator to increase the gain without increasing the settling time.

10.12. A system has the transfer function

$$G_x(s) = \frac{K_x}{(s^2 - 2s + 2)(s + 10)}$$

The roots of the closed-loop system must have $\zeta = 0.5$. Add a lead compensator, with two zeros and two poles, to produce a stable system. Determine $C(s)/R(s)$. Is this a practical compensator? Explain.

10.13. In the figure the servomotor has inertia but no viscous friction. The feedback through the accelerometer is proportional to the acceleration of the output shaft. (*a*) When $H(s) = 1$, is the servo system stable? (*b*) Is the system stable if $K_x = 10$, $K_y = 2$, and $H(s) = 1/(2s)$? (*c*) Determine another $H(s)$ that causes the system to be stable. Show that the system is stable with the value of $H(s)$ selected by sketching the transfer function $\mathbf{C}(j\omega)/\mathbf{E}(j\omega)$ and the root locus.

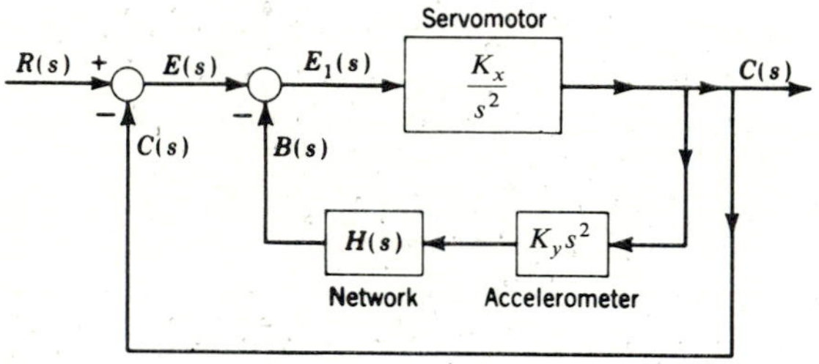

10.14. For the system shown, find (*a*) the open-loop transfer function $C(s)/E(s)$; (*b*) ω_n and ζ for the open-loop system and for the closed-loop system $C(s)/R(s)$.

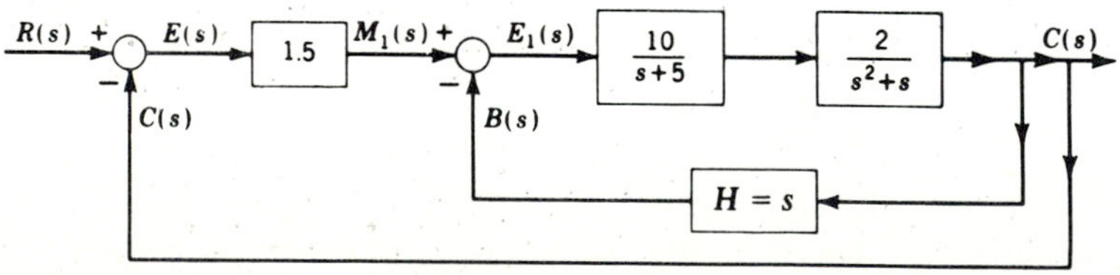

10.15. The accompanying schematic shows a method for maintaining a constant rate of discharge from a water tank by regulating the level of the water in the tank. The relationships governing the dynamics of the flow into and out of the tank are

$$Q_1 - Q_2 = 16\frac{dh}{dt} \qquad Q_2 = 4h \qquad Q_1 = 10\theta$$

where Q_1 = volumetric flow into tank
Q_2 = volumetric flow out of tank
h = pressure head in tank
θ = angular rotation of control valve

The motor damping is much smaller than the inertia. The system shown is stable only for very small values of system gain. Therefore, feedback from the control valve to the amplifier is proposed. Show conclusively which of the following feedback functions would produce the best results: (*a*) feedback signal proportional to control-valve position; (*b*) feedback signal proportional to rate of change of control-valve position; (*c*) feedback signal comprising a component proportional to valve position and a component proportional to rate of change of valve position.

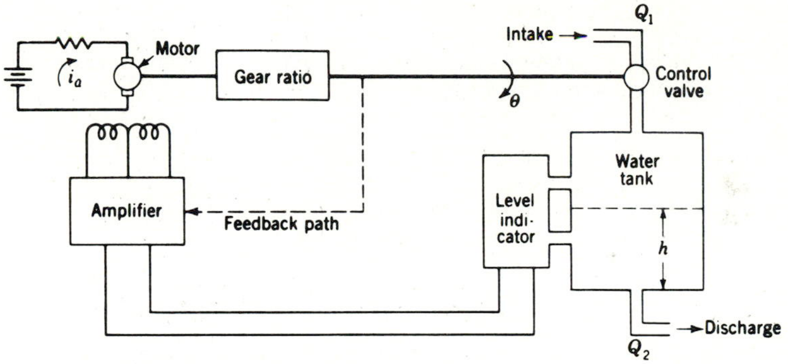

10.16–10.17. The block diagram shows a simplified form of roll control for an airplane. Overall system specifications with a step input are t_s = 1.0 s and $1 < M_p \leq 1.3$. A_h is the gain of an amplifier in the H_2 feedback loop. G_2 is an amplifier of adjustable gain with a maximum value of 100. Restrict b between 15 and 50.

$$G_3(s) = \frac{1.7}{(1 + 0.25s)(1 + s)} \qquad H_2(s) = 0.2A_h\frac{s + a}{s + b} \qquad G_4(s) = \frac{1}{s}$$

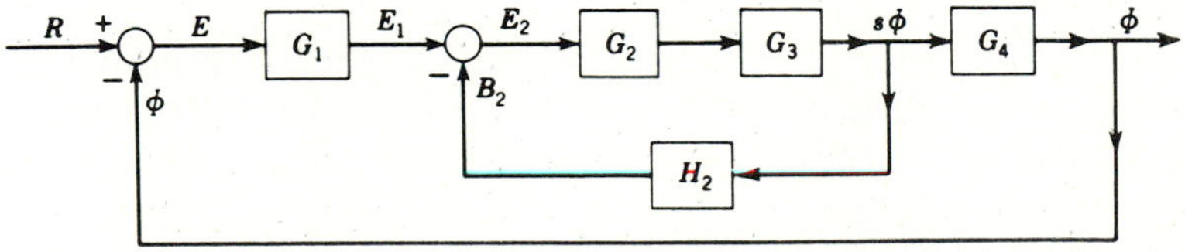

10.16. (*a*) By use of the root locus, method 1, with $G_1(s) = 1$, determine the parameters A_h, a, and b in $H_2(s)$ to meet the overall system specifications. (*b*) Design $G_1(s)$ to improve the value of K_1 of part (*a*) by a factor of 5 while maintaining the desired overall system specifications. Determine all roots of the characteristic equation for Φ/R after compensation. (*c*) For the values obtained in part (*a*), determine T_s and M_p for the inner loop. Compare these values with those given in part (*a*) of Prob. 10.17(*a*).

10.17. (*a*) By use of the root locus, method 2, determine the parameters A_h, a, and b to meet the inner-loop specifications M_p = 1.05 and T_s = 1.0 s. Select G_2 so that $\dot{\phi}(t)$

follows a step input $e_1(t) = u_{-1}(t)$. Determine the necessary values of A_h and G_2. (b) Design $G_1(s)$ to improve the value of K_1, resulting from part (a), by a factor of 5 and also to meet the overall system specifications.

10.18. For the feedback-compensated system of Fig. 10.29,

$$G_x(s) = \frac{K_x}{s(s+1)} \qquad H(s) = \frac{s(s+a)}{s+4}$$

(a) Use method 1 to determine $C(s)/R(s)$ using $T_s = 2$ s (2 percent criterion) and $\zeta = 0.707$ for the dominant roots. (b) Design a cascade compensator which yields the same dominant roots as part (a). (c) Compare the gains of the two systems.

10.19. A unity-feedback system has

$$G_x(s) = \frac{K_x}{s(s+6)}$$

(a) For $\zeta = 0.5$, determine K_x and $C(s)/R(s)$. (b) Minor-loop and cascade compensation are to be added (see Fig. 10.29) with the amplifier replaced by $G_c(s)$:

$$G_c(s) = \frac{As}{s+b} \qquad H(s) = \frac{K_h s}{s+a}$$

Use the value of K_x determined in part (a). By method 1 of the root-locus design procedure, determine values of a, b, K_h, and A to produce dominant roots having $\zeta = 0.5$ and $\sigma = -8$. Only positive values of K_h and A are acceptable. To minimize the effect of the third real root, it should be located near a zero of $C(s)/R(s)$.

10.20. For $G(s)$ given in Prob. 10.3, add feedback rate (tachometer) compensation. Use root-locus method 1 or 2 as assigned. Design the tachometer so that an improvement in system performance is achieved while maintaining a $\zeta = 0.5$ for the dominant roots. Compare with the results of Prob. 10.3. *Note*: Use Fig. 10.29.

10.21. Refer to Fig. 10.29. It is desired that the dominant poles of $C(s)/R(s)$ have $\zeta = 0.5$ and $T_s = 1$ s. Using method 1, with $A = 5$, determine the values of b and K_t to meet these specifications. Be careful in selecting b.

$$G_x = \frac{4(s+7)}{s(s+4)^2} \qquad H(s) = \frac{K_t s}{s+b}$$

10.22. Use minor-loop feedback to compensate the basic system of Prob. 10.1 to achieve the same specification. Compare the cascade and feedback performance.

10.23. Repeat Prob. 10.22 for the system of Prob. 7.9, comparing the results with those of Prob. 10.5.

10.24. Repeat Prob. 10.2 using a feedback compensator.

10.25.

$$G_x(s) = \frac{K_x}{s(s^2 + 4.2s + 14.4)} \qquad H(s) = 1$$

The specifications for the closed-loop performance with a step input are $1 < M_p \le 1.123$, $t_p \le 1.6$ s, $t_s \le 3.0$ s, $K_1 \ge 1.5$. Show that these specifications can be achieved without the use of a compensator.

10.26. An interesting control problem occurs when an open-loop Type 0 system $G_x(s)$ is lightly damped and increased gain makes the closed-loop system even more lightly damped as shown in figure a.

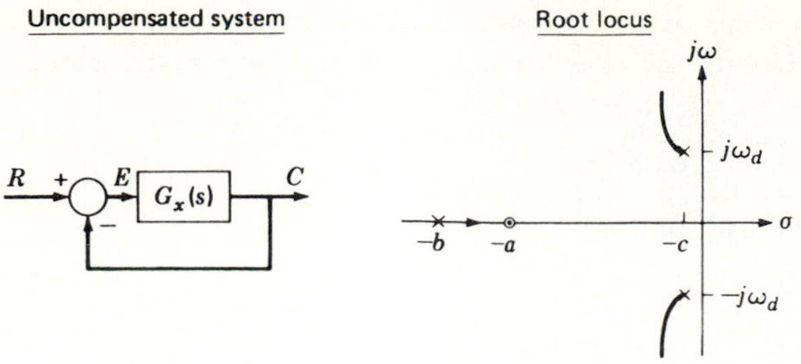

(a)

$$G_x(s) = \frac{K_x(s+a)}{(s+b)(s+c+j\omega_d)(s+c-j\omega_d)}$$

Rate feedback (see figure b) is used to increase the damping ratio and move the root locus to the left. A forward-path lag compensator, Eq. (10.23), is also used to increase K_0. The closed-loop system is shown in figure b:

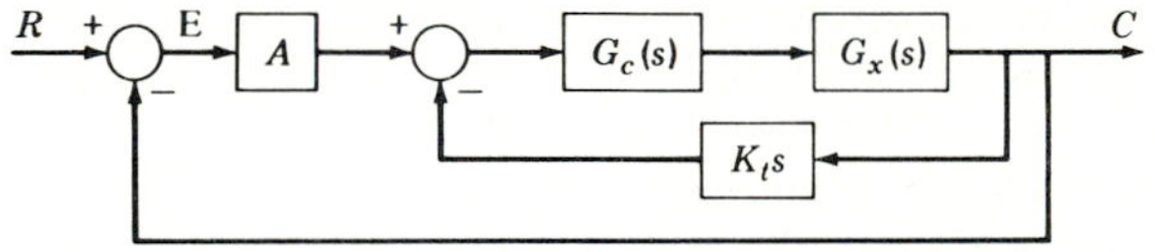

(b)

The closed-loop system can be redrawn in two different ways, one for root-locus computation (figure c) and one for computation of the step-error coefficient $K_0 = \lim_{s\to\infty} G(s)$ (figure d). The design procedure is to first select the value of A/K_t and the lag compensator, and then to draw the root locus. From the root locus the desired roots are selected, and the corresponding value of the static loop sensitivity K is used to evaluate K_0. If it is not acceptable, change $G_c(s)$ or A/K_t and repeat the design process. Use the following steps to achieve the desired design

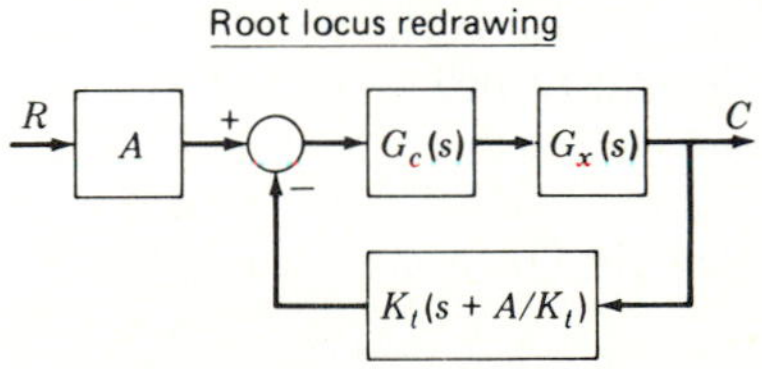

(c)

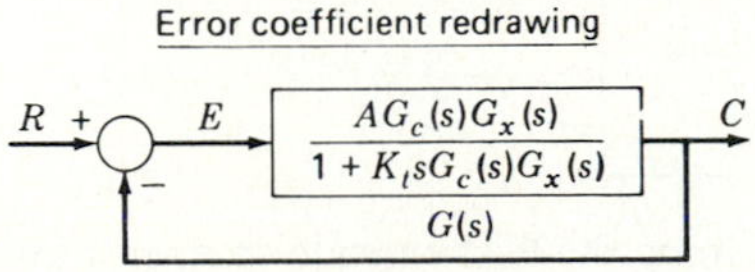

(d)

1. Obtain the root locus of the uncompensated system up to $K_x = 34{,}000$. For $K_x = 25{,}000$ determine the closed-loop zeros, closed-loop poles, and step-error coefficients.

Case	a	b	c	ω_d
1	50	100	10	50
2	50	100	5	25
3	75	100	15	75
4	75	100	12	60

2. Select the pole and zero for the lag compensator with $\alpha = 10$. Let the compensator zero be 0.1 times the left-half-plane nearest the origin, excluding the pole at the origin.
3. The zero $-A/K_t$ for the rate feedback should draw the root locus to the left. Place this zero to the left of the pole $-b$ (try $-2b$). Obtain the root locus for the compensated system.
4. The goal is to make the compensated step-error coefficient at least 20 times the uncompensated step-error coefficient with a damping ratio of $\zeta = 0.7$ for the dominant complex-conjugate pole pair.
5. With $\zeta = 0.7$ for the dominant complex-conjugate pair, determine the closed-loop zeros, closed-loop poles, and static-loop sensitivity. Then compute the step-error coefficient.
6. If the compensated step-error coefficient is at least 20 times the uncompensated step-error coefficient when $\zeta = 0.70$ for the dominant complex-conjugate pair, you are finished. If not, place the rate feedback zero farther to the left and return to step 3.

10.27. Repeat Prob. 10.21, with A unspecified, for

$$G_x(s) = \frac{2(s+6)}{s(s+4)^2}$$

10.28. Given the system of Fig. 10.29 with

$$G_x(s) = \frac{K_x(s+8)}{s(s+1)} \qquad H(s) = \frac{K_h(s+0.1)}{s+10}$$

(a) The specification for the inner loop is that $c(t)_{ss} = 1$ for $i(t) = u_{-1}(t)$. Determine the value of K_h in order to satisfy this condition. *Hint:* Analyze $C(s)/I(s)$. (b) It is desired that the dominant poles of $C(s)/R(s)$ be exactly at $-4 \pm j4$. By use of method 1, with K_h having the value determined in part (a), determine the values of A, K_m, and K_x that yield these dominant poles.

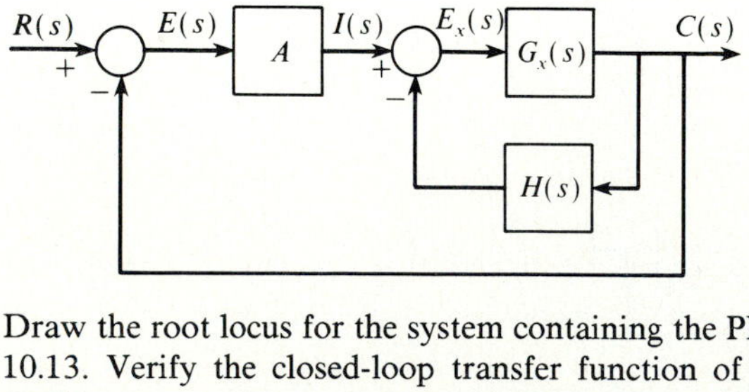

10.29. Draw the root locus for the system containing the PID cascade compensator of Sec. 10.13. Verify the closed-loop transfer function of Eq. (10.47) and the figures of merit.

Chapter 11

11.1. The control system of Prob. 8.4 is to have an $M_m = 1.16$. (*a*) Add a lag compensator, with $\alpha = 10$, so that this value of M_m can be obtained. (*b*) Add a lead compensator with $\alpha = 0.1$. (*c*) Add a lag-lead compensator with $\alpha = 10$. Indicate the time constants of the compensator in each case. Compare the results obtained by using each type of compensator. Also, compare the results of this problem with those of Prob. 10.3.

11.2. A control system has the forward transfer function

$$G_x(s) = \frac{K_x}{s^2(1 + 0.2s)} \qquad K_2 = 1$$

The closed-loop system is to be made stable by adding a compensator G_c and an amplifier A in cascade with $G_x(s)$. An $M_m = 1.5$ is desired with $\omega_m \approx 1.4$ rad/s. (*a*) What kind of compensator is needed? (*b*) Select an appropriate α and T for the compensator. (*c*) Select the necessary value of amplifier gain A. (*d*) Plot the compensated curve. (*e*) Compare the results of this problem with those obtained in Prob. 10.2.

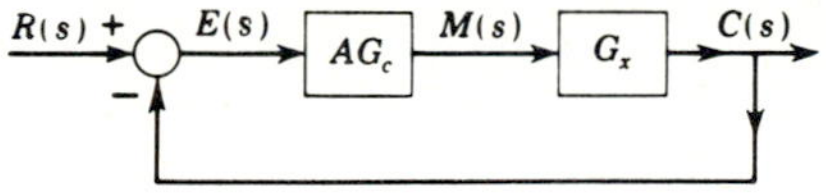

11.3. A control system with unity feedback has a forward open-loop transfer function

$$G_x(s) = \frac{K_x}{s(1 + 0.5s)(1 + 0.1s)^2}$$

(*a*) Find the gain K_x for 45° phase margin, and determine the corresponding phase-margin frequency ω_ϕ. (*b*) For the same phase margin as in (*a*), it is desired to increase the phase-margin frequency to a value of $\omega_\phi = 3.0$ with the maximum possible improvement in gain. To accomplish this, a lead compensator is to be used. Determine the values of α and T that will satisfy these requirements. For these values of α and T, determine the new value of gain. (*c*) Repeat part (*b*) with the lag-lead compensator of Fig. 11.16. Select an appropriate value for T. With $G_c(s)$ inserted in cascade with $G_x(s)$, find the gain needed for 45° phase margin. (*d*) Show how the compensator has improved the system performance, i.e., determine M_p, t_p, t_s, and K_1 for each part of the problem.

11.4. (*a*) Using the $G_x(s)$ of Prob. 11.3 with $K_x = 1$, determine ω_{ϕ_x} and γ_x for the basic system. (*b*) Design a cascade lead compensator, using $\alpha = 0.1$, in order to achieve an $\omega_\phi = 2$ rad/s, for this value of γ. Determine the required value of T and A. (*c*) For the basic (uncompensated) unity-feedback control system it is desired to increase K_m by a factor of approximately 10 while maintaining $\gamma = \gamma_x$. Design a cascade compensator that yields this increase in K_m. Specify the values of α and T so that $\left|\angle G_c(j\omega)\right| \le 0.6°$ at $\omega = \omega_{\phi_x}$. Determine the values of A and K_m for the compensated system. (*d*) It is desired that a phase margin γ of 45° and a phase-margin frequency $\omega_\phi = 3.0$ rad/s be achieved by use of the lag-lead compensator

$$G_c'(s) = \frac{(1 + 100s)(1 + 0.5s)}{(1 + 1000s)(1 + 0.5\alpha s)}$$

Determine the values of α and A that yield these values of γ and ω_ϕ for the compensated system. Determine the ramp error coefficient of the compensated system. (e) It is desired, for the feedback compensated system of Fig. 11.20, that $\gamma = 45°$ at $\omega_\phi = 3$ rad/s and the phase margin be determined *essentially* by the characteristics of the feedback unit. It is also specified for the compensated system that $K_1 = 100\ \text{s}^{-1}$ and that $\gamma = 45°$ occur at *exactly* $\omega = \omega_\phi = 3$. Initially, by use of the straight-line approximation method, determine the values of A, K_H, and α for the feedback compensator

$$H(s) = \frac{K_H(s+0.1)}{s+10/\alpha} \qquad 0.1 \le \alpha \le 10$$

Then by trial and error, with the use of a CAD program, select an α and adjust K_H until $C(s)/I(s)$ is stable and yields $\gamma = 45°$. (f) It is desired, for the compensated system of Fig. 11.20, that $\gamma = 45°$ and the phase margin be determined *essentially* by the characteristics of a feedback unit. Determine if the following proposed feedback unit will result in a stable system. If a stable response can be achieved, determine the value of A that yields the desired value of γ and the corresponding value of ω_ϕ.

$$H(s) = \frac{K_H s^2}{1+s/2} \qquad K_H = 100$$

Note: For all parts determine the figures of merit: M_p, t_p, t_s, K_m, and M_m, and the actual values of ω_ϕ and γ that are achieved.

11.5. The mechanical system shown has been suggested for use as a compensating component in a mechanical system. (a) Determine whether it will function as a lead or a lag compensator by finding $X_2(s)/F(s)$. (b) Sketch $\mathbf{X}_2(j\omega)/\mathbf{F}(j\omega)$.

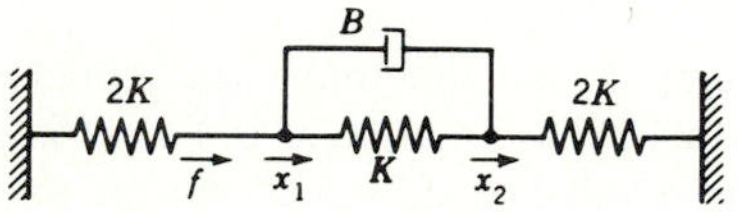

11.6. A unity-feedback control system has the transfer function

$$G(s) = \frac{K}{(1+s)^3} G_c(s)$$

(a) What is the value of the gain of the basic system for an $M_m = 1.12$? (b) Design a cascade compensator $G_c(s)$ that will increase the step-error coefficient by a factor of 8 while maintaining the same M_m. (c) What effect does the compensation have on the closed-loop response of the system? (d) Repeat the design using a minor-loop feedback compensator.

11.7. A unity-feedback control system has the transfer function

$$G(s) = G_c(s)G_1(s)$$

where $G_1(s) = 1/s^2$. It is desired to have an $M_m = 1.4$ with damped natural frequency of about 10.0 rad/s. Design a compensator $G_c(s)$ that will help to meet these specifications.

11.8. A system has

$$G_2(s) = \frac{10(1+0.316s)}{(1+0.1s)(1+0.01s)} \qquad G_3(s) = 1$$

For the control system shown in Fig. (a), determine the cascade compensator $G_c(s)$ required to make the system meet the desired open-loop frequency-response characteristics shown in Fig. (b). What is the value of M_m for the compensated system?

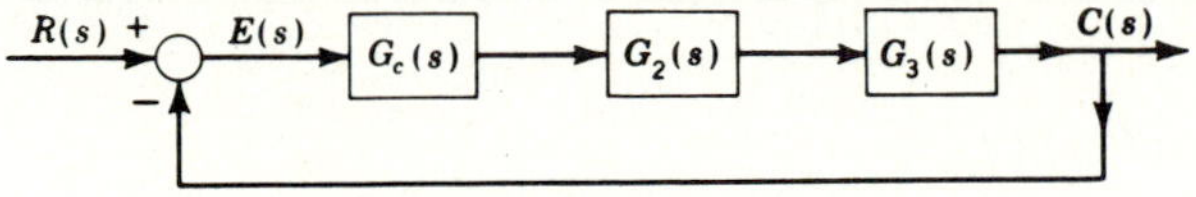

(a)

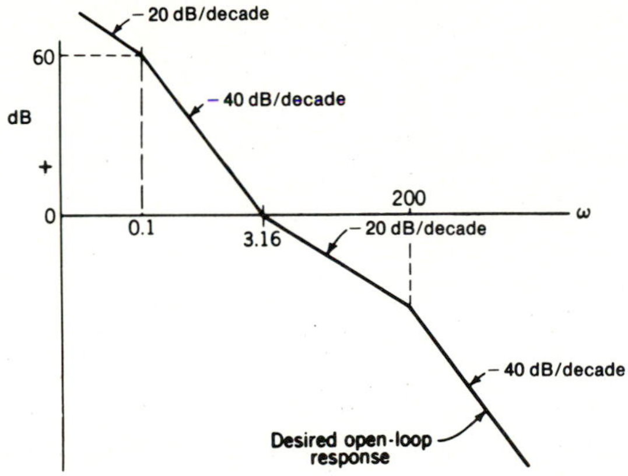

(b)

11.9. A system has the transfer function

$$G_x(s) = \frac{K_1(1 + 4s)}{s(1 + 0.5s)(s + 0.5s + s^2)}$$

(a) For $M_m = 1.26$ determine K_1 and ω_m for the original system. (b) Add a lag compensator with $\alpha = 10$ and determine T, K_1, and ω_m for the same M_m. (c) Add a lead compensator with $\alpha = 0.1$ and determine T, K_1, and ω_m for the same M_m.

11.10. Repeat Prob. 10.1, using $M_m = 1.16$ as the basis of design. Compare the results.

11.11. Repeat Prob. 10.16 but solve by the use of the logarithmic plots. *Note*: Incorporate G_4 into the minor loop. Limit G_1 and G_2 to less than 100.

11.12–11.18. Analyze the 0-dB crossing slope of the straight-line Bode plot of Lm **C**/**E** for system stability.

11.12. A system has

$$G_1(s) = 3 \qquad G_2(s) = 10 \qquad G_3(s) = \frac{1}{1+s} \qquad G_4(s) = \frac{1}{s} \qquad H_1(s) = \frac{s}{1+s}$$

System specifications are $K_1 = 30\ \text{s}^{-1}$, $\gamma \geq +50°$, and $\omega_\phi = 3$ rad/s. Using approximate techniques, (a) determine whether the feedback-compensated system satisfies all the specifications; (b) find a cascade compensator to be added between $G_1(s)$ and $G_2(s)$, with the feedback loop omitted, to produce the same $\mathbf{C}(j\omega)/\mathbf{E}(j\omega)$ as in part (a); (c) determine whether the system of part (b) satisfies all the specifications.

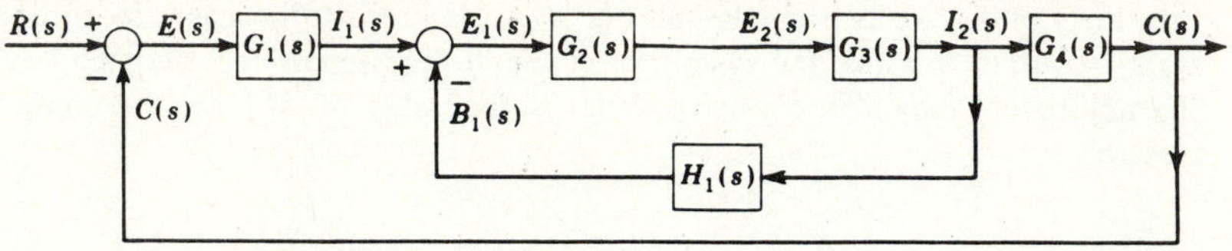

11.13. In the figure of Prob. 11.12,

$$G_1(s) = A_1 \qquad G_2(s) = A_2 \qquad G_3(s) = \frac{1}{(1 + 0.1s)(1 + 0.01s)} \qquad G_4(s) = \frac{1}{s}$$

$H_1(s)$ is a passive network. Choose A_1, A_2, and $H_1(s)$ to meet the desired open-loop response characteristic shown in the figure of Prob. 11.8 for $\omega < 120$.

11.14. For the feedback-compensated system of Fig. 11.20

$$G_x(s) = \frac{10}{s(1+s)} \qquad H(s) = \frac{K_t s^2}{1 + Ts}$$

(*a*) For the original system, without feedback compensation, determine the values of A, K_1, and ω_ϕ for a phase margin of 45°. (*b*) Add the minor-loop compensator $H(s)$ and determine, for a phase margin of 45°, $\omega_\phi = 2$ rad/s, and $K_1 = 100\ \text{s}^{-1}$, the corresponding values of A, K_t, and T. The specifications are to be dictated by $H(s)$. (*c*) Compare the results.

11.15. A servo with tachometric feedback can be represented as shown in Fig. (*a*). $G_x(s)$ represents an ideal servomotor which develops a torque $T = K_T E_1$. The total inertia on the output shaft is J_0, and there is negligible viscous friction. The torque-speed characteristics are shown in Fig. (*b*) for several values of E_1. $H_1(s)$ represents a tachometric generator which develops voltage $B = K_t Dc$ and $G_3(s) = C(s)/E(s)$. (*a*) Find, in terms of system constants, $G_x(s)$, $H_1(s)$, and $G_x(s)$. (*b*) Sketch the polar plots of $\mathbf{G}_x(j\omega)$, $\mathbf{H}_1(j\omega)$, and $\mathbf{G}_3(j\omega)$.

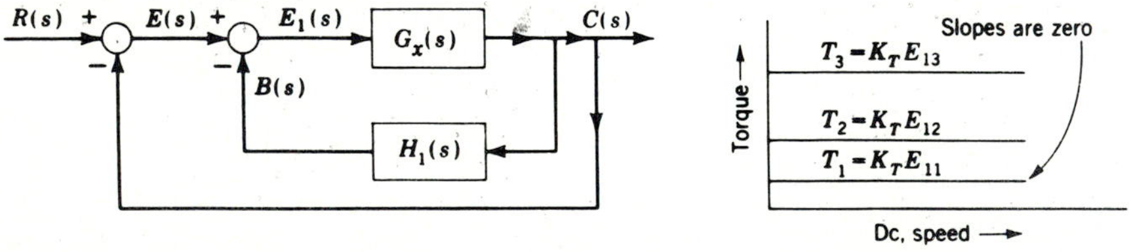

(*a*) (*b*)

11.16. The system represented in the figure is an ac voltage regulator. (*a*) Calculate the response time of the compensated system, when the switch S is closed, to an output disturbance. (*b*) Compare this with the value obtained from the uncompensated system. Comment upon any difference in performance.

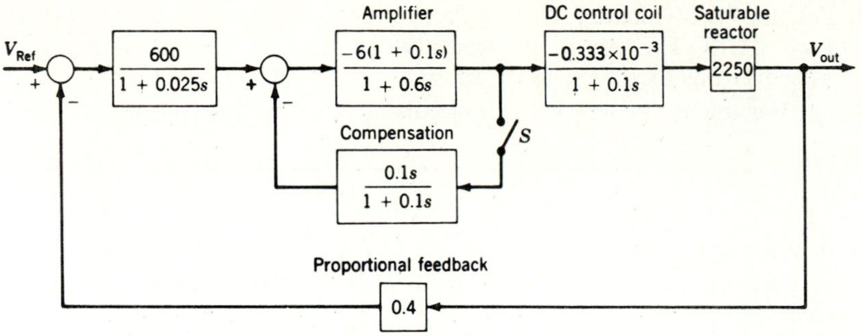

11.17. The simplified pitch-attitude-control system used in a space-launch vehicle is represented in the block diagram. (*a*) Determine the upper and lower gain margins. (*b*) Is the closed-loop system stable?

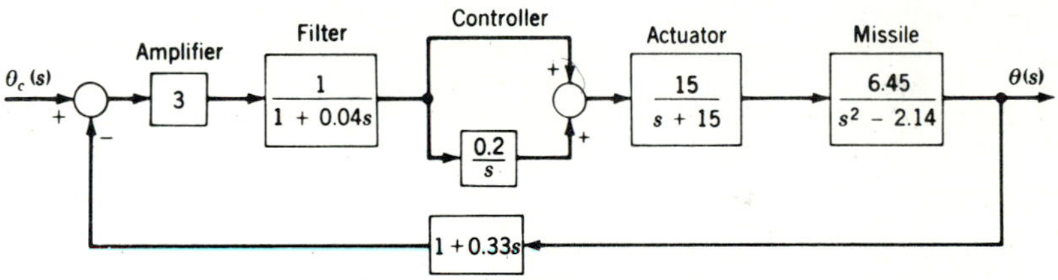

11.18. Repeat Prob. 10.20 using the log plots for $M_m = 1.16$. In determining the value of K_t to use, first determine the range of values it can have in order to have an intersection between Lm $\mathbf{G}_x$ and Lm[$1/\mathbf{H}$] and to maintain a stable open-loop $C(s)/I(s)$ transfer function. A root-locus sketch for $G_x(s)H(s) = -1$ reveals that $C(s)/I(s)$ can have right-half plane poles. Thus, for a stable $C(s)/I(s)$ the 0 dB crossing slope of Lm $\mathbf{G}_x\mathbf{H}$ must be greater than -40 dB/dec. Compare with the results in Probs. 10.3 and 11.1. Obtain the time response. *Hint:* Since $\angle 1/\mathbf{H}(j\omega) = -90°$, then, for $\gamma < 90°$, the new ω_ϕ is close to the intersection of $1/\mathbf{H}(j\omega)$ and $\mathbf{G}(j\omega)$. *Note*: For Prob. 10.20 (1), for $K_t = 1$, the system is unstable if the value of K is set at the value obtained for the uncompensated system to meet the desired value of M_m.

11.19. For the plant $G_x(s)$ of Prob. 10.26, design a lag-lead compensator given by Eq. (11.16). (*a*) Without compensation, determine the gain K_0 to obtain a phase margin of 45°. Determine the phase-margin frequency ω_ϕ. (*b*) Select T_1 for the lag portion of $G_c(s)$ with $\alpha = 10$ so that it contributes 4° at ω_ϕ. (*c*) Select T_2 for the lead portion of $G_c(s)$ with $\alpha = 10$. For a Type 0 system, an empirical location for the compensator zero is at $-2\omega_n$, where ω_n is associated with the complex poles of $G_x(s)$. (*d*) Adjust the gain A of the compensated system $G_c(s)G_x(s)$ for a phase margin of 45°. (*e*) Sketch the root locus of the compensated system. (*f*) Obtain the time response with a unit step input.

11.20. For the plant $G_x(s)$ of Prob. 10.26, add a pole at the origin to make the plant transfer function Type 1. Repeat Prob. 11.19, where the zero of the lead portion of $G_c(s)$ is located at $-0.5\omega_n$.

11.21. For the Type 0 plant of $G_x(s)$ of Prob. 10.26, change to $b = 2$ for all cases. Repeat Prob. 11.19. The zero of the lead portion of $G_c(s)$ may be located at $-0.5\omega_n$ whenever this location is to the left of the real pole $-b$.

11.22. (a) For the uncompensated unity-feedback control system where

$$G_x(s) = \frac{K_1}{s\left(\frac{s}{4} + 1\right)\left(\frac{s}{8} + 1\right)}$$

determine, for $\gamma = 50°$, K_{1_x}, and ω_{ϕ_x}. (b) Assume K_{1_x} has the value determined in part (a) and is not adjustable. For the feedback system of Fig. 11.20 the feedback compensator is given by $\mathbf{H}(j\omega) = K_h\mathbf{H}'(j\omega)$, where $K_h = K_t/T$ and $\mathbf{H}'(j\omega) = (j\omega T)^2/(1 + j\omega T) = \mathbf{H}'(ju) = (ju)^2/(1 + ju)$, *where* $u = \omega T$. Draw plots of $\mathrm{Lm}\,[1/\mathbf{H}'(ju)]$ and $\underline{/1/\mathbf{H}'(ju)}$. Analyze the 0 dB crossing slope of $\mathrm{Lm}\,\mathbf{C}(j\omega)/\mathbf{E}(j\omega)$ for system stability. (c) It is desired, for the compensated system, that $\gamma = 50°$ and the phase-margin frequency $\omega_\phi = 3\omega_{\phi_x}$ be determined *essentially* by the characteristics of the feedback unit. Using straight-line approximations determine, as a first trial, the values of T, K_t, and A that yield this value of γ and the corresponding value of ω_ϕ. (d) Compare the values of M_p, M_m, ω_m, t_p, t_s, and K_1 of part (c) with part (a). For the graphical value of A, what are the actual values of ω_ϕ and γ of part (c)?

Chapter 12

12.1. For the control system of Prob. 10.25 the desired specifications are $1 < M_p \le 1.15$, $t_s \le 2$ s, $t_p \le 1.6$ s, and $K_1 \ge 2.5\ \mathrm{s}^{-1}$. If these specifications cannot be achieved by the basic system, then (a) synthesize a desired control ratio that satisfies these specifications. (b) Using the Guillemin-Truxal method (see Sec. 12.3), determine the required compensator $G_c(s)$. (c) Using state feedback, determine $\mathbf{k}_c$.

12.2. Determine the required passive cascade compensator $G_c(s)$ for each unity-feedback system. The desired closed-loop time response for a step input is achieved by each system having the respective control ratio:

$$(1)\quad M(s) = \frac{20(s + 3)}{(s^2 + 2s + 5)(s + 2)(s + 6)}$$

$$(2)\quad M(s) = \frac{30(s^2 + 1.9s + 4.9025)}{(s^2 + 0.9088s + 2.6928)(s^2 + 2.132s + 5.26945)}$$

Given: System (1)

$$G_x(s) = \frac{10}{s(s^2 + 4s + 8)} \quad \text{and} \quad G_c(s) = \frac{A}{s + b}$$

System (2)

$$G_x(s) = \frac{10}{s(s^2 + 2s + 5)} \quad \text{and} \quad G_c(s) = \frac{A(s^2 + as + b)}{(s + c)(s + d)}$$

Compare M_p, t_p, and t_s for the desired and actual $M(s)$.

12.3. Determine a passive cascade compensator for a unity-feedback system. The desired closed-loop time response for a step input is achieved by the system having the control ratio

$$M(s) = \frac{5.6(s + 4)}{(s^2 + 4s + 8)(s + 2.8)} \qquad G_x(s) = \frac{2}{s(s + 3)} \qquad G_c(s) = \frac{A(s + a)}{s + b}$$

Compare M_p, t_p, t_s, and K_1 for the desired and the actual systems.

12.4. Determine a cascade compensator for a unity-feedback system. The desired closed-loop time response is the same as for Prob. 12.2. The open-loop transfer function is $G_x(s) = 10/s(s+2)(s+4)$. Determine the simplest possible approximate transfer functions that may yield the desired time response.

12.5. For the nonunity-feedback control system of Fig. 10.3,

$$G_x(s) = \left[\frac{K_x \prod_{i=1}^{w}(s+a_i)}{\prod_{j=1}^{n}(s+b_j)}\right]_{n \geq w}$$

$$M(s) = \frac{G_x(s)}{1+G_x(s)H(s)} = \left[\frac{K' \prod_{m=1}^{w'}(s-z_m)}{\prod_{k=1}^{n'}(s-p_k)}\right]_{n' \geq w'}$$

Determine whether a feedback compensator $H(s)$ (other than just gain) can be evaluated by use of the Guillemin-Truxal method such that the order of its denominator is equal to or greater than the order of its numerator.

12.6. Determine a cascade compensator by the Guillemin-Truxal method where

$$G_x(s) = \frac{4(s+2)}{s(s+1)(s+5)} \qquad M(s) = \frac{210(s+1.5)}{(s+1.75)(s+16)(s^2+3s+11.25)}$$

12.7. Repeat Prob. 4.17, where, for disturbance rejection, it is desired that $y(t)_{ss} = 0$ for $D(s) = D_o/s$ with $a_1 = 0$. For this problem $Y(s)$ represents the output of a nonunity-feedback control system, where $G_x(s) = K_x G'(s) = N_1/D_1$ is fixed (the degree of the numerator is w_1 and the degree of the denominator is n_1) and $H(s) = K_h H'(s) = N_2/D_2$ is to be determined (the degree of the numerator is w_2 and the degree of the denominator is n_2). *Hint:* First determine $Y(s)$ in terms of N_1, N_2, D_1, and D_2 and the degree of its numerator and denominator in terms of w_1, w_2, n_1, and n_2. For case (2) of Prob. 4.17, consider $n = w + 1$ and $n > w + 1$.

12.8. The figure below represents a disturbance rejection control system, where

$$G_x(s) = \frac{2}{s(0.5s+1)}$$

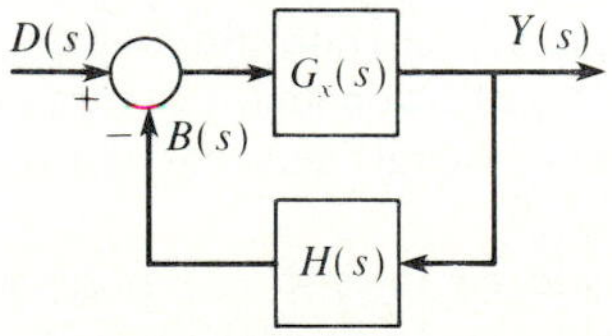

(a) The following feedback unit

$$H(s) = \frac{K_H(s+4)^2}{(s+a)(s+b)} \qquad K_H > 0$$

is proposed to meet the specifications of $y(t)_{ss} = 0$ for $d(t) = u_{-1}(t)$, $|y(t_p)| \leq 0.004$, and $\text{Lm}\,|\mathbf{Y}(j\omega)/\mathbf{D}(j\omega)| \leq -48$ dB $= -\text{Lm}\,\alpha_p$ for $0 \leq \omega \leq 4$ rad/s. Select values of a and b that will satisfy these specifications. (b) *Graphically*, by use of straight-line asymptotes, determine a value of K_H that meets these specifications with the specified $H(s)$. (c) For your value of K_H, determine if the system is stable by use of Routh's stability criterion. Correlate this with the 0 dB crossing slope of $\text{Lm}\,\mathbf{G}_x\mathbf{H}$. ($d$) Obtain a plot of $y(t)$ vs. t and determine α_p and t_p. *Note*: Adjust K_H so that $|y(t_p)| = \alpha_p$.

12.9. For the disturbance rejection control system of Prob. 12.8

$$G_x(s) = \frac{200}{s(s+2)} \qquad \text{and} \qquad H(s) = H_x(s)H_y(s)$$

where

$$H_x(s) = \frac{200}{s+200} \qquad H_y(s) = K_y\frac{N_y(s)}{D_y(s)} = \frac{K_y(s-z_1)(s-z_2)(s-z_3)}{s(s-p_1)(s-p_2)}$$

Determine the feedback compensator $H_y(s)$ such that the control system can satisfy the following specifications for a unit step disturbance input $d(t)$.

$$|y(t_p)| \le 0.002 = \alpha_p \qquad t_p \le 0.15 \text{ s} \qquad y(t)_{ss} = 0$$

and

$$\text{Lm}\left|\frac{\mathbf{Y}(j\omega)}{\mathbf{D}(j\omega)}\right| \le -51 \text{ dB} = -\text{Lm}\,\alpha_p$$

within the bandwidth of $0 < \omega \le 10$ rad/s. Analyze the 0 dB crossing slope of the straight-line Bode plot of $\text{Lm}\,\mathbf{G}_x\mathbf{H}$ for system stability and correlate this to a root-locus sketch of $G_x(s)H(s) = -1$.

Chapter 13

13.1. For a single-input single-output system the open-loop matrix state and output equations are $\dot{\mathbf{x}} = \mathbf{A}\mathbf{x} + \mathbf{b}u$ and $y = \mathbf{c}^T\mathbf{x}$. Derive the overall closed-loop control ratio for a state-variable feedback system with $u = \mathbf{k}^T\mathbf{x}$:

$$\frac{Y(s)}{R(s)} = \mathbf{c}^T[s\mathbf{I} - (\mathbf{A} - \mathbf{b}\mathbf{k}^T)]^{-1}\mathbf{b}$$

13.2. For the system described by

$$D^2y + 6Dy + 5y = 2u$$

(a) draw the simulation diagram. (b) Write the matrix state and output equations for the simulation diagram. (c) Rewrite the state equation and the output equation in uncoupled form. (d) From part (c) determine whether the system is completely controllable and/or observable.

13.3. Determine whether the following systems are completely (a) observable and (b) controllable. (c) Obtain the transfer functions. (d) Determine how many observable states and how many controllable states are present in each system. (e) Are the systems stable?

(1) $$\dot{\mathbf{x}} = \begin{bmatrix} 1 & 1 \\ 2 & 0 \end{bmatrix}\mathbf{x} + \begin{bmatrix} 1 \\ 0 \end{bmatrix}\mathbf{u} \qquad \mathbf{y} = [1 \quad 2]\mathbf{x}$$

(2) $$\dot{\mathbf{x}} = \begin{bmatrix} 0 & 1 & 0 \\ 0 & 0 & 1 \\ 0 & -1 & -2 \end{bmatrix}\mathbf{x} + \begin{bmatrix} 0 \\ 1 \\ 1 \end{bmatrix}\mathbf{u} \qquad \mathbf{y} = [0 \quad 1 \quad 1]\mathbf{x}$$

(3) $\dot{\mathbf{x}} = \begin{bmatrix} 1 & 0 & 0 \\ 0 & 0 & 1 \\ 0 & -2 & -1 \end{bmatrix}\mathbf{x} + \begin{bmatrix} 1 \\ 0 \\ 1 \end{bmatrix}\mathbf{u} \qquad \mathbf{y} = [1 \quad 0 \quad 1]\mathbf{x}$

(4) $\dot{\mathbf{x}} = \begin{bmatrix} 0 & 1 & 0 \\ 0 & 0 & 1 \\ -6 & -11 & -6 \end{bmatrix}\mathbf{x} + \begin{bmatrix} 1 \\ 1 \\ 0 \end{bmatrix}\mathbf{u} \qquad \mathbf{y} = [1 \quad 1 \quad 0]\mathbf{x}$

13.4. A single-input single-output system is described by the transfer function

$$G(s) = \frac{s^3 + 3s^2 + 5s + 8}{s^4 + 7s^3 + 14s^2 + 8s}$$

(*a*) Derive the phase-variable state equation. Draw the block diagram. The $\mathbf{A}_c$ matrix is singular. How is this interpreted physically? (*b*) Represent the system in terms of canonical variables. Draw the corresponding block diagram. (*c*) Is the system observable and controllable?

13.5. A control system is to use state-variable feedback. The plant has the given transfer function. The system is to meet the following requirements: (1) it must have zero steady-state error with a step input, (2) dominant poles of the closed-loop control ratio are to be $-2 \pm j2$, (3) the system is to be stable for all values $A > 0$, and (4) the control ratio is to have low sensitivity to increase in gain A. A cascade element may be added to meet this requirement.

$$G_x(s) = \frac{A}{s(s+1)}$$

(*a*) Draw the block diagram showing the state-variable feedback. Use the form illustrated in Fig. 13.14. (*b*) Determine $H_{eq}(s)$. (*c*) Find $Y(s)/R(s)$ in terms of the state-variable feedback coefficients. (*d*) Determine the desired control ratio $Y(s)/R(s)$. (*e*) Determine the necessary values of the feedback coefficients. (*f*) Sketch the root locus for $G(s)H_{eq}(s) = -1$ and show that the system is insensitive to variations in A. (*g*) Determine $G_{eq}(s)$ and K_1. (*h*) Determine M_p, t_p, and t_s with a step input.

13.6. A control system is to use state-variable feedback. The plant has the given transfer function. The system is to meet the following requirements: (1) it must have zero steady-state error with a step input, (2) the dominant poles of the closed-loop control ratio are supposed to be $-1 \pm j1$, (3) the system is to be stable for all values $A > 0$, and (4) the control ratio is to have low sensitivity to increase in gain A and contains a zero at -1.4. A cascade element may be added to meet this requirement.

$$G_x(s) = \frac{A(s+2)}{s(s+3)(s+5)}$$

(*a*) Determine the desired control ratio $Y(s)/R(s)$. (*b*) Draw the block diagram showing the state-variable feedback. Use the form illustrated in Fig. 13.16 where the pole -1 in the figure is changed to -3 and the transfer function $1/(s+5)$ is replaced by $(s+1.4)/(s+5)$ for this problem. (*c*) Determine $H_{eq}(s)$. (*d*) Find $Y(s)/R(s)$ in terms of the state-variable feedback coefficients. (*e*) Determine the necessary values of the feedback coefficients. (*f*) Sketch the root locus for $G(s)H_{eq}(s) = -1$ and show that the system is insensitive to variations in A. (*g*) Determine $G_{eq}(s)$ and K_1. (*h*) Determine M_p, t_p and t_s with a step input.

13.7. Design a state-variable feedback system for the given plant $G_x(s)$. The desired complex dominant roots are to have a $\zeta = 0.425$. For a unit step function the approximate specifications are $M_p = 1.2$, $t_p = 0.15$ s, and $t_s = 0.3$ s. (a) Determine a desirable $M(s)$ that will satisfy these specifications. Use steps 1 to 6 of the design procedure given in Sec. 13.3 to determine **k**. Draw the root locus for $G(s)H_{eq}(s) = -1$. From it show the "good" properties of state feedback. (b) Obtain $y(t)$ for the final design. Determine the values of the figures of merit and the ramp-error coefficient.

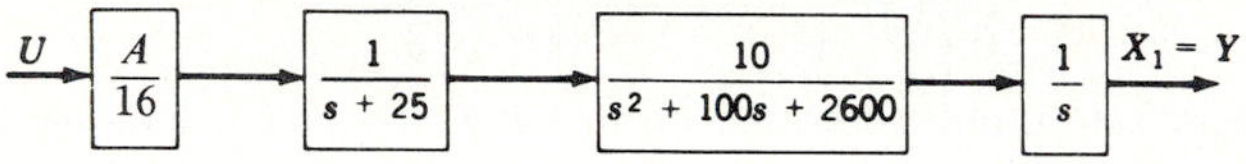

13.8. For the plants of Prob. 10.3 design a state-variable feedback system utilizing the phase-variable representation. The specifications are that dominant roots must have $\zeta = 0.5$, a zero steady-state error for a step input, and $t_s \le 1/3$ s. (a) Determine a desirable $M(s)$ that will satisfy the specifications. Use steps 1 to 6 of the design procedure given in Sec. 13.3 to determine **k**. Draw the root locus for $G(s)H_{eq}(s) = -1$. From it show the "good" properties of state feedback. (b) Obtain $y(t)$ for the final design. Determine the values of the figures of merit and the ramp-error coefficient.

13.9. Design a state-variable feedback system that satisfies the following specifications: (1) For a unit step input $M_p = 1.15$ and $T_s \le 1$ s. (2) The system follows a ramp input with zero steady-state error. (a) Determine $y(t)$ for a step and for a ramp input. (b) For the step input determine M_p, t_p, and t_s. (c) What are the values of K_p, K_v, and K_a?

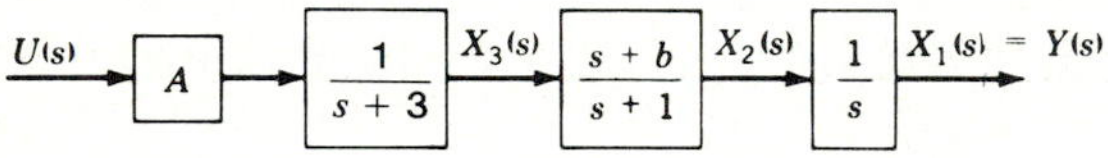

13.10. A closed-loop system with state-variable feedback is to have a control ratio of the form

$$\frac{Y(s)}{R(s)} = \frac{A(s+2)(s+0.5)}{(s^2+2s+2)(s+1)(s-p_1)(s-p_2)}$$

where p_1 and p_2 are not specified. (a) Find the two necessary relationships between A, p_1, and p_2 so that the system has zero steady-state error with both a step and a ramp input. (b) With $A = 100$, use the relationships developed in (a) to find p_1 and p_2. (c) Use the system shown in Fig. 13.17 with $Z_i = -0.5$. Determine the values of the feedback coefficients. (d) Draw the root locus for $G_x(s)H_{eq}(s) = -1$. Does this root locus show that the system is insensitive to changes of gain A? (e) Determine the time response with a step and with a ramp input. Discuss the response characteristics.

13.11.

$$M(s) = \frac{K_G(s+2)(s+b)}{(s^2+2s+2)(s+a)} \qquad b > 0$$

The desired specifications for a stable system are $e(t)_{ss} = 0$ for step and ramp inputs. Determine: (a) the required condition to satisfy the specifications for a step input; (b) the required condition to satisfy the specification for a ramp input. If this specification can be satisfied, give the range of values that a and b can have. (c) Make an appropriate selection of b in order to minimize its effect on $y(t_p)$.

13.12. For the plants shown design a state-variable feedback system that satisfies the following specifications: dominant roots must have a $\zeta = 0.53$, a zero steady-state error for a step input, and $T_s = 0.8$ s. Determine M_p, t_s, t_p, and K_1.

(1)
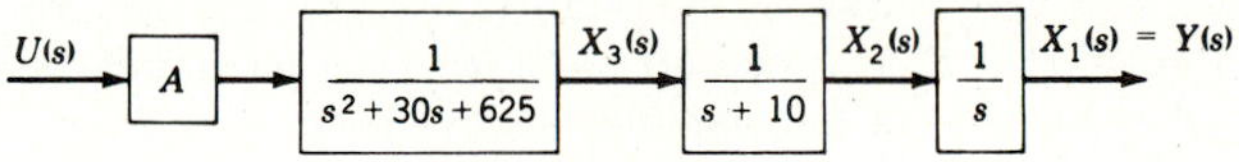

(2)
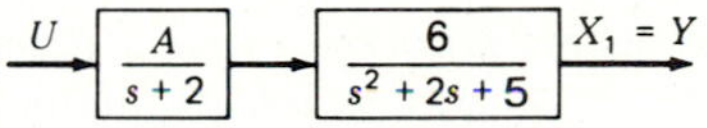

13.13. For the system of Fig. 13.14 it is desired to improve t_s from that obtained with the control ratio of Eq. (13.93) while maintaining an underdamped response with a smaller overshoot and faster settling time. A desired control ratio that yields this improvement is

$$\frac{Y(s)}{R(s)} = \frac{1.428(s + 1.4)}{(s + 1)(s^2 + 2s + 2)}$$

The necessary cascade compensator can be inserted only at the output of the amplifier A. (a) Design a state-variable feedback system that yields the desired control ratio. (b) Draw the root locus for $G(s)H_{eq}(s)$ and comment on its acceptability for gain variation. (c) For a step input determine M_p, t_p, t_s, and K_1. Compare with the results of Examples 1 to 3 of Sec. 13.10.

13.14. Derive Eqs. (13.49) and (13.50).

13.15. For Example 3, Sec. 13.10, determine $G_{eq}(s)$ and the system type. Redesign the state-variable feedback system to make it Type 2.

13.16. The desired control ratio $M(s)_T$ is to satisfy the following requirements: the desired dominant poles are given by $s^2 + 4s + 8$; there are no zeros; for a step input $e(t)_{ss} = 0$; and the system is insensitive to variations in gain A. Determine an $M(s)_T$ that achieves these specifications for the plant

$$G_x(s) = \frac{A(s + 3)}{s(s + 1)(s + 2)(s + 4)}$$

13.17. Repeat Prob. 13.16 with the added stipulation that $e(t)_{ss} = 0$ for a ramp input. A cascade compensator may be added to achieve the desired specifications. A zero may be required in $M(s)_T$ to achieve $e(t)_{ss}$ for a ramp input. Compare the figures of merit with those of Prob. 13.16.

13.18. (a) For the system shown derive the transfer function $Y(s)/U(s)$. (b) If $u(t) = u_{-1}(t)$, sketch the responses of the system variables $x(t)$, $z(t)$, and $y(t)$. (c) A linear system is said to be stable if, in response to a bounded input, the system produces a bounded output. Is this system stable? Is it observable? Discuss your answer briefly.

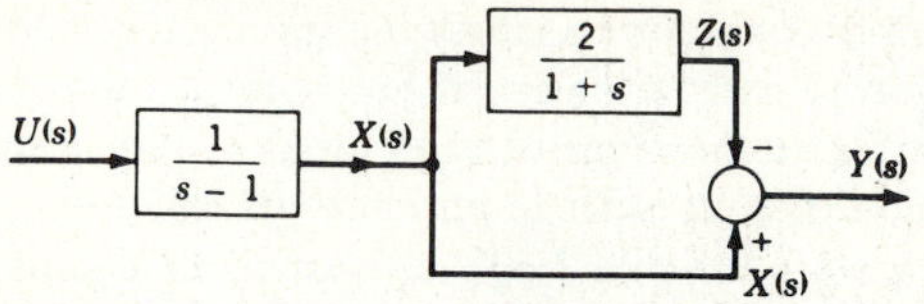

13.19. For the plant $G_x(s) = 2/s(s+1)$ it is desired to design a state-variable feedback control system that satisfies the following specifications: (1) The dominant poles of $M(s)$ are $s_{1,2} = -1 \pm j1$ and $s_3 = -1.5$; (2) $M(s)$ has a zero at -2.0; (3) $e(t)_{ss} = 0$ for $r(t) = R_0 u_{-1}$; and (4) $|\mathbf{S}_A^M(j\omega)|$ is a minimum over the range $0 \le \omega \le \omega_b$. A is the gain of the amplifier which has $U(s)$ as its input. Do the following: (a) Specify an $[M(s)]_T$ that satisfies the above specifications. (b) Determine if it is possible to achieve $[M(s)]_T$ using only the plant and the amplifier A in the final design. If your answer is no, specify the transfer function(s) of the unit(s) that must be added to the system to achieve the desired system performance.

13.20. Given

$$M(s) = \frac{A(s+a)^2}{(s^2+2s+2)(s+2)(s+b)}$$

(a) For $b = 1.5$ it is desired that $e(t)_{ss} = 0$ for both step and ramp inputs. Determine the values of A and a that will satisfy these specifications. (b) With A, a, and b unspecified, determine $G_{eq}(s)$. (c) For part (b) determine the condition that must be satisfied for $G_{eq}(s)$ to be a Type 1 transfer function.

Chapter 14

14.1. A unity-feedback control system has the following forward transfer function:

$$G(s) = \frac{K(s+a)}{s^m(s+b)(s^2+cs+d)}$$

The nominal values are $K = 20$, $a = 6$, $b = 3$, $c = 4$, and $d = 8$. (a) With $m = 1$, determine the sensitivity with respect to (1) K, (2) a, (3) b, (4) c, (5) d. (b) Repeat with $m = 0$.

14.2. The range of values of a is $3 \ge a \ge 1$, $M_p \le 1.10$ for a unit step input, and $T_s = \le 1$ s. Design a state-variable feedback system that satisfies the given specifications and is insensitive to variation of a. Assume that all states x_i are accessible and are fed back directly through the gains k_i. Obtain plots of $y(t)$ for $a = 1$, $a = 2$, and $a = 3$ for the final design. Compare M_p, t_p, t_s, and K_1. Determine the sensitivity functions for variation of A and a (at $a = 1, 2$, and 3).

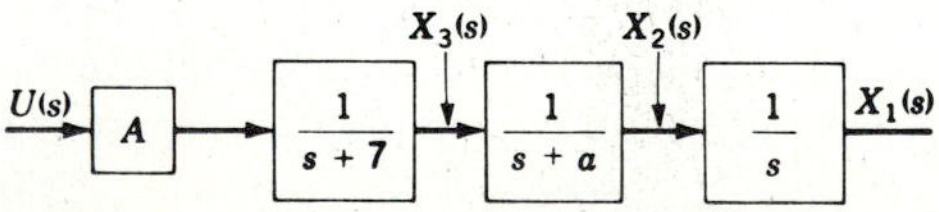

14.3. Repeat Prob. 14.2 but with state X_3 inaccessible.

14.4. Determine the sensitivity function for Example 1 in Sec. 13.10 for variations in A. Insert the values of k_i in Eq. (13.92) and then differentiate with respect to A to obtain S_A^M.

14.5. For Example 2, Sec. 13.10, obtain $Y(s)/R(s)$ as a function of K_G [see Fig. 13.16 with $a = 1$]. Determine the sensitivity function $S_{K_G}^M$.

14.6. For the state-variable feedback system shown the zero term a varies about the nominal value of 5. The feedback gains are $k_1 = 0.4$ and $k_2 = -0.2$. With $b = 2$, determine $S_{\delta=a}^M(s)$ when (a) the system is to be implemented as shown and (b) the system's structure is modified to minimize the sensitivity function. Evaluate $\mathbf{S}_{\delta=5}^{\mathbf{M}}(j0)$ for each case and compare.

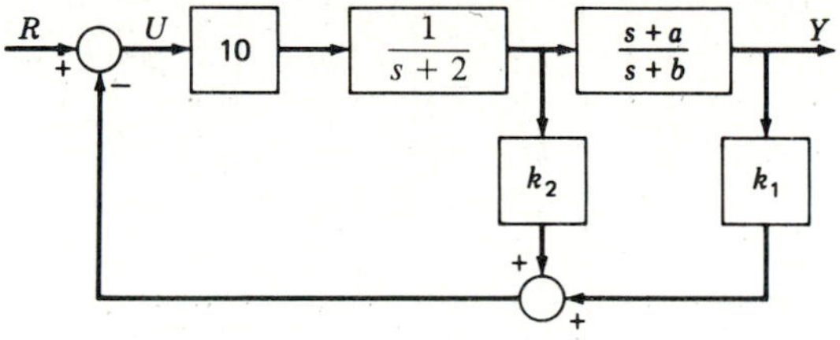

14.7. Repeat Prob. 14.6 where $a = 5$ = constant and b varies about the nominal value of 2.

14.8. The desired control ratio for a tracking system is

$$M(s)_T = \frac{K_G(s + 1.4)}{(s + 1)(s^2 + 2s + 2)(s - p)}$$

This control ratio must satisfy the following specifications: $e(t)_{ss} = 0$ for $r(t) = R_0 u_{-1}(t)$, and $|\mathbf{S}_{K_G}^M(j\omega)|$ must be a minimum over the range $0 \le \omega \le \omega_b$. ($a$) Determine the values of K_G and p that satisfy these specifications. The transfer function of the basic plant is

$$G_x(s) = \frac{K_x(s + 2)}{s(s + 1)(s + 5)} = \frac{K_x(s + 2)}{s^3 + 6s^2 + 5s}$$

A controllable and observable state-variable feedback control system is to be designed for this plant so that it produces the desired control ratio above. (b) Is it possible to achieve $M(s)_T$ using only the plant in the final design? If not, what simple unit(s) must be added to the system to achieve the desired system performance? Specify the parameters of these unit(s). Using phase-variable representation, determine $\mathbf{k}^T$ for the state-variable feedback control system. Assume appropriate values to perform the calculations.

14.9. For the state-feedback system shown, all the states are accessible, the parameters a and b vary, and sensors are available that can sense all state variables. The values of k_1, k_2, k_3, and A are known for a desired $M(s)_T$ based upon nominal values for a and b. It is desired that the system response $y(t)$ be insensitive to parameter variations. (a) By means of a block diagram indicate an implementation of this

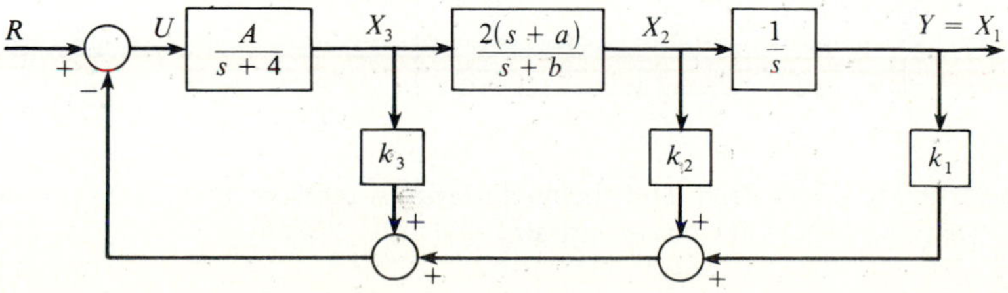

design to minimize the effect of parameter variations on $y(t)$ by use of physically realizable networks. (b) Specify the required feedback transfer functions, ignoring sensor dynamics.

14.10. For the state-feedback control system of Prob. 14.9 with $a = 5$, $b = 1$, and a and b do not vary, the specifications are as follows: $e(t)_{ss} = 0$ for $r(t) = u_{-1}(t)$, $|\mathbf{S}_A^M(j\omega)|$ must be a minimum over the range $0 \le \omega < \omega_b$, and the dominant poles of $M(s)_T$ are given by $s^2 + 2s + 2$. Determine (a) an $M(s)_T$ that achieves these specifications and (b) (the values for K_G, k_1, k_2, and k_3 that yield this $M(s)_T$.

Chapter 15

15.1. Obtain the isocline equation for each of the following differential equations. Draw sufficient isoclines to draw trajectories starting from the point indicated. Determine the kind of singularity and the stability in each case.

(a) $\ddot{x} + 3\dot{x} + 2x = 0$	$x_1 = 0,\ x_2 = 5$
(b) $\ddot{x} + 10\dot{x} + 9x = 0$	$x_1 = 0,\ x_2 = 6$
(c) $\ddot{x} + 5\dot{x} + 6x = 6$	$x_1 = 0,\ x_2 = 5$
(d) $\ddot{x} - 3\dot{x} + 3x = 0$	$x_1 = 4,\ x_2 = 4$
(e) $\ddot{x} + 4\dot{x} + 64x = 0$	$x_1 = 6,\ x_2 = 0$
(f) $\ddot{x} - 8\dot{x} + 17x = 34$	$x_1 = 0,\ x_2 = 4$
(g) $\dot{x} - 4x = 0$	$x_1 = 2,\ x_2 = 1$

15.2. Construct a phase portrait using the isocline method for the following second-order differential equations. Determine the kind of singularity and the stability for each case.

(a) $\ddot{x} + \dot{x}/\lvert\dot{x}\rvert + 2x = 0$	(b) $\ddot{x} + 4\dot{x} + x^2 = 0$
(c) $\ddot{x} + 2\dot{x}^2 + x = 0$	(d) $\ddot{x} + 2\dot{x}\lvert\dot{x}\rvert + 2x = 0$
(e) $\ddot{x} + \dot{x} + 8x\lvert x\rvert = 0$	(f) $\ddot{x} + \dot{x} - 4x\lvert x\rvert = 0$

15.3. Write the state equations for Prob. 15.1 and convert to normal form. Draw the trajectory in the normal plane. For Eqs. (c) and (f) it is simpler to change variables to move the equilibrium point to the origin.

15.4. The nonlinearity has a *dead-zone* characteristic, as shown. Outside the dead-zone region, $-0.1 \le E \le 0.1$, the output E_1 of the nonlinearity is proportional to the input E. For $r(t) = 0$, determine the isocline equation for each region of the phase plane. Draw the phase portrait.

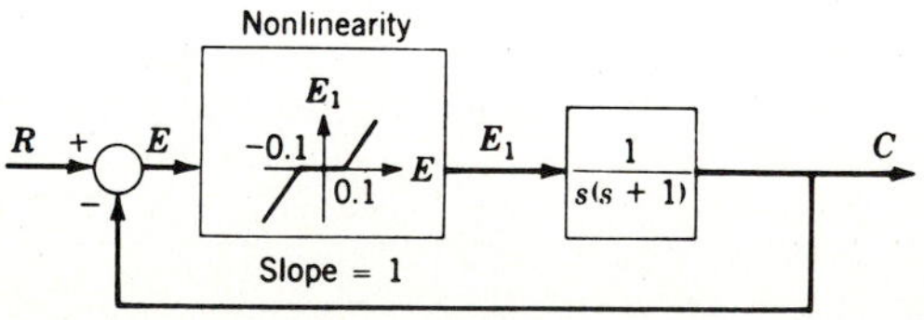

15.5. In the figure for Prob. 15.4 the nonlinearity is replaced by each of the characteristics shown. Draw the phase portrait.

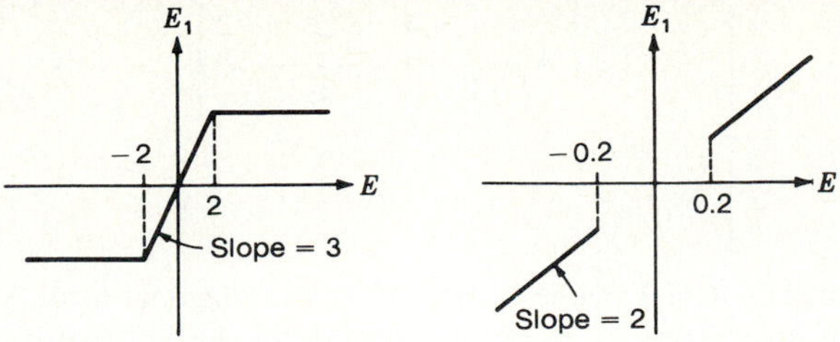

(1) (2)

15.6. The nonlinearity in the block diagram of Prob. 15.4 is replaced by a three-position relay with the characteristic shown. (*a*) For $b = 0.2$ determine the isocline equations for each region: $E > b$, $-b \le E \le b$, $E < -b$. (*b*) Draw several trajectories. (*c*) Repeat for $b = 0.3$.

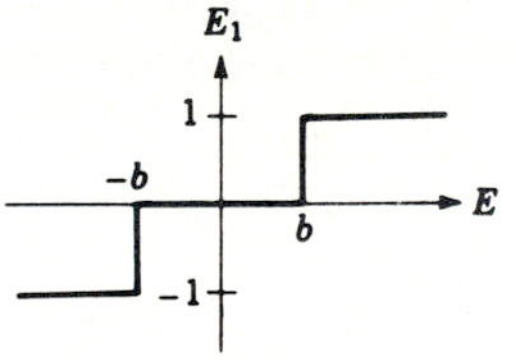

15.7. Use the Jacobian matrix to determine the stability in the vicinity of the equilibrium points for the system described by:

(*a*) $\dot{x}_1 = x_1 + x_1^2 - x_1x_2 + u_1$

$\dot{x}_2 = x_1 - x_2 + u_1u_2$ where $\mathbf{u}_0 = [0 \quad 1]^T$

(*b*) $\dot{x}_1 = 4x_1 - x_1x_2$

$\dot{x}_2 = x_1^2 - x_2$

(*c*) $\dot{x}_1 = -x_1^2 + x_2$

$\dot{x}_2 = 4x_1 - 2x_2$

For parts (*b*) and (*c*), determine the equation for the slope of the phase-plane trajectories.

15.8. For the systems described by the state equations

(1) $\dot{x}_1 = x_1 + x_1^2 - 3x_1x_2 + u_1 \qquad \dot{x}_2 = x_1 - x_2 + u_1u_2 \qquad \mathbf{u}_0 = [1 \quad 0]^T$

(2) $\dot{x}_1 = x_2 \qquad \dot{x}_2 = -x_1 - 3x_2 + 0.25x_3^2 \qquad \dot{x}_3 = x_1x_3 - x_3$

(3) $\dot{x}_1 = x_2 \qquad \dot{x}_2 = -x_2 - 2\sin x_1 + 1$

(*a*) Determine the equilibrium points for each system. (*b*) Obtain the linearized equations about the equilibrium points. (*c*) Determine whether the system is asymptotically stable or unstable in the neighborhood of its equilibrium points. (*d*) Identify the type of equilibrium points.

15.9. Determine the definiteness of the following quadratic forms, $V(\mathbf{x}) = \mathbf{x}^T\mathbf{A}\mathbf{x}$, where

$$(a)\ \mathbf{A} = \begin{bmatrix} 10 & 6 & 4 \\ 0 & 1 & 0 \\ -2 & 0 & 4 \end{bmatrix} \qquad (b)\ \mathbf{A} = \begin{bmatrix} 0 & 1 & 0 \\ 0 & 0 & 1 \\ -10 & -5 & -6 \end{bmatrix}$$

(c) $\mathbf{A} = \begin{bmatrix} 4 & 4 & 0 \\ 3 & 1 & 4 \\ 0 & -2 & 2 \end{bmatrix}$ (d) $\mathbf{A} = \begin{bmatrix} -4 & 2 & -2 \\ 0 & -1 & 0 \\ 0 & 0 & -1 \end{bmatrix}$

(e) $\mathbf{A} = \begin{bmatrix} 2 & 2 & 0 \\ 5 & -1 & 3 \\ 0 & 1 & 2 \end{bmatrix}$

15.10. (a) Transform the matrix **A** of Example 7, Sec. 15.4, into diagonal form. Use this matrix to verify the definiteness of **A**. (b) Determine the Jordan matrix for the matrix **A** of Example 2, Sec. 15.4. Use this matrix to verify the definiteness of **A**.

15.11. Given

$$(1)\quad \dot{\mathbf{x}} = \begin{bmatrix} 0 & 1 & 0 \\ 0 & 0 & 1 \\ -K & -5 & -6 \end{bmatrix}\mathbf{x} \qquad (2)\quad \dot{\mathbf{x}} = \begin{bmatrix} 0 & 1 & 0 \\ 0 & 0 & 1 \\ -K & -1 & -2 \end{bmatrix}\mathbf{x}$$

and $$V(\mathbf{x}) = 6Kx_1^2 + 2Kx_1x_2 + 41x_2^2 + 12x_2x_3 + x_3^2$$

Show that $V(\mathbf{x})$ remains positive definite for the same range of K for which $dV(\mathbf{x})/dt$ is negative semidefinite for each state equation. Also show that global asymptotic stability exists.

15.12.

$$\dot{\mathbf{x}} = \begin{bmatrix} 0 & 1 & 0 \\ 0 & -2 & 1 \\ -K & -1 & 0 \end{bmatrix}\mathbf{x} \qquad V(\mathbf{x}) = 2Kx_1^2 + 2Kx_1x_2 + x_2^2 + x_3^2$$

(a) Determine the range of K for which $\dot{V}(\mathbf{x})$ is negative semidefinite, while also assuring that $V(\mathbf{x})$ is positive definite. (b) Determine whether this system has global asymptotic stability.

15.13. Given

$$\dot{\mathbf{x}} = \begin{bmatrix} -a & K \\ -1 & 0 \end{bmatrix}\mathbf{x} \qquad a > 0$$

Use the Liapunov function of Eq. (15.69) which is positive definite with $p_1 > 0$ and $p_2 > 0$. Determine the range of values of K for global asymptotic stability.

15.14. Use the method of Sec. 15.7 to determine the range of values of K that yields a positive definite matrix **P** in the Liapunov function $V(\mathbf{x}) = \mathbf{x}^T\mathbf{P}\mathbf{x}$, thus assuring asymptotic stability for:

(a) $\mathbf{A} = \begin{bmatrix} -1 & 0 & 1 \\ 2 & -2 & 0 \\ K & 0 & -2 \end{bmatrix}$ $\mathbf{N} = \begin{bmatrix} 0 & 0 & 0 \\ 0 & 2 & 0 \\ 0 & 0 & 0 \end{bmatrix}$

(b) $\mathbf{A} = \begin{bmatrix} -K & 1 & 0 \\ -11 & 0 & 1 \\ -6 & 0 & 0 \end{bmatrix}$ $\mathbf{N} = \begin{bmatrix} 0 & 0 & 0 \\ 0 & 0 & 0 \\ 0 & 0 & 2 \end{bmatrix}$

(c) $\mathbf{A} = \begin{bmatrix} 0 & 1 & 0 \\ 0 & 0 & 1 \\ 0 & -5 & -6 \end{bmatrix}$ $\mathbf{N} = \begin{bmatrix} 0 & 0 & 0 \\ 0 & 0 & 0 \\ 0 & 0 & 2 \end{bmatrix}$

(d) $\mathbf{A} = \begin{bmatrix} -1 & K \\ 0 & -2 \end{bmatrix}$ $\mathbf{N} = \begin{bmatrix} 1 & 0 \\ 0 & 0 \end{bmatrix}$ where $K \neq 0$

(e) For part (a) change a_{33} to $+4$.

15.15. Determine the bound on the settling time for $\mathbf{x}(t)$ to reach and remain within the region $x_1^2 + x_2^2 = 0.02$ for the system represented by

$$A = \begin{bmatrix} 0 & 1 \\ -4 & -2 \end{bmatrix} \qquad \mathbf{x}(0) = \begin{bmatrix} 0 \\ 2 \end{bmatrix}$$

15.16. For the quadratic form $V(\mathbf{x}) = \mathbf{x}^T\mathbf{Q}\mathbf{x}$, where $\mathbf{Q}$ is a symmetric matrix, prove that

$$\frac{dV(\mathbf{x})}{d\mathbf{x}} = 2\mathbf{Q}\mathbf{x}$$

15.17. Determine the equilibrium conditions for $\dot{x}_1 = x_2$ and $\dot{x}_2 = 0$.

Chapter 16

16.1. For the example of Sec. 13.8, determine the feedback coefficients and the gain A in order to yield the response corresponding to a third-order Butterworth form (Table 16.3), with $\omega_0 = 4.642$. Compare the time responses.

16.2.

$$\dot{\mathbf{x}} = \begin{bmatrix} 0 & 1 \\ -2 & -3 \end{bmatrix}\mathbf{x} + \mathbf{B}\mathbf{u} \qquad \mathbf{B}_1 = \begin{bmatrix} 10 \\ 0 \end{bmatrix} \qquad \mathbf{B}_2 = \begin{bmatrix} 10 & 0 \\ 0 & 10 \end{bmatrix} \qquad \mathbf{Z} = \mathbf{I}$$

$$\mathbf{y} = [1 \quad 0]\mathbf{x} \qquad \mathbf{Q}_a = \begin{bmatrix} 1 & 0 \\ 0 & 0 \end{bmatrix} \qquad \mathbf{Q}_b = \begin{bmatrix} 1 & 0 \\ 0 & 1 \end{bmatrix}$$

(a) Find the feedback coefficient matrix for this system for each matrix $\mathbf{B}$. Obtain solutions for each matrix $\mathbf{Q}$ indicated. (b) For the matrix $\mathbf{B}_1$ compare the time responses with a unit step input. [See Prob. 13.1 for $Y(s)/R(s)$.]

16.3. Find the feedback coefficient matrix for the system and PI indicated:

$$\mathbf{A} = \begin{bmatrix} 0 & 1 & 0 \\ 0 & 0 & 1 \\ -6 & -11 & -6 \end{bmatrix} \qquad \mathbf{B} = \begin{bmatrix} 0 \\ 0 \\ 10 \end{bmatrix} \qquad \mathbf{y} = [1 \quad 2 \quad 0]\mathbf{x}$$

$$\text{PI} = \int_0^\infty \left(x_1^2 + 0.01x_2^2 + 0.01x_3^2 + zu^2\right) dt$$

(a) $z = 1$, (b) $z = 10$, (c) $z = 0.1$. (d) Compare the time responses of the system with a step input for each value of z.

16.4. For the example in Sec. 16.9 determine $G(s)H_{\text{eq}}(s)$. Verify that the plot of $\mathbf{G}(j\omega)\mathbf{H}_{\text{eq}}(j\omega)$ is outside the unit circle centered at -1 (see Fig. 16.12).

16.5. Use the results of the example in Sec. 13.8 and the method of Sec. 16.11 [Eq. (16.79)] to determine the approximate value of $\mathbf{h}$. This requires the phase-variable representation of the system. Plot the straight-line curve for Lm $\mathbf{G}_K(j\omega)$. Show whether or not all the corner frequencies are below ω_ϕ.

16.6. For the example of Sec. 16.9, verify that the third root of the characteristic equation is $s_3 \approx -100$.

16.7. A linear-regulator system is to be designed using the method of Sec. 16.9. Use $z = 1$, $\mathbf{h}^T = [1 \quad 0 \quad 0] = \mathbf{c}^T$ [so that $G_K(s) = G(s)$], and $\omega_\phi = 2\omega_{cf}$, where ω_{cf} is the highest corner frequency of $\mathbf{G}(j\omega)$. (a) Determine the value of K_G required. (b) Using the phase-variable representation, determine the feedback coefficients.

$$G(s) = \frac{K_G}{s(s+1)(s+2)}$$

16.8. Repeat Prob. 16.7 using $\mathbf{h}^T = [1 \quad 1 \quad 2]$ and $\omega_\phi = 10\omega_{cf}$, where ω_{cf} is the highest corner frequency of $\mathbf{G}_K(j\omega)$.

16.9. For the system shown with $a = 2$, $b = 3$, and $c = 4$, determine $\mathbf{k}_p$ by use of the Bode diagram. Values of K_G equal to 8 or 100 are proposed, with $z = 1$ and $\mathbf{h}_p^T = [1 \quad 0 \quad 0]$. Give your reason for selecting one value of K_G over the other. Compare with the Riccati solution.

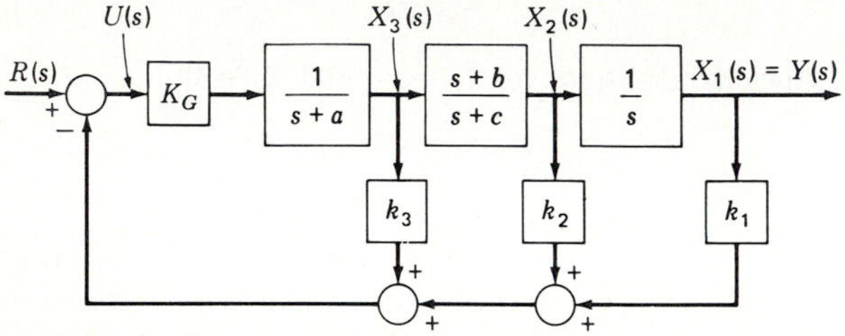

16.10. Repeat Prob. 16.9 with $a = 10$, $b = 4$, and $c = 2$. The values of K_G are 8 or 1600.

16.11. Repeat Probs. 16.9 and 16.10 with $\mathbf{h}_p^T = [1 \quad 0.1 \quad 0.1]$. Select $\omega_\phi \geq 20\omega^*$.

16.12. In the state-variable control system of Prob. 16.9 the states x_2 and x_3 are inaccessible, and noise is present throughout. (*a*) Derive a minor-loop compensator that permits a physical *realization of this system*. (*b*) State reasons for the minor-loop compensator chosen.

16.13. Repeat Probs. 16.7 to 16.11 by the RSL method. Use the values of K_G obtained in those problems.

16.14. A state feedback system is to be designed for a plant whose transfer function is

$$G(s) = \frac{K_G(s+4)}{s(s+2)(s+10)}$$

which is to satisfy

$$\mathrm{PI} = \int_0^\infty (\mathbf{x}^T\mathbf{Q}\mathbf{x} + u^2)\, dt$$

where $\mathbf{Q} = \mathbf{h}\mathbf{h}^T$. The RSL for this system, for $\mathbf{h}^T = [2 \quad 0 \quad 0]$, $K_G = 200$, and phase-variable representation, yields the following roots:

$$\pm 4.0203 \pm j4.5895 \text{ and } \pm 10.630$$

Determine the value of $\mathbf{k}$.

16.15. Use the system in the example of Sec. 16.9 with $A = 50$ and represent it in terms of phase variables. Use Eq. (16.79) to evaluate $\mathbf{k}$. Transform to physical variables to obtain $\mathbf{k}_p$. Compare with the values obtained in Sec. 16.9.

16.16. Repeat Prob. 16.11 using Eq. (16.79) to evaluate $\mathbf{k}$ in terms of phase variables. Transform to physical variables to obtain $\mathbf{k}_p$. Compare with the values obtained in Prob. 16.11.

16.17. By the method of Sec. 16.9 determine $\mathbf{k}^T$, where $\mathbf{h}^T = [9 \quad 6 \quad 1 \quad 0]$, $z = 1$, and $\omega_\phi = 10\omega^*$ for the plant

$$G(s) = \frac{K_G(s+3)^2}{s^2(s+1)(s+2)}$$

Hint: Determine $\mathbf{c}$.

16.18. Given

$$G(s) = \frac{K_G(s+4)}{s(s+2)(s+10)} \qquad \omega_\phi = 5\omega^* \qquad \mathbf{h} = \mathbf{c} \qquad \text{and} \qquad z = 1$$

Determine if a minimum to Eq. (16.22) exists. If it does, apply the method of Sec. 16.9 to determine $\mathbf{k}^T$, K_G, and ω_ϕ.

16.19. By the method of Sec. 16.11, assuming $K_G = 1600$ and $\mathbf{h}_c^T = [4 \quad 0 \quad 1]$, determine k_c for Prob. 16.10.

16.20. Find the feedback coefficient matrix for the following SISO system

$$\mathbf{A} = \begin{bmatrix} 0 & 1 \\ 0 & 0 \end{bmatrix} \qquad \mathbf{b} = \begin{bmatrix} 0 \\ 10 \end{bmatrix} \qquad \mathbf{Q} = \begin{bmatrix} 1 & 0 \\ 0 & 1 \end{bmatrix} \qquad z = 1$$

Chapter 17

17.1. Given $G(s) = K_G/(s+1)(s+2)(s+3)$. The desired characteristic polynomial is $(s^2 + 2s + 2)(s + \alpha)$. ($a$) Under the hfg condition and with $z = 1$ determine the $\mathbf{k}_c$ required to produce an equivalent Type 1 system. Find $G_{\text{eq}}(s)$ and verify the system type. (b) Using the $\mathbf{k}_c$ and K_G obtained in (a), use the method of Sec. 16.11 to determine $\mathbf{h}_c$. Is the system optimal? (c) Determine $y(t)$ with a step and with a ramp input. (d) For the physical configuration shown, find $\mathbf{k}_p$.

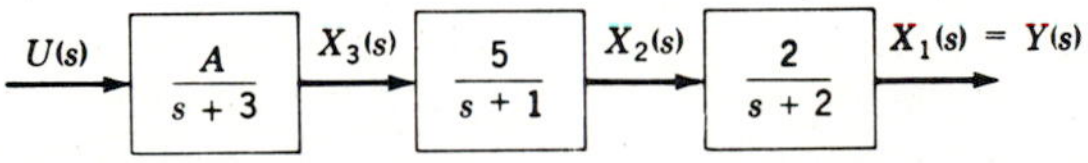

17.2. Repeat Prob. 17.1(a)–(c) for $G(s) = K_G/(s+1)(s+3)(s+4)$.

17.3. Repeat Prob. 17.1(a)–(c) for $G(s) = K_G/s(s+0.1)(s+0.5)$.

17.4. Given $G(s) = K_G(s+3)(s+4)/[s(s+1)^2(s+2)]$. ($a$) Determine the value of $\mathbf{k}$ under hfg conditions that yields zero steady-state error for the three standard type inputs using $z = 1$. Find $G_{\text{eq}}(s)$ and determine the system type. (b) Utilizing the value of $\mathbf{k}$ and K_G determined in (a), use the design method of Sec. 16.11 to determine the value of $\mathbf{h}$. Does this value of $\mathbf{h}$ yield an optimal performance of the system? (c) Determine $y(t)$ for a unit step input and determine M_p, t_p, and t_s. Also determine the error $e(t)$ for ramp and parabolic inputs. (d) The physical configuration of this system is shown. Determine the value of $\mathbf{k}_p$ that is required to implement the design values of part (a).

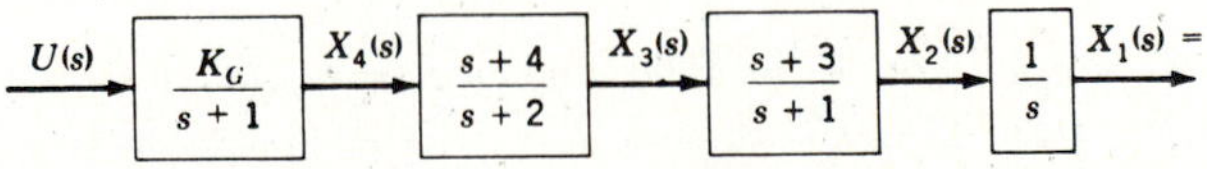

17.5. Repeat Prob. 17.4 for the following plant:

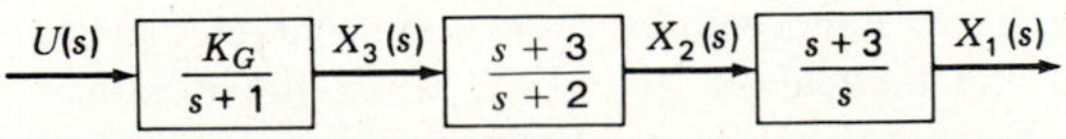

17.6. Repeat parts (a) through (c) of Prob. 17.4 for the plant of Prob. 17.5 and for $z = 0.5$ and $z = 2$.

17.7. The plant of Prob. 17.1 is to be optimized utilizing the PI of Eq. (17.7) with $z = 1$. The desired performance is $M_p = 1.04$, $t_p = 2$ s, and $e(t)_{ss} = 0$ with a unit step input. (*a*) Determine the minimum value of K_G required to use the nomogram of Fig. 17.3. (*b*) For this K_{min} determine the matrix **Q** from the nomogram, the **k** that results, and the actual $y(t)$. Compare the actual values of M_p, t_p, and t_s with those of the nomogram. Does the comparison indicate that a higher value of K_G is required? (*c*) For $z = 0.5$ and $z = 2$ determine the value of q_1 required to achieve $e(t)_{ss} = 0$ with a unit step input and the new value of K_G while maintaining the same value of K_{min}. (*d*) For each z determine **k** and $y(t)$ with a unit step input. Use the values of q_2 and q_3 obtained in part (*b*) and the appropriate value of q_1. (*e*) Compare the values of M_p, t_p, and t_s for all three values of z.

17.8. Utilize the plant of Prob. 16.1, the PI of Eq. (17.7), and the matrix

$$\mathbf{Q} = \begin{bmatrix} 1 & & \mathbf{0} \\ & 0 & \\ \mathbf{0} & & 0 \end{bmatrix}$$

Select $z = 0.5$, 1, or 2 and a set of values of K_G. Determine the corresponding **k** and $y(t)$. Each person in a class can be assigned to a different value of z and K_G. For each value of z, plot the values of M_p, t_p, and t_s vs. K_G using the format of Fig. 17.4. If Probs. 17.1 and/or 17.3 have been worked, compare the values of M_p, t_p, and t_s for the different **Q** matrices used.

17.9. A plant is to be optimized by use of the quadratic cost function with $z = 1$. The desired response is $M_p = 1.1$, $t_p = 0.6$ s, and $e(t)_{ss} = 0$ with a unit step input. (*a*) Using Fig. 17.6 and a computer program, determine the numerator coefficient a that will satisfy these specifications. (*b*) Determine the feedback coefficients and $y(t)$.

$$G(s) = \frac{K_G(s+a)}{s(s+1)(s+5)}$$

17.10. Repeat Prob. 17.9 for $z = 0.5$ and $z = 2$ and maintain the same value of K_{min}. In each case reevaluate K_G and q_1 and use the values of q_2 and q_3 obtained in Prob. 17.9. For each z determine **k** and $y(t)$ with a unit step input. Compare the values of M_p, t_p, and t_s for all three values of z.

17.11. State-variable feedback is to be used to design a system having the plant transfer function

$$G(s) = \frac{K_G(s+10)}{s(s^2+4s+4)}$$

Assume that $K_x = 500$. For $K_G = 1000$ and $r(t) = u_{-1}(t)$ it is desired that $e(t)_{ss} = 0$, $M_p \approx 1.055$, $t_p \approx 0.67$ s, and $t_s \approx 1$. The system is to be optimal according to Eq. (17.7) with $z = 1$. (*a*) Determine the matrix **Q** that satisfies the desired performance specification (see Fig. 17.6). (*b*) Determine the value of the feedback coefficient k_1 and the approximate value of k_3. (*c*) Determine the approximate value of the nondominant root of the characteristic equation. (*d*) Determine k_2, using the method of Sec. 17.8, when the dominant roots are $-3.973 \pm j3.974$. (*e*) Verify the values k_i obtained and the desired time-response characteristics by using a computer solution to Eq. (17.7).

17.12. A controllable system having the plant transfer function shown is to be designed to satisfy Eq. (17.7). Find the values of q_1 and k_1 to satisfy $e(t)_{ss} = 0$ for a step input for $z = 1$.

$$G(s) = \frac{10}{(s+1)(s+2)^2}$$

17.13. Use the design procedure of Secs. 17.11 and 16.11. (a) Determine **h** and **k** for $z = 1$.

$$G(s) = \frac{K_G}{s^2(s+1)^2} \qquad M(s)_T = \frac{200}{(s+2)(s^2+1.4117s+1)(s+100)}$$

(b) Use a computer program to determine **k** for the **h** determined in part (a). (c) Determine $y(t)$ with a unit step input for each **k** and compare with that for $M(s)$.

17.14. Repeat Prob. 17.13 with

$$(1)\ G_x(s) = \frac{K_G(s+0.5)}{s^2(s+1)^2} \qquad M(s)_T = \frac{200}{(s^2+2s+2)(s+100)}$$

$$(2)\ G_x(s) = \frac{K_G(s+3)}{s(s+1)^2(s+5)} \qquad M(s)_T = \frac{160(s+2)}{(s^2+2s+2)(s+1.6)(s+100)^2}$$

Note: For plant (2), determine if $M(s)_T$ can be achieved with $G_x(s)$ alone. If not, what must be done to the system to achieve $M(s)_T$?

17.15.

$$G_x(s) = \frac{400(s+2)(s+1)}{s^3(s+3)(s+5)} = \frac{400(s^2+3s+2)}{s^3+8s^4+15s^3} \qquad z = 1$$

The desired control ratio is to have no zeros, and the only dominant poles are to be $-1 \pm j1$. Use the method of Sec. 17.11 to solve for **h**. (Specify the desired control ratio.) Then use Eq. (16.79) to determine **k**. *Note*: The system is to use only the basic plant $G_x(s)$. Assume that your **h** satisfies Eq. (16.28).

17.16. Design a state feedback control system for the given plant that achieves the same figures of merit as does $M(s)$. Use the design procedures of Secs. 16.11 and 17.11.

$$G_x(s) = \frac{K_G(s+2)}{s^4} \qquad M(s) = \frac{8}{s^2+4s+8}$$

(a) Determine an $M_T(s)$ that achieves the design objectives for the control system whose plant is $G_x(s)$. (b) For $z = 1$ determine **h** and **k**.

17.17. The desired control ratio $M(s)_T$ for a state-variable feedback system having the basic plant

$$G_x(s) = \frac{K_G(s+3)}{s(s+1)^2(s+5)} = \frac{K_G(s+3)}{s^4+7s^3+11s^2+5s} \qquad z = 1$$

is (1) to have no zeros, (2) to have the dominant poles $-1 \pm j1$, and (3) to yield $e(t)_{ss} = 0$ for a step input. Use the method of Sec. 17.11 to determine **k**. (a) Specify $M(s)_T$. (b) Solve for **h**. (c) Use Eq. (17.59) to determine **k**.

17.18. A controllable and observable state-variable feedback control system, utilizing the given basic plant $G(s)$, is to be designed by the method of Sec. 17.11.

$$G(s) = \frac{K_G(s+2)(s+1)}{s(s+3)^2(s+5)^2}$$

The system must (1) satisfy the quadratic PI (with $z = 1$) and (2) yield $e(t)_{ss} = 0$ for both a step and a ramp input. Also, (3) the control ratio may contain only one zero at $-2 + j0$ and must have the dominant poles $-4 \pm j4$. (*a*) Specify $[Y(s)/R(s)]_T$. (*b*) Determine **h**. (*c*) By use of Eq. (16.79) determine **k**. (*d*) Determine $e(t)_{ss}$ for step and ramp inputs for the $\mathbf{k}_c$ of part c.

Chapter 18

18.1. The states are inaccessible for the plant of Fig. 14.2. (*a*) Synthesize a full-order observer which generates estimates of the plant states $\hat{\mathbf{x}}$. Select suitable values of the observer eigenvalues. (Use $\sigma(\mathbf{A} - \mathbf{LC}) = \{-5, -10, -20\}$.) (*b*) Compute the feedback matrix **K** which assigns the set of closed-loop eigenvalues $\{-0.708 + j0.706, -0.708 - j0.706, -100\}$. (*c*) Compute the state responses with the input $r(t) = u_{-1}(t)$, $\mathbf{x}(0) = [1 \quad 1 \quad 0]^T$, and the observer initially quiescent. Use the values $p_1 = -1$, $p_2 = -5$, and $A = 10$.

18.2. Repeat Prob. 18.1 for the plant represented by

$$\dot{\mathbf{x}} = \begin{bmatrix} 0 & 1 \\ -2 & -3 \end{bmatrix}\mathbf{x} + \begin{bmatrix} 0 \\ 1 \end{bmatrix}\mathbf{u} \qquad \mathbf{y} = [1 \quad 0]\mathbf{x}$$

Assign the set of closed-loop plant eigenvalues as $\{-4, -6\}$. The initial plant states are $\mathbf{x}(0) = [1 \quad -1]^T$ and the observer is initially quiescent.

18.3. Repeat Prob. 18.2 for the plants of Prob. 5.19 with $\mathbf{y} = [1 \quad 2]\mathbf{x}$.

18.4. A SISO system represented by the state equation $\dot{\mathbf{x}}_p = \mathbf{A}_p\mathbf{x}_p + \mathbf{b}_p\mathbf{u}$ has the matrices

$$\mathbf{A}_p = \begin{bmatrix} -1 & 0 & 0 \\ -2 & -2 & -2 \\ -1 & 0 & -3 \end{bmatrix} \qquad \mathbf{b}_{p_1} = \begin{bmatrix} 1 \\ 0 \\ 1 \end{bmatrix} \qquad \mathbf{b}_{p_2} = \begin{bmatrix} 1 \\ 1 \\ 0 \end{bmatrix}$$

For each $\mathbf{b}_p$: (*a*) Determine controllability; (*b*) find the eigenvalues; (*c*) find the transformation matrix **T** which transforms the state equation to the control canonical (phase-variable) form; (*d*) determine the state feedback matrix $\mathbf{k}_p^T$ required to assign the eigenvalue spectrum $\sigma(\mathbf{A}_{cl}) = \{-4, -5, -6\}$.

18.5. For the system of Prob. 18.4, determine the feedback matrix **K** which assigns the closed-loop eigenvalue spectrum $\sigma(\mathbf{A}_{cl}) = \{-1 + j1, -1 - j1, -2\}$.

18.6. $\dot{\mathbf{x}} = \mathbf{A}\mathbf{x} + \mathbf{B}\mathbf{u}$

$$(1)\ \mathbf{A} = \begin{bmatrix} -1 & 0 & 0 \\ 1 & -2 & 0 \\ -1 & 0 & -4 \end{bmatrix} \qquad \mathbf{B} = \begin{bmatrix} 0 & 1 \\ -1 & 0 \\ 1 & 0 \end{bmatrix}$$

$$(2)\ \mathbf{A} = \begin{bmatrix} -1 & 0 & 0 \\ -2 & -2 & -2 \\ -1 & 0 & -3 \end{bmatrix} \qquad \mathbf{B} = \begin{bmatrix} 1 & 0 \\ 0 & 2 \\ 0 & 2 \end{bmatrix}$$

Determine (*a*) controllability and (*b*) the control invariants. Transform the state equation to the generalized control canonical form and determine (*c*) the required transformation matrices **T** and **F** and (*d*) the required feedback matrix **H**. (*e*) Determine the state feedback matrix **K** to assign the closed-loop eigenvalue spectrum $\sigma(\mathbf{A}_{cl}) = \{-4, -5, -6\}$. (*f*) Obtain $\mathbf{A}_{cl}$. (*g*) Compute the time response with zero input and $\mathbf{x}(0) = [1 \quad 0 \quad 1]^T$.

18.7. For the state equation in Prob. 18.6, identify the permissible subspaces in ker $\mathbf{S}(\lambda)$ in which the eigenvectors must be located. Do not use specific values of λ. (See Sec. 19.2.)

18.8. Repeat Prob. 18.6 for

$$(1)\ \mathbf{A} = \begin{bmatrix} 1 & 1 & 0 \\ 1 & 0 & 1 \\ 0 & 0 & 1 \end{bmatrix} \qquad \mathbf{B} = \begin{bmatrix} 0 & 1 \\ 1 & 0 \\ 0 & 1 \end{bmatrix}$$

$$(2)\ \mathbf{A} = \begin{bmatrix} 1 & 0 & 0 \\ 0 & 1 & 0 \\ 0 & -2 & 1 \end{bmatrix} \qquad \mathbf{B} = \begin{bmatrix} 1 & 0 \\ 0 & 1 \\ 1 & 0 \end{bmatrix}$$

Find $\mathbf{A}_{cl} = \mathbf{A} + \mathbf{BK}$.

18.9. Repeat Prob. 18.7 with the state equation matrices of Prob. 18.8.

18.10. $\dot{\mathbf{x}} = \mathbf{Ax} + \mathbf{Bu}$, where

$$(1)\ \mathbf{A} = \begin{bmatrix} -2 & 1 & 0 & 0 \\ -2 & 0 & 1 & 0 \\ 0 & 1 & 1 & 0 \\ 0 & 0 & 0 & 1 \end{bmatrix} \qquad \mathbf{B} = \begin{bmatrix} 0 & 0 \\ 0 & 0 \\ 1 & 1 \\ 0 & 1 \end{bmatrix}$$

$$(2)\ \mathbf{A} = \begin{bmatrix} 1 & 1 & 0 & 0 \\ 0 & 1 & 0 & 0 \\ 1 & 0 & 0 & 1 \\ 0 & 1 & 0 & 1 \end{bmatrix} \qquad \mathbf{B} = \begin{bmatrix} 1 & 0 \\ 1 & 0 \\ 0 & 0 \\ 0 & 1 \end{bmatrix}$$

(*a*) Determine controllability. Transform the state equation to the generalized control canonical form and determine (*b*) the control invariants, (*c*) the required transformation matrices **T** and **F**, (*d*) the required feedback matrix **H**, and (*e*) the feedback matrix **K** which assigns the closed-loop eigenvalue spectrum

$$\sigma(\mathbf{A}_{cl}) = \{-5 + j2, -5 - j2, -6, -10\}$$

(*f*) Obtain the time response of the closed-loop system with zero input and the initial conditions $\mathbf{x}(0) = [1 \quad 0 \quad 0 \quad 1]^T$. *Hint:*

$$\begin{bmatrix} \mathbf{M}_1 & \mathbf{0} \\ \mathbf{0} & \mathbf{M}_2 \end{bmatrix}^{-1} = \begin{bmatrix} \mathbf{M}_1^{-1} & \mathbf{0} \\ \mathbf{0} & \mathbf{M}_2^{-1} \end{bmatrix}$$

18.11. Repeat Prob. 18.7 with the state equation matrices of Prob. 18.10.

18.12. Repeat Prob. 18.4 with

$$\mathbf{A}_p = \begin{bmatrix} 0 & 1 & 0 \\ 0 & -1 & -3 \\ 0 & 0 & -5 \end{bmatrix} \qquad \mathbf{b}_p = \begin{bmatrix} 0 \\ 2 \\ 2 \end{bmatrix} \qquad \sigma(\mathbf{A}_{cl}) = \{-1 + j1, -1 - j1, -2\}$$

18.13. Repeat Prob. 18.6 with

$$\mathbf{A} = \begin{bmatrix} 1 & 1 & 1 \\ 0 & 0 & 1 \\ 0 & -2 & 1 \end{bmatrix} \qquad \mathbf{B} = \begin{bmatrix} 1 & 1 \\ 0 & 0 \\ 3 & 0 \end{bmatrix} \qquad \sigma(\mathbf{A}_{cl}) = \{-1 + j1, -1 - j1, -4\}$$

18.14. Repeat Prob. 18.7 with the matrices of Prob. 18.13.

18.15. Given

(1) $$\mathbf{M}^{-1} = \begin{bmatrix} -0.5 & 1 & 0.5 \\ 1 & 0 & 0 \\ 0.5 & 0 & -0.5 \end{bmatrix} \qquad \mathbf{A} = \begin{bmatrix} 1 & 1 & 0 \\ 1 & 0 & 1 \\ 0 & 0 & 1 \end{bmatrix}$$

$\gamma_1 = 2$, and $\gamma_2 = 1$. Determine the transformation matrix **T** which converts **A** to GCCF where $\mathbf{x} = \mathbf{T}\mathbf{z}$. Do *not* perform the transformation on **A**.

(2) $\dot{\mathbf{x}}_p = \mathbf{A}_p\mathbf{x}_p + \mathbf{B}_p\mathbf{u}$

$$\mathbf{A}_p = \begin{bmatrix} 1 & 4 \\ -5 & 3 \end{bmatrix} \qquad \mathbf{B}_p = \begin{bmatrix} 1 \\ 0 \end{bmatrix} \qquad \mathbf{C}_p^T = [1 \quad 0]$$

Design a state feedback control law $u = r - \mathbf{K}_p\mathbf{x}_p$ which assigns the closed-loop eigenvalue spectrum $\sigma(\mathbf{A}_{cl}) = \{-5, -7\}$.

Chapter 19

19.1. For the system of Prob. 18.7, assign the eigenvalue spectrum $\sigma(\mathbf{A}_{cl}) = \{-4, -5, -6\}$. (*a*) Select linearly independent eigenvectors from the subspaces determined by the results of Prob. 18.7. (*b*) Determine the required state feedback matrix **K**. (*c*) Obtain the time response with $\mathbf{x}(0) = [1 \quad 0 \quad 1]^T$. Compare with the results of Prob. 18.6.

19.2. Repeat Prob. 19.1 for the system of Prob. 18.8, using the results of Prob. 18.9.

19.3.

$$\dot{\mathbf{x}} = \mathbf{A}\mathbf{x} + \mathbf{B}\mathbf{u} \qquad \mathbf{A} = \begin{bmatrix} -1 & 0 & 0 \\ -2 & -2 & -2 \\ -1 & 0 & -3 \end{bmatrix} \qquad \mathbf{B} = \begin{bmatrix} 1 & 0 \\ 0 & 2 \\ 0 & 1 \end{bmatrix}$$

(*a*) Determine the uncontrollable mode. (*b*) Identify the permissible subspace in $\ker \mathbf{S}(\lambda)$ in which the eigenvector must be located. (*c*) Assign the eigenvalue spectrum $\sigma(\mathbf{A}_{cl}) = \{-2, -4, -5\}$. (*c*) Compute $\mathbf{A}_{cl}$.

19.4. Repeat Prob. 19.1 for the system of Prob. 18.10, using the results of Prob. 18.11 with $\sigma(\mathbf{A}_{cl}) = \{-1, -2, -3, -4\}$ and $\mathbf{x}(0) = [1 \quad 0 \quad 0 \quad 1]^T$.

19.5. Repeat Prob. 19.1 for the system of Prob. 18.13, using the results of Prob. 18.14.

19.6. The system represented by

$$\dot{\mathbf{x}} = \begin{bmatrix} 1 & 6 & -3 \\ -1 & -1 & 1 \\ -2 & 2 & 0 \end{bmatrix}\mathbf{x} + \begin{bmatrix} 1 \\ 1 \\ 1 \end{bmatrix}\mathbf{u} \qquad \mathbf{y} = [1 \quad 0 \quad 1]\mathbf{x}$$

is required to follow a step input $r(t) = u_{-1}(t)$. (*a*) Determine the composite system matrices. (*b*) Is this system controllable? (*c*) Determine the feedback matrix $\overline{\mathbf{K}}$ required to assign the eigenvalue spectrum $\sigma(\overline{\mathbf{A}} + \overline{\mathbf{B}}\overline{\mathbf{K}}) = \{-2, -3, -4, -5\}$. (*d*) For the initially quiescent system compute the time response and show that $y(t)_{ss}$ does follow $r(t)$.

19.7. Repeat Prob. 19.6 for the system of Prob. 18.4 and $\mathbf{y} = [1 \quad 0 \quad 1]\mathbf{x}$.

19.8. Repeat Prob. 19.6 for the system of Prob. 18.12 and $\mathbf{y} = [1 \quad 1 \quad 0]\mathbf{x}$.

19.9. Repeat Prob. 19.6 for the system of Prob. 18.6 with $r(t) = [1 \quad -1]^T$, $\mathbf{y} = \begin{bmatrix} 1 & 0 & 1 \\ 0 & 1 & 1 \end{bmatrix}\mathbf{x}$, and $\sigma(\overline{\mathbf{A}} + \overline{\mathbf{B}}\overline{\mathbf{K}}) = \{-4, -5, -6, -7, -8\}$.

19.10. Repeat Prob. 19.6 for the system of Prob. 18.8 with $\mathbf{r}(t) = [1 \quad -1]^T$, $\sigma(\overline{\mathbf{A}} + \overline{\mathbf{B}}\overline{\mathbf{K}}) = \{-2, -3, -4, -5, -6\}$, and

$$\mathbf{y} = \begin{bmatrix} 0 & 1 & 1 \\ 1 & 1 & 0 \end{bmatrix}\mathbf{x}$$

19.11. Repeat Prob. 19.6 for the system of Prob. 19.3 with $r(t) = [1 \quad -1]^T$, $\sigma(\overline{\mathbf{A}} + \overline{\mathbf{B}}\overline{\mathbf{K}}) = \{-2, -4, -5, -6, -7\}$, and

$$\mathbf{y} = \begin{bmatrix} 0 & 1 & 0 \\ 1 & 0 & 1 \end{bmatrix}\mathbf{x}$$

19.12. Repeat Prob. 19.6 for the systems of Prob. 18.10 with $r(t) = u_{-1}(t)$ and $\mathbf{y} = [1 \quad 0 \quad 1 \quad 0]\mathbf{x}$.

19.13. Repeat Prob. 19.6 for the system of Prob. 18.13 with $r(t) = u_{-1}(t)$, $\mathbf{y} = [0 \quad 0 \quad 1]\mathbf{x}$, and $\sigma(\overline{\mathbf{A}} + \overline{\mathbf{B}}\overline{\mathbf{K}}) = \{-1 + j1, -1 - j1, -4, -6\}$.

19.14. Synthesize an observer for the system of Prob. 18.6 with $\mathbf{y} = [1 \quad 1 \quad 1]\mathbf{x}$. Use $\sigma(\mathbf{A} - \mathbf{LC}) = \{-8, -9, -10\}$.

19.15. Synthesize an observer for the system of Prob. 19.6. Use the observer eigenvalues $\sigma(\mathbf{A} - \mathbf{LC}) = \{-10, -11, -12\}$.

19.16. Synthesize an observer for the system of Prob. 18.4 with

$$\mathbf{y} = [1 \quad 1 \quad 1]\mathbf{x} \qquad \text{and} \qquad \sigma(\mathbf{A} - \mathbf{LC}) = \{-8, -9, -10\}.$$

19.17. Expand the simulation diagram of Fig. 16.7*a* to represent the tracking system of Eq. (19.44) and the feedback control law of Eq. (19.48).

19.18. For the system given below the specified closed-loop eigenvalue spectrum is $\sigma(\mathbf{A} + \mathbf{BK}) = \{-2, -3\}$. Find $\ker \mathbf{S}(\lambda_i)$.

$$\mathbf{A} = \begin{bmatrix} 1 & 1 \\ 1 & 0 \end{bmatrix} \qquad \mathbf{B} = \begin{bmatrix} 1 & 0 \\ 0 & 1 \end{bmatrix}$$

19.19. Assume for the plant of Prob. 18.8(2) that

$$\ker \mathbf{S}(-2) = \left\{ \begin{bmatrix} 2 \\ 0 \\ 1 \\ \hdashline 1 \\ 2 \end{bmatrix}, \begin{bmatrix} 0 \\ 1 \\ 2 \\ \hdashline 0 \\ 1 \end{bmatrix} \right\},$$

$$\ker \mathbf{S}(-3) = \left\{ \begin{bmatrix} 1 \\ 1 \\ 1 \\ \hdashline 0 \\ 2 \end{bmatrix}, \begin{bmatrix} 0 \\ 2 \\ 0 \\ \hdashline 1 \\ 1 \end{bmatrix} \right\}$$

$$\ker \mathbf{S}(-4) = \left\{ \begin{bmatrix} 0 \\ 0 \\ 1 \\ \hdashline 2 \\ 0 \end{bmatrix}, \begin{bmatrix} 1 \\ 1 \\ 0 \\ \hdashline -1 \\ 1 \end{bmatrix} \right\}$$

It is desired that the mode $(\ )e^{-2t}$ not appear in $x_1(t)$ and that the mode $(\ )e^{-3t}$ not appear in $x_1(t)$ and $x_3(t)$. (*a*) Select the eigenvectors that produce this result. (*b*) Form a permissible matrix $\mathbf{V}$.

19.20. A control system described by

$$\mathbf{A} = \begin{bmatrix} 1 & 1 & 0 & 0 \\ 0 & 1 & 1 & 0 \\ 1 & 0 & 0 & 1 \\ 0 & 0 & 0 & 1 \end{bmatrix} \qquad \mathbf{B} = \begin{bmatrix} 1 & 0 \\ 1 & 0 \\ 0 & 0 \\ 0 & 1 \end{bmatrix} \qquad \mathbf{C} = \begin{bmatrix} 1 & 1 & 0 & 0 \\ 0 & 1 & 0 & 1 \\ 1 & 0 & 0 & 0 \end{bmatrix}$$

is to be designed as a tracking system. The forcing function is $\mathbf{r}(t) = [2 \quad -1]^T$, where $y_1(t)$ tracks $r_1(t)$ and $y_2(t)$ tracks $r_2(t)$. (*a*) Determine the following matrices: $\overline{\mathbf{A}}$, $\overline{\mathbf{B}}$, $\overline{\mathbf{C}}$, **E**, and **F**. (*b*) Specify an acceptable eigenvalue spectrum for the closed-loop tracking control system, i.e., specify $\sigma(\overline{\mathbf{A}} + \overline{\mathbf{B}}\overline{\mathbf{K}})$.

19.21. Given

$$\mathbf{A} = \begin{bmatrix} 1 & 0 & 0 \\ 0 & 1 & 0 \\ 0 & -1 & -1 \end{bmatrix} \qquad \mathbf{B} = \begin{bmatrix} 1 & 0 \\ 0 & 1 \\ 1 & 0 \end{bmatrix}$$

It is desired to assign the closed-loop eigenvalue spectrum $\sigma(\mathbf{A} + \mathbf{BK}) = \{\lambda_1, \lambda_2, \lambda_3\}$, where λ_1, λ_2, and λ_3 are all different from the open-loop eigenvalue spectrum $\sigma(\mathbf{A}) = \{-0.5 + j\sqrt{3}/2, -0.5 - j\sqrt{3}/2, 0\}$. (*a*) Is it possible to achieve the desired closed-loop eigenvalues? If so, state why. (*b*) By the method of Example 2 of Sec. 19.2, determine $\ker \mathbf{S}(\lambda_i)$.

Chapter 20

20.1. Verify the values of the set of zeros S_2 as given by Eq. (20.60) for the system of Sec. 20.6.

20.2. For the system of Prob. 18.6, the output matrix is

$$\mathbf{C} = \begin{bmatrix} 0 & 1 & 0 \\ 0 & 0 & 1 \end{bmatrix}$$

(*a*) Design a PI controller. (*b*) Determine the asymptotic transfer function. (*c*) Determine the time responses with $\mathbf{r}_1(t) = [1 \quad 0]^T$ and $\mathbf{r}_2(t) = [0 \quad 1]^T$. Use a CAD program for the solution of this problem.

20.3. For the system of Prob. 18.10(1), the output matrix is

$$\mathbf{C} = \begin{bmatrix} 1 & 0 & 1 & 0 \\ 0 & 0 & 0 & 1 \end{bmatrix}$$

(*a*) Design a PI controller. (*b*) Determine the transmission zeros. (*c*) Determine the asymptotic transfer function. (*d*) Determine the closed-loop time responses with $\mathbf{r}_1(t) = [1 \quad 0]^T$ and $\mathbf{r}_2 = [0 \quad 1]^T$.

20.4. The state and output equations of a system are

$$\dot{\mathbf{x}} = \begin{bmatrix} 0 & 0 & 1 \\ 1 & -1 & 1 \\ -2 & 0 & 2 \end{bmatrix}\mathbf{x} + \begin{bmatrix} 0 & 0 \\ 1 & 1 \\ 3 & 0 \end{bmatrix}\mathbf{u}$$

$$\mathbf{y} = \begin{bmatrix} 0 & -1 & 1 \\ 1 & 1 & 0 \end{bmatrix}\mathbf{x}$$

(*a*) Design a PI controller. (*b*) Determine the asymptotic transfer function. (*c*) Determine the time responses with $\mathbf{r}_1(t) = [1 \quad 0]^T$ and $\mathbf{r}_2(t) = [0 \quad 1]^T$. Use a CAD program for the solution of this problem.

20.5. Repeat Prob. 20.4 with the output matrix

$$\mathbf{y} = \begin{bmatrix} 0 & 1 & 0 \\ 1 & 0 & 0 \end{bmatrix}\mathbf{x}$$

20.6. Repeat Prob. 20.3 with the output matrix

$$\mathbf{y} = \begin{bmatrix} 1 & 0 & 0 & 0 \\ 0 & 0 & 0 & 1 \end{bmatrix} \mathbf{x}$$

20.7. Derive Eq. (20.25) by inserting Eq. (20.24) into Eq. (20.20) and taking the limit as $g \to \infty$.

20.8. Show that Eq. (20.47) can be put into the form of Eq. (20.48). *Hint:* Premultiply the matrix $[\lambda \mathbf{I}_m + g\mathbf{B}_2\mathbf{K}_1\mathbf{C}_2]$ by $\mathbf{C}_2^{-1}\mathbf{C}_2$ and factor to the form $[\mathbf{C}_2^{-1}(\ )\mathbf{C}_2]$. Then apply the property on the inverse of a product of matrices given in Sec. 4.14. Note that the eigenvalues are not changed by these matrix operations, and therefore Eq. (20.44) yields the same eigenvalues as Eq. (20.43).

Chapter 21

Problems 21.1 through 21.3 are intended to give the reader experience in using a trial-and-error procedure, with the aid of straight-line Bode plots, in synthesizing a compensator $G_c(s)$ that will satisfy the performance specifications for the closed-loop system.

21.1. The plant for a unity-feedback control system is

$$P(s) = \frac{K_x}{s(s+5)} \qquad K_x = 10$$

(a) The desired closed-loop performance specifications are: $K_1 = 25\ \text{s}^{-1}$ and $\gamma = 40°$. Design a lag compensator $G_c(s)$ [see Eq. (10.3)] that will achieve these closed-loop system specifications. Determine the values of α, T, and A of $G_c(s)$, where $1 < \alpha \leq 100$. (b) Adjust K_x of the basic system to yield $K_1 = 25\ \text{s}^{-1}$ and determine the corresponding value of the phase-margin frequency ω_{ϕ_x}. (c) Repeat the design by use of a lead compensator, i.e., with K_x fixed, design $G_c(s)$ such that $K_1 = 25\ \text{s}^{-1}$ and $\gamma = 40°$ at $\omega_\phi = \omega_{\phi_x}$ of the basic system.

21.2. For the plant of Prob. 21.1 it is desired that $K_1 = 25\ \text{s}^{-1}$, $\gamma = 40°$, and $\omega_\phi = 15$ rad/s. Design a lag-lead compensator $G_c(s)$ that meets these closed-loop system performance specifications. *Hint:* Adjust K_x of the basic system to yield the desired value of K_1. Based upon the values of $\text{Lm}\,\mathbf{P}(j15)$ and $\angle \mathbf{P}(j15)$, for this value of K_x, determine the location of the poles and zeros of $G_c(s)$ with the aid of a straight-line Bode plot that satisfies the desired performance specifications.

21.3. For the plant of Prob. 21.1 it is desired that $K_1 = 25\ \text{s}^{-1}$, $\gamma = 40°$, $\omega_\phi = 15$ rad/s, the gain margin $a = 3$ (9.54 dB), and $\angle \mathbf{L}(j\omega) \geq -140°$ for $\omega \leq \omega_\phi$. Design a lag-lead compensator $G_c(s)$ that will achieve these closed-loop system specifications. Locate the poles and zeros of $G_c(s)$ with the aid of a straight-line Bode plot.

21.4. A design has been made for the following figure:

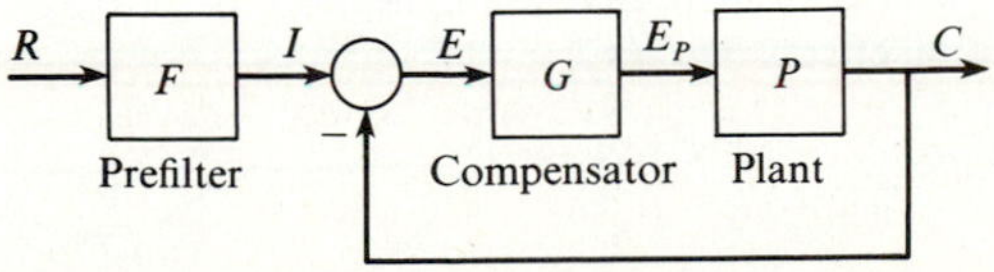

where the transfer functions $P(s)$, $F(s)$, and $G(s)$ are known. It turns out to be more convenient to implement the system by using the following figure:

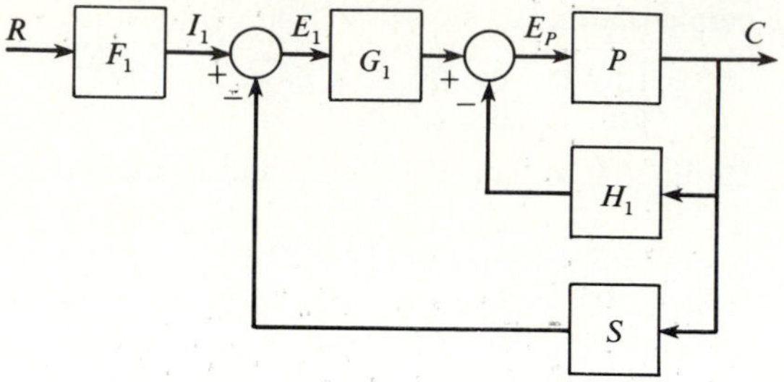

$H_1(s)$ is a tachometer and $S(s)$ is a sensor, and these transfer functions are known. Find $F_1(s)$ and $G_1(s)$ as functions of the known transfer functions $F(s)$, $G(s)$, $H_1(s)$, and $S(s)$ so that the two figures are equivalent. *Note*: the loop transmissions for both control systems must be identical.

21.5. A system is described by the differential equations

$$\ddot{y} - A\dot{z} = r \quad (1) \qquad \text{and} \qquad Bz + D\dot{y} + r = 0 \quad (2)$$

where r is the input and y is the output. (a) Draw the SFG of the system. (b) Determine the loop transmission function.

21.6. The signal flow graph for a control system is

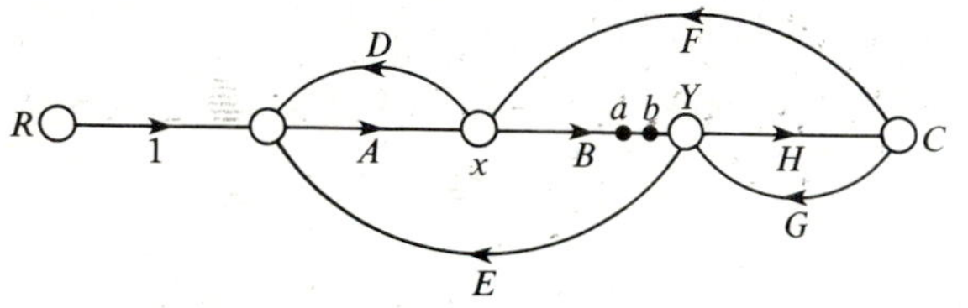

(a) Determine the transfer function C/R. (b) Redraw the flow graph into the fundamental feedback (single loop) flow graph in which B is the reference, i.e.,

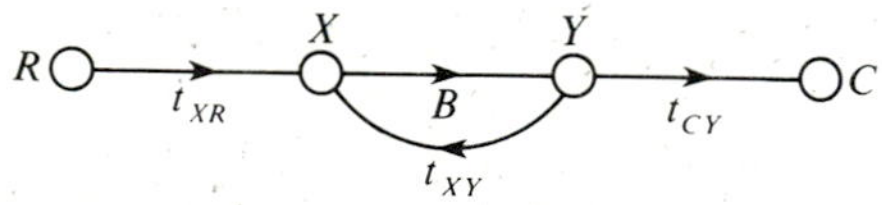

Identify the transfer functions t_{XR}, t_{YS}, and t_{CY} in terms of A, B, C, etc. *Hint:* First determine t_{XR} and t_{XY} by letting $B = 0$, i.e., break open the branch between points a and b in Fig. (a).

21.7. In Fig. (a) below $P(s) = 1/s$ and $G_1(s) = 1/(s + 2)$. In Fig. (b) below $P(s) = 1/s$ and $H(s) = s/(s + 1)$. (a) Find $G_2(s)$ so that $C(s)/R(s)$ is the same for both configurations. (b) Find $S_P^T(s)$, system sensitivity to variations in P, for each configuration. If these are different, which one yields smaller values of $|\mathbf{S}_P^T(j\omega)|$ over a range of finite values of ω? (c) Which configuration is the best for minimizing the parameter variation effects on $C(s)$?

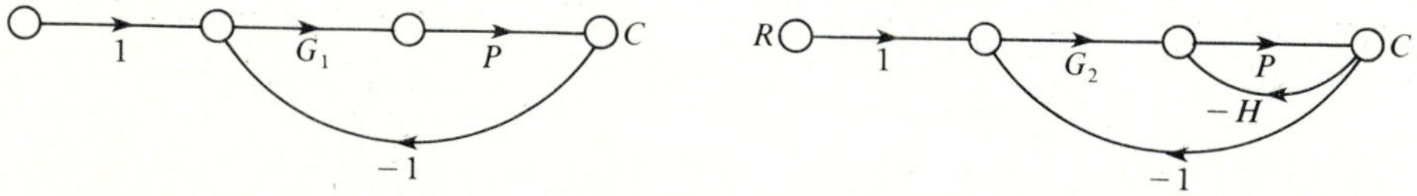

(a) (b)

21.8. For the control system shown in Fig. (a), $P(s) = 10/(s+1)$ and $d(t) = u_{-1}(t)$. It is desired that $|\lim_{t\to\infty}[c(t)]| \le 10^{-3}$. Also, the maximum value of $|c(t)| < 0.35$ is reached within 1 s, and $c(t)$ must not exceed this value. Design $G(s)$ so that these specifications are satisfied, where $\lim_{\omega\to\infty} \mathbf{G}(j\omega) = 0$ and $|\mathbf{G}(j\omega)|$ must be as small as possible for all ω.

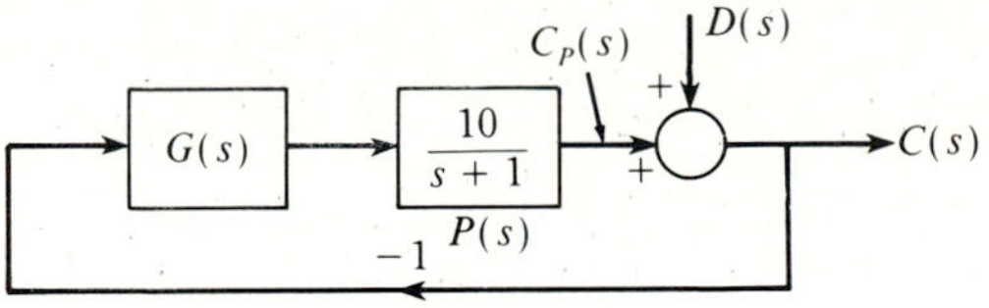

(a)

Hint: It is best to design the system from analyzing $c(t) = d(t) + c_p(t)$, where $C_p(s) = -\{L(s)/[1 + L(s)]\}D(s)$. Figure ($b$) shows the desired straight-line approximations of the upper and lower bounds on $c_P(t)$.

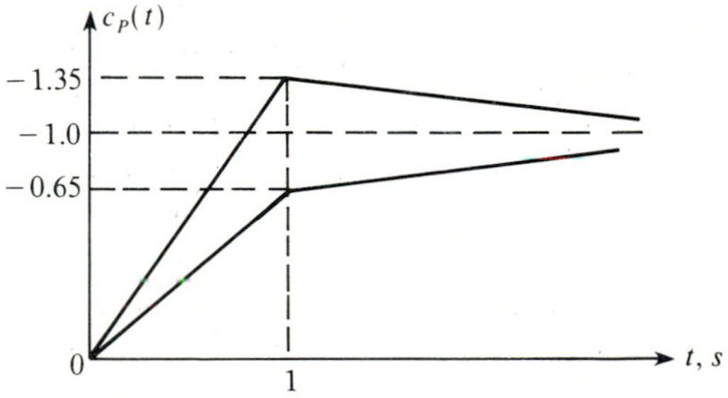

(b)

21.9. For the design example of Sec. 21.20: (a) Obtain the templates. (b) Draw $\text{Lm}\,\mathbf{L}_0$ vs. $\angle\mathbf{L}_0$ on a NC. (c) Obtain the data in order to draw the plots of Fig. 21.32. (d) Determine a suitable $F(s)$. (e) Obtain the four response characteristics for $T_R(s)$ for the points A, B, C, and D. Are the desired time-response specifications achieved at each point?

21.10. For the control system shown, $P(s) = Ka/s(s+a)$ and the parameter variations shown in Fig. (b) are $K_x \le K \le K_y$ and $a_w \le a \le a_z$. (*Note*: Parameter variation values are to be supplied by the instructor. A typical set is $1 \le K \le 10$ and $2 \le a \le 12$.) (a) Design $G(s)$ and $F(s)$ so that the following specifications are met:

For T_R: $T_U \to M_p = 1.2$ ($\text{Lm}\,M_m = 1.58$ dB) and $t_s = 1.8$ s

$T_L \to$ overdamped response and $t_s = 1.8$ s

For T_D: $c(t_p) \le 0.1$ where $t_p = 60$ ms, $c(\infty) = 0$, and $r(t) = d(t) = u_{-1}(t)$.

(b) For the design achieved, for each point A, B, C, and D in Fig. (b), obtain $c(t)$ due to $r(t) = u_{-1}(t)$ and $d(t) = 0$. Repeat for $d(t) = u_{-1}(t)$ and $r(t) = 0$.

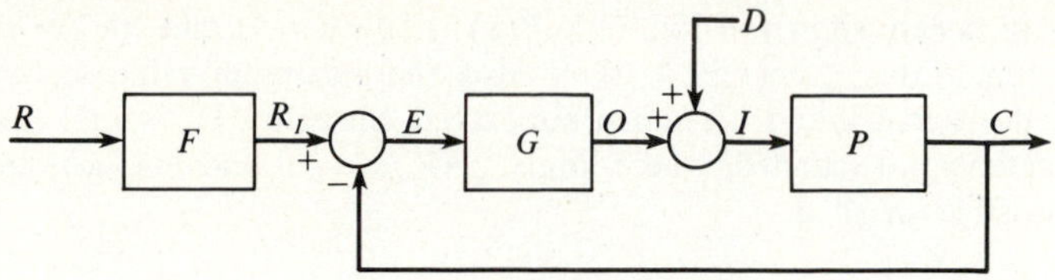

(a)

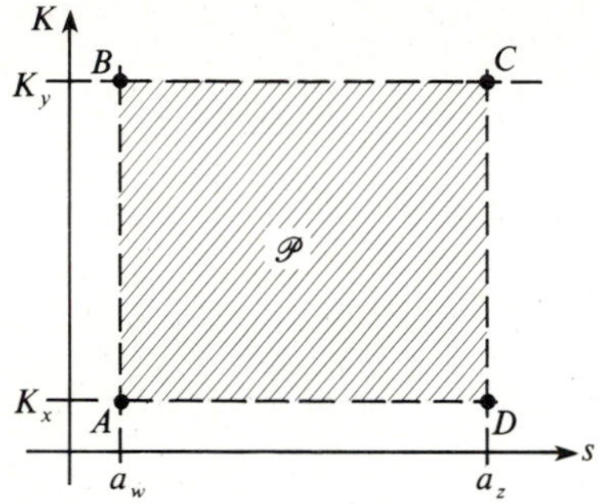

(b)

21.11. In Prob. 21.4 assume that $P = P_A P_p$, where P_A represents an actuator or a power amplifier with no uncertainty and P_p represents a plant with uncertainty. One designer made a QFT design with $P = P_A P_p$ that satisfied the system performance specifications. Another designer forgot to include the actuator in his QFT design, i.e., he assumed $P = P_p$. Is it necessary for him to redo his design?

Chapter 22

22.1. By use of Eqs. (22.13) and (22.14) verify the following z transforms in Table 22.1: (a) entry 4, (b) entry 7, (c) entry 10, (d) entry 13.

22.2. Determine the inverse z transform, in closed form, of

$$(a)\ F(z) = \frac{z(z+1)}{(z-1)(z^2-z+1)} \qquad (b)\ F(z) = \frac{-10(11z^2-15z+6)}{(z-2)(z-1)^2}$$

22.3. Repeat Prob. 22.2 but use the power-series method to obtain the open form for $f(nT)$.

22.4. Determine the initial and final values of the z transforms given in (a) Prob. 22.2, (b) Eq. (22.17), (c) Eq. (22.34), and (d) Eq. (22.43), i.e., solve for $c(0)$ and $c(\infty)$, where $R(z) = z/(z-1)$.

22.5. Let $F(z) = C(z)/E(z)$ in Prob. 22.2. Determine the difference equation for $c(nT)$ by use of the translation theorem of Sec. 22.3.

22.6. For the sampled-data control system of Fig. 22.10, $G(s)$ is given by Eq. (22.15), where $a = 2$ and T is unspecified. (a) Determine $G(z)$ and $C(z)/R(z)$. (b) Assume a sufficient number of values of T between 0.01 and 10 s, and for each value of T determine the maximum value that K can have for a stable response. (c) Plot K_{max} vs. T and indicate the stable region. (d) Obtain the root locus for this system with $T = 0.1$ s. (e) Use Eq. (22.19) to locate the roots on the root locus which have a $\zeta = 0.707$ and determine the corresponding value of K. (f) For a unit step input determine $c(nT)$, M_P, t_p, and t_s.

22.7. The sampled-data control system of Prob. 22.6 is to be compensated in the manner shown in Fig. 22.12. Determine a digital compensator that reduces T_s of Prob. 22.6, by one-half. Determine M_p, t_p, t_s, and K_m.

22.8. For $T = 1$ s obtain the PCT system of Prob. 22.6. Obtain a root locus for this pseudosystem and compare it to the root locus for a unity-feedback continuous-time control system whose forward transfer function is $K/s(s+2)$. This problem illustrates the degradation in system stability that results in converting a continuous-time system into a sampled-data system.

22.9. Repeat Prob. 22.8 for $T = 0.04$ s.

22.10. Given

$$G(s) = \frac{K(s+2)(s+10)}{s(s+1)(s+2 \pm j2)(s+10 \pm j20)}$$

and $T = 0.04$ s. Map each pole and zero of $G(s)$ by use of Eq. (22.50) and $z = e^{sT}$. Compare and analyze the results with respect to Fig. 22.17 and the warping effect.

22.11. Repeat Prob. 22.10, where $T = 0.01$ s, for:
(a) $s + 1000$
(b) $s^2 + 2000s + 1{,}002{,}500 = s + 1000 \pm j50$

22.12. The poles and zeros of the following desired control ratio of the PCT system have been specified:

$$\left[\frac{C(s)}{R(s)}\right]_M = \frac{K(s+7)}{(s+5 \pm j10)(s+6)}$$

(a) What must be the value of K in order for $y(t)_{ss} = r(t)$ for a step input? (b) A value of $T = 0.1$ s has been proposed for the sampled-data system whose design is to be based upon the above control ratio. Do all the poles and zeros lie in the good Tustin approximation region of Fig. 22.17? If the answer is no, what should be the value of T so that all the poles and zeros do lie in this region?

22.13†. For the basic system of Fig. 22.18, where $G_x(s) = K_x/s(s+5)$: (a) Determine M_p, t_p, t_s, and K_1 for $\zeta = 0.7$ and $T = 0.01$ s by the DIR method. (b) Obtain the PCT control system of part a and the corresponding figures of merit. (c) Generate a digital compensator $D_c(z)$ (with no ZOH) that reduces the value of t_s of part a by approximately one-half for $\zeta = 0.7$ and $T = 0.01$ s. Do first by the DIG method (s plane) and then by the DIR method. Obtain M_p, t_p, and t_s for $c(t)$ of the PCT and $c^*(t)$ (for both the DIG and the DIR designs) and K_m. Draw the plots of $c^*(t)$ vs. kT for both designs. (d) Find a digital compensator $D_c(z)$ (with no ZOH) that increases the ramp error coefficient of part a, with $\zeta = 0.7$ and $T = 0.01$ s with a minimal degradation of the transient-response characteristics of part a. Do first by the DIG method (s plane) and then by the DIR method. Obtain M_p, t_p, and t_s for $c(t)$ of the PCT and $c^*(t)$ for both the DIG and the DIR designs and K_m. Draw a plot of $c^*(t)$ vs. kT for both designs. (e) Summarize the results of this problem in tabular form and analyze. (f) Obtain the frequency-response plots of parts a to d and evaluate.

22.14. Repeat Prob. 22.13 for $G_x(s) = K_x/[s(s+1)(s+5)]$.

---

† For part b use 4-decimal-digit accuracy; for parts c and d use 4-decimal-digit accuracy, then repeat using 8-digit accuracy. Compare the results.

ANSWERS TO SELECTED PROBLEMS

Chapter 2

2.1. (*a*) *Loop equations*:

(1) $e(t) = (LD + R_1 + R_2)i_1 - (LD + R_2)i_2$

(2) $0 = -(LD + R_2)i_1 + \left(LD + R_2 + R_3 + \dfrac{1}{CD}\right)i_2$

(*b*) *Node equations*:

(3) $\left(\dfrac{1}{R_1} + \dfrac{1}{R_3} + \dfrac{1}{LD + R_2}\right)v_a - \dfrac{1}{R_3}v_b = \dfrac{1}{R_1}e$

(4) $-\dfrac{1}{R_3}v_a + \left(\dfrac{1}{R_3} + CD\right)v_b = 0$

(*c*) Let $x_1 = v_b$, $\dot{x}_1 = Dv_b$, $x_2 = i_L$, $\dot{x}_2 = Di_L$, $i_C = C\,Dv_b = C\dot{x}_1$, $i_1 = i_C + i_L$, and $R = R_1 + R_3$

$$\mathbf{A} = \begin{bmatrix} \dfrac{-1}{RC} & \dfrac{-R_1}{RC} \\ \dfrac{R_1}{RL} & \dfrac{-(R_1R_2 + R_2R_3 + R_3R_1)}{RL} \end{bmatrix} \qquad \mathbf{b}^T = \begin{bmatrix} \dfrac{1}{RC} & \dfrac{R_3}{RL} \end{bmatrix}$$

(*d*) $G(D) = \dfrac{v_b}{e}$

$$= \frac{(LD + R_2)}{CL(R_1 + R_3)D^2 + [(R_1R_2 + R_2R_3 + R_3R_1)C + L]D + (R_1 + R_2)}$$

2.2. (*a*) $\left(\mathbf{R}_1 + \dfrac{1}{C_1D}\right)i_i - \dfrac{1}{C_1D}i_2 = E$

$$-\frac{1}{C_1D}i_1 + \left(LD + R_2 + \frac{1}{C_1D} + \frac{1}{C_2D}\right)i_2 = 0$$

(*b*) Node 1: $\left(C_1D + \dfrac{1}{R_1} + \dfrac{1}{R_2}\right)v_1 - \dfrac{1}{R_2}v_2 = \dfrac{1}{R_1}E$

Node 2: $-\dfrac{1}{R_2}v_1 + \left(\dfrac{1}{R_2} + \dfrac{1}{LD}\right)v_2 - \dfrac{1}{LD}v_3 = 0$

Node 3: $-\dfrac{1}{LD}v_2 + \left(C_2D + \dfrac{1}{LD}\right)v_3 = 0$

(*c*) $x_1 = v_{c_1},\ x_2 = v_{c_2},\ x_3 = i_L$

$$\dot{\mathbf{x}} = \begin{bmatrix} -\dfrac{1}{R_1C_1} & 0 & -\dfrac{1}{C_1} \\ 0 & 0 & \dfrac{1}{C_2} \\ \dfrac{1}{L} & -\dfrac{1}{L} & -\dfrac{R_2}{L} \end{bmatrix}\mathbf{x} + \begin{bmatrix} \dfrac{1}{R_1C_1} \\ 0 \\ 0 \end{bmatrix} E$$

2.3. Circuit 2: (*b*)

$$\dot{\mathbf{x}} = \begin{bmatrix} -\dfrac{R_1 + R_2}{L} & -\dfrac{1}{L} \\ \dfrac{1}{C} & 0 \end{bmatrix}\mathbf{x} + \begin{bmatrix} \dfrac{1}{L} \\ 0 \end{bmatrix}\mathbf{u}$$

2.7. (*b*) $(M_1D^2 + B_1D + K_1)x_1 - (B_1D + K_1)x_2 = 0$

$-(B_1D + K_1)x_1 + [M_2D^2 + (B_1 + B_2)D + (K_1 + K_2)]x_2 = f$

(*c*) $x_3 = \dot{x}_1,\ x_4 = \dot{x}_2$

$$\dot{\mathbf{x}} = \begin{bmatrix} 0 & 0 & 1 & 0 \\ 0 & 0 & 0 & 1 \\ -\dfrac{K_1}{M_1} & \dfrac{K_1}{M_1} & -\dfrac{B_1}{M_1} & \dfrac{B_1}{M_1} \\ \dfrac{K_1}{M_2} & -\dfrac{K_1 + K_2}{M_2} & \dfrac{B_1}{M_2} & -\dfrac{B_1 + B_2}{M_2} \end{bmatrix}\mathbf{x} + \begin{bmatrix} 0 \\ 0 \\ 0 \\ \dfrac{1}{M_2} \end{bmatrix} f$$

(*d*) $$G = \frac{x_1}{f} = \frac{B_1D + K_1}{M_1M_2D^4 + (M_1B_1 + M_1B_2 + M_2B_1)D^3 + (M_1K_1 + M_1K_2 + M_2K_1 + B_1B_2)D^2 + (B_1K_2 + B_2K_1)D + K_1K_2}$$

2.8. (*c*) $$G(D) = \frac{y}{f} = \frac{BD + K_2}{D^4 + 2BD^3 + KD^2 + (K_1 + K_3)BD + (K_aK_b - K_2^2)}$$

where $K_a = K_1 + K_2,\ K_b = K_2 + K_3,\ K = K_a + K_b$

(*d*) Let $x_1 = x_a,\ x_2 = \dot{x}_1,\ x_3 = x_b$, and $x_4 = \dot{x}_3;\ u = f$, and $y = x_3$

Thus $\dot{\mathbf{x}} = \mathbf{Ax} + \mathbf{Bu}$ and $\mathbf{y} = \mathbf{c}^T\mathbf{x}$

$$\mathbf{A} = \begin{bmatrix} 0 & 1 & 0 & 0 \\ -K_a & -B & K_2 & B \\ 0 & 0 & 0 & 1 \\ K_2 & B & -K_b & -B \end{bmatrix} \qquad \mathbf{B} = \begin{bmatrix} 0 \\ 1 \\ 0 \\ 0 \end{bmatrix} \qquad \mathbf{c}^T = [0 \quad 0 \quad 1 \quad 0]$$

2.12. (*a*) $(LD + R)i = e,\ K_i i = f,\ \dfrac{x_a}{l_1} = \dfrac{x_b}{l_2}$

$(M_2D^2 + B_2D + K_2)x_b = f_2$

$(M_1D^2 + B_1D + K_1)x_a + \dfrac{l_2}{l_1}f_2 = f = k_i i$

Let $\dfrac{l_2}{l_1} = a,\ \dfrac{M_1}{a} + aM_2 = b,\ \dfrac{B_1}{a} + aB_2 = c$, and $\dfrac{K_1}{a} + aK_2 = d$

$(bD^2 + cD + d)x_b = K_i i$

(*b*) There are only three independent states:
$x_1 = x_b$, $x_2 = \dot{x}_b$, $x_3 = i$, and $u = e$

$$\dot{x}_1 = x_2$$

$$\dot{x}_2 = -\frac{d}{b}x_1 - \frac{c}{b}x_2 + \frac{K_i}{b}x_3$$

$$\dot{x}_3 = \frac{R}{L}x_3 + \frac{1}{L}u$$

2.13. Traction force $= MD^2x + BDx$

2.14. (*b*) (1) $T = (J_1D^2 + B_1D + K_1)\theta_1 - K_1\theta_2$
(2) $0 = -K_1\theta_1 + [J_2D^2 + B_2D + (K_1 + K_2)]\theta_2 - K_2\theta_3$
(3) $0 = -K_2\theta_2 + [J_3D^2 + B_3D + (K_2 + K_3)]\theta_3$
(*d*) Let $x_1 = \theta_1$, $x_2 = D\theta_1$, $x_3 = \theta_2$, $x_4 = D\theta_2$, $x_5 = \theta_3$, $x_6 = D\theta_3$, and $u = T$.

$$\dot{\mathbf{x}} = \begin{bmatrix} 0 & 1 & 0 & 0 & 0 & 0 \\ -\frac{K_1}{J_1} & -\frac{B_1}{J_1} & \frac{K_1}{J_1} & 0 & 0 & 0 \\ 0 & 0 & 0 & 1 & 0 & 0 \\ \frac{K_1}{J_2} & 0 & -\frac{K_1 + K_2}{J_2} & -\frac{B_2}{J_2} & \frac{K_2}{J_2} & 0 \\ 0 & 0 & 0 & 0 & 0 & 1 \\ 0 & 0 & \frac{K_2}{J_3} & 0 & -\frac{K_2 + K_3}{J_3} & -\frac{B_3}{J_3} \end{bmatrix}\mathbf{x} + \begin{bmatrix} 0 \\ \frac{1}{J_1} \\ 0 \\ 0 \\ 0 \\ 0 \end{bmatrix}\mathbf{u}$$

2.20. (*a*)

$$\left[\frac{C_1ad}{(a+b)(c+d)} + D\right]y_1 = \frac{C_1b}{a+b}x_1$$

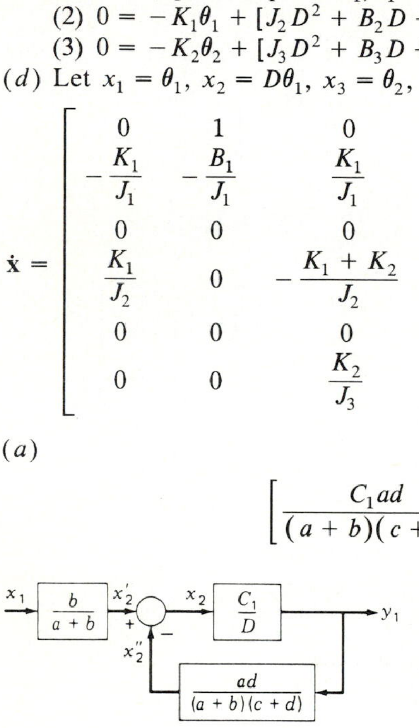

2.23. $K_Tv_1(t) = JL_a\,D^3\theta_m + (JR_a + BL_a)\,D^2\theta_m + (BR_a + K_TK_b)\theta_m$

Chapter 3

3.2. $\theta_3(t) = -16.4 + 2t + 21.7e^{-0.22t} - 4.81e^{-0.48t}$
$+ 0.852e^{-0.389t}\sin(0.744t - 59°) + 0.113e^{-0.411t}\sin(1.19t - 227°)$

3.5. (*b*) $c(t) = -1 + 0.5t + e^{-t} + 0.5e^{-t}\sin t$

3.6. $\omega_m(t) \approx 6.35\cos(5t + 11.5°) - 56.52e^{-0.5t}\sin 0.289t$

3.9. (*a*) Physical variables: $x_1 = v_c$ and $x_2 = i_L$

$$\mathbf{A} = \begin{bmatrix} -10^4 & -10^6 \\ 0.1 & -465 \end{bmatrix} \qquad \mathbf{b} = \begin{bmatrix} 10^4 \\ 0 \end{bmatrix} \qquad \mathbf{c}^T = [1 \quad 0]$$

(*b*) $v_C(t) = m_1A_1e^{m_1t} + m_2A_2e^{m_2t}$, where $m_1 = -492.5$, $m_2 = -9970$, $A_1 = 1.25$, and $A_2 = -50.2$

3.10. (*b*) $x(t) = 3 + 1.5e^{-3t} - 4.5e^{-t}$
(*c*) $x(t) = 0.18 + 0.237t + 0.242e^{-0.5t}\sin(2t + 3.98)$

3.15.

$$\Phi(t) = \begin{bmatrix} -e^{-2t} + 2e^{-4t} & 2e^{-2t} - 2e^{-4t} \\ -e^{-2t} + e^{-4t} & 2e^{-2t} - e^{-4t} \end{bmatrix} \qquad \mathbf{x}(t) = \begin{bmatrix} \frac{1}{2} - 3e^{-2t} + \frac{9}{2}e^{-4t} \\ \frac{3}{4} - 3e^{-2t} + \frac{9}{4}e^{-4t} \end{bmatrix}$$

$$y(t) = x_1 = \tfrac{1}{2} - 3e^{-2t} + \tfrac{9}{2}e^{-4t}$$

3.18. (*a*) $\mathbf{x}(t)_h = \begin{bmatrix} 2e^{-3t} - e^{-4t} \\ -6e^{-3t} + 4e^{-4t} \end{bmatrix}$ (*b*) $\mathbf{x}(t) = \begin{bmatrix} 1 - 2e^{-3t} + 2e^{-4t} \\ 6e^{-3t} - 8e^{-4t} \end{bmatrix}$

3.19. (1) (*a*) $\lambda_1 = -1, \lambda_{2,3} = -1 \pm j1$

$$\Phi(t) = \mathbf{Z}_1 e^{-t} + \mathbf{Z}_2 e^{-t}e^{jt} + \mathbf{Z}_3 e^{-t}e^{-jt}$$

Chapter 4

4.1. (*a*) $f(t) = 10e^{-t}(t - \sin t)$ (*b*) $f(t) = 1 - 1.3c^{-t} + 1.14e^{-2t}\sin(3t + 164.74°)$

4.2. (*c*) $x(t) = 1 - 1.047e^{-t} + 0.0673e^{-t}\sin(t + 135°)$

4.4. (*b*) $(4.35 + 0.683e^{-10t} - 5.01e^{-0.965t} - 1.87 \times 10^{-3}e^{-239t})10^{-3}$

4.5. (*d*) $F(s) = \dfrac{1}{s^2} - \dfrac{3}{4s} + \dfrac{2}{3(s+1)} + \dfrac{1}{12(s+4)}$

(*h*) $F(s) = \dfrac{1}{s} + \dfrac{20}{s+3} + \dfrac{10.6\angle 172°}{s+3-j1} + \dfrac{10.6\angle -172°}{s+3+j1}$

4.6. (*b*) $x(t) = 3 + 3e^{-3t} - 6e^{-t}$

4.7. (*b*) $(s^2 + 2s + 5)X(s) = \dfrac{2(s^2+5)}{s}$ $\quad x(t) = 2 - 2e^{-t}\sin 2t$

(*d*) $(s^3 + 4s^2 + 14s + 20)X(s) + 4s^2 + 15s + 52 = \dfrac{5}{s^2 + 25}$

4.8. (*a*) Prob. 4.1(*a*): 0, Prob. 4.1(*b*): 1
(*b*) Prob. 4.2(*a*): 0, Prob. 4.2(*c*): 1, Prob. 4.2(*e*): 2

4.11. (1) (*b*) $x(t) = 2.5 + 0.5e^{-4t} + 3.162e^{-t}\sin(t + 251.6°)$
(*c*) $x(t) = 0.25t - 0.1875 - 0.0125e^{-4t} + 0.22361e^{-t}\sin(t + 116.6°)$

4.12. (1)(*a*)

$$\mathbf{X}(s) = \begin{bmatrix} \dfrac{0.2}{s} - \dfrac{0.2}{s+5} \\ \dfrac{0.4}{s} + \dfrac{0.6}{s+5} \end{bmatrix}$$

(*b*) $G(s) = \dfrac{1}{(s+1)(s+5)}$

4.15. (1) $\sigma(\mathbf{A}) = \{-1, -1 + j1, -1 - j1\}$

4.19. (*a*) $Q(\lambda) = \lambda^4 - 1 = (\lambda^2 + 1)(\lambda^2 - 1)$

$$\mathbf{A}^{-1} = \begin{bmatrix} -1 & 0 & 0 & 1 \\ 1 & 0 & 0 & 0 \\ 1 & 1 & 1 & -1 \\ 0 & 0 & 1 & 0 \end{bmatrix}$$

Chapter 5

5.5. $G(s) = \dfrac{-60s - 4}{(s + 10)(s^2 + 0.13s + 8.004)}$

5.12. (2) $\lambda_1 = -1, \lambda_2 = -2, \lambda_3 = -3, \lambda_4 = -4$
(*a*) $y = [10 \quad 0 \quad 5 \quad 0]\mathbf{x}$
(*b*) $y = [2.5 \quad -16 \quad 27.5 \quad -15]\mathbf{x}$

5.14. For system (2):
(*b*) $\lambda_1 = -1, \lambda_2 = -2, \lambda_3 = -3$

(*c*) $\mathbf{T} = \begin{bmatrix} 1 & 1 & 1 \\ 5 & 4 & 3 \\ 6 & 3 & 2 \end{bmatrix}$

(*d*) $\dot{\mathbf{z}} = \begin{bmatrix} -1 & 0 & 0 \\ 0 & -2 & 0 \\ 0 & 0 & -3 \end{bmatrix}\mathbf{z} + \begin{bmatrix} \frac{3}{2} \\ 0 \\ -\frac{3}{2} \end{bmatrix}\mathbf{u} \qquad \mathbf{y} = [1 \quad 1 \quad 1]\mathbf{z}$

(*e*) The mode with $\lambda = -2$ is uncontrollable

5.18.

$$\Lambda_m = \begin{bmatrix} -1.5 & 0.5\sqrt{3} & 0 \\ -0.5\sqrt{3} & -1.5 & 0 \\ 0 & 0 & -2 \end{bmatrix}$$

5.19. (2) (*a*) $G(s) = \dfrac{s}{(s+2)(s+1)}$ (*c*) $\mathbf{x}(t) = \begin{bmatrix} e^{-t} - e^{-2t} \\ 1 - 2e^{-t} + e^{-2t} \end{bmatrix}$

5.20. (1) (*a*) $\mathbf{T} = \begin{bmatrix} 8 & 6 & 1 \\ 4 & 1 & 0 \\ -2 & 2 & 1 \end{bmatrix}$ (*b*) $\mathbf{T} = \begin{bmatrix} 1 & 0 & 0 \\ 1 & 1 & 0 \\ -1 & -1 & 1 \end{bmatrix}$

5.23. (1) (*a*) $\lambda_1 = -1, \lambda_2 = -1, \lambda_3 = -2$
(*b*) *Hint*: Use $[\lambda_i \mathbf{I} - \mathbf{A}]\mathbf{v}_{i+1} = \mathbf{v}_i$ for multiple eigenvalues

$$\mathbf{v}_1 = \begin{bmatrix} 1 \\ -1 \\ 1 \end{bmatrix} \qquad \mathbf{v}_2 = \begin{bmatrix} -2 \\ 1 \\ 0 \end{bmatrix} + \alpha\mathbf{v}_1 \qquad \mathbf{v}_3 = \begin{bmatrix} 1 \\ -2 \\ 4 \end{bmatrix}$$

where α is any real number. $\mathbf{v}_2$ is called a *generalized eigenvector*.

5.24. (1) $\mathbf{T} = \begin{bmatrix} 2 & 1 & 2 \\ -1 & 0 & 1 \\ 1 & 0 & -3 \end{bmatrix}$ $\sigma(\mathbf{A}) = \{-1, -2, -3\}$

5.30. (*a*) $\sigma(\mathbf{A}) = \{-1, -2, -3\}$ (*b*) Use the Vandermonde matrix

Chapter 6

6.2. (*b*) $0 < K < 1$; $s_{1,2} = \pm j1$ (*c*) $-20 < K < \infty$; for $K = -20$, $s_1 = 0$
(*e*) $-6 < K < 60$; $s_{1,2} = \pm j3.317$ for $K = 60$
(*f*) $0 < K < 8.486$; $s_{1,2} = \pm j1.520$

6.3. (*a*) $(s+3)(s^2+3s+3)$ (*c*) $(s^2+3s+3)(s^2-s+2)$
(*i*) $s^2+6s+10)(s^2-4s+40)(s+1)$

6.4. (*a*) None (*c*)2 (*i*)2 (*f*) None

6.10. (*a*) For (1): $K_p = \infty$, $K_v = 1.6$, $K_a = 0$ (*b*) For $r_1(t)$: (1) $e(t)_{ss} = 1.25$, (2) $e(t)_{ss} = 0$

6.13. (*g*) $\infty, -6, 0$; (*j*) $\infty, \infty, -3$ (*i*) $\infty, -50, 0$

6.14. (*g*): (*a*) 0, (*b*) $-\frac{1}{3}$, (*c*) ∞; (*j*): (*a*) 0, (*b*) 0, (*c*) $-\frac{2}{3}$

6.20. $K_H = (K_G c_0 - a_0)/K_G c_0 > 0$

6.26. (*a*) $0 < K < 9/8$ (*b*) $e(t)_{ss} = 160/9$

Chapter 7

7.3. (*c*) For $K > 0$, real-axis branches: none; four branches; $\gamma = \pm 45°, \pm 135°$; angles of departure: $\pm 45°, \pm 135°$; imaginary-axis crossing: $\pm j1$ with $K = 0$. Branches are straight lines and coincide with the asymptotes.

(*e*) Real-axis branch: 0 to $-\infty$; five branches; $\gamma = \pm 36°, \pm 108°, 180°$; $\sigma_0 = -\frac{8}{5}$; no breakaway or break-in points; angles of departure: $-90°$, $-26°$; and the imaginary-axis crossing: $\pm j0.9$ with $K = 20.6$. Locus crosses asymptotes at $\pm 36°$.

7.5. (1) (*a*) For $K > 0$, real-axis branches: 0 to -50 and -100 to $-\infty$; three branches; $\gamma = \pm 60°, 180°$; $\sigma_0 = -50$; breakaway point: -21.1 ($K_1 = 9.5$); and imaginary-axis crossing: $\pm j70.7$ ($K_1 = 150$).

(*b*) $K_1 = 150$ (*c*) $K_1 = 25.92$

(*d*) $e(t) = -1.29e^{-16.67t}\sin(28.9t + 223.9°) + 0.102e^{-116.7t}$

(*e*)

ω	M	α, deg	ω	M	α, deg
0	1	0	22.85	1.1323	−63.4
5	1.01	−11.2	25	1.126	−71.9
10	1.039	−23.2	30	1.053	−92.5
15	1.083	−36.8	35	0.908	−112.3
20	1.123	−52.9	41	0.7078	−132.0

7.9. Poles: $s_1 = -1$, $s_2 = -5$, $s_{3,4} = -3 \pm j2$

(*a*) For $K > 0$, real-axis branch: -1 to -5; four branches; $\gamma = \pm 45°, \pm 135°$; angles of departure: $-180°, 0°, -90°, 90°$; imaginary-axis crossing: $\pm j3$ with $K = 340$; break-in and breakaway points coincide at $s = -3$. For $K < 0$, real-axis branches: -5 to $-\infty$ and -1 to $+\infty$; $\gamma = 0°, \pm 90°, 180°$; angles of departure: $0°, 180°, 90°, -90°$; imaginary-axis crossing: $s = 0$ with $K = -320$.

(*b*) $-320 > K > 340$ (*c*) $K = 16$

7.10. (1) (*a*) For $K = -8K_1$; four branches; $\gamma = 0°, \pm 120°$ for $K_1 > 0$; $\gamma = \pm 60°$, $180°$ for $K_1 < 0$; $\sigma_0 = -3$; break-away points: -0.323 for $K = -1.02$, -3.244 for $K = -1.613$, -1.48 for $K = 0.375$; break-in point: 1.72 for $K = -138.2$; imaginary-axis crossings; $\pm j0.627$ with $K = -5.28$, $\pm j4.54$ with $K = 136.2$.

(*b*) Stable for $0 < K_1 < 0.66$

7.13. (*a*) For $K > 0$; six branches; real-axis branches: -3.9 to $-\infty$; $\gamma = \pm 36°, \pm 108°$, $\pm 180°$; angles of departure: $140°, -94.5°, 2.5°$; imaginary crossing: $\pm j17.1$ with $K = 1.16 \times 10^6$; break-in point: -17.5.

(*b*) 2.22

7.14. (1) (*a*) $A = 1.44$

Chapter 8

8.1. (*b*) (1) $\mathbf{G}(j0+) = \infty\underline{/-90°}$, $\mathbf{G}(j\infty) = 0\underline{/-270°}$, $\omega_x = \pm\sqrt{12.5}$, $\mathbf{G}(j\omega_x) = -10/7.5$

(2) $\mathbf{G}(j0+) = \infty\underline{/-90°}$, $\mathbf{G}(j\infty) = 0\underline{/-360°}$

8.4. (1) (*b*) $\mathbf{G}(j0+) = \infty\underline{/-90°}$, $\mathbf{G}(j\infty) = 0\underline{/0°}$, $\omega_x = 12.5$, $\omega_y = \pm 39.4$

8.6. (*a*) It must have an initial slope of -40 dB/decade

(*b*) The phase-angle curve approaches an angle of 180° as $\omega \to 0$

8.9. Case II: $G(s) = \dfrac{50(1 + 0.2s)}{s(1 + 2s)(1 + 0.0625s)}$
correction at $\omega = 4$ is 1.76 dB.

8.10. (*a*) $K_0 = 10$ (*b*) $K_2 = 0.1$

8.11. (*a*) Transfer function given in Prob. 8.2(3)

8.14. (*a*) $N = 0$, $Z_R = 0$, stable (*c*) $N = -2$, $Z_R = 2$, unstable

8.15. (*a*) For stability $K_3 > \frac{1}{2}$

8.17. (*a*) Stable for $K > -1$ (*e*) stable for $0 < K < 1.5$
(*h*) stable for $K > \sqrt{2}$ (*i*) stable for $-1 < K < 5/3$

8.18. (1) (*a*) $K_1 = 7.61$, $\omega_\phi = 6.34$ (*b*) $K_1 = 4.35$, $\omega_\phi = 4$
(*c*) maximum $K_1 = 308/9$
(2) (*a*) $K_0 = 16.65$, $\omega_\phi = 6.6$ (*b*) $K_0 = 6.91$, $\omega_\phi = 3.7$
(*c*) maximum $K_0 = \infty$

8.24. (*d*) (2) $K_0 \simeq 7.15$ ($\omega_x \simeq 1.2$)

Chapter 9

9.1. (2) $K_1 = 15$, $M_m = 1.35$, $\omega_m = 16.2$, $\omega_\phi = 13.64$, $\omega_c = 28.9$
(4) $K_1 = 0.5714$, $M_m = 1.346$, $\omega_m = 0.61$, $\omega_\phi = 0.52$, $\omega_c = 1.24$

9.3. $M_m = 1.29$, $K_1 = 20$, $\omega_m = 11.50$

9.6. (1) (*a*) $M_m = 1$, $\omega_m = 0$ (*b*) $M_m = 1.31$, $\omega_m = 6.75$, $K_1 = 7.61$

9.9. (1) Original system is unstable; for $M_m = 1.12$, $K_1 = 3.06$, $\omega_m = 2.38$, $\gamma = 56.8°$
(3) $M_m = 3.27$ at $\omega_m = 1.44$. The minimum achievable is $M_m = 2.34$, $K_2 = 8$, $\omega_m = 3.4$, $\gamma = 35.2°$ and $\omega_\phi = 3.60$

9.11. (1) $K_1 = 0.467$, $\omega_m = 0.495$

9.13. (1) $M_p = 1.163$, $T_p = 3.628$ s, $T_s = 8.0$ s
(2) $M_p = 1.854$, $T_p = 0.315$ s, $T_s = 8.0$ s
(3) $M_p = 1.163$, $T_p = 36.28$ s, $T_s = 80.0$ s

Chapter 10

10.3, 10.20.

System (1)	Dominant Roots	Other roots	K_1, s^{-1}	T_p, s	T_s, s	M_o
Basic	$3.57 \pm j6.19$	$-16.4 \pm j19.2$	5.23	0.507	1.12	0.166
Lag-compensated: $\alpha = 10$, $T = 10$	$-3.54 \pm j16.13$	$-16.4 \pm j19.2$ -0.10176	53.2	0.54	1.55	0.19
Lead-compensated: $\alpha = 0.1$, $T = 0.1$	$-9.6 \pm j16.6$	-11.5 -99.4	6.7	0.324	0.273	0.0025
Lag-lead-compensated: $\alpha = 10$, $T_1 = 10$, $T_2 = 0.1$	$-9.6 \pm j16.8$	-0.101 -11.3 -99.4	67.8	0.324	0.268	0.0071
Tachometer-compensated: $A = 15$, $K_t = 1$	$-8.8 \pm j15.17$	$-11.2 \pm j9.85$	6.34	0.36	0.26	0.1105

10.7. (*a*) $a = 2$, $b = 1.333$ (*b*) $a = 1.5$ (*c*) lag compensator

10.8. (*a*) $|\sigma| \geq 4/T_s = 3.33$; $\zeta = 0.5$ and $|\sigma| = 3.5$ yields $a = 5$ and $b = 2$
(*b*) $A = 0.576$ and $\alpha = 2.5$
(*c*) lag, original $K_1 = 3.28$, new $K_1 = 4.71$

10.14. $G(s) = \dfrac{30}{s(s^2 + 6s + 25)}$ $\dfrac{C(s)}{R(s)} = \dfrac{30}{(s + 1.695)(s^2 + 4.305s + 17.703)}$

10.20. (1) One possible solution is $A/K_t = 15$, $KK_t = 4578$, dominant roots $-8.8 \pm j15.22$, other roots $-11.22 \pm j9.85$, $K_1 = 6.34$. See the solution to Prob. 10.3 for a comparison with cascade compensation. *Note*: $K_t = 1$.

10.23. With the desired roots $s = -1.5 \pm j2.45$, $K = 68.43$, $A = 1.406$, and $K_t = 0.343$, $M_p = 0.647$, $t_p = 1.86$ s, $t_s = 2.47$ s, $K_0 = 1.51$.

10.25. With $K_x = 22$ the specifications are met

Chapter 11

11.1. For system (1): Basic system: $K_1 = 5.23$, $\omega_m = 5.10$
Lag-compensated: $K_1 = 47.5$, $\omega_m = 4.35$, $T_1 = 4$
Lead-compensated: $K_1 = 9.69$, $\omega_m = 14.4$, $T_2 = 0.1$
Lag-lead-compensated: $K_1 = 95.7$, $\omega_m = 14.0$, $T_1 = 4$, $T_2 = 0.1$

11.5. (*a*) $G(s) = \dfrac{1}{8K}\left[\dfrac{1 + Ts}{1 + \alpha Ts}\right]$, where $T = \dfrac{B}{K}$ and $\alpha = 0.5$

11.12. (*a*) At $\omega = 3$: $\mathbf{G}(j\omega) = \mathbf{C}(j\omega)/\mathbf{E}(j\omega) = 1\angle{-122^\circ}$, thus $K_1 = 30$, $\omega_\phi = 3$, and $\gamma \geq 50^\circ$

(*b*) $G_c(s) = \dfrac{1.05(1 + s)^2}{(1 + 10s)(1 + 0.1s)}$

11.18. For system (1) tachometer-compensated with $A = 1.15$, $K_t = 1/15$, $K_1 = 6.42$, $\omega_m = 7.26$, $M_p = 1.037$, $t_p = 0.58$ s, $t_s = 0.736$ s

11.20. Case 1: (*a*) $K_1 = 14.875$, $\omega_{\phi_x} = 46.1$, $M_p = 1.17$, $t_p = 0.313$ s, $t_s = 1.61$ s
(*b*) $\alpha = 10$, $T_1 = 20$ (*c*) $T_2 = 0.0392$ (*d*) $A = 5.94$
(*f*) $M_p = 1.0048$, $t_p = 1.5$ s, $t_s = 0.463$ s, $K_1 = 88.35$

Chapter 12

12.2. (1) $G_c(s) = \dfrac{2(s + 3)}{s + 6}$
With this cascade compensator $M_p = 1.081$, $t_p = 1.9$ s, $t_s = 3.65$ s, and $K_1 = 1.25$. The specified $M(s)$ yields $M_p = 1.127$, $t_p = 1.95$ s, $t_s = 3.9$ s, and $K_1 = 1.36$

12.4. (1) As a first trial use

$$G_c \approx \frac{2(s + 3)}{s + 5.4} \qquad \begin{matrix} M_p = 1.16, t_p = 2.3 \text{ s} \\ T_s = 4.9 \text{ s}, K_1 = 1.39 \end{matrix}$$

12.8. (*a*) $a = 0$ for $c(t)_{ss} = 0$; make pole $-b$ nondominant: $b = 100$
(*b*) $K_H = 6250$
(*c*) Characteristic equation: $s^4 + 102s^3 + 12740s^2 + 100000s + 200000 = 0$. Routh stability criterion reveals that all roots lie in the left-half s plane. A plot of Lm $\mathbf{G}_x\mathbf{H}$ reveals that its final slope crosses the 0-dB axis with a -12-dB/octave slope, indicating a possibility of a large overshoot for $c(t_p)$. The sys-

tem poles are: $p_{1,2} = -46.94 \pm j149.1$, $p_3 = -4.426$, and $p_4 = -3.701$. $|c(t_p)| = 0.001487 < 0.004$, $t_p = 0.24$ s, $t_s > 0.7$ s

Chapter 13

13.3. (1) (*a*) Observable, (*b*) controllable

$$(c)\ G(s) = \frac{s+4}{s^2 - s - 2}$$

(*d*) all states, (*e*) unstable

(4) (*a*) Not observable, (*b*) controllable

$$(c)\ G(s) = \frac{2(s+5.5)}{(s+2)(s+3)}$$

(*d*) two observable modes; the eigenvalues are $-1, -2, -3$, (*e*) stable

13.5. (*a*) Add a cascade compensator $G_c(s) = 1/(s+1)$

(*b*) $H_{eq}(s) = k_3 s^2 + (k_3 + k_2)s + k_1$

(*c*) A desired control ratio is $\dfrac{Y(s)}{R(s)} = \dfrac{A}{(s^2 + 4s + 8)(s + 25)}$

(*e*) with $A = 200$, $\mathbf{k}^T = [1 \quad 0.135 \quad 0.4]$

13.8. For system (1) (*a*) $M(s) = \dfrac{702{,}000}{(s + 12 \pm j21)(s + 16)(s + 75)}$

$$k_1 = 1,\ k_2 = 0.108,\ k_3 = 4.34 \times 10^{-3},\ k_4 = 1.07 \times 10^{-4}$$

(*b*) $y(t) = 1 + 0.974e^{-12t}\sin(21t + 142.6°) - 1.627e^{-16t} + 0.036e^{-75t}$

$$t_s = 0.205 \text{ s},\ M_p = 1.1018,\ t_p = 0.250 \text{ s},\ K_1 = 8.56$$

13.12. For plant (1) the zeros of $G(s)H_{eq}(s)$ are selected as $z_{1,2} = -5 \pm j10$ and $z_3 = -30$. Then

$$M(s) = \frac{163{,}000}{(s + 4.93 \pm j7.99)(s + 36.82 \pm j22.25)}$$

$$c(t) = 1 + 1.14e^{-4.93t}\sin(7.99t + 219.7°) + 0.111e^{-36.82t}\sin(22.25t - 81.3°)$$

$$k_1 = 1,\ k_2 = \tfrac{1}{30},\ k_3 = \tfrac{3}{750},\ k_4 = \tfrac{2}{375}$$

$$t_p = 0.44 \text{ s},\ M_p = 1.14,\ t_s = 0.866 \text{ s}$$

13.16. $M(s)_T = \dfrac{-8p(s+3)}{(s+3)(s^2+4s+8)(s-p)}$

13.18. (*a*) $\dfrac{Y(s)}{U(s)} = \dfrac{1}{s+1}$ (*b*) $y(t) = 1 - e^{-t}$ (*c*) bounded, unobservable

Chapter 14

14.1. For $m = 1$: (*a*) $S_K^M(s)|_{K=20} = \dfrac{s^4 + 7s^3 + 20s^2 + 24s}{s^4 + 7s^3 + 20s^2 + 44s + 120}$

14.6. (*a*)

$$S_\delta^M(s) = \frac{0.5}{(s+5)^2}\left(\frac{s^2 + 4s + 4}{s^2 + 6s + 20}\right) \qquad \text{where } \delta = \delta_n = a = 5$$

14.8. (*a*) $p = -14$, $K_G = 20$ (*b*) No, $G_c = \dfrac{A(s+a)}{(s+b)(s+c)}$

Chapter 15

15.1. Let $x_1 = x$ and $x_2 = \dot{x}_1$; equilibrium points: (*a*) $x_1 = x_2 = 0$, (*b*) $x_1 = x_2 = 0$, (*c*) $x_1 = 1$, $x_2 = 0$

$$(a)\ x_2 = \frac{-2x_1}{3+N} \text{ stable node} \quad (c)\ x_2 = \frac{-6x_1+6}{5+N} \text{ stable node}$$

$$(f)\ x_2 = \frac{-17x_1+34}{N-8} \text{ unstable focus} \quad (g)\ x_2 = \frac{4x_1}{N} \text{ saddle point}$$

15.2. Let $x_1 = x$ and $x_2 = \dot{x}_1$:

$$(c)\ N = \frac{-2x_2^2 - x_1}{x_2} \text{ vortex}$$

$$(d)\ N = \frac{2(-x_2|x_2| - x_1)}{x_2} \text{ vortex}$$

15.4. Let $x_1 = e$, $x_2 = \dot{e}$:

(1) For $e < -0.1$: $N = \dfrac{-x_2 - (x_1 + 0.1)}{x_2}$

Equilibrium point is $x_1 = -0.1$, $x_2 = 0$

(2) For $-0.1 < e < 0.1$: $N = -1$

(3) For $e > 0.1$: $N = \dfrac{-x_2 - (x_1 - 0.1)}{x_2}$

Equilibrium point is $x_1 = 0.1$, $x_2 = 0$

15.8. For system (2) (*a*) Three equilibrium points are

$$\mathbf{x}_a = \begin{bmatrix} 0 \\ 0 \\ 0 \end{bmatrix} \quad \mathbf{x}_b = \begin{bmatrix} 1 \\ 0 \\ 2 \end{bmatrix} \quad \mathbf{x}_c = \begin{bmatrix} 1 \\ 0 \\ -2 \end{bmatrix}$$

(*b*) For $\mathbf{x}_b$, $\mathbf{J}_{\mathbf{x}_b} = \begin{bmatrix} 0 & 1 & 0 \\ -1 & -3 & 1 \\ 2 & 0 & 0 \end{bmatrix}$

(*c*) $\mathbf{x}_b$ is unstable

15.9. (*a*) Positive definite (*b*) indefinite

15.12. (*a*) $V(\mathbf{x})$ is positive definite for $0 < K < 2$

$\dot{V}(\mathbf{x}) = (2K - 4)x_2^2$ is negative semidefinite for $K \le 2$

(*b*) $\dot{V}(\mathbf{x}) = 0$ only at the equilibrium point $x = 0$; therefore globally asymptotically stable

15.14. (*a*)

$$\mathbf{P} = \begin{bmatrix} \dfrac{-4/3}{K-2} & \dfrac{-4}{K-12} & \dfrac{20/3}{(K-2)(K-12)} \\ \dfrac{-4}{K-12} & 0.5 & \dfrac{-1}{K-12} \\ \dfrac{20/3}{(K-2)(K-12)} & \dfrac{-1}{K-12} & \dfrac{10/3}{(K-2)(K-12)} \end{bmatrix} \quad \textbf{PD} \text{ for } K < 2$$

Chapter 16

16.2. (*a*) Using $\mathbf{B}_1$, $\mathbf{Q}_a$, and $z = 1$ gives

$$\mathbf{P}_a = \begin{bmatrix} 0.098504 & 0.007423 \\ 0.007423 & 0.001556 \end{bmatrix} \quad \mathbf{k} = \begin{bmatrix} 0.98504 \\ 0.07423 \end{bmatrix}$$

Using $\mathbf{B}_2$, $\mathbf{Q}_b$, and $z = 1$ gives

$$\mathbf{P}_b = \begin{bmatrix} 0.10044 & -0.0023396 \\ -0.002396 & 0.074152 \end{bmatrix} \qquad \mathbf{K}_b = \frac{\mathbf{P}_b}{10}$$

$(b)\ \dfrac{Y(s)}{R(s)} = \dfrac{10(s+3)}{(s+3.076)(s+9.774)}$

$y(t) = 0.998 + 0.0369e^{-3.076t} - 1.0347e^{-9.774t}$

$M_p \approx 1.001,\ t_p \approx 0.65$ s, $t_s = 0.354$ s

16.7. $\mathbf{G_K}(s) = \mathbf{h}^T\mathbf{\Phi}(s)\mathbf{b} = G(s)$
Using $\omega_\phi = 2\omega_{cf} = 4$ yields $K_G = 64$. Matching coefficients of $1 + G(s)H_{eq}(s) = T_s(s)$ yields $\mathbf{k}^T = [1 \quad 0.4688 \quad 0.07813]$

16.10. $\mathbf{G_K}(s) = \mathbf{h}^T\mathbf{\Phi}(s)\mathbf{b} = G(s) = \dfrac{K_G(s+4)}{s(s+10)(s+2)}$; $\omega_\phi > \omega^* = 10$;
$K_G = \omega_\phi^2 > 100$; thus use $K_G = 1600$ and $\omega_f = 40$
$\mathbf{k}_p^T = [1 \quad 0.03375 \quad -0.00375]$

16.13. (For Prob. 16.7) the RSL with $K_G = 64$ yields roots
$s_1 = -4.2143,\ s_{2,3} = -2.0984 \pm j3.2838$
Characteristic equation: $s^3 + 8.4111s^2 + 32.8733s + 64 = 0$
$\mathbf{k}^T = [1 \quad 0.4824 \quad 0.08455]$
(For Prob. 16.8): the RSL with $K_G = 2$ yields roots: $s_{1,2} = -0.3472 \pm j0.5585$, $s_3 = -4.624$

Chapter 17

17.1. $G(s) = \dfrac{K_G}{(s+1)(s+2)(s+3)} = \dfrac{K_G}{s^3 + 6s^2 + 11s + 6}$

(a) From the RSL, $K_{\min} = 314$. Use $K_G = 400$
From Eq. (17.28), $q_{11} = 1 - (\frac{6}{400})^2 \approx 1 \approx h_{11}$
From Eq. (17.27), $k_1 = 1 - \frac{6}{400} = 0.985$

$$G_{eq}(s) = \frac{400}{s[s^2 + (6 + K_Gk_3)s + (11 + K_Gk_2)]}$$

$\mathbf{k}^T = [0.985 \quad 0.9775 \quad 0.49]$, $\alpha = 200$

(b) $\mathbf{h}^T = [1 \quad 1.01 \quad 0.4951]$
Hint: Use Eq. (17.74) for a first estimate of h_3:
$k_3 \approx \sqrt{q_3} = h_3$
When Eq. (16.28) is used, the system is optimal

(c) $y(t) = 1 + 1.42e^{-t}\sin(t + 224.7°) - 0.5 \times 10^{-4}e^{-200t}$
$M_p = 1.0432,\ t_p = 3.15$ s, $t_s = 4.22$ s

17.7. (a) $K_G = 400$ (see Prob. 17.1)
(b) Use $q_1 = 1$ and $q_2 = 0.008$. Then from Fig. 17.3, for $M_p = 1.04$ and $t_p \approx 1.93$ s, $t_s = 2.4^+$ s, $q_3 = 0.03$. From a computer solution of the algebraic Riccati equation,
$\mathbf{k}^T = [0.98511 \quad 0.58277 \quad 0.16704]$
$y(t) = 1 + 1.4667e^{-1.7176t}\sin(1.678t + 222.9°) - 0.001259e^{-69.381t}$
Actual values: $M_p = 1.04,\ t_p = 1.90^-$ s, $t_s \approx 2.47$ s
(c) For $z = 0.5$, $q_1 = 0.499775$, $K_G = 282.842$
(d) For $z = 0.5$, $\mathbf{k}^T = [0.97879 \quad 0.68756 \quad 0.23425]$, $M_p = 1.0388$, $t_p = 2.28$ s, $t_s = 2.95$ s

17.9. (*a*) First trial values obtained from Fig. 17.6 are $a = 9$, $q_3 = 0.06$. From the RSL, use $K_G = 300$. Then, $q_1 = c_0^2 = 81$. Also, $q_2 = 0.008$ for Fig. 17.6. Use a computer solution of the algebraic Riccati equation to evaluate **k**, the roots of the characteristic equation, and the time response.

(*b*) $\mathbf{k}^T = [9 \quad 2.2062 \quad 0.254067]$

$s_{1,2} = -4.2778 \pm j4.2842$

$s_3 = -73.662$

$y(t) = 1 - 1.0608e^{-4.2778t} \sin(4.2842t + 263.73°) + 0.0545e^{-73.662t}$

$M_p = 1.064$, $t_p = 0.58$ s, $t_s = 0.89$ s

Note: For $M_p = 1.06$ a solution using $a = 10$ is given in Table 17.6

17.13. (*a*) DDE: $(s^2 + 1.417s + 1)(s + 2) = 0$

$\mathbf{h}^T = [1 \quad 1.917 \quad 1.7085 \quad 0.5]$

From Eq. (16.79), $\mathbf{k}^T = [1 \quad 1.927 \quad 1.72267 \quad 0.507085]$

(*b*) $\mathbf{k}^T = [1 \quad 1.9269 \quad 1.7226 \quad 0.50704]$

(*c*) $\lambda_{1,2} = -0.70489 \pm j0.70572$, $\lambda_3 = -2$, $\lambda_4 = -99.99$

$y(t) = 1 - 1.9395e^{-0.70489t} \sin(0.70572t + 195.83°) - 0.47114e^{-2t} + 0.0^+ e^{-99.99t}$

$M_p = 1.0352$, $t_p = 5.17$ s, $t_s = 6.46$ s

17.14. Plant (1); (*a*) DDE: $(s^3 + 2.5s^2 + 3s + 1) = 0$; $\mathbf{h}^T = [0.5 \quad 1.5 \quad 1.25 \quad 0.5]$ for $z = 1$;

$\mathbf{k}^T = [0.5 \quad 1.505 \quad 1.26 \quad 0.50250]$

(*b*) $\mathbf{k}^T = [0.5 \quad 1.505 \quad 1.26 \quad 0.50254]$

(*c*) $\lambda_{1,2} = -1.0019 \pm j1.0061$, $\lambda_3 = -0.49602$, $\lambda_4 = -100$

Computer values of the model: $M_p = 1.0432$, $t_p = 3.13$ s, $t_s = 4.225$ s

17.15. $\mathbf{h}^T = [2 \quad 5 \quad 5 \quad 2.5 \quad 0.5]$; $\mathbf{k}^T = [2 \quad 5.01 \quad 5.025 \quad 2.4875 \quad 0.4925]$

Chapter 18

18.2. Using the set of observer eigenvalues $\{-15, -20\}$; (*a*) $\mathbf{L} = [32 \quad 202]^T$
(*b*) $\mathbf{K} = [-22 \quad 7]$

18.4. (1) (*a*) Completely controllable, (*b*) $\sigma(\mathbf{A}_p) = \{-1, -2, -3\}$,

(*c*) $\mathbf{T} = \begin{bmatrix} 6 & 5 & 1 \\ -6 & -4 & 0 \\ 0 & 2 & 1 \end{bmatrix}$ (*d*) $\mathbf{k}_c^T = [114 \quad 63 \quad 9]$, $\mathbf{k}_p^T = [31 \quad 12 \quad -22]$

$$\mathbf{A}_{\text{cl}} = \begin{bmatrix} -32 & -12 & 22 \\ -2 & -2 & -2 \\ -32 & -12 & 19 \end{bmatrix}$$

18.7. (1)

$$\ker \mathbf{S}(\lambda) = \text{span}\left\{ \begin{bmatrix} 1 \\ 0 \\ 0 \\ \hline 1 \\ 1+\lambda \end{bmatrix}, \begin{bmatrix} 0 \\ -(4+\lambda) \\ 2+\lambda \\ \hline (2+\lambda)(4+\lambda) \\ 0 \end{bmatrix} \right\}$$

18.8. (1) (*a*) Controllable, (*b*) $\gamma_1 = 2$, $\gamma_2 = 1$,

(*c*) $\mathbf{T} = \begin{bmatrix} 1 & 0 & 1 \\ -1 & 1 & 0 \\ 0 & 0 & 1 \end{bmatrix}$ $\quad \mathbf{F} = \mathbf{I}$

(*d*) $\mathbf{H} = \begin{bmatrix} 1 & 1 & 2 \\ 0 & 0 & 1 \end{bmatrix}$

(*e*) $Q_1(\lambda) = (\lambda + 4)(\lambda + 5) = \lambda^2 + 9\lambda + 20,\ Q_2(\lambda) = \lambda + 6$

$\mathbf{\Gamma} = \begin{bmatrix} -20 & -9 & 0 \\ 0 & 0 & -6 \end{bmatrix} \qquad \mathbf{K} = \begin{bmatrix} -31 & -10 & 29 \\ 0 & 0 & -7 \end{bmatrix}$

(*f*) $\mathbf{A}_{\text{cl}} = \begin{bmatrix} 1 & 1 & -7 \\ -30 & -10 & 30 \\ 0 & 0 & -6 \end{bmatrix}$

18.9. (1)

$$\ker \mathbf{S}(\lambda) = \text{span}\left\{\begin{bmatrix} 1 \\ -1+\lambda \\ 0 \\ \hdashline -1-\lambda+\lambda^2 \\ 0 \end{bmatrix}, \begin{bmatrix} 0 \\ 1-\lambda \\ 1 \\ \hdashline -1+\lambda-\lambda^2 \\ -1+\lambda \end{bmatrix}\right\}$$

18.10. (1) (*a*) Controllable (*b*) $\gamma_1 = 3,\ \gamma_2 = 1$

(*c*) $\mathbf{T} = \begin{bmatrix} 1 & 0 & 0 & 0 \\ 2 & 1 & 0 & 0 \\ 2 & 2 & 1 & 0 \\ 0 & 0 & 0 & 1 \end{bmatrix} \qquad \mathbf{F} = \begin{bmatrix} 1 & -1 \\ 0 & 1 \end{bmatrix}$

(*d*) $\mathbf{H} = \begin{bmatrix} 4 & 1 & -1 & 0 \\ 0 & 0 & 0 & 1 \end{bmatrix}$

(*e*) $Q_1(\lambda) = \lambda^3 + 16\lambda^2 + 89\lambda + 174,\ Q_2(\lambda) = \lambda + 10$

$\mathbf{\Gamma} = \begin{bmatrix} -174 & -89 & -16 & 0 \\ 0 & 0 & 0 & -10 \end{bmatrix}$

$\mathbf{K} = \begin{bmatrix} -28 & -60 & -15 & 11 \\ 0 & 0 & 0 & -11 \end{bmatrix}$

18.11. (1)

$$\ker \mathbf{S}(\lambda) = \text{span}\left\{\begin{bmatrix} 1 \\ 2+\lambda \\ 2+2\lambda+\lambda^2 \\ 0 \\ \hdashline -4-\lambda+\lambda^2+\lambda^3 \\ 0 \end{bmatrix}, \begin{bmatrix} 0 \\ 0 \\ 0 \\ 1 \\ \hdashline 1-\lambda \\ -1+\lambda \end{bmatrix}\right\}$$

Chapter 19

19.1. (1) (*b*)

$$\mathbf{K} = \mathbf{Q}\mathbf{V}^{-1} = \begin{bmatrix} 1 & 3 & 8 \\ -3 & 0 & 0 \end{bmatrix}\begin{bmatrix} 1 & 0 & 0 \\ 0 & 1 & 2 \\ 0 & -3 & -4 \end{bmatrix}^{-1} = \begin{bmatrix} 1 & 6 & 1 \\ -3 & 0 & 0 \end{bmatrix}$$

$$\mathbf{A}_{\text{cl}} = \mathbf{A} + \mathbf{BK} = \begin{bmatrix} -4 & 0 & 0 \\ 0 & -8 & -1 \\ 0 & 6 & -3 \end{bmatrix}$$

(c)

$$e^{\mathbf{A}_{cl}t} = \begin{bmatrix} e^{-4t} & 0 & 0 \\ 0 & -2e^{-5t} + 3e^{-6t} & -e^{-5t} + e^{-6t} \\ 0 & 6e^{-5t} - 6e^{-6t} & 3e^{-5t} - 2e^{-6t} \end{bmatrix}$$

$$\mathbf{x}_{zi}(t) = e^{\mathbf{A}_{cl}t}\mathbf{x}(0) = \begin{bmatrix} e^{-4t} \\ -e^{-5t} + e^{-6t} \\ 3e^{-5t} - 2e^{-6t} \end{bmatrix}$$

19.3. Rank $\mathbf{M}_c$ = Rank $[\mathbf{B} \quad \mathbf{AB}] = 2 < n$. There is one uncontrollable mode, e^{-2t}

19.6. (a)

$$\overline{\mathbf{A}} = \begin{bmatrix} \mathbf{A} & \mathbf{0} \\ -\mathbf{C} & \mathbf{0} \end{bmatrix} = \left[\begin{array}{ccc|c} 1 & 6 & -3 & 0 \\ -1 & -1 & 1 & 0 \\ -2 & 2 & 0 & 0 \\ \hline -1 & 0 & -1 & 0 \end{array}\right] \qquad \overline{\mathbf{B}} = \begin{bmatrix} \mathbf{B} \\ \mathbf{0} \end{bmatrix} = \begin{bmatrix} 1 \\ 1 \\ 1 \\ 0 \end{bmatrix}$$

$$\overline{\mathbf{C}} = [\mathbf{C} \quad \mathbf{0}] = [1 \quad 0 \quad 1 \quad 0]$$

(b) Rank $\overline{\mathbf{M}}_c$ = Rank $\begin{bmatrix} \mathbf{B} & \mathbf{A} \\ \mathbf{0} & -\mathbf{C} \end{bmatrix} = 4$

(c)

$$\overline{\mathbf{K}} = \overline{\mathbf{Q}}\overline{\mathbf{V}}^{-1} = [0 \quad 8 \quad 100 \quad 135] \begin{bmatrix} 1 & 4 & 10 & 0 \\ 0 & -3 & -28 & -30 \\ 1 & 2 & -6 & -15 \\ 1 & 2 & 1 & -3 \end{bmatrix}^{-1}$$

$$= [-27 \quad -20.667 \quad 33.667 \quad -6.667]$$

$$\mathbf{A}_{cl} = \overline{\mathbf{A}} + \overline{\mathbf{B}}\overline{\mathbf{K}} = \begin{bmatrix} -26 & -14.667 & 30.667 & -6.667 \\ -28 & -21.667 & 34.667 & -6.667 \\ -29 & -18.667 & 33.667 & -6.667 \\ -1 & 0 & -1 & 0 \end{bmatrix}$$

(d) $y(t) = 1 - 10e^{-2t} + 13.3e^{-3t} - 1.67e^{-4t} - 2.67e^{-5t}$

19.14. (1) (a)

$$\left\{ \begin{bmatrix} 10 + 6\lambda + \lambda^2 \\ 4 + 5\lambda + \lambda^2 \\ 2 + 3\lambda + \lambda^2 \\ 8 + 14\lambda + 7\lambda^2 + \lambda^3 \end{bmatrix} \right\}$$

Use $\sigma(\mathbf{A} - \mathbf{LC}) = \{-8, -9, -10\}$

$\mathbf{L}^T = [100.8 \quad -67.2 \quad -13.6]$

(c)

$$\mathbf{A} - \mathbf{LC} = \begin{bmatrix} -101.8 & -100.8 & -100.8 \\ 68.2 & 65.2 & 67.2 \\ 12.6 & 13.6 & 9.6 \end{bmatrix}$$

Chapter 20

20.4. $\sigma(\mathbf{A}) = \{-1, 1 + j1, 1 - j1\}$. Plant is unstable

1. Rank $\mathbf{M}_c = 3$, Rank $\mathbf{M}_0 = 3$, therefore the plant is completely controllable and observable

2. Rank $\begin{bmatrix} \mathbf{B} & \mathbf{A} \\ \mathbf{0} & \mathbf{C} \end{bmatrix} = 5$. The augmented system is controllable
3. Since **CB** has full rank, there is one transmission zero, $\lambda = -1$
4. Use $\mathbf{\Sigma} = \mathbf{I}$. $\mathbf{K}_1 = (\mathbf{C}_2\mathbf{B}_2)^{-1}\mathbf{\Sigma} = \begin{bmatrix} 1/3 & 1/3 \\ -1/3 & 2/3 \end{bmatrix}$
5. Let $\lambda_0 = -2$
6. $\mathbf{\Gamma}(\lambda) = \text{diag}\left\{\dfrac{g}{\lambda + g} \quad \dfrac{g}{\lambda + g}\right\}$

20.5. Plant is unstable
1. Plant is controllable and observable
2. Augmented system is controllable
3. Rank $\mathbf{C}_2\mathbf{B}_2 = \text{Rank}\begin{bmatrix} 1 & 1 \\ 0 & 0 \end{bmatrix} = 1$

   (*a*) Using $d_1 = d_2 = 0$

$$\mathbf{B}^* = \begin{bmatrix} 0 & 0 \\ 0 & 1 \end{bmatrix} \qquad \mathbf{M} = \begin{bmatrix} 0 \\ m \end{bmatrix}$$

   (*b*) $\mathbf{F}_2 = \mathbf{C}_2 + \mathbf{M}\mathbf{A}_{12} = \begin{bmatrix} 1 & 0 \\ 0 & m \end{bmatrix}$, $\mathbf{F}_2^{-1} = \begin{bmatrix} 1 & 0 \\ 0 & 1/m \end{bmatrix}$

   (*c*) $\mathbf{C}_2\mathbf{F}_2^{-1} = \begin{bmatrix} 1 & 0 \\ 0 & 0 \end{bmatrix}$

   (*d*) $\mathbf{F}_1 = [0 \quad 1]^T$

4. Using $\mathbf{\Sigma} = \mathbf{I}$, $\mathbf{K}_1 = (\mathbf{F}_2\mathbf{B}_2)^{-1}\mathbf{\Sigma} = \begin{bmatrix} 0 & (1/3)m \\ 1 & (-1/3)m \end{bmatrix}$
5. Use $\lambda_0 = -2$
6.

$$\mathbf{\Gamma}(\lambda) = \begin{bmatrix} \dfrac{g}{\lambda + g} & 0 \\ 0 & \dfrac{1/m}{\lambda + 1/m} \end{bmatrix}$$

Chapter 21

21.1. (*a*) $\alpha = 3.548$, $T = 2$, $A = 12.5$
(*b*) For the basic system: $K_x = 125$, $\omega_{\phi_x} = 10.6$ rad/s
For $\gamma = 40°$: $\alpha = 0.3816$, $T = 0.04359$, $A = 2.6205$

21.6. (*a*) By Mason's rule obtain

$$\frac{C}{R} = \frac{ABH}{1 - AD - BHF - HG - ABE + ADHG}$$

(*b*) $t_{XR} = \dfrac{X}{R} = \dfrac{A}{1 - AD}$, $t_{XY} = \dfrac{X}{Y} = \dfrac{HF + EA}{(1 - HG)(1 - AD)}$

$t_{CY} = \dfrac{C}{Y} = \dfrac{H}{1 - HG}$

Chapter 22

22.3. (*a*) $z^{-1} + 3z^{-2} + 4z^{-3} + 3z^{-4} + \cdots$

22.4. (*a*) Entry 4: $f(0) = 0$, $f(\infty) = \infty$
(*c*) $c(0) = 0$, $c(\infty) = 1$

22.5. (*a*) $c(nT) = e[(n-1)T] + e[(n-2)T] + 2c[(n-1)T] - 2c[(n-2)T] + c[(n-3)T]$

INDEX